10th International Conference and Workshop on Numerical Simulation of 3D Sheet Metal Forming Processes (NUMISHEET 2016)

Journal of Physics: Conference Series Volume 734

Bristol, United Kingdom
4 – 9 September 2016

Editors:

Rui P.R. Cardoso
Jeong Whan Yoon
Robert E. Dick
Eduardo Souza Neto
J.M.A. Cesar de Sa
O.B. Adetoro

ISBN: 978-1-5108-2913-8
ISSN: 1742-6588

Printed from e-media with permission by:

Curran Associates, Inc.
57 Morehouse Lane
Red Hook, NY 12571

Some format issues inherent in the e-media version may also appear in this print version.

Printed by Curran Associates, Inc. (2016)

Institute of Physics
Dirac House, Temple Back
Bristol BS1 6BE UK

Phone: 44 1 17 929 7481
Fax: 44 1 17 920 0979

techtracking@iop.org

Additional copies of this publication are available from:

Curran Associates, Inc.
57 Morehouse Lane
Red Hook, NY 12571 USA
Phone: 845-758-0400
Fax: 845-758-2633
Email: curran@proceedings.com
Web: www.proceedings.com

Table of contents

Volume 734

August 2016

Part A - Benchmarks

Editors: Rui P.R. Cardoso, Jeong Whan Yoon and Robert E. Dick

Part B – General Papers

Editors: Rui P.R. Cardoso, Eduardo Souza Neto, J.M.A Cesar de Sa and O.B. Adetoro

Paper

Journal of Physics: Conference Series **734** (2016) 011001 doi:10.1088/1742-6596/734/1/011001

NUMISHEET 2016: The 10th International Conference and Workshop on Numerical Simulation of 3D Sheet Metal Forming Processes:

Part A - Benchmark Problems and Results
Part B - General Papers

Bristol, United Kingdom
4-9 September 2016

Editors
Part A: Rui P.R. Cardoso, Jeong Whan Yoon and Robert E. Dick
Part B: Rui P.R. Cardoso, Eduardo Souza Neto, J.M.A Cesar de Sa and O.B. Adetoro

All papers have been peer reviewed

Sponsoring Organizations
AutoForm
ESI
JSTAMP

Conference Organizers
University of the West of England, UK
University of Porto, Portugal
Swansea University, UK

Journal of Physics: Conference Series **734** (2016) 011001 doi:10.1088/1742-6596/734/1/011001

Editors

Rui P.R. Cardoso
Brunel University London
Department of Mechanical, Aerospace & Civil Engineering
Kingston Lane, Uxbridge, UB8 3PH
United Kingdom
E-mail: Rui.Cardoso@brunel.ac.uk

Eduardo Souza Neto
Swansea University
College of Engineering
Fabian Way, Swansea, SA1 8EN
United Kingdom
E-mail: e.desouzaneto@swansea.ac.uk

J.M.A. Cesar de Sa
University of Porto
Faculty of Engineering
Rua Dr Roberto Frias, Porto, 4200-465
Portugal
E-mail: cesarsa@fe.up.pt

O.B. Adetoro
University of the West of England
Department of Engineering, Design & Mathematics
Coldharbour Lane, Bristol, BS16 1QY
United Kingdom
E-mail: Mayo.Adetoro@uwe.ac.uk

Jeong Whan Yoon
Deakin University
School of Engineering
Geelong Waurn Ponds, VIC 3220
Australia
E-mail: j.yoon@deakin.edu.au

Robert E. Dick
Alcoa Technical Center
100 Technical Drive, Alcoa center, PA 15069
United States
E-mail: Robert.Dick@alcoa.com

Journal of Physics: Conference Series **734** (2016) 011001 doi:10.1088/1742-6596/734/1/011001

Conference Chairmen

Rui Cardoso (Chair)	Brunel University London, UK
Eduardo Souza Neto (co-Chair)	Swansea University, UK
Jose Cesar de Sa (co-Chair)	FEUP, University of Porto, Portugal
Mayo Adetoro (co-Chair)	University of the West of England, UK

Conference Steering Committee

Chung, K.	South Korea
Gelin, J.C.	France
Hora, P.	Switzerland
Pourboghrat, F.	USA
Stoughton, T.B.	USA
Wagoner, R.H.	USA
Yang, D.Y.	South Korea
Yoon, J.W. (Chair of the Steering Committee)	Australia

Local Organising Committee

Adetoro, O.B.	University of the West of England, UK
Amali, R.	University of the West of England, UK
Butuc, C.	University of Aveiro, Portugal
Cardoso, R.	Brunel University London, UK
Cesar Sa, J.	FEUP, University of Porto, Portugal
Dashwood, R.	University of Warwick, UK
Hazra, S.	University of Warwick, UK
Neto, E.S.	Swansea University, UK
Santos, A.	FEUP, University of Porto, Portugal
Teixeira, P.	INEGI, Portugal
Vincze, G.	University of Aveiro, Portugal
Wen, P.H.	Queen Mary, University of London, UK

International Scientific Committee

Aliabadi, F.	UK
Alfano, G.	UK
Aretz, H.	Germany
Bahai, H.	UK
Balan, T.	France
Banabic, D.	Romania
Barlat, F.	South Korea
Cao, J.	USA
Cardoso, R.	UK
Cazacu, O.	USA
Cesar de Sa, J.	Portugal
Chung, K.	South Korea
de Borst, R.	UK

Dick, R.E.	USA
Gelin, J.C.	France
Green, D.	Canada
Habraken, A.M.	Belgium
Hora, P.	Switzerland
Iadicola, M.	USA
Khan, A.	USA
Kim, H.Y.	South Korea
Kuwabara, T.	Japan
Korkolis, Y.	USA
Kyriakides, S.	USA
Lee, M.G.	South Korea
Manach, P.	France
Massoni, E.	France
Meinders, V.T.	Netherlands
Menezes, L.	Portugal
Narasimhan, K.	India
Ponthot, J.P.	Belgium
Pourboghrat, F.	USA
Reese, S.	Germany
Stoughton, T.B.	USA
Thuillier, S.	France
Tisza, M.	Hungary
Wagoner, R.H.	USA
Worswick, M.	Canada
Wu, P.	Canada
Yang, D.Y.	South Korea
Yoon, J.W.	Australia
Yoshida, F.	Japan
Volk, W.	Germany

Benchmark Committee Chairs

Cardoso, R.	Brunel University London, UK
Yoon, J.W.	Deakin University, Australia
Dick, R.E.	Alcoa, USA

Benchmark 1 Committee - Fracture prediction after Cup Drawing and Expansion

Watson, M.	Crown Packaging, UK
Huang, H.	Crown Packaging, UK
Lockley, A.	Crown Packaging, UK
Dick, R.E.	Alcoa, USA
Cardoso, R.	Brunel University London, UK
Santos, A	FEUP, University of Porto, Portugal

Journal of Physics: Conference Series **734** (2016) 011001 doi:10.1088/1742-6596/734/1/011001

Benchmark 2 Committee - Springback of a Jaguar Land Rover aluminium panel

Allen, M.	Jaguar Land Rover, UK
Oliveira, M.	University of Coimbra, Portugal
Hazra, S.	University of Warwick, UK
Adetoro, O.B.	University of the West of England, UK
Das, A.	University of Warwick, UK

Benchmark 3 Committee - Springback of an Al-Mg alloy in warm forming conditions

Manach, P.Y.	University of Bretagne-Sud, IRDL, France
Laurent, H.	University of Bretagne-Sud, LIMATB, France
Coer, J.	University of Bretagne-Sud, LIMATB, France
Jegat, A.	University of Bretagne-Sud, LIMATB, France
Yoon, J.W.	Deakin University, Australia

PREFACE

The NUMISHEET conference series have been established as a world-class forum through which new intellectual ideas and technologies in the area of sheet metal forming simulation are exchanged. Previous NUMISHEET conferences have given enormous contributions to industry and academia in what regards the development of new methods and ideas for the numerical simulation of sheet metal forming processes. Previous NUMISHEET conferences were held in: Zurich (Switzerland, 1991), Isehara (Japan, 1993), Dearborn (USA, 1996), Besancon (France, 1999), Jeju Island (South Korea, 2002), Detroit (USA, 2005), Interlaken (Switzerland, 2008), Seoul (South Korea, 2011) and Melbourne (Australia, 2014).

The NUMISHEET 2016 conference will be held in Bristol, UK. It features technical, keynote and plenary sessions and mini-symposiums in diverse sheet metal forming areas including the recently introduced incremental sheet forming and electromagnetic forming, as well as new prominent numerical methods such as IsoGeometric Analysis and meshless methods for sheet analysis. NUMISHEET 2016 will have eight academic plenary lectures delivered by worldwide recognised experts in the areas of sheet metal forming, material modelling and numerical methods in general. Also, NUMISHEET 2016 will have three industrial plenary lectures which will be addressed by three different companies with strong businesses in sheet metal forming processes: AutoForm, Crown Technology and Jaguar Land Rover.

One of the most distinguishing features of NUMISHEET conference series is the industrial benchmark sessions, during which numerical simulations of industrial sheet formed parts are compared with experimental results from the industry. The benchmark sessions provide an extraordinary opportunity for networking, for the exchange of technologies related to sheet metal forming and for the numerical validation of sheet metal forming codes/software. Three benchmark studies have been organised in NUMISHEET 2016: BM1) *"Benchmark 1 – Failure Prediction after Cup Drawing, Reverse Redrawing and Expansion"*, for both an aluminium AA 5352 and steel TH330 alloys, where the major objectives are predicting the earing profile of the cup and the failure point after complex nonlinear strain paths; BM2) *"Benchmark 2 – Springback of a Jaguar Land Rover Aluminium Panel"*, where it is intended to study the thickness distribution and the springback profiles of a component of a Jaguar car at three particular cross-sections; BM3) *"Benchmark 3 – Springback of an Al-Mg alloy in warm forming conditions"*, where the main objective is to predict the springback of a Al-Mg alloy after warm forming conditions.

The conference proceedings will include a standard proceedings for regular and keynote papers (Part B) and a separated proceedings for the industrial benchmarks (Part A). Both proceedings will be published by the Institute of Physics (IOP) and will be included in SCOPUS database for a wide scientific dissemination. The benchmarks book proceedings, Part A, will be available through a printed paperback book and an electronic version and the proceedings book for the regular papers, part B, will be available through a USB pen drive. Both proceedings books will also be available online through the Institute of Physics publishers' platform.

The NUMISHEET 2016 international conference is organising two special issues in two prestigious peer-reviewed international journals: 1) The International Journal of Solids and Structures; 2) The European Journal of Computational Mechanics. The NUMISHEET 2016 conference delegates are thus invited to submit an extended version of their papers for these two special issues, which will have a peer reviewed process as usual before publication in the journals.

Bristol, United Kingdom, July 15, 2016
NUMISHEET2016 Chairs
Rui P.R. Cardoso, E. Souza Neto, J.M.A. Cesar de Sa, O.B. Adetoro

ACKNOWLEDGEMENT

The NUMISHEET conference has a particular characteristic of joining experts from academia and industry for the discussion of numerical methods around the simulation of 3D sheet metal forming processes. The organising committee would like to express their gratitude to the many distinguished individualities that contributed enormously to the success of this conference. We gratefully acknowledge:

The conference participants that are taking part in this prestigious conference either by presenting scientific papers or by participating with their simulations in the benchmark case studies;

Professors: Ferri Aliabadi (Imperial College London, UK), Rene De Borst (University of Sheffield, UK), Anne-Marie Habraken (University of Liege, Belgium), Akhtar Khan (University of Maryland, USA), Stelios Kyriakides (University of Texas, USA), Eugenio Onate (Technical University of Catalunya, Spain), Rade Vignjevic (Brunel University London, UK) and Jeong Whan Yoon (Deakin University, Australia) for accepting to deliver the plenary lectures and for all their support to this conference;

The Industrial Plenary Speakers: Dr Kevin Ambrose (Crown Technology, UK), Dr Bart Carleer (AutoForm), Dr Neil Small and Mr Matt Stanton (Jaguar Land Rover, UK) for accepting to deliver these important Industrial Plenary sessions, giving more of an industrial flavour to this conference;

The organisers of the mini-symposiums: MS1 - "Forming defects, material and shape instabilities", P. Manach, J.B. Kim and S. Thuillier"; MS2 - "Challenges in characterization and modelling of Titanium materials", O. Cazacu and B. Revil-Baudard; MS3 - "Formability in Sheet Metal Forming Processes", X. Zhu, S. Li and T. B. Stoughton; MS4 - "Meshless Method in Manufacturing", E. Barbieri and P.H. Wen; MS5 - "Isogeometric Analysis of Structures with Applications to Shells and Contact/Impact", S. Hartmann, Y. Bazilevs and D. Benson; MS6 - "Material Modelling for Sheet Metal Forming", F. Barlat, J.W. Yoon; MS7 – "Incremental sheet forming processes", R.J. Alves de Sousa, A. K. Behera and F. A. Lora;

The University of the West of England, Bristol, UK, for hosting this conference and for facilitating administrative resources such as contracting and administrative staff. In particular, we would like to thank Professor Glenn Lyons for his fantastic support to this conference;

The sponsors of this conference: AutoForm (Gold Sponsor), ESI Group (Bronze Sponsor) and JSTAMP (Bronze Sponsor) for the financial support given;

The organisers of the three benchmarks in general for their various contributions on the preparation of these challenging case studies about sheet metal forming.
Regarding benchmark 1 (BM1), we would like to acknowledge the contributions of Crown Technology, from Martin Watson and his research team composed by Helen Huang, Andrew Lockley and Rosie Moffat for the coordination and dedication to this benchmark. We would like also to thank Alcoa Technical Center, USA, particularly Robert E. Dick, for the cup testing

for the different alloys, the material characterisation of the aluminium alloy and, most importantly, for his wisdom on sheet metal forming processes. We are also very grateful to the support of Prof. Abel Santos from University of Porto, Portugal and Dr. Gabriela Vincze from the University of Aveiro, Portugal, for their contributions in the discussion of the benchmark and on the material testing for the material characterisation of the different material parameters.

Regarding benchmark 2 (BM2), we would like first of all to thank Jaguar Land Rover for the industrial benchmark study on springback prediction of a complex industrial sheet metal formed part. We would like also to acknowledge Dr. Marta Oliveira from the University of Coimbra, Portugal, for her guidance and expertise, which turned out to be fundamental for the success of this benchmark. We are also very grateful to Dr. Sumit Hazra from Warwick University, UK, and his team for the discussions and for the 3D measurements of the thickness distribution and springback profiles at the different cross-sections.

Regarding benchmark 3 (BM3), we are really grateful to Professor Pierre Manach and his team at the University of Bretagne Sud and IRDL institute in France. Professor Pierre Manach and his team performed all of the necessary material testing and characterisation for warm forming conditions and conducted as well all the experiments required for the success of this benchmark;

The Steering Committee for their approval for the organisation of this important conference in Bristol UK. In particular, we are thankful for the guidance and mentorship given by the Chair of the Steering Committee, Professor Jeong Whan Yoon from Deakin University, Australia;

Miss Claire Garland from the Institute of Physics for her excellent support on the construction of the website and for the professionalism in dealing with several different requests from conference delegates;

Miss Vivien Calway of the University of the West of England, UK, for efficiently managing all of the secretarial tasks that were thrust upon her throughout the preparations for the conference.

We also greatly appreciate the devoted services of everybody involved somehow in this conference but are not explicitly recognised in any printed form.

Bristol, United Kingdom, July 15, 2016
NUMISHEET2016 Chairs
Rui P.R. Cardoso, E. Souza Neto, J.M.A. Cesar de Sa, O.B. Adetoro

Journal of Physics: Conference Series **734** (2016) 011001 doi:10.1088/1742-6596/734/1/011001

Conference Sponsors

Gold Sponsors

AUTOFORM
Forming Reality

Bronze Sponsors

esi
get it right®

JSTAMP

Journal of Physics: Conference Series **734** (2016) 011001 doi:10.1088/1742-6596/734/1/011001

Benchmark Sponsors

Benchmark 1

CROWN

ALCOA

Brunel University London

U.PORTO
FEUP FACULDADE DE ENGENHARIA
UNIVERSIDADE DO PORTO

Benchmark 2

JAGUAR

LAND ROVER

Brunel University London

THE UNIVERSITY OF WARWICK

UWE BRISTOL University of the West of England

Benchmark 3

Université Bretagne Sud
ubs

iRDL
Institut de Recherche Dupuy de Lôme

Journal of Physics: Conference Series **734** (2016) 011001 doi:10.1088/1742-6596/734/1/011001

Conference Organizers

UWE
University of the West of England
BRISTOL

Swansea University
Prifysgol Abertawe

U. PORTO
FEUP FACULDADE DE ENGENHARIA
UNIVERSIDADE DO PORTO

Brunel University London

Journal of Physics: Conference Series **734** (2016) 011002 doi:10.1088/1742-6596/734/1/011002

Peer review statement

All papers published in this volume of *Journal of Physics: Conference Series* have been peer reviewed through processes administered by the proceedings Editors. Reviews were conducted by expert referees to the professional and scientific standards expected of a proceedings journal published by IOP Publishing.

Benchmark 1 – Failure Prediction after Cup Drawing, Reverse Redrawing and Expansion

Part A: Benchmark Description

Martin Watson[a], Robert Dick[b], Y. Helen Huang[a], Andrew Lockley[a], Rui Cardoso[c], Abel Santos[d]

[a]*Crown Packaging, Oxfordshire OX12 9BP, UK*
[b]*Alcoa Technical Center, PA 16069-0001, USA*
[c]*Brunel University London, Uxbridge, UB8 3PH, UK*
[d]*FEUP, University of Porto, 4200-465, Portugal*

Abstract. This Benchmark is designed to predict the fracture of a food can after drawing, reverse redrawing and expansion. The aim is to assess different sheet metal forming difficulties such as plastic anisotropic earing and failure models (strain and stress based Forming Limit Diagrams) under complex nonlinear strain paths. To study these effects, two distinct materials, TH330 steel (unstoved) and AA5352 aluminum alloy are considered in this Benchmark. Problem description, material properties, and simulation reports with experimental data are summarized.

Keywords: Fracture, cup drawing, redraw, expansion, FLD, anisotropy

1. INTRODUCTION

The numerical simulation of cupping processes is fundamental for the can making industry. It allows the prediction of many different sheet metal defects and instabilities that significantly affect the efficient production of these parts. These defects include thinning from cup drawing, earing from plastic anisotropy, and damage and fracture from different combinations of strain paths, e.g. drawing and expansion. Due to the earing profile after cup drawing and reverse redrawing, it is usually necessary to perform a trimming operation to bring the cup to a uniform height prior to conducting other challenging sheet forming operations. The numerical simulation will not include this trimming operation.

In sheet metal forming, the Forming Limit Diagram (FLD) is commonly used to predict material failure in forming operations. The sheet material failure or tearing can be triggered by different loading conditions or strain paths ranging from uniaxial tension to plane strain and bi-axial tensile loading. The FLD has been integrated in many different finite element commercial packages for the simulation of sheet metal forming processes, as well as many academic research codes.

More recently there have been intensive studies on the development and use of strain-path independent forming limit diagrams such as the stress-based FLD and the polar FLD. Despite the fact that strain-based FLD has been used with some success in the past, it has been experimentally verified that the material tearing point is strain-path dependent and so the use of strain based forming limit diagrams can lead to erroneous predictions of failure in sheet metal forming analysis. The stress-based FLD can be obtained from a mapping from the strain-based FLD and enables strain-path independence for the prediction of material tearing in sheet metal forming simulations.

In this benchmark study, a strain-based FLD is provided for both the AA5352 and TH330 materials. Other forms of FLD may be used.

Journal of Physics: Conference Series **734** (2016) 022001 doi:10.1088/1742-6596/734/2/022001

Shell elements, solid elements, or solid-shell elements are recommended for this benchmark with careful control of the incremental punch stroke, with sufficient number of elements in the mesh to reproduce the curvature of the dies and capture plastic strain accurately. The analysis in this benchmark is highly nonlinear, including double sided contact with anisotropy. It is recommended that a simple isotropic material model (such as von-Mises yield function) be used before attempting an advanced anisotropic material model.

This benchmark study has the main objective of predicting the failure point after drawing, reverse redrawing and expansion. Different challenging outputs will be required:

i) prediction of earing after the reverse redrawing operation due to the plastic anisotropy of the material;
ii) prediction of the thickness profile after the reverse redrawing operation;
iii) prediction of the failure point after the expansion operation, using either strain-based, stress-based and/or polar FLD.

2. DESCRIPTION OF FORMING OPERATIONS

This section contains a description of the cup-forming and die expansion operations for this benchmark.

Cup forming is a two-stage process of drawing and reverse redrawing which occur sequentially during a single stroke of the drawing operation punch.

2.1 Drawing operation

- Drawing speed: 400.0 mm/sec (15.748 in/sec).
- Blank holder force between pressure pad and die: 21.1 kN or 4743.5 lbf. (For a quarter model, divide by 4.)
- Friction coefficient (recommended): 0.03

Journal of Physics: Conference Series **734** (2016) 022001 doi:10.1088/1742-6596/734/2/022001

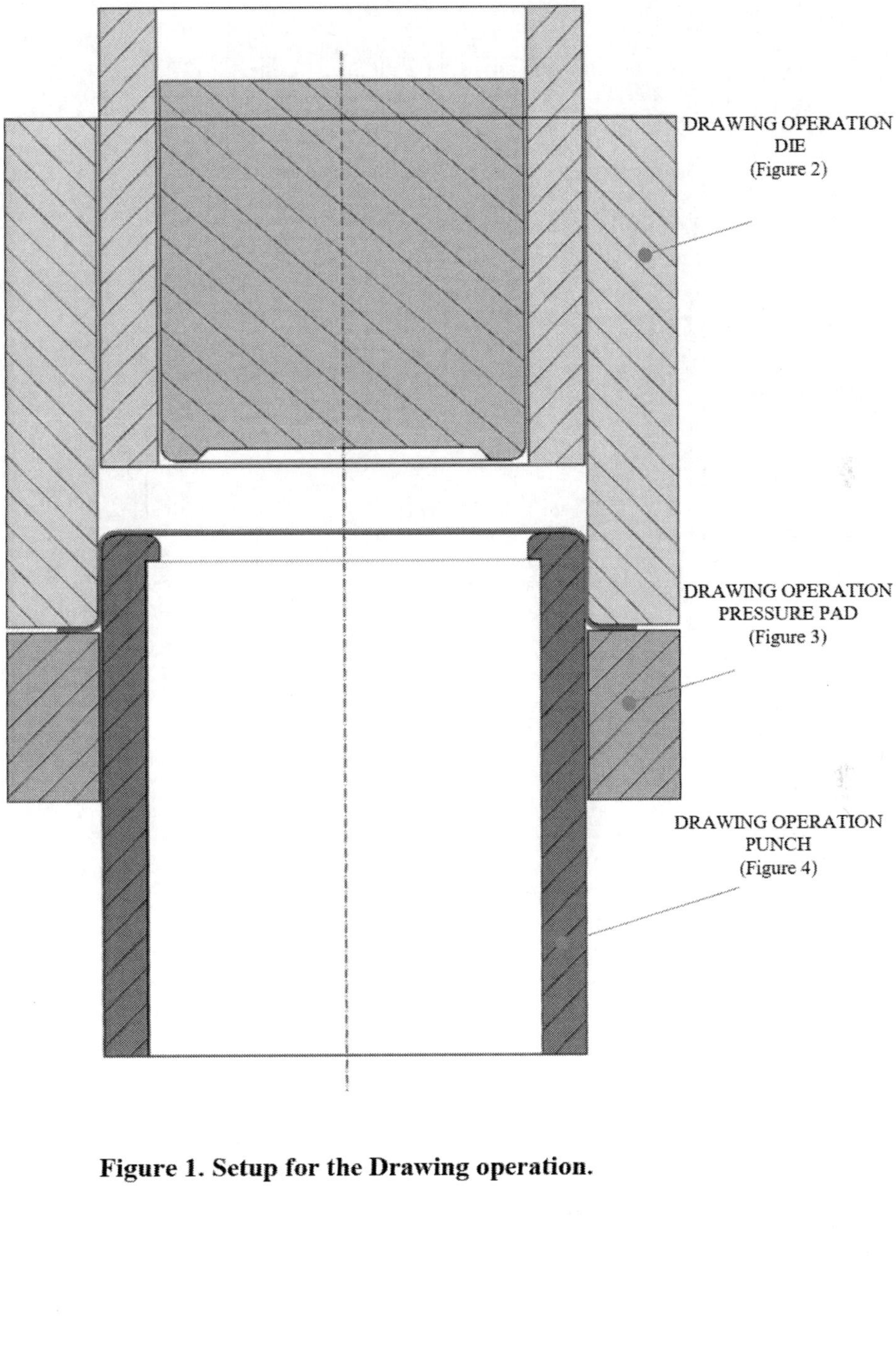

Figure 1. Setup for the Drawing operation.

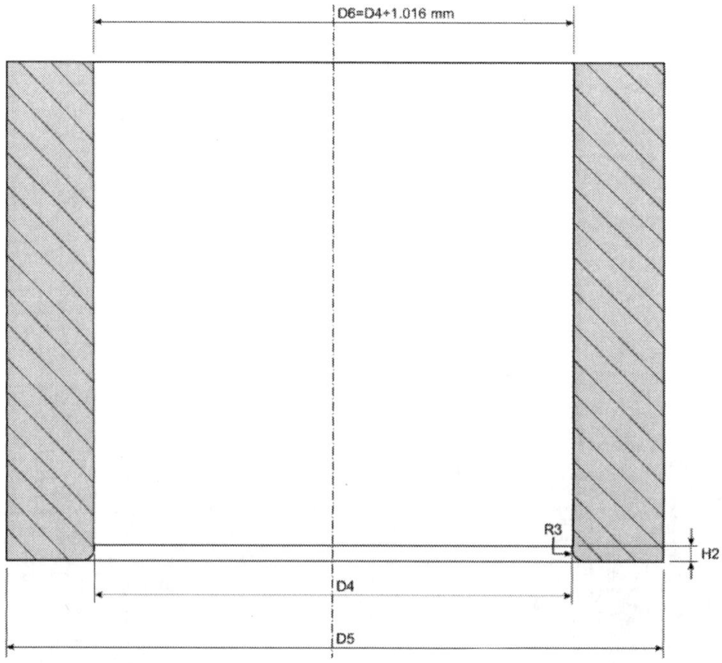

Parameter	D4	D5	R3	H2
mm	117.445	162.911	2.54	6.35
inch	4.6238	6.4138	0.1	0.25

Figure 2. Drawing operation – Die.

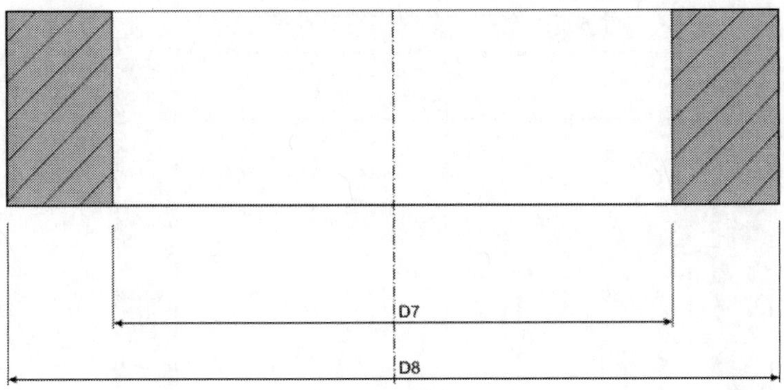

Parameter	D7	D8
mm	117.856	162.839
inch	4.64	6.411

Figure 3. Drawing operation – Pressure Pad.

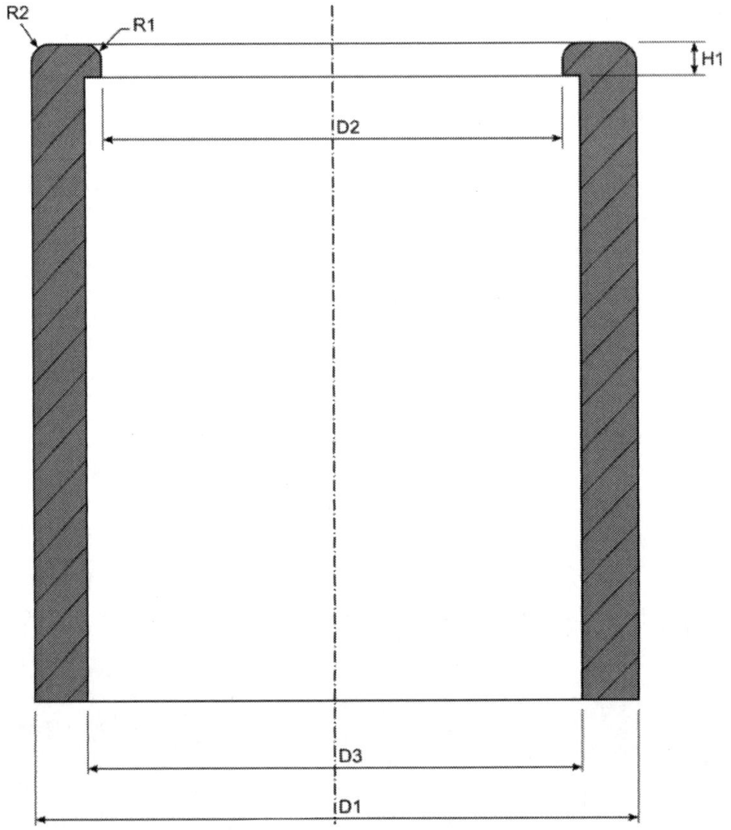

Parameter	D1	D2	D3	R1	R2	H1
mm	116.84	88.489	96.291	2.54	4.064	6.35
inch	4.6	3.4838	3.791	0.1	0.16	0.25

Figure 4. Drawing operation – Punch.

2.2 Reverse Redrawing operation

- Reverse redrawing occurs after the cup is fully drawn.
- Figure 5 shows a schematic representation of the tools used in the reverse redrawing stage, with the cup part-formed.
- The geometry and dimensions for the pressure pad and reverse redrawing punch are shown in Figures 6 and 7 respectively.
- The forming speed remains at 400.0 mm/sec (15.748 in/sec).
- Reverse redrawing displacement defined as zero when where the reverse redrawing punch contacts the drawn cup.
- Blank holder force: 16.6 kN or 3731.8 lbf. For a quarter model, divide by 4.
- Friction coefficient (recommended): 0.03

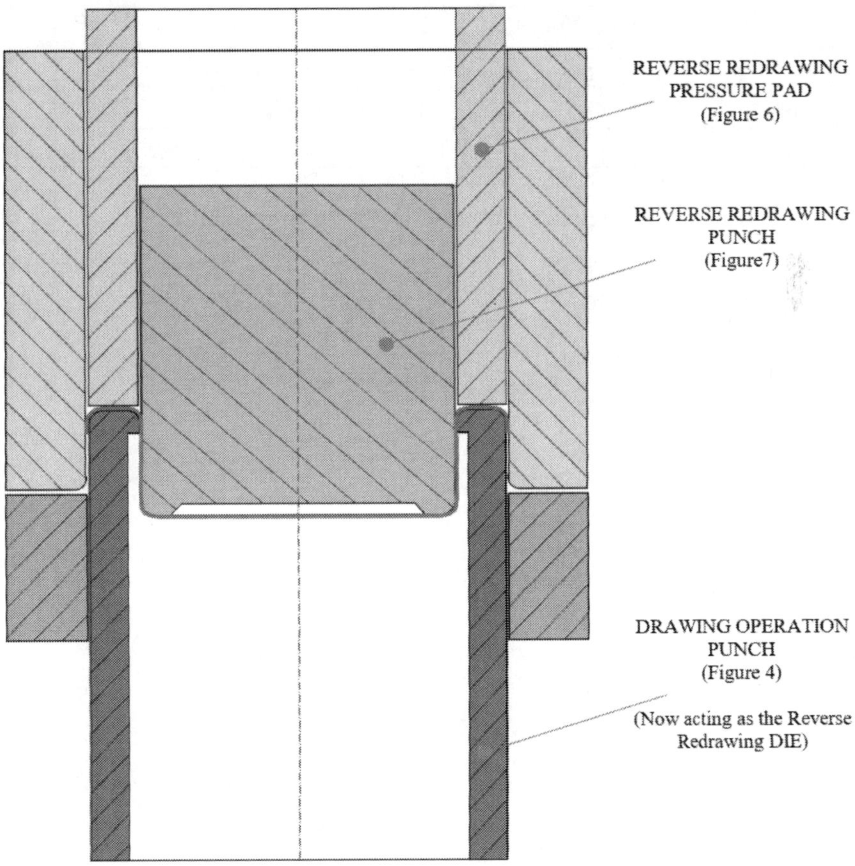

Figure 5. Setup for the Reverse Redrawing operation.

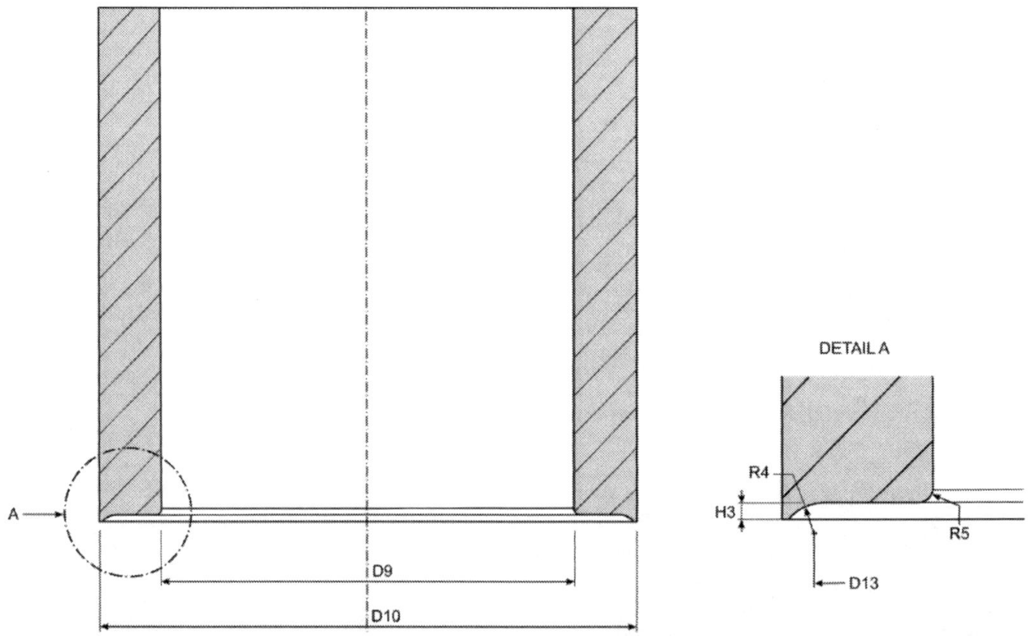

Parameter	D9	D10	D13	R4	R5	H3
mm	89.408	116.713	108.712	4.572	1.524	1.524
inch	3.52	4.595	4.28	0.18	0.06	0.06

Figure 6. Reverse Redrawing operation – Pressure Pad.

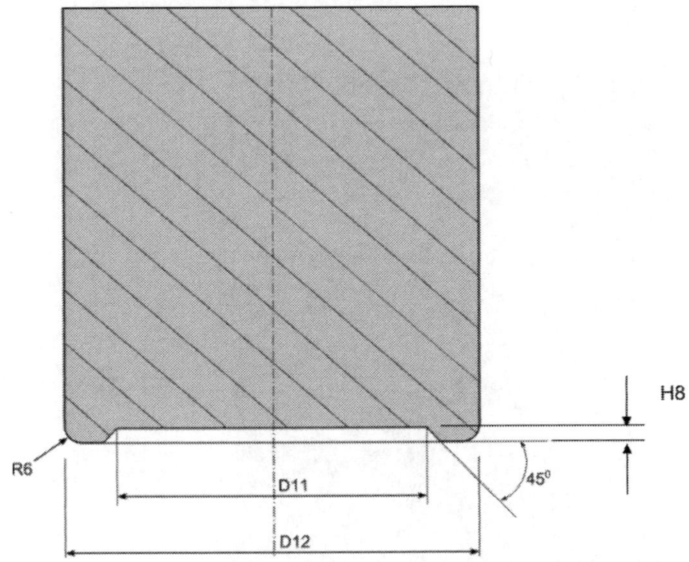

Parameter	D11	D12	R6	H8
mm	65.532	87.859	4.064	3.05
Inch	2.58	3.459	0.16	0.12

Figure 7. Reverse Redrawing operation – Punch.

2.3 Die Expansion operation

Practical experiment has shown that the force required to expand the steel cup to fracture will cause metal to flow into the base of the cup. This causes wrinkles in the base and reduces the height of the cup. To prevent this metal flow, the base was clamped using an inner clamp plate bolted through the centre of the cup base as shown in Figure 8.

The benchmark participants shall decide how to define the numerical boundary conditions representing experimental clamping conditions. The same clamp arrangement is used for both materials - steel and aluminium.

- The cup sits in the cup support tool.
- The clamp plate is bolted through the centre of the cup to hold the material securely in the base of the cup to prevent material flow.
- The cup has a 9 mm diameter hole drilled through the centre of the base.
- The expansion punch is driven into the cup until the point of fracture.
- A vent on the punch prevents internal air pressure build up.
- Punch speed is 200 mm per minute (3.33 mm/sec)
- Friction coefficient (recommended): 0.03

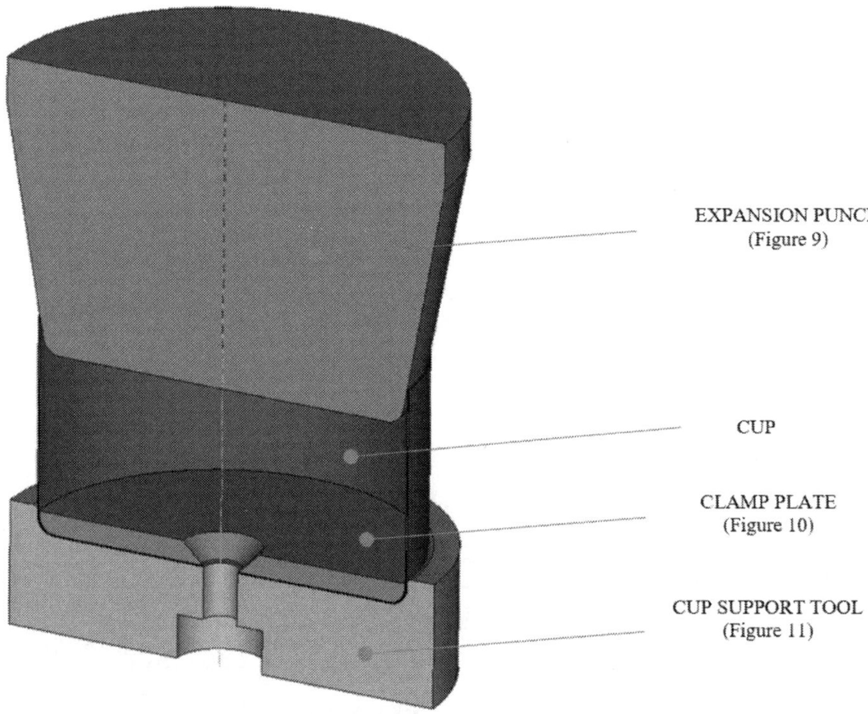

Figure 8. Practical setup for the die expansion operation.

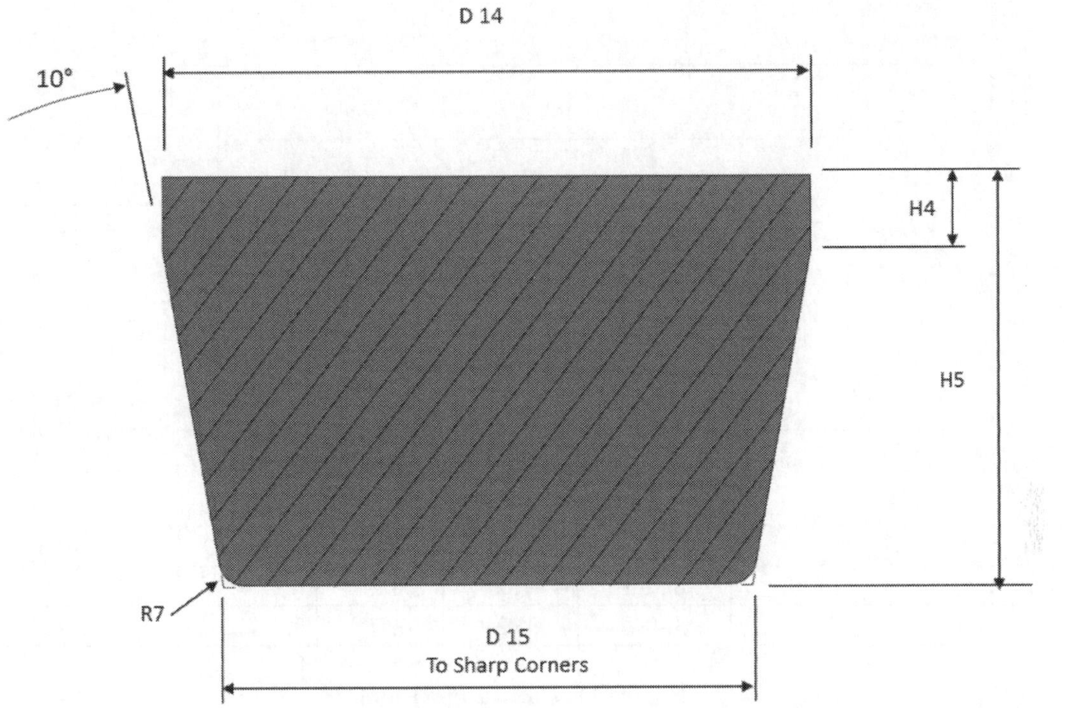

Parameter	D14	D15	H4	H5	R7
mm	104.43	85.04	12.0	66.50	4.0
inch	4.111	3.348	0.472	2.618	0.157

Figure 9. Expansion operation – Punch.

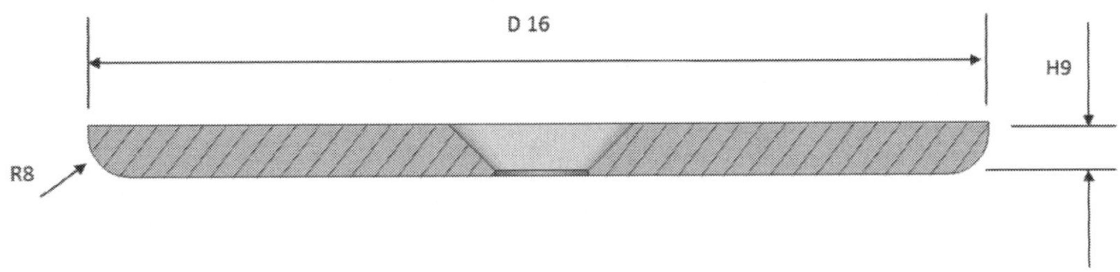

Parameter	D16	R8	H9
mm	87.77	3.86	5.00
inch	3.456	0.152	0.197

Figure 10. Clamp plate

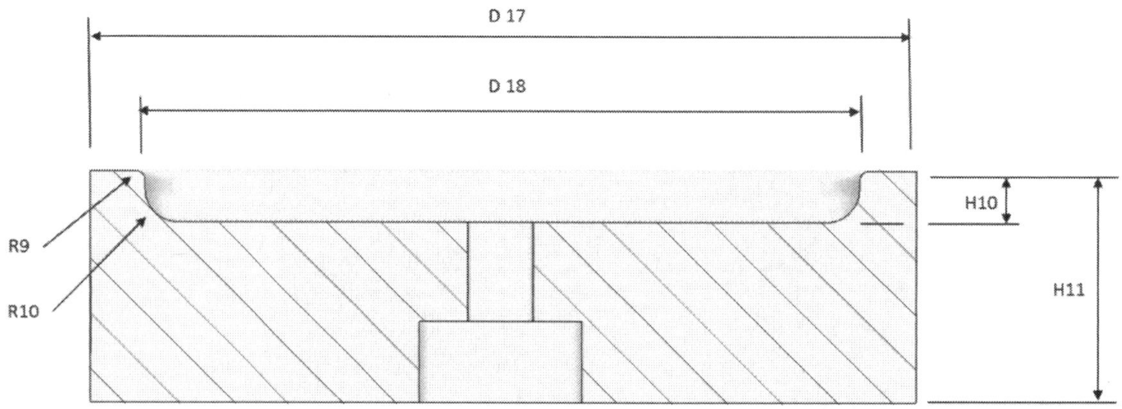

Parameter	D17	D18	R9	R10	H10	H11
mm	100	88.43	1.00	4.28	6.50	28.50
inch	3.947	3.481	0.039	0.169	0.256	1.122

Figure 11. Cup support tool

3. BLANK MATERIALS

- AA5352 aluminium and TH330 steel are considered for this benchmark.
- Blank diameter for both materials: 162.9664 mm / 6.416 in
- Thickness: 0.279 mm / 0.0110 in for AA5352; 0.270 mm / 0.0106 in for TH330
- Material properties: see tables in Section 5 of this document.
- The tooling geometry shown in Figures 1 to 11 is used for both materials.

4. BENCHMARK REPORT

All results are expected to be reported using the benchmark report template, which can be downloaded from the conference website, and when completed, uploaded to the website at a later date to submit the entry.

The primary metric of the benchmark will be the degree of correlation with the location, punch depth, and the plane strain tensor components of the onset of failure. The additional data reported will be used to understand the origin and significance of factors contributing to the success or failure of the correlation obtained with respect to the primary metric.

4.1 General Description

1) Benchmark participant: name, affiliation, address, email and phone number
2) Simulation software: name of the FEM code, general aspects of the code, basic formulations, element/mesh technology, type of elements, number of elements, contact property model and friction formulation
3) Simulation hardware: CPU type, CPU clock speed, number of cores per CPU, main memory, operating system and total CPU time
4) Material model: Yield function/Plastic potential, Hardening rule and Stress-Strain Relation, strain-based, stress-based or Polar Forming Limit Diagram (FLD) used
5) Remarks

4.2 Simulation results required for AA5352 aluminium and for TH330 steel

1) Cup height "h" (mm) after the reverse redrawing operation, measured from the centre point of the lower surface of the blank around the circumference from the rolling direction (0°) to 360°, reported in 5° increments or less.
2) Plot of punch load (kN) vs punch stroke (mm) for the cup forming operation (both drawing and reverse redrawing). The zero punch stroke is defined at the position when the punch makes initial contact with the blank with no interaction forces.
3) The strain history (Major and Minor principal strains) of the upper surface of the blank as a function of the drawing punch stroke (drawing and reverse redrawing) and the expansion punch stroke. This should be reported for the element at the leading edge of the cup at 0°, 45° and 90° (points 1, 2 and 3 in Figure 12.)
4) Thickness profile at a height of 20mm and 45mm from the base of the cup, after the reverse redrawing operation, measured around the circumference from the rolling direction (0°) to 360°, reported in 5° increments or less.
5) The expansion punch stroke at onset of failure (fracture or necking), and the location of the onset of failure, "*Point 4*"; reported as angle θ from the rolling direction and the height of the

cup at this location, measured from the base of the cup. Zero expansion punch stroke is defined at the position when the punch makes initial contact with the cup with no interaction forces.

6) The strain history (Major and Minor principal strains) as a function of punch stroke for the predicted onset failure point (point 4).

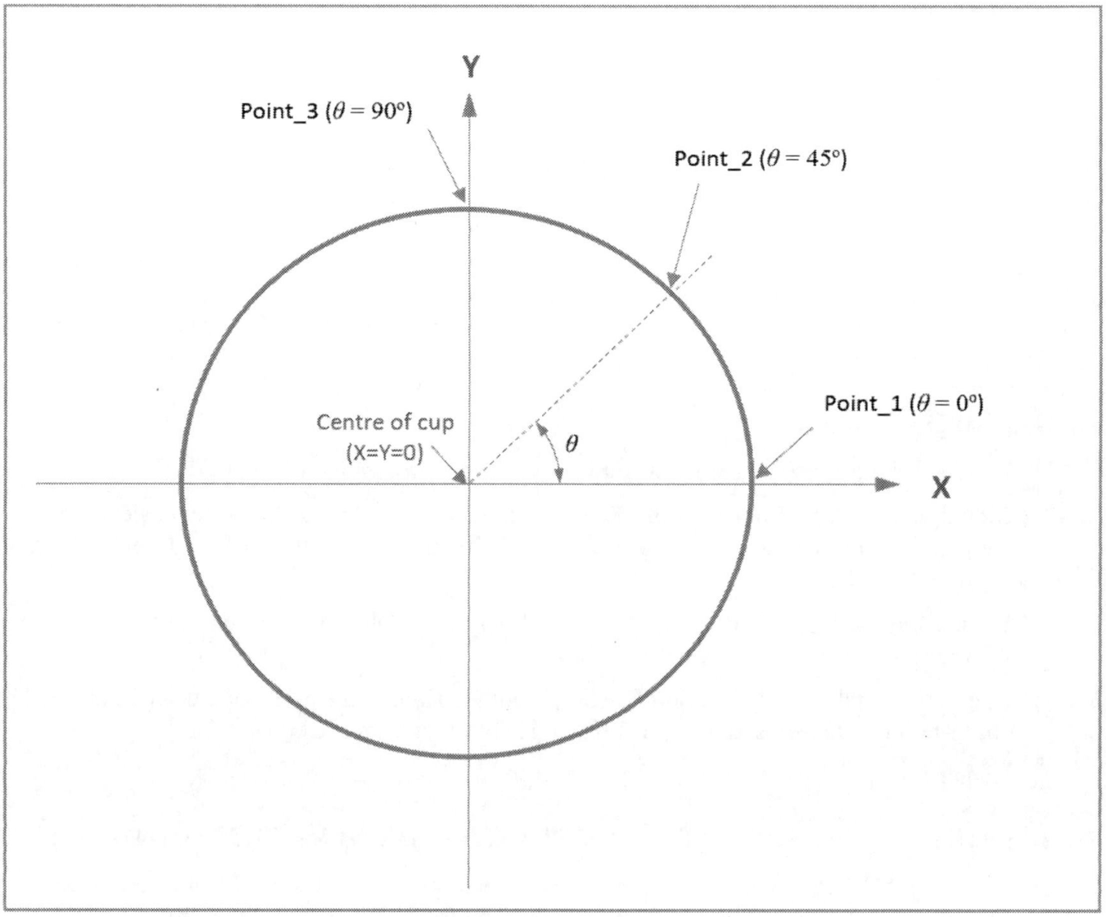

Figure 12. Schematic view of the location of the required points on the cup.

5. MATERIALS CHARACTERIZATION

5.1 Aluminum AA 5352

TABLE 1.1. Elastic mechanical properties

Sample	Density	Young's modulus	Poisson's ratio
	g/cm^3	GPa	
AA5352	2.72	68.5	0.33

Table 1.2. Uniaxial Tension Test Data

Test Direction	YS	UTS	Engineering Strain (%)		r value
	MPa	MPa	Uniform	Total	
0°	197.59	239.29	8.219	11.23	0.535
15°	193.39	238.45	8.964	12.16	0.465
30°	196.62	236.21	9.516	12.62	0.655
45°	193.94	231.54	9.580	13.09	1.105
60°	197.03	234.19	9.772	15.09	1.415
75°	196.89	235.38	11.602	17.96	1.595
90°	198.74	239.04	11.032	16.11	2.270

Table 1.3. Uniaxial Tension Test Data

Test Direction	Swift $\sigma = k(\varepsilon_0 + \bar{\varepsilon}^p)^n$			Voce $\sigma = A - B \cdot \exp(-C \cdot \bar{\varepsilon}^p)$			Hollomon $\sigma = K(\bar{\varepsilon}^p)^n$		Ludwik $\sigma = YS + K(\bar{\varepsilon}^p)^n$	
	k	ε0	n	A	B	C	n	K	n	K
	MPa			MPa	MPa			MPa		MPa
0°	376.59	0.01334	0.1594	285.41	91.88	15.5271	0.1036	333.28	0.6433	293.01
15°	373.76	0.01564	0.1640	283.34	91.05	14.5781	0.1030	326.66	0.6426	283.76
30°	365.19	0.01598	0.1552	283.55	88.44	14.0703	0.1018	326.11	0.6372	264.13
45°	347.94	0.01529	0.1439	276.53	84.11	14.2729	0.0951	313.89	0.6341	244.62
60°	350.22	0.01896	0.1448	278.71	83.53	13.7658	0.0943	315.99	0.6153	240.19
75°	359.58	0.01694	0.1535	286.96	90.77	12.0959	0.0967	321.79	0.6457	256.66
90°	365.67	0.01675	0.1537	286.88	89.46	13.7037	0.0992	328.26	0.6448	272.27

TABLE 1.4. Equal Biaxial Tension Test Data

t, mm	Maximum		Swift $\sigma = k(\varepsilon_0 + \bar{\varepsilon}^p)^n$			Voce $\sigma = A - B \cdot \exp(-C \cdot \bar{\varepsilon}^p)$			Hollomon $\sigma = K(\bar{\varepsilon}^p)^n$	
	σ	ε	K	ε_0	n	A	B	C	n	K
	MPa		MPa			MPa	MPa			MPa
0.279	304.24	0.1200	378.43	0.0006	0.10120	314.53	81.74	17.1051	0.1000	377.53

TABLE 1.5. Disk Compression Test Data

Alloy/Temper	r_{BX}
AA 5352	0.62

TABLE 1.6a. Material Constants for Yield Function CPB06 (a=2.0)

C_{11}	C_{12}	C_{13}	C_{22}	C_{23}	C_{33}	C_{44}	C_{55}	C_{66}	k
0.77037	-0.32531	-0.342559	1.273241	0.13047	2.175042	1	1	1.329576	-0.414982

TABLE 1.6b. Material Constants for Yield Function CPB06ex2 (a=3.0)

C_{11}	C_{12}	C_{13}	C_{22}	C_{23}	C_{33}	C_{44}	C_{55}	C_{66}	k
0.212707	-0.015974	0.307484	0.924971	0.78806	1.817972	1	1	0.73172	-0.283913
C'_{11}	C'_{12}	C'_{13}	C'_{22}	C'_{23}	C'_{33}	C'_{44}	C'_{55}	C'_{66}	k'
1.07512	0.207383	0.014346	1.220567	0.086062	0.06489	1	1	0.946897	-0.075132

TABLE 1.7 Material Constants for Yield Function Yld2004-18p (a=8.0)

A1	A2	A_3	A_4	A_5	A_6
0.6402	0.7462	0.5014	0.0636	1.0439	0.2379
A_7	A_8	A_9	A_{10}	A_{11}	A_{12}
0.9089	0.9140	0.8654	0.6105	1.0386	1.1726
A_{13}	A_{14}	A_{15}	A_{16}	A_{17}	A_{18}
1.3065	0.3464	1.4753	0.9812	0.9410	1.0763

TABLE 1.8. Material Constants for Yield Function Yld2000-2d (a=8.0)

Sample	α_1	α_2	α_3	α_4	α_5	α_6	α_7	α_8
AA5352	0.6547	1.2883	0.9050	0.8641	0.9710	0.6820	1.0091	1.2084

TABLE 1.9. Material Constants for Yield Function Yld89 r-value based (m=8.0)

Sample	a	c	h	p
AA5352	1.0162	0.9838	0.7086	0.8961

TABLE 1.10. Material Constants for Yield Function Yld89 stress value based (m=8.0)

Sample	a	c	h	p
AA5352	0.3401	1.6599	1.0052	1.0315

TABLE 1.11. Material Constants for Yield Function Hill, r-value based (1948)

Sample	F	G	H	L	M	N
AA5352	0.3071	1.3029	0.6971	3.0	3.0	2.5841

TABLE 1.12. Material Constants for Yield Function Hill, stress-based (1948)

Sample	F	G	H	L	M	N
AA5352	0.7782	0.7572	1.2428	3.0	3.0	3.4687

TABLE 1.13. Forming Limit Curve for AA5352

Major True Strain Ɛ1	Minor True strain Ɛ2
0.14	-0.038
0.109	-0.012
0.09	0.005
0.076	0.012
0.078	0.026
0.11	0.092

Journal of Physics: Conference Series **734** (2016) 022001 doi:10.1088/1742-6596/734/2/022001

5.2 Steel TH330 unstoved.

TABLE 2.1. Elastic mechanical properties

Sample	Density	Young's modulus	Poisson's ratio
	g/cm^3	GPa	
TH330 Steel	7.8	205	0.3

TABLE 2.2. Uniaxial Tension Test Data

Test Direction	YS	UTS	Engineering Strain (%)		r value
	MPa	MPa	Uniform	Total	
0°	258.87	366.87	15.881	29.042	1.4492
15°	258.25	369.08	15.835	29.239	1.3734
30°	254.36	367.66	15.357	27.086	1.3012
45°	255.25	371.02	15.665	29.505	1.2664
60°	251.47	368.25	15.126	28.824	1.3350
75°	250.33	368.49	16.104	30.521	1.4434
90°	249.42	367.08	15.771	29.347	1.5100

TABLE 2.3. Uniaxial Tension Data

Test Direction	Swift $\sigma = k(\varepsilon_0 + \bar{\varepsilon}^p)^n$			Voce $\sigma = A - B \cdot \exp(-C \cdot \bar{\varepsilon}^p)$			Hollomon $\sigma = K(\bar{\varepsilon}^p)^n$		Ludwik $\sigma = YS + K(\bar{\varepsilon}^p)^n$	
	k	ε_0	n	A	B	C	n	K	n	K
	MPa			MPa	MPa			MPa		MPa
0°	575.58	0.00500	0.16200	506.63	250.51	7.6342	0.14946	568.01	0.35952	334.14
15°	573.69	0.00484	0.15941	505.96	251.22	7.7497	0.14800	566.95	0.35276	334.01
30°	576.85	0.00463	0.15867	508.46	254.19	7.8542	0.14786	570.49	0.34961	338.03
45°	577.59	0.00442	0.15561	514.28	259.86	7.5711	0.14551	571.63	0.34524	339.66
60°	576.02	0.00424	0.15600	512.92	260.80	7.5685	0.14639	570.29	0.34404	340.17
75°	575.59	0.00414	0.15723	511.69	263.13	7.6440	0.14792	570.11	0.34006	342.34
90°	575.45	0.00445	0.15567	512.98	259.61	7.5040	0.14625	570.02	0.34493	338.23

TABLE 2.4. Equal Biaxial Tension Test Data

t, mm	Maximum		Swift $\sigma = k(\varepsilon_0 + \bar{\varepsilon}^p)^n$			Voce $\sigma = A - B \cdot \exp(-C \cdot \bar{\varepsilon}^p)$			Hollomon $\sigma = K(\bar{\varepsilon}^p)^n$	
	σ	ε	K	ε_0	n	A	B	C	n	K
	MPa		MPa			MPa	MPa			MPa
0.270	579.86	0.4642	666.94	0.00775	0.17050	575.54	253.87	7.4950	0.1480	650.48

TABLE 2.5. Disk Compression Test Data

Alloy/Temper	r_{BX}
TH330 steel	0.984

TABLE 2.6. Material Constants for Yield Function CPB06 (a=2.0)

C_{11}	C_{12}	C_{13}	C_{22}	C_{23}	C_{33}	C_{44}	C_{55}	C_{66}	k
0.68903	-0.10408	-0.16354	0.79676	0.1931	-0.60226	1.0	1.0	0.85596	-0.1693

TABLE 2.7 Material Constants for Yield Function Yld2004-18p (a=6.0)

A1	A2	A3	A4	A5	A6
0.8226	0.3634	0.9005	0.8555	0.5851	0.6972
A7	**A8**	**A9**	**A10**	**A11**	**A12**
0.9704	0.9994	0.8117	1.1570	1.3698	1.1071
A13	**A14**	**A15**	**A16**	**A17**	**A18**
0.9483	1.1485	1.0410	1.0426	1.0273	1.1368

TABLE 2.8. Material Constants for Yield Function Yld2000-2d (a=6.0)

Sample	α1	α2	α3	α4	α5	α6	α7	α8
TH 330	1.0352	0.9875	0.8241	0.9284	0.9345	0.8019	0.9867	1.0585

TABLE 2.9. Material Constants for Yield Function Yld89, r-value based (m=6.0)

Sample	a	c	h	p
TH 330	0.8068	1.1932	0.9917	0.9570

TABLE 2.10. Material Constants for Yield Function Yld89, stress-based (m=6.0)

Sample	a	c	h	p
TH 330	0.5275	1.4725	0.9843	0.9793

TABLE 2.11. Material Constants for Yield Function Hill, r-value based (1948)

Sample	F	G	H	L	M	N
TH 330	0.7837	0.8167	1.1833	3.0	3.0	2.8262

TABLE 2.12. Material Constants for Yield Function Hill, stress-based (1948)

Sample	F	G	H	L	M	N
TH 330	0.7646	0.8267	1.1733	3.0	3.0	3.0543

Journal of Physics: Conference Series **734** (2016) 022001 doi:10.1088/1742-6596/734/2/022001

TABLE 2.13. Forming Limit Curve for TH330

Major True Strain $\varepsilon 1$	Minor True strain $\varepsilon 2$
0.541	-0.270
0.205	0.000
0.284	0.213
0.310	0.310

Benchmark 1 – Failure Prediction after Cup Drawing, Reverse Redrawing and Expansion

Part B: Material Characterisation

Martin Watson[a], Robert Dick[b], Y. Helen Huang[a], Andrew Lockley[a], Rui Cardoso[c], Abel Santos[d]

[a]*Crown Packaging, Oxfordshire OX12 9BP, UK*
[b]*Alcoa Technical Center, PA 16069-0001, USA*
[c]*Brunel University London, UB8 3PH, UK*
[d]*FEUP, University of Porto, 4200-465, Portugal*

Abstract. This report summarises the material testing on the two metals used in this benchmark study, AA5352 aluminium alloy and TH330 steel (unstoved). In order to characterise the material anisotropic plastic properties and to specify failure criteria, the tests carried out were uniaxial tensile, bulge and disc compression tests, as well as the tests for obtaining forming limit curves. The parameters / material constants required for various material models are also described in this report.

Keywords: material characterisation, uniaxial tensile, bulge, disc compression, anisotropy, r-value, FLC

1. INTRODUCTION

Two metals were selected for Benchmark 1 in Numisheet 2016; AA5352 aluminium alloy supplied by Alcoa USA and TH330 steel (unstoved) obtained from Tata Steel. The thickness is 0.279 mm / 0.0110 inch for AA5352, and 0.270 mm / 0.0106 inch for TH330. This report provides a description of the material testing carried out on both metals for characterising the material properties and for specifying failure criteria. Data fitting to obtain the material constants of various material models is also documented in this report.

2. MATERIAL TESTS

2.1 Uniaxial tensile tests

Uniaxial tensile tests were carried out on both metals at 15 degree increments starting from the rolling direction as zero degrees to 90 degrees to the rolling direction on the plane of a metal sheet, i.e. 7 tests on each metal. For aluminium AA5352, the tests were conducted at University Aveiro, Portugal, based on ASTM norm E8M Standard and at Crown Technology Centre based on ISO 6892-1:2009 Standard. A single tensile test was conducted at the SP Technical Research Institute of Sweden using Digital Image Correlation (DIC) analysis and the results compared well with those obtained from mechanical test. Similarly, the tests on steel TH330 were performed at Tata Steel according to ISO 6892-1:2009 Standard and repeated at University Aveiro and at Crown. The test data were used to obtain the mechanical properties of the material such as yield stress, hardening coefficient, uniform elongation and r-value, etc.

The r-values were determined between 2% and 20% strain or between 2% and uniform elongation when the uniform elongation was lower than 20%.

Journal of Physics: Conference Series **734** (2016) 022001 doi:10.1088/1742-6596/734/2/022001

2.2 Bulge tests

Thin sheet metal of such small gauge as used in this Benchmark study usually splits at a strain of approximately 10% for Aluminium and 20% for steel in a uniaxial test. In order to obtain the material data at higher strains and to reduce discrepancies on data extrapolation of hardening curves, hydraulic bulge tests are commonly used to extend the tensile test data. The tests were performed at University of Porto, Portugal, and at Alcoa for AA5352. For TH330 the tests were conducted at both Tata Steel and University of Porto.

Figure 1 shows the setup of the bulge test system used in the University of Porto, as well as the measurement and variables used to determine the stress and strain. The full detailed description of the theory for stress and strain measurement can be found in [1], [2] and [3]. The equibiaxial stress is determined by

$$\sigma_b = \frac{p\rho}{2t} \tag{1}$$

where p is hydraulic pressure and t is the sheet metal thickness. The radius of curvature ρ is calculated by

$$\rho = \frac{(D_{cv}/2)^2 + h^2}{2h} - \frac{t}{2} \tag{2}$$

where D_{cv} is the diameter defined by the spherometer and h is the difference between the spherometer support and displacement transducer as illustrated in Figure 1. A correction is performed for half thickness of the sheet, since the calculation is done at the external surface of the cap. The current thickness (t) of the sample can be obtained through Eq. 3, knowing the initial thickness (t_0) and the current thickness strain ε_t.

$$t = t_0 \cdot \exp(-\varepsilon_t) \tag{3}$$

Concerning thickness strain, the following equation is used:

$$\varepsilon_t = -2 \cdot \ln\left(\frac{D_{st}}{D_{st0}}\right) \tag{4}$$

where D_{st} and D_{st0} are also defined in

Figure 1.

The bulge test on steel TH330 carried out in Tata Steel followed procedure ISO Standard 16808:2014. Figure 2 shows the results of the bulge test in which 3D digital image correlation (DIC) was applied to determine the in-plane strains and the bulge height. True stress and strain were reported in all the bulge tests.

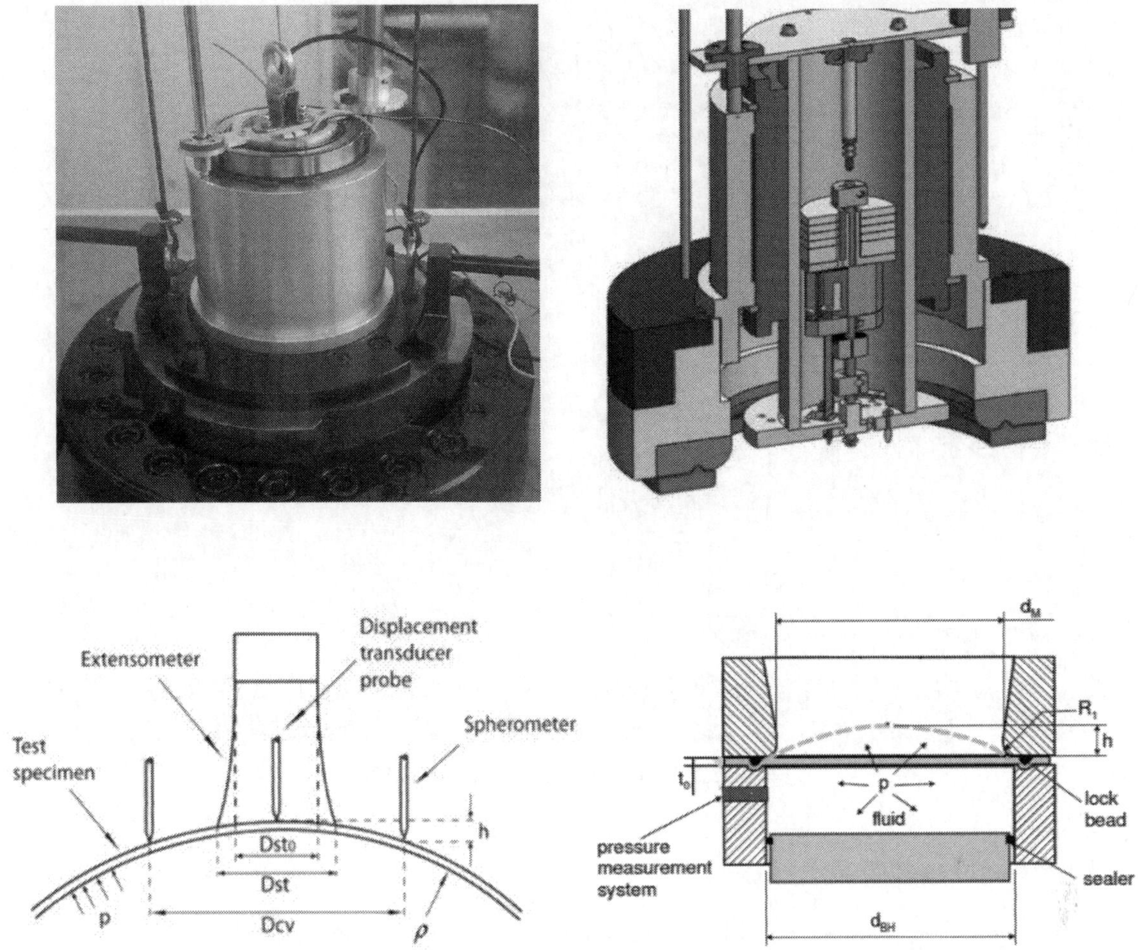

Figure 1: Hydraulic bulge test system and variables for stress-strain determination

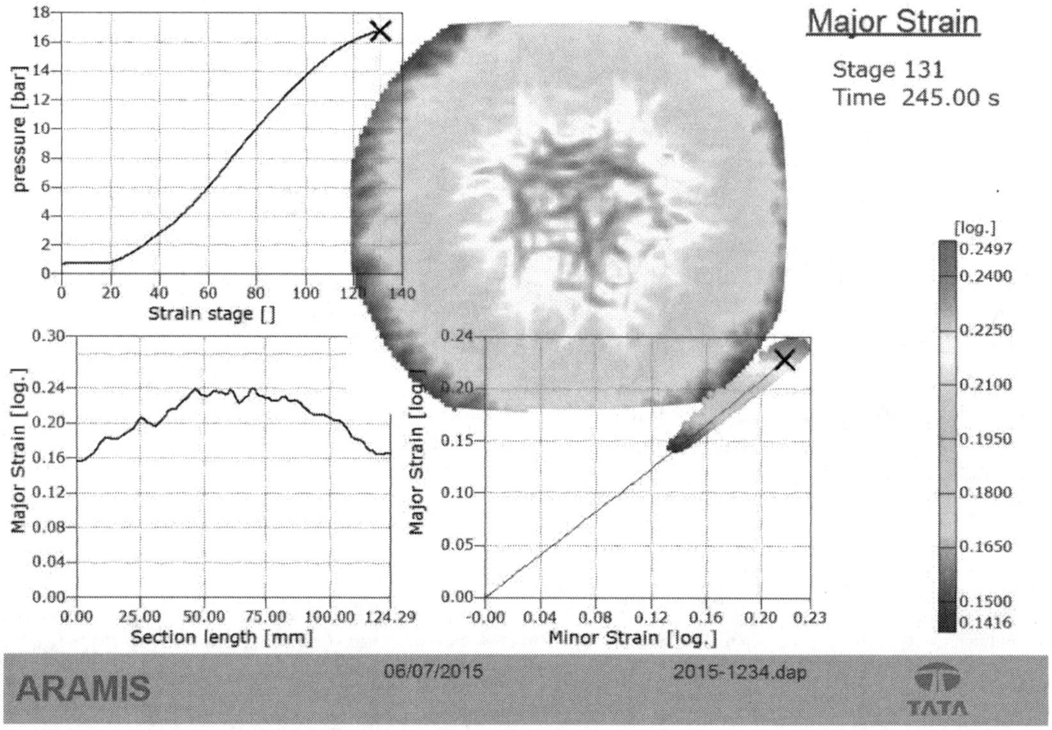

Figure 2. Bulge test result for TH330

2.3 Disk compression tests

For anisotropic yield models, the biaxial r value (r_b) is required and obtained from disk compression tests (**Figure 3**). For steel TH330, the tests were conducted at CERTETA with Prof. Dorel Banabic and at the University of Porto. For aluminium AA5352, the tests were conducted at the University of Porto and at Alcoa.

Different disk dimensions were used in the various test labs, ranging from 9 mm to 15 mm in diameter. Teflon foil and lubricant were used in both sides of samples to minimise friction effect.

The true total strains along the rolling direction of the metal and the transverse direction were measured and calculated. The biaxial r value (r_b) is independent of the amount of thinning of the disk and determined using

$$r_b = \frac{\sum_{i=1}^{N} \varepsilon_{RD,i}^{(p)} \varepsilon_{TD,i}^{(p)}}{\sum_{i=1}^{N} \varepsilon_{RD,i}^{(p)} \varepsilon_{RD,i}^{(p)}}. \tag{5}$$

where $\boldsymbol{\varepsilon}_{\boldsymbol{RD}}^{(\boldsymbol{p})}$ and $\boldsymbol{\varepsilon}_{\boldsymbol{TD}}^{(\boldsymbol{p})}$ are the plastic parts of the logarithmic strain of the i-th specimen (i = 1, …, N) in the rolling and transverse direction, respectively [4].

Figure 3. Disk compression test setup

2.4 Forming Limit Curve (FLC)

The Forming Limit Curve for aluminium AA5352 was provided by Alcoa USA based on the concept of the forming limit diagram (FLD) first introduced by [5] and [6]. This diagram is usually plotted on axes representing the major (ε_1) and the minor (ε_2) strains in the plane of a sheet and corresponds to the maximum admissible local strains achievable just prior to the occurrence of visible defects in the sheet metal like necking and fracture.

The experimental technique to determine the FLD involves subjecting specimens of the considered sheet metal to different in-plane strain states, by simple tensile testing or stretching over a hemispherical punch and is known as the Nakajima test (Figure 4). Different deep draw and stretch forming conditions (from biaxial tension to plane strain to uniaxial tension) are achieved in the sheet metal by varying the specimen width. The tests on the AA5352 material followed the ISO 12004-2:2008 standard and used a 101.6 mm ball traveling at 1.5mm/s with strains being measured using an Aramis 5M system.

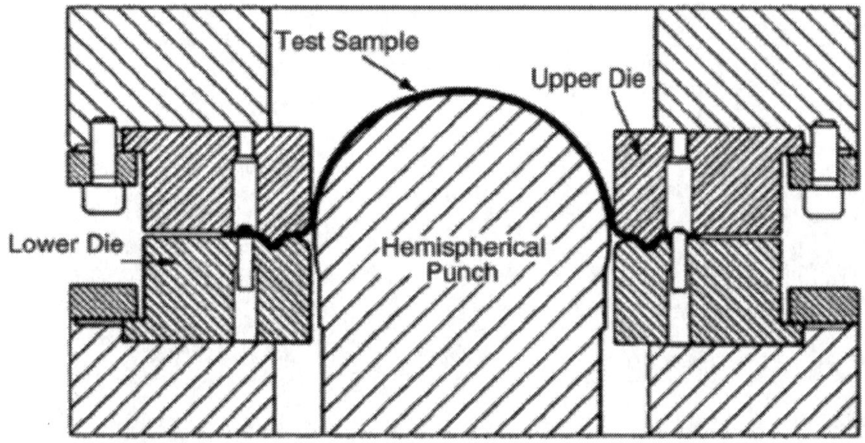

Figure 4. A schematic view of Nakajima test [7]

The Forming Limit Curve for steel TH330 was provided by Tata Steel using the Abspoel & Scholting method [8] in which four points are chosen to define the FLC, i.e., uniaxial tension necking point (TE), plane strain point (PS), intermediate biaxial stretch point (IM) and equi-biaxial stretch point (BI) (see Figure 5). The major and minor strains of these four points can be calculated based on the mechanical properties of an A_{80} tensile bar (total elongation and r-values in transverse direction plus the minimum A_{80} of the three directions, i.e., transverse, diagonal or longitudinal to rolling direction).

The uniaxial tension local necking points were obtained using an MTS 300 test bench with a GOM Aramis optical strain measurement system. The samples were measured transverse to the rolling direction. The other characteristic points were derived from Nakazima tests according to ISO12004-2: 2008, performed on an Erichsen model 145/60 laboratory press. The samples were measured transverse to the rolling direction. All measured strains in the Nakazima tests were corrected to the mid-plane. Figure 5 shows the fracture test data and the derived forming limit curve for TH330.

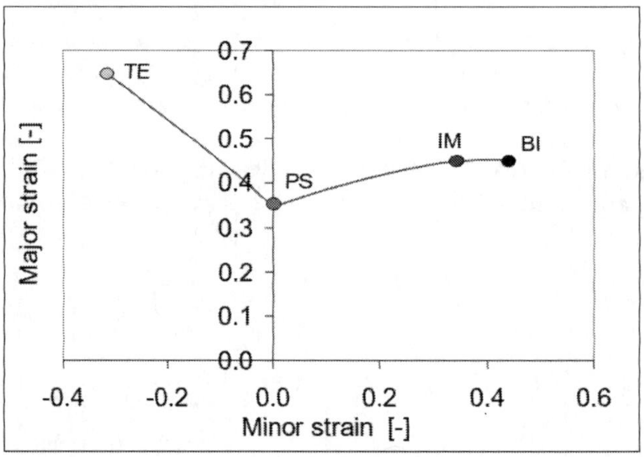

Figure 5. The points chosen for defining FLC [8]

The three different notched tensile fracture experiments were performed on an MTS 300 test bench with a GOM Aramis optical strain measurement system, where the specimens are "gridded" with a random speckle pattern (Figure 6). The deformation is filmed with two cameras and with the digital image correlation (DIC) method and the 3D coordinates of the speckle pattern are converted into strain components or principal strains contour plots. For every Aramis optical strain measurement, the used facet size is 19 pixels by 19 pixels with a step size of 9 pixels by 9 pixels, which results into an overlap of 53%.

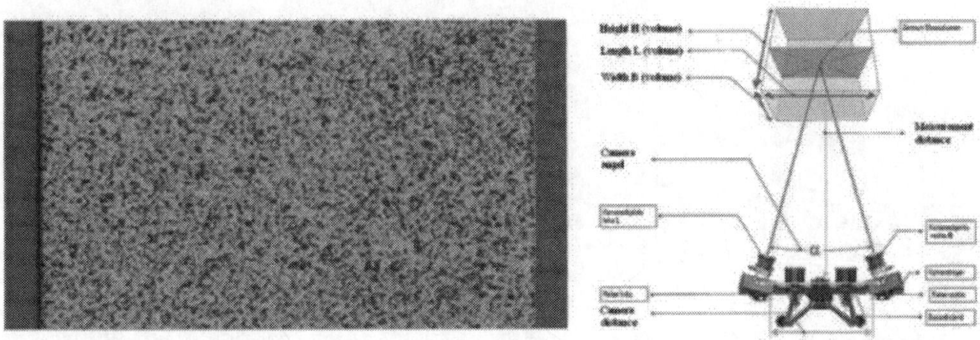

Figure 6. a) Speckle pattern, b) GOM Aramis system

The geometry of the three different notched tensile specimens are shown in Figure 7. The obtained spatial resolution with the GOM Aramis optical measurement system is approximately 0.3 mm for the notched tensile specimens. A remark should be made here regarding the edge of the notched tensile fracture experiments, the edges are milled and polished with a P1000 sand paper. A stack of sheets was magnetically clamped and milled into shape using a die, because a single sheet of the TH330 material is too thin to mill to these geometries.

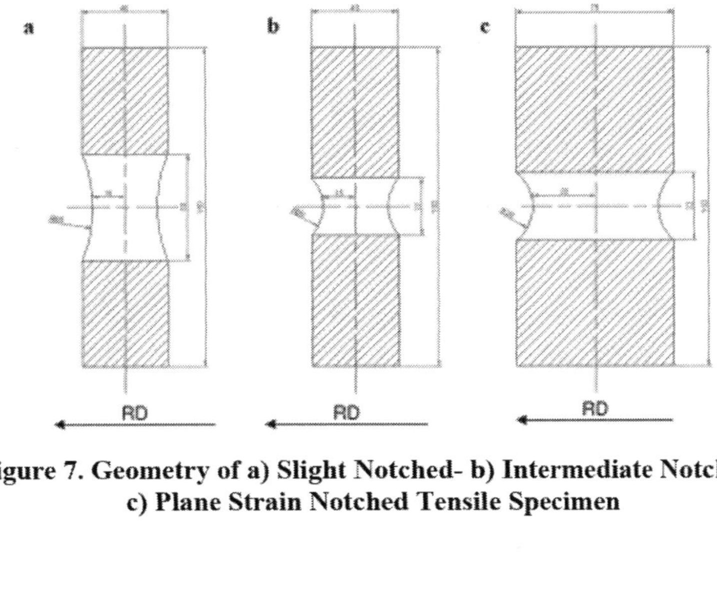

Figure 7. Geometry of a) Slight Notched- b) Intermediate Notched- c) Plane Strain Notched Tensile Specimen

The Nakajima fracture experiments were performed according to ISO12004-2: 2008, on an Erichsen model 145/60 laboratory press using a GOM Aramis optical strain measuring system (Figure 8). The obtained spatial resolution for the Nakajima experiments with the GOM Aramis optical measurement system is approximately 0.4 mm.

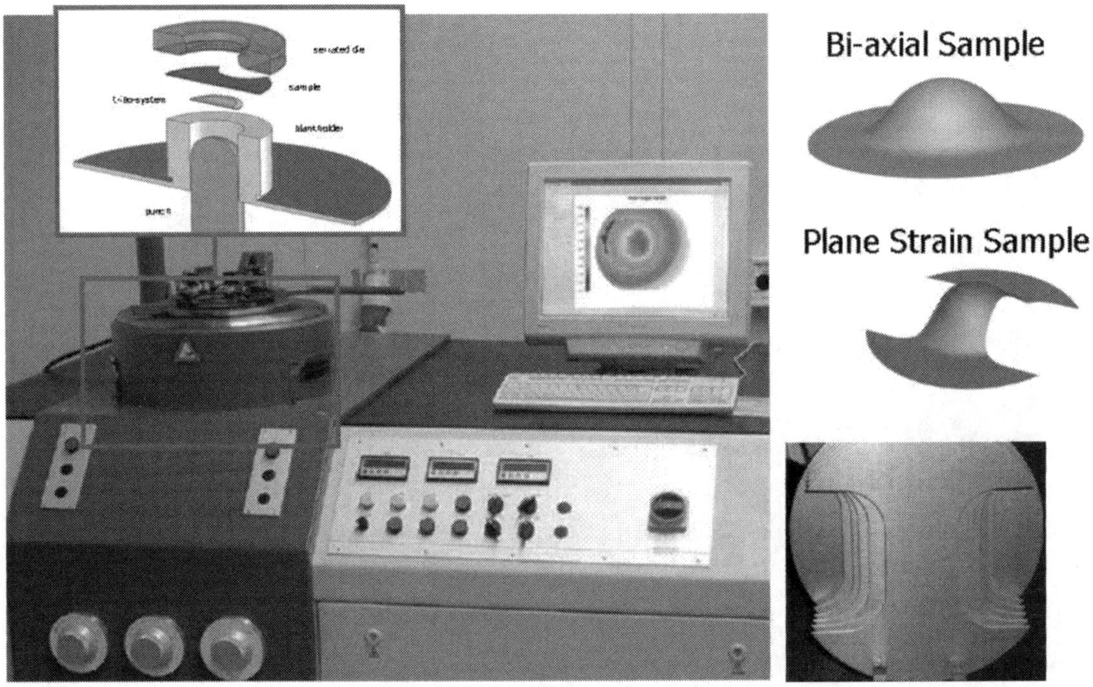

Figure 8. Test Setup Nakajima Experiments

The in-plane fracture experiments were performed on the MTS 300 test bench and the geometry is shown in Figure 9. Normally, the groove is milled at half the thickness of the sheet, but the TH330 material is approximately 0.27 mm thick. Therefore a stack of sheet were glued with DP490 (3M Scotch-weld). The glued strips are placed in a holder and the stacked block is compressed with a PST hydraulic press at a load of 10 kN for 3 hours for the adhesive to cure. The specimen is left at room temperature for 12 hours to acquire the full bonding strength.

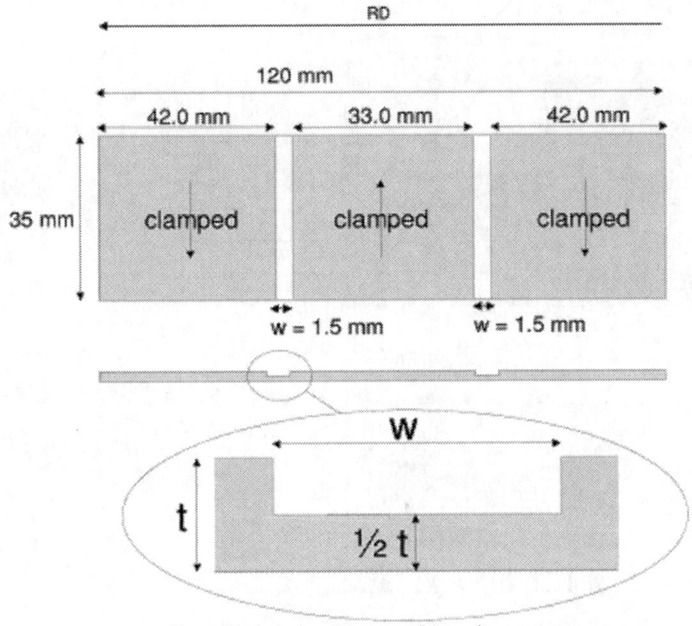

Figure 9. Geometry In-Plane Shear Fracture Specimen

For the in-plane shear fracture test the maximum shear angle is measured with a macroscope (Figure 10). The front of the in-plane shear fracture test sample has a groove with a width of 1.5 mm, where 3 scribes are made on the back. One scribe is in the middle of the sample and the two other scribes are made at 5 mm from the edge of the sample, see top right picture in Figure 10. After the macroscope has been calibrated and the in-plane shear fracture test sample is properly aligned, the maximum shear angle can be measured with the Leica measuring software as shown in the pictures on the left side of Figure 10.

The maximum shear angle at fracture is then converted into principal fracture strains. All the fracture experiments were measured transverse to the rolling direction and on the surface of the specimen. Figure 11 shows the fracture test data and the derived Forming Limit Curve (FLC) for TH330.

Journal of Physics: Conference Series **734** (2016) 022001 doi:10.1088/1742-6596/734/2/022001

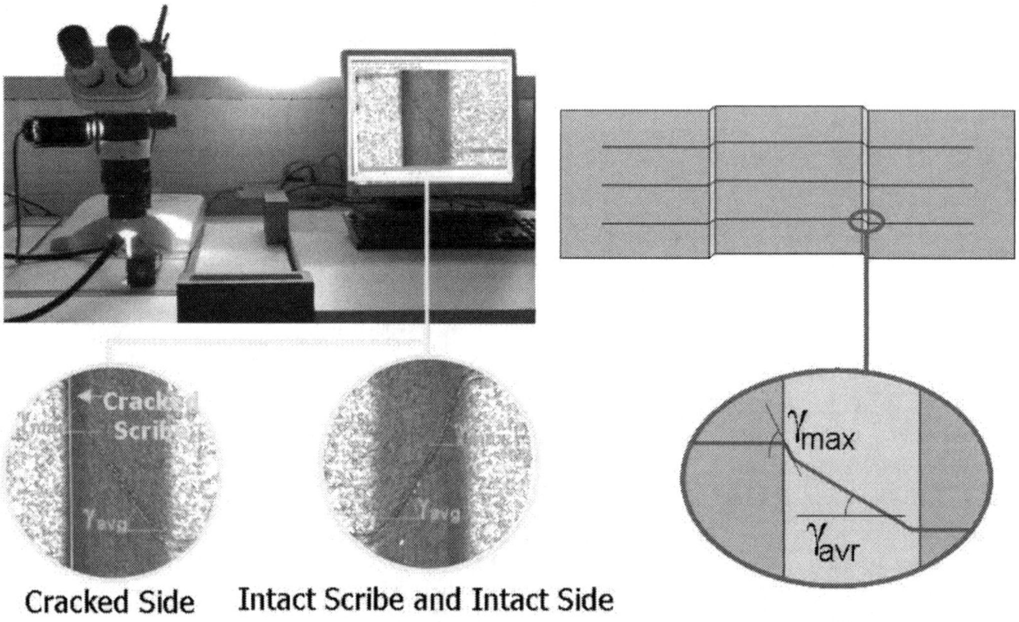

Figure 10. Maximum Shear Angle Measurements with a macroscope

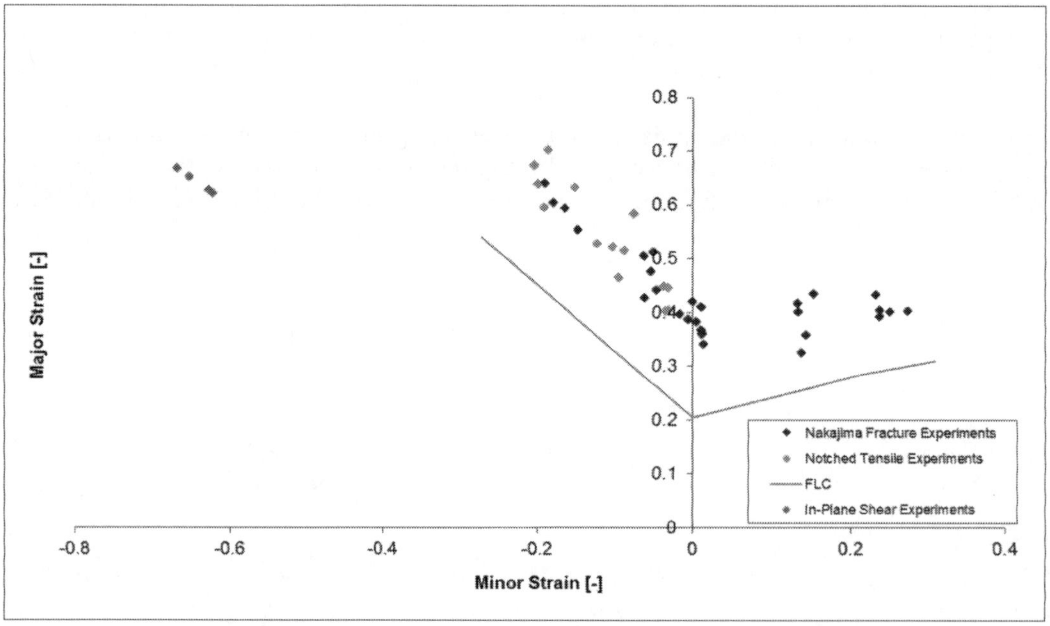

Figure 11. The FLC of TH330 and the fracture experiment data

3. DATA FITTING

3.1 Stress-strain curves and hardening coefficients

The raw data obtained from the uniaxial tensile tests were mainly load-deformation datasets. The engineering stress and strain data as well as the corresponding true stress-strain data were calculated based on the dimensions of individual specimen depending on different standards used in the tests. The yield stress was determined as proof stress at 0.1% plastic strain from the true stress-strain curve and averaged from all the sample data of each test.

Uniaxial tensile tests gave maximum uniform strains of approximately 10% and 16% for aluminium and steel respectively. The bulge test data were scaled such that they overlay the uniaxial test data and plastic work is conserved. This scaled bulge test data was used to extend the uniaxial test data beyond the point of maximum uniform strain to higher levels of strain [9]. The extended true stress-strain curves were used for data fitting and determination of material constants. Parameters for the following hardening functions were calculated using a least square fitting algorithm applied to the data.

Swift law: $$\sigma = k(\varepsilon_0 + \bar{\varepsilon}^p)^n \quad (6)$$

Voce law: $$\sigma = A - B \cdot \exp(-C \cdot \bar{\varepsilon}^p) \quad (7)$$

Hollomon's equation: $$\sigma = K(\bar{\varepsilon}^p)^n \quad (8)$$

Ludwik's equation: $$\sigma = YS + K(\bar{\varepsilon}^p)^n \quad (9)$$

The obtained material constants for the above four hardening laws are shown in Table 1.3 for aluminium AA5352 and Table 2.3 for steel TH330 in Part A.

Table 1.4 and 2.4 in Part A are the constants obtained from the equal biaxial test data only for AA 5352 and TH330, respectively.

3.2 Yield Functions

Yld2004-18p

The Yld2004-18p model requires an exponent, *a*, and 18 coefficients to be defined according to the procedure described in [10]. *a* is defined based on the crystal system of the alloy. For BCC alloys including steels, 6 is used, and for FCC alloys including Aluminium alloys 8 is used. The 18 coefficients were obtained by an iterative process which matches the material model behaviour to various properties determined from practical tests or assumed. These material properties and their origins are summarised in TABLE 1. The flow stress ratio is a ratio relative to a reference flow stress, in this case that of the 0° uniaxial test.

TABLE 1: Summary of measured and assumed properties used for calibration of the Yld2004-18p model.

Test	Flow stress ratio	r-value
Uniaxial (x7 @ 0°, 15°, 30°, 45°, 60°, 75°, 90°)	Y	Y
Disk compression		Y
Bulge	Y	
Assumptions	**Flow stress ratio**	**r-value**
45° tension in TD-ND plane	1	
45° tension in ND-RD plane	1	
Simple shear in TD-ND plane	$1/(1+2^{a-1})^{1/a}$	
Simple shear in ND-RD plane	$1/(1+2^{a-1})^{1/a}$	

The iterative process to determine the 18 coefficients was driven by an optimisation algorithm. For each iteration, a new set of 18 parameters was generated, and the values of each property in TABLE 1 predicted by the material model with this set of parameters was calculated by a program. The calculated and measured values were then used to calculate an error term for each property:

$$err = \left(\frac{p_{pr}}{p_{ex}} - 1\right)^2, \tag{10}$$

where p_{pr} is the predicted value of a property and p_{ex} is the experimentally determined value of a property. The error term for each property is multiplied by a weighting factor and summed to calculate the objective function:

$$E = \sum_p w_p err_p, \tag{11}$$

Where w_p are weighting functions. The optimisation algorithm selects a new set of 18 coefficients for the next iteration based on previous coefficient sets and objective functions.

Yld2000-2d

The Yld2000-2d model requires an exponent, a, and 8 coefficients to be defined according to the procedure described in [11]. a is defined based on the crystal system of the alloy. For BCC alloys including steels, 6 is used, and for FCC alloys including Aluminium alloys 8 is used. The 8 coefficients for each material given in the benchmark description were obtained by numerical solution of a set of 8 nonlinear simultaneous equations based on the experimental data summarised in TABLE 2.

TABLE 2: Summary of experimental data used in the calibration of the Yld2000-2d model.

Test	Flow stress ratio	r-value
Uniaxial (x7 @ 0°, 15°, 30°, 45°, 60°, 75°, 90°)	Y	Y
Disk compression		Y
Bulge	Y	

CPB06

The CPB06 model requires an exponent, a, a strength differential parameter, k, and 7 anisotropy coefficients to be defined. The value of a is taken as 2, and the strength differential and anisotropy parameters are determined according to an optimisation similar to that used for the Yld2004-18p model. The error function in this case also requires terms relating to the error in the predicted compressive yield stress, but since no experimental data was available in compression, the biaxial yield stress determined from the bulge test was taken as the uniaxial compressive stress at each angle:

$$\sigma_\theta^C = \sigma_b^T; \in [0,90], \tag{12}$$

Where superscripts C and T denote compression and tension respectively, a subscript b represents equibiaxial loading and a subscript θ represents the angle of uniaxial loading.

CPB06ex2

The CPB06ex model requires an exponent, a, two strength differential parameters, k and k', and 14 anisotropy coefficients to be defined. The value of a is taken as 3, and the strength differential and anisotropy parameters are determined according to an optimisation similar to that used for the Yld2004-18p model. The error function in this case also requires terms relating to the error in the predicted compressive yield stress, but since no experimental data was available in compression, the biaxial yield stress determined from the bulge test was taken as the uniaxial compressive stress at each angle, as in Eq. 12.

Hill

Stress based

Hill's 1948 yield function [12] may be written in a coordinate system where x is parallel to the rolling direction, y is parallel to the transverse direction and z is normal to the sheet as:

$$F\left(\frac{\sigma_{yy}}{\sigma_0}-\frac{\sigma_{zz}}{\sigma_0}\right)^2+G\left(\frac{\sigma_{zz}}{\sigma_0}-\frac{\sigma_{xx}}{\sigma_0}\right)^2+H\left(\frac{\sigma_{xx}}{\sigma_0}-\frac{\sigma_{yy}}{\sigma_0}\right)^2+2L\left(\frac{\sigma_{yz}}{\sigma_0}\right)^2+2M\left(\frac{\sigma_{zx}}{\sigma_0}\right)^2 + 2N\left(\frac{\sigma_{xy}}{\sigma_0}\right)^2=2 \tag{13}$$

σ_0 is the reference yield stress, which is taken as the value of yield stress for uniaxial tension in the rolling direction. Yield stress ratios, $\frac{\sigma_\alpha}{\sigma_0}$, are calculated using the Swift data for the relevant test given in the benchmark description, and a plastic strain determined such that the plastic strain energy is the same in each case. From these yield stress ratios, the components in the material coordinate system, $\frac{\sigma_{ij}}{\sigma_0}$, may be derived. By substituting these values of yield stress ratios from the 0°, 45°, 90° and bulge tests, the following system of equations is obtained:

$$G+H=2\left(\frac{\sigma_0}{\sigma_0}\right)^2 \tag{14}$$

$$\frac{F}{2}+\frac{G}{2}+N=4\left(\frac{\sigma_0}{\sigma_{45}}\right)^2 \tag{15}$$

$$F+H=2\left(\frac{\sigma_0}{\sigma_{90}}\right)^2 \tag{16}$$

$$F+G=2\left(\frac{\sigma_0}{\sigma_b}\right)^2 \tag{17}$$

These equations were then solved for F, G, H and N, while values for L and M were fixed at 3.

r-value based

The definition of σ_0 as the yield stress for uniaxial tension in the rolling direction means that:

$$G + H = 2 \tag{18}$$

r-values are provided in the benchmark description for uniaxial tests loaded at 0°, 45° and 90° to the rolling direction. These r values may be expressed in terms of incremental plastic strain components as:

$$r_0 = \frac{d\varepsilon_{yy}^{pl}}{d\varepsilon_{zz}^{pl}} \tag{19}$$

$$r_{90} = \frac{d\varepsilon_{xx}^{pl}}{d\varepsilon_{zz}^{pl}} \tag{20}$$

The flow rule for Hill's 1948 yield function is:

$$d\boldsymbol{\varepsilon}^{pl} = d\lambda \frac{df}{d\boldsymbol{\sigma}} \tag{21}$$

Substituting this expression into the expression for the r-values gives:

$$r_0 = \frac{H}{G} \tag{22}$$

$$r_{90} = \frac{H}{F} \tag{23}$$

There are now 3 equations which may be solved for F, G and H. The 45° uniaxial test was used to determine N.

$$r_{45} = \frac{d\varepsilon^{pl,135}}{d\varepsilon_{zz}^{pl}} \tag{24}$$

Where a superscript on ε denotes the angle from the rolling direction at which the strain is measured, with:

$$d\varepsilon^{135} = \frac{1}{2}d\varepsilon_{11} + \frac{1}{2}d\varepsilon_{11} - d\gamma_{12} \tag{25}$$

Leads to:

$$N = G\left(r_{45} + \frac{1}{2}\right)\left(1 + \frac{r_x}{r_y}\right) \tag{26}$$

Yld89

Stress based

The Yld89 [13] yield function is:

$$f_{yld89} = a|k_1 + k_2|^M + a|k_1 - k_2|^M + c|2k_2|^M = 2\sigma_0^M \tag{27}$$

With $M=8$ for aluminium alloys, $M=6$ for steel and:

$$k_1 = \frac{\sigma_{11} + h\sigma_{22}}{2} \tag{28}$$

$$k_2 = \sqrt{\left(\frac{\sigma_{11} - h\sigma_{22}}{2}\right)^2 + p^2\sigma_{12}^2} \tag{29}$$

Yield stress values are calculated from the Swift data given for uniaxial and bulge tests, with a plastic strain determined such that the plastic strain energy is the same in each case. σ_0 is the reference yield stress, which is taken as the value of yield stress for uniaxial tension in the rolling direction. By substituting the stress components on the LHS of the yield function with values corresponding to the yield stresses in 0°, 45° and 90° uniaxial tests and the bulge test, a system of 4 equations is formed:

$$a + c = 2 \tag{30}$$

$$a\left|\frac{(1+h)\sigma_{45}}{4} + \sqrt{\left(\frac{(1-h)\sigma_{45}}{4}\right)^2 + p^2\frac{\sigma_{45}^2}{4}}\right|^M + a\left|\frac{(1+h)\sigma_{45}}{4} - \sqrt{\left(\frac{(1-h)\sigma_{45}}{4}\right)^2 + p^2\frac{\sigma_{45}^2}{4}}\right|^M + c\left|\sqrt{(1-h)\sigma_{45} + p^2\sigma_{45}^2}\right|^M = 2\sigma_0^M \tag{31}$$

$$h = \frac{\sigma_0}{\sigma_{90}} \tag{32}$$

$$a = \frac{2\sigma_0^M}{(1+h^M)\sigma_b^M} \tag{33}$$

This set of equations was solved for a, c, h and p.

r-value based
Based on the flow rule:

$$d\varepsilon^{pl} = d\lambda\frac{df_{yld89}}{d\sigma}, \tag{34}$$

incompressibility during plastic flow:

$$\varepsilon_{xx} + \varepsilon_{yy} + \varepsilon_{zz} = 0 \tag{35}$$

and the yield function:

$$f_{yld89} = a|k_1 + k_2|^M + a|k_1 - k_2|^M + c|2k_2|^M = 2\sigma_0^M, \tag{36}$$

the following relationships between Lankford's coefficients and a and h may be derived:

$$a = 2 - 2\sqrt{\frac{r_0}{1+r_0}\frac{r_{90}}{1+r_{90}}} \tag{37}$$

$$h = \sqrt{\frac{r_0}{1+r_0}\frac{1+r_{90}}{r_{90}}} \tag{38}$$

The use of the yield stress for uniaxial tension in the rolling direction as the reference stress leads to:

$$a + c = 2 \tag{39}$$

The predicted r-value for uniaxial loading at 45° from the rolling direction is given by:

$$r_{45} = \frac{2m\sigma_0^m}{\left(\frac{\partial f_{yld89}}{\partial \sigma_0} + \frac{\partial f_{yld89}}{\partial \sigma_{90}}\right)\sigma_{45}} - 1 \tag{40}$$

This equation was used iteratively to find p such that predicted and measured values of r_{45} matched.

4. ACKNOWLEDGEMENTS

The authors wish to thank many people for significant contributions they made to the material characterisation for this benchmark study including:
Frank Schouten, Romke Boterman, Gerard Krusemeijer, Jan Heijne, Michael Abspoel, Marc Scholting, Marcel Lansbergen, Henri Kwakkel and Martin Sadhinoch , Tata Steel.
Prof. Dorel Banabic, CERTETA.
Dr Gabriela Vincze, University Aveiro, Portugal.
Torsten Sjögren, SP Technical Research Institute of Sweden.

5. REFERENCES

[1] H. Campos, A. D. Santos, B. Martins, K. Ito, N. Mori and F. Barlat, "Hydraulic bulge test for stress-strain curve determination and damage calibration for Ito-Goya model," in *11th World Congress on Computational Mechanics (WCCM XI), J. Oliver and A. Huerta (Eds)*, Barcelona, Spain, 2014.

[2] H. Campos, A. D. Santos, R. Amaral and J. B. Pacheco, "Desenvolvimento de um Sistema Experimental de Caraterização de Materiais Metálicos Usando o Ensaio de Expansão Biaxial Hidráulico - Development of an Experimental System for Characterization of Metalic Materials Using Hydraulic Bulge Test," *Revista da Associação Portuguesa de Análise Experimental de Tensões (ISSN 1646-7078), APAET Mecânica Experimental, Vol 25,* pp. 93-100, 2015.

[3] L. C. Reis, M. C. Oliveira, A. D. Santos and J. V. Fernandes, "On The Determination Of The Work Hardening Curve Using The Bulge Test," *International Journal of Mechanical Sciences, Vol. 105,* pp. 158-181, 2015.

[4] F. Barlat, J. C. Brem, J. W. Yoon, K. Chung, R. E. Dick, D. J. Lege, F. Pourboghrat, S. -H. Choi and E. Chu. , "Plane stress yield function for aluminum alloy sheets — part 1: theory," *International Journal of Plasticity,* vol. 19, pp. 1297-1319, 2003.

[5] S. P. Keeler and W. A. Backofen, "Plastic instability and fracture in sheets stretched over rigid punches," *ASM Trans. Quart.,* vol. 56, pp. 25-48, 1964.

[6] G. M. Goodwin, "Application of strain analysis to sheet metal forming problems in the press," *SAE paper No. 680093,* 1968.

[7] M. Turkoz, H. S. Halkaci, O. Yigit, M. Dilmec and F. Ozturk, "A new method for determining limit strains of materials that show post-uniform elongation behavior," *Proc Institute of Mechanical Engineers Part B: J Engineering Manufacture,* vol. 228, no. 3, pp. 450-457, 2014.

[8] M. Abspoel, M. E. Scholting and J. M. M. Droog, "A new method for predicting Forming Limit Curves from mechanical properties," *Journal of Materials Processing Technology,* pp. 759-769, 2013.

[9] M. Sigvant, K. Mattiasson, H. Vegter and P. Thilderkvist, "A viscous pressure bulge test for the determination of a plastic hardening curve and equibiaxial material data," *Int J Mater Form,* pp. 2:235-242, 2009.

[10] F. Barlat, H. Aretz, J. Yoon, M. Karabin, J. Brem and R. Dick, "Linear transformation-based anisotropic yield functions," *International Journal of Plasticity,* pp. 1009-1039, 2005.

[11] F. Barlat, J. Brem, J. Yoon, K. Chung, R. Dick, S. Choi, F. Pourboghrat, E. Chu and D. Lege, "Plane stress yield function for aluminum alloy sheets–part 1: theory," *Int. J. Plasticity,* vol. 19, p. 1297, 2003.

[12] R. Hill, "A theory of the yielding and plastic flow of anisotropic metals," *Proceedings of,* no. 193, p. 281–297, 1948.

[13] F. Barlat and J. Lian, "Plastic behaviour and stretchability of sheet metals (Part I): A yield function for orthotropic sheet under plane stress conditions," *International Journal of Plasticity,* vol. 5, pp. 51-56, 1989.

[14] O. Cazacu, B. Plunkett and F. Barlat, "Orthotropic yield criterion for hexagonal close packed metals," *International Journal of Plasticity,* vol. 22, p. 1171–1194, 2006.

Benchmark 1 – Failure Prediction after Cup Drawing, Reverse Redrawing and Expansion

Part C: Physical Tryout

Martin Watson[a], Robert Dick[b], Y. Helen Huang[a], Andrew Lockley[a], Rui Cardoso[c], Abel Santos[d]

[a]*Crown Packaging, Oxfordshire OX12 9BP, UK*
[b]*Alcoa Technical Center, PA 16069-0001, USA*
[c]*Brunel University London, UB8 3PH, UK*
[d]*FEUP, University of Porto, 4200-465, Portugal*

Abstract: The practical experiments and measurements of BM1, "Failure Prediction after Cup Drawing, Reverse Redrawing and Expansion" are summarised, including details of the equipment used, process conditions, measurement methods and the results obtained.

Keywords: Cup, cupping, expansion die, ear height, wall thickness.

1. OUTLINE OF EXPERIMENTS AND MEASUREMENTS

A two-stage drawing and reverse redrawing process was used to produce cups in both steel and aluminium.

Cup wall heights were measured at 1° increments circumferentially to produce an ear profile for the entire circumference of the cups.

Cup wall thicknesses were measured at 10° increments circumferentially and at 20mm, 45mm and 50mm from the base of each cup to produce a map of wall thickness.

Cups were expanded to fracture using a lubricated, vented conical die tool. The vent holes in the tool ensured that there was no internal air pressure increase as the conical die entered the cup.

Axial displacement and axial load were recorded at approximately 0.01mm increments during the die expansion-to-fracture testing.

2. EXPERIMENTAL FORMING TOOLS & PROCESS CONDITIONS

Details of all tooling are given in Part A: Benchmark report, along with the displacements, blank holding forces, velocities and friction coefficients.

2.1 Draw – Reverse Redraw Cupping Process

Cups were made using Alcoa's Draw and Reverse Redraw process at the Alcoa Technical Center, Pittsburgh, USA.

Figure 1: Alcoa Press

Figure 2: Draw, Reverse-Redraw tool

Figure 3: Typical aluminium cup

Figure 4: Typical steel cup

Journal of Physics: Conference Series **734** (2016) 022001 doi:10.1088/1742-6596/734/2/022001

2.2 Die Expansion to Fracture

The die expansion-to-fracture experiment was conducted at Crown's Corporate Technology Centre in Wantage, UK. Cups were driven downwards over a lubricated conical tool while both displacement and load were measured. The conical tool had holes drilled to vent the cup during die expansion.

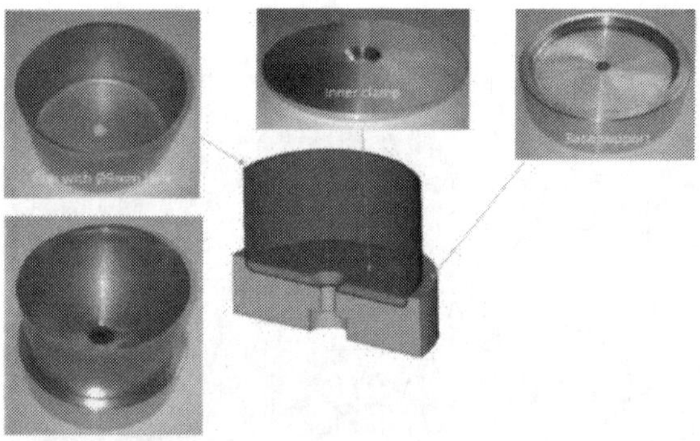

Figure 5: Cup mounting details.

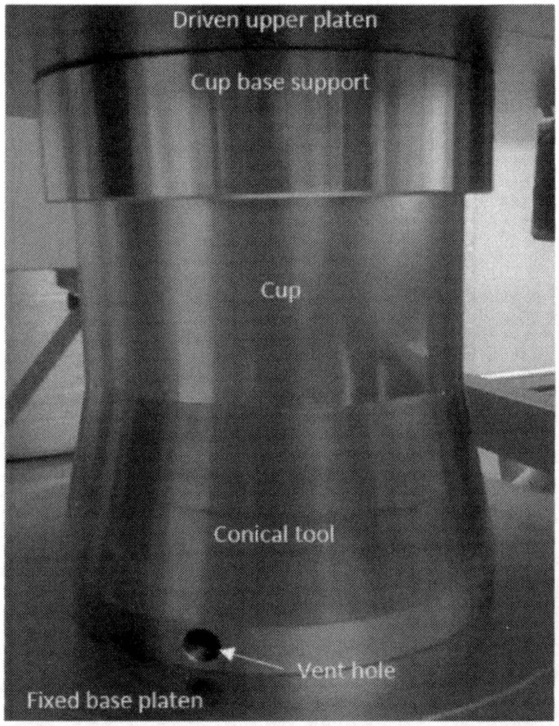

Figure 6: Die expansion of cup

Figure 7: Typical aluminium cup fracture

Figure 8: Typical steel cup fracture

Journal of Physics: Conference Series **734** (2016) 022001 doi:10.1088/1742-6596/734/2/022001

3. MEASUREMENT

3.1 Cup height and ear profile

Cup heights were measured at Crown's Corporate Technology Centre in Wantage, UK.

Cups were clamped and rotated through 360°.

A sprung-loaded probe measured the vertical position of the edge at 1 degree increments during the rotation, producing an ear profile for the cup.

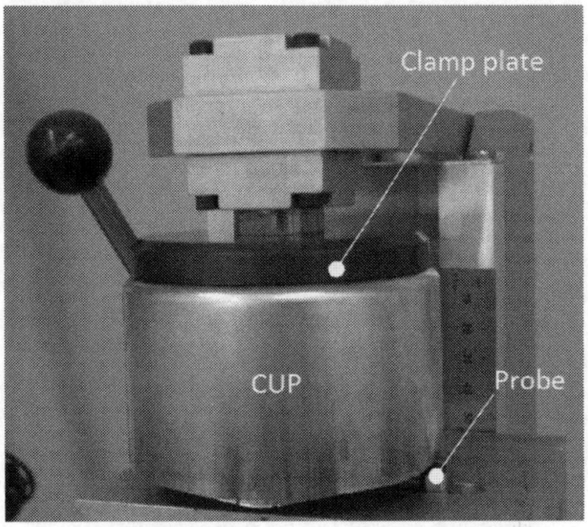

Figure 9: Ear measurement

Steel cups (averaged data)

- 4 ears
- Mean height was 52.61mm
- Amplitude (max – min) was 1.14mm.

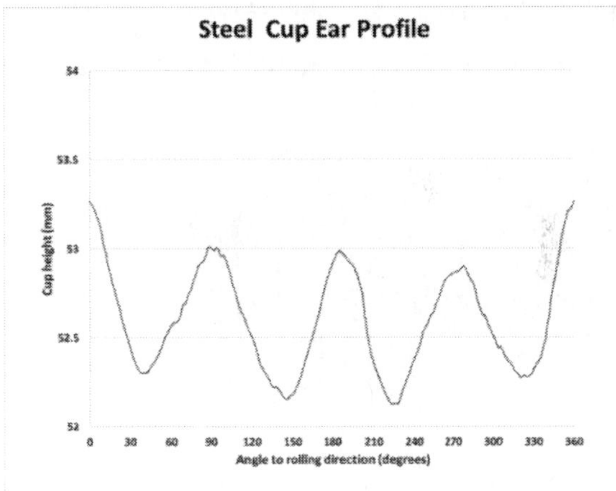

Figure 10: Steel cup ear profile

Journal of Physics: Conference Series **734** (2016) 022001 doi:10.1088/1742-6596/734/2/022001

Aluminium cups (averaged data)

- 8 ears.
- Mean height was 52.05mm
- Amplitude (max – min) was 4.13mm.
- The metal in the ears at 0° and 180° was severely thinned due to the blank holder load being concentrated on these last two ears during the forming process.

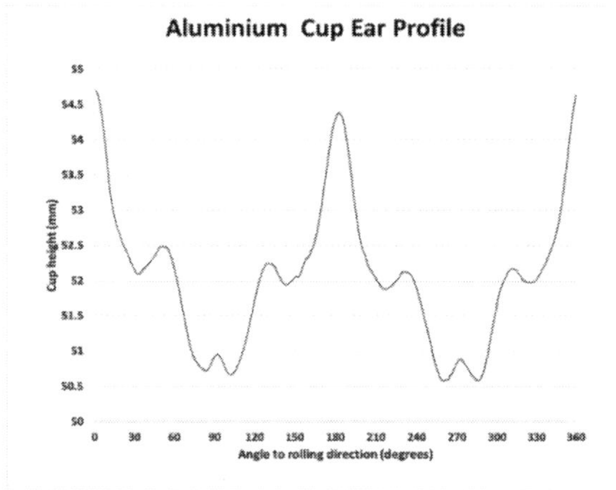

Figure 11: Aluminium cup ear profile

Comment on aluminium cup ear profile

For the aluminium cups, the upper region of the ears at 0° and 180° were significantly thinner than the rest of the cup, as shown in the images in figure 12 below.

This thinning is believed to be due to the blank holding force exerted on these two ears in the final stages of cup forming.

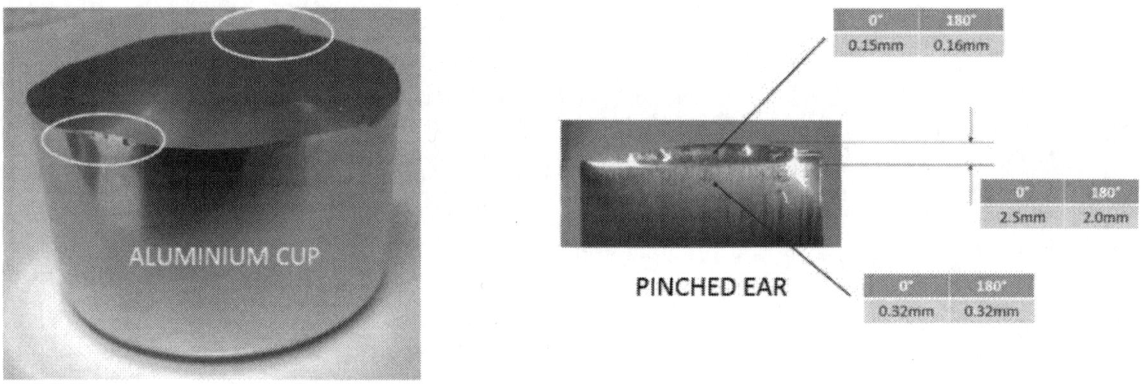

Figure 12: Aluminium cups – pinched ears

3.2 Cup wall thickness

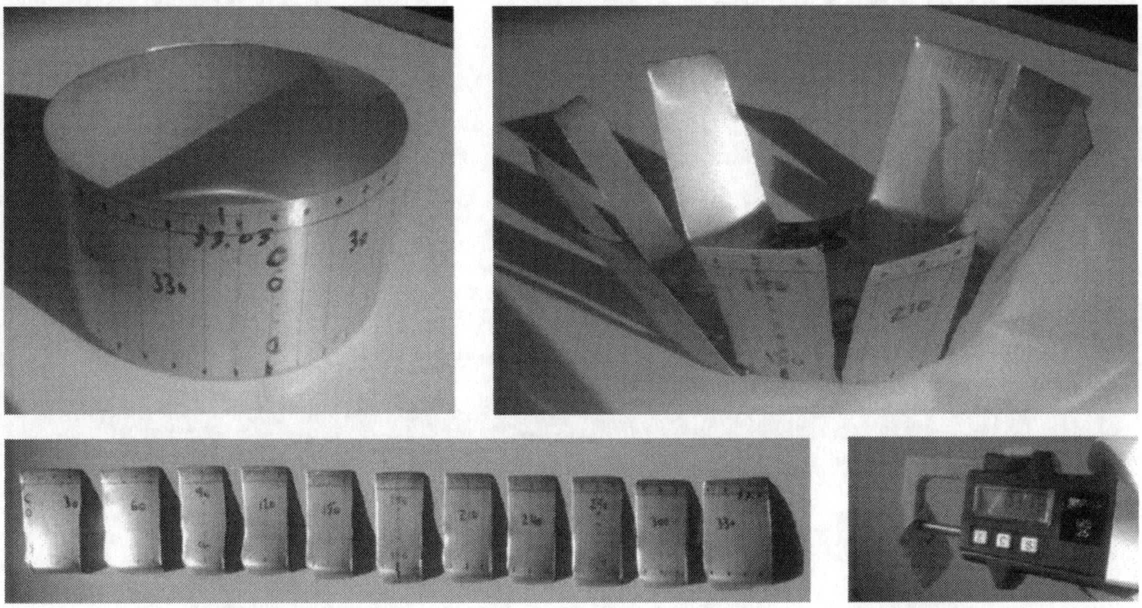

Figure 13: Sample preparation and measurement of wall thickness

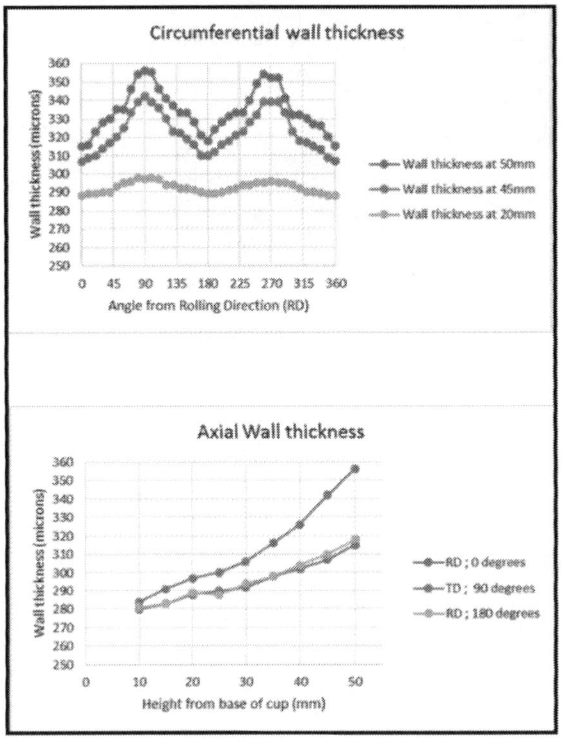

Figure 14: Aluminium cup wall thickness

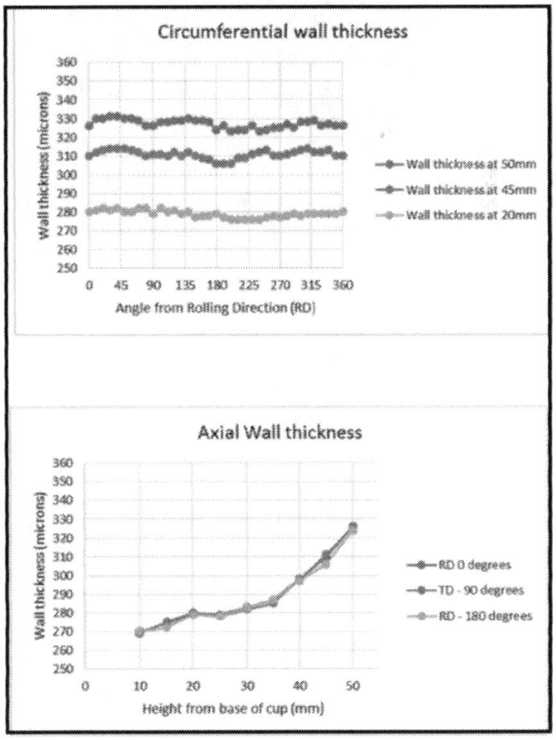

Figure 15: Steel cup wall thickness

Journal of Physics: Conference Series **734** (2016) 022001 doi:10.1088/1742-6596/734/2/022001

3.3. Die expansion to fracture

Load and displacement data were recorded at approx. 0.01mm travel increments.

Steel cups (averaged data)

- Load at fracture = 8.55 kN
- Travel at fracture = 30.2mm
- Fracture angle = 42°
- Banding observed all around the free edge.

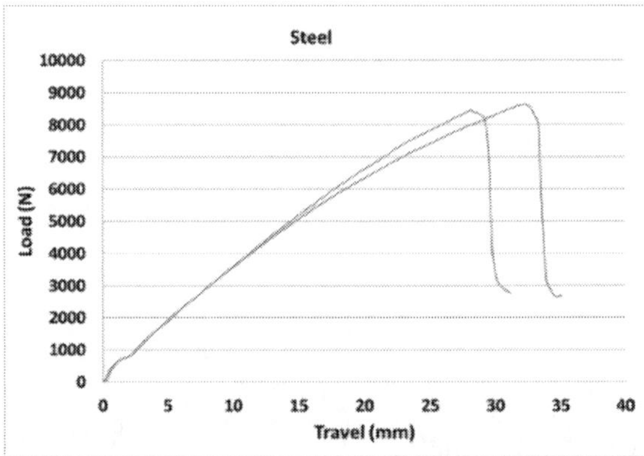

Figure 16: Load to fracture (Steel)

Figure 17: Steel cup after fracture

Journal of Physics: Conference Series **734** (2016) 022001 doi:10.1088/1742-6596/734/2/022001

Aluminium cups (averaged data)

- Load at fracture = 2.76 kN
- Travel to fracture = 16.4mm
- Fracture angle = 96°

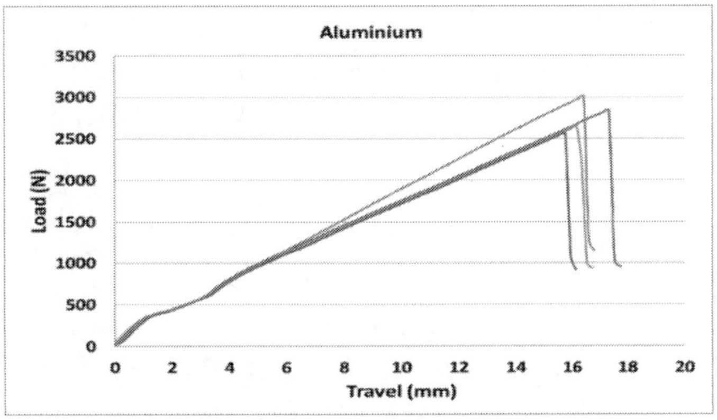

Figure 18: Load to fracture (Aluminium)

Figure 19: Aluminium cup after fracture

Benchmark 1 – Failure Prediction after Cup Drawing, Reverse Redrawing and Expansion

Part D: Responses

Martin Watson[a], Robert Dick[b], Y. Helen Huang[a], Andrew Lockley[a], Rui Cardoso[c], Abel Santos[d]

[a]*Crown Packaging, Oxfordshire OX12 9BP, UK*
[b]*Alcoa Technical Center, PA 16069-0001, USA*
[c]*Brunel University London, UB8 3PH, UK*
[d]*FEUP, University of Porto, 4200-465, Portugal*

1. Experimental data : BM1_00	
Name	**Martin Watson, Robert Dick, Y. Helen Huang, Andrew Lockley, Rui Cardoso, Abel Santos**
Prepared by	Benchmark 1 Committee
Email	NumisheetBenchmark1@uwe.ac.uk

Journal of Physics: Conference Series **734** (2016) 022001 doi:10.1088/1742-6596/734/2/022001

1. Benchmark Participant : BM1_01	
Name	**Toshiro Amaishi1*, Masahi Arai1, Yuko Watanabe1**
Affiliation	1JSOL Corporation
Address	Tosabori Daibiru Building, 2-2-4, Tosabori Nishi-ku, Osaka 550-0001, Japan
Email	* amaishi.toshirou@jsol.co.jp
Phone number	'+81-6-4803-5820
Fax number	'+81-6-6225-3517

2. Simulation Software	
Name of the FEM code	JSTAMP/NV (solver: LS-DYNA)
General aspect of the code	Integrated sheet metal forming simulation system
Basic formulations	Dynamic Explicit
Element/Mesh technology	
Number of elements	29430 (for Blank), Quarter model
Type of elements	Solid element, reduced integration
Contact property model	Penalty Mehtod, Surface to Surface
Friction formulation	Coluomb's friction law

3. Simulation Hardware	
CPU Type	Intel Xeon E5-1650
CPU clock speed	3.5GHz
Number of cores per CPU	6
Main memory	16.0G
Operating system	Windows 7 Professional

Total CPU time	AA5352: 2 hours 31 min. for drawing, 4 hours 28 min. for reverse redrawing, 2 hours 10 min. for die expansion TH330: 2 hours 8 min. for drawing, 3 hours 35 min. for reverse redrawing, 2 hours 4 min. for die expansion

4. Describe the material model used for each material		
Material	**AA5352 (aluminium)**	**TH330 (steel)**
Yield Function/ Plastic Potential	Yoshida (sixth order polynomial model)	Yoshida (sixth order polynomial model)
Hardening Rule (e.g. Isotropic, kinematic)	Kinematic	Kinematic
Stress-Strain Relation (e.g. Swift, Voce)	Yoshida-Uemori Kinematic hardening model	Yoshida-Uemori Kinematic hardening model
Failure model	Strain based FLD	

1. Benchmark Participant : BM1_02	
Name	**P.D. Carvalho, P.D. Barros, D.M. Neto, M.C. Oliveira, J.L Alves and L.F. Menezes**
Affiliation	Departament of Mechanical Engineering, University of Coimbra
Address	Polo II, Rua Luís Reis Santos, 3030-788 Coimbra, Portugal
Email	pedro.barros@dem.uc.pt
Phone number	+351 239 790 700
Fax number	+351 239 790 701

2. Simulation Software	
Name of the FEM code	DD3IMP
General aspect of the code	Static fully implicit
Basic formulations	Updated Lagrangian formulation with associated flow rule
Element/Mesh technology	
Number of elements	11552
Type of elements	Isoparametric 3D brick elements with selective reduced integration technique
Contact property model	Rigid tools modelled by 7550 Nagata patches, Augmented lagrangian method
Friction formulation	Coulomb friction law

3. Simulation Hardware	
CPU Type	Intel® Core™ i7-2600K
CPU clock speed	3.4 GHz
Number of cores per CPU	4 cores
Main memory	8 GB RAM
Operating system	Windows 7 Professional (64-bit)

Total CPU time	13 hours (AA5352 aluminium); 22 hours (TH330 steel)

4. Describe the material model used for each material		
Material	**AA5352 (aluminium)**	**TH330 (steel)**
Yield Function/ Plastic Potential	Cazacu and Barlat 2001	Cazacu and Barlat 2001
Hardening Rule (e.g. Isotropic, kinematic)	Isotropic	Isotropic
Stress-Strain Relation (e.g. Swift, Voce)	Voce	Swift
Failure model	Maximum element thinning rate	

1. Benchmark Participant : BM1_03	
Name	**Kai Oide1*, Takaya Kobayashi1, Yasuko Mihara1, Hideo Takizawa2**
Affiliation	1Mechanical Design & Analysis Corporation, 2Nippon Institute of Technology
Address	AXIS Chofu 2F, 1-40-2 Fuda Chofu Tokyo, JAPAN
Email	*oide@mech-da.co.jp
Phone number	81(42)482-1539
Fax number	81(42)482-5106

2. Simulation Software	
Name of the FEM code	Abaqus 3DEXPERIENCE R2016x
General aspect of the code	Dynamic Implicit in Drawing and Reverse redrawing operation Static Implicit with artificial damping method in Die expansion operation
Basic formulations	Updated Lagrangian formulation with associated flow rule
Element/Mesh technology	
Number of elements	16000 elements
Type of elements	4-node quadrilateral shell element with reduced integration : S4R
Contact property model	Penalty method
Friction formulation	Coulomb friction model

3. Simulation Hardware	
CPU Type	Intel® Xeon® CPU X5650
CPU clock speed	2.66GHz
Number of cores per CPU	6 cores
Main memory	48.0 GB RAMM

Journal of Physics: Conference Series **734** (2016) 022001 doi:10.1088/1742-6596/734/2/022001

Operating system	Windows 7 Professional
Total CPU time	average total CPU time is 586868[sec]

4. Describe the material model used for each material		
Material	**AA5352 (aluminium)**	**TH330 (steel)**
Yield Function/ Plastic Potential	YLD2004	YLD2004
Hardening Rule (e.g. Isotropic, kinematic)	Isotropic	Isotropic
Stress-Strain Relation (e.g. Swift, Voce)	Voce	Voce
Failure model	MSFLD	

1. Benchmark Participant : BM1_04	
Name	**Kai Oide1*, Takaya Kobayashi1, Yasuko Mihara1, Hideo Takizawa2**
Affiliation	1Mechanical Design & Analysis Corporation, 2Nippon Institute of Technology
Address	AXIS Chofu 2F, 1-40-2 Fuda Chofu Tokyo, JAPAN
Email	*oide@mech-da.co.jp
Phone number	81(42)482-1539
Fax number	81(42)482-5106

2. Simulation Software	
Name of the FEM code	Abaqus 3DEXPERIENCE R2016x
General aspect of the code	Dynamic Implicit in Drawing and Reverse redrawing operation Static Implicit with artificial damping method in Die expansion operation
Basic formulations	Updated Lagrangian formulation with associated flow rule
Element/Mesh technology	
Number of elements	16000 elements
Type of elements	4-node quadrilateral shell element with reduced integration : S4R
Contact property model	Penalty method
Friction formulation	Coulomb friction model

3. Simulation Hardware	
CPU Type	Intel® Xeon® CPU X5650
CPU clock speed	2.66GHz
Number of cores per CPU	6 cores
Main memory	48.0 GB RAMM

Operating system	Windows 7 Professional
Total CPU time	average total CPU time is 586868[sec]

4. Describe the material model used for each material		
Material	**AA5352 (aluminium)**	**TH330 (steel)**
Yield Function/ Plastic Potential	YLD2000	YLD2000
Hardening Rule (e.g. Isotropic, kinematic)	Isotropic	Isotropic
Stress-Strain Relation (e.g. Swift, Voce)	Voce	Voce
Failure model	MSFLD	

1. Benchmark Participant : BM1_05	
Name	**Kai Oide1*, Takaya Kobayashi1, Yasuko Mihara1, Hideo Takizawa2**
Affiliation	1Mechanical Design & Analysis Corporation, 2Nippon Institute of Technology
Address	AXIS Chofu 2F, 1-40-2 Fuda Chofu Tokyo, JAPAN
Email	*oide@mech-da.co.jp
Phone number	81(42)482-1539
Fax number	81(42)482-5106

2. Simulation Software	
Name of the FEM code	Abaqus 3DEXPERIENCE R2016x
General aspect of the code	Dynamic Implicit in Drawing and Reverse redrawing operation Static Implicit with artificial damping method in Die expansion operation
Basic formulations	Updated Lagrangian formulation with associated flow rule
Element/Mesh technology	
Number of elements	16000 elements
Type of elements	4-node quadrilateral shell element with reduced integration : S4R
Contact property model	Penalty method
Friction formulation	Coulomb friction model

3. Simulation Hardware	
CPU Type	Intel® Xeon® CPU X5650
CPU clock speed	2.66GHz
Number of cores per CPU	6 cores
Main memory	48.0 GB RAMM

Operating system	Windows 7 Professional
Total CPU time	average total CPU time is 586868[sec]

4. Describe the material model used for each material		
Material	**AA5352 (aluminium)**	**TH330 (steel)**
Yield Function/ Plastic Potential	YLD89 stress based	YLD89 stress based
Hardening Rule (e.g. Isotropic, kinematic)	Isotropic	Isotropic
Stress-Strain Relation (e.g. Swift, Voce)	Voce	Voce
Failure model		

1. Benchmark Participant : BM1_06	
Name	**Kai Oide1*, Takaya Kobayashi1, Yasuko Mihara1, Hideo Takizawa2**
Affiliation	1Mechanical Design & Analysis Corporation, 2Nippon Institute of Technology
Address	AXIS Chofu 2F, 1-40-2 Fuda Chofu Tokyo, JAPAN
Email	*oide@mech-da.co.jp
Phone number	81(42)482-1539
Fax number	81(42)482-5106

2. Simulation Software	
Name of the FEM code	Abaqus 3DEXPERIENCE R2016x
General aspect of the code	Dynamic Implicit in Drawing and Reverse redrawing operation Static Implicit with artificial damping method in Die expansion operation
Basic formulations	Updated Lagrangian formulation with associated flow rule
Element/Mesh technology	
Number of elements	16000 elements
Type of elements	4-node quadrilateral shell element with reduced integration : S4R
Contact property model	Penalty method
Friction formulation	Coulomb friction model

3. Simulation Hardware	
CPU Type	Intel® Xeon® CPU X5650
CPU clock speed	2.66GHz
Number of cores per CPU	6 cores
Main memory	48.0 GB RAMM

Operating system	Windows 7 Professional
Total CPU time	average total CPU time is 586868[sec]

4. Describe the material model used for each material		
Material	**AA5352 (aluminium)**	**TH330 (steel)**
Yield Function/ Plastic Potential	Hill stress based	Hill stress based
Hardening Rule (e.g. Isotropic, kinematic)	Isotropic	Isotropic
Stress-Strain Relation (e.g. Swift, Voce)	Voce	Voce
Failure model	MSFLD	

Journal of Physics: Conference Series **734** (2016) 022001 doi:10.1088/1742-6596/734/2/022001

1. Benchmark Participant : BM1_07	
Name	**Hariharasudhan Palaniswamy, Subir Roy**
Affiliation	Altair Engineering
Address	1820 E Big Beaver road, Troy, MI, 48085, USA
Email	hpalaniswamy@altair.com, subir@altair.com
Phone number	248-614-2400
Fax number	248-614-2411

2. Simulation Software	
Name of the FEM code	HyperForm - RADIOSS
General aspect of the code	Commercial nonlinear finite element software
Basic formulations	Forming (Explicit), Springback(Implicit)
Element/Mesh technology	
Number of elements	810934
Type of elements	Shell element - QEPH formulation
Contact property model	Penaltly based contact formulations
Friction formulation	Coulomb's Law

3. Simulation Hardware	
CPU Type	Laptop with Inter Core I7 processor
CPU clock speed	2.5Ghz
Number of cores per CPU	4 cores with 2 threads each, 3 cores with 2 threads used for calculation
Main memory	16GB
Operating system	Windows 7

Total CPU time	Forming 1: 25 min 52 secs, Forming 2 : 39 mins 42 secs, Expansion :

4. Describe the material model used for each material		
Material	**AA5352 (aluminium)**	**TH330 (steel)**
Yield Function/ Plastic Potential	Barlat 3 parameter model	HILL 1948 Material model
Hardening Rule (e.g. Isotropic, kinematic)	Combined hardening rule	Combined hardening rule
Stress-Strain Relation (e.g. Swift, Voce)	Hollmon hardening law	Hollmon hardening law
Failure model		

1. Benchmark Participant : BM1_08	
Name	**Jan SLOTA, Marek, SISER**
Affiliation	Technical University of Kosice
Address	Masiarska 74, 040 01 Kosice, Slovakia
Email	jan.slota@tuke.sk, marek.siser@tuke.sk
Phone number	'+421556023545
Fax number	

2. Simulation Software	
Name of the FEM code	PamStamp 2G, v2015.1
General aspect of the code	Dynamic explicit code
Basic formulations	
Element/Mesh technology	
Number of elements	blank: 5472 elements
Type of elements	shell/membrane
Contact property model	accuracy
Friction formulation	Coulomb friction model, u=0.03

3. Simulation Hardware	
CPU Type	Intel XEON E5
CPU clock speed	2.60GHz
Number of cores per CPU	8
Main memory	16 GB
Operating system	Win 8.1 Pro 64-bit

Total CPU time	

4. Describe the material model used for each material		
Material	**AA5352 (aluminium)**	**TH330 (steel)**
Yield Function/ Plastic Potential	Barlat2000	Orthotropic Hill48
Hardening Rule (e.g. Isotropic, kinematic)	Isotropic	Isotropic
Stress-Strain Relation (e.g. Swift, Voce)	Swift/Krupkowski	Swift/Krupkowski
Failure model		

Table of figures

Figure 28: (AA5352) Forming limit diagram and principal strain plot at point of first failure for cup draw, redraw and expansion operations.

Figure 29: (AA5352) Forming limit diagram and principal strain plot at point of first failure for cup draw, redraw and expansion operations.

Figure 30: (AA5352) Forming limit diagram and principal strain plot at point of first failure for cup draw, redraw and expansion operations.

Figure 31: (AA5352) Forming limit diagram and principal strain plot at point of first failure for cup draw, redraw and expansion operations.

Figure 32: (AA5352) Forming limit diagram and principal strain plot at point of first failure for cup draw, redraw and expansion operations.

Figure 33: (TH330) Forming limit diagram and principal strain plot at point of first failure for cup draw, redraw and expansion operations.

Figure 34: (TH330) Forming limit diagram and principal strain plot at point of first failure for cup draw, redraw and expansion operations.

Figure 35: (TH330) Forming limit diagram and principal strain plot at point of first failure for cup draw, redraw and expansion operations.

Figure 36: (TH330) Forming limit diagram and principal strain plot at point of first failure for cup draw, redraw and expansion operations.

Figure 37: (TH330) Forming limit diagram and principal strain plot at point of first failure for cup draw, redraw and expansion operations.

Figure 38: (AA5352) Position of point of first failure during expansion test.

Figure 39: (AA5352) Stroke of expansion punch at first failure.

Figure 40: (TH330) Position of point of first failure during expansion test.

Figure 41: (TH330) Stroke of expansion punch at first failure.

Journal of Physics: Conference Series **734** (2016) 022001 doi:10.1088/1742-6596/734/2/022001

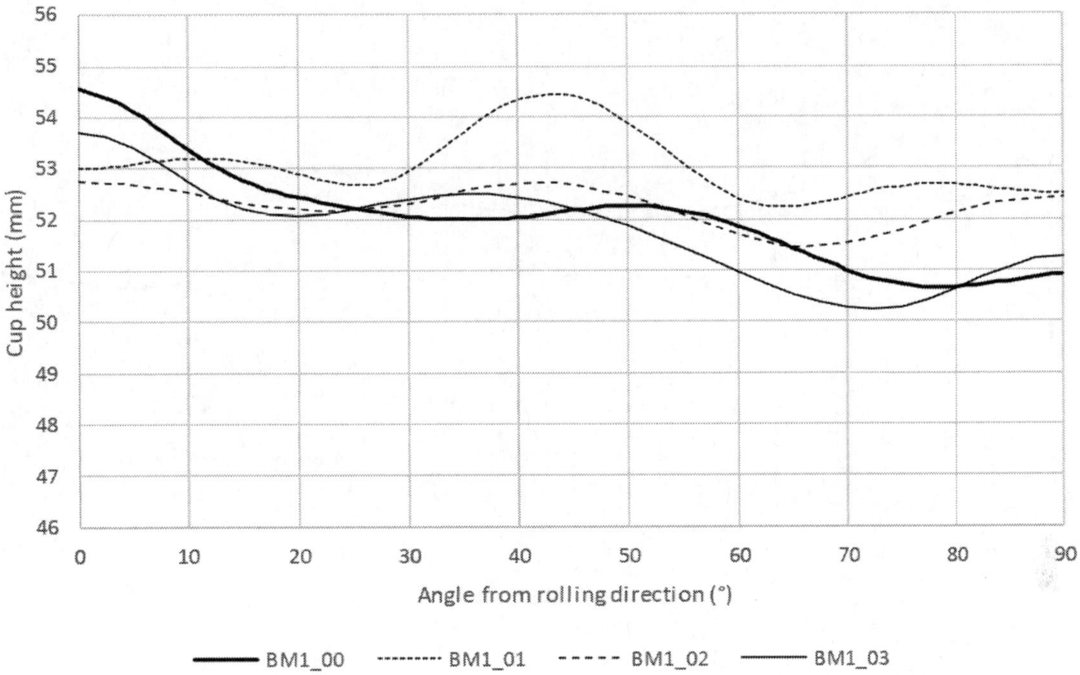

Figure 1: (AA5352) Cup height vs angle from rolling direction after the reverse redraw operation.

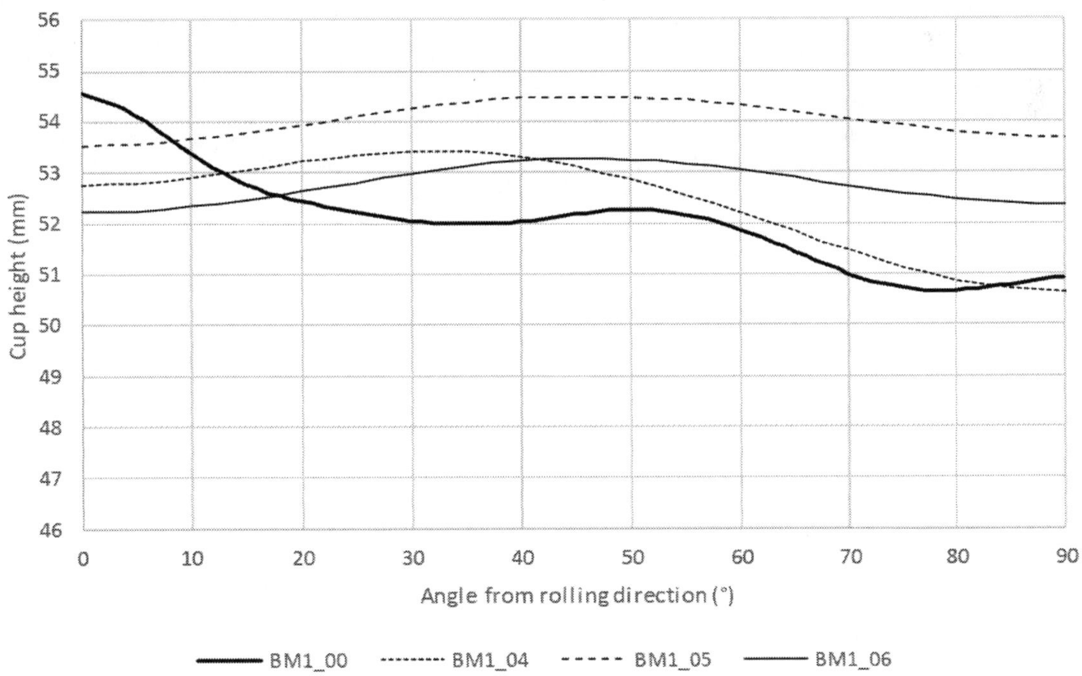

Figure 2: (AA5352) Cup height vs angle from rolling direction after the reverse redraw operation.

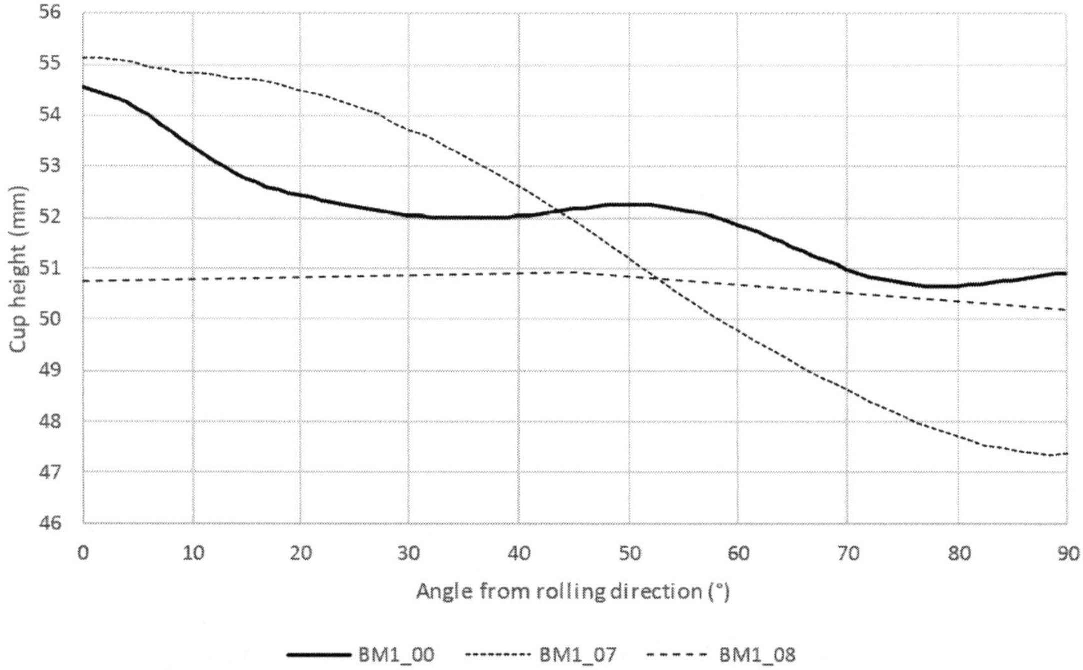

Figure 3: (AA5352) Cup height vs angle from rolling direction after the reverse redraw operation.

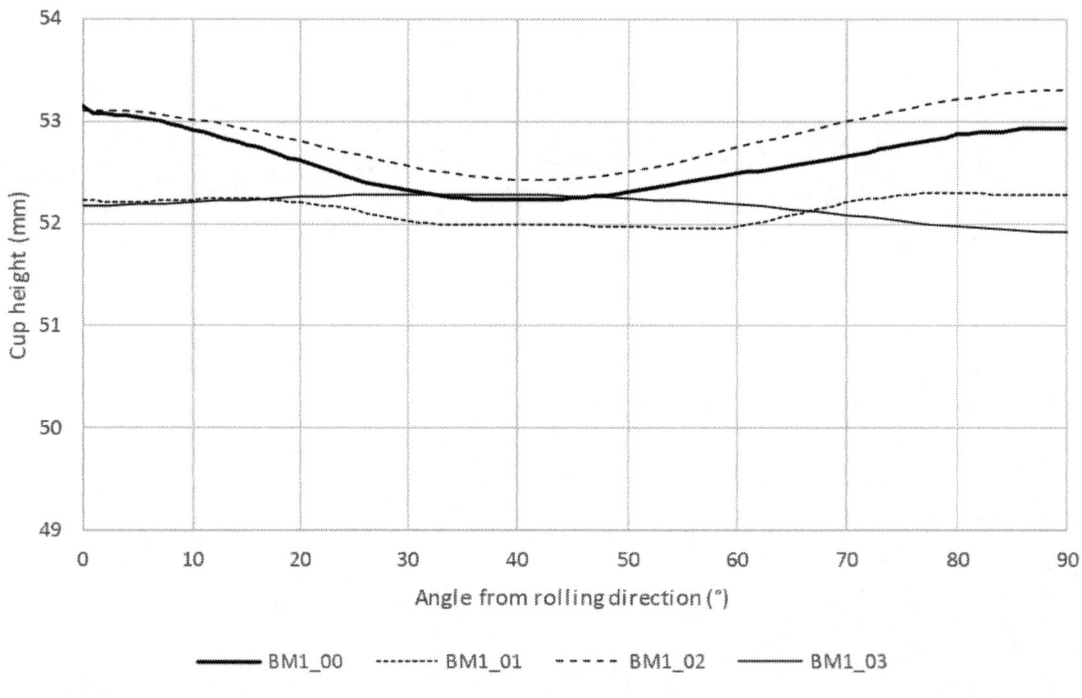

Figure 4: (TH330) Cup height vs angle from rolling direction after the reverse redraw operation.

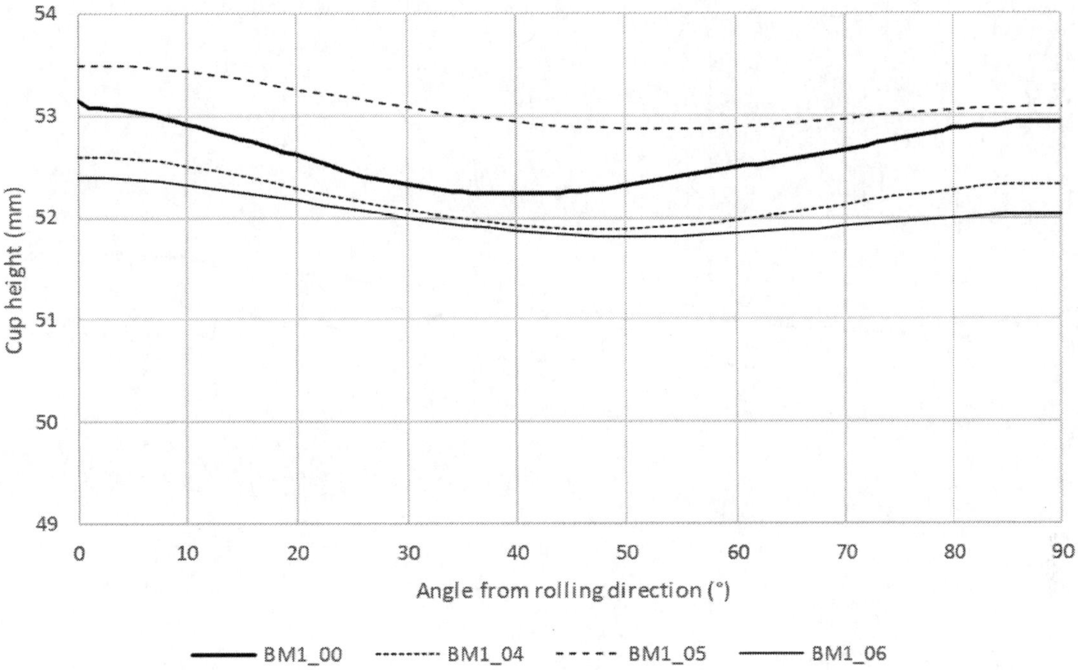

Figure 5: (TH330) Cup height vs angle from rolling direction after the reverse redraw operation.

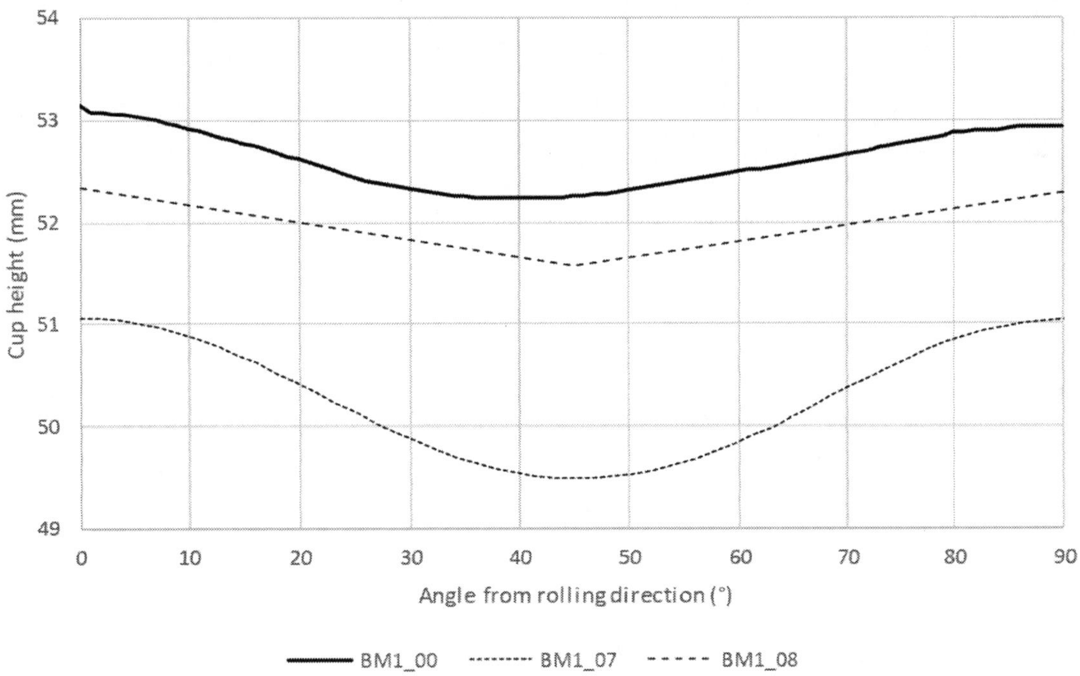

Figure 6: (TH330) Cup height vs angle from rolling direction after the reverse redraw operation.

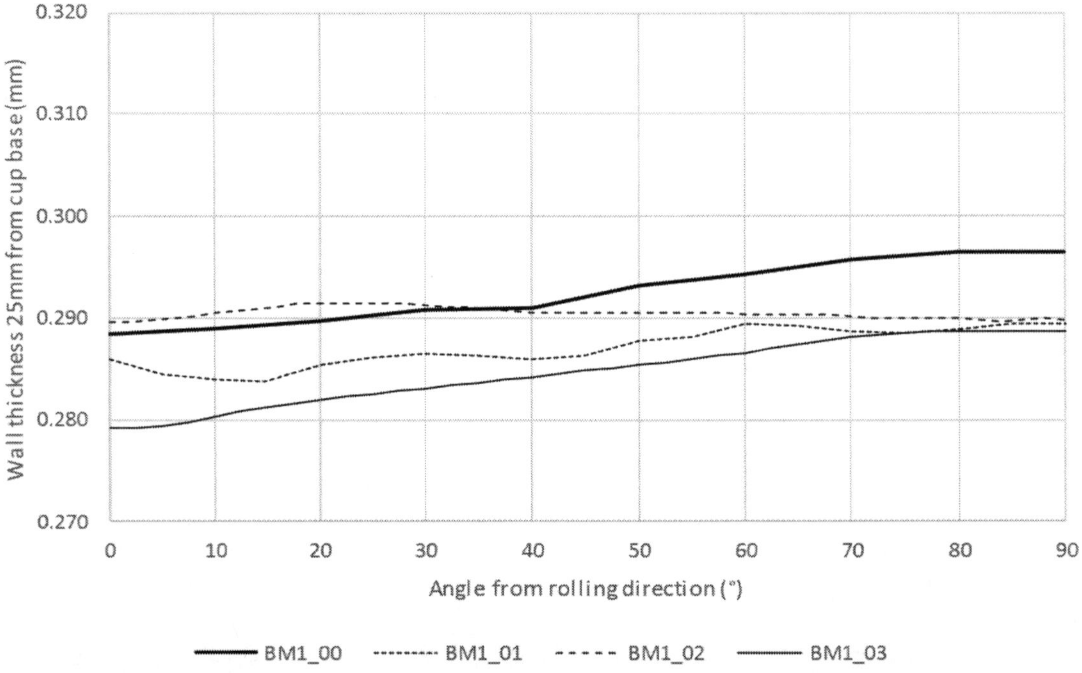

Figure 7: (AA5352) Wall thickness vs angle from rolling direction 25mm above cup base.

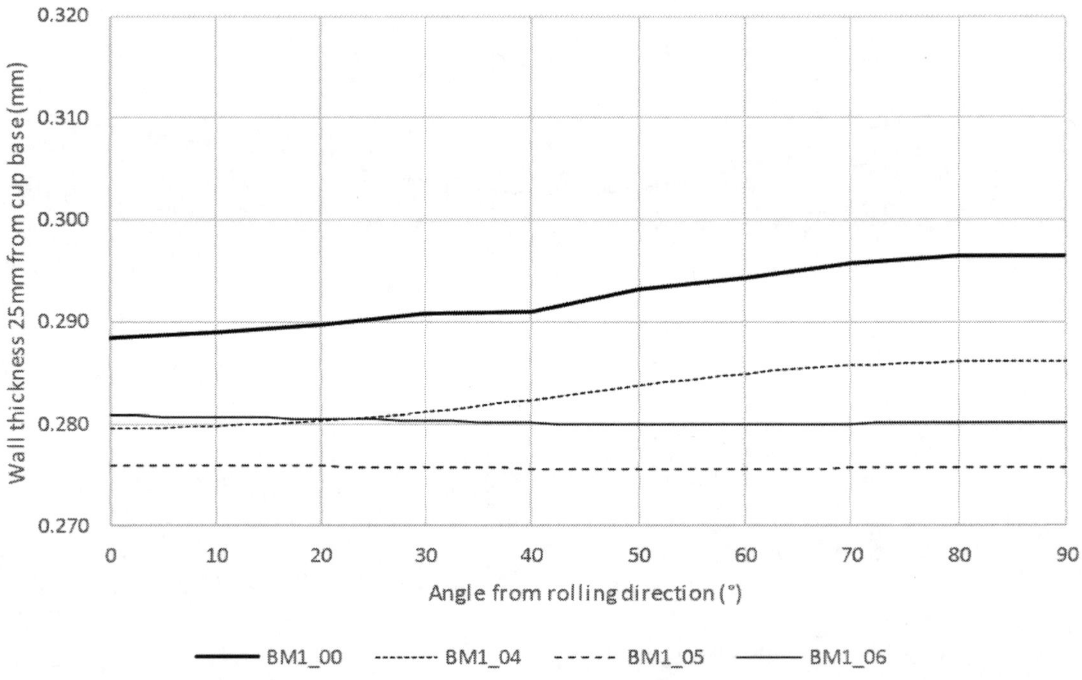

Figure 8: (AA5352) Wall thickness vs angle from rolling direction 25mm above cup base.

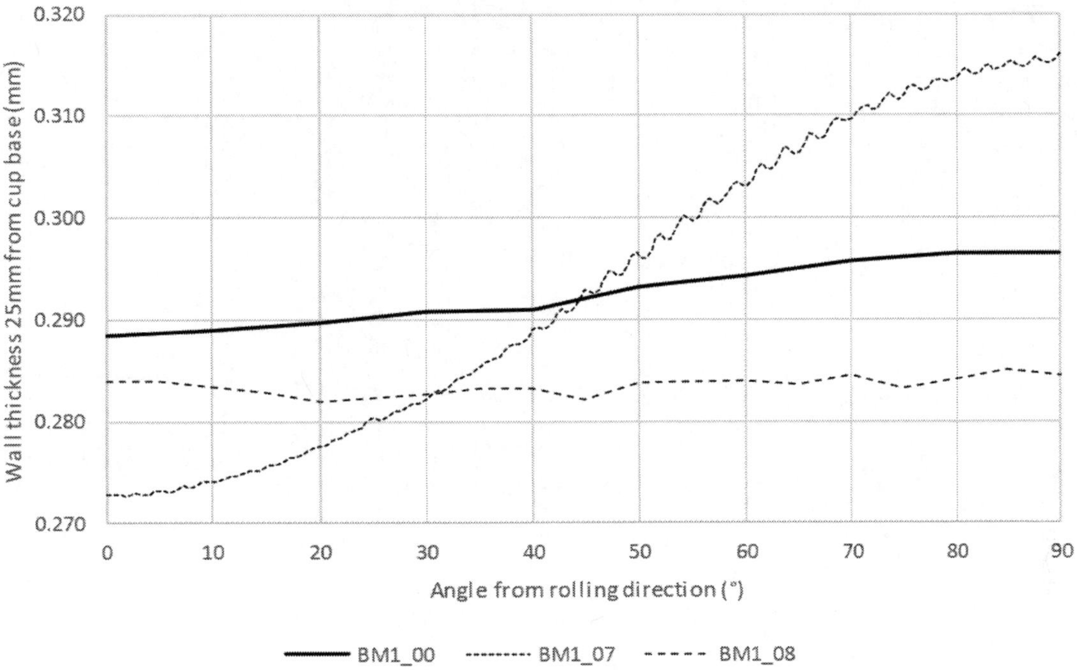

Figure 9: (AA5352) Wall thickness vs angle from rolling direction 25mm above cup base.

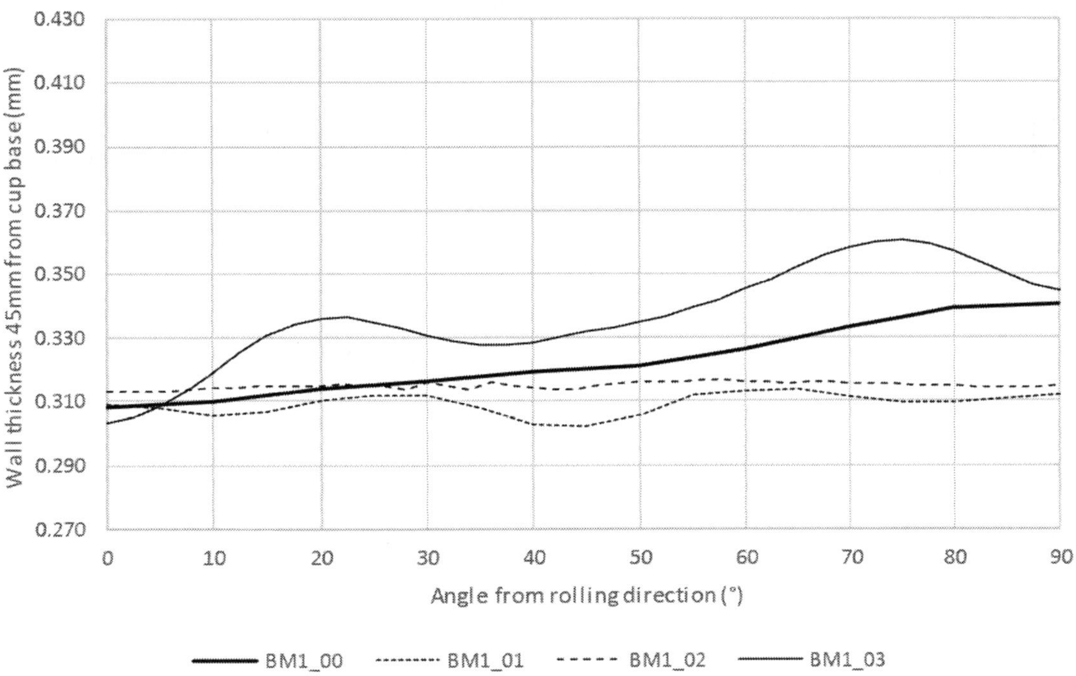

Figure 10: (AA5352) Wall thickness vs angle from rolling direction 45mm above cup base.

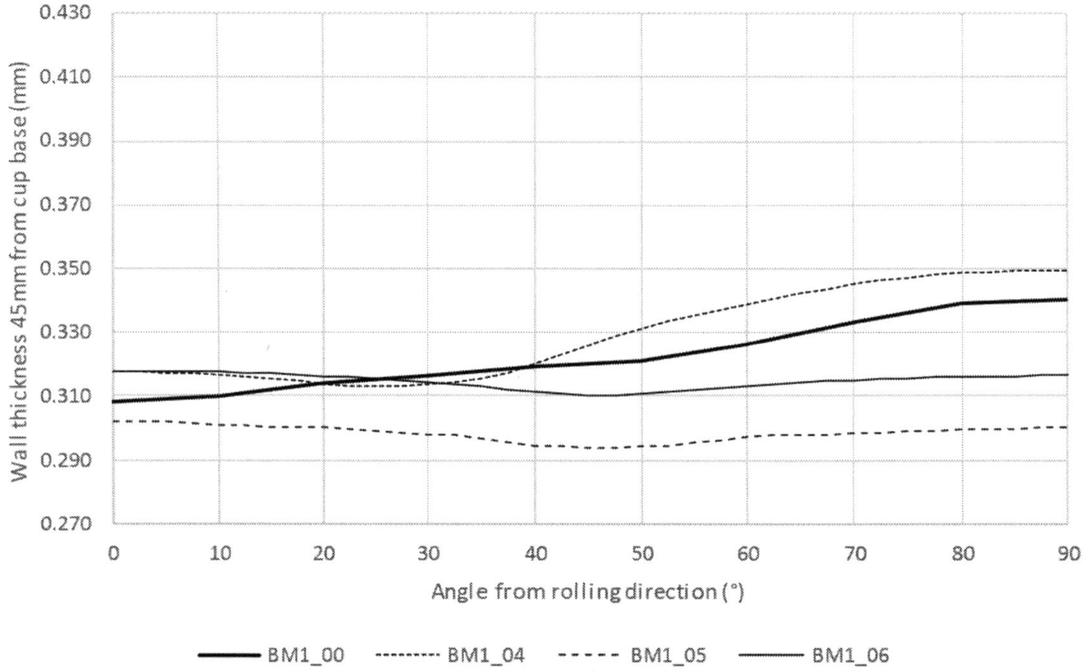

Figure 11: (AA5352) Wall thickness vs angle from rolling direction 45mm above cup base.

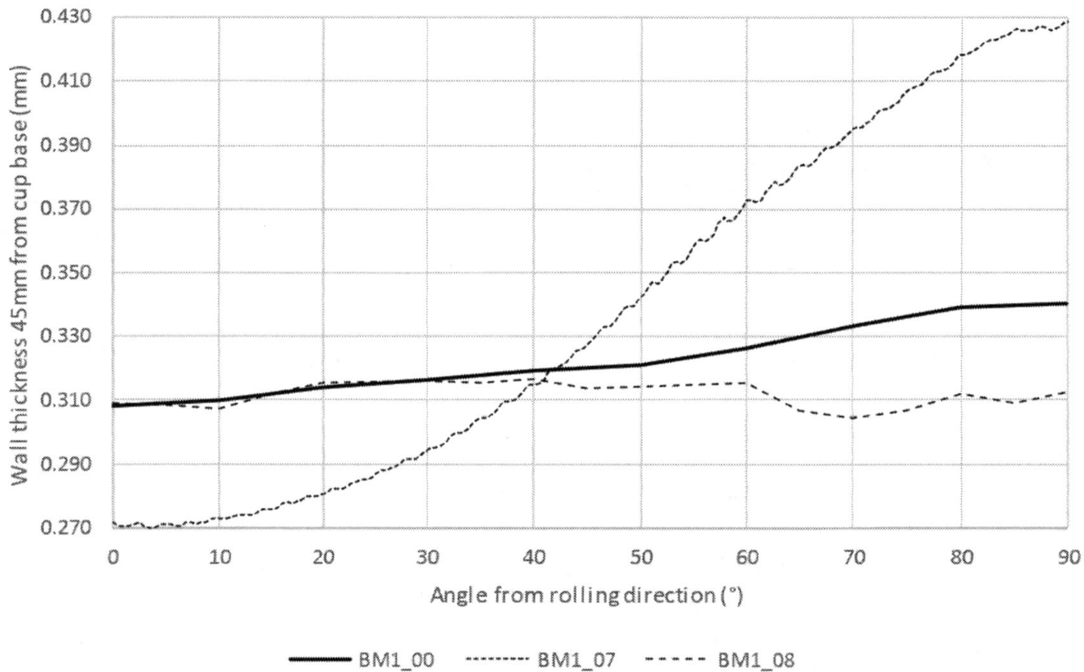

Figure 12: (AA5352) Wall thickness vs angle from rolling direction 45mm above cup base.

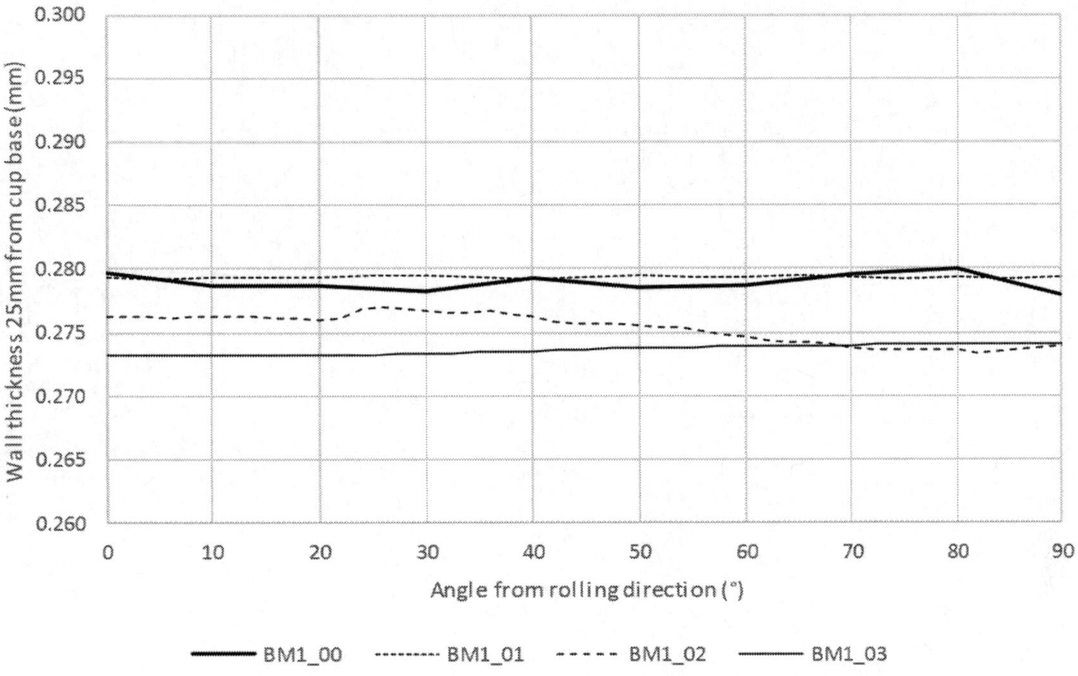

Figure 13: (TH330) Wall thickness vs angle from rolling direction 25mm above cup base.

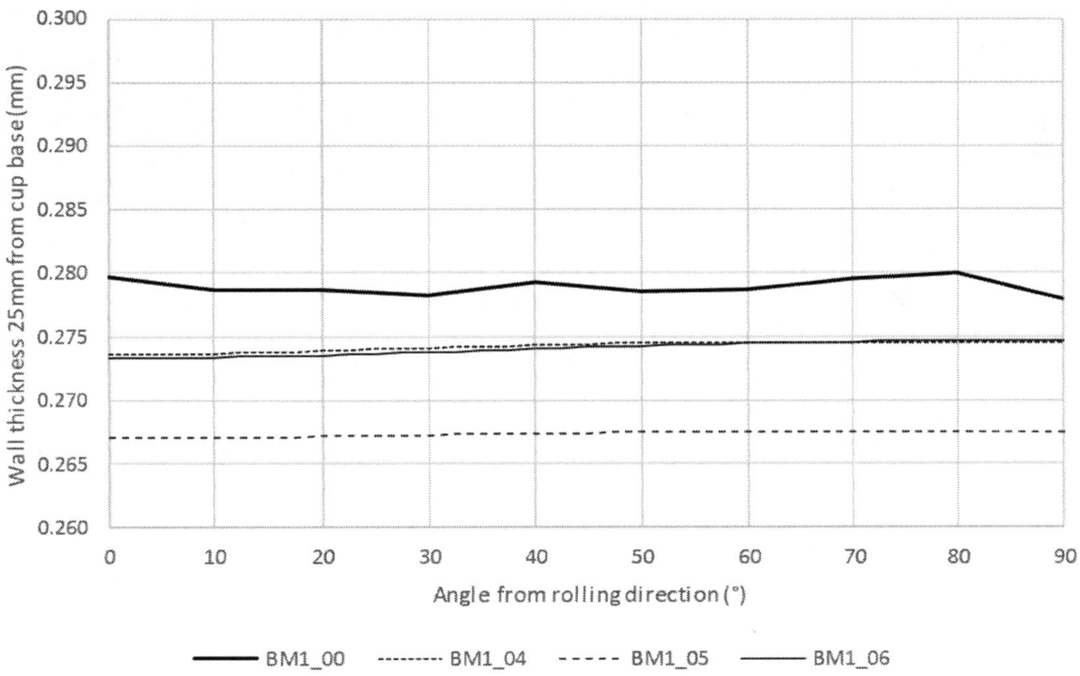

Figure 14: (TH330) Wall thickness vs angle from rolling direction 25mm above cup base.

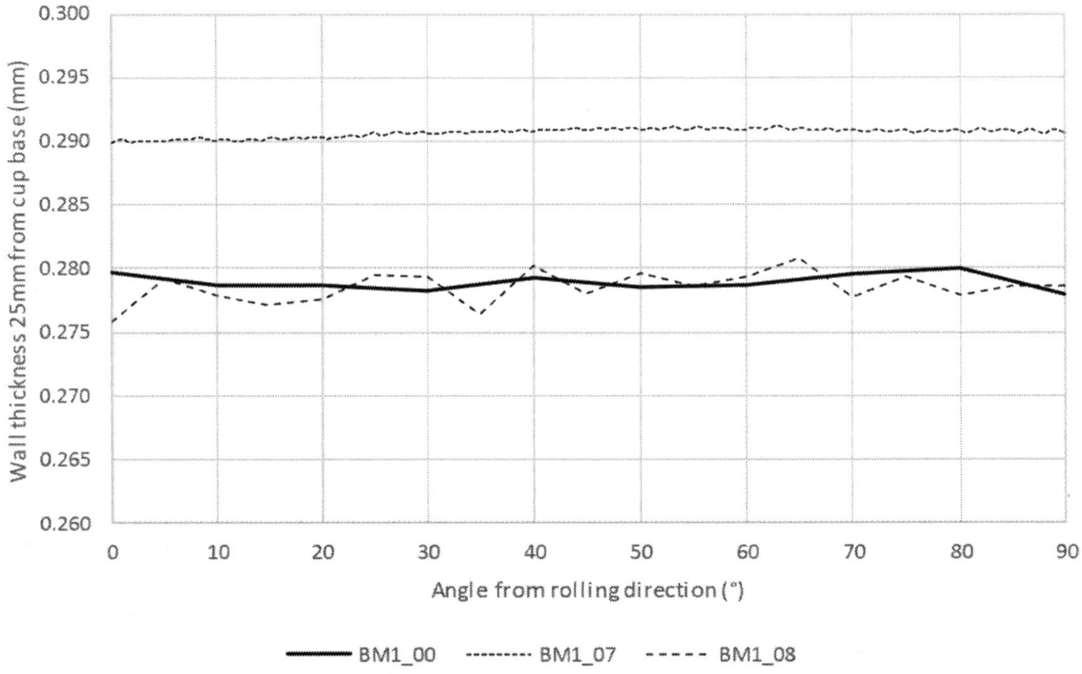

Figure 15: (TH330) Wall thickness vs angle from rolling direction 25mm above cup base.

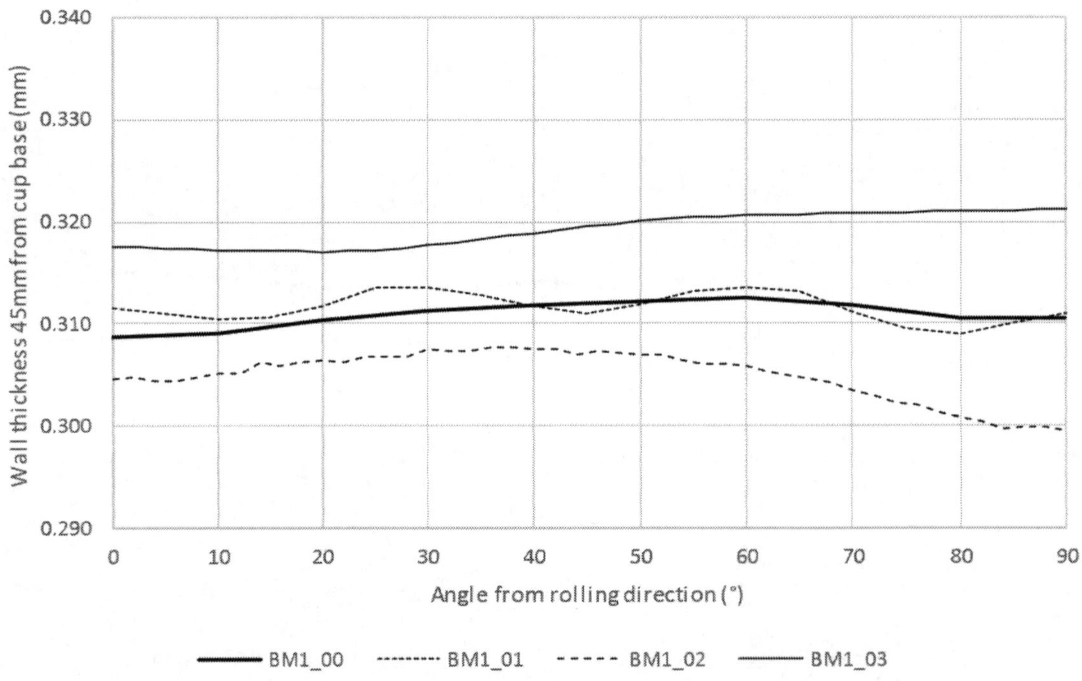

Figure 16: (TH330) Wall thickness vs angle from rolling direction 45mm above cup base.

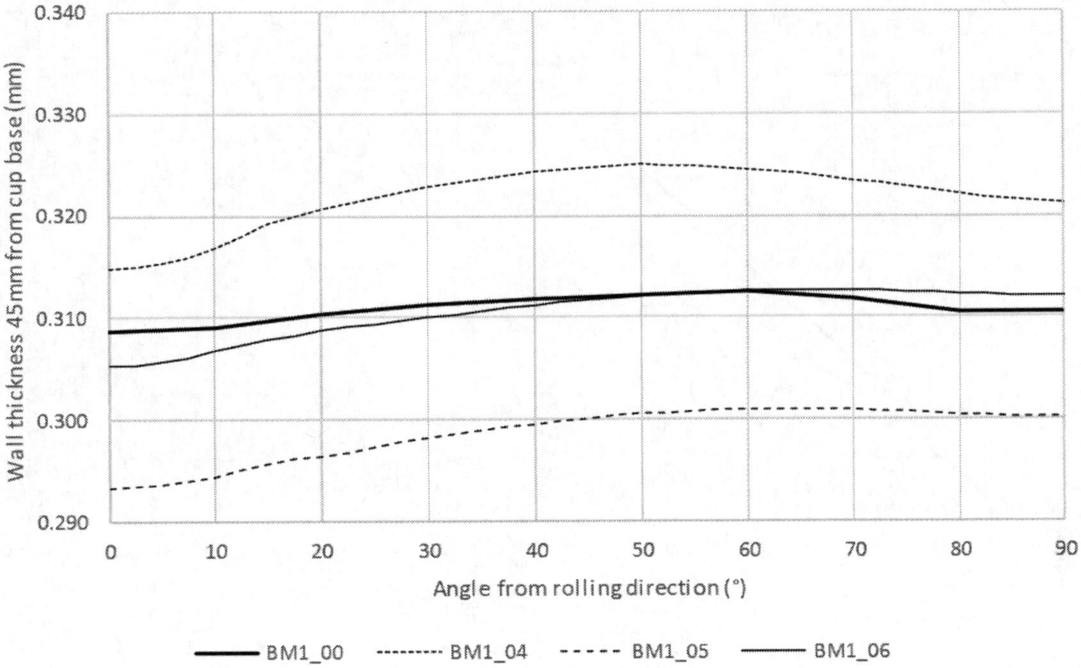

Figure 17: (TH330) Wall thickness vs angle from rolling direction 45mm above cup base.

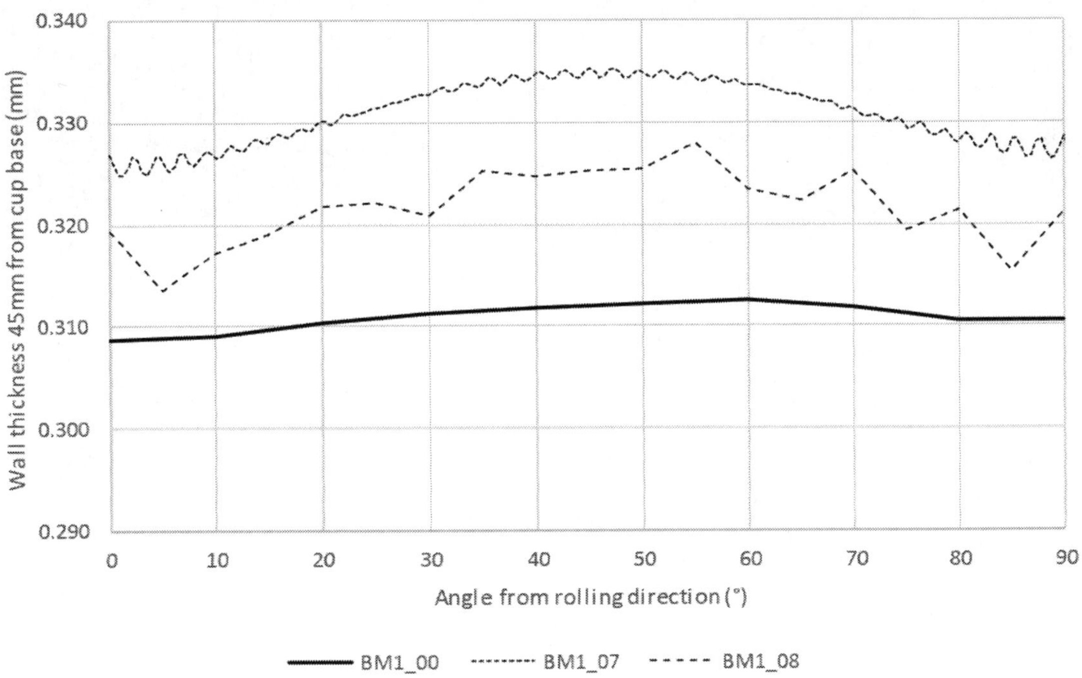

Figure 18: (TH330) Wall thickness vs angle from rolling direction 45mm above cup base.

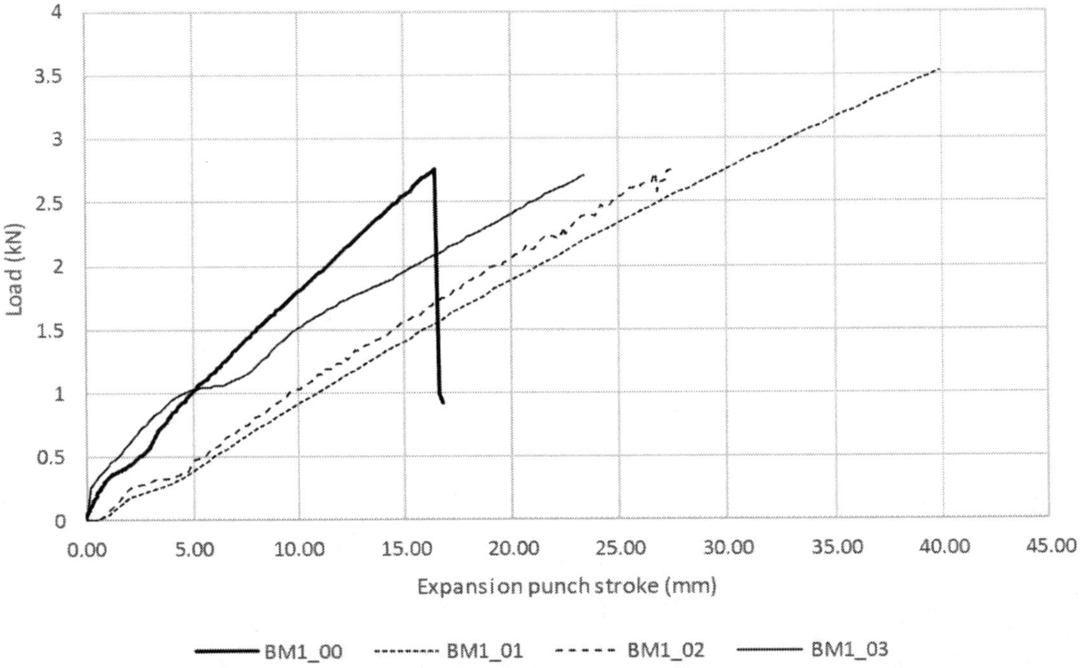

Figure 19: (AA5352) Load vs expansion punch stroke.

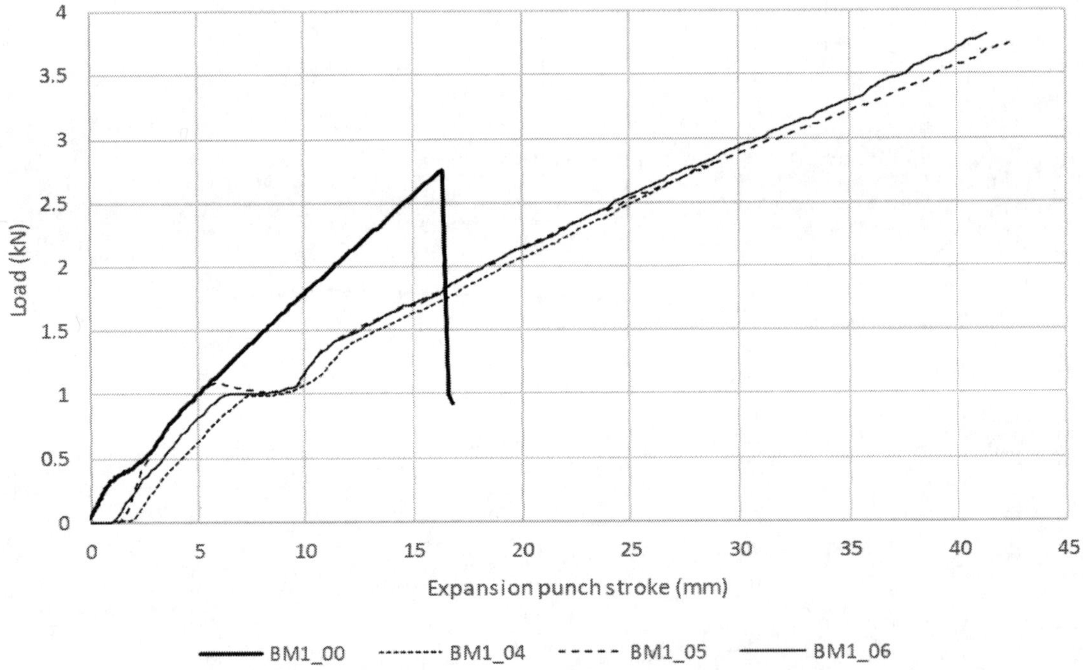

Figure 20: (AA5352) Load vs expansion punch stroke.

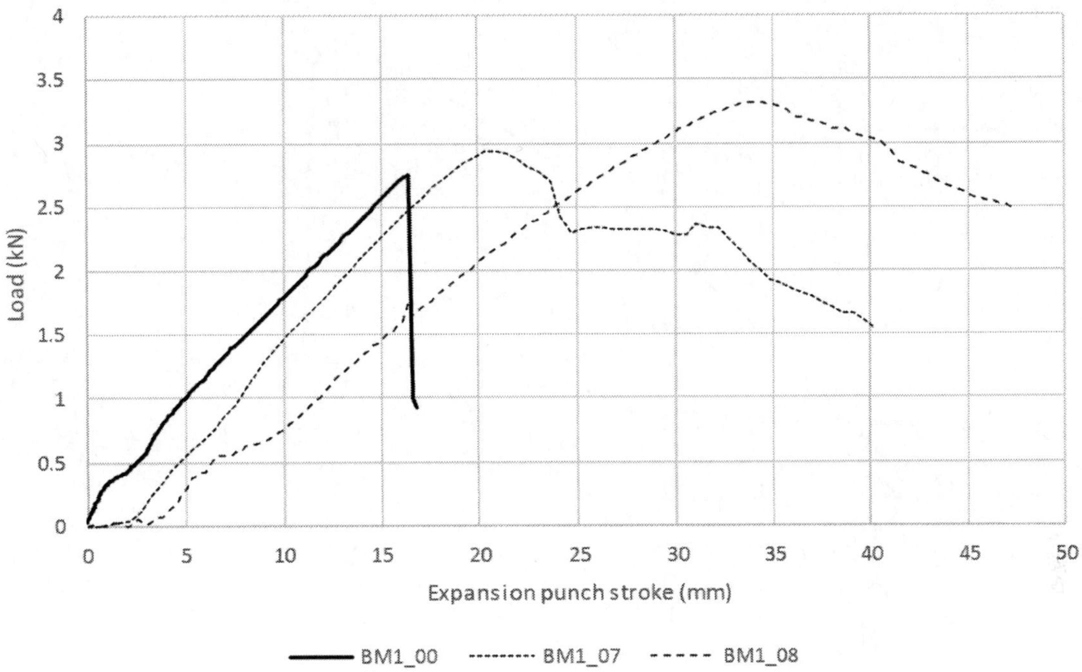

Figure 21: (AA5352) Load vs expansion punch stroke.

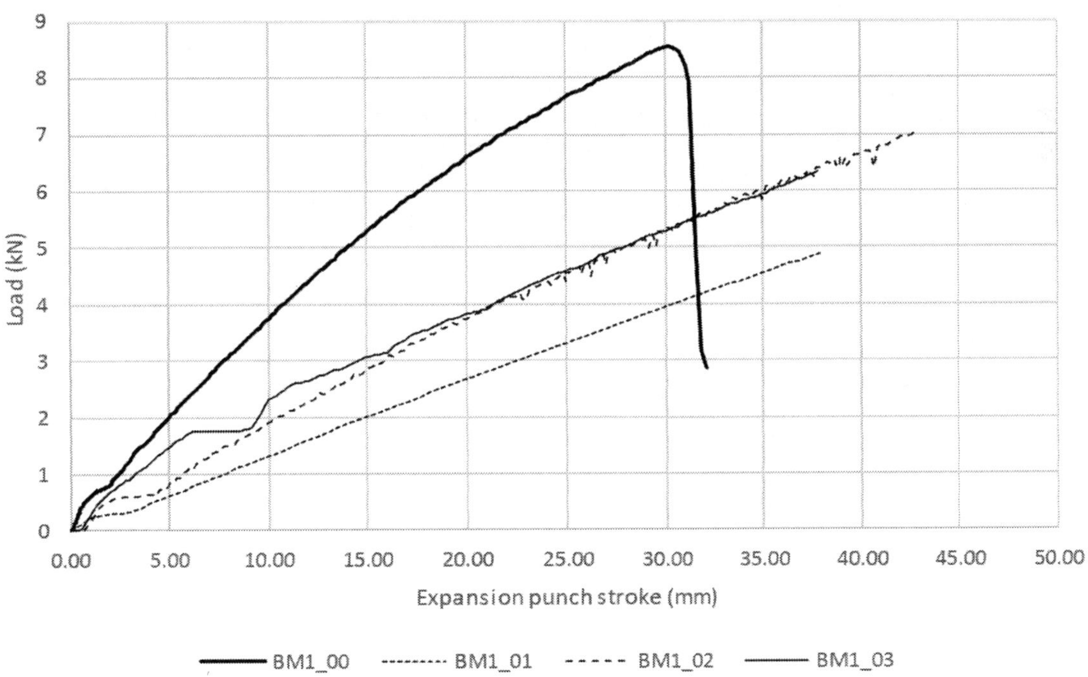

Figure 22: (TH330) Load vs expansion punch stroke.

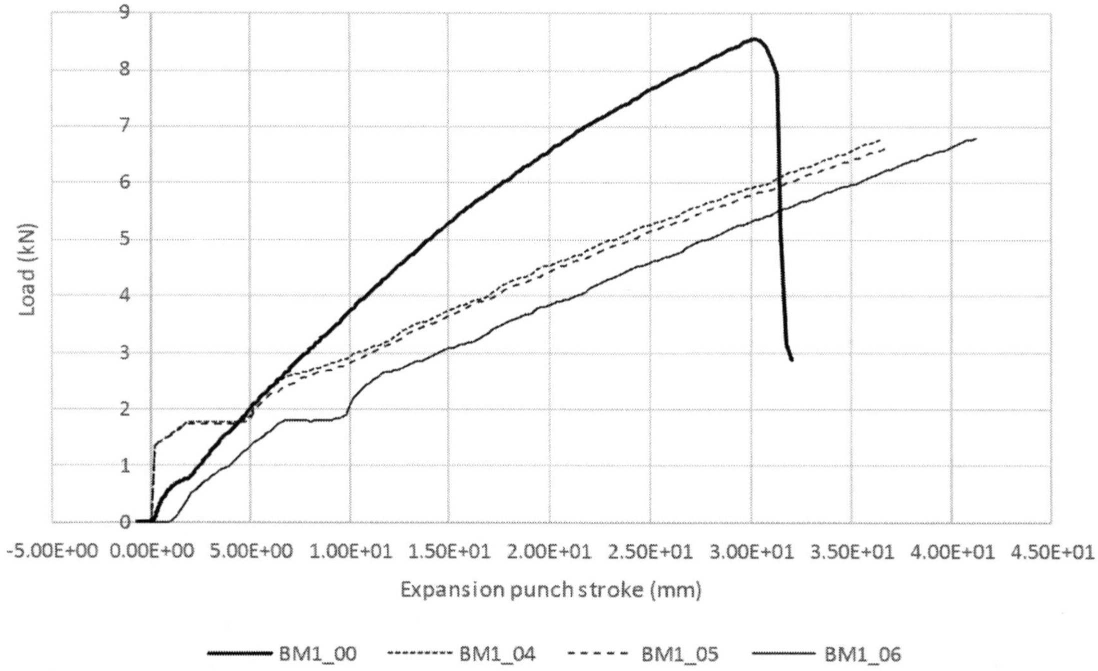

Figure 23: (TH330) Load vs expansion punch stroke.

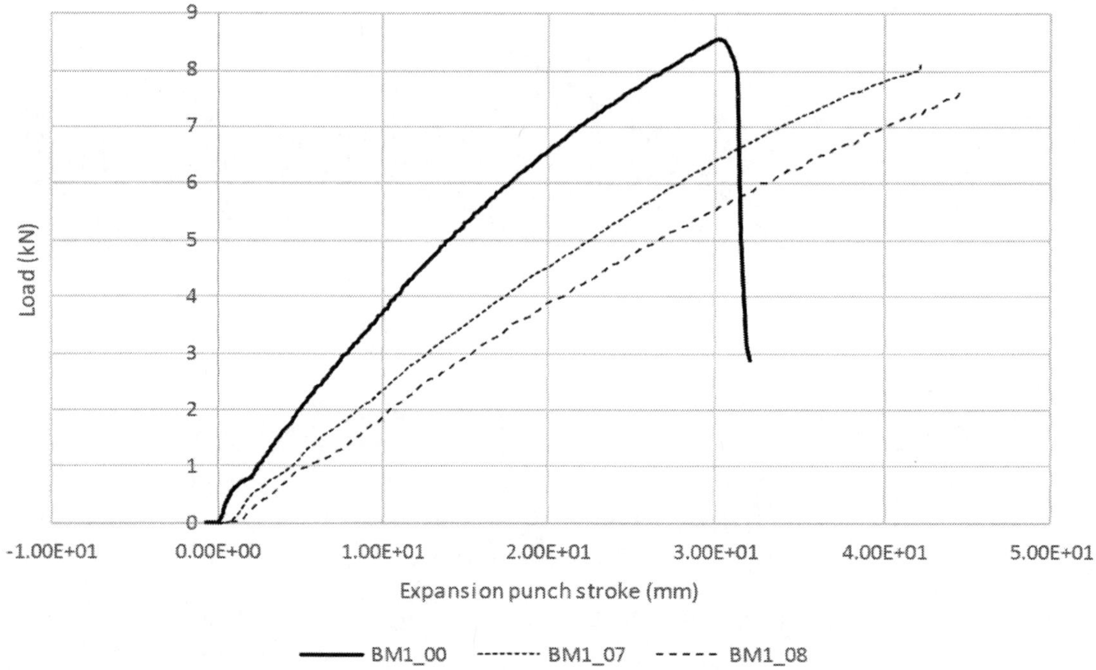

Figure 24: (TH330) Load vs expansion punch stroke.

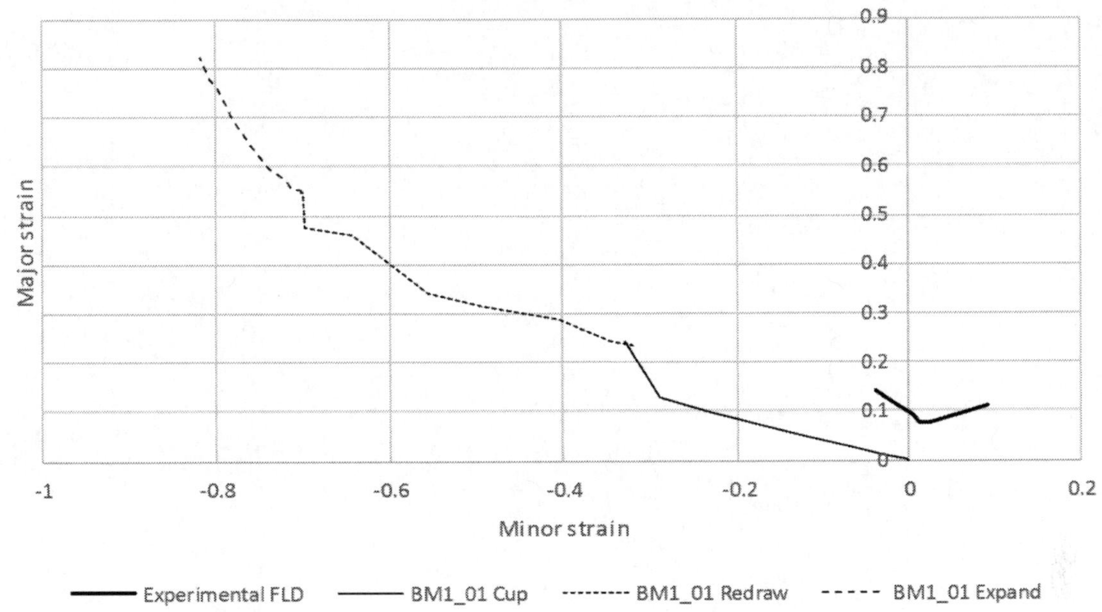

Figure 25: (AA5352) Forming limit diagram and principal strain plot at point of first failure for cup draw, redraw and expansion operations.

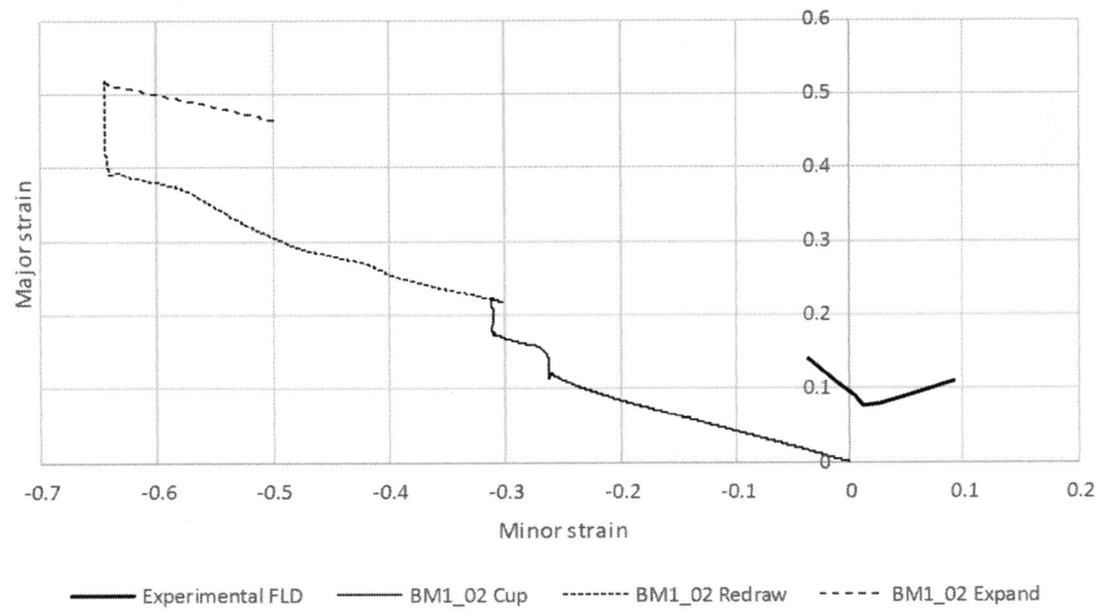

Figure 26: (AA5352) Forming limit diagram and principal strain plot at point of first failure for cup draw, redraw and expansion operations.

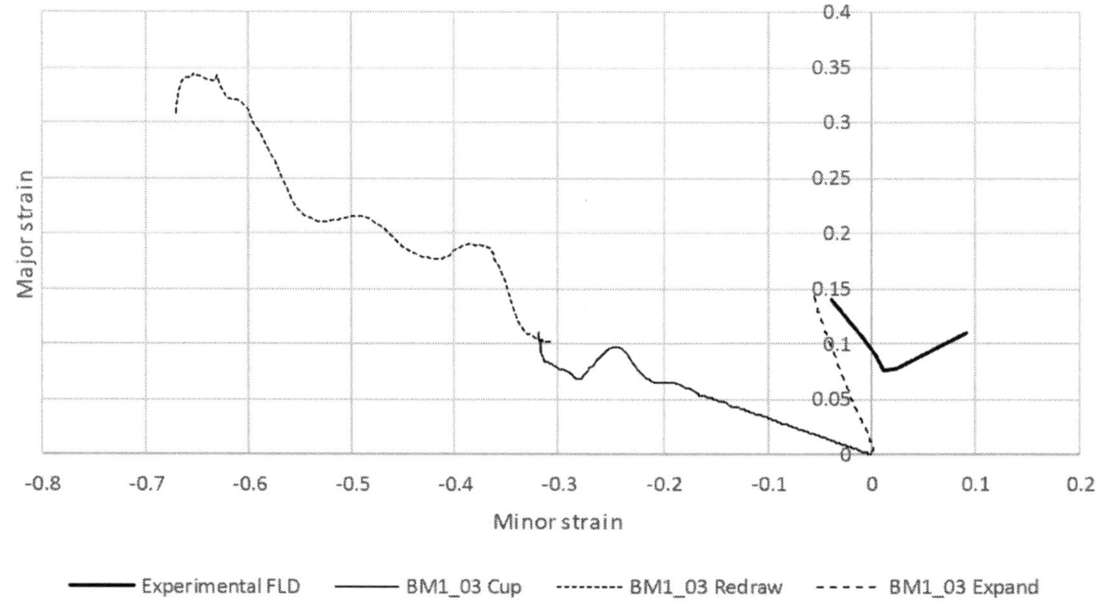

Figure 27: (AA5352) Forming limit diagram and principal strain plot at point of first failure for cup draw, redraw and expansion operations.

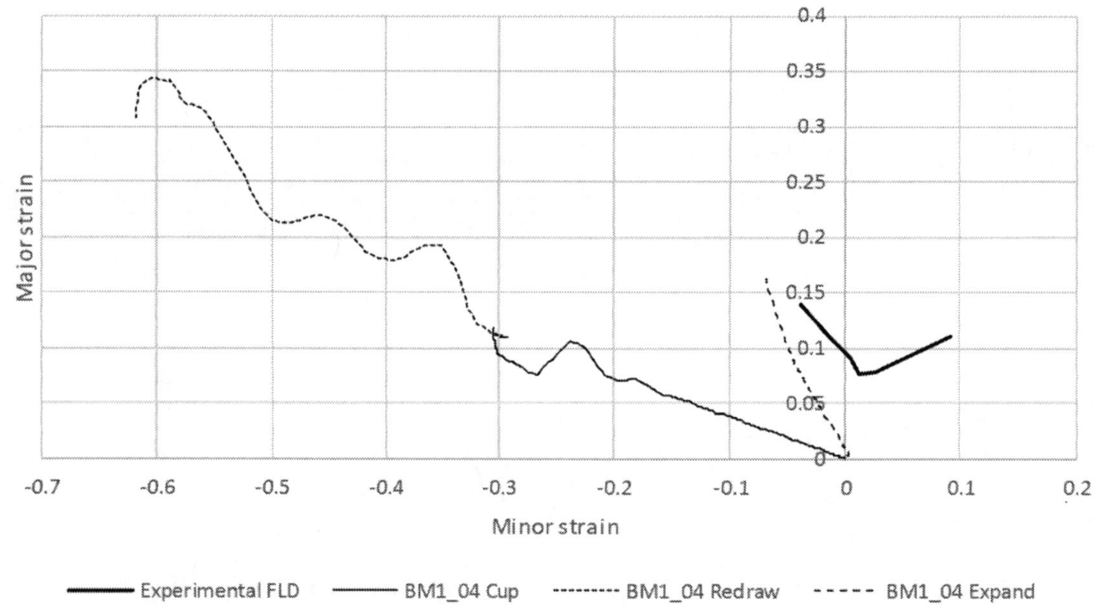

Figure 28: (AA5352) Forming limit diagram and principal strain plot at point of first failure for cup draw, redraw and expansion operations.

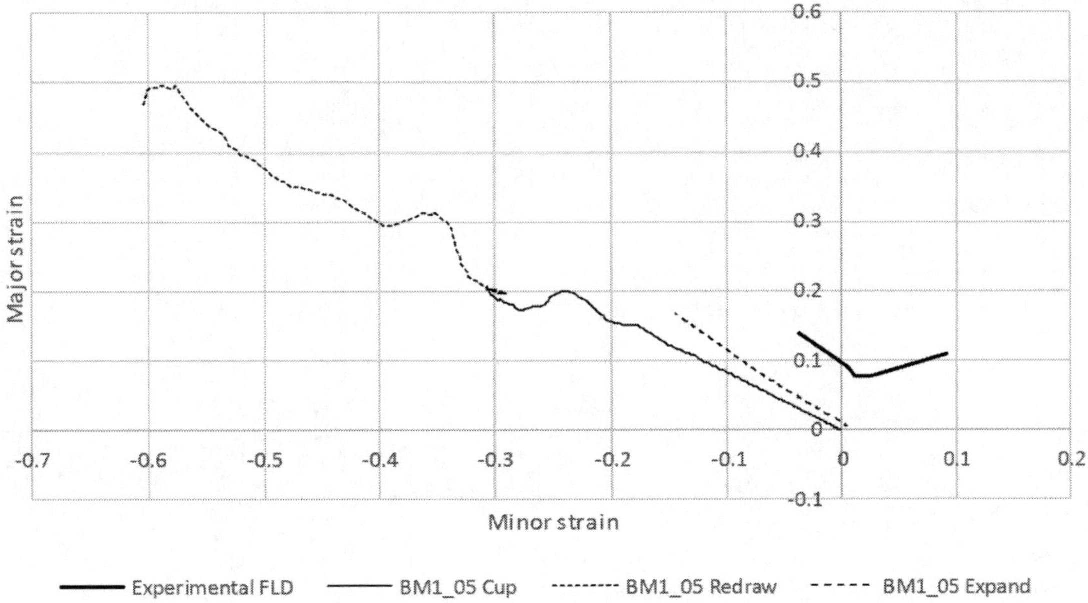

Figure 29: (AA5352) Forming limit diagram and principal strain plot at point of first failure for cup draw, redraw and expansion operations.

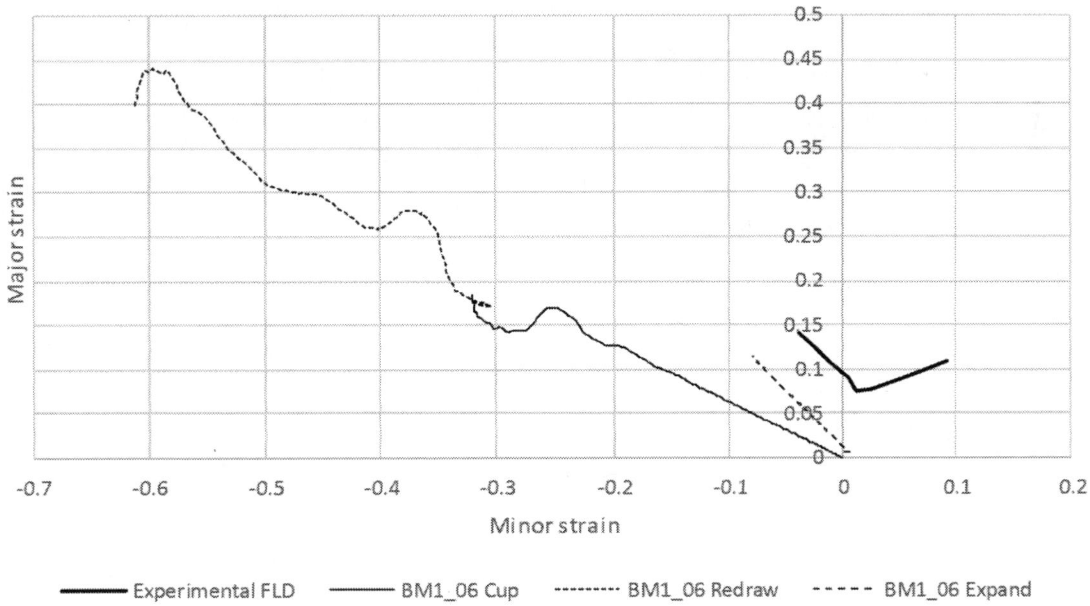

Figure 30: (AA5352) Forming limit diagram and principal strain plot at point of first failure for cup draw, redraw and expansion operations.

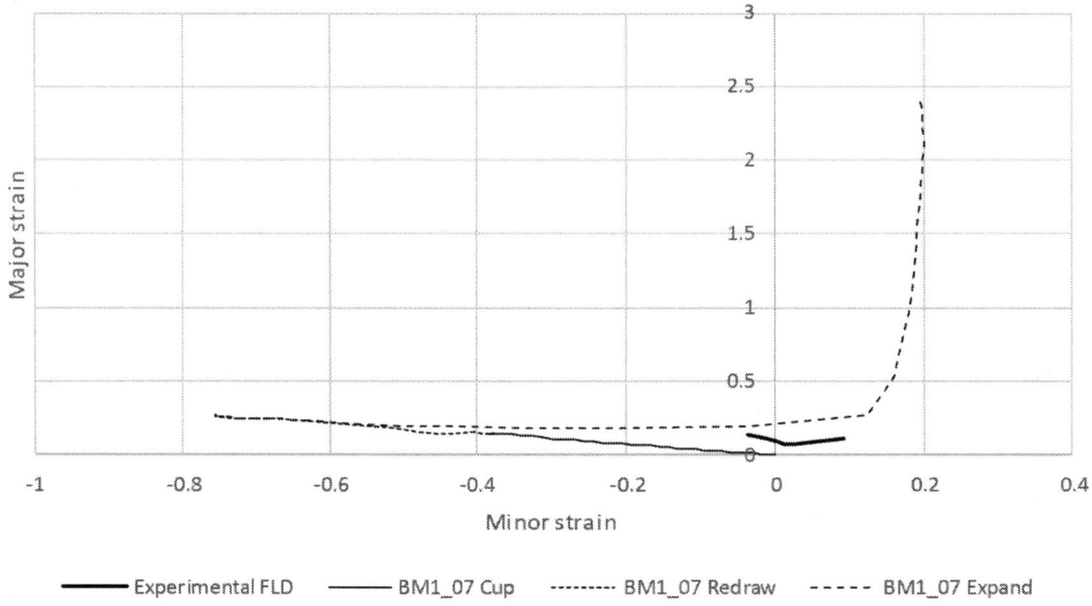

Figure 31: (AA5352) Forming limit diagram and principal strain plot at point of first failure for cup draw, redraw and expansion operations.

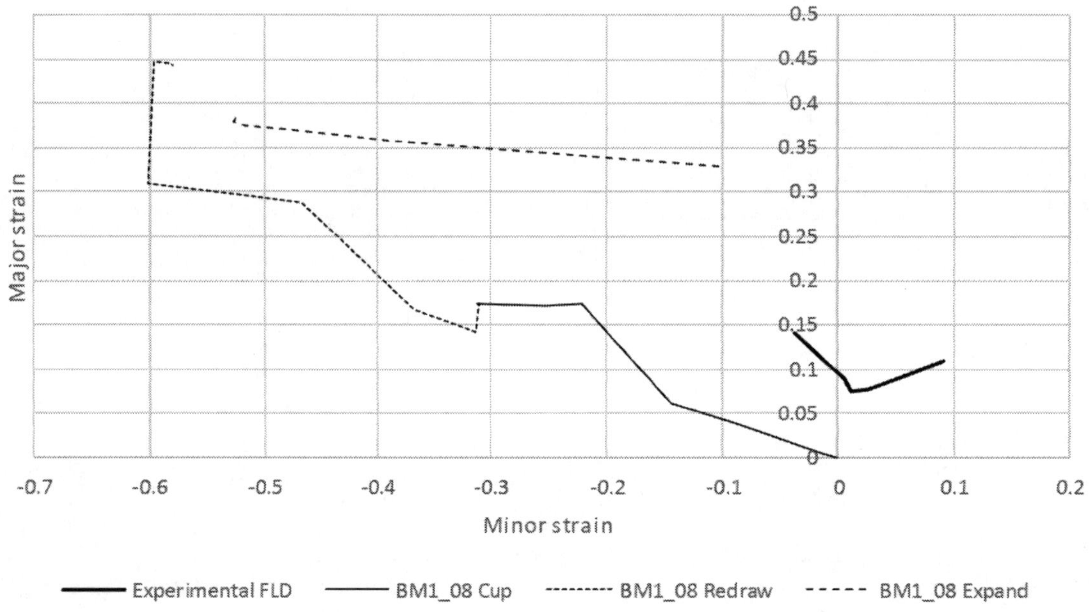

Figure 32: (AA5352) Forming limit diagram and principal strain plot at point of first failure for cup draw, redraw and expansion operations.

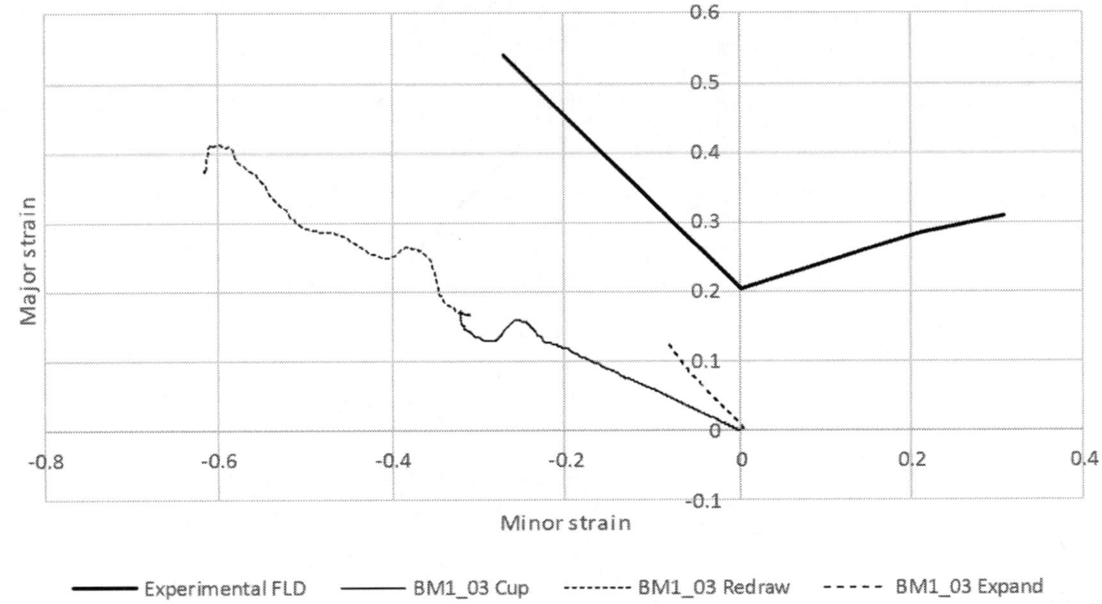

Figure 33: (TH330) Forming limit diagram and principal strain plot at point of first failure for cup draw, redraw and expansion operations.

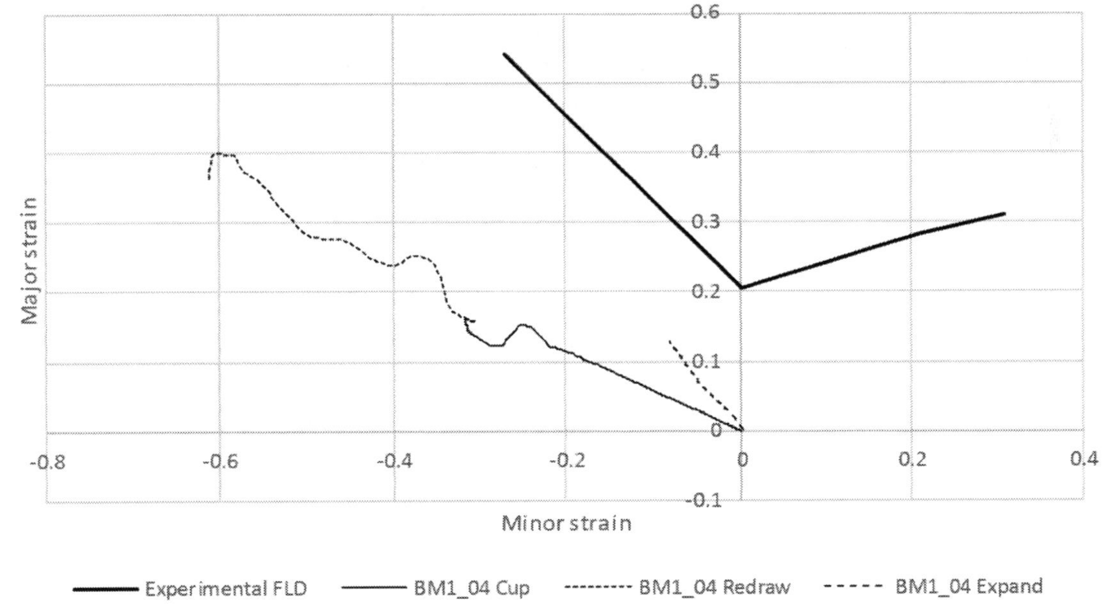

Figure 34: (TH330) Forming limit diagram and principal strain plot at point of first failure for cup draw, redraw and expansion operations.

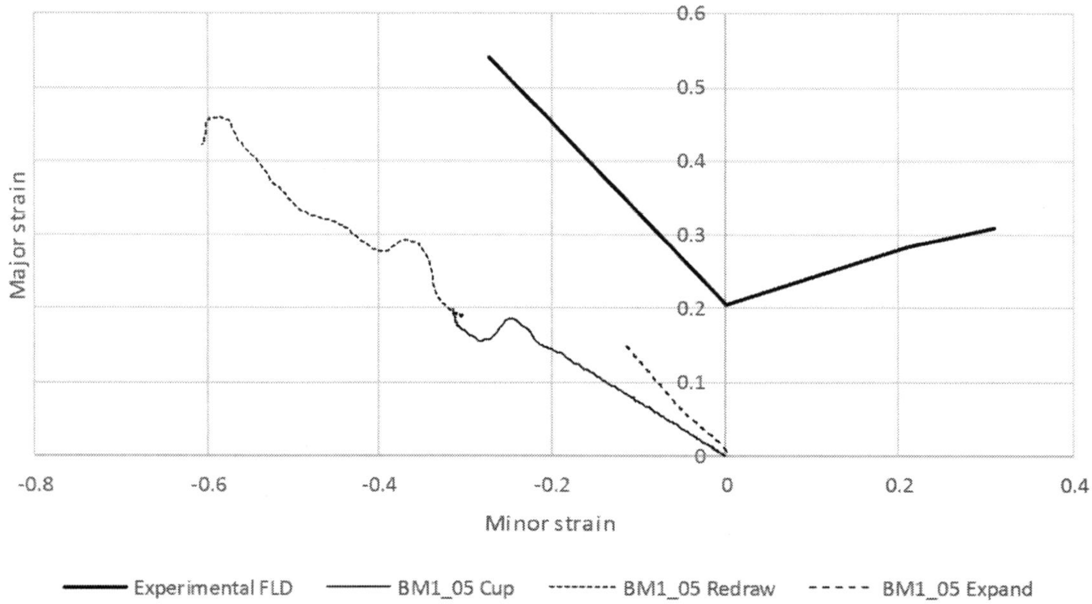

Figure 35: (TH330) Forming limit diagram and principal strain plot at point of first failure for cup draw, redraw and expansion operations.

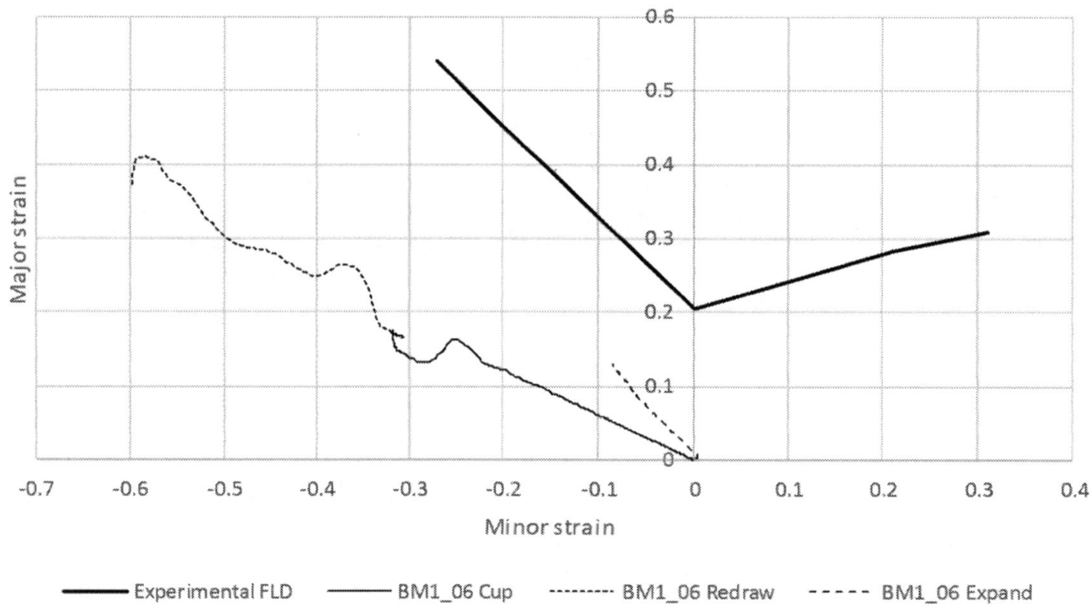

Figure 36: (TH330) Forming limit diagram and principal strain plot at point of first failure for cup draw, redraw and expansion operations.

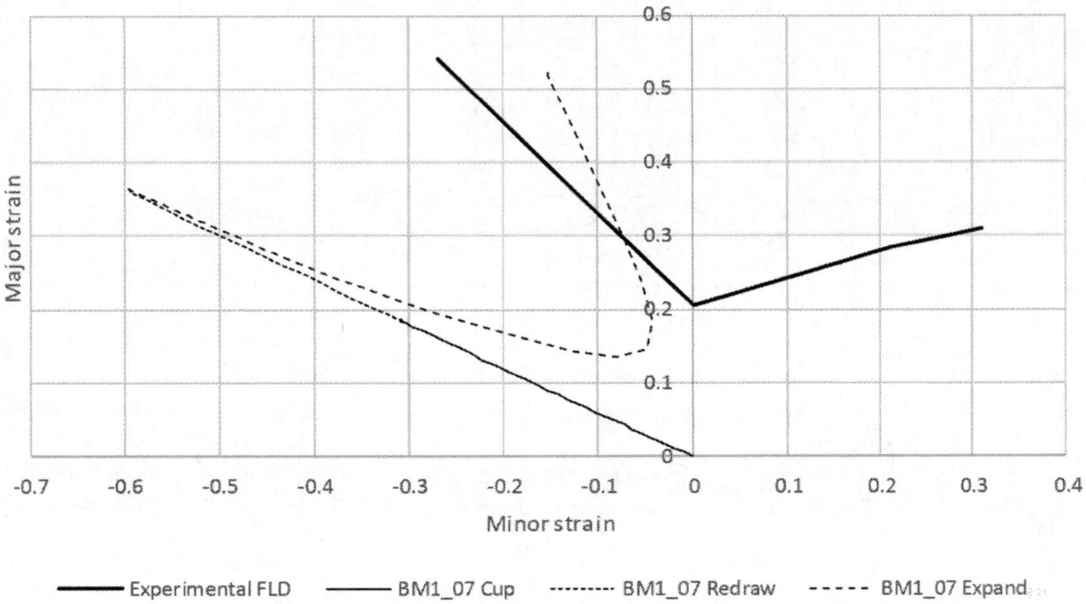

Figure 37: (TH330) Forming limit diagram and principal strain plot at point of first failure for cup draw, redraw and expansion operations.

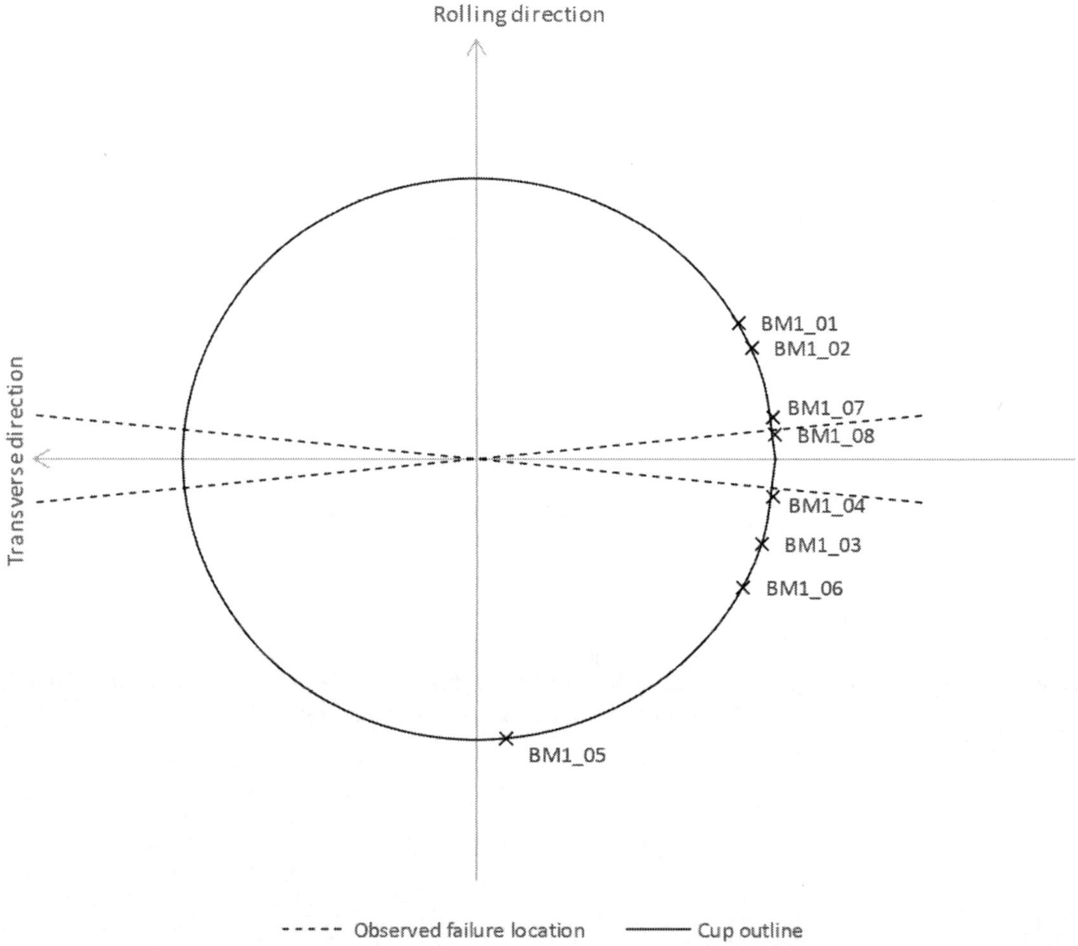

Figure 38: (AA5352) Position of point of first failure during expansion test.

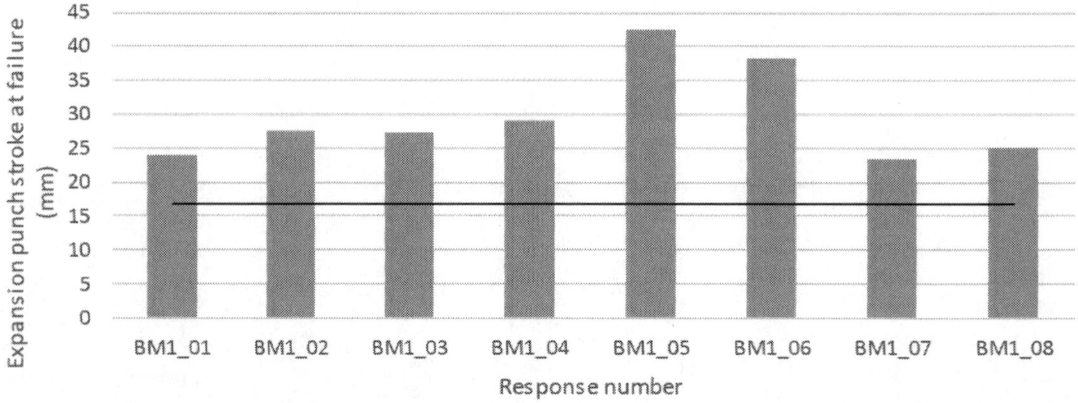

Figure 39: (AA5352) Stroke of expansion punch at first failure.

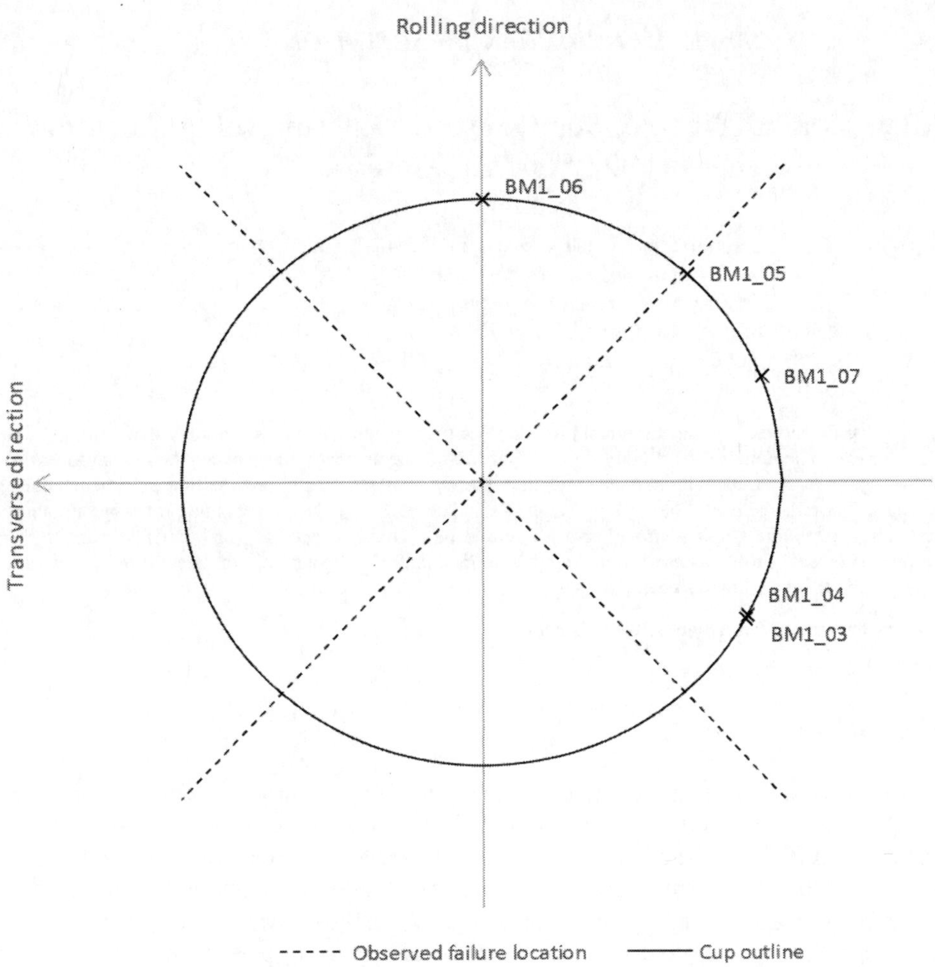

Figure 40: (TH330) Position of point of first failure during expansion test.

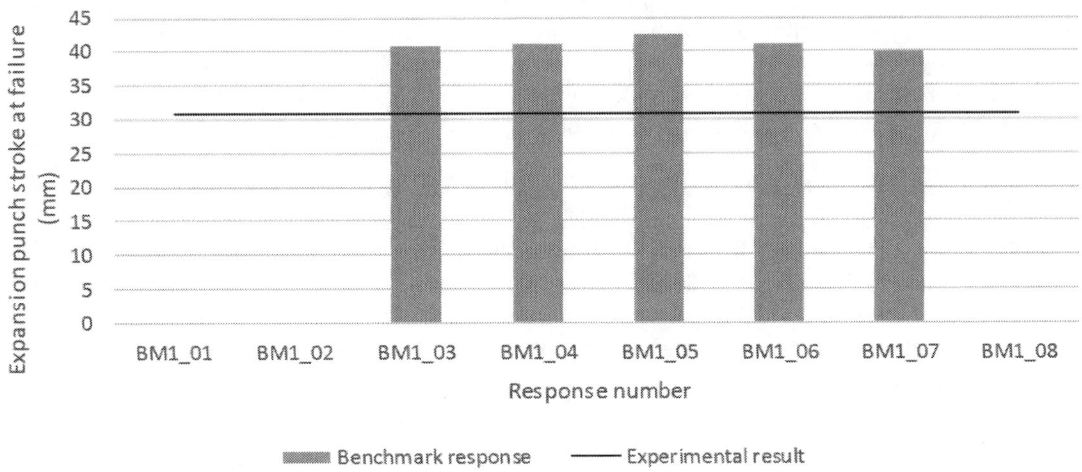

Figure 41: (TH330) Stroke of expansion punch at first failure.

Benchmark 2 – Springback of a Jaguar Land Rover Aluminium Panel

Part A: Benchmark Description

Martin Allen[a], Marta Oliveira[b], Sumit Hazra[c], Oluwamayokun Adetoro[d], Abhishek Das[c] and Rui Cardoso[e]

[a] *Jaguar Land Rover, Coventry CV3 4LF, UK*
[b] *University of Coimbra, Coimbra 3030-790, Portugal*
[c] *University of Warwick, Coventry CV4 7AL, UK*
[d] *University of the West of England, Bristol BS16 1QY, UK*
[e] *Brunel University London, Uxbridge, UB8 3PH, UK*

Abstract. The aim of this benchmark is the numerical prediction of the springback of an aluminium panel used in the production of a Jaguar car. The numerical simulation of springback has been very important for the reduction of die try outs through the design of the tools with die compensation, thereby allowing for the production of dimensionally accurate complex parts at a reduced cost. The forming stage of this benchmark includes one single forming operation followed by a trimming operation. Cross-sectional profiles should be reported at specific (provided) sections in the part before and after springback. Problem description, tool geometries, material properties, and the required simulation reports are summarized in this benchmark briefing.

Keywords: Forming, Trimming, Springback, Plastic Anisotropy

1 INTRODUCTION

Springback is one of the most important problems for the sheet metal forming industry due to the strong geometrical deviations which occurs through elastic recovery after forming. These deviations can lead to many manufacturing difficulties such as joining parts together into a more complex assembly. Springback is influenced by the forming operations and the degree of constraints imposed by the geometry of the part but it is also strongly dependent on the material properties of the blank sheet. For aluminum, springback behaviour is more complex because of its strong plastic anisotropy and low Young's modulus. Consequently, inaccurate material models can lead to major or unexpected deviations in the prediction of springback.

The main objective of this benchmark is to predict the springback of a single stage formed panel, assess the influence of material models and quantify the influence of different numerical modelling techniques that affect springback prediction. Numerical techniques includes the finite elements used, integration rules, implicit or explicit code analysis, contact and friction models and the use of emerging techniques such as isogeometric analysis and meshless methods.

The kinematic hardening effect of bending and unbending deformation through the different die radius and curvatures of the tools can significantly influence the nature and prediction of panel's springback. The springback prediction of different loading/unloading forming operations requires the use of appropriate kinematic and/or combined kinematic/isotropic hardening models, together with sophisticated flow rules and yield functions. Cyclical shear tests for different levels of pre-strains were therefore performed for the material characterisation of the kinematic/isotropic hardening (the Bauschinger effect) for this benchmark study and the measured shear strain-stress curves are summarised in the attached excel file "Cyclical_Shear.xls".

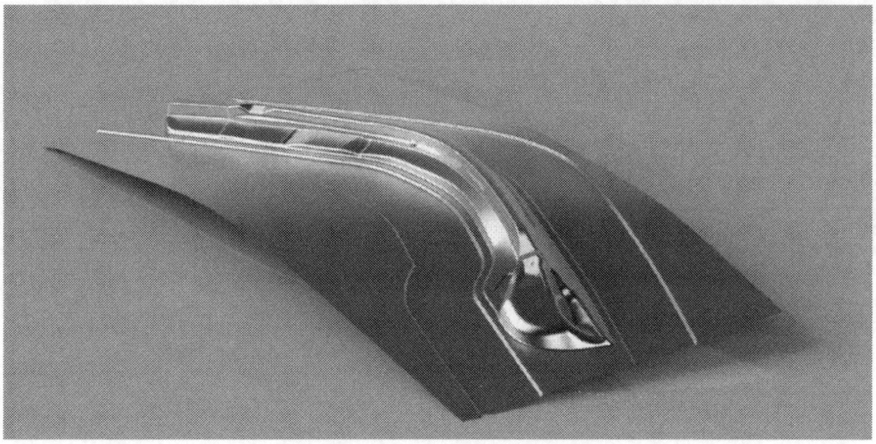

Figure 1. Die face.

2 BLANK MATERIAL

The blank material to be used in this benchmark is the aluminium alloy (AA6451-T4) with thickness t = 3.0 mm. The elastic mechanical properties are given in Table 1.

Table 1. Elastic mechanical properties

Sample	Density, ρ ($g.cm^{-3}$)	Young's modulus, E (GPa)	Poisson's ratio, ν
AA6451-T4	2.7	70.0 GPa	0.3

The uniaxial tensile yield stress and r-values are given in Table 2.

Table 2. Uniaxial Tension Test Data

Test Direction	YS, σ_{yld} (MPa)	r-value
0°	151.28	0.62
45°	171.2	0.33
90°	163.6	0.8

The equal biaxial tensile yield stress and the biaxial r-value are given in Table 3.

Table 3. Equal Biaxial Tension Test Data

σ_b (MPa)	r-value, r_b
153.6	0.55

The material constants for the hardening curve at 0 degrees from the rolling direction (RD) are described in Table 4 for the Voce hardening law.

Table 4. Hardening curve

Voce		
A, (MPa)	B, (MPa)	C
359.093260	196.310139	9.374256

The Voce hardening curve gives a better fitting to the experimental results at 0 degrees from RD. The material constants for Barlat's Yld2000-2d yield function are provided in Table 5 with the eight anisotropy coefficients and the material constants for Barlat's Yld89 yield function are provided in Table 6.

Table 5. Material Constants for Yield Function **Yld2000-2d** (a = 8.0)

Sample	α_1	α_2	α_3	α_4	α_5	α_6	α_7	α_8
AA6451-T4	1.065173	0.841891	0.960059	0.958652	1.034037	1.027112	0.838988	0.877033

Table 6. Material Constants for Yield Function **Yld89** (m = 8.0)

Sample	a	c	h	p
AA6451-T4	1.3033	0.9556	0.9247	0.8465

Cyclical shear mechanical tests were conducted (with the specimen at 0 degrees from RD) for different pre-strains so that a full characterization of the kinematic and/or combined kinematic/isotropic hardening can be conducted effectively for the numerical simulation of the springback of the aluminium panel. The plots for the shear stress vs shear strain for the different pre-strain levels are shown in Figure 2. The excel file "Cyclical_Shear.xls" with the full data for the cyclical shear tests is available on the website of the conference.

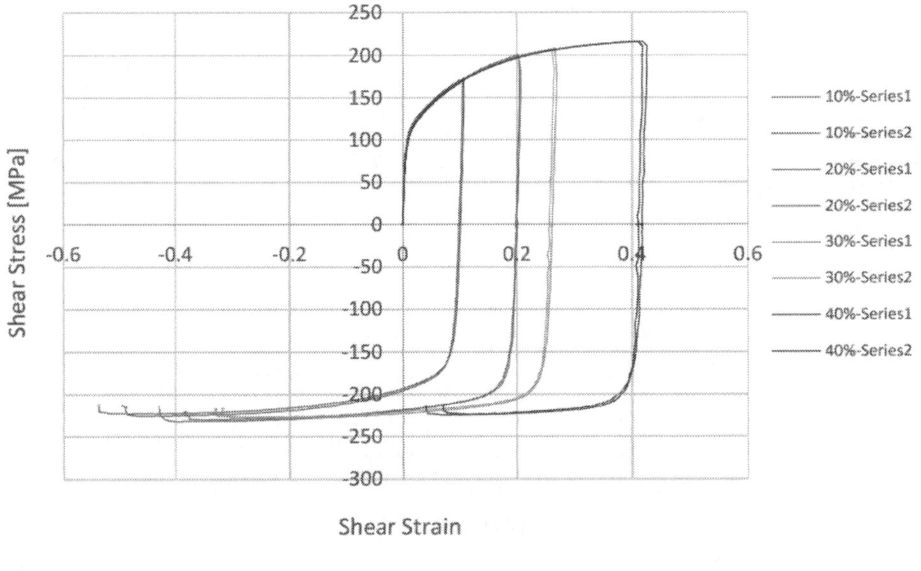

Figure 2. Experimental results for the cyclical shear tests on the AA6451-T4 aluminium alloy.

The rolling direction is specified schematically in Figure 3, with the rolling direction making an angle of 87^0 with the global x-axis.

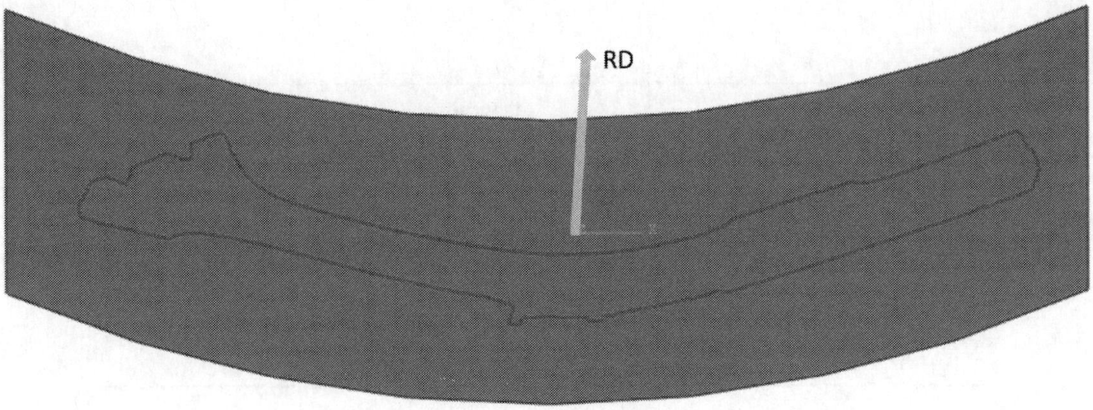

Figure 3. Rolling direction.

3 SIMULATING THE FORMING OPERATION

The simulation involves three operations: forming, trimming and springback. The drawing occurs continuously in a single action process during which the die moves at 100 $mm.s^{-1}$. The CAD geometries for the blank, the lower punch, the upper die and the binder, as well as a mesh for the punch, die and blank holder are provided. The parts/tools are provided in their corresponding orientation and position in the global axis and the forming direction is aligned to the global *z*-axis, whilst no symmetry plane exists as shown in Figure 4. Participants should not move the tool position in the *x-y* plane.

The indicative values for the coefficient of friction to be used in the forming operations are: i) 0.08 for Pam-Stamp and LS-DYNA; ii) 0.14 for AutoForm.

The lower punch, binder and upper die are illustrated in Figure 4. Only one blank material (3 *mm* thick) is investigated in this benchmark, properties of which are given in the previous section. The required simulation boundary conditions are given in Table 7.

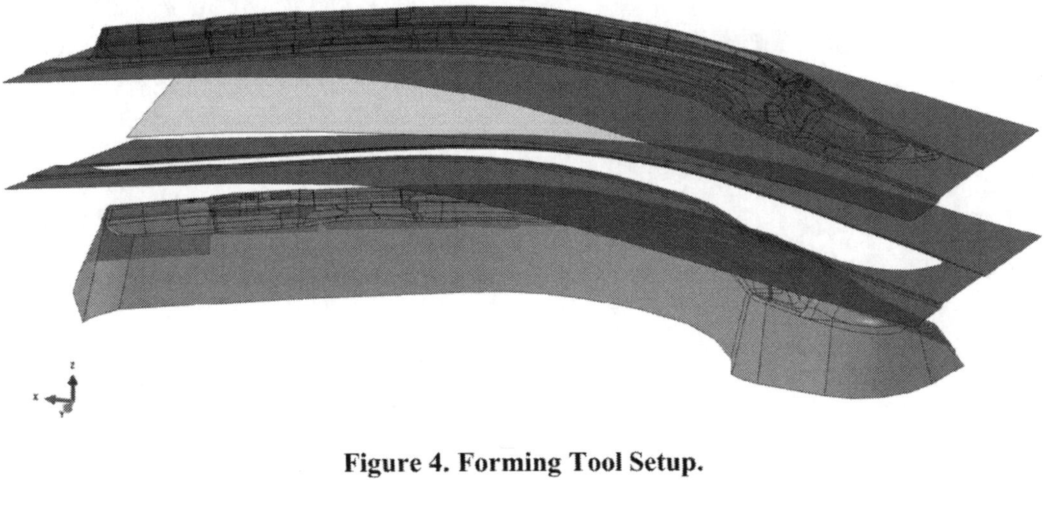

Figure 4. Forming Tool Setup.

Table 7. Boundary Conditions

Part	Operation			
	Binder Closure	**Forming**	**Trimming**	**Springback**
Die	Z-Disp: −300 *mm*	Z-Disp: −200 *mm*	Clamped	Refer to Section 4.3 (SB Analysis)
Binder	Clamped	BHF: 1900 *kN*	Clamped	
Blank	Free	Free	Free	
Punch	Clamped	Clamped	Clamped	

4 FORMING ANALYSIS

4.1 Tool moving directions and force:

4.1.1 Binder Closure

Lower Punch: stationary
Upper Die: moving (z-direction), see Table 7
Binder: stationary

4.1.2 Forming

Lower Punch: stationary
Upper Die: moving (z-direction), see Table 7
Binder: loading (z-direction), see Table 7

4.1.3 Blank holding force

The blank holding force is defined in table 7. It should be applied after the binder has been moved into position.

4.2 Trimming

The trim line is illustrated in Figure 5 (the red line/edge) and it is provided in the attached IGES file.

Figure 5. Trim line on the formed part.

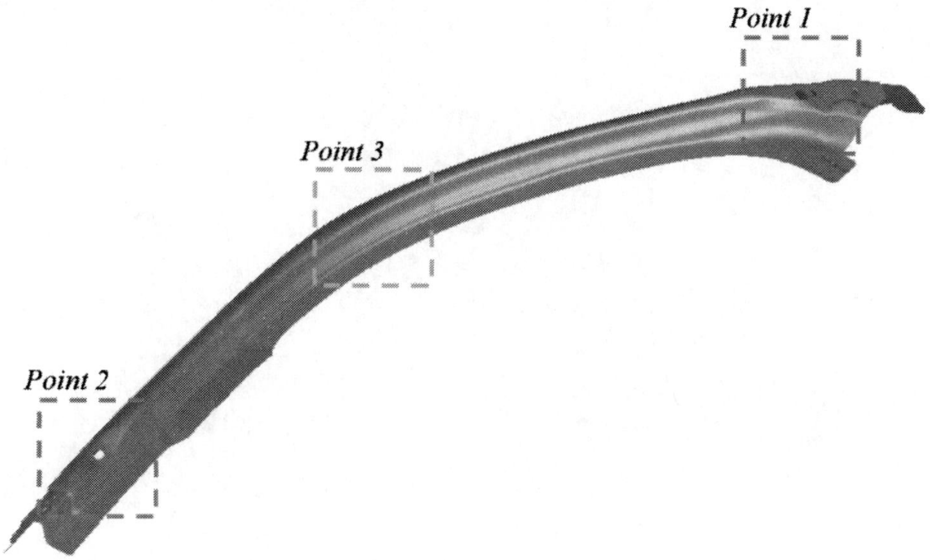

(a) Location of boundary conditions for springback analysis

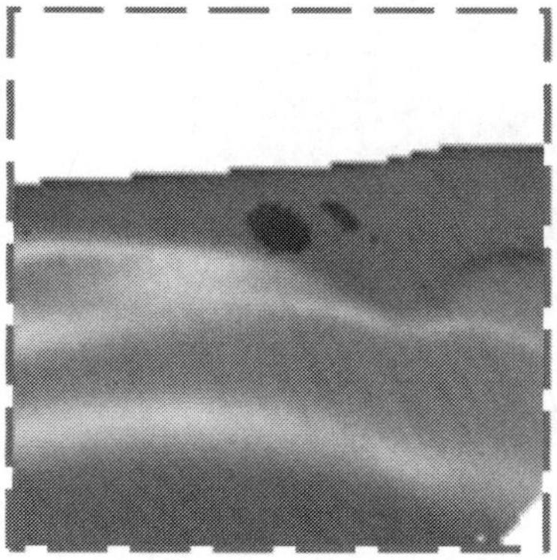

(b) Point 1 – Pin BC

(c) Point 2 – Slot BC

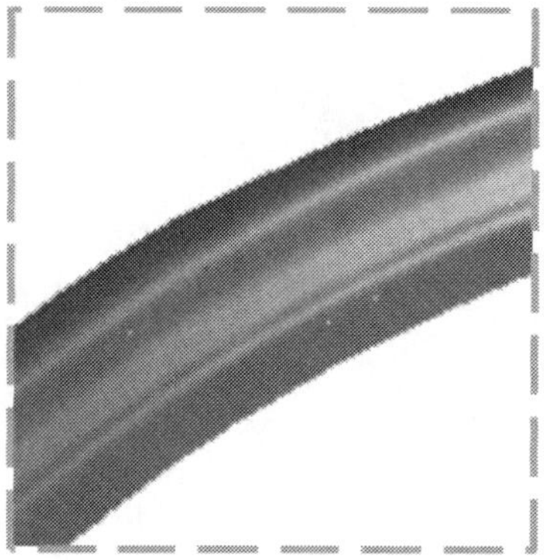

(d) Point 3 – Simply Supported

Figure 6. Springback BC locations.

4.3 Springback Analysis

The locations of the boundary conditions (BCs) to be defined for springback analysis simulation are depicted in Figure 6. A 3-2-1 locating configuration will be used for part measurement. Points 1 and Point 2 correspond to the centre of the holes shown in Figures 5 and 6.

4.3.1.1 ***Point 1 – Pin BC (all dimensions in mm)***

The blank is restrained in all global translation directions, *X*, *Y*, *Z* at Point 1 with the coordinates, (-749.3, 75.5, 206.2).

4.3.1.2 ***Point 2 – Slot (all dimensions in mm)***

A local coordinate system is to be defined and restrained in translation directions, y', z'. The coordinates of the origin (Point 2) of the local coordinate system is (711.0, 83.8, 220.0) and the vector defining the free x' local axis is (30.0, 10.0, 0.1).

4.3.1.3 ***Point 3 – Simply Supported (all dimensions in mm)***

The blank is restrained in global translation direction, *Z* at Point 3 with the coordinates, (-68.7, -46.5, 193.4).

4.4 Simulation Files

CAD geometry (IGES) files are provided for the die face, binder, blank, punch and the trim line. The trim lines are indicated by lines in the IGES file.

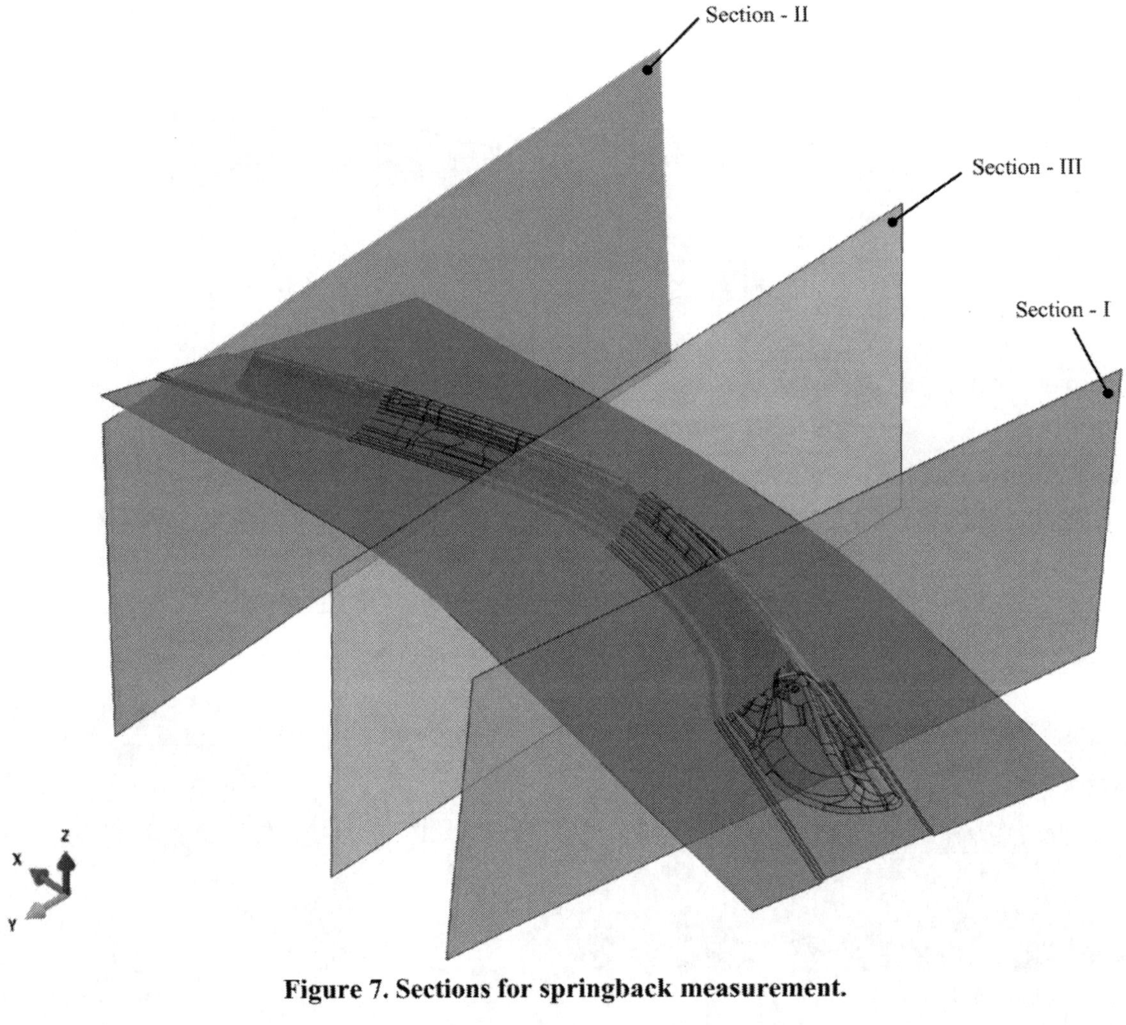

Figure 7. Sections for springback measurement.

Journal of Physics: Conference Series **734** (2016) 022002 doi:10.1088/1742-6596/734/2/022002

5 BENCHMARK REPORT

The due date for benchmark submission is listed on the website. All results are to be reported using the benchmark report template which can be downloaded from the conference website.

5.1 General Description

- Benchmark participant: name, affiliation, address, email and phone number.
- Simulation software: name of the FEM code, general aspects of the code, basic formulations, element/mesh technology, type of elements, number of elements, contact property model and friction formulation.
- Simulation hardware: CPU type, CPU clock speed, number of cores per CPU, main memory, operating system, a breakdown of CPU time for the three stages and analysis methods adopted (e.g. explicit or implicit) for each operation.
- Material model: Yield function/Plastic potential, Hardening rule and Stress-Strain Relation, strain-based.
- Delegate's remarks on the results template.

Table 8. Section normal vectors

Plane	x	y	z
Section I	- 0.985572	0.100936	- 0.135870
Section II	- 0.997984	- 0.062806	0.009108
Section III	-0.998390	-0.044252	-0.035492

5.2 Simulation Results Required

The following information are requested from your simulation:

- Die stroke (*mm*) vs. total punch force (*kN*) from the simulation during forming, reported for at least every 5 *mm* of die movement.
- Blank thickness after forming at Sections I, II and III (as shown in Figure 7). The sections are provided as IGES files and the normal vectors of these sections are provided in Table 8, whilst the origin points coincide with the points defined in sections 4.3.1.1, 4.3.1.2 and 4.3.1.3, respectively. Local in-plane axes are defined for each section as described in figures 8, 9 and 10 and Table 9 and, together with the normal vectors from Table 8, they form a right-handed local coordinate system that should be used for the report of the blank thickness after forming.
- Profiles of the formed sheet at Sections I, II and III, taken of the punch-side surface for two different instants: (i) end of the forming operation and (ii) after trimming and springback. The profiles should be plotted in graphs with local coordinate system defined by local axes described schematically in figures 8, 9 and 10 and Table 9 and the normal vectors from Table 8. The origin of these coordinate systems are the BC points defined in section 4.3.1.1, 4.3.1.2 and 4.3.1.3., respectively.
- As an option, the part after springback can be reported in the form of a geometric (*.stl) file. The committee will report the springback results from correlation with the real part after springback. This will be carried out by aligning the springback result to the measured data by using the same three BC points from section 4.3 – Springback Analysis.

Table 9. Local axes for the plot of springback profiles

Local axis	x	y	z
$\vec{X}_I$	-0.099951436	-0.994896865	-0.013781841
$\vec{Y}_I$	-0.136593212	0.0	0.990627223
$\vec{X}_{II}$	0.062801814	-0.998026018	0.0
$\vec{Y}_{II}$	0.00908191	0.000571486	0.999958595
$\vec{X}_{III}$	0.04432748	-0.999017054	0.0
$\vec{Y}_{III}$	-0.035464739	-0.001573602	0.999369689

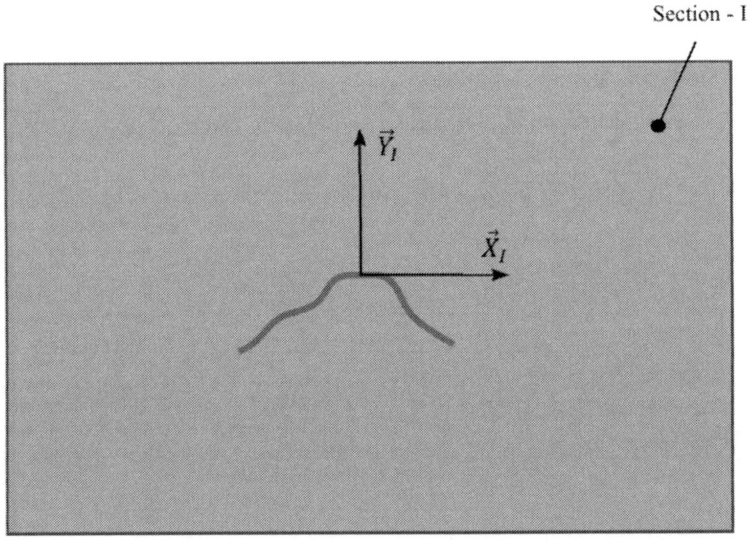

Figure 8. Local coordinate system $\vec{X}_I$ - $\vec{Y}_I$ for the report of springback profile at section I.

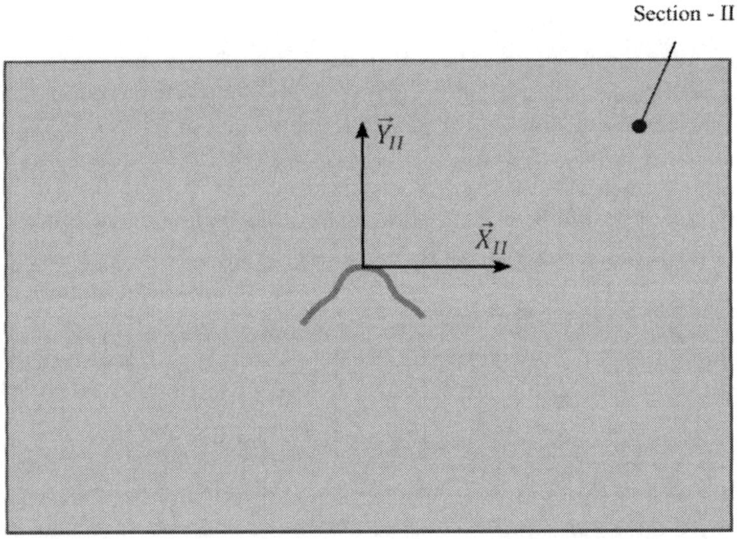

Figure 9. Local coordinate system $\vec{X}_{II}$ - $\vec{Y}_{II}$ for the report of springback profile at section II.

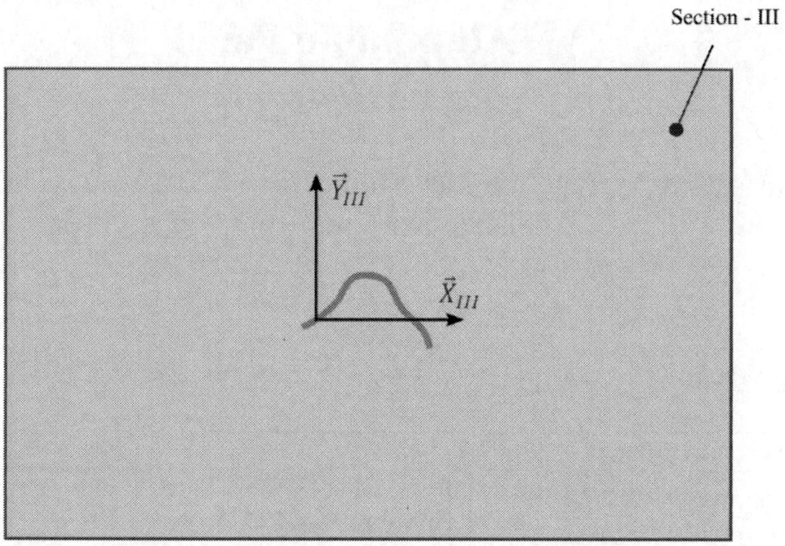

Figure 10. Local coordinate system $\vec{X}_{III}$ - $\vec{Y}_{III}$ for the report of springback profile at section III.

Journal of Physics: Conference Series **734** (2016) 022002 doi:10.1088/1742-6596/734/2/022002

Benchmark 2 – Springback of a Jaguar Land Rover Aluminium Panel

Part B: Responses

Martin Allen[a], Marta Oliveira[b], Sumit Hazra[c], Oluwamayokun Adetoro[d], Abhishek Das[c] and Rui Cardoso[e]

BM2-00

1. Benchmark Participant	
Name	**Martin Allen, Marta Oliveira, Sumit Hazra, Oluwamayokun Adetoro, Abhishek Das, Rui Cardoso**
Prepared by	Benchmark-2 Committee

BM2-01

1. Benchmark Participant	
Name	**Jan NOVY, Takayuki OGAWA**
Affiliation	ESI Group (MECAS ESI, ESI Japan)
Address	Brojova 16 33201 Plzen, Czech Republic; 6-14-1 Nishi-Shinjuku, Shinjyuku-ku, Tokyo 1600-0023, Japan
Email	jno@esi-group.com, tog@esi-group.com
Phone number	+420724269068; +81-3-6381-8494
Fax number	+420 377 432 930; +81-3-6381-8494

2. Simulation Software	
Name of the FEM code	Pam-Stamp
General aspect of the code	Dynamic explicit(forming), Static implicit (gravity, springback)
Basic formulations	Updated Lagrangian formulation with associated flow rule, Barlat2000 yield function, Yoshida kinematic hardening
Element/Mesh technology	
Number of elements	175,582 (After stamping), 75,084 (after trimming)
Type of elements	explicit solution:Belytschko-Tsay shell , implicit solution: Batoz Q4 gamma shell
Contact property model	explicit solution: non-linear penalty contact, implicit solution: contact 54
Friction formulation	Standard Coulomb friction

3. Simulation Hardware	
CPU Type	Intel Xeon CPU E5645 approach 1, Xeon e5-2650 approach 2
CPU clock speed	2.40GHz approach 1, 2.6GHz approach 2
Number of cores per CPU	12 approach 1, 8 approach 2
Main memory	48 GB approach 1, 64 GB approach 2
Operating system	Linux
Total CPU time	17 hours approach 1, 27 hours approach 2

4. Describe the material model used for each material	
Material	AA6451-T4
Yield Function/ Plastic Potential	**Yld2000-2D - the parameters for Yld2000-2D used as provided**
Hardening Rule (e.g. Isotropic, kinematic)	**Kinematic hardening**
Stress-Strain Relation (e.g. Swift, Voce)	**Yoshida-Uemori (Y-U): Cyclic shear data "Cyclical.xls" were transered into stress-strain curves. And then, Y-U parameters are evaluated from them by**

5. Remarks

There were used 2 approaches of computation:

Approach 1:

Gravity - Holding - Stamping - Trimming&springback using locked nodes of model

Approach 2:

OP20 (Gravity-Holding-Stamping-Springback) - OP30 (Holding-Trimming-Springback)-Fixure(Clamping)
Since trimming dies and fixtures are not provided by organizer, those shapes are estimated from the specifications and provided CAD data.
For more please check video:Fixture.avi
There are submitted two result only in STL

BM2-02

1. Benchmark Participant	
Name	**HARSH R AGRAWAL**
Affiliation	M.Tech Student (First Year) (Branch - Steel Technology), Department of Metallurgical Engineering & Materials Science, IIT Bombay, Powai , Mumbai (India)
Address	Hostel No. 4, Room No. 332, Indian Institute of Technology Bombay, Powai, Mumbai (Maharashtra, India) (PIN - 400076)
Email	harshagrawal1993@gmail.com
Phone number	9770892835
Fax number	+91 22 2572 6975

2. Simulation Software	
Name of the FEM code	PAM-STAMP 2012.2
General aspect of the code	Dynamic Explicit (for Holding/Stamping), Static Implicit (springback after trimming)
Basic formulations	Updated Lagrangian formulation with associated flow rule, Barlat 2000 (Yld 2000-2D), Isotropic Hardening, Tabulated data for hardening curve following Voce Equation
Element/Mesh technology	
Number of elements	Number of blank elements = 4856 (initial mesh), 188528 (after mesh refinements at the end of stamping stage)
Type of elements	Type of blank elements = 4-node Belytschko-Tsay shell, reduced integration, hour glass control, 5 integration points through thickness.
Contact property model	Accurate Contact
Friction formulation	Standard Coulomb friction, value is 0.08 which is constant at all blank-tool interface

3. Simulation Hardware	
CPU Type	Intel® Core™ i7-3770 CPU @ 3.40 GHz
CPU clock speed	3.4 GHz
Number of cores per CPU	1 Core
Main memory	16 GB
Operating system	64-bit Operating System
Total CPU time	Total time = 25 hours [Binder closure (explicit) = 3.25 hours, Forming (Explicit) = 21.5 hours, Trimming-Springback (Implicit) = 0.25 Hours]

4. Describe the material model used for each material	
Material	AA6451-T4
Yield Function/ Plastic Potential	**Barlat 2000 or Yld 2000-2D**
Hardening Rule (e.g. Isotropic, kinematic)	**Isotropic Hardening**
Stress-Strain Relation (e.g. Swift, Voce)	**Tabulated data following Voce Equation**

5. Remarks
Not Applicable In this Case.

BM2-03

1. Benchmark Participant	
Name	**T.F. Neves, D.M. Neto, M.C. Oliveira, J.L Alves and L.F. Menezes**
Affiliation	Departament of Mechanical Engineering, University of Coimbra
Address	Polo II, Rua Luís Reis Santos, 3030-788 Coimbra, Portugal
Email	diogo.neto@dem.uc.pt
Phone number	+351 239 790 700
Fax number	+351 239 790 701

2. Simulation Software	
Name of the FEM code	DD3IMP
General aspect of the code	Static fully implicit
Basic formulations	Updated Lagrangian formulation with associated flow rule
Element/Mesh technology	
Number of elements	141476
Type of elements	Isoparametric 3D brick elements with selective reduced integration technique
Contact property model	Rigid tools modelled by 132719 Nagata patches, Augmented lagrangian method
Friction formulation	Coulomb friction law

3. Simulation Hardware	
CPU Type	Intel® Core™ i7-5930K
CPU clock speed	3.5 GHz
Number of cores per CPU	6 cores
Main memory	64 GB RAM
Operating system	Windows 10 Professional (64-bit)
Total CPU time	284 hours (forming) 11 hours (trimming)

4. Describe the material model used for each material	
Material	**AA6451-T4**
Yield Function/ Plastic Potential	**Barlat 91**
Hardening Rule (e.g. Isotropic, kinematic)	**Armstrong–Frederick kinematic hardening**
Stress-Strain Relation (e.g. Swift, Voce)	**Voce law**

5. Remarks

BM2-04

1. Benchmark Participant	
Name	**Jun Hu, Fadi Abu-Farha**
Affiliation	Clemson University
Address	4 Research Drive, Greenville, SC 29607
Email	junh@g.clemson.edu | fadi@clemson.edu
Phone number	1-859-489-2926
Fax number	1-864-283-7225

2. Simulation Software	
Name of the FEM code	LS-DYNA
General aspect of the code	Forming: dynamic explicit; springback: static implicit
Basic formulations	Updated Lagrangian formulation with associated flow rule
Element/Mesh technology	
Number of elements	80657
Type of elements	Fully integrated shell element (ELFORM=16)
Contact property model	Surface to surface contact
Friction formulation	Coulomb friction

3. Simulation Hardware	
CPU Type	Intel Xeon64
CPU clock speed	
Number of cores per CPU	8 SMP double-precision
Main memory	16GB
Operating system	Scientific Linux 6
Total CPU time	12 hours 22 minutes

4. Describe the material model used for each material	
Material	**AA6451-T4**
Yield Function/ Plastic Potential	**Hill1948-3R, associated flow rule**
Hardening Rule (e.g. Isotropic, kinematic)	**Yoshida-Uemori model (isotropic + nonlinear kinematic hardening rule)**
Stress-Strain Relation (e.g. Swift, Voce)	**Voce**

5. Remarks

Journal of Physics: Conference Series **734** (2016) 022002 doi:10.1088/1742-6596/734/2/022002

BM2-05

1. Benchmark Participant	
Name	**Kaiping Li, Yang Hu**
Affiliation	FCA US LLC
Address	800 Chrysler Dr, Auburn Hills, MI48326, USA
Email	Kaiping.li@fcagroup.com, yang.hu@fcagroup.com
Phone number	(1)2485766176, (1)2487015302
Fax number	(1)2485767190

2. Simulation Software	
Name of the FEM code	AutoForm R6
General aspect of the code	
Basic formulations	
Element/Mesh technology	
Number of elements	117081
Type of elements	Shell element
Contact property model	N/A
Friction formulation	constant (0.14 as Instructed)

3. Simulation Hardware	
CPU Type	Working Station with 8 Cpus
CPU clock speed	N/A
Number of cores per CPU	48
Main memory	32 GB
Operating system	LINUX
Total CPU time	Forming: 2 hours & Springback: 1min

4. Describe the material model used for each material	
Material	**AA6451-T4**
Yield Function/ Plastic Potential	**Barlat**
Hardening Rule (e.g. Isotropic, kinematic)	**Isotropic**
Stress-Strain Relation (e.g. Swift, Voce)	**Combined Swift/Hockett-Sherby**

5. Remarks

Journal of Physics: Conference Series **734** (2016) 022002 doi:10.1088/1742-6596/734/2/022002

BM2-06

1. Benchmark Participant	
Name	**Kaiping Li, Yang Hu**
Affiliation	FCA US LLC
Address	800 Chrysler Dr, Auburn Hills, MI48326, USA
Email	Kaiping.li@fcagroup.com, yang.hu@fcagroup.com
Phone number	(1)2485766176, (1)2487015302
Fax number	(1)2485767190

2. Simulation Software	
Name of the FEM code	LS-DYNA
General aspect of the code	
Basic formulations	
Element/Mesh technology	
Number of elements	560000
Type of elements	Fully integrated shell element
Contact property model	FORMING_ONE_WAY_SURFACE_TO_SURFACE + Penalty
Friction formulation	constant (0.08 as instructed)

3. Simulation Hardware	
CPU Type	HPC
CPU clock speed	N/A
Number of cores per CPU	48
Main memory	N/A
Operating system	LINUX
Total CPU time	Forming: 23 hours & Springback: 17mins

4. Describe the material model used for each material	
Material	**AA6451-T4**
Yield Function/ Plastic Potential	**M36: Barlat89**
Hardening Rule (e.g. Isotropic, kinematic)	**Isotropic**
Stress-Strain Relation (e.g. Swift, Voce)	**Swift**

5. Remarks

BM2-08

1. Benchmark Participant	
Name	**1Rongfeng Liu , 2Ninshu Ma**
Affiliation	1 2JSOL Corporation
Address	117F KS Bulding 2-18-25, Marunouchi Naka-ku, Nagoya 460-0002, Japan 2Tosabori Daibiru Building, 2-2-4, Tosabori Nishi-ku, Osaka 550-0001, Japan
Email	rongfeng.liu@jsol.co.jp, ma.ninshu@jsol.co.jp
Phone number	+81-52-202-8181
Fax number	+81-52-202-8172

2. Simulation Software	
Name of the FEM code	JSTAMP/NV
General aspect of the code	Integrated sheet metal forming simulation system
Basic formulations	Forming: Dynamic Explicit(LS-DYNA); Spribingback : Static implicit(LS-DYNA)
Element/Mesh technology	
Number of elements	Solid Blank: 1398855 / Die shell: 150884 /Holder shell:87704/Punch shell:97480
Type of elements	Constant stress solid element with 8 nodes
Contact property model	Penalty Mehtod, Node to Surface
Friction formulation	Coluomb's friction law, fricition coefficient m=0.08

3. Simulation Hardware	
CPU Type	Xeon E5-2670
CPU clock speed	2.60GHz
Number of cores per CPU	16 Core
Main memory	64G
Operating system	CentOS5.8
Total CPU time	37533 seconds(10 hours 25 min. 33 sec.) for 1420998 cycles

4. Describe the material model used for each material	
Material	**AA6451-T4**
Yield Function/ Plastic Potential	**Hill48**
Hardening Rule (e.g. Isotropic, kinematic)	**Yoshida-Uemori Kinematic hardending model**
Stress-Strain Relation (e.g. Swift, Voce)	**Swift**

Journal of Physics: Conference Series **734** (2016) 022002 doi:10.1088/1742-6596/734/2/022002

BM2-09

1. Benchmark Participant	
Name	**Jan NOVY, Takayuki OGAWA**
Affiliation	ESI Group (MECAS ESI, ESI Japan)
Address	Brojova 16 33201 Plzen, Czech Republic; 6-14-1 Nishi-Shinjuku, Shinjyuku-ku, Tokyo 1600-0023, Japan
Email	jno@esi-group.com, tog@esi-group.com
Phone number	+420724269068; +81-3-6381-8494
Fax number	+420 377 432 930; +81-3-6381-8494

2. Simulation Software	
Name of the FEM code	Pam-Stamp
General aspect of the code	Dynamic explicit(forming), Static implicit (gravity, springback)
Basic formulations	Updated Lagrangian formulation with associated flow rule, Barlat2000 yield function, Yoshida kinematic hardening
Element/Mesh technology	
Number of elements	175,582 (After stamping), 75,084 (after trimming)
Type of elements	explicit solution:Belytschko-Tsay shell , implicit solution: Batoz Q4 gamma shell
Contact property model	explicit solution: non-linear penalty contact, implicit solution: contact 54
Friction formulation	Standard Coulomb friction

3. Simulation Hardware	
CPU Type	Intel Xeon CPU E5645 approach 1, Xeon e5-2650 approach 2
CPU clock speed	2.40GHz approach 1, 2.6GHz approach 2
Number of cores per CPU	12 approach 1, 8 approach 2
Main memory	48 GB approach 1, 64 GB approach 2
Operating system	Linux
Total CPU time	17 hours approach 1, 27 hours approach 2

4. Describe the material model used for each material	
Material	**AA6451-T4**
Yield Function/ Plastic Potential	**Yld2000-2D - the parameters for Yld2000-2D used as provided**
Hardening Rule (e.g. Isotropic, kinematic)	**Kinematic hardening**
Stress-Strain Relation (e.g. Swift, Voce)	**Yoshida-Uemori (Y-U): Cyclic shear data "Cyclical.xls" were transered into stress-strain curves. And then, Y-U parameters are evaluated from them by**

5. Remarks
There were used 2 approaches of computation: **Approach 1:** **Gravity - Holding - Stamping - Trimming&springback using locked nodes of model** **Approach 2:** **OP20 (Gravity-Holding-Stamping-Springback) - OP30 (Holding-Trimming-Springback)-Fixure(Clamping)** **Since trimming dies and fixtures are not provided by organizer, those shapes are estimated from the specifications and provided CAD data.** **For more please check video:Fixture.avi** **There are submitted two result only in STL**

BM2-10

1. Benchmark Participant	
Name	**Albert Forgas**
Affiliation	Quantech ATZ
Address	C/Gran Capità 2/4 08034 Barcelona, Spain
Email	aforgas@stampack.com
Phone number	+342047083
Fax number	

2. Simulation Software	
Name of the FEM code	Stampack V7.1.2
General aspect of the code	Finite Element Method
Basic formulations	Explicit Formability, Implicit Springback
Element/Mesh technology	
Number of elements	93099
Type of elements	Hexahedra Special Integration
Contact property model	Penalty Method
Friction formulation	Coluomb

3. Simulation Hardware	
CPU Type	Intel Core i7-3770
CPU clock speed	3,40 GHz
Number of cores per CPU	8 threads
Main memory	16 GB
Operating system	Win 7 64 Bit
Total CPU time	3 Hour 35 Min 20 Sec

4. Describe the material model used for each material	
Material	**AA6451-T4**
Yield Function/ Plastic Potential	**Yoshida Uemori**
Hardening Rule (e.g. Isotropic, kinematic)	**Isotropic**
Stress-Strain Relation (e.g. Swift, Voce)	**Voce**

5. Remarks
Variable young modulus

Journal of Physics: Conference Series **734** (2016) 022002 doi:10.1088/1742-6596/734/2/022002

BM2-11

1. Benchmark Participant	
Name	**Albert Forgas**
Affiliation	Quantech ATZ
Address	C/Gran Capità 2/4 08034 Barcelona, Spain
Email	aforgas@stampack.com
Phone number	+342047083
Fax number	

2. Simulation Software	
Name of the FEM code	Stampack V7.1.2
General aspect of the code	Finite Element Method
Basic formulations	Explicit Formability, Implicit Springback
Element/Mesh technology	
Number of elements	59638
Type of elements	Basic Shell Triangle
Contact property model	Penalty Method
Friction formulation	Coluomb

3. Simulation Hardware	
CPU Type	Intel Core i7-3770
CPU clock speed	3,40 GHz
Number of cores per CPU	8 threads
Main memory	16 GB
Operating system	Win 7 64 Bit
Total CPU time	0 Hour 40 Min 10 Sec

4. Describe the material model used for each material	
Material	**AA6451-T4**
Yield Function/ Plastic Potential	**Yoshida Uemori**
Hardening Rule (e.g. Isotropic, kinematic)	**Isotropic**
Stress-Strain Relation (e.g. Swift, Voce)	**Voce**

5. Remarks
Variable young modulus

BM2-12

1. Benchmark Participant	
Name	**SARIN BABU THOKALA**
Affiliation	STADCO LTD
Address	Queens way, Hortonwood, Telford - TF1 7LL, UK
Email	S.Thokala@stadco.co.uk
Phone number	0044 (0) 787 554 6247 or 0044 (0) 772 481 6535
Fax number	0044 (0) 1952 222050

2. Simulation Software	
Name of the FEM code	AutoForm^plus R6
General aspect of the code	Stamping Simulation
Basic formulations	Static Implicit
Element/Mesh technology	
Number of elements	Form - 205000, After Trim -88000
Type of elements	Triangular elastic plastic shell, 11 Integration points through thickness
Contact property model	
Friction formulation	Coulomb friction (0.14)

3. Simulation Hardware	
CPU Type	Intel®Core™i7-2760QM CPU @ 2.40GHz
CPU clock speed	2.40GHz
Number of cores per CPU	4
Main memory	32.0GB
Operating system	Windows 7 Professional
Total CPU time	01Hour:20Mins:32Sec

4. Describe the material model used for each material	
Material	**AA6451-T4**
Yield Function/ Plastic Potential	**Barlat -1989**
Hardening Rule (e.g. Isotropic, kinematic)	**Isotropic hardening**
Stress-Strain Relation (e.g. Swift, Voce)	**Aproximation (Combined Swift - Hockett-Sherby formulation)**

5. Remarks

BM2-13

1. Benchmark Participant	
Name	**Jan Slota, Marek Siser**
Affiliation	Technical University of Kosice
Address	Masiarska 74, 040 01 Kosice, Slovakia
Email	jan.slota@tuke.sk, marek.siser@tuke.sk
Phone number	
Fax number	

2. Simulation Software	
Name of the FEM code	PAM-Stamp 2015.1
General aspect of the code	Explicit
Basic formulations	
Element/Mesh technology	
Number of elements	82000
Type of elements	Quadrilateral
Contact property model	Accurate
Friction formulation	0.08

3. Simulation Hardware	
CPU Type	Intel Xeon CPU E5 2670
CPU clock speed	2,6 GHz
Number of cores per CPU	6, total 12 cores
Main memory	16 GB
Operating system	Win 8.1 64 bit
Total CPU time	14:30 hod.

4. Describe the material model used for each material	
Material	**AA6451-T4**
Yield Function/ Plastic Potential	**Barlat2000**
Hardening Rule (e.g. Isotropic, kinematic)	**Isotropic**
Stress-Strain Relation (e.g. Swift, Voce)	**Krupkowski/Swift**

5. Remarks

BM2-14

1. Benchmark Participant	
Name	**Hariharasudhan Palaniswamy, Subir Roy**
Affiliation	Altair Engineering
Address	1820 E Big Beaver road, Troy, MI, 48085, USA
Email	hpalaniswamy@altair.com, subir@altair.com
Phone number	248-614-2400
Fax number	248-614-2411

2. Simulation Software	
Name of the FEM code	HyperForm - RADIOSS
General aspect of the code	Commercial nonlinear finite element software
Basic formulations	Forming (Explicit), Springback(Implicit)
Element/Mesh technology	
Number of elements	810934
Type of elements	Shell element - QEPH formulation
Contact property model	Penaltly based contact formulations
Friction formulation	Coulomb's Law

3. Simulation Hardware	
CPU Type	HPC Cluster
CPU clock speed	2.50GHz
Number of cores per CPU	13 Node, 24 cores per node. 24 cpu's used for the simulation
Main memory	128 GB of RAM per core
Operating system	Linux
Total CPU time	Forming: 16118 Secs, Trimming: 0 Sec, springback : 390 Secs

4. Describe the material model used for each material	
Material	AA6451-T4
Yield Function/ Plastic Potential	**Barlat 3 parameter model**
Hardening Rule (e.g. Isotropic, kinematic)	**Combined hardening rule**
Stress-Strain Relation (e.g. Swift, Voce)	**Voce hardening law**

5. Remarks

Journal of Physics: Conference Series **734** (2016) 022002 doi:10.1088/1742-6596/734/2/022002

BM2-15

1. Benchmark Participant	
Name	[1]Yasuyoshi Umezu , [1]Toshiro Amaishi, [2]Wan-Jin Chung
Affiliation	[1] JSOL Corporation , [2] Seoul National Univerisity of Science & Technology
Address	[1]Tosabori Daibiru Building, 2-2-4, Tosabori Nishi-ku, Osaka 550-0001, Japan [2]232 Gongneung-ro, Nowon-gu, Seoul, 01811, korea
Email	umezu.yasu@jsol.co.jp, amaishi.toshirou@jsol.co.jp, wjchung@seoultech.ac.kr
Phone number	+81-6-4803-5820
Fax number	+81-6-6225-3517

2. Simulation Software	
Name of the FEM code	ASTAMP(for Forming), JOH/NIKE(for Spring back)
General aspect of the code	Press Simulation Softrware Optimized for GPGPU
Basic formulations	Forming:Dynamic Explicit, Spring back:Static Implicit
Element/Mesh technology	
Number of elements	466544 (for Blank)
Type of elements	Quadrilateral Belytschko-Tsay and C0 Triangular
Contact property model	Penalty Method, Node to Surface
Friction formulation	Coulomb's friction law

3. Simulation Hardware	
CPU Type	GPGPU(TESLA-K20)
CPU clock speed	706MHz (TESLA-K20)
Number of cores per CPU	2496 cores (TESLA-K20)
Main memory	5Gb (TESLA-K20)
Operating system	Windows 7 Professional
Total CPU time	15226 sec (4 hours 14 min 46sec) for 194124 Binder&Forming steps, 551 sec for Springback

4. Describe the material model used for each material	
Material	AA6451-T4
Yield Function/ Plastic Potential	Hill 48
Hardening Rule (e.g. Isotropic, kinematic)	Isotropic Hardening
Stress-Strain Relation (e.g. Swift, Voce)	Voce

5. Remarks
Binder Closure and Forming steps are not separated in this calculation, 60% of computation times was required for Binder Closure and 40% for Forming.

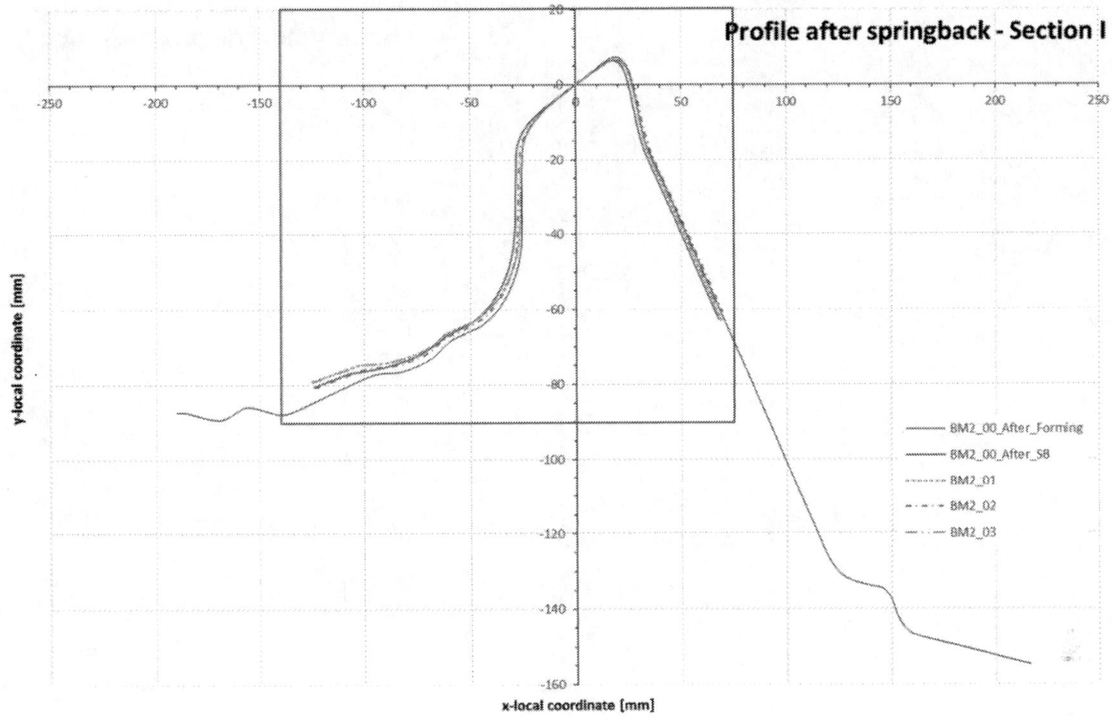

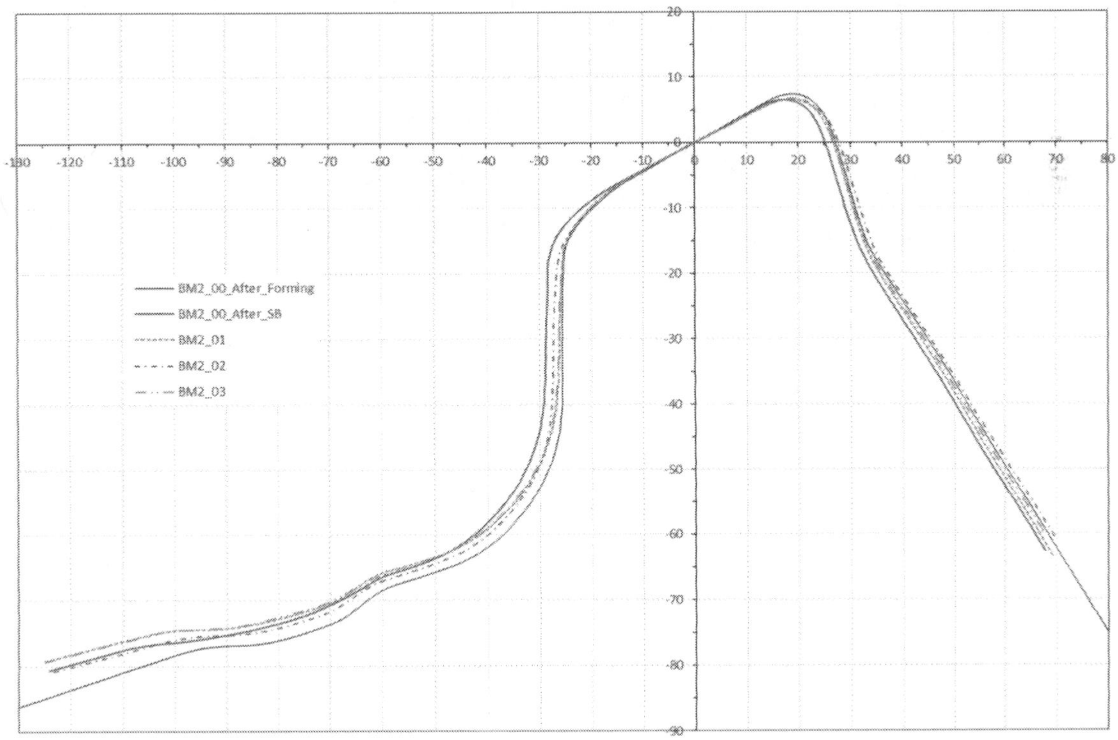

Figure 6.1. Profile after springback for Section I: BM2_01, BM2_02, BM2_03.

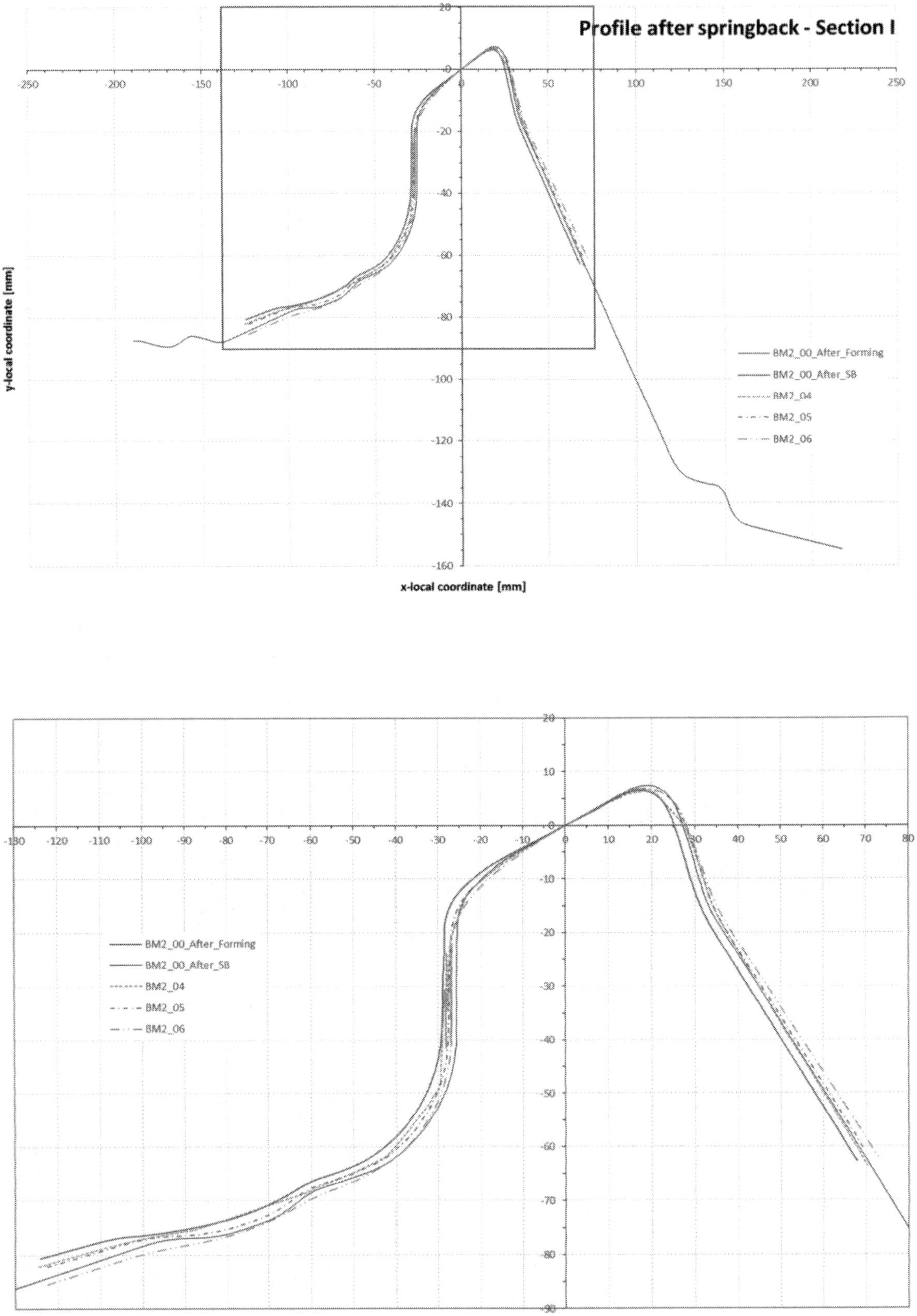

Figure 6.2. Profile after springback for Section I: BM2_04, BM2_05, BM2_06.

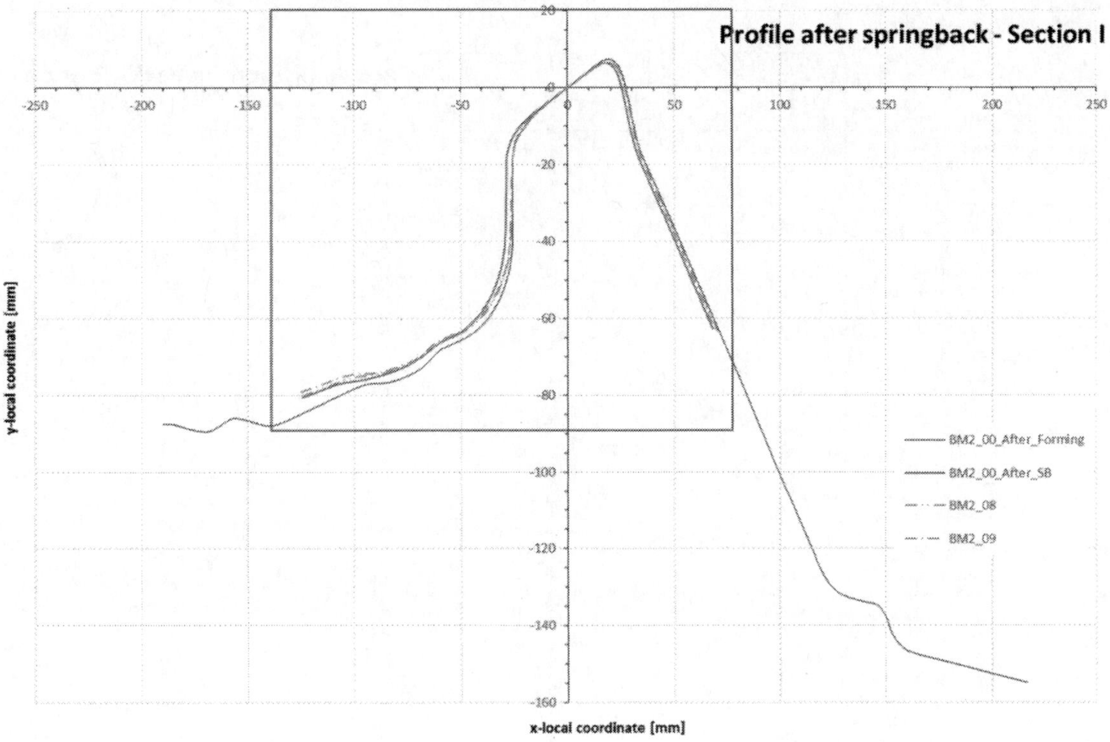

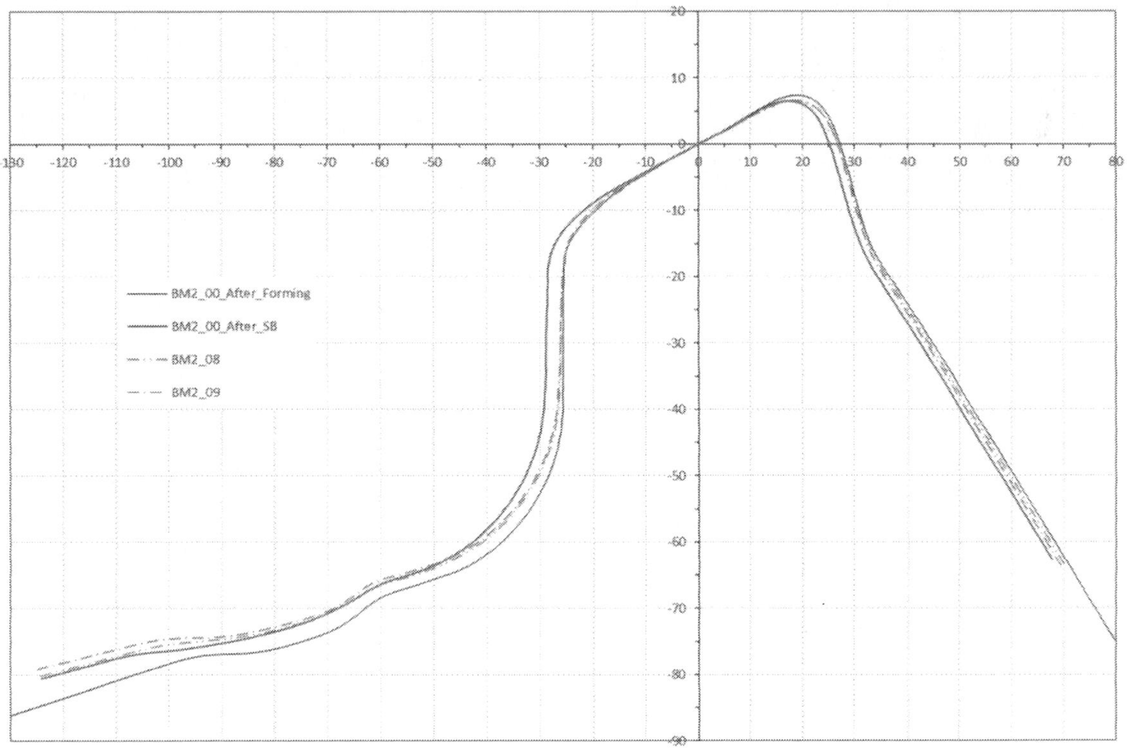

Figure 6.3. Profile after springback for Section I: BM2_08, BM2_09.

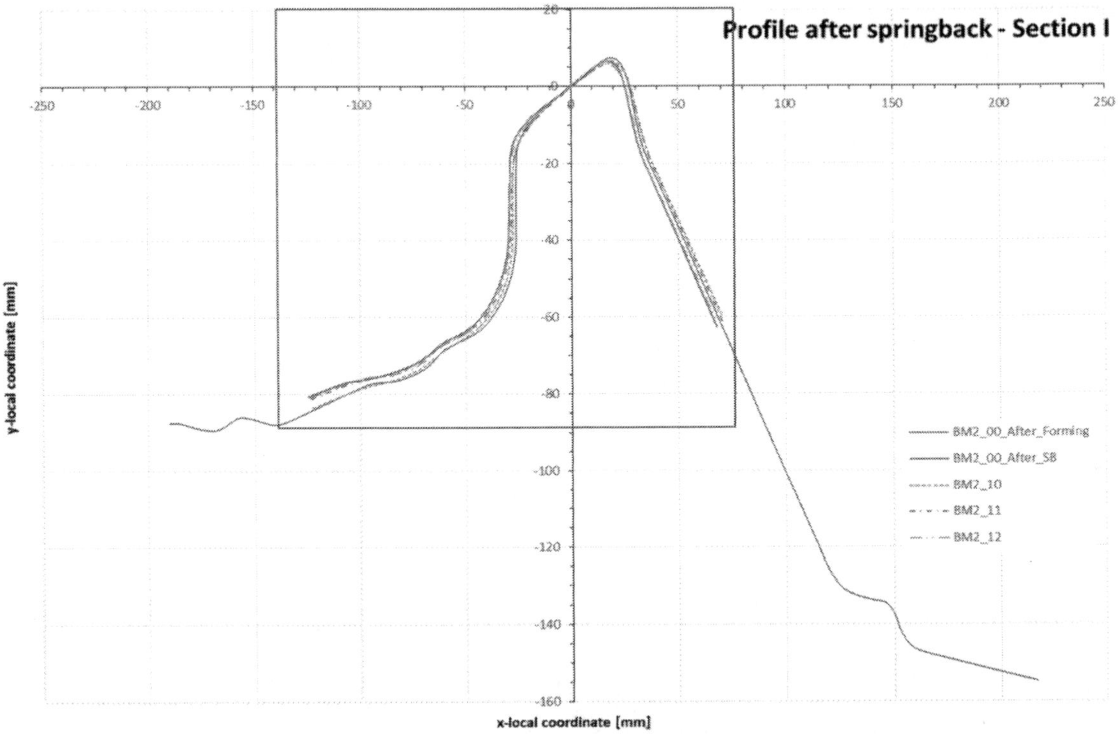

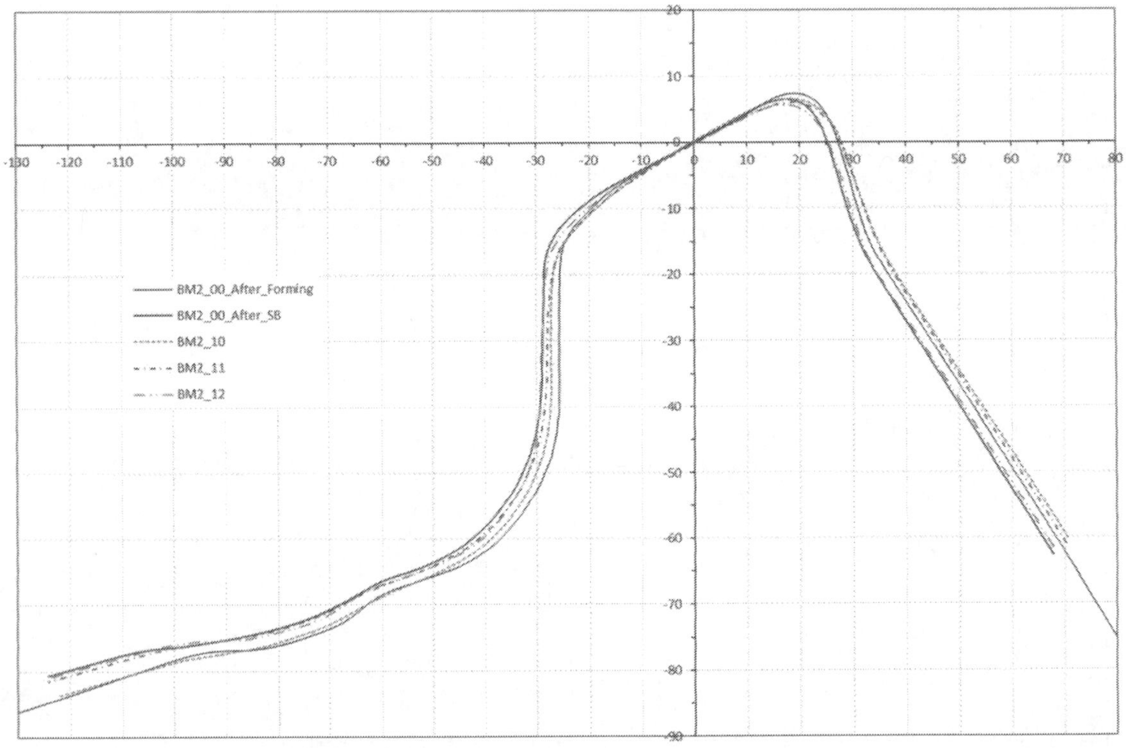

Figure 6.4. Profile after springback for Section I: BM2_10, BM2_11, BM2_12.

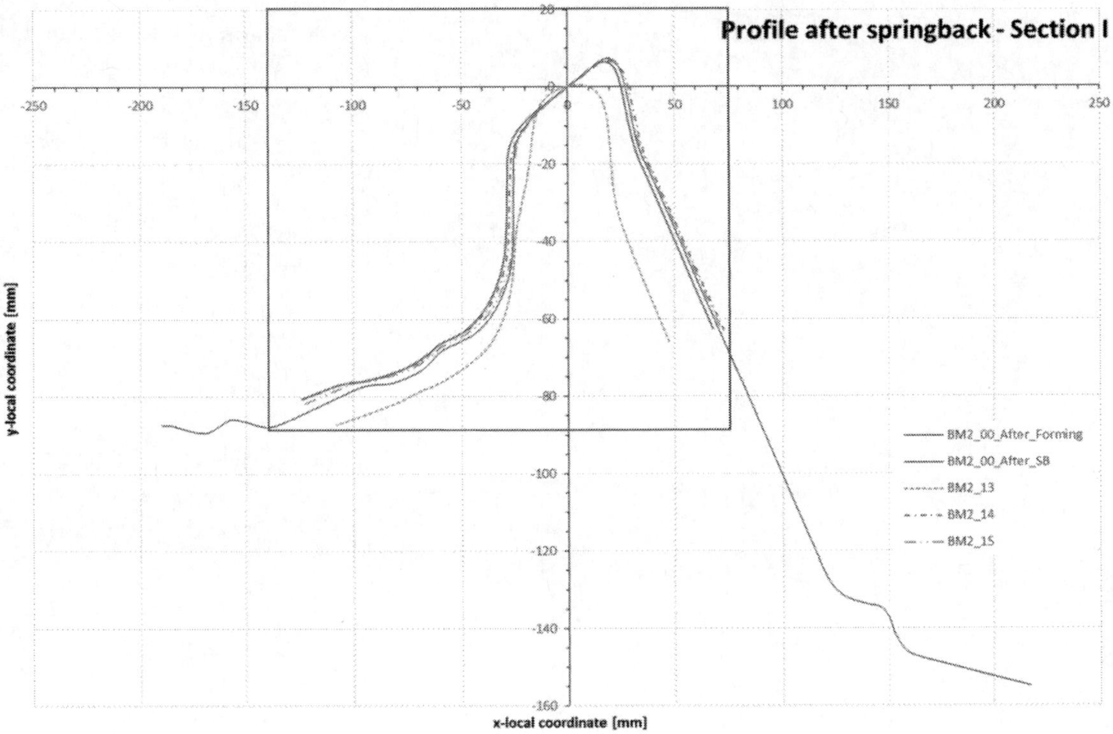

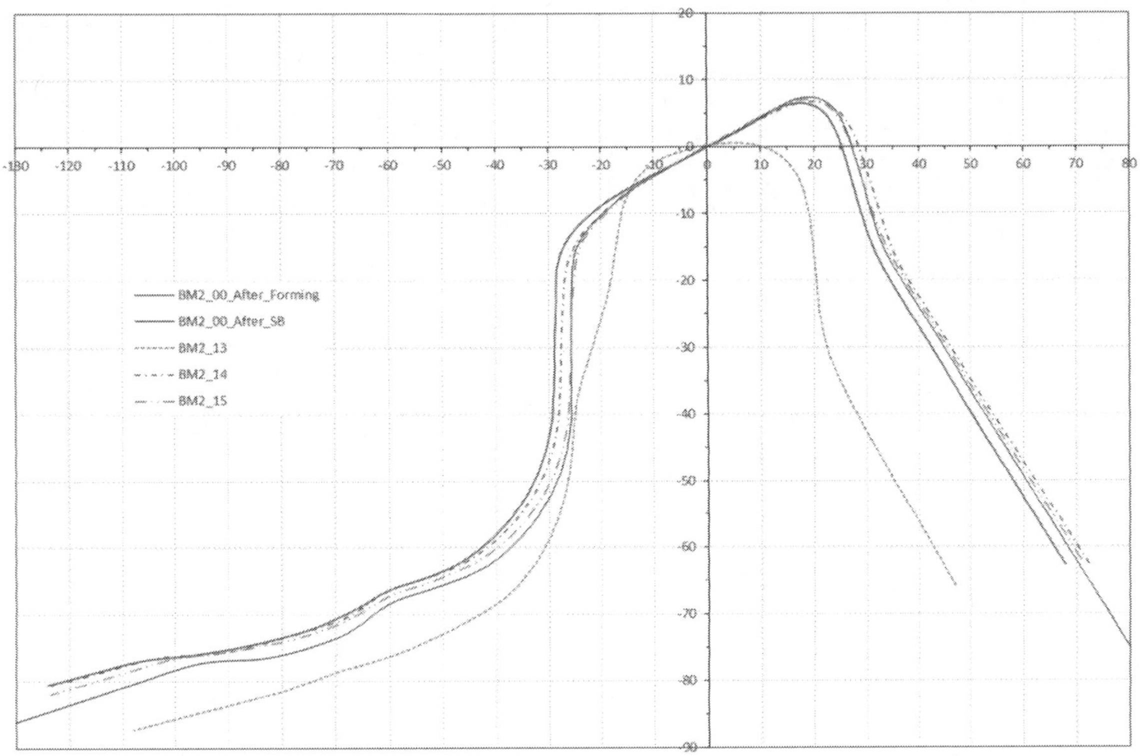

Figure 6.5. Profile after springback for Section I: BM2_13, BM2_14, BM2_15.

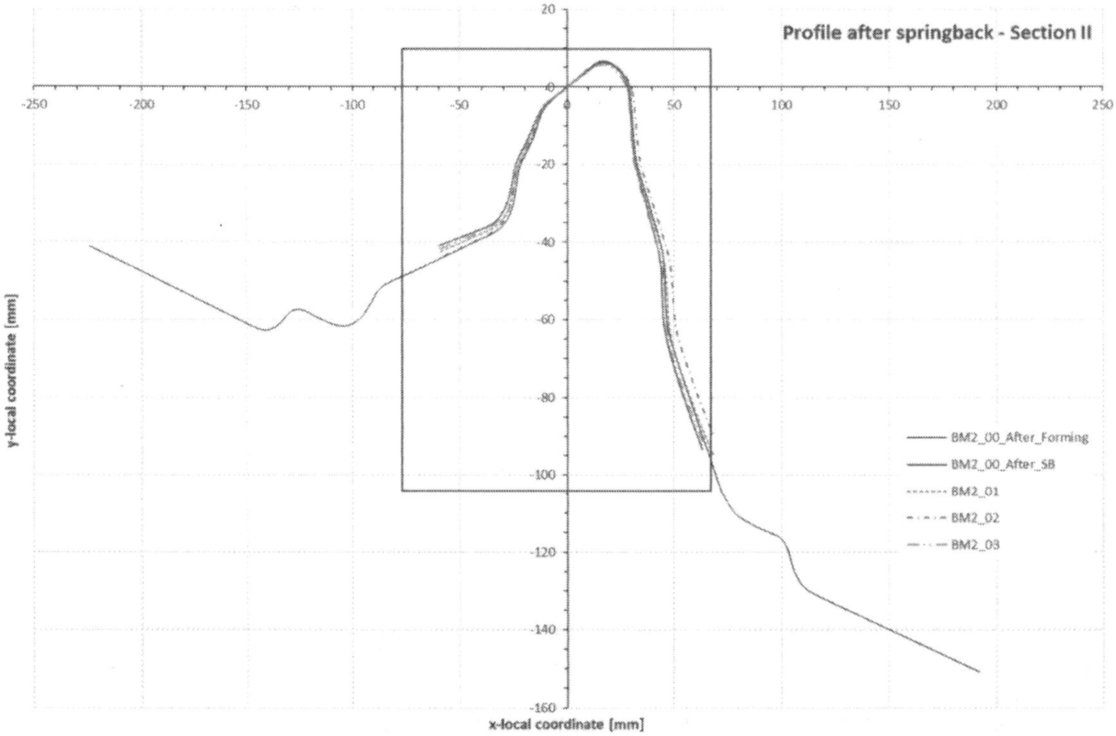

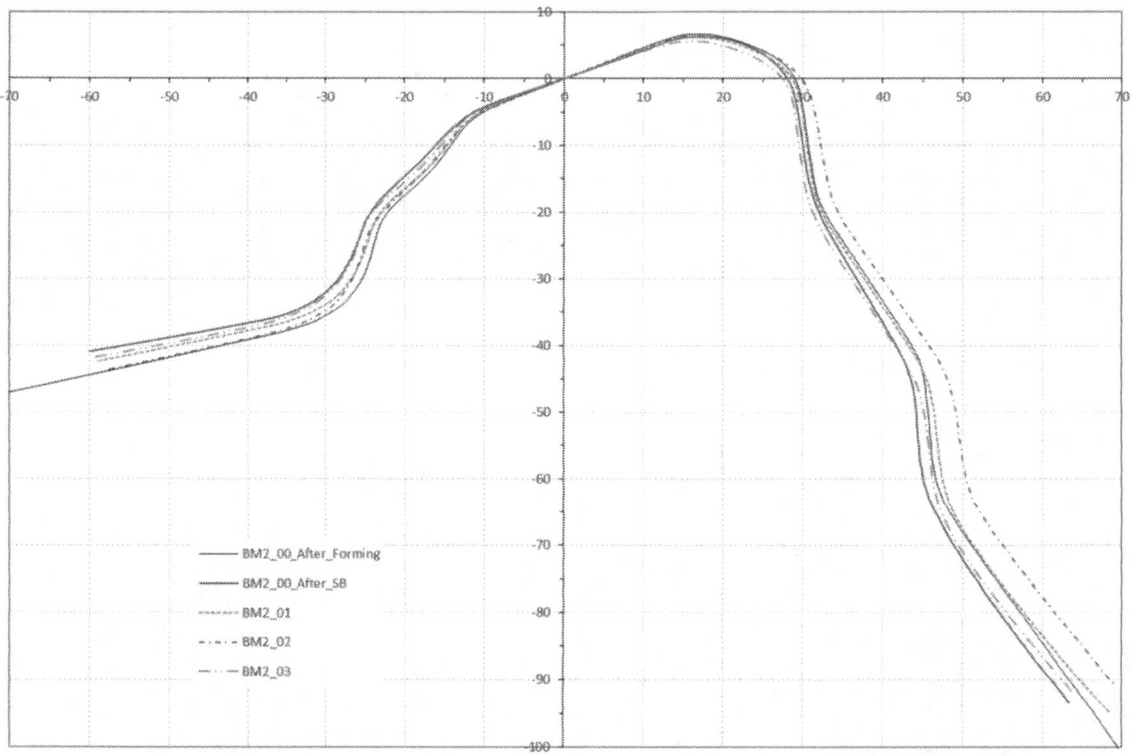

Figure 6.6. Profile after springback for Section II: BM2_01, BM2_02, BM2_03.

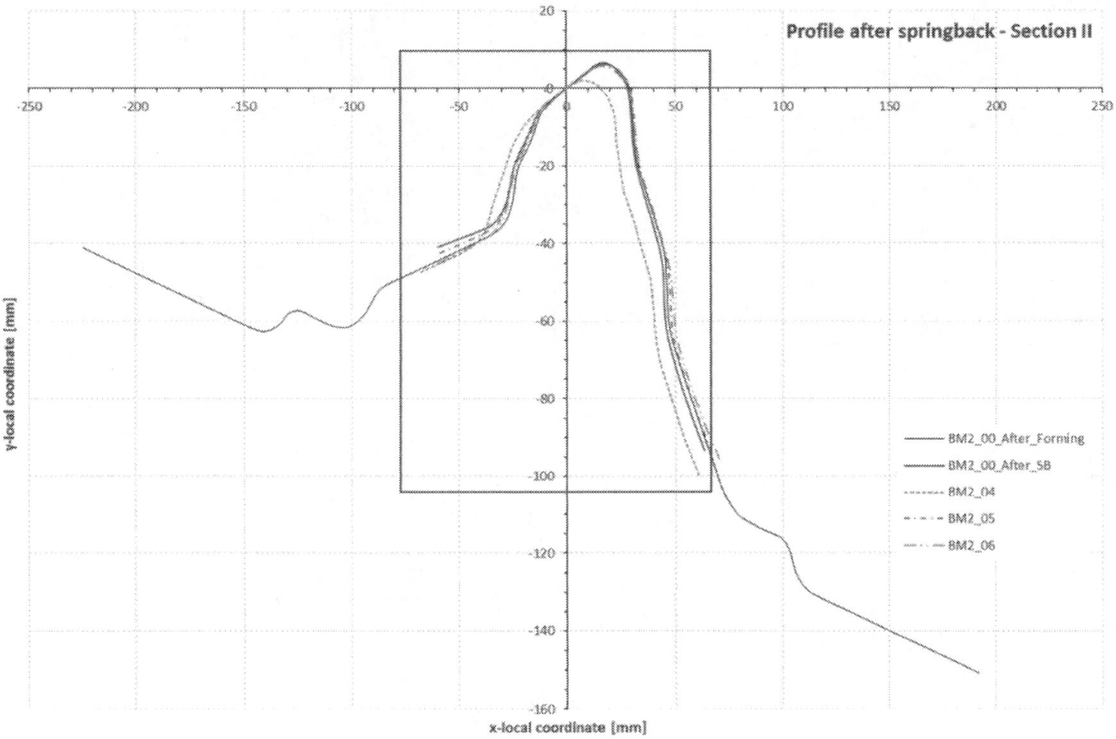

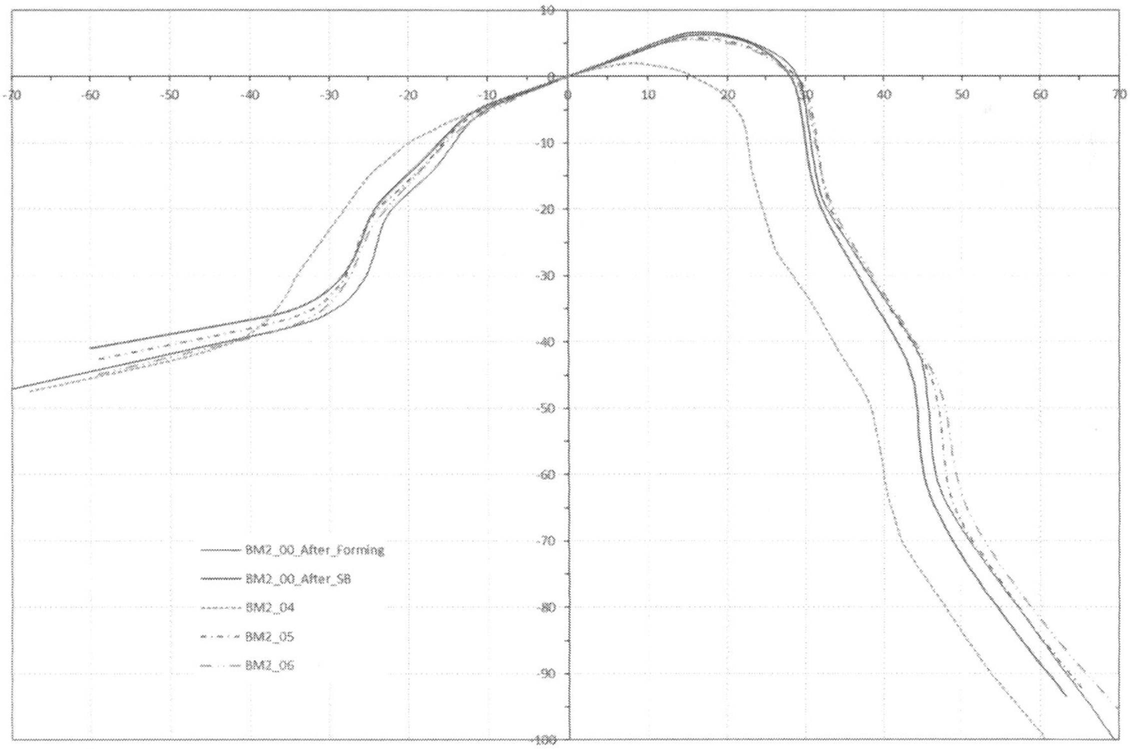

Figure 6.7. Profile after springback for Section II: BM2_04, BM2_05, BM2_06.

Journal of Physics: Conference Series **734** (2016) 022002 doi:10.1088/1742-6596/734/2/022002

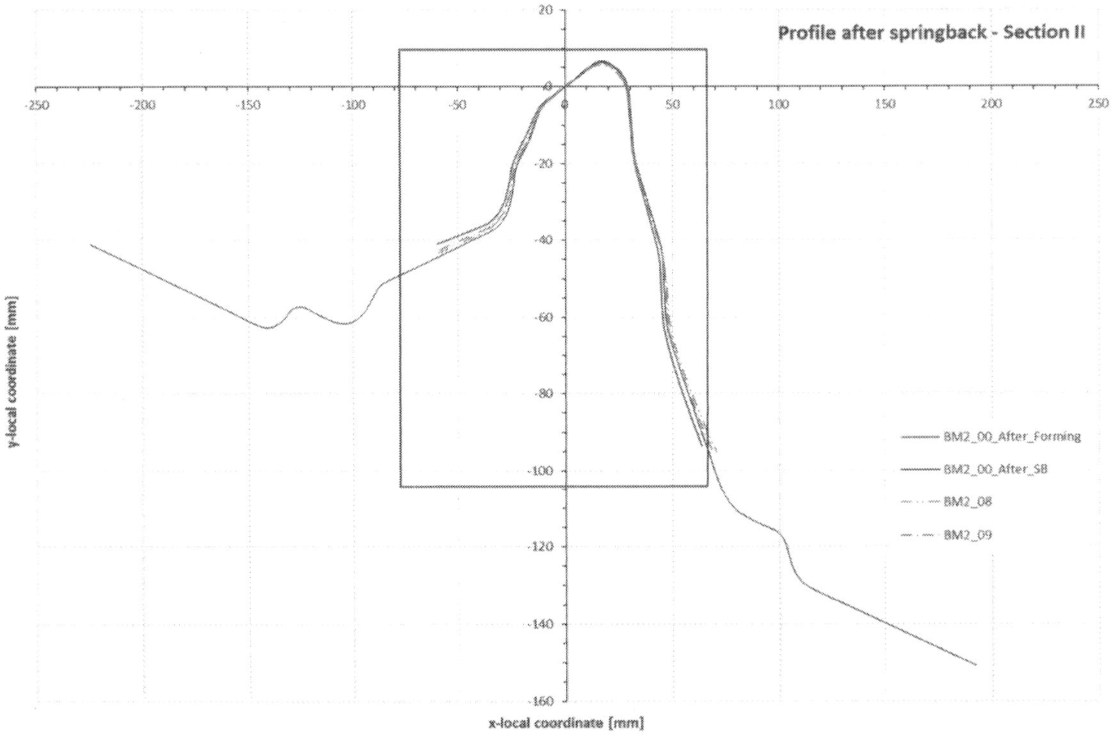

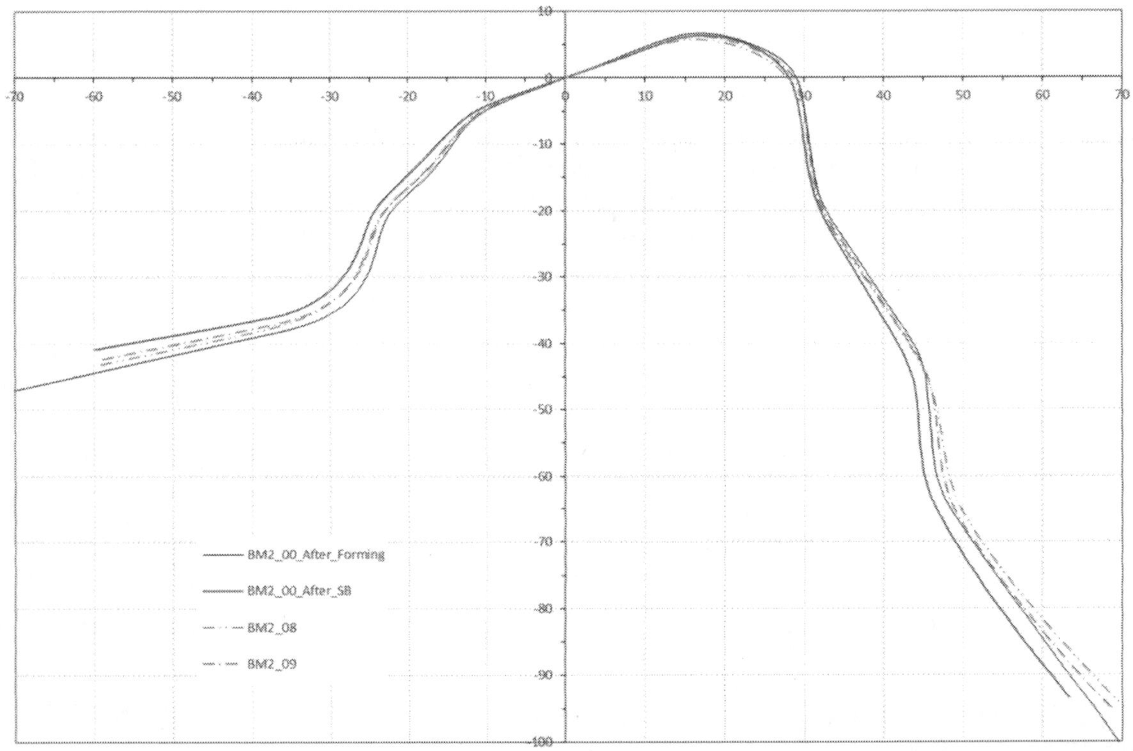

Figure 6.8. Profile after springback for Section II: BM2_08, BM2_09.

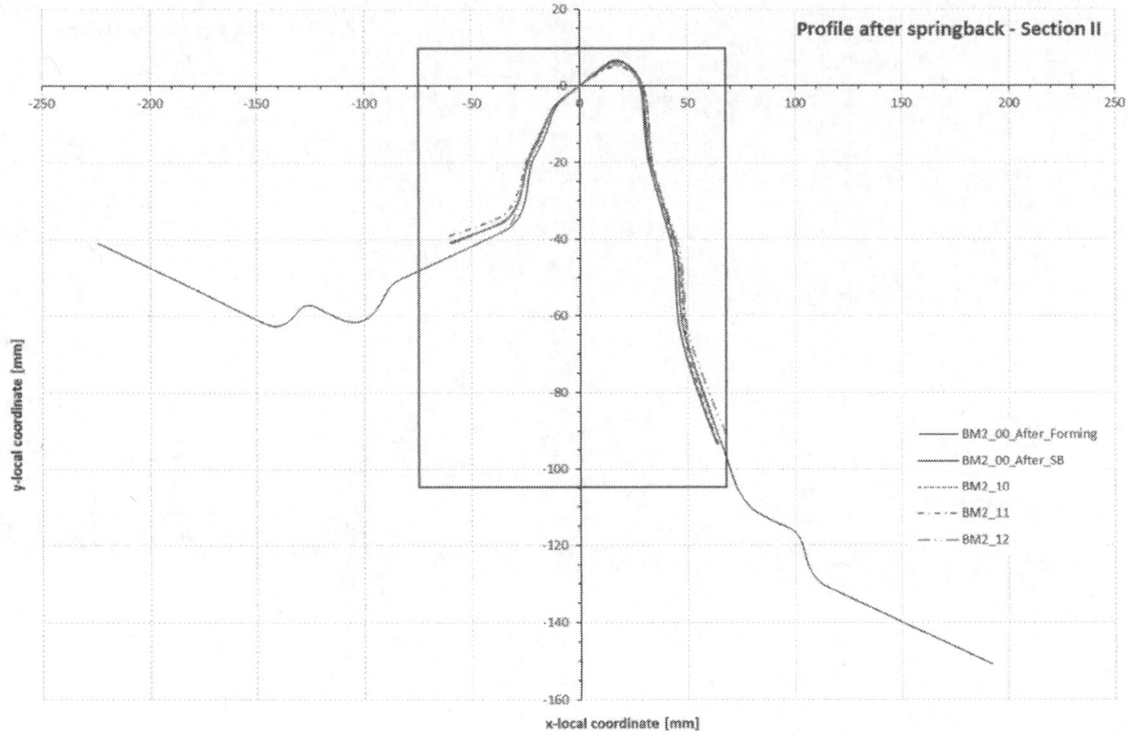

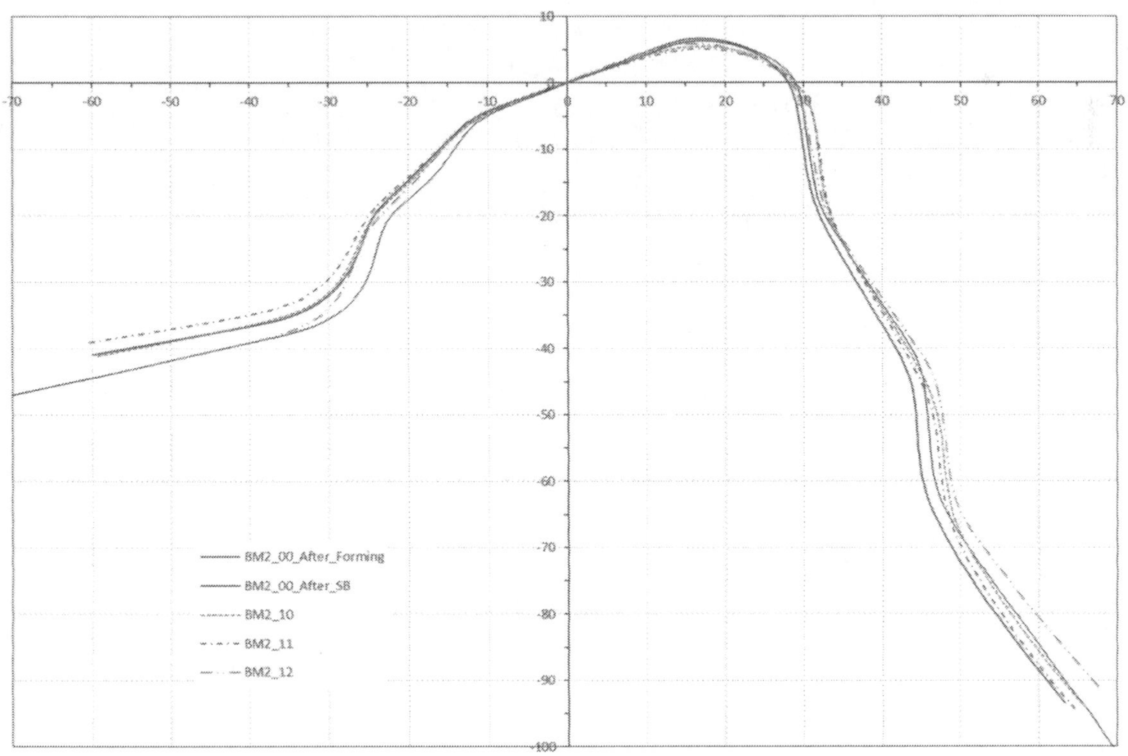

Figure 6.9. Profile after springback for Section II: BM2_10, BM2_11, BM2_12.

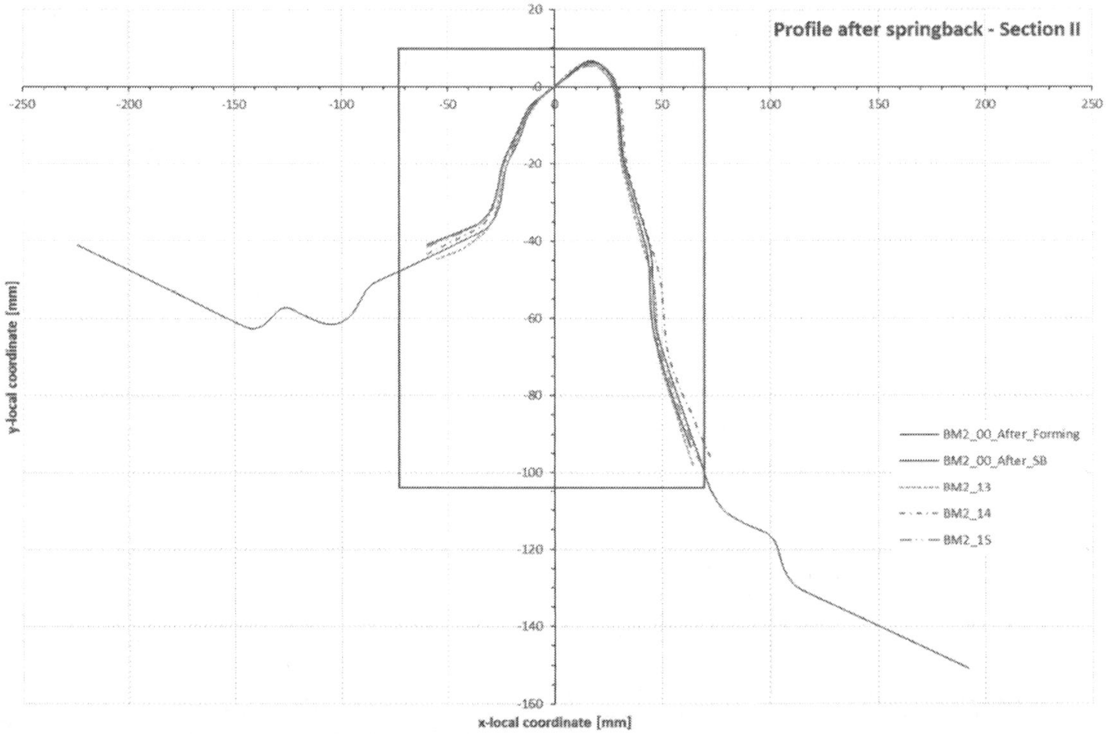

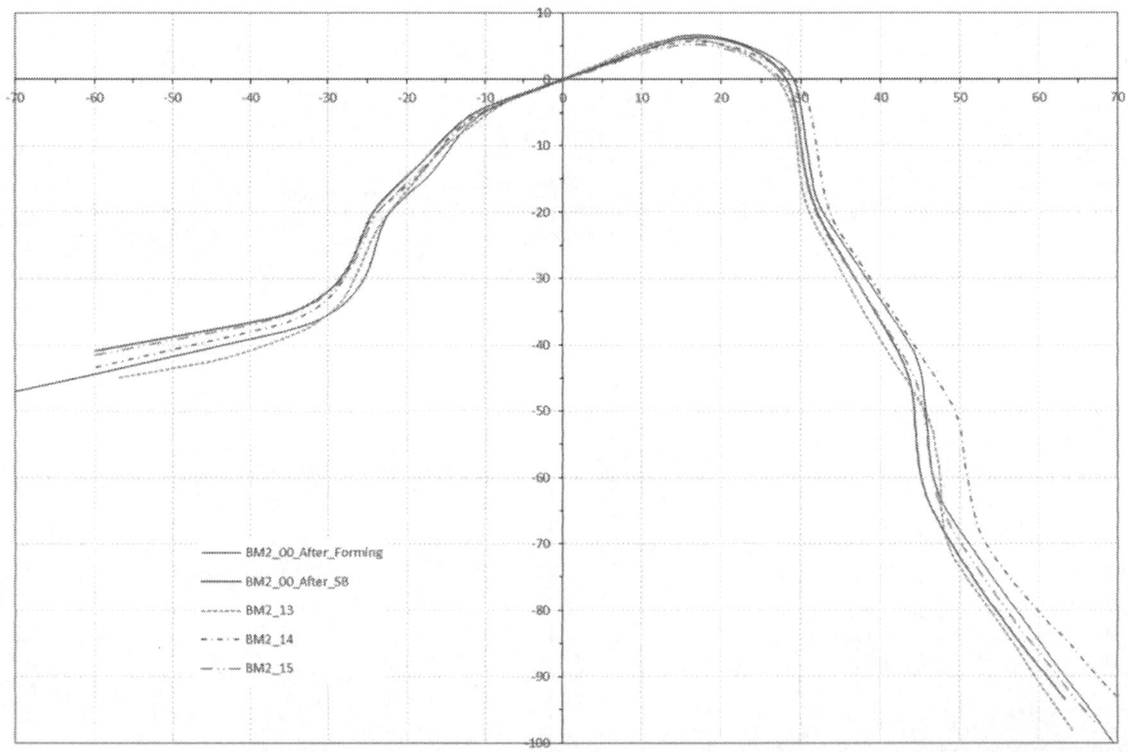

Figure 6.10. Profile after springback for Section II: BM2_13, BM2_14, BM2_15.

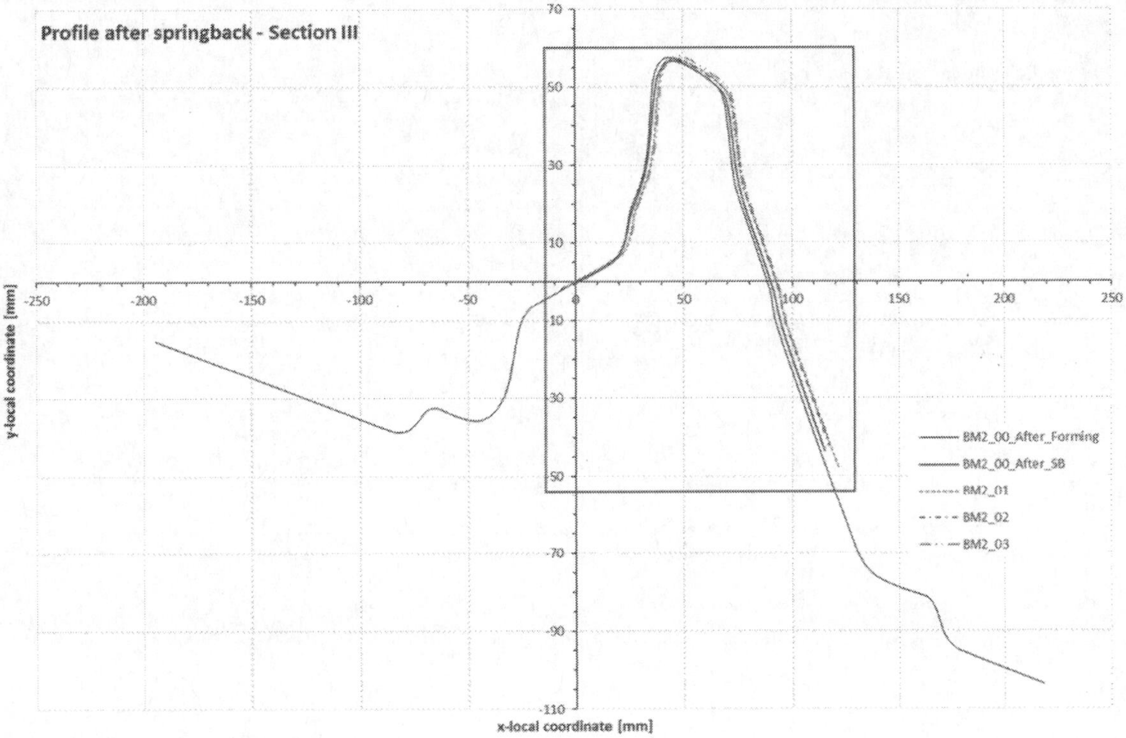

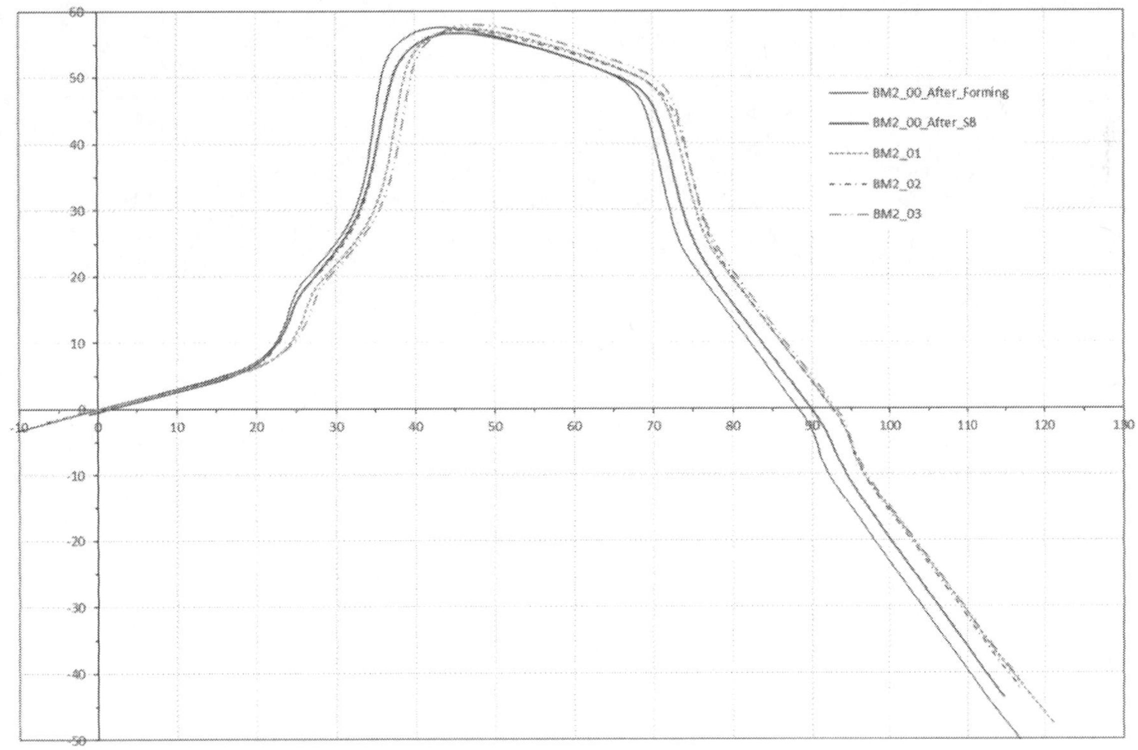

Figure 6.11. Profile after springback for Section III: BM2_01, BM2_02, BM2_03.

Journal of Physics: Conference Series **734** (2016) 022002 doi:10.1088/1742-6596/734/2/022002

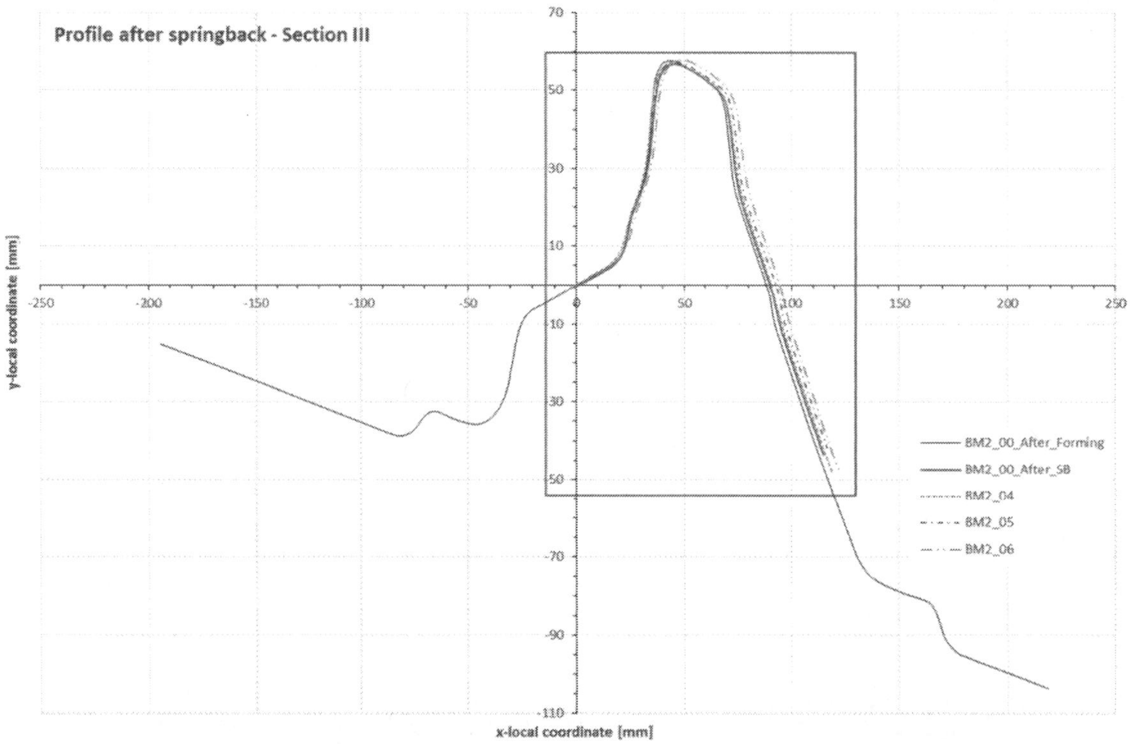

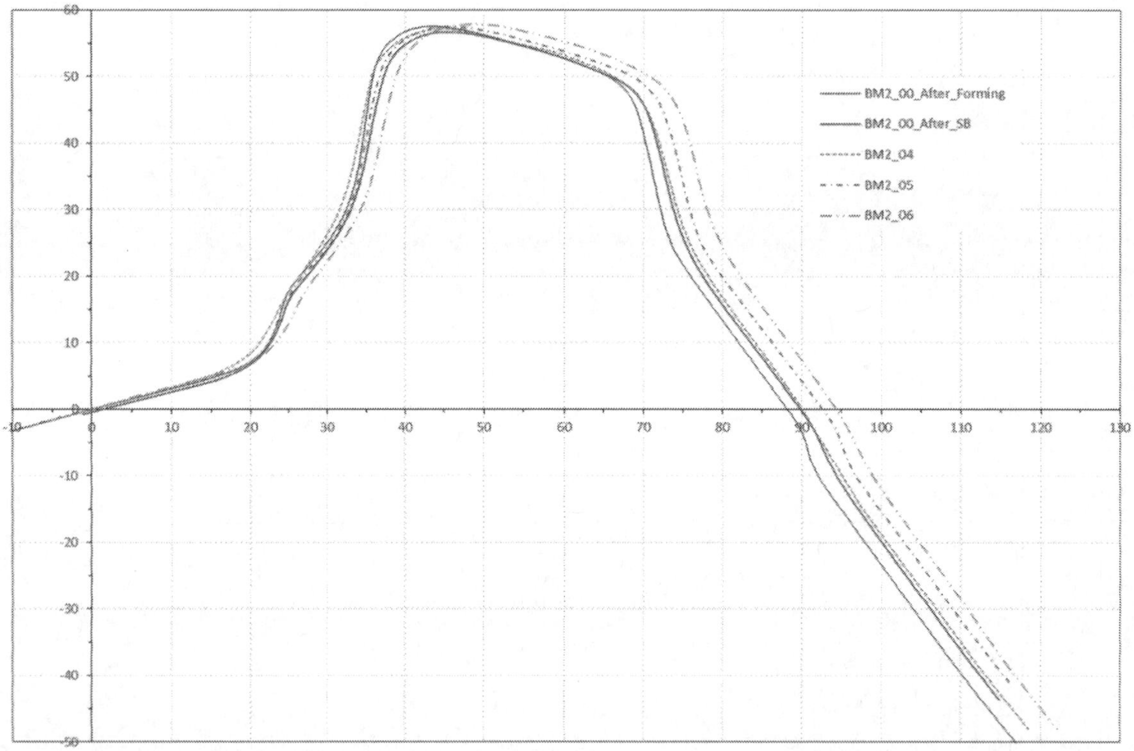

Figure 6.12. Profile after springback for Section III: BM2_04, BM2_05, BM2_06.

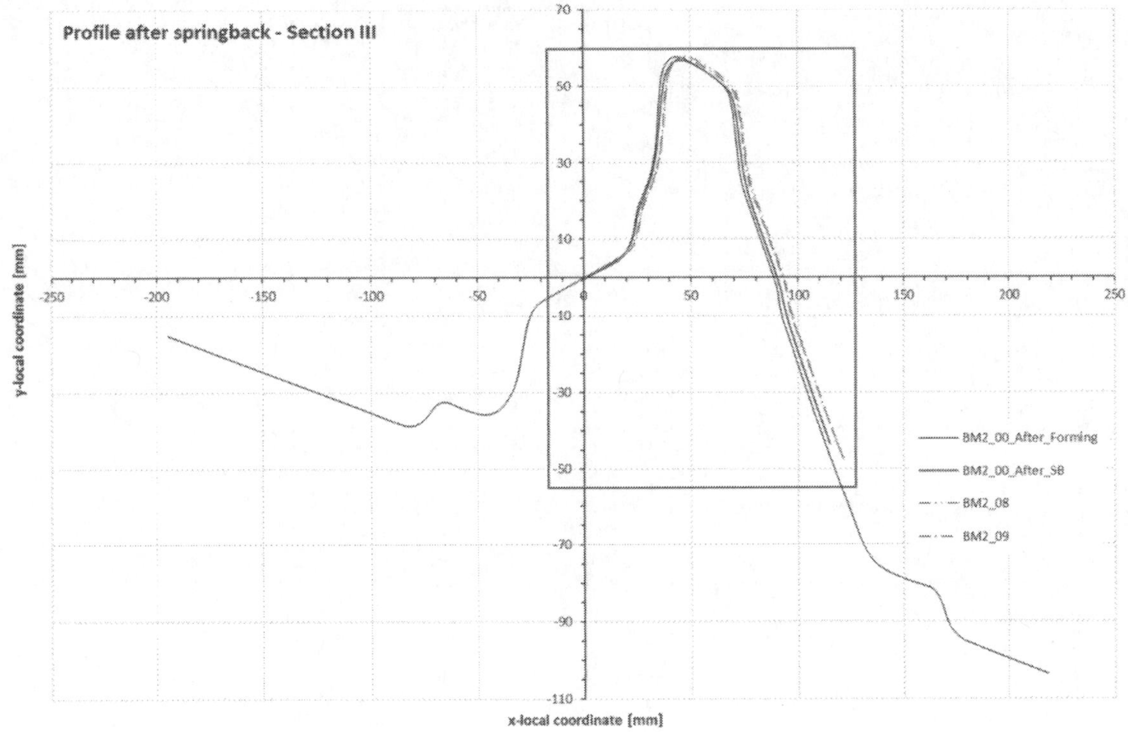

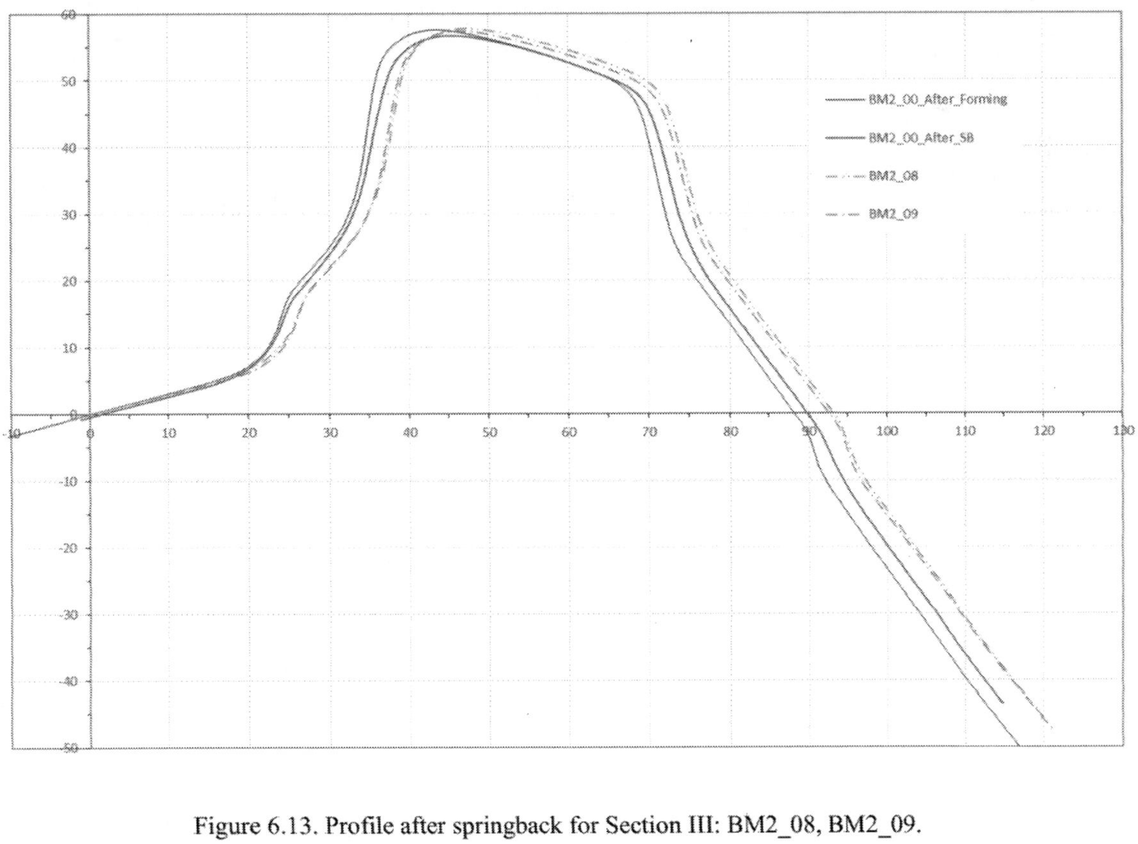

Figure 6.13. Profile after springback for Section III: BM2_08, BM2_09.

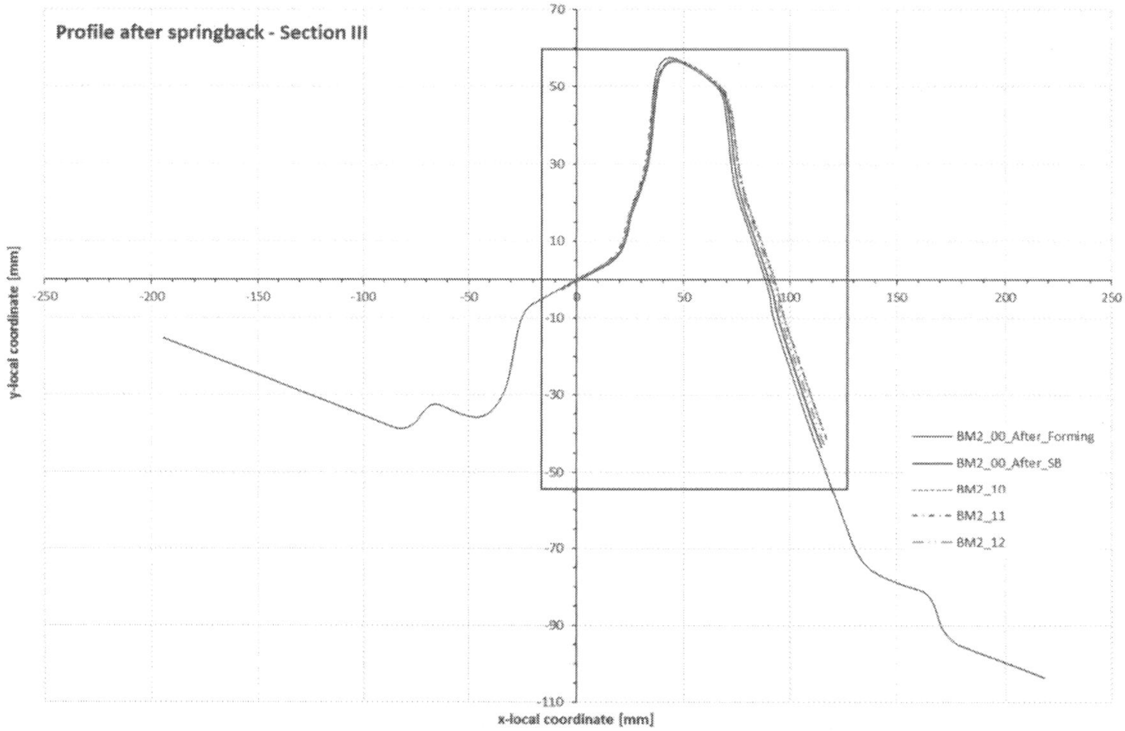

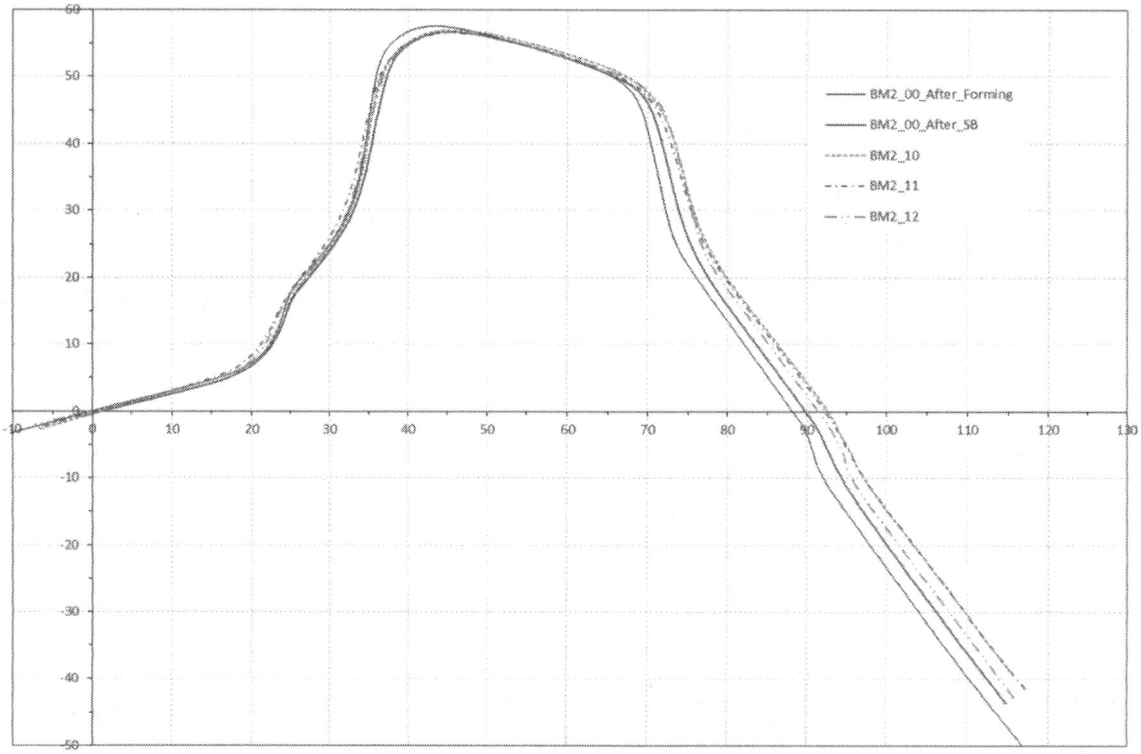

Figure 6.14. Profile after springback for Section III: BM2_10, BM2_11, BM2_12.

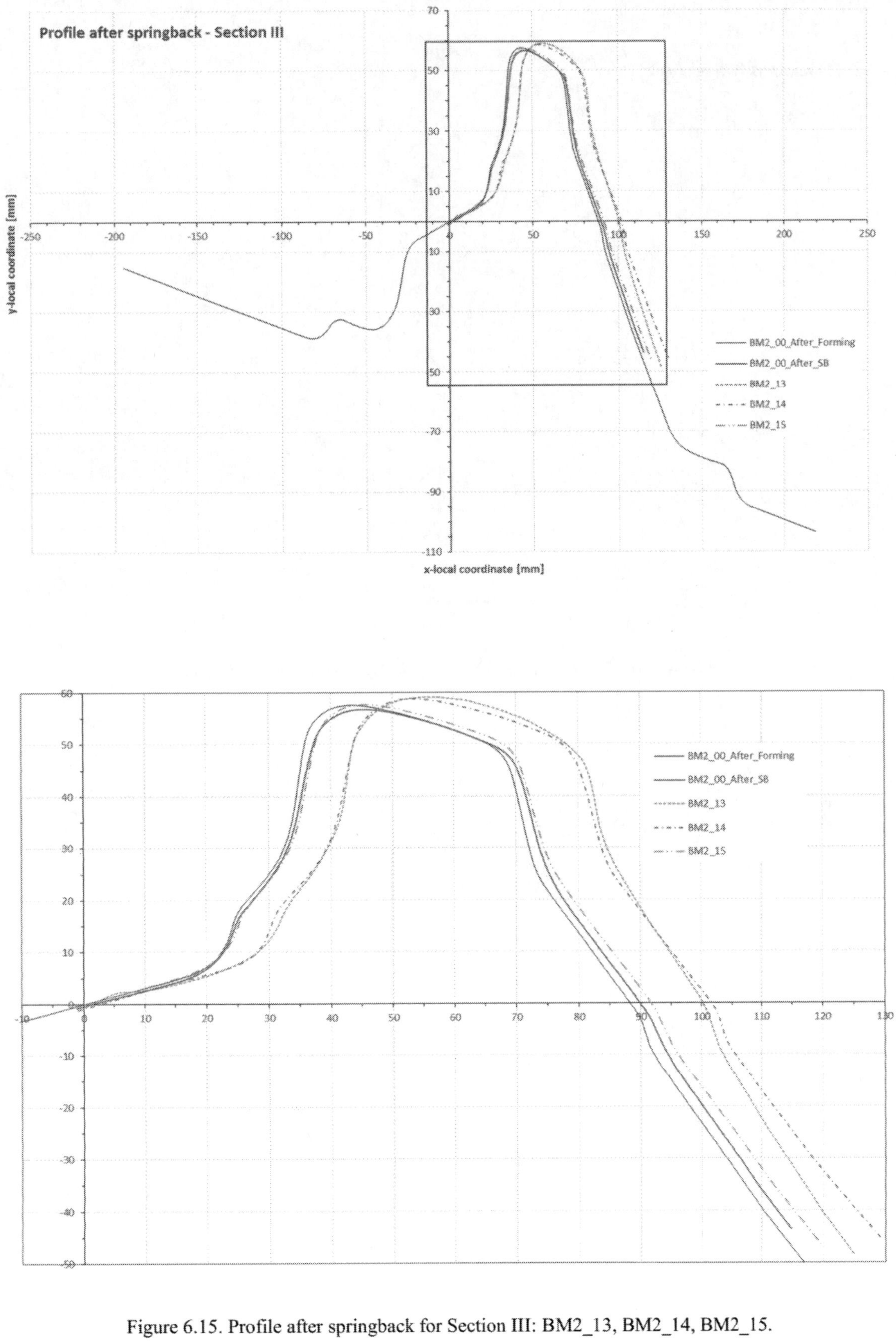

Figure 6.15. Profile after springback for Section III: BM2_13, BM2_14, BM2_15.

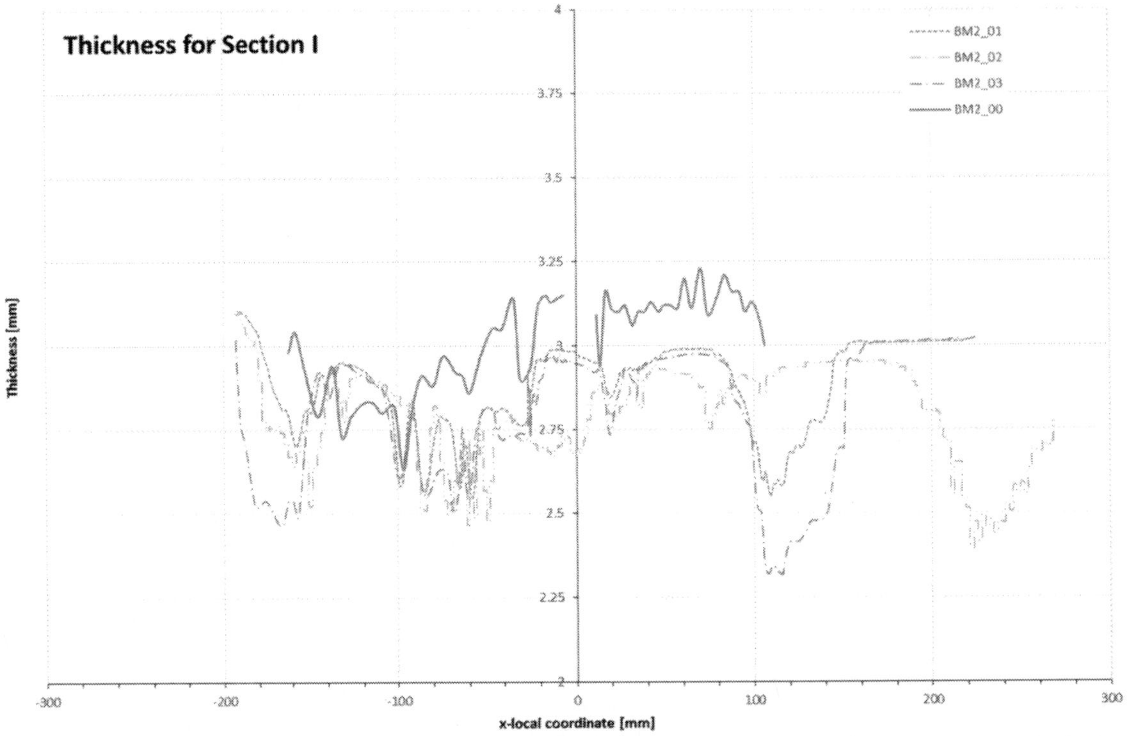

Figure 6.16. Thickness for Section I: BM2_01, BM2_02, BM2_03.

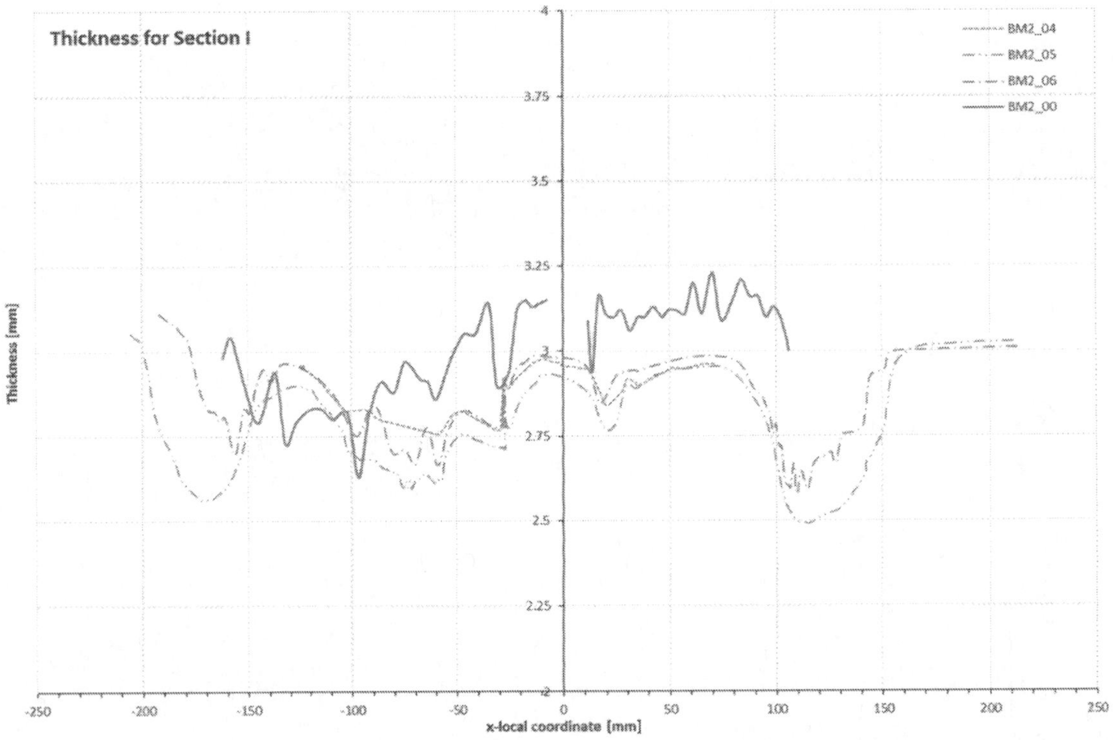

Figure 6.17. Thickness for Section I: BM2_04, BM2_05, BM2_06.

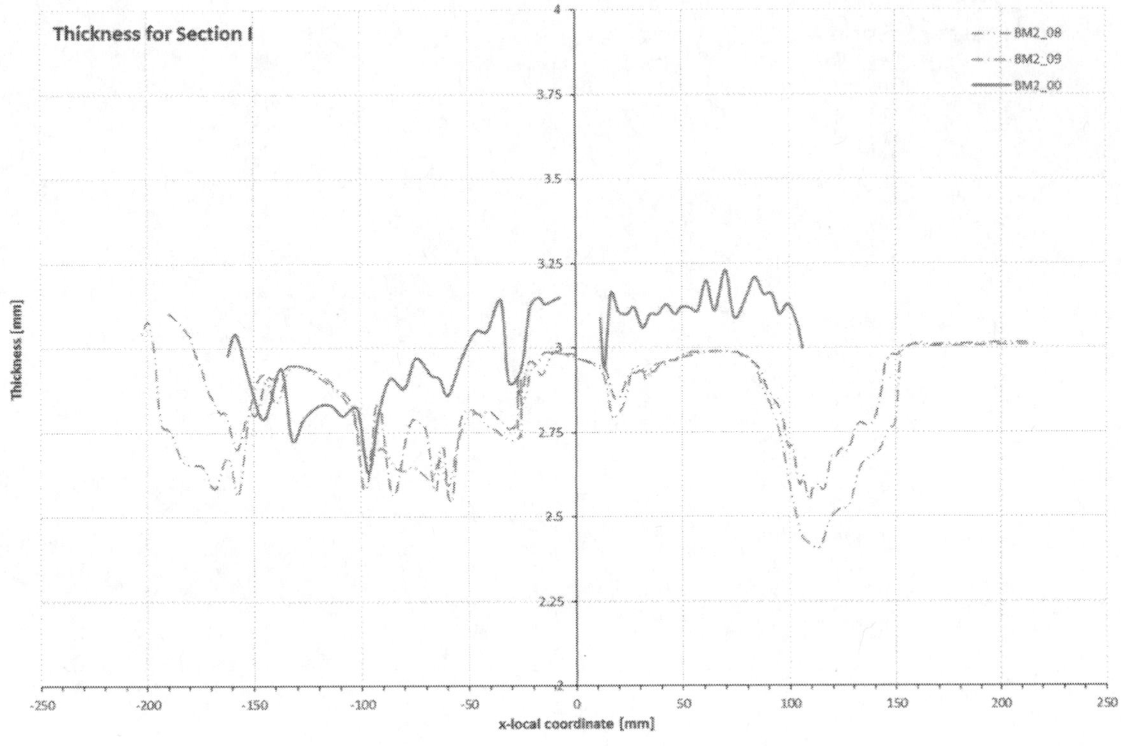

Figure 6.18. Thickness for Section I: BM2_08, BM2_09.

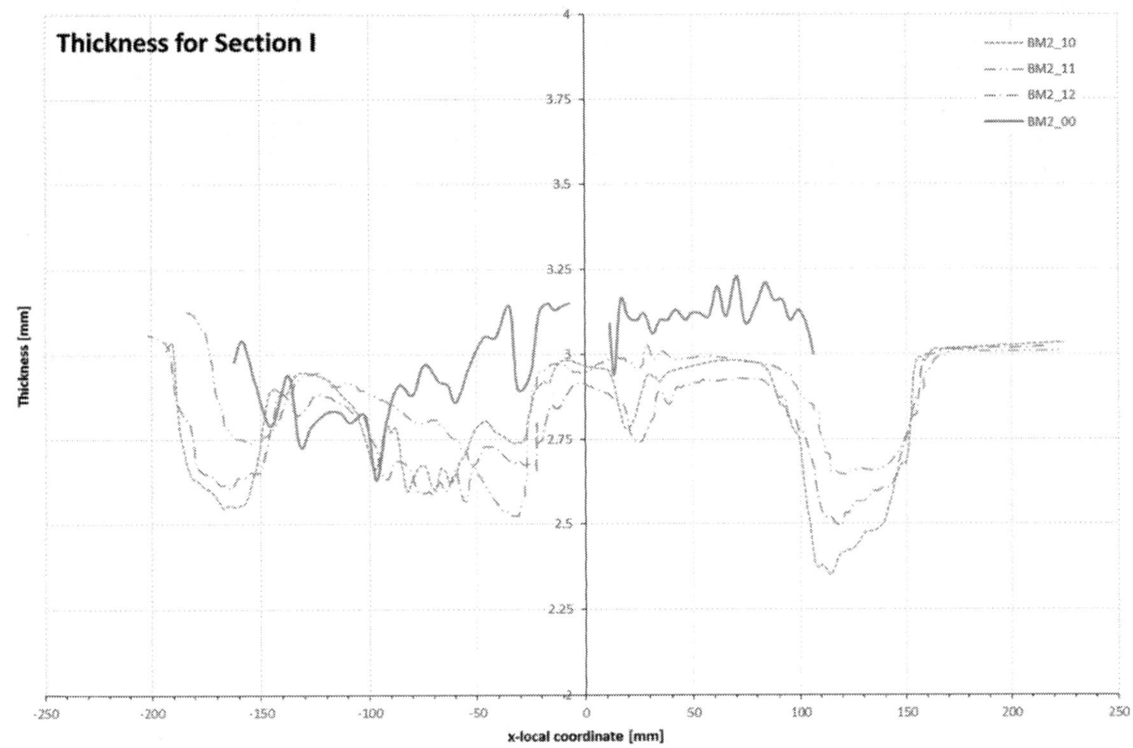

Figure 6.19. Thickness for Section I: BM2_10, BM2_11, BM2_12.

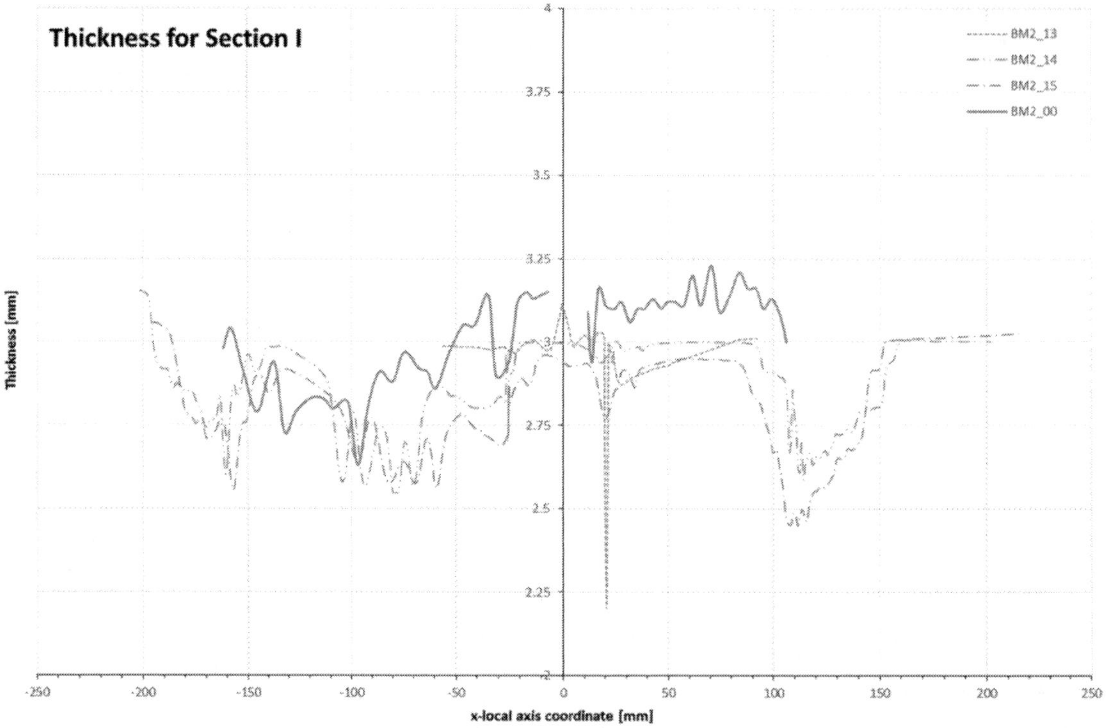

Figure 6.20. Thickness for Section I: BM2_13, BM2_14, BM2_15.

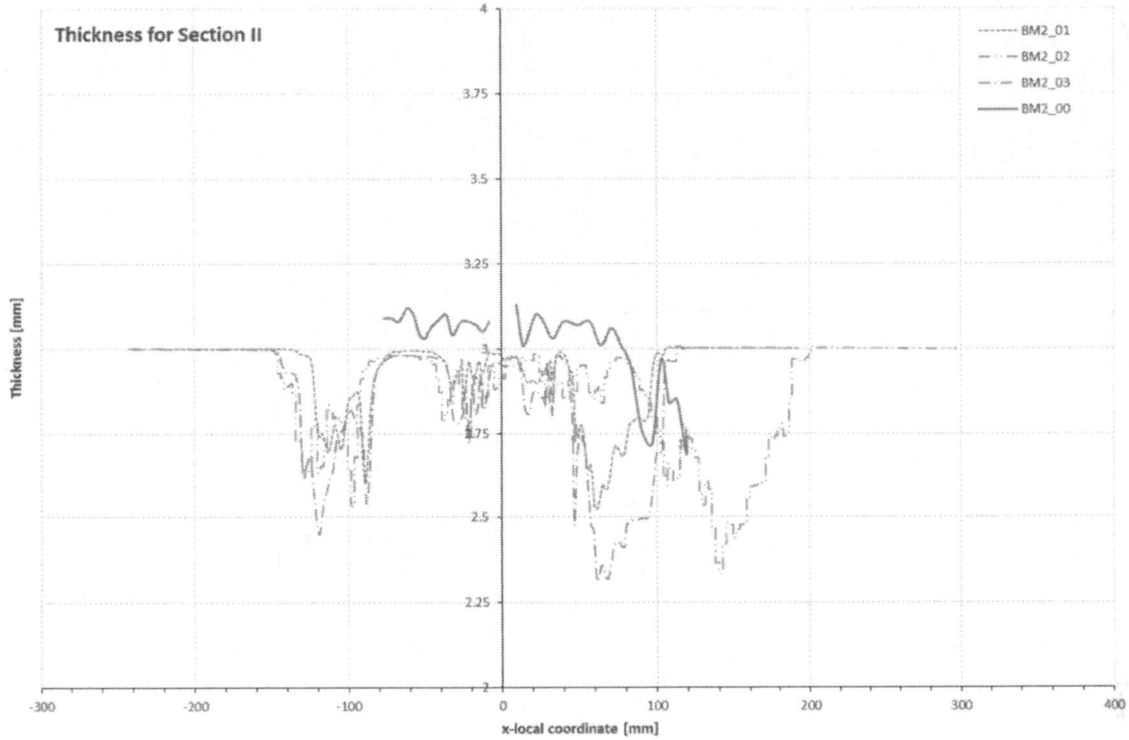

Figure 6.21. Thickness for Section II: BM2_01, BM2_02, BM2_03.

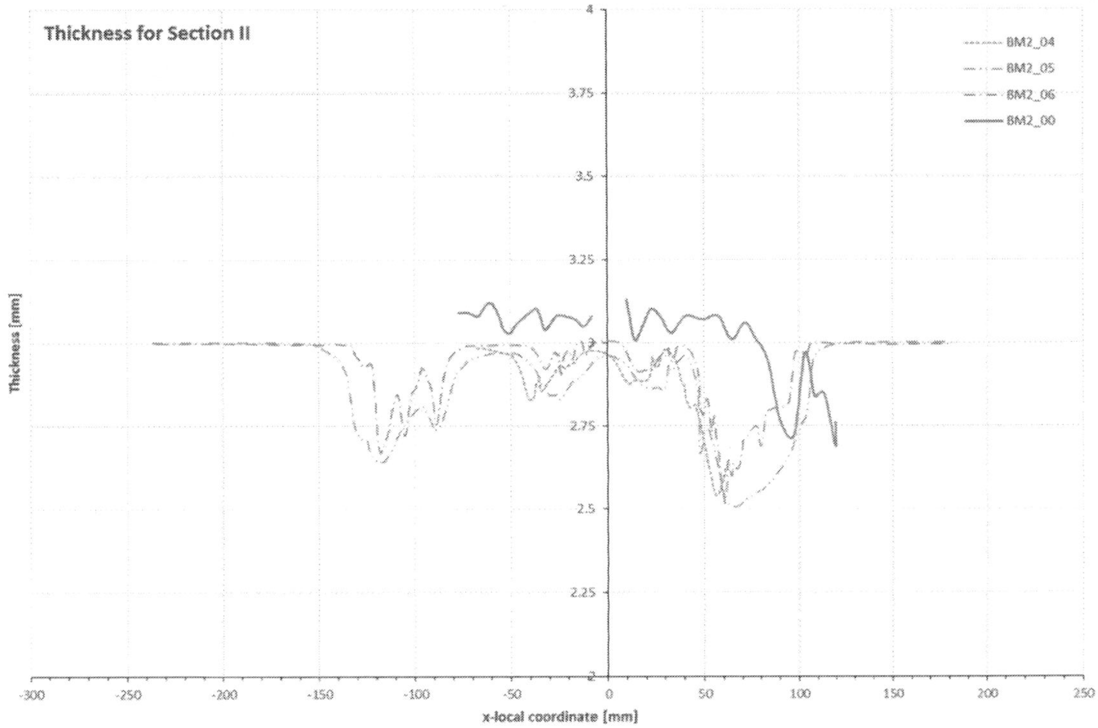

Figure 6.22. Thickness for Section II: BM2_04, BM2_05, BM2_06.

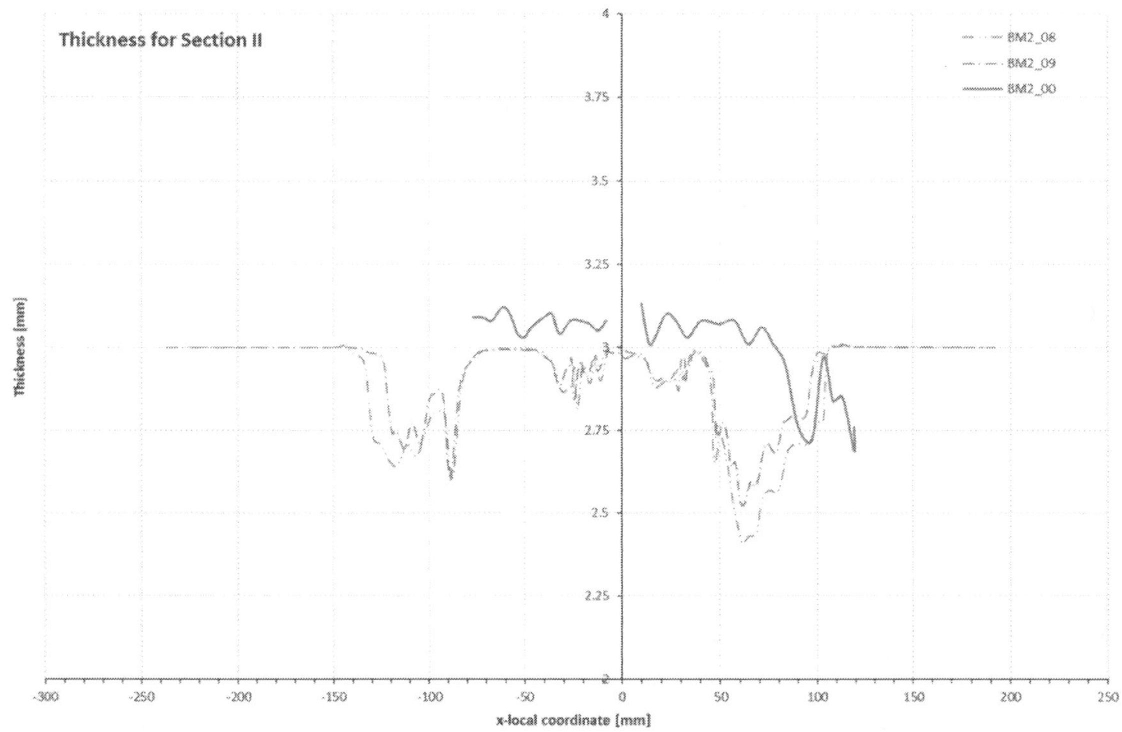

Figure 6.23. Thickness for Section II: BM2_08, BM2_09.

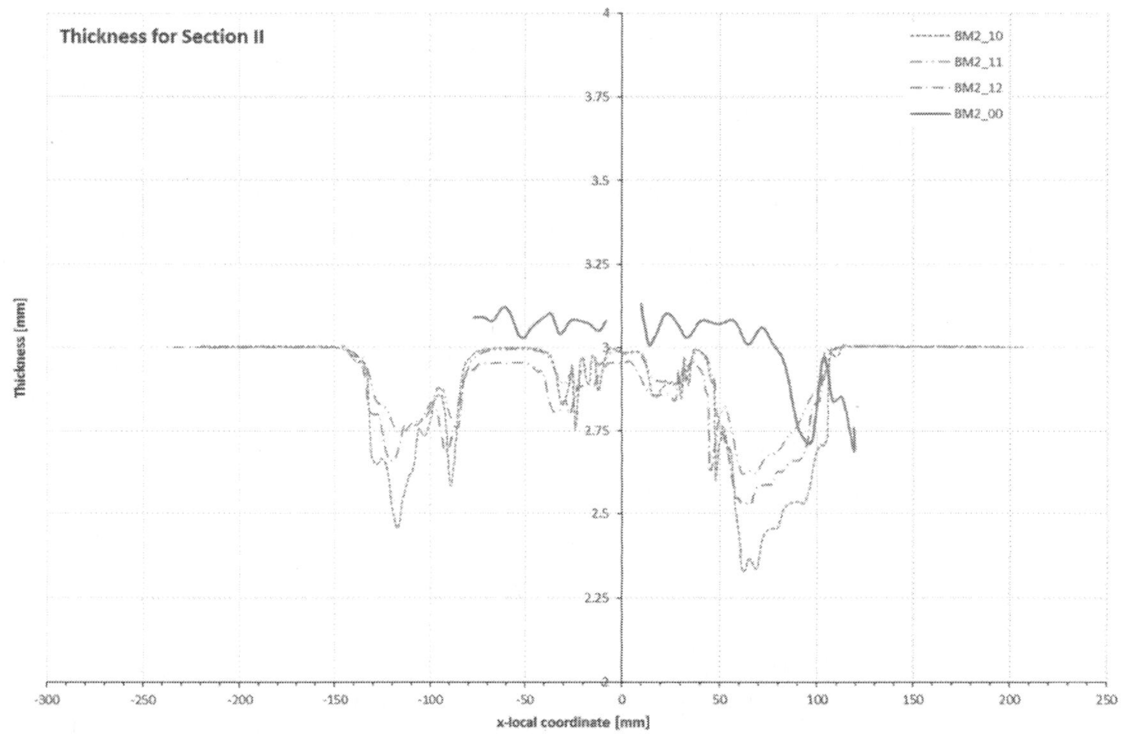

Figure 6.24. Thickness for Section II: BM2_10, BM2_11, BM2_12.

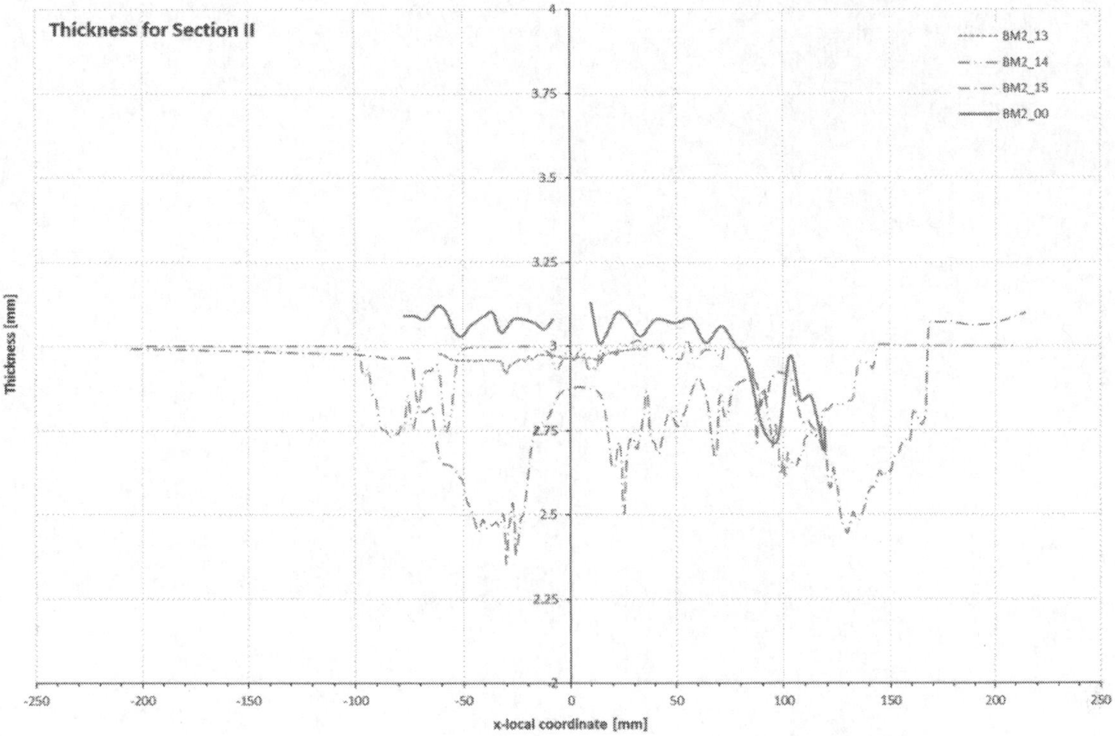

Figure 6.25. Thickness for Section II: BM2_13, BM2_14, BM2_15.

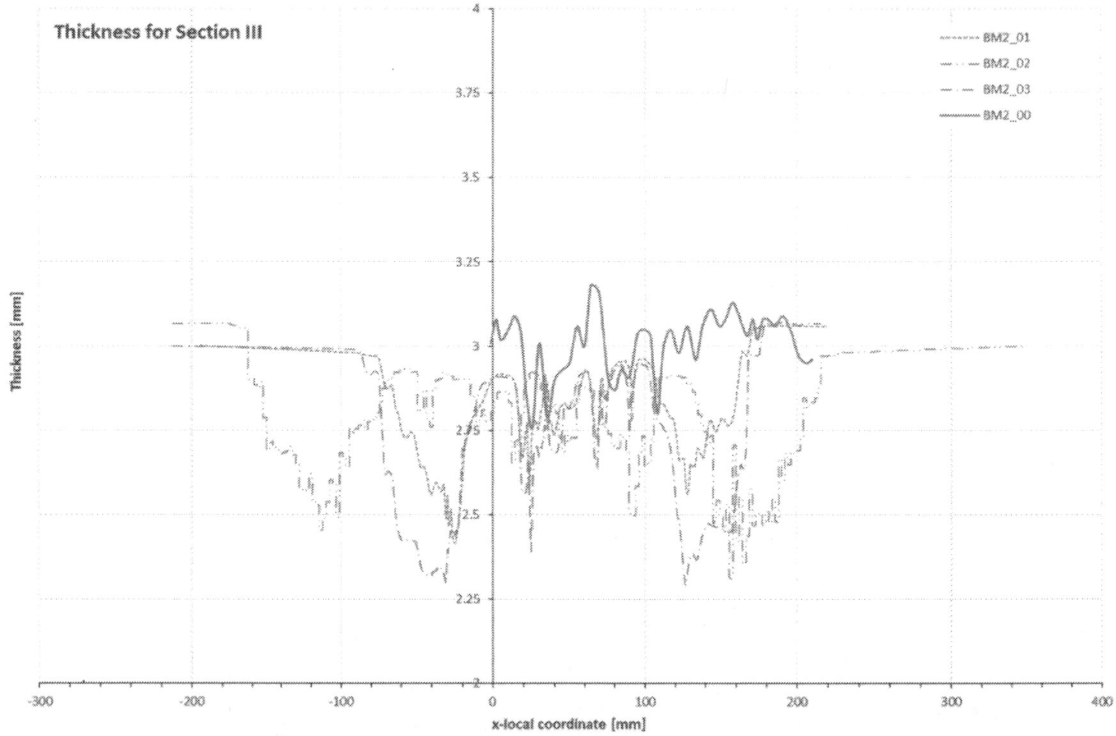

Figure 6.26. Thickness for Section III: BM2_01, BM2_02, BM2_03.

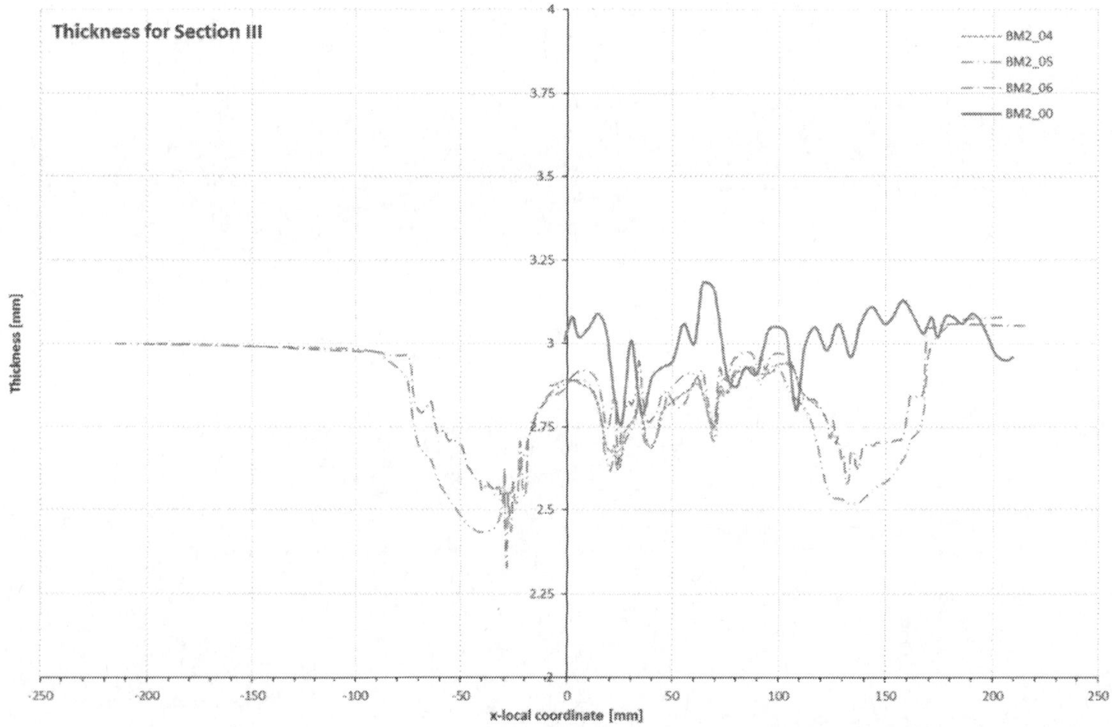

Figure 6.27. Thickness for Section III: BM2_04, BM2_05, BM2_06.

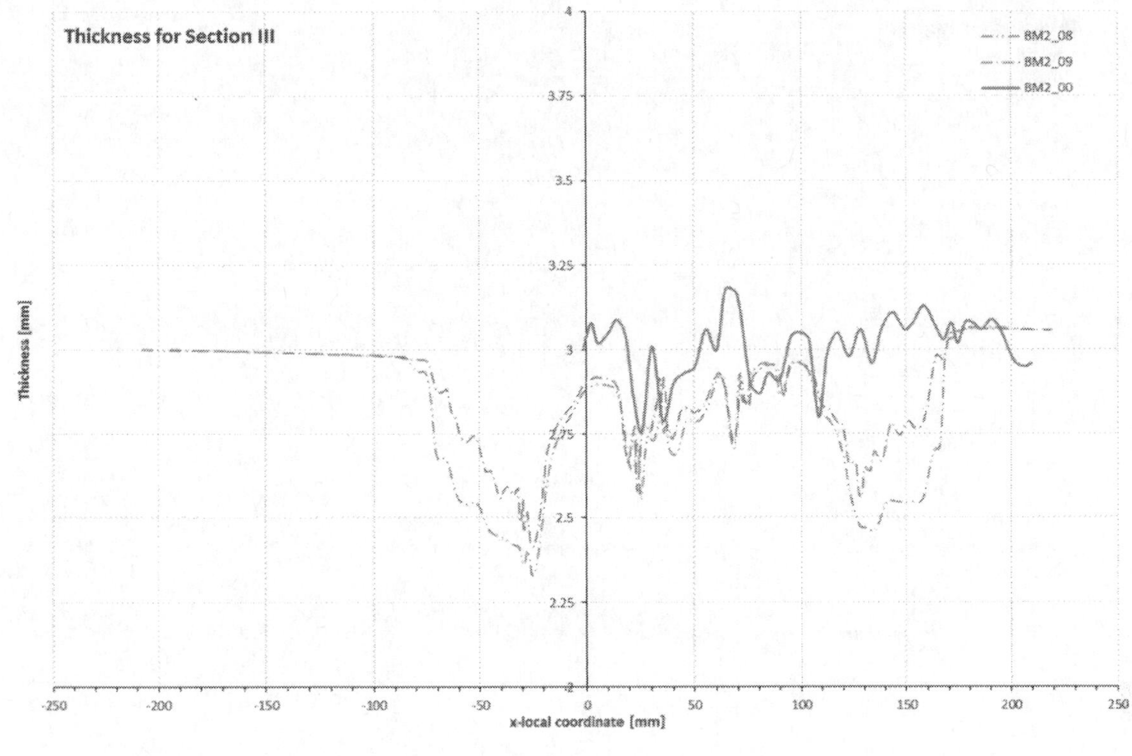

Figure 6.28. Thickness for Section III: BM2_08, BM2_09.

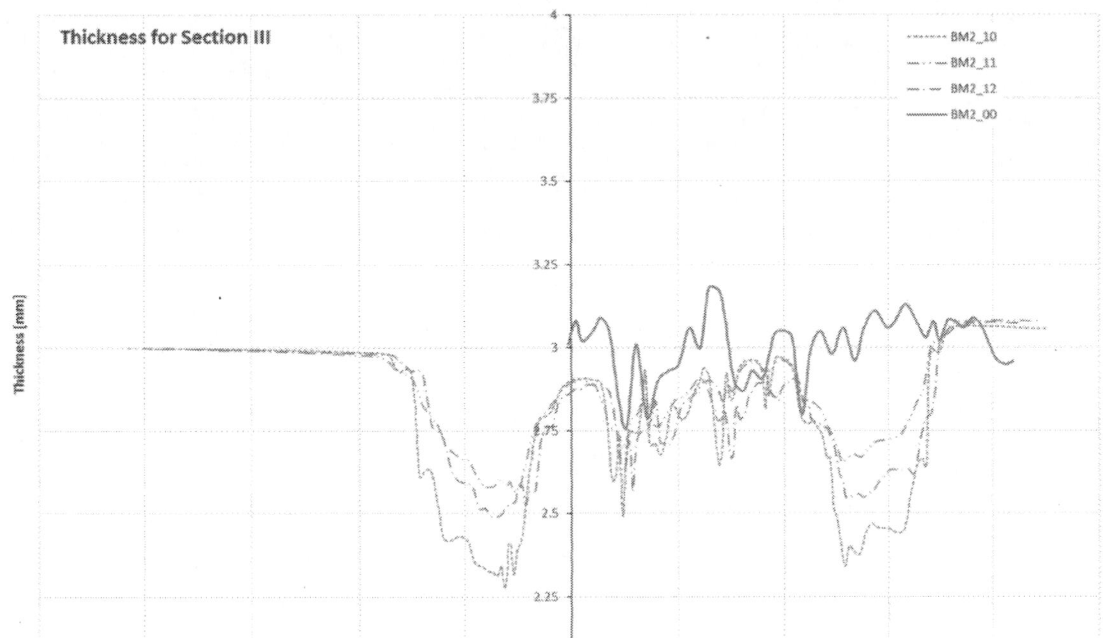

Figure 6.29. Thickness for Section III: BM2_10, BM2_11, BM2_12.

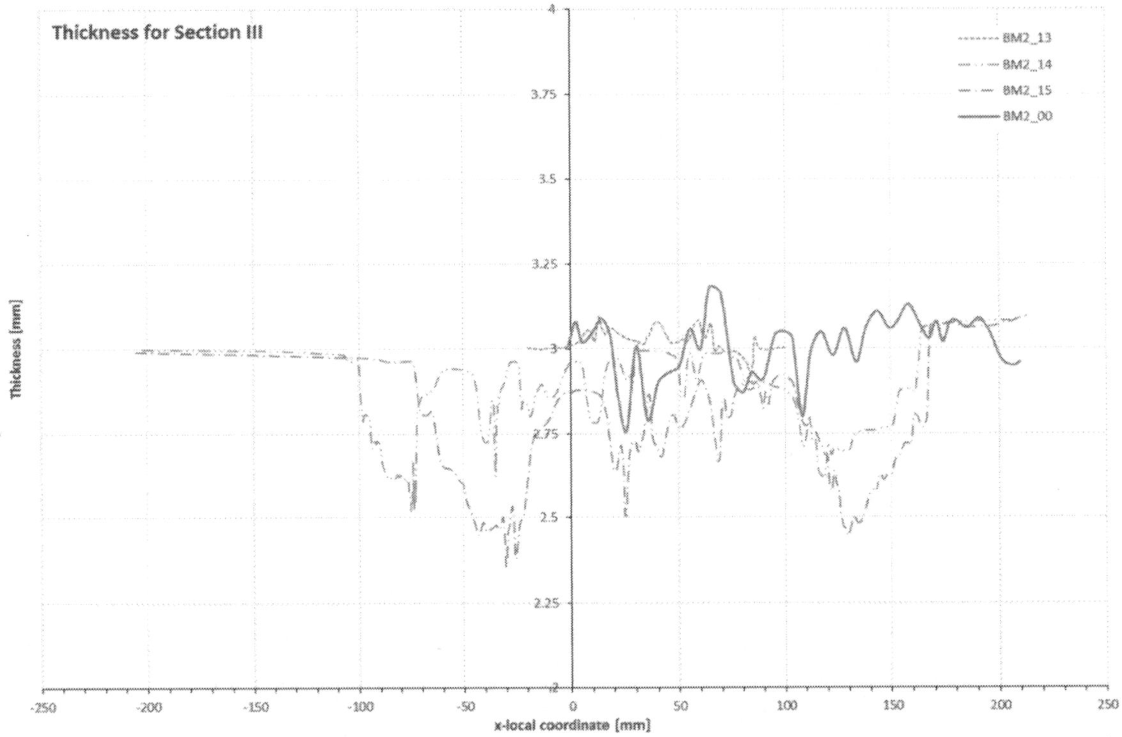

Figure 6.30. Thickness for Section III: BM2_13, BM2_14, BM2_15.

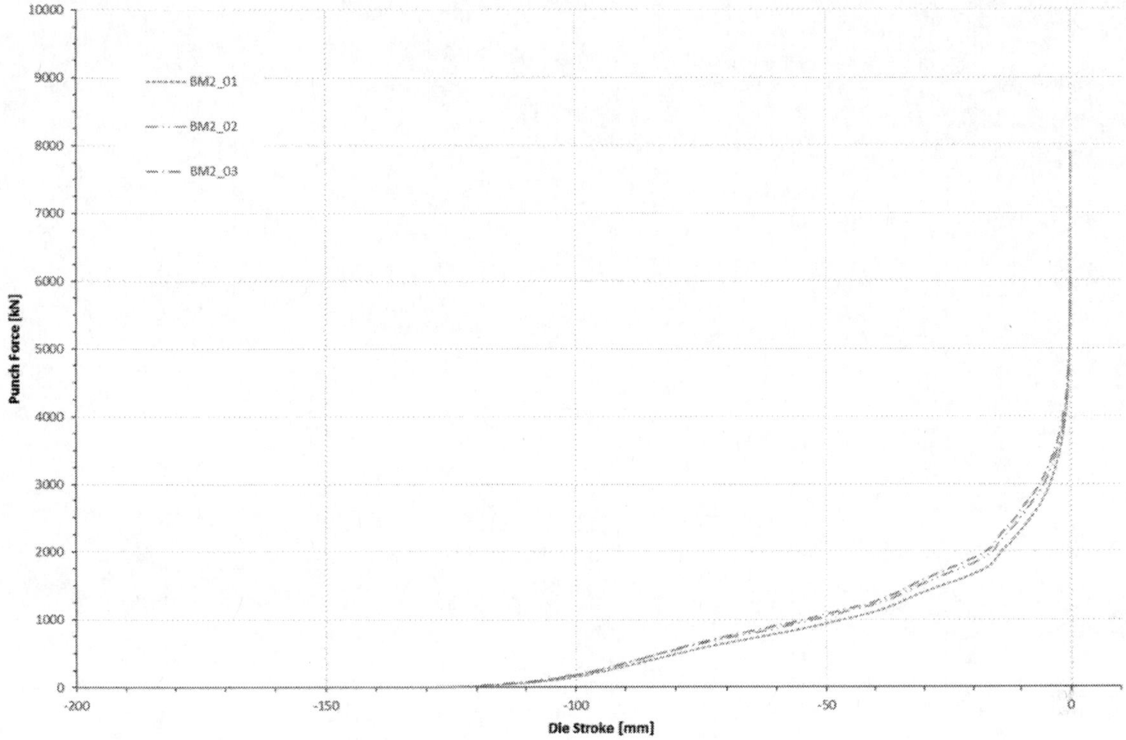

Figure 6.31. Punch Force: BM2_01, BM2_02, BM2_03.

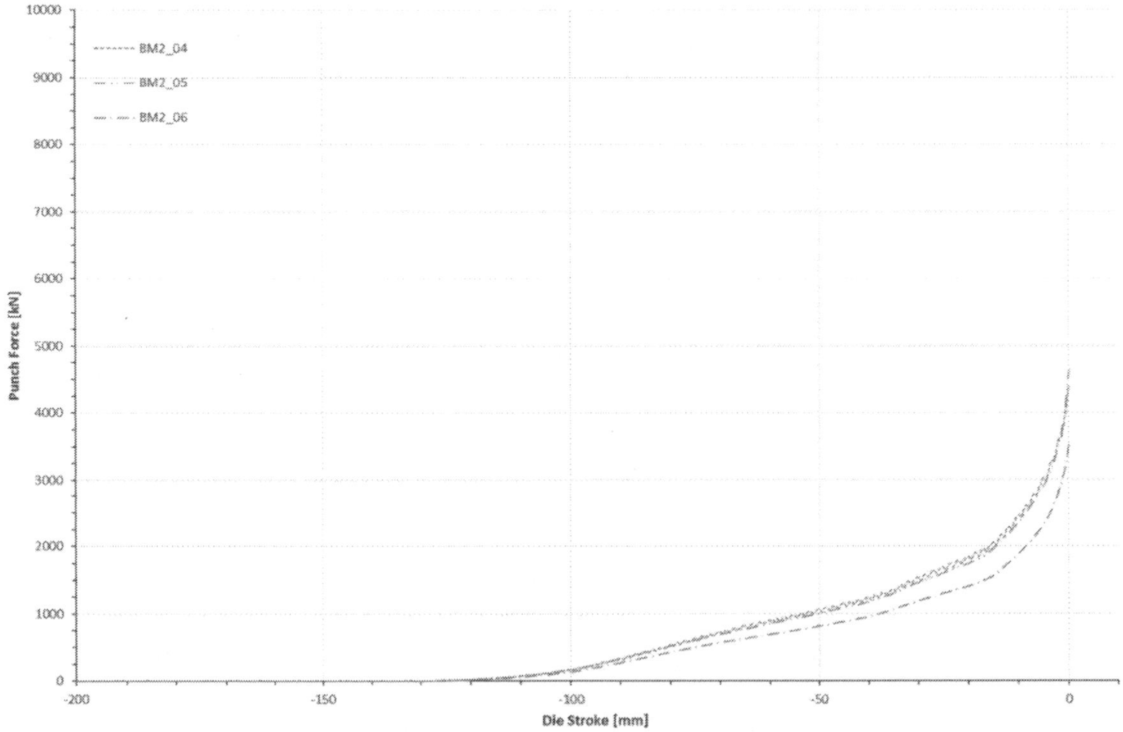

Figure 6.32. Punch Force: BM2_04, BM2_05, BM2_06.

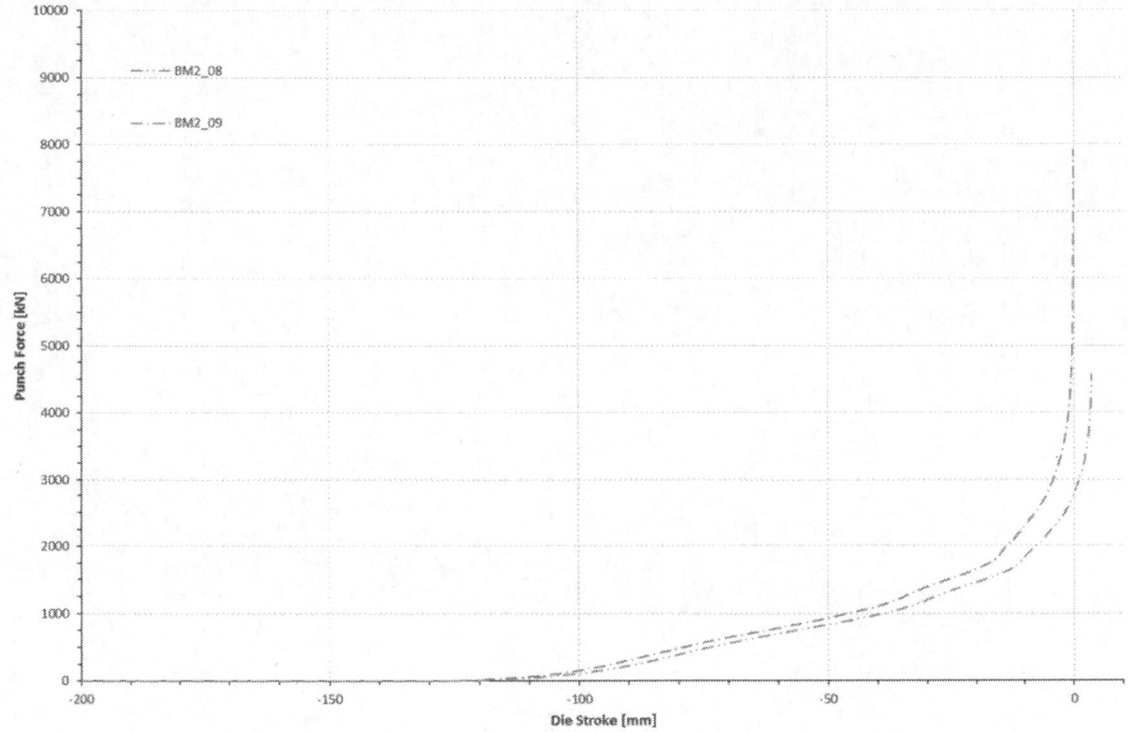

Figure 6.33. Punch Force: BM2_08, BM2_09.

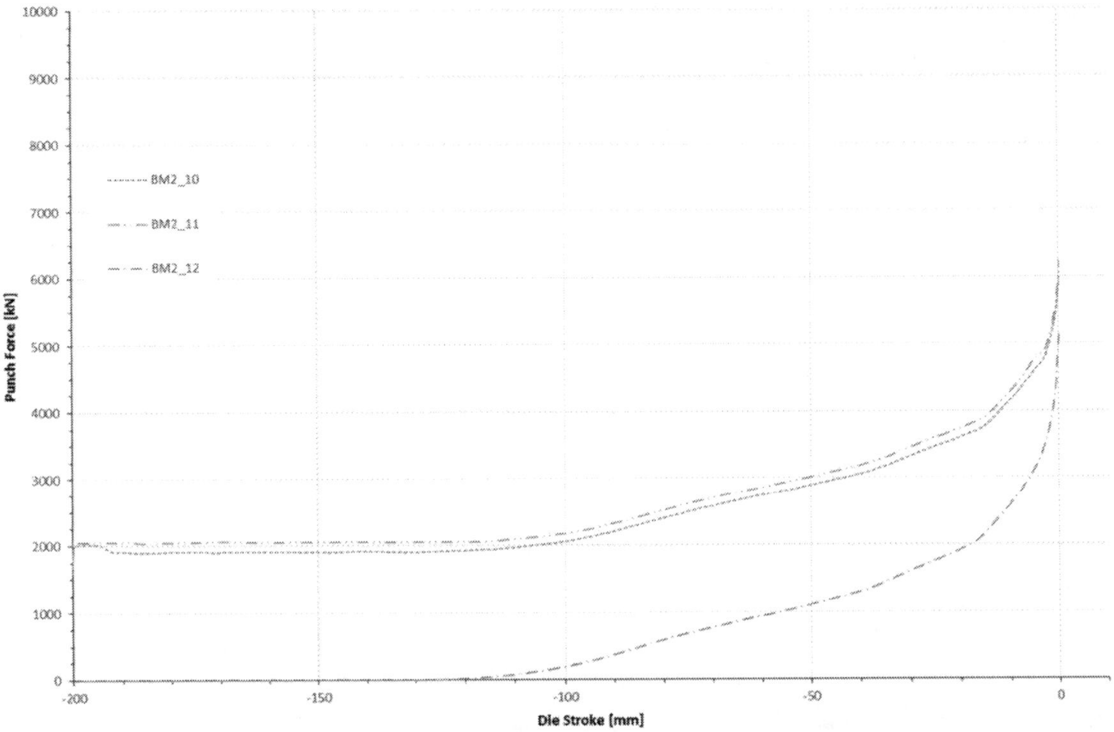

Figure 6.34. Punch Force: BM2_10, BM2_11, BM2_12.

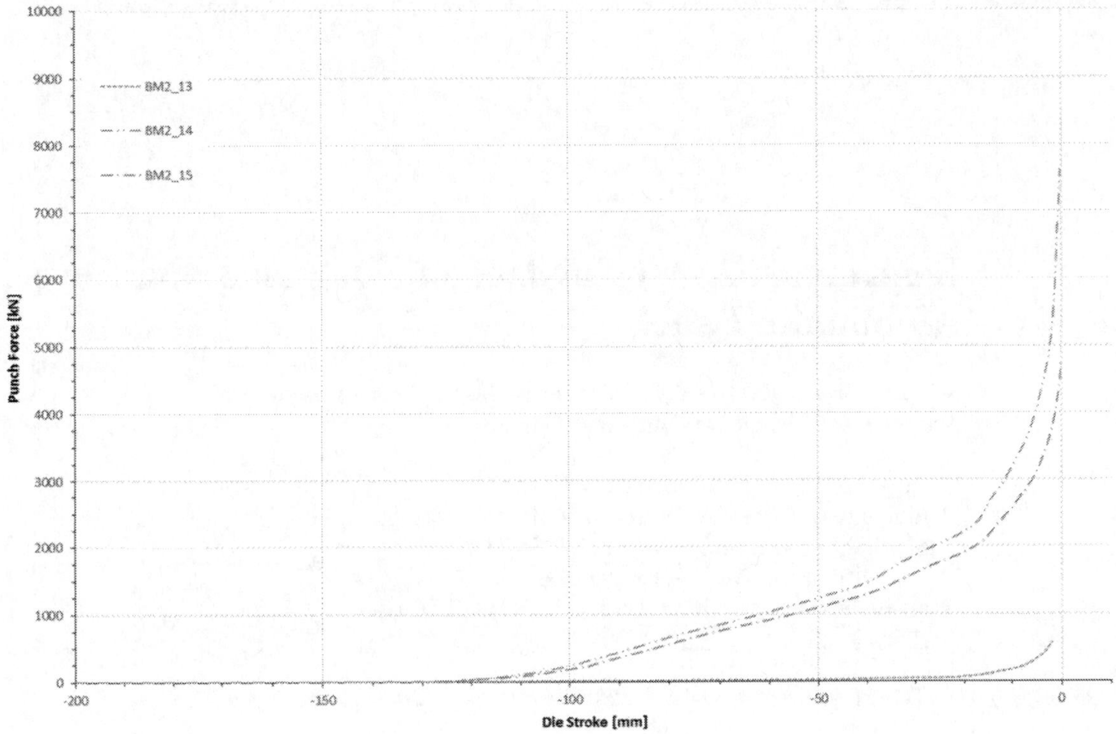

Figure 6.35. Punch Force: BM2_13, BM2_14, BM2_15.

Benchmark 2 – Springback of a Jaguar Land Rover Aluminium Panel

Martin Allen[1], Marta Oliveira[2], Sumit Hazra[3], Oluwamayokun Adetoro[4], Abhishek Das[3] and Rui Cardoso[5]

[1] Jaguar Land Rover, Coventry CV3 4LF, UK
[2] University of Coimbra, Coimbra 3030-790, Portugal
[3] University of Warwick, Coventry CV4 7AL, UK
[4] University of the West of England, Bristol BS16 1QY, UK
[5] Brunel University London, Uxbridge, UB8 3PH, UK

CORRIGENDUM TO: M Allen *et al* 2016 *J. Phys.: Conf. Ser.* **734** 022002

The editor would like to add additional material that was omitted from the original paper. The introduction of the new material results in all of the figures appearing after the new material being renumbered, the figures are not being overwritten. The new material and all renumbered figures are as follows:

BM2-16

1. Benchmark Participant	
Name	**Bart Carleer, Dave Ling, Igor Burchitz**
Affiliation	AutoForm Engineering B.V.
Address	Industrieweg 2, 2921 LB Krimpen aan den Ijssel, The Netherlands
Email	igor.burchitz@autoform.nl
Phone number	0031 180 668 255
Fax number	

2. Simulation Software	
Name of the FEM code	AutoForm^plus R6
General aspect of the code	Static Implicit
Basic formulations	
Element/Mesh technology	
Number of elements	Initial number of elements - 31555 Final number of elements due to adaptive mesh refinement - 213120
Type of elements	Triangular elastic plastic shell, 11 integration points through thickness
Contact property model	Penalty method
Friction formulation	Coulomb friction

3. Simulation Hardware	
CPU Type	Intel Core i7-5960X
CPU clock speed	3.0 GHz
Number of cores per CPU	8 cores used to run a simulation
Main memory	64 GB
Operating system	Windows 7 Pro
Total CPU time	Elapsed Time - 23 minutes 13 seconds

4. Describe the material model used for each material	
Material	**AA6451-T4**
Yield Function/ Plastic Potential	BBC Model (Banabic 2005). R-values are based on raw tensile test data provided by the organizing committee upon request
Hardening Rule (e.g. Isotropic, kinematic)	Isotropic hardening
Stress-Strain Relation (e.g. Swift, Voce)	Combined Swift - Hockett-Sherby formulation based on raw tensile test data provided by the organizing committee upon request

5. Remarks
Although boundary conditions were requested for analysis of springback, real measurement fixture was used in this submission. The main goal was to have a better comparison to reality. Simulated fixture included two pilots, supporting clamps and one double sided clamp. These elements were used to represent pin support, slot support and the simple clamp used in the real fixture.

BM2-17

1. Benchmark Participant	
Name	**Bart Carleer, Dave Ling, Igor Burchitz**
Affiliation	AutoForm Engineering B.V.
Address	Industrieweg 2, 2921 LB Krimpen aan den Ijssel, The Netherlands
Email	igor.burchitz@autoform.nl
Phone number	0031 180 668 255
Fax number	

2. Simulation Software	
Name of the FEM code	AutoForm^plus R6
General aspect of the code	Static Implicit
Basic formulations	
Element/Mesh technology	
Number of elements	Initial number of elements - 31555 Final number of elements due to adaptive mesh refinement - 211850
Type of elements	Triangular elastic plastic shell, 11 integration points through thickness
Contact property model	Penalty method
Friction formulation	Coulomb friction

3. Simulation Hardware	
CPU Type	Intel Core i7-5960X
CPU clock speed	3.0 GHz
Number of cores per CPU	8 cores used to run a simulation
Main memory	64 GB
Operating system	Windows 7 Pro
Total CPU time	Elapsed Time - 24 minutes 44 seconds

4. Describe the material model used for each material	
Material	**AA6451-T4**
Yield Function/ Plastic Potential	BBC Model (Banabic 2005). R-values are based on raw tensile test data provided by the organizing committee upon request
Hardening Rule (e.g. Isotropic, kinematic)	Kinematic hardening considering early re-plastification, transient softening and work hardening stagnation formulated under plane stress condition
Stress-Strain Relation (e.g. Swift, Voce)	Combined Swift - Hockett-Sherby formulation based on raw tensile test data provided by the organizing committee upon request

5. Remarks
In this submission, springback analysis was performed with boundary conditions requested in the benchmark briefing.

BM2-18

1. Benchmark Participant	
Name	**Bart Carleer, Dave Ling, Igor Burchitz**
Affiliation	AutoForm Engineering B.V.
Address	Industrieweg 2, 2921 LB Krimpen aan den Ijssel, The Netherlands
Email	igor.burchitz@autoform.nl
Phone number	0031 180 668 255
Fax number	

2. Simulation Software	
Name of the FEM code	AutoForm^plus R6
General aspect of the code	Static Implicit
Basic formulations	
Element/Mesh technology	
Number of elements	Initial number of elements - 31555 Final number of elements due to adaptive mesh refinement - 211460
Type of elements	Triangular elastic plastic shell, 11 integration points through thickness
Contact property model	Penalty method
Friction formulation	Pressure dependent coefficient of friction

3. Simulation Hardware	
CPU Type	Intel Core i7-5960X
CPU clock speed	3.0 GHz
Number of cores per CPU	8 cores used to run a simulation
Main memory	64 GB
Operating system	Windows 7 Pro
Total CPU time	Elapsed Time - 25 minutes 58 seconds

4. Describe the material model used for each material	
Material	AA6451-T4
Yield Function/ Plastic Potential	BBC Model (Banabic 2005). R-values are based on raw tensile test data provided by the organizing committee upon request
Hardening Rule (e.g. Isotropic, kinematic)	Kinematic hardening considering early re-plastification, transient softening and work hardening stagnation formulated under plane stress condition
Stress-Strain Relation (e.g. Swift, Voce)	Combined Swift - Hockett-Sherby formulation based on raw tensile test data provided by the organizing committee upon request

5. Remarks
Main goal was to investigate influence of friction on springback prediction of the part. Pressure dependent friction was described by a power law, i.e. Reference Pressure – 4MPa; Pressure Exponent – 0.85; Reference friction coefficient – 0.12. In this submission, springback analysis was performed with boundary conditions requested in the benchmark briefing.

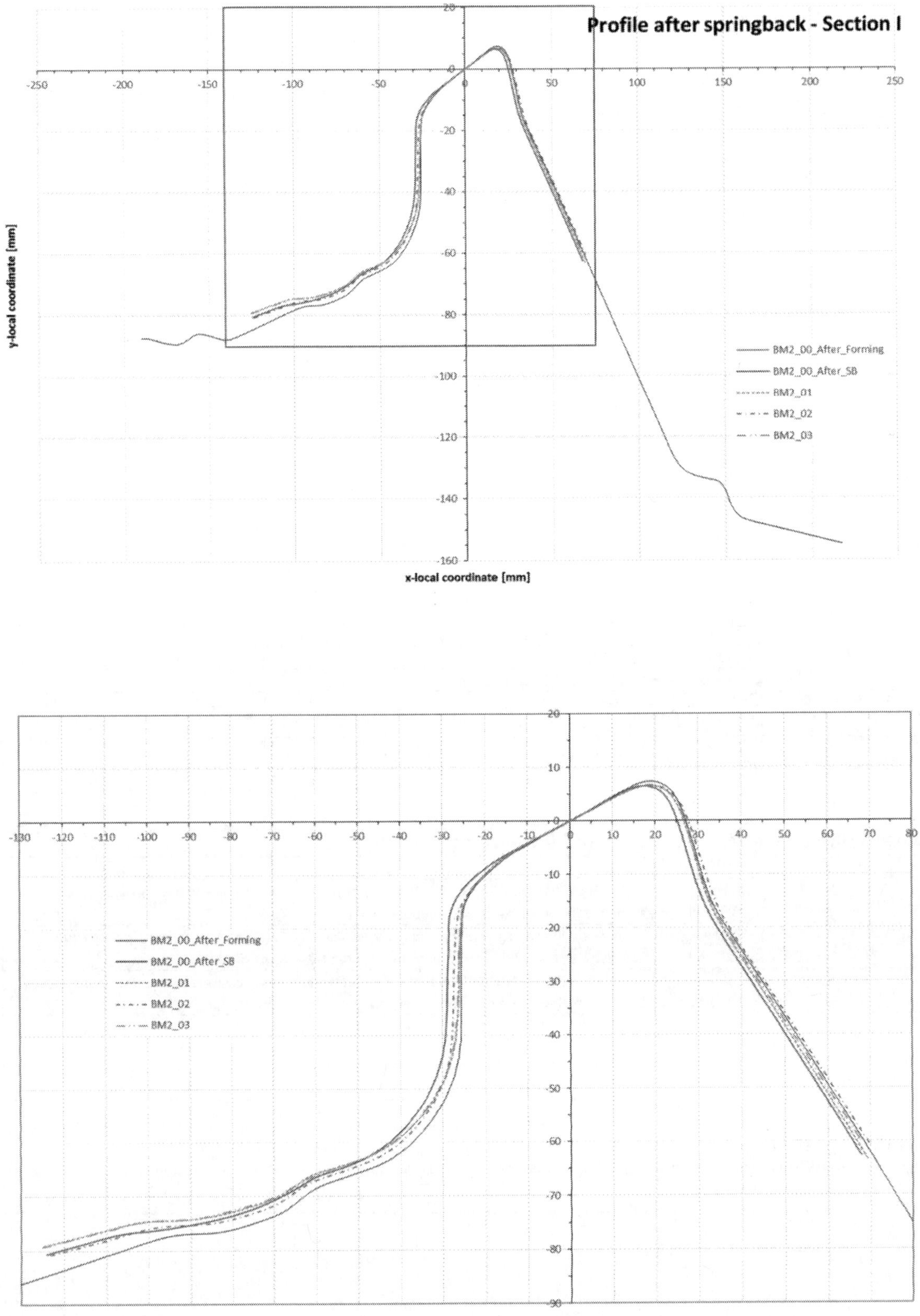

Figure 6.1. Profile after springback for Section I: BM2_01, BM2_02, BM2_03.

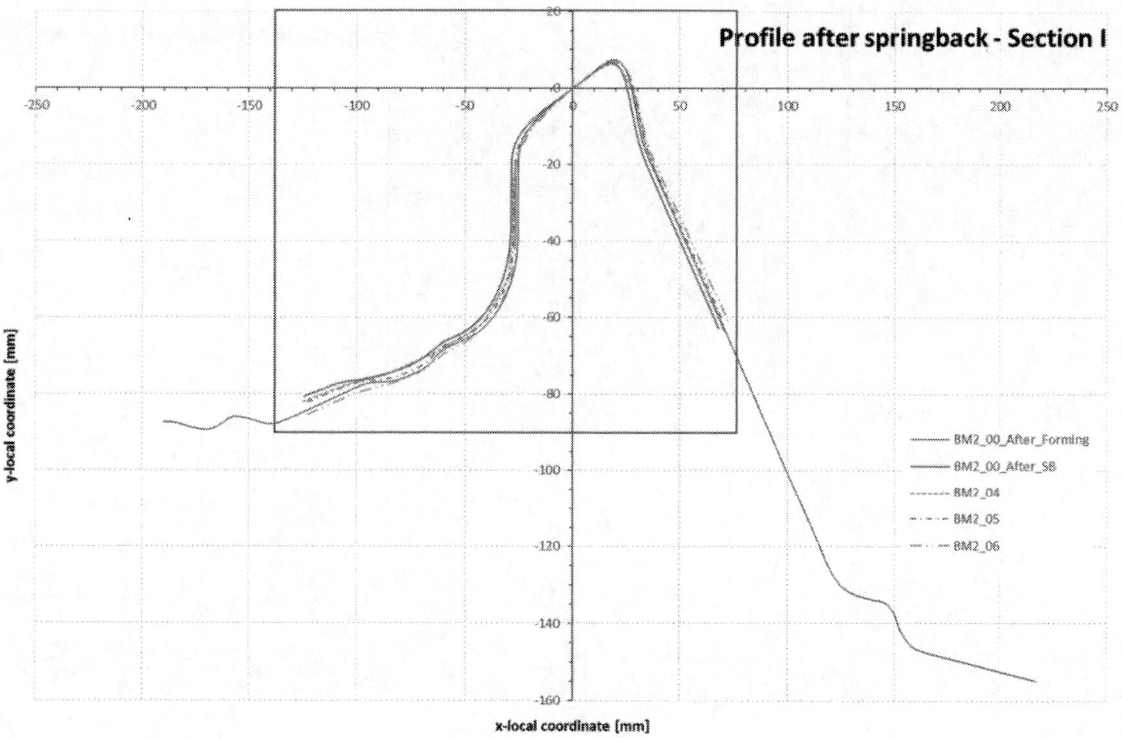

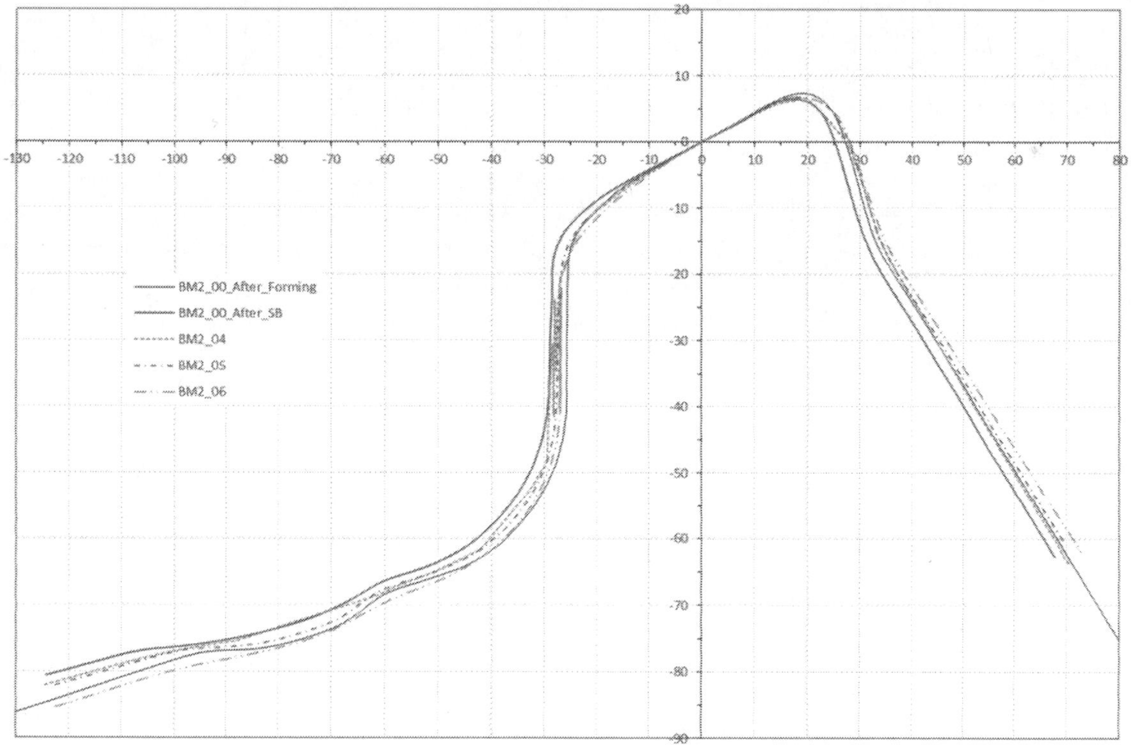

Figure 6.2. Profile after springback for Section I: BM2_04, BM2_05, BM2_06.

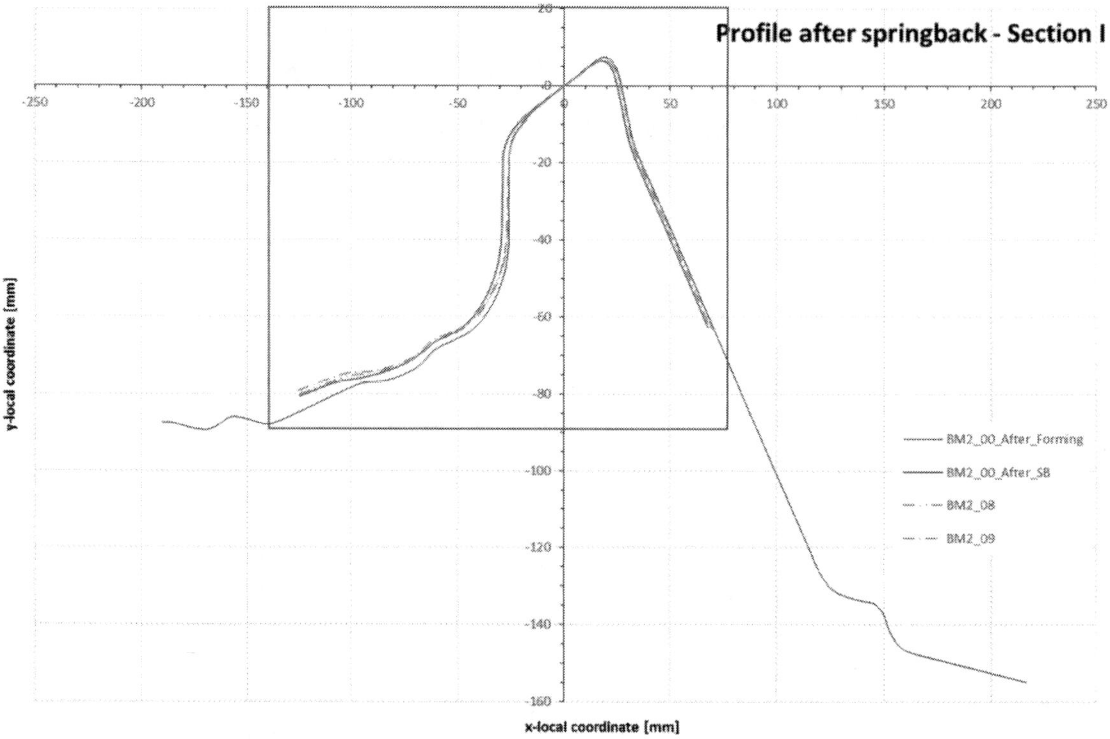

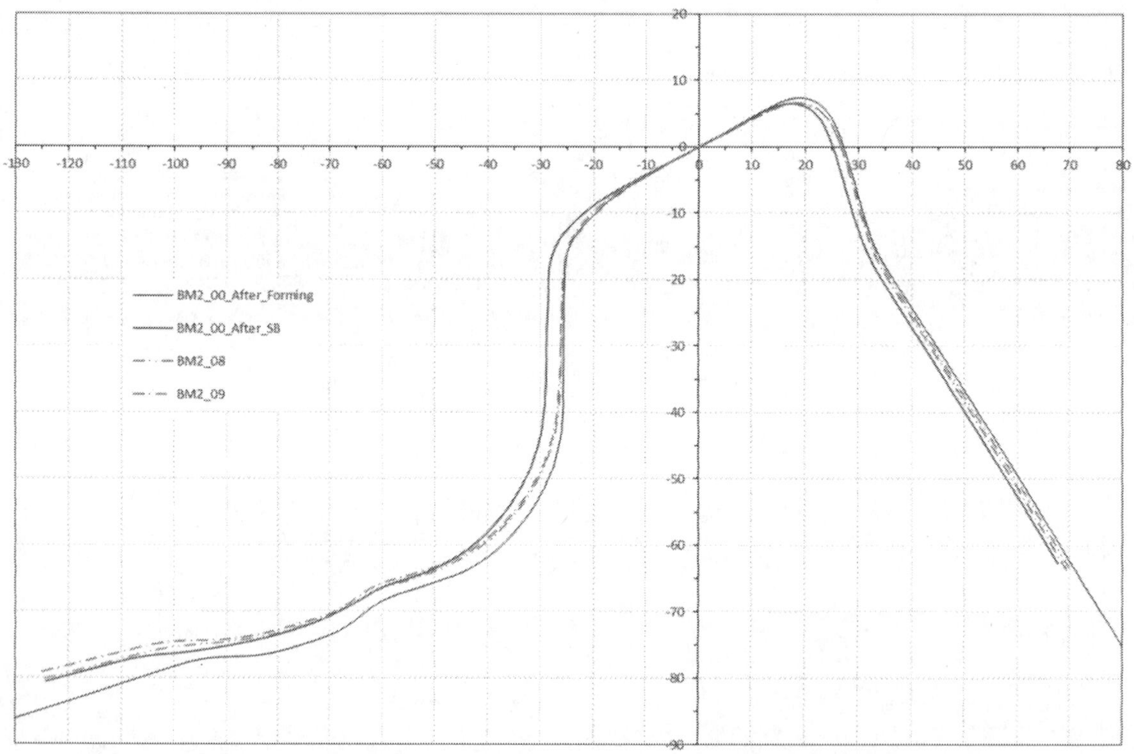

Figure 6.3. Profile after springback for Section I: BM2_08, BM2_09.

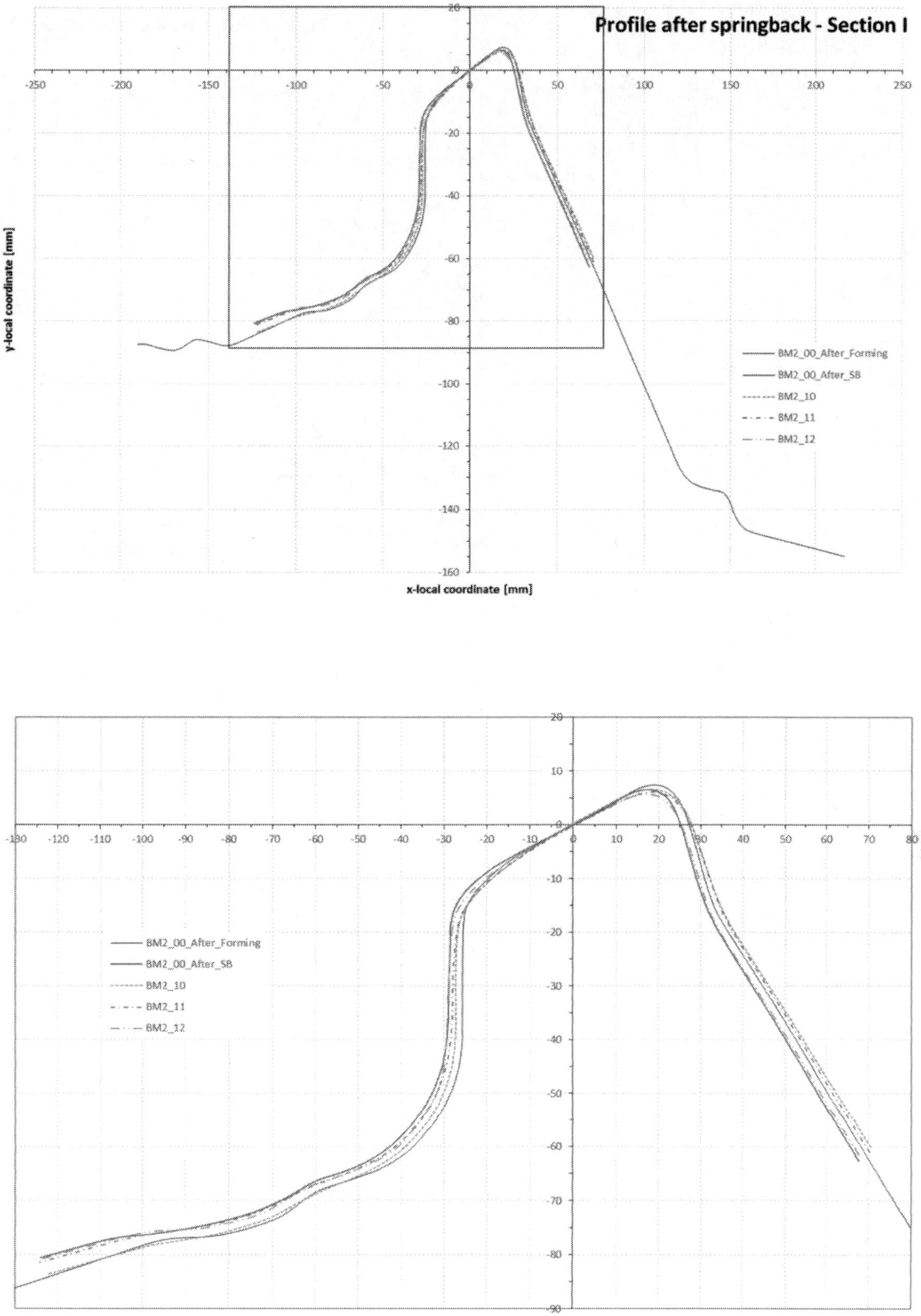

Figure 6.4. Profile after springback for Section I: BM2_10, BM2_11, BM2_12.

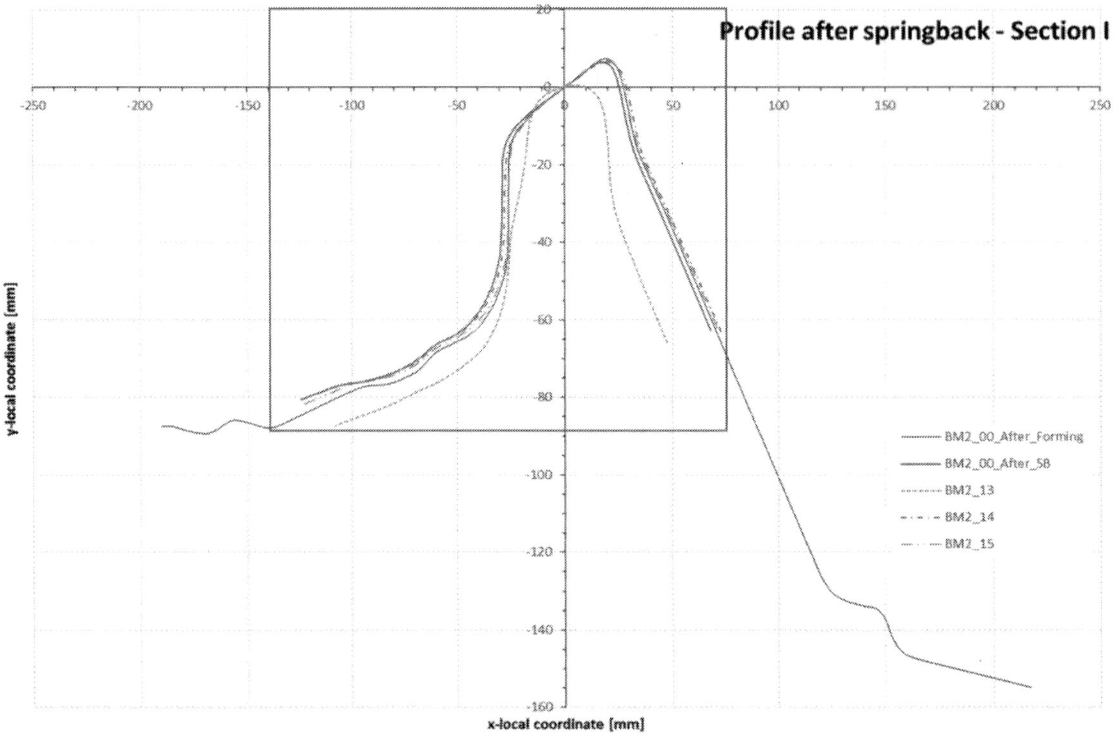

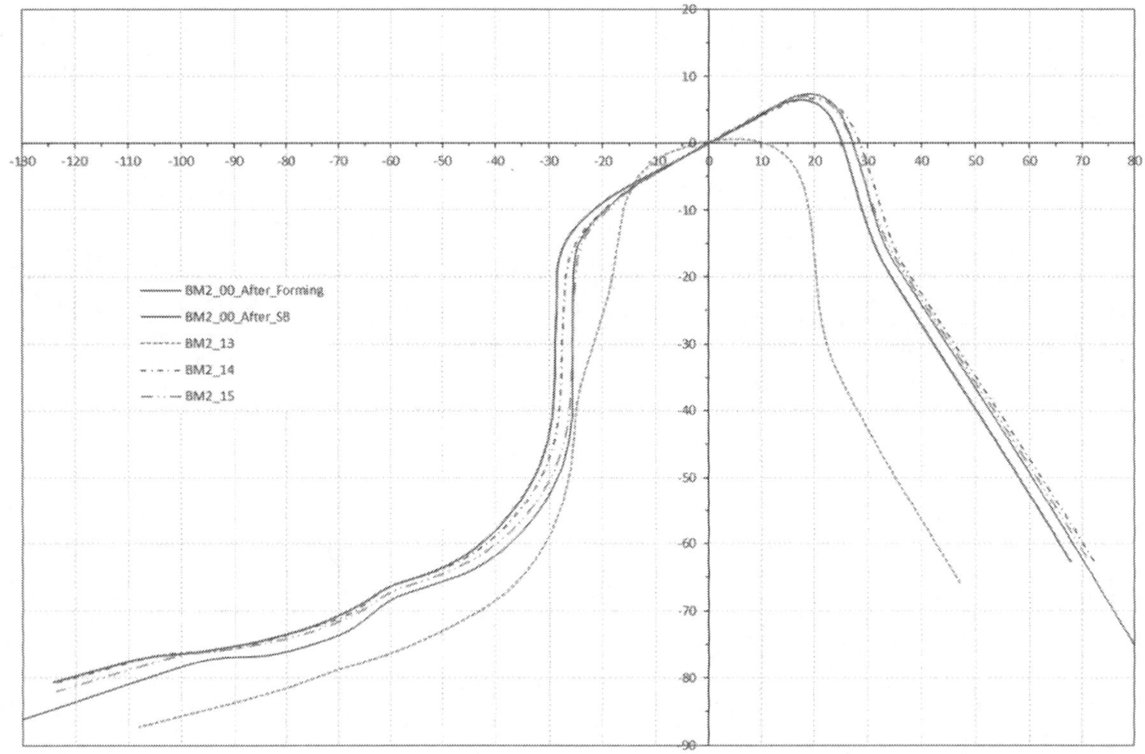

Figure 6.5. Profile after springback for Section I: BM2_13, BM2_14, BM2_15.

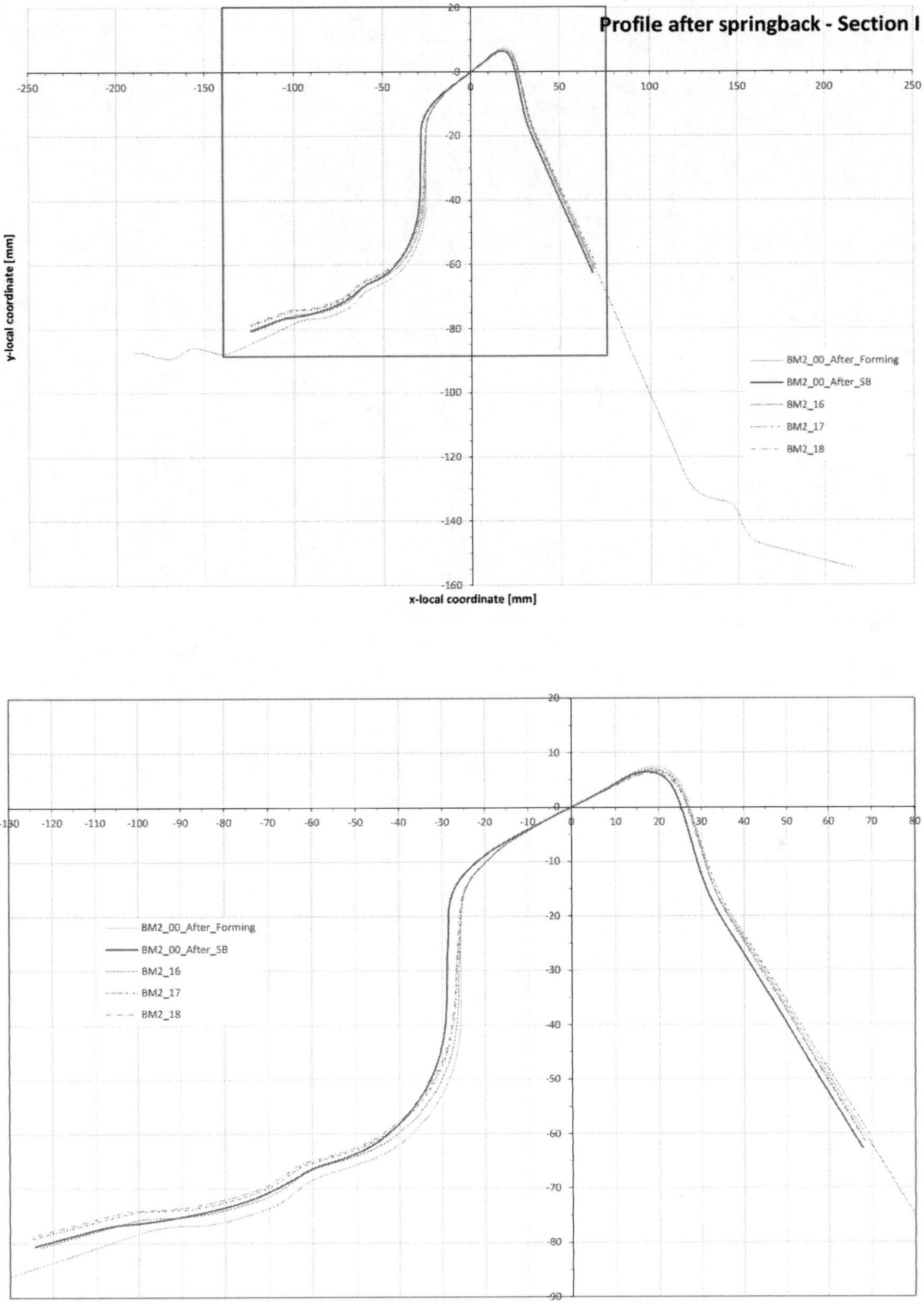

Figure 6.6. Profile after springback for Section I: BM2_16, BM2_17, BM2_18.

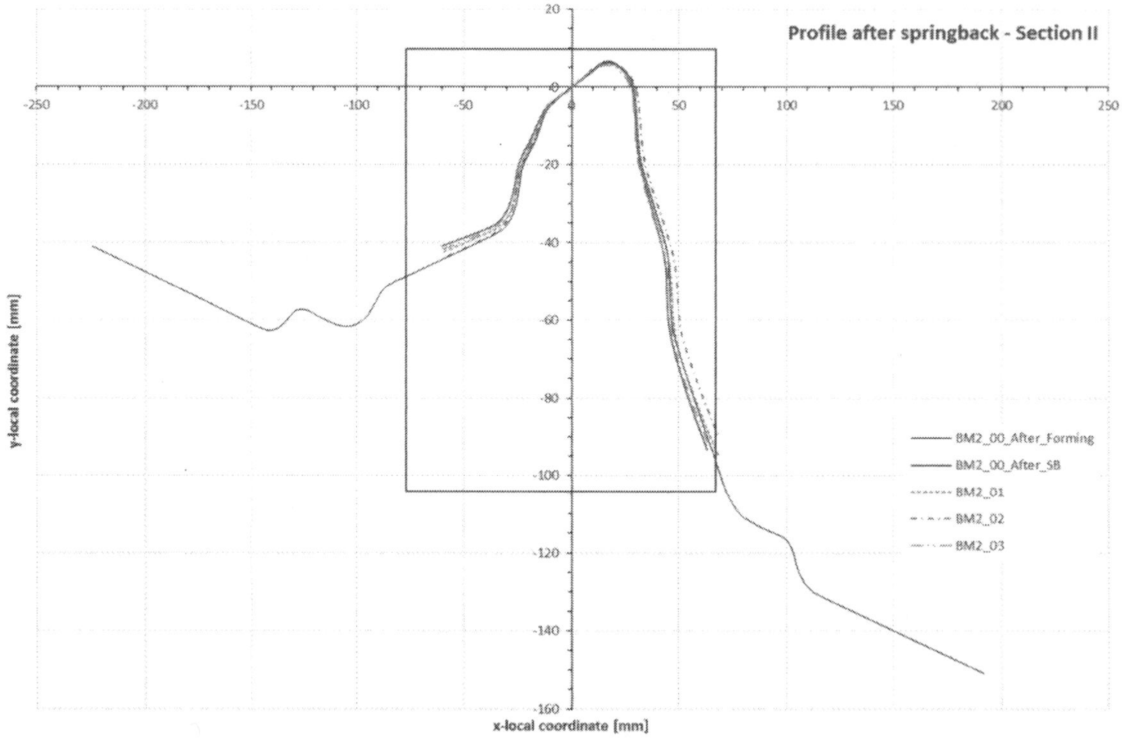

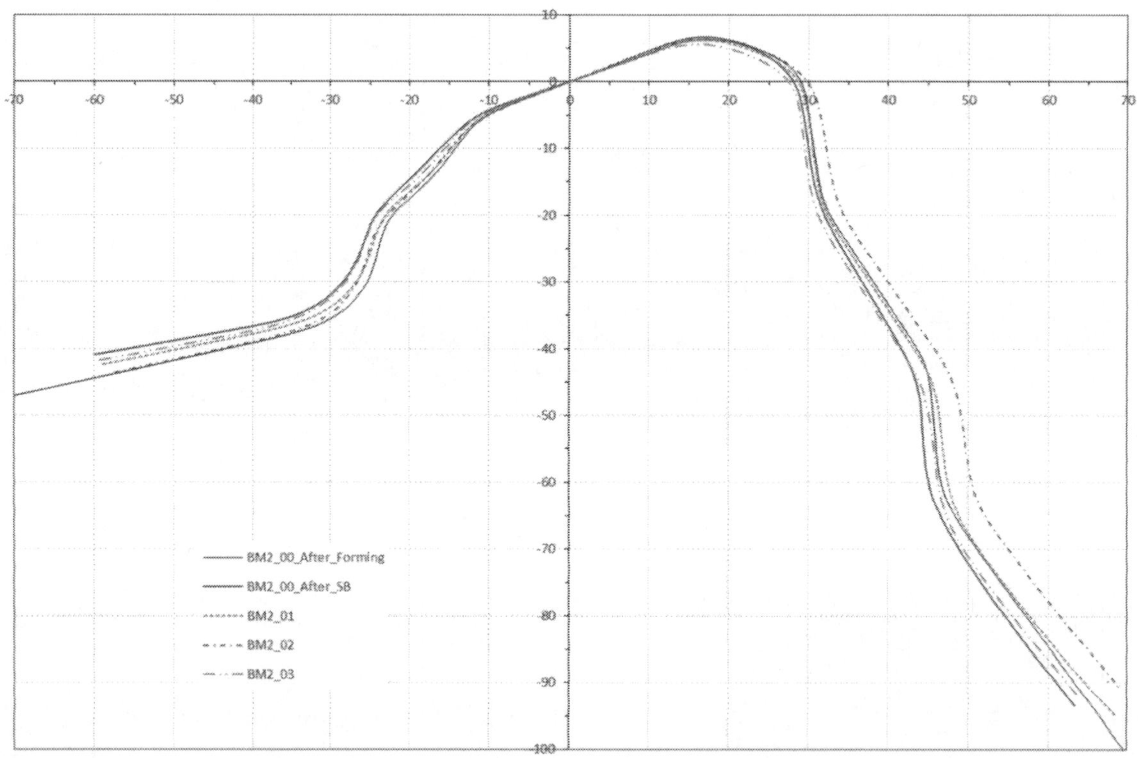

Figure 6.7. Profile after springback for Section II: BM2_01, BM2_02, BM2_03.

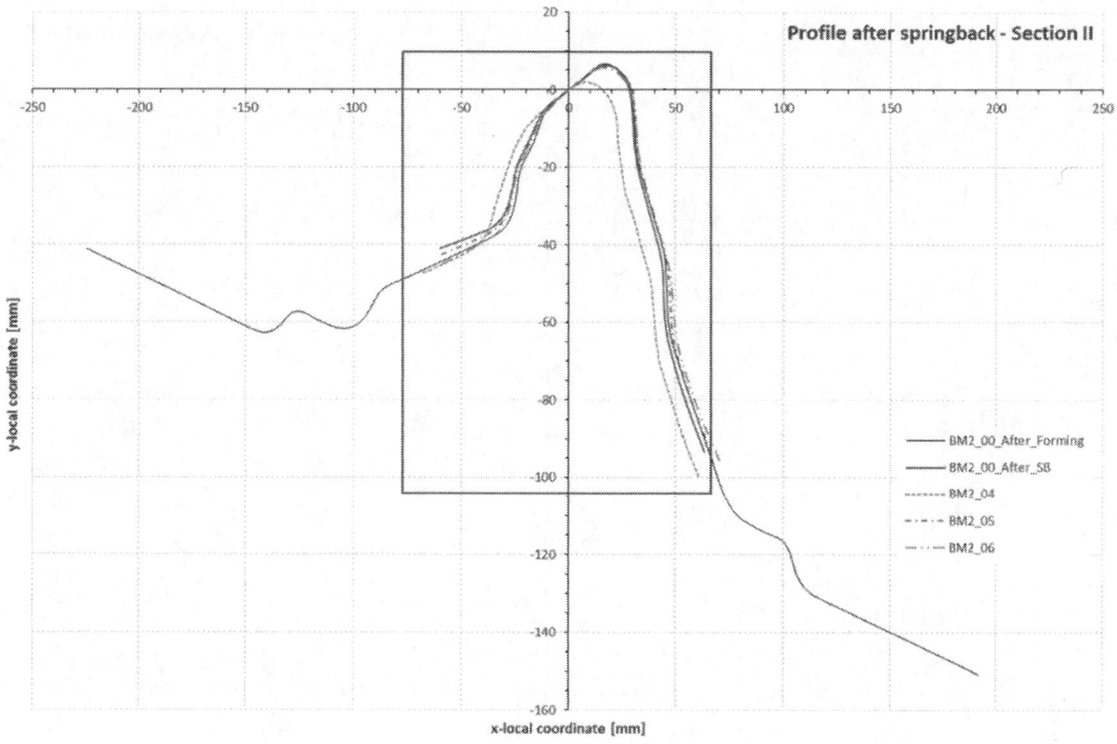

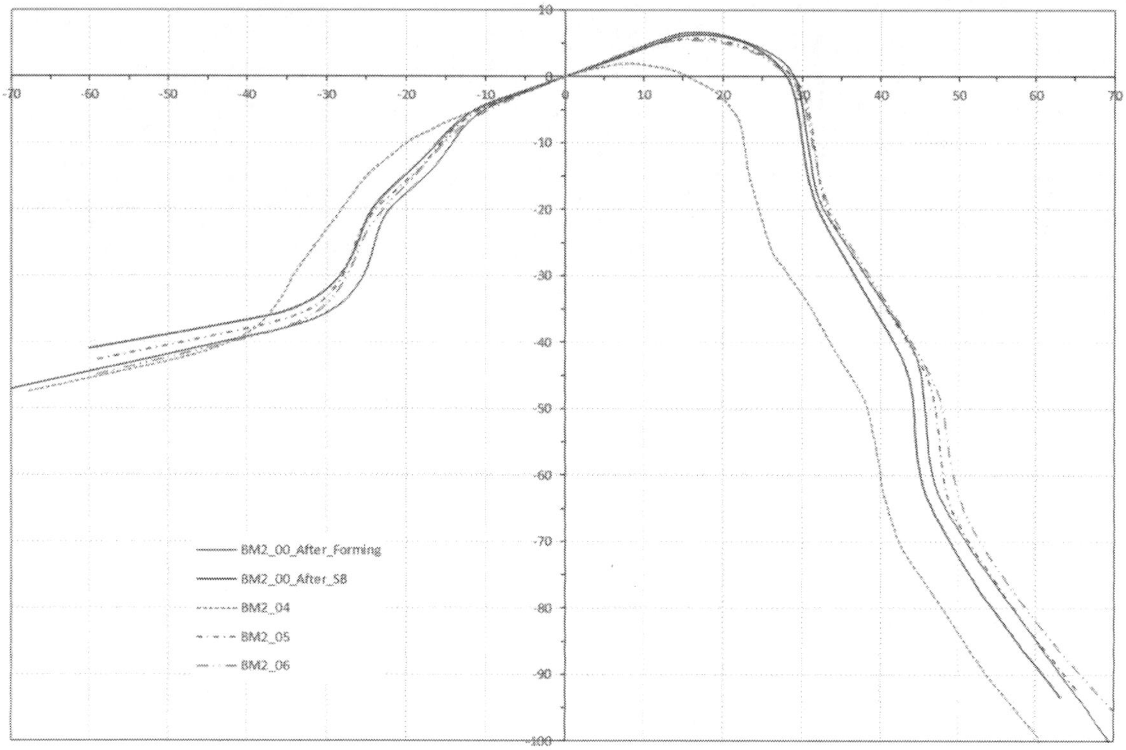

Figure 6.8. Profile after springback for Section II: BM2_04, BM2_05, BM2_06.

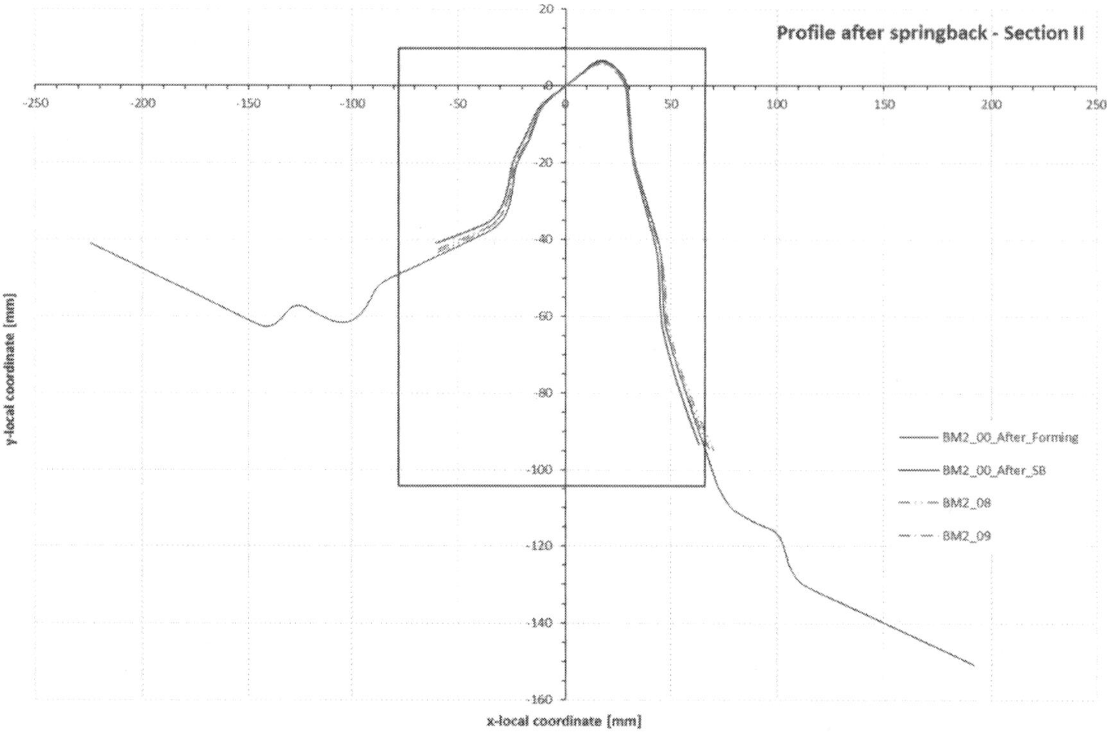

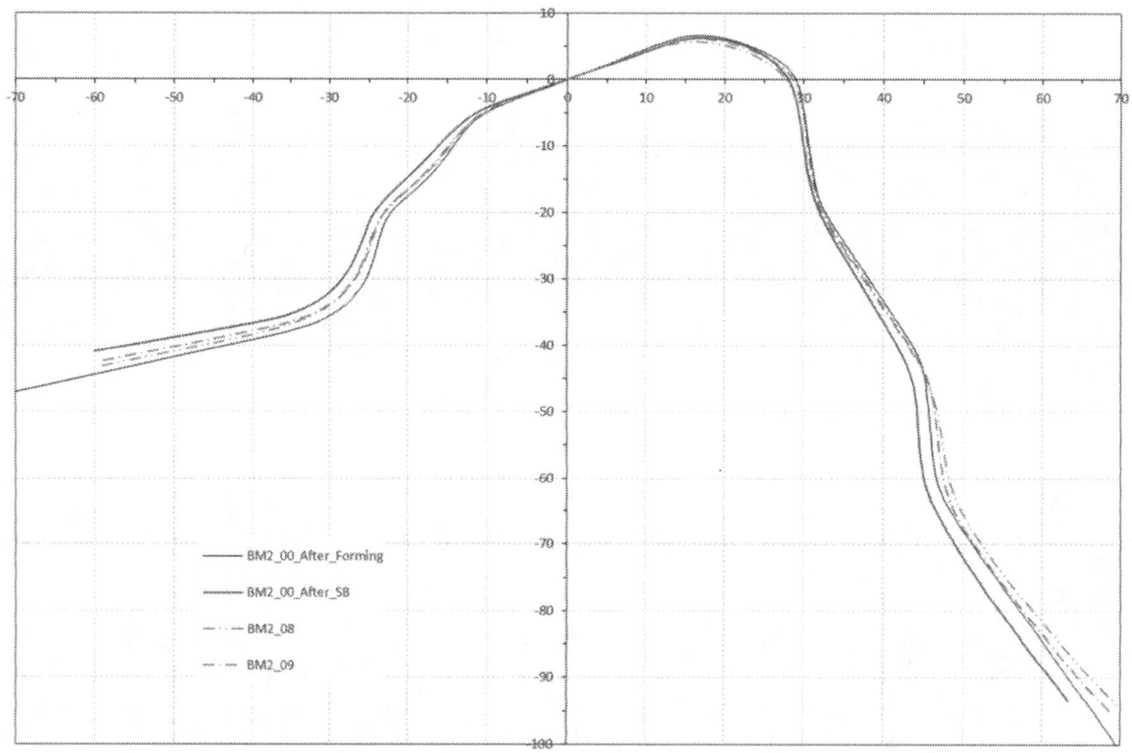

Figure 6.9. Profile after springback for Section II: BM2_08, BM2_09.

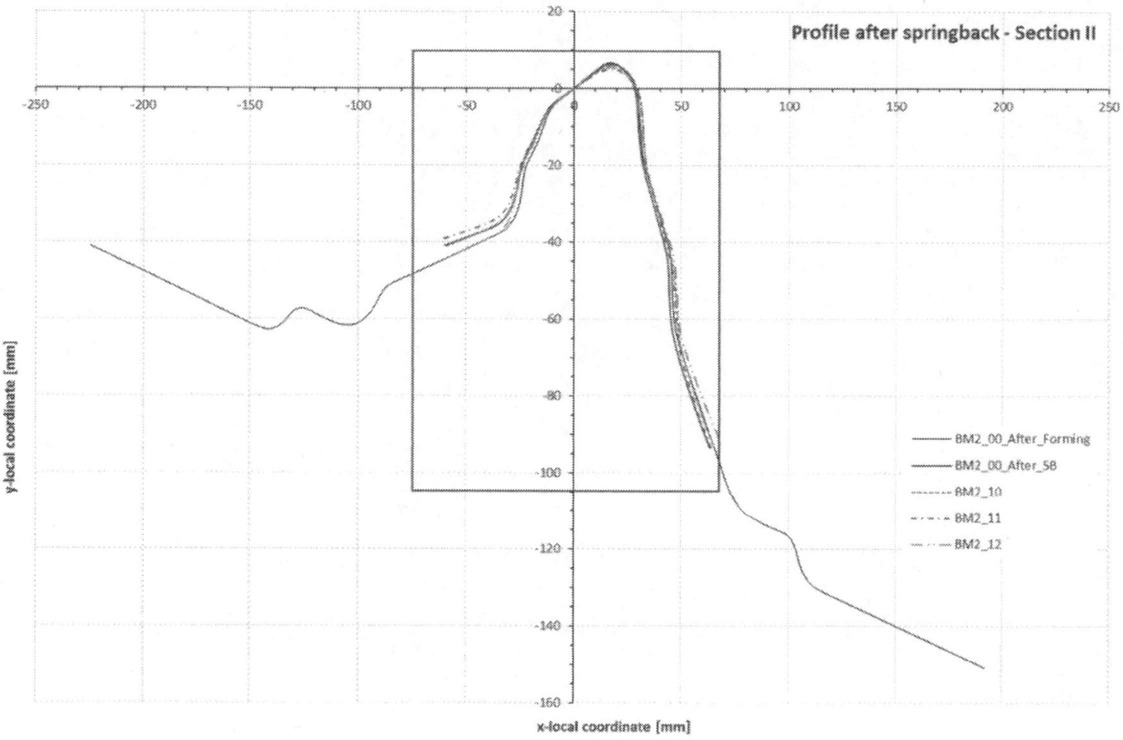

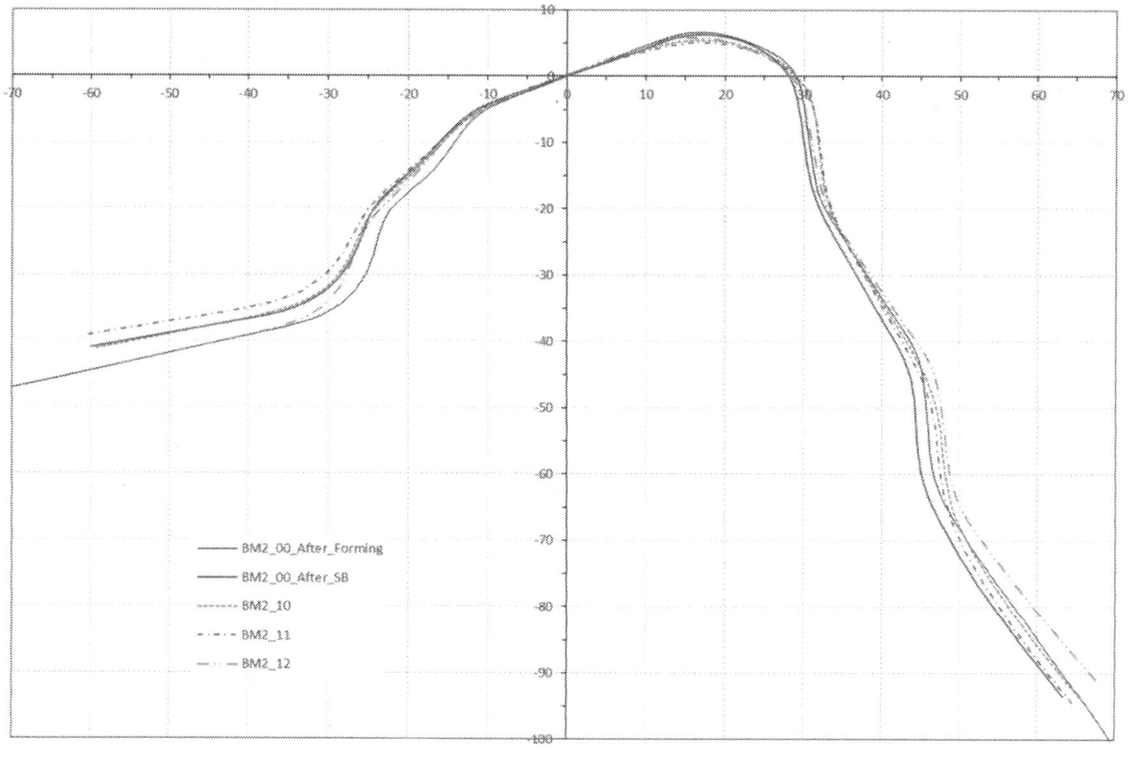

Figure 6.10. Profile after springback for Section II: BM2_10, BM2_11, BM2_12.

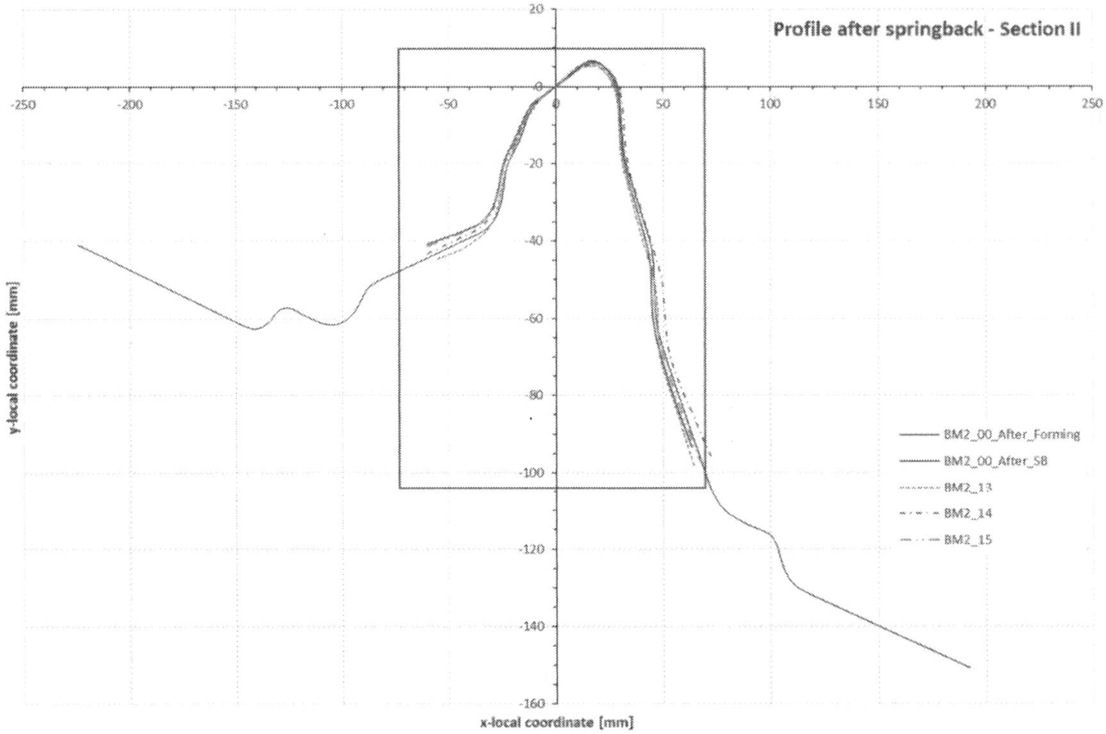

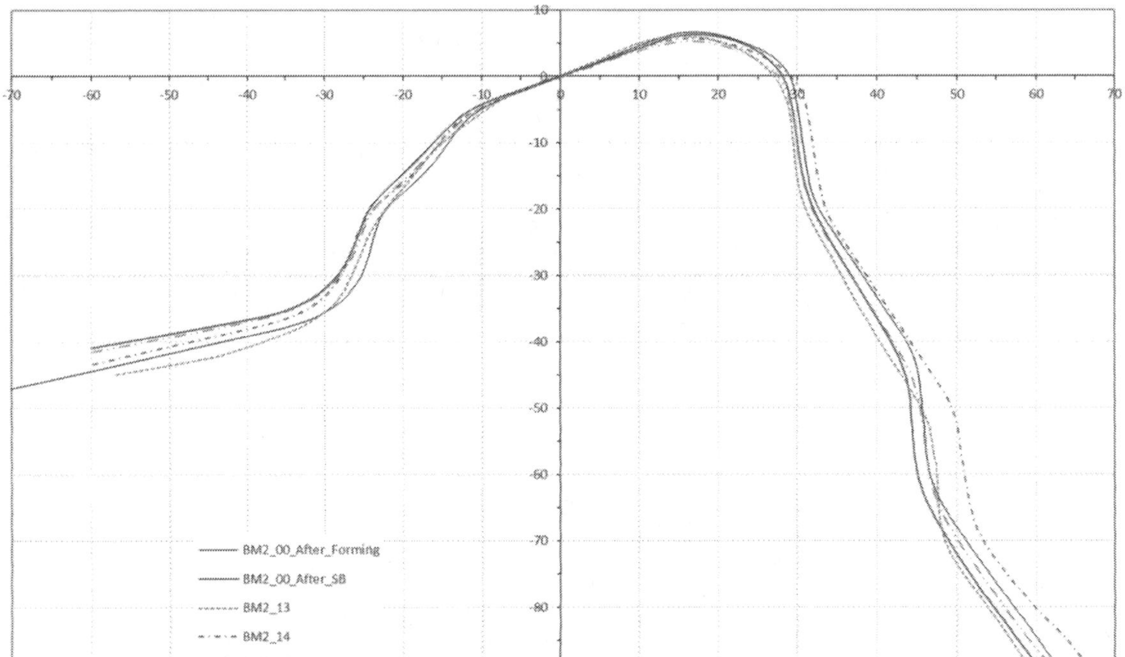

Figure 6.11. Profile after springback for Section II: BM2_13, BM2_14, BM2_15.

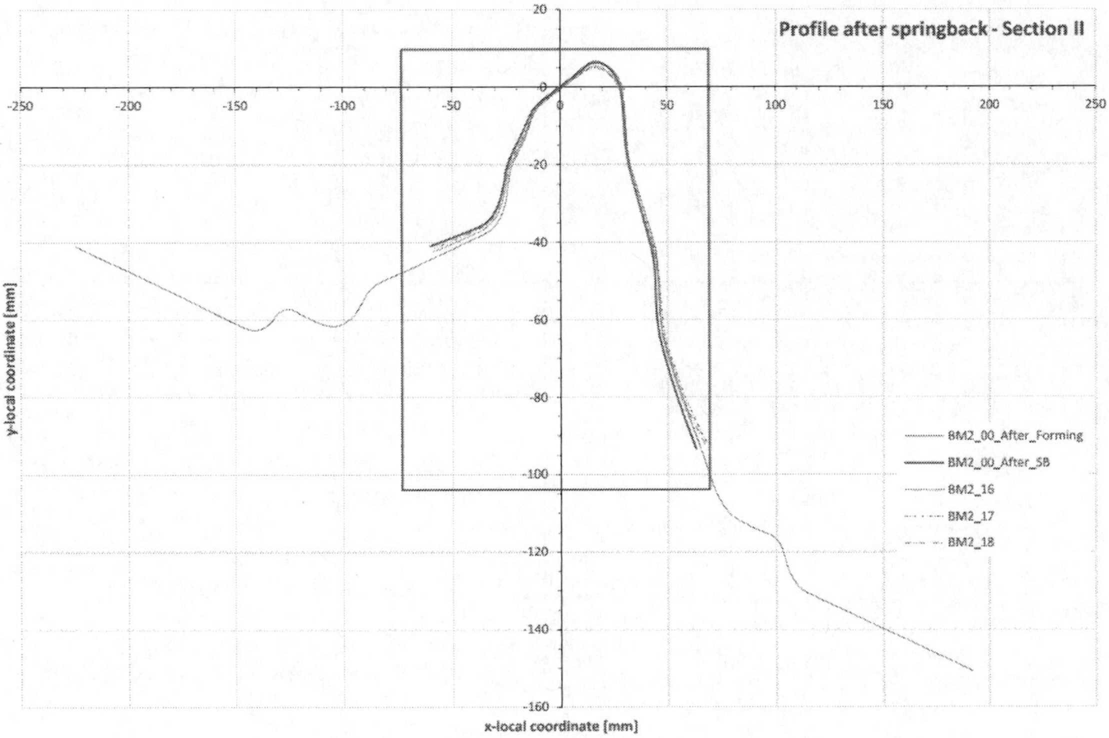

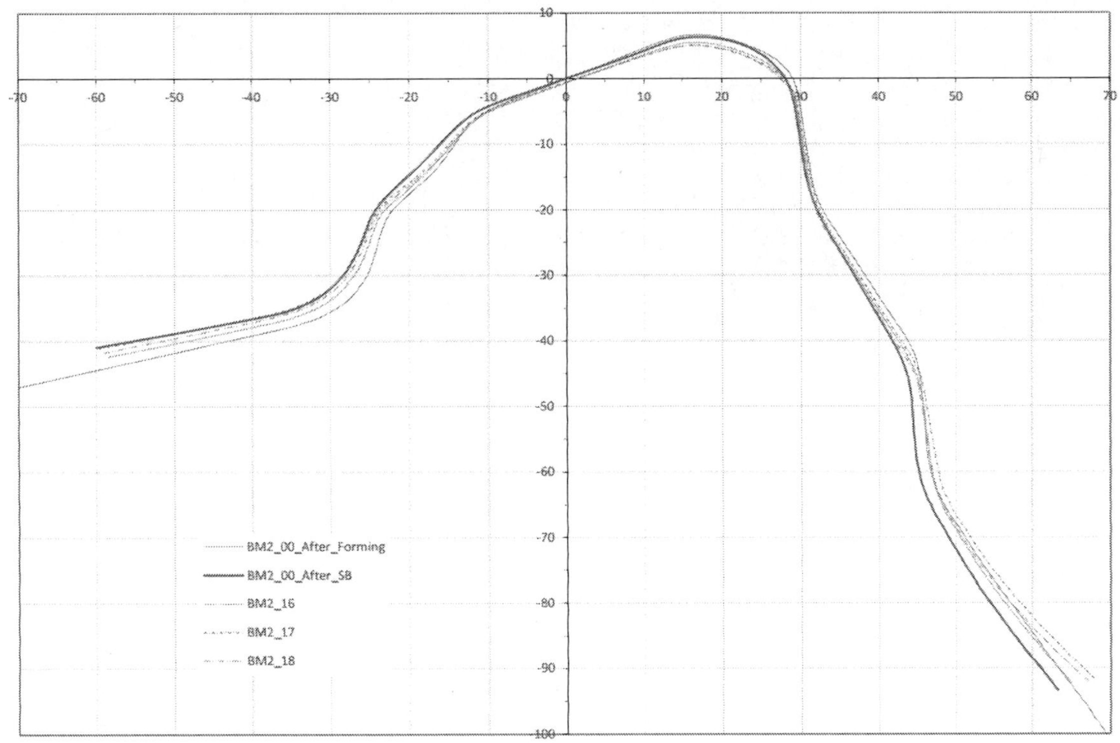

Figure 6.12. Profile after springback for Section II: BM2_16, BM2_17, BM2_18.

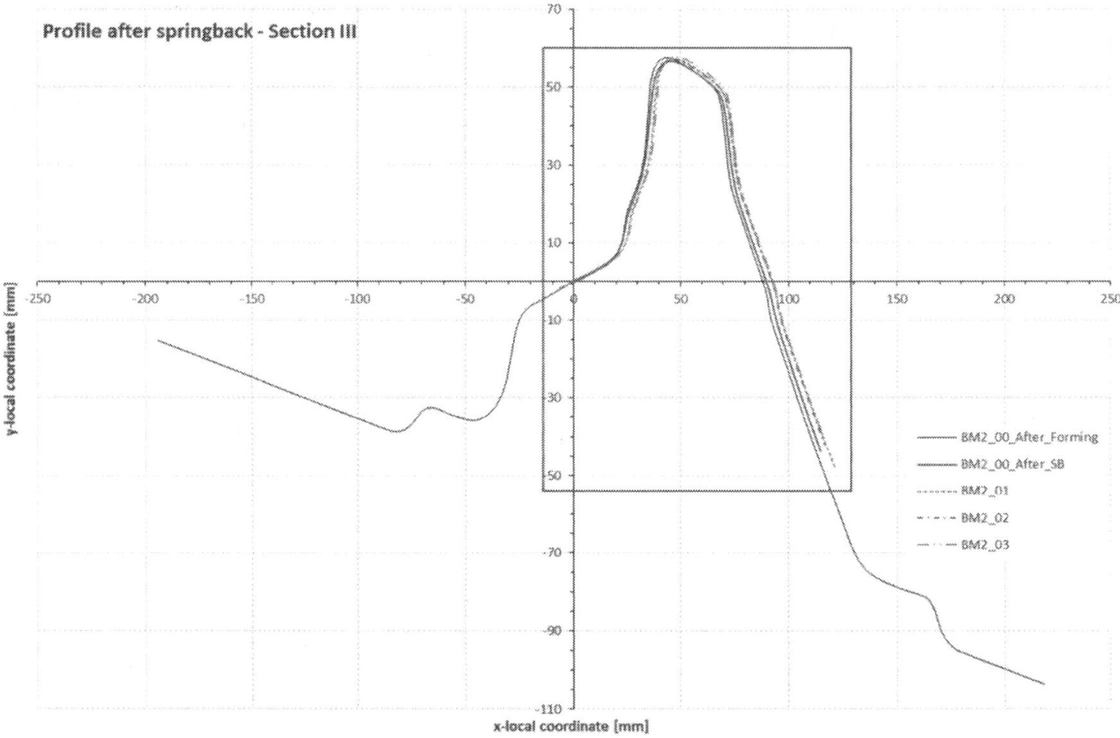

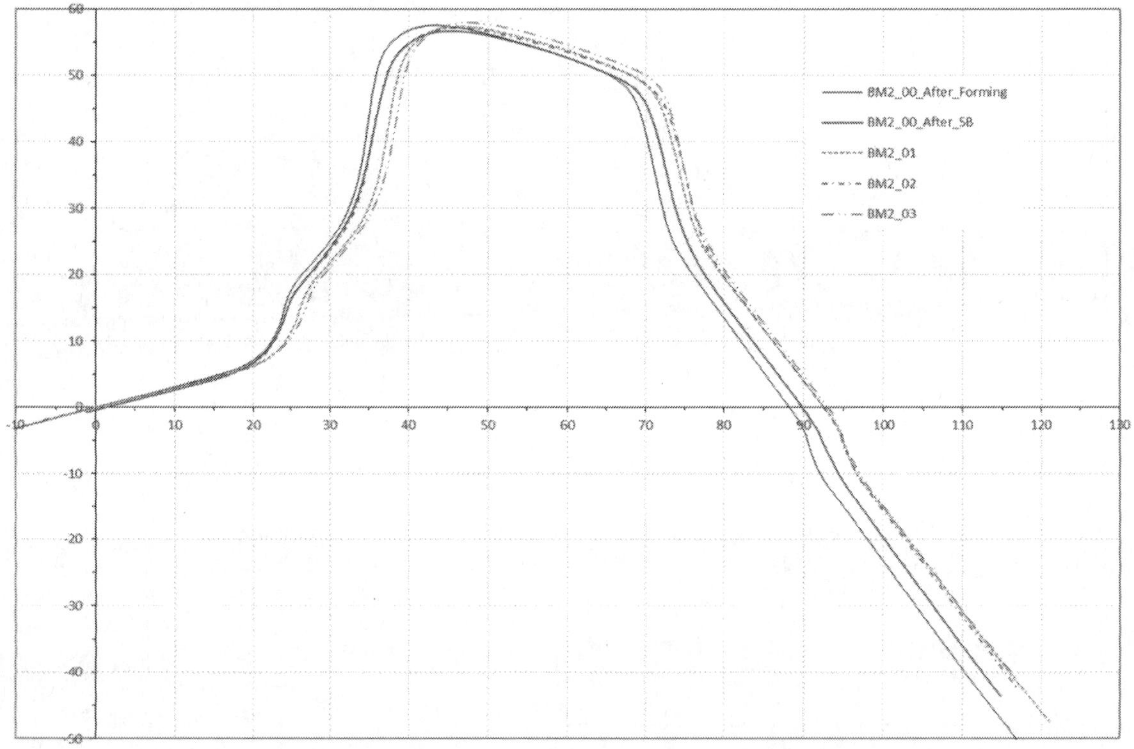

Figure 6.13. Profile after springback for Section III: BM2_01, BM2_02, BM2_03.

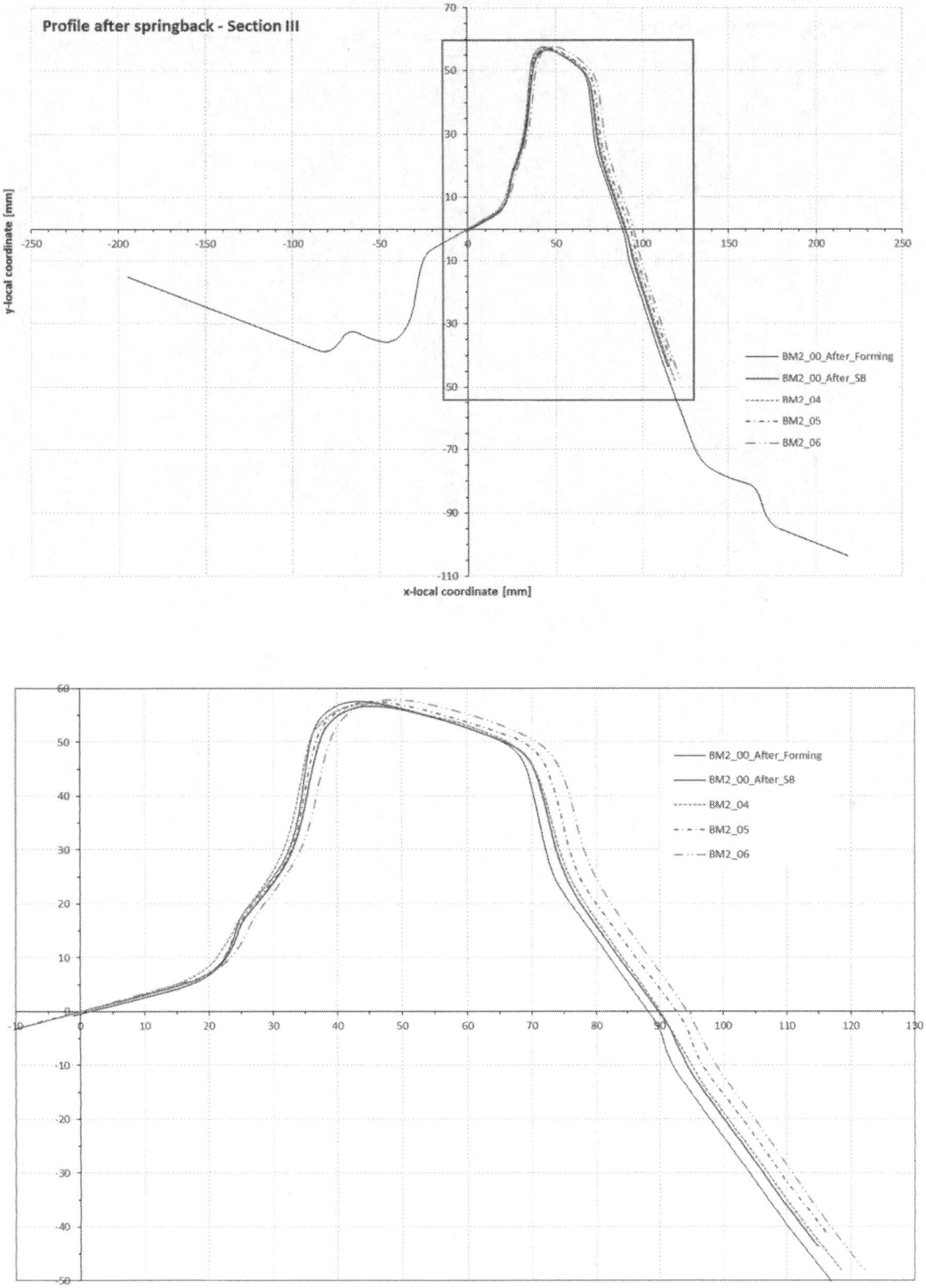

Figure 6.14. Profile after springback for Section III: BM2_04, BM2_05, BM2_06.

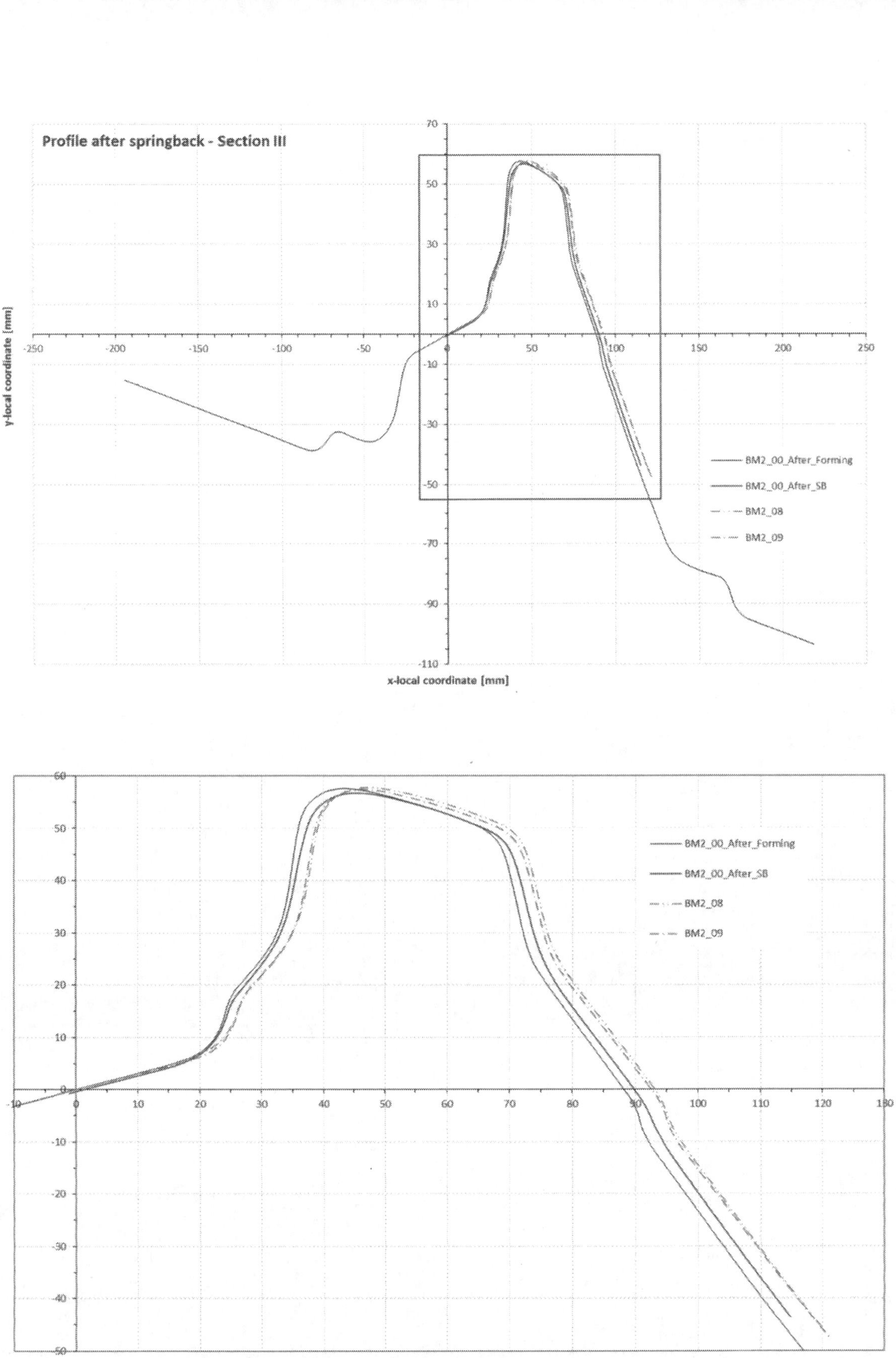

Figure 6.15. Profile after springback for Section III: BM2_08, BM2_09.

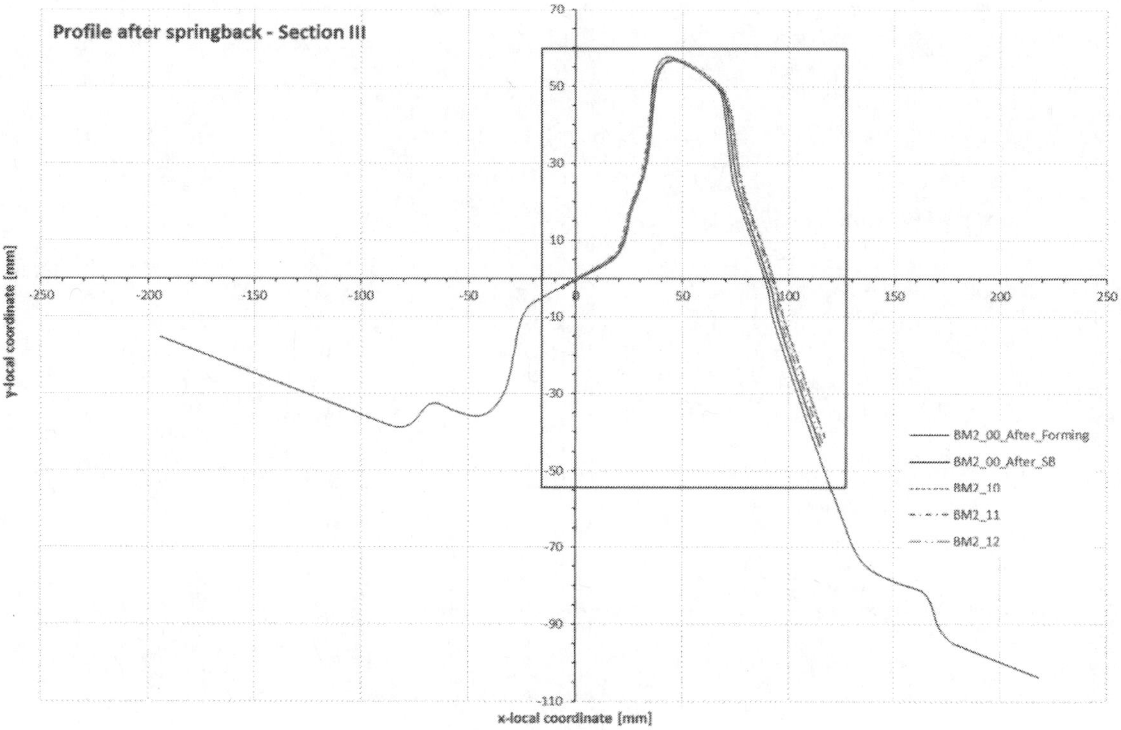

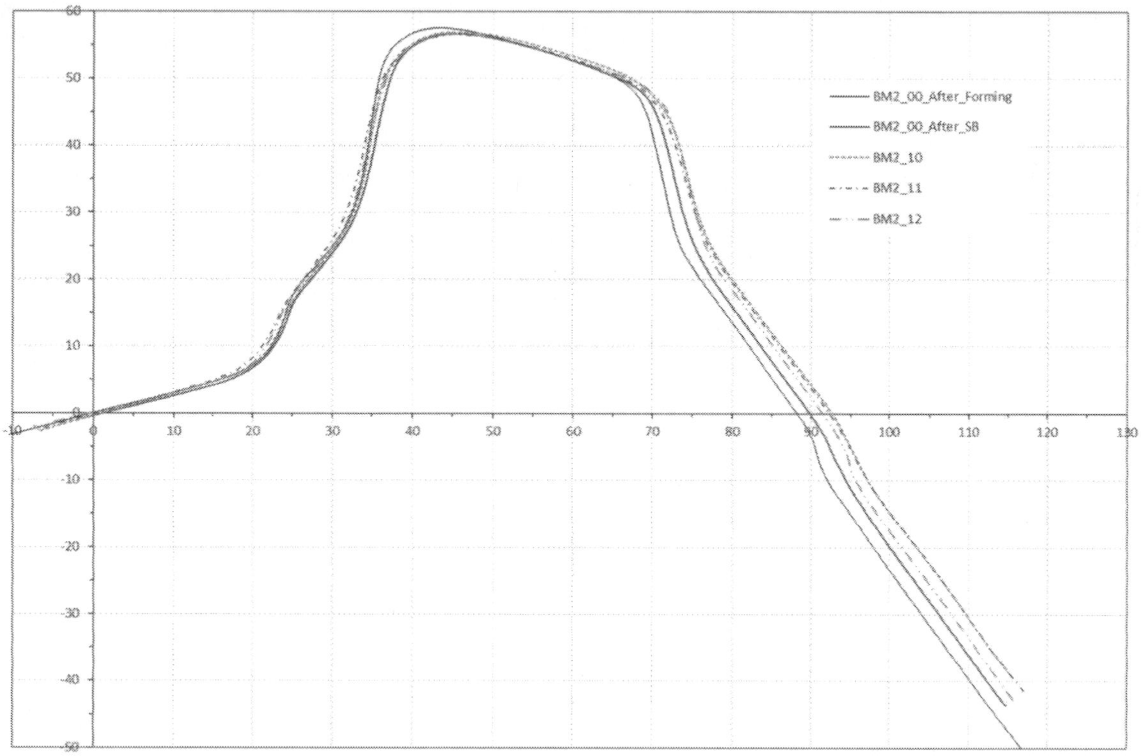

Figure 6.16. Profile after springback for Section III: BM2_10, BM2_11, BM2_12.

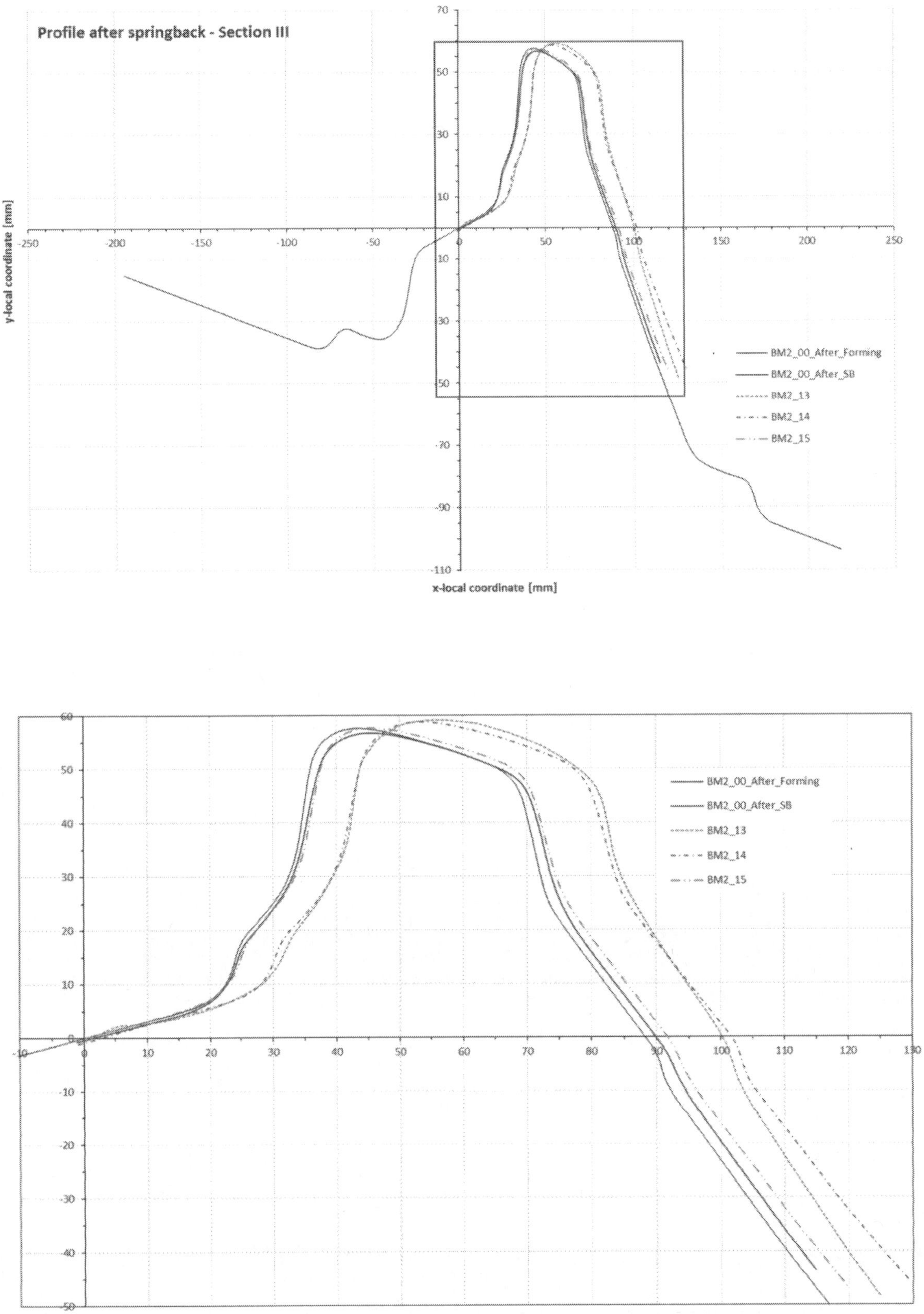

Figure 6.17. Profile after springback for Section III: BM2_13, BM2_14, BM2_15.

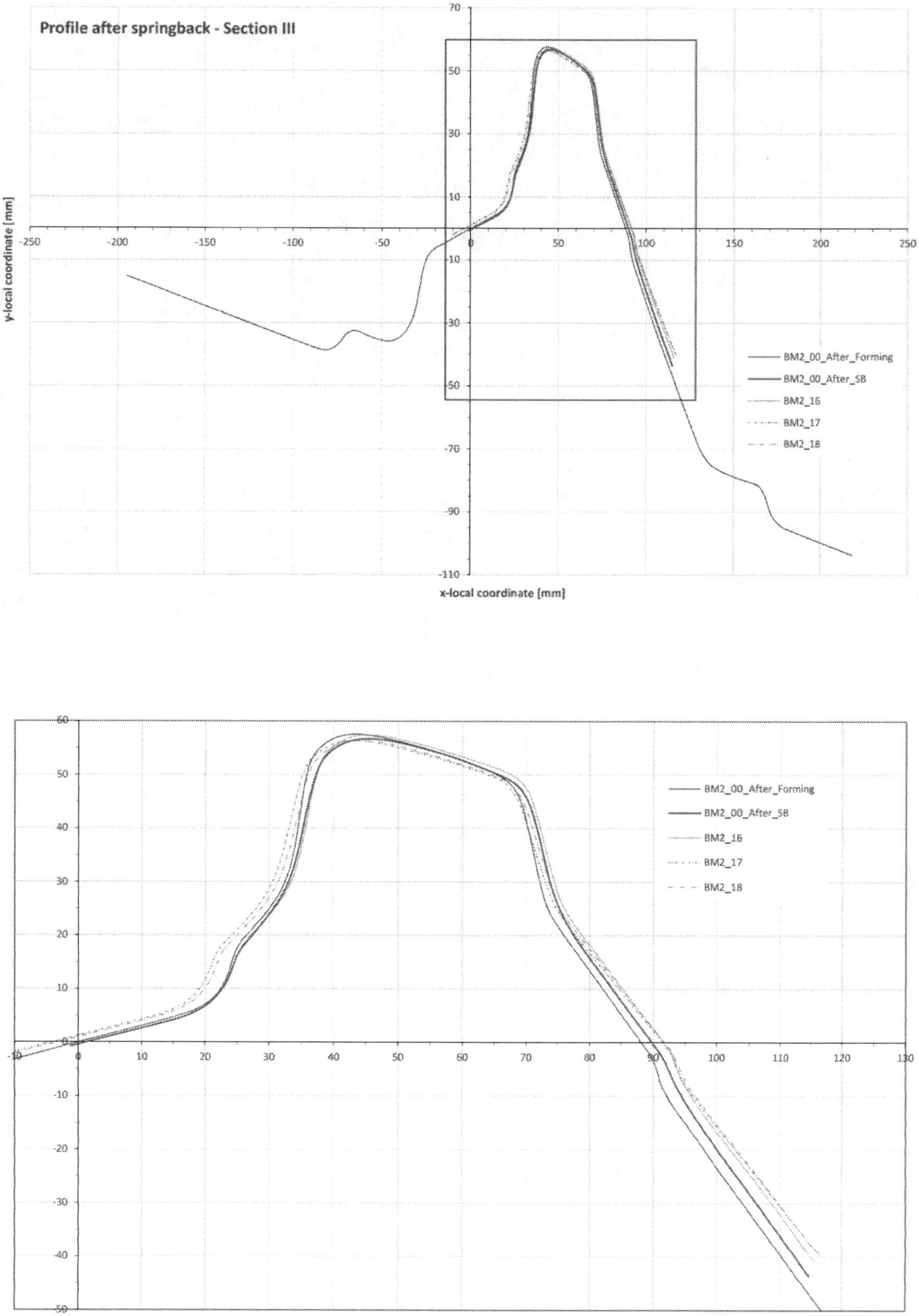

Figure 6.18. Profile after springback for Section III: BM2_16, BM2_17, BM2_18.

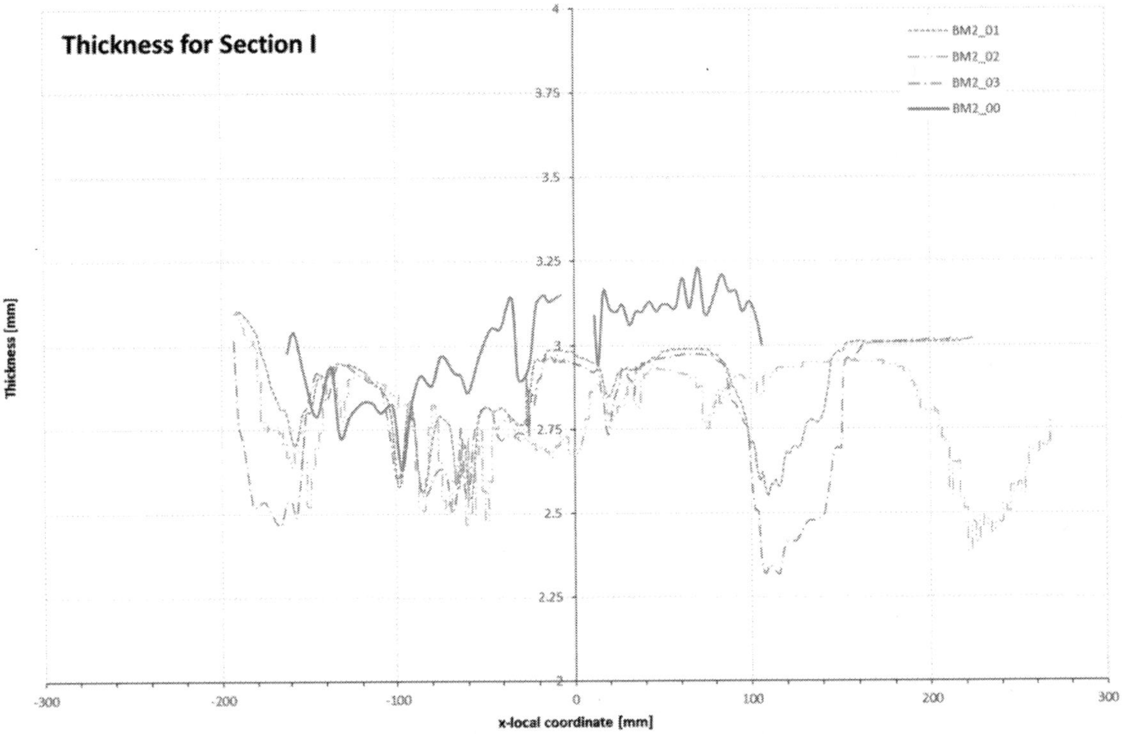

Figure 6.19. Thickness for Section I: BM2_01, BM2_02, BM2_03.

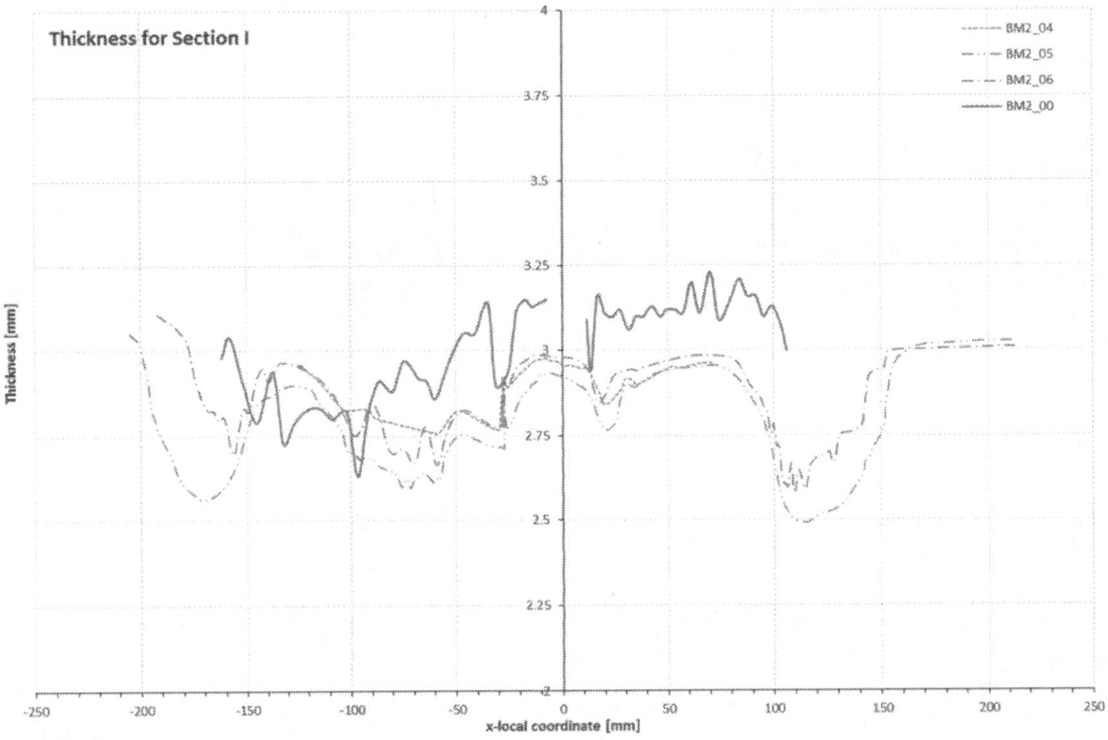

Figure 6.20. Thickness for Section I: BM2_04, BM2_05, BM2_06.

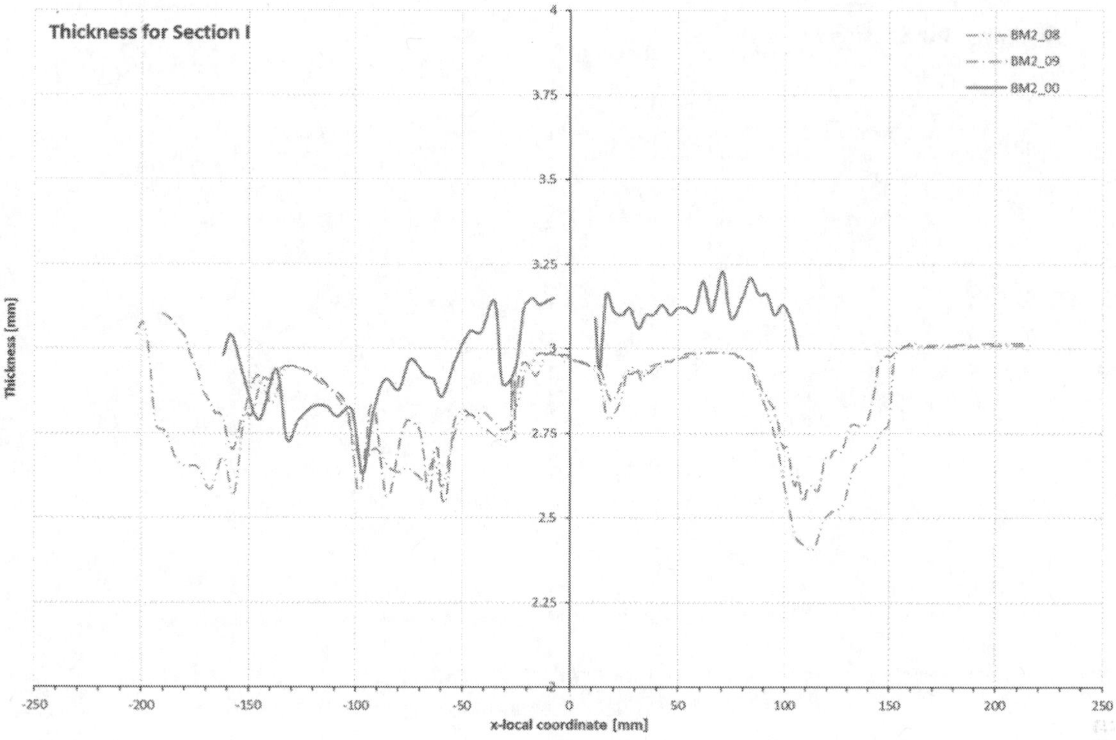

Figure 6.21. Thickness for Section I: BM2_08, BM2_09.

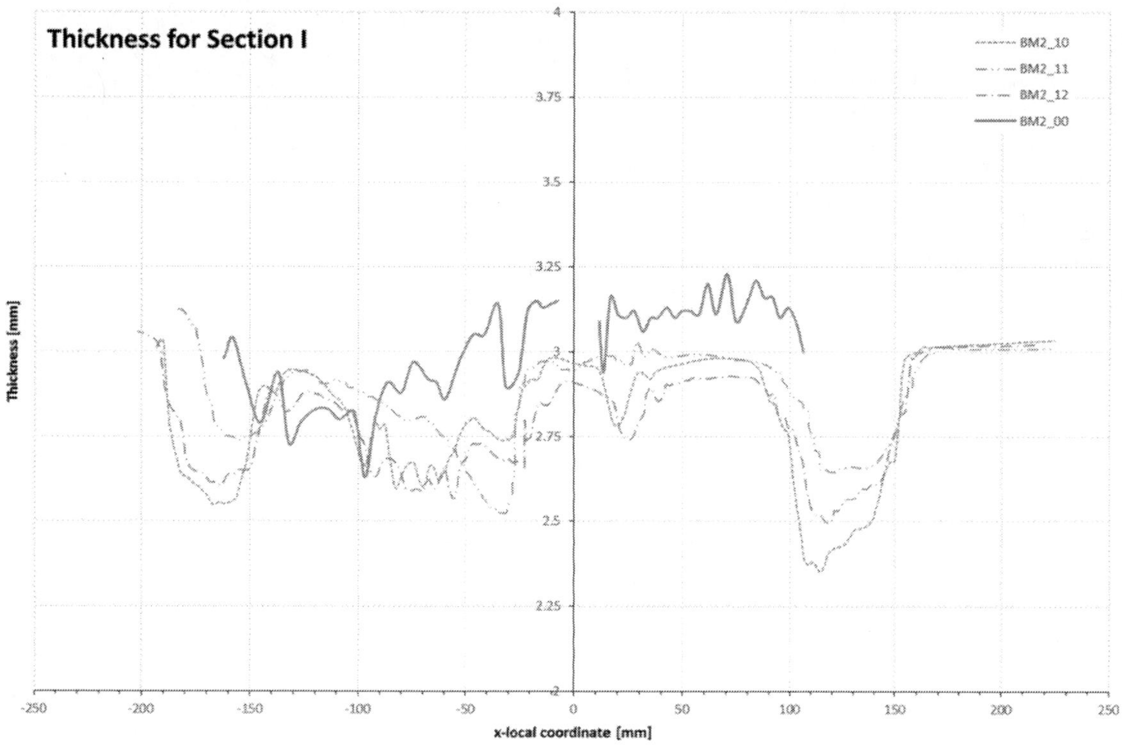

Figure 6.22. Thickness for Section I: BM2_10, BM2_11, BM2_12.

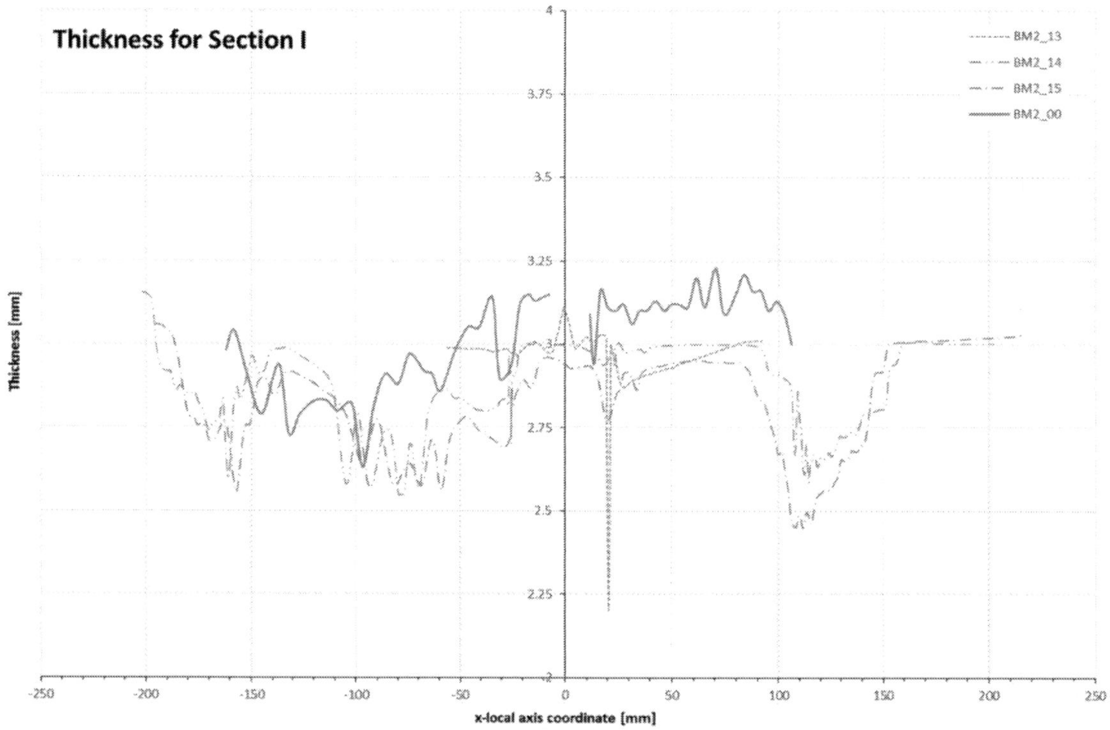

Figure 6.23. Thickness for Section I: BM2_13, BM2_14, BM2_15.

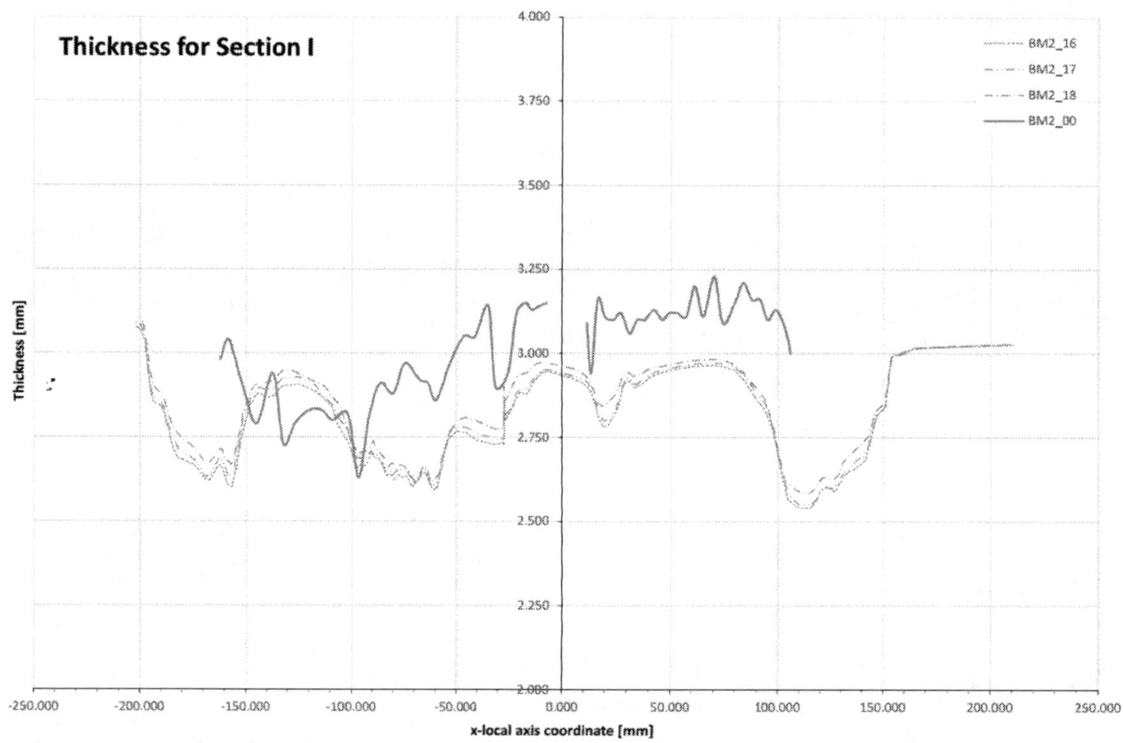

Figure 6.24. Thickness for Section I: BM2_16, BM2_17, BM2_18.

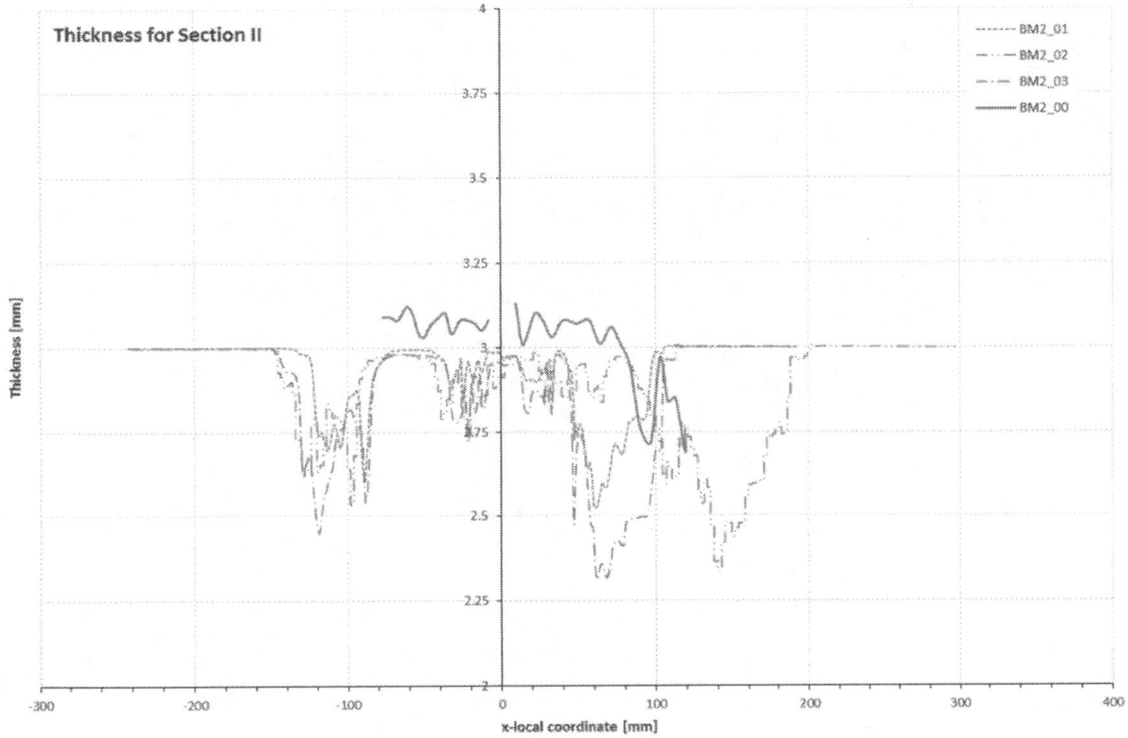

Figure 6.25. Thickness for Section II: BM2_01, BM2_02, BM2_03.

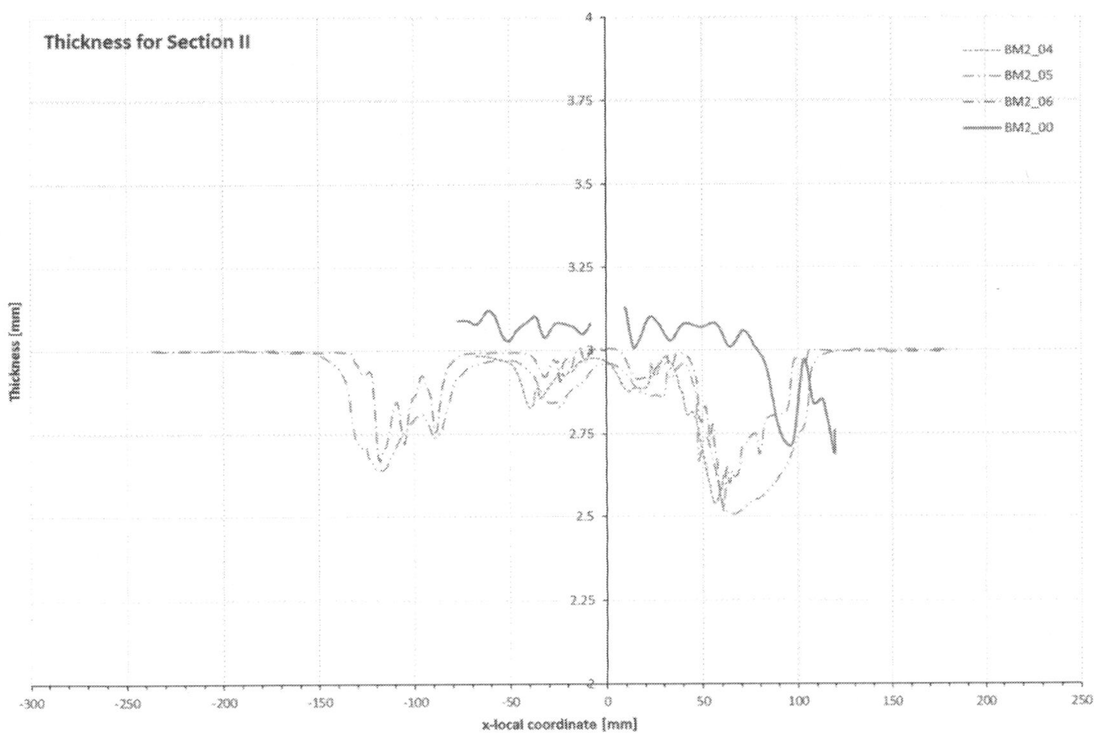

Figure 6.26. Thickness for Section II: BM2_04, BM2_05, BM2_06.

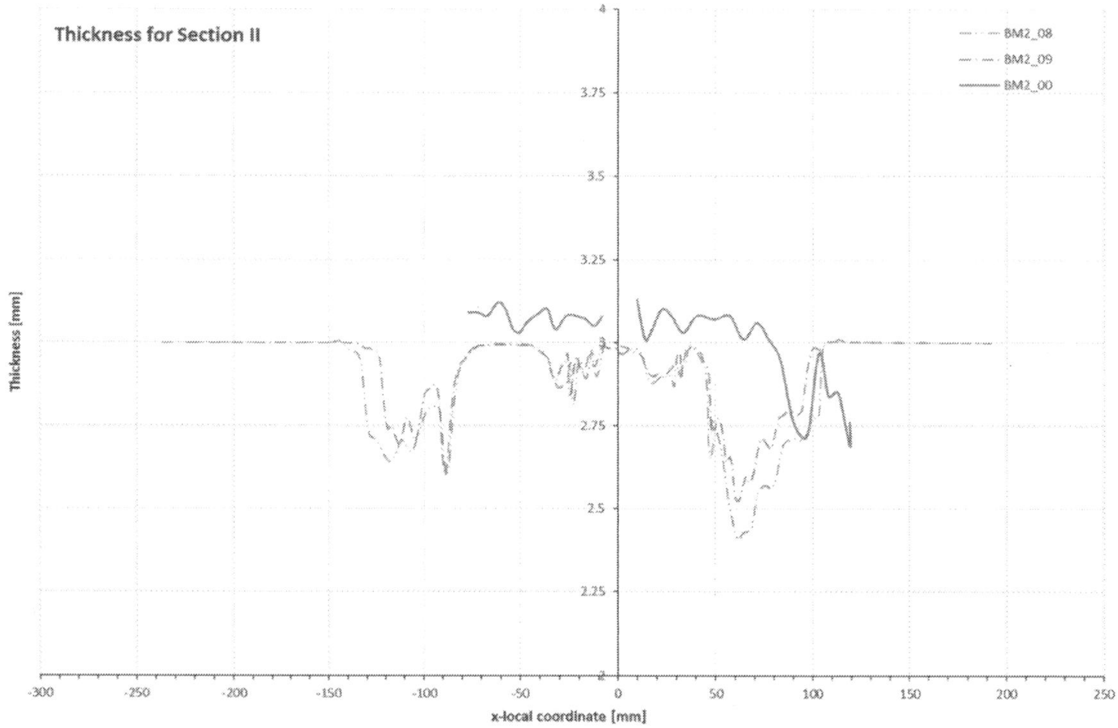

Figure 6.27. Thickness for Section II: BM2_08, BM2_09.

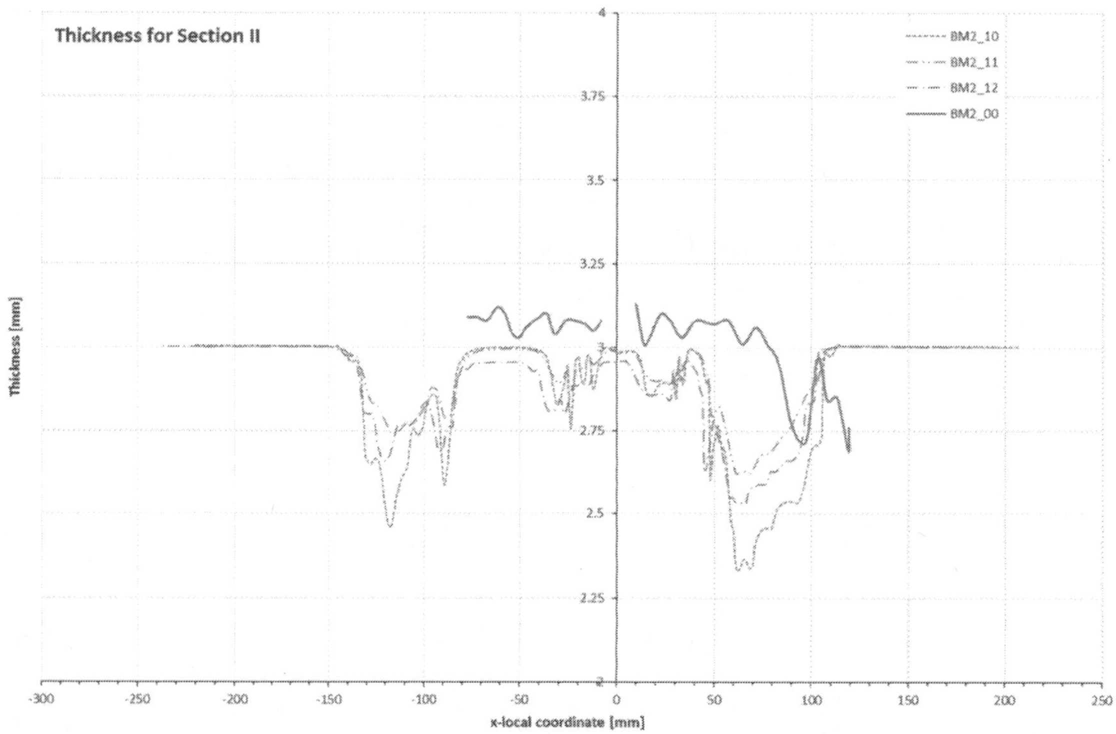

Figure 6.28. Thickness for Section II: BM2_10, BM2_11, BM2_12.

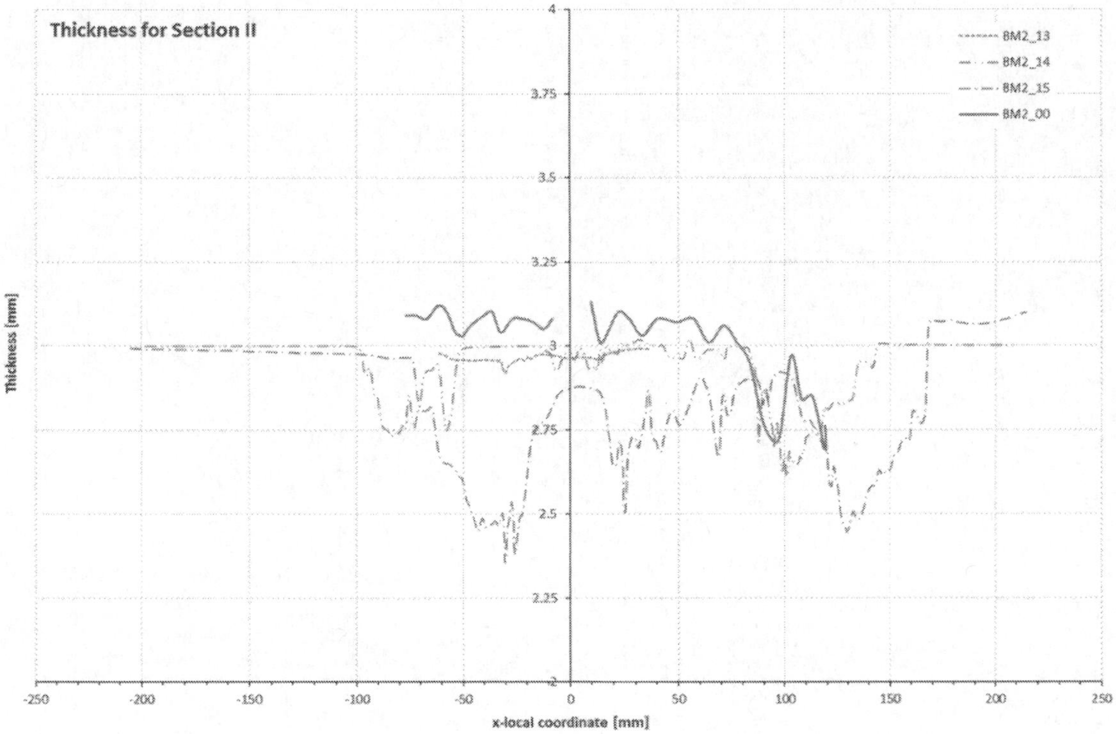

Figure 6.29. Thickness for Section II: BM2_13, BM2_14, BM2_15.

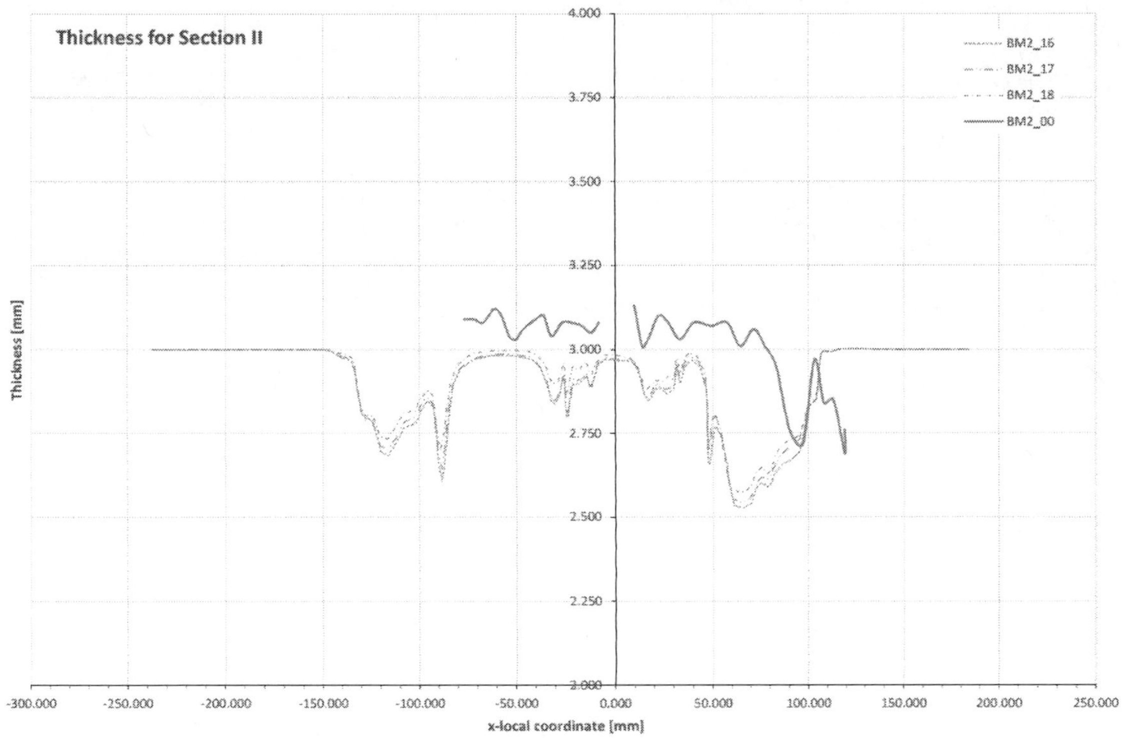

Figure 6.30. Thickness for Section II: BM2_16, BM2_17, BM2_18.

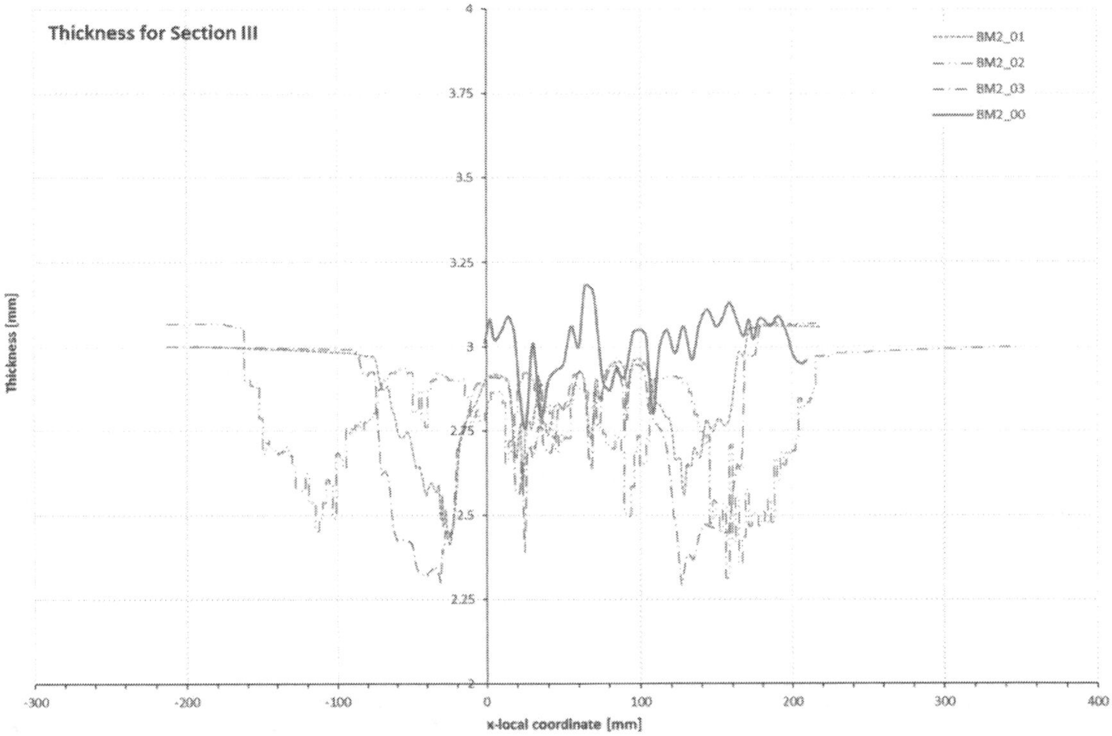

Figure 6.31. Thickness for Section III: BM2_01, BM2_02, BM2_03.

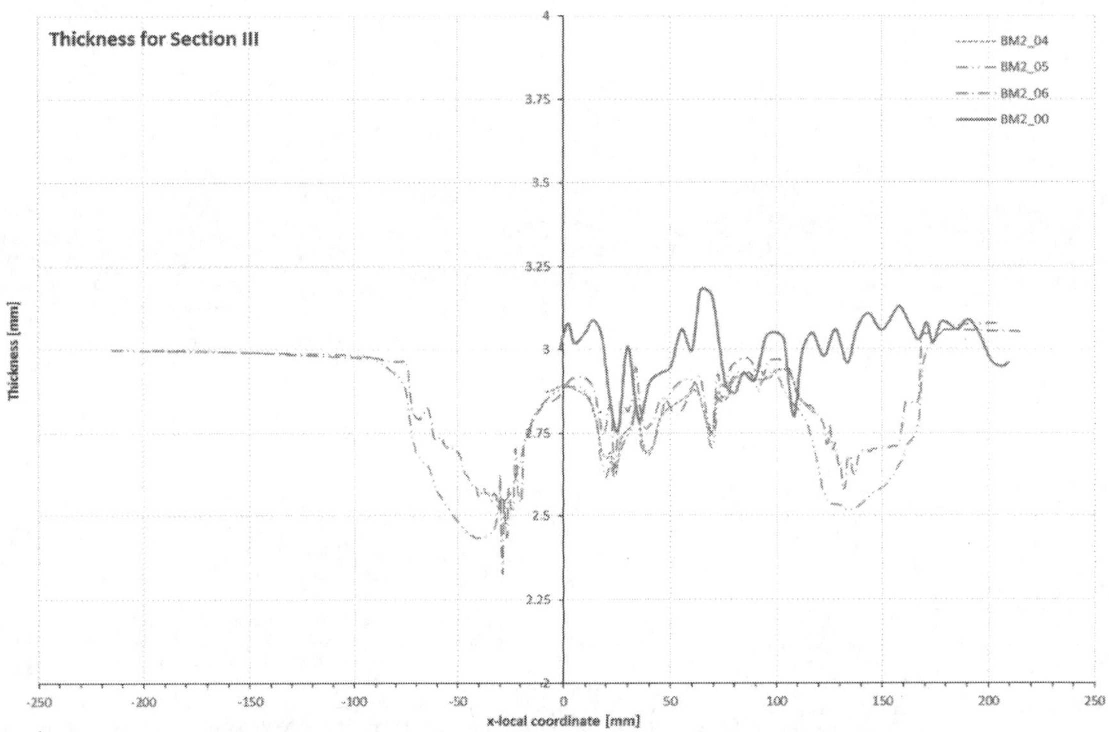

Figure 6.32. Thickness for Section III: BM2_04, BM2_05, BM2_06.

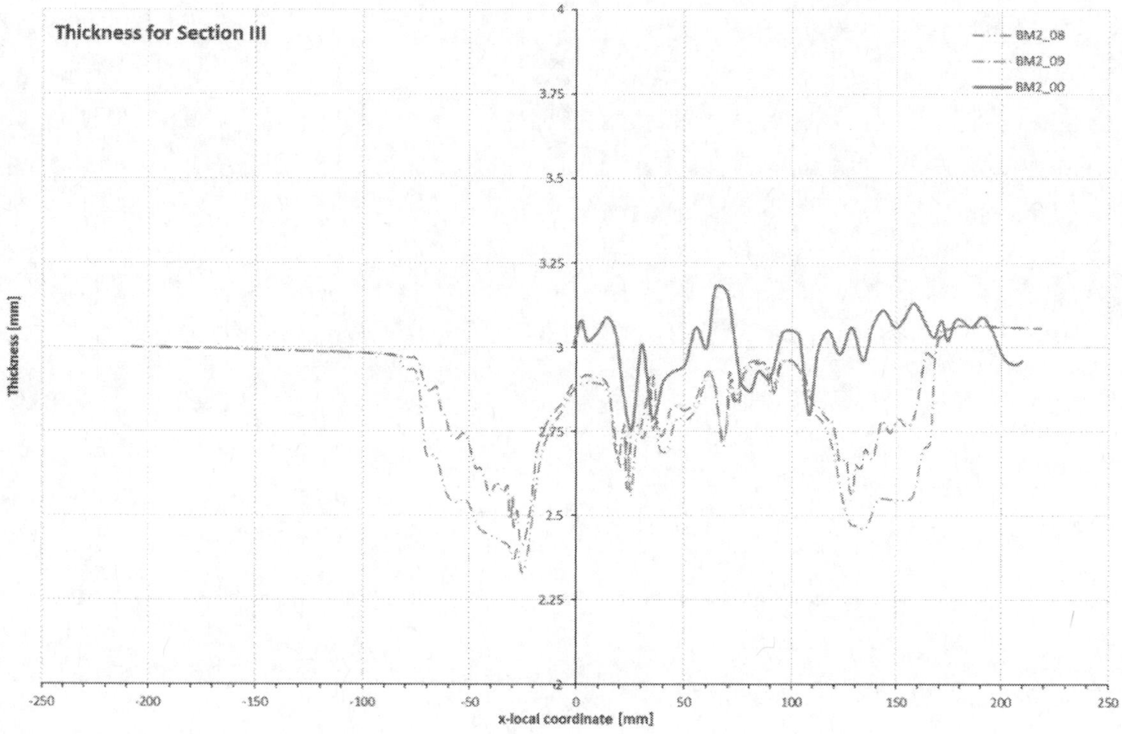

Figure 6.33. Thickness for Section III: BM2_08, BM2_09.

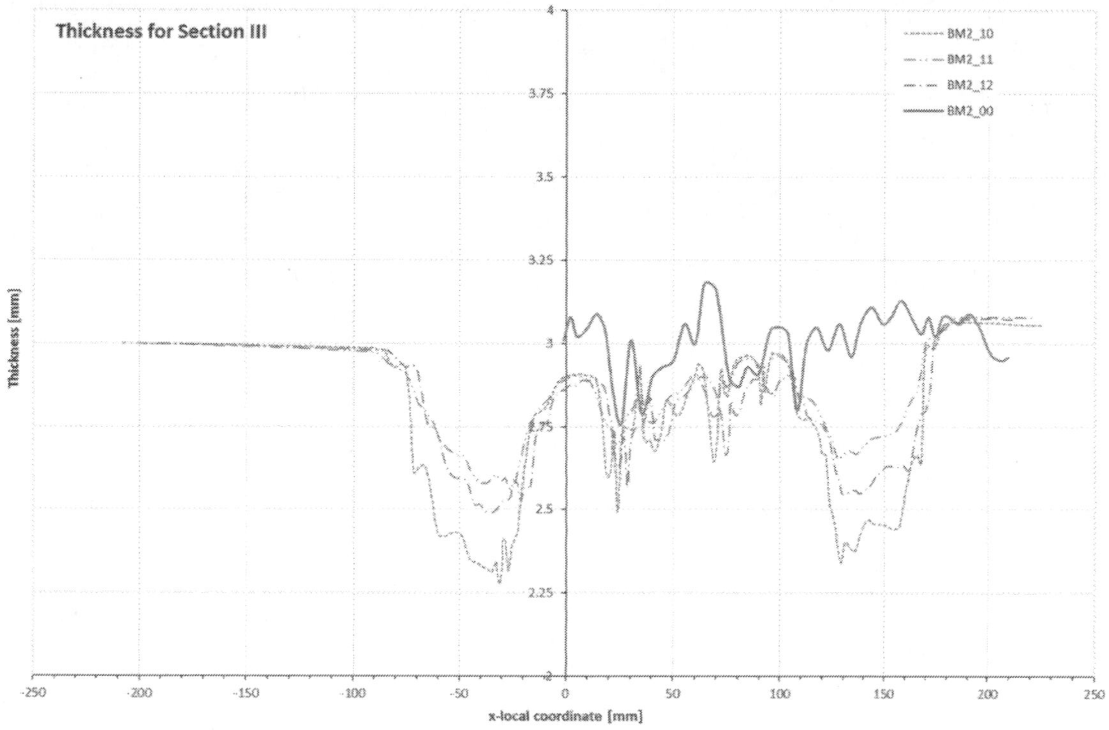

Figure 6.34. Thickness for Section III: BM2_10, BM2_11, BM2_12.

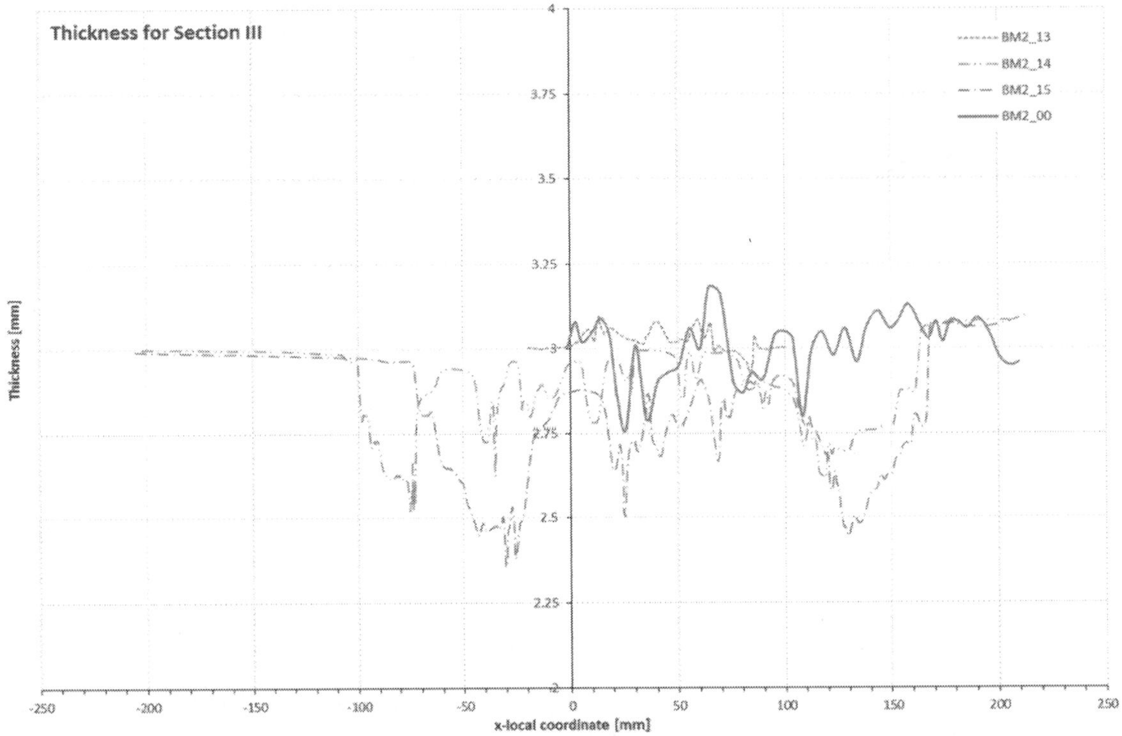

Figure 6.35. Thickness for Section III: BM2_13, BM2_14, BM2_15.

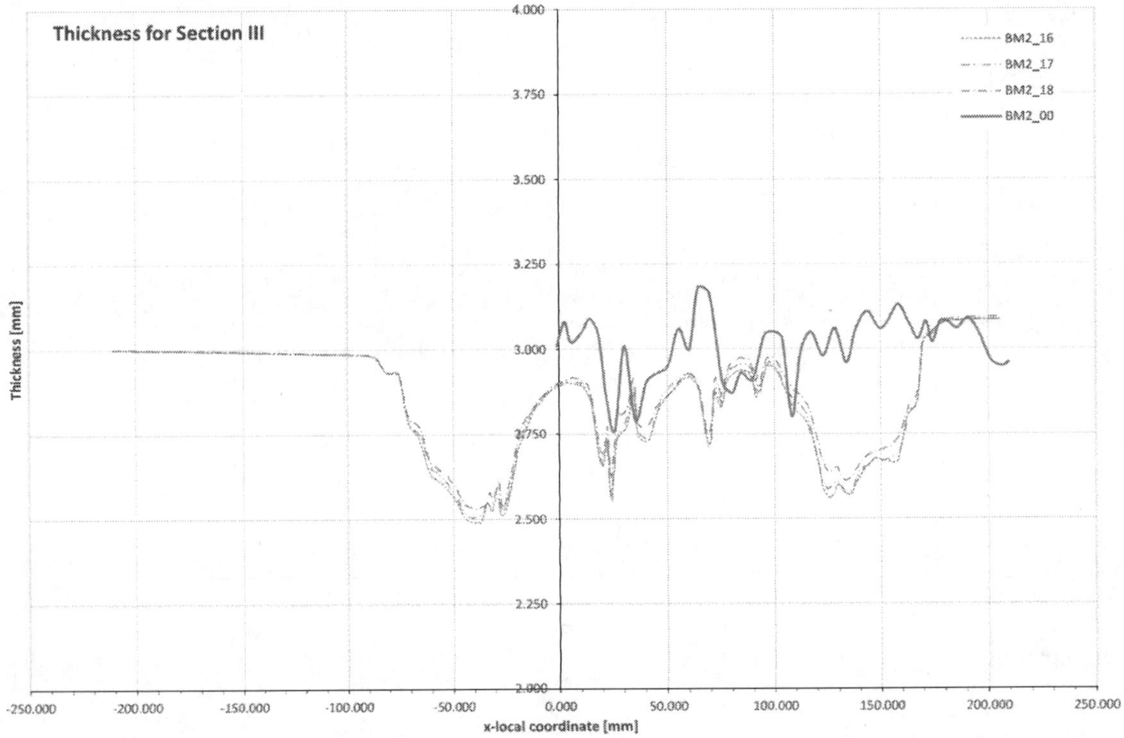

Figure 6.36. Thickness for Section III: BM2_16, BM2_17, BM2_18.

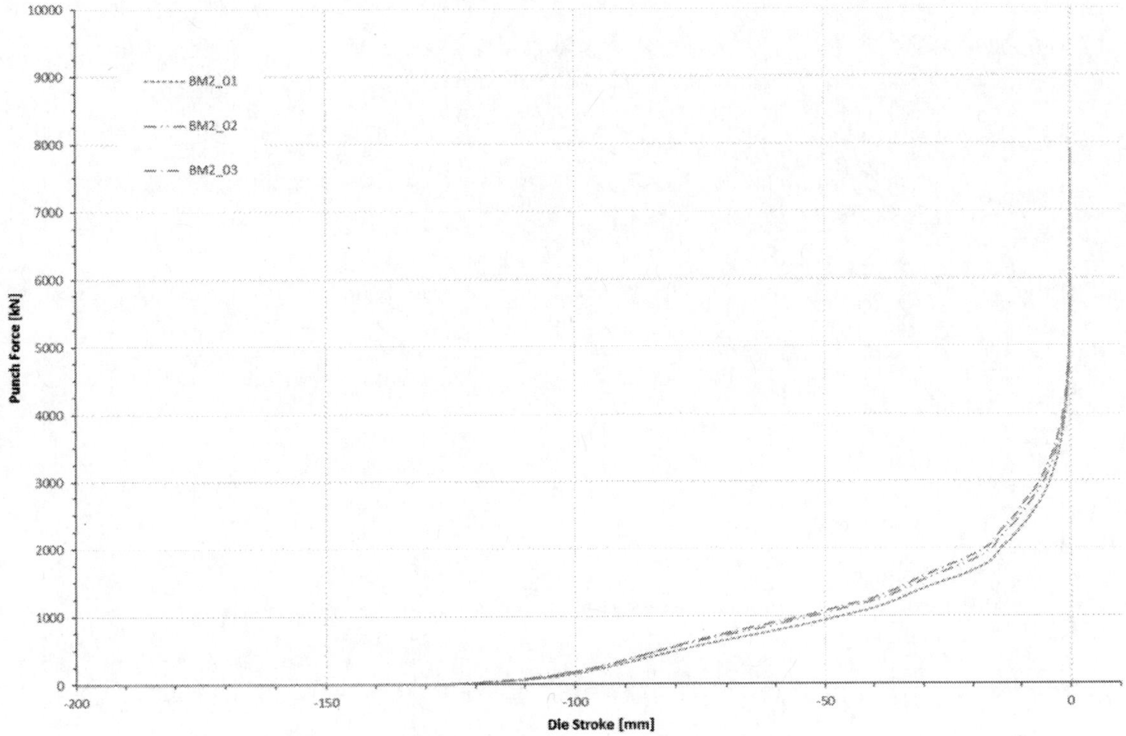

Figure 6.37. Punch Force: BM2_01, BM2_02, BM2_03.

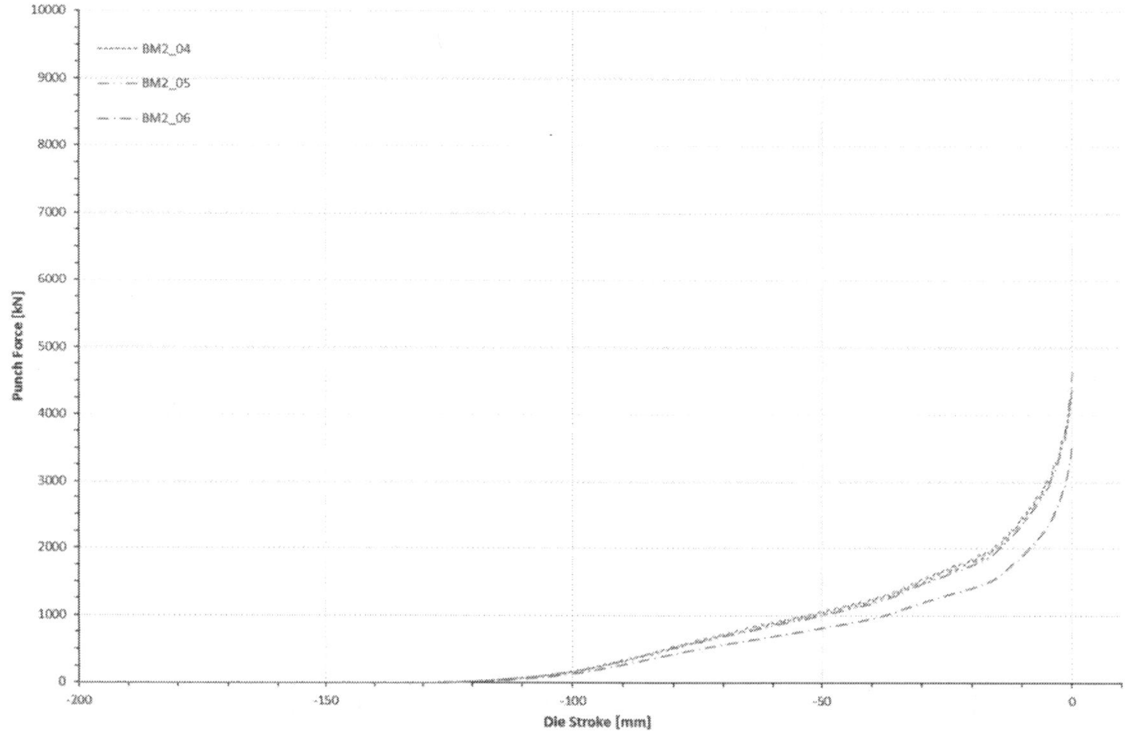

Figure 6.38. Punch Force: BM2_04, BM2_05, BM2_06.

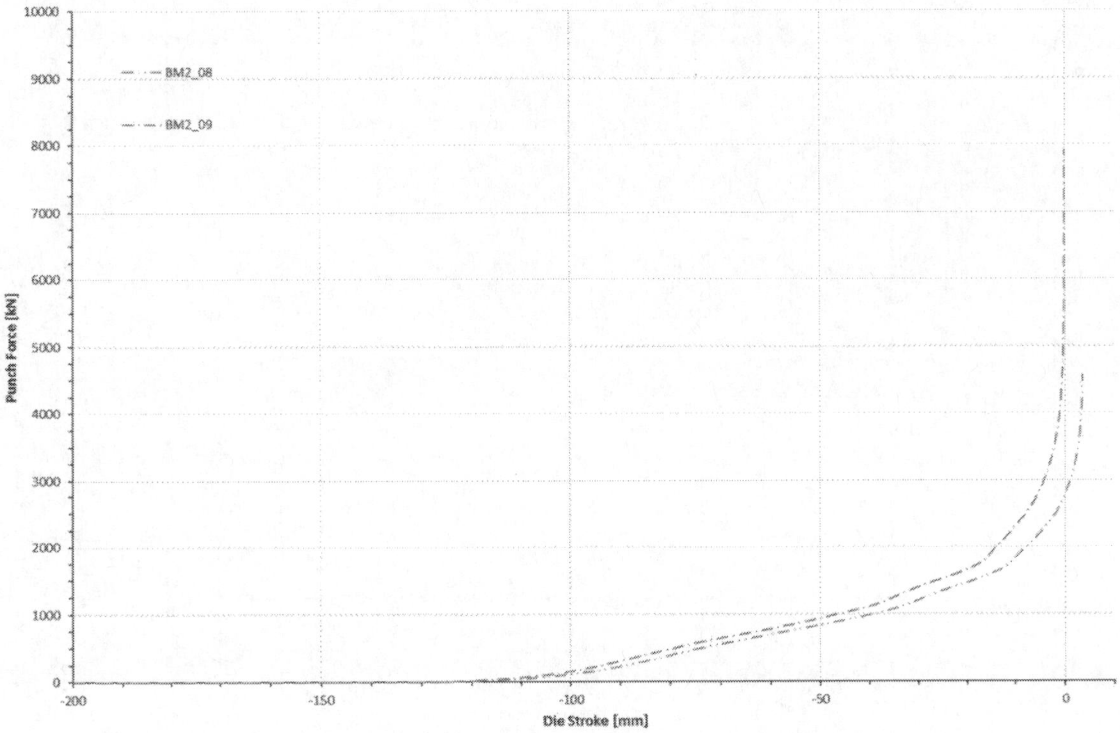

Figure 6.39. Punch Force: BM2_08, BM2_09.

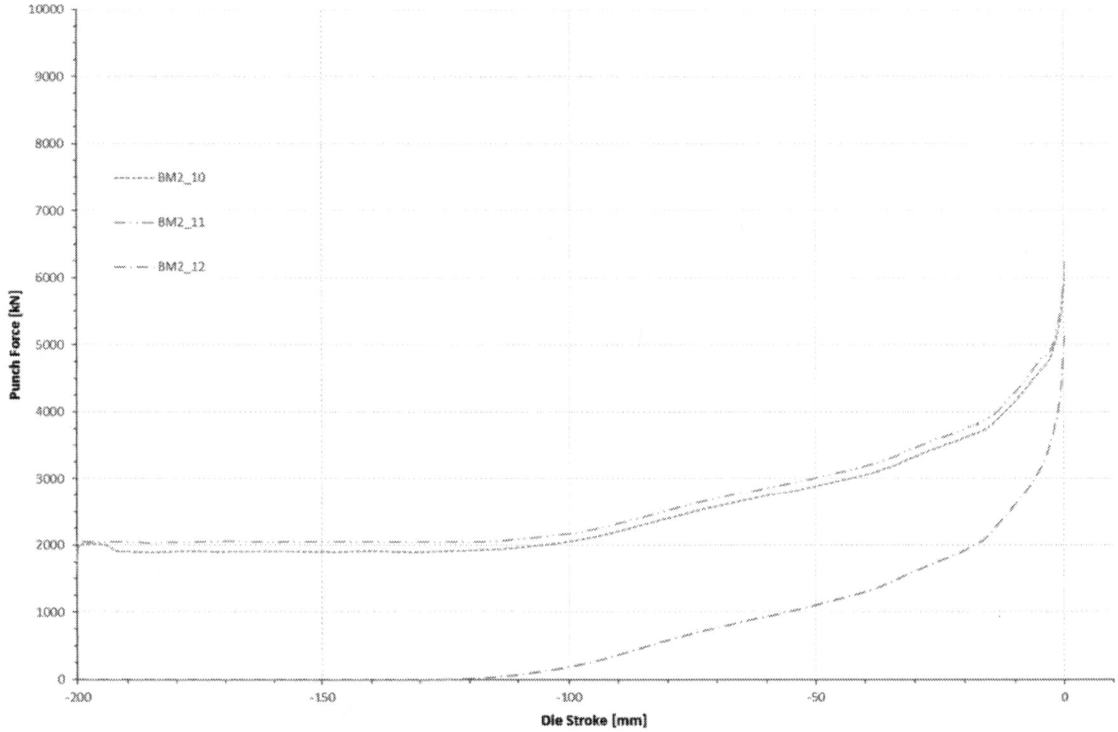

Figure 6.40. Punch Force: BM2_10, BM2_11, BM2_12.

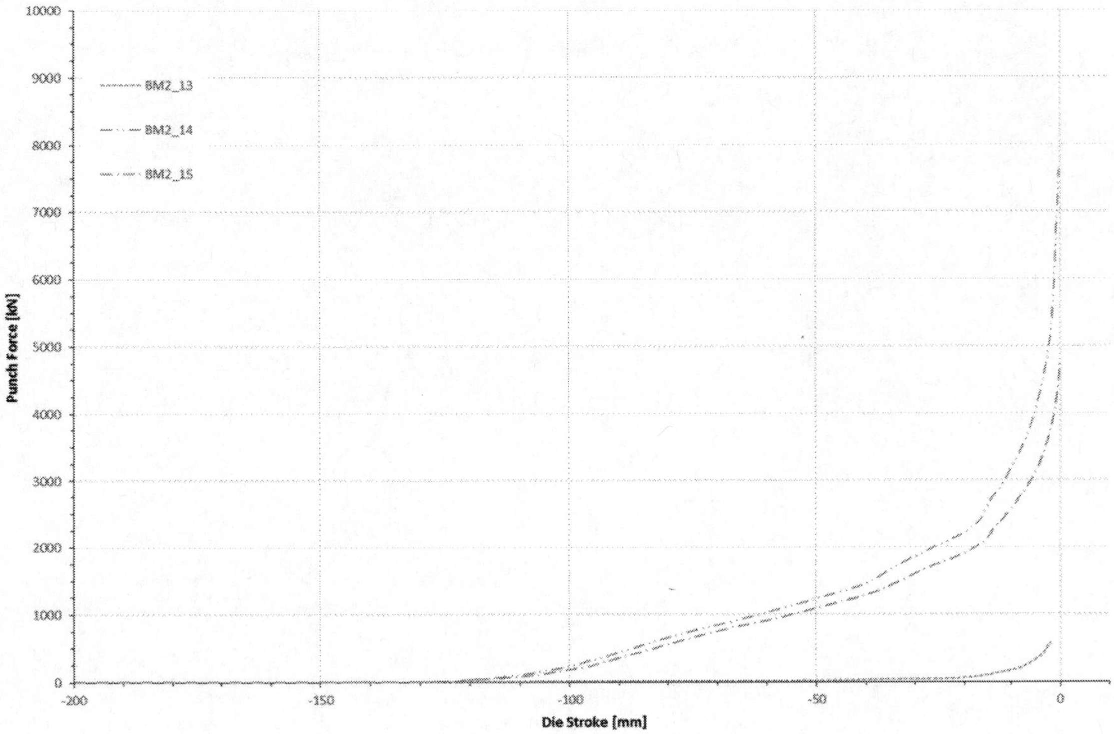

Figure 6.41. Punch Force: BM2_13, BM2_14, BM2_15.

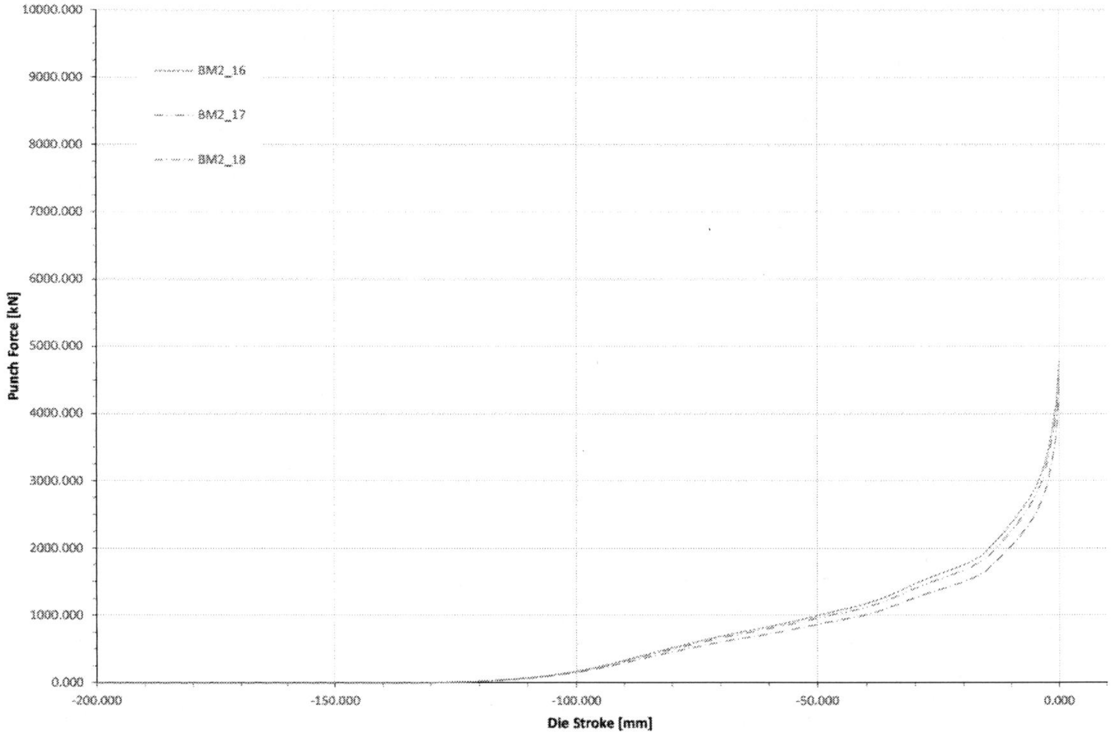

Figure 6.42. Punch Force: BM2_16, BM2_17, BM2_18.

Benchmark 3 – Springback of an Al-Mg alloy in warm forming conditions

Pierre-Yves Manach[a], Jérémy Coër[a], Anthony Jégat[a]
Hervé Laurent[a], Jeong Whan Yoon[b]

[a]*Univ. Bretagne Sud, FRE CNRS 3744, IRDL, Lorient, France*
[b]*Daekin University, Geelong, Australia*

Abstract. Accurate prediction of springback is a long-standing challenge in the field of warm forming of aluminium sheets. The objective of this benchmark is to predict the effect of temperature on the springback process through the use of the split-ring test [1] with an Al-Mg alloy. This test consists in determining the residual stress state by measuring the opening of a ring cut from the sidewall of a formed cylindrical cup. Cylindrical cups are drawn with a heated die and blank-holder at temperatures of 20, 150 and 240°C. The force-displacement response during the forming process, the thickness and the earing profiles of the cup as well as the ring opening and the temperature of the blank are used to evaluate numerical predictions submitted by the benchmark participants. Problem description, material properties, and simulation reports with experimental data are summarized.

Keywords: Deep drawing, warm conditions, Al-Mg alloy, springback

1. INTRODUCTION

Nowadays, aluminium alloys are increasingly used in the automotive industry, since they allow weight reduction in body-in-white. However, the large springback that occurs after aluminium alloy sheets have been formed at room temperature is one of the main reasons why this material has not been more widely used. In order to overcome this issue, good results on the stamping process are obtained for aluminium alloys when the temperature is elevated up to an intermediate temperature, below the recrystallisation temperature. This process is called warm forming, that promoted a great interest during the last few years, especially with the 5xxx series (Al–Mg alloys). The warm forming process has now become a widely used alternative to the classical forming processes performed at room temperature.

The aim of this benchmark is to investigate experimentally springback tests performed on an AA5086 alloy under warm forming conditions, such as to serve as a reference to compare the results obtained by numerical simulations. An experimental setup has been designed to perform the deep drawing of a cylindrical cup by heating the tools separately. Indeed, several authors have shown that the formability increases when selective localized heating strategies are applied to the forming tools, causing an inhomogeneous distribution of the temperature in the blank. Therefore, the aim is to confirm this improvement of the formability and to study the effects of warm forming conditions on residual stresses and springback. For that purpose, the springback is determined by measuring the opening of a ring cut from the sidewall of a drawn cylindrical cup (see Fig.1). This is the so-called split-ring test that was first presented by Demeri et al. [1]. It provides a simple and effective way of predicting the forming and springback properties of alloys based on experimental measurements.

Cylindrical cups are drawn with a heated die and blank-holder at temperatures of 20 (room temperature), 150 and 240°C. The force-displacement response during the forming process, the thickness and the earing profiles of the cup as well as the ring opening and the temperature of the blank are used to evaluate numerical predictions submitted by the benchmark participants.

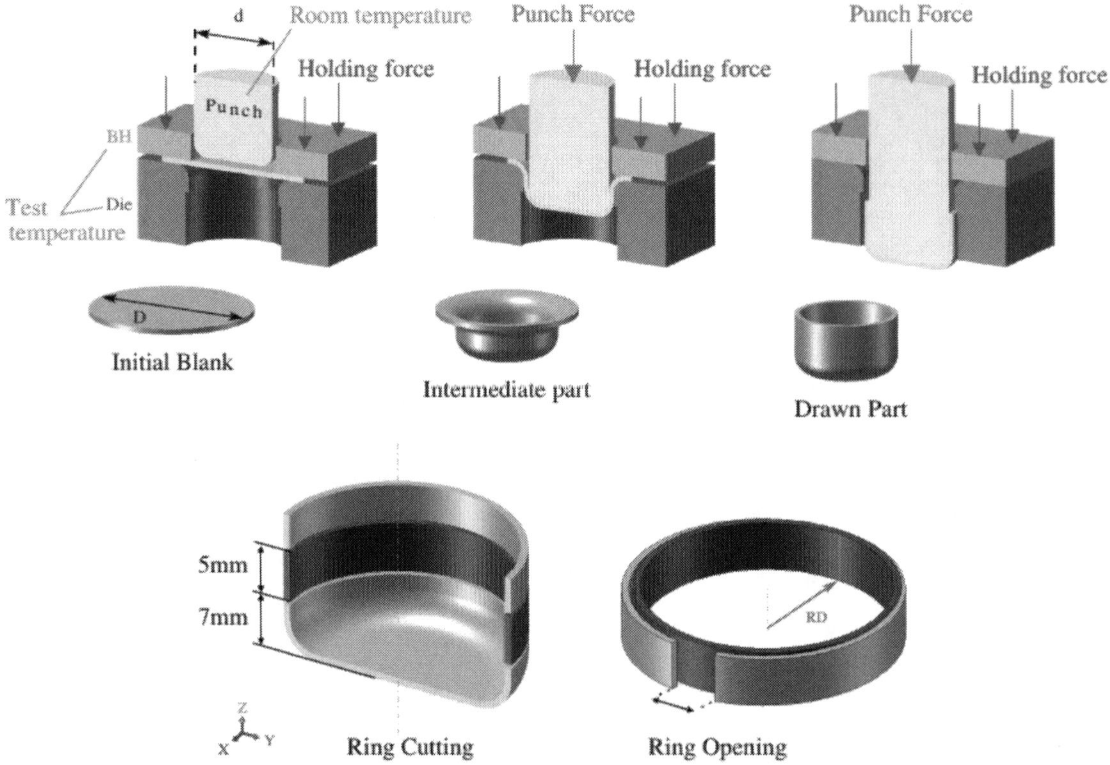

Figure 1: Main steps in the split-ring test

To simulate this process, a temperature-dependent anisotropic constitutive model is required for the material. The parameters of hardening models and strain rate dependency can be identified using data given in uniaxial tensile and shear tests at various temperatures and strain rates as well as biaxial expansion test, in order to account for temperature, viscous effects and anisotropy in a coupled thermo-mechanical constitutive law. Shell elements, solid elements, or solid-shell elements are recommended for this benchmark with careful control of the incremental punch stroke, with sufficient number of elements in the mesh to reproduce the curvature of the die and to capture plastic strain accurately. The analysis in this benchmark is highly non-linear, including thermal, viscosity and anisotropy. It is recommended to use a simple isotropic material model (such as von-Mises yield function) before attempting an advanced anisotropic material model.

This benchmark study has the main objective of predicting springback after warm forming, cutting and opening. Different challenging outputs will be required as a function of forming temperature:

i) Prediction of earing after the warm forming operation due to the plastic anisotropy of the material;
ii) Prediction of thickness profiles for several orientation to the rolling direction after the warm forming operation;
iii) Prediction of springback through ring opening;
iv) Prediction of punch force-punch stroke and temperature of the blank evolutions.

2. DESCRIPTION OF FORMING OPERATIONS

This section contains a description of the warm forming, cutting and opening operations for this benchmark.

2.1 Drawing operation

The benchmark is based on a paper presented in [2]. Cylindrical cup forming tests (Swift tests) are carried out on a Zwick/Roell Amsler BUP 200 sheet metal testing machine. A diagram of the deep-drawing procedure is presented in Fig.1. The blanks can be heated between the die and the blank-holder up to 240°C. Heating is obtained using electrical rods embedded both in the die and in the blank-holder. Axial water input and output channels are machined into the punch that allow controlling the temperature of the punch. An ejector located inside the punch is used to eject the cup from the punch at the end of the forming process. Type K thermocouples (TC) are used to control the temperature of the blank, the punch, the die and the blank-holder.

The geometry of the tools is given in Fig.2. The material is a rolled sheet of AA5086 aluminium alloy of 0.8 mm thick. The circular blank has an initial diameter of D=60 mm. At the beginning of the test, an oil lubricant (Jelt Oil) is applied manually on both sides of the blanks. To fully draw the cup, a punch displacement of 32.5 mm is imposed with a constant punch travel speed of 5 mm/s. The punch force, the punch stroke and the blank-holder force are recorded during the test.

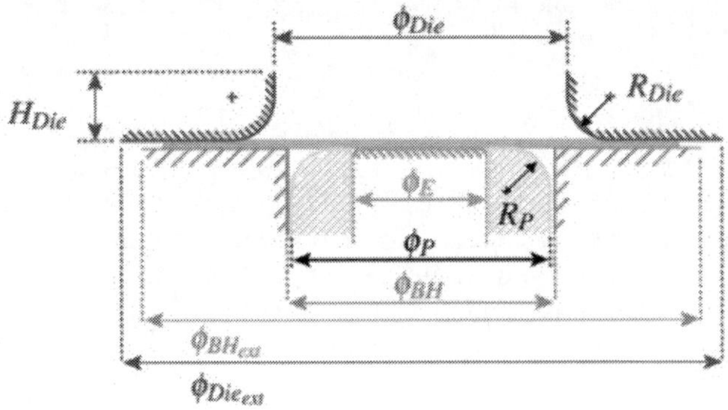

Tool Dimensions [mm]

ϕ_{Die}	R_{Die}	H_{Die}	ϕ_P	R_P
35,25	5	8,75	33	5

ϕ_{BH}	$\phi_{BH_{ext}}$	$\phi_{Die_{ext}}$	ϕ_E
33,6	70	80	20

Figure 2: Dimensions of the tools used in the Swift-cupping test

All the tools are axisymmetric. The blank-holder force at the beginning of the deep drawing operation is set to 5 kN, and this force is maintained until the cup is fully drawn. Heating is applied using electrical heating rods as shown in Fig.3. The set of tools used includes:

i) A draw die composed of two inserts containing a resistance coil and a copper plate;
ii) A blank-holder machined with a suitable upper annular insert for placing the heating rods;
iii) The punch and the internal ejector;
iv) A base used to support previous parts, connected to the BUP 200 machine.

The positions of thermocouples used to control the temperature of the punch, the die and the blank-holder are shown in Fig. 3. In the die, the thermocouple (TC-Die) is located on a diameter $\phi_{TC\text{-}Die}$ = 38.4 mm, 1 mm from the contact surface. In the blank-holder, TC-BH is located on a diameter $\phi_{TC\text{-}BH}$ = 47.3 mm, 1 mm from the contact surface and for the punch, TC-Punch is located under the ejector, on a diameter $\phi_{TC\text{-}Punch}$ = 16 mm, 1.5 mm from the contact surface. The evolution of the temperature of the tools is supposed homogeneous around the circumference and as those of the thermocouples, and is given as a function of the punch stroke in the file **BM3_Process.xlsx** for each temperature. The resulting temperature of the blank is measured on the side in contact with the die, at a location of $\phi_{TC\text{-}Blank}$ = 11.5 mm for the orientation of 22.5° to the rolling direction.

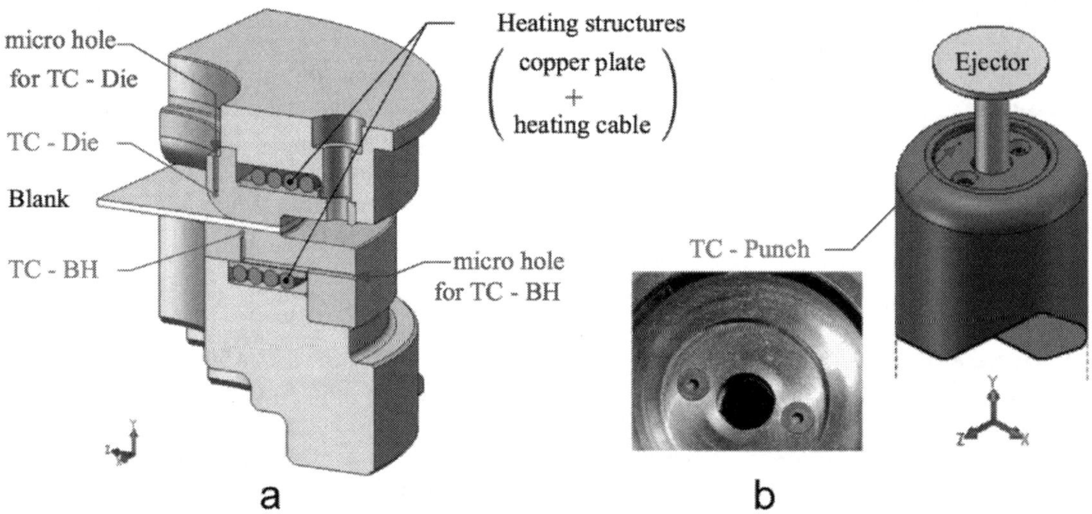

Figure 3: Heating parts in the tools and location of the thermocouples (TC)
(a) Die and BH, (b) Punch and ejector

2.2 Springback

Rings are cut from the sidewall of a formed cylindrical cup and split perpendicularly to the circle plane, in the rolling direction (RD). The cutting and splitting operations are carried out using a wire electro-erosion machine. Ring gap measurements are performed along the straight line connecting the two ends of the split rings (see Fig.1) in order to characterize residual stress state and to measure the springback effect.

2.3 Tool Materials

- Punch: XC38CrMoV5 Tool steel, 58-60 HRC, 2-4 Finish working surfaces
- Die: XC38CrMoV5 Tool steel, 58-60 HRC, 2-4 Finish working surfaces
- Blank-holder: XC38CrMoV5 Tool steel, 58-60 HRC, 2-4 Finish working surfaces

2.3 Experimental Measurements

The distributions of the thickness of the cup are measured in several directions (rolling direction RD, transverse direction TD, diagonal direction DD) from the center to the outer diameter every 1 mm in curvilinear distance, using a 3-D measuring machine. The curvilinear distance corresponds to the length of the average fiber of the cup. The earing profiles of the cups are also recorded and the cup height is plotted from the bottom of the cup as a function of the angular position (every 5°) to the RD. The punch force-displacement curves as well the temperature of the tools and the blank are recorded during the forming test. To help participants, the evolutions of temperatures of the tools as well as the temperature of the blank during the forming process are given in **BM3_Process.xlsx** file (see Fig.4). For example, these temperatures may be used to estimate the contact heat transfer coefficient. Thus, the participants can evaluate the relevance of their thermomechanical simulations on the temperature of the blank. Rings of 5 mm high are cut 7 mm from the bottom of the cups (see Fig.1), perpendicularly to the revolution axis of the cup. Ring gap measurements are performed along the straight line connecting the two ends of the split rings in order to evaluate the springback.

Time (s)	F - BH (kN)	Tc-Punch (°C)	Tc-BH (°C)	Tc-Blank (°C)	Tc-Die (°C)
0.000	5.258	23.860	23.563	23.662	23.364
0.005	5.265	23.860	23.563	23.662	23.265

Figure 4: Experimental data contained in the file **BM3_Process.xlsx**. For each temperature (RT, 150°C and 240°C), from columns left to right: time (s), blank-holder force (kN), Temperatures of the tools and the blank (Punch, Blank-holder, Blank and Die)

3. BLANK MATERIAL

The material used is sampled from a rolled sheet of 0.8 mm thick AA5086-H111 aluminium alloy. This material presents, at least at room temperature, the Portevin-Le Châtelier (PLC) effect. This effect is no more present for temperatures above 200°C. For the material parameters required in the constitutive models, the material is characterized under different conditions (temperature and strain rate) and strain paths (tensile, shear, bulge). Uniaxial tensile tests are performed under isothermal conditions at room temperature (20), 150 and 240°C. Tensile tests at room temperature are performed on a hydraulic Instron 8803 machine while the tests at 150 and 240°C are carried out with a Gleeble 3500 testing machine where the specimen is heated by Joule effect [3]. On this last machine, a constant crosshead velocity is difficult to achieve and the strain rate is thus non linear. For all the tensile tests, the participants can calculate the strain rate by fitting the time-strain signal. For the strain rate effects, a decade has been imposed between two consecutive tests, denoted by x1, x10 and x100.

Monotonous and reverse shear tests for several values of pre-strains are provided in order to evaluate Bauschinger effect and therefore kinematic hardening. Shear tests are performed with a tensile machine, using a specific shear device placed in a heating furnace [4]. But due to experimental considerations, it was not possible to reach temperatures higher than 150°C. Shear samples have been machined at dimensions: $60x15mm^2$ and the shear width is constant equal to 3mm (see [4] for details). Finally, biaxial tests are carried out in a hydraulic bulge test setup only at room temperature. The material data necessary to identify the influence of temperature, anisotropy and strain rate is given in Section 5 and in the Excel file **AA5086-H111.xlsx**.

4. BENCHMARK REPORT

All results are expected to be reported using the benchmark report template **BM3_Report.xlsx**, which can be downloaded from the conference website, and when completed, uploaded to the website at a later date to submit the entry. The report file contains the following informations:

4.1 General description

1) Benchmark participant: name, affiliation, address, email and phone number
2) Simulation software: name of the FEM code, general aspects of the code, basic formulations, element/mesh technology, type of elements, number of elements, contact property model and friction formulation
3) Simulation hardware: CPU type, CPU clock speed, number of cores per CPU, main memory, operating system and total CPU time
4) Material model: Yield function/Plastic potential, Hardening rule and Stress-Strain Relation, and heat transfer model
5) Remarks

4.2 Simulation results

1) Earing profiles plotted through Cup height (h mm) after the warm forming operation for each temperature (RT, 150, 240°C), measured from the lower surface to the upper edge of the cup around the circumference starting from the rolling direction (0°) to 360°, reported every 5°
2) Plot of punch load (kN) vs punch stroke (mm) during the cup forming operation for each temperature (RT, 150, 240°C). The zero punch stroke is defined as the position when the punch makes initial contact with the blank with no interaction forces
3) Blank surface temperature as a function of punch stroke on the side in contact with the die, during the test for RT, 150 and 240°C. The temperatures of the tools should also be given for each test temperature
4) Thickness distribution (mm) vs curvilinear distance from the cup center after the forming operation, in the rolling direction (0°), transverse direction (90°) and diagonal direction (45°) for each temperature (RT, 150, 240°C). For the curvilinear distance, the medium thickness should be considered. The experimental values are the average between the four quarters of the cup
5) The ring opening (mm) for each temperature (RT, 150, 240°C) measured as the straight line connecting the two ends of the ring. As the ring may be slightly conical, the distance should be measured at the mid-height of the ring.

5. MATERIAL CHARACTERIZATION

Table 1. Elastic mechanical properties

Sample	Density	Young's modulus	Poisson's ratio
	g/cm3	GPa	
AA5086	2.70	71.7	0.31

Table 2. Uniaxial tension test data[1]

Test orientation	YS MPa	UTS MPa	% Elongation	r value
RD – 20°C	138.5	267.4	22.15	0.71
DD – 20°C	135.0	258.4	28.90	1.08
TD – 20°C	138.8	256.8	23.90	0.73
RD – 150°C	148.4	246.1	29.50	0.63
DD – 150°C	139.7	232.9	31.80	0.97
TD – 150°C	142.6	237.1	23.90	0.66
RD – 240°C	119.4	150.4	39.25	0.60
DD – 240°C	115.7	141.7	40.70	0.88
TD – 240°C	114.9	142.0	36.60	0.67

Table 3. Thermal properties of AA5086-H111

Material	AA5086-H111
Thermal expansion coefficient	2.2×10^{-5}
Specific heat (J/kg.°C)	900
Thermal conductivity (W/m.°C)	220
Inelastic heat fraction (%)	100

[1] Stress-strain curves are provided in the Excel file AA5086-H111.xlsx for several strain rates and temperatures. Equal Biaxial Tension Test Data and Reversed shear stress data are given directly in the Excel file.

Table 4. Mechanical and thermal properties of the tools

Tools	XC38CrMoV5
Density (kg/m³)	8150
Young modulus (GPa)	215
Poisson's ratio	0.3
Thermal expansion coefficient	1.19×10^{-5}
Specific heat (J/kg.°C)	500
Thermal conductivity (W/m.°C)	25.
Contact heat transfer coefficient (W/m² °C)	To be estimated from the temperature of the tools
Die and BH temperature (°C)	20, 150, 240
Blank-Holder force (kN)	5
Punch speed (mm/s)	5
Friction coefficient (recommended)	0.09

6. REFERENCES

[1] M.Y. Demeri, M. Lou, M.J. Saran. A benchmark test for springback simulation in sheet metal forming, In Society of Automotive Engineers, Inc., Volume 01-2657 (2000)
[2] H. Laurent, J. Coër, P.Y. Manach, M.C. Oliveira, L.F. Menezes. Experimental and numerical studies on the warm deep drawing of an Al-Mg alloy, International Journal of Mechanical Sciences, 93 (2015) 59-72
[3] J. Coër, C. Bernard, H. Laurent, A. Andrade Campos, S. Thuillier. The effect of temperature on anisotropy properties of an aluminium alloy, Experimental Mechanics, 51 (2010) 1185-1195
[4] J. Coër, P.Y. Manach, H. Laurent, L.F. Menezes, M.C. Oliveira. Piobert-Lüders plateau and Portevin-Le Chatelier effect in an Al-Mg alloy in simple shear, Mechanics Research Communications 48 (2013) 1-7

7. RESULTS

General description BM01

A. Benchmark participant	
Name	Bart Carleer, Alper Güner, Thomas Brenne
Affiliation	AutoForm Engineering
Address	Emil-Figge-Strasse 76-80, 44227, Dortmund, Germany
Email	alper.guener@autoform.de
Phone number	0049 231 9742 278
Fax number	0049 231 9742 322

B. Software/Hardware	
Name of the FEM code	AutoForm^plus R7
General aspect of the code	Static Implicit
Basic formulations	
Element/Mesh technology	
Number of elements	26000
Type of elements	Triangular elastic plastic shell, 11 integration points through thickness
Contact property model	Penalty method
Friction formulation	Coulomb friction
CPU type	
CPU clock speed	Intel Core i7-4910MQ - 2.90 GHz
Number of cores per CPU	8
Main memory	16 GB
Operating system	Windows 7 Professional
Total CPU time	Elapsed Time - RT: 08:49 , 150°C: 09:16, 240°C: 09:20 (minute:second)

C. Describe the material model	
Name of the material model	Barlat
Yield Function/Plastic Potential	Barlat89
Hardening Rule	Strain rate dependent isotropic hardening with temperature dependent r-values
Flow rule (Associated/Non Associated)	Associated
Heat transfer model	

D. Remarks
Temperature of blankholder and die are modeled as constant. Punch and sheet temperature change is calculated through thermo-solver. Plotted punch temperatures are measured directly on the surface of the punch. So there is no delay due to depth of measurements or air gap.

General description BM02

A. Benchmark participant	
Name	Bart Carleer, Alper Güner, Thomas Brenne
Affiliation	AutoForm Engineering
Address	Emil-Figge-Strasse 76-80 44227 Dortmund
Email	alper.guener@autoform.de
Phone number	0049 231 9742 278
Fax number	0049 231 9742 322

B. Software/Hardware	
Name of the FEM code	AutoForm^plus R7
General aspect of the code	Static Implicit
Basic formulations	
Element/Mesh technology	
Number of elements	26000
Type of elements	Triangular elastic plastic shell, 11 integration points through thickness
Contact property model	Penalty method
Friction formulation	Coulomb friction
CPU type	
CPU clock speed	Intel Core i7-4910MQ - 2.90 GHz
Number of cores per CPU	8
Main memory	16 GB
Operating system	Windows 7 Professional
Total CPU time	Elapsed Time - RT: 08:14 , 150°C: 09:00 and 240°C: 08:56 (minute:second)

C. Describe the material model	
Name of the material model	BBC Model
Yield Function/Plastic Potential	Banabic 2005 (BBC 2005)
Hardening Rule	Strain rate dependent isotropic hardening with temperature dependent r-values
Flow rule (Associated/Non Associated)	Associated
Heat transfer model	

D. Remarks
Temperature of blankholder and die are modeled as constant. Punch and sheet temperature change is calculated through thermo-solver. Plotted punch temperatures are measured directly on the surface of the punch. So there is no delay due to depth of measurements or air gap.

Journal of Physics: Conference Series **734** (2016) 022003 doi:10.1088/1742-6596/734/2/022003

General description BM03

A. Benchmark participant	
Name	Bart Carleer, Alper Güner, Thomas Brenne
Affiliation	AutoForm Engineering
Address	Emil-Figge-Strasse 76-80 44227 Dortmund
Email	alper.guener@autoform.de
Phone number	0049 231 9742 278
Fax number	0049 231 9742 322

B. Software/Hardware	
Name of the FEM code	AutoForm^plus R7
General aspect of the code	Static Implicit
Basic formulations	
Element/Mesh technology	
Number of elements	26000
Type of elements	Triangular elastic plastic shell, 11 integration points through thickness
Contact property model	Penalty method
Friction formulation	Coulomb friction
CPU type	
CPU clock speed	Intel Core i7-4910MQ - 2.90 GHz
Number of cores per CPU	8
Main memory	16 GB
Operating system	Windows 7 Professional
Total CPU time	Elapsed Time - RT: 08:36 , 150°C: 09:18 and 240°C: 08:41 (minute:second)

C. Describe the material model	
Name of the material model	BBC Model
Yield Function/Plastic Potential	Banabic 2005 (BBC 2005)
Hardening Rule	Strain rate dependent isotropic hardening with temperature dependent r-values and temperature dependent Young's Modulus
Flow rule (Associated/Non Associated)	Associated
Heat transfer model	

D. Remarks
Temperature of blankholder and die are modeled as constant. Punch and sheet temperature change is calculated through thermo-solver. Plotted punch temperatures are measured directly on the surface of the punch. So there is no delay due to depth of measurements or air gap.

Journal of Physics: Conference Series **734** (2016) 022003 doi:10.1088/1742-6596/734/2/022003

General description BM04

A. Benchmark participant	
Name	Martin Holeček, David Lorenz
Affiliation	ESI Group
Address	Brojova 16, 326 00 Czech Republic
Email	martin.holecek@esi-group.com, david.lorenz@esi-group.com
Phone number	00420 37 7432959
Fax number	00420 37 7432930

B. Software/Hardware	
Name of the FEM code	PAM Stamp 2015.1
General aspect of the code	Dynamic explicit
Basic formulations	Updated Lagrangian formulation with associated flow rule
Element/Mesh technology	
Number of elements	6500
Type of elements	Belytschko-Tsay underintegrated shell element
Contact property model	Non-linear penalty contact
Friction formulation	Coulomb friction
CPU type	
CPU clock speed	Intel Xeon E5410 @ 2.33 GHz
Number of cores per CPU	8
Main memory	16 GB
Operating system	Linux
Total CPU time	3 hours

C. Describe the material model	
Name of the material model	Vegter-Lite
Yield Function/Plastic Potential	Hocket-Sherby
Hardening Rule	isotropic
Flow rule (Associated/Non Associated)	Associated
Heat transfer model	

D. Remarks

General description BM05

A. Benchmark participant	
Name	Kaimin Yan, Gang Fang
Affiliation	Tsinghua University
Address	Room A807, Lee Shau Kee Sci.& Tech Building, Dept.of Mechanical Engineering, Tsinghua University, Beijing, 100084,China
Email	fangg@tsinghua.edu.cn
Phone number	+86-10-62782694
Fax number	+86-10-62770190

B. Software/Hardware	
Name of the FEM code	AutoForm plus R6
General aspect of the code	Implicit; Iterative Solver
Basic formulations	3-node Shells
Element/Mesh technology	
Number of elements	Adaptive mesh refinement: 6514 initially; 5721 at the end.
Type of elements	elastic shell (gravity) ; EP shell (drawing, cutting & springback)
Contact property model	Penalty method with automatically selected penalty; Tool stiffness = 50
Friction formulation	Coulomb: mu=0.09
CPU type	
CPU clock speed	4 GHz
Number of cores per CPU	8
Main memory	16GB
Operating system	Windows 10 education
Total CPU time	RT: one minute and eleven seconds; 150°C: one minute and twenty four seconds; 240°C: one minute and twenty three seconds

C. Describe the material model	
Name of the material model	AA5086-H111-2016-02-25
Yield Function/Plastic Potential	von Mises/von Mises
Hardening Rule	Isotropic Hardening
Flow rule (Associated/Non Associated)	Associated
Heat transfer model	HTC to tool equals to 5.000 mW/(mm²K)

D. Remarks
A negative strain rate sensitivity of the material AA5086 was observed, however, the software AutoForm doesn't allow flow curves with a negative strain rate sensitivity. As a compromise, the fitted flow stresses at room temperature at strain rate 0.001 and 0.1, as well as at 150°C at strain rate 0.001 and 0.01 were slightly modified compared to the given data, so that higher strain rates lead to higher stress values at the same temperature

General description BM06

A. Benchmark participant	
Name	Byron Bonyoung Ghoo, Yasuyoshi Umezu, and Yuko Watanabe
Affiliation	JSOL Corporation
Address	Harumi Center Building, 2-5-24 Harumi, Chuo-ku, Tokyo 104-0053, JAPAN
Email	ghoo.bonyoung@jsol.co.jp
Phone number	+81-80-3023-7012
Fax number	+81-3-5859-6035

B. Software/Hardware	
Name of the FEM code	JSTAMP (Solver: LS-DYNA)
General aspect of the code	Nonlinear Implicit(heat transfer & springback) / Explicit(mechanical forming) Coupling
Basic formulations	Thermo-Elasto-Viscoplastic Material, Planar Anisotropy, Kinematic Hardening
Element/Mesh technology	
Number of elements	Blank(4,638) for Thermo-mechanical / Tools(16,915) for Themal Analysis Only
Type of elements	Nonlinear Thermal 12node Shell Formulation (Fully Integrated)
Contact property model	Surface to Surface Contact / Penalty Method
Friction formulation	Coulomb's Law of Friction
CPU type	
CPU clock speed	3.2GHz
Number of cores per CPU	8 Cores
Main memory	16GB
Operating system	Windows 7 Professional (64bit)
Total CPU time	Drawing (2 hours 10 minutes) / Cooling (4 minutes) / Springback (1 minute)

C. Describe the material model	
Name of the material model	MAT_3-PARAMETER_BARLAT (MAT036) / MAT_KINEMATIC_HARDENING_BARLAT2000 (MAT242)
Yield Function/Plastic Potential	Barlat and Lian [1989] for Non-isothermal / Barlat et al. [2003] for Isothermal Analyses
Hardening Rule	Isotropic for Non-isothermal / Non-linear Kinematic (Yoshida-Uemori) for Isothermal Analyses
Flow rule (Associated/Non Associated)	Associated Elastic Viscoplastic Thermal (Stress-Strain-Strainrate-Temperature)
Heat transfer model	Transient Analysis by Diagonal Scaled Conjugate Gradient Iterative Method

D. Remarks
We add a special sheet (2.4) that shows an isothermal analysis result for room temperature case in the report (referred as BM06b). The isothermal analysis for room temperature case was conducted using Yoshida-Uemori hardening and Barlat 2000 2d yield function for more accurate springback result. Non-isothermal analyses were conducted using the planar anisotropic yield function, Barlat 89, for the room temperature, 150C, and 240C cases respectively.

Journal of Physics: Conference Series **734** (2016) 022003 doi:10.1088/1742-6596/734/2/022003

General description BM07

A. Benchmark participant	
Name	P.M. Cunha, J.M.P. Martins, D.M. Neto, M.C. Oliveira, J.L Alves and L.F. Menezes
Affiliation	Departament of Mechanical Engineering, University of Coimbra
Address	Polo II, Rua Luís Reis Santos, 3030-788 Coimbra, Portugal
Email	joao.pmartins@dem.uc.pt
Phone number	+351 239 790 700
Fax number	+351 239 790 701

B. Software/Hardware	
Name of the FEM code	DD3IMP
General aspect of the code	Static fully implicit
Basic formulations	Updated Lagrangian formulation with associated flow rule
Element/Mesh technology	
Number of elements	30237
Type of elements	Isoparametric 3D brick elements with selective reduced integration technique
Contact property model	Rigid tools modelled by 390 Nagata patches, Augmented lagrangian method
Friction formulation	Coulomb friction law
CPU type	
CPU clock speed	3.5 GHz
Number of cores per CPU	4 cores
Main memory	16 GB RAM
Operating system	Windows 8 Professional (64-bit)
Total CPU time	6 hours (RT); 8 hours (150ºC and 240ºC)

C. Describe the material model	
Name of the material model	Elastoplastic
Yield Function/Plastic Potential	Barlat 91 (RT); Hill'48 (150ºC and 240ºC)
Hardening Rule	Voce + Frederick and Armstrong (RT); Hockett-Sherby (150ºC and 240ºC)
Flow rule (Associated/Non Associated)	Associated
Heat transfer model	Classical Fourier's law for heat conduction

D. Remarks
For the thermal problem, a constant temperature was assumed for the rigid tools.

Journal of Physics: Conference Series **734** (2016) 022003 doi:10.1088/1742-6596/734/2/022003

General description BM08

A. Benchmark participant	
Name	Hervé Laurent
Affiliation	Université Bretagne Sud
Address	iRDL, Rue de Saint Maudé, 56100 Lorient (France)
Email	herve.laurent@univ-ubs.fr
Phone number	0033 297874570
Fax number	0033 297874572

B. Software/Hardware	
Name of the FEM code	Abaqus-6.14
General aspect of the code	Standard Implicit
Basic formulations	Coupled Temperature-Displacement
Element/Mesh technology	
Number of elements	15936
Type of elements	C3D8RT
Contact property model	Surface to surface finite sliding
Friction formulation	Isotropic Coulomb friction coefficient 0.09
CPU type	
CPU clock speed	3.60 GHz
Number of cores per CPU	4
Main memory	16 GB
Operating system	3.16.0-4-amd64 GNU/Linux
Total CPU time	55227 s

C. Describe the material model	
Name of the material model	H. Laurent et al. / International Journal of Mechanical Sciences 93 (2015) 59–72
Yield Function/Plastic Potential	Hill 1948
Hardening Rule	Hockett–Sherby hardening model and a power law strain rate dependency
Flow rule (Associated/Non Associated)	Associated
Heat transfer model	Fully transient coupled thermal-stress analysis

D. Remarks
The ring opening is perfomed as in the reference: H. Laurent et al., International Journal of Mechanical Sciences 93 (2015) 59–72, with a ring of 7mm height instead of 5mm, located at 8mm from the cup bottom instead of 7mm

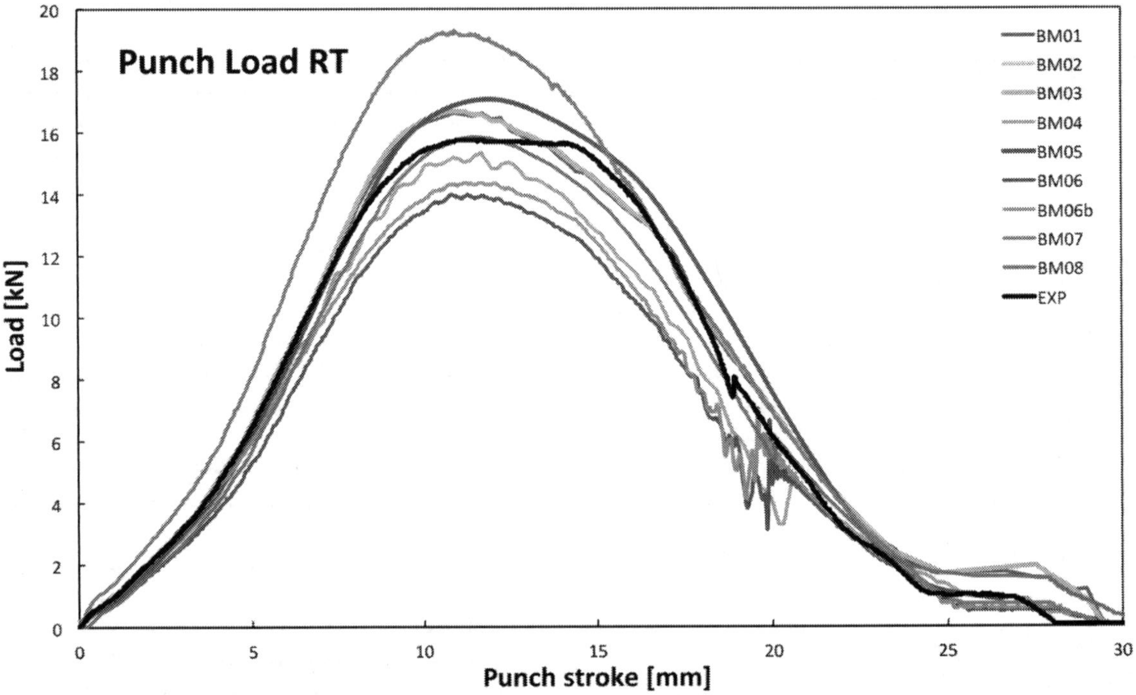
Punch Load RT
Load [kN]
Punch stroke [mm]
0
2
4
6
8
10
12
14
16
18
20
0
5
10
15
20
25
30
BM01
BM02
BM03
BM04
BM05
BM06
BM06b
BM07
BM08
EXP

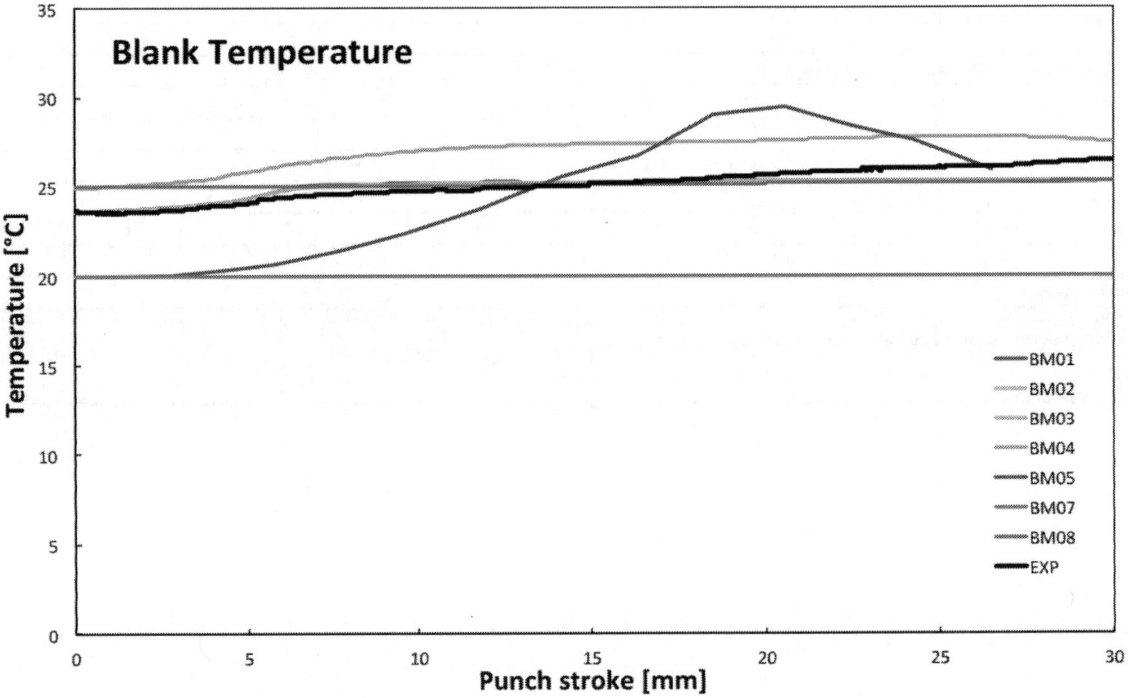
Blank Temperature
Temperature [°C]
Punch stroke [mm]
0
5
10
15
20
25
30
35
0
5
10
15
20
25
30
BM01
BM02
BM03
BM04
BM05
BM07
BM08
EXP

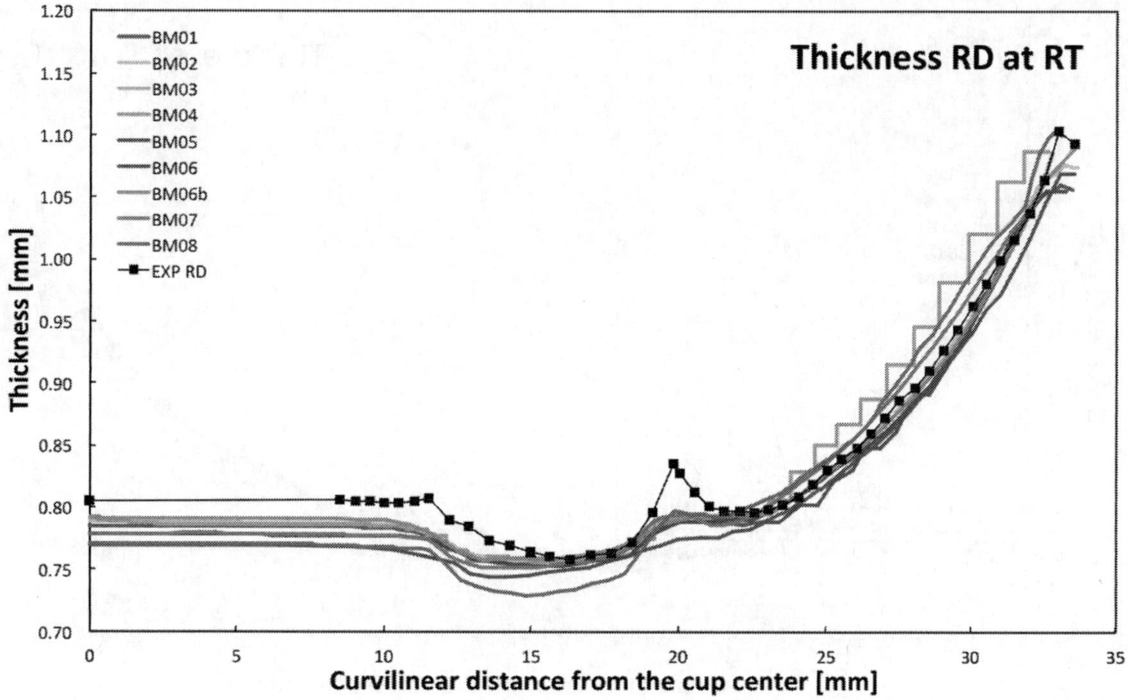
Thickness RD at RT
BM01
BM02
BM03
BM04
BM05
BM06
BM06b
BM07
BM08
EXP RD
Thickness [mm]
1.20
1.15
1.10
1.05
1.00
0.95
0.90
0.85
0.80
0.75
0.70
0
5
10
15
20
25
30
35
Curvilinear distance from the cup center [mm]

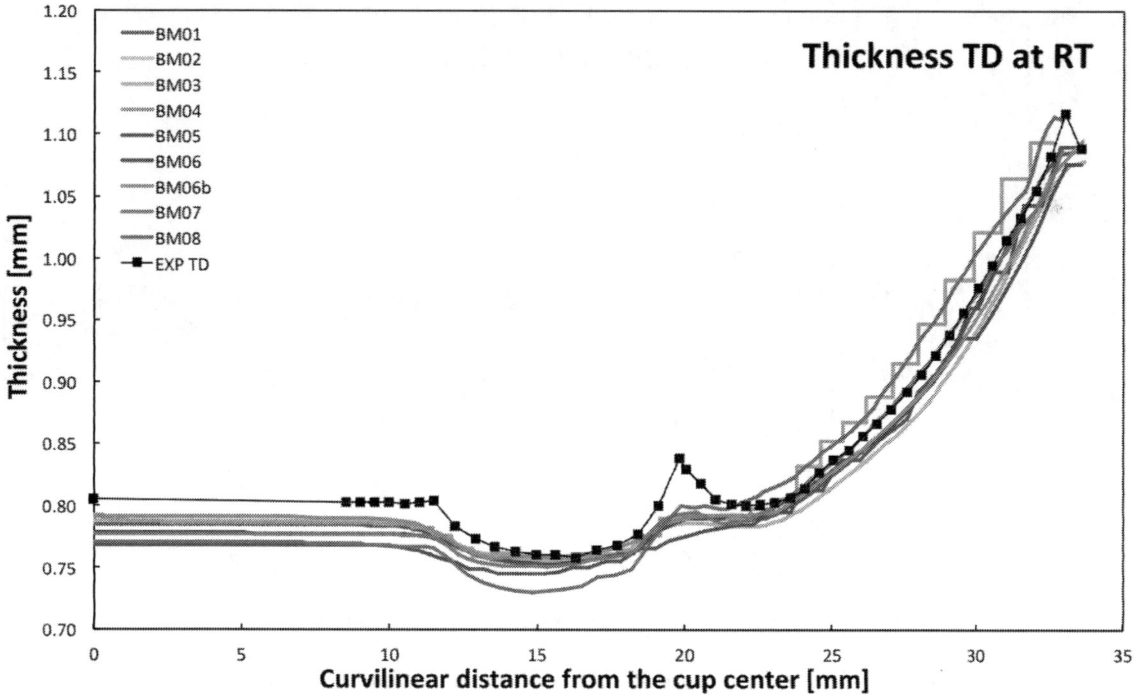
Thickness TD at RT
BM01
BM02
BM03
BM04
BM05
BM06
BM06b
BM07
BM08
EXP TD
Thickness [mm]
1.20
1.15
1.10
1.05
1.00
0.95
0.90
0.85
0.80
0.75
0.70
0
5
10
15
20
25
30
35
Curvilinear distance from the cup center [mm]

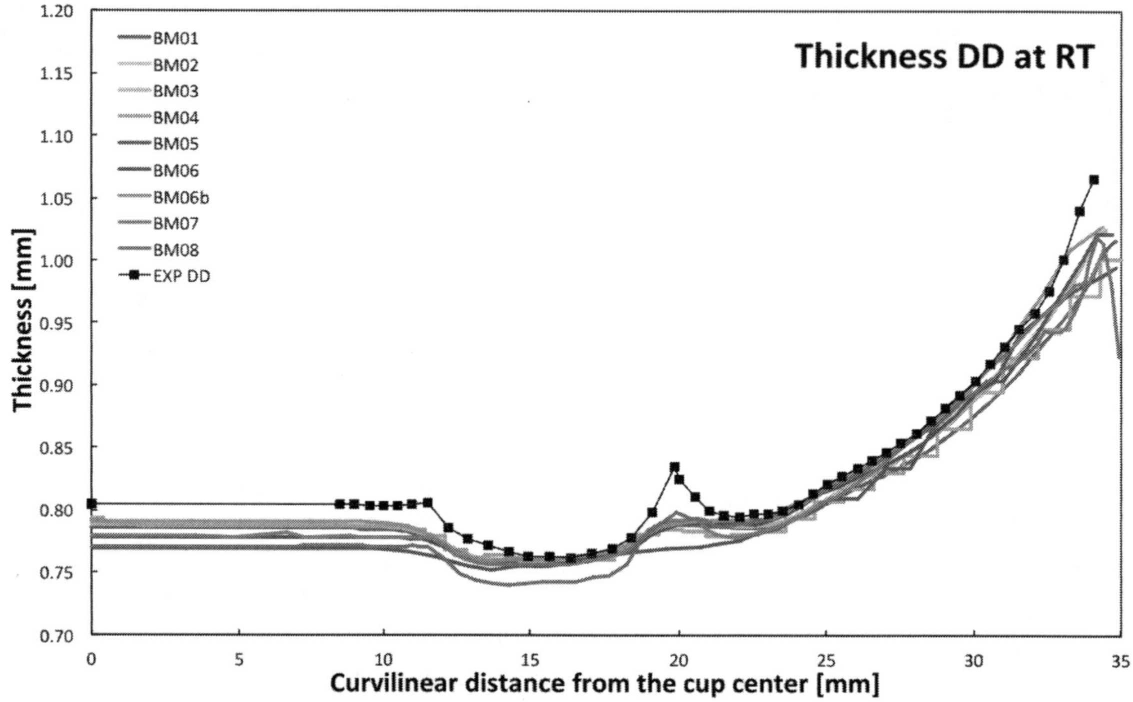
Thickness DD at RT
BM01
BM02
BM03
BM04
BM05
BM06
BM06b
BM07
BM08
EXP DD
Thickness [mm]
Curvilinear distance from the cup center [mm]
1.20
1.15
1.10
1.05
1.00
0.95
0.90
0.85
0.80
0.75
0.70
0
5
10
15
20
25
30
35

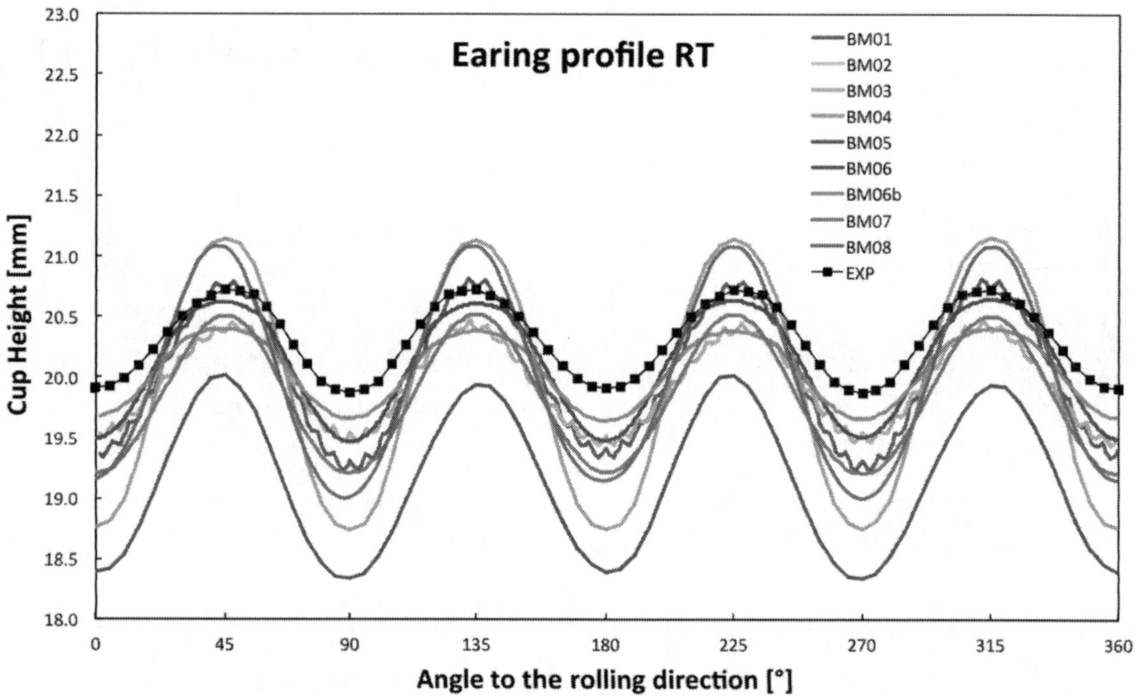
Earing profile RT
BM01
BM02
BM03
BM04
BM05
BM06
BM06b
BM07
BM08
EXP
Cup Height [mm]
Angle to the rolling direction [°]
23.0
22.5
22.0
21.5
21.0
20.5
20.0
19.5
19.0
18.5
18.0
0
45
90
135
180
225
270
315
360

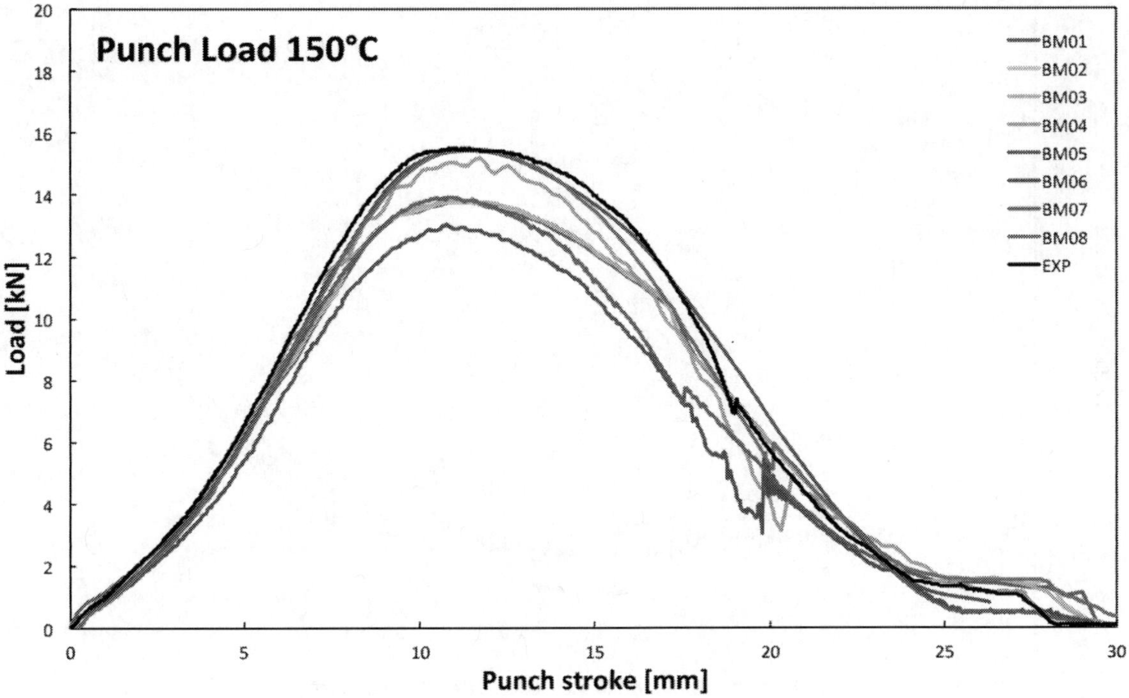
Punch Load 150°C
Load [kN]
Punch stroke [mm]
BM01
BM02
BM03
BM04
BM05
BM06
BM07
BM08
EXP
0
2
4
6
8
10
12
14
16
18
20
0
5
10
15
20
25
30

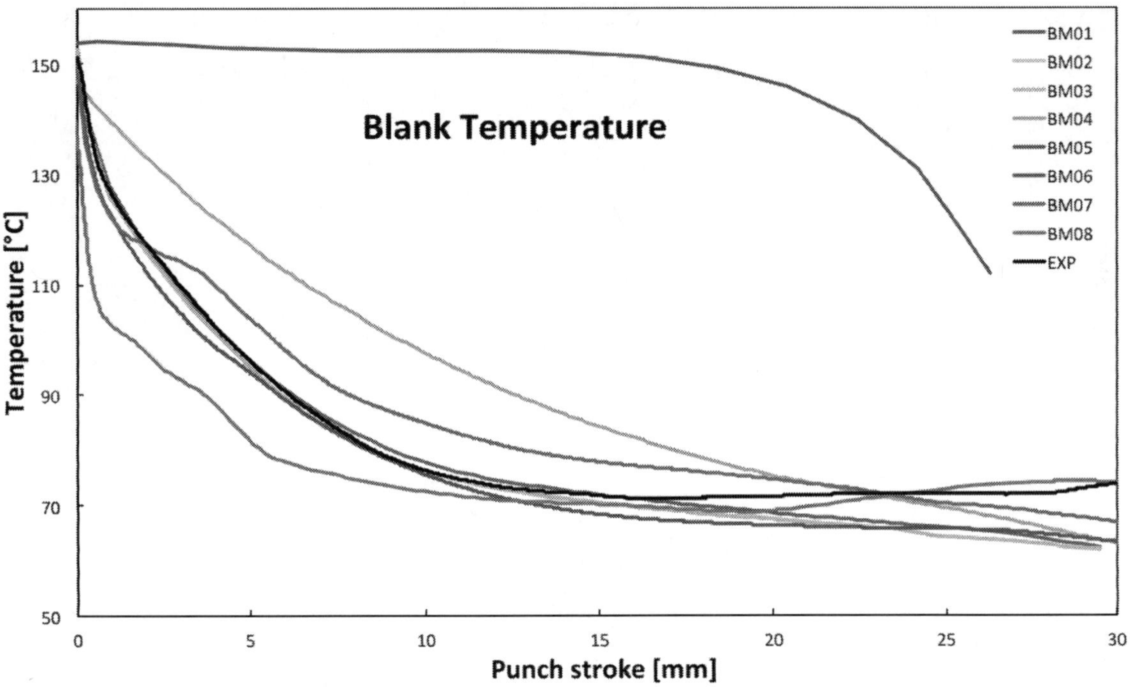
Blank Temperature
Temperature [°C]
Punch stroke [mm]
BM01
BM02
BM03
BM04
BM05
BM06
BM07
BM08
EXP
50
70
90
110
130
150
0
5
10
15
20
25
30

Journal of Physics: Conference Series **734** (2016) 022003 doi:10.1088/1742-6596/734/2/022003

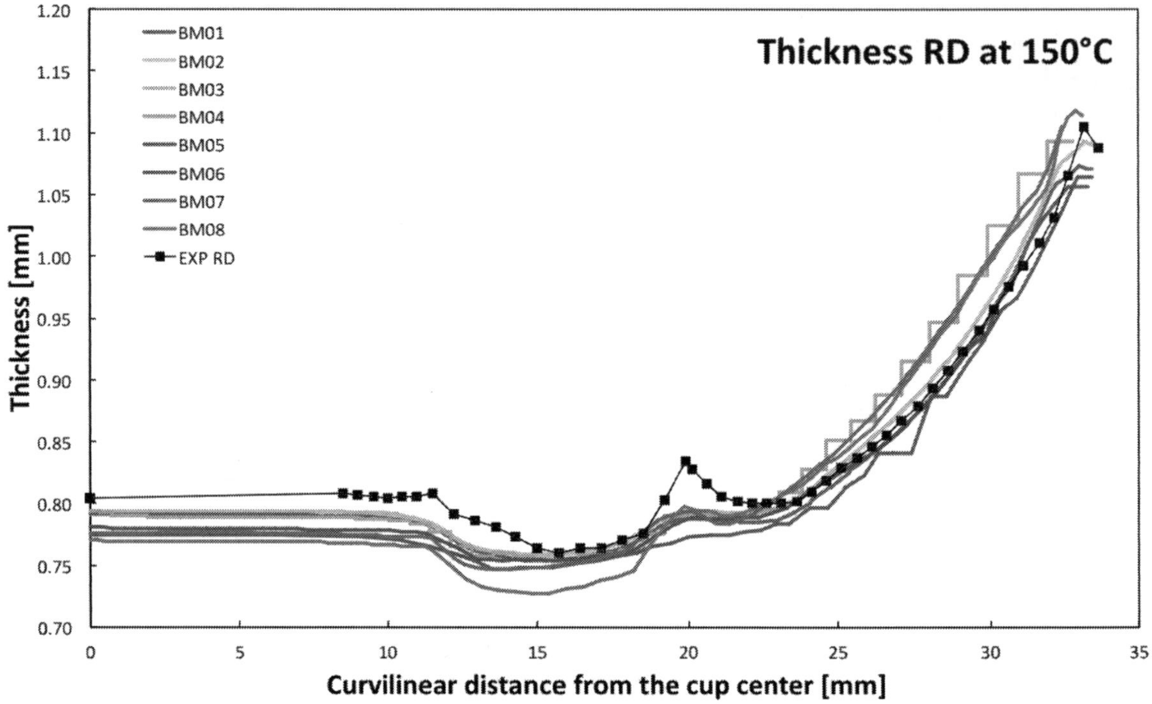

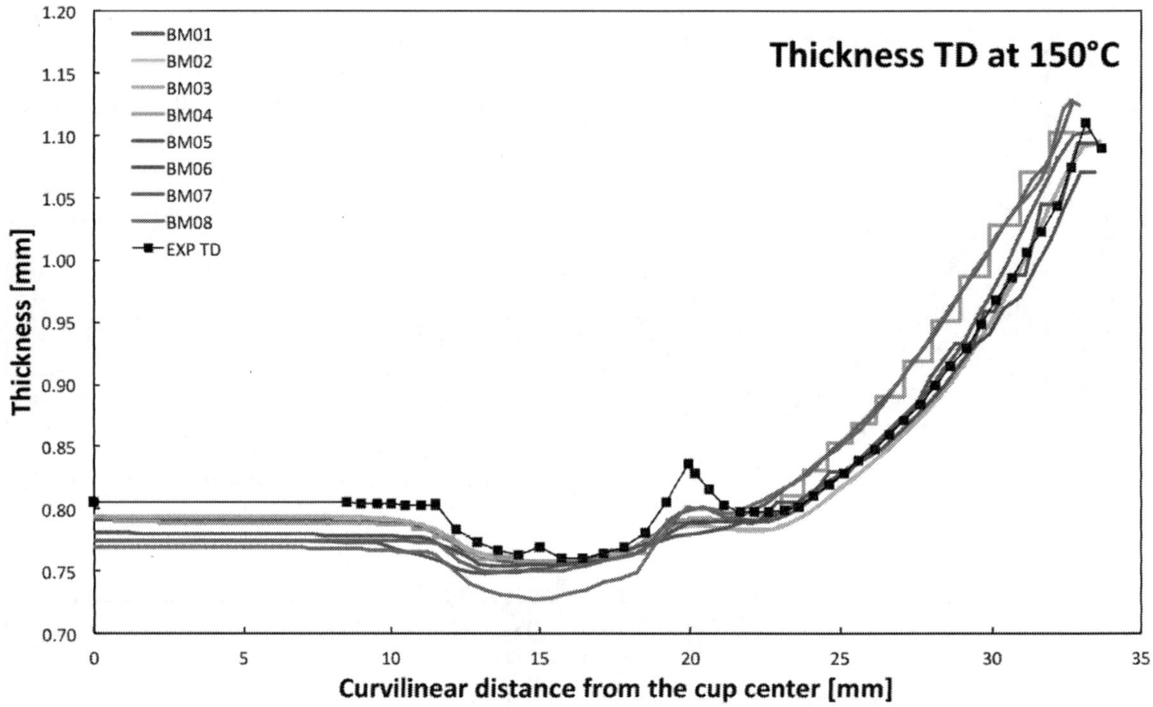

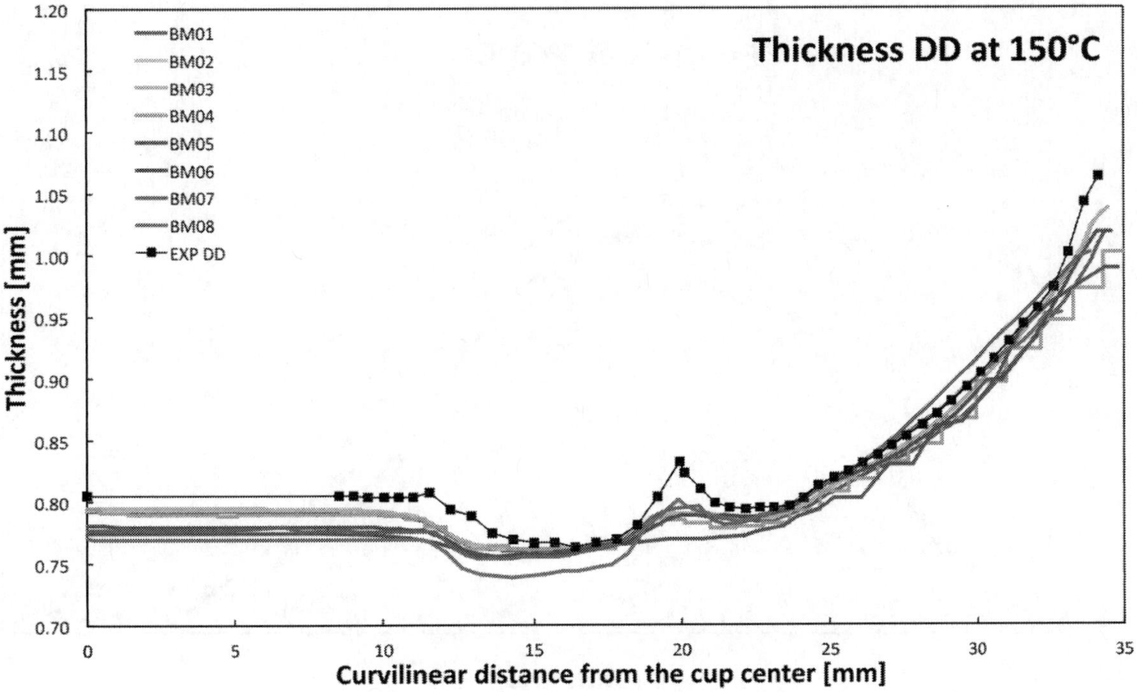
Thickness DD at 150°C
BM01
BM02
BM03
BM04
BM05
BM06
BM07
BM08
EXP DD
Thickness [mm]
1.20
1.15
1.10
1.05
1.00
0.95
0.90
0.85
0.80
0.75
0.70
0
5
10
15
20
25
30
35
Curvilinear distance from the cup center [mm]

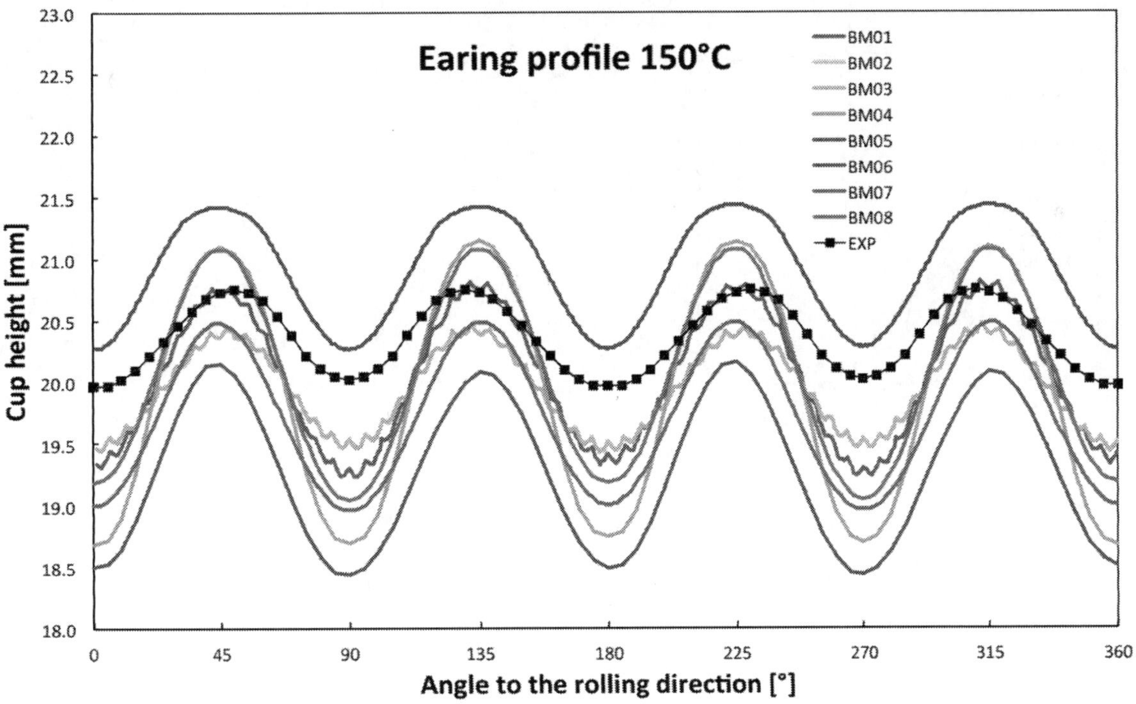
Earing profile 150°C
BM01
BM02
BM03
BM04
BM05
BM06
BM07
BM08
EXP
Cup height [mm]
23.0
22.5
22.0
21.5
21.0
20.5
20.0
19.5
19.0
18.5
18.0
0
45
90
135
180
225
270
315
360
Angle to the rolling direction [°]

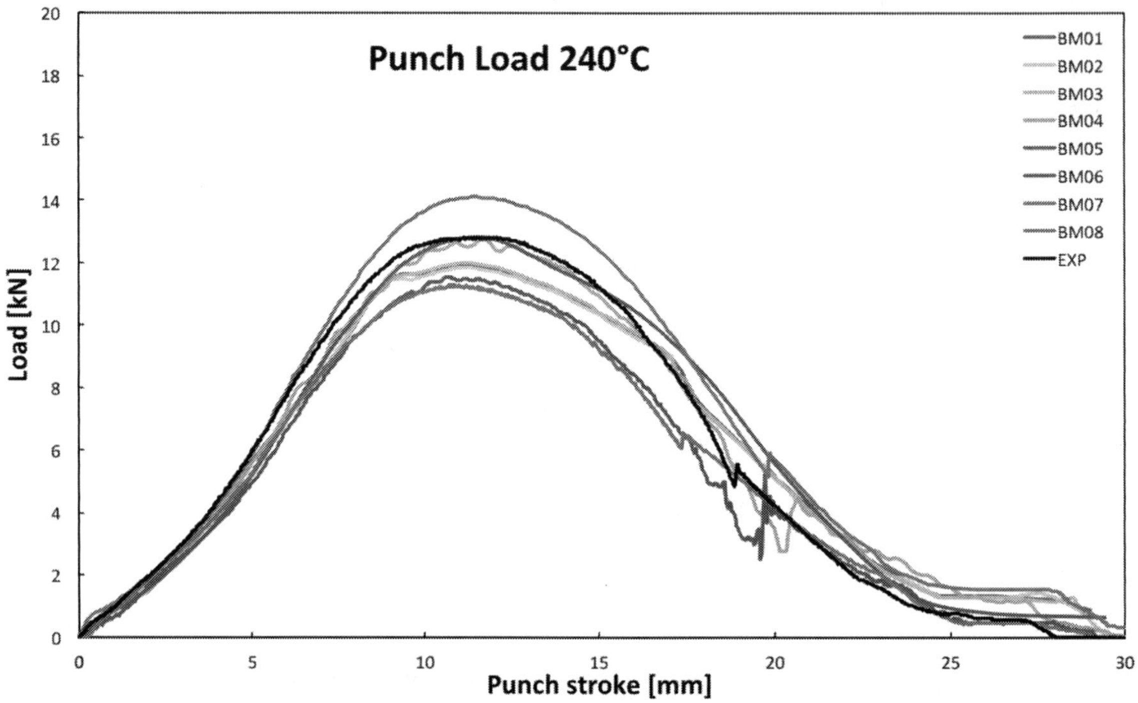
Punch Load 240°C
Load [kN]
Punch stroke [mm]
BM01
BM02
BM03
BM04
BM05
BM06
BM07
BM08
EXP

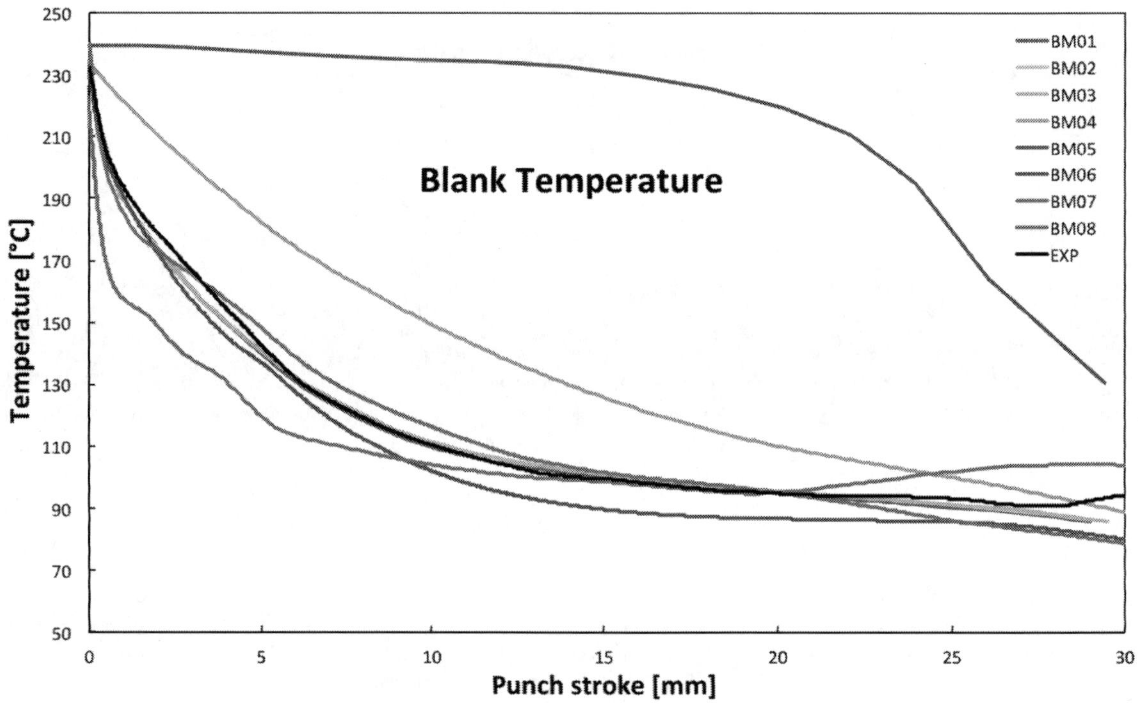
Blank Temperature
Temperature [°C]
Punch stroke [mm]
BM01
BM02
BM03
BM04
BM05
BM06
BM07
BM08
EXP

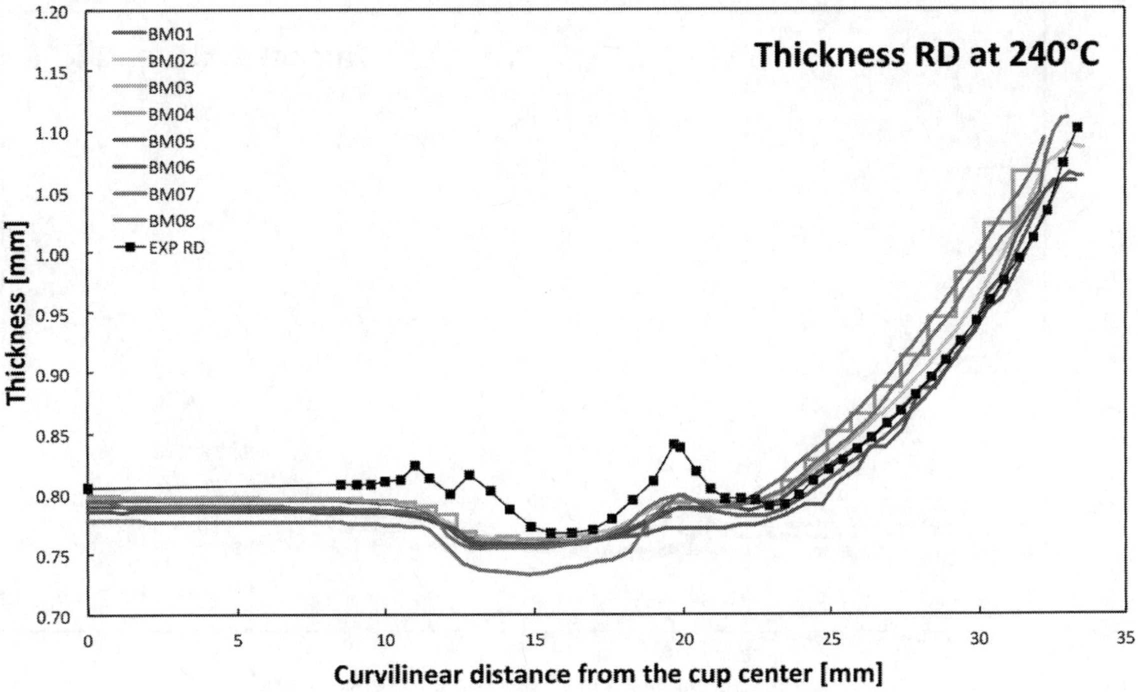
Thickness RD at 240°C
BM01
BM02
BM03
BM04
BM05
BM06
BM07
BM08
EXP RD
Thickness [mm]
1.20
1.15
1.10
1.05
1.00
0.95
0.90
0.85
0.80
0.75
0.70
0
5
10
15
20
25
30
35
Curvilinear distance from the cup center [mm]

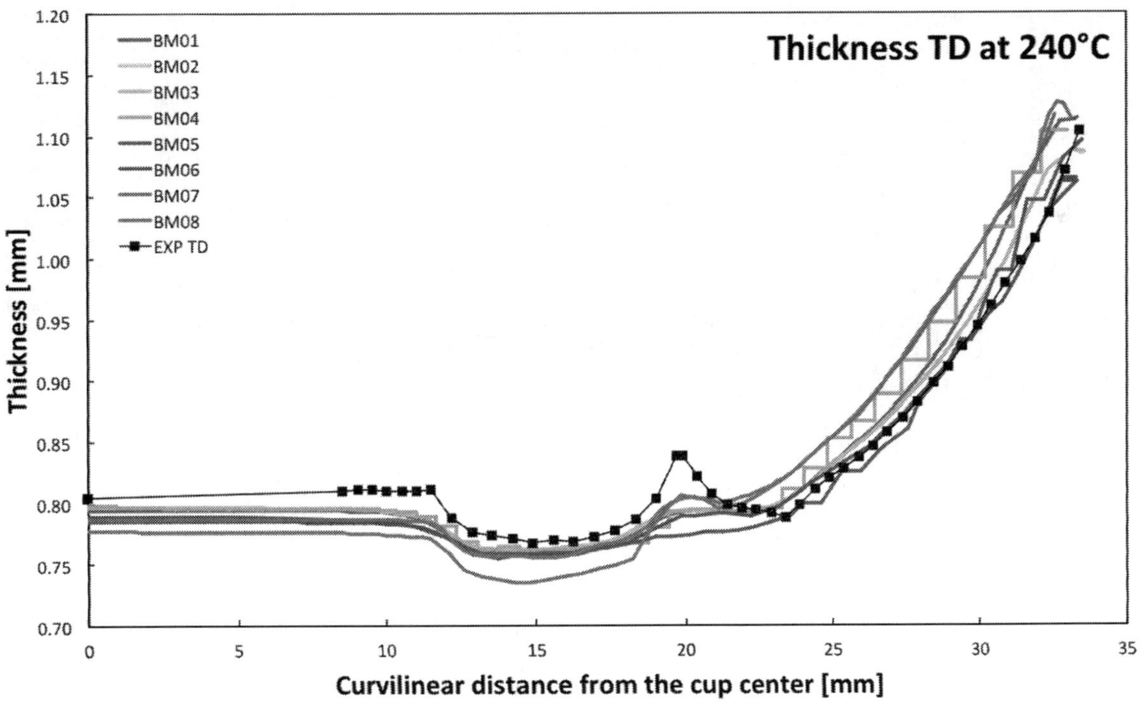
Thickness TD at 240°C
BM01
BM02
BM03
BM04
BM05
BM06
BM07
BM08
EXP TD
Thickness [mm]
1.20
1.15
1.10
1.05
1.00
0.95
0.90
0.85
0.80
0.75
0.70
0
5
10
15
20
25
30
35
Curvilinear distance from the cup center [mm]

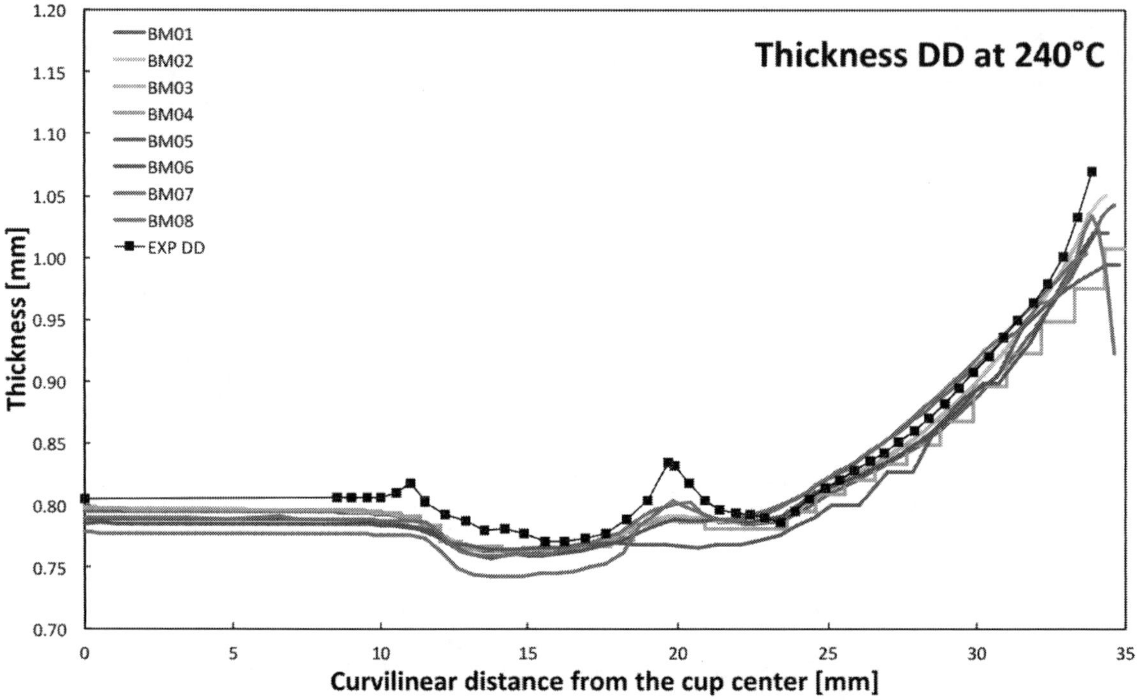
Thickness DD at 240°C
BM01
BM02
BM03
BM04
BM05
BM06
BM07
BM08
EXP DD
Thickness [mm]
1.20
1.15
1.10
1.05
1.00
0.95
0.90
0.85
0.80
0.75
0.70
0
5
10
15
20
25
30
35
Curvilinear distance from the cup center [mm]

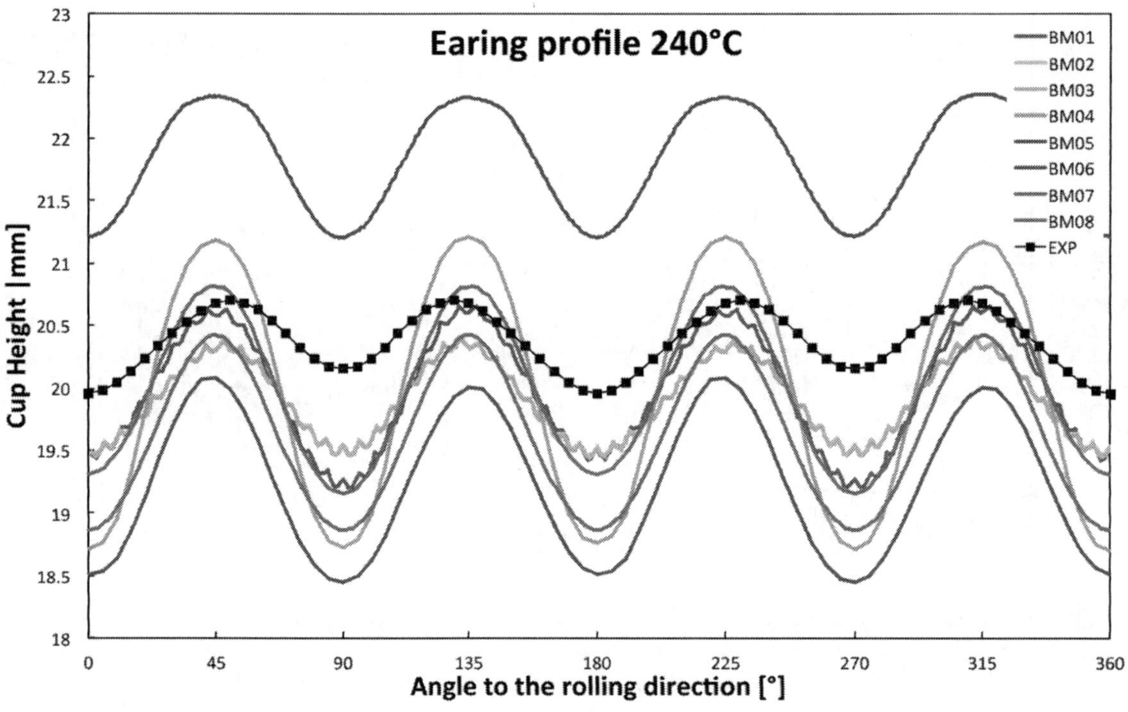
Earing profile 240°C
BM01
BM02
BM03
BM04
BM05
BM06
BM07
BM08
EXP
Cup Height [mm]
23
22.5
22
21.5
21
20.5
20
19.5
19
18.5
18
0
45
90
135
180
225
270
315
360
Angle to the rolling direction [°]

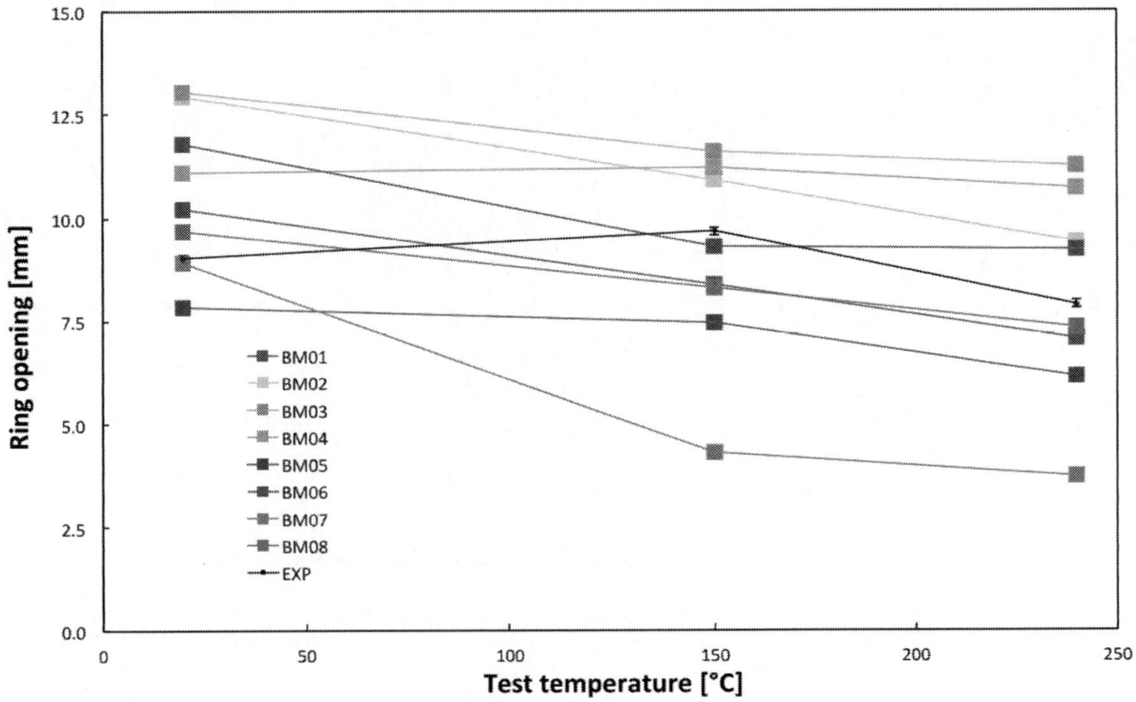
15.0
12.5
10.0
7.5
5.0
2.5
0.0
Ring opening [mm]
0
50
100
150
200
250
Test temperature [°C]
BM01
BM02
BM03
BM04
BM05
BM06
BM07
BM08
EXP

Modeling strategy for clinched joints in assemblies

A Breda, S Coppieters and D Debruyne

Departement of Materials Engineering, KU Leuven technology campus Ghent, Gebroeders De Smetstraat 1, Ghent 9000, Belgium

E-mail: andreas.breda@kuleuven.be

Abstract. Clinching is a mechanical joining technique which involves severe local plastic deformation of two or more metal sheet parts resulting in a permanent mechanical interlock. Today, it is a reliable joining technique used in automotive, HVAC and general steel constructions whilst still gaining interest. As it is not computationally feasible to include detailed sub models of these type of joints in FE simulations of real-life clinched assemblies, this paper proposes a methodology to represent these connections with simplified elements. In order to calibrate the parameters governing the equivalent model, a simple shear lap and pull-out test is used. This methodology is applied to clinched configurations and validated using a modified Arcan test in which both shear and pull-out loads are considered.

1. Introduction

For both economical and environmental point of view, lightweight constructions have gained more interest in recent years. The need to join dissimilar, coated or hard to weld lightweight materials, led to rapid development of mechanical joining techniques such as clinched joints. Clinching is a mechanical joining technique that involves severe local plastic deformation of two or more sheet metal parts, resulting in a permanent mechanical interlock.

The forming process and the mechanical performance of a single clinched joint have been extensively investigated by finite element simulations [1, 8, 9]. The clinched region is a complex shaped zone where the material state varies from point to point. If a structure or assembly contains many such joints, it is unrealistic because of the computational costs and means to build a numerical model containing a huge number of detailed sub-models. The goal in this work is to obtain a methodology to replace the complex full-scale clinch model in numerical simulations for quasi-static loading. The focus is on the global force displacement response up to maximum force. Therefore a relatively simple connector element can be used. Simplified models are already successfully used in other joining techniques such as bolts, spot welds, self piercing rivets etc. Today, however, the possibilities in this field for clinched joints have not yet been thoroughly investigated.

2. Methodology

2.1. Experimental reference case

DC01 sheet metal was selected because it has excellent deep drawing properties, is widely available as sheet metal and is therefore ideal to use as clinching material. The material properties of DC01 sheet metal (thickness 1 mm) were obtained by means of an uni-axial tensile test. This material was joined with the Non Cutting Single Stroke (NCSS) clinch technology

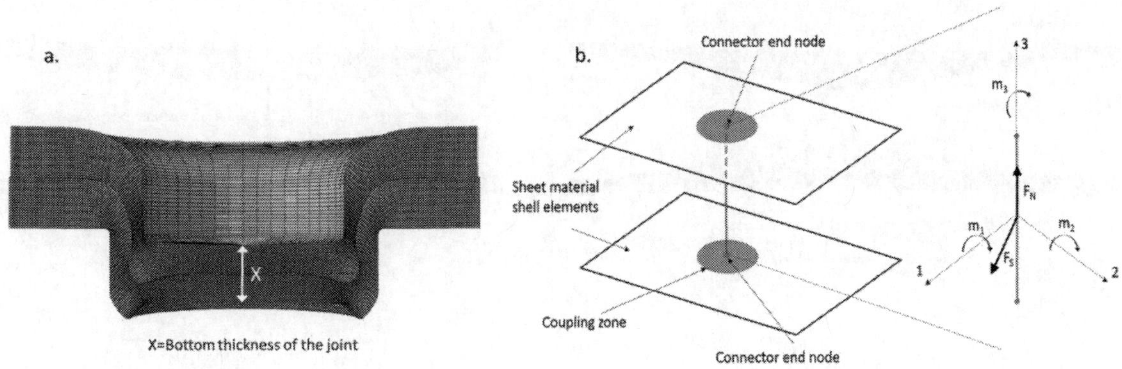

Figure 1. Principle of equivalent modeling: a. full scale model b. equivalent model

using an extensible die. The average clinch diameter was 8 mm with a bottom thickness of X=0.55 mm at the base of the joint. In order to calibrate the numerical models, a reference case is necessary. Therefore a simple pull-out and shear lap experiment are performed on a clinched specimen. These experiments will be used to calibrate the equivalent model.

2.2. Equivalent model representation

For modeling clinched joints a connector formulation is used. This formulation has been proven to be optimal in terms of computational cost compared to a virtual element formulation [6]. The connector formulation consists of a beam like connection with a local coordinate system and 6 degrees of relative motion (U1,U2,U3,UR1,UR2,UR3), all in which elastic and plastic behaviour can be introduced. The end nodes of this connector are coupled to a coupling zone onto the shell elements which acts as the clinched joint influence area on the sheet material (Fig. 1 b.). To couple the end nodes with the influence zone, Weyer S et al. [5] and Grujicic M et al. [4], suggested to use a structural distributed coupling. This coupling divides the moments and forces acting on the connector over the elements of the influence area which can cause local deformation of the coupling zone. As a result, the model will have a lower stiffness compared with the experimental results (Fig. 2 a.). It can be concluded that the influence zone of the clinched joint can be represented as a very stiff zone inside the sheet material which restricts the sheet deformation in this zone. The use of a kinematic coupling is therefore a good approximation for a clinched joint (Fig. 2 b.). The radius of the influence zone is assumed to be equal to the clinch joint maximum radius.

2.3. Connector elastic behaviour calibration

Uncoupled elastic behaviour will be used. This means the spring stiffness's D_{ii} are defined independently for each load direction. Following equation is used to calculate the elastic response of the connector:

$$F_i = D_{ii} \cdot u_i \tag{1}$$

To calibrate the connector spring stiffness's D_{11}, D_{22} and D_{33}, respectively the two shear and the normal directions of the connector, a rigid stiffness is initially assumed as connector section in both numerical models. Through comparison with the experimental force displacement curves of the shear lap and pull-out tests, the connector stiffness is calibrated. Due to the limited bending of the clinched joint, D_{44} and D_{55} (bending around the radial axes) are assumed to be

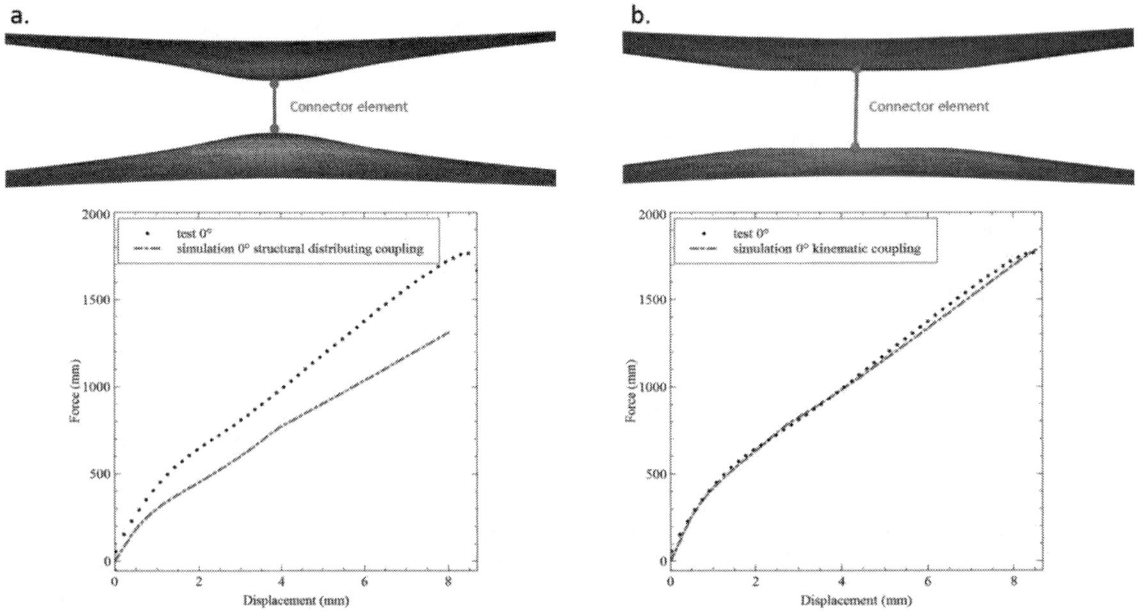

Figure 2. Coupling methods: a. structural distributing coupling b. kinematic coupling

rigid. The bending around the normal axis, D_{66}, is assumed to be free as the joint can rotate around the normal axis with limited amount of resistance.

2.4. Connector plastic behaviour calibration

To describe the plastic behaviour of the connector element, a independent force displacement based hardening law is defined for each of the main load directions (shear and normal) exerting on the connector. For mixed-mode load conditions, the connector acts as the resultant of both hardening laws. These hardening laws are calibrated using the experimental results of the reference pull-out and shear lap tests. The equivalent yielding force can directly be derived from matching the elastic connector simulation with the experimental reference results (equation 2 and 4). The plastic displacements however, retrieved from the experimental reference cases, need scaling factors $K_{u,N}$ and $K_{u,S}$ to determine the correct displacements at the connector end nodes (equation 3 and 5). The experimentally measured plastic displacement components require to be scaled to the plastic displacement at the connector end nodes. Indeed, it is physically impossible to obtain the latter values directly from the reference experiments, and, consequently, a scaling factor K needs to be introduced. The specimen geometry and type of test determine the size of the scaling factor. By matching the simulations with the reference case experiments, the scaling factors and hardening laws are determined. The resulting connector hardening laws are independent of the specimen geometry and size.

For the normal direction:

$$F_N = f_3 \tag{2}$$

$$u_N^{pl} = u_3^{pl} \cdot K_{u,N} \tag{3}$$

For the shear direction:

$$F_S = f_1 = f_2 \tag{4}$$

$$u_S^{pl} = u_1^{pl} \cdot K_{u,S} \tag{5}$$

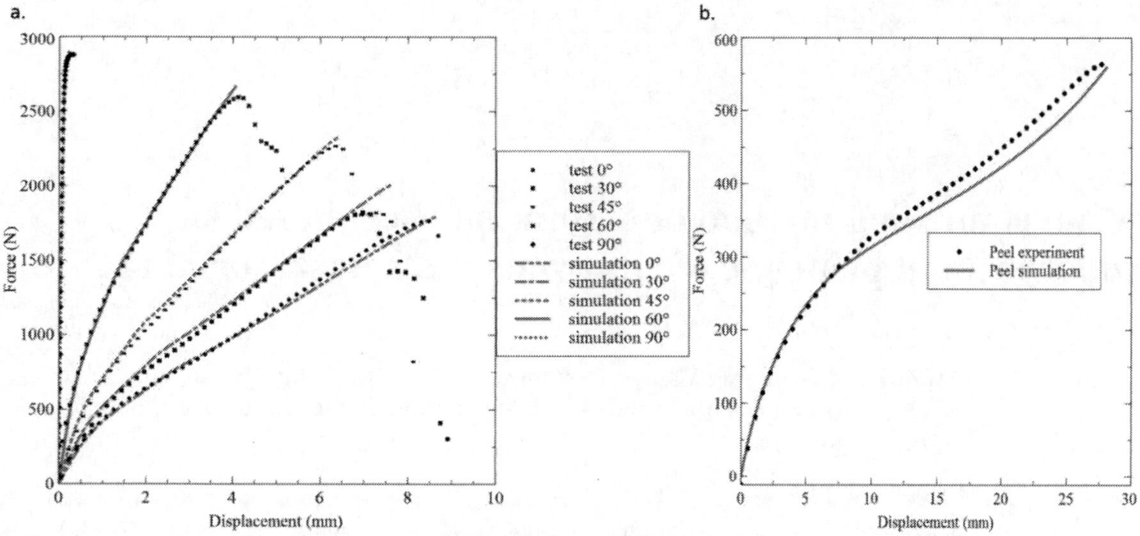

Figure 3. Results: a. DC01 Arcan test results b. DC01 Peel test results

3. Validation

In practice, the clinched joint will be exposed to a combination of shear and pull-out loads. In order to validate the calibrated joint behaviour, mixed-mode load conditions need to be applied on the joint. To achieve that, a modified Arcan test [7, 8] was performed for 3 different load cases (30°, 60°, 45°). Our in-house produced modified Arcan fixture was mounted in a single axis tensile machine. This test has been widely accepted as validation method for equivalent models [2, 3, 5]. Digital Image Correlation (DIC) [10] was used to track the displacements of the Arcan fixture. The simulation results of the Arcan test (Fig. 3 a.) show that the presented methodology for clinched joints gives a accurate prediction of the global force displacement behaviour of the clinched specimen up to maximum force. Additionally, the influence of bending moments was investigated using a peel test on a clinched specimen (Fig. 3 b.). Comparing the results of the simulation with the experiment, it can be concluded that the moments acting onto the joint have no substantial influence on the intrinsic deformation behaviour of the joint as the equivalent formulation does not include bending moments and is still able to reproduce the force displacement response.

References

[1] He X 2010 *Int. J. Adv. Manuf. Tech.* **48** Iss: pp 607-12
[2] Langrand B, Patronelli L, Deletombe E, Markiewicz E, Drazétic P 2002 *Aerosp. Sci. Technol.* **6** pp 343 - 54
[3] Hanssen A G, Olovsson L,Porcaro R, Langseth M 2010 *Eur. J. Mech. A-Solid.* **29** pp 484 - 95
[4] Grujicic M, Snipes J, Ramaswami S, Abu-Farha F 2014 *Multidiscipline Model. Mater. Struct.* **10** pp 631 - 658
[5] Weyer S, Hooputra H, Zhou F 2006 Modeling of self-piercing rivets using fasteners in crash analysis *ABAQUS users conference paper*
[6] Bérot M, Malrieu M, Bay F 2014 *Eng. Computation.* **31** pp 453 - 66
[7] Porcaro R, Hanssen A G, Langseth M, Aalberg A 2005 *Int. J. Solids. Struct.* **43** pp 5110 - 31
[8] Coppieters S, Van Hecke R, Cooreman S, Sol H, Van Houtte P, Debruyne D 2013 *Int. J. Mater. Form.* **6** pp 437 - 51
[9] Coppieters S, Cooreman S, Lava P, Sol H, Van Houtte P, Debruyne D 2011 *Int. J. Mater. Form.* **4** pp 429 - 40
[10] MatchID 2015 *www.matchidmbc.com*

A novel forming method of aluminum sheet based on superposition principle of electromagnetic local forming

Anlin Long, Wenping Wang[1], Chunping Fang, Xiangdong Wu and Min Wan
School of Mechanical Engineering and Automation, Beihang University, Beijing 100191, P R China

E-mail: anlinlong@buaa.edu.cn

Abstract: A novel forming method aimed at large size aluminum sheet is introduced. In this method, aluminum sheet is clamped between a common flat spiral electromagnetic actuator and a punch matrix. Driven by pulsed electromagnetic force and restrained by punch matrix, the blank sheet will form local shallow dome accordingly. Moving actuator along sheet surface and triggering the pulsed power generator in sequence results in dome matrix that is uniformly distributed. Superposition of those spherical domes leads to macroscopically curved figuration of aluminum sheet. This paper demonstrates the newly proposed method experimentally, which verifies its feasibility. The mechanism analysis of this method is also presented using a simplified analytical model. The results show that this novel method is feasible and can be explained by the proposed mechanism well.

1. Introduction

Electromagnetic forming process is one of high speed forming processes. Compared to conventional quasi-static forming operations, researches show that formability of metallic materials such as steel [1], aluminum alloy [2], magnesium alloy [3] and titanium alloy [4] is dramatically improved due to the high speed of this forming process. Furthermore, it can inhibit springback of workpiece significantly. More detailed advantages are concluded by Psyk V et al. [5].

A lot of researchers are interested in electromagnetic forming field because of those advantages. However, limited by electromagnetic actuator and capacitor bank, they lack research on electromagnetic forming of large scale workpiece. Cui X et al. [6] proposed a novel electromagnetic incremental forming for aluminum alloy sheet and tube. The results show that this newly proposed technology is feasible for the manufacturing of large parts.

This work also aims at forming large scale workpiece using pulsed electromagnetic force. A novel forming method of aluminum sheet based on the superposition principle of electromagnetic local forming is proposed. A regular electromagnetic coil and punch matrix that is used in multi-point forming technology [7] are adopted in this method. Different from the former work [6], no macroscopic deformation appears during the forming procedure. The specimen becomes cambered only if it is released from clamps after this procedure.

This paper demonstrates the novel forming method using detailed experimental procedure. Deflections and several sectional profiles of the specimens are measured too. A simplified qualitative analysis model is then adopted to illustrate the mechanism of this newly proposed forming method.

[1] Corresponding author: Wenping Wang, Ph. D., Associate Professor, School of Mechanical Engineering and Automation, Beihang University, Beijing 100191, P. R. China, E-mail: wwp_2003@163.com

2. Experiments

2.1. Experimental procedure
As shown in figure 1, during experimental procedure, the clamps fasten specimens and keep them horizontal. Actuator and punch matrix are then positioned along the length direction of sheets synchronously. Once they are positioned correctly, the pulsed power generator will be triggered and the electromagnetic local forming at that position is accomplished. Moving those tools to next position and triggering the generator again, and so forth. The synchronous feeding distance of actuator and punch matrix is input into PLC by a Human Machine Interface and detected by a grating scale, respectively.

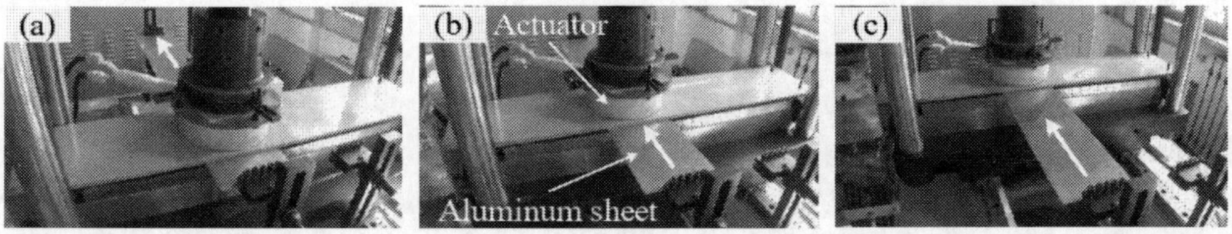

Figure 1. Positioning procedure of experiments. Beginning position (a), middle position (b) and ending position (c). The thicker arrows mark the positioning direction.

2.2. Experimental parameters
The 2xxx series of aluminum alloy are widely used in aerospace field because of their nice comprehensive properties. Thus 2524-T3 aluminum alloy sheets were then investigated in this work. Dimensions of the specimens were set to be 900mm×150mm×1.6mm. Table 1 shows more detailed parameters. Because effective outer diameter of the actuator (149mm) and width of the sheets (150mm) are almost the same, the actuator and punches were not moved along the width direction of sheets. They were kept symmetry with specimen in that direction during the forming procedure.

Table 1. The detailed parameters of forming experiments.

No.	Discharging voltage (V)	Feeding distance (mm)	Cushion thickness (mm)	Epoxy sheet thickness (mm)
1	2500	30	2	10
2	5500	30	8	10

3. Results and discussions
Figure 2 shows the figuration of specimens. The clamping allowances of these specimens were trimmed away. Thus only the deformed part of each specimen is exhibited, which is in the size of 600mm×150mm×1.6mm. The specimen number is in accordance with experiment number listed in table 1. An undeformed sheet is inserted to illustrate deformation of specimen vividly. According to this picture, these two specimens bend towards different directions. Specimen 1 bends towards the punches side while specimen 2 bends towards the coil side. The maximum deflections of specimens are amplified and demonstrated on the right side. It shows that specimen 1 has smaller deflection which approximately equals 9.5mm. And specimen 2 approximately reaches 15.5mm.

In order to analyse forming mechanism, detailed surface features are then taken into consideration. Specially, some profiles are extracted along several certain paths. A hand held laser scanner is used to obtain the three dimensional coordinate data of these specimens in STL format. Importing the files into commercial 3D CAD software CATIA, the reconstitution models and extracting paths are demonstrated in figure 3. The paths are located in the surfaces adjacent to coil. A shallow dome matrix can be observed on specimen 2 in case of careful observation.

Based on the extracted point cloud data, two sectional profiles are then fitted by high order curves. The maximum deviation between point cloud and fitted curve is set to be 0.03mm because of the measuring accuracy of laser scanner. Figure 4 shows the porcupine curvature analysis of these profiles. The profile number is in accordance with path number shown in figure 3.Comb length represents the curvature magnitude at that point. And the combs are shown in outward normal direction of each sub-segment. Some extremum values are labelled.

Macroscopically, profile 1 is a concave curve while profile 2 is a convex curve. However, the wavy envelop of combs means that these profiles are not smooth enough but locally wavy. For profile 1, sub-segments such as 1a, 1c, 1e, 1g and 1i have smaller curvature than the adjacent segment. On the contrary, for profile 2, sub-segments such as 2a, 2c, 2e, 2g and 2i have larger curvature than the adjacent segment. It should be noticed that the macroscopic bending direction of each profile is in accordance with sub-segment that has larger curvature. Generally, profile 1 has smaller curvature than profile 2, which means it is smoother.

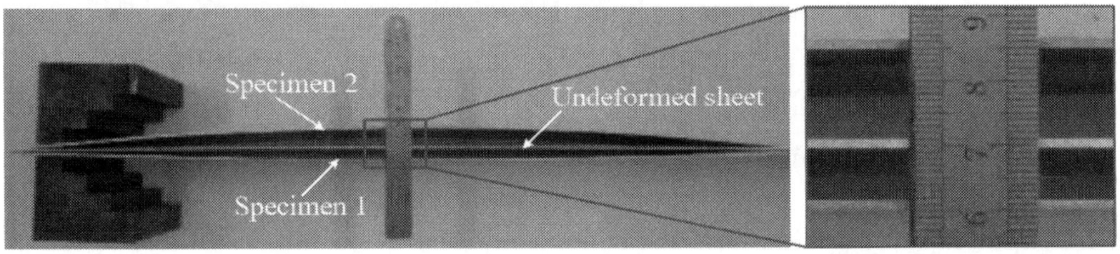

Figure 2. Figuration of the specimens. These specimens bend towards contrary directions and reach different deflection magnitude.

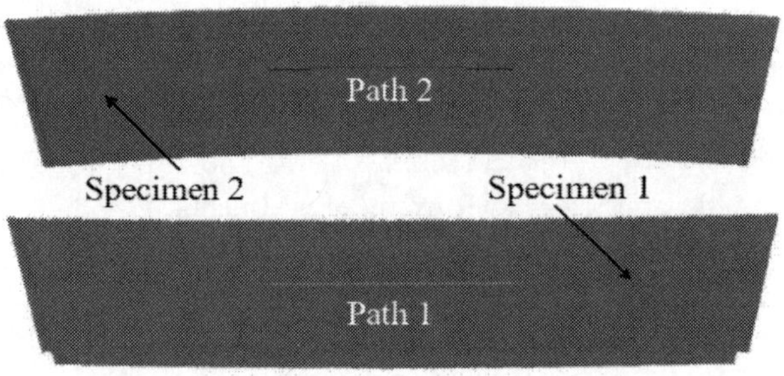

Figure 3. This figure shows the reconstitution models of specimens and the corresponding extracting path. Shallow dome matrix can be observed in specimen 2.

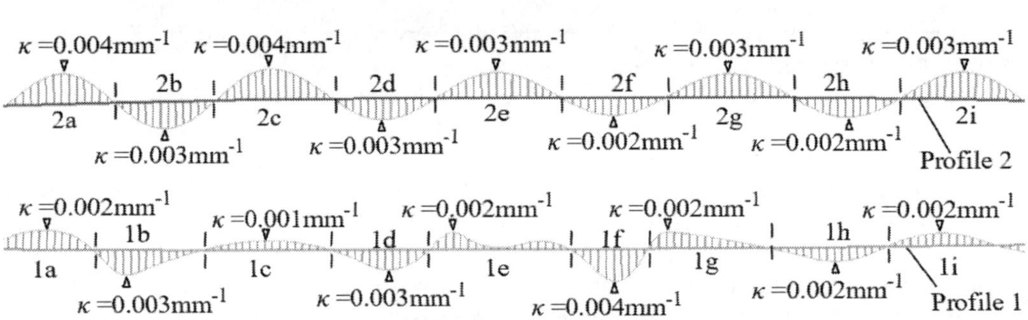

Figure 4. Porcupine curvature analysis of the profiles.

Base on the extracted sectional profiles, a simplified schematic mechanism diagram of this method is demonstrated in figure 5. It shows a part of the cross section of aluminum sheet during forming procedure. The rubber cushion is omitted. Furthermore, this section is divided into several subsections. Subsection 1 and subsection 2 bend towards contrary directions. The neutral surface radii of these subsections are represented by R and r, respectively. It should be noticed that curvature κ is the reciprocal of radius R or r.

According to theory of plates and sheets, smaller radius leads higher bending stress. It means that subsection is easier to generate larger plastic deformation magnitude. The forming mechanism depends on the relative magnitude relationship between R and r. In case that R is bigger than r, subsection 1 will generate smaller plastic deformation while subsection 2 gets a bigger one. When the symmetric boundary condition is released, this section will bend towards the punches side. Superposition of these local bending lead macroscopic deformation of the specimen, which also bends towards the punches side. Similarly, if R is smaller than r, the specimen will bend towards the coil side.

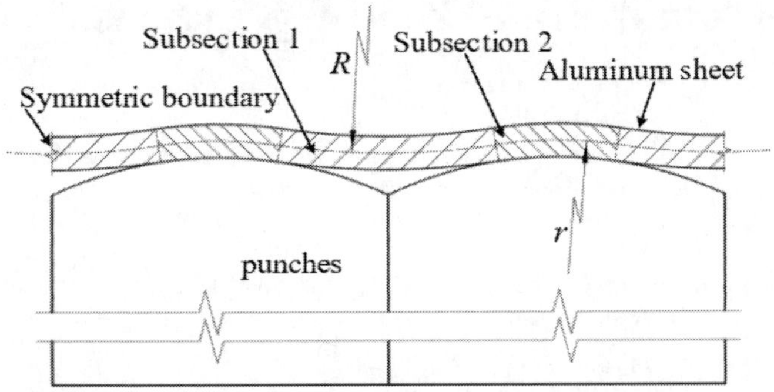

Figure 5. A simplified schematic mechanism diagram of this novel forming method.

This simplified qualitative model can be popularized to actual situation. During the forming procedure, sheets are driven by electromagnetic force and restrained by punch matrix. Because of spherical end of punch, local bulging with rigid punch is achieved. On the other side, a local electromagnetic bulging is also generated. They result in shallow domes when the bulging stress is large enough for plastic deformation. The final figuration of specimen is determined by superposition of these local domes and relative magnitude relationship between these neutral surface spherical radii. Based on the figurations of specimens shown in figure 2 and detailed information demonstrated in figure 4, the forming mechanism of this novel method is verified to be correct.

4. Conclusions

Based on the experiments carried out and the simplified analysis model, this forming method is feasible. Thus a novel forming method of aluminum sheet is proposed. It extends the application of electromagnetic force. Discharging voltage of capacitor bank and rubber thickness will determine the figuration of specimen. The effects of such parameters on specimen figuration and its surface integrity are prepared for further research.

5. Acknowledgments

This work was supported by the National Basic Research Program of China (973 Program, Grant No. 2011CB012804) and the National Natural Science Foundation of China (Grant No. 51275026). The Authors wishing to acknowledge assistance of these programs for the financial support.

6. References

[1] Seth M, Vohnout V J and Daehn G S 2005 *J. Mater. Process. Technol.* **168** 390-400
[2] Golovashchenko S F 2007 *J. Mater. Eng. Perform* **16** 314-20
[3] Xu J, Yu H, Cui J and Li C 2013 *Mat. Sci. Eng.* A **569** 150-58
[4] Li F, Mo J, Li J, Huang L and Zhou H 2013 *Mater. Des.* **52** 337-44
[5] Psyk V, Risch D, Kinsey B L, Tekkaya A E and Kleiner M 2011 *J. Mater. Process. Technol.* **211** 787-829
[6] Cui X, Li J, Mo J, Fang J, Zhou B, Xiao X and Feng F 2016 *J. Mater. Process. Technol.* **233** 79-88
[7] Li M, Cai Z, Sui Z and Yan Q 2002 *J. Mater. Process. Technol.* **129** 333-38

Hencky's model for elastomer forming process

A A Oleinikov[1] **and A I Oleinikov**[2,3]

[1] Faculty of informational technologies, mathematics and physics, Amur State University of Humanities and Pedagogy, Komsomolsk-na-Amure, 681000, Russia
[2] Research and Technology Center, Central Aerohydrodynamic Institute, Zhukovsky, MR, 140180, Russia
[3] Department of Aeromechanics and Flight Engineering, Moscow Institute of Physics and Technology, MR, 140180, Zhukovsky, Russia

E-mail: andriy.oleinikov@mail.ru

Abstract. In the numerical simulation of elastomer forming process, Henckys isotropic hyperelastic material model can guarantee relatively accurate prediction of strain range in terms of large deformations. It is shown, that this material model prolongate Hooke's law from the area of infinitesimal strains to the area of moderate ones. New representation of the fourth-order elasticity tensor for Hencky's hyperelastic isotropic material is obtained, it possesses both minor symmetries, and the major symmetry. Constitutive relations of considered model is implemented into MSC.Marc code. By calculating and fitting curves, the polyurethane elastomer material constants are selected. Simulation of equipment for elastomer sheet forming are considered.

1. Introduction

The greatest problem of machine building consists in permanent upgrading of the manufacturing methods. The main idea behind the process of equipment development is to replace the die forming, that is based on rigid dies and punches, by flexible dies and punches variable in forming. In the equipment, that is considered, the die and punch are formed as a system of coaxial stem rods with the blank being secured between them. The high points of the rods, contacting to the panel blank are defined by Henckys elasticity shaped as hemispheres in unloading state. Such equipment ensures die forming simultaneously for both principal curvatures without development of cylindrical stiffness and allows parts with double curvature to be manufactured within one process pass. The paper deals with simulation of equipment actuator operation of industrial equipment for the pressure treatment of materials, including elastomer forming process.

It is well known, that application of complex material models that are efficient in all the range of elastomers deforming needs accurate experiment definition and huge work of parameters searching for description of experimental curves. It is shown, that the Hencky's isotropic hyperelastic material model provide good approximation of elastomers strains up to 50 % of scale.

2. Hencky's isotropic hyperelastic constitutive material model

Currently theoretical fundamentals of the hyperelastic theory are well developed, but there are many open questions yet, like constructing more efficient forms of equations in terms of their usage in the applications. In the present work new approach for formulation of constitutive relations of Hencky's isotropic hyperelastic material is suggested. This approach is stated on tensors representations through eigenprojections of right Cauchy–Green strain tensor $\mathbf{C}$.

2.1. Basic kinematics
Introduce the velocity vector $\mathbf{v}$ and the velocity gradient $\mathbf{l} \in \mathcal{T}^2$

$$\mathbf{v} = \dot{\mathbf{x}}, \quad \mathbf{l} \equiv \operatorname{grad} \mathbf{v} \equiv \mathbf{r}_F = \dot{\mathbf{F}} \cdot \mathbf{F}^{-1}. \tag{1}$$

Introduce the symmetric Eulerian strain rate tensor $\mathbf{d} \in \mathcal{T}^2_{\text{sym}}$ and the Lagrangian counterpart $\mathbf{D}$ of the tensor $\mathbf{d}$ called the rotated strain rate tensor $\mathbf{D} = \mathcal{R}_{R^T}(\mathbf{d})$

$$\mathbf{d} \equiv \operatorname{sym} \mathbf{l} = \frac{1}{2}(\mathbf{l} + \mathbf{l}^T), \quad \mathbf{D} = \frac{1}{2}(\dot{\mathbf{U}} \cdot \mathbf{U}^{-1} + \mathbf{U}^{-1} \cdot \dot{\mathbf{U}}), \quad \mathbf{d} = \mathbf{R} \cdot \mathbf{D} \cdot \mathbf{R}^T. \tag{2}$$

Let $\mathbf{H} \in \mathcal{T}^2$ is a Lagrangian tensor. Enter the Lagrangian D-rate of the tensor $\mathbf{H}$ [1]:

$$\mathbf{H}^D \equiv \dot{\mathbf{H}} - \boldsymbol{\Omega}^D \cdot \mathbf{H} + \mathbf{H} \cdot \boldsymbol{\Omega}^D, \tag{3}$$

where

$$\boldsymbol{\Omega}^D \equiv \boldsymbol{\Psi}_r(\mathbf{U}, \mathbf{D}) = \sum_{i \neq j = 1}^{m} r(z) \mathbf{U}_i \cdot \mathbf{D} \cdot \mathbf{U}_j. \tag{4}$$

Here $\boldsymbol{\Psi}_r(\mathbf{U}, \mathbf{D})$ — skew-symmetric tensor function, where $z \equiv \lambda_i/\lambda_j$, and scalar-valued function $r(z)$ evaluated in this way:

$$r(z) \equiv \begin{cases} \frac{2z}{1-z^2} + \frac{1}{\ln z}, & \text{if } z \neq 1, \\ 0, & \text{if } z = 1 \; [= \lim\limits_{z \to 1} (\frac{2z}{1-z^2} + \frac{1}{\ln z}). \end{cases} \tag{5}$$

Notice, that the D-rate of the tensor $\mathbf{H}$ possesses remarkable property

$$\mathbf{E}^{(0)\,D} = \mathbf{D}. \tag{6}$$

2.2. Hyperelastic Constitutive Relations
Let consider constitutive relations of the Hencky's hyperelastic isotropic material model in form

$$\bar{\boldsymbol{\tau}} = \lambda (\operatorname{tr} \mathbf{E}^{(0)}) \mathbf{I} + 2\mu \mathbf{E}^{(0)}, \tag{7}$$

where $\bar{\boldsymbol{\tau}}$ is Noll's (Lagrangian) stress tensor, tensor $\mathbf{E}^{(0)}$ called the right logarithmic strain tensor (right (Lagrangian) Hencky strain tensor) [2], λ, μ — Lame constants. Or [3]

$$\bar{\boldsymbol{\tau}} = \mathbb{C}^E : \mathbf{E}^{(0)}, \tag{8}$$

where $\mathbb{C}^E \in \mathcal{T}^4_{\text{sym}}$ — Lagrangian tensor:

$$\mathbb{C}^E \equiv \lambda \mathbb{C}_{\text{I}} + \mu(\mathbb{C}_{\text{II}} + \mathbb{C}_{\text{III}}) = \lambda \mathbb{C}_{\text{I}} + 2\mu \mathbb{S}. \tag{9}$$

For usage in applications, we express the second Piola–Kirchhoff stress tensor $\mathbf{S}^{(2)}$ through the eigenprojections of the right Cauchy–Green strain tensor $\mathbf{C}$, principal stretches λ_i express through the eigenvalues μ_i of the tensor $\mathbf{C}$ ($\mu_i = \lambda_i^2$ $(i = 1, \ldots, m)$).

Constitutive relations of the Hencky's isotropic hyperelastic medium we enter in form

$$\mathbf{S}^{(2)} = \sum_{i=1}^{m} \tau_i/\mu_i \mathbf{C}_i. \tag{10}$$

Enter constitutive relations of the Hencky's material in velocity:

$$\dot{\bar{\boldsymbol{\tau}}} = \mathbb{C}^E : \dot{\mathbf{E}}^{(0)} \quad \Leftrightarrow \quad \bar{\boldsymbol{\tau}}^D = \mathbb{C}^E : \mathbf{D}, \tag{11}$$

where $\bar{\boldsymbol{\tau}}^D$ identify the objective Lagrangian corotational D-rate of the tensor $\bar{\boldsymbol{\tau}}$ (see (3)) [1, 4]. From (11) we derive:

$$\dot{\mathbf{S}}^{(2)} = \mathbb{C} : \dot{\mathbf{E}}^{(2)}, \tag{12}$$

where

$$\mathbb{C} \equiv \sum_{i,j=1}^{m} \frac{\lambda}{\mu_i \mu_j} \mathbf{C}_i \otimes \mathbf{C}_j + \sum_{i=1}^{m} \frac{2(\mu - \tau_i)}{\mu_i^2} \mathbf{C}_i \overset{\text{sym}}{\otimes} \mathbf{C}_i + \sum_{i \neq j=1}^{m} \frac{2}{\mu_i - \mu_j} \left(\frac{\tau_i}{\mu_i} - \frac{\tau_j}{\mu_j} \right) \mathbf{C}_i \overset{\text{sym}}{\otimes} \mathbf{C}_j. \tag{13}$$

The proof of the fact, that derived elasticity tensor possesses both minor symmetries, and also the major symmetry is very useful in applications. This properties have major theoretical meaning in incremental formulations of hyperelasticity, when formulate variational principles for equations, stated with velocities, and criteria of uniqueness and stability of the solutions for this equations.

3. Application of the finite element method for solving of motion equations of deformable bodies of Hencky's isotropic hyperelastic material

It is well known, that performing a finite element analysis (FEA) on a hyperelastic material is difficult due to nonlinearity, large deformations, and material instability. In this paper Lagrangian formulation of Hencky's isotropic hyperelastic material constitutive relations is implemented into FEA software package MSC.Marc, that is utilized for the simulations, based on different hyperelastic material models. This implementation is realized for three-dimensional isoparametric finite elements with usage of Fortran-code of user's subroutine hypela2.f.

In all testing calculations we used constants of the Hencky's material, which are close to constants of the real elastomers: $\nu = 0,45$, $E = 3,37$ MPa, $\rho = 2000$ kg/m^3. Here ν — Poisson's ratio, E — Young's modulus, ρ — mass density of the rod's material. Young's modulus and Poisson's ratio are associated with Lame constants following way:

$$\lambda = \frac{E\nu}{(1+\nu)(1-2\nu)}, \quad \mu = \frac{E}{2(1+\nu)}. \tag{14}$$

Let consider a simple shear problem. This problem is testing problem to approbation elastic and nonelastic material models in terms of large deformations. Solution of this problem, that is derived with usage of the FEA software package MSC.Marc, gives us good agreement for components of Cauchy stress tensor with exact solution. We consider a cube with edge length 0,1 m, which is undergoing simple shear. Finite element model is represented by one eight-nodal three-dimensional isoparametric finite element. Derived solution for stress tensor component s_{12} coincide with solution, that is derived in [5, 6]. Equilibrium points, that are derived in numerical solution, with pinpoint accuracy lay on curves of such relations of exact solution. The stress tensor component s_{12} have maximum point, which have coordinates [5, 6]:

$$\gamma_m = 3,0177171, \quad s_{12,m} = 4\mu/\gamma_m = 1,32548684\,\mu. \tag{15}$$

Cauchy stress tensor components subject to parameter γ in Cartesian coordinate system:

$$s_{11} = \frac{\mu\gamma \ln \mu_1}{\sqrt{4+\gamma^2}}, \quad s_{22} = -s_{11}, \quad s_{12} = \frac{2\mu \ln \mu_1}{\sqrt{4+\gamma^2}}. \tag{16}$$

Component of stress tensor s_{22} gives us important information for researching of Pointing's effect. Pointing's effect is that, at free bending of rod with circular cross-section in axial direction, in terms of large strain, length of axial line of rod is changing. Simple shear problem is two-dimensional analogue of three-dimensional problem of bending torsion of elastic rod.

4. Simulation of equipment actuator operation

The computation model corresponds to the specially conducted experiment on forming of a 600 × 400 × 10 mm flat sheet blank (V95pch2 material). In this experiment, each lower rod moved uniformly with its own speed, so that desired blank bending was achieved at the final time. Owing to blank forming symmetry, the computation model was built for 1/4 of the plate at each instant of rod travel. Results of using software support and simulation of the WP geometry in forming under elastoplastic conditions are further provided.

The first series of computations ensured simulations of the technical procedure of forming through blank bending with upward camber provided by the downward travel of the upper rods. Fig. 1 (a, b) shows the blank position during its bending at the times $t = 10$, and 20, respectively. The time $t = 10$ corresponds to the maximum forming (anticipatory working profile), and $t = 20$ corresponds to the result obtained after springback. A three-dimensional stress analysis is performed at each time instant by the finite element method in the MSC.Marc package with allowance for the geometrical nonlinearity of the DBM equations relative to the default configuration with elastoplastic material performance corresponding to the V95pch2 alloy at 20°C. The shear stress intensity distribution on the upper blank surface at $t = 10$ is shown Fig. 1. Thus, three-dimensional simulations of actuator operation allows simulating the work piece performance and variable tooling geometry, and also reproducing the operation of moving equipment parts.

a) b)

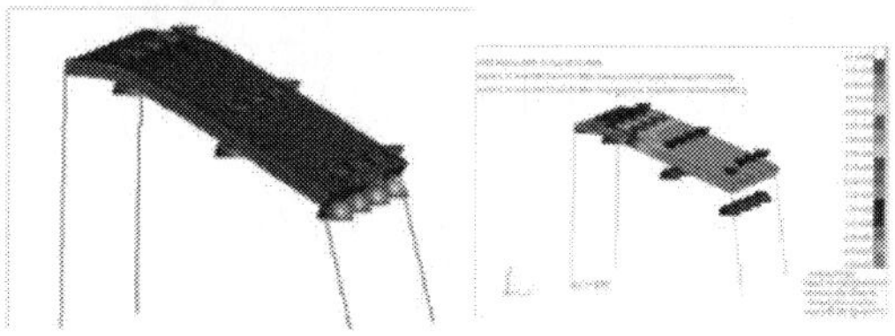

Figure 1. Positions of the blank during its bending by the upper rods: a) at the times $t = 10$ (anticipatory WP); b) at the time $t = 20$ (unloaded condition).

References

[1] Xiao H, Bruhns O T, Meyers A 1998 *Arch. Mech.* **50** 1015–1045.
[2] Curnier A and Rakotomanana L 1991 *Eng. Trans.* **39** 461–538.
[3] Korobeynikov S N 2008 *J. Elast.* **93** 105–140.
[4] Korobeynikov S N 2011 *Acta Mech.* **216** 301–332.
[5] Xiao H, Bruhns O T, Meyers A 1997 *Acta Mech.* **124** 89–105.
[6] Xiao H, Bruhns O T, Meyers A 1997 *J. Elast.* **47** 51–68.

Systematic investigation of geometrical parameters' influence on the appearance of surface deflections in sheet metal forming

A Weinschenk and W Volk

Institute of Metal Forming and Casting, Technical University Munich, Walther-Meissner-Strasse 4, 85748 Garching, Germany

E-mail: annika.weinschenk@tum.de

Abstract. Surface deflections occur during springback, which follows deep drawing. They highly affect the visual appearance of outer skin components and are, therefore, undesirable. In this work, the influence of the part geometry on the shaping of surface deflections is investigated. The geometrical parameters of an exemplary component are varied and existing surface deflections are detected. For this, a component consisting of a multiple curved surface with an inserted door handle hollow is used, and AA6016, with a sheet thickness of 1.0 mm, as well as DC06, with a sheet thickness of 0.7 mm, are chosen. After the simulations are performed in AutoForm plus R6^{TM}, a virtual stone, Three-Point Gauging and the analysis of curvatures of the part before and after springback are used to detect surface deflections.

1. Introduction

Surface deflections are small geometrical deviations from the ideal surface of a part. They distribute the course of the light and are, therefore, often visible after painting [1]. Various specifications of the dimension of surface deflections exist in the literature. Thus, a depth in the range of 10 µm up to several hundreds of µm can be found [2]. In contrast to the small size of the depth, the area of the surface deflection is much bigger. Values between 30 mm and 50 mm are indicated for length and width of the area [1].

Surface deflections occur due to springback during the opening of a deep drawing tool and can often be found near strong geometrical changes on large, flat parts with little thinning [2]. In this area, the part has lower stiffness and lower plastic strains compared to the surrounding area and is, therefore, especially prone to springback [3]. The springback calculation is sufficiently accurate to map such phenomena [4]. On curved parts, surface deflections occur mainly because of elastic forming during relief at the end of the forming process. Internal stresses, which remain in the part if the material cannot relax uniformly, are responsible for the occurrence of surface deflections on flat surfaces [5]. The position and visibility of surface deflections depend on several parameters. These are geometrical parameters such as curvature [6] and depth of hollows on the part, process parameters, like blank holding force as well as material parameters such as yield stress, r-value and sheet rolling direction [7].

In a virtual environment, such as AutoForm plus R6^{TM}, methods like stoning and Three-Point gauging can be used to detect surface deflections [5].

2. Objective and Approach

The formation of surface deflections depends on a sheet metal part's geometry and the stress state. This study focuses on the geometry. The influence of a sheet metal part's surface curvatures on the depth of surface deflections is systematically investigated.

A part geometry with a door handle hollow is chosen for the study. At first, a parametric CAD construction is created in CATIA V5^{TM}. For each set of geometrical parameters, a finite element model is created in AutoForm plus R6^{TM}. While the geometry of each finite element model is different, the settings of all the other parameters are the same. In post-processing, surface deflections are detected by a stoning method implemented in AutoForm plus R6^{TM}. The aluminium alloy AA6016, with a sheet thickness of 1.0 mm, and the steel DC06, with a sheet thickness of 0.7 mm, are chosen for the investigation.

3. Tool design and finite element model

The deep drawing tool consists of a two-part die, a blank holder, and a punch, as can be seen in Figure 1. The blank holder and the outer die remain constant for the parameter study. The dimensions of the punch and the inner part of the die are varied. The size of the blank is 800 mm x 1000 mm. A circumferential lock drawbead prevents draw-in of the material.

The part generated by the deep drawing tool consists of a single convex curved profile in x direction and a multiple curved profile in y direction, as shown in Figure 2. The profile in y direction is based on five tangentially merged radii. Four of these can be modified for the investigation. In addition, there is a door handle hollow with an entrance radius D whose dimension can be varied. The range of all parameters is limited by geometric constraints.

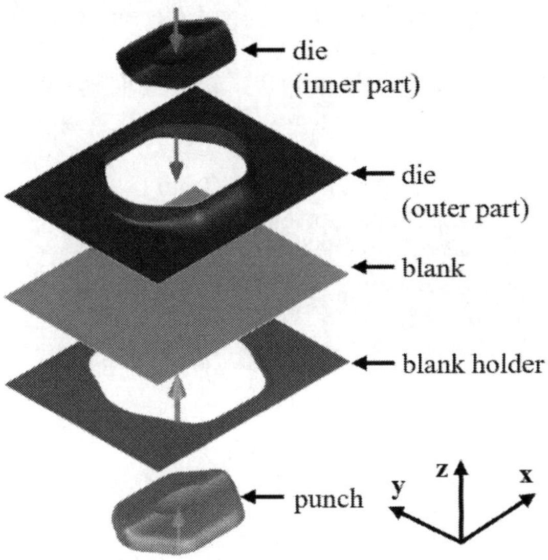

Figure 1. Surfaces of the deep drawing tool for the finite element simulation.

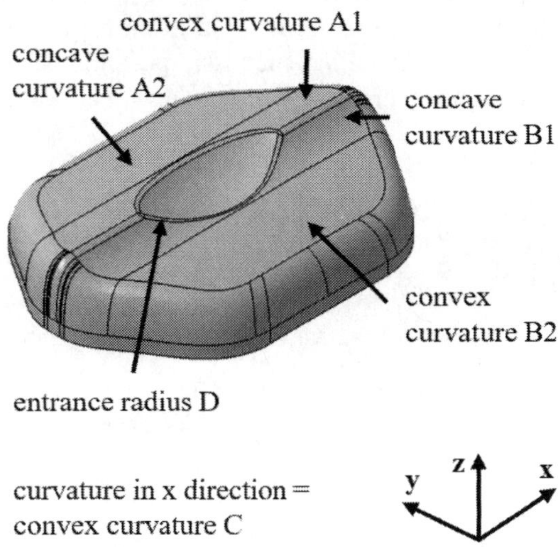

Figure 2. Punch of the deep drawing tool with variable parameters.

The clearances between the components of the tool depend on the used sheet thickness. The offset between blank holder and outer part of the die is always equal to the sheet thickness. When AA6016 is used, the offset between the punch and inner part of the die is 1.05 mm to take variations between material batches into account. For DC06 an offset of 0.75 mm is chosen.

The finite element model is built in AutoForm plus R6^{TM}, as described in Table 1.

Table 1. Settings for the finite element simulation in AutoForm plus R6^{TM}.

Option	Setting
geometries	The tool components and the blank of Figure 1 are used.
symmetry	The yz plane is used as a plane of symmetry.
operations	First: a deep drawing operation, second: free springback is performed.
anisotropy	The x axis is parallel to the rolling direction of the sheet.
lubrication	Drawing oil is used and, therefore, the friction coefficient is set to 0.05.
force	A constant force of 1300 kN is applied to the blank holder.
drawbead	An analytical constant line drawbead is used to represent the lock drawbead.
element type	Blank: An elastic plastic shell element with 11 layers is used.
meshing	Blank: An initial element size of 5.00 mm is used. Adaptive mesh refinement is allowed and restricted by a maximum refinement level of 3. Therefore, the minimum size of elements is 0.62 mm.

4. Evaluation and results

This initial configuration is used for the parametric study: A1 = 1/200 mm^{-1}, A2 = 1/600 mm^{-1}, B1 = 1/75 mm^{-1}, B2 = 1/450 mm^{-1}, C = 1/35000 mm^{-1}, and D = 10 mm.

The one-factor-at-a-time method is used to obtain an overview of the influence of the parameters on the depth of surface deflections. Virtual stones at various lengths are used to detect surface deflections on the upper side of each part. The length of a stone resulting in the lowest surface deflection is determined. For each part, the stone is moved in x direction, because this is the direction of the minimum principal curvature.

The results show that it is possible to reduce the depth significantly by a suitable selection of the parameters. Figure 3 shows an example for which a stone length of 150 mm was used.

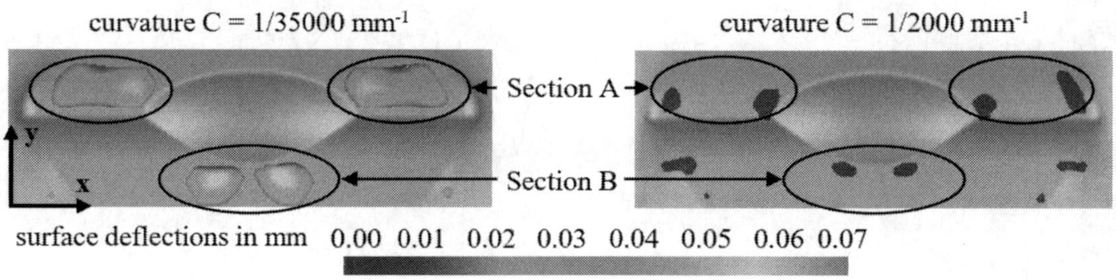

Figure 3. DC06: Detection of surface deflections on the part with a stone length of 150 mm

For DC06, the initial configuration of C = 1/35000 mm^{-1} results in a clearly visible surface deflection with a depth of 0.053 mm in Section A and 0.06 mm in Section B, while a larger curvature of C = 1/2000 mm^{-1} results in a depth of 0.006 mm in Section A and 0.003 mm in Section B. The relationship between C and depth is shown in detail in Figure 4. When C increases and the other parameters remain constant, the Gaussian curvature increases. Hence, the local stiffness of the part increases and the risk of surface deflections decreases.

As [2] pointed out, surface deflections occur near strong geometrical changes. Therefore, these should be avoided. As one can see in Figure 5, the depth decreases with an increase of D. With a radius of 28 mm or higher, the surface deflection in Section B disappears.

It was found that A2 has a small effect on the depth because it does not directly affect the areas with surface deflections. Within the given geometric constraints, a minimum depth of

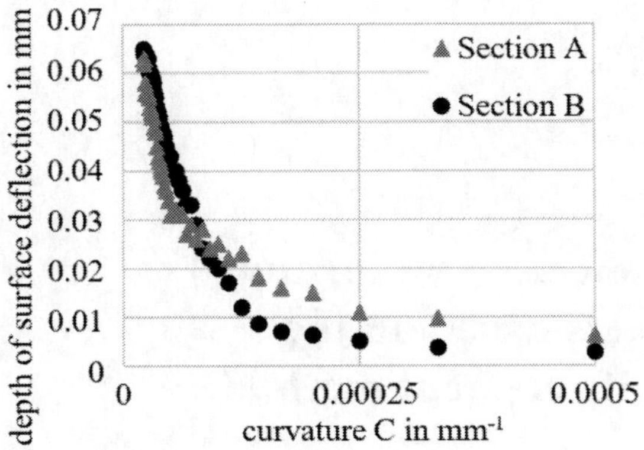

Figure 4. DC06: Influence of curvature C on depth of surface deflection

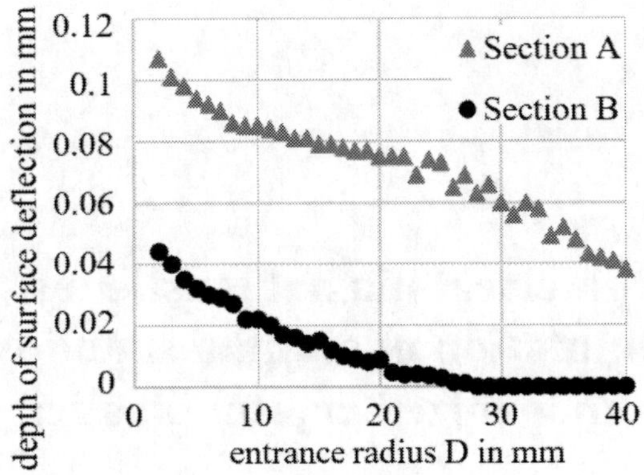

Figure 5. AA6016: Influence of entrance radius D on depth of surface deflection

0.048 mm in Section A, and 0.056 mm in Section B was detected for DC06. For AA6016, the initial configuration leads to 0.085 mm in Section A and 0.022 mm in Section B. An increase in A1 results in a decrease in depth in Section A. $1/100\ \mathrm{mm}^{-1}$ leads to 0.047 mm for AA6016 and 0.02 mm for DC06. By variation of A2, only a minimal depth of 0.084 mm in Section A and 0.017 mm in Section B can be achieved. With the variation of B1, the depth of the surface deflection in Section B can be reduced to 0.01 mm for AA6016 and for DC06 it even reaches 0 mm. B2 also has a high impact on the depth in Section B. It is possible to reduce the depth significantly.

5. Future work

In the next step, the interactions between the parameters and the influence of stress and strain on surface deflections will be investigated. In addition, robustness analyses will be carried out and for a selected set of parameters an experimental validation will be performed.

Acknowledgments

We extend our sincere thanks to the European Research Association for Sheet Metal Working and the German Federation of Industrial Research Associations for financing our investigations as part of the program for the furthering of the Industrial Collective Research by the Federal Ministry for Economic Affairs and Energy.

References

[1] Hazra A, Williams D, Roy R, Aylmore R and Allen M 2011 Predicting the occurrence of cosmetic defects in automotive skin panels *AIP Conf. Proc.* **1353** 253-257

[2] Fukumura M, Yamasaki Y, Inage D and Fujita T 2004 Finite element simulation of surface defects in the automobile door outer panel *AIP Conf. Proc.* **712** 1917-22

[3] Anderson A 2005 Evaluation and visualization of surface defects a numerical and experimental study on sheet-metal parts *AIP Conf. Proc.* **778** 113-118

[4] Muthler A, Duester A, Volk W, Wagner M, Rank E 2006 High order thin-walled solid finite elements applied to elastic spring-back computations *Comput. Methods Appl. Mech. Engrg.* **195** 5377-89

[5] Schoenbach T and Avermann M 2013 Simulation based method of surface deflect detection *IDDRG Proc.* 243-248

[6] Port A, Thuillier S and Manach P 2009 Occurrence and numerical prediction of surface defects during flanging of metallic sheets *Int. Journal of Material Forming* **3** 215-223

[7] Yonemura S, Uenishi A, Kosugi S, Yoshida T, Hiwatashi S and Nakagawa J 2013 Advanced technologies related to simulation and evaluation for performance of automotive exterior panels *Nippon Steel Technical Report* **103** 11-17

Journal of Physics: Conference Series **734** (2016) 032005 doi:10.1088/1742-6596/734/3/032005

Numerical biaxial tensile test for sheet metal forming simulation of aluminium alloy sheets based on the homogenized crystal plasticity finite element method

A Yamanaka[1], Y Ishii[2], T Hakoyama[3], P Eyckens[4], and T Kuwabara[1]

[1] Division of Advanced Mechanical Systems Engineering, Institute of Engineering, Tokyo University of Agriculture and Technology, 2-24-16, Naka-cho, Koganei-shi, Tokyo, 184-8588, Japan
[2] Faculty of Engineering, Tokyo University of Agriculture and Technology, 2-24-16, Naka-cho, Koganei-shi, Tokyo, 184-8588, Japan
[3] JSPS Research Fellow, Tokyo University of Agriculture and Technology, 2-24-16, Naka-cho, Koganei-shi, Tokyo, 184-8588, Japan
[4] Department of Metallurgy and Materials Engineering, Katholieke Universiteit Leuven, Kasteelpark, Arenberg 44, 3001, Leuven, Belgium

E-mail: a-yamana@cc.tuat.ac.jp

Abstract. The simulation of the stretch forming of A5182-O aluminum alloy sheet with a spherical punch is performed using the crystal plasticity (CP) finite element method based on the mathematical homogenization theory. In the simulation, the CP constitutive equations and their parameters calibrated by the numerical and experimental biaxial tensile tests with a cruciform specimen are used. The results demonstrate that the variation of the sheet thickness distribution simulated show a relatively good agreement with the experimental results.

1. Introduction

In order to improve the accuracy of finite element simulations of sheet metal forming processes, it is essential to check whether the material models and parameters used in the simulations adequately describe the plastic deformation behavior of sheet metals under a multiaxial stress state or not. Recently, the authors have developed the numerical biaxial tensile test (NBT) based on the crystal plasticity finite element (CPFE) method and the mathematical homogenization method [1]. Furthermore, NBTs of a A5182-O aluminum alloy sheet were performed. The results demonstrated that NBT enables us to calibrate the material models and parameters by comparing the contours of equal plastic work calculated by NBT with those measured by the biaxial tensile test using a cruciform specimen [2].

In this paper, the CPFE simulation of the stretch forming of the A5182-O aluminum alloy sheet with a spherical punch is presented. In the forming simulation, the data of crystallographic texture in the sheet measured experimentally and the material parameters calibrated by NBT are used. In order to verify our methodology, the results of the forming simulation are compared with the corresponding experimental results.

2. Crystal plasticity model

NBT and the stretch forming of A5182-O aluminum alloy sheet are performed using the CP constitutive equations proposed by Peirce et al. [3] and the two-scale CPFE method proposed by Nakamachi et al. [4]. In this study, the following constitutive equations are used.

$$\hat{\sigma}_{ij} = D^e_{ijkl}\dot{\epsilon}_{kl} - \sum_\alpha R^{(\alpha)}_{ij}\dot{\gamma}^{(\alpha)} \tag{1}$$

$$R^{(\alpha)}_{ij} = D^e_{ijkl}P^{(\alpha)}_{kl} + W^{(\alpha)}_{ik}\sigma_{kj} - \sigma_{ik}W^{(\alpha)}_{kj} \tag{2}$$

Here, $\hat{\sigma}_{ij}$ represents the objective rate of the Cauchy stress tensor, D^e_{ijkl} is the elastic modulus tensor and $\dot{\varepsilon}_{ij}$ is the strain rate tensor. $P^{(\alpha)}_{ij}$ and $W^{(\alpha)}_{ij}$ are the Schmid and the spin tensors for the αth slip system, respectively. In this study, twelve {111}<110> slip systems in a face-centered cubic crystal are taken into account. The plastic shear strain rate on each slip system is given by the following equation [5]:

$$\dot{\gamma}^{(\alpha)} = \dot{\gamma}^{(\alpha)}_0 sgn\left(\tau^{(\alpha)}\right)\left|\frac{\tau^{(\alpha)}}{g^{(\alpha)}}\right|^{\frac{1}{m}} \tag{1}$$

where $\dot{\gamma}^{(\alpha)}_0$ is the reference plastic shear strain rate, $\tau^{(\alpha)}$ is the resolved shear stress for αth slip system, m is the strain rate sensitivity factor. $g^{(\alpha)}$ is the critical resolved shear stress and its evolution is described as

$$g^{(\alpha)} = \sum_\beta h(\gamma)\left|\gamma^{(\beta)}\right| \tag{2}$$

where $h(\gamma)$ is the hardening coefficient matrix. In this study, we use the following equation [4]:

$$h(\gamma) = h_0 nC\left\{C(\gamma_0 + \gamma)\right\}^{n-1} \tag{3}$$

where δ is the Kronecker delta. q is a parameter describing the self and latent hardenings and assumed to be $q = 1$ in this study. h_0 is the initial hardening coefficient, n is the hardening exponent and C is the hardening coefficient. γ_0 is the initial plastic shear strain. γ is the accumulated plastic shear strain. In this study, the initial resolved shear stress, τ_0, and the material parameters, h_0, n, C and γ_0, identified by NBT [2] are used.

As similar to Nakamachi et al. [4], in order to simulate the macroscopic deformation of the sheet during the forming on the basis of the data of crystallographic texture, the mathematical homogenization method proposed by Guedes et al. [6] is used. By using this homogenization method, the principals of virtual work for both macroscopic and microscopic (crystalline) scales are derived simultaneously. Furthermore, the dynamic explicit solver and a parallel computing technique are used for accelerating the simulation. The detail of the formulation can be found in the literature [4].

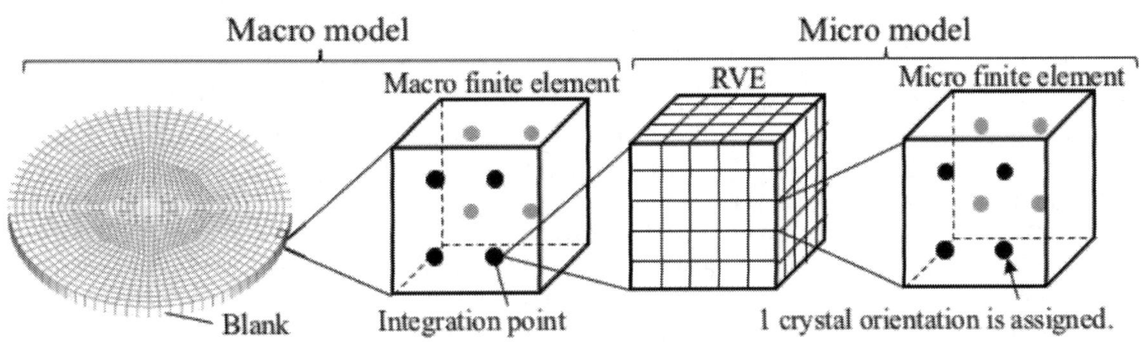

Figure 1. Finite element models used in the CPFE simulation of the stretch forming.

Journal of Physics: Conference Series **734** (2016) 032005 doi:10.1088/1742-6596/734/3/032005

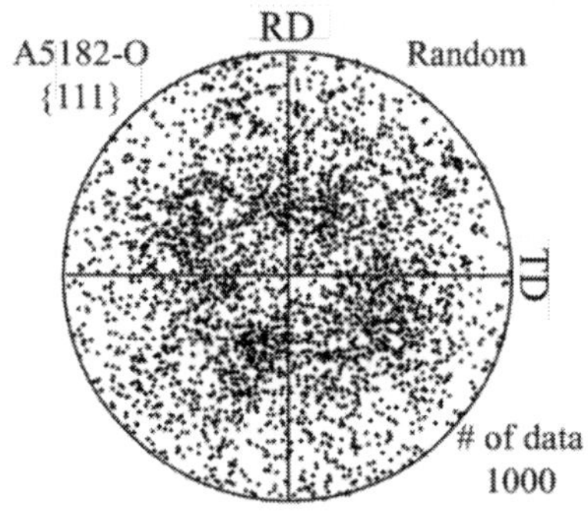

Figure 2. {111} pole figure of the initial crystal orientation [2].

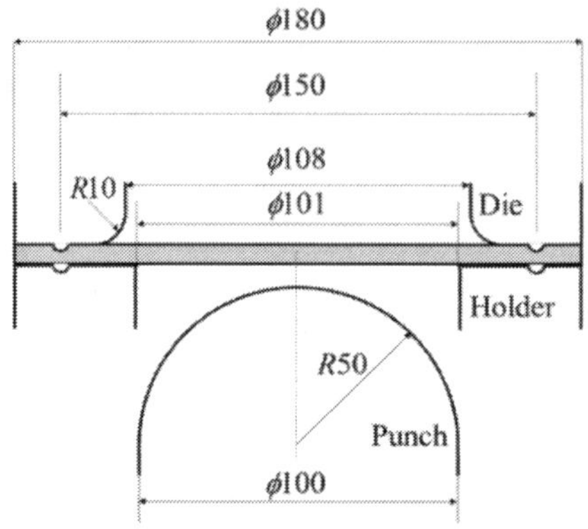

Figure 3. Dimensions of tools.

3. Simulation of stretch forming of A5182-O aluminium alloy sheet using a spherical punch

3.1. Simulation model and condition

Figure 1 shows the finite element models for the CPFE simulation of the stretch forming of A5182-O aluminum alloy sheet with a spherical punch. The macro and micro models describes a blank and a representative volume element (RVE) of polycrystalline aggregate in the sheet, respectively. The number of finite elements for the blank is 1083. One orientation data is assigned to each integration point in the RVE. The number of finite element for the RVE is 125. Therefore, 1000 initial crystal orientations are taken into account at each integration point in a finite element consisting of the macro model. The initial crystal orientations are randomly sampled from the orientation data measured by electron backscattered diffraction (EBSD). Figure 2 shows {111} pole figure of the initial crystal orientations. Note that the same initial orientations were used in NBT [2]. The initial critical resolved shear stress and the material parameters for A5182-O aluminum alloy sheet validated by NBT [2] are as follows: $\tau_0 = 43.5$ MPa, $h_0 = 120$ MPa, $n = 0.24$, $C = 17$, and $\gamma_0 = 0.1$.

Figure 3 shows the size and geometry of tools used in this study. In the experiment, the punch speed was set to be 6 mm/min. and the maximum forming height was 30 mm. In the simulation, the friction coefficient between the blank and the punch is assumed to be 0.1. Instead of modeling the bead in the simulation explicitly, the nodal displacement at 75 mm away from the center of the blank is fixed to be zero.

3.2. Simulation results

Figure 4 shows the distributions of the plastic strain along the through-thickness direction at different forming heights, h. It is seen that the strain localization occurs at the center of the blank at h = 10 mm. As the forming height is increased, the region where shows the maximum plastic strain moves to a little far away from the center of the blank due to the friction between the punch and the blank. Figure 5 shows the calculated and measured distributions of the sheet thickness along the rolling direction (RD) at h = 10, 20, 25 and 30 mm. The experimental results were reported by Kawaguchi et al. [7]. It is shown that the calculated thickness distribution at h = 10 mm shows very good agreement with the experimental. As shown in the experimental results, the distance between the center of the blank and the minimum thickness position is increased with increasing h. Although the present simulation can qualitatively predict this tendency, the minimum sheet thickness predicted by the CPFE simulation underestimates the experimental results. In order to improve the simulation results obtained in this study, we need to investigate the effects of initial crystal orientations and the friction coefficient.

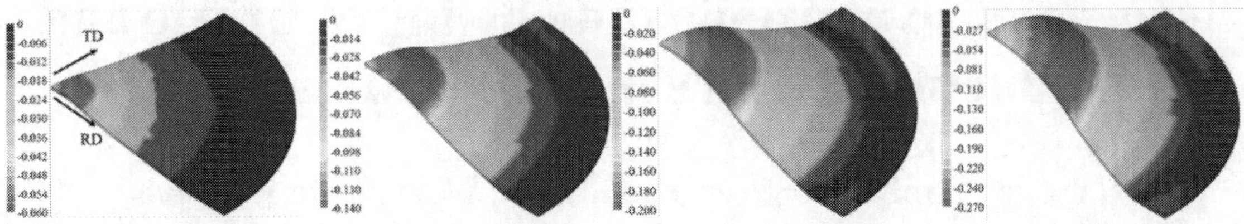

(a) h = 10 mm (b) h = 20 mm (c) h = 25 mm (d) h = 30 mm

Figure 4. Distributions of the plastic strain along the through-thickness direction calculated by the CPFE simulation.

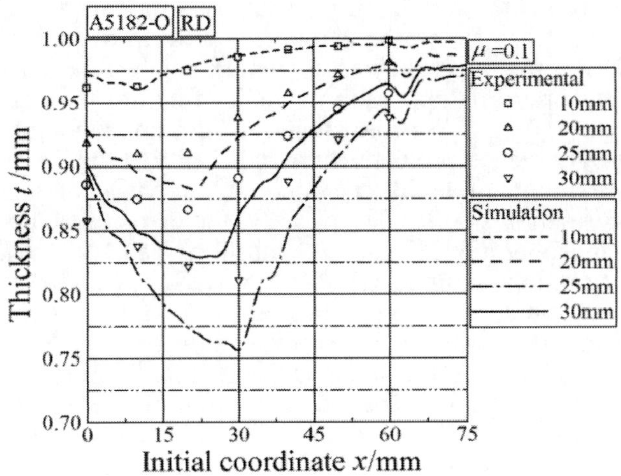

Figure 5. Calculated distributions of the sheet thickness along the RD compared with the experimental results [7].

4. Conclusion

The stretch forming of A5182-O aluminum alloy sheet with a spherical punch was simulated using the CPFE method based on the mathematical homogenization theory. In order to perform the forming simulation accurately, we used the material parameters calibrated by NBT. It was demonstrated that the simulated variation of the sheet thickness shows relatively good agreement with the experimental results.

Acknowledgement

This study was financially supported by Grant-in-Aid for Young Researcher (A) (No. 25709002) from Japan Society for the Promotion of Science (JSPS).

References

[1] Hashimoto K, Yamanaka A, Kawaguchi J, Sakurai T and Kuwabara T 2015 *J. Jpn. Inst. Light Met.* **65** 196.

[2] Kuwabara T, Ikeda S and Kuroda K 1998 *J. Mater. Proc. Technol.* **80-81** 517.

[3] Pierce D, Asaro R J amd Needleman A 1983 *Acta Metall.* **30** 1087.

[4] Nakamachi E, Tam N N and Morimoto H 2007 *Int. J. Plasticity* **23** 450.

[5] Pan J and Rice J 1992 *Int. J. Solid Struct.* 59 973.

[6] Guedes J M and Kikuchi N 1990 *Comput. Methods Appl. Mech. Engng.* **83** 143.

[7] Kawaguchi J, Kuwabara T, Sakurai T and Noguchi O 2014 *Abstracts of the 127th JILM Symposium (Tokyo)* (Tokyo: The Japan Institute of Light Metals) p 169.

Isogeometric phase-field modeling of brittle and ductile fracture in shell structures

Marreddy Ambati*[1], Josef Kiendl[1], and Laura De Lorenzis[1]

[1] Institut für Angewandte Mechanik
Technische Universität Braunschweig, Germany

Phase-field modeling of brittle and ductile fracture is a modern promising approach that enables a unified description of complicated failure processes (including crack initiation, propagation, branching, merging), as well as its efficient numerical treatment [1-4]. In the present work, we apply this approach to model fracture in shell structures, considering both thin and thick shells. For thin shells, we use an isogeometric Kirchhoff-Love shell formulation [5-6], which exploits the high continuity of the isogeometric shape functions in order to avoid rotational degrees of freedom, i.e., the shell geometry is modeled as a surface and its deformation is fully described by the displacements of this surface. For thick shells, we use an isogeometric assumed natural strain (ANS) solid shell formulation [7], i.e., a 3D solid formulation enhanced with the ANS method in order to alleviate geometrical locking effects. According to the discretization of the structural formulations, an isogeometric basis is also used for the phase-field. While the phase-field fracture formulation for solid shells is basically the same as for standard solids, some reformulation is necessary for thin shells, accounting for the interaction of stresses devoted to membrane and bending deformation. We test both formulations on several numerical examples and perform comparisons of the results obtained by the two methods to each other as well as to reference solutions, which confirm the validity and applicability of the presented methods.

References

[1] M. Ambati, T. Gerasimov, L. De Lorenzis. A review on phase-field models of brittle fracture and a new fast hybrid formulation. *Computational Mechanics*, 55:383–405 (2014).

[2] T. Gerasimov, L. De Lorenzis. A line search assisted monolithic approach for phase-field computing of brittle fracture. *Computer Methods in Applied Mechanics and Engineering*, http://dx.doi.org/10.1016/j.cma.2015.12.017 (2016).

[3] M. Ambati, T. Gerasimov, L. De Lorenzis. Phase-field modeling of ductile fracture. *Computational Mechanics*, 55:1017-1040 (2015).

[4] M. Ambati, R. Kruse, L. De Lorenzis. A phase-field modeling of ductile fracture at finite strains and its experimental verification. *Computational Mechanics*, DOI 10.1007/s00466-015-1225-3 (2015).

[5] J. Kiendl, K.-U. Bletzinger, J. Linhard, R. Wüchner. Isogeometric shell analysis with Kirchhoff-Love elements. *Computer Methods in Applied Mechanics and Engineering*, 198:3902–3914 (2009).

[6] J. Kiendl, M.C. Hsu, M.C.H. Wu, A. Reali. Isogeometric Kirchhoff-Love shell formulations for general hyperelastic materials. *Computer Methods in Applied Mechanics and Engineering*, 291:280–303 (2015).

[7] J.F. Caseiro, R.A.F. Valente, A. Reali, J. Kiendl, F. Auricchio, R.J. Alves de Sousa. On the Assumed Natural Strain method to alleviate locking in solid-shell NURBS-based finite elements. *Computational Mechanics*, 53:1341–1353 (2014).

ACCURATE DRAWBEAD MODELING IN STAMPING SIMULATIONS

M. Sester[1], I. Burchitz[2], E. Saenz de Argandona[3], F. Estalayo[4], B. Carleer[5]

[1]AutoForm Development, Zürich, Switzerland
[2]AutoForm Engineering, Krimpen a/d IJssel, Netherlands
[3]Mondragon University, Mondragon, Spain
[4]Matrici, S. Coop, Bilbao, Spain
[5]AutoForm Engineering, Dortmund, Germany

E-mail: matthias.sester@autoform.ch

Abstract. An adaptive line bead model that continually updates according to the changing conditions during the forming process has been developed. In these calculations, the adaptive line bead's geometry is treated as a 3D object where relevant phenomena like hardening curve, yield surface, through thickness stress effects and contact description are incorporated. The effectiveness of the adaptive drawbead model will be illustrated by an industrial example.

1. Introduction

Drawbeads are frequently used in stamping tools to control the material flow. The geometrical modeling of drawbeads results in very long computation times. To overcome this problem equivalent drawbead models are applied. Stoughton proposed a model based on bending around an effective radius [1]. This model resulted in an effective description but also showed limitations. This paper describes several extensions to improve the accuracy of the equivalent drawbead model.

2. Adaptive equivalent drawbead model

2.1. Geometry and strain. The distribution of longitudinal strain depends on the sheet's bending shape. In our model, the bending shape in a partially penetrated round bead is described with five sections of constant curvature (*Figure 1*). The Stoughton model [1] employs three sections (1, 3 and 5 in our notation) with an approximated "effective radius", and two straight sections in between.

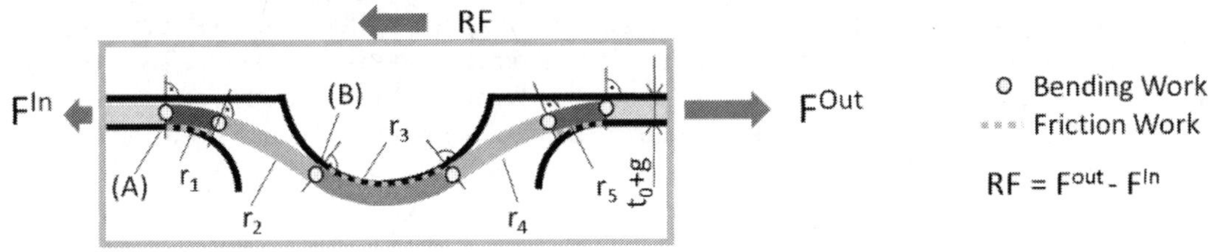

Figure 1: Bending shape with 5 sections of constant curvature $1/r_i$. Definition of restraining force RF.

The longitudinal strain change between two sections is given by the expression $\Delta\epsilon_{11} = \Delta\varepsilon_0 + \Delta\kappa\zeta$, where $\Delta\varepsilon_0$ is the mid-plane stretch increment, ζ is a natural coordinate in the sheet thickness direction, and $\Delta\kappa$ is the curvature increment $1/r_i - 1/r_{i-1}$ between current and previous section [2]. This can be interpreted as superposition of a membrane strain and a bending strain around the geometric middle.

2.2. *Material model.* An incremental elastoplastic material description defines the longitudinal stress σ_{11} from strain increments $\Delta\varepsilon_{11}$. Plasticity is modelled with Hill'48 yield condition and plane strain assumption. The equivalent plastic strain increment is given by $\Delta\varepsilon_{eq} = \frac{1+\bar{r}}{\sqrt{1+2\bar{r}}}\,\Delta\varepsilon_{11}$ with the normal anisotropy coefficient $\bar{r}$. The hardening curve $\sigma_{eq}(\varepsilon_{eq})$ relates the accumulated equivalent plastic strain $\varepsilon_{eq} = \int d\varepsilon_{eq}$ to the equivalent stress σ_{eq}. Finally, σ_{11} is computed from σ_{eq} and σ_{33} via Hill'48.

The thickness stress σ_{33} is given by a differential equation $d\sigma_{33} = (\sigma_{11} - \sigma_{33})\ dr/r$ which arises from a force balance of an infinitesimal thin layer with radius r. This equation is integrated numerically, starting at the outer surface with the natural boundary condition $\sigma_{33} = 0$. Basically, the Bauschinger effect should be essential for drawbeads due to the inherent bending/unbending [2]. We found taking into account the through thickness stress σ_{33} in the Hill48 yield surface to be more relevant than using mixed hardening.

2.3. *Section forces and section moments.* For each constant curvature section introduced above, axial forces and bending moments are evaluated from current normal stress σ_{11} and current sheet thickness t. The thinning Δt is directly determined from $\Delta\varepsilon_0$ via plastic incompressibility. The axial force is $F = \int_{-t/2}^{+t/2} \sigma_{11}\ dt$ and the bending moment is $M = \int_{-t/2}^{+t/2} \sigma_{11}\ t\ dt$. A numerical integration with N=25 integration points of equal distance is employed in our implementation.

2.4. *Thinning iteration.* The curvature change $\Delta\kappa$ is given directly by the assumed bending shape of the sheet strip, but the mid-plane stretch increment $\Delta\varepsilon_0$ and the thinning Δt are not known a priori. In our model, it follows from a procedure we call "thinning iteration". As in [2], we assume that the total curvature change $\Delta\kappa$ occurs over an infinitesimal small length along which the bending work $V_i = \int_{-\frac{t}{2}}^{+\frac{t}{2}} \left(\int_0^{\Delta\varepsilon} \sigma\Delta\varepsilon\right)\ dt$ is dissipated. The virtual work equation $F_{(i)}dx_{(i)} = F_{(i-1)}dx_{(i-1)}\ +\ V_{(i)}$ is used as a conditional equation for the unknown sheet thickness t and is solved iteratively. The virtual displacements $dx_{(i)}$, $dx_{(i-1)}$, the membrane strain $\Delta\varepsilon_0$ and the sheet thickness t are related via continuity and incompressibility.

2.5. *Iteration of bending shape.* In many early drawbead models, the geometry of the sheet is predetermined [1],[2],[3]. Our approach uses geometric constraints and moment equilibrium to adapt the bending form to geometry, material state and process conditions. Such an iteration of the bending shape was suggested by Sanchez and Weinmann [4]. They applied it in the framework of a finite difference scheme. We apply bending iteration to the free sections between die and bead in order to determine the unknowns r_1, r_2, r_4 and r_5 (*Figure 1*). The "entry iteration" of radius r_1 and r_2 is discussed now. Left to section 1, the sheet is straight, and right to section 2, it matches the bead. At section boundaries, normals are identical. All combinations of r_1 and r_2 that fulfil this continuity condition and do not lead to tool penetrations are called "geometric admissible states". The vertical tool gap $t_0 + g$ is crucial for the description of the bending form, as already pointed out by Yellup and Painter [3]. Among all geometric admissible solutions, the final solution is defined from balance of moments: For both sections of the unsupported region between die and bead, the moments with respect to point (A) or point (B), respectively, are evaluated. Using this reference, in the moment balance only known internal section forces and moments enter. This gives two conditions for the two unknowns r_1 and r_2 which are fulfilled by the bending iteration. In the same way, the bending form between bead radius and die *exit* radius is described.

2.6. *Forces.* The equivalent restraining force (RF) and equivalent uplift force (UF) are computed from the section forces and section moments defined in chapter 2.3. First, the "lever forces" $F^{Le,In}$ and $F^{Le,Out}$ between upper tool and sheet are derived (Figure 2). Lever forces are tool reactions that balance internal bending moments. They are an important element since they dominate the uplift force and contribute via frictional forces to the restraining forces. The lever force at die exit radius follows from the known section moment M of the last section and the lever arm L_e with help of a moment balance $M = F^{Le,Out}L_e$ around point (D). For the lever arm L_e a geometric assumption from a circular

Journal of Physics: Conference Series **734** (2016) 032007 doi:10.1088/1742-6596/734/3/032007

segment formula is used. At the entry radius a similar relation holds.

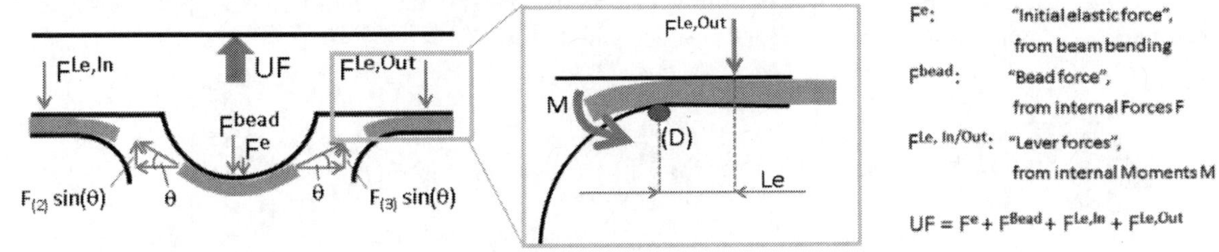

Figure 2: Definition of bead force, lever forces and total uplift force UF.

For a given bending shape, in each section the axial forces are known from the numerical solution of the virtual work equation, see chapter 2.4. The axial forces are solved section by section, and frictional forces are added where applicable. The Coulomb friction law $F^{\mu} = \mu N$ is applied for point contact, so the lever force contributes $F^{\mu} = \mu F^{Le}$. Euler's law is applied for the line contact along the bead radius, increasing F^- to $F^+ = F^- e^{2\mu\theta}$ [1]. The equivalent restraining force is defined as the axial force increase between entry and exit, $RF = F^{Out} - F^{In}$. The vertical uplift force between sheet and the upper tool is the sum of the lever forces $F^{Le,In}$ and $F^{Le,Out}$ and the vertical part of the internal forces $F_{(2)}$ and $F_{(3)}$. For shallow beads, the initial elastic force F^e from elastic beam bending is added to UF, as in [1]. Finally, the uplift force *is* $UF = F^e + F^{Le,In} + F^{Le,Out} + F_{(2)} sin\theta + F_{(3)} sin\theta$.

2.7. *Generalization and application to double beads.* The simplest case is a single bead with no entry force applied. In general, the actual state at drawbead entry (entry force, sheet thickness, plastic strain and strain hardening) enters the model as a starting condition for the first section. This allows a straightforward application of the model to general situations, including also double beads, without additional assumptions. Note that the model nowhere does make use of fitting parameters.

2.8. *Verification of uplift forces for single beads*

At Mondragon University, three sheet materials and thicknesses were examined for a typical industrial round bead geometry. In Figure 3 the bead geometry is shown; measured and modelled uplift forces are plotted over bead height. The results clearly demonstrate the important role the lever forces play in our model. Without lever force, the uplift force is roughly a factor two smaller than measured. The full model reproduces the measured data well.

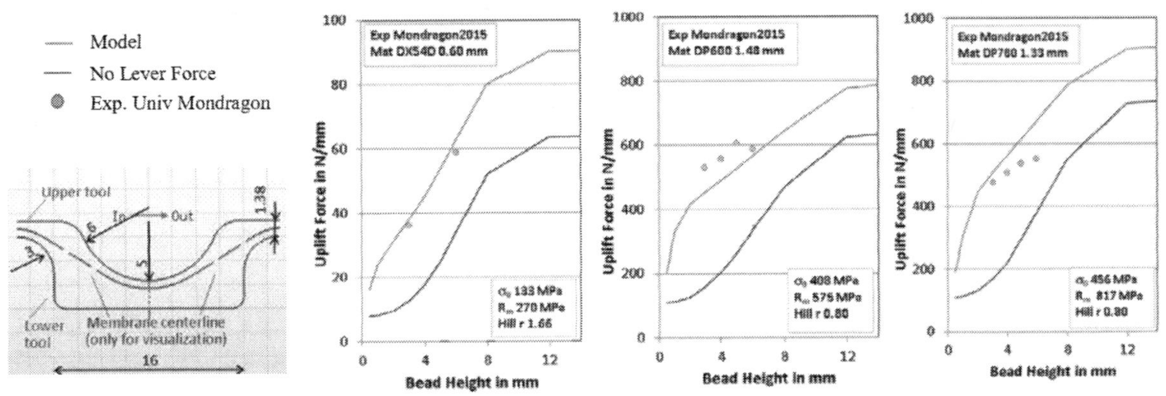

Figure 3: Example bead geometry (DP780 1.33 mm, height 5 mm). Measured uplift forces (points) by Mondragon University, modelled uplift forces (lines) with and without contribution of lever forces.

3. Application to an industrial example

Performance of the proposed equivalent drawbead model is validated using real example – underbody cross member, see Figure 4. This part was used as Numisheet 2005 benchmark, see [5].

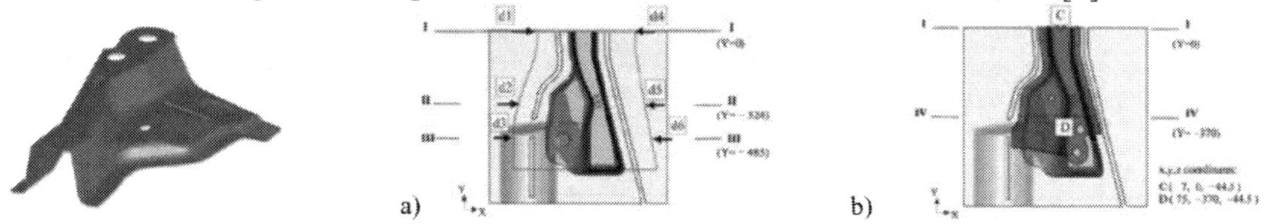

Figure 4: Underbody cross member. Sections to measure: a) draw-in; b) strains and part shape, [5].

Blank material is aluminum alloy AL5182-O. Results of two simulations are compared with real measurements. One simulation is run using the proposed adaptive line bead model and the other using geometric beads. All other process conditions as described in the conference proceedings.

Total upper die force predicted by two simulations is similar, i.e. about 1700kN. Figures 5-6 show comparison of material draw-in, true thickness strain after forming and part's shape after springback. Although results calculated with geometric beads are slightly closer to experimental measurements, adaptive line bead model gives a good prediction of typical variables used to validate a process. CPU time is about 2.5 times less compared to the geometric beads and therefore the proposed line bead model can be considered as an efficient way to incorporate drawbead effects in the simulation.

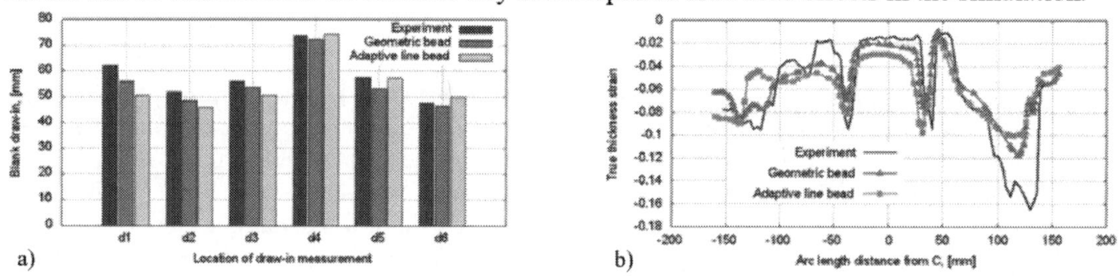

Figure 5: Simulation results: a) material draw-in after drawing; b) true thickness strain in section I

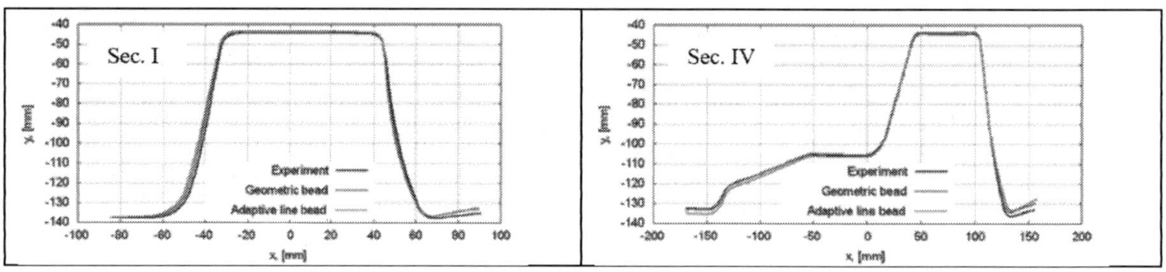

Figure 6: Springback profiles at sections I and IV.

References

[1] Stoughton T.B.: Model of drawbead forces in sheet metal forming. *Proceedings of 15th Biennal Congress of IDDRG, Dearborn, MI, 1988*

[2] Matiasson K., Bernspang L: Drawbead modelling in sheet metal stamping simulation. *Proceedings of 4th Numisheet Conference, Besancon, France, 1999*

[3] Yellup J.M., Painter M.J: The prediction of strip shape and restraining force for shallow drawbead systems. *Jounal of Applied Metalworking, Vol. 4, No. 1, pp 30-38, 1985*

[4] Sanchez L.R., Weinmann K.J.: An analytical and experimental study of the flow of sheet metal between cirular drawbeads. *Jounal of Engineering for Industry, Vol. 118, pp 45-54, 1996*

[5] L.M. Smith, F. Pourboghrat, J.W. Yoon, and T.B. Stoughton, editors. *Proc. of 6th Numisheet Conf.*, Detroit, MI, 2005

Shape optimization for contact problems based on isogeometric analysis

Benjamin Horn, Stefan Ulbrich
TU Darmstadt, Germany
E-mail: bhorn@mathematik.tu-darmstadt.de

Abstract. We consider the shape optimization for mechanical connectors. To avoid the gap between the representation in CAD systems and the finite element simulation used by mathematical optimization, we choose an isogeometric approach for the solution of the contact problem within the optimization method. This leads to a shape optimization problem governed by an elastic contact problem. We handle the contact conditions using the mortar method and solve the resulting contact problem with a semismooth Newton method. The optimization problem is nonconvex and nonsmooth due to the contact conditions. To reduce the number of simulations, we use a derivative based optimization method. With the adjoint approach the design derivatives can be calculated efficiently. The resulting optimization problem is solved with a modified Bundle Trust Region algorithm.

Introduction

There are different works on shape optimization for contact problems with standard finite elements e.g. [1] or shape optimization based on isogeometric analysis without contact e.g. [2]. In this work we present a optimization scheme for mechanical connectors based on isogeometric analysis. In the first part we describe the underlying system of partial differential equations, which represents the physical model. After that we give a short introduction in isogeometric analysis. The third part deals with the presentation of the shape optimization problem and the used optimization algorithm. Followed by an exemplary numerical result. Finally we give a short outlook on future research.

1. Physical Model

We describe the structural mechanic behavior of the connection through a system of partial differential equations (PDE). For simplicity reasons we assume that only one part of the connection is elastic. The other part is assumed to be rigid. Further we consider frictionless contact. This setting is called Signorini problem. The elastic body Ω is fixed on Γ_D and a surface load f^s acts on Γ_N and a volume load f^v on the whole body. The potential contact zone Γ_C includes the a-priori unknown contact zone, where $\partial\Omega = \Gamma_N \cup \Gamma_D \cup \Gamma_C$. The gap between Ω and the obstacle is denoted by g. One can formulate the Signorini Problem as a constrained convex quadratic optimization problem. It can be shown that for Problem 1 exists a unique solution. For further details we refer to [3, 4].

Problem 1 (Weak Formulation of the frictionless Signorini Problem)
Find $y \in \mathcal{K} := \{v \in H^1_D(\Omega) : Bv - g \leq 0 \text{ on } \Gamma_C\}$ *such that*

$$J_{pe}(y) = \min_{v \in \mathcal{K}} J_{pe}(v) := \underbrace{\frac{1}{2}\int_\Omega \varepsilon(v) : \mathcal{C}\varepsilon(v)\ dx}_{a(v,v)=:\langle Av,v\rangle_{H^1_D(\Omega)}} - \underbrace{\left(\int_\Omega f^v \cdot v\ dx + \int_{\Gamma_N} f^s \cdot v\ dS(x)\right)}_{f(v)=:\langle F,v\rangle_{H^1_D(\Omega)}}, \qquad (1)$$

where B *denotes the normal trace operator and* $H^1_D(\Omega) := \{v \in H^1(\Omega) : v = 0 \text{ on } \Gamma_D\}$. *The objective function* J_{pe} *denotes the potential energy.*

For the following we denote the saddle point formulation of Problem 1 by

$$C(y, \lambda, u) := \begin{pmatrix} A(u)y - F(u) + B^*(u)\lambda \\ \lambda - \max(0, \mu + \alpha^{-1}(B(u)y - g)) \end{pmatrix} = 0, \qquad (2)$$

where u denotes the design of Ω, y represents the displacement and $\lambda \in H^{\frac{1}{2}}(\Gamma_C)'$ constitutes the Lagrange multipliers of the contact conditions, since we apply a mortar method [5, 6, 7]. The state equation is solved by a semismooth Newton method. To allow for a convergence theory of the semismoth Newton method in function space, we introduce a regularization parameter α, see [8].The state equation (2) is the nonsmooth reformulation of the optimality system of Problem 1. For a more detailed derivation of the nonsmooth reformulation we refer to [8]. To discretize the state equation we use the approach of isogeometric analysis.

2. Isogeometric Analysis

The approach of isogeometric analysis follows an isoparametric idea. The characteristic is that we use the basis functions of the geometry description as basis functions for the solution space not like it is common in isoparametric settings the other way around. This gives us the advantage of a direct link from the CAD geometry to the finite element simulation and additionally that we can directly perform the simulation on the correct geometry, without the need to approximate it. In particular for contact mechanics the smooth boundaries lead to better results compared to standard finite elements [9]. Since in isogeometric analysis NURBS are used both as geometry description and ansatz function, we shortly summarize the basic concept of NURBS.

Definition 1 (NURBS curve) *For a degree* p*, a set of d-dimensional control points* $\{P_1, \dots, P_n\}$, $d \geq 1$, $n \geq p+1$, *a set of real valued weights* $\{w_1, \dots, w_n\}$, *with* $w_i \geq 0$ *and an open knot vector* $\Xi := \{\bar{\xi}_1, \dots, \bar{\xi}_{n+p+1}\}$, *with* $\bar{\xi}_i \in [0,1]$ *and* $\bar{\xi}_1 = \dots = \bar{\xi}_{p+1} = 0$ *and* $\bar{\xi}_{n+1} = \dots = \bar{\xi}_{n+p+1} = 1$,
the resulting NURBS curve is defined as

$$G(\xi) = \sum_{i=1}^{n} R_i^p(\xi) P_i, \text{ with } R_i^p(\xi) := \frac{w_i N_i^p(\xi)}{\sum_{j=1}^n w_j N_j^p(\xi)},$$

where the basis functions $N_i^p : [0,1] \to \mathbb{R}$ *are defined through the recurrence formula due to Cox, deBoor and Mansfield [10].*

Through the knot vector Ξ one can not only control the geometry of the curve but also the differentiability. With the concept of multiple knots, e.g. $\bar{\xi}_i = \dots = \bar{\xi}_{i+k-1}$, one reduces the smoothness at $\bar{\xi}_i$ from C^{p-1} to C^{p-k}, where the case of C^{-1} denotes a point of discontinuity of the considered curve. In this work we assume an open knot vector. As a result the NURBS curve is interpolatory at both ends. A detailed introduction in NURBS theory is given in [10].

Definition 2 (NURBS solid) *A d-dimensional NURBS solid* $V(\xi,\eta,\zeta)$, *with* $(\xi,\eta,\zeta)\in[0,1]^3$ *of degree* $(p,q,r)\in\mathbb{N}_0^3$ *is defined as the tensor product of three NURBS curves defined as in Definition 1,*

$$V(\xi,\eta,\zeta)=\sum_{i=1}^{n}\sum_{j=1}^{m}\sum_{k=1}^{l}R_{i,j,k}^{p,q,r}(\xi,\eta,\zeta)P_{i,j,k},$$

$$\text{with } R_{i,j,k}^{p,q,r}(\xi,\eta,\zeta):=\frac{w_{i,j,k}N_i^p(\xi)M_j^q(\eta)L_k^r(\zeta)}{\sum_{a=1}^{n}\sum_{b=1}^{m}\sum_{c=1}^{l}w_{a,b,c}N_a^p(\xi)M_b^q(\eta)L_c^r(\zeta)},\quad P_{i,j,k}\in\mathbb{R}^d.$$

3. Optimization problem

The aim of our optimization scheme is to find a shape Ω for a connector such that the stiffness is maximized under different constraints. One way to do this is to minimize the compliance of the considered connection. The compliance can be defined as

$$J_c(y,u)=\int_{\Omega(u)}f\cdot y\,dx+\int_{\Gamma_N(u)}q\cdot y\,dS(x). \tag{3}$$

To determine subgradients with respect to the shape parameters, we use an adjoint approach. Since we know that Problem 1 has a unique solution for a given domain Ω, we can define a nonsmooth solution operator $S:U\to(\mathcal{K}\times H^{\frac{1}{2}}(\Gamma_C)^*)$ that maps each shape $u\in U$ of Ω to the corresponding unique solution of (2). Thereby we can define the reduced form of (3) by

$$j_c(u):=J_c(u,S(u)).$$

It can be checked that in points of strict complementarity, i.e., if the contact forces do not vanish in all contact nodes, the shape gradient can be obtained by

$$j_c'(u)=\frac{\partial}{\partial u}J_c(S(u),u)+p^\top\frac{\partial}{\partial u}C(S(u),u), \tag{4}$$

where the adjoint p is defined through the adjoint equation

$$\frac{\partial}{\partial(y,\lambda)}C(\lambda,y,u)^\top p=-\frac{\partial}{\partial(y,\lambda)}J_c(y,\lambda,u).$$

If strict complementarity is violated, a perturbation argument justifies to compute an approximate subgradient by (4). The shape optimization problem has the following form:

$$\min_{u\in U_{ad}}\ j_c(u), \tag{5}$$

with the set of admissible designs $U_{ad}:=\{u\in U:\ h(u)=0,\ g(u)\le 0,\ \underline{u}\le u\le\overline{u}\}$. The differentiable functions $h(\cdot)$ and $g(\cdot)$ can describe different constraints like volume, distance or curvature restrictions. The resulting optimization problem (5) is nonsmooth and nonconvex. We therefore apply a nonconvex Bundle Trust Region (BTR) method, as shown in [11]. In general the BTR method generates a cutting plane model to approximate the objective function from below. For this purpose one calculates a subgradient in the current point and adds the plane defined through this subgradient to the bundle of planes already generated. Then the optimization problem is solved over the cutting plane model. If the step satisfy a sufficient descend condition the step is called essential step and will be performed. If not it is called null step. A null step is an indicator for a bad approximation of the real objective function by the cutting plane model and so it has to be improved by adding an additional cutting plane and start the optimization again. In [11] a BTR method for the unconstrained case is proposed. We have modified the inner loop of the algorithm such that the calculated steps are always within the feasible set.

4. Numerical Results

Now we give a short application of the algorithm. We assume the following scenario. The bar has the size of 12 mm×3 mm×1.5 mm. From the top there is a homogeneous down acting force applied, see upper detached plane in Fig. 1. We additionally assume gravity. The bar is in contact with a rigid obstacle at the outer 2 mm, see the lower two gray planes in Fig. 1. For an improved visibility the Neumann boundary is plotted with a constant offset towards the bar. The colour in Fig. 1 represents the von Mises stress. The displacement is scaled by a factor of 100. As objective function we choose the compliance (3). The tri-cubic reference mesh consisting

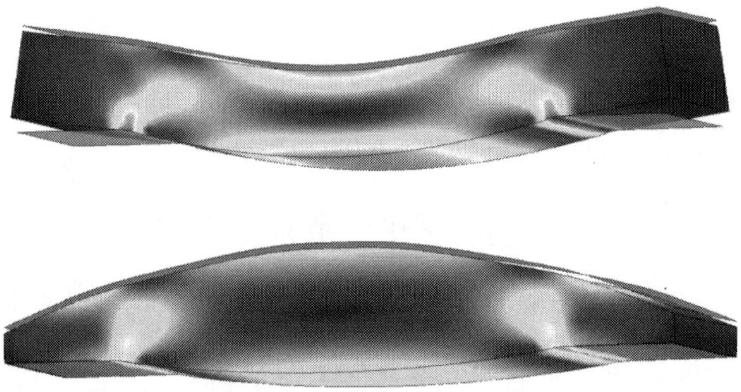

Figure 1: Comparison of the start (top) and the optimized (bottom) geometry of the connector.

of two elements and 80 control points. The mesh has been refined through knot insertion up to 1200 elements and 4362 control points. The degrees of freedom in this example have been the 20 control points of the upper surface in the reference mesh. There was no change in the x and y direction allowed. Under the constraint of constant volume, a minimal hight of 0.5 mm and a maximal height of 10 mm we could reduce the compliance about 30.57 % within 13 iterations.

Outlook and Acknowledgment

The next steps of our work are to consider multibody contact and an extension of the current physical model due to nonlinear elasticity and large deformations. This work was supported by the German Research Foundation (Deutsche Forschungsgemeinschaft, DFG) Research Grant CRC 666 Subproject A2.

References

[1] Beremlijski P, Haslinger J, Kočvara M and Outrata J 2002 *SIAM J. Optim* **13** 561–587
[2] Wall W A, Frenzel M A and Cyron C 2008 *Comput. Methods Appl. Mech. Engrg.* **197** 2976–2988
[3] Bratzke D 2015 *Optimal Control of Deep Drawing Processes based on Reduced Order Models* Ph.D. thesis TU-Darmstadt
[4] Kikuchi N and Oden J T 1988 *Contact Problems in Elasticity: A Study of Variational Inequalities and Finite Elemend Methods* (*Studies in Applied and Numerical Mathematics* vol 8) (SIAM)
[5] Hüeber S and Wohlmuth B I 2005 *Comput. Methods Appl. Mech. Engrg.* **194** 3147–3166
[6] Wohlmuth B I and Krause R H 2003 *SIAM J. Sci. Comput.* **25** 324–347
[7] Wohlmuth B I 2000 *SIAM J. Numerical Analysis* **38** 989–1012
[8] Ulbrich M, Ulbrich S and Koller D 2013 A multigrid semismooth newton method for contact problems in linear elasticity Tech. rep. Technische Universität Darmstadt, Technische Universität München
[9] Wriggers P, Temizer İ and Hughes T J R 2011 *Computer Methods in Applied Mechanics and Engineering* **200** 1100–1112
[10] Piegl L and Tiller W 1997 *The NURBS Book* (Springer)
[11] Schramm H and Zowe J 1992 *SIAM J. Optim.* **2** 121–152

Numerical Tool Path Optimization for Conventional Sheet Metal Spinning Processes

Benedikt Rentsch[a,*], Niko Manopulo[a] and Pavel Hora[a]

[a] *ETH Zurich, Institute of Virtual Manufacturing, CLA F9, Tannenstrasse 3, 8092 Zurich, Switzerland*

Abstract. To this day, conventional sheet metal spinning processes are designed with a very low degree of automation. They are usually executed by experienced personnel, who actively adjust the tool paths during production. The practically unlimited freedom in designing the tool paths enables the efficient manufacturing of complex geometries on one hand, but is challenging to translate into a standardized procedure on the other. The present study aims to propose a systematic methodology, based on a 3D FEM model combined with a numerical optimization strategy, in order to design tool paths. The accurate numerical modelling of the spinning process is firstly discussed, followed by an analysis of appropriate objective functions and constraints required to obtain a failure free tool path design.

Keywords: INCREMENTAL SHEET FORMING; OPTIMIZATION; FEM

INTRODUCTION

This study presents and discusses a numerical optimization strategy for the first forward pass tool path of an industrial aluminum component. A 3D FE-model is introduced, based on which a multitude of tool paths are processed. The results are captured in the form of objective functions and their corresponding values are used to generate meta-models. On one hand these meta-models can be used to find a constrained optimum, on the other for a global analysis of the process.

LS-DYNA 3D Model & Material Characterization

The simulations are conducted using the commercial FEM software LS-DYNA. Figure 1 displays the setup of the 3D model. Involved components are: Mandrel (red), DuraspinR23 roller (blue), Tailstock (yellow) and blank (green).

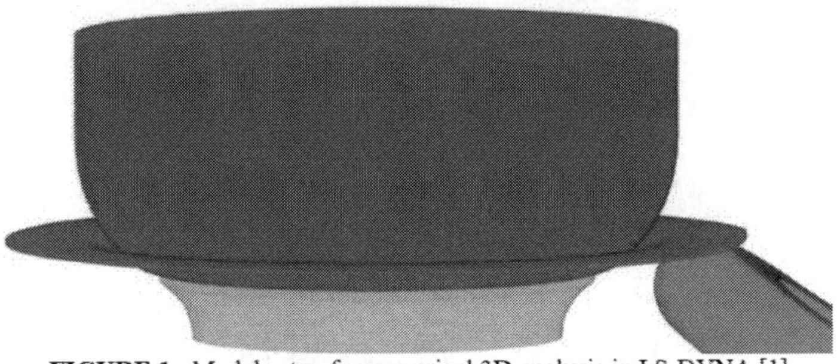

FIGURE 1. Model setup for numerical 3D analysis in LS-DYNA [1]

All components are meshed using quad shell elements. The tools are modeled as undeformable rigid structures. For the blank, the Belytschko-Lin-Tsay shell element formulation with five integration points in thickness direction is applied. The material properties are characterized by an elasto-plastic material model which is based on an arbitrary stress versus plastic strain curve with the von Mises yield criterion and isotropic hardening behavior. The material data is determined by conducting uniaxial tensile tests under quasi-static conditions and hydraulic bulge tests, in order to retrieve material hardening characteristics for plastic strains up to twice the uniform elongation. For extrapolation of the hardening behavior the Ghosh-hardening model is found to be suitable:

$$\bar{\sigma} = k(\epsilon_0 + \bar{\epsilon})^n + C \tag{1}$$

* *Corresponding Author*

where $\bar{\sigma}$ and $\bar{\epsilon}$ are equivalent stress and strain. The Parameters k, ϵ_0, n and C are identified as 132.35, 0.0027, 0.29 and -12.83. The experimental data and the results of the fitting procedure are shown in Figure 2.

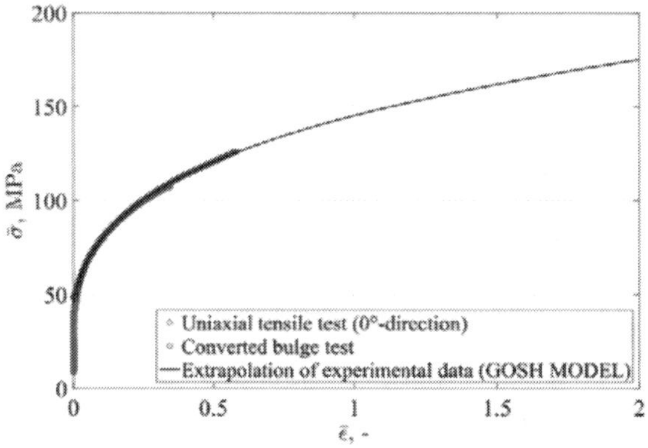

FIGURE 2. Al EN AW-1050A H111 Experimental Data and Ghosh-model fit [1]

The presented FE-model provides an efficient method to map conventional metal spinning process. A more detailed discussion and validation can be found in [1].

Meta-Model & Optimization

In this section a meta-model based methodology to design tool paths is discussed. The process is executed with the optimization tool LS-OPT 5.2 in combination with MATLAB R2016a (Figure 3). To make the analysis most efficient, an iterative strategy is chosen. This enables sequential adding of points until a minimum meta-model accuracy is attained. The iterations are structured into the sub steps: a) Sampling, b) Preprocessing, c) Simulation, d) Postprocessing, e) Build Meta-models and f) Optimization.

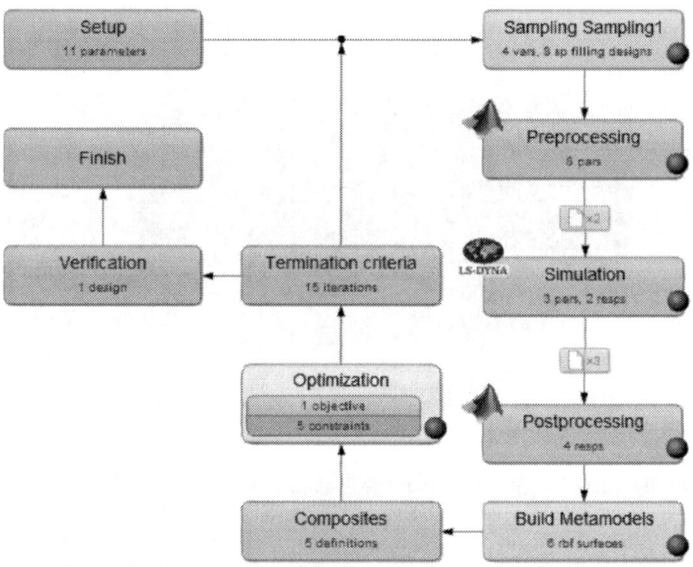

FIGURE 3. Iterative strategy for meta-model based optimization in LS-OPT 5.2

The content of the sub steps is shortly discussed:

a) A space filling sampling design has been chosen, it is obtained by maximizing the minimum distance between any two points. With this method the points are evenly distributed in the parameter space.
b) In the preprocessing step, the tool path load curve is generated based on the parametrized template shown in Figure 4.

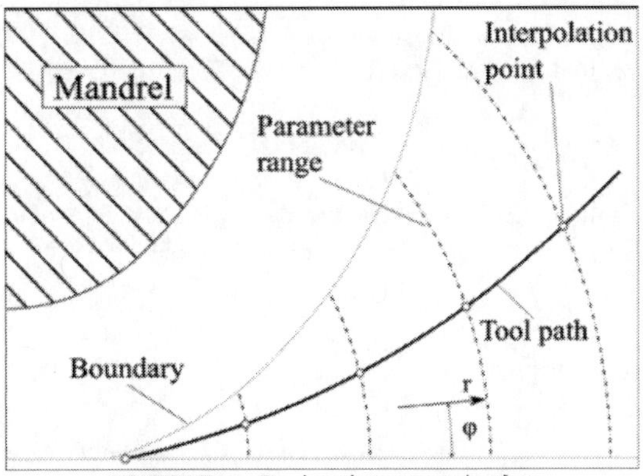

FIGURE 4. Tool path parametrization

Five interpolation points whose coordinates are defined by their radial and angular components (r, φ) are used. This allows the points to move on radial paths (parameter range represented by the dashed lines). The data between the points is interpolated using a Piecewise Cubic Hermite Interpolating Polynomial (PCHIP).
c) The simulation introduced in the previous section is run with the settings for the given iteration.
d) In the step postprocessing, the simulation results are extracted and the objective functions are evaluated as system responses. The main objectives are maximizing the forming and minimizing wrinkle formation.
e) Combining the system responses of all iterations, meta-models are built using neural networks with radial basis functions.
f) Based on the metal-models a constrained optimum is found.

Results

In each iteration eight points are added. After 15 iterations the meta-model accuracy does not significantly increase anymore and the cycle is interrupted. Based on the final configuration, the search for a constrained global optimum is started. Aiming to maximize the forming in in the first forward tool path without wrinkling failure, the result illustrated in Figure 5 arises.

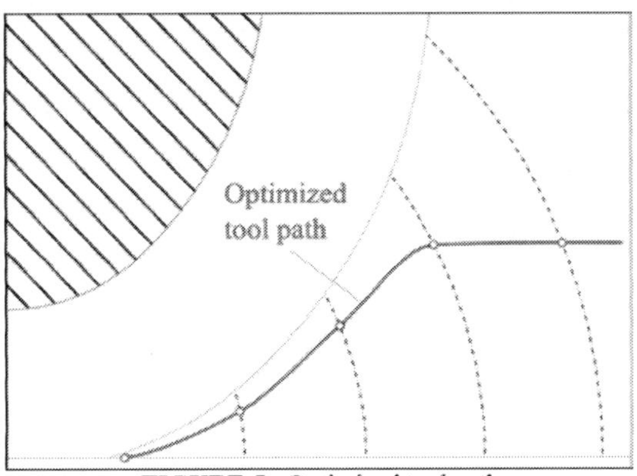

FIGURE 5. Optimized tool path

The introduced methodology provides an effective strategy to design failure free tool paths for the conventional sheet metal spinning process. The method can also be chained and used to design multi-pass processes.

ACKNOWLEDGMENTS

The present work is carried out in the scope of Project Nr. 17231.1 PFIW-IW funded by the Swiss Federal Commission for Technology and Innovation (CTI).

REFERENCES

1. B Rentsch, N Manopulo, P Hora. "Numerical modelling, validation and analysis of multi-pass sheet metal spinning processes". *International journal of material forming.* 2015, submitted for 2nd review

The effect of welding line heat-affected-zone on the formability of tube hydroforming process

Cheng-Kai ChiuHuang, Cheng-En Hsu, Ping-Kun Lee

Product Application Research Section, China Steel Corporation, 1, Chung Kang Road, Kaohsiung, Taiwan

Email:185066@mail.csc.com.tw

Abstract. Tube hydroforming has been used as a lightweight design approach to reduce CO_2 emission for the automotive industry. For the high strength steel tube, the strength and quality of the welding line is very important for a successful tube hydroforming process. This paper aims to investigate the effect of the welding line's strength and the width of the heat-affected zone on the tube thinning during the hydroforming process. The simulation results show that both factors play an important role on the thickness distribution during the tube expansion.

1. Introduction

Tube hydrofoming process has been applied extensively in the automotive body structure [1]. During the process, a tube is first put in the lower die cavity. After upper die closing, two cylinders are sealed on the both side of the tube. Then high pressure (>1000 bar) water is pumped into the tube to make tube expand to the die shape [2]. Most simulation studies focus on the thickness/wrinkle/burst prediction during tube-bending, preforming and hydroforming process [3-4]. Usually, in the finite element simulation model, a uniform mechanical property without considering the welding line's strength is assumed. However, during the tube making process and die-tryout, we found that the difference of mechanical property between the welding line and non-welding line area plays an important role in the tube thinning and accuracy of simulation prediction. In this study, we use finite element software PAM-STAMP 2G to investigate the effect of the welding line's mechanical strength and heat-affected-zone's width on the tube thinning behavior during hydroforming.

2. Model setup

An automotive chassis twist beam (see Figure 1) is selected as the simulation model in this study. In the middle is a V-shape section. On both end, the expansion ratio is around 11% by choosing a round tube of 113 mm outside diameter. The tube grade is 590 MPa thickness is 3.5 mm. The software Pam-Stamp 2G is used to build up the simulation model as shown in Figure 2. Young's modulus is 210 GPa and Poisson's ratio is 0.3. The total element number is 7056. The water pressure is multi-stage linearly increased from 0 to 1500 bar. The axial feeding distance is multi-stage linearly increased from 0 to 26 mm on both tube end.

During the electric resistance welding (ERW) tube making process, the mechanical properties of welding line will be changed by adjusting the heat treatment parameters. Usually, the strength of the welding line is higher than the other area of the material. In this study, we consider three different welding line's strength level during the simulation. Three cases of welding line are shown in Figure 3.

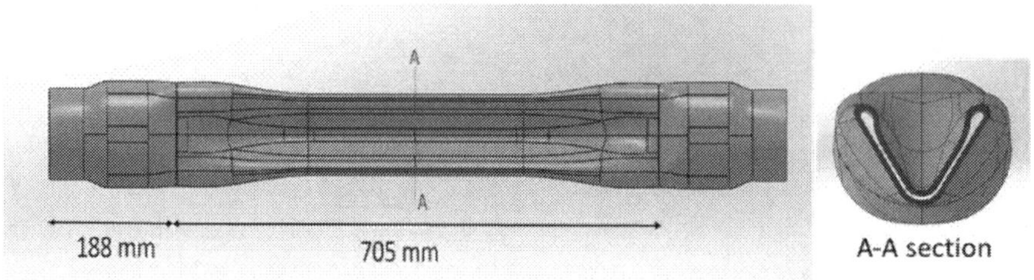

Figure 1. Automotive chassis twist beam structure

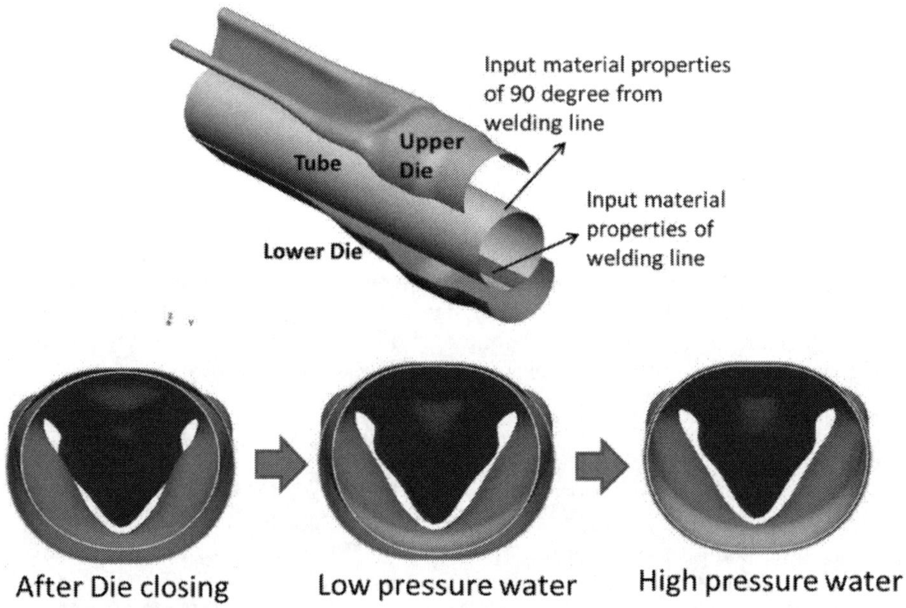

Figure 2. Tube profile change during hydroforming

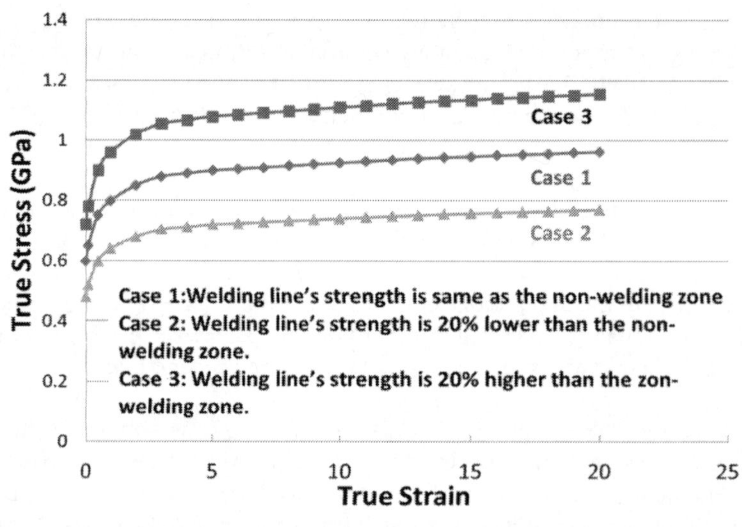

Figure 3. Flow-Stress curves for different welding line's strength

3. Simulation results and discussions

During the tube making process, the welding line's strength and the range of heat affecting zone (HAZ) will be affected by using different online annealing parameters. In the following, we discuss how the welding line's material property and the width of HAZ affect the tube thinning during hydroforming.

3.1. The effect of welding line's material property on the tube thinning

Figure 4(a) shows the material thinning of the welding line after hydroforming when we consider different strength of the welding line. The welding line including the HAZ is assumed to be 25mm. Case 1 is the case when we incorporate a uniform material property (i.e. welding line's strength is same as the non-welding line material). It could be observed that the thinning is very uniform through the width of the welding line, ranging from 4.8% to 6.5%. If the strength of the welding line is 20% lower than the non-welding line area, during hydroforming the thinning of the welding line become more significant as shown in Figure 4(b). Thinning of the welding line increases to range from 19.2% to 26.1% because now the welding line is a soft zone compared to the non-welding area. Figure 4(c) is the case when the welding line's strength is 20% higher than the non-welding line material. It could be observed that the thinning of the welding line is further improved compared to the case 1 and case 2. The main reason is that during tube expansion, the soft area of the tube will be mainly stretched. This result shows that during the heat treatment of tube making process, the mechanical property of the welding line plays an important role in the tube thinning and fracture location. A local thinning on the welding line during hydrofoming should be avoided.

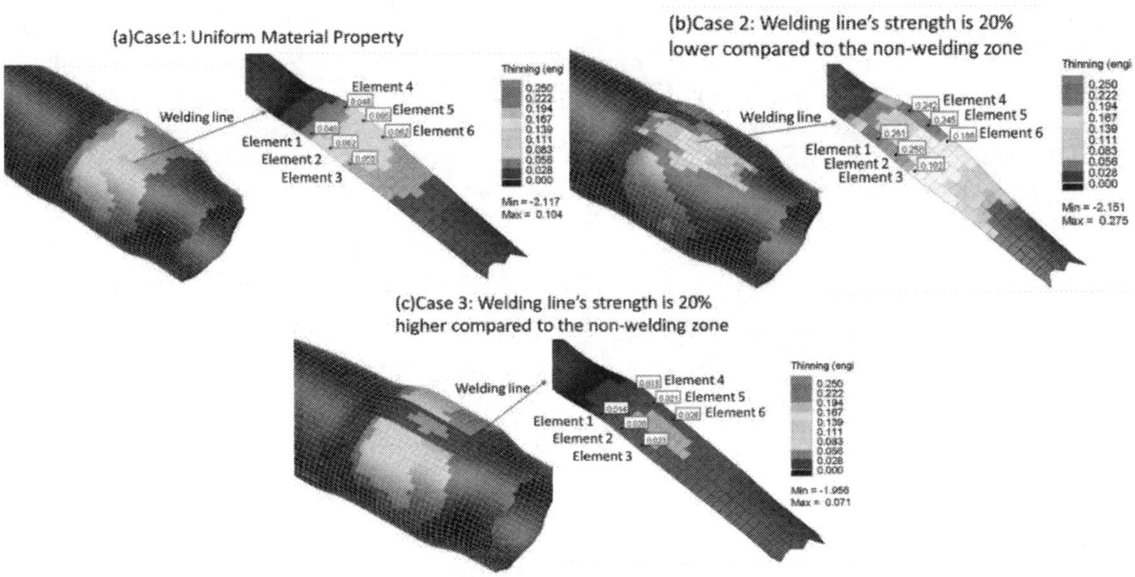

Figure 4. The effect of welding line material property on the tube thinning

3.2. The effect of width of heat-affected-zone on the tube thinning

In this section, we discuss the effect of the width of heat-affected-zone (HAZ) on the tube thinning during hydroforming. Three different cases (HAZ=25, 45, 60 mm) are considered in this study and we all assume the welding line's strength is 20% higher than the non-welding area. When the HAZ is 25mm, after tube expansion the thinning of the HAZ's edge is between 7.4~9.5% as shown in Figure 5(a). When the HAZ is increased to 45 mm and 60 mm, the thinning at the same element has been changed to the range of 6% to 8.1% and 5.4% to 7.3%, respectively. This result indicates that with the increasing of the HAZ, the local thinning of the welding line could be improved, thus lower the risk of bursting during hydroforming.

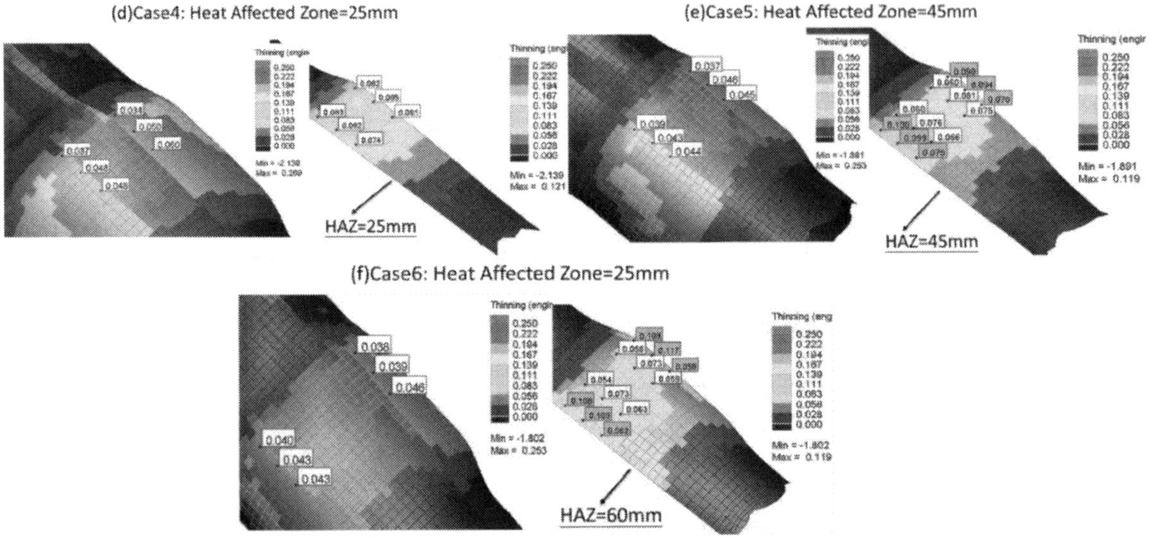

Figure 5. The effect of width of heat-affected-zone on the tube thinning

4. Conclusions

In this paper, we investigate the effect of welding line's mechanical properties and the width of heat-affected-zone on the tube thinning during hydroforming. The simulation results show that the local thinning of the welding line could be improved by (1) Increase the strength of the welding line and (2) Increase the range of the heat-affected-zone. The results also suggest that in the finite element simulation, a model incorporating the welding line and non-welding's line's material properties should have a better prediction of the tube thickness distribution.

References

[1] Ahmetoglu M, Altan T 2000 Tube hydroforming: state-of-the-art and future trends *J. Mater. Process. Technol.* **98** 25-33

[2] Koc M, Altan T 2001 An overview of the Tube Hydroforming Technonogy *J. Mater. Process. Technol.* **108** 384-393

[3] Teng B, Li K, Yuan S 2013 Optimization of loading path in hydroforming T-shape using fuzzy control algorithm *Int. J. Adv. Manuf. Technol.* **69** 1079-1086

[4] Manikandan G, Mathan G, Verma R, Rout D 2013 Elimination of Bursting Defect in Tube Hydroforming Using Finite Element Analysis *J Machin. Form. Technol.* **5** 195-206

Journal of Physics: Conference Series **734** (2016) 032011 doi:10.1088/1742-6596/734/3/032011

Inverse modelling approach in 3-point bending for elasto-plastic material parameter identification of thin spring steel

C Mertin, K Huse and G Hirt

Institute of Metal Forming, RWTH Aachen University, Intzestr. 10, 52056 Aachen, Germany

mertin@ibf.rwth-aachen.de

Abstract. Under process conditions such as bending of flat wire made from high strength spring steel, the occurring strains are many times higher than the maximum strains determined from uniaxial tensile tests. To determine the elasto-plastic material behaviour of high strength spring steel (X10CrNi18-8), an inverse modelling approach using a simple testing method is presented. A 3-point bending test with the resulting force-displacement measurements is used for the inverse analysis. The inverse approach is used for determining the Young's modulus and hardening parameters of the Ludwik-Hollomon's law for bending of high strength spring steel. FE simulations with the optimised material data meet the experimentally measured punch forces during bending. The optimised material data considerably enhances the springback prediction.

1. Introduction

For the electrical industry, springs made from high strength flat wire with distinct elastic properties are formed by multi-stage bending processes. The used materials, i.e. spring steel, are characterised by a brittle material behaviour in terms of high yield stress combined with low ultimate elongation. The occurring strains in bending processes of those materials are many times higher than the maximum strains determined in commonly used standard tests, e.g. tensile tests. Furthermore, the stress profile during bending differs from that in tensile tests. In order to determine the material properties for the required level of strain in the Finite Element Method (FEM) of bending processes, an inverse modelling approach by means of a 3-point bending test is presented.

The basic principle of inverse modelling is to optimise input parameters iteratively by minimising a cost function, e.g. least squares method, which evaluates the difference between the experimental and the simulated response of a physical system. One advantage of this inverse approach is that the material parameters are determined under similar load conditions as in the manufacturing of the components. Yoshida et al. and Eggertsen et al. determined the Bauschinger effect in cyclic bending tests using an inverse approach [1,2]. Antonelli et al. used air-bending tests to determine elasto-plastic material parameters of various steel grades without depicting details of the inverse algorithm [3]. Mendiguren et al. determined material data in v-bending of two sheet materials with thicknesses of 1 mm and 1.5 mm and compared it to material data from tensile test. The results show that material data from tensile test are not adequate to predict the punch force in v-bending [4].

This research discusses an inverse modelling approach for the material parameter identification of high strength spring steel. The inverse algorithm is used to optimise elasto-plastic material parameters by minimising iteratively the deviation of the measured and simulated punch forces. The determined material behaviour is evaluated by comparing the simulated springback angles to experimental results and simulations using extrapolated tensile test data.

2. Material and methods

2.1. Materials

The analysed material is a flat wire made from high strength spring steel (X10CrNi18-8) with a thickness of 0.5 mm and a width of 4.8 mm. Figure 1 a) shows a stress-strain diagram from spring steel determined in tensile test. The ultimate elongation is 2.4 % with an elastic amount of approximately 1 %. The tensile strength is about 1900 N/mm². Due to the steep rise of the engineering stress, the initial yield strength is hard to determine from tensile test data, while $R_{p0.2}$ overestimates the initial yield strength.

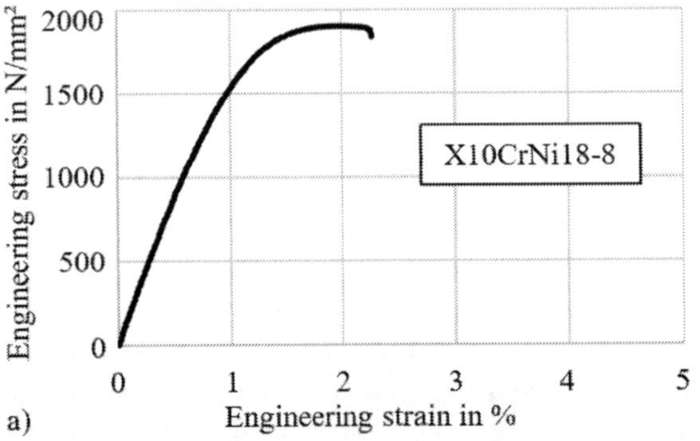

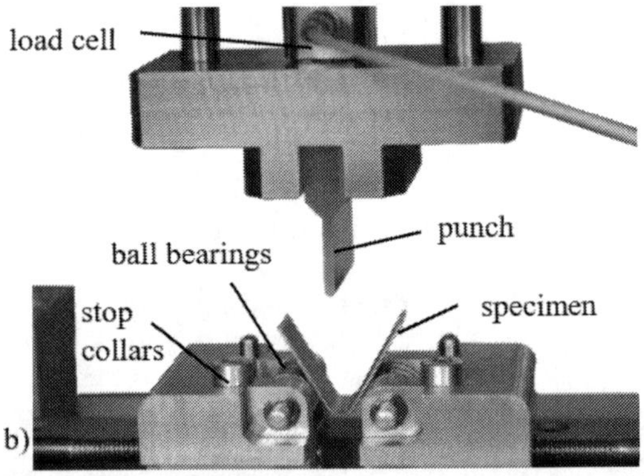

Figure 1. a) Stress-strain diagram from high strength spring steel determined in tensile test; b) experimental set-up of the 3-point bending test used for the inverse approach.

2.2. 3-point bending experiment

The 3-point bending tests were performed on a servo-hydraulic testing machine. Figure 1 b) shows the experimental set-up of the 3-point bending test. The punch radius is 1 mm and moves with a constant velocity of 1 mm/s. The die consists of ball bearings with a radius of 5.5 mm to ensure a minimal friction between die and specimen. For all experiments, a die width of 20 mm and a tool path of 10 mm are used. The force is measured at the centre and vertically to the surface of the specimen. Stop collars ensure a correct positioning of the specimen. The force is measured with an accuracy of ±1 N.

2.3. Finite Element model

The 3-point bending test is simulated with the FE code LS-DYNA. The implicit solver is used for both forming simulation and springback simulation. The tools are considered as rigid bodies. The sheet is discretised using fully-integrated shell elements, which are accurate in springback prediction [5]. The tool displacement is described by the measured displacement-curve from the servo-hydraulic testing machine to minimise the error during the acceleration and deceleration.

3. Inverse modelling approach

3.1. Material model

The Young's modulus basically describes the elastic material behaviour. A sensitivity analysis showed that the Poisson ratio has no considerable effect during bending and therefore, a constant value from literature of 0.3 is used. The plastic material behaviour in terms of the true stress – true strain curve is described by the Ludwik-Hollomon equation (1) and assuming the von Mises yield criterion:

$$\sigma(\varphi) = \sigma_0 + K\varphi^n \tag{1}$$

where σ is the yield stress, φ is the plastic strain, K is a material constant and n is the hardening parameter.

3.2. Inverse algorithm and parameter identification

For the inverse modelling approach, the FE-model was coupled with an optimisation algorithm which is available in the commercial software MATLAB®. The least squares method combined with the Levenberg-Marquardt algorithm is used. Boundary conditions are set to avoid local minima and stabilise the identification algorithm. Force-displacement curves from the 3-point bending experiments are used as reference values. The cost function of the least squares method is defined as:

$$c(\boldsymbol{\rho}) = \sum_{i=1}^{n} (F_i^{exp} - F_i^{num}(\boldsymbol{\rho}))^2 \tag{2}$$

where F^{exp} and F^{num} are the measured and simulated bending forces, $\boldsymbol{\rho}$ is the target vector of the material parameters and n is the number of reference values.

The proposed identification strategy is used to optimise the elastic material parameter and plastic material parameters separately. The Young's modulus (E) is calculated using the elastic region of the force-displacement curve in the beginning of the experiment. The hardening behaviour in terms of the Ludwig-Hollomon parameters are determined using the complete data set from the force-displacement curves. The target vectors $\boldsymbol{\rho}_{elast}$ and $\boldsymbol{\rho}_{plast}$ for the separated approach are defined as:

$$\boldsymbol{\rho}_{elast} = [E], \quad \boldsymbol{\rho}_{plast} = [\sigma_{yo}, K, n] \tag{3}$$

4. Results and discussion

Table 1 shows the start values used for the optimisation algorithm and the calculated parameters of the target vectors $\boldsymbol{\rho}_{elast}$ and $\boldsymbol{\rho}_{plast}$. For the optimisation of the elastic material behaviour, a total of 7 iterations, which result in 16 FE-simulations, are needed. The plastic material parameters are determined within 20 iterations, which are 84 simulations using the Levenberg-Marquardt algorithm. Using the start values, the cost function $c(\boldsymbol{\rho})$ of the least squares method of the target vector $\boldsymbol{\rho}_{plast}$ shows a sum of 642. The optimised material parameters of $\boldsymbol{\rho}_{plast}$ lower the result of the cost function to 33. The comparatively high end value of the least squares results from noise of sliding contacts between the sheet and the die in the FE-model.

Table 1. Calculated material parameters from high strength spring steel.

Material parameter	Unit	Start value	Determined value
E	N/mm²	140,000	170,660
$\boldsymbol{\sigma_0}$	N/mm²	1,250	1,157
K	-	1,000	939
n	-	0.10	0.134

The calculation of the Young's modulus is stable by using a minimum of four reference values during the elastic loading to avoid errors resulting from noise of the force measurement. The optimised value of the Young's modulus is independent from the choice of the start value. The optimised values of the Ludwik-Hollomon parameters are dependent on the start values due to correlation between the parameters. Start values of extrapolated tensile test data support the optimisation with respect to material data that meet the bending experiment. Furthermore, a mutual balancing of the parameters is minimised. However, the result of $c(\boldsymbol{\rho})$ is not necessarily the global optimum. An iterative adjustment of the start values is necessary to avoid local minima and to receive a global minimum in the least squares method. Using underestimated values from extrapolated tensile test data and overestimated ones facilitate the choice of start values.

Figure 2 shows the force-time curve of the 3-point bending experiment compared to simulations with material data from tensile tests and those from the inverse modelling approach. Material data from tensile tests overestimate the bending force in 3-point bending of spring steel by up to 10 %. During bending, a characteristic stress distribution occurs in the cross section of the bending zone – compression to tension. Material data from the inverse modelling approach meet the experimental data. To model the load conditions during bending of high strength flat wire, enhanced material data for numerical investigations should be used.

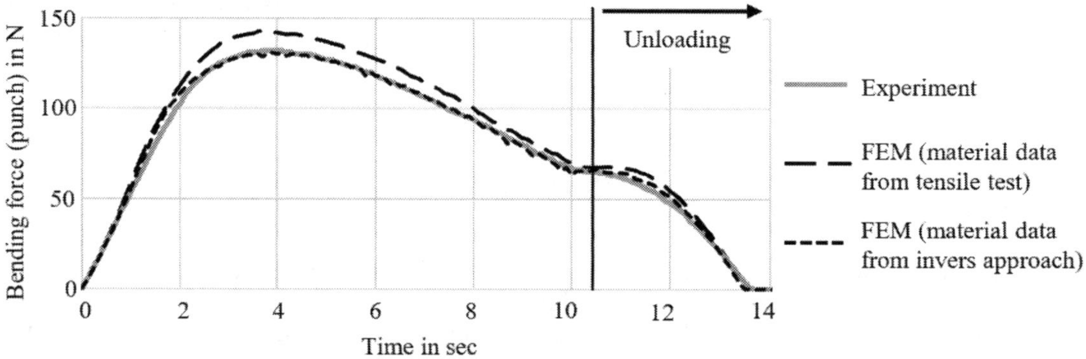

Figure 2. Force-time curves of the bending experiment compared to those from FE-simulations with material data from tensile test and material data from the inverse modelling approach.

To evaluate the quality of the optimised material data, springback angles from the bending experiment are compared to those from FE-simulations using material data from tensile tests and material data from the inverse modelling approach. The experimental and simulated springback angle of the bending tests are summarised in table 2. The bending angle of 128° results from a tool path of 10 mm. Using material data from the inverse modelling approach improves the accuracy of the springback prediction in the FEM of bending by approximately 2°.

Table 2. Experimental and simulated springback angles of high strength spring steel.

	Bending angle	Angle after springback	Deviation to the experiment
Experiment	128°	93.7°	-
FEM (material data from tensile test)	128°	90.0°	3.7°
FEM (material data from inverse approach)	128°	92.2°	1.5°

5. Conclusion

The inverse modelling approach in 3-point bending is suitable to determine the elasto-plastic material behaviour in terms of the Young's modulus and Ludwik-Hollomon law of high strength flat wire made from spring steel. The introduced inverse modelling approach is sensitive to the choice of start values. Start values of extrapolated data from tensile tests improve the quality of the calculation. Simulations using the optimised material data meet the measured punch forces of the 3-point bending experiments. Furthermore, the optimised material data enhance considerably the accuracy of the springback prediction in the FEM during bending of spring steel.

Acknowledgments

The authors gratefully would like to acknowledge the support obtained from the Phoenix Feinbau GmbH & Co. KG for the depicted research.

References

[1] Yoshida F, Urabe M and Toropov V V 1998 *Int. J. Mech. Sci.* **40** 237–49
[2] Eggertsen P A and Mattiasson K 2010 *Eng. Comput* **26** 159–70
[3] Antonelli L, Salvini P, Vivio F and Vullo V 2007 *J. Mat. Pro. Tech.* **183** 127–39
[4] Mendiguren J, Galdos L, Sáenz de Argandoña E and Silvestre E 2012 *Key Eng. Mat.* **504** 889–94
[5] Maker B N and Zhu X 2001 *Livermore Software Technology Corp.* 1–11

Fracture Prediction for High-strength Steel Sheet Subjected to Draw-bending Using Forming Limit Stress Criterion

C Sekiguchi[1], T Hakoyama[2], T Kuwabara[3] and H Fukiharu[4]

[1] JSPS Research Fellow (Doctoral course student), Graduate School, Department of Mechanical Systems Engineering, Tokyo University of Agriculture and Technology, Tokyo, Japan
[2] JSPS Research Fellow(PD), Institute of Engineering, Tokyo University of Agriculture and Technology, Tokyo, Japan
[3] Division of Advanced Mechanical Systems Engineering, Institute of Engineering, Tokyo University of Agriculture and Technology, Tokyo, Japan
[4] Engineering Technology Division, CAE Technology Department, JSOL Corporation, Tokyo, Japan

E-mail: sekiguchi@st.go.tuat.ac.jp

Abstract. A fracture criterion for sheet metals subjected to draw-bending is investigated using the concept of the forming limit stress criterion. The test material used is a 1.0-mm-thick high-strength steel sheet with a tensile strength of 590MPa. The specimen undergoes bending-unbending under tension when passing over the die profile. The drawing speed was set to 5-100 $\mathrm{mm \cdot s^{-1}}$. The magnitude of true stress σ_{DB} when a specimen fractured has been precisely determined. Moreover, multiaxial tube expansion tests of the test material are performed to measure the forming limit stress σ_{PT} of the test material under plane-strain tension. It is found that σ_{DB} is larger than σ_{PT} by 2.8-6.3 %. Therefore, it is concluded that the forming limit stress criterion is effective as a fracture criterion in draw-bending.

1. Introduction

Draw-bending is one of the typical deformation modes in sheet metal forming. It causes serious thickness reduction to sheet metals and very often leads to fracture. Therefore, it is crucial to establish a fracture criterion for sheet metals subjected to draw-bending. A forming limit curve (FLC) [1] is commonly used for the prediction of sheet metal fracture. However, experimental results [2] and numerical analyses [3] suggest that FLCs notably depend on the strain paths. On the other hand, a forming limit stress curve (FLSC) is reported to be independent of strain paths [4]. Yoshida and coworkers measured the forming limit strains and forming limit stresses of an extruded aluminum alloy tube [5] and a steel tube [6], using the servo-controlled, combined tension-internal pressure testing apparatus [7], [8]. They found that forming limit stresses in stress space lie on a single curve, irrespective of the loading paths, when the observed equivalent stress-equivalent plastic strain curves for given linear and combined (two-stage) stress paths are on a single curve. Saito and Kuwabara [9] compared the forming limit stress and the draw-bending fracture of a 590 MPa dual-phase steel sheet . However, the drawing speed was 1 mm/s, much slower than those in industrial press forming processes.

The objective of this study is to clarify whether the forming limit stress criterion can be applied to the fracture prediction of the sheet metals subjected to draw-bending with industrial forming speeds. A high-speed draw-bending testing machine has been developed to measure the fracture stresses of a 590 MPa high strength steel sheet subjected to draw-bending. The draw-bending fracture stresses are compared to the forming limit stress under plane strain tension measured using the multiaxial tube expansion testing method [10].

2. Experimental method

2.1. Test material

The test material used in this study was a 1.0 mm-thick high-strength steel sheet (JSC590R) with a tensile strength of 590MPa. The work hardening characteristics and r-values at 0 and 90° to the rolling direction (RD) of the sample are listed in table 1. Hereafter, the RD, TD and the thickness directions of the sample are defined as the x-, y- and z-axes, respectively.

Table 1. Material properties of the test material.

Tensile direction [°]	$\sigma_{0.2}$ [MPa]	c^{a} [MPa]	n^{a}	α^{a}	r-value[b]
0 (RD)	464	1043	0.177	0.0089	0.71
90 (TD)	486	1037	0.169	0.0094	1.02

[a] Approximated using $\sigma = c(\alpha + \varepsilon^{p})^{n}$ for $0.002 \le \varepsilon^{p} \le 0.093$.
[b] Measured at nominal strain $\varepsilon_{N} = 0.1$.

2.2. Draw-bending test

Draw-bending experiments were carried out to experimentally determine the forming limit stress of the sheet metal subjected to draw-bending. Figure 1(a) shows a schematic diagram of the servo-controlled high-speed draw-bending testing machine developed in this study. The hydraulic cylinder A draws the specimen while the hydraulic cylinder B applies a back tensile force, which is kept constant by a relief valve, to the specimen. Thus, the specimen is subjected to draw-bending as it passes over the die profile. The drawing speed V_{DB} was chosen to be 5, 50 and $100\,\mathrm{mm \cdot s^{-1}}$. The die profile radius R was 4mm. The RD of the test sample was taken to be parallel to the drawing direction.

Figure 1(b) shows the dimensions of the specimen used in this study. A square grid of 2×2 mm^2 was printed on the outer surface of the specimen for strain measurement. The specimen was first bent along a bending line and then a drawing force was applied. The cross-sectional area of the draw-bent section decreases with the progress of drawing as the sheet edges have an arc shape. As a result, the back-tension for making the specimen fracture at the exit of the tool profile could be easily determined.

The strain components were measured at the positions adjacent to the localized necking area, 2 mm distant from the fracture surface, as shown in figure 2(a). The width strains were measured using the deformation of the square grids. The specimens were cut along the white lines by a wire electrical discharge machine for the measurement of thickness. The draw-bending stress σ_{DB} at which the specimen fractured was calculated using the cross-sectional area and the drawing force F measured by a load cell. Figure 2(b) shows the division of the cross-sectional area. It was found that the specimens fractured under plane strain tension at S_1 to S_5, while they fractured under uniaxial tension at S_6 and S_7. From this observation σ_{DB} was determined as follows:

$$\sigma_{DB} = \frac{F - \sigma_0 (S_6 + S_7)}{S_1 + S_2 + S_3 + S_4 + S_5}. \qquad (1)$$

2.3. Multiaxial tube expansion test

Multiaxial tube expansion tests were performed to measure the forming limit stress σ_{PT} for the test material under plane strain tension. The servo-controlled multiaxial tube expansion testing machine [10] was used. An axial load and internal pressure were applied to a tubular specimen by a hydraulic

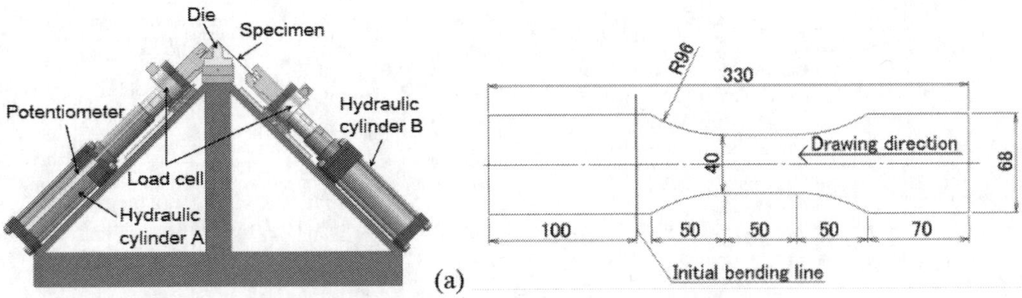

Figure 1. (a) Experimental apparatus for the high-speed draw-bending test. (b) A schematic diagram of a specimen used for the draw-bending test (dimensions are in mm).

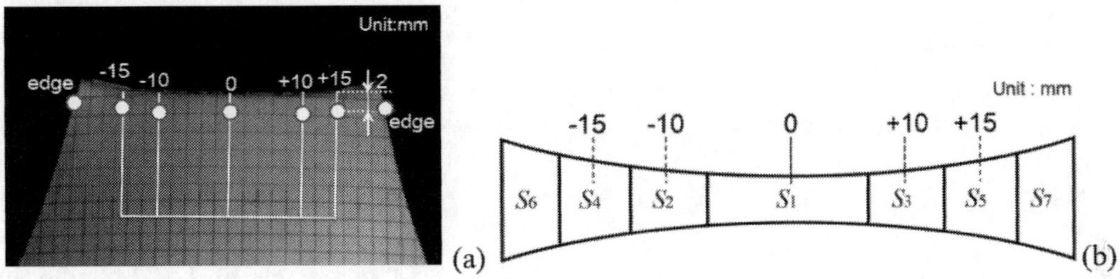

Figure 2. (a) The positions of strain measurement for determining σ_{DB}. (b) Division of the cross-sectional area for determining σ_{DB}. Dimensions are in mm.

cylinder and pressure booster, respectively. Tubular specimens with an inner diameter of 53.9 mm and a length of 230 mm were fabricated by bending the sheet sample into a cylindrical shape and laser-welding the sheet edges together. Linear stress paths were applied to the tubular specimens; the true stress ratio was chosen to be $\sigma_x : \sigma_y = 2:1$. The equivalent plastic strain rates were approximately constant at 5×10^{-4} s^{-1}. Details of the testing apparatus and testing method are given in [10].

3. Results and discussion

Figure 3 shows the distribution of strain components in the width direction of the fractured specimens; $\varepsilon_z^p \approx -0.3$, $-0.04 \le \varepsilon_w^p \le -0.01$, and $0.3 \le \varepsilon_l^p \le 0.35$ for a range of $-15 \le y \le 15$ mm. The drawing speed had little effect on the strain distribution within the limit of the experimental conditions. Therefore, it was concluded that the deformation mode of the fractured specimens was close to a plane strain tension for a range of $-15 \le y \le 15$ mm.

Figure 4 shows the measured results of σ_{DB} and σ_{PT}. The values of σ_{DB} is larger than σ_{PT} by 2.8-6.3 %. Therefore, σ_{PT} can be used effectively as a fracture criterion for draw-bending with a margin of safety. The strain rates under draw-bending with $V_{DB} = 5$, 50 and 100 mm·s^{-1} are approximately 0.2, 2.5 and 4.9 s^{-1}, respectively, while the average strain rate under multiaxial tube expansion test was 5×10^{-4} s^{-1}. σ_{PT} is estimated to increase to 844, 853 and 856 MPa for strain rates of 0.2, 2.5 and 4.9 s^{-1}, respectively, assuming the constitutive model proposed by Kuroda et al. [11], which takes into account the strain rate effect on the flow stress. Thus, the values of σ_{PT} become much closer to the measured values of σ_{DB}. Therefore, it is concluded that the predictive accuracy of σ_{DB} is improved by using σ_{PT} as a fracture criterion and by taking into account the strain rate effect in evaluating σ_{PT}.

Kim et al. [12] investigated the shear fracture of high strength steel sheets under draw-bending, and concluded that deformation-induced heating has a dominant effect on the occurrence of shear failure. For the material used in this study, however, the concept of forming limit stress criterion seems to work well, even without taking account of temperature effect, as a practical method for predicting the draw-bending fracture.

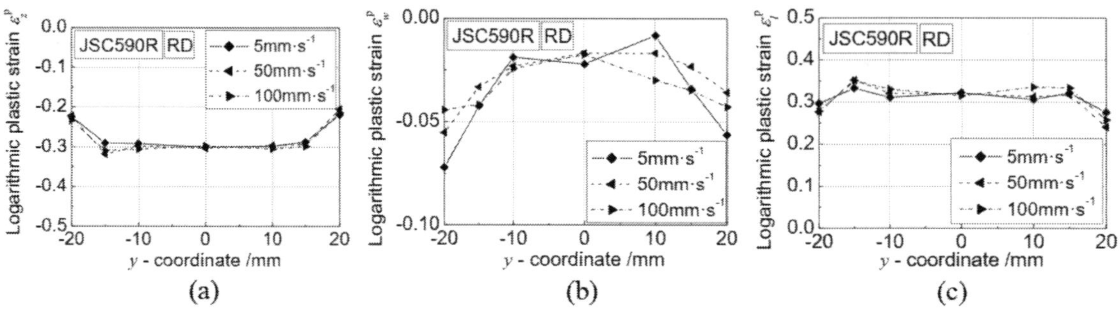

Figure 3. Distributions of the forming limit strains (logarithmic plastic strains) in the width direction (y-direction): (a) thickness strain ε_{t}^{p}, (b) width strain ε_{w}^{p}, and (c) longitudinal strain ε_{l}^{p}.

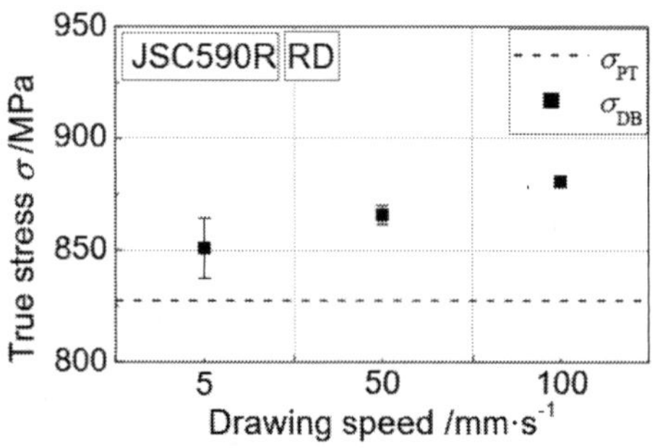

Figure 4. Comparison of the forming limit stresses for the sheet sample under draw-bending (σ_{DB}) and plane strain tension (σ_{PT}).

4. Concluding remarks

The forming limit stress σ_{PT} under plane strain tension can be effectively used as a forming limit stress criterion for the draw-bending of the high-strength steel sheet with a margin of safety. The predictive accuracy of σ_{DB} is further improved by taking into account the effect of the strain rate on the magnitude of σ_{PT}.

Acknowledgements
This work was partially supported by Grant-in-Aid for JSPS Fellows (28 8450).

References

[1] Keeler SP and Backofen WA 1963 *ASM TRANS Q* **56** 25-48
[2] Nakajima K, Kikuma T and Hasuka K 1968 *Yawata Technical Report* **264** 8517-30
[3] Kuroda M and Tvergaard V 2000 *Int. J. Mech. Sci.* **42** 867-87
[4] Arrieux R, Bedrin C and Boivin M 1982 *Proceedings 12th International Deep Drawing Research Group Congress* 61-71
[5] Yoshida K, Kuwabara T, Narihara K and Takahashi S 2005 *Int. J. Forming Processes* **8-SI** 283-298
[6] Yoshida K and Kuwabara T 2007 *Int. J. Plasticity* **23** 1260-84
[7] Kuwabara T, Ishiki M, Kuroda M and Takahashi S 2003 *Journal de Physique* IV **105** 347-54
[8] Kuwabara T, Yoshida K, Narihara K and Takahashi S 2005 *Int. J. Plasticity* **21** 101-17
[9] Saito M and Kuwabara T 2013 *Proceedings of the ASME 2013 International Mechanical Engineering Congress & Exposition (IMECE2013)*
[10] Kuwabara T and Sugawara F 2013 *Int. J. Plasticity* **45** 103-18
[11] Kuroda M, Uenishi A, Yoshida H and Igarashi A 2006 *Int. J. Solids. Struct.* **43** 4465-83
[12] Kim J, Sung J, Piao K and Wagoner RH 2011 *Int. J. Plasticity* **27** 1658-76

A comparison of strain calculation using digital image correlation and finite element software

M Iadicola[1] and D Banerjee[1]
[1]Material Measurement Laboratory, NIST, Gaithersburg, Maryland, USA
E-mail: Dilip.Banerjee@nist.gov

Abstract. Digital image correlation (DIC) data are being extensively used for many forming applications and for comparisons with finite element analysis (FEA) simulated results. The most challenging comparisons are often in the area of strain localizations just prior to material failure. While qualitative comparisons can be misleading, quantitative comparisons are difficult because of insufficient information about the type of strain output. In this work, strains computed from DIC displacements from a forming limit test are compared to those from three commercial FEA software. Quantitative differences in calculated strains are assessed to determine if the scale of variations seen between FEA and DIC calculated strains constitute real behavior or just calculation differences.

1. Introduction

Digital image correlation (DIC) data are being extensively used for many forming applications including constitutive law calibration, benchmark calibration, and for comparisons with finite element analysis (FEA) simulated results. The most challenging comparisons are often in the area of strain localizations just prior to material failure. This is because limit strains produce an inhomogeneous strain field prior to imminent failure. There is not a consensus on what constitutes "raw" data in DIC measurement and how estimates of errors and uncertainties in "raw" measurands (e.g., shape or displacement) affect the comparison of DIC to FEA. Smoothing related to DIC analysis parameters used for matching and the effective gauge length used for strain calculation can affect the limit of spatial resolution and appropriateness of comparison to FEA. For example, strain measurement uncertainty increases when the virtual gauge length (over which strain is ascertained) decreases. While qualitative comparisons can be misleading, quantitative comparisons are difficult because of insufficient information about the type of strain output from FEA software. This is because FEA software often do not specify how exactly strains are computed for a given type of element. Additionally, each software computes strain differently for a given type of element i.e. shell etc. In order to understand this, three benchmark problems are constructed and results obtained from three commercial FEA software are compared with manual computation performed using properties of element shape functions and known displacements at nodes. Finally, strains computed from DIC displacements from a Marciniak [1] forming limit test (just prior to failure) are compared to those from three commercial FEA software. Quantitative differences in calculated strains are assessed to determine if the scale of variations seen between FEA and DIC calculated strains constitute real behavior or just calculation differences.

2. Overview of DIC system and strain calculation procedure

Although strains are frequently reported, the basic level of output from stereo digital image correlation measurements is an array of 3D initial shape positions (X,Y,Z) and the associated array of displacements (U,V,W) in that same coordinate system. In the measured Marciniak test reported here, a pair of 5MP CCD grey-scale lab grade cameras with 35 mm focal length lenses are used to acquire at 5 frames/s. The average magnification of the system is approximately 22.4 pixels/mm. Correlations were done over squares of 19 pixels (≈0.85 mm) with a raster step of 7 pixels (≈0.31 mm). The DIC manufacturer software [**2**] calculates strains using neighboring data points and applies a Gaussian weighted smoothing with a diameter of five step points (≈ 1.25 mm here) centered on the point of interest decaying to a 10 % weight at the edges. Details of the software calculation are not known. Assuming an effective strain gauge length can be approximated as (Step Size x (Smoothing Diameter -1)+Subset Size) / Average Magnification, the gauge length is approximately 2.10 mm for the DIC software with smoothing and 1.16 mm without smoothing. To compare the DIC software and typical FEA calculations to manually calculated strains, the DIC data were organized into four node elements. For the manual calculations, the elements are defined in a local (ξ, η) coordinate system where linear shape functions are used to determine the displacement gradients in (ξ, η)

space that are transformed back to physical space. Using these gradients the Lagrangian strains are calculated including the second order terms at any points of interest in the element. From these values the principal strains and true Hencky strains (ε_{xx}, ε_{yy}, ε_{xy}) are calculated at each point of interest. This formulation of strain does not include the effects of an element that is not initially flat. Although the specimen is nominally flat to begin the test, the measured values do not exactly match a flat X-Y plane. To correct for this each element was rotated flat as part of the manual strain calculation. This flattening procedure induced small strains in the calculation on the order of 10^{-6} strain. Calculations for strains at the integration points were done on each individual element using the lower left node as the reference point for rotation to flat. Strain at each node was calculated separately on each of the four elements that neighbor that node (after the afore mentioned element rotation), and then the strains were averaged at that point weighted by the area of each contributing element. Missing elements due to lack of data at a node point were not included in the average calculation.

3. FEA analysis overview and software used in this study – description of elements used

Three different problems were considered: (A) a 2-element uniaxial (strictly planar) deformation problem, (B) a 2-element by 2-element patch from the high strain band (see Fig. 3), and (C) the entire area near the high strain band at about 0.4 s prior to fracture in Marciniak test (see Fig. 3). FEA study was conducted using three commercially available software: ABAQUS, ANSYS, and LS-Dyna[i] [3-5]. Based on our selection of problems two different types of elements were used, either 4-noded plane strain or 4-noded shell/membrane element. For Problem A plane strain elements were used since there was no variation in Z-coordinate among the nodes and the strains were functions of planar coordinates alone and the out-of-plane normal and shear strains are equal to zero. Plane strain elements are defined in the X–Y plane, in which all loading and deformation occur. Shell elements are used to model structures in which dimension in one direction is significantly smaller than the other dimensions. Typically, shell elements use this condition to discretize a domain by defining the geometry at a reference surface, where the thickness is defined through the section property definition. Traditional shell elements have displacement and rotational degrees of freedom. Membrane elements (and shell elements with a membrane option) are used for Problems B & C. These elements are akin to shell elements and are essentially surface elements that transmit in-plane forces only (no moments) with no bending stiffness. Element formulations vary among commercial software. In this work, efforts were made to use comparable elements in FEA software to solve the three problems. The elements used are: a) ANSYS (PLANE182 and SHELL181 with membrane option), ABAQUS (continuum plane strain CPE4 and membrane M3D4), and LS-Dyna (Shell element with plane strain and fully integrated option). Although, both reduced and fully integrated elements have been used, only results from fully integrated elements are reported here for brevity.

4. Benchmark problems and discussion of results

Problem A, the 2-element model (N_1-N_2-N_4-N_3 and N_4-N_6-N_5-N_3), was constructed using quadrilateral elements (Fig. 1). The left 2 nodes (N_1 and N_2) were constrained in X, and Y directions. A 0.1 mm displacement in the X-direction was applied to nodes N_3 and N_4 and a 0.3 mm displacement in the X-direction was applied to nodes N_5 and N_6. ANSYS Plane182, ABAQUS CPE4 and M3D4, and LS-Dyna Shell (with plane strain option) elements were used to compute the true normal and shear strains at integration points. Note that each fully integrated elements [6] has 4 integration points that are numbered starting from lower left corner and are incremented in counterclockwise manner in each element. Table 1 lists ε_{xx} computed using the procedure described in Section 2 above (hereafter called "manual calculation") and those computed using the three commercial FEA software. The differences between strains obtained with manual calculation and from each software are listed under the "Difference column". Similar results are obtained for the ε_{yy}, and ε_{xy} but are not shown for brevity. It is clear from this table that although all FEA software produce reasonable values, ABAQUS membrane element is the most consistent with the manual calculation. A similar exercise was conducted on a 2 element by 2 element patch from the Marciniak test just before onset of failure (Fig. 3), Problem B. This patch is shown in Fig. 2. For this exercise, the displacement values at each node were used as boundary conditions in a simple static analysis. Strains computed at each integration point were compared to those computed manually. In addition,

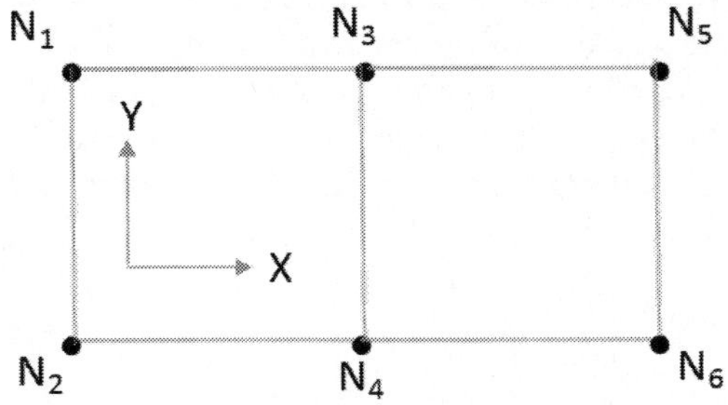

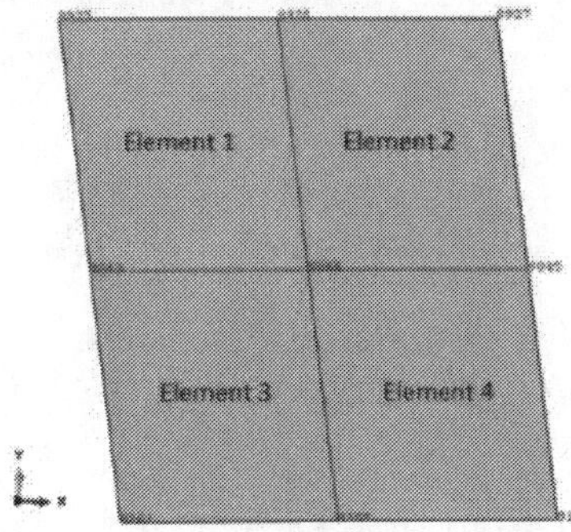

Fig. 1 Problem A and 2 Problem B.

average nodal strains computed at the common node in the center of the patch are compared. Table 2 shows the true strain values at the integration points. Only results from element 4 are shown here. Again,

Table 1 Problem A ε_{xx} at integration points

	int	Manual	ABAQUS - CPE4		ABAQUS-M3D4		ANSYS-Plane182		LS-Dyna-full integration	
	pnts	calculation	Value	Difference	Value	Difference	Value	Difference	Value	Difference
Element 1	1	0.095348	0.095269	-7.90E-05	0.095349	6.08E-07	0.095142	-2.07E-04	0.095369	1.70E-05
	2	0.095348	0.095269	-7.90E-05	0.095349	8.17E-09	0.095142	-2.07E-04	0.095234	-1.18E-04
	4	0.095349	0.095268	-8.10E-05	0.095349	-3.15E-08	0.095141	-2.08E-04	0.095382	2.90E-05
	3	0.095349	0.095268	-8.10E-05	0.095349	-6.31E-07	0.095141	-2.08E-04	0.095253	-1.00E-04
Element 2	1	0.182323	0.182319	-4.00E-06	0.182323	-4.61E-07	0.181814	-5.09E-04	0.182340	1.70E-05
	2	0.182323	0.182319	-4.00E-06	0.182323	-4.61E-07	0.181814	-5.09E-04	0.182290	-3.30E-05
	4	0.182323	0.182319	-4.00E-06	0.182323	-4.68E-07	0.181814	-5.09E-04	0.182340	1.70E-05
	3	0.182323	0.182319	-4.00E-06	0.182323	-4.68E-07	0.181814	-5.09E-04	0.182310	-1.30E-05

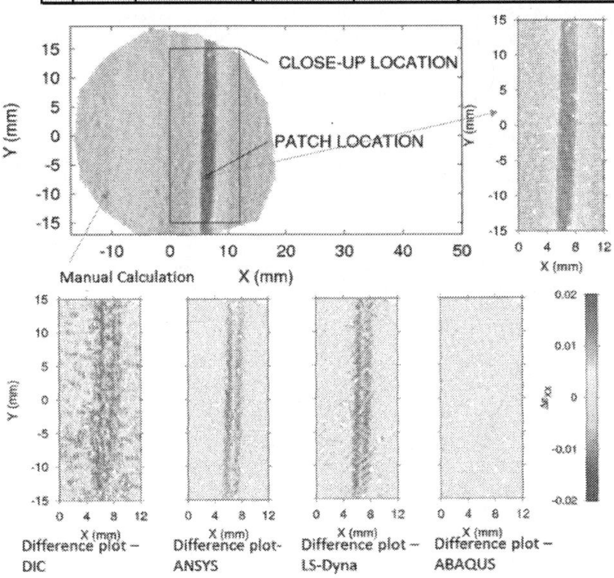

Fig. 3. Manual and FEA computed ε_{xx} difference plots in a high strain band prior to fracture in a Marciniak test.

ABAQUS with membrane elements provide the best match to the manual calculation. Both LS-Dyna and ANSYS values are similar and the difference between manual calculation results and values predicted by these software are on the order of 10^{-3}. Note that ε_{xy} are tensor shears in manual calculation which is half the value in vector or Voigt notation in FEA. Table 3 shows the results at the central node, 9044. It is clear that ABAQUS predicted nodal values are closest to manual calculation values. ANSYS predicts slightly larger differences and predicts somewhat better values than LS-Dyna. There are many assumptions in the averaging technique and it is not clear how the nodal strain values are averaged by each commercial FEA software. Finally, the nodal results are compared in Fig. 3 for the actual forming limit test just prior to failure. Here, manual average ε_{xx} are shown in the upper portion of Fig. 3, while the bottom portion shows the difference between this calculation with DIC software (smooth), ANSYS membrane 181, LS-Dyna shell, and ABAQUS membrane elements. Gaussian smoothing applied to the manual calculated strains (not shown here) resulted in very similar results to the DIC software (left most figure in the bottom of Fig.3). In this scale of +/- 0.02 strain difference, the DIC smoothing option shows somewhat large strain difference from those obtained with manual calculation. While FEA software used shows better agreement with manual calculation in general (excepting near the high strain band), ABAQUS membrane element provides the best match with the manual calculation. Both ANSYS and LS-Dyna show similar behavior and show strain up to about 0.02 strain below the manual calculation at the strain band, which is somewhat large difference. The present exercise makes the point that reporting of strain values should be made with a better description of how the strains are calculated, what types of elements (along with integration scheme) are used, and what type of averaging scheme is used for obtaining nodal values from element integration point values.

Table 2. Problem B results at integration points.

Element 4		**Manual calculation**		**ANSYS-mem181-full int**			**Difference**		
Int Pt	**Exx**	**Eyy**	**Exy**	**Exx**	**Eyy**	**Exy**	**Exx**	**Eyy**	**Exy**
1	0.361250	-0.023102	-0.001579	0.36577	-2.11E-02	2.66E-03	4.52E-03	1.97E-03	5.82E-03
2	0.361250	-0.006998	-0.001845	0.3554747	-5.94E-03	-5.19E-03	-5.78E-03	1.06E-03	-1.51E-03
3	0.361330	-0.004073	0.004426	0.3557066	-6.08E-03	8.29E-03	-5.61E-03	-2.01E-03	-5.60E-04
4	0.361320	-0.020195	0.004531	0.3659433	-2.12E-02	1.64E-02	4.61E-03	-1.03E-03	7.35E-03
				ABAQUS-M3D4-full int			**Difference**		
1	0.361250	-0.023102	-0.001579	0.361250	-0.023102	-0.003144	0.00E+00	2.00E-07	1.37E-05
2	0.361250	-0.006998	-0.001845	0.361253	-0.006998	-0.003672	3.00E-06	-1.00E-08	1.76E-05
3	0.361330	-0.020195	0.004531	0.361335	-0.020195	0.009076	5.00E-06	3.00E-07	1.42E-05
4	0.361320	-0.004073	0.004426	0.361325	-0.004073	0.008869	5.00E-06	2.00E-08	1.71E-05
				LS-Dyna-membrane-full int			**Difference**		
1	0.361250	-0.023102	-0.001579	0.354880	-0.016713	-0.039607	-6.37E-03	6.39E-03	-3.64E-02
2	0.361250	-0.006998	-0.001845	0.350370	-0.001411	-0.041182	-1.09E-02	5.59E-03	-3.75E-02
3	0.361330	-0.020195	0.004531	0.351100	-0.001625	-0.041257	-1.02E-02	1.86E-02	-5.03E-02
4	0.361320	-0.004073	0.004426	0.355600	-0.016928	-0.039682	-5.72E-03	-1.29E-02	-4.85E-02

Table3. Problem B results at central node (9044).

Nodal strains at the central node (9044)

	Manual	ANSYS-membrane 181		ABAQUS-M3D4		LS-Dyna-membrane	
Strains	Value	Value	Difference	Value	Difference	Value	Difference
ε_{xx}	0.36711	0.36720	-8.948E-05	0.367076	3.35E-05	0.36003	7.08E-03
ε_{yy}	-0.025112	-0.02433	-0.0007775	-0.025218	1.05E-04	-0.02031	-4.80E-03
ε_{xy}	-0.00819	-0.01227	-4.11E-03	-0.016534	1.54E-04	0.00033	-1.67E-02

5. References

[1] ISO 2008, "Metallic materials-Sheet and strip-Determination of forming-limit curves-Part 2: Determination of forming-limit curves in the laboratory", ISO12004-2:2008(E).
[2] VIC3D (v2010), http://www.correlatedsolutions.com/.
[3] ABAQUS 13.6 software, Dassault Systemes, http://www.3ds.com/.
[4] ANSYS Mechanical Release12.1, ANSYS Inc., http://www.ansys.com/.
[5] LS-Dyna-971, Livermore Software Technology Corporation, http://www.lstc.com/.
[6] Zienkiewicz O C, Taylor, R L and Zhu J Z 2013, *The Finite Element Method: Its Basis and Fundamentals*, 7th Edition (Oxford: Butterworth-Heinemann).

[i] Certain commercial software or materials are identified to describe a procedure or concept adequately. Such identification is not intended to imply recommendation, endorsement, or implication by NIST that the software or materials are necessarily the best available for the purpose.

Process Windows for Sheet Metal Parts based on Metamodels

D. Harsch [a,*], J. Heingärtner [a], D. Hortig [b] and P. Hora [c]

[a] *inspire AG – ivp (Institute of Virtual Manufacturing), Technoparkstrasse 1, 8005 Zürich, Switzerland, harsch@inspire.ethz.ch, www.inspire.ethz.ch*

[b] *Daimler AG, Käsbrünnlestr., 71059 Sindelfingen, Germany, dirk.hortig@daimler.com, www.daimler.com*

[c] *ETH Zurich, Institute of Virtual Manufacturing, Tannenstrasse 3, 8092 Zürich, Switzerland, pavel.hora@ivp.mavt.ethz.ch, www.ivp.ethz.ch*

Abstract. Achieving robust production of deep drawn sheet metal parts is challenging. The fluctuations of process and material properties often lead to robustness problems. Numerical simulations are used to validate the feasibility and to detect critical regions of a part. To enhance the consistency with the real process conditions, the measured material data and the force distribution are taken into account. The simulation metamodel contains the virtual knowledge of a particular forming process, which is determined based on a series of finite element simulations with variable input parameters. Based on the metamodels, process windows can be evaluated for different parameter configurations. This helps improving the operating point search, to adjust process settings if the process becomes unstable and to visualize the influence of arbitrary parameters on the process window.

Keywords: Deep Drawing; Numerical Simulation; Metamodeling; Process Optimisation; Process Window

INTRODUCTION

Scattering of process parameters and material properties as well as the increasingly higher tolerance requirements often lead to robustness problems during production [1], [2]. The factors influencing robustness are not systematically measured and online action is limited to a manual intervention on a trial-and-error basis. Thus the quality of the outcome is strongly correlated to the experience of the staff and eventual corrections on the tools are costly. Other circumstances, such as the reduction of material thickness, complex geometries with sharp radii and the flexible choice of the press, limit the process stability even further.

In order to increase the significance and accuracy of the simulations, a variety of different measures are taken into account, such as the use of more accurate material models based on experiments, the consideration of the press construction and the implementation of digitised tool geometries [3]. The purpose of this work is to evaluate the influences of the process settings on the critical regions of the part. A methodological approach based on simulation metamodels is developed to find a suitable operating point, particularly by means of two-dimensional process windows. Thus the influence of a particular process parameter can be visualised and the understanding of a specific process can be improved.

PROCESS DESCRIPTION

The part presented in this work is the trunk lid of the actual B-class of Daimler, which can show several imperfections during production, such as splits, wrinkling and insufficient stretching, which can be predicted well in simulation (see Figure 2a). However, skid impact lines and sink marks are not easily detectable, because of missing meaningful, virtual result variables.

To enhance the accuracy of the material model of the DC06 steel, various experiments are used, such as tensile experiments, Bulge-Tests and Nakajima experiments. The evaluation of the material model with all its details is shown in [3]. Another improving adjustment is considering the coordinates of the cushion quills in the press construction. The acting forces from the die cushions are transferred via cushion quills to the blank holder (see Figure 1) [3].

In simulation it is common practice to use initial pressures, which are distributed homogeneously. As the tools are assumed to be rigid in AutoForm, the deflection of any tool is locked. Because of the equilibrium condition all acting forces can be reduced to one total force with a specific force application point. These simulative adjustments do not influence the simulation result [3]. This measure allows in simulation to redistribute the acting forces inhomogeneously on the binder surface by moving the force application point, in order to better map the set values on serial press.

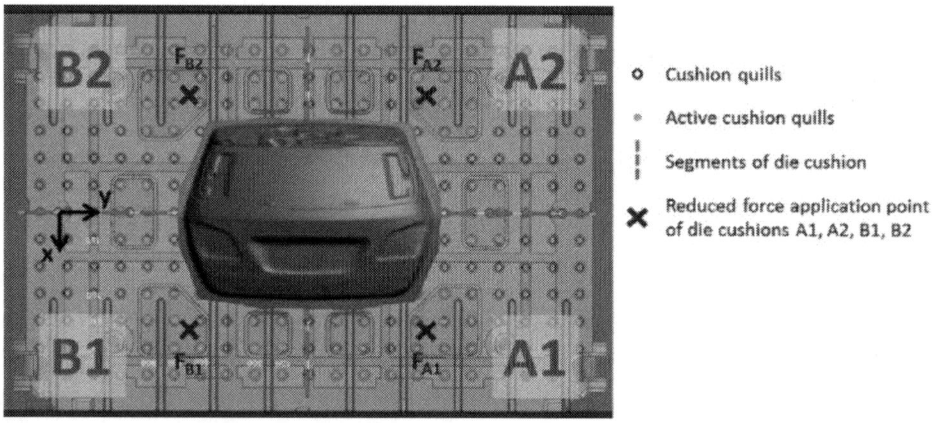

Figure 1: Binder suspension on cushion quills and segments of die cushion [3]

Once the tools have been milled, the surfaces have to be processed by hand in tool tryout. Regions, where splits, wrinkles, sink marks or skid impact lines occur, are then reworked. Furthermore, most of the efforts are put into the tryout of the drawbeads to adjust the draw-in of the blank, which is done in many loops. The influences of the adjustments made on part quality are checked after each loop in the tryout press. Therefore tool geometries after tool tryout vary strongly from the designed tool geometries of the process plan. Because of large differences between the simulations coming from the process engineering and the simulation with digitised tool geometries (with identical simulation settings), the tools are digitised after the first tryout loop with a GOM ATOS measurement system. Thus the major problems are remedied, but the complex fine-tuning is still pending.

The simulation is then built up and calculated in AutoForm R3.1. The total binder force and the coordinates of the force application point are implemented in AutoForm with the function *Columns (User def.)*. In case of the presented part, the eccentricity in y-direction can be neglected because of symmetry.

PROCESS WINDOWS

Finding a suitable operating point in the field of simulation design is often challenging. Many parameters, such as binder forces, force distributions, blank size and position are influencing the process distinctively. In many cases it is unclear how strong a specific parameter is influencing the determined quality criteria, especially when taking multiple design parameters into account. In process engineering the simulation parameters are first changed iteratively, which is strongly depending on the user's understanding of the process behaviour. Thus, the quality of the outcome and the required time for iterative optimisation correlates with the experience of the process engineer.

For more complex and larger car body shell parts, for example sidewall frames, the process parameters are optimised based on variant simulations. Depending on the number of independent variation parameters, the simulations are calculated based on a design of experiment. Then, for each detected quality criterion a metamodel can be evaluated, which allows to analyse the influences and sensitivities of the varied parameters on the specific quality criterion. Moreover, in the current state of the art different process windows can be displayed, such as one- and two-dimensional process windows based on conservative minimum/maximum analysis, convex hulls and multi-dimensional hypercubes parallel to the axes. For each of these simplification methods some information gets lost [4] [5].

For the trunk lid in this work four design parameters are optimised over 30 simulations: The total binder force, the x-coordinate of the force application point (to allow different force distributions), the force on the die inlet and the x-coordinate of the blank position. The die inlet is preventing skid impact lines. Because of the symmetry of the part, the y-coordinates of the forces and of the blank position stay constant. After simulating in AutoForm the simulation results are imported in Matlab. For each defined quality criterion a metamodel is evaluated and validated with the *leave-one-out cross validation*. The used types of metamodels are based on the *Response Surface Methodology*, *Kriging method* or *Radial Basis Functions*. The polynomial degree is limited to a quadratic base model without interaction coefficients. Each metamodel is limited by an upper or lower limit, e.g. *Max. Failure* in AutoForm (ratio between the simulated major strain and the major strain from FLC for a given minor strain [7]) at 0.8. By keeping the die inlet force and the coordinates of blank position constant at a predefined value, the contour lines of the metamodels can then be projected in the plane of the remaining parameters "total binder force – force distribution" (see Figure 2b).

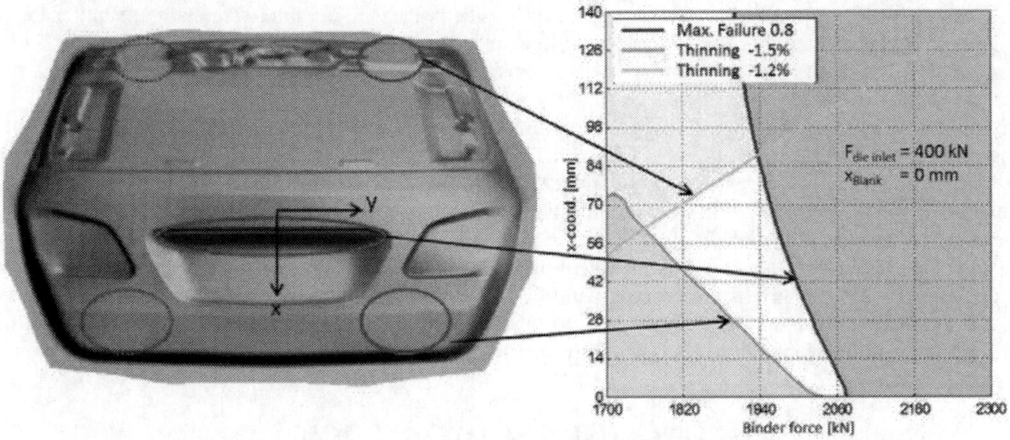

Figure 2: Definition of quality criteria (a) and their process limit in the process window (b)

The light-blue and light-green coloured regions in Figure 2b illustrate the parameter combinations, where the corresponding quality criterion regarding stretching is limiting the process. In the dark-blue region the *Max. Failure*-criterion exceeds the predefined limit of risk of splits. The remaining white triangle represents the valid process window. It shows the amount of force necessary to avoid insufficient stretching and the optimal force application point for which all of the predefined quality criteria are fulfilled. In the shown process window above this point lies around 1900 kN and 56 mm.

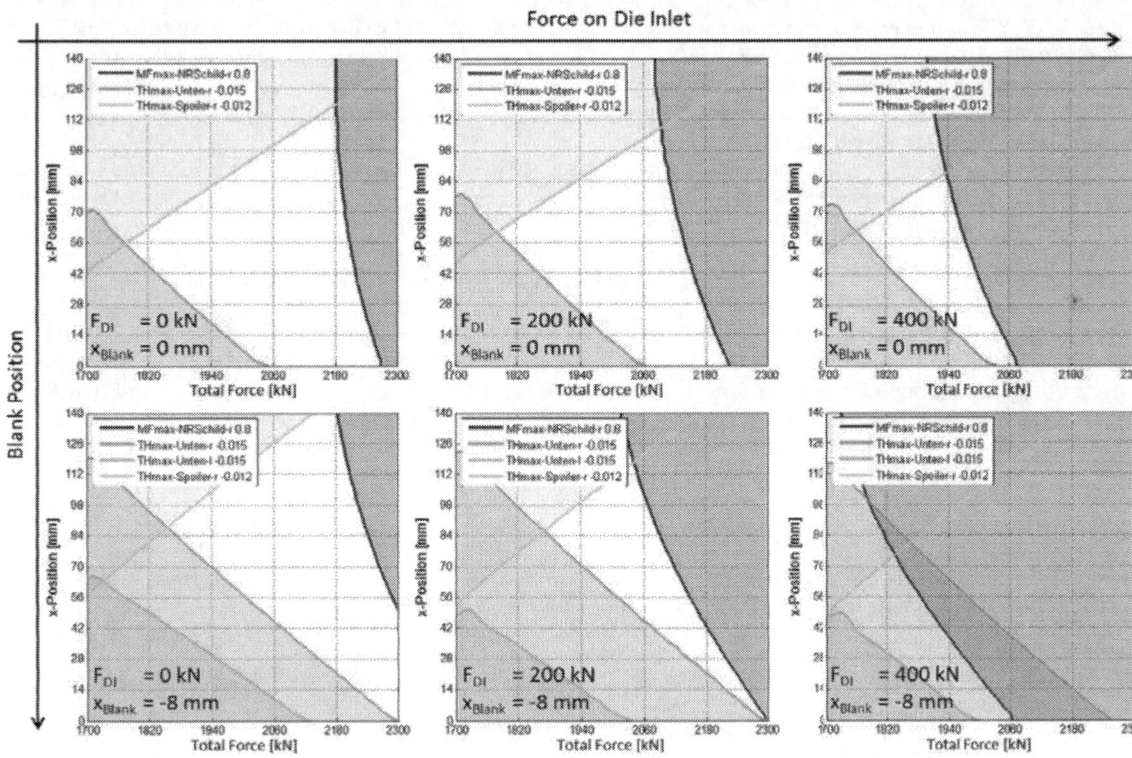

Figure 3: Set of process windows with different parameter settings

By repeating this method for different die inlet forces and blank positions a set of different process windows can be created (see Figure 3). The process windows in the upper row have a constant blank position at the nominal value of 0 mm and in the lower row of minus 8 mm. Considering the windows column by column the die inlet force stays constant at 0, 200 and 400 kN respectively. It becomes obvious, that the die inlet force does not influence the stretching at all, whereas the failure criterion is affected significantly. The higher the die inlet force is, the smaller the process window gets, which means the robustness could be influenced negatively. On the other hand the die inlet avoids skid impact lines. Therefore a suitable compromise needs to be found, which reduced the skid impact lines to a minimum but keeps the process in a safe zone.

By moving the blank out of the nominal position about 8 mm in spoiler direction, the process boundary of the stretching in the lower part region moves upward right. This can be explained by the fact that less material is in contact with the lower binder region and consequently the material is held back less. To counteract this, the force in the near binder region should be increased, either by moving the force application point in positive x-direction (redistribution of binder force) or by raising the global binder force. Both of these measures are intended to increase the corresponding retention force on the blank. In general the movement of the blank position out of its nominal value does not enhance the size of process window. Indeed the process window is even getting smaller. Actually, with an eccentric blank position and a maximal die inlet force of 400 kN, no binder force and force distribution combination can fulfil all required limit values of the quality criteria, which means no process window would be available for this parameter combination.

CONCLUSION AND OUTLOOK

The presented methodological approach allows to visualize two-dimensional process windows as a function of two variation parameters containing predefined quality criteria. By changing additional parameters gradually and showing the resulting process windows in rows (third parameter) and column by column (fourth parameter), four-dimensional process windows can be illustrated. On one hand the changing of the size of the process window can be seen at one glance, on the other hand the individual effects of the process design or noise variables on the quality criteria can be observed. Thus this way of illustrating process boundaries provides a spatial feeling about the process window. Moreover the process behaviour of a complex part can easily be visualized and the process understanding can be enhanced. The illustrated process windows have been compared with the real tryout steps that have been done on the presented tool. As far as a comparison is possible, a very good accordance of the tendency between the virtual tryout and the real behaviour of the part could be determined [8].

Skid impact lines and sink marks cannot be predicted very well. Therefore novel result variables should be developed, in order to improve the prediction accuracy.

Furthermore the methodological approach can also be used for draw-in regions. Each draw-in measuring point can be defined as a normal quality criterion and limited by a boundary value. Based on the calculated metamodels an inline feedback control can be realised [6].

ACKNOWLEDGMENTS

The authors are very grateful to the CTI (The Swiss Innovation Promotion Agency) for the financial support of this work within the project 17366.1 PFIW-IW. Also thanks to Daimler AG for providing the material, for digitising the tools and for the great opportunity to perform experiments on the press. The project partner AutoForm Engineering GmbH and GOM International AG are also gratefully acknowledged for their contribution on the project results.

REFERENCES

[1] D. Hortig. *Experiences with the Robustness of sheet metal forming processes*. Forming Technology Forum 2011, Proceedings, Zürich, 2011.

[2] P. Hora, J. Heingärtner, N. Manopulo, L. Tong. *On the way from an Ideal Virtual Process to the Modelling of the Real Stochastic*. Forming Technology Forum 2011, Proceedings, Zürich, 2011

[3] D. Harsch, J. Heingärtner, D. Hortig, P. Hora. *Virtual tryout planning in automotive industry based on simulation metamodels*. The International Deep Drawing Research Group, Proceedings, 2016

[4] C. Annen. *Entwicklung einer neuen Methode zur Ermittlung und Visualisierung von robusten Prozessfenstern in der Blechumformung*. Diss. ETH Nr. 20573, ISBN 978-3-906031-35-4, 2012.

[5] N. Stander, W. Roux, T.Goel, T. Eggleston, and K. Craig. *LS-OPT User's Manual*. Livermore Software Technology Corporation, pp. 9-11, 2009.

[6] P. Fischer, D. Harsch, J. Heingärtner, Y. Renkci, P. Hora. *Inline feedback control for deep drawing applications*. The International Deep Drawing Research Group, Proceedings, 2016

[7] AutoForm. *Helpviewer 11.5.4*. AutoForm Engineering GmbH, Version AutoForm R3.1, 2011.

[8] N. Schelhammer, Mercedes-Benz Cars Operation – Press line 3 & 4 (TF/PW4), Sindelfingen; *Private communications*, 2014.

Numerical determination of the onset of local necking using time dependent evaluation method and dynamic material parameters

D Jocham and W Volk

Institute of metal forming and casting, Technische Universität München, Walther-Meißner-Straße 4, 85748 Garching, Germany

David.Jocham@utg.de

Abstract. The forming limit curve (FLC) is a valid instrument for the evaluation of failure in sheet metal processes. However, its experimental evaluation is challenging, in particular for modern lightweight sheet metals, in which the failure occurs without an evident necking transition. Therefore, the numerical analysis can represent a valid alternative for the investigation of the onset of necking phenomena. Prerequisite for realistic failure prediction are an accurate material characterization for high strain levels and a stable and coherent numerical model. Within this paper, an approach for the determination of forming limits by using the time dependent evaluation method is investigated and an analysis of the material sensitivity on the simulation results is performed. The results are discussed for mild steel DX56 and first suggestions for the improvement of the simulation input data are derived.

1. Introduction

To predict the formability in sheet metal forming process simulation the forming limit curve is state of the art. Herein local instability due to local necking in thickness direction is the predominant failure mechanism. In ISO 12004-2 [1] the experimental procedure to create the forming limit curve in form of the Nakajima test [2] and an evaluation method, the so called cross section method, are standardized. The principle of the cross section method is the fitting of a parabola curve through the strain values along a cross section, which is perpendicular to the occurring crack. The limit strains are determined as the values of the curve at the location of the crack. Merklein et al. [3] showed that this method leads to reliable results for mild steels but shows weaknesses for bainitic steel, due to the immediate fracture without necking. The evaluation method is also not suitable for aluminum and steel grades, which show more than one necking zone, because the parabola fit gets inaccurate. To overcome this disadvantages different authors proposed time dependent evaluation methods, which are more physically motivated [3-6]. The principle is to use the thinning rate as a physical quantity. The remaining plastic deformation concentrates in small shear bands. This leads to high thinning rates, while the thinning rates outside of the shear bands nearly vanish. By using the time dependent evaluation method the onset of local necking can be determined directly in the necking zone. Furthermore, with this method the limit strains can be examined easily in finite element simulations [7,8]. For the numerical determination of forming limits the material, e.g. hardening, strain rate sensitivity, temperature sensitivity and anisotropic behavior, has to be described as precise as possible to avoid high deviations between simulation and experimental data. The material properties also interact strongly with each other, whereby simulation results in

IOP Publishing
doi:10.1088/1742-6596/734/3/032015

relation to experimental data can be misinterpreted. In this paper the influence on the resulting forming limits for different stress states by using dynamic material parameters is investigated.

2. Experimental and numerical setup

2.1. Material characterisation

The material properties of the mild steel DX56 with thickness 0.8 mm were recorded in hydraulic bulge tests and in tensile tests with tactile and optical measurement systems. The usage of optical measurement enables the determination of the material behavior directly in the necking zone [9]. Tactile measurement is only suitable until yield strength. The results of the tensile tests are the Lankford coefficient in different directions, the elastic modulus and the yield curve at specific strain rates for different strain levels depending on the measurement system. By performing a hydraulic bulge test, data for extrapolation of the yield curve for equivalent plastic strain up to 0.7 is possible. Jocham et al. [10] detected the strain rate sensitivity under biaxial stress in the hydraulic bulge test. The described procedure was applied on the mild steel DX56. Figure 1 displays the dynamic behavior of the strain rate sensitivity in dependence on the true strain and the load case. The correlation can be described with the equation, proposed by Jocham et al. [10], to display the dynamic hardening. Further the Lankford coefficients in 0°, 45° and 90° to the rolling direction were determined in tensile tests until 0.8 true strain, whereby a significant decrease was observed, see figure 2. Nowadays in most simulation cases this dynamic behavior of the strain rate sensitivity and anisotropy is not taken into account, though the strain dependence of the Lankford coefficients is commonly known.

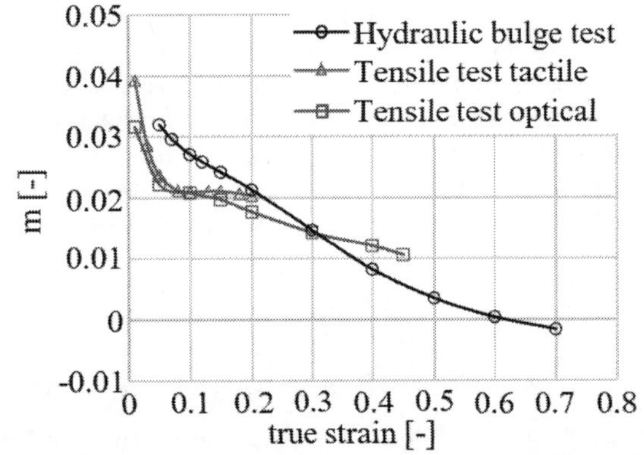

Figure 1. Dynamic strain rate sensitivity of mild steel DX56 determined in hydraulic bulge and tensile tests.

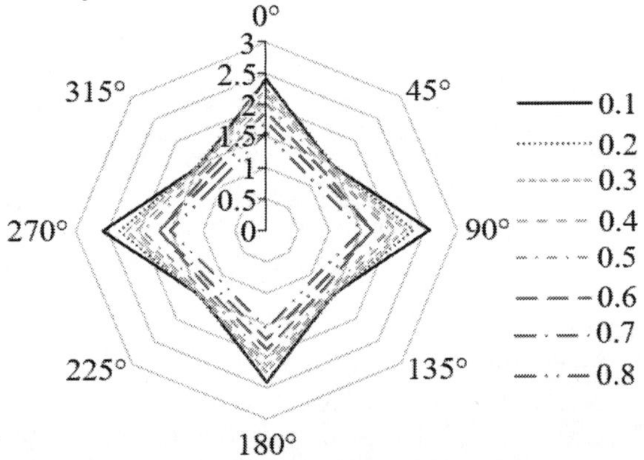

Figure 2. Dynamic Lankford parameters of mild steel DX56 for directions 0°, 45° and 90° to the rolling direction and different true strains.

2.2. Simulation and material model

For the simulation the software Abaqus™ 6.12-3 supplied by Dessault Systèmes was utilized. The simulation model of the Nakajima test consists of a rigid spherical punch with 50 mm radius, a rigid blank holder and rigid die with edge radius of 5 mm. The specimen is elasto-plastic and consists of shell elements with element size of 0.1 mm in the evaluation area, which was shown as best suited in a sensitivity analysis. 5 different specimens with widths of 40, 70, 100, 120 mm and one full sample were used. To create the same condition as in the experiment the lubrication system in form of an elastic pad was implemented. The contact between punch and pad as well as the contact between pad and specimen were frictionless. The clamping force was neglected and replaced with displacement boundaries in the area of the blank holder. The reference punch velocity was 1 mm/s. For the description of the hardening behavior of the steel grade DX56 a combination of the approaches by Swift and by Hockett-Sherby was used, see equation (1). To take into account the additional hardening through the increase of strain rate in the necking zone the Johnson-Cook model was implemented, see equation (2). The parameter $C_{J\text{-}C}$

can be calculated directly from the m-value from figure 1. The anisotropy was displayed by the Hill48 yield locus. The material parameters of a conventional determined material card are displayed in table 1.

$$\sigma_0(\varepsilon)=(\alpha)\cdot\left(A-(A-B)\cdot e^{-C\cdot\varepsilon^D}\right)+(1-\alpha)\cdot\left(E\cdot(\varepsilon+F)^G\right) \quad (1)$$

$$\sigma(\varepsilon,\dot{\varepsilon})=\sigma_0(\varepsilon)\cdot\left(1+C_{J\text{-}C}\cdot\ln\left(\frac{\dot{\varepsilon}}{\dot{\varepsilon}_0}\right)\right) \quad (2)$$

Table 1. Conventional determined material model.

α [-]	A [MPa]	B [MPa]	C [-]	D [-]	E [MPa]	F [-]	G [-]
0.35	415.9	147.4	4.44	0.78	518.9	0.00877	0.2685
$C_{J\text{-}C\ \varepsilon=15\%}$ [-]	$\dot{\varepsilon}_0$ [s^{-1}]	$r_{0°\ \varepsilon=8\text{-}12\%}$ [-]	$r_{45°\ \varepsilon=8\text{-}12\%}$ [-]	$r_{90°\ \varepsilon=8\text{-}12\%}$ [-]	Elastic modulus [GPa]	Density [kg/m³]	Poisson's ratio [-]
0.0204	0.00025	2.543	1.73	2.525	210	7850	0.3

3. Experimental and numerical results

The simulations were terminated at the strain rate level, which occurred in the experiments before crack. For the time dependent evaluation method equal output parameters as in the experiments were applied. The output frequency was 10 Hz. For the stable line fit 40-20 stages and for the instable line fit 3 stages before termination were selected. In figure 3 the results of the experiments and simulations with different material parameters are compared.

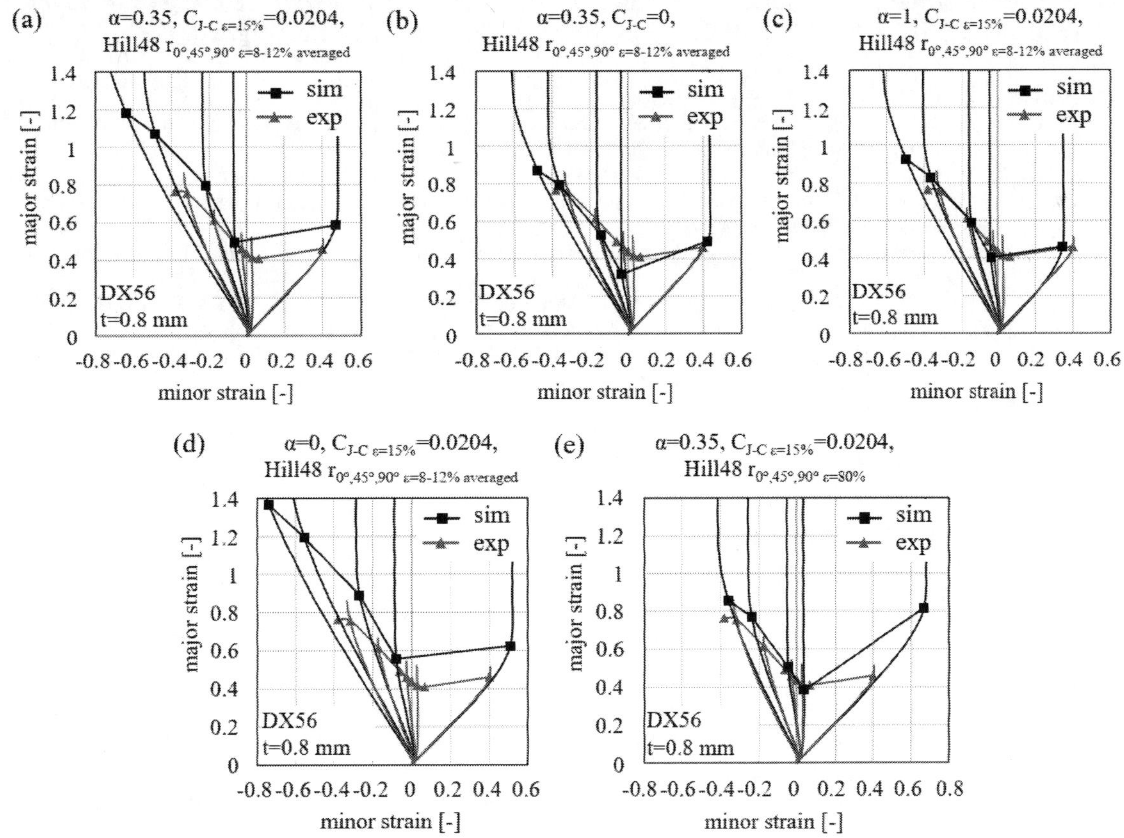

Figure 3. Simulation and experimental results for different material models.

The mild steel DX56 shows a high strain rate sensitivity and anisotropy for low strain levels, whereby a high deviation between experimental and simulation results occurs, see figure 3a. Also the strain paths do not match. By neglecting the strain rate sensitivity, see figure 3b, or by using a 100% Hockett-Sherby extrapolation approach, as you can see in figure 3c, the FLCs fit each other quite well. Nevertheless the strain paths do not fit in both cases. In contrast to this, the implementation of the 100% Swift approach leads to a higher deviation compared to the FLC resulting by a conventionally determined material model, see figure 3d. The variation of the extrapolation approach do not describe the real physical material behavior, which was characterized in tensile and hydraulic bulge tests. First by considering Lankford coefficients at higher true strain levels the strain paths fit quite well, see figure 3e. With this setup the biaxial limit strains increase in figure 3e compared to figure 3a. This depends on the nonexistent alteration of the texture under biaxial strain condition.

4. Conclusions

In this paper the influence of the dynamic strain rate sensitivity and Lankford coefficients on the results of a Nakajima simulation was investigated. By applying the time dependent evaluation method the forming limits were determined. The strain rate sensitivity of the mild steel DX56 shows a high variation between uniaxial and biaxial load case. Further the strain rate sensitivity decreases with increasing true strain. The same behavior shows the anisotropy, described by Lankford coefficients and determined in tensile tests. By neglecting the strain rate sensitivity, which corresponds to high strain levels, the resulting forming limit curve in simulation fits the experimental data quite well. Nevertheless the interpolation points and corresponding strain paths are different. The strain paths can be displayed correctly by taken decreasing Lankford coefficients into account. In future investigations the dynamic strain rate sensitivity and anisotropy will be implemented in the simulation model by using subroutines. This enables an accurate modelling of the material parameters, which were measured already in bulge and tensile tests. Furthermore the influence of different yield loci will be investigated.

References

[1] ISO copyright office 2006 Metallic materials-sheet and strip determination of forming limit curves-part 2: determination of forming limit curves in laboratory *ISO/DIS 12004-2*

[2] Nakajima K, Kikuma T, Asaku K 1968 Study on the formability of steel sheet *Yawata Tech Rep* **264**

[3] Merklein M, Kuppert A, Geiger M 2010 Time dependent determination of forming limit diagrams *Annals CIRP* **59(1)** pp 295–298 DOI 10.1016/j.cirp.2010.03.001

[4] Volk W 2006 New experimental and numerical approach in the evaluation of the FLD with the FE-method *Proc FLC-Zurich* pp 26–30

[5] Eberle B, Volk W, Hora P 2008 Automatic Approach in the Evaluation of the Experimental FLD with a full 2D Approach Based on a time Depending Method *Numisheet Part A* pp 279–284

[6] Volk W, Hora P 2010 New algorithm for a robust user-independent evaluation of beginning instability for the experimental FLC determination *Int. J. Material Forming* **10** DOI 1007/s12289-010-1012-9

[7] Wu X 2013 Benchmark 1 – nonlinear strain path forming limit of a reverse draw *part C: benchmark analysis. AIP Conf Proc* **1567(1)** pp 39–176 DOI 10.1063/1.4849979

[8] Gaber C, Jocham D, Weiss H A, Böttcher O, Volk W 2016 Evaluation of non-linear strain paths using Generalized Forming Limit Concept and a modification of the Time Dependent Evaluation Method *Int. J. Material Forming* DOI 10.1007/s12289-016-1283-x

[9] Hoffmann H, Vogl C 2003 Determination of true stress-straincurves and normal anisotropy in tensile tests with optical strain measurement *CIRPAnnals Manuf Technol* **52** pp 217–220 DOI 10.1016/S0007-8506(07)60569-1

[10] Jocham D, Norz R, Volk W 2016 Strain rate sensitivity of DC06 for high strains under biaxial stress in hydraulic bulge test and under uniaxial stress in tensile test *Int. J. Material Forming* DOI 10.1007/s12289-016-1293-8

Acknowledgments

We extend our sincere thanks to the European Research Association for Sheet Metal Working and the German Federation of Industrial Research Associations for financing our investigations as part of the program for the furthering of the Industrial Collective Research by the Federal Ministry for Economic Affairs and Energy. The authors also want to thank their industrial partners for their support throughout the project.

Numerical Simulation of the Roll Forming Process of Aluminum Folded Micro-channel Tube

Tianxia Zou[a], Ning Zhou [a], Yinghong Peng[a], Ding Tang[a], Dayong Li[a]*

[a]*School of Mechanical Engineering, Shanghai Jiao Tong University, Shanghai 200240, China*

**Corresponding author: 86-21-34206313, dyli@sjtu.edu.cn*

Abstract. Micro-channel tube is the most important component of flat tube heat exchangers. The folded micro-channel tube is made of clad aluminum sheet through roll forming process, and has great advantage in the aspect of corrosion resistance over extruded tube. The folded tube's sub-millimeter channel size as well as tight dimensional precision requirement brings great challenge to roll forming process design. In this paper, the finite element model of the whole roll forming process of a ten-channel tube is established by using ABAQUS/Explicit. The deformation at different forming stands are investigated and compared with experiment. The hydraulic pressure test is carried out on the developed tube and its pressure bearing capacity is evaluated.

Keywords: Folded micro-channel tube; Finite element simulation; Roll forming; Pressure bearing capacity

INTRODUCTION

Micro-channel tube is a kind of heat exchanger tube with a row of sub-millimeter-diameter channels, and has been widely used in automobile air condition systems[1-6]. Most micro-channel tubes are extruded ones, as shown in Figure 1(a), i.e. they are formed by hot extrusion of cylindrical aluminum alloy ingot through a very small extrusion die, with an extrusion ratio over than 500[7-8]. The folded tube is a novel micro-channel tube that is formed by roll forming of thin aluminum plates, as shown in Figure1(b). It can be made of cladding aluminum plates, which consists of a 3XXX core aluminum alloy and a 4XXX aluminum alloy[9-10] on both sides, as shown in Figure 2. Therefore, the folded tube has great advantage in the aspect of corrosion-resistancy than extruded tube[11], and has great potential, as a substitute for extruded tube, in parallel flow heat exchangers, especially those used under harsh conditions.

A representative folded tube shown in Figure 1(b) is 16mm in width and 1.8mm in thickness. The requirement on high-precision overall dimension should be guaranteed for assembly with fins and header pipes in manufacturing heat exchangers. The sub-millimeter channel size poses challenge to roll forming of the folded tube. Moreover, reduction of wall thickness should be as small as possible to ensure the pressure bearing and anti-corrosion performance of the tube in future service.

(a) (b)

FIGURE 1. Micro-channel tube: (a) extruded tube; (b) folded tube

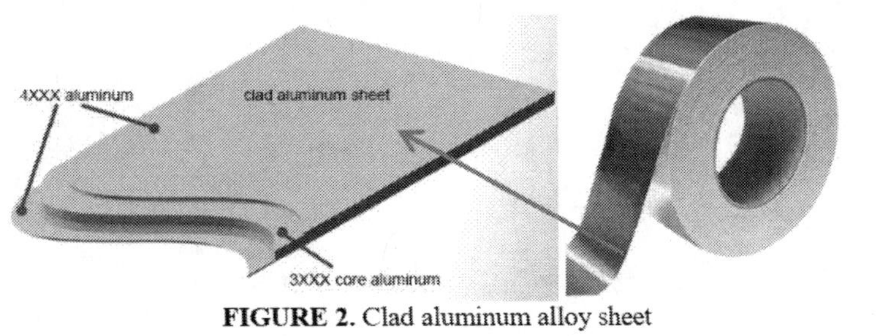

FIGURE 2. Clad aluminum alloy sheet

In the present work, the roll forming process of a ten-channel folded tube is numerically investigated by using the finite element method. First, the uniaxial tension test of clad aluminum sheet is carried out and a Swift type hardening model is established. A three dimensional finite element model of the whole roll forming process is then built by using ABAQUS/Explicit. The deformation behaviors of the clad aluminum sheet at different forming steps is analyzed. The experiment is carried out, and deformation configurations at different forming stages are measured and compared with simulated results. Finally, the formed folded tube is evaluated by pressure bearing test.

PATTERN DESIGN

As shown in Figure 3, roll forming process of folded micro-channel tube can be divided into three stages: the trapezoidal grooves on both hands of the plate section are formed at the first stage (from step 1 to step 5); at the second stage (from step 6 to step 14), parts of sheet with trapezoidal grooves are folded to form the first four micro channels; The rest of six channels, as well as the two semi-circular arcs of micro-channel tube, are formed by folding the sheet once more at the stages from step 15 to step 22.

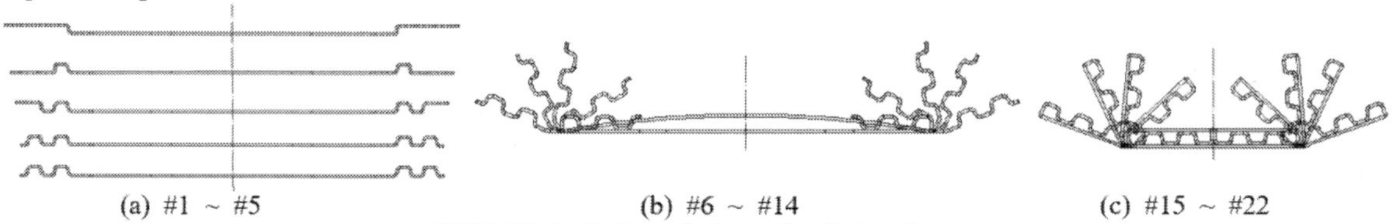

FIGURE 3. Pattern design of roll forming

FINITE ELEMENT MODEL

The clad aluminum sheet studied in current work is an AA4343/AA3003/AA4343 sandwich sheet, in which AA3003 is an Al-Mn base alloy and AA4343 an Al-Si base alloy. The thickness of the sheet is 0.265mm. The uniaxial tension test of the clad aluminum sheet is carried out. The Von Mises associated plasticity model with the exponential hardening is used to describe the stress-strain response, as below.

$$\sigma = K(\varepsilon + \varepsilon_0)^n \tag{1}$$

where K and n are material parameters of the exponential hardening model.

Based on data fitting of the material test, the mechanical properties of clad aluminum sheet are obtained as listed in Table 1, where E, σ_s and v are elastic modulus, yield stress and Poisson's ratio respectively.

TABLE 1. Mechanical properties of clad aluminum sheet

Parameters	E (Gpa)	σ_s(Mpa)	v	ε_0	K(Mpa)	N
Values	65.5	180.7	0.33	0.00494	296.8	0.09336

The three dimensional finite element model of the roll forming process is established by using ABAQUS/Explicit[12], as shown in Figure 4. Only a half of aluminum sheet is modeled due to the symmetry of geometry, boundary and loading conditions. According to the pattern design, the plate is formed in a folded tube after experiencing deformation at total 22 forming stands. Upper and lower rollers are installed on each forming stand, as shown in Figure 4(a). The aluminum sheet is discretized by 3D stress elements (C3D8R) with different sizes according to various forming radii, as shown in Figure 4(b). The total element number is 135000 and the minimum element size is 0.03mm. The dimension of metal sheet and forming parameters are listed in Table 2. All forming tools are modeled as analytical rigid surfaces. The contact between tools and aluminum sheet is set as a strict "master-slave" algorithm. The penalty function model with friction coefficient of 0.1 is employed.

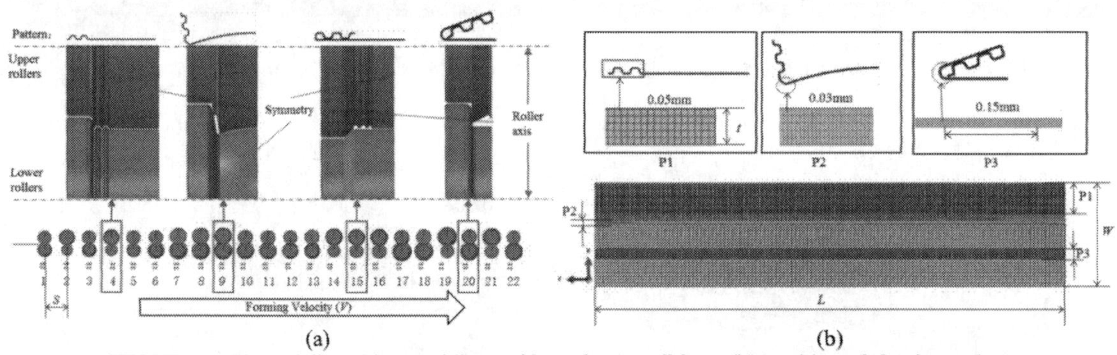

FIGURE 4. FE modeling: (a) assembling and boundary conditions; (b) meshing of aluminum sheet

TABLE 2. Sheet dimension and forming parameters

Parameters	L(mm)	W(mm)	t (mm)	V (mm/s)	S(mm)
Values	120	26.5	0.26	2000	100

Journal of Physics: Conference Series **734** (2016) 032016 doi:10.1088/1742-6596/734/3/032016

NUMERICAL SIMULATION RESULTS

In order to verify the finite element simulation, the comparison between FE simulation and experimental results is carried out. Some forming results in both of experiment and simulation are compared in figure 5. In the first five steps, the edges of the aluminum sheet are gradually formed into two trapezoidal grooves, as shown in Figure 5(a). Figures 5(b) and (c) show the second stage of the forming sequence and Figures 5(d) and (e) show the third forming stage. The final shapes of the folded tube are demonstrated in Figure 5(f). As can be seen from the figures, simulation results show good consistency in the geometric configurations comparing with the experimental ones.

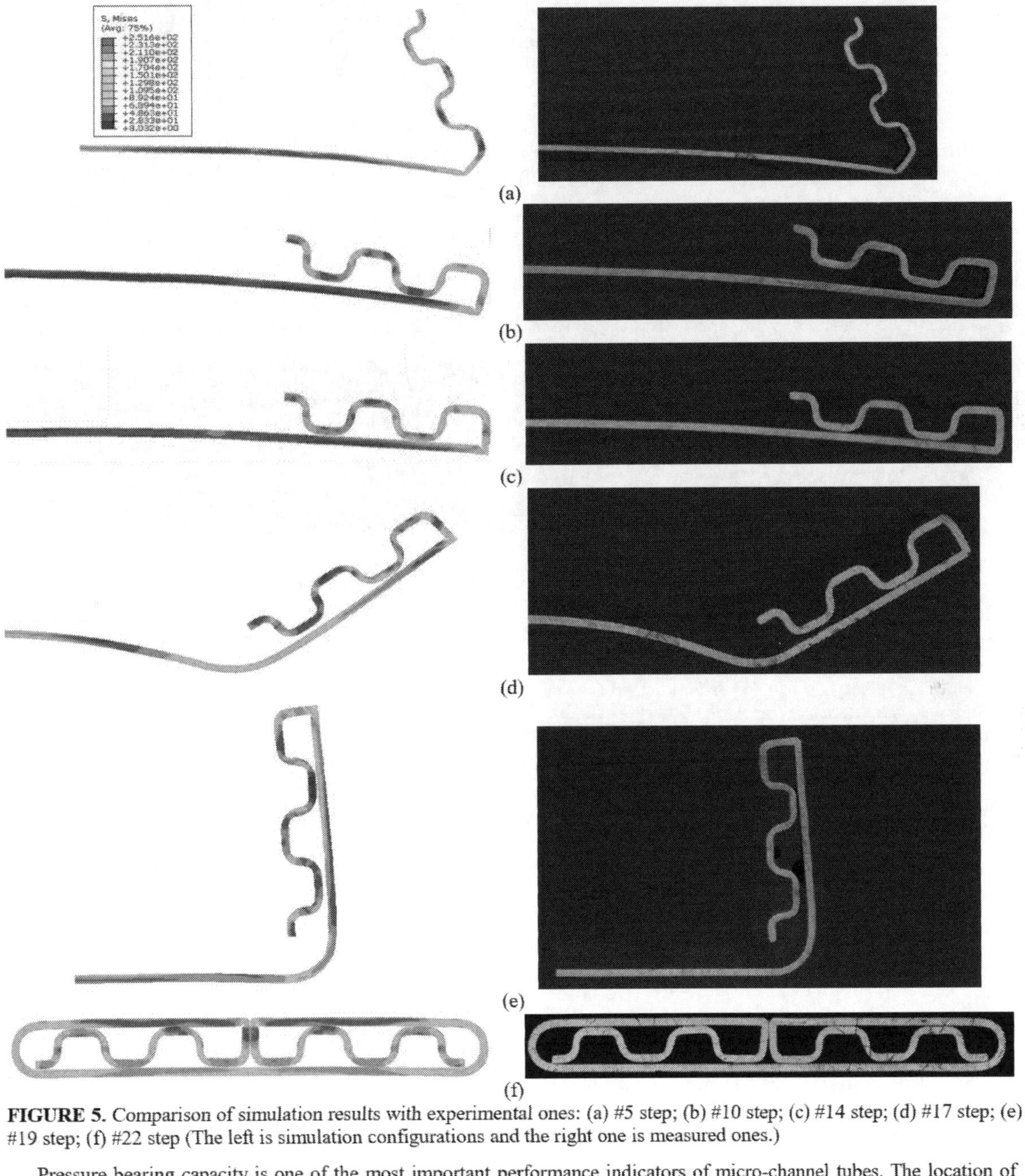

FIGURE 5. Comparison of simulation results with experimental ones: (a) #5 step; (b) #10 step; (c) #14 step; (d) #17 step; (e) #19 step; (f) #22 step (The left is simulation configurations and the right one is measured ones.)

Pressure bearing capacity is one of the most important performance indicators of micro-channel tubes. The location of burst failure is always at the weaknesses of the micro-channel tube. Therefore, wall thickness reduction of the folded tube in roll forming process directly affects its bearing capacity. Generally, the smaller bending radius will lead to more severe thickness reduction. As can be seen from the forming pattern, small bending radii are at the corner of trapezoidal grooves which is formed during the first five steps. Therefore, it is necessary to evaluate the thickness reduction of sheet wall at these bending corners. The thickness reductions at the eight corners in experiment and simulation are compared in Figure 6. The thickness reduction shows a decreasing trend from the center to the edge and the most server thickness reduction (23%) appears at the first bending corner. The simulation results agree with the experimental ones very well.

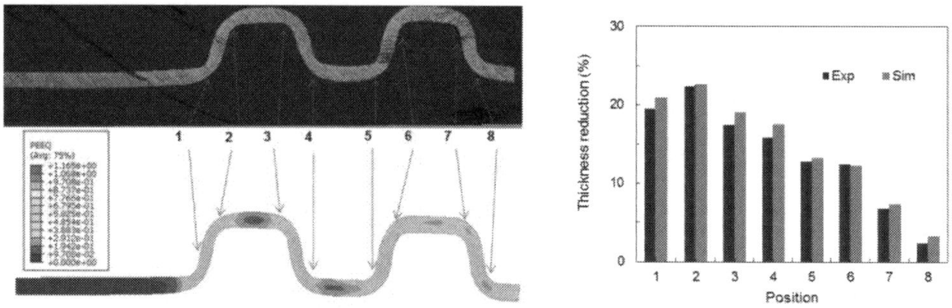

(a) (b)

FIGURE 6. Compared results of thickness reduction between experiment and simulation: (a) forming shape; (b) thickness reduction

In order to evaluate the pressure bearing capacity of folded micro-channel tube, NOCOLOK brazing[13] and hydraulic pressure tests are carried out. The folded tube is firstly heated to 605 °C in SECO/WARWICK Nitrogen protection brazing furnace, then held for 3~5 minutes at this temperature. The welded micro-channel tube is obtained after air cooling, as shown in Figure 7(a). Hydraulic pressure tests show that leak failure occurs at the top center of the tube when the pressure increases to about 18MPa. The tube cross section after hydraulic pressure test is shown in Figure 7(b), which indicates that the fracture areas are at the bending corners of the trapezoidal grooves and corresponds to the positions with thickness reduction.

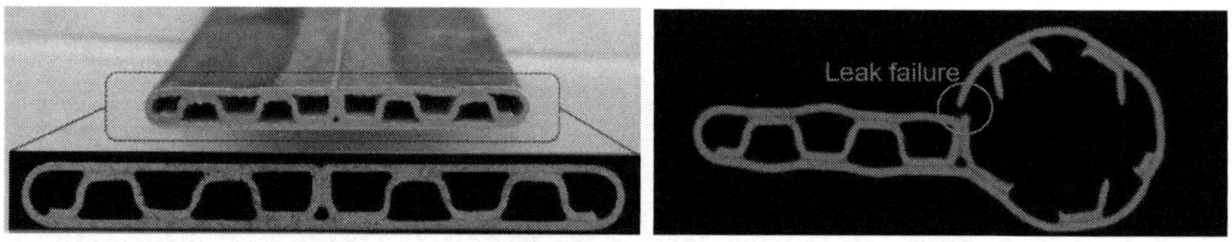

(a) (b)

FIGURE 7. Hydraulic pressure test: (a) brazed folded tube; (b) folded tube after hydraulic pressure test

CONCLUSIONS

Numerical simulation of the roll forming process of a clad aluminum folded micro-channel tube is carried out. The deformation characteristics of the sheet during the forming processes are analyzed. The simulation results show good consistency with the experimental ones in the geometric configurations and thickness reductions. In addition, the pressure bearing capacity of the folded micro-channel tube is evaluated by the hydraulic pressure test. The failure mode indicates that the bending corners of the trapezoidal grooves formed in the first five steps are the weakness of the folded tubes. Future work will focus on the pattern design optimization for alleviating the wall thickness reduction.

ACKNOWLEDGMENTS

Authors would like to acknowledge the financial support of Ministry of Science and Technology of China (No. 2012BAF01B06).

REFERENCES

1. T. Ren, G. Ding, T. Wang, et al., Applied Thermal Engineering, 59, 660-674(2013).
2. J. Pettersen, A. Hafner, G. Skaugen, Int. J. Refrig.-Rev. Int. Froid, 21, 180-193(1998).
3. Z.Z. Qi, Y. Zhao, J.P. Chen, Int. J. Refrig.-Rev.Int. Froid, 33, 301-312(2010).
4. C.Y. Park, P.S. Hrnjak, Int. J. Refrig.-Rev. Int. Froid, 31, 822-831(2008).
5. B. Dai, M.Li, C. Dang, et al., Int J Heat Mass Tran, 70, 114-118(2014).
6. G. Vamadevan, F.F. Kraft, J Mater Process Tech, 181, 30-33(2007).
7. Q.Zhang, D.Tang, D. Li, Y. Peng, J Mater Process Tech, 210, 1876–1884(2010).
8. D.Tang, Q.Zhang, D. Li, Y. Peng, J Mater Process Tech, 214, 2777–2783(2014).
9. S.Tierce, N.Pebere, C.Blanc, et al., Corrosion Science, 49, 4581-4593(2007).
10. X.Han, B.Shao, K. Zuo, et al., Trans. Nonferrous Met. Soc. China, 26, 658-664(2016).
11. W. Wei, K. Xu, Y. Liang, et al., CIESC Journal, 64(7), 2369-2375(2013).
12. H., Karlson, S. Inc, ABAQUS/Explicit: User's Manual (2001).
13. J. Liu, T. Worlitz, SAE Technical Paper Series #950117, Detroit, Michigan (1995).

Journal of Physics: Conference Series **734** (2016) 032017 doi:10.1088/1742-6596/734/3/032017

Failure Investigation for QP Steel Sheets under uniaxial and Equal-Biaxial Tension Conditions

Danqing Zou[1,2] , Shuhui Li[1,2,*], Ji He[1,2], Ronggao Cui[1,2]

[1] Shanghai Key Laboratory of Digital Manufacture for Thin-walled Structures, Shanghai Jiao Tong University, Shanghai 200240, PR China

[2] State Key Laboratory of Mechanical System and Vibration, Shanghai Jiao Tong University, Shanghai 200240, PR China

E-mail: lishuhui@sjtu.edu.cn

Abstract. The Quenching and Partitioning (QP) steel sheet is new generation material to induce phase transformation for plasticity in forming vehicle parts. The phase transformation is strongly stress state dependent behavior in experiments, which should affect the failure timing and limit strain in forming processes. In this paper, Nakajima test with QP980 and DP1000 steel sheets under equal-biaxial loading condition is performed for failure behavior. X-ray diffraction (XRD) is adopted to obtain the volume fraction of retained austenite (f_A). Digital Image Correlation (DIC) is used to record the surface strain field and its evolution during equal-biaxial tension deformation. The same level Dual Phase (DP) steel is also employed for the purpose of comparison. The results show that phase transformation in QP steel gives small impact on failure strain under equal biaxial tension condition which is contradicted with our understanding. It suggests that failure behavior under uniaxial tension of QP980 is strongly phase transformation dependent. But it shows almost independent under equal biaxial tension condition.

1 Introductions

The Quenching and Partitioning (QP) steel sheet is new generation material with excellent elongation for forming the vehicle safety part. The austenite-to-martensite phase transformation during material deformation provides excellent ductility for forming processes. The martensitic transformation kinetics has been widely studied with different advanced material characterization methods such as XRD, TEM, SEM and EBSD. The published experimental results show that the phase transformation behaviour depends on the stress state applied on the material. In sheet metal forming process, forming limits are employed as failure criterion for predicting material split in simulation. Nakajima test is performed on sheet metal specimens with different shapes for the purpose of obtaining forming limits under specific strain path. Obviously, certain stress state will be applied on the sheet metal specimen during the Nakajima test with accordingly strain path. The micro structure evolution of the austenite-to-martensite phase transformation is expected to impact the macro mechanical behavior as well as its failure limits consequently.

Due to the length limitation of the paper, we will briefly summarize the research work for austenite-to-martensite phase transformation behaviour. Recently, experimental study [1] indicates that the transformation-induced plasticity (TRIP) effect occurs in QP steel, which enhances the strain hardenability and delays the necking of materials [2-5]. Beese and Mohr [6] produce a comprehensive set of data that illustrates the effect of stress state on the martensitic transformation kinetics in an austenitic stainless steel. They found that the martensitic transformation depends on both stress triaxiality and Lode angle. Compare to the research of phase transformation kinetics in steel, limited effort has been made to understand the detailed relationship of austenite-to-martensite phase transformation and the failure behaviour of sheet steels.

In this paper, Nakajima test with QP980 and DP1000 steel sheets under equal-biaxial loading condition is performed for failure behavior. X-ray diffraction (XRD) is adopted to obtain the volume fraction of retained austenite (f_A). Digital Image Correlation (DIC) is used to record the surface strain field and its evolution during equal-biaxial tension deformation. Different micro structures but similar strength level steels (QP980 and DP1000) are compared with each other. Interesting results are obtained and discussed in detail.

2 Mechanical Experiments

2.1 Test materials

The cold-rolled QP980 steel sheets provided by the Baosteel Group Corp. in Shanghai China with the thickness of 1.2mm are adopted in experiments. The provided QP980 has a multi-phase microstructure of retained austenite (RA), martensite (M) and ferrite (F) as shown in Fig.1. For the comparison purpose, the commercial DP980 sheets steels are employed for comparative study. The microstructure of DP980 (1.0mm) used in this study is martensite and ferrite.

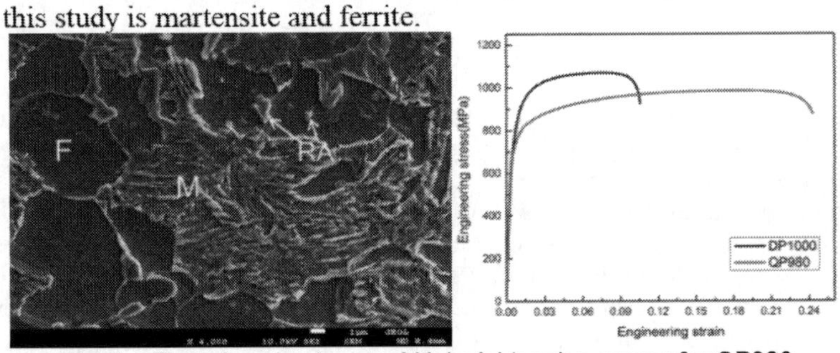

Fig. 1 *The microstructure and Uniaxial tension curves for QP980*

2.2 Uniaxial tension tests

Uniaxial tension tests for QP980 and DP1000 under static deformation condition are performed with Zwick Roell tensile testing machine at room temperature. The engineering stress and strain curves of these two materials are presented in Fig. 1.The elongation between these two different materials is very different due to the phase transformation-induced plasticity (TRIP) effect from the austenite-to-martensite transformation. The failure limit strain between the QP980 and DP1000 is about 12% to 17% under uniaxial stress state condition (or uniaxial strain path condition). The retained austenite gives significant impact on sheet metals failure under uniaxial tension. Micro structures of these two materials are critical factors for the discrepant ductility. Based on this phenomenon, it suggests that TRIP effect enhances failure limit strain for QP steel and also provides similar strength.

2.3 Equal biaxial tension tests

The Nakajima test is performed for equal biaxial tension test of both QP980 and DP1000 sheets. The Zwick Roell BUP400 forming limit testing equipment with regular radius 50mm punch is adopted to tension the sheet metal as shown in Fig.2.

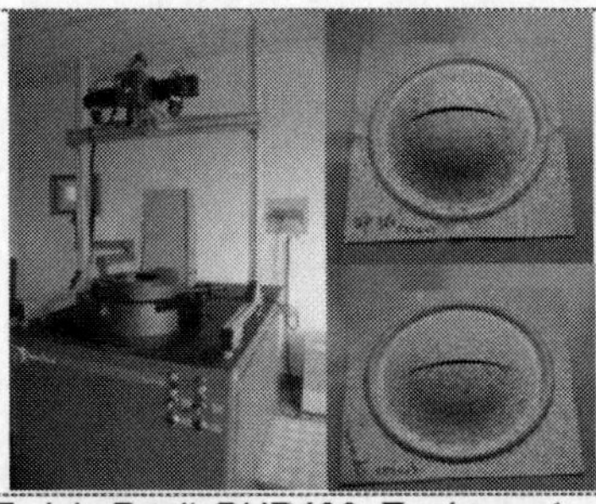

Fig. 2 Zwick Roell BUP400 Equipment and tested samples

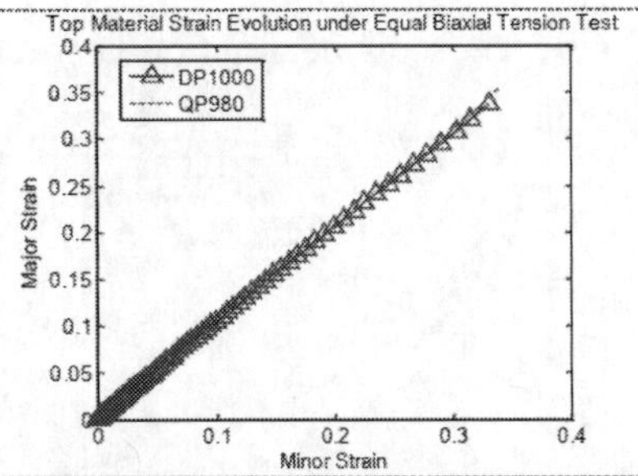

Fig. 3 The strain evolution of the punch top material for both QP980 and DP1000.

Digital Image Correlation (DIC) is set up for strain computing and evolution recording. When testing with the complete disk to obtain balanced-biaxial deformation, rectangular shaped Teflon plastic about 1 mm thick was used. Viscous oil is painted uniformly on the Teflon which provides as close to a frictionless condition as is possible. We tested all specimens using a fixed punch velocity of 4.5 mm/s, irrespective of the different material samples. The binder force is fixed as 350kN for all tests. The 12Hz sampling frequency is employed to take speckle pictures of tests due to the limitation of used DIC. The tested samples are presented in Fig.2.

The fracture occurs at the center area of the tested sample for both QP980 and DP1000 which suggest that the friction is almost zero and the material on the punch top is almost deformed under equal biaxial tension state. The arbitrary material point can be selected on strain calculating image for strain evolution and strain field of the top surface.

The strain evolution of the punch top surface material at top point is shown in Fig.3 for both QP980 (1.2mm) and DP1000 (1.0mm). Similar strain evolution can be achieved for both sheets even with different thickness.

In Fig.3, the end point is the last frame before the material split which can be calculated the surface strain of the top material. A little lower limit strain of DP980 than QP980 can be obtained in this figure. If consider the factor that QP980 steel sheet (1.2mm) is a little thicker than DP980 steel sheet (1.0mm), then it may conclude that almost equal formability of these two material under equal biaxial tension loading can be achieved. The experimental results show that TRIP effect gives no impact on sheet metal's formability under equal biaxial tension loading. The material's mechanical behavior shows insensitive phenomenon to its micro stricture which is inconsistent with our understanding. This phenomenon suggests failure limits of QP steel material shows very different dependent phenomenon on the austenite-to-martensite phase transformation behaviour.

3 Deformation induced martensitic transformation

To further studied the micro structure evolution of the QP steel and explain discrepant phenomenon between failure behaviour under uniaxial and equal biaxial tension, XRD method is used to determine the volume fraction of retained austenite during different deformation stages. The similar deformation process is applied on the QP980 sheet metal.

The XRD experiment was performed at the Thin-walled Structures Laboratory of Shanghai Jiao Tong University. Fig. 4 shows the experimental setup used for the experiments. The detailed volume fraction of retained austenite has been obtained through the XRD method. A size of 10×10 mm area material has been detected in this equipment. The previous study [7] results show that the initial volume fraction of retained austenite is very stable and insensitive to the environmental temperature conditions.

The volume fraction of retained austenite under incremental uniaxial tension and equal biaxial tension are shown in Fig.4. The M.Radu model [8] has been used to fit the experimental results. The detailed expression is:

$$f_A = 0.1061 * exp(-(3.43 * \eta + 1.05) * (\varepsilon_{eq}^{p})^{0.78}) \qquad (1)$$

Where f_A is the volume fraction of retained austenite, η is stress triaxiality and ε_{eq}^{p} is equivalent plastic strain.

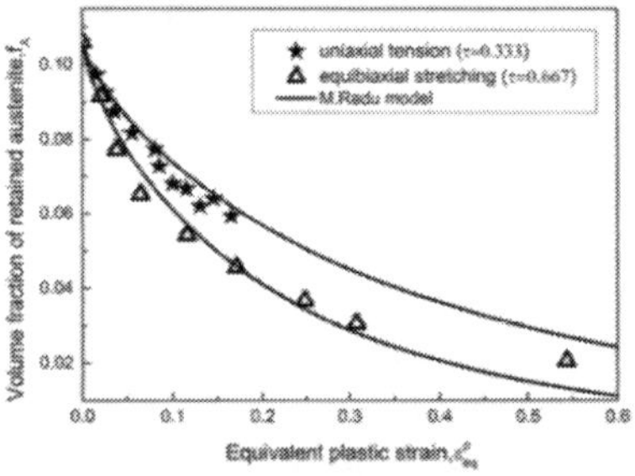

Fig. 4 *The evolution of volume fraction of retained austenite as true strain*

Based on the experimental results of retained austenite during incremental deformation, obvious difference is observed between the loading process of uniaxial and equal biaxial tension on QP980 sheet metals. The volume fraction of retained austenite keeps reducing in deformation process which is consistent with our expectation. Moreover, under biaxial tensile tests, the retained austenite transformed to martensite faster than that of uniaxial tensile tests. Therefore, the further question is proposed: if deformation induced martensitic deformation is promoted under equal biaxial tension than uniaxial tension for this QP980 steel sheet, why the QP980 steel sheet show higher failure limit strain under uniaxial tension but almost identical failure limit strain under equal-biaxial tension when compared with the dual-phase DP1000 steel

4 CONCLUSIONS

Both QP980 and DP1000 are tested with uniaxial and equal biaxial tension loading. XRD method has been used to obtain the volume fraction of retained austenite during different incremental deformation stages. The top surface strain field and its evolution were recorded by the DIC system. Experimental results show that the austenite-to-martensite phase transformation induced TRIP effect impacts materials ductility under uniaxial but has little effect on its formability under equal biaxial tension. It suggests that the failure limit of QP980 sheet metals shows obviously TRIP effect (or phase transformation effect) dependent phenomenon under uniaxial tension loading. But the failure strain under equal biaxial tension shows almost TRIP effect independent phenomenon. It is still elusive to answer the contradictory phenomenon of failure behaviour of QP980 sheet metal in this paper. The physical explanation should be continuously searched in future.

5 KNOWLEDGEMENT

Financial support from China Postdoctoral Science Foundation funded project (2014M560330) and the National Natural Science Foundation of China (Grant No. U1360203) are gratefully acknowledged.

References

[1] E. D Moor, S. Lacroix, A. Clarke, J. Penning, J. Speer, 2008 *Metall. Mater. Trans. A.* **39** pp 2586-2595.
[2] P. Jacques, Q. Furnemont, F. Lani, T. Pardoen, F. Delannay, 2007 *Acta Mater.* **55** pp 3681-3693.
[3] B. Fu, W.Y. Yang, Y.D. Wang, L.F. Li, Z.Q. Sun, Y. Ren, 2014 *Acta Mater.* **76** pp 342-354.
[4] H.N. Han, C.-S. Oh, G. Kim, O. Kwon, 2009 *Mater. Sci. Eng. A.* **499** pp 462-468.
[5] N. Tsuchida, Y. Tomota, 2000 *Mater. Sci. Eng. A.* 285 **pp** 346-352.
[6] A.M Beese, D. Mohr, 2011 *Acta Mater.* 59 **pp** 2589-2600.
[7] S.H. Li, D.Q. Zou, J. He, C. Xia, 2015 *Steel Res. Int.* **86** pp:1-10.
[8] M. Radu, J. Valy, A.F. Gourgues, F.L. Strat, A. Pineau, 2005 *Scripta Mater.* **52** pp 525-530.

Modeling and Simulating Material Behavior during Hot Blank – Cold Die (HB-CD) Stamping of Aluminium Alloy Sheets

Nan Zhang, Fadi Abu-Farha*
Clemson University - International Center for Automotive Research (CU-ICAR)

* Corresponding Author. Tel.: +1 864 283 7231; E-mail: fadi@clemson.edu

Key Words: *Deformation Mechanisms; Non-Isothermal Material Deformation; Hot Stamping Process*

Abstract:

Hot blank – cold die (HB-CD) stamping, non-isothermal hot stamping, of aluminium alloy sheets offers great opportunities for high production rates at low cost, while overcoming limited material formability issues. Yet developing an accurate model that can describe the complex material behavior over the wide ranging conditions of HB-CD stamping (temperatures ranging between 25 and 350 °C) is challenging. Moreover, validation of the developed models under transient conditions is problematic. This work presents he results of a comprehensive characterization, material modeling, FE simulation and experimental validation effort to capture the behavior of an aluminium alloy sheet during HB-CD stamping. In particular, we highlight the integration between temperature measurements (thermography) and strain measurements (digital image correlation) for the accurate validation of model predictions of non-isothermal material deformation.

Journal of Physics: Conference Series **734** (2016) 032019 doi:10.1088/1742-6596/734/3/032019

Modification of Yoshida-Uemori Model with Consideration of Transformation-Induced Plasticity Effect

Jun Hu[1], Lay Knoerr[2], Fadi Abu-Farha[1]*
[1]Clemson University - International Center for Automotive Research
[2]ThyssenKrupp Steel NA

* Corresponding Author. Tel.: +1 864 283 7231; E-mail: fadi@clemson.edu

Abstract:

Transformation-induced plasticity (TRIP) assisted steels possess improved strain hardening behavior and resistance to necking that are favorable for automotive body applications. However, the TRIP effect causes complex springback behavior of these steels that can hardly be predicted by existing constitutive models for other steels. In this work, the functions in the original Yoshida-Uemori model describing isotropic and kinematic hardening were modified by adding new parameters that can represent the TRIP effect. Cyclic tension/compression experiments were performed on a selected TRIP-steel grade, and the results were used to calibrate the modified model. The modified model was coded via user subroutine into a commercial FE solver. The springback predictions were compared with actual try-out stamping experimental results for highlighting the improvement of predictions with the modified model.

A Triaxial Failure Diagram to predict the forming limit of 3D sheet metal parts subjected to multiaxial stresses

F Rastellini [1,2,*], G Socorro [1], A Forgas [1] and E Oñate [1,2,3]

[1] Quantech ATZ S.A., c/ Gran Capitán 2-4, Barcelona, 08034, Spain.
[2] Centre Internacional de Mètodes Numèrics en Enginyeria (CIMNE).
[3] Universitat Politècnica de Catalunya (UPC), c/ Jordi Girona 3, Barcelona, Spain.

[*] fernando@quantech.es

Abstract: Accurate prediction of failure and forming limits is essential when modelling sheet metal forming processes. Since traditional Forming Limit Curves (FLCs) are not valid for materials subjected to triaxial loading, a new failure criterion is proposed in this paper based on the stress triaxility and the effective plastic strain accumulated during the history of material loading. Formability zones are identified inside the proposed Triaxial Failure Diagram (TFD). FLCs may be mapped into the TFD defining a new Triaxial Failure Curve, or it can be defined by triaxial failure experiments. Several TFD examples are validated and constrasted showing acceptable accuracy in the numerical prediction of forming failure/limit of 3D thick sheet parts.

1. Motivation

Since the classical Forming Limit Curves (FLCs) are only valid for in-plane loading cases, a new approach is required to appropriately render failure curves for materials subjected to multiaxial loads.

The Forming Limit Diagram (FLD) was initially introduced by Keller and Backofen [1] and later improved by Goodwin [2]. Currently, the FLD is widely used in sheet metal forming numerical simulation as a practical and useful criterion to predict the formability limit, based on the in-plane principal strain components.

The fracture loci of metal sheets may be depicted by the Fracture Forming Limit line (FFL) originally proposed by Atkins [3] and by the Shear Fracture Forming Limit line (SFFL). The FFL is related to the reduction in sheet thickness that originates crack opening by tensile stresses (mode I), while the SFFL is related to the distortion that originates crack opening by in-plane shear stresses (mode II).

Whilst the FLC is used for evaluating the formability of the principal strain components under in-plane loading, this criterion is not suitable to predict neither failure nor forming limits when triaxial loads are involved, such in hydro-forming, ironing processes or thick-sheet forming where multiaxial stresses arise.

Due to all these limitations, it is necessary to include the stress triaxility effect into the formability failure models employed in the numerical simulation of metal forming processes.

2. Formability failure detection under multiaxial loading

Several researches have considered the effect of the stress triaxiality in their fracture failure models. It is worth mentioning the work of Johnson-Cook [4], Lemaitre [5] and Wierzbicki [6].

2.1. Formability failure model including shear fracture

The Modified Mohr-Coulomb (MMC) model was proposed by Bai and Wierzbicki [7] as a ductile fracture model. Subsequently, the same model was successfully applied by Li and Wierzbicki [8]-[9] to predict cracked areas during deep drawing processes.

The MMC fracture model [7]-[10] can predict the ductile fracture damage evolution (D) in terms of the equivalent plastic strain $\bar{\varepsilon}$, the stress triaxiality η and the normalized Lode angle $\bar{\theta}$:

$$D = \int_0^{\varepsilon} \frac{d\bar{\varepsilon}}{f(\eta, \bar{\theta})} \qquad \eta = \frac{\sigma_m}{\bar{\sigma}} \qquad \bar{\theta} = 1 - \frac{6}{\pi}\theta \tag{1}$$

where σ_m is hydrostatic stress and $\bar{\sigma}$ is equivalent stress.

2.2. Proposal of a Triaxial Failure Diagram

An innovative TFD is proposed in this paper including a 3D failure criterion, together with the identification of valid safety zones. The equivalent plastic strain at failure is related in terms of the stress triaxiality. The FLC points are mapped to the new diagram, and extra assumptions are taken to close it for high values of volumetric stresses. The Figure 3 illustrates an example of FLD and TFD safety zones. The names for each safety zone are taken from the industrial jargon currently employed in the FLD for thin sheets: Fail, Thinning, Risk, Safe, Low stretch, Soft compression, Compression, Strong compression.

An internal damage variable should be used for each material point through the simulation aiming to consider the non-linear path evolution and possible changes of the triaxiality values. When fracture damage variable reaches the unity, then the TFC envelope is also reached and failure arises.

2.3. Failure envelope mapping from FFLD to TFD

The failure envelope conversion mapping may be done from the FFLD to TFD space. In the case that complete FFLC (Fracture Failure Limit Curve) is not available; the FLC curve may be used instead.

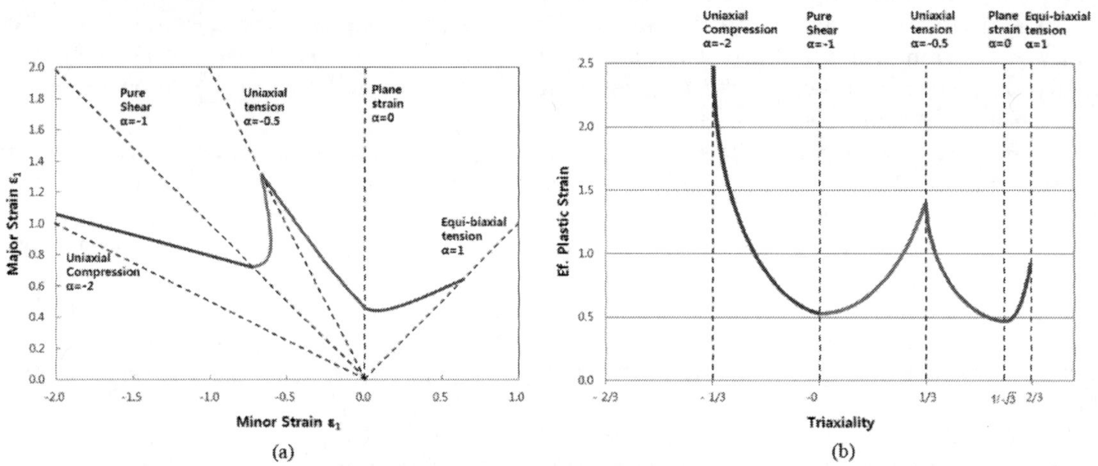

(a) (b)

Figure 1. Conversion mapping from FFLC to TFC. (a) FFLD space. (b) TFD space.

The best would be to perform triaxial testing in the lab, in order to obtain the Triaxial Failure Curve (TFC) envelope. Although, under lack of tests, the FFLC-TFC mapping maybe done as a quite good approximation. The elasto-plastic material parameters are employed. This mapping can only be done for the plane stress condition; i.e. from triaxiality = –0.66 up to triaxiality = +0.66. Outside this range, some closure lines must also be considered, under lack of additional testing data.

Usually the pure shear failure point is not provided in the FLC curve, so this value may be incorporated as an input parameter from the end-user, aiming to enrich the obtained TFC. Proper consideration of the pure shear failure is essential to capture the crack in many simulation tests.

Even though the mapping is done under plane stress condition, the mapped curve to TFD can be employed in solid elements subjected to multiaxial loading, as a really good failure criterion for 3D solids. The Figure 1 shows the MMC failure model [10] converted from the complete FFLC, including pure shear fracture, into the TFD space.

3. Numerical example validation

A complex industrial process is taken in order to validate the capability of the TFD criterion to predict the failure or the feasibility of metal forming components.

The Stampack^{TM} explicit solver [11] is employed to perform the numerical simulations. Two models are created: the first one employing triangular shell elements [12] with FLC as failure criterion; and the second one employing hexahedral 3D solid elements [13] with TFD failure criterion.

The shear hydro-bending experiment made by Han et al [14] is taken as validation example to execute the numerical simulation analysis, for both shell and 3D solid models.

3.1. Process and material description

The Figure 2 depicts the schematic principle of the shear hydro-bending forming process. During the filling step, the internal pressure is applied. Secondly, the left and right punches provide the proper feeding speed while top and bottom dies are moved down the stroke distance, 40 mm in this example.

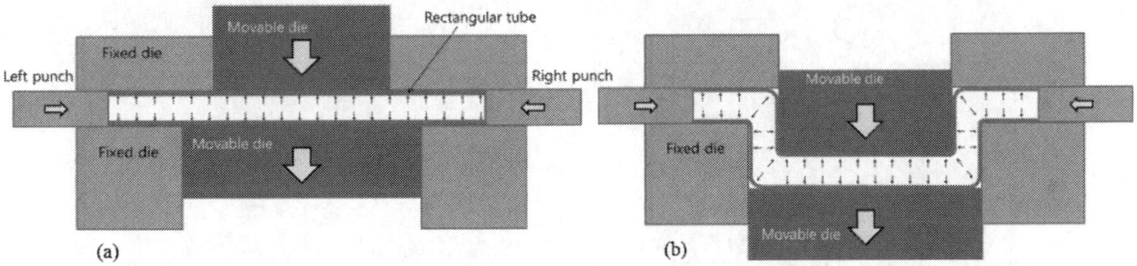

Figure 2. Principle of shear hydro-bending process. (a) Filling stage. (b) Tools movement stage.

Han et al [14] performed the experimental tests under several process conditions: different tool radius, different internal pressures, different axial feeding speeds and different transversal strokes. Nevertheless, in the present analysis, the numerical simulations are studied only for following specific conditions: feeding ratio of 1.2 and internal pressure of 0.6 σ_s. The process defects, such as cracking and wrinkling, of the formed tube do not appear for these parameters [14]. The material employed in the experiment is the 5A02 aluminum alloy, which properties are listed in the Table 1.

Table 1. Material properties of aluminium alloy 5A02.

Properties	Values
Young Modulus	75 GPa
Poisson ratio	0.30
Yield stress (σ_s)	76 MPa
Hardening constant(K)	336.99 Mpa
Hardening exponent(n)	0.254
Tensile strength	184 Mpa
Elongation	22.8 %

The Figure 3 illustrates the FLD and TFD employed during the numerical simulation of the part.

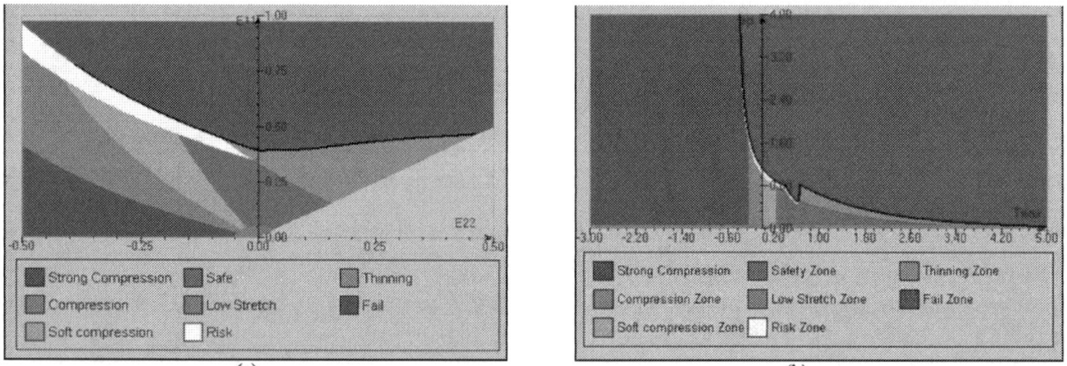

(a) (b)

Figure 3. Safety zones and formability curves for AA 5A04 material. (a) FLD. (b) TFD.

3.2. *Comparison with physical part*

The numerical simulation results are validated against the physical industrial part produced and published by Han et al [14] and Hong-lae et al [15]. The Figure 4 shows the final shape result obtained by simulation and laboratory. The difference in shape is minimal, but more remarkable aspect is that the part is shown feasible when employing 3D solid elements.

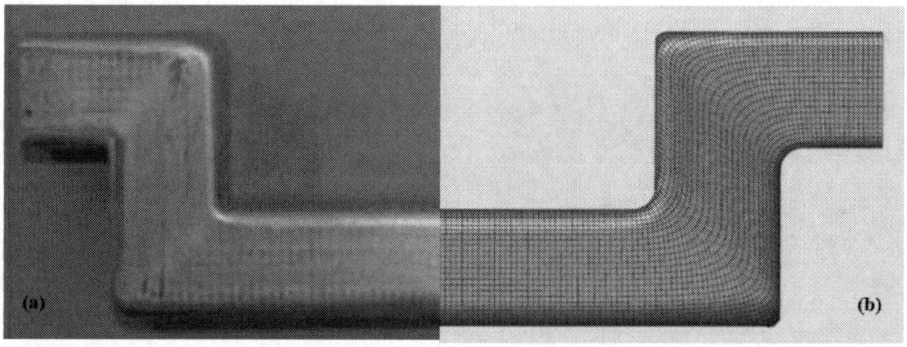

Figure 4. Shear hydro-bending final shape. (a) Physical part (b) Simulation with 3D solid elements.

3.3. *Formability analysis given by shell and 3D solid solutions*

The Figure 5 shows the postprocess contour fill for the FLD map and safety zones at the end of the forming modelled with shell elements. It can be seen that values over 100% is found in the FLD map meaning that failure is reached for the red colored areas, also marked as failed in safety zone plot.

The Figure 6 displays the contour fill for the TFD map and safety zones obtained with the solid solution, employing MMC triaxial failure. The failure map values below 100% indicate that the onset of fracture was not reached; fact that is also confirmed in TFD safety zone plot.

The Figure 7 provides the formability graph for both shell and solid solutions, where material points are shown to overpass the FLC, while all points remain below the TFC in the 3D solid analysis.

The differences obtained when comparing shell and solid analysis, reflects that FLC is no longer valid as a trusted criterion when multiaxial loads are involved. In contrast with the triaxility approach that provides more reliable results.

The accuracy of the TFD methodology is validated in this analysis of tube hydro-forming process. The result of the 3D solid model is successfully validated against the physical part. Only the new TFD approach is able to confirm the feasibility of this part.

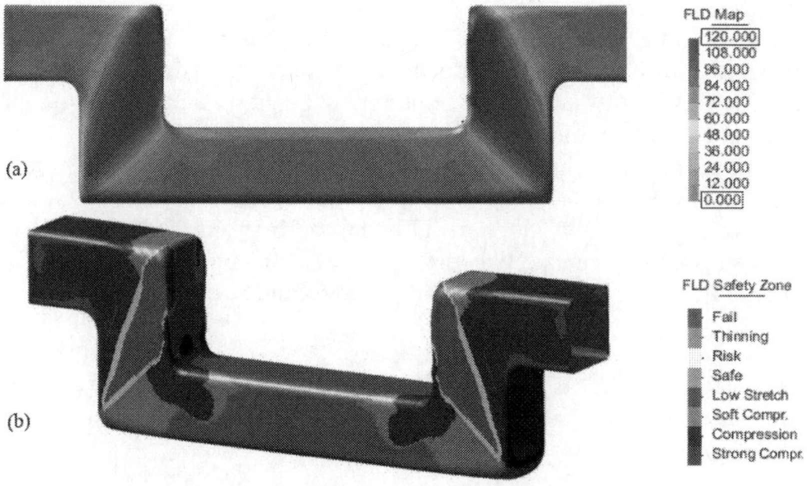

Figure 5. Post-process contour plots from shell elements solution. (a) FLD map. (b) FLD safety zones.

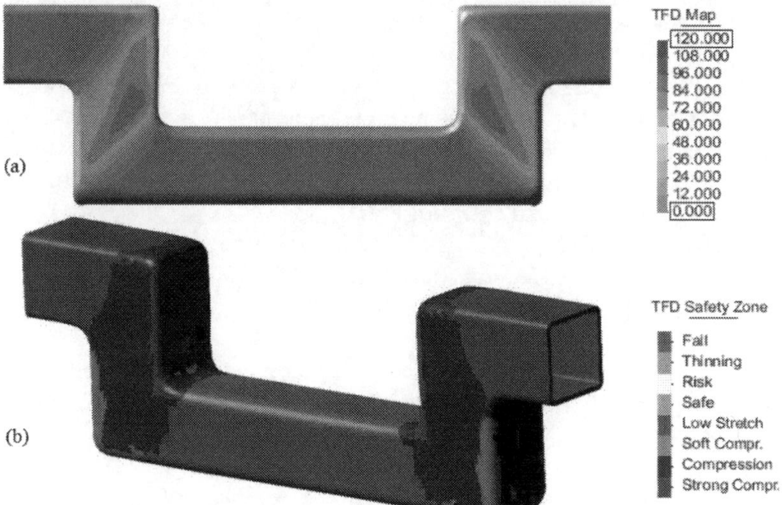

Figure 6. Post-process contour plots from 3D solid elements solution. (a) TFD map. (b) TFD safety zones.

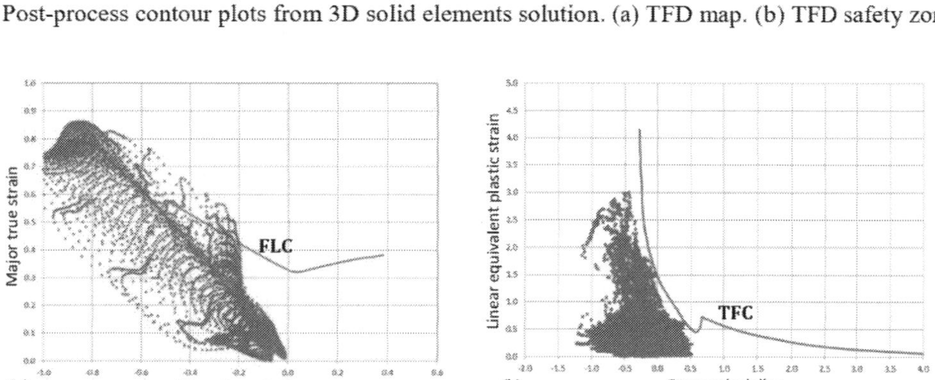

Figure 7. Formability points distribution graph. (a) FLC from shell solution. (b) TFC from solid solution.

4. Concluding remarks

A novel Triaxiality Failure Diagram is proposed as 3D failure criterion, which also allows the identification of valid safety zones. The equivalent plastic strain at failure is related in terms of the stress triaxiality. The FLC points are mapped to the new diagram, and extra assumptions are taken to close it for high values of volumetric stresses.

Numerical simulations of tube hydro-forming are performed employing an explicit solution. Shell and 3D solid models are created to compare both criteria. Results are validated against the physical part produced by the industry. Only the new TFD approach is able to confirm the feasibility of the part. The use of stress-based criteria, like the proposed one, seems to be very convenient for the numerical failure prediction of industrial parts subjected to multiaxial loading.

Acknowledgments

We want to thank Dr. Pablo Vargas who contributed with the results achieved in this research work. We also acknowledge the financial contribution received by the Catalonian government through AGAUR funding granted to Gerardo Socorro.

References

[1] Keeler S, Backofen W. 1963. Plastic instability and fracture in sheets stretched over rigid punches. *ASM TRANS Q* **56** (1), 25–48.
[2] Goodwin G M. 1968. Application of strain analysis to sheet metal forming in the press shop. *SAE paper 680093*.
[3] Atkins A G. 1996. Fracture in forming. *J. Mat. Processing Technology*, **56**, 609-618.
[4] Johnson G R, Cook W H. 1985. Fracture characteristics of three metals subjected to various strains, strain rates, temperatures and pressures. *Engineering Fracture Mechanics*, **21**: 31-48.
[5] Lemaitre J. 1985. A continuous damage mechanics model for ductile fracture. *Journal of Engineering Materials and Technology*, **107**: 83–89.
[6] Wierzbicki T, Bao Y, Lee Y, Bai Y. 2005. Calibration and evaluation of seven fracture models. *International Journal of Mechanical Sciences*, **47**, 719–743.
[7] Bai Y, Wierzbicki T. 2010. Application of extended Mohr-Coulomb criterion to ductile fracture. *International Journal of Fracture* **161** (1), 1–20.
[8] Li Y, Wierzbicki T. 2009. Mesh-size study of ductile fracture by non-local plasticity model. *Proceeding of SEM Annual Conference and Exposition on Experimental and Applied Mechanics*. Paper No. 387.
[9] Li Y, Wierzbicki T. 2010. Prediction of plane strain fracture of AHSS sheets with post-initiation softening. *International Journal of Solids and Structures* **47** (17), 2316–2327.
[10] Li Y, Luo M, Gerlach J, Wierzbicki T. 2010. Prediction of shear-induced fracture in sheet metal forming. *Journal of Materials Processing Technology* **210**, 1858–1869.
[11] STAMPACK^{TM} Reference manual, Version 7.1, Quantech ATZ (www.Stampack.com).
[12] Flores F G, and Oñate E. 2005. Improvements in the membrane behavior of the three node rotation-free BST shell triangle using an assumed strain approach. *Computer Methods in Applied Mechanics and Engineering*, **194**(6), 907–932.
[13] Rastellini F, Díaz G, Forgas A, and Socorro G. 2011. Cutting Strategy for the Numerical Simulation of Multi-Stage Forming of Thick Metal Sheets. *American Institute of Physics Conference Series*. **1383**: 406–412.
[14] Han C, Xu Y, Wang Y, Zang C, Yuan S. 2012. Shear hydro-bending of 5A02 aluminum alloys rectangular tubes. *Transactions of Nonferrous Metals Society of China* **22,** 382–388.
[15] Hong-lae K, Forgas A, Jungsun K, Soontak K and Hotaek K. 2013. Numerical Prediction of the Fracture Using Triaxial Forming Limit in Tube Hydro Forming. *Proceedings of the 6th International Conference on Tube Hydroforming TUBEHYDRO2013*. Jeju, Korea. 243-249.

On the fracture prediction of 304L stainless steel sheets utilizing different hardening models

S A Dizaji[1], H Darendeliler[1], B Kaftanoğlu[2]

[1]Department of Mechanical Engineering, Middle East Technical University, 06800, Ankara, Turkey
[2]Department of Manufacturing Engineering, Atilim University, 06836, Ankara, Turkey
E-mail: abbasnegzad@gmail.com, hdarende@metu.edu.tr, bilgin.kaftanoglu@atilim.edu.tr

Abstract: Fracture prediction is one of the challenging problems in sheet metals. Forming limit curves at fracture (FLCF), as a tool to determine fracture in sheet metal processes, are obtained through the use of numerical analyses. As one of the approaches, the ductile fracture criteria (DFCs) represent the fracture initiation of the sheets formed by different loading histories. In this study, the effects of three different hardening models on different DFCs to predict the fracture for stainless steel 304L have been investigated. The results show that most of DFCs work better in the region $\varepsilon_2 < 0$ especially with the kinematic hardening model. However, for the region $\varepsilon_2 > 0$ where the stretching conditions are dominant, none of them could precisely estimate the fracture initiation.

1. Introduction

Forming limit curves (FLCs) which were first drawn experimentally in 1960s have been used as an effective tool to analyze the behavior of the sheet metal for years. It is reported by many researchers that the strain paths can produce considerable variations in the forming limits [1-2]. Ductile fracture criterion as one of the less dependent methods to strain-path, has been employed extensively in the recent years to draw the forming limit curves at fracture (FLCF). But there are many parameters affecting the efficiency of the DFCs and Dizaji et al [3] show that the hardening rule is one of those factors.

In this study, the effects of three hardening models; namely, isotropic, kinematic and combined hardening were investigated for different ductile fracture criteria. For this purpose, ductile fracture criteria proposed by Freudenthal [4], Cockroft and Latham [5], Oh et al. [6], Ayada et al. [7] and Brozzo et al. [8] have been used. Zeigler-Prager [9] equation and the Chaboche-Zeigler formulation with five constants [10] have been utilized in the kinematic hardening and the combined hardening models, respectively. All of the criteria and models were implemented to a commercial FE software using user subroutines [11]. The FEM results have been compared with experimental data for Nakazima tests to recognize the DFC and the corresponding hardening model which predict the fracture better in stainless steel material SS304L. Also a deep drawing process with square blank was utilized to compare the DFCs predictions.

2. Experimental tests and Numerical models

The explicit solver of the commercial finite element code ABAQUS has been used in all numerical analyses of the deformation processes. All tools were assumed as rigid and the mesh for the sheets were created by using an 8-node linear brick, reduced integration elements. The friction coefficients, based on Coulomb law, have been taken as 0.05 for lubricated interfaces and 0.13 for dry interfaces in the simulations. The penalty contact algorithm has been utilized to model the interaction between the surfaces. To have a quasi-static conditions in the dynamic explicit solutions, the kinetic energy is kept less than 10 percent of the total internal energy [11].

The properties of the stainless steel SS304L that were obtained by employing uniaxial tensile test (UTT) and are presented in table 1. To calculate the combined hardening constants, the half cycle data shown in table 1 were used as input for the numerical analyses. Then, with some numerical manipulation, the required material constants were found as shown in table 2.

Table 1. Material properties of SS304L steel.

Parameter	$E\ (GPa)$	ν	$\sigma_Y^0\ (MPa)$	r	$K\ (MPa)$	n
Value	194	325	339	1.0125	1196	0.32

Table 2. The parameters of Chaboche-Zeigler combined hardening model.

Parameter	γ	C	σ_Y^0	b	Q
Value	1732.6	120.7	339	5.32	347.54

3. Results and Discussion

The required constants for each empirically formulated phenomenological uncoupled DFCs used in this study, are shown in table 3 for different hardening rules.

Table 3. Criteria constants for SS304L material.

Constants	$C_{Freu.}$	$C_{Cock.}$	$C_{Broz.}$	C_{Oh}	C_{Ayada}
Isotropic Hardening	922.5	989.5	0.996	0.985	0.371
Kinematic Hardening	843	872	0.901	0.855	0.313
Combined Hardening	903.4	995.6	1.02	1.007	0.392

To evaluate the reliability of each implemented DFC on the sheet forming process the Nakazima tests were carried out using the SS304L stainless steel. The comparison of the strain values obtained by the simulations for different DFCs with the experimental results are presented in figures 1-3.

It is observed that the FLCF curves of all ductile fracture criteria are almost linear lines with a negative slope predicting the fracture better when $\varepsilon_2 < 0$; while for the region $\varepsilon_2 > 0$, all of the criteria have inaccurate estimations except for the certain deformation states. In the region $\varepsilon_2 < 0.1$, FLCF curves obtained by DFCs of Ayada et al. and Brozzo et al. are closer together and have better predictions among all DFCs for all hardening models. In this region, the best results are obtained by the DFC of Ayada et al. for kinematic hardening rule and DFC of Brozzo et al. for isotropic and combined hardening rules. The better results of Ayada et al. and Brozzo et al. can be attributed to the consideration of both mean stress and equivalent plastic strain simultaneously in their formulations [2]. For the region $\varepsilon_2 > 0.1$ the predictions of all DFCs with all hardening models deviate from the experimental results. However, the predictions of Freudenthal, Cockroft and Latham, and Oh et al. are better for equi-biaxial conditions especially with isotropic hardening rule. Fracture initiation in square cup drawing, as a process which has complicated deformation history, was also predicted by various DFCs together with the three hardening models. For this purpose, 1 mm thick, 80 × 80 mm blanks made of SS304L steel were drawn by using an 40 × 40 mm punch as shown in figure 4.

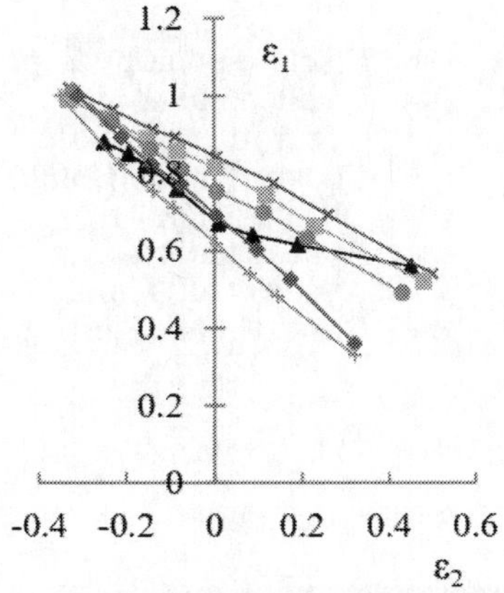

Figure 1. Nakazima test results rule using piecewise isotropic hardening model.

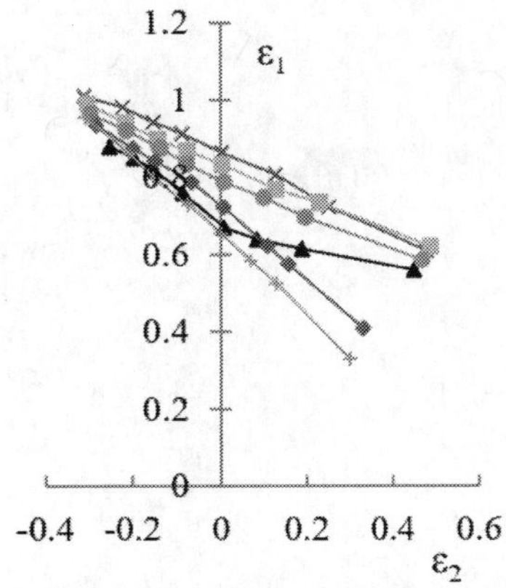

Figure 2. Nakazima test results using Zeigler-Prager kinematic hardening model.

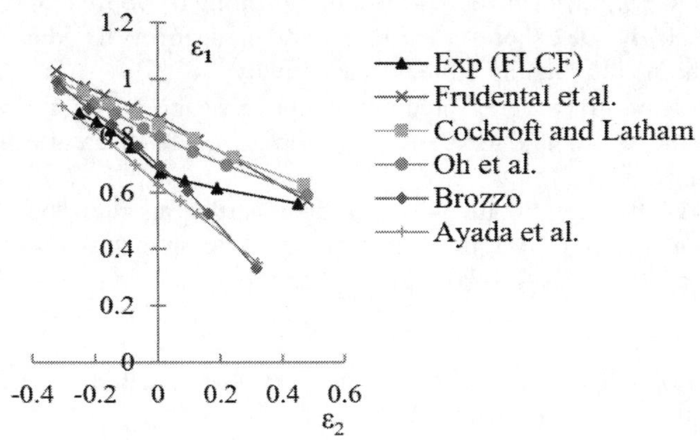

Figure 3. Nakazima test results for combined hardening that uses Chaboche- Zeigler equation.

In figure 5, each point in the diagram stands for the major and minor principal strains in the critical element of the mesh in which the fracture is initiated by using specific DFC and hardening model. The deformation histories of the points have not been shown to avoid confusedly diagram.

It is observed by figure 5 that fracture initiation is better predicted by the DFC of Oh et al. with isotropic hardening between all DFCs. Also the DFCs of Cockroft and Latham with isotropic hardening and DFC of Brozzo et al. with kinematic hardening rule are the other DFCs which have closer estimations. These results can also be remarked based on the figures 1-3 in the region that fracture has been initiated ($0.2 < \varepsilon_2 < 0.3$).

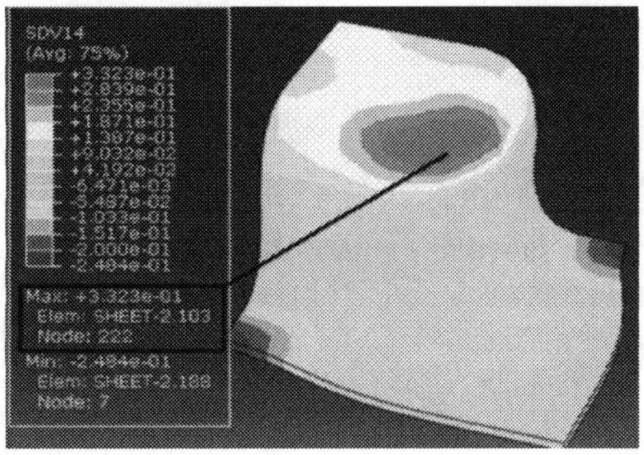

Figure 4. Distribution of the variable SDV14 indicating the element where fracture has occurred at 21.3 mm of cup height using DFC of Ayada et al. with kinematic hardening model.

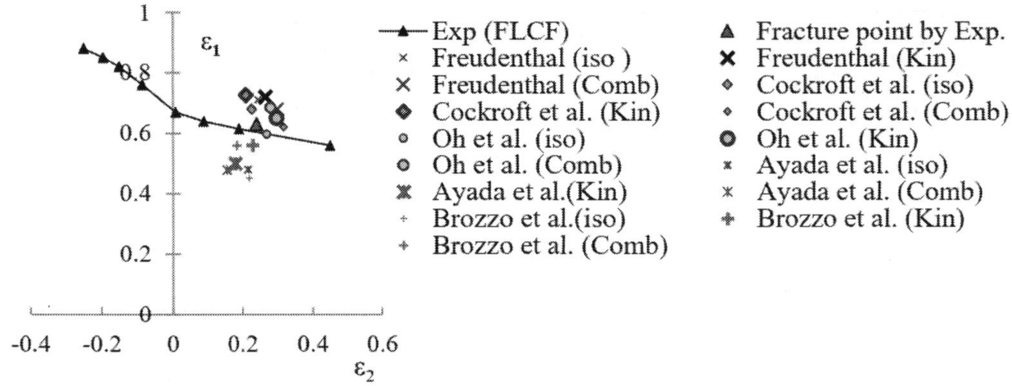

Figure 5. Comparison of different DFCs results for square cup drawing.

4. Conclusions

The effects of hardening rules on the reliability and applicability of different ductile fracture criteria were studied and it is shown that they have significant influences in the forming of SS304L sheets.

1. The best predictions were obtained by DFCs of Ayada et al. and Brozzo et al. when $\varepsilon_2 < 0$ especially with kinematic and combined hardening models, respectively.
2. For $\varepsilon_2 > 0$, although DCF of Oh et al. gives better predicitions, none of the DFCs are accurate enough. Especially DFCs of Ayada et al. and Brozzo et al. should be used cautiously with stretching dominant deformations.
3. The DFCs of Oh et al. and Cockroft and Latham with isotropic hardening rule and DFC of Brozzo et al. with kinematic hardening rule have shown the better fracture predictions between all DFCs in square cup drawing as expected from the Nakazima test results.

5. Acknowledgements

The principal author willingly thanks TÜBİTAK (The Scientific and Technological Research Council of Turkey) for supporting this study under the grant BIDEB2215.

References

[1] Yoshida K, Kuwabara T, Kuroda M 2007 Path-dependence of the forming limit stresses in a sheet metal *Int. J. Plasticity* **23** 361-84.

[2] Hongsheng, L, Yuying, Y, Zhongqi Y, Zhenzhong S, Yongzhi W 2009 The application of a ductile fracture criterion to the prediction of the forming limit of sheet metals *J. of Mat. Proc. Tech.* **209** 5443–7.

[3] Dizaji S A, Darendeliler H, Kaftanoğlu B, 2014 Effect of hardening models on different ductile fracture criteria in sheet metal forming *Int. J. Mat. Forming*, doi: 10.1007/s12289-014-1188-5.

[4] Freudenthal, FA 1950 *The inelastic behavior of engineering materials and structures* (New York, Wiley).

[5] Cockroft MG, Latham DJ 1968 Ductility and workability of metals *J. Inst. Metall.* **96** 33–9.

[6] Oh SI, Chen CC, Kobayashi S 1979 Ductile fracture in axisymmetric extrusion and drawing *J. Eng. Ind. Trans.* ASME, **101** 36–44.

[7] Ayada M, Higashino T, Mori K 1984 *Central bursting in extrusion of inhomogeneous materials* Proc. of the 1st ICTP. Adv. Tech. of Pals. Tokyo, Japan. **1** 553–8.

[8] Brozzo P, Deluca B, Rendina R. 1972 *A new method for the prediction of formability limits of metal sheets. In: Sheet Metal Forming and Formability*, Proc. DDRG Amsterdam, Netherlands.

[9] Prager W 1956 A new method of analyzing stress and strains work-hardening plastic solids *J. of Applied Mech.*, **23** 493-6.

[10] Lemaitre J, Chaboche L 1990 *Mechanic of solid materials* (U.K., Cambridge university press).

[11] Hibbit, Karlsson, Sorensen Inc. 2002 *ABAQUS/Explicit user's manual*, Version 6.3.

Influence of material modeling on earing prediction in cup drawing of AA3104 aluminum alloy sheet

H Fukumasu[1], T Kuwabara[2] and H Takizawa[3]

[1] Mitsubishi Aluminum Co., Ltd., 85, Hiramatsu, Susono, Shizuoka, 410-1127, Japan
[2] Division of Advanced Mechanical Systems Engineering, Tokyo University of Agriculture and Technology, 2-24-26, Naka-cho, Koganei, Tokyo, 184-8588, Japan
[3] Mechanical Engineering Dept., Nippon Institute of Technology, 4-1 Gakuendai Miyashiro, Saitama, 345-8501, Japan
Email: fukumasu@malco.co.jp

Abstract. In-plane biaxial tension and combined tension-compression tests are carried out for AA3104 aluminum alloy sheets. Linear stress paths are applied to cruciform specimens to measure the contours of plastic work in the stress space and the directions of plastic strain rates at each stress path. Coefficients $\alpha_1 - \alpha_8$ and exponent M of the Yld2000-2d yield function are determined to minimize the mean square error of the analytical yield locus from a measured work contour. The values of the weighting coefficients in the evaluation of the error are varied to check the effect of a specific stress state on the earing behavior. The effects of the combinations of the weighting coefficients on the accuracy of earing prediction in the cup drawing process are discussed.

1. Introduction

For aluminum beverage can body stock, a low level of earing during cup drawing is required. A sheet forming simulation, especially for earing prediction in the cup drawing process, is required to avoid difficulties in the drawing and ironing process. In the simulation, it is necessary to use a material model capable of accurately reproducing the deformation behavior of the aluminum sheet.

Determining the anisotropic coefficients of yield functions by a biaxial test is effective for obtaining accurate material models. Kuwabara et al. [1] developed a biaxial tensile testing method using a cruciform specimen for sheet metals. It is necessary to consider the effect of shear stress σ_{xy}, where x and y are the axes of in-plane plastic anisotropy, because the axis of principal stress is not always identical with the axis of plastic anisotropy. Yoon et al. [2] predicted the earing profile very accurately by employing a material model that reproduced the r-value and the yield stress directionalities precisely. In the cup drawing process, the stress state in the flange area is the in-plane tension-compression combined state. Therefore, it is necessary to reproduce the deformation behavior at the 2nd and 4th quadrants on the 2D stress plane accurately in order to predict the earing precisely. However, no studies have yet reported the plastic deformation behavior of aluminum beverage can body stock measured under the in-plane combined tension-compression stress states when the axes of principal stresses are not identical with the axes of anisotropy.

In this work, in-plane uniaxial tension, biaxial tension and combined tension-compression tests [3] were carried out for an AA3104 aluminum alloy sheet to measure the successive contours of plastic work in the stress space and the directions of plastic strain rates under linear stress paths. The angles made by the axes of principal stresses and the axes of plastic anisotropy were varied in uniaxial tension and combined tension-compression tests to measure the experimental data including shear stress σ_{xy}. Anisotropic coefficients $\alpha_1 - \alpha_8$ and exponent M of the Yld2000-2d yield function [4] were determined to minimize the mean square error of the analytical yield locus from a measured work contour. The values of the weighting coefficients in the evaluation of the error were varied to check the effect of a specific stress state on the earing behavior. The effects of the combinations of weighting coefficients on the accuracy of the earing prediction in the cup drawing process are discussed.

2. Experimental

2.1. Test material
An AA3104 aluminum alloy sheet for an aluminum can body of 0.4 mm thickness was used in this work. The uniaxial tensile mechanical properties in the rolling direction are listed in Table 1. Hereinafter, the rolling, transverse and thickness directions are defined as the x, y and z axes, respectively.

Table 1. Tensile properties measured parallel with rolling direction.

$\sigma_{0.2}$ / MPa	c^a / MPa	n^a	α^a	r^b
258.9	629.3	0.303	0.0504	0.420

[a] Approximated using $\sigma = c(\alpha + \varepsilon^p)^n$ for a logarithmic plastic strain range of $0.002 \leq \varepsilon^p \leq 0.035$.
[b] Measured at uniaxial nominal strain ε_N=0.04.

2.2. Procedure of biaxial tensile tests
Biaxial tensile tests were employed using cruciform specimens, the geometry of which is standardized in ISO 16842 [5]. Seven linear stress paths, $\sigma_x : \sigma_y$ = 4:1, 2:1, 4:3, 1:1, 3:4, 1:2, and 1:4, were applied to the cruciform specimens. Combined tension-compression tests were also performed. Six linear stress paths, $\sigma_x : \sigma_y$ =1:-2, 1:-1, 2:-1, -1:2, -1:1, and -2:1, were applied, and the angles θ made by the axes of principal stresses and the axes of plastic anisotropy were varied as 0, 22.5, 45, 67.5, and 90°.

The contour of plastic work associated with ε_0^p=0.003 in the stress space was determined in order to quantitatively evaluate the biaxial deformation behavior of the test material, where ε_0^p is the logarithmic plastic strain measured in the uniaxial tensile test along the RD, and was employed as a reference datum for the plastic work per unit volume.

2.3. Cup drawing test
A cup drawing test was carried out using a cylindrical punch of 33 mm diameter. The punch and die clearance was set to avoid cup wall ironing. The blank holding force was 5 kN. Beef tallow was applied to both sides of the blanks as a lubricant. The earing profile of the drawn cups was measured.

3. Finite element simulation
Finite element (FE) simulations using LS-DYNA v971 R6.1.2 were carried out for the cup drawing test. Due to the orthotropic anisotropy of the sheet, one quarter of a circular blank was analyzed. All dies and punch were assumed to be analytical rigid bodies. Fully integrated shell elements were employed for the circular blank model. A friction coefficient of 0.05 was employed for all contacts in the model. The Yld2000-2d yield function was used in the FE simulations. Anisotropic coefficients $\alpha_1 - \alpha_8$ and exponent M of Yld2000-2d were determined to minimize the mean square error of the analytical yield locus from the measured work contour. Combinations of weighting coefficients of evaluation function (1) were changed for four cases, (a) to (d) shown in Table 2, to check the effect of a specific stress state on the earing behavior.

$$f(\alpha_i, M) = \sum_{j=1}^{N} w_{j,\sigma}\left(\delta_{j,C} - \delta_{j,M}\right)^2 + \sum_{k=1}^{O} w_{k,\beta}\left(\beta_{k,C} - \beta_{k,M}\right)^2 \quad (1)$$

Here, $w_{j,\sigma}$ and $w_{k,\beta}$ are the weighting coefficients for the stress points forming a work contour and the directions of plastic strain rates $\delta_{j,C}$ and $\delta_{j,M}$ are the distances from the origin to the calculated and measured stress points, and $\beta_{k,C}$ and $\beta_{k,M}$ are the calculated and measured directions of the plastic strain rates. The calculations of the plastic strain rates are based on the normality rule.

Table 2 Experimental data for calculating parameters of the Yld2000-2d yield function.

	Uniaxial test				Biaxial test		
	$\sigma_0, \sigma_{45}, \sigma_{90},$	$\sigma_{15}, \sigma_{30}, \sigma_{60}, \sigma_{75}$	$r_0, r_{45}, r_{90},$	$r_{15}, r_{30}, r_{60}, r_{75},$	Quad. I	Quads. II, IV $\sigma_{xy} = 0$	Quads. II, IV $\sigma_{xy} \neq 0$
(a)	X		X		X [a]		
(b)	X		X		X		
(c)	X	X	X	X	X	X	
(d)	X	X	X	X	X	X	X

[a] Equibiaxial tensile properties (σ_b, r_b), only.

4. Results and discussion

Figure 1 compares the measured deformation behavior of the test material and the behavior predicted by using Yld2000-2d (a)–(d). Figure 1(a) reproduces the yielding behavior under uniaxial tension precisely, whereas the reproducibility in the biaxial stress state and, in particular, that at the 2nd and 4th quadrants on the 2D stress plane is not accurate. In (b), no significant differences are observed from (a). In (c), the reproducibility is improved at the 2nd and 4th quadrants, whereas the accuracy under uniaxial tension has decreased. In (d), the reproducibility at the 2nd and 4th quadrants is the best in this study, whereas the accuracy under uniaxial tension has decreased further.

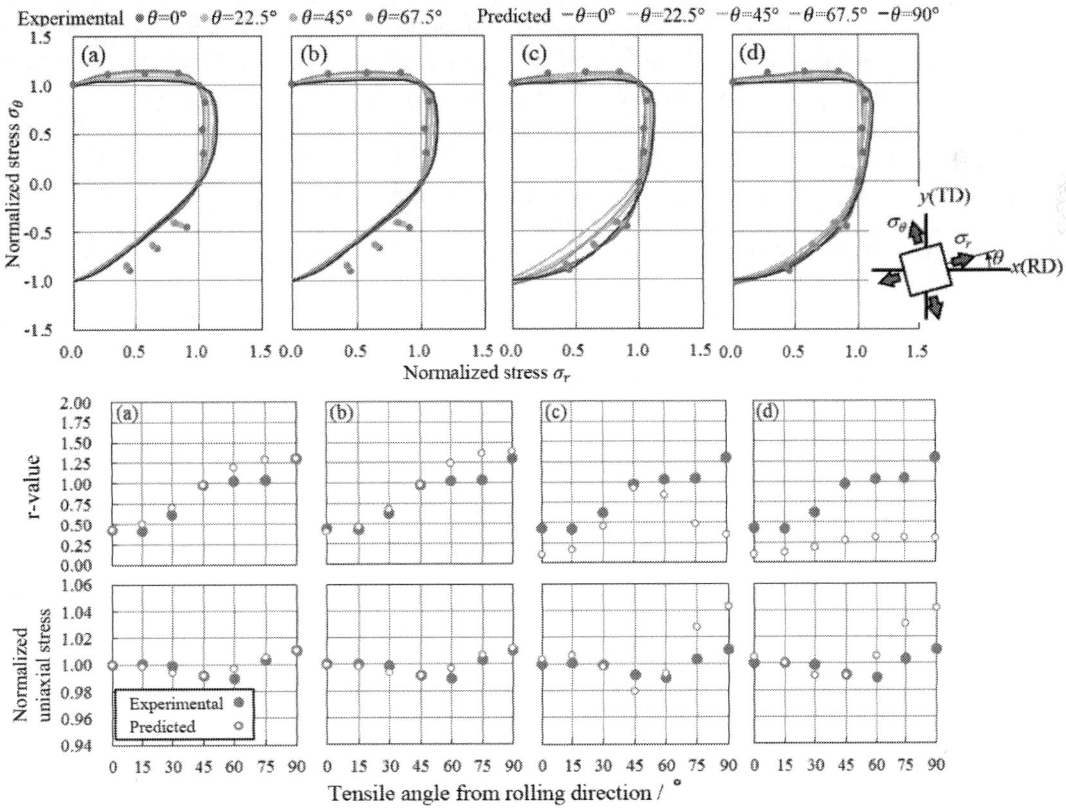

Figure 1 The deformation behavior measured experimentally and predicted by Yld2000-2d (a)–(d).

Figure 2 compares the measured cup height with the heights predicted using the Yld2000-2d yield function. The cup height is normalized by the average cup height. Figure 2(d), which reproduces the yielding behavior at the 2nd and 4th quadrants in the stress space, provides the closest agreement with the measurement. In a previous study [6] of the relation between the shape variation of the contour of plastic work according to θ and the predicted earing profile, it was reported that the cup height becomes relatively higher when the shape of the work contour at the 2nd and 4th quadrants is relatively shrunk. In our study, the same feature is observed in (a)–(d); namely, the shape of the work contour at the 2nd and 4th quadrants is relatively shrunk at $\theta = 45°$, and the predicted earing profile indicates four ears at 45° from the rolling direction. Furthermore, the relation between the earing amplitude and the magnitude of the expanded/shrunk difference of the shape of the work contour at the 2nd and 4th quadrants is observed; namely, the low magnitude of the expanded/shrunk difference results in the low amplitude of the earing in (a) and (b); on the other hand, the high magnitude of the expanded/shrunk difference results in the high amplitude of the earing in (c).

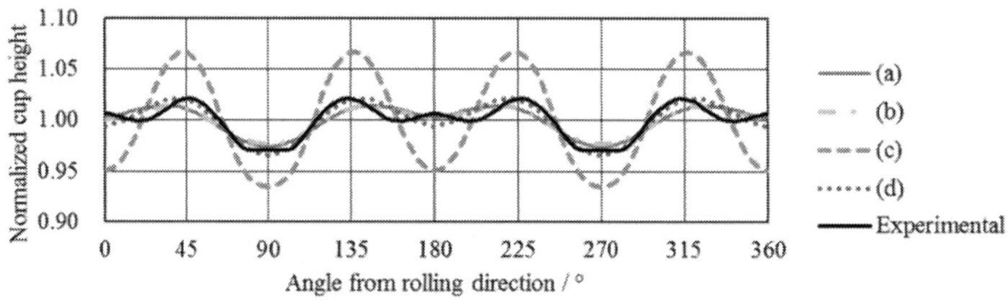

Figure 2 Comparison of the measured cup height with those predicted by Yld2000-2d (a)–(d).

5. Conclusion

In-plane uniaxial tension, biaxial tension and combined tension-compression tests were carried out for AA3104 aluminum alloy sheet, and the contour of plastic work and the directions of the plastic strain rates (r-values) at ε_0^p=0.003 were precisely measured. The predicted earing profile changed depending on the focused stress state used to determine the anisotropy coefficients of the Yld2000-2d yield function. In this work, case (d), which reproduces the yielding behavior at the 2nd and 4th quadrants of the stress space, provides the closest agreement with the measurement. It was observed that the magnitude of the difference in the expanding/shrinking rate of the work contour shapes at the 2nd and 4th quadrants affects the earing amplitude.

References

[1] Kuwabara T, Ikeda S and Kuroda T 1998 Measurement and analysis of differencial work hardening in cold-rolled steel sheet under biaxial tension *J. Mater. Process Technol.* **80-81** 517-23

[2] Yoon J W, Barlat F, Dick R E and Karabin M E 2006 Prediction of six or eight ears in a drawn cup based on a new anisotropic yield function *Int. J. Plasticity* **22** 174-93

[3] Kuwabara T, Horiuchi Y, Uema N and Ziegelheimova J 2007 Material testing method of applying in-plane combined tension-compression stresses to sheet specimen *J. Japan Soc. Technol. Plasticity* **48** 630-34

[4] Barlat F, Brem J C, Yoon J W, Chung K, Dick R E, Lege D J, Pourboghrat F, Choi S H and Chu E 2003 Plane stress yield function for aluminum alloy sheets – Part 1 : Theory *Int. J. Plasticity* **19** 1297-319

[5] ISO 16842: 2014 Metallic materials −Sheet and strip −Biaxial tensile testing method using a cruciform test piece

[6] Fukumasu H, Hakoyama T, Kuwabara T and Takizawa H 2015 Effect of the yielding surface on earing prediction in cup drawing process for 3104 aluminum alloy sheets *Proc. The 129th Conf. Japan Inst. Light Met.* 177-178

Journal of Physics: Conference Series **734** (2016) 032023 doi:10.1088/1742-6596/734/3/032023

Large-strain cyclic response and martensitic transformation of austenitic stainless steel at elevated temperatures

H Hamasaki[1], T Nakano[1] E Ishimaru[2] and F Yoshida[1]

[1] Deparmtment of Mechanical Science and Engineering, Hiroshima University, 1-4-1, Kagamiyama, Higashi-Hiroshima, Hiroshima, 739-8527 Japan

[2] Nippon Steel & Sumikin Stainless Steel Corporation, 3434, Ooaza-shimada, Hikari, Yamaguchi, 743-8550 Japan

E-mail: hamahiro@hiroshima-u.ac.jp

Abstract. Cyclic tension-compression tests were carried out for austenitic stainless steel (SUS304) at elevated temperatures. The significant Bauschinger effect was found in the obtained stress-strain curve. In addition, stagnation of deformation induced martensitic transformation was observed just after stress reversal until the equivalent stress reached the maximum value in the course of experiment. The constitutive model for SUS304 at room temperature was developed, in which homogenized stress of SUS304 was expressed by the weighed summation of stresses of austenite and martensite phases. The calculated stress-strain curves and predicted martensite volume fraction were well correlated with those experimental results.

1. Introduction

Austenitic stainless steel shows extensive workhardening even at large strain owing to the precipitation of hard α′-martensite phase by plastic deformation. The constitutive equations (e.g. Stringfellow et al [1]. and Geijselares et al. [2]) for an austenitic stainless steel with which martensitic transformation kinetics are incorporated were suggested. In the present study, large cyclic tension compression tests for an austenitic stainless steel (SUS304) were conducted at various temperatures (from 293 to 473K). The stress-strain curves as well as martensite volume fraction were measured during the deformation. From the obtained results, it was found that austenite phase show extensive workhardening at high temperature even though martensitic transformation did not take place. Additionally, stagnation of martensitic transformation due to the Bauschinger effect was found. The macroscopic constitutive model for the present material was proposed and it was found that the calculation results well predicted the stress-strain curves and martensitic volume fraction observed by the experiment at 293K.

2. Cyclic tension-compression test of austenitic stainless steel at elevated temperature

The material used for this study is an austenitic stainless steel type SUS304. The material was machined in the shape of tension-compression test specimen shown in Figure 1(a), and solution treated at 1223K for 5 minutes. Cyclic tension-compression apparatus is shown in Figure 1(b). A specimen was covered by the thermostatic chamber and specimen was warmed by the heaters installed in the chamber. Uniaxial tension and cyclic tests were conducted at every 10K from 293 to 473K. The temperature error was found to be within 1K in every experiment.

(a) (b)

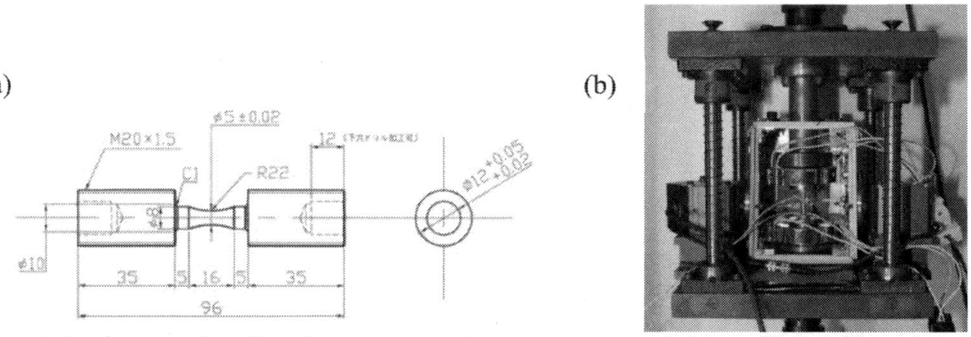

Figure 1. (a) Schematic of tension-compression specimen, and (b) cyclic tension-compression test apparatus.

3. Constitutive model for austenitic stainless steel

Let us assume the stress of austenitic stainless steel $\boldsymbol{\sigma}$ is expressed by the following rule of mixture;

$$\boldsymbol{\sigma} = f_m \boldsymbol{\sigma}_m + (1 - f_m)\boldsymbol{\sigma}_a \tag{1}$$

where subscripts m and a denote martensite and austenite, respectively, and f_m is the volume fraction of martensite. The objective rates of martensite and austenite stresses are given as;

$$\overset{\circ}{\boldsymbol{\sigma}}_m = \boldsymbol{C}^e_m \dot{\boldsymbol{\varepsilon}}^e = \boldsymbol{C}^{ep}_m \dot{\boldsymbol{\varepsilon}} \quad , \quad \overset{\circ}{\boldsymbol{\sigma}}_a = \boldsymbol{C}^e_a \dot{\boldsymbol{\varepsilon}}^e = \boldsymbol{C}^{ep}_a \dot{\boldsymbol{\varepsilon}} \tag{2}$$

where $\boldsymbol{C}^e$ and $\boldsymbol{C}^{ep}$ are the elastic and epasto-plastic tangent moduli, and $\dot{\boldsymbol{\varepsilon}}$ is the strain rate tensor.

A linear isotropic hardening model was assumed for martensite as;

$$F_m = \bar{\sigma}_m - \left(Y_m + H_m \bar{\varepsilon}^p_m\right) \tag{3}$$

where $\bar{\sigma}_m$ and $\bar{\varepsilon}^p_m$ are the equivalent stress and equivalent plastic strain of martensite, respectively, and Y_m and H_m are the material constants. By assuming the von Mises yield criteria and associative flow rule, one can have the constitutive relationship of martensite as the following equation.

$$\overset{\circ}{\boldsymbol{\sigma}}_m = \left[\boldsymbol{C}^e_m - \frac{\boldsymbol{C}^e_m \dfrac{\partial F_m}{\partial \boldsymbol{\sigma}_m} \otimes \dfrac{\partial F_m}{\partial \boldsymbol{\sigma}_m} \boldsymbol{C}^e_m}{\dfrac{\partial F_m}{\partial \boldsymbol{\sigma}_m} \boldsymbol{C}^e_m \dfrac{\partial F_m}{\partial \boldsymbol{\sigma}_m} + H_m}\right] \dot{\boldsymbol{\varepsilon}} \tag{4}$$

On the other hand, we assumed Yoshida-Uemori kinematic hardening rule [3] for austenite phase since the strong Bauschinger effect was observed in cyclic test results.

(Yiled function) $$F_a = \phi(\boldsymbol{\sigma}_a, \boldsymbol{\alpha}) - Y_a \tag{5}$$

(Kinematic hardening) $$\boldsymbol{\alpha} = \boldsymbol{\alpha}_* + \boldsymbol{\beta} \tag{6}$$

$$\overset{\circ}{\boldsymbol{\alpha}}_* = C\left\{\frac{a}{Y_a}(\boldsymbol{\sigma}_a - \boldsymbol{\alpha}) - \sqrt{\frac{a}{\bar{\alpha}_*}}\boldsymbol{\alpha}_*\right\}\dot{\bar{\varepsilon}}^p_a \tag{7}$$

$$\overset{\circ}{\boldsymbol{\beta}} = m\left\{\frac{b}{Y_a}(\boldsymbol{\sigma}_a - \boldsymbol{\alpha}) - \boldsymbol{\beta}\right\}\dot{\bar{\varepsilon}}^p_a \tag{8}$$

$$a = a_0 + R \tag{9}$$

$$\dot{R} = m(R_{sat} - R)\dot{\bar{\varepsilon}}^p_a \tag{10}$$

Similar to the case of martensite phase, von Mises yield criteria and associative flow rule were assumed then constitutive equation of austenite phase was derived as the following equation.

$$\overset{\circ}{\boldsymbol{\sigma}}_a = \left[\frac{\boldsymbol{C}_a^e \frac{\partial F_a}{\partial \boldsymbol{\sigma}_a} \otimes \frac{\partial F_a}{\partial \boldsymbol{\sigma}_a} \boldsymbol{C}_a^e}{\frac{\partial F_a}{\partial \boldsymbol{\sigma}_a} \boldsymbol{C}_a^e \frac{\partial F_a}{\partial \boldsymbol{\sigma}_a} + \frac{Ca + mb}{Y_a} \frac{\partial F_a}{\partial \boldsymbol{\sigma}_a} (\boldsymbol{\sigma}_a - \boldsymbol{\alpha}) - \frac{\partial F_a}{\partial \boldsymbol{\sigma}_a} \left(C \sqrt{\frac{a}{\bar{\alpha}_*}} \boldsymbol{\alpha}_* + m\boldsymbol{\beta} \right)} \right] \dot{\varepsilon} \qquad (9)$$

From the experimental observation, it was found that the martensitic transformation was induced by the maximum stress imposed in austenite phase. Hence the martensite volume fraction was given by the following form here;

$$f_m = 1 - \exp\left[-\beta_{oc} \left\{ 1 - \exp\left(-\alpha_{oc} \bar{\sigma}_{a,\max} \right)^{n_{oc}} \right\} \right] \qquad (10)$$

where α_{oc}, β_{oc} and n_{oc} are material constants. It should be mentioned that the equation (10) is the empirical equation just to approximate the relationship between the martensite volume fraction and austenite stress obtained by the experiment. For an accurate model applicable to multiaxial stress state, further investigation such as biaxial tension with variety of stress ratio are essential. In addition to that, it was observed from the experiments that parameters in equation (10) could be different in tension and compression, therefore, α_{oc}, β_{oc} and n_{oc} were determined from monotonic tension and compression tests separately, and the following assumptions were used for the subsequent calculations:

i) when starting from tension, parameters for tension were used throughout calculation
ii) when starting from compression, parameters for compression were used throughout calculation
iii) for cyclic test with ±5% of strain, parameters for tension were used throughout calculation

Material parameters used for the calculation are shown in Table 1 and 2.

Table 1. Parameters for martensitic evolution.

	α_{oc}	β_{oc}	n_{oc}
Tension	0.0010	19.2	8.2
Compression	0.0012	4.1	6.1

Table 2. Parameters for hardening model.

Ya /MPa	a_0 /MPa	C_1	C_2	b /MPa	m	R_{sat} /MPa	h	Y_m /MPa	H_m /MPa
200	150	100	150	4.0	2.1	950	0.25	1750	10

4. Results

Figure 2(a) and (b) are the experimentally obtained stress-strain curves and f_m vs. plastic strain curves for each temperatures. The martensitic transformation becomes moderate with increase of test temperature. Since stress are almost the same below 8% of strain at any temperature, stress was not affected by temperature up to 353K. Therefore, one can conclude that the decrease in stress at large strain as the increase of temperature was caused by less precipitation of martensite which supposed to be harder than the austenite phase.

Figure 3(a) and (b) are the comparisons of stress-strain and f_m vs. plastic strain curves at 293K between experiment and simulation results, respectively. It is clear that the proposed model successfully predicted both stress-strain curve and martensite volume fraction, especially stagnation of martensitic transformation, were well predicted.

5. Conclusions

From the cyclic tension-compression tests for an austenitic stainless steel, we obtained the following findings:

(i) deformation induced martensitic transformation would be minor at higher temperature,

(ii) martensitic transformation stagnates due to the Bauschinger effect even if plastic strain accumulated.

Based on the experimental observations, constitutive model for austenitic stainless steel was proposed. The calculated results well predicted both cyclic stress-strain curves and fm vs. plastic strain curves obtained from experiments.

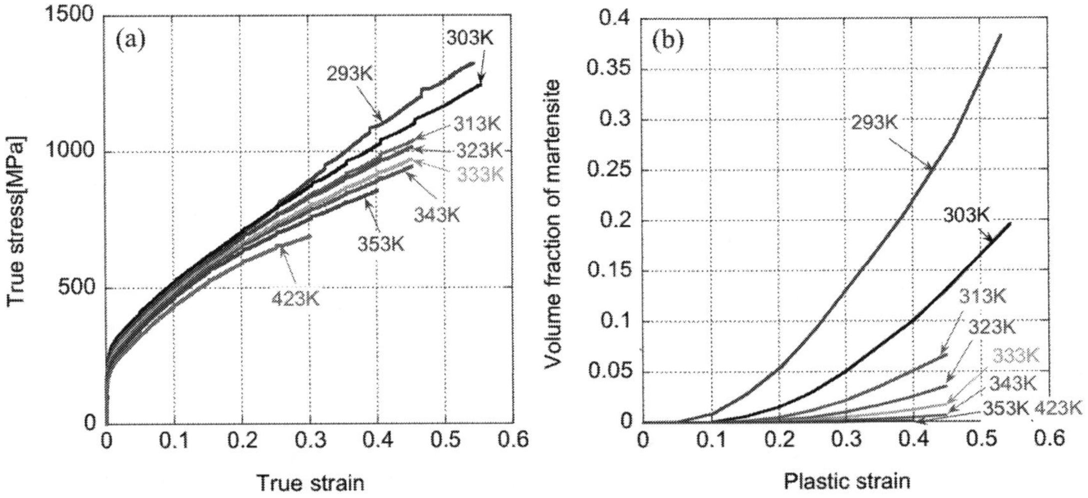

Figure 2. (a) Stress-strain curves and (b) volume fraction of martensite vs. plastic strain curves obtained from experiments at various temperature

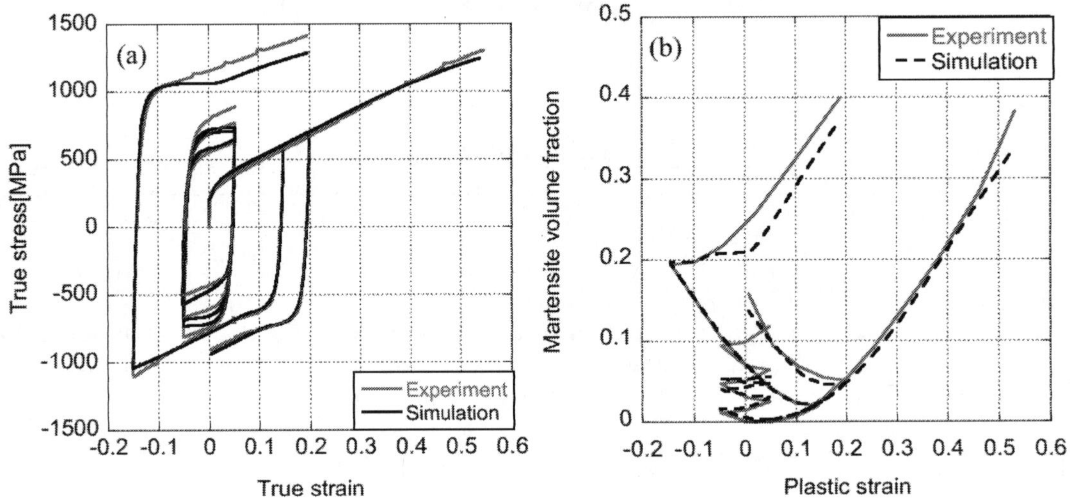

Figure 3. Comparisons of (a) stress-strain and (b) fraction of martensite vs. plastic strain curves between experiments and calculation results using the proposed model.

References

[1] Stringfellow R G, Parks D M and Olson G B 1992 *Acta Metall. Mater.* **40** 1703.

[2] Geijselaers H J M, Hilkhuijsen P, Bor T C and van den Boogaard 2015 *Mater. Sci. Eng. A* **631** 166.

[3] Yoshida F and Uemori T *Int. J. Plast.* **18** 661.

Prediction of part shape and associated material properties in hot-press forming using finite element analysis

Hwigeon Kim[1], Jinjin Ha[2], Myoung-Gyu Lee[3], Frederic Barlat[1]

[1]Graduate Institute of Ferrous Technology, Pohang University of Science and Technology, Pohang, 790-784, South Korea
[2]Mechanical Engineering Department, University of New Hampshire, Durham, NH, 03824, USA
[3]Department of Materials Science and Engineering, Korea University, Seoul, 136-713, South Korea

Abstract. The hot-press forming of a U-channel was conducted on a boron-steel blank. The die consisted of two separate parts in order to perform the partial quenching process. The cold die was initially at 25 °C while the heated die was set to five different temperatures, namely, 25, 120, 220, 320 and 400 °C. The cooling temperature history, Vickers hardness and springback of the channel were measured. A thermo-mechanical-metallurgical model, which accounts for the prior austenite deformation effect, was successfully implemented in the LS-DYNA explicit solver to simulate the hot-press forming process under partial quenching conditions. The predicted and experimental results were compared and found in reasonable agreement.

1. Introduction

The hot-press forming (HPF) process has drawn considerable attention in the automobile industry in order to produce ultra high-strength automotive components. This technology significantly contributes to lower the car weight, reduce gas emission and improve fuel efficiency. The HPF process consists of high temperature forming of a boron steel sheet (22MnB5) followed by die quenching. Although a full martensitic microstructure in the final product increases the strength, limited ductility results in lower crash performance. For this reason, the partial quenching method focuses to secure both the structural strength and ductility as in-service properties by tailoring the microstructure in press-hardened products. The purpose of this work is to simulate HPF and partial quenching of a 22MnB5 steel sheet sample.

2. Experiments

2.1. The blank and tool design

An S-rail shaped, U-draw bending die was designed to conduct HPF experiments with a partial quenching (PQ-HPF) method. The measurement of the temperature history during the whole process was recorded and further experiments were performed after the process to characterize hardness and springback of the partial quenched HPF parts. The geometry of the product is depicted in Fig. 1. In order to achieve PQ-HPF, the die was separated into three independent blocks with a 3 mm air gap between each other. Fig. 1(a) indicates that only the block corresponding to Zone (1) was heated up (heated die) while (2) and (3) were initially set to room temperature (cooled die). A 22MnB5 boron steel sheet sample provided by POSCO was investigated in this article.

2.2. Experiments with partial quenching die

In order to produce partially quenched products, the block corresponding to Zone (1) was heated up to five different temperatures, namely, 25, 120, 220, 320 and 400 °C. The others two blocks were kept to room temperature. This method allows variations of the cooling rate in the different zones of the channel. This results in gradients of microstructure, and mechanical properties such as strength, ductility and hardness. Fig. 2 summarizes the different steps of the experimental procedure. Before HPF, a 22MnB5 blank with an optimized shape was put into a furnace at 930 °C and kept for 300 seconds for full austenitization. Then, the blank was transferred and set on the press die within 10 seconds. The forming was conducted during 2 seconds, during which the material was still fully austenitic, and die quenched during 10 seconds. Finally, the die was opened to let the blank cool down in the air. In this work, the zone in the blank, which contacted the cooled die zone underwent rapid cooling, leading to a final martensite microstructure. In contrast, the zone in the blank in contact with heated die did not cool down as rapidly and resulted in a mixture of martensite and softer phases. A thermocouple was attached to the blank in this zone only to monitor the temperature history during the whole process. After PQ-HPF, hardness tests were conducted in each channel only near the heated zones.

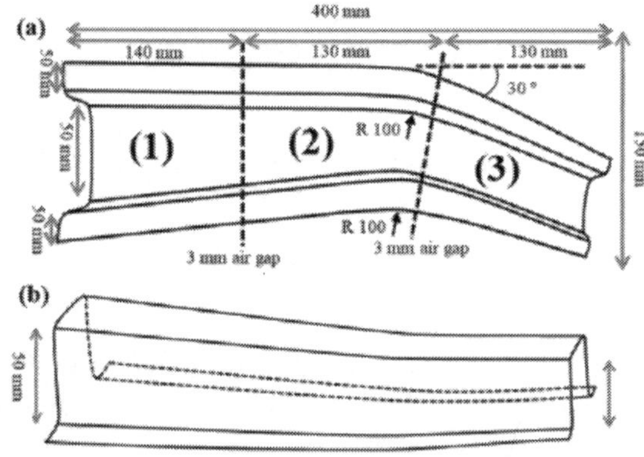

Figure 1. (a) Geometry of channel from rear view; (1) Section in contact with heated die; (2) & (3) Section in contact with cooled die), (b) Schematics of side view.

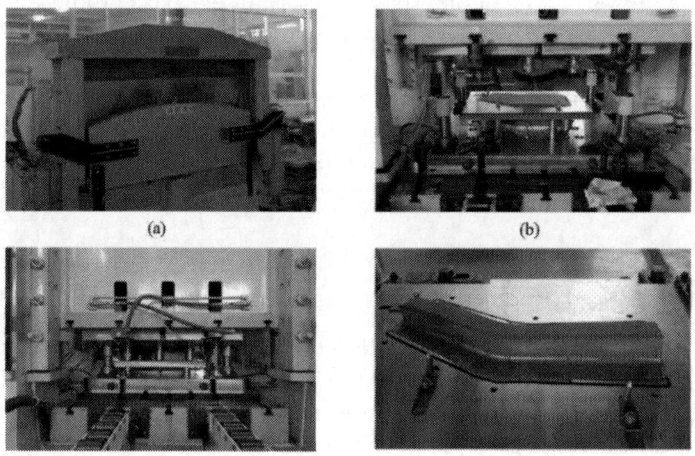

Figure 2. (a) Heating of blank in furnace (300 sec); (b) Transferring heated blank to die (10 sec); (c) Forming (2 sec) and die quenching (10 sec) and; (d) Air cooling (180 sec).

3. Verification of FE-user subroutine model

3.1. Implementation of user-subroutine for LS-Dyna

An approach, developed in-house by Bok [1] and tested successfully in the Abaqus user-subroutine interface, was successfully implemented in the LS-Dyna explicit solver. This new code allows coupled thermo-mechanical-metallurgical simulations and includes two main calculations, i.e., stress-strain field and phase evolution, since thermal aspects are taken care off by LS-Dyna. For the stress-strain field, von-Mises plasticity with isotropic hardening was combined with a modified Johnson-Cook law for the austenite phase and Swift for other phases. Three different strains were considered as a result of phase transformations, namely, thermal, volumetric and transformation plasticity strain. For the phase evolution, semi-empirical models were used. The Continuous Cooling Transformation (CCT) diagram was calculated using the Time Temperature Transformation (TTT) data. Since the steels transform under a continuously decreasing temperature field, a proper treatment to convert the incubation from isothermal to non-isothermal conditions is required [1].

3.2. Finite Element (FE) simulation

Finite Element (FE) simulations of the PQ-HPF of a U-draw bending were conducted to predict the temperature history, hardness and final shape of the product after springback. Fig. 3 shows the FE meshes for the tools and blank. The process conditions were similar to that of the real process. The cooled die was fixed as 25 °C and the heated die was set to 5 different temperatures, i.e., 25, 120, 220, 320 and 400 °C. Note that, since the blank transfer was not included in the simulations, the initial blank temperature was that measured at the beginning of the forming step. The heat transfer coefficients, which depend on the temperature difference between the blank and tools [2], were also adjusted for each condition. The heat transfer coefficients, measured in [2], ranged from 2500 to 8000 W/m^2 °C in this work.

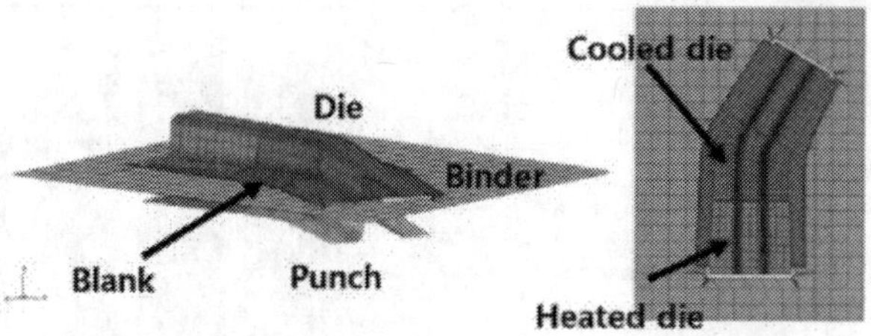

Figure 3. FE meshes of blank and tool for hot-press forming of U-channel.

3.3. Results

First of all, it is important to capture the temperature history in the blank because it is directly connected to the phase transformation, which mainly affects the mechanical property of the products. The experimental and predicted temperature histories are shown in Figs. 4(a) ~ (c). During the forming and quenching stages, the predicted temperatures are in good agreement with the experimental results. Only for an initial die at 400 °C in the air cooling stage (see the green line in Fig. 4(c)), a significant difference is observed because the latent heat was neglected in the FE model. The in-house user-subroutine cannot control the latent heat in the explicit LS-Dyna code. Actually the drastic temperature increase in the experiment indicates that bainite phase transformation occurred for this low cooling rate. In spite of this drawback, Fig. 4(d) demonstrates that the predicted hardness after air cooling was in quite good agreement with the experimental value for each die temperature. Thus, it is likely that neglecting the latent heat does not affect the predicted mechanical properties of the final products because the high cooling rate is the most important feature to capture in this work. Fig. 5 shows the predicted and experimental shapes of the channels after unloading. In this work, the springback was very low for all conditions in both simulations and experiments.

4. Conclusion

The present work aimed at predicting the material properties and shape in partial quenching HPF using a thermo-mechanical-metallurgical model in a FE analysis. The main conclusions of the present work are summarized below:

- The reliability of the model was successfully demonstrated, by the good agreement between the predicted and experimental temperature history, hardness and springback.

- Although most of the experimental and FE simulation results were in good agreement, the temperature history was not well predicted with the FE simulation for the heated die at 400 °C. This is because a significant amount of bainite transformation occurred due to the lower cooling rate but the corresponding latent heat was neglected in the simulation. However, this did not affect the predicted mechanical properties of the final product.

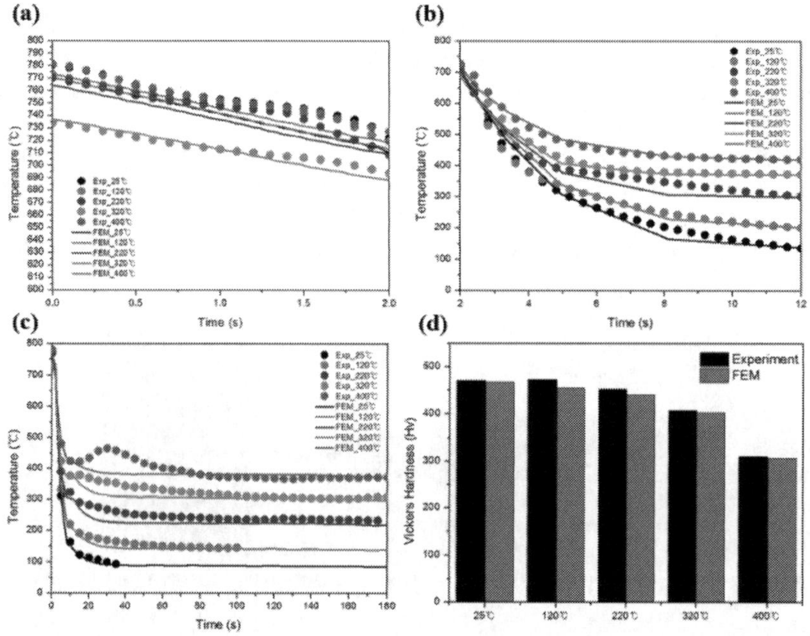

Figure 4. Experimental and predicted temperature history in the blank (a) During forming (0 ~ 2 sec); (b) During die quenching (2 ~ 12 sec); (c) During whole process (0 ~ 180 sec). (d) Experimental and simulated Vickers hardness after whole process.

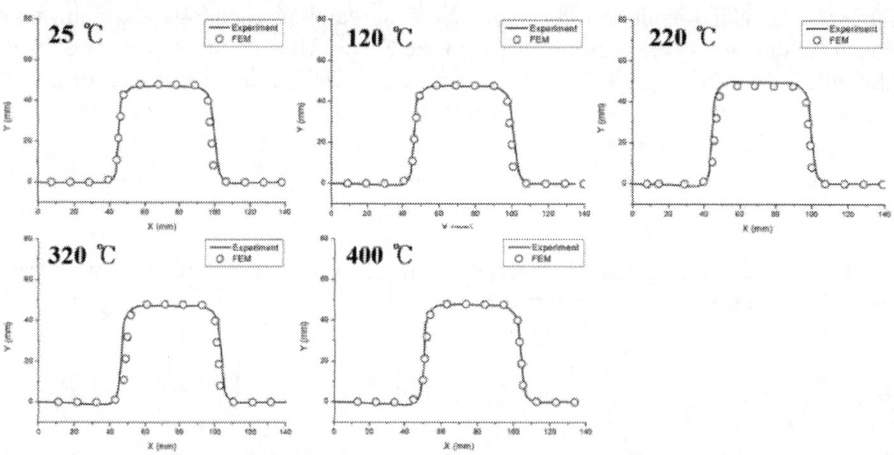

Figure 5. Springback prediction and comparison with experimental results.

Acknowledgements
The authors greatly appreciate the support by POSCO. This work was also sponsored by the National Research Foundation of Korea (NRF) Grant funded by the MSIP (No. 2014R1A2A11052889).

References

[1] H.H. Bok, J.W. Choi, F. Barlat, D.W Seo and M.G Lee 2014 *Int. J. Plasticity* **58** 154-183.
[2] H.R. Kim 2016 *Master Thesis, POSTECH, Korea.*

Development of new hole expansion testing method

Hyunok Kim[1], Jianhui Shang[1], Kevin Beam[2], Anoop Samant[2], Cliff Hoschouer[3], and Jim Dykeman[4]

[1] EWI Forming Center,
1250 Arthur E. Adams Drive, Columbus, Ohio 43221, USA

[2] KTH Parts Industries Inc.; K-Group Technical R&D Center,
8205 Business Way, Plain City, OH 43064

[3] SHILOH,
5580 Wegman Drive, Valley City, Ohio 44280, USA

[4] Honda R&D of Americas, Inc.,
21001 State Route 739, Raymond, Ohio 43067, USA

E-mail: hkim@ewi.org

Abstract. This paper introduces a new hole expansion (HE) testing method that could be more relevant to the edge cracking problem observed in stamping advanced high strength steel (AHSS). The new testing method adopted a large hole diameter of 75 mm compared to the standard hole diameter of 10 mm. An inline monitoring system was developed to visually monitor the hole edge cracking during the test and synchronize the load-displacement data with the recorded video for capturing the initial crack. A new hole expansion testing method was found to be effective in evaluating the edge cracking by considering the effects of material properties and trimming methods. It showed a much larger difference, up to 11%, of the HE ratio between DP980 and TRIP780 compared to the standard HE testing method giving less than a 2% difference.

1. Introduction

Edge cracking occurs more frequently in stamping AHSS than other ductile steels and aluminum alloys. Major reasons for edge cracking are:

- Reduced local ductility (i.e. residual stress and poor edge quality) during blanking or punching operations
- Inadequate clearance between the trim die and punch
- Worn-out trim tool or die.

Edge cracking is difficult to predict using analytical or numerical methods. It increases scrap rate in the stamping production. The ISO standard hole expansion testing method is available to evaluate the local formability of the sheet material [1]. Unfortunately, the ISO standard testing method with a 10-mm initial diameter of the hole often gives highly inconsistent test results with the same material depending on the hole preparation methods and interpretation procedure. From these impractical limitations, the material supplier and automotive original equipment manufacturers (OEM) need better testing methods [2]. This paper introduces a new hole expansion testing method to reliably evaluate the edge cracking and to correlate the test results with the industrial stampings that have edge cracking problems.

2. Approach and Results

2.1. Technical Approach

Three major steps should be considered to reliably evaluate and predict the edge cracking of sheet materials as shown in Figure 1. First, the consistent edge quality of the hole should be obtained. Second, the testing method should be simple enough to accurately measure or interpret the onset of edge cracking. Third, the test data should include useful inputs or reference for finite element (FE) simulations to predict edge cracking in stamping. The ISO standard hole expansion testing provides general guidelines on how to conduct hole expanding testing [1]. However, there are many variations in punching a 10-mm hole under various punch-die clearance, concentricity of the hole position, and interpretation of the through-thickness crack. This could easily result in different edge qualities and different hole expansion ratios from the same material. In addition, a 10-mm hole is not large enough to clearly differentiate the local ductility change of different materials by considering the statistical deviations of the data. The new testing method starts with punching the hole with the dedicated tooling as shown in Figure 2.

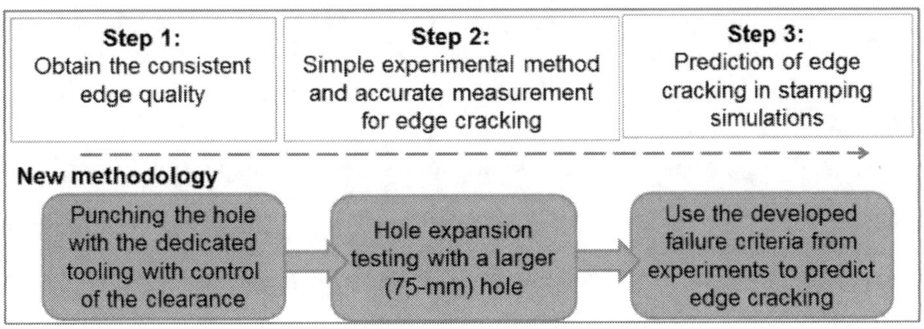

Figure 1. Strategy in evaluating and predicting edge cracking.

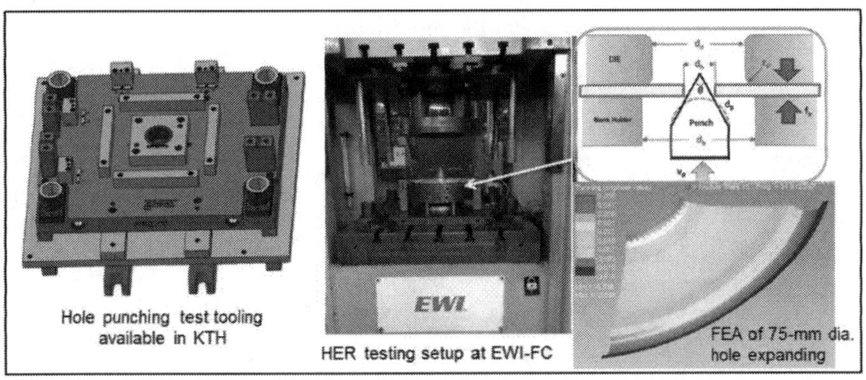

Figure 2. A new hole expansion testing.

A 75-mm hole diameter was determined through FE simulations to maximize the sensitivity of the hole edge quality and material properties in the hole expansion ratio. The effects of the hole size on hole expansion testing were thoroughly investigated and addressed by other researchers [3]. To monitor and capture the onset of edge cracking, a visual monitoring system was developed. This system is synchronized with the load cell and laser displacement data as shown in Figure 3. This allows for unlimited play back and review of the video file to ensure that the onset of edge cracking is found.

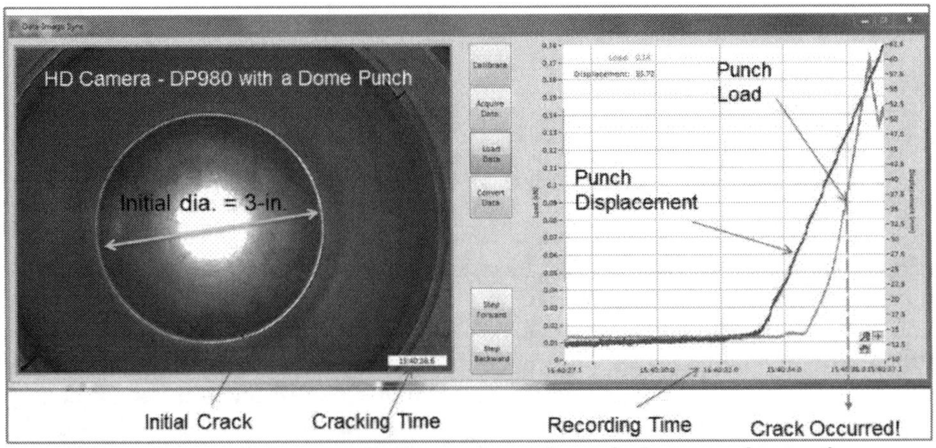

Figure 3. A real-time monitoring system of new hole expansion testing.

2.2. Results

Three different sheet materials were tested with the new hole expansion testing as well as the ISO standard hole expansion testing. Table 1 summarizes the detailed testing conditions used for new hole expansion testing. Regarding the ISO standard hole expansion testing, only punched hole samples were tested. Three experiments were repeated for each testing condition.

Table 1. Experimental conditions for the new hole expansion testing.

Sheet Materials	Punch Types	Hole Edge Preparation Methods
Al 5182-O (1.0-mm thickness) TRIP780 (1.4-mm thickness) DP980 (1.0-mm thickness)	Conical punch and dome punch (diameter=152.4-mm and 60-degree cone angle)	Punching die and water-jet cutting

Figure 4 compares the hole expansion ratios obtained with the new testing method. The following major findings can be summarized:

- Al 5182 showed less sensitivity to the edge cracking between the water-jet cut and punching methods.
- The edge-cracking sensitivity to the trimming methods was found to be different with the strength and microstructure of the materials. TRIP780 showed 20% differences between punching and water-jet, DP980 showed about 10% difference, and Al showed negligible difference.
- The 75-mm punched hole samples still showed large variations of the HER values (i.e. 10~25% for different materials), although a fresh punching tool was used.
- The ISO standard hole expanding tests gave the hole expansion ratio of 15.32% for DP980 and 17.28% for TRIP780 with the punched 10-mm initial hole.

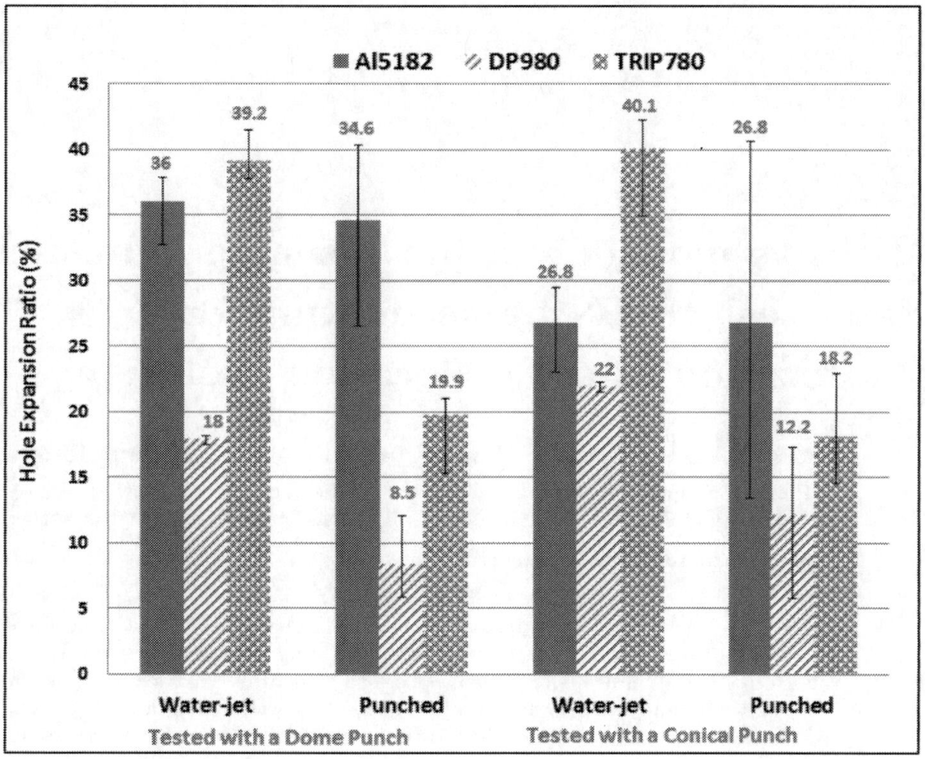

Figure 4. Results of the hole expansion testing at EWI Forming Center

3. Conclusions

The new hole expansion test procedure was found to be more effective than standard testing methods to evaluate the edge cracking by considering the effects of material properties and trimming methods. The ISO standard test with 10-mm hole size gave only 1.96% (i.e. about 200 μm) difference between DP980 and TRIP780 samples with the punched hole edge conditions. The new testing method gave 11.4% (8.55 mm) difference between DP990 and TRIP780 with the dome punch, and 6% (4.5 mm) difference with the conical punch. The punch shape changes the strain path on the hole edge. For example, a dome punch gives the uniaxial tension while the conical punch gives near the plane strain condition according to the strain measurements and FE predictions.

Acknowledgements

This project was conducted with support from the state of Ohio, Ohio Development of Services Agency (ODSA) on Advanced Manufacturing Program (Grant No. TECG20141159). The author would also like to thank Dr. Taylan Altan, Emeritus Professor, at The Ohio State University for his technical advice on this research program.

References

[1] ISO/TS 16630: 2003 (E) Method of hole expanding test

[2] Shih H, Hsiung C K and Wendt B 2014 Optimal production trimming process for AHSS sheared edge stretchability improvement *SAE International* 2014-01-0094

[3] Akinobu I., Masaki U, Toru I 2013 Press forming analysis contributing to the expansion of high strength steel sheet applications *JFE Technical Report* **18**

Constitutive modeling of hot horming of austenitic stainless steel 316LN by accounting for recrystallization in the dislocation evolution

H. Kooiker[1,2], E.S. Perdahcioğlu[2] and A.H. van den Boogaard[2]
[1] Philips Consumer Lifestyle, Drachten, the Netherlands
[2] University of Twente, Department of Applied Mechanics, Enschede, the Netherlands
E-mail: harm.kooiker@philips.com

Abstract. Hot compression test data taken from Zhang [1] of metastable austenitic stainless steel AISI 316LN over a range of strain rates and temperatures shows typical dynamic recovery and recrystallization behavior. It is proposed to model this behavior by incorporating not only the hardening and recovery into the Bergström dislocation evolution equation, but also the recrystallization. It is shown that the initial mechanical response before recrystallization can be accurately represented by assuming that the mean free path evolves as the microstructure evolves from homogeneously spaced dislocations to cell-pattern. Results show that this novel continuum mechanical model can predict the observed behavior, showing a good match to the experimental data and capturing the transition from recrystallization to (almost) no recrystallization.

1. Introduction

Metastable austenitic stainless steels are widely used in industry due to their excellent corrosion resistance and formability. During hot forming, the stresses introduced when deforming are considerably lower and thus the resulting residual stresses and springback. In order to design optimal and safe hot-formed components and products a good understanding and constitutive description of the material behavior is vital and an important feature of the behavior during hot forming is dynamic recovery and dynamic recrystallization.

Several authors have presented continuum mechanical models that predict the flow stress evolution of metallic materials experiencing dynamic recrystallization during hot forming, however most of these models fall in one of two classes. The first one employs the Arrhenius equation to link strain rate, deformation temperature and flow stress, the second one treats dynamic recrystallization similar to static recrystallization and is based on Avrami kinetics.

The Arrhenius-based models start out with assuming a phenomenological relation between the flow stress and Zener-Holloman parameter Z and are supplemented with 4th to 7th-order polynomial relations between strain and model parameters to be able to describe stress-strain relations. The recrystallization models based on Avrami kinetics describe recrystallization as a fractional process lowering the stress towards a steady state. In this model, recrystallization evolves from zero to one, suggesting that recrystallization is finished. However, it is commonly recognized that multiple peak versus single peak recrystallization is due to successive waves of recrystallization, where in single peak the waves are in rapid succession and in multiple peak, recrystallization is (almost) finished before the next wave starts [2].

In this research a new continuum mechanical model is proposed to account for single-peak recrystallization in a more physically based manner. It is based on the assumption that at the saturation stress an equilibrium exists between hardening on one hand and recovery and recrystallization on the other. Experimental curves of hot compression of AISI 316LN are taken from Zhang et al. [1] and used to calibrate the proposed model.

2. Modeling of Hot Forming

The constitutive equations for the new model are based on the decomposition of the flow stress into a work hardening stress σ_w, which accounts for dislocation-dislocation interactions, and a strain and strain rate independent stress σ_0. The strain and strain rate independent stress is assumed to relate to stresses in the atomic lattice of which the temperature dependence is assumed to be similar to the temperature dependence of the shear modulus [3].

$$\sigma_f(\dot{\varepsilon}, T) = \sigma_0(T) + \sigma_w(\dot{\varepsilon}, T) = \mu(T)\left(\sigma_o + \alpha\mu_{ref} bM\sqrt{\rho(\dot{\varepsilon}, T)}\right) \tag{1}$$

Where $\mu(T)$ reperesent the temperature dependence of the shear modulus with respect to the reference value μ_{ref} and is taken linear in this work:

$$\mu(T) = \frac{\mu_{ref} - C_5 T}{\mu_{ref}} \tag{2}$$

The work hardening part will be described in two subsections. In the first subsection the Taylor equation and the corresponding dislocation density evolution are described. Furthermore the relation between the model parameters and strain rate and temperature is explained. In the second subsection the evolution of mean free path with strain is elaborated.

2.1. Taylor and Bergström equation

The work hardening of the model is based on the evolution of the statistical average mobile dislocation density ρ (henceforth dislocation density) and is described by the Taylor equation:

$$\sigma_w = \alpha\mu bM\sqrt{\rho} \tag{3}$$

Where α, μ, b and M are a material constant, temperature dependent shear modulus, the length of the Burgers vector and the Taylor orientation factor. To complete the relation between stress and strain a supplementary equation is needed to describe the evolution of the dislocation density with strain. A novelty of the current model is the incorporation of dynamic recrystallization directly into the Bergström dislocation evolution. After all, the effect of recrystallization is to nucleate and grow new dislocation-free grains into the dislocated substructure, thereby removing dislocations from the structure [4,5] and oppose work hardening. The modified Bergström equation reads:

$$\frac{d\rho}{d\varepsilon} = h\sqrt{\rho} - f\rho - R_{rate}(\rho - \rho_0) \tag{4}$$

In this equation $h\sqrt{\rho}$ describes the hardening by storage (immobilization) of mobile dislocations and is linked to the mean free path of dislocations, f governs the dynamic recovery and is dependent on temperature and strain rate due to its thermally activated nature [6].

$$f(\dot{\varepsilon}, T) = f_0 + C_1 Z^{m_1} \tag{5}$$

Now it must be noted that R_{rate} does not describe the state of recrystallization, it describes the current recrystallization *rate*. As discussed in the introduction, at the steady state, equilibrium exists between dislocation generation and annihilation by recovery *and* recrystallization. This means that at the steady state, recrystallization is ongoing and not finished. In this research, a simple mathematical formulation is used to describe the expected recrystallization rate explicitly. It captures both the sigmoidal trend and the constant rate of recrystallization which is expected at the steady state stress.

$$R_{rate} = \frac{A}{1 + \exp\{n(\varepsilon_c - \varepsilon)\}} \tag{6}$$

Here A determines the saturation recrystallization rate (and thus influences the steady state stress), n determines the rate at which the saturation recrystallization rate is reached and ε_c is linked to the critical strain for the onset of recrystallization by shifting the sigmoid over the strain-space. The dependence of ε_c on strain rate and temperature is mirrored from the Bergström approach for recovery, Eq. 5. For A the constant is omitted to allow the recrystallization rate to approach zero at low temperatures and high strain rates, n is taken as a constant.

2.2. Evolution of mean free path on hardening behavior

In a recent paper by Angella [7], focusing on the hardening and recovery behavior of AISI 316L, it was found by means of TEM analysis that annealed AISI 316L starts out with homogeneously spaced dislocations which, upon straining, arrange themselves into a cell pattern. One of the fundamental hypotheses of the family of Bergström and similar Kocks-Mecking hardening models is the principle of similitude, which states that the dislocation structure is self similar, i.e. it scales with the inverse of the applied stress [8]. Obviously, the evolution from homogeneously spaced dislocations to cell-pattern violates the principle of similitude.

To account for this behavior, Angella assumes that the mean free path of dislocations is a state variable (next to the dislocation density) and evolves from an initial value to a saturation value. In the model presented in this paper (for the similar material AISI 316LN) the same strategy is used and a simple evolution equation is presented for the parameter controlling the hardening.

$$\frac{dh}{d\varepsilon} = \frac{K}{\sqrt{\rho}}(h_s - h) \tag{7}$$

In the results it will be shown that the behavior of AISI 316LN is not well represented by the Bergström model alone and that Eq. 7 offers a significant improvement in capturing the hardening behavior. To complete this model a relation is needed to define the rate at which h evolves from the initial value to the saturation value depending on strain rate and temperature. This is achieved through a similar dependence of K on the Zener-Holloman parameter as in Eq. 5.

3. Results

A least squares fitting routine was used to obtain the parameter values shown in Table 1. The model, experimental results and the merit of the mean free path evolution are shown in Figure 1.

Table 1: Model parameters obtained by least squares optimization (α, b, and M were taken constant). Units: σ_0 and μ_{ref} in MPa, Q in J/mol, b in mm, ρ_0 in mm^{-2} and h_0 and h_s in mm^{-1}, the rest is dimensionless.

Taylor		Bergström		$\frac{dh}{d\varepsilon}$		f		Z	
α	0.5	ρ_0	4.63e5	h_0	1.71e6	f_0	15.06	Q	3.75e5
b	2.8e-7	h	—	h_s	2.67e5	C_1	7.68e4		
M	3	f	—	K	—	m_1	-0.25		
σ_0	42	R_{rate}	—						
K		A		n		ε_c		$\mu(T)$	
K_0	4.40e5	C_3	2.1e8	n	15.40	ε_{c0}	2e-3	μ_{ref}	8.67e4
C_2	1.5e11	m_3	-0.56			C_4	0.16	C_5	-32.25
m_2	-0.47					m_4	3.48e-2		

4. Discussion and Conclusion

The results show that the presented constitutive equations are capable of accurately capturing the behavior of AISI 316 LN over a broad range of hot working temperatures and strain rates. This includes the transition from pronounced recrystallization at 'high temperatures - low strain rates' to negligible recrystallization at 'low temperature - high strain rates'. The material constants have logical dependencies on temperature and strain rate through simple power law relations with the Zener-Holloman parameter, similar to the one suggested by Bergström for dynamic recovery. Furthermore, from Figure 1, it is clear that the evolution of mean free path improves the accordance between model and experiment significantly. The hardening saturation value is reached around 5% strain, however the effect on the fit between measurement and model is significant up to large strains.

Some discrepancies between experiment data and model still exist, e.g. at 900°C, $\dot{\varepsilon} = 10^{-1}s^{-1}$ the experimental curve shows more softening then predicted, furthermore there seems to be pronounced recrystallization at 1200°C, $\dot{\varepsilon} = 10^{-3}s^{-1}$ which is not predicted by the model. Upon close examination it seems that the yield stress at these high temperature-low strain rate conditions is softened *below* the initial yield stress. It is hypothesized that at these extreme conditions concurrent grain coarsening is taking place.

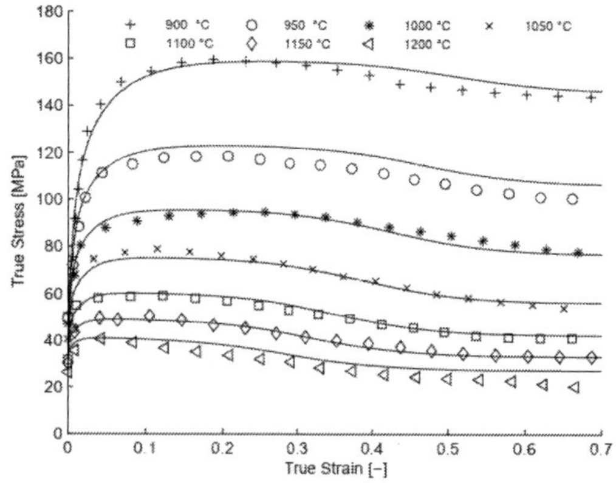

(a) strain rate $10^{-3}s^{-1}$

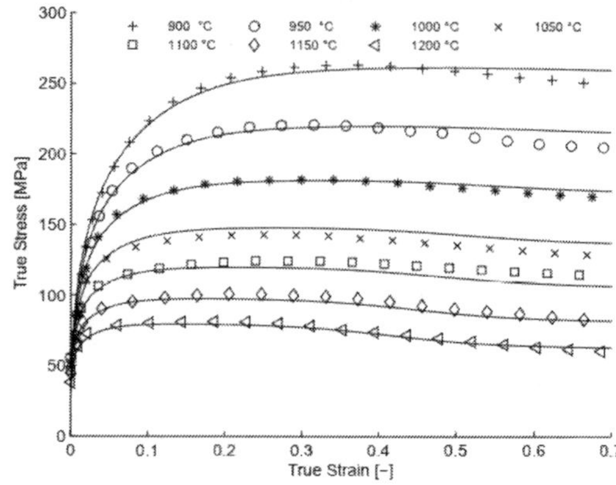

(b) strain rate $10^{-1}s^{-1}$

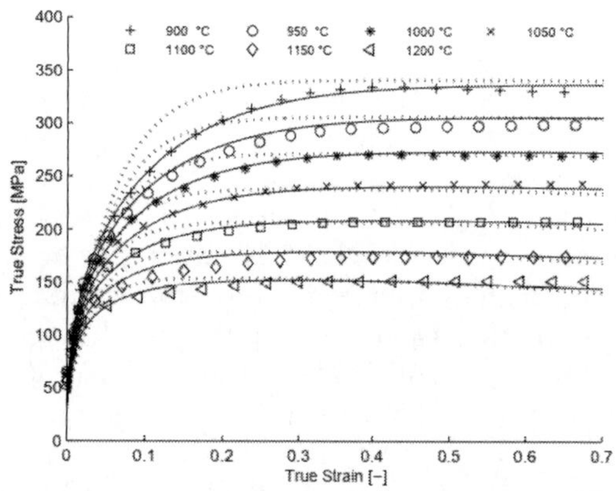

(c) strain rate $10^{1}s^{-1}$

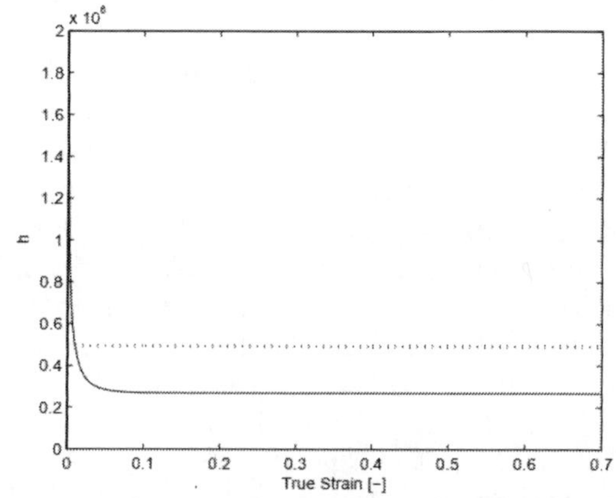

(d) Evolution of h, *with* and *without* mean free path evolution (solid vs dotted)

Figure 1: True stress vs true strain curves ($\dot{\varepsilon} = 10^{-3}s^{-1}$ - $10^{1}s^{-1}$, T = 900°C - 1200°C), model results shown in solid line and dotted line, respectively *with* and *without* evolving mean free path, subfigure (d) shows the evolution of h *with* and *without* mean free path evolution in the model

References

[1] Zhang L, Feng X, Wang X, and Liu C 2014 *On the constitutive model of nitrogen-containing austenitic stainless steel 316LN at elevated temperature*, *PLos ONE* **9(11)**
[2] Brown A A and Bammann D J 2012 *Int. Journal of Plasticity* **32-33** 17-35
[3] Boogaard A H van den, Bolt P J 2003 *VII Int. Conf. on Computation Plasticity, COMPLAS*
[4] Humphreys F J and Hatherly M 2004 *Recrystallization and Related Annealing Phenomena* second edition (Elsevier Ltd.)
[5] McQueen H J and Tegart W J 1975 *Sci. Am.* **4** 116-25
[6] Bergström Y 2010, *A theory for the temperature and strain-rate dependences of dislocation re-mobilisation* (URL http://www.plastic-deformation.com/paper3.pdf)
[7] Angella G, Donnini R, Maldini M and Ripamonti D 2014 *Materials Science and Engineering* A **594** 381-8
[8] Kocks U F and Mecking H 2003 *Progress in Materials Science* **48** 171-273

Texture evaluation in ductile fracture process by neutron diffraction measurement

H Sunaga [1], M Takamura [1], Y Ikeda [1], Y Otake [1], T Hama [2], M Kumagai [3], H Suzuki [4] and S.Suzuki [5]

Center for Advanced Photonics, RIKEN, Wako, Saitama 351-0198, Japan
Research Reactor Institute, Kyoto University, Kumatori, Osaka 590-0494, Japan
Faculty of Engineering, Tokyo City University, Setagaya, Tokyo 158-8857, Japan
Materials Sciences Research Center, Japan Atomic Energy Agency, Tokai, Ibaraki 319-1195, Japan
Faculty of Science and Engineering, Waseda University, Shinjuku, Tokyo 169-8555, Japan

sunaga@riken.jp

Abstract. A neutron diffraction measurement was performed to reveal microstructural aspects of the ductile fracture in ferritic steel. The diffraction patterns were continuously measured at the center of the reduced area while a tensile specimen was loaded under tension until the end of the fracture process. The measurement results showed that the volume fraction of (110)-oriented grains increased when the texture evolved as a result of plastic deformation. But the mechanism of texture evolution may be changed during necking, decreasing an increase rate of the volume fraction.

1. Introduction

Fracture is one of the most crucial forming defects encountered in manufacturing processes. The classic phenomena of ductile fracture are observed in a tensile specimen subjected to a large amount of plastic deformation showing necking or a reduced area particularly near the fracture itself.

Understanding of the microstructural phenomena occurring during post-uniform elongation will help define numerous criteria expected to be necessary for the establishment of accurate simulation technology[1]. This should reduce the preparation time of the manufacturing processes carried out by trial-and-error method, and enable the realization of high quality and low-cost products with a short delivery period.

Neutron diffraction can be used to measure texture evolution[2]. It has an advantage that the neutrons penetrate more deeply into the bulk than the conventional X-ray diffraction. The neutron diffraction technique has been extended to investigate the evolution of lattice strain and internal stress[3-5]. Therefore, it is used to obtain data under various conditions such as loading or changing temperature by in situ measurements.

The deformation behavior of a stainless steel is investigated under tensile loading without any interruption to apply a load or displacement. The transformation behavior or volume fraction of the retained austenite under the enhancement of uniform elongation is studied to increase the work-hardening resulting in delaying the onset of necking[6]. The local deformation behavior occurring during post-uniform elongation, which takes place before fracture, is still poorly understood. In this study, a neutron diffraction measurement was performed to reveal the texture evolution in ferritic steel during elongation until the end of the fracture process.

2. Experimental procedure

The neutron diffraction measurement was performed using the time-of-flight-type engineering neutron diffractometer TAKUMI at the Material and Life Science Experimental Facility (MLF) of Japan Proton Accelerator Research Complex (J-PARC).

The material used in this study was JIS-SS400, which is a rolled steel used in general structures. The dimensions of the JIS-SS400 steel specimen used in this study are given in Figure 1. The shoulders of the specimen used for gripping are machined to have dimensions fully consistent with the specifications of the tensile testing machine. The most important part of the specimen is the middle section. The cross-sectional area of the middle section was designed to be smaller than that of the other sections, because necking and ductile fracture were expected to occur in the middle section, for which a width of 10mm was determined to provide sufficient area for the diffraction of the incident beam.

The tensile testing machine was oriented at an angle of 45° to the incident beam. The diffraction patterns both parallel and transverse to the loading axis were measured simultaneously by using two detector banks which were oriented at angle of ±90° to the incident beam as shown in Figure 2. A movable crosshead was controlled to move at a constant speed (0.005 mm/s) and the tensile testing machine itself moved to the opposite direction to the crosshead movement to allow the incident beam to be diffracted in the middle section of the specimen during the tensile loading. The neutron data were collected continuously throughout the entire tensile test when the JIS-SS400 steel specimen was loaded under uniaxial tension until the fracture occurred.

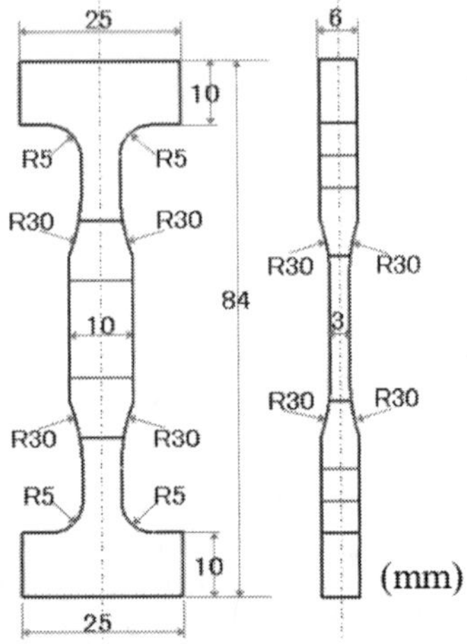

(a) Dimensions of a tensile specimen

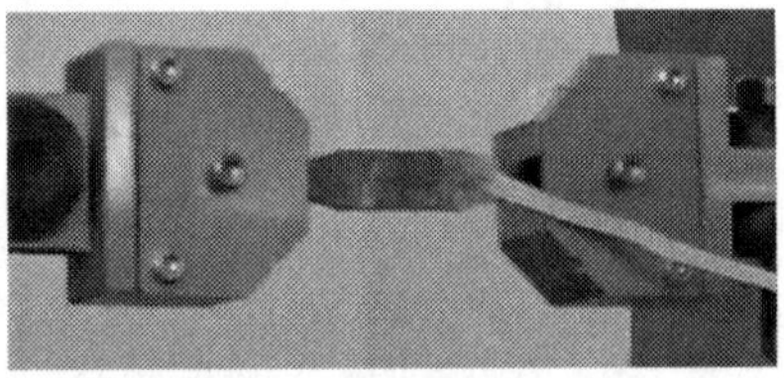

(b) Specimen fixed to the tensile machine

Figure 1. Tensile specimen used in this study

3. Results and discussion

Figure 3 shows the tensile load vs crosshead displacement obtained by the tensile test during the neutron

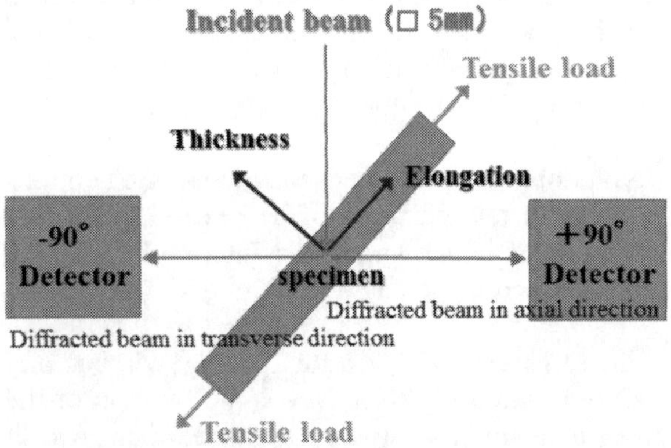

Figure 2. Schematic view of neutron diffraction measurement

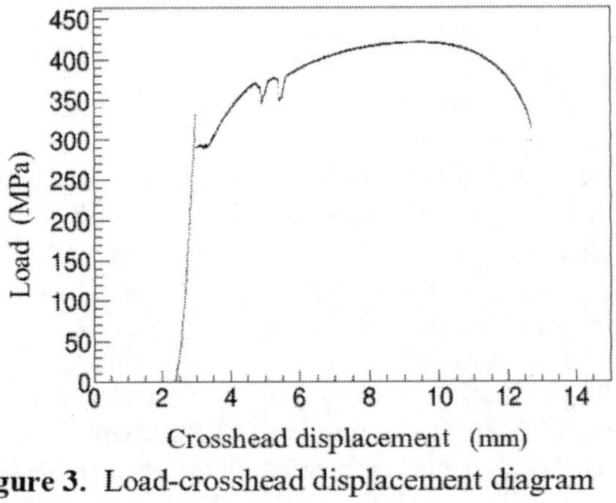

Figure 3. Load-crosshead displacement diagram

diffraction measurement. Figure 4 shows the changes in the thickness and width at the most reduced section of the specimen where the incident beam was diffracted during the tensile loading, as shown in Figure 5. It was measured with the photographs taken from the front and side of the specimen during the tensile test, which was performed under the same condition as the diffraction measurement. Both the width and thickness decrease at the same rate relative to the initial size, and the maximum load corresponds to the point at which the deformation starts to localize in the middle of the specimen. It was also confirmed that neutron diffraction could measure the material elongation during necking and ductile fracture, because necking occurred in an area of the square equal to the incident beam size (□5 mm).

The neutron data were sliced every 30 sec to calculate the diffraction patterns in each detector bank. Figure 6 shows the diffraction patterns for the axial direction parallel to the loading axis taken before deformation. Here, we discuss the texture evolution based on changes in the peak intensity of the 110, 200, 211 and 220 reflections during deformation. The peak intensities were normalized by a sampling volume which was estimated from a change in thickness at the center of the specimen.

Figure 7 shows the change in the diffraction intensity in the axial direction as a function of crosshead displacement. Initially, the intensity of 110 increased while those of 200 and 211 decreased with increasing applied load. A transition from an increase to a decrease in the intensity occurred for 110 prior to the fracture.

The volume fraction of (110)-oriented grains increased when the texture evolved as a result of plastic deformation. But the mechanism of texture evolution may be changed during necking, decreasing an increase rate of the volume fraction.

Figure 8 shows the change in the diffraction intensity in the transverse direction as a function of crosshead displacement. In contrast of the result for the axial direction, the intensity of all the lattice planes including 110 decreased monotonically. No behavior

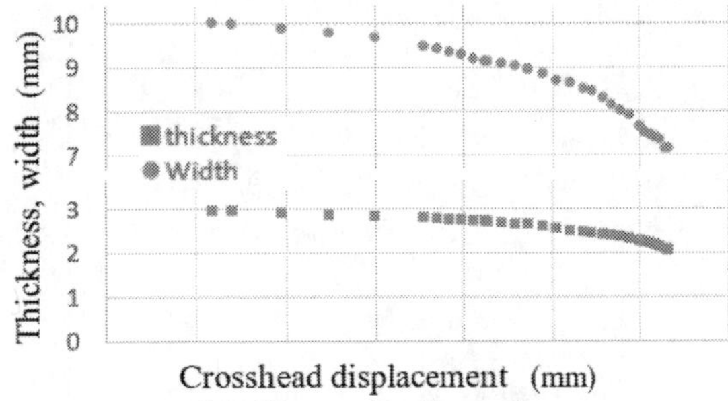

(a) Change in dimensions

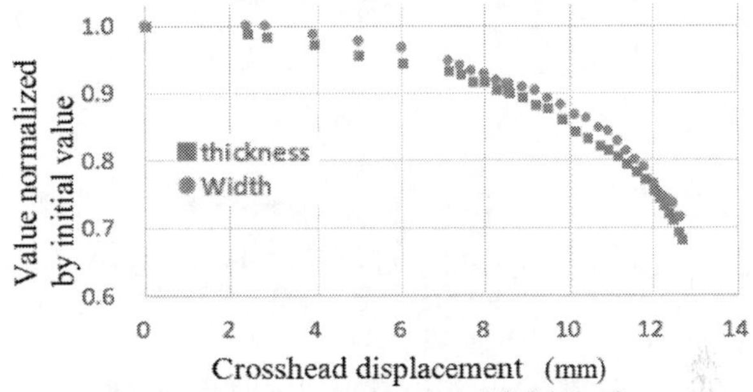

(b) Change relative to initial length

Figure 4. Changes in the thickness and width at the most reduced section of the specimen

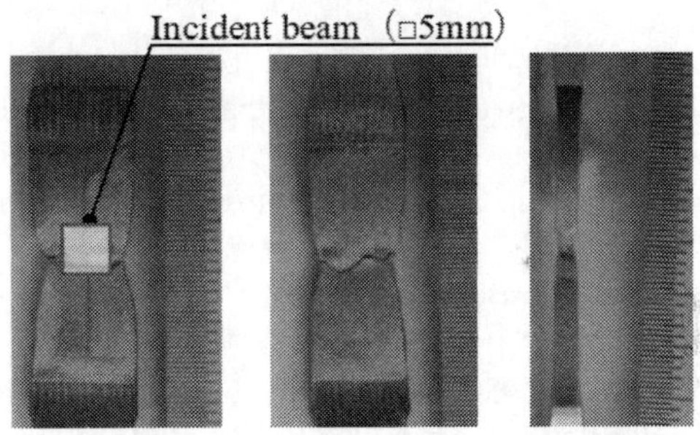

(a) Front view (b) Back view (c) Side view

Figure 5. Photographs of specimen after fracture

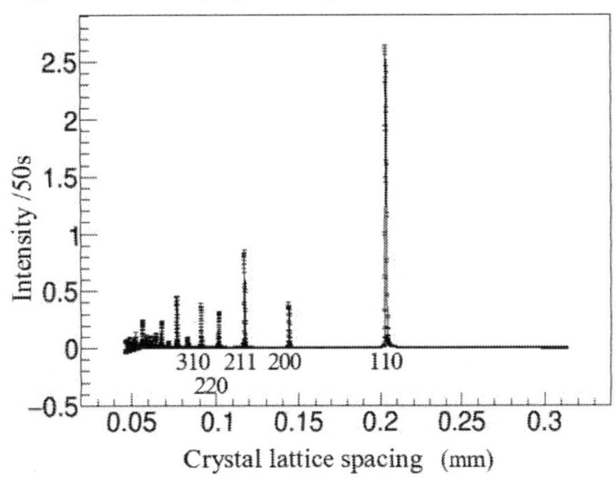

Figure 6. Diffraction pattern in the axial direction taken before deformation

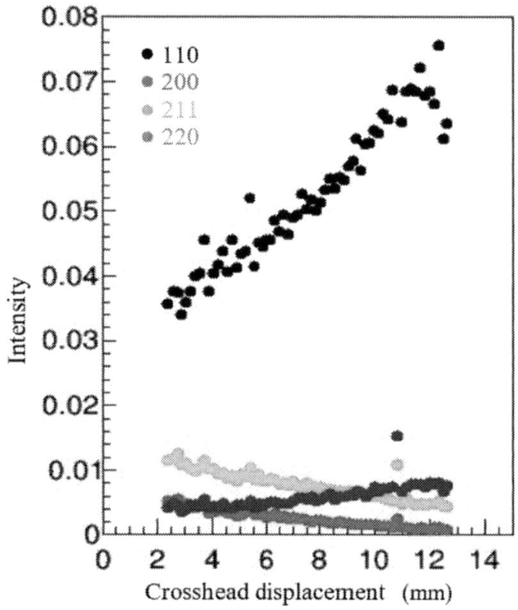

Figure 7. Change in diffraction intensity in the axial direction

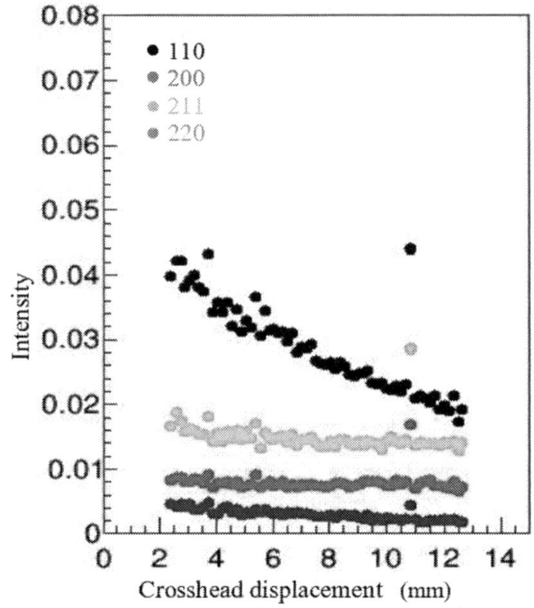

Figure 8. Change in diffraction intensity in the transverse direction

that could be ascribed to the effect of fracture was observed.

4. Conclusion
The local deformation behavior that occurs before fracture was studied by neutron diffraction measurement. The results of a change in the peak intensity of the diffraction pattern supposed that the mechanism of the texture evolution may be changed during necking.

Acknowledgments
This work has been partially supported by Photon and Quantum Basic Research Coordinated Development Program from the Ministry of Education, Culture, Sport, Science and Technology, Japan. Furthermore, this work has been supported by JSPS KAKENHI Grant Numbers 25289265 and 25420078. The authors also wish to acknowledge Dr. Yujiro Ikeda of Japan Atomic Energy Agency for his beneficial advice and comments.

[1] M. Takamura, T. Ozaki, Y. Miyoshi, H. Sunaga, S. Takahashi, Proc. 10th Int. Conf. Techn. Plast. (ICTP 2011), (2011), 591-595.
[2] B. Clausen, D. W. Brown, C. N. Tomé, L. Balogh and S. C. Vogel, Proceedings of the 31st Risø International Symposium on Materials Science: Challenges in materials science and possibilities in 3D and 4D characterization techniques (2010) 11.
[3] S. Morooka, Y. Tomota, Y. Adachi, S. Morita and T. Kamiyama, Advanced Materials Research, 15-17 (2006) 912
[4] S. Harjo, J. Abe, K. Aizawa, W. Gong and T. Iwahashi, JPS Conf. Proc. 1, 014017(2014) .
[5] H.Wang, B.Clausen, C.N.Tome, P.D.Wu, Acta Materialia, 61 (2013) 1179-1188
[6] Y.Tomota, H.Tokuda, Y.Adachi, M.Wakita, N.Minakawa, A.Moriai, Y.Morii, Acta Materialia, 52 (2004) 5737-5745

Development of the subroutine library 'UMMDp' for anisotropic yield functions commonly applicable to commercial FEM codes

Hideo Takizawa[1,5], Toshihiko Kuwabara[2], Kai Oide[3] and Junji Yoshida[4]

[1] Mechanical Engineering Dept., Nippon Institute of Technology, 4-1 Gakuendai Miyashiro, Saitama, 345-8501, Japan
[2] Division of Advanced Mechanical Systems Engineering, Tokyo University of Agriculture and Technology, 2-24-26, Naka-cho, Koganei, Tokyo, 184-8588, Japan
[3] Mechanical Design and Analysis Co., 1-40-2, Fuda, Chofu, Tokyo, 182-0024, Japan
[4] Dept. of Civil and Environmental Engineering, University of Yamanashi, 4-3-11, Takeda, Kofu, Yamanashi, 400-8511, Japan
[5] htaki@nit.ac.jp

Abstract. Numerous types of yield functions have been proposed to describe the shape of a realistic yield surface. Major commercial finite element codes include few anisotropic functions. Alternatively, the codes allow users to implement material models through user-subroutines. We develop the Unified Material Model Driver for Plasticity (UMMDp) subroutine library, which enables users to implement an arbitrary yield function easily. In this paper, the framework of the UMMDp is presented and its availabilities is shown through examples of sheet metal forming analyses.

1. Introduction

Yield functions generally represent the properties of plasticity in a numerical simulation. The yield functions under the associated flow rule describe not only relationship between the stress components in the plastic state, but also the direction of the plastic strain increment. Therefore, employment of appropriate anisotropic yield function, which can quantitatively express the realistic plasticity, is required for accurate prediction. Numerous types of yield functions have been proposed for this purpose so far in the past research [1]. But, general-purpose FE codes include few classical anisotropic yield functions. Alternatively, most FE codes allow users to implement their own material models as user-subroutines. However, it is not easy for the standard user to write subroutines by themselves, because detailed knowledge of continuum mechanics and numerical procedures is required to write subroutines appropriately.

The present paper describes the development of an easy-to-use user-subroutine library for anisotropic yield functions, which is widely applicable to major commercial FE codes.

2. Elasto-plastic constitutive model in FE codes

The classical FE codes for elasto-plastic deformation use a hypoelastic material model, and it is used for stress integration and calculation of the consistent tangent matrix. Although the above-mentioned roles of the material model are identical for all FE codes, detailed applications of subroutines in those codes are only discussed in closed communities such as user's conferences held by the software vendor. In order to promote industrial use of anisotropic yield functions, a universal easy-to-use subroutines package is needed to implement various yield functions for those codes.

3 Stress integration and consistent tangent matrix

In the backward Euler scheme, all of the constitutive equations are satisfied in the increment n+1. The state in which the stress $\{\sigma_{n+1}\}$ is located on the isotropically hardened yield surface is described as

$$\sigma_e(\sigma_{n+1}) - \sigma_Y(p_{n+1}) = 0\,. \tag{1}$$

where σ_e is an arbitrary yield function, $p_{n+1} = p_n + \Delta p$ is the equivalent plastic strain, and $\sigma_Y(p)$ is the hardening function driven by p. By assuming the associated flow rule, the plastic strain increment $\{\Delta\varepsilon^{\mathrm{p}}\}$ can be described as follows in the Voigt notation:

$$\{\Delta\varepsilon^{\mathrm{p}}\} = \Delta p \left.\frac{\partial\sigma_e}{\partial\{\sigma\}}\right|_{n+1} \equiv \Delta p\{N_{n+1}\} \tag{2}$$

The direction of the plastic strain increment coincides with the outward normal to the yield surface of increment n+1. Here, $\{N\}$ is the 1st-order partial differential of σ_e with respect to the stress. If the elastic deformation is small enough, the additive decomposition of the total strain increment can be applied as $\{\Delta\varepsilon\} = \{\Delta\varepsilon^{\mathrm{e}}\} + \{\Delta\varepsilon^{\mathrm{p}}\}$. Thus, the updated stress can be written as

$$\{\sigma_{n+1}\} = \{\sigma_n\} + [C^{\mathrm{e}}]\{\Delta\varepsilon^{\mathrm{e}}\} = \{\sigma_n\} + [C^{\mathrm{e}}]\{\Delta\varepsilon\} - \Delta p[C^{\mathrm{e}}]\{N_{n+1}\} \equiv \{\sigma^{\mathrm{Try}}\} - \Delta p[C^{\mathrm{e}}]\{N_{n+1}\}, \tag{3}$$

where, $[C^{\mathrm{e}}]$ is the matrix of Hooke's law, and $\{\sigma^{\mathrm{Try}}\}$ is the stress if the strain increment is all elastic.

The stress increment $\{\Delta\sigma\}$ and equivalent plastic strain increment Δp are solved to satisfy equations (1) and (3). First, the residual scalar g_1 and vector $\{g_2\}$ are defined as

$$g_1 = \sigma_e(\sigma_{n+1}) - \sigma_Y(p_n + \Delta p) \quad \text{and} \quad \{g_2\} = \{\sigma_{n+1}\} - \{\sigma^{\mathrm{Try}}\} + \Delta p[C^{\mathrm{e}}]\{N_{n+1}\}, \tag{4}$$

respectively. These equations are linearized to obtain corrections $\{\delta\Delta\sigma\}$ and $\delta\Delta p$ by Newton-Raphson iteration, as follows:

$$g_1 + \{N\}^{\mathrm{T}}\{\delta\Delta\sigma\} - \frac{\partial\sigma_Y}{\partial p}\{\delta\Delta p\} = 0\,, \quad \{g_2\} + [E]\{\delta\Delta\sigma\} + \delta\Delta p[C^{\mathrm{e}}]\{N\} = 0 \tag{5}$$

where [I] is the unit matrix, and $[E] = [\mathrm{I}] + \Delta p[C^{\mathrm{e}}][\partial\{N\}/\partial\{\sigma\}^{\mathrm{T}}]$. Solving the above equations as simultaneous equations, $\delta\Delta p$ and $\{\delta\Delta\sigma\}$ can be obtained as

$$\delta\Delta p = \frac{g_1 - \{N\}^{\mathrm{T}}[E]^{-1}\{g_2\}}{\{N\}^{\mathrm{T}}[E]^{-1}[C^{\mathrm{e}}]\{N\} + \dfrac{\partial\sigma_Y}{\partial p}} \quad \text{and} \quad \{\delta\Delta\sigma\} = -[E]^{-1}\{\{g_2\} + \delta\Delta p[C^{\mathrm{e}}]\{N\}\}\,. \tag{6}$$

The equations are updated iteratively with $\{\Delta\sigma\} + \{\delta\Delta\sigma\}$ and $\Delta p + \delta\Delta p$ until g_1 and $\{g_2\}$ converge.

The consistent tangent matrix that conforms to the algorithm can be obtained as the relationship between the perturbations of $\{\Delta\sigma\}$ and $\{\Delta\varepsilon\}$. The total differentials of equations (1) and (3) are

$$\{N\}^{\mathrm{T}}\{\delta\Delta\sigma\} - \frac{\partial\sigma_Y}{\partial p}\delta\Delta p = 0 \text{ and } \{\delta\Delta\sigma\} = [C^{\mathrm{e}}]\{\delta\Delta\varepsilon\} - [C^{\mathrm{e}}]\left\{\delta\Delta p\{N\} + \Delta p\frac{\partial\{N\}}{\partial\{\sigma\}^{\mathrm{T}}}\{\delta\Delta\sigma\}\right\}, \tag{7}$$

respectively. Using the above-mentioned $[E]$ and $[e] = [E]^{-1}[C^{\mathrm{e}}]$, the relationship between the perturbations of $\{\Delta\sigma\}$ and $\{\Delta\varepsilon\}$ can be obtained as follows:

$$\{\delta\Delta\sigma\} = \left[[e] - [e]\{N\}\{N\}^{\mathrm{T}}[e]\cdot\left(\{N\}^{\mathrm{T}}[e]\{N\} + \frac{\partial\sigma_Y}{\partial p}\right)^{-1}\right]\{\delta\Delta\varepsilon\} \tag{8}$$

From the above, the updated stress and the consistent tangent matrix can be calculated only if the yield function σ_e and its 1st- and 2nd-order partial differentials, $\{N\}$ and $[\partial\{N\}/\partial\{\sigma\}^{\mathrm{T}}]$, respectively, are given.

4. Framework of the UMMDp subroutine library

As mentioned above, the roles of the material user-subroutines are identical for all FE codes, and the numerical procedures in the user-subroutines are independent of the type of yield functions. Based on these two features, the material model can be unified as a common numerical procedure, externalizing the variety of FE codes and the variety of yield functions.

Figure 1 shows the framework of the Unified Material Model Driver for Plasticity (UMMDp). The core that executes the common numerical procedure is called from particular user-subroutines of FE codes. Each user-subroutine acts as a 'plug-in', which connects the UMMDp core to the FE codes. In the right branch, various yield functions are modularized as subroutines, in which σ_e and its differentials are calculated as the arguments. The various yield functions were implemented in the UMMDp: Hill's 1948 and 1990, Gotoh's bi-quadratic (2D), Hu's bi-quadratic (3D), Yoshida's 6th-order polynomial, Bralat's Yld89, Yld2000-2d, and Yld2004, Banabic's BBC2005 and BBC2008, Cazacu's CPB2006, Karafills-Boyce, and Vegter's spline. Furthermore, as shown in the left branch, various hardening rules, including kinematic hardening, are modularized for future expansion.

5. Working example of the UMMDp

As an example, FEA for a hole-expansion test is performed using the UMMDp. In this test, the material anisotropy is emphasized in the thickness distribution around the hole. Figure 2 shows the experimental set-up of the test [2]. Five yield functions Yld2000-2d [3], Yld2004 [4], CPB2006 [5], Yoshida's 6th-order polynomial [6], and BBC2005 [7] are used in the simulations. The coefficients of these functions are determined based on the results of biaxial tensile tests using cruciform specimens [8]. Figure 3 shows the yield loci together with experimental plots of the normalized plastic work

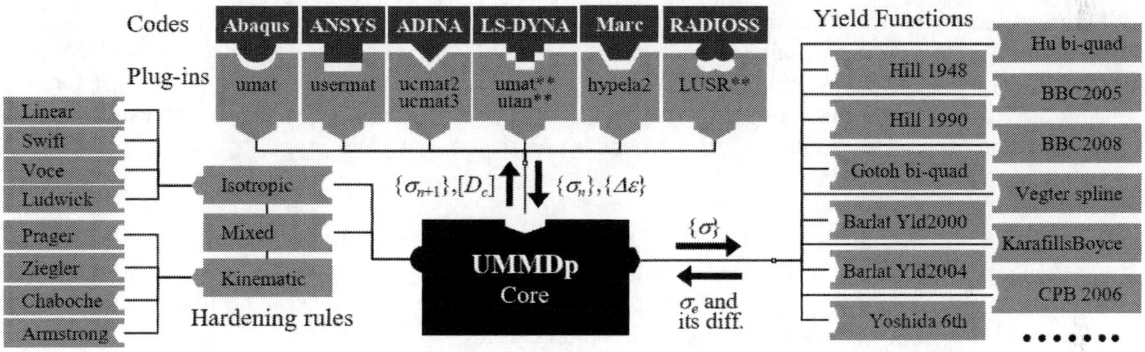

Figure 1. Framework of UMMDp (Unified Material Driver for Plasticity).

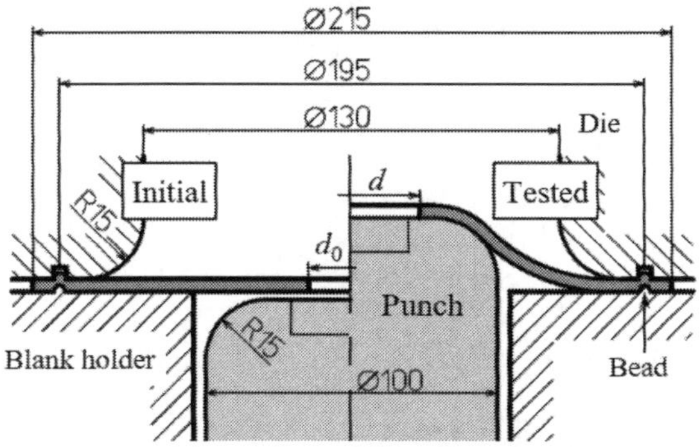

Material	SPCD (JIS G3141)
Thickness	0.65mm
Young's modulus	E=200GPa
Poisson's ratio	ν=0.3
Hardening curve	$\sigma_Y = 541(p+0.004)^{0.25}$ MPa

Figure 2. The experimental set-up for the hole expansion test and material properties.

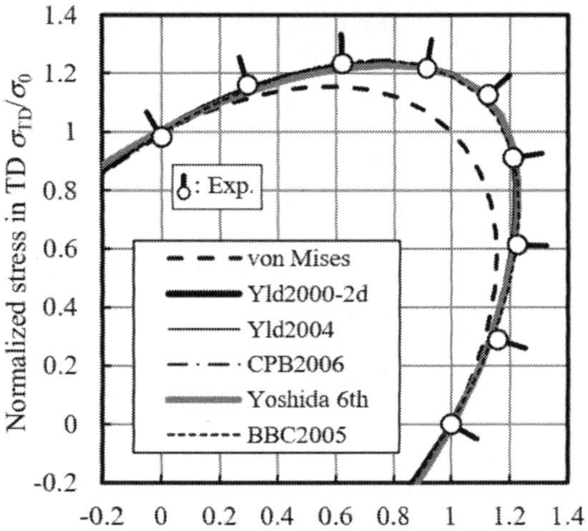

Figure 3. Yield loci with experimental plots.

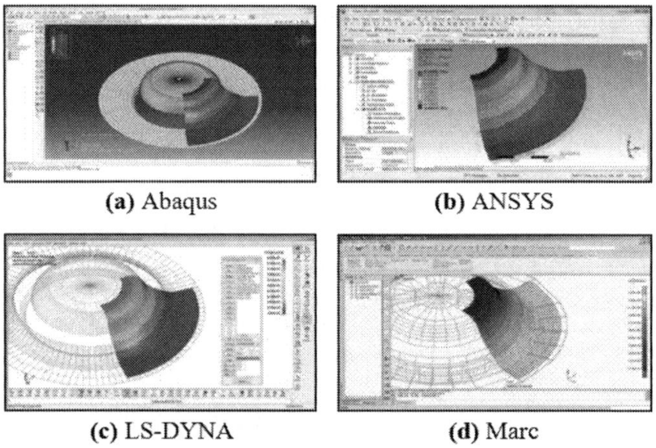

Figure 4. Post processing of thickness distributions calculated by various FE codes (Yld2004).

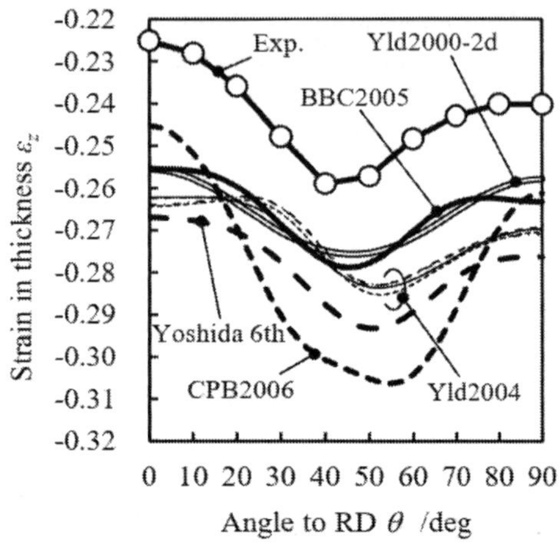

Figure 5. Distributions of thickness strain along the edge of the hole.

contour in which the axes of anisotropy coincide with that of the principal stress. The yield loci of all functions are in agreement with the experimental results.

Figure 4 shows the results of FEA performed using four commercial FE codes (Abaqus, ANSYS, LS-DYNA, and Marc) in conjunction with the same UMMDp. Figure 5 shows the circumferential distributions of the thickness strain along the edge of the hole. The results obtained using each FE code with Yld2004 were very similar. Also figure 4 shows the simulation results of various yield functions using Marc. The results differed depending on the type of yield function used in analyses.

6. Conclusion

A universal user-subroutine library for various anisotropic yield functions was developed. This library is available for major FE codes and enables users to easily implement their yield functions. Verification of the library is performed using an isotropic hardening model. The library is currently being expanded to include various kinematic hardening rules.

Acknowledgments

All programs were developed by participants of the Material Modeling Working Group of the Japan Association for Nonlinear CAE (JANCAE). The authors would like to thank all of the people who supported the present project.

References

[1] Banabic D, Barlat F, Cazacu O and Kuwabara T 2010 *Int. J. Mater. Form.* **3** 165
[2] Kuwabara T and Ichikawa K 2015 *Ro. J. Techn. Sci. Appl. Mechanics* **60** 63
[3] Barlat F, Brem J C, Yoon J W, Chung K, Dick R E, Lege D J, Pourboghrat F, Choi S-H and Chu E 2003 *Int. J. Plasticity* **19** 1297
[4] Barlat F, Aretz H, Yoon J W, Karabin M E, Brem J C and Dick R E 2005 *Int. J. Plasticity* **21** 1009
[5] Cazacu O, Plunkett B and Barlat F 2006 *Int. J. Plasticity* **22** 1171
[6] Yoshida F, Hamasaki H and Uemori T 2013 *Int. J. Plasticity* **45** 119
[7] Banabic D, Aretz H, Comsa D S and Paraianu L 2005 *Int. J. Plasticity* **21** 493
[8] ISO-16842 2014 Metallic materials −Sheet and strip −Biaxial tensile testing method using a cruciform test piece

Journal of Physics: Conference Series **734** (2016) 032029 doi:10.1088/1742-6596/734/3/032029

Advanced constitutive modeling of AHSS sheets for application to springback prediction after U-draw double stamping process

Jisik Choi[1], Jinwoo Lee[2], Myoung-Gyu Lee[3], Frederic Barlat[1]

[1]Graduate Institute of Ferrous Technology, Pohang University of Science and Technology, Pohang, 790-784, Republic of Korea
[2]Materials deformation department, KIMS, Changwon, 51508, Republic of Korea
[3]Department of Materials Science and Engineering, Korea University, Seoul, 136-713, South Korea

Abstract. The reduction of springback for a U-shaped channel using a double drawing process was investigated. In this test, the punch strokes of the 1st and 2nd stamping steps were controlled and each followed by unloading. The simulations were conducted using kinematic and distortional hardening models, which were implemented into a finite element (FE) code to describe the Bauschinger effect and its associated anisotropic hardening effects during strain path change. In addition to the usual mechanical characterization tests, in-plane compression-tension experiments were conducted on DP980 and TWIP980 to determine the constitutive parameters pertaining to load reversal. Experimental and FE simulated results of the channel shape were compared for both materials in order to understand the effect of anisotropic hardening under non-proportional loading on springback.

1. Introduction

Advanced high strength steels (AHSS) have replaced conventional steels for automotive parts to reduce the body weight and improve the crash performance. However, the higher strength and thinner design of the parts manufactured with AHSS sheets result in significant springback increase. Thus, tool adjustments for a springback-free design should be optimized based on trial-and-error, which usually increases the process design time and cost.

In order to understand and predict the mechanics of springback for AHSS, finite element (FE) approaches have been frequently used. An accurate prediction of springback based on FE simulation can significantly facilitate a successful achievement of the optimum tool design. In order to increase the reliability of the springback estimation using FE simulations, an accurate elasto-plasticity constitutive model must be used. Specifically, the Yoshida-Uemori (YU) constitutive model has been extensively used for springback simulation. It is based on a two-surface kinematic hardening approach and can predict most of the complex hardening behaviors under load reversal [1]. Recently, a new approach was proposed as an alternative to kinematic hardening, known as a homogenous yield function based, anisotropic hardening (HAH) model [2].

Automotive parts are generally produced using multi-stage forming processes in industry. In this case, since blanks undergo strain path change several times, it is difficult to simulate accurate predictions of springback with conventional constitutive descriptions. In this work, the U-draw double stamping process was investigated for two different AHSS sheets with various blank holding force (BHF) and strokes. Different types of hardening models were considered to predict springback in this work.

2. Experiment

To evaluate the performance of U-draw double stamping predictions, two AHSS sheet samples provided by POSCO were investigated; i.e., a dual-phase steel (DP 980) and a twinning-induced plasticity steel (TWIP 980), both of which with a nominal ultimate tensile strength of 980 MPa. The size of the sheet specimen was 350 (length), 45 (width) and 1.2 mm (thickness).

Continuous compression-tension (CT) tests were conducted to determine the hardening behavior during load reversal. The testing device was designed with an anti-buckling system where a constant normal force is applied to prevent sheet buckling during compression. The experimental data were calibrated by considering the effects of the clamping force and friction. All the tests were performed with a quasi-static, constant strain rate of $10^{-3}/s$.

For U-draw double stamping, an equipment recently modified, Fig. 1(a), was employed. The punch width was 50 mm and the clearance was 1.5 mm. The punch velocity and total stroke were set to 70 mm/min and 70 mm, respectively. For conventional (one stroke) U-draw bending, three different blank holder forces, BHF, (10, 25 and 50 kN) were selected but for double stamping, only the 1st stroke was varied (30, 50 and 70 mm). The BHF and 2nd stroke were fixed to 50 kN and 70 mm, respectively. An oil, BW-70PSB, was used for lubrication. The method of springback measurement is illustrated in Fig.1(b). There are three black dash lines, which are defined unambiguously: bottom line, middle line (sidewall region) and upper line (flange region). Based on these three lines, the springback indicators θ_1, θ_2 and ρ were calculated using MATLAB.

(a)

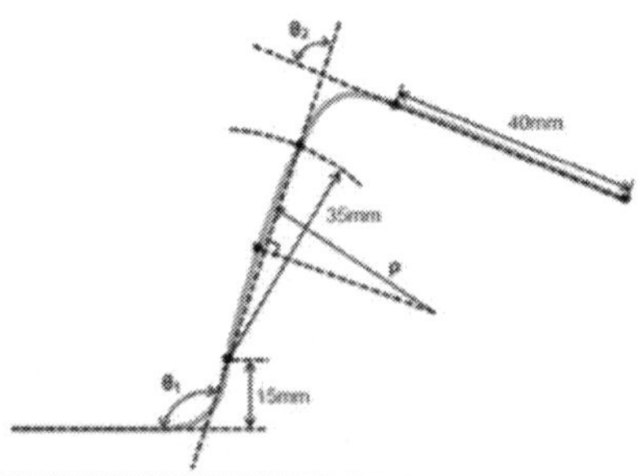

(b)

Figure 1. (a) U-draw stamping apparatus, (b) springback measurement

3. Constitutive model

The Swift hardening law was adopted for isotropic hardening. In addition, to consider the elastic modulus degradation when sheets are unloaded after prior deformation, the chord modulus concept [3] was considered. A non-quadratic yield function, Yld2000-2d [4], was employed to describe the plastic anisotropy of the sheet metals during forming.

3.1. Isotropic-kinematic hardening: Yoshida-Uemori model

The Yoshida–Uemori (YU) model is based on a nonlinear, two-surface, kinematic hardening law. In this study, the YU model was combined with Yld2000-2d to investigate the effect of both the yield surface and hardening on the springback. In the following paragraph, a summary of the YU model is presented, but the details of the model can be found in [1]. The relative kinematic motion $(\boldsymbol{\alpha}_*)$ of the yield surface is

$$\boldsymbol{\alpha}_* = \boldsymbol{\alpha} - \boldsymbol{\beta} \tag{1}$$

$$\overset{\circ}{\boldsymbol{\alpha}}_* = \mathrm{C}\left[\left(\frac{\mathrm{a}}{\mathrm{Y}}\right)(\boldsymbol{\sigma} - \boldsymbol{\alpha}) - \sqrt{\frac{\mathrm{a}}{\bar{\alpha}_*}}\boldsymbol{\alpha}_*\right]\dot{\mathrm{p}} \tag{2}$$

where $\boldsymbol{\alpha}$ is the back stress tensor and $\boldsymbol{\beta}$ is the center of the bounding surface. From Eq. (1), the evolution of $\boldsymbol{\alpha}_*$ is assumed to be given by Eq. (2). $\dot{\mathbf{p}}$ is the effective strain rate and $\mathbf{C}$ a material constant controlling the kinematic hardening rate. ($^{\circ}$) stands for time derivation with an objective rate. Y is the size of the yield surface.

3.2. Distortional anisotropic hardening: HAH model

In this work, the HAH model was also coupled with Yld2000-2d to represent the anisotropy under linear strain paths. The HAH yield condition can be expressed as

$$\bar{\sigma}(\mathbf{s}) = \left[\phi^q(\mathbf{s}) + f_1^q\left|\hat{\mathbf{h}}:\mathbf{s} - |\hat{\mathbf{h}}:\mathbf{s}|\right|^q + f_2^q\left|\hat{\mathbf{h}}:\mathbf{s} + |\hat{\mathbf{h}}:\mathbf{s}|\right|^q\right]^{\frac{1}{q}} = \sigma(\bar{\varepsilon}) \tag{3}$$

where $\phi(\mathrm{s})$ can be any homogeneous anisotropic yield function of first degree, and $\hat{\mathbf{h}}$ is the microstructure deviator, which is associated with the deformation history in a continuum sense. The superscript "^" implies a normalized value, ":" denotes the double dot product, $\mathbf{s}$ is the stress deviator and q is a constant exponent. The detailed HAH formulation was reported in [2].

4. FE simulation

For the YU model, the parameters were determined using a nonlinear least-square method in MATLAB to fit both the monotonic and reverse stress–strain curves obtained from the compression-tension test. In the HAH model, the Swift hardening coefficient were obtained by fitting the monotonic flow curve. In order to determine the anisotropic hardening parameters, a single element FE simulation was performed iteratively under compression-tension loading to determine the optimized hardening coefficients, k_i

For the FE simulations, the 2D draw-bending springback model proposed in the Numisheet'93 benchmark was considered. ABAQUS/explicit was employed for the forming process simulations and ABAQUS/implicit for springback analysis. The element type was a 4-node shell element with reduced integration (S4R). A half model of the specimen only was considered in the simulations. All the numerical conditions, such as type and number of elements, integration scheme, boundary conditions, etc., remained identical as in the benchmark problem. The FE simulations were conducted with four material descriptions for each case.

5. Results

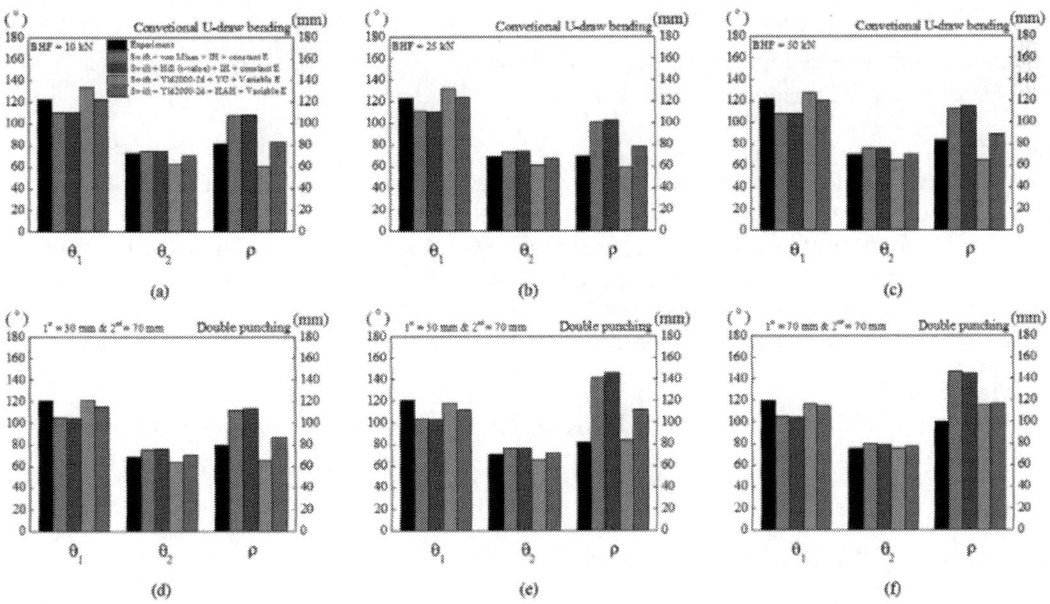

Figure 2. Springback results for θ_1, θ_2 and ρ for DP980: (a) conventional U-draw when BHF = 10 kN, (b) conventional U-draw when BHF = 25, (c) conventional U-draw when BHF = 50 kN, (d) double stage 1st stroke 30 mm, (e) double stage 1st stroke 50 mm and (f) double stage 1st stroke 70 mm

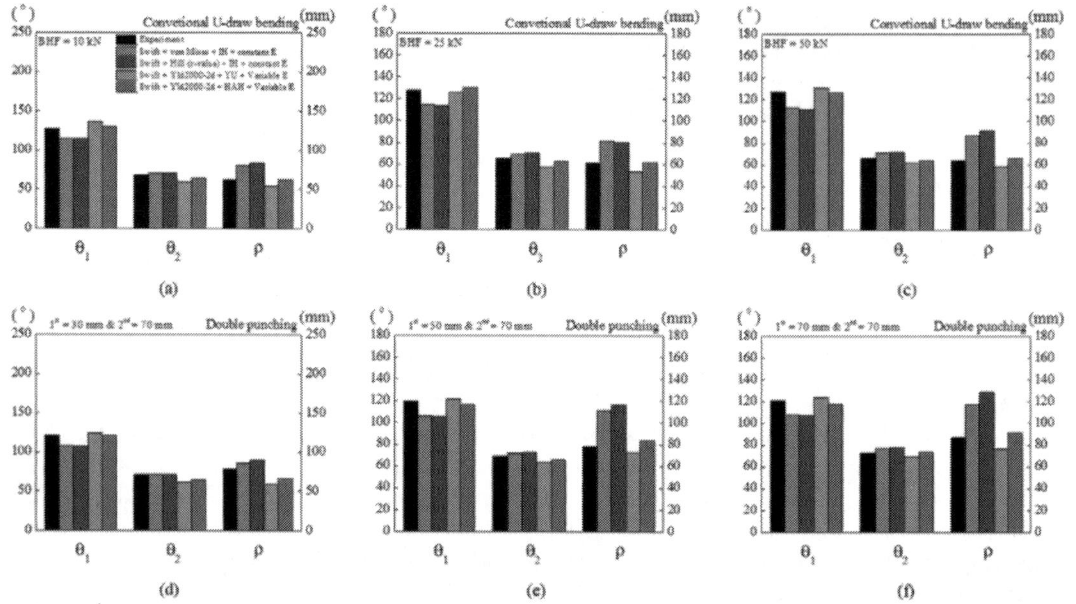

Figure 3. Springback results for θ_1, θ_2 and ρ for TWIP980: (a) conventional U-draw when BHF = 10 kN, (b) conventional U-draw when BHF = 25, (c) conventional U-draw when BHF = 50 kN, (d) double stage 1st stroke 30 mm, (e) double stage 1st stroke 50 mm and (f) double stage 1st stroke 70 mm

Figs. 2-3 represent the springback predictions after conventional U-draw bending (a-c) and double stage stamping (e-f) for DP980 and TWIP980, respectively, as well as the experimental results. Two advanced hardening models lead to better predictions of the experiments compared to the results using isotropic hardening. After double stage stamping, the springback prediction accuracy deteriorates but the results obtained by the two advanced models are still in reasonable agreement with the experiments. It can be shown that this is because the two advanced models can describe complex hardening behaviors under strain path changes. Interestingly, the influence of the anisotropic yield function, using either Hill or von Mises, is very minor for springback.

6. Discussion

The internal stress predicted by the anisotropic hardening formulation (YU & HAH) is lower than that from isotropic hardening because the two advanced models can describe the complex flow behavior after load reversal. In turn, this results in a smaller amount of springback. On the other hand, the elastic modulus degradation contributes to larger amounts of springback. Eventually, these opposite elastic-plastic relations lead to an accurate prediction of springback. This work confirms that, in order to obtain accurate predictions, both elastic and plastic hardening descriptions must be considered simultaneously.

Acknowledgement

The authors appreciate the support from POSCO. MGL acknowledges the support from ERC grant (No.2012R1A5A1048294).

References

[1] F. Yoshida and T. Uemori 2003 Int. J. Mech. Sci. **45** 1687
[2] F. Barlat et al 2011 Int. J. Plast **27** 1309
[3] F. Yoshida et al 2002 Int. J. Plast **18** 633
[4] F. Barlat et al 2003 Int. J. Plast **19** 1297

Numerical assessment of residual formability in sheet metal products: towards design for sustainability

Javad Falsafi*, Emrah Demirci, Vadim.V. Silberschmidt

The Wolfson School of Mechanical, Electrical and Manufacturing Engineering, Loughborough University, LE11 3TU, Loughborough, UK

*falsafi.javad@gmail.com

Abstract. A new computational scheme is presented to addresses cold recyclability of sheet–metal products. Cold recycling or re-manufacturing is an emerging area studied mostly empirically; in its current form, it lacks theoretical foundation especially in the area of sheet metals. In this study, a re-formability index was introduced based on post-manufacture residual formability in sheet metal products. This index accounts for possible levels of deformation along different strain paths based on Polar Effective Plastic Strain (PEPS) technique. PEPS is strain-path independent, hence provides a foundation for residual formability analysis. A user-friendly code was developed to implement this assessment in conjunction with advanced finite-element (FE) analysis. The significance of this approach is the advancement towards recycling of sheet metal products without melting them.

1. Introduction

The environmental challenges have been considered as a serious problem across all industries, and reduction of energy and material consumption is a pressing matter for them. Conventional recycling saves significant amount of energy, however it involves melting process. High melting points of metals and additional processes required to get final products, turn it yet into an energy intensive process. Recycling metal waste without melting has potential for additional level of energy saving [1].

The possibility of cold recycling of sheet metals has been studied by some researchers. Takano et al. [2] showed that cold recycling of sheet metal can be accomplished using incremental forming, in which strain localization can be almost inhibited. Tekkaya et al. [3] analysed remanufacturing of a contoured sheet metal part using a hydro-forming technology. They showed that material inhomogeneity induced by a primary forming process could be ignored.

The main challenge in the cold recycling/remanufacturing process is varying and inhomogeneous material behaviour as a result of the primary forming process. Therefore, a very important requirement for re-manufacturing is the identification of residual formability across the material. As a matter of fact, the theoretical basis of remanufacturing has been barely addressed in the literature. The present research was an effort to bridge this gap by looking at the problem from a point of view of formability.

Forming limit diagrams (FLD) have been such an effective and widely accepted tool in sheet metal industries. A significant limitation of characterizing formability with the FLC was highlighted experimentally by Graf et al. [4] and it was concluded that no unified curve in a strain space could represent the forming limit of a material. Stoughton and Yoon [5] proposed an alternative approach as a remedy to the aforementioned difficulty. They modified the idea proposed by Zeng et al. [6] and employed effective plastic strain as a metric to evaluate formability. They suggested a PEPS diagram and demonstrated the insensitivity of their technique to a strain path using experimental data. Assuming a secondary life and additional forming stages for a formed product required multistage/multiple strain path formability assessment technique, therefore PEPS took the central stage in present research.

With an outlook for sustainable manufacturing, this study aimed at introducing an index that quantifies reformability of a pre-formed material on a scale of 0 to 1, that respectively denotes fully damaged and a healthy material. This index is strain-path-sensitive and provides an insight into the allowable formability for different strain paths.

2. Reformability index

Stoughton and Yoon [5] pointed out that, basically, two parameters carry nonlinear paths information on, first, β which is the ratio of the principal strain rates; and second, effective plastic strain $\bar{\varepsilon}_p$, which is defined by a time integral of a function of the principal strain rates. Assuming that principal strains, $\varepsilon_1(t)$ and $\varepsilon_2(t)$ are parameters accounting for the strain history, and r_m is the plastic anisotropy, the following relations are the components of the *PEPS* diagram:

$$\bar{\varepsilon}_p(t) = \frac{1+r_m}{\sqrt{1+2r_m}} \int_0^t \sqrt{\dot{\varepsilon}_1{}^2(t') + \dot{\varepsilon}_1{}^2(t') + \frac{2r_m}{1+r_m} \dot{\varepsilon}_1(t')\dot{\varepsilon}_1(t')dt'}, \tag{1}$$

$$\beta = \frac{\dot{\varepsilon}_2(t)}{\dot{\varepsilon}_1(t)}. \tag{2}$$

In order to plot diformation in a polar diagram of the $\theta = tan^{-1}(\beta)$, is measured with reference to vertical axis. Figure 1 illustrates PEPS diagram and how a multi-path operation can be examined against the limit curve.

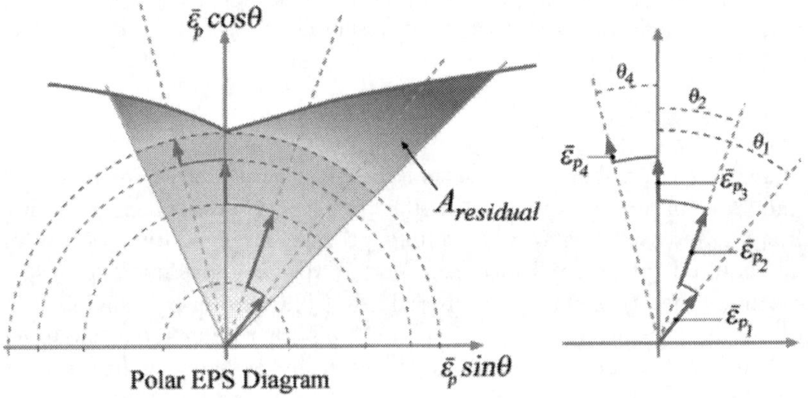

Figure 1. Graphical illustrations of Polar EPS diagram showing deformation with four different consecutive paths in which the progression of forming stages leads in loss of formability. In the proposed approach $A_{residual}$ is the basis for residual formability calculation.

2.1. *Strain-path-sensitive reformability index*

The technique proposed in this study is to calculate the area under the forming-limit curve and enclosed between the stress triaxiality $\eta = 2/3$ and $\eta = 1/3$, as a quantitative measure of original formability. Every stage of deformation exploit part of this area, and the residual formability can be described as the reduced area under the limit curve as shown in Figure 1. direction of the subsequent strain path is also an important parameter in describing the residual formability. Figure 2 illustrates that depending on strain-path the residual formability varies.

In the *PEPS* approach, the circle passing through the end point of the effective strain vector is the locus of the starting point for next vector. This implies that after the deformation process a circular sector, consists of incremental deformations, would be taken off from the original formability. Therefore the reformability index can be formulated as:

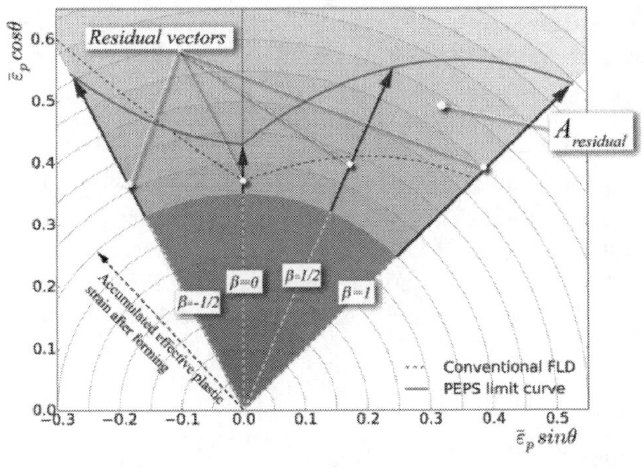

Figure 2. The ratio of residual area to original area; (A_r/A_0), does not address the residual formability in different strain path, therefore $\Phi = (A_r/A_0)^{B(\beta)}$ was introduced.

$$\Phi = \left(\frac{A_r}{A_0}\right)^{B(\beta)}, \tag{3}$$

$$A_0 = \frac{1+r_m}{\sqrt{1+2r_m}} \int_{\frac{-\pi}{8}}^{\frac{\pi}{4}} \sqrt{{\varepsilon_{1FLD}}^2 + {\varepsilon_{2FLD}}^2 + \frac{2r_m}{1+r_m} \varepsilon_{1FLD}\varepsilon_{2FLD}}\, d\theta\,, \tag{4}$$

$$A_r = A_0 - \frac{1+r_m}{\sqrt{1+2r_m}} \int_{\frac{-\pi}{8}}^{\frac{\pi}{4}} \int_0^t \sqrt{{\dot{\varepsilon}_1}^2(t') + {\dot{\varepsilon}_1}^2(t') + \frac{2r_m}{1+r_m} \dot{\varepsilon}_1(t')\dot{\varepsilon}_1(t')dt'}\, d\theta = A_0 - \bar{\varepsilon}_p \left(3\frac{\pi}{8}\right). \tag{5}$$

$$B(\beta) = C_3\beta^3 + C_2\beta^2 + C_1\beta^2 + C_0\,. \tag{6}$$

2.1.1. Calculating $B(\beta)$ constants. The exponent B is defined to modulate the reformability index to account for the strain-path effect. B is introduced as polynomial function of the strain path β.To calculate the associated constants, applying log function to both sides of the Eq. (3) gives:

$$B(\beta) = \frac{log\left(\frac{A_r}{A_0}\right)}{log\ (\Phi_\beta)}, \tag{7}$$

where A_r/A_0 is the residual to original area ratio, and Φ_β is the residual to original vector ratio for relevant strain path (see Figure 2). Four data values, are required to calculate C_0 to C_3. It is suggested to assume a total effective plastic strain equivalent to 70% of the limit curve in $\beta = 0$ direction, and workout $B(1)$, $B(1/2)$, $B(0)$ and $B(-1/2)$ from Eq.7. A set of four simultaneous equations are then obtained to calculating the polynomial constants.

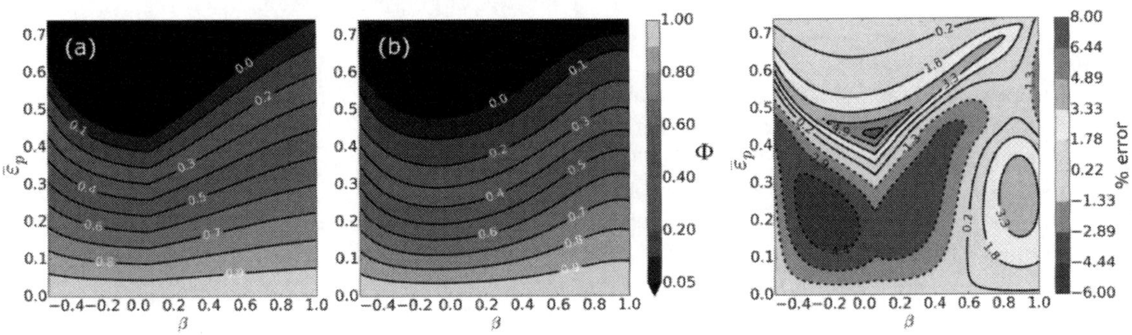

Figure 3. The strain-path sensitive reformability index (a) actual, (b) predicted and error.

A sample $B(\beta)$ for the *PEPS* curve is presented in Figure 3, in which "predicted" Φ index is shown together with the "actual" Φ obtained from vector ratios in every strain path angle. The error contour plot shows a good agreement between the model presented in Eq. (3) and the actual distribution. Higher order polynomial for $B(\beta)$ could improve the predicted Φ.

3. REFORMAP

Implementing this technique using the finite element provides an element by element assessment of reformability index across the part. A code called REFORMAP was developed for mapping out the reformability index downstream the FE simulation. Figure 4 shows a typical quarter model of rectangular deep drawn part. The presented FEA analysis is based on the analytical forming limit prediction. Once the result file is imported into REFORMAP, it helps visualizing the distribution of the reformability index and some additional analysis.

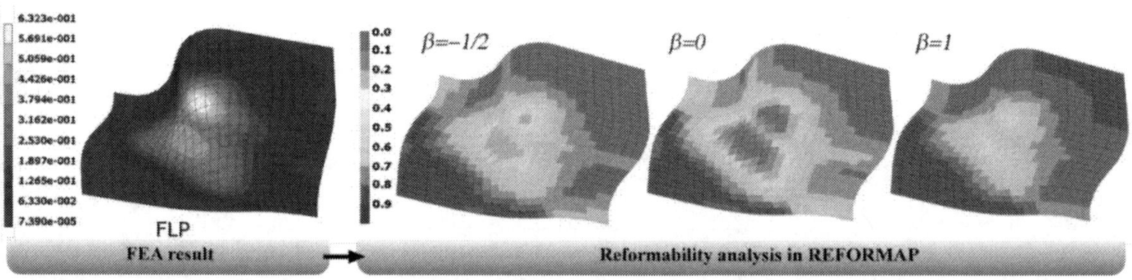

Figure 4. The results from FEA quarter deep drawing simulation was imported to REFORMAP for reformability analysis.

4. Conclusion

In this paper a strain-path-dependent index was proposed to quantify the reformability of deformed sheet on a scale from 0 to 1, denoting fully damaged and a healthy material respectively. This scheme was introduced based on the path independent Polar EPS. The proposed model was established and formulated in this paper, together with a way to obtain the relevant constants. The primary advantages of the PEPS are robustness against strain path change, and no dependence on the flow curve and any saturation or softening in the material model.

The presented model is suitable to be incorporated in numerical platforms. An in-house code called REFORMAP was developed to carry out reformability analysis downstream the FE simulation of forming process. The REFORMAP GUI provides three-dimensional visualization of the model as well as colour map of the re-formability index and other relevant outputs. This re-formability assessment implies the possibility of cold-recycling and contributes to design for recycling.

References

[1] Cooper DR, Gutowski TG 2015 Journal of Industrial Ecology
[2] Takano H Kitazawa K Goto T 2008 Int J of Machine Tools and Manufacture **48**(3) 477
[3] Tekkaya E Franzen V Trompeter M 2008 15.SFU 187
[4] Graf A Hosford W 1993 Metall Trans A**24** 2503
[5] Stoughton TB Yoon JW 2012 International Journal of Solids and Structures **49**(25) 3616
[6] Zeng D Chappuis L Xia ZC Zhu X 2008 SAE Int J of Mater and Manuf **1** 809

A Grain Structure Model based on Voronoi polygon of Non-oriented Electrical Steel in Blanking Process

Zhe Wang[1,2], Shuhui Li[1,2], Liang Dong[1,2], Ji He[1,2]

[1] State Key Laboratory of Mechanical System and Vibration, Shanghai Jiao Tong University, Shanghai 200240, PR China
[2] Shanghai Key Laboratory of Digital Manufacture for Thin-walled Structures, Shanghai Jiao Tong University, Shanghai 200240, PR China

E-mail: lishuhui@sjtu.edu.cn

Abstract. World-wide there is a trend to develop higher permeability grades, thin thickness and coarse grain of non-oriented electrical steels, a core function material of motors. Blanking is the most popular technique for producing the motor laminations. However, the deformation of material is significantly influenced by grain size. In this paper, Voronoi polygon is used for generate the random microstructures of the studied non-oriented electrical steel. Finite Element (FE) model considering grain size is thus established to analysis the blanking process. The material behaviour of grains is derived from the widely accepted surface layer model. Compared to the conventional model without considering the grain size, the novel model shows good matching with the experimental results.

1. Introduction

Due to the energy and environmental protection, non-oriented electrical steel with higher permeability grade, thin thickness (0.3-0.6mm) and coarse grain (50-150μm) becomes the major product. In order to producing the motor laminations efficiently, blanking is the most popular manufacturing technique. Plenty of prior studies show that the deformation is significantly influenced by grain size. Therefore, how to study the blanking process with considering grain size is of great importance.

The grain size effect on polycrystal material is generally expressed by the famous surface layer model. Peng et al. [1] and Fu et al. [2] used the surface layer model to study the flow and deformation behaviours, which is widely accepted in many micro-scaled forming processes. In order to modeling the micro grain structures of polycrystal materials, Cao et al. [3] and Gao et al. [4] adopted the Voronoi method for investigating the plastic deformation in microforming process.

Grain size effect in blanking process was mainly investigated via experiments. Xu et al. [5] innovatively investigated the grain size effects in micro-blanking test and established a grain size phenomenological model of blanking process. However, grain size effect in blanking process using FE method need be paid more attention.

Therefore, this paper aims at addressing this issue via development of a grain structure model. To realize this thought, the Voronoi polygon and surface layer model are used to establish the grain structure model. The simulation and experiment results are compared and the proposed grain model is verified.

2. Grain Structure Modelling

2.1. Material

High-grade non-oriented electrical steel B30AHV1500 (Baosteel Co., Ltd, Shanghai, China) is selected for the study. The chemical composition is shown in Table 1. With the help of Baoteel Company, some samples in different anneal stages (800℃,900℃,950℃,1000℃) are obtained. The metallographic images across the sheet thickness section and in-plain section are shown in Fig.1.

Table 1. Chemical composition of the investigated non-oriented electrical steel.

Element	Si	Al	Fe
Mass Fraction(%)	3.3	0.7	96.0

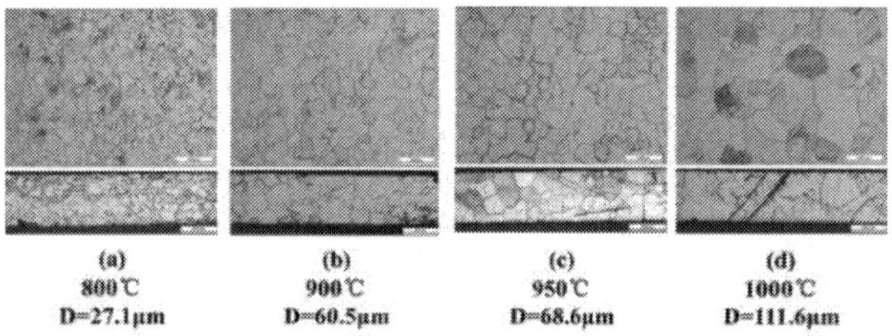

Figure. 1. Metallographic structure of non-oriented electrical steel samples under different anneal stage: (a)800℃ (b)900 ℃ (c) 950℃ (d)1000 ℃.

According to the hypothesis of surface layer model, the polycrystalline metal sheet is considered to consist of surface and internal grains, the flow stress can be represented as follows:

$$\begin{cases} \sigma_{sm}(\varepsilon) = \eta\sigma_s(\varepsilon) + (1-\eta)\sigma_i(\varepsilon) \\ \sigma_s(\varepsilon) = \mathrm{m}\tau_R(\varepsilon) \\ \sigma_i(\varepsilon) = M\tau_R(\varepsilon) + k(\varepsilon)\cdot D^{-1/2} \\ \eta = N_s/N \approx D/t \end{cases} \tag{1}$$

In Eq.(1) $\sigma_{sm}(\varepsilon)$ and N are the flow stress and grain number of the specimen, $\sigma_s(\varepsilon)$ and N_s are the flow stress and number of the surface grains, $\sigma_i(\varepsilon)$ is the flow stress of internal grains, D is the grain size, t is the thickness of metal sheet, $\tau_R(\varepsilon)$ is the critical shear resolved stress of single crystal, $k(\varepsilon)$ is the locally intensified stress needed to propagate general yield stress across the polycrystal grain boundaries, m and M are the orientation factors of the surface and internal grains. M is known as 3.06 and m is known as 2.0 for BCC material based on Taylor's Model [2].

To identify the flow stress curves of the surface and internal grains, the flow curves of the different samples at room temperature are first obtained by uniaxial tensile tests, then the hypothetic equation of $\tau_R(\varepsilon)$ and $k(\varepsilon)$ can be calculated using the least square fitting method. Fig.2 shows the results of surface layer model.

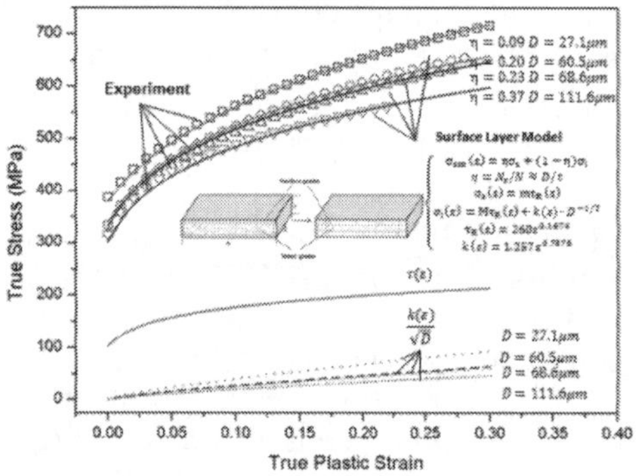

Figure 2. Results of surface layer model

2.2. FE model

In this paper, according to the average grain size, the Voronoi polygon is generated via a code in software Matlab to model the grain structure. Through the software interface, the geometric model was introduced into ABAQUS/CAE using python language (Fig.3). Different material properties of grains are assigned to geometric parts according to the hypothesis of surface layer model. A compared conventional model without grain structure is also established.

The blanking process is assumed to be plane strain situation and the finite element code ABAQUS/Standard is used to simulate the blanking operation in 2D. The blanking region is set to 0.30 mm×0.30 mm and the mesh of the region is defined very dense (element size: 0.005 mm). The right of blanking region is constrained against movement in any direction and the left is symmetrically constrained. 4-node bilinear plane strain quadrilateral elements with reduced integration (CPE4R) and hourglass control are used. A coulomb friction model is used to represent the contact between the plate and the tools with a friction coefficient value equal to 0.1. The punch and die are assumed to be rigid bodies. The punch-die clearance is set to 0.045 mm, which is same with the experiment. The Arbitrary Lagrangian Euleurian (ALE) technique is used to avoid element distortion in the sheared section. Then the FE model based on the Voronoi polygon is established (Fig.3b).

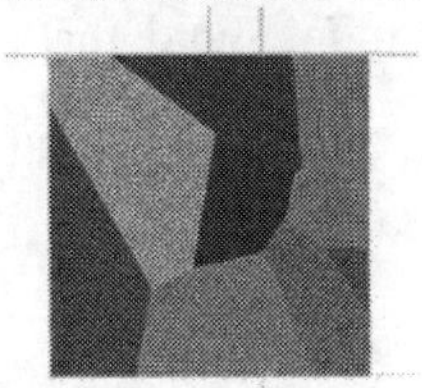

Figure 3. FE model based on the Voronoi polygon

2.3. *Fracture Strain*

In order to simulate the fracture of blanking process, Digital Image Correlation (DIC) method is adopted to measure the fracture strain more precisely. Fig.4a shows the strain distribution just when the crack occurs and the equivalent fracture strain is 0.241. From Fig. 4b, the fracture mode of the studied non-oriented electrical steel is transgranular fracture. Therefore, it is assumed that the fracture strain in uniaxial tensile test can also be suitable in blanking process. In the FE model, element suicides when the strain reaches the critical fracture strain.

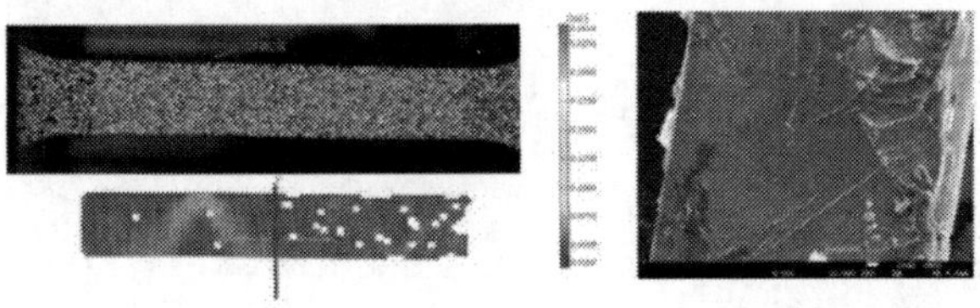

(a) (b)

Figure 4. Fracture analysis: (a) DIC test; (b) SEM image.

3. Results and discussion

3.1. *Verification of FE model*

Grain structure model based on Voronoi polygon are first analysed by uniaxial tensile simulation. By controlling the displacement of the edge in the FE model, the total elongation of the whole grain structure model and the total reaction force on the edge are respectively calculated. Then the flow curve of the whole grain structure model can be obtained.

Fig.5 shows the comparison between the simulation and experiment. The simulation results coincide well with the experimental data, which verifies the fine feasibility of the Grain structure model.

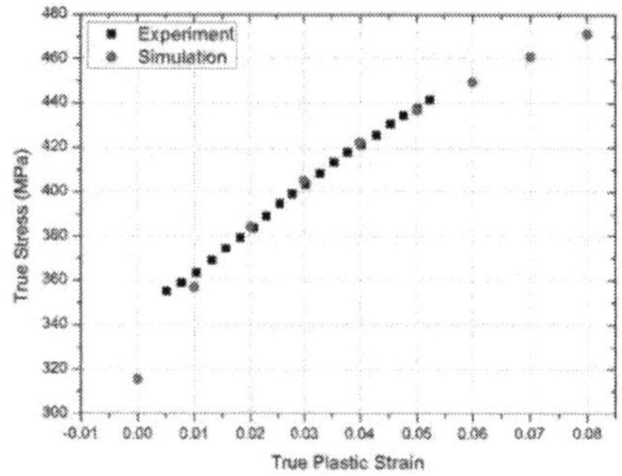

Figure 5. Comparison between the simulation and experiment

3.2. *Blanking process*

The simulation of blanking process is conducted using the grain structure model and the conventional model. The simulation and experiment results are shown in Fig.6, it is found that the fracture mechanism is transgranular fracture due to the coarse grain. The blanked edge is flat and without tensile fracture. The grain structure model can clearly predict the real blanked edge and the conventional model is not suit for the situation.

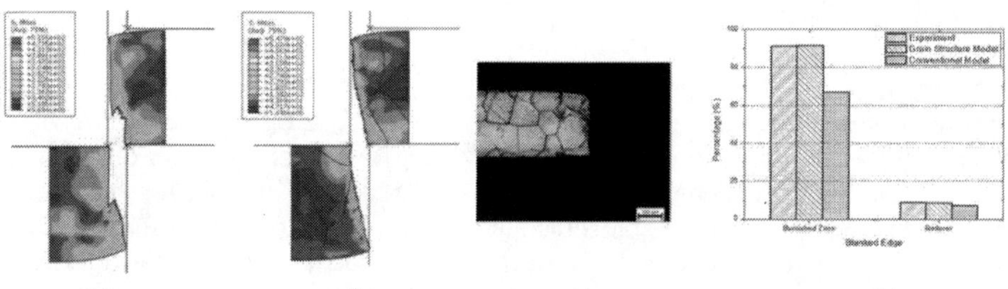

(a) (b) (c) (d)

Figure 6. Simulation and experiment results: (a) Conventional Model (b) Grain Structure Model; (c)Experiment result; (d)Distribution of blanked edge

4. Conclusion

In this paper, the FE simulation and experiments are conducted to investigate the blanking process. It can be concluded that when the scale of grain size is close to the punch-die clearance, the conventional FE model is no longer suit for blanking prediction. A grain structure model based on Voronoi polygon can be adopted in FE simulation. The good match with experiment results demonstrates the feasibility of the novel model.

5. Acknowledgements

The authors gratefully acknowledge the support provided by the National Natural Science Foundation of China (Grant No. U1360203). The authors are grateful to Baosteel Company for providing experimental materials. Thanks to other colleagues for their help in this study too.

References

[1] L Peng, X Lai, H.J Lee, J.H Song, J Ni 2009 *Mater. Sci. Eng. A.* **526** pp 93–99
[2] J.Q. Ran, M.W. Fu, W.L. Chan, 2013. *Int. J. Plast.* **41**, pp 65–81.
[3] Cao J., Zhuang W., Wang S., Lin J., 2010 *Int. J. Adv. Manuf. Technol.* **47** pp 981–991.
[4] Z Gao, L Peng, P Yi, X Lai, 2015 *J. Mater. Process. Technol.* **219** pp 28–41
[5] J Xu , B Guo, C Wang, D Shan, 2012 *Int. J. Mach. Tools Manuf.* **60** pp 27-34.

A NUMERICAL ANALYSIS ON THE LOCAL DEFORMATION OF A SPACER GRID STRUCTURE FOR NUCLEAR FUEL CELLS

Myung-Geun Jang[1], Geum Ju Na[1], Hyunho Shin[2] and Jong-Bong Kim[1]

[1] Dept. Automotive. Eng. SeoulTech. Nowon-gu, Seoul, 01811, Korea
[2] Dept. Mat. Eng., Gangneung-Wonju Nat. Univ., Gangwon-do, 25457, Korea
Corresponding author: Jong-Bong Kim, jbkim@seoulutech.ac.kr

Abstract. The result of a preliminary numerical investigation on local deformation characteristics of a multi-layered spacer-grid structure with five guide tubes is reported based on implicit finite element analysis. For the numerical analysis, displacements of top and bottom cross sections of each guide tube in a single-layer model were constrained while a lateral displacement was imposed on the single layer. Unlike the impact hammer test that is generally employed to characterize the deformation characteristics of the space-grid structure, the buckling phenomenon occurs locally in this study; it takes place at the inner grids around each tube and the degree of bucking is more apparent for tubes near the lateral surface where the lateral displacement was imposed.

1. Introduction

Multi-layered spacer grid structures (Fig. 1) are generally employed for a nuclear fuel cell assembly. In order to investigate the deformation characteristics of a spacer-grid assembly, a hammer impact test [1] has been carried out extensively. This test simulates an accident in which all of the fuel cell assembly is pushed in one direction. After this test, almost all of investigated space-grid structures in the literature [1-3] showed the buckling phenomenon across the whole grid structure. However, studies on the deformation characteristics when only a single layer was subjected to deformation are not easily found. In this regard, this study considers the case when only a single layer received a lateral load and reports the deformation characteristics of a spacer-grid assembly with five guide tubes (see Fig. 2).

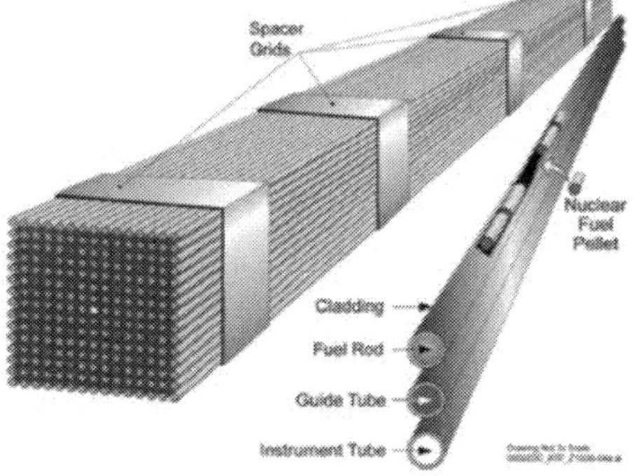

Fig. 1 An example of multi-layered spacer grid [4].

2. Results and Discussions

Fig.2 shows a single layer of grid structure with guide tubes in a multi-layered structure. The welding region and spring and dimple shapes are also shown in Fig. 2 (b) and (c), respectively. Such a layer is repeated to form more than 10 layers. For computational efficiency, we analyzed one layer of the structure via constraining the displacements of the nodes on top and bottom cross sections of each guide tube. External displacement (marked as S in Fig. 2(a)) was imposed on a lateral surface.

The application of a lateral force (artificial imperfection) on local grid surfaces surrounding each tube was informative for achieving convergence in numerical analysis (not shown). Zircaloy-4 and stainless steel (SUS 304L [5]) are used for the spacer grid and guide tubes, respectively. The mechanical properties for the materials are shown in Table 1.

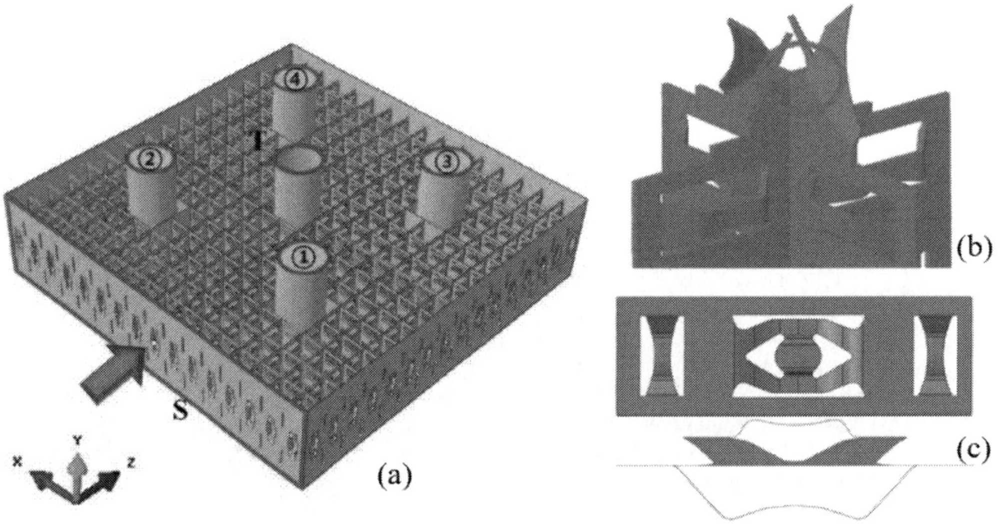

Fig. 2 (a) Assembled shape of one layer of spacer grids and guide tubes, (b) detailed view of a welding region, and (c) spring and dimple shapes.

Table 2. Mechanical properties of Zirconium and stainless steel (SUS 302).

	Zircaloy-4	stainless steel (SUS 304L)
Young's modulus	113.7 GPa	200 GPa
Poisson's ratio	0.29	0.3
Stress-strain relation	$\sigma = 157{+}543\varepsilon^{0.163}$	$\sigma = 310{+}1000\varepsilon^{0.65}$

Fig. 3 shows the deformed shape of the single layer and the von-Mises stress distribution. As can be seen in this figure, the buckling phenomenon occurs locally in this study; it takes place at the inner grid surfaces around each tube, which phenomenon is different from deformed shapes of many spacer-grid structures after the hammer impact test. Although this study does not compare the different tube regions, the degree of bucking was more apparent for the tube region near the lateral surface where the lateral displacement was imposed.

Fig. 4 shows the reaction force on each guide tubes. The center guide tube contact to the spacer grid firstly and support most of the force at the early stage of deformation. The other four guide tubes contact to the spacer grid almost at the same time. The maximum reaction forces are almost the same for the center and front tubes (number 1 and 2). However, the reaction force on the rear tubes (number 3 and 4) are smaller than the reaction forces on the other tubes.

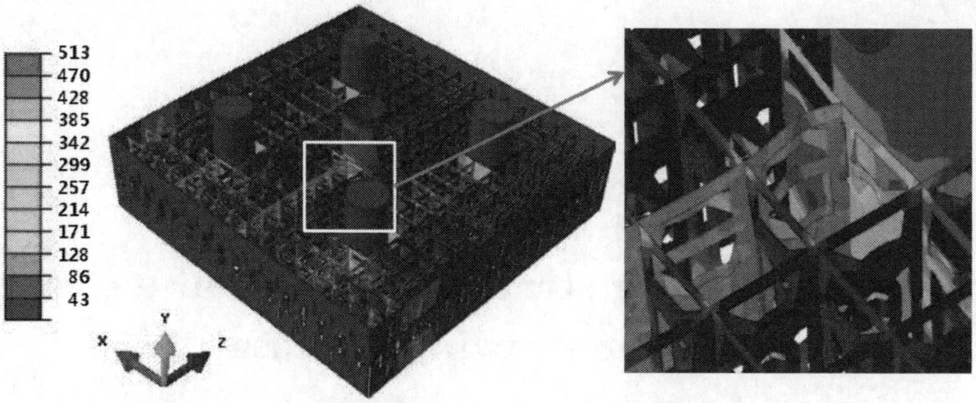

Fig. 3 Deformed shape of the spacer grid and von-Mises stress distribution.

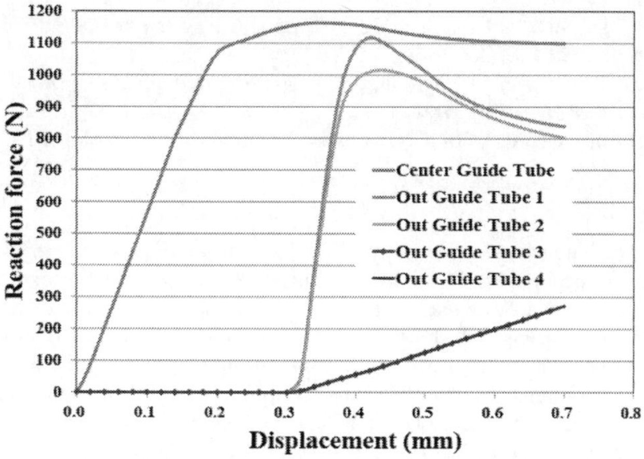

Fig. 4 Reaction force on each guide tubes.

3. Conclusion

Deformation of a multi-layered spacer grid structure by a lateral load on a single layer was analyzed numerically via an appropriate boundary condition. Unlike the hammer impact test, the buckling phenomenon occurred locally (at the inner grids around each tube). The reaction forces on each guide tubes are investigated. It was shown that the reaction forces on the center and front tubes are higher than the reaction forces on the rear tubes.

References

[1] Song K-N, Lee S-H and Lee S-B 2008 *Int. J. Modern Physics B* vol 22 p 1228
[2] Park K-J, Kang B-S, Song K-N, and Park G-J 2003 *J. Nucl. Sci. Technol.* vol 40 p 989–997.
[3] Jung M-J, Song H-G and Park K-B 1992 *J. Kor. Nucl. Soc.* vol 24 p 111–120.
[4] https://sanonofresafety.org/nuclear-waste
[5] Zhang X, Lu J and Shi S 2011, *Mech. of Advanced Mat. Struct.* Vol 18 p 572-577.

Acknowledgments

This work was supported by the Radioactive Waste Management Technology Program of the Korea Institute of Energy Technology Evaluation and Planning (KETEP), granted financial resource from the Ministry of Science, ICT and Future Planning, Republic of Korea. (No. NRF- 2016M2A8A5904254).

A staggered coupling strategy for the finite element analysis of warm deep drawing process

J M P Martins[1], P M Cunha[1], D M Neto[1], J L Alves[2], M C Oliveira[1], H Laurent[3] and L F Menezes[1]

[1] CEMUC, Department of Mechanical Engineering, University of Coimbra, Polo II, Rua Luís Reis Santos, 3030-788 Coimbra, Portugal

[2] CMEMS, Department of Mechanical Engineering, University of Minho, Campus de Azurém, 4800-058 Guimarães, Portugal

[3] Univ. Bretagne Sud, FRE CNRS 3744, IRDL, F-56100 Lorient, France

E-mail: joao.pmartins@dem.uc.pt

Abstract. The thermomechanical finite element analysis of warm forming processes enables an improved comprehension of the process parameters affecting the material formability. However, the thermal and mechanical coupling problem is still a challenge from the computational standpoint. A staggered strategy for the thermomechanical coupling problem is presented in this study, which is based on an isothermal split approach and allows the treatment of the two problems separately. The exchange of information between the mechanical and the thermal problem is performed to achieve a compromise between computational cost and accuracy. The proposed algorithm was implemented in DD3IMP in-house finite element code. Its performance is analysed and compared with a classical strategy commonly employed for solving thermomechanical problems.

1. Introduction

Warm forming is one of the most promising technologies and it has been studied for many years, with increasing attention of the scientific community over recent years. This is a consequence of its advantages, namely the increase of the ductility and decrease of the springback effect in the sheet metal forming of aluminium alloys [1].

The finite element analysis has been playing an important role in the understanding of the deformation mechanism and a great effort has been done to improve the confidence on its results, particularly in the material modelling. Another important aspect, for a successful analysis, is the coupling strategy chosen to handle the dependence of the mechanical behaviour on the thermal behaviour and *vice-versa*. The resultant coupled thermomechanical problem can be approached by monolithic or by staggered algorithms. The monolithic approach treats both problems (thermal and mechanical) in a single system of equations. Despite its unconditional stability, it leads to a large and non-symmetric system of equations and requires a high computational cost for the resolution phase [2]. The staggered approach was designed to reduce the computational costs via partitioning the thermomechanical problem into two sub-problems, thermal and mechanical. Thus, this strategy yields two smaller systems of equations. Within the staggered approach, two methodologies can be applied, the isothermal split and the adiabatic split [3]. The last one has been proposed to circumvent the conditional stability of the isothermal split. However, in case of metal forming processes, this unstable performance is not expected, since it only occurs for values of the thermal expansion coefficient which are non-physical for metalic materials [4].

In this paper, a new coupling strategy is proposed based on the isothermal split. The mechanical problem is solved at constant temperature and the thermal problem is solved for a fixed configuration. This algorithm was designed with the objective of enhancing the computational performance of the classical algorithms based on isothermal split, without compromising the accuracy.

2. Thermomechanical coupling

A brief description of a classical algorithm based on isothermal split and the proposed algorithm is presented. Both were implemented in the in-house finite element code DD3IMP. Concerning the formulation and solution strategy used for each problem (mechanical and thermal), a detailed description can be founded in [5, 6].

2.1. Implicit algorithm

The classical algorithm presented here, is the so-called iterative or implicit one [7, 8], which was designed to attempt an higher accuracy [7]. It consists of an iterative procedure for the mechanical and thermal problems, in each time increment. Thus, the information between problems is interchanged in each iteration until attaining the convergence criterion. This algorithm may be summarized as follows: for each increment, it starts with the resolution of the mechanical problem for a fixed temperature field from the last increment. After that, the thermal problem is solved and subsequently the temperature-dependent material properties are updated. This procedure is repeated until attaining the convergence criterion.

2.2. Proposed algorithm

The proposed algorithm can be divided into two phases for each time increment. The first phase is a predictor one, where an explicit trial solution for the temperature and displacement fields is determined. The second is a corrector phase, where the trial solution is improved by an implicit method. Therefore, first the thermal problem is solved based on an explicit approach and assuming the boundary conditions of the last increment. After that, the temperature-dependent material properties are updated with the trial temperature field, and the mechanical problem is solved, based on an explicit approach. Then, the corrector phase is performed, where the mechanical problem is solved first using the previously obtained displacement field, which corresponds to the initial solution for the Newton-Raphson method. This is corrected until the equilibrium state is attained. Finally, the thermal problem is solved using an implicit approach, for the current mechanical equilibrium configuration.

3. Warm deep drawing example

In order to compare the two algorithms previously described, the warm deep drawing of circular AA5754-O aluminium alloy blank was chosen, which is inspired by the work of Laurent et al. [9]. In this work, the deep drawing tools were assumed as rigid and at constant temperature. Regarding the mechanical behaviour of the aluminium alloy, the Hockett-Sherby hardening law and the Hill'48 anisotropic yield criterion were adopted. However, the strain rate dependency was neglected. The mechanical and thermal material parameters, as well as the friction coefficient, are the ones described in [9].

One fourth of the model was simulated, with a total of 9068 tri-linear hexahedron finite elements (2 layers through the thickness) for the blank. Selective reduced integration was adopted for the mechanical problem and full integration for the thermal one. The initial temperature of the blank, the blank holder and the die was 200°C, whereas the temperature of the punch was 145°C. These thermal conditions imposed a non-homogeneous temperature distribution on the blank, which affects the mechanical behaviour of the material.

3.1. Results and discussion

The punch force/displacement curves are presented in Figure 1 a), for both algorithms. The punch force attains a maximum value for 11.6 kN, for a punch displacement of 10.6 mm. The two algorithms attain the same value and an identical evolution for the curve. The plastic strain distribution is also presented in Figure 1 b) and c) and once again identical results were attained for the two algorithms.

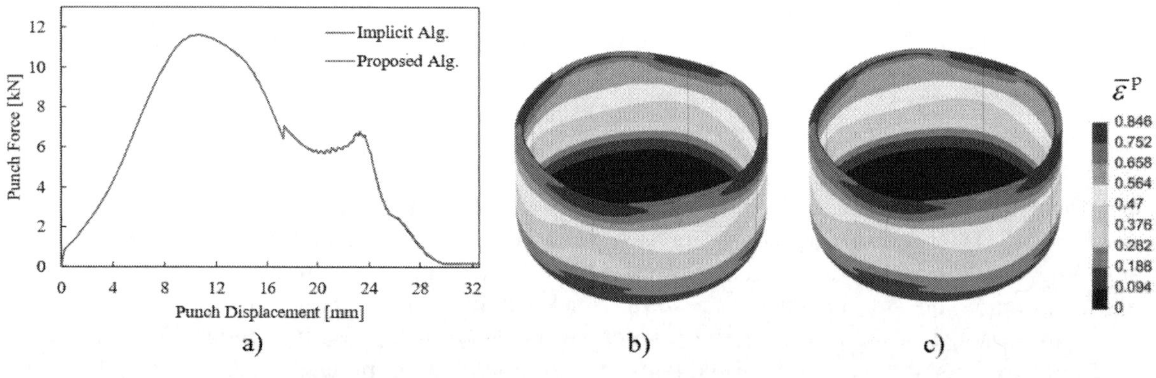

Figure 1. Comparison between the proposed and the implicit algorithms: a) punch force evolution, equivalent plastic strain distribution for the b) proposed and c) implicit algorithm. Note that the evolution of the punch force curves is almost coincident for the two algorithms.

Figure 2 a) shows the temperature evolution for two different nodes: the first is located under the blank holder in the beginning of the process, while the other is located under the punch.

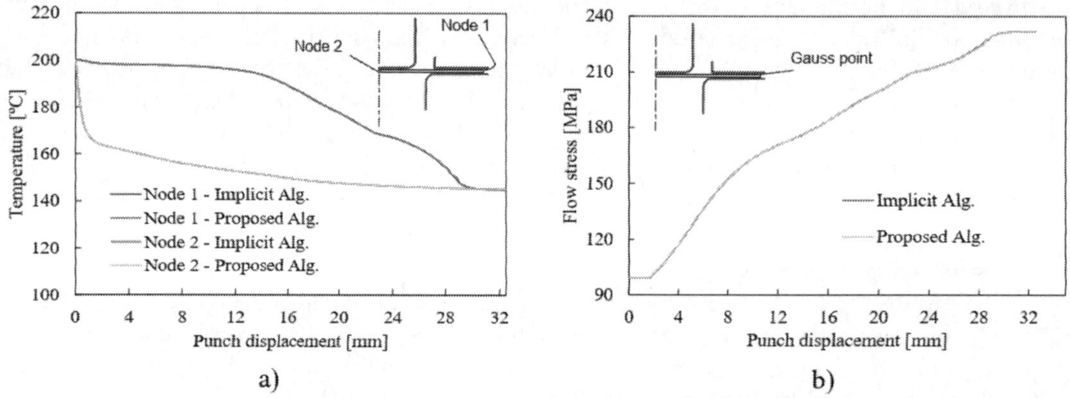

Figure 2. Comparison between the proposed and the implicit algorithms: a) temperature evolution for two nodes and b) flow stress evolution for a gauss point. Note that in both cases, the evolution of the curves is almost coincident for the two algorithms.

The second node attains lower values of temperature quicker than the first one, due to the contact with the punch, although both nodes achieved the punch temperature at the end of the process. Figure 2 b) presents the flow stress evolution for a gauss point located under the blank holder in the beginning of the process. The temperature and the flow stress evolution are equal for both algorithms, revealing a good accuracy of the proposed algorithm.

The numerical simulations were carried out on a computer machine equipped with an Intel Core i74700HQ Quad-Core processor (2.6 GHz). The computational performance of both algorithms is presented in Table 1. The simulation required 536 and 524 increments for the implicit and the proposed algorithms, respectively. Although both algorithms present identical results in terms of accuracy, the computational cost was 32% lower for the proposed algorithm.

Table 1. Computational performance of both algorithms.

	Implicit algorithm	Proposed algorithm
N° increments	536	524
Computational Time [s]	8538	5731

4. Conclusions

In this article, a new staggered algorithm for thermomechanical coupling, based on an isothermal split methodology was presented. This algorithm was designed to achieve a better compromise between computational cost and accuracy. For the sake of comparison and to demonstrate the main advantage of the proposed algorithm, a classical algorithm based on isothermal split was also presented. This classical algorithm, in contrary to the proposed one, is only focused on the accuracy of the results. Both algorithms have been tested on a warm deep drawing example, for which the temperature distribution have a strong effect on the material behaviour. The results show good accuracy of the proposed algorithm and a significant decrease of the computational time, when compared with the classical algorithm.

Acknowledgments

The authors gratefully acknowledge the financial support of the Portuguese Foundation for Science and Technology (FCT) under projects with reference UID/EMS/00285/2013, PTDC/EMS-TEC/0702/2014 (POCI-01-0145-FEDER-016779) and PTDC/EMS-TEC/6400/2014 (POCI-01-0145-FEDER-016876) by UE/FEDER through the program COMPETE2020. The second author is also grateful to the FCT for the Postdoctoral grant SFRH/BPD/101334/2014.

References

[1] Ghosh M, Miroux A, Werkhoven R J, Bolt P J and Kestens L A I 2014 Warm deep-drawing and post drawing analysis of two Al-Mg-Si alloys *J. Mater. Process. Technol.* **214** 756-766

[2] Adam L, Ponthot J P 2002 Numerical simulation of viscoplastic and frictional heating during finite deformation of metal. Part I: Theory *J. Eng. Mech.* **128** 1215-1221

[3] Armero F and Simo J C 1992 A new unconditionally stable fractional step method for non-linear coupled thermomechanical problems *Int. J. Numer. Meth. Eng.* **35** 737766

[4] Armero F and Simo J C 1993 A priori stability estimates and unconditionally stable product formula algorithms for nonlinear coupled thermoplasticity *Int. J. Plasticity* **9** 749-782

[5] Oliveira M C , Alves J L and Menezes L F 2008 Algorithms and strategies for treatment of large deformation frictional contact in the numerical simulation of deep drawing process *Arch. Comput. Meth. Eng.* **15** 113162

[6] Martins J M P, Alves J L, Neto D M, Oliveira M C and Menezes L F 2016 Numerical analysis of different heating systems for warm sheet metal forming *Int. J. Adv. Manuf. Technol.* **83** 897909

[7] Vaz Jr M, Muoz-Rojas P A and Lange M R 2011 Damage evolution and thermal coupled effects in inelastic solids *Int. J. Mech Sci.* **53** 387-398

[8] Erbts P and Dster A 2012 Accelerated staggered coupling schemes for problems of thermoelasticity at finite strains *Comput. Math. Appl.* **64** 2408-2430

[9] Laurent H, Coër J, Manach P Y, Oliveira M C and Menezes L F 2015 Experimental and numerical studies on the warm deep drawing of an AlMg alloy *Int. J. Mech Sci.* **93** 59-72

Journal of Physics: Conference Series **734** (2016) 032034 doi:10.1088/1742-6596/734/3/032034

Roll levelling semi-analytical model for process optimization

E. Silvestre[1], D. Garcia,[2]L. Galdos[3], E. Saenz de Argandoña[3], J. Mendiguren[3*]

[1] KONIKER, Forming and Assembly technological center, San Andres 20, 20500, Mondragon, Gipuzkoa, Spain

[2] FAGOR ARRASATE S. Coop., Apdo. 18, San Andrés 20, 20500, Mondragón, Gipuzkoa, Spain

[3] Mechanical and Manufacturing Department, Mondragon Unibertsitatea, Loramendi 4, 20500, Mondragon, Gipuzkoa, Spain

*jmendiguren@mondragon.edu

Abstract. Roll levelling is a primary manufacturing process used to remove residual stresses and imperfections of metal strips in order to make them suitable for subsequent forming operations. In the last years the importance of this process has been evidenced with the apparition of Ultra High Strength Steels with strength > 900 MPa. The optimal setting of the machine as well as a robust machine design has become critical for the correct processing of these materials. Finite Element Method (FEM) analysis is the widely used technique for both aspects. However, in this case, the FEM simulation times are above the admissible ones in both machine development and process optimization. In the present work, a semi-analytical model based on a discrete bending theory is presented. This model is able to calculate the critical levelling parameters i.e. force, plastification rate, residual stresses in a few seconds. First the semi-analytical model is presented. Next, some experimental industrial cases are analyzed by both the semi-analytical model and the conventional FEM model. Finally, results and computation times of both methods are compared.

1. Introduction

The tightening of the CO2 emissions policies, together with the increase of safety requirement, is challenging the automotive sector to be competitive in car weight and price. In order to the challenge, new high strength materials are been incorporated to the market such as third generation steels i.e. Fortiform1050, Nippon1180, or the high strength aluminium alloys. However, these new materials lead to important manufacturing issues such as early fractures, high forming forces and high springback. One of the key processes to assure stability of the manufacturing line is the roll levelling process [1]. After the rolling process, the sheet usually has shape defects [2] and internal residual stresses that could lead to manufacturing quality issues down the line. In order to get rid of those defects and aiming at homogenising the through thickness stress profile the sheet is subjected to an iterative bending-unbending process.

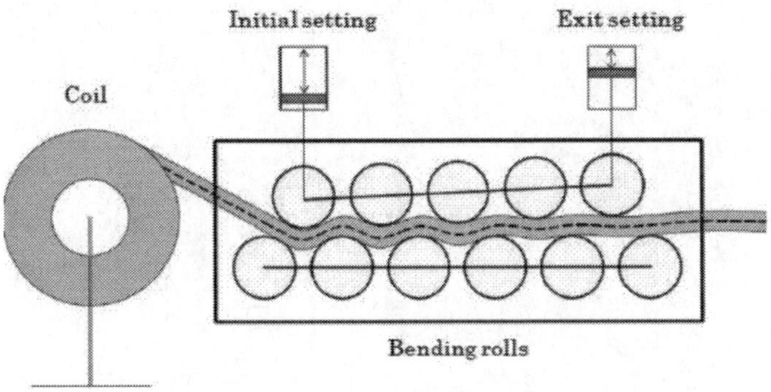

Figure 1. Roll levelling process, where the sheet is subjected to an iterative bending-unbending process.

Multiple works have been conducted on the analysis of the levelling process and on the correct set up of the rolls for an optimal process development. In this regard, previous researchers developed numerical finite element method (FEM) models to predict the evolution of the material properties during the levelling process [1-7]. The main advantage of this models is the capability of prediction prior process that allows the correct set-up of the machine or even close loop control systems.

The main disadvantage of the FEM models is the high computational time required that exceeds the admissible ones in both machine development and process optimization. In the present work, a semi-analytical model based on a discrete bending theory is presented. This model is able to calculate the critical levelling parameters i.e. force, plastification rate, residual stresses in a few seconds. First the semi-analytical model is presented. Then, the performance of model compared to the FEM one is evaluated in terms of accuracy in torque and computational time.

2. Roll levelling semi-analytical model

The developed semi-analytical model is based on a model proposed by Mendiguren et al. [8] in the framework of bending and springback modelling. As in Figure 2 it can be shown, first the thickness of the sheet is discretizated in a finite number of points (N = 2000 in this case). Then, once that the curvature that will take the sheet is known, a perfect bending under plane strain conditions is assumed and an elasto-plastic formulation [9] is applied to each discrete point. It has to be noted the necessity of anisotropic hardening models to be able to correctly represent the behaviour of the material during the bending-unbending process [10]. Finally if the section value is needed the integral through the thickness is numerically performed.

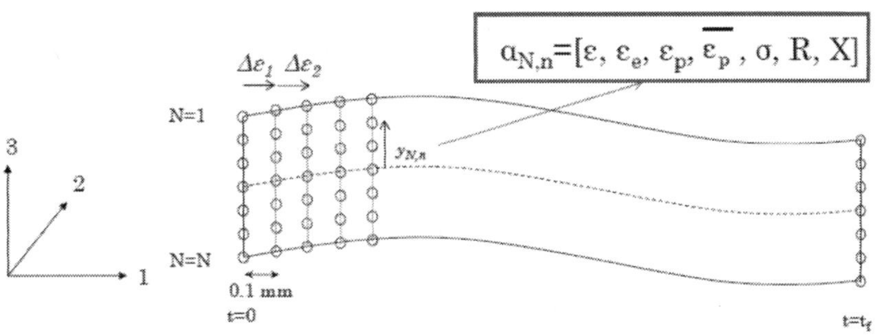

Figure 2. Semi analytical model definition inspired in [2].

Previous works have stated that that the contact position between the rolls and the sheet are not at the Summit of the rolls, but there is a contact angle regarding the summit of the roll, as shown in Figure 3. Previous research has shown that this contact angle is strength material and thickness dependent [2, 11].

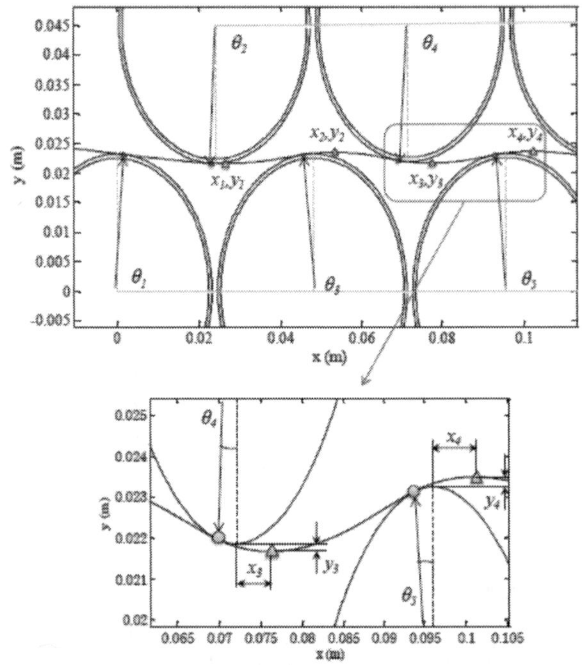

Figure 3. Offset of the maximum and minimum points of the sheet.

A neuronal network has been developed to predict the bending line based on the thickness and material strength. This neuronal network has been trained with more than 300 FEM simulations in a wide range of thicknesses and material strengths.

The process of the semi-analytical model works as follow:

- First, using the neuronal network, based on the material strength and the thickness the bending line is obtained
- Next, the sheet is discretizated in sections (one each 0.1 mm) and the sections are discretizated in 2000 points
- Then, knowing the strain increment of each point, from section to section, a plane strain elastic predictor-plastic correct algorithm is conducted
- Finally, the desired values of residual stress profile, levelling forces, torques or plastification rate are calculated

Figure 4 shows the through thickness stress gradient evolution calculated using the semi-analytical model from de coil to the exit of the leveller. During this process, it is taken into account the coil radius reduction, the springback after uncoiling and the springback when the sheet leaves the leveller.

3. Model results

In table 1 the average error on the calculated total torque between the FEM model [1, 7] and the semi-analytical model are shown. In the same way, the calculated times are shown as a reference.

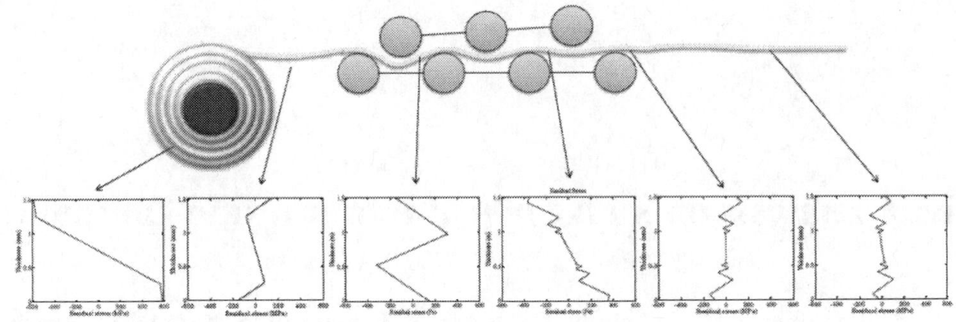

Figure 4. Uncoiling process simulation where the residual stresses profile can be monitored.

Table 1. Performance of the numerical models.

Model	Average error of total torque	Computational time
FEM	29.2 %	5 h
Semi-analytical	16.7 %	18 s

4. Conclusions

In this work a semi-analytical model of levelling has been developed. This model is able to predict the basic levelling variables in less than 30 s with a similar accuracy compared to a 3h FEM model. The correct performance of this model will allow the close loop control process implementation on future levelling machines.

Acknowledgments

The authors gratefully acknowledge the financial support of the Basque Government received within the research project "Leungor" and “Thirdform”. Furthermore, the authors would like to thank the Fagor Arrasate company for their technical support and the scientific exchange.

References

[1] J. Morris, S. Hardy, A. Lees and J. Thomas 2001 Ironmaking & Steelmaking Vol 28, N 1, 44-52
[2] E. Silvestre 2015 Sheet metal roll levelling optimization by means of advanced numerical models and development of new concepts for last generation materials. Mondragon Unibertsitatea.
[3] K. Park and S. Hwang. 2002. On-line flatness measurement of large steel plates using Moiré topography. Technical research centre of Fineland.
[4] L. Steinwender, A. Kainz, K. Krimpelstatter and K. Zeman.2010 Computational analysis of the curvature distribution and power losses of metal strip in tension leveller. IOP Publishing.
[5] L. Madej 2011. *CIRP Annals-Manufacturing Technology* V 60 N 1 291-294
[6] U. Muller, H. Krambeer and A. Wolff. 2013 Optimization of final plate flatness by set-up coordination for subsequent manufacturing process (Final plaate flatness). European Commission.
[7] M. Gruber, M. Oligschlager and G. Hirt. 2014. *Advanced materials research* V 1018. 207-214
[8] J. Mendiguren, J.J. Trujillo, F. Cortes and L. Galdos. 2013. *International journal of mechanical sciences* V 77. 57-64
[9] J. Hu, Z. Marciniak and J. Duncan. 202. Mechanics of Sheet Metal Forming. Second ed. Engla: Butterworth-Heinemann.
[10] E. Silvestre, J. Mendiguren, L. Galdos and E. Saenz de Argandoña. 2015. *International journal of mechanical sciences* V 101-102. 10-20
[11] J. Petruska, T. Navrat and F. Sebek. 2012. *Advanced materials research* V 586. 289-393

Sensitivity analysis on an AC600 aluminum skin component

J. Mendiguren[1*], J. Agirre[1], E. Mugarra[1], L. Galdos[1], E. Saenz de Argandoña[1]

[1] Mechanical and Manufacturing Department, Mondragon Unibertsitatea, Loramendi 4, 20500, Mondragon, Gipuzkoa, Spain

*jmendiguren@mondragon.edu

Abstract. New materials are been introduced on the car body in order to reduce weight and fulfil the international CO2 emission regulations. Among them, the application of aluminum alloys is increasing for skin panels. Even if these alloys are beneficial for the car design, the manufacturing of these components become more complex. In this regard, numerical simulations have become a necessary tool for die designers. There are multiple factors affecting the accuracy of these simulations e.g. hardening, anisotropy, lubrication, elastic behavior. Numerous studies have been conducted in the last years on high strength steels component stamping and on developing new anisotropic models for aluminum cup drawings. However, the impact of the correct modelling on the latest aluminums for the manufacturing of skin panels has been not yet analyzed. In this work, first, the new AC600 aluminum alloy of JLR-Novelis is characterized for anisotropy, kinematic hardening, friction coefficient, elastic behavior. Next, a sensitivity analysis is conducted on the simulation of a U channel (with drawbeads). Then, the numerical an experimental results are correlated in terms of springback and failure. Finally, some conclusions are drawn.

1. Introduction

The weight reduction trend is driving the automotive industry design and development. Clear example of that trend are the nowadays widely used in BIW press hardening and HSS steels aimed at reducing the weight of the primary structure while maintaining if not increasing the safety standards [1]. Traditionally, the outer skins have been designed in soft thin mild steels [2]. However, the trend in recent times, mainly pushed by some OEM, is to introduce aluminium alloys in those components in order to reduce weight [3]. Although the aluminium is an ideal material from a design point of view, it poses some manufacturing issues at die design level. High anisotropy and high springback, associated to its low elastic modulus, together with a limited formability leads to extensive work on the optimization of the stamping tooling [4].

In this work a sensitivity analysis has been conducted on the simulation of an AC600 aluminium roof panel. The basic idea is to be able to asses which are the critical material parameters to be taken into account in order to accurately predict the forming of the component. First, the material characterization is shown. Next, simulations with different material models are conducted. Finally, the springback and formability results are analysed and come conclusions are drawn.

2. Material and characterization

Material behaviour for stamping simulations is mainly divided in three aspects: elastic behaviour, plastic yielding and hardening. The elastic behaviour and the yielding can be analysed by combining standard tensile test and loading-unloading tests, Fig. 1 [5]. The hardening on the other hand can be evaluated using different experimental methodologies such as tension-compression or shear test methods [6]. Another critical input for stamping simulations is the friction coefficient definition. In this case a Strip Drawing tests have been conducted to characterize the contact bahavior between the sheet and the die material.

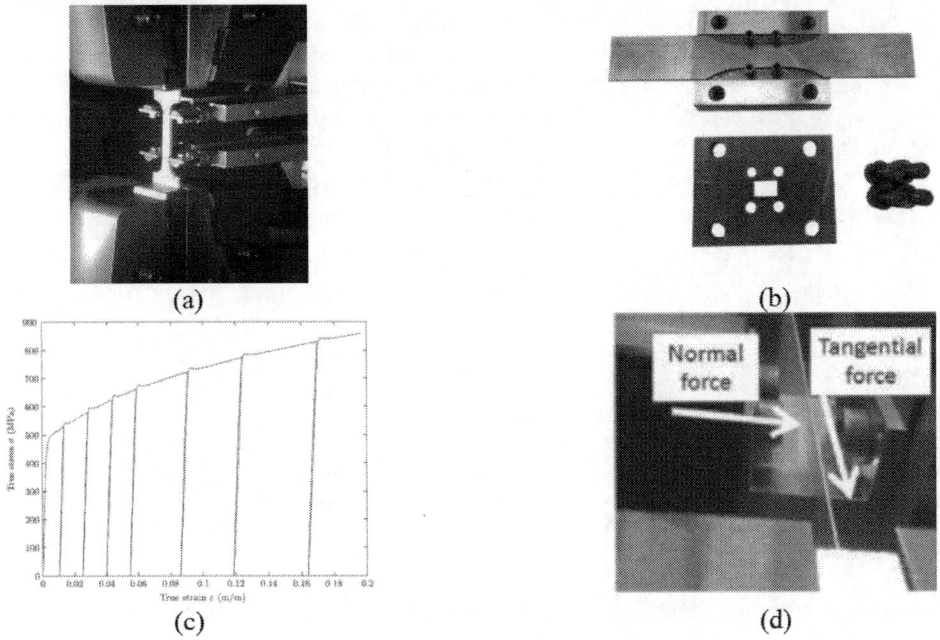

Figure 1. Material advanced characterization: a) r values measurement with DIC, b) tension compression tests, c) loading-unloading test and d) Strip drawing test.

Table 1 shows the anisotropy values and the yielding stresses (Rp02) of the analysed AC600 material. It should be noted that the elastic modulus decrease and the anisotropic hardening have been taken from the software database.

Table 1. AC600 material anisotropy values.

Sample	RD	45D	TD
Yield stress Rp02	152 MPa	144 MPa	150 MPa
r value	0.636	0.344	0.802

The Strip Drawing test shows an average coefficient of 0.22 in contrast to the 0.14 proposed by the aluminium supplier.

3. Roof panel simulation

Figure 2 shows the roof panel simulation conducted on this work in a single action press and with blankholder and drawbeads.

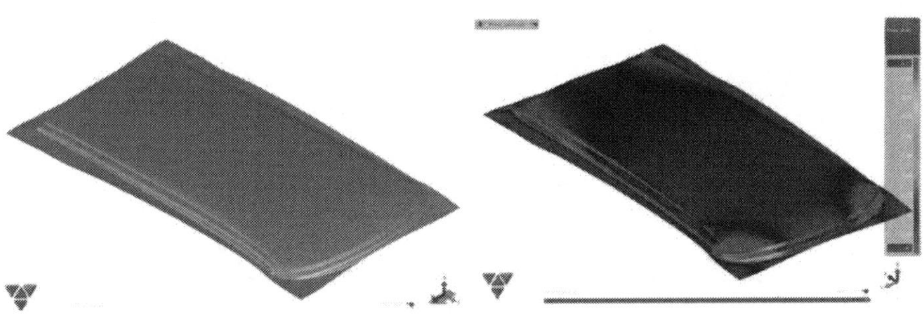

Figure 2. Uncoiling process simulation where the residual stresses profile can be monitored.

Different simulations have been conducted in this study. First, isotropic hardening (IH) and mixed isotropic-kinematic hardening (IKH) have been compared. Next, the yielding definition has been analysed from a Barlat 1989 model supposing 0.6 of anisotropy coefficients (Barlat*) to a Barlat 1989 using the real coefficients of Table 1 to a BBC model where all yielding stresses and anisotropy coefficients are used on the definition. Then, the elastic modulus (constant or variable) has been analysed and finally the influence of the correct definition of the friction coefficient.

Table 2. Sensitivity analysis model definition

Model	Hardening		Yielding			Elastic modulus		Friction	
	IH	IKH	Barlat*	Barlat	BBC	CT	Variable	0.14	0.22
1	X			X		X		X	
2	X		X			X		X	
3	X				X	X		X	
4		X		X		X		X	
5	X			X			X	X	
6	X			X		X			X

4. Results and discussion

Figure 3 shows the transverse springback profile in a XZ plane for the models shown in Table 2.

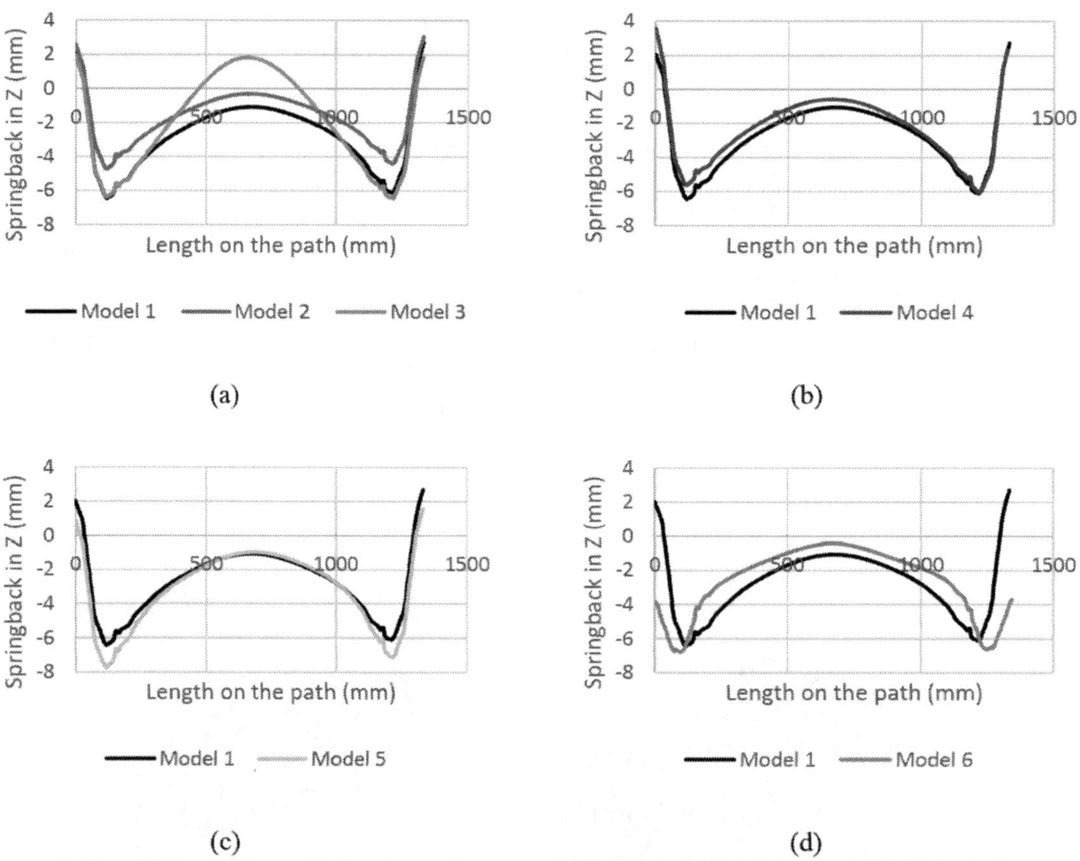

Figure 3. Sensitivity of the springback results to the different material inputs: a) Yield function definition, b) Hardening definition, c) Elastic behaviour definition and d) Friction coefficient

Figure 4 shows the forming limit diagram of the different models.

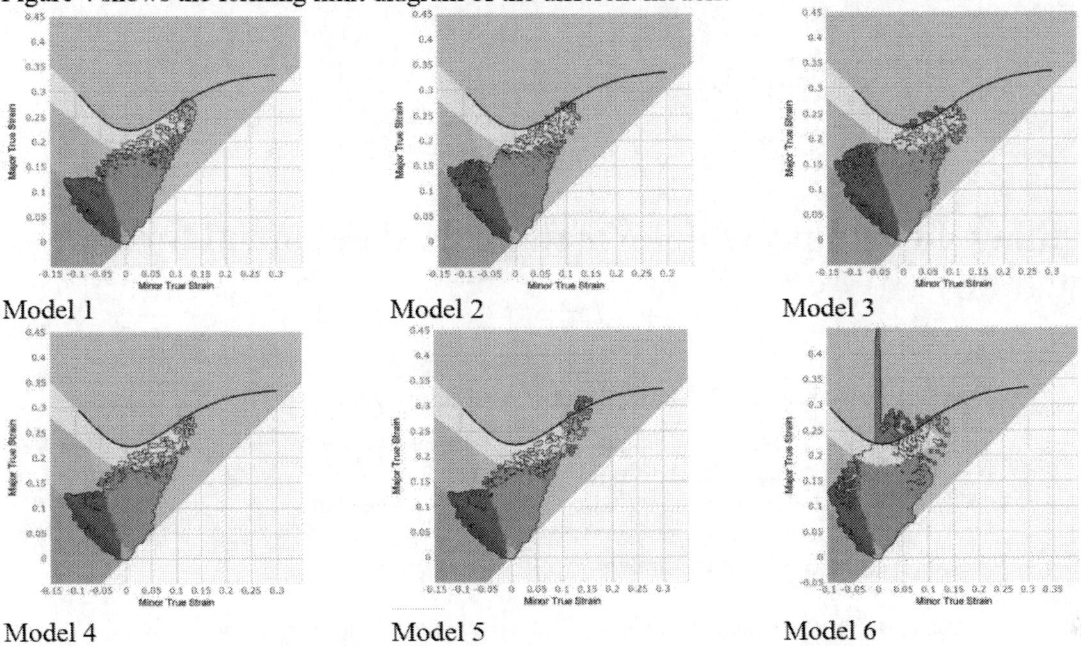

Figure 4. Formability results of the different models.

5. Conclusions

From the present study it can be shown that for the actual component the yield criteria definition and the friction coefficient are the critical parameters in terms of springback and formability respectively. Differences of almost 4 mm have been found between the use of the Barlat 1989 model and the BBC model while differences of 2 mm have been derived from the correct definition of the anisotropy coefficients. In the case of this component, small influence of the elastic behaviour definition and hardening definition has been found. The friction coefficient on the other hand shows a critical impact on formability where increasing the coefficient the sheet is stretched and therefore the formability reduced.

It has to be noted that these are particular results for this specific component and that a different component could show a different trend.

References

[1] S.L. Costa, J.P. Mendonça, N. Peixinho 2016 Material and Design V 91 pp 143-154
[2] B. Zillmann, M.F.X. Wagner, S. Schmaltz, E. Schmidl, T. Lampke, K. Willner, T. Halle 2015 International Journal of Solids and Structures V 66 pp111-120
[3] J. Kang, Y. Chen, D. Sigler, B. Carlson, D.S. Winkinson 2016 Engineering Failure Analysis In press.
[4] S. Suttner, M. Merklein 2016 Journal of Material Processing Technology V 235 pp 121-133
[5] J. Mendiguren, F. Cortés, X. Gómez, L. Galdos 2015 Material Science and Engineering A. V 634, pp 147-152
[6] E. Silvestre, J. Mendiguren, L. Galdos, E. Saenz de Argandoña 2015 International Journal of Mechanical Sciences V 101-102, pp 10-20

Including die and press deformations in sheet metal forming simulations

Johan Pilthammar[1,2,3], Mats Sigvant[1,2] and Sharon Kao-Walter[2]

[1] Volvo Cars, Dept. 81153 Stamping CAE & Die Development, Olofström, Sweden
[2] Blekinge Institute of Technology, Valhallavägen, 371 41 Karlskrona, Sweden
[3] Author to whom any correspondence should be addressed

E-mail address: johan.pilthammar@volvocars.com

Abstract. Structural analysis, in Abaqus, of a stamping die and subsequent morphing of the tool surfaces in AutoForm were performed to improve a sheet metal forming simulation. First, the tool surfaces of the XC90 rear door inner were scanned. They were not matching when the die was unloaded and could therefore not give any satisfying results in sheet metal forming simulations. Scanned surface geometries were then added to a structural FE-model of the complete stamping die and some influential parts of the production press. The structural FE-model was analysed with Abaqus to obtain the structural deformations of the die. The calculated surface shapes were then transferred to AutoForm where a forming simulation was performed. Results from the different sheet metal forming simulations were compared to measured draw in curves and showed a substantial increase in accuracy and ability to analyse dies in running production when the morphed surfaces were used.

1. Introduction

A correct representation of the forming surfaces is of great importance when sheet metal forming simulations are performed for stamping dies in running production. Scanning of the tool surfaces is one way of getting data for the forming simulations. However, it is well known that the tool surfaces deforms during the drawing of the blank. This will influence the process conditions and final part geometry. These deformations are hard to predict in standard forming simulations where the tool surfaces are represented as rigid 2D surfaces. AutoForm, which is used for sheet metal forming simulations in this paper, has some capabilities for modelling of the blankholder deformations through its different binder models.

Previous works have been done regarding elastic tool deformations and its influence on the forming process and simulations. One example and one of the most extensive works [1], by R A Lingbeek describes different methods for analysis and virtual rework of tool structures and surfaces. Another example is [2] where a method for compensation against elastic tool deformations is outlined and demonstrated by simulations. Most of the previous work in the area is performed on small industrial tools or experimental tools. It is noted in several of the works that more work is required, especially simulations and validations for dies on an industrial scale. Many suggested methods combines the structural analysis and the sheet metal in the same FE-model. The focus of this work is instead two separate models where information is transferred between them.

Volvo Cars has also performed previous work on structural analysis and optimization of stamping dies. There seems to be a large potential in this area. One thesis was performed at Volvo Cars in 2007

[3]. The other one is running during the spring 2016 and is about stress analysis and subsequent optimization steps for reducing stress levels and increasing stiffness of a blankholder.

2. Initial simulations

A production run with a die, XC90 rear door inner, was followed under real conditions. The die was initially of interest in a project regarding friction between Volvo Cars and Triboform [4-7]. It was expected and verified that the surfaces in contact with the sheet deviated from the original CAD model. The tool surfaces that deviated the most was the shape of the blankholder. It was also concluded that the surfaces needed to be deformed virtually to be able to run the forming simulation. No settings in AutoForm for the binder models were able to compensate for the difference in surface shapes. The best result obtained with the scanned surfaces can be seen in figure 1.

The die material is GGG70L with a chrome plated die and punch. The blankholder is polished and laser hardened. The blank material is VDA239-CR4 GI with a thickness of 0.7 mm and Fuchs 4107 as prelube (1.0 g/m^2). The simulations are done in AutoForm^{plus} R6.0 with friction models from the TriboForm FEM Plug-In based on measurements of the die and sheet material. The BBC2005 material model is used for all sheet metal forming simulations [8].

Figure 1. (a) XC90 rear door inner. (b) Blank shape compared to draw-in, red curve, after forming simulation with scanned surfaces.

3. Structural analysis and Sheet metal forming simulation

To be able to virtually deform the tool surfaces, a structural FE-model was created in Hypermesh 13.0 and solved with Abaqus 6.14. The entire tool was modelled as elastic bodies and solved with a dynamic quasi-static method. The ram was assumed to be rigid and an elastic model of the cushion was included in the model. The model was created from the original CAD model. The scanned surfaces were imported into Hypermesh in stl-format and the nodes on the blankholder and die surfaces were morphed with the scanned surfaces as target. This enabled solving of the structural model with the scanned tool surfaces.

The purpose of this structural simulation was to obtain the correct surface geometries for the loaded blankholder. The calculated shapes of the surfaces were then used during a new forming simulation. This will be valid under the assumption that the change in tool deformations are relatively small after the blankholder has been closed. The punch was not included in the structural FE-analysis since no forming was done in that simulation. In order to further reduce the model size the steel sheet was replaced with a contact surface between the blankholder and the die. The contact surface was given the same thickness as the sheet, 0.7mm.

The resulting surfaces from the Abaqus simulation have the correct global shape of the loaded die. However, the mesh is way too coarse to be used in the sheet metal forming simulations. This was solved in AutoForm by morphing the already imported scanned surfaces of the blankholder and the die. In AutoForm morphing is done by using guide curves. They describe from where and where to the surfaces should be morphed. Catia V5 was used to create guide curves on the original scan surfaces and on the deformed surfaces from the structural FE-simulation. The curves have exactly the same shape if they are projected onto the xy-plane. The difference can only be seen in the levels and shape in the z-direction.

Position of the guide curves and difference between morphed and unmorphed surfaces can be seen in figure 2.

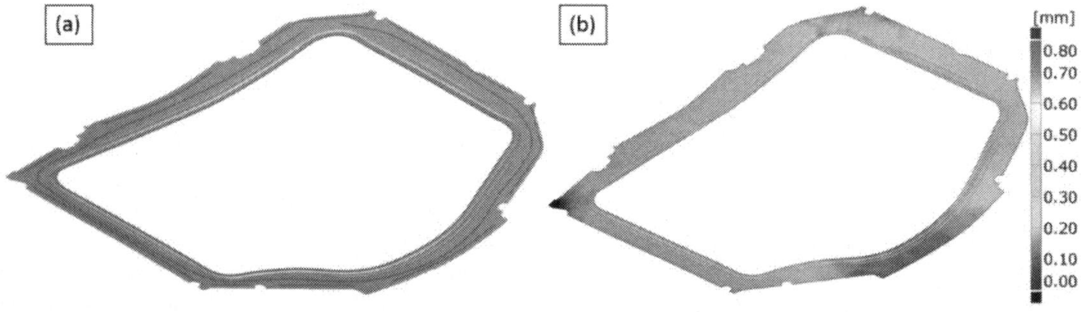

Figure 2. (a) Blankholder surface with position of guide curves for morphing. (b) Difference between scanned and morphed blankholder surface.

For the proposed method to work the blankholder shape must be kept constant during the forming simulation. This is obtained by setting a high value for the tool stiffness in a binder model with support type Force Controlled. Exactly what level that should be used is not certain but this simulation shows good results when the tool stiffness is in the interval 5000 – 30000 MPa/mm. The best value has to be determined by analysing more models.

4. Results

The result from the structural FE-simulation showed that the blankholder deformed up to 0.7 mm in Z direction which can be seen in figure 3. The surfaces of the die and blankholder that are in contact with the sheet was exported as stl-files from Hyperview. Other parts of the die and the cushion in the press also showed deformations but they are much smaller than the deformations of the blankholder.

The sheet metal forming simulations with the morphed surfaces shows a very large improvement compared to the initial simulation with scanned surfaces. The results can be seen in figure 4. The draw-in is now matching the measured curved very well except for minor deviations in some areas. The deformed surfaces together with TriboForm´s advanced friction model are shown to be of great importance if final part properties, e.g. shape, strains, wrinkles et.c. should be predicted with high accuracy. This paper has focused on the importance of tool surface shapes and how to incorporate them into sheet metal forming simulations. The importance of an accurate friction model and the influence on other results such as strains and wrinkles for this die are described and discussed in [5-7].

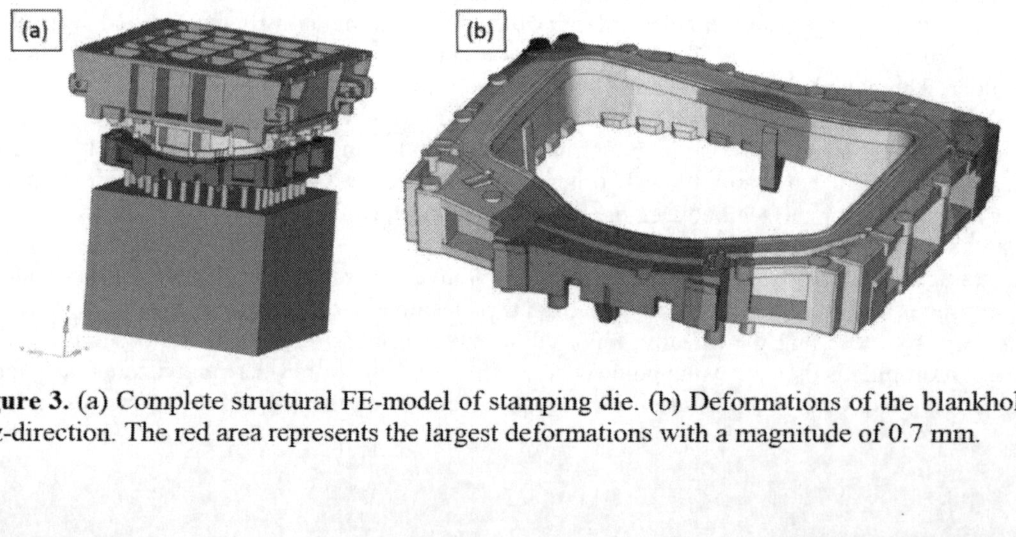

Figure 3. (a) Complete structural FE-model of stamping die. (b) Deformations of the blankholder in z-direction. The red area represents the largest deformations with a magnitude of 0.7 mm.

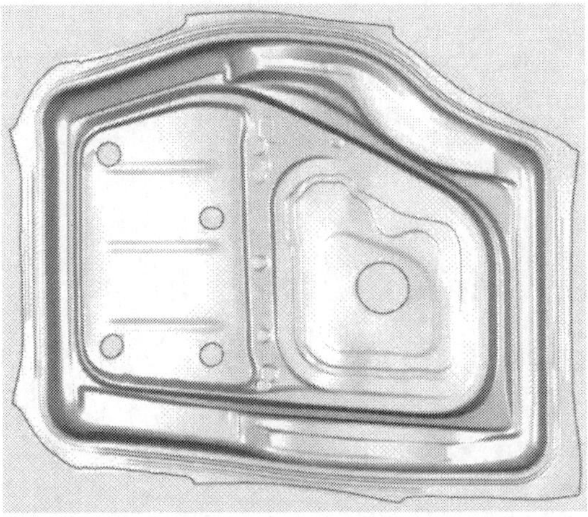

Figure 4. Geometry after final simulation with correct surface geometries and a stiff blankholder model in AutoForm. The red curves represent measured draw-in on a real blank.

5. Discussion and future work

The ability to virtually deform a die and include the resulting geometry in forming simulations is of high importance. One example has been demonstrated in this paper with good results. However, there needs to be further investigations into efficient workflow and how to set parameters in the simulation sofwares to best represent the blankholder conditions. This can be done by simulating more dies, blankholder designs and lubrication systems. For some dies there will probably be deformations during the forming operation. It might be handled by applying an average deformation based on different points in time during the drawing of the blank. Another benefit of the method is that the total simulation time of both models is very short compared to some other methods for incorporating die deformations, e.g. combining both structural and forming simulations in the same model.

The method can be used when support is requested from running production. This link between structural analysis and sheet metal forming will also enable methods for virtual design and rework of stamping dies. Structural optimization should be used for reducing the die deformations together with methods for virtual spotting. That would give an optimum design of the blankholder.

References

[1] Lingbeek R 2008 *Virtual Tool Reworking New Strategies in Die Design using Finite Element Forming Simulations* (Netherlands Enschede: Print Partners Ip-skamp)

[2] Groβmann K, Wiemer H, Hardtmann A, Penter L and Kriechenbauer S 2009 Adjusting the contact surface of forming tools in order to compensate for elastic deformations during the process *Proc. 7th European LS-DYNA Conference*

[3] Nilsson A and Birath F 2007 Topology optimization of a stamping die *Proc. NUMIFORM '07*

[4] Sigvant M, Hol J, Chezan T and van den Boogard T 2015 Friction modeling in sheet metal forming simulations *Proc. FTF 2015 (Zürich, Switzerland)*

[5] Sigvant M, Pilthammar J, Hol J, Wiebenga J H, Chezan T, Carleer B and van den Boogard T 2016 Friction and lubrication modeling in sheet metal forming simulations of a Volvo XC90 inner door *Proc. IDDRG 2016 (Linz, Austria)*

[6] Sigvant M, Pilthammar J, Hol J, Wiebenga J H, Chezan T, Carleer B and van den Boogard T 2016 Friction and lubrication modeling in sheet metal forming simulations of the Volvo XC90 inner door *Proc. Numisheet 2016 (Bristol, UK)*

[7] Sigvant M, Pilthammar J, Hol J, Wiebenga J H, Chezan T, Carleer B and van den Boogard T 2016 Friction in sheet metal forming: Simulations of the Volvo XC90 inner door *Proc. FTF 2016 (Ohlstadt/Garmisch-Partenkirchen, Germany)*

[8] Banabic D, Carleer B, Comsa D S, Kam E, Krasivskyv A, Mattiasson K, Sester M, Sigvant M and Zhang X 2010 *Sheet Metal Forming Processes, Constitutive Modelling and Numerical Simulation* (Berlin Heidelberg: Springer Verlag)

Lagrangian and Arbitrary Lagrangian Eulerian Simulations of Cold Roll Forming Processes

Yanick Crutzen, Romain Boman, Luc Papeleux, Jean-Philippe Ponthot

LTAS-Computational Mechanics, University of Liège, 9 Allée de la Découverte, Liège, B-4000, Belgium

E-mail: Y.Crutzen@ulg.ac.be, R.Boman@ulg.ac.be, L.Papeleux@ulg.ac.be, JP.Ponthot@ulg.ac.be

In the manufacturing industry, cold roll forming has achieved a renewed interest since high-yield-strength materials have been brought to the market. The primary reason is that it is a convenient way for forming materials with high mechanical strength getting rid of the limitations particular to the classical deep drawing process for such materials (e.g., the limited press power, the sheet tearing, etc.). Besides, roll forming enables a high productivity, close geometric tolerances and complex shaped sections.

Finite element simulation of the roll forming process is regarded as an essential tool for the early design and optimization stages of a roll forming mill. However, such simulations are generally incredibly time-consuming, limited to some simple cases and to the pre-cut forming method (see Fig. 1).

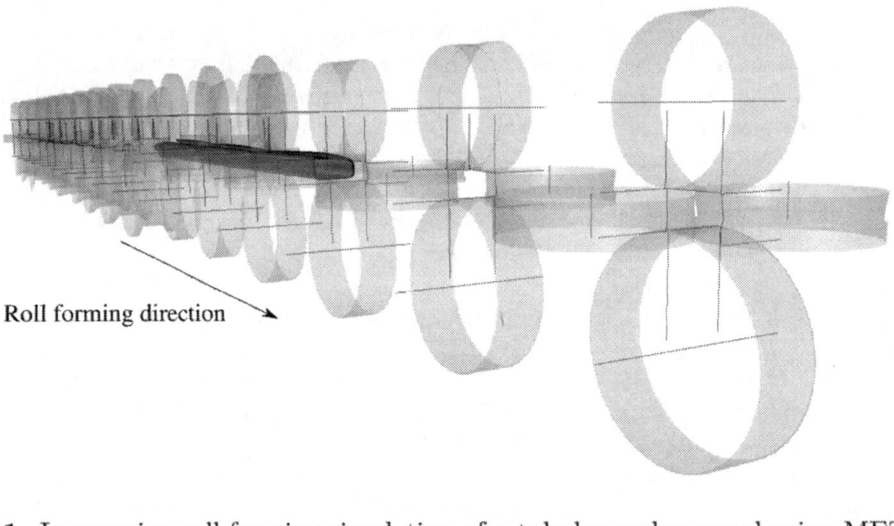

Figure 1. Lagrangian roll forming simulation of a tubular rocker panel using METAFOR

This paper presents an original contribution based on the Arbitrary Lagrangian Eulerian formulation (ALE) [1, 2] to simulate continuous roll forming operations. In contrast to the classical Lagrangian approach, the ALE formalism, which consists in decoupling the motion of the material and the mesh, has the capability to simulate the continuous process for the whole roll forming line (see Fig. 2) while keeping CPU cost under control. This is indeed the case especially when the algorithm has been parallelized and the mesh, which can be maintained fixed in the forming direction, has been optimized. Taking advantage from this approach, the mesh is only refined in the close vicinity of the forming tools for accurate modeling of contact and bending. The numerical simulation starts with an initial guess on the sheet geometry hopefully leading to the steady state. This initial geometry is built by the interpolation from the flower diagram of the mill simplifying the beginning of the simulation in terms of contact conditions.

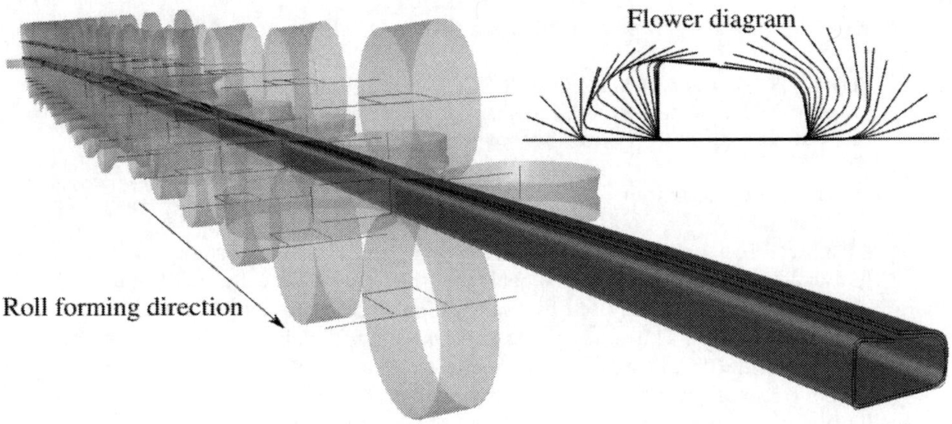

Figure 2. ALE roll forming simulation of a tubular rocker panel using METAFOR

In this work, a validation of both Lagrangian and ALE models is performed, comparing numerical results to some experimental data for a U-channel forming. Furthermore, advantages of the ALE formalism, especially its ease of use, are highlighted with the simulation of a tubular rocker panel on a 16-stand forming mill (see Fig. 2), which is a complex industrial mill.

These developments have been implemented in METAFOR [3, 4], an in-house finite element code of the Aerospace & Mechanical Engineering Department of the University of Liège, Belgium.

References

[1] J. Donea, A. Huerta, J.-P. Ponthot and A. Rodríguez-Ferran, *Encyclopedia of Computational Mechanics, Volume 1*, chapter 14, John Wiley & Sons, Ltd, 2004.

[2] R. Boman and J.-P. Ponthot, Continuous Roll Forming Simulation Using Arbitrary Lagrangian Eulerian Formalism. *Key Engineering Materials* 473:564–571, 2011.

[3] J.-P. Ponthot, *Traitement unifié de la Mécanique des Milieux Continus solides en grandes transformations par la méthode des éléments finis (in French)*. PhD thesis, Université de Liège, Liège, Belgium, 1995.

[4] METAFOR Website, http://metafor.ltas.ulg.ac.be/.

Journal of Physics: Conference Series **734** (2016) 032038 doi:10.1088/1742-6596/734/3/032038

Lubrication study for Single Point Incremental Forming of Copper

Kishore Jawale[1], José Ferreira Duarte[2], Ana Reis[1] and M B Silva[3]

[1] UTAF, INEGI, Campus da FEUP, 4200 465 Porto, Portugal
[2] Mechanical Department, FEUP, 4200 465 Porto, Portugal
[3] IDMEC, IST, 1049-001 Lisboa, Portugal

kishore.jawale@gmail.com

Abstract. In conventional machining and sheet metal forming processes, in general, lubrication assists to increase the quality of the final product. Similarly it is observed that there is a positive effect of the use of lubrication in Single point incremental forming, namely in the surface roughness. This study is focused on the investigation of the most appropriate lubricant for incremental forming of copper sheet. The study involves the selection of the best lubricant from a range of several lubricants that provides the best surface finishing. The influence of the lubrication on other parameters such as the maximum forming angle, the fracture strains and the deformed profile are also studied for Copper.

1. Introduction

Recent studies have shown the application of single point incremental forming (SPIF) in medical [1] and industrial field [2]. Despite the fact that the SPIF lead time is high when compared with conventional sheet metal forming, its application for prototyping and job production can be justified. In this process the sheet metal blank is clamped firmly on the rig, the blank is then formed incrementally by a SPIF tool controlled by CNC. SPIF uses an universal tool i.e. a ball or hemispherical ended tool eliminating the need for the customised die used in conventional sheet metal forming. Lubrication has been primarily used in SPIF to reduce tool wear [3,4] and recently used in order improve the surface finish [5].

It is interesting to note that even though some studies [3,5] have shown that the surface finish improves when lubrication is used, other [6] shows negative effect of lubrication on formability. Thus it is important that the effect of lubrication on formability of the material should be studied. In this study formability is assessed on the basis of maximum forming angle, fracture strains and deformed geometry.

2. Material and Methodology

2.1 Material

The material used for this study is Copper. Copper finds a wide application in sheet metal industry in particular in the home appliances industry. Copper has a relative good ductility. Although for direct comparison of all the lubricants no spring-back compensation strategies were used in the experimental work. The material properties are presented in table 1. The preparation of the blanks involved electrochemically etching of grid with circles of 2.5mm diameter in order to evaluate the strains. It can be observed from table 1 that the material anisotropy is almost 1.

2.2 SPIF tool and machine.
A 5 axis CNC machining centre and ball ended forming tool with 12 mm of diameter were used to carry out the experiments. The blank was firmly clamped on the rig as shown in figure 1. The geometry selected for the SPIF test was the benchmark truncated conical geometry, as shown in figure 1, which starts with constant angle of 30° and later increases constantly.

Table 1: Material properties of Copper

Engineering stress (MPa)		True stress (MPa)		A [%]	Young's Modulus E (MPa)	Normal anisotropy
σ_Y	σ_{UTS}	σ_Y	σ_{UTS}			
208.1	261.7	208.9	331.3	26.7	114343.8	0.98

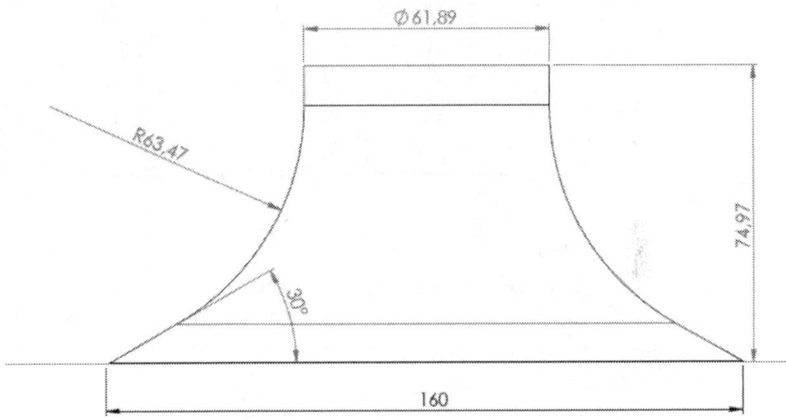

Figure 1: a) Experimental setup b) Truncated conical geometry (dimension in mm)

2.3 *Lubrication*
The process parameters and the tool path were kept the same throughout the experimental work. Six lubrication conditions were used as presented in table 2. For each lubrication condition two specimen were produced.

Table 2: List of Lubrication used

Lubrication condition	Lubricant reference	Chemical composition	Base	State	Viscosity (mm/s^2) at 40°C
1	Copaslip	15% hydrated Mg silicate + 15% Copper flakes + 15% hydrocarbon polymer	Mineral oil	Paste	96.2
2	AS40	40% MoS_2 + 15% graphite powder + 25% EP lithium grease	Petroleum oil	Paste	na
3	Weicon Ni Special	20% Ni powder + 10% Calciumhydroxide	Mineral oil	Paste	na
4	Weicon montage	20% ceramic powder + 10% Aluminiumpulver	Mineral oil	Paste	185
5	Castrol Magnaglide D68	Mineral oil	na	Oil	68
6	no Lubrication	na	na	Na	na

3. Results

3.1 Surface Roughness

Table 3 presents the average surface roughness obtained in the experimental tests. It was observed that all lubrication conditions reduced surface roughness. Lubrication conditions 3 and 5 provided an optimum surface roughness although when compared, lubrication condition 5 is more economical. It is also observed that lubrication state 2 is highly not recommended as it results in stains on the finished part. Stains might be a result of the reaction of MoS_2 with copper [7].

Table 3: Effect of lubrication on Surface roughness

Lubrication condition	Quantity of Lubrication (gm)	Ra (μm)
1	25.81	0.6950
2	31.93	0.5667
3	37.69	0.4216
4	45.06	0.4350
5	18.96	0.3933
6	-	0.7200

3.2 Strain analysis

The true fracture strains were calculated using circle grid analysis and the results obtained are presented in table 4. As the conical geometry results to in-plane strains; only major strains are used for analysis. It can be inferred from the observation that there is no significant effect of lubrication on the resultant fracture strains.

3.3 Maximum forming angle

The maximum forming angle was calculated based on the depth at which the fracture occurred and comparing the equivalent angle for the same depth in the CAD geometry. Here, as can be seen in table 4, there is no significant effect of lubrication on the maximum forming angle. However quantitatively, it can be stated that the lubrication has negligible positive effect on the maximum forming angle.

Table 4: Fracture strains and maximum forming angle

Lubrication state	Average fracture strain	Maximum forming angle (ψ in deg)
1	1.34	76.08
2	1.34	76.46
3	1.35	76.85
4	1.34	76.35
5	1.35	77.38
6	1.34	75.99

3.4 Profile deviation

As the geometry formed in this study is axi-symmetrical and anisotropy is negligible, it was analysed only one profile along the depth of the final geometry. Here it was observed that there was no significant effect of lubrication on all the profiles. Figure 2, presents the theoretical geometry from CAD and the experimental profiles obtained from several lubrication conditions. It can be stated that the formed geometries are significantly deviated from the CAD geometry due to spring-back effect.

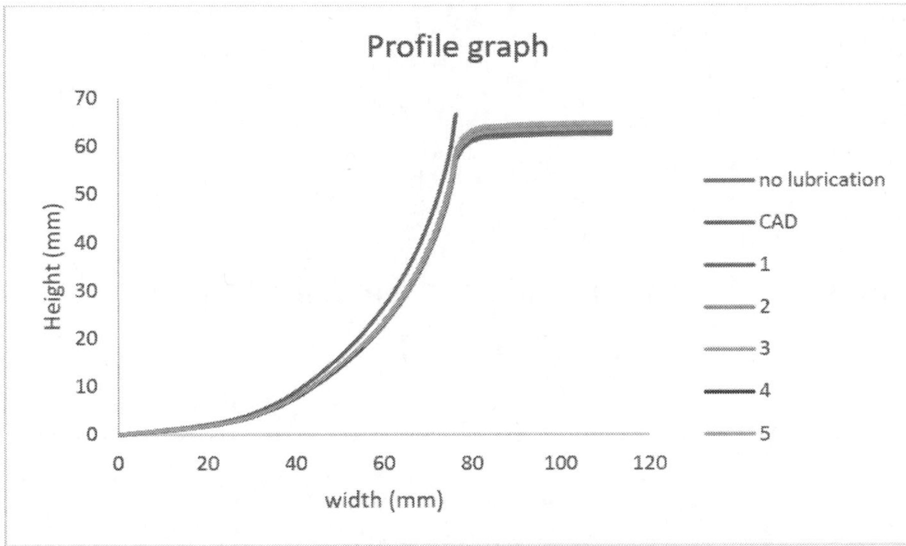

Figure 2: Profile deviation

4. Conclusion

From this study it can be concluded that the mineral oil based lubricants are economical and an optimum choice for incremental forming of Copper, having a significant and positive effect on surface finish. It can also be concluded that the lubrication conditions, in the case of Copper, do not affect the formability.

5. Acknowledgement

Author(s) gratefully acknowledge the funding of SciTech - Science and Technology for Competitive and Sustainable Industries, R&D project cofinanced by Programa Operacional Regional do Norte ("NORTE2020"), through Fundo Europeu de Desenvolvimento Regional (FEDER) and *Fundação para a Ciência e a Tecnologia of Portugal* and LAETA – UID/EMS/50022/2013.

6. References

[1] Araújo R, Teixeira P, Montanari L, Reis A, Silva M B and Martins P A 2013 Single point incremental forming of a facial implant. *Prosthet. Orthot. Int.* **0** 1–10

[2] Cristino V a, Montanari L, Silva M B, Atkins a G and Martins P a F 2014 Fracture in hole-flanging produced by single point incremental forming *Int. J. Mech. Sci.* **83** 146–54

[3] Hussain G, Gao L, Hayat N, Cui Z, Pang Y C and Dar N U 2008 Tool and lubrication for negative incremental forming of a commercially pure titanium sheet *J. Mater. Process. Technol.* **203** 193–201

[4] Zhang Q, Xiao F, Guo H, Li C, Gao L, Guo X, Han W and Bondarev A B 2010 Warm negative incremental forming of magnesium alloy AZ31 Sheet: New lubricating method *J. Mater. Process. Technol.* **210** 323–9

[5] Azevedo N G, Farias J S, Bastos R P, Teixeira P, Davim J P, Jose R and Sousa A De 2015 Lubrication Aspects during Single Point Incremental Forming for Steel and Aluminum Materials *Int. J. Precis. Eng. Manuf.* **16** 1–7

[6] Kim Y . and Park J . 2002 Effect of process parameters on formability in incremental forming of sheet metal *J. Mater. Process. Technol.* **130-131** 42–6

[7] Kalikhman V L, Gladchenko E P, Duksina A G and Krupina A G 1971 X-ray diffraction investigation of the reactions of some Mo, W, Nb, and Ta chalcogenides with copper, silver, nickel, and cobalt *Sov. Powder Metall. Met. Ceram.* **10** 475–7

Drawing simulation by static implicit analysis with the artificial damping method

K Oide[1], Y Mihara[1], T Kobayashi[1], H Takizawa[2], T Amaishi[3], Y Umezu[3]

[1] Mechanical Design & Analysis Corporation, Fuda Chofu City Tokyo, 182-0024 Japan
[2] Nippon institute of technology, Miyashiro-machi Minamisaitama-gun Saitama Pref., 345-8501 JAPAN
[3] JSOL Corporation, Tosabori Nishi-ku Osaka, 550-0001 Japan

E-mail: oide@mech-da.co.jp

Abstract. Wrinkling during draw is typically a local instability problem. When the structural instability is localized, there will be a local transfer of strain energy from one part of the structure to neighboring parts, and global solution methods, which is typically represented by the arc length method, may not work. So, this type of problems has to be solved either dynamically or with the artificial damping. On the other hand, the essential nature of the buckling behavior can be regarded as a static problem, even though it may be possible to raise some side issues due to the inertia effect. In this study, we traced the local buckling behavior of anisotropic elasto-plastic thin shells in Numisheet2014 BM4 using the artificial damping method.

1. Introduction

Advanced can manufacturing technology and cost control efforts have resulted in a consistent reduction of the net metal weight used, which has led to beverage cans with thinner sidewalls, reduced neck diameters, and smaller base diameters. However, small-diameter cans increase the likelihood of wrinkling during production. Wrinkling during redraw is typically a local instability problem that may be triggered by a small, local disturbance. When the structural instability is localized, there will be a local transfer of strain energy from one part of the structure to neighboring parts, and global solution methods, typified by the arc length method, may not work. This class of problems has to be solved either dynamically or statically with the aid of artificial damping. The latest general purpose finite element codes provide automatic mechanisms for stabilizing unstable quasi-static problems through the automatic addition of viscous damping to the model. When local instability occurs, the deformation rate of that portion begins to increase and, consequently, locally released strain energy is dissipated due to the appended artificial damping effect. In the phenomena of the real cases, the energy generated by the artificial damping is equivalent to the energy released as the inertia effect. Namely, replacing the kinetic energy with the artificial damping energy makes it possible for the unstable phenomenon to be analyzed stably in order to obtain the equilibrium solution within the static analysis.

In this study, we traced the local buckling behavior of anisotropic elasto-plastic thin shells in Numisheet2014 BM4[2] using the artificial damping method.

2. Simulation model and result

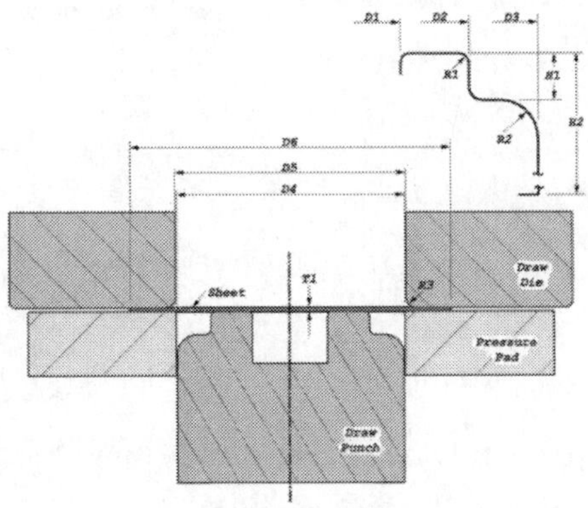

Figure 1. Schematic view of tools for the drawing process.

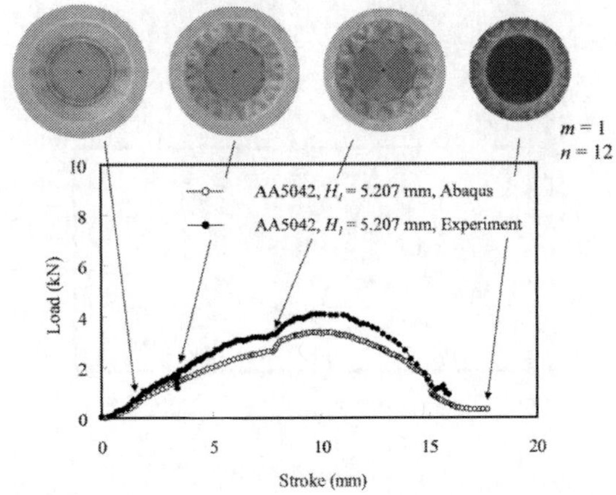

Figure 2. Load-stroke diagram during the drawing process.

A schematic view of tools used for the drawing process, including their dimensions, is shown in Figure 1. The drawing occurs continuously during one punch stroke, and the punch speed is 140 mm/sec. The recommended friction coefficient is 0.03, and the blank holding force is given as 8900 N. AA5042 aluminum alloy ($t = 0.2083$mm) is considered as sheet materials. This material is assumed to have the YLD2000 anisotropic yield criteria[1].

The anisotropic elasto-plastic material behaviors were implemented in Abaqus by using the UMAT user-defined material subroutine. In this study, we traced the local buckling behavior of anisotropic elasto-plastic thin shells using the artificial damping method. The artificial damping method is a technique used to achieve total energy balance in static analysis, compensating for strain energy changes due to local elastic instability. Abaqus provides an automatic mechanism for stabilizing unstable quasi-static problems through the automatic addition of volume-proportional damping to the model, while the artificial damping method or its comparable quasi-static capabilities are available in other codes. Via Eq.(1), viscous forces are added to the global equilibrium equations, as shown in Eq.(2):

$$\boldsymbol{F}_v = c\boldsymbol{M}^*\boldsymbol{v} \tag{1}$$

$$\boldsymbol{P} - \boldsymbol{Q} - \boldsymbol{F}_v = \boldsymbol{0} \tag{2}$$

where $\boldsymbol{M}^*$ is an artificial mass matrix calculated with unity density, c is a damping factor, $v = \Delta u/\Delta t$ is the vector of nodal velocity, Δu is the vector of incremental displacement, Δt is the increment of time (which may or may not have a physical meaning in the context of the problem being solved), P is the total applied load, and Q is the internal force. The automatic seamless simulation provided good agreement with the experimentally observed buckling behavior, which covered the entire drawing process well, as shown in Figure 2.

3. Eigenvalue analysis

Using implicit codes, it makes it possible to calculate eigenmodes at any midpoint in the analysis. Figure 3 shows the results from the eigenvalue analysis for AA5042 of $H_1 = 5.207$ mm. The eigenvalue analysis was performed at the very early stage of the punching process (2 mm of stroke). The magnitude of this punch stroke is small enough to assume a nearly uniform

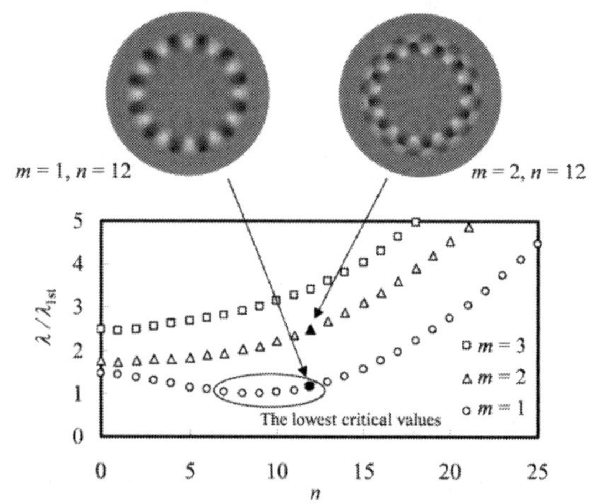

Figure 3. Relationship between the normalized eigenvalue and the number of circumferential full-waves n.

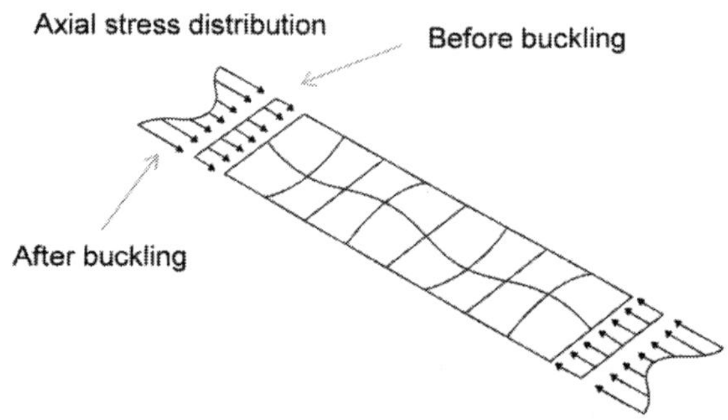

Figure 4. Analogy of buckling behavior compressed rectangular plate

circumferential stress distribution on the shell. The generated plastic strain at this point is around 1-2%.

The curves in Figure 3 represent the relationship between the normalized eigenvalue and the number of circumferential full-waves n, where m is the number of radial half-waves. In the vicinity of the lowest critical value, multiple eigenmodes with different numbers of waves (n=9-12 for m=1) exist. The proximity of the eigenmodes must have impaired the reproducibility of experiments, where our history analysis of the punching process provides a unique buckling mode with n=12.

The buckled area on the shell can be regarded as a circular plate with a hole at the center and with the inner and outer boundaries clamped. Timoshenko[3] suggested that such a circular plate, when subjected to radial loading, buckles in multiple waves along the circumference. For the ratio of the inner and outer radius approaching unity, the buckling conditions for a compressed ring are analogous to those of a long, compressed rectangular plate. In the case of the rectangular plate, a typical deformation mode indicates several half-waves in the direction of compression but only one half-wave in the perpendicular direction (in Figure 4). The lengths of the half-waves are supposed to approach the width of the plate, because a buckled plate subdivides approximately into squares. A single eigenvalue will be extracted for this deformation mode. However, manufacturing experience indicates that the real process conditions, such as the mechanical properties of the aluminum sheet, the tooling geometry, and the contact conditions (including lubrication), cause multiple buckling modes with different numbers of circumferential waves in close proximity. This suggests that the effective width of the buckled area is influenced by the above-mentioned factors. In fact, the conference organizer reported that the number of circumferential full-waves was found to be n=13 in their benchmark experiment, whereas our analyses provides n=12.

4. Conclution

In this paper, we traced the local buckling behavior of anisotropic elasto-plastic thin shells in Numisheet2014 BM4[2] using the static implicit analysis with artificial damping method. We estimated the buckling mode by eigenvalue analysis and the effective width of the buckled area by the basic buckling theory for shell. However the relationship between the eigenmodes obtained from the linear eigenvalue analysis and the deformation modes actually observed in the buckling process requires further investigation.

References

[1] Barlat, F., et al., 2003, Plane stress yield function for aluminum alloy sheets-part 1: theory, International Journal of Plasticity, 19, 1297-1319.
[2] Dick, R., et al., 2014, Benchmark 4, The 9th International Conference and Workshop on Numerical Simulation of 3D Sheet Metal Forming Processes (NUMISHEET2014), Melbourne, http://www.numisheet2014.org/
[3] S.T. Timoshenko, J.M. Gere, 1961, Theory of Elastic Stability, 2nd ed., McGraw-Hill, New York, pp.389-392.

Investigation of High Speed Friction Test for Aluminum Alloys

K Ooki[1] and S Takahashi[2]

[1] Graduate student, Nihon University, Narashino, Japan
[2] Professor, Nihon University, Narashino, Japan

E-mail:takahashi.susumu32@nihon.u.ac.jp

Abstract. To shorten the development stage of automobiles, FEM simulation has been applied. It was important to increase the accuracy of the sheet metal simulation results. The friction coefficient between the sheet metal and dies the greatly affected the simulation results. Therefore, apparatus for measuring the friction coefficient with a specific press forming speed (300 mm/s) has been developed. The materials of the sheet metals and dies were aluminum alloys and die steel respectively. It was found that the friction was affected by the difference between the velocity of the sheet metal and that of the dies.

1. Introduction

Automobile makers are reducing emissions from automobiles as a countermeasure against global warming. Therefore, the weight of the vehicle bodies has to be reduced. Although aluminum alloys and high tensile steel sheets are applied for this purpose, their formability is low. To shorten the development stage of sheet metal forming dies numerical simulation is applied. Accurate simulation results can be obtained by using accurate stress-strain curves, yield functions, friction coefficients, and so on. There are many investigation regarding biaxial tension and compression behavior for sheet metal [1-6]. One of the key parameters for obtaining accurate sheet metal forming simulation is friction coefficient. Friction is measured at the speed performed by conventional mechanical testing machine. The sliding speed of the blank on the die during sheet metal forming is about 300mm/s. The testing speed is not fast enough to investigate friction behavior at sheet metal forming speed. Therefore it is difficult to observe friction phenomena at high speed.

In this study, the friction behavior between the sheet metal and dies with respect to the sheet metal forming speed was investigated. A high speed friction test apparatus was developed. To perform the high speed friction test, a servo press was used. Aluminum alloy sheets was used as the blank material and tool material was used for the die. The wide range of testing speed was performed for investigating the influence of the testing speed on the friction.

High speed friction test method

1.1. Experimental apparatus

To perform friction tests at high speed, new apparatus was developed. A servo press (SDE-2025, 2000 kN, Amada Co., Ltd.) was used to move the dies for the friction test, because the slide of the servo press can be used to control the speed. Figure1 shows the movement of the experimental apparatus of the high speed friction test before and after the experiment. To increase the speed of the slide of the servo press, it needs to be moved some distance. Therefore, the minimum distance between the flange of the chuck

Journal of Physics: Conference Series **734** (2016) 032040 doi:10.1088/1742-6596/734/3/032040

and the upper part of the lower flange to achieve test speed (500 mm/s), was 50 mm. Both sides of the specimen were in contact with the dies. To apply compression pressure to the specimen with the dies, a hydraulic hand pump and a jack (TWA - 0.3 Osaka Jack. Co., Ltd. and EF10S1.1 Osaka Jack . Co., Ltd.) were used. Two load cells (LUK-A 20kN Kyowa Electronic Instruments Co., Ltd.) were located on the other side of the hydraulic jack for measuring the compression pressure. The drawing (friction) force was measured by the load cell (LUK-A 100 kN Kyowa Electronic Instruments Co., Ltd.), which is located in the upper frame attached to the press slide. The material of the dies is SKD11. The hardness and surface roughness after heat treatment are Ra=0.30 - 0.55 μm and HRC59, respectively. The contact area between the dies and the specimen is 234 mm^2.

1.2. Specimen
The material of the specimen was A5023-O with a dull finish. The dimensions are 300 × 30 ×1 mm. The specimens were machined in the rolling and transverse directions.

2.3 Experimental conditions
Wash oil (R303P, Sugimura Chemical Industrial Co., Ltd) was used as lubricant. The oil was applied to both sides of the specimen prior to testing. The coating amount was 0.1 - 0.15 g. The compression load applied to the dies was changed in the range 1.1 - 1.9 kN. The pressure was 5 -10 MPa. The drawing speed of the specimen was changed in the range 5 - 500 mm/s.

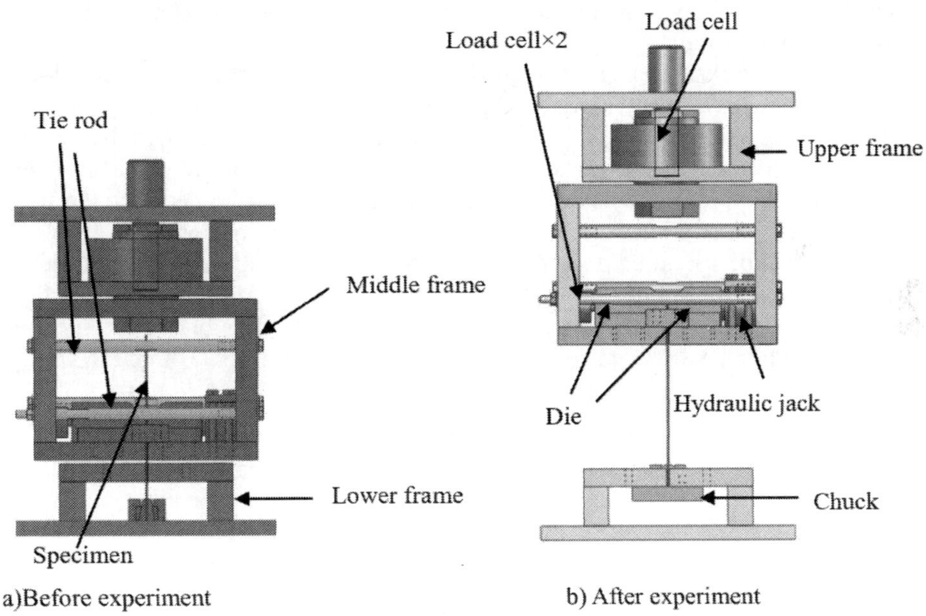

Figure 1. Movement of experimental apparatus of high speed friction test.

2. Test results and discussion

3.1 Drawing force
The drawing force during the experiment is shown in figure 2. The test conditions were 5 MPa (compression pressure), and 500 mm/s (drawing speed). It was found that there was electrical noise when measuring the force, so it was difficult to evaluate the friction coefficient from the result.
To reduce the noise, the location of ground for the measuring devices was investigated. The most appropriate location of ground was the bolster of the press. Figure 3 shows the results after reducing

noise. The range of noise before and after changing the location of ground is ±250 N and ±15 N, respectively. The reduction rate of the noise is about 95%. In the subsequent test, the appropriate location of ground was used.

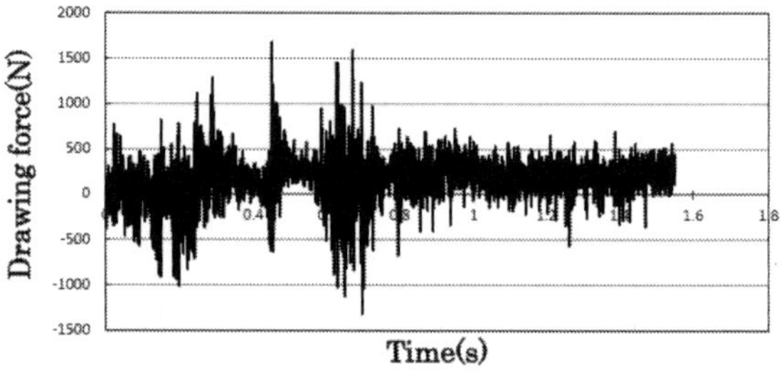

Figure 2. Drawing force during experiment with noise.

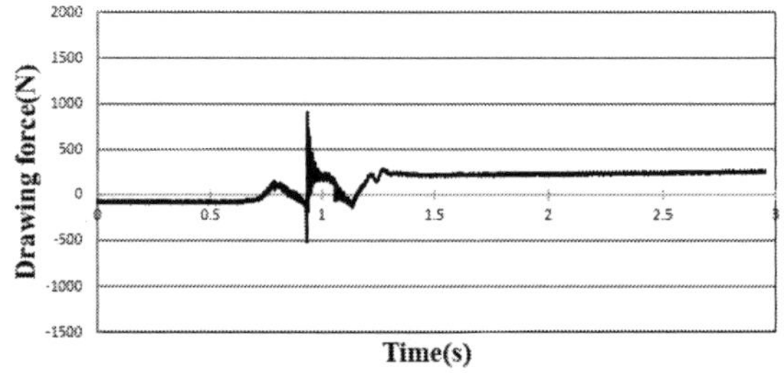

Figure 3. Drawing force during experiment without noise.

3.2 Friction coefficient

The relationship between the friction coefficient and time in the case of a pressure of 5 MPa and drawing speed of 500 mm/s during the evaluation stage in figure 3 is shown in figure 4. To reduce the oscillation of the results, a moving average method was applied. The friction coefficient calculated using the average of the values in figure 4 is 0.09.

Figure 5 indicates the relationship between the friction coefficient and pressure. Figure 5 shows the relationship between friction coefficient and pressure under the influence of drawing speed. It indicates that a faster drawing speed leads to a lower friction coefficient. It can be said that if the viscosity of the lubricant is high enough, the faster drawing speed can reduce the movement of lubricant between specimen and dies towards the outside of the specimen. That helps to decrease the friction coefficient. Figure 5 b) shows the relationship between friction coefficient and pressure under the influence of machining direction of specimen. It indicates that the friction coefficient in the transverse direction is lower than that in the rolling direction, and the direction of scratch in the rolling direction coincided with the drawing direction. In this case, lubricant could easily flow out of the specimen. This makes causes a larger friction coefficient than that of the case in the transverse direction.

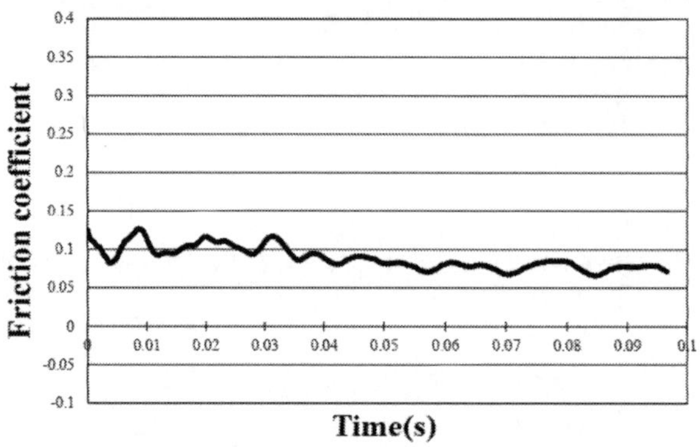

Figure 4. Friction coefficient during experiment (Pressure 5MPa Drawing speed 500mm/s).

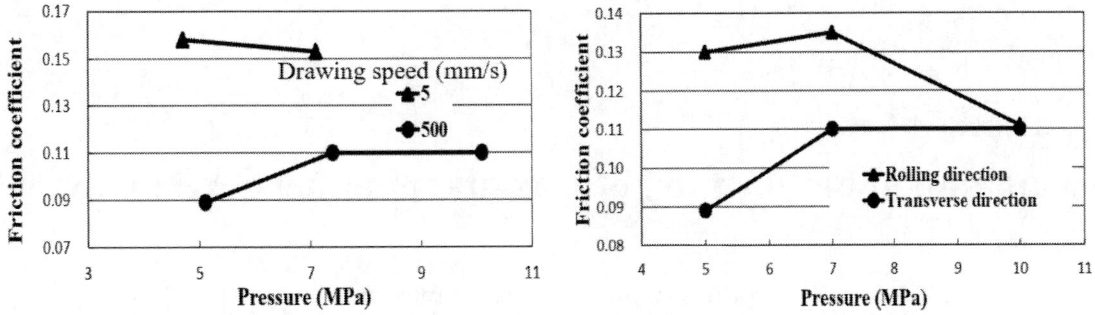

a) Changing drawing speed (Rolling direction) b) Changing machining direction (Drawing speed 500mm/s)

Figure 5. Relationship between friction coefficient and pressure.

3. Conclusions

1) To reduce the electrical noise of the measured results, the location of ground of the measurement device is important.

2) A faster drawing speed can result in a lower friction coefficient.

3) The friction coefficient of the specimen in the rolling direction is greater than that of the specimen in the transverse direction. It can be said that a more accurate sheet metal forming simulation can be obtained by taking into account the changing friction and moving direction of the material.

Acknowledgements
The authors acknowledge the support from the Amada Foundation

References

[1] Hoferlin E, Van Bael A, Van Houttec P, Steyaert G, De Mare C, The design of a biaxial tensile test and its use for the validation of crystallographic yield loci, *Modelling Simul. Mater. Sci. Eng.*, 8(2000) 432-3

[2] Green DE, Neale KW, MacEwen SR, Makinde, R. Perrin, A.; Experimental investigation of the biaxial behaviour of an aluminum sheet, *Intl. J. Plasticity*, 20(2004) 1677-1706

[3] Wu X-D, Wan M, Zhou X-B, Biaxial tensile testing of cruciform specimen under complex loading, *J. Materials Processing Technology*, 168(2005) 181-83

[4] Nagayasu T, Takahashi S, Kuwabara T, Development of compact biaxial tensile testing apparatus using conventional compression testing machine and evaluation of the test reslts, Conference proceeding of IDDRG2010, (2010) 593-602

[5] Takahashi S, Nagayasu T, Kuwabara T, Improvement of compact biaxial tensile testing apparatus using conventional tensile testing machine and evaluation of the test results, IDDRG2011, CD-ROM

[6] Hanabusa Y, H Takizawa, Kuwabara T, Numerical verification in biaxial stress tests using cruciform specimen, J. *The Japan Society for Technology of Plasticity*, 52-601 (2011) 282-87

Journal of Physics: Conference Series **734** (2016) 032041 doi:10.1088/1742-6596/734/3/032041

Evaluation of hot forming effects mapping for CAE analyses

L Knoerr[1], T Faath[2], J Dykeman[3] and S Malcolm[4]

[1,2]thyssenkrupp Steel NA, 22355 W. 11 Mile Road, Southfield, MI 48033 USA;
[3,4]Honda R&D Americas, Inc. 21001 State Route 739, Raymond, OH 43067 USA.

lay.knoerr@thyssenkrupp.com
timo.faath@thyssenkrupp.com
jdykeman@oh.hra.com
smalcolm@oh.hra.com

Abstract. Hot forming has grown significantly in the manufacturing of structural components within the vehicle Body-In-White construction. The superior strength of press hardened steels not only guarantee high resistance to deformation, it also brings a significant weight saving compared to conventional cold formed products. However, the benefit of achieving ultrahigh strength with hot stamping, comes with a reduction in ductility of the press hardened part. This will require advanced material modeling to capture the predicted performances accurately. A technique to optically measure and map the thinning distribution after hot stamping has shown to improve numerical analysis for fracture prediction. The proposed method to determine the forming effects and mapping to CAE models can be integrated into the Vehicle Development Process to shorten the time to production.

1. Introduction

The demands for improved vehicle safety and fuel economy have intensified with the development of Advanced to Ultra High Strength Steels. However, increased strength has come at the expense of ductility and formability, making parts more challenging to be defect free and meet required dimensional tolerances. To overcome this, ultrahigh strength parts above 1200MPa in tensile strength are typically hot formed by austenitizing Manganese Boron steel blanks in a furnace to 880-950° C, subsequently forming and quenching in a cooled die. Structural components manufactured in this process can attain tensile strength of up to 1500MPa to 2000MPa, with significantly lower thicknesses. Furthermore, the heat treatment can be tailor-tempered to make crash relevant components that can absorb impact energy and have high deformation resistance. A functional component where intrusion resistance is desired, has martensitic hot formed microstructure that exhibits very little ductility and strain to failure. This behavior can be difficult to predict accurately in a crash performance analysis, where uniform part thickness is assume and a basic material model with no inherent damage accumulation criterion specified. This paper evaluates the benefits of validating the numerical analysis of a hot formed hat profiled section part under quasi-static three pointing bending with the experimental study, by incorporating thickness mapping and damage accumulation criterion. The approach of gridding the blanks to optically measuring the strains after hot stamping, and mapping the thickness over to the finite element mesh model is fully addressed here.

2. Experimental Study

The validation study of the hot formed part under three point bending loading condition was conducted in conjunction with an automotive OEM (Honda of America). The hat profiled parts were hot formed in the thyssenkrupp Steel Europe Research facility in Dortmund, Germany, under series-like conditions. The material used is MBW®1500 (22MnB5) steel grade with Aluminum Silicon (AS) coating of 1.5 mm thickness. The geometrical dimensions of the hat profiled part are shown in figure 1 with a 1.6 mm welded back plate of a High Strength Low Alloy (HSLA) material grade. The assembled part is subjected to a three point bending loading condition in the OEM's universal testing machine, instrumented to acquire the load and stroke curves of the test setup as illustrated in figure 2.

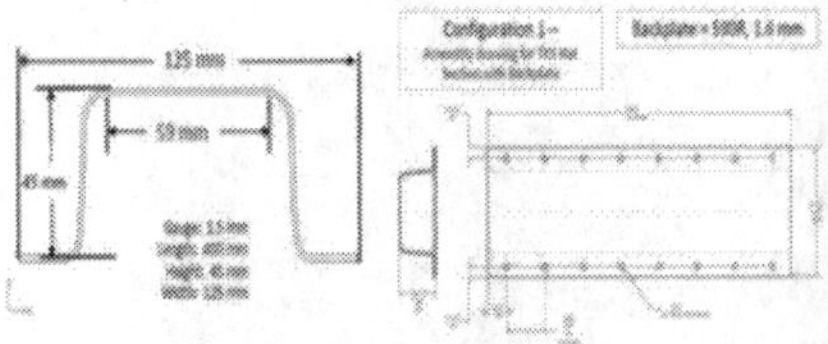

Figure 1. Hat profiled dimensions.

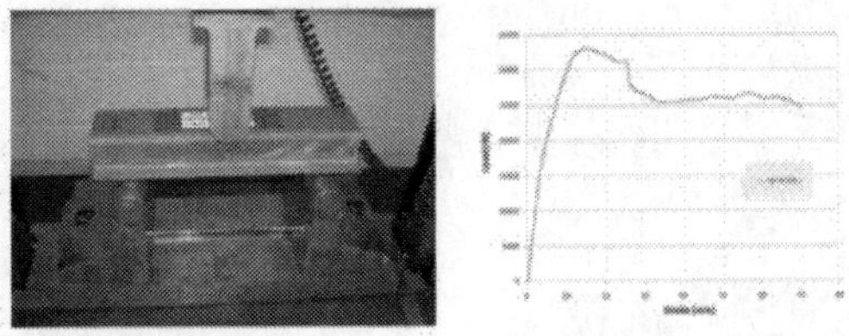

Figure 2. Three Point Bending test setup.

2.1. Material Test Specimens

The material selected to investigate the hot forming effect on the crash performance is of the Manganese-Boron steel grade and requires at least two tests, such as uniaxial and plane strain tension for material model calibration. The uniaxial tension test conforms to the Japan Industrial Standard (JIS) Z 2241 under quasi-static strain rate of 0.003/s. The geometry of the test specimen is undersized of 32 mm gauge length and width of 6 mm. The dog bone shaped specimens were cut from material blanks after heat treatment, and prepared with speckled pattern for strain measurement using the Digital Image Correlation (DIC) method as shown in figure 3 (a). Another test specimen configured for plane strain loading as shown in figure 3(b) to evaluate the parameters for fracture model calibration.

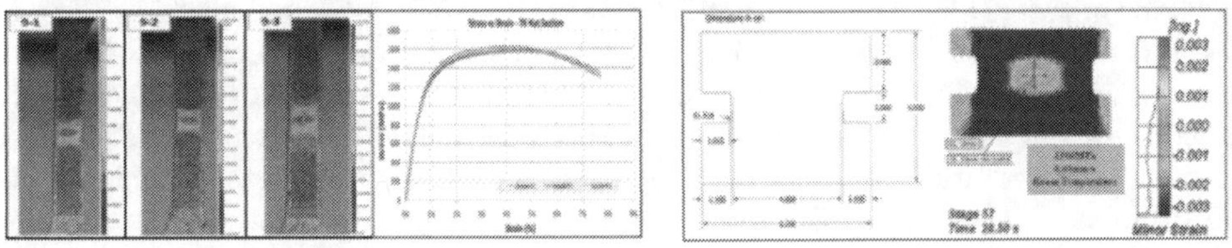

Figures 3. (a) JIS uniaxial coupon level test. **(b)** Plane strain coupon test.

2.2 Blank Preparation

The rectangular blanks developed for the hot forming of the hat-profiled part is 1.5 mm thick, width of 200 mm and 400 mm in length. The blanks are gridded for optical strain measurements after hot forming using the GOM Argus system. The gridding process as shown in figure 4(a), etched on the blank (b) utilized a patented varnish developed by Sikora et al at thyssenkrupp [1] that will not fade in the hot furnace. The hot stamped part (c) has visible grid markings for strain measurements and mapping over to a numerical model for crash performance evaluation.

Figures 4. (a) Gridding process. **(b)** Etched blank. **(c)** After hot stamping.

2.3 Experimental Three Point Bending

The bending experiments were conducted on a universal test machine at Honda Research Laboratory with a punch speed of 5 mm/s. Figure 5 illustrates the hat profiled assembled part set on two stationary roller fulcrums at equidistance from the moving hemi-sectional punch of 50 mm radius and 115 mm width under a certain deformation load. The punch load transmitted thru' the load cell and the stroke via a LVDT (Linear Variable Displacement Transducer), provided the signals to the data logger which synchronized and filtered the noise to output a smooth force-stroke curve as depicted in figure 2 to the maximum distance of 70 mm. The fracture modes of the assembled parts at maximum stroke are displayed in figure 6. The raw force-stroke data acquired correlates to the predicted results of the virtual three point bending simulation using an advanced fracture model for the validation study.

Figure 5. Hat profiled assembly under load.

Figure 6. Fractured parts at maximum stroke.

3. Numerical Study

A comprehensive numerical study that incorporates the thinning distribution of a hot stamped component under three point bending, was performed using an advanced material model that can reliably predict fracture under different loading conditions. The constitutive model was formulated on the basis of shear-modified Gurson model [2], by simplifying the governing equations and implementing them in the shell mechanics framework [5, 6, 7]. The shell-based formulation is the only effective approach to analysing large scaled ductile problems, such as vehicle crashworthiness assessment. The calibrated model, was conducted using only two coupons level tests: uniaxial and plane strain tension. The initial analyses conducted with assumed uniform thickness of hat profiled mesh model was compared with thickness mapping from the actual optically measured part. Excellent correlation between the experimental and predicted results was observed on the mesh model with the thickness mapped, validating the facture prediction methodology utilized.

3.1. Thickness Mapping

The gridded hat profiled part shown in figure 7 is optically measured following the procedural method outlined in GOM Argus system's manual [3]. The measured 2D strain tensors are processed in the system's software to compute the thinning distribution which is the more profound effect of hot forming with the limited residual ductility on the finished part. The thickness/thinning mapping workflow as outlined in the manual, includes initial best fit alignment of the measured to the mesh part, then projection followed by interpolation of the nodal thickness values to the mesh model. The resultant mesh model of the mapped distributed thickness values as shown here in figure 8 can be plotted and exported in various commercial finite element solvers' formats.

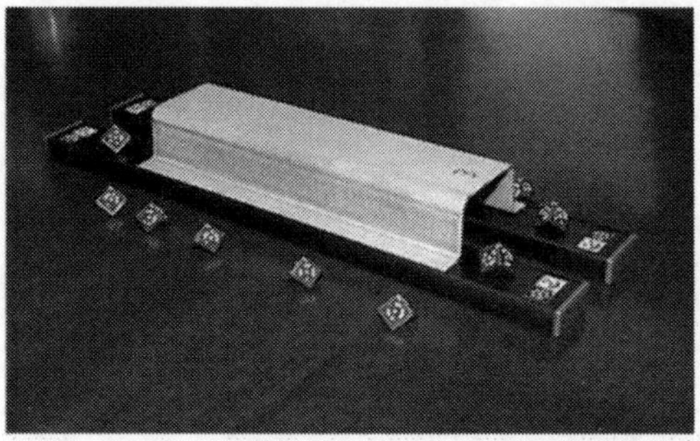

Figure 7. Optical measurement with Argus.

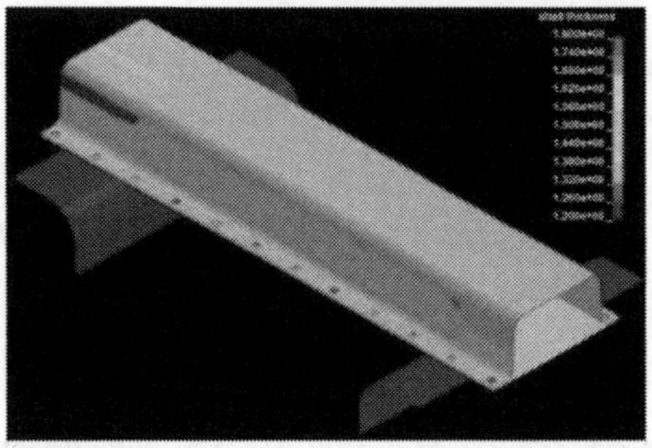

Figure 8. Thickness mapped model.

3.2. Material Characterization and Model Calibration

Calibration of the ductile fracture constitutive model used in the simulation requires that effective material characterisation after localized necking to failure is performed. This is necessary to account for the influence of multi-axial stress states on ductile fracture. This process can be extensive, considering the numerous coupon level tests typically required for full calibration of ductile fracture models. In this study, only two coupon level tests were used for model calibration. This was possible due to the micromechanically motivated constitutive formulation that explicitly accounts for underlying ductile fracture mechanisms [5, 6, 7]. The resulting fracture locus is as shown in figure 9. Despite limited datasets, notable validation is accomplish as discussed in the following section.

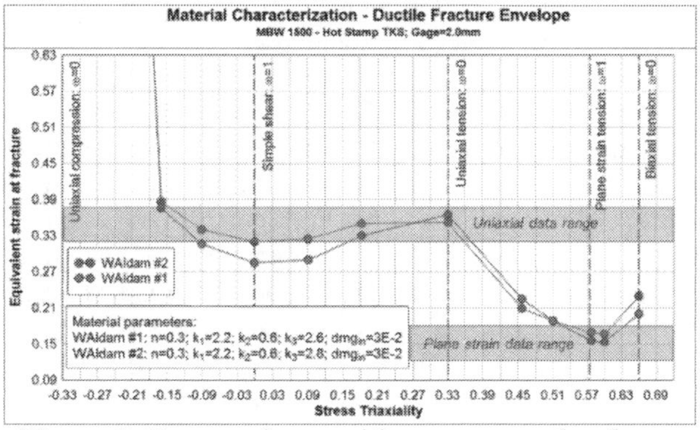

Figure 9. Fracture locus with two coupon level tests

3.3 Fracture Prediction Methodology

The fracture simulation methodology used here allows for reliable prediction of fracture initiation and propagation in thin sheets under multiaxial stress states. The model is based on an advanced dilatational plasticity model with a scalar damage variable governing softening behavior [4, 5]. The elasto-plastic damage criterion for ductile metals is fundamental in capturing the macroscopic response under large strain and varying stress triaxiality. Consequently, accurate experimental material characterization and systematic calibration of the model parameters are imperative to enhance fracture prediction capability. An approach to automate the iterative calibration process was accomplished using custom build software that includes a user-friendly Graphical User Interface [6]. The constitutive equations governing the model formulated in the framework of shell mechanics are suitable for large structural simulation like crashworthiness for which the use of solid elements is time and cost prohibitive. The described formulation as referenced earlier was developed and implemented by principals mentioned in [7, 8]. It is also available as a user-material in ABAQUS and validated against several large-scale shell deformation case studies.

3.4 Analysis Validation

A calibrated model obtained as outlined in the method described above, was utilized for the quasi-static three point bending simulation of a hot formed hat profiled part which is representative of an impact

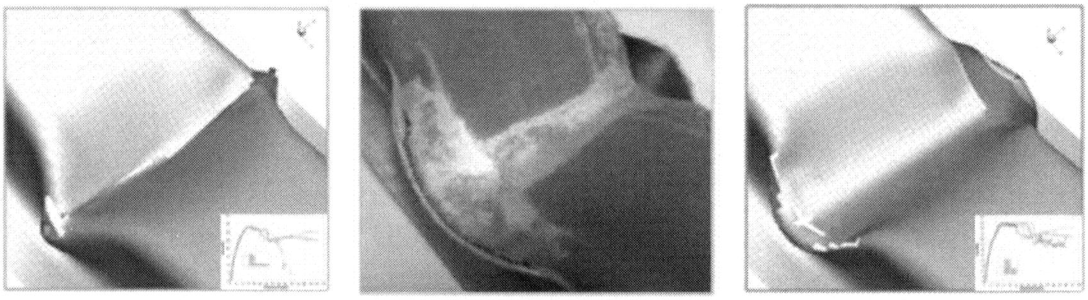

Figures 10. (a) Uniform thickness (b) Actual Part (c) Mapped thickness

beam under side crash loading condition. Two analyses case scenario of the part mesh model were performed, one with the uniform part thickness assuming no forming effects, and the other with thickness/thinning mapping over the mesh part model. The predicted results for the two case study at maximum stroke are displayed in figures 10(a) and (c), as compared to the actual experimental part (b). The results clearly show the close validation of the latter case with the thickness mapped model that captures the fracture mode behaviour and the force-stroke curve more precisely. A detailed evaluation of the fracture mode revealed that the test part was under complex multiaxial stress states of bending of nonlinear strain paths with dominant crack initiated by in plane shear deformation. Overall, excellent correlation with the experimental test results was achieved.

4. Conclusion

Hot forming with its ability to stamp lightweight components to meet the demands for stringent fuel efficiency and vehicle safety has proliferated since the past decade as the preferred manufacturing process. The resultant ultrahigh strength low ductility however, has prompted application of these manufactured components to be very selective like bumper beams, roof rails, side impact beams etc. Early concept design to evaluate vehicle structural, fatigue and crash assessments can be improved by utilizing advanced constitutive material models that can accurately predict fracture initiation and propagation as investigated in this study. The process of material selection based on forming and performance requirements can be incorporated with a modular software to commercial numerical solvers that calibrates the model parameters for the prediction methodology utilized. Mapping techniques of forming effects from simulation or actual measurements can be integrated seamlessly in the vehicle development process to improve the prediction capability of any numerical analysis.

References

[1] Sikora S et al 2012 *United States Patent No. US 8,215,143,B2*

[2] Nahshon K and Hutchinson J W 2008 Modifications of the Gurson Model for shear failure *European Journal of Mechanics A/Splids 27 1-17*

[3] GOM Optical Measuring Techniques 2009 FEA Comparison *Argus Manual Chapter 7 v6-2*

[4] Voyiadjis G Z and Woelke P 2006 Elasto-Plastic and damage analysis of plates and shells *Int. J Solids Structures 2008 Springer -Verlag*

[5] Woelke P, Voyiadjis G Z and Perzyna P 2006 Elasto-plastic finite element analysis of shells with damage due to microvoids *Int. J Num. Meth. Eng Vol 68 930 pp 338-380*

[6] Woelke P, Bentowitz B, Hiriyur B, Knoerr L and Hutchinson J W 2014 Ductile Fracture Prediction for Automotive Applications *Int. Automotive Body Congress*

[7] Woelke P and Abboud N 2012 Modeling fracture in large scale shell structures *J of Mechanics and Physics of Solids 60 12 2044-2063*

[8] WAimat Suite: www.waimat.com

Knowledge Based Cloud FE simulation – data-driven material characterization guidelines for the hot stamping of aluminium alloys

Ailing Wang, Yang Zheng, Jun Liu, Omer El Fakir, Marc Masen, Liliang Wang*

Department of Mechanical Engineering, Imperial College London, London, SW7 2AZ, UK

E-mail address: liliang.wang@imperial.ac.uk

Abstract. The Knowledge Based Cloud FEA (KBC-FEA) simulation technique allows multi-objective FE simulations to be conducted on a cloud-computing environment, which effectively reduces computation time and expands the capability of FE simulation software. In this paper, a novel functional module was developed for the data mining of experimentally verified FE simulation results for metal forming processes obtained from KBC-FE. Through this functional module, the thermo-mechanical characteristics of a metal forming process were deduced, enabling a systematic and data-driven guideline for mechanical property characterization to be developed, which will directly guide the material tests for a metal forming process towards the most efficient and effective scheme. Successful application of this data-driven guideline would reduce the efforts for material characterization, leading to the development of more accurate material models, which in turn enhance the accuracy of FE simulations.

1. Introduction

Material property characterisation is an essential requirement of FE simulations, as their accuracy is greatly enhanced with a detailed definition of the material's deformation behaviour, input in the form of either discrete data known as 'look-up tables', or a set of constitutive equations describing their flow stresses. Thus, sensible selection of the conditions at which the material properties are characterised, through tensile or formability testing for example, is needed. Currently, tensile testing conditions are usually selected based on experience as opposed to the real requirements of the actual application of the material. Strain rates in the range of 10^{-5} – 10^{-2} s^{-1} are used to study superplastic forming [1, 2], strain rates of 10^{-1} – 1 s^{-1} are used for steel sheet metal forming [3], and strain rates of 10^{-4} –10^{-2} s^{-1} are utilised in the study of low-cycle fatigue [4]. Large variations in temperature ranges also exist across different applications, such as in warm and hot stamping of aluminium alloys, where a range of 150 – 525 °C [5-7] is used. To enhance the utility of the material data input, the selected test matrix must be targeted according to the specific application of the material and its associated conditions of strain rate and temperature. In the present research, an innovative data driven approach was developed by data mining existing experimentally verified FE simulation results obtained from the Knowledge Based Cloud FEA (KBC-FEA) technique. By tracking the deformation history of each element, the real thermo-mechanical conditions in a hot stamping process were identified.

Extraction of the data was achieved through the development of a functional module that operates on the KBC-FEA platform, the structure for which is shown in Figure 1. The module, 'Strategy', would provide analysis of the data collected from the experimentally verified FE simulation results, and output a material and forming technique specific tensile test guide.

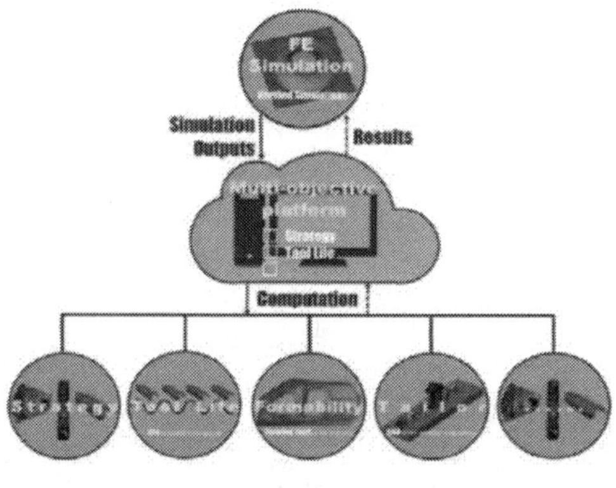

Figure 1. Structure of KBC-FEA system

2. Methodology: Data-driven approach for material characterization

As a FE simulation orientated multi-objective platform, experimentally verified simulations from KBC-FEA can be used to construct a database from historical FE simulations to provide a comprehensive quantification of temperature and strain rate conditions. In this paper, the 'Strategy' module is implemented to study the deformation characteristics of AA6082 under hot stamping conditions.

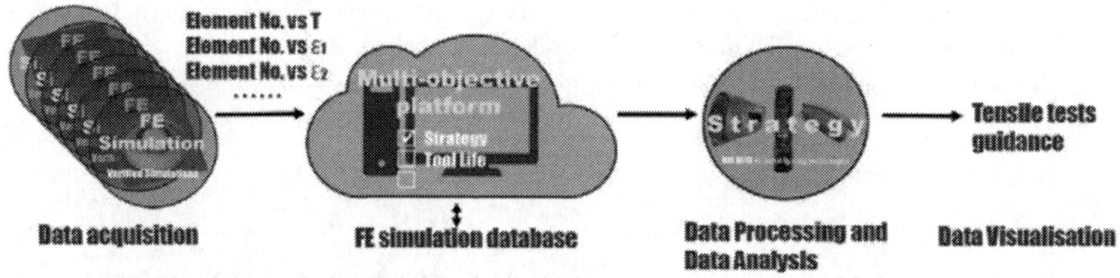

Figure 2. KBC-FE simulation technique – 'Strategy' implementation process

The 'Strategy' module works in three stages: data acquisition, data processing and data analysis, as shown in Fig. 2. Data is acquired through the multi-objective KBC-FEA platform, which can be accessed via www.smartforming.com. This is an online portal system developed at Imperial College enabling engineers to share their knowledge and sophisticated models. The case study in this paper is based on a database for hot stamping simulations of lab scale dome tests. Simulations were conducted at different forming conditions, such as different forming speeds and blank holding forces. The database consists of the thermo-mechanical history of all elements in all stages of each simulation to capture the characteristics of the deformation process. Data from each simulation is stored as a separate entity and managed in the central cloud data system, and is then processed and analysed in the functional module. To establish a uniaxial tensile test guide for the hot stamping technique, evolutions of temperature and strain rate are selected from each simulation file. The first step involves filtering the raw data to distinguish between elements from different regions, *e.g.* the blank holding area, die cavity *etc.* In this study, the elements in the blankholder region are removed since plastic deformation in this region is very low, and does not represent the characteristics of hot stamping. Due to the possibility of each simulation file having different numbers of elements, the data could potentially be distorted, thus a weighting strategy is applied to adjust the proportional relevance of each simulation to generate representative data. Single and bivariate parameter distributions are then generated for data analysis.

3. Results and discussion: thermo-mechanical characteristics of the hot aluminium stamping technology

Hot stamping of the dome shapes exhibits complicated temperature, strain rate and strain path histories. Figure 3 shows the temperature and strain rate distributions for AA6082 dome shaped parts formed under hot stamping conditions. The temperature distribution is found to be negatively skewed as the majority of elements undergoing hot stamping tend to have higher temperatures. 66% of the elements were formed in the 400-450 °C temperature range. The peak of the distribution indicates the most frequent occurrence, and corresponds to the initial forming temperature of 450 °C used in the lab scale dome tests. The minimum temperature was found to be around 210 °C, while the maximum was 455°C, which may have been caused by frictional heat. The strain rate distribution resembles a normal distribution, and although the peak region of strain rates of 0.1-1 s^{-1} are commonly tested, strain rates higher than 1 s^{-1} are also encountered during hot stamping, occupying 29% of the distribution, where the maximum strain rate was around 30 s^{-1}. Single parameter distributions provide valuable insights into the lower and upper limits, as well as the variation of the conditions encountered in hot stamping processes.

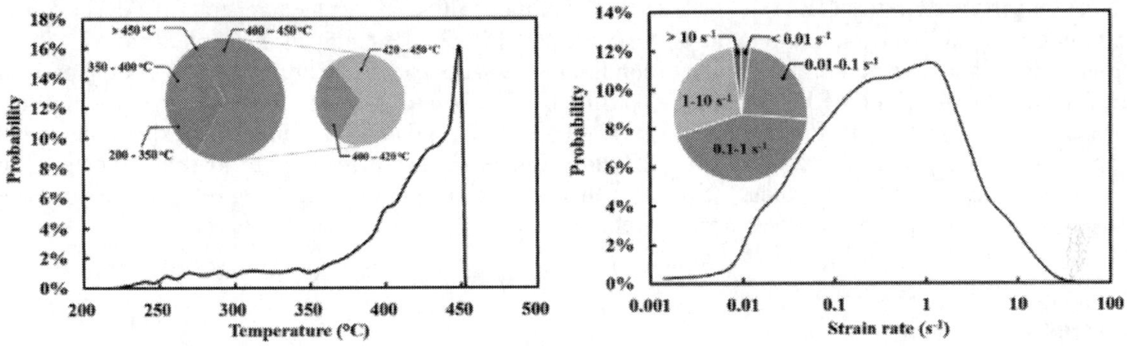

Figure 3. Temperature and strain rate distribution for AA6082 formed under hot stamping conditions

To facilitate the design of tensile test guidelines, a bivariate distribution incorporating the relationship between temperature and strain rate was investigated. This offers greater benefits as tensile tests are conducted at conditions based on both parameters. The relationship between temperature and strain rate obtained from the dome shape forming simulations is visualised in Figure 4, and provides the tensile test boundaries for the hot stamping of dome shapes from AA6082. The maximum and minimum values are the same as those identified in the single parameter distribution. However, higher strain rates (> 5 s^{-1}) were found to be only associated with high temperatures between 350 – 450 °C; this phenomenon is expected as at higher temperatures, the material is softer and hence more plastic deformation occurs leading to higher strain rates. 71.5 % of the elements deformed in the temperature and strain rate range of 400 – 455 °C and 0.01 – 10 s^{-1} respectively. For all temperatures below 350 °C, a maximum strain rate of 5 s^{-1} would be sufficient, which occurred for 13.5% of the elements. It is therefore critical to conduct tensile tests that cover the lower and upper limits, as well as to focus on specific combinations of conditions with the greatest probabilities.

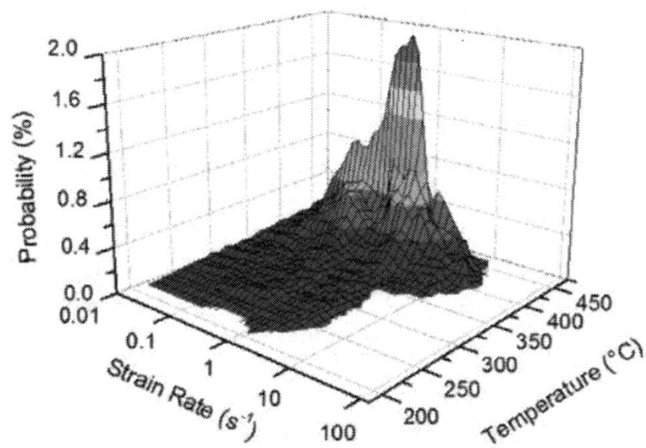

Figure 4. Temperature and strain rate distribution for AA6082 formed under hot stamping conditions

4. Conclusion

The parametric study utilised in the 'Strategy' functional module provides a viable approach for determining test matrices for tensile tests and the conditions to be investigated. The module integrates data from different simulation parts used in the KBC-FEA technique, and numerically analyses the temperature and strain rate distributions to reduce uncertainties in the critical input used in simulating the mechanical behaviour. It has captured dominant temperature and strain rate conditions the blank undergoes during a hot stamping process and detected the trends in the hot stamping of AA6082. The boundaries and trends identified will allow researchers to conduct more focused and targeted tensile tests, reducing the efforts of experiments, and improving efficiency and productivity. As the KBC-FEA database expands, more material and forming technique specific tensile test guides could be generated, with opportunities to develop more strategic analyses to gain insights into various forming techniques.

References

[1] Z. Horita, M. Furukawa, M. Nemoto, A.J. Barnes and T. G. Langdon, *Acta Materialia,* **48** (2000), pp. 3633-3640.
[2] H. Watanabe, T. Mukai, M. Kohzu, S. Tanabe and K. Higashi, *Acta Materialia,* **47** (1999), pp. 3753-3758
[3] A. Turetta, S. Bruschi and A. Ghiotti, *J. Mater. Process. Technol.,* **177** (2006), pp. 396-400.
[4] S. Hong and S. Lee, *Int. J. Fatigue,* **26** (2004), pp. 899-910.
[5] D. Li and A. Ghosh, *Materials Science and Engineering: A,* **352** (2003), pp. 279-286.
[6] M. S. Mohamed, A. D. Foster, J. Lin, D. *Int. J. Mach. Tools Manuf.,* **53** (2012) pp. 27-38.
[7] O. El Fakir, L. Wang, D. Balint, J. P. Dear, *Int. J. Mach. Tools Manuf.,* **87** (2014), pp. 39-48.

Comparison of Constitutive Relationships of Tubes Established Using Uniaxial Tensile Tests and Tube Hydroforming Experiments

Jing Li and Lianfa Yang*

Faculty of Mechanical & Electrical Engineering, Guilin University of Electronic Technology, Guilin, Guangxi, 541004, China.
Corresponding author: y-lianfa@163.com

Abstract. The aim of the present paper is to evaluate existing constitutive models and to fitting hardening laws of SS304 tubes for the accurate prediction of the deformation behaviors of the tubes in hydroforming. Uniaxial tensile test (UTT) and free hydro-bugling (FHB) experiments were conducted on SS304 tubes, and a hi-speed three-dimensional (3D) digital image correlation (DIC) system was applied to obtain the deformation data of the samples. Eight constitutive relationships of the tubes were then established by fitting the equivalent stress and strain data with the four existing constitutive models of Hollomon, Ghosh, Voce and Ghosh/Voce, and the fitting accuracy of the obtained constitutive relationships were analyzed and compared. The results show that Ghosh/Voce model holds the highest accuracy in describing the deformation behaviors of the tubes in UTT and FHB, followed by the Ghosh model and then the Hollomon model. The Voce model holds the lowest accuracy. A distinct discrepancy between the constitutive relationships obtained using UTT and FHB experiments are observed in present research conditions.
Keywords. Tube; Constitutive relationship; Uniaxial tensile test; Free hydro-bugling

1. Introduction

Tube hydroforming (THF) is an advanced near net-shape technology of hollow parts with complex cross-sections. It has been widely used in automotive and aircraft industries owing to its improvement of structural stiffness and weight reduction in products [1]. An accurate constitutive relationship is essential for the analyses of the forming process of THF. Constitutive relationship is an equation to describe the relationship between the equivalent stresses and strains of metallic materials under plastic deformation. Uniaxial tensile test (UTT) is widely used to establish the constitutive relationship of metals in plastic deformation. However, free hydro-bugling (FHB) experiments have increasingly been adopted to establish the constitutive relationships of metallic tubes for THF, due to their consistent biaxial stress state. For instance, N H Wang et al. [2] fitting the polynomial hardening law for tubes based on the plastic incremental theory by conducting UTT and THF. T Hakoyama et al. [3] fitting the hardening law of multiaxial tubes using the Ghosh model by conducting FHB experiments.

The aim of this paper is to fitting hardening laws of SS304 tubes using UTT and FHB experiments, respectively. In the experiments, a hi-speed three-dimensional (3D) digital image correlation (DIC) system was applied to obtain the deformation data of the samples. Eight constitutive expressions were then obtained by fitting the equivalent stress and strain data with four existing constitutive models of Hollomon, Ghosh, Voce and Ghosh/Voce. The obtained constitutive relationships were analyzed and compared.

2. Constitutive models

Diversified constitutive models exist at present. Among them, Hollomon and Ghosh model are both power functions. Hollomon model, a special form of Ghosh model, was put forward by Hollomon [4] in 1944. Hollomon model is commonly used to describe the stress-strain relation of steels and nonferrous metals in plastic deformation. Voce model is an exponential function that was put forward by Voce [5] in 1948. The Ghosh/Voce model, a combination model of the power function and the exponential function, have been used to describe the stress-strain relationship in the deformation process of metal materials [6, 7].

In the present research, four typical models including Hollomon, Ghosh, Voce, Ghosh/Voce, as shown in Tab. 1, are selected to investigate based on the features of THF.

Table 1. The constitutive models and the values of the material parameters obtained by fitting the data of UTT and FHB

Constitutive models	Expressions of the models	Material Parameters	Values based on UTT (Fitting precision)	Values based on FHB (Fitting precision)
Hollomon	$\sigma = k\varepsilon^n$	k, n	k=1594, n=0.4491 (R^2=0.9987)	k=1540,n=0.4517 (R^2=0.9838)
Ghosh	$\sigma = a(\varepsilon_0+\varepsilon)^b - c$ $\varepsilon_0 = \sqrt[b]{\sigma_y / a}$	a, b, c, σ_y	a=1767,b=0.3871,c=203.1 (R^2=0.999, $RMSE$= 6.638) σ_y=360MPa	a=1733,b=0.51,c=105.6 (R^2=0.9845) σ_y=360MPa
Voce	$\sigma = a - be^{-c\varepsilon}$	a, b, c	a=-4829, b=-5140, c=-0.3707 (R^2=0.9168)	a=-9022,b=-9346,c=-0.2147 (R^2=0.9692)
Ghosh/Voce	$\sigma = k(Ghosh)+(1-k)(Voce)$	k	k=1.017(R^2=0.999, $RMSE$=6.512)	k=0.9955 (R^2=0.985)

(Tips: R^2, the determination coefficient, indicate the fitting accuracy, and a smaller value of $|R^2\text{-}1|$ indicates a higher fitting precision. *RMSE*, the root-mean-square error, judge the deviation between the fitted values and the experimental values, and a bigger value of $|RMSE\ \text{-}1|$ indicates a higher fitting precision.)

3. Descriptions of experiments

3.1. Uniaxial Tensile Test (UTT)

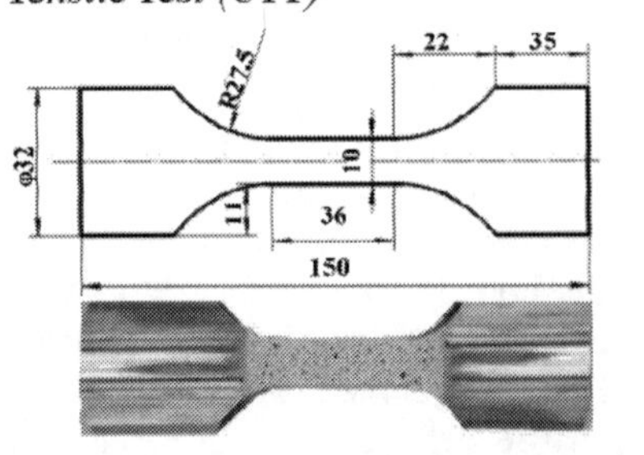

Figure 1. Dimensions of the specimen for UTT

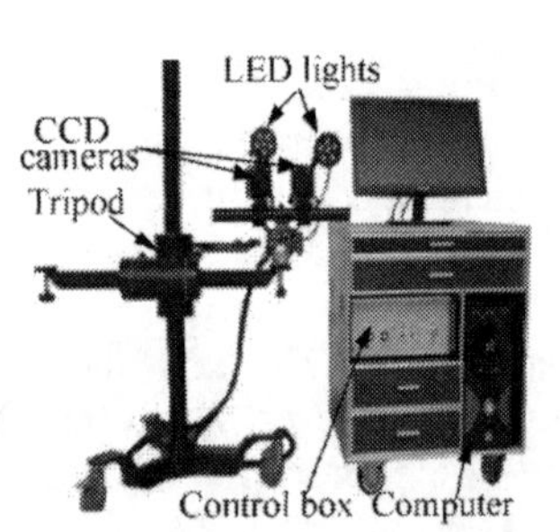

Figure 2. Hi-speed 3D DIC system

SS304 tubes with the out diameter of 32 mm and wall thickness of 0.6 mm were adopted in present experiments. The specimens for the UTT were shown in Fig. 1. The UTT were performed on an electronic universal testing machine, a model of WDW-100E, made by Time Group Inc. The testing machine holds the following specifications: maximum capacity of 100 KN, moving speed range of 0.005-500 mm/min, etc. The UTT were carried out with the tension speed 6 mm/min at room temperature.

A hi-speed 3D DIC system, as shown in Fig. 2, was applied to obtain the deformation data of the specimens. The images of the deformation zone of the specimens were synchronously captured by the CCD cameras. The deformation data of 3D displacement, strain and other parameters were calculated by the analysis system based on the images. The effective stresses and strains were then obtained by dealing with the deformation data using the method described in reference [2]. Finally, four constitutive expressions, as shown in Tab. 1, were achieved by fitting the effective stresses and strains in Matlab.

3.2. Free hydro-bugling (FHB) experiments

Tubular specimens, with the initial length L=110 mm and the bulging length l= 50 mm, for FHB were made from the SS304 tubes. The FHB experiments were conducted on the THF system, as shown in Fig. 3. It consisted of a hydraulic power generation system, a hi-speed 3D DIC system and a FHB device.

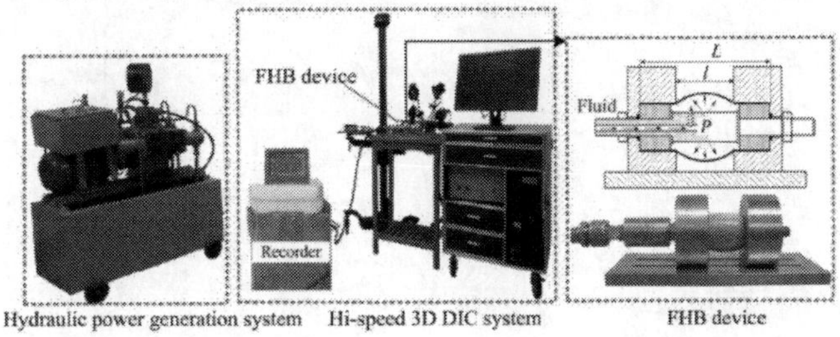

Figure 3. Configuration of self-developed THF system

In the pressure supply system, required hydraulic fluid with a certain pressure for the hydro-bugling was created. The hydraulic fluid was then led into the tubular specimen to bulge it to crack. During the entire process of bugling, the hydraulic pressure was recorded by a sensor, and the deformation data on the bulging zone l were synchronously captured by the hi-speed 3D DIC system.

The effective stress and strain were determined by the methods as described in reference [2]. After the equivalent stress and strain in FHB process were obtained, four constitutive models, as shown in Tab. 1, were achieved by fitting the effective stresses and strains in the software Matlab.

4. Results and discussion

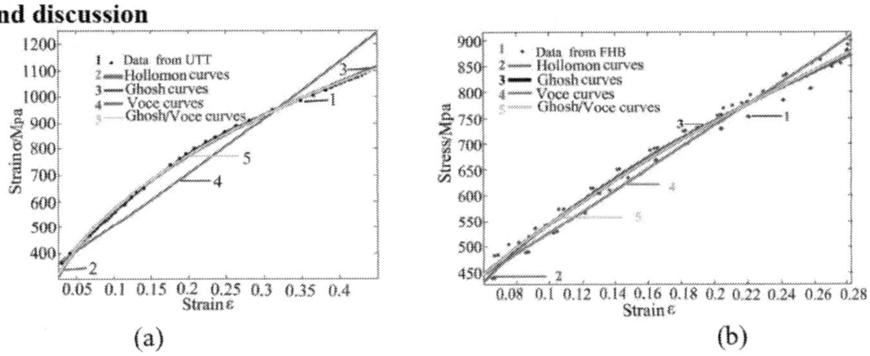

Figure 4. Constitutive curves of tubes: (a) based on data from UTT, and(b) based on data from FHB

The eight constitutive curves plotted according to the constitutive expressions in Tab. 1 are shown in Fig. 4. In view of the degree of the approximation of the fitting curves to the experimental data, it is obvious to observe that Ghosh/Voce model holds the best approximation, followed by the Ghosh model and the Hollomon model, and the Voce model holds the worst approximation on either UTT or FHB cases.

Moreover, with respect to fitting accuracy, compared with the constitutive curves based on FHB, the constitutive curves based on UTT are closer to their own experimental data.

The constitutive curves based on the three constitutive models and the experimental data from the UTT and FHB are shown in Fig. 5. In this section, a relative deviation $\delta=(\sigma_{UTT}-\sigma_{FHB})/\sigma_{FHB}$ is introduced to indicate the maximum difference between the stress σ_{UTT} obtained from UTT and the stress σ_{FHB} obtained from FHB. In the case of the Hollomon model, as illustrated in Fig. 5(a), a relative deviation δ=3.81% happens when the strain arrives 0.33, and δ=3.68% when the strain comes to 0.5, the end of the deformation. In the case of the Ghosh model, as illustrated in Fig. 5(b), a relative deviation 4.69% appears when the strain comes to 0.15, despite a relative deviation of 38.33% at the start of the deformation. In the case of the Ghosh/Voce model, as illustrated in Fig. 5(c), a relative deviation 4.90% appears when the strain comes to 0.15, despite a relative deviation of 39.43% at the start of the deformation.

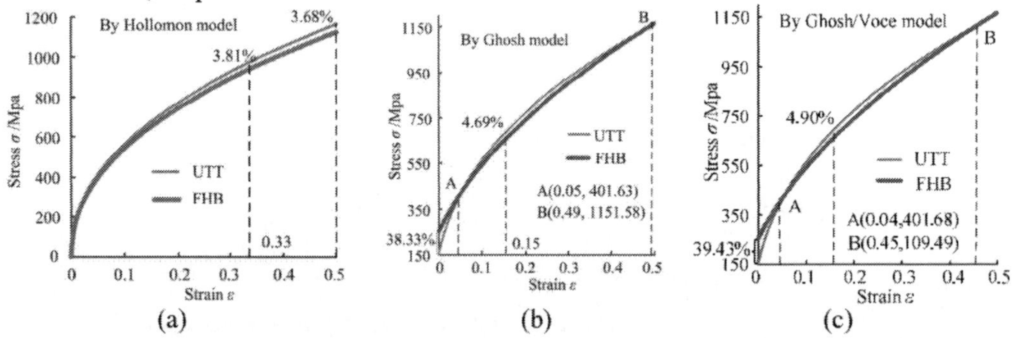

Figure 5. Comparison of the constitutive curves based on the experimental data between the UTT and FHB and the constitutive models of: (a)Hollomon, (b) Ghosh, and (c) Ghosh/Voce

5. Conclusions

Based on the analyses mentioned above, the following conclusions can be drawn from the study:

(1) Among Hollomon, Ghosh, Voce, Ghosh/Voce constitutive models, the deformation behaviors of SS304 tubes in UTT and FHB can best be described by the Ghosh/Voce model, followed by the Ghosh model and the Hollomon model, but worst by Voce model.

(2) With respect to of fitting accuracy, compared with the constitutive curves based on FHB, the constitutive curves based on UTT are closer to their own experimental data.

(3) There are distinct difference between the constitutive curves based on UTT and FHB. The maximum differences of the two constitutive curves based on UTT and FHB are 3.81%, 4.69% and 4.90% as for using the Hollomon, Ghosh and Ghosh/Voce models, respectively.

6. References

[1] Tao Z N, Yang L F, and Mao X C 2012 *Hydraulics Pneumatics & Seals*. **3** 20
[2] Wang N H and Yang L F 2015 *Forging & Stamping Technology*. **40** 133
[3] Hakoyama T and Kuwabara T 2015 *Key Engineering Materials*. **651-653** 9
[4] Hollomon J H 1944 *Trans of ASM*. **32** 123
[5] Voce E 1948 *Journal of the Institute of Metals*. **74** 537
[6] Kim J H, Kim D, Han H N *et al* 2013 *Materials Science and Engineering: A*. **559** 222
[7] Kim, Lee S, Jinwoo *et al* 2013 *Journal of Materials Processing Technology*. **11** 1929

Acknowledgements

The authors gratefully acknowledge the support of the National Natural Science Foundation of China (grant number 51271062), Guangxi Natural Science Foundation (grant number 2013GXNFAA019305).

Springback Compensation Process for High Strength Steel Automotive Parts

M. Fatih Onhon

Ford Otosan Tool&Die Department,Ihsaniye-Kocaeli,Turkey

fonhon@ford.com.tr

Abstract. This paper is about an advanced stamping simulation methodology used in automotive industry to shorten total die manufacturing times in a new vehicle project by means of benefiting leading edge virtual try-out technology.

1. Introduction

We are talking about a springback compensation process in the automotive tool&die industry.Usually one part can only be produced through four operations to its final shape.To perform four operations,you need four dies and four presses which are building together a pressline.

Neither springback prediction nor its compensation is given considering only one forming operation.All existing operations of the part should be taken into account to reach desired results.For that reason we developed a methodology which we call as stamping simulation system.Accurate springback prediction is the key point of that system.All other following processes are depending on the success of prediction.The main goal of virtual try-out engineering in the industry is always to decrease die manufacturing costs and delivery times.If your prediction is not enough accurate than you can not decrease the costs and times because you would have to recut the dies at the try-out.

2. Components of Stamping Simulation System

Stamping simulation system is a die process work flow including all aspects of stamping an automotive part within satisfying tolerances. The basic goal of that system is to solve all issues of the part during the die process phase and will have not any need to touch the dies at the try-out.

In the figure 1 you can see the general phases of a die manufacturing project and the above mentioned die process phase during which we implement the stamping simulation system.It begins with formability studies and ends with a virtual try-out step.One of the interesting things of that system is that we are also reporting and reviewing the die process output with the management with help of a virtual inspection report which is corresponding to the ISIR (Initial Sample Inspection Report).ISIR is the name of try-out measurement reports of the stamped parts used for part buy-off.In the figure 2 steps of stamping simulation system are showed in detail.

Journal of Physics: Conference Series **734** (2016) 032044 doi:10.1088/1742-6596/734/3/032044

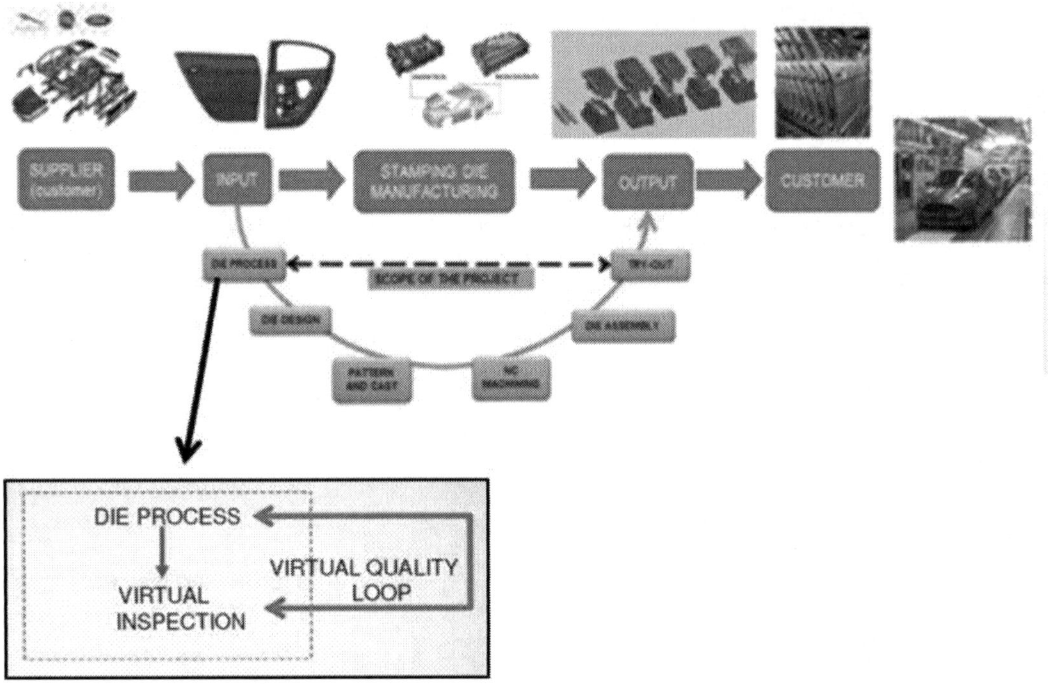

Figure 1. Die manufacturing phases and virtual try-out concept

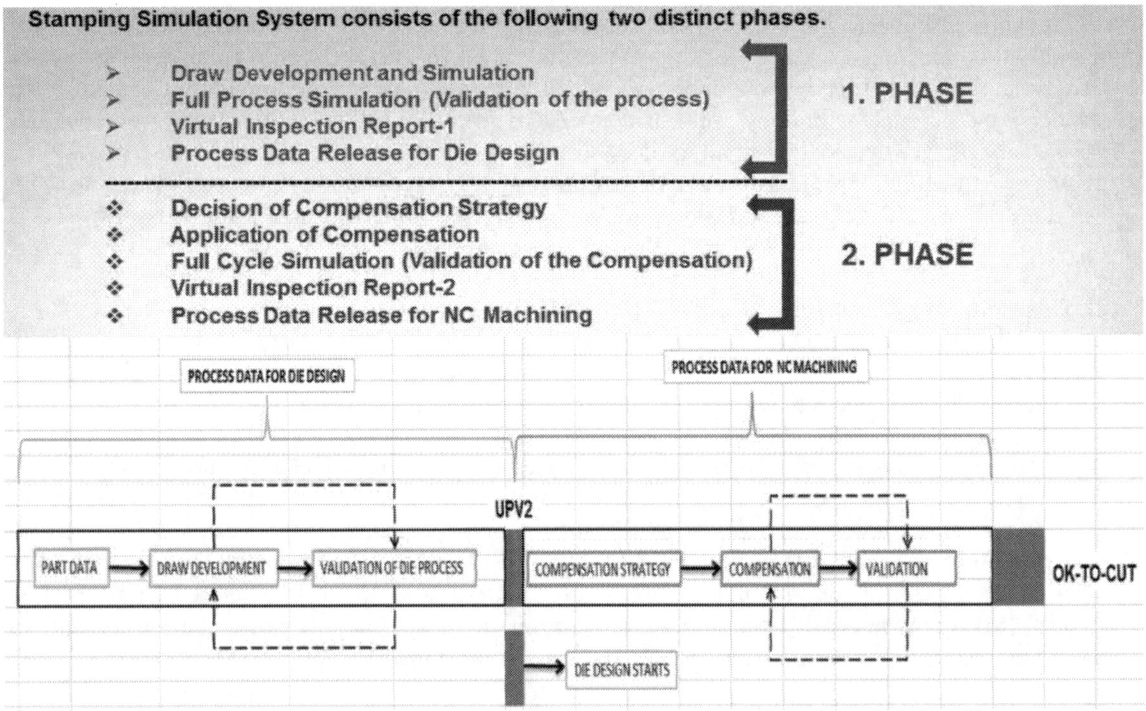

Figure 2. Components and phases of the stamping simulation system

Journal of Physics: Conference Series **734** (2016) 032044 doi:10.1088/1742-6596/734/3/032044

2.1. Draw Development

Draw development is the most important phase where binder and addenda surfaces are developed for a part forming the draw die surface together.

In the figure 3 some of the main post-processing parameters are listed which must be checked during draw development.Before any springback compensation, minimizing opportunities of it should be carefully studied.Minimizing springback is the best compensation method.We are using only Autoform as FEA software for all kind of simulations.

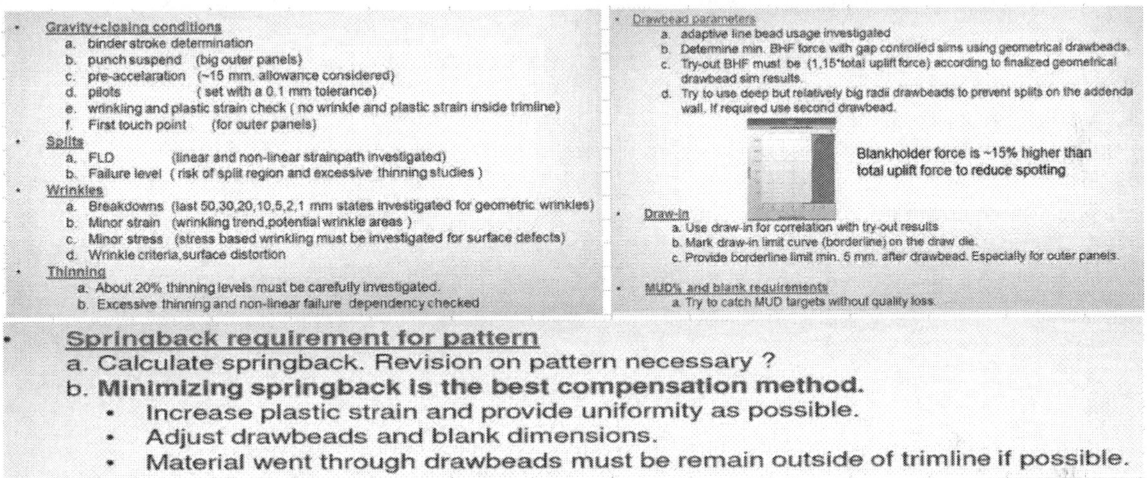

Figure 3. Details of Draw development

2.2. Springback Compensation

For a satisfying springback compensation there should be some strategies you can follow.Compensation is seen as a process because all operations of the part must be considered and will be affected because springback differs after each operation.What we want is that stamped part would be within %90 inside the satisfying tolerance limit.In the figure 4 several compensation strategies are showed which we apply according to the springback behavior of the part.

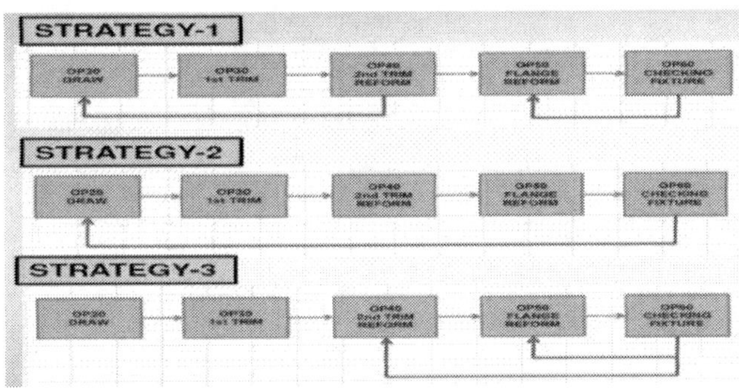

Figure 4. Compensation strategy samples

2.3. *Virtual Try-Out*

Virtual try-out is a final validation purpose advanced level simulation.If something not satisfying with the results more than one iteration (virtual quality loops) can be done until you achieve expected results.In the figure 5 parameters of a such simulation is listed.

- Simulation with morphed surfaces and all operations.
- Springback calculation after each operation.
- Fitting of blank checked and studied for all operations. (gravity+locating)
- First trim tool developed according to springback after draw.
- Used tool structures are taken from die design (pad relief taken into account)
- Forces similar with die design (gas springs, pad force, air cushions etc.)
- Trim operations are performed with pad as the reality.
- Geometric drawbeads are used. (Mesh size must be small accordingly)

Figure 5. Items of the virtual try-out simulation

3. A High Strength Steel Case Study

One part from our projects is given as an example below.Its material is HSLA380 and it has a thickness of 1 mm. There is a special point on the springback prediction of HSS parts.You must consider kinematic hardening which has a big effect on the springback results.The below part is a good example for that.In the figure 6 comparison can be seen between the virtual part and real stamped part.All phases mentioned in the above sections are applied during compensation process.

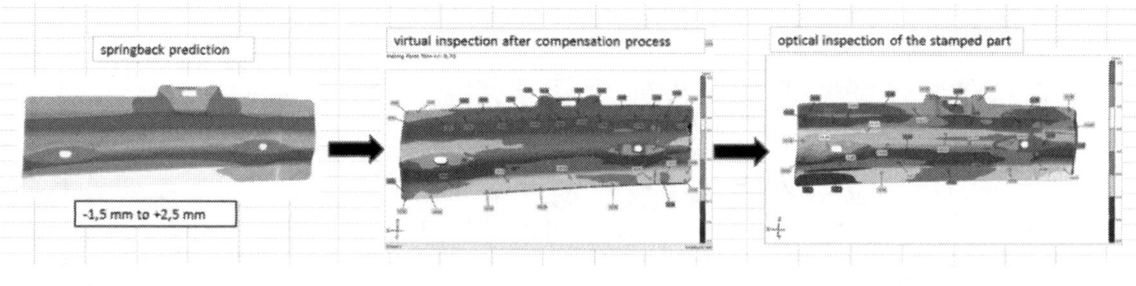

Figure 6. Results of a HSS part

Figure 7. Stamped part on the checking fixture

Influence of Yield Condition on the Accuracy of Earing Prediction for Steel Sheets

Marco Gösling

BILSTEIN GmbH & Co. KG, Im Weinhof 36, D-58119 Hagen, Germany

marco.goesling@bilstein-kaltband.de

Abstract. This paper is dealing with the material modelling of steel sheets and is focused on the input parameters for a correct earing prediction. The cause of earing is the anisotropy of the rolled sheet which is usually modelled by a yield criterion. In a first study earing predictions with the Hill'48 yield criterion and with the Barlat'2000 yield criterion are conducted for different steel grades between 200 and 800 MPa yield strength. A comparison of the results shows that the Barlat'2000 yield criterion leads in almost all cases to a better earing prediction. In a second study the measurements for the Barlat'2000 law were analysed, to find the main parameter influencing the accuracy in earing prediction. The results of this study show that it is not affected by the biaxial measurements, but by the yield strength in 45° regarding to rolling direction.

1. Introduction

Forming simulations by Finite Element Analysis are well-established in the development process of sheet metal parts in order to predict cracks, wrinkles, springback and press forces. One of the main influencing variables on the simulation quality is the material model.

Sheet metals usually exhibit a plastic anisotropy influenced by previous processes like hot rolling, cold rolling and annealing. One manifestation of the plastic anisotropy is a phenomenon called earing. Earing occurs during deep drawing of a cylindrical cup and can be described as a periodic variation of wall height around the diameter of the finished cup (see figure 1). In accordance to DIN EN 1669 the anisotropy can be described by the Mean Ear Height h_e and the Earing in percent Z.

Figure 1. Earing Test.

In figure 2 the correlation between earing and anisotropy can be seen by means of the planar anisotropy. An increasing anisotropy is leading to an increasing earing.

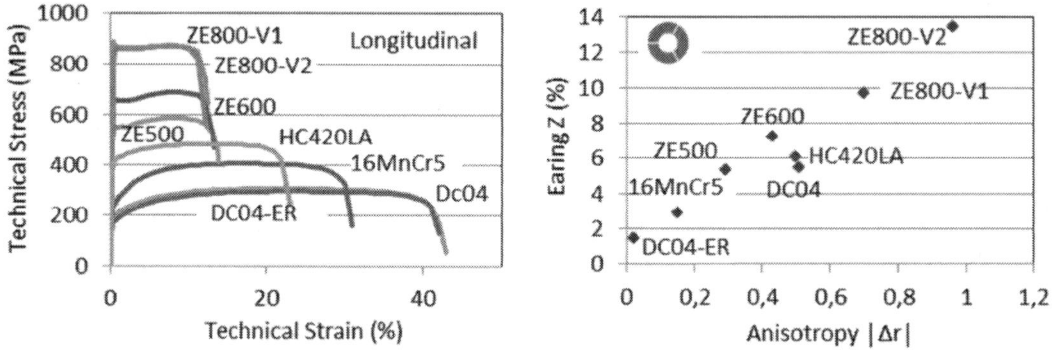

Figure 2. Materials, Earing and Anisotropy.

All materials, used in this study, are cold rolled single-phase steels: mild steels (DC04), steels for case hardening (16MnCr5) and high strength low-alloy steels (HC420LA and ZE-grades). BILSTEIN's ZE-grades are non-standard high strength low-alloy steels, which are usually customized to the specific strip application. This is demonstrated with the help of two variants of a ZE800. These materials are adjusted to different directional characteristics because the application, respectively the position in forming press regarding to rolling direction is different. A similar case is the material DC04-ER. This is a so called "Earing Reduced" mild steel. This material is adjusted to nearly isotropic behavior due to a combination of cold rolling and heat treatment strategy.

Many authors show for aluminum alloys that the accuracy of earing prediction is depending to the chosen yield criterion [1], [2], [3]. For this reason, the influence of the yield criterion on the accuracy of earing prediction is analysed also for steel sheets. The Hill'48 criterion [4] is compared to the Barlat'2000 criterion [5].

2. Earing Prediction

PAM-STAMP 2G (V2012.2) was used to model the cup drawing test. The blank was modelled by shell elements and the tools by rigid bodies. The contact between tools and blank was modelled by a nonlinear penalty contact with coulombs friction. The friction value was calibrated by comparing the measured punch force to simulated punch forces.

The influence of the yield condition to earing prediction can be seen in figure 3, where earing predictions for the Hill'48 and the Barlat'2000 yield criterion are compared.

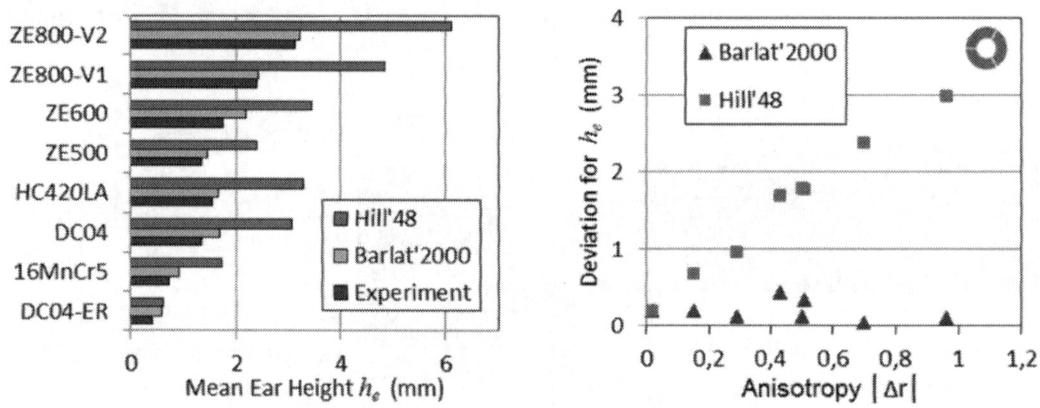

Figure 3. Influence of Yield Condition

Both yield conditions can describe the influence of anisotropy to earing. Both models predict bigger ears for materials with higher anisotropy. But a comparison of the results shows that the Barlat'2000 yield criterion leads in almost all cases to a much better earing prediction. The deviation between simulation and experiment for the Barlat'2000 criterion is only small. For the Hill'48 criterion the deviation tends to increase with increasing anisotropy.

3. Input parameters

One difference of the used yield criteria are the input parameters, respectively the measurements for characterization. The input parameters for the Hill'48 criterion are the r-values (r_{00}, r_{45}, r_{90} from tensile tests). The Baralat'2000 criterion requires the r-values, the yield stresses (σ_{00}, σ_{45}, σ_{90} from tensile tests), the biaxial yield stress σ_{bi} (from hydraulic bulge tests) and the biaxial anisotropy r_{bi} (from disc compression tests). Further it is possible to choose the exponent M. The influence to earing prediction of these input parameters was analysed by using Barlat'2000 models with different input parameter sets (see Table 1).

Table 1. Parameter Settings.

	r_{00}	r_{45}	r_{90}	σ_{00}	σ_{45}	σ_{90}	σ_{bi}	r_{bi}	M
Model M6	M	M	M	M	M	M	M	M	6
Model M2	M	M	M	M	M	M	M	M	2
Model Bi	M	M	M	M	M	M	H	H	6
Model ST	M	M	M	M	M	H	H	H	6
Model SD	M	M	M	M	H	M	H	H	6
Model Si	M	M	M	M	H	H	H	H	6
Model Hi	M	M	M	M	H	H	H	H	2

M: Measurement; H: Prediction by Hill'48

The input parameters for material model calibration were switched between measurements and predictions from Hill'48 model. The exponent was changed between M=2 and M=6. The results of these simulations are displayed in figure 4.

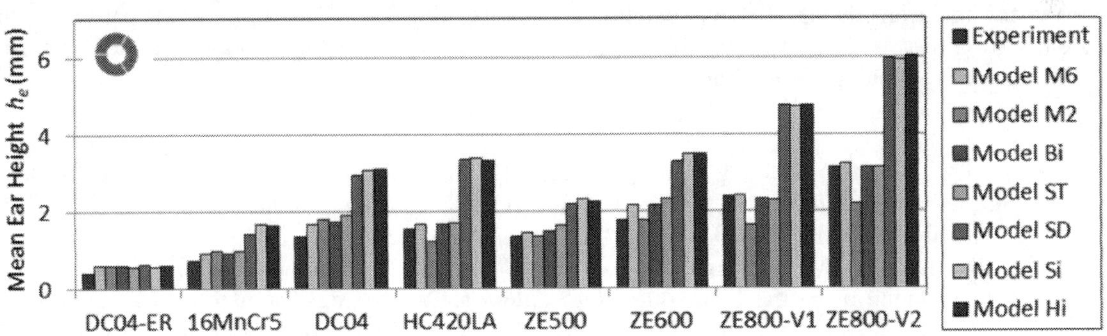

Figure 4. Influence of Input Parameters to Earing Prediction.

Comparing model M6 with Bi shows that the results are changing only slightly when biaxial measurements are taken into account. This means that one of the main differences (biaxial measurements) between Barlat'2000 and Hill'48 criterion has nearly no effect to earing prediction. Comparing model M6, M2, Bi and ST with SD, Si and Hi shows that the prediction accuracy is improved strongly when the yield criterion includes the yield strength σ_{45}. The parameter σ_{45} seems to be the main parameter causing the improvement in earing prediction.

In figure 5 the predictions of the yield criteria for material characteristics to the measurements of the material characterization are compared.

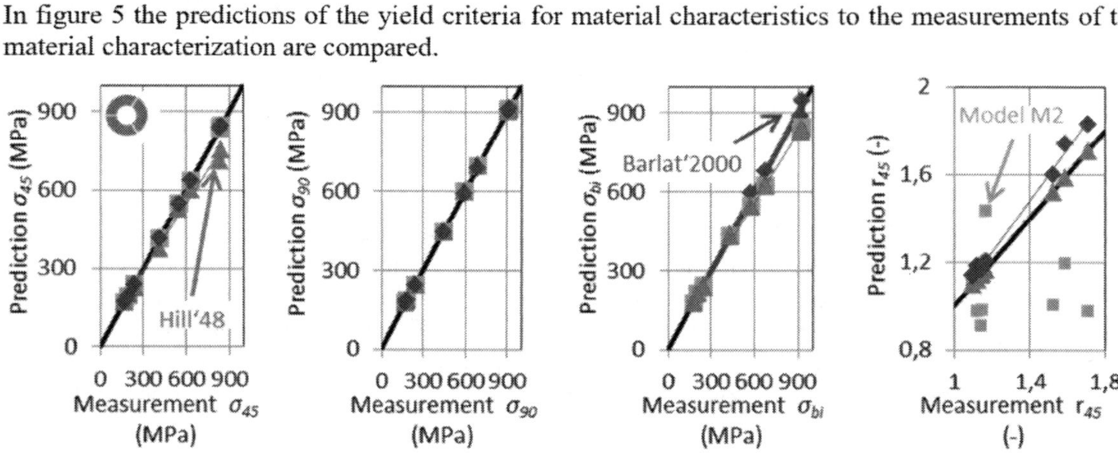

Figure 5. Predictions for material characteristics

As expected the Barlat'2000 predictions for σ_{45} are better than the Hill'48 predictions. The predictions for σ_{bi} are also better. But as mentioned before, this improvement in the material model has only slightly effects on earing prediction. The effect of σ_{90} on earing prediction is small, because both used yield criteria predicts σ_{90} very well. Note that this does not mean that σ_{90} has no influence on earing prediction. It means only that both yield criteria are suitable to predict σ_{90}. The influence of the exponent M on earing prediction seems to be unclear. This results from the model calibration. An exponent M=2 for the Barlat'2000 criterion is not suitable to predict all measured material characteristics (e.g. r_{45} for Barlat'2000).

4. Conclusion

For the Hill'48 criterion the prediction accuracy of earing is decreasing with increasing anisotropy. So, it is useful to apply a more complex yield criterion for materials with high anisotropy. The prediction accuracy of the Barlat'2000 criterion is much better than the Hill'48 model. This improvement is caused by considering σ_{45} in the yield criterion. The biaxial measurements σ_{bi} and r_{bi} do not improve the results. This means that the better prediction accuracy of the Barlat'2000 criterion is caused by the flexibility to meet both r_i and σ_i from tensile tests and not by the more complex material characterization (biaxial measurements).

References

[1] Yoon, J.W.; Barlat, F.; Chung, K.; Pourboghrat F.; Yang, D.Y.: Influence of initial back stress on the earing prediction of drawn cups for planar anisotropic aluminum sheets, Journal of Materials Processing Technology 80–81, pp. 433–437, 1998.

[2] Vrh, M.; Halilovic, M.; Starman, B.; Stok, B.; Comsa, D.S., Banabic, D.: Earing Prediction in Cup Drawing using the BBC2008 Yield criterion, Proceedings of international Conference on Numerical Simulation of 3D Sheet Metal Forming Processes, AIP Conf. proc. 1383 pp. 142- 149, 2011.

[3] Soare, S.; Banabic, D: About the mechanical data required to describe the anisotropy of thin sheets to correctly predict the earing of deep-drawn cups. Proceedings of ESAFORM Conference, 2008.

[4] Hill, R.: A theory of the yielding and plastic flow of anisotropic metals. Proc. Roy. Soc. A193, pp. 281-297, 1948.

[5] Barlat, F.; Brem, J.C.; Yoon, J.W.; Chung, K.; Dick, R.E.; Choi, S.H.; Pourboghrat, F.; Chu, E.; Lege, D.J.: "Plane stress yield function for aluminum alloy sheets", Int. J. Plasticity, 19, pp. 1297-1319, 2003.

Remapping algorithms: application to trimming operations in sheet metal forming

D M Neto[1], C M A Diogo[1], T F Neves[1], M C Oliveira[1], J L Alves[2] and L F Menezes[1]

[1] CEMUC, Department of Mechanical Engineering, University of Coimbra, Polo II, Rua Luís Reis Santos, 3030-788 Coimbra, Portugal
[2] CMEMS, Department of Mechanical Engineering, University of Minho, Campus de Azurém, 4800-058 Guimarães, Portugal

E-mail: diogo.neto@dem.uc.pt

Abstract. Most of sheet metal forming processes comprise intermediate trimming operations to remove superfluous material. These operations are required for subsequent forming operations. On the other hand, the springback is strongly influenced by the trimming operations that change the part stiffness and the stress field. From the numerical point of view, this involves the geometrical trimming of the finite element mesh and subsequent remapping of the state variables. This study presents a remapping method based on Dual Kriging interpolation, specifically developed for hexahedral finite elements, which has been implemented in DD3TRIM in-house code. Its performance is compared with the one of the Incremental Volumetric Remapping method, using the split-ring test to highlight their advantages and limitations. The numerical simulation of the forming processes is performed with DD3IMP finite element solver.

1. Introduction

Typically, the finite element simulation of multi-stage sheet metal forming processes requires the modification of the blank geometry, which is the consequence of trimming operations. Therefore, some remapping procedure is required to transfer the state variables from the old mesh to the new one, including the ones evaluated at the integration points, within each finite element [1]. The number of state variables depends on the constitutive model adopted, but usually includes the current flow stress, the stress field, the plastic strain, etc.

The remapping methods can be divided into four groups [2]. The first group refers to pointwise interpolation and extrapolation methods, where the variables are transferred from the old mesh to the new one using a function that interpolates/extrapolates the variables. The second group refers to the area/volume weighted averaging methods, which use the area/volume intersection between the old and the new mesh to define weighting factors, expressing the contribution of each finite element from the old mesh to the new one. The third group refers to mortar element methods, which are general techniques for projecting data at interfaces between non-conforming subdomains [3]. From a mathematical point of view, this method comprises the minimization of a weighted residual, where the weight functions are usually chosen from the space spanned by the basis functions of the mortar side. The last group refers to specialized methods, which are designed for specific applications [4].

This study presents and compares two distinct remapping methods for finite element meshes composed by linear hexahedral elements, namely the Incremental Volumetric Remapping (IVR) [5] and the Dual Kriging (DK) [6]. Both methods were implemented in the in-house code DD3TRIM, which was specifically developed to perform trimming and remapping operations involved in the multi-stage sheet metal forming processes. In order to assess the performance of each remapping method, the springback of an aluminum alloy is quantified, using the split-ring test example.

2. Remapping algorithms

The accuracy and the computational performance are key points of any remapping algorithm. In fact, the remapping operation introduces errors in the new mesh, due to the approximation of the state variables. In order to try to control and reduce the error arising from the remapping operations, several authors point out some desirable characteristics. The method should be self-consistent, guarantee the locality and avoid spurious local extreme values. Accordingly, the IVR and the DK methods are described in detail, where the old and the new finite element meshes are denoted by donor and target mesh, respectively [2,7].

2.1. *Incremental volumetric remapping*

The IVR method is based on the concept that the value of each state variable, at the region corresponding to a Gauss volume (an eight part of the standard brick element), is equal to the state variable of the correspondent Gauss point. Therefore, the variable value assigned to a given Gauss point of the target mesh element is evaluated based on a weighted average of the values of the donor mesh [5]. Hence, the value of the state variable α in a Gauss point *i* of the target mesh is given by:

$$\alpha_i = \sum_{j=1}^{ng} w_{ij}\alpha_j, \quad \text{with} \quad w_{ij} = \frac{1}{v_i^{\text{total}}}\sum_{k=1}^{nl^3} v_{ij}^k, \tag{1}$$

where *ng* denotes the number of donor Gauss volumes that contributes for the target Gauss volume *i*. w_{ij} is the fraction of the target Gauss volume *i*, contained within the donor Gauss volume *j*, and α_j is the value of the state variable in the donor Gauss volume *j*. The target Gauss volume is partitioned in *nl* equal subdivisions in each direction (see Figure 1 (c)), in order to approximate the fraction of the target Gauss volume contained in each donor Gauss volume (see Figure 1 (d)). The weight function w_{ij} is the fraction between the summation of the elementary volumes v_{ij}^k of the target Gauss volume *i*, contained in the donor Gauss volume *j*, and the total volume of the target Gauss volume *i*, denoted by v_i^{total}.

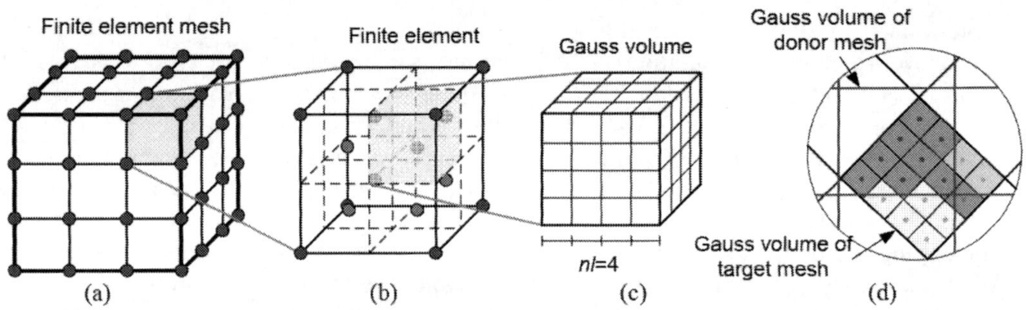

Figure 1. Schematic representation of the IVR method: (a) finite element mesh; (b) subdivision of each element into 8 Gauss volumes; (c) subdivision of the target Gauss volume; (d) intersection between target and donor Gauss volumes.

2.2. *Dual-Kriging method*

The Dual Kriging method is applied in the present study to remap the state variables between different finite element meshes. This method provides an explicit parametric interpolation, allowing to evaluate the value of the state variables in any point (Gauss points of the target mesh). In order to reduce the computational effort, only the donor Gauss points neighboring to the target Gauss point under analysis are selected and used [6]. The general form of the DK model is decomposed into the sum of two terms:

$$\alpha(\mathbf{x}) = d(\mathbf{x}) + f(\mathbf{x}) = (d_1 + d_2 x + d_3 y + d_4 z) + \sum_{i=1}^{n} b_i K(h_i), \tag{2}$$

where $d(\mathbf{x})$ denotes the global trend function (polynomial) and $f(\mathbf{x})$ represents the fluctuation or deviation function, which captures the local deviations from the trend function. The fluctuation depends on the n donor Gauss points used to calculate the state variables in the target Gauss point. Indeed, it is constructed using n parameters b_i, which are weighted by the generalized covariance function $K(h_i)$ associated with each donor Gauss point. The accuracy of the DK method is affected by the covariance function selected. In the present study, it is simply the Euclidean distance between the target Gauss point and the donor Gauss point. The n parameters b_i and the four parameters d_i (the last term $d_4 z$ disappears for 2D applications) are determined by solving a system of n+4 linear equations, which has to be solved only once for each target Gauss point. Then, the interpolated value of all state variables is computed.

For each target Gauss point, the set of donor Gauss points located in its region of influence must be selected. Since the sheet metal forming simulation is the main application of the proposed remapping method, the selection method is carried out by layers through the thickness, i.e. only the donor Gauss points located in the same layer of the target Gauss point are considered. Accordingly, the DK method is applied in a 2D framework, where each donor Gauss point is projected into a plane defined by the finite element of the donor mesh containing the target Gauss point. Considering a structured discretization, nine is the maximum number of donor Gauss points selected (based on the mesh connectivity).

3. Trimming example

The split-ring test was used to measure the effect of the remapping method on the springback prediction. The cylindrical cup forming (fully draw) was carried using the aluminum alloy AA5754-O [8]. The opening diameter of the die is 35.25 mm and the punch diameter is 33 mm, while the shoulder radius of both the die and the punch is 5 mm. The circular aluminum blank presents 60 mm of diameter and 1 mm of thickness. The residual stress state is evaluated by measuring the opening of a ring (7 mm high) cut from the sidewall of the formed cylindrical cup (8 mm from the bottom).

The finite element simulation of the forming operation was performed with the in-house static implicit finite element code DD3IMP [9], while the ring cut and splitting was performed with the in-house code DD3TRIM [10]. The plastic behavior of the sheet is described by the isotropic work hardening (voce law) and isotropic yield criterion (von Mises). Only half model is simulated, allowing to perform the discretization of the blank with 30.237 solid finite elements (3 layers through the thickness). The von Mises stress distribution in the cylindrical cup is presented in Figure 2 (a), showing the position of the ring cutting. The von Mises stress distribution in the ring, obtained with the IVR and DK remapping methods is presented in Figure 2 (b) and (c), respectively. Both methods provide identical results for the state variables, in agreement with the variable distribution before the cutting and remapping operations (see Figure 2).

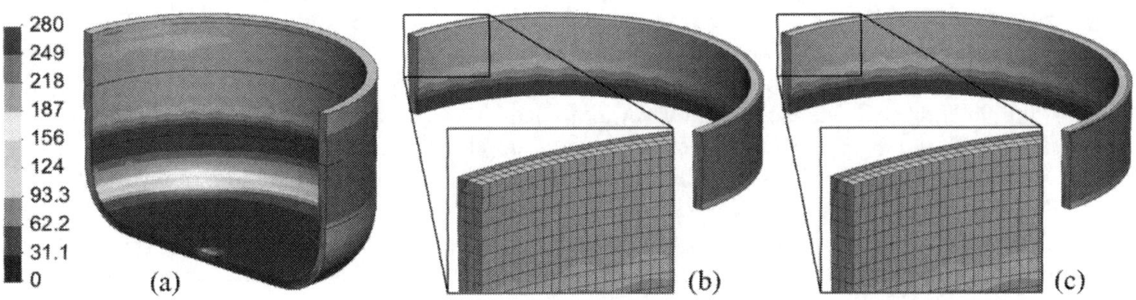

Figure 2. von Mises stress distribution in the: (a) formed cylindrical cup; (b) ring before splitting using the IVR method; (c) ring before splitting using the DK method.

The springback of the ring after cutting and splitting operations is strongly affected by the stress state in the ring, which is dictated by the remapping algorithm adopted. Since both remapping methods (IVR and DK) provide identical stress states in the ring (see Figure 2), the obtained ring opening after springback is similar for both methods (approximately 4.3 mm). The ring gap measurements were performed along the straight line connecting the two ends of the split ring. Regarding the computational performance of each remapping procedure, the computational cost is about 300 seconds using the IVR method, while the DK method requires less than 2 seconds. The calculations were carried out on a computer equipped with an Intel Core i7–4770K Quad-Core processor (3.5 GHz).

4. Conclusions

This study presents the application of the Dual Kriging (DK) interpolation method as remapping scheme, which is compared with the Incremental Volumetric Remapping (IVR) method, both in terms of accuracy and computational performance. The split-ring test is the example considered, which involves the cut of a ring from the sidewall of a cylindrical cup and consequent splitting operation, for springback evaluation. Both remapping methods provide results with a good level of accuracy, specifically the stress field, which presents a strong impact in the springback prediction. Thus, the predicted ring opening is identical in both remapping methods. On the other hand, the computational cost of the DK method is significantly lower (about 100 times) than the IVR method.

Acknowledgments

The authors gratefully acknowledge the financial support of the Portuguese Foundation for Science and Technology (FCT) under the projects with reference UID/EMS/00285/2013, PTDC/EMS-TEC/0702/2014 (POCI-01-0145-FEDER-016779) and PTDC/EMS-TEC/6400/2014 (POCI-01-0145-FEDER-016876) by UE/FEDER through the program COMPETE2020. The first author is also grateful to the FCT for the Postdoctoral grant SFRH/BPD/101334/2014.

References

[1] Rashid M M 2002 Material state remapping in computational solid mechanics *Int. J. Numer. Methods Eng.* **55** 431–50

[2] Jiao X and Heath M T 2004 Common-refinement-based data transfer between non-matching meshes in multiphysics simulations *Int. J. Numer. Methods Eng.* **61** 2402–27

[3] Arbogast T, Cowsar L C, Wheeler M F and Yotov I 2006 Mixed Finite Element Methods on Nonmatching Multiblock Grids *SIAM J. Numer. Anal.*

[4] Chung S W and Kim S J 2003 A remeshing algorithm based on bubble packing method and its application to large deformation problems *Finite Elem. Anal. Des.* **39** 301–24

[5] Baptista A J 2005 Application of the Incremental Volumetric Remapping Method in the Simulation of Multi-Step Deep Drawing Processes *AIP Conference Proceedings* vol 778 (AIP) pp 173–8

[6] Diogo C M A, Neto D M, Oliveira M C, Alves J L and Menezes L F 2016 Incremental Volumetric and Dual Kriging remapping methods *Eng. Comput.*

[7] Perić D, Hochard C, Dutko M and Owen D R J 1996 Transfer operators for evolving meshes in small strain elasto-placticity *Comput. Methods Appl. Mech. Eng.* **137** 331–44

[8] Laurent H, Coër J, Manach P Y, Oliveira M C and Menezes L F 2015 Experimental and numerical studies on the warm deep drawing of an Al–Mg alloy *Int. J. Mech. Sci.* **93** 59–72

[9] Menezes L F and Teodosiu C 2000 Three-dimensional numerical simulation of the deep-drawing process using solid finite elements *J. Mater. Process. Technol.* **97** 100–6

[10] Barros P D, Baptista A J, Alves J L, Oliveira M C, Rodrigues D M and Menezes L F 2013 Trimming of 3D solid finite element meshes: sheet metal forming tests and applications *Eng. Comput.* **31** 237–57

Non-destructive Texture Measurement of Steel Sheets with Compact Neutron Source "RANS"

M Takamura[1], Y Ikeda[1], H Sunaga[1], A Taketani[1], Y Otake[1], H Suzuki[2], M Kumagai[3], T Hama[4] and Y Oba[5]

Center for Advanced Photonics, RIKEN, Wako, Saitama 351-0198, Japan
Materials Sciences Research Cente, Japan Atomic Energy Agency, Tokai, Ibaraki 319-1195, Japan
Faculty of Engineering, Tokyo City University, Setagaya, Tokyo 158-8857, Japan
Graduate School of Energy Science, Kyoto University, Sakyo-ku, Kyoto 606-8501, Japan
Research Reactor Institute, Kyoto University, Kumatori, Osaka 590-0494, Japan

takamura@riken.jp

Abstract. Neutron diffraction is well known to be a useful technique for measuring a bulk texture of metallic materials taking advantage of a large penetration depth of the neutron beam. However, this technique has not been widely utilized for the texture measurement because large facilities like a reactor or a large accelerator are required in general. In contrast, RANS (Riken Accelerator-driven Compact Neutron Source) has been developed as a neutron source which can be used easily in laboratories. In this study, texture evolution in steel sheets with plastic deformation was successfully measured using RANS. The results show the capability of the compact neutron source for the analysis of the crystal structure of metallic materials, which leads us to a better understanding of plastic deformation behavior.

1. Introduction

Sheet forming technologies for high strength steel sheets are becoming more and more important in automotive body parts with the aim of weight reduction, but their use urgently requires further improvement by analyzing the behavior of plastic deformation. On the other hand, recent advances in numerical analyses have enabled the practical application of advanced sheet metal forming simulation technology using finite-element methods (FEM), e.g., simulations of shearing processes performed by authors [1]. It is obvious that the crystal plasticity FEM is one of the most powerful numerical tools to model microscopic deformation mechanisms and to predict macroscopic plastic behavior [2], which is often performed in conjunction with texture measurement techniques such as Electron backscatter diffraction (EBSD) or X-ray diffraction (XRD). However, these methods can provide information only from localized area or surface with the thickness of μ m or 10 μ m.

In contrast, neutron diffraction is well known to be a useful technique to quantitatively measure microstructural factors of metals [3] such as texture in bulk-average taking advantage of a large penetration depth of the beam, which is strongly related to its formability in metal forming processes, bringing advanced understanding of the mechanical behavior of metals by investigating the relationship with macroscopic behavior. Such studies using neutron diffraction typically require a neutron engineering diffractometer [4-11] installed in large experiment facilities such as a reactor and a large accelerator to obtain high flux neutrons. Therefore, we have only few chances in a year to use their facilities because of high-level competition of the beam time. However, investigations of plastic deformation require much more opportunities to measure the texture anywhere at any time like commercial X-ray equipment available at their own locations.

In contrast, Riken Accelerator-driven Compact Neutron Source (RANS) [12, 13] has been developed as a neutron source which can be used easily in laboratories. If we can evaluate microstructural factors easily by neutron diffraction using a compact neutron source at our own laboratory, it is expected to achieve effective analyses of metal forming processes more efficiently. Authors have recently shown diffraction patterns of ferritic steels obtained by RANS and evaluation results of retained austenite volume fraction [14]. In this study, the neutron diffraction experiments are carried out using RANS and plastically deformed mild steel samples to clarify a capability of the compact neutron source for the analysis of the texture evolution, which leads us to a better understanding of plastic deformation behavior.

2. Experimental method

Figure 1 shows an entire view of RANS [12-14], which consists of a proton accelerator, a neutron production target and instruments for the neutron experiment. Protons are accelerated with a proton liniac to 7 MeV, and injected to a beryllium (Be) target [15]. Neutrons with the maximum energy of about 5 MeV are generated via the Be (p,n) reaction. The fast neutrons are moderated in a polyethylene moderator, and the thermal neutrons with approximately 0.01 eV (0.1 nm in wavelength), which is a suitable energy for the diffraction experiment, can be extracted from the moderator surface. Neutrons with 10^4 $s^{-1}cm^2$ in flux are provided to the camera box installed at approximately 5 m far away from the moderator.

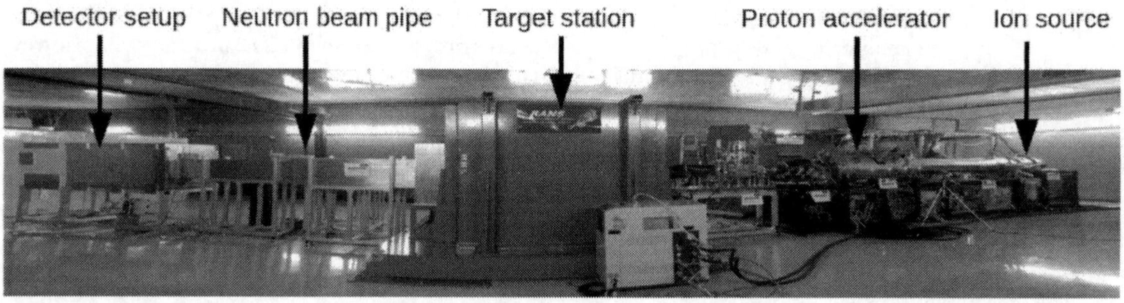

Figure 1. Entire view of RANS. This is about 15 m in total length, consists of an ion source, a proton accelerator, a target station, a neutron beam pipe and a camera box.

Figure 2 shows the experimental setup for measuring diffraction from steel samples [14]. A neutron detector, which consists of a ZnS(Li) scintillator and a position sensitive photo multiplier tube RPMT, were installed inside the camera box. Diffraction patterns were measured based on the time-of-flight (TOF) principle. Energy range from 0.05 to 0.5 nm in wavelength is used to measure the diffraction pattern from the steel in this study.

The sample used here was IF (interstitial free) steel with the thickness of 1.0 mm, which was plastically deformed by 10 % in uniaxial compression along the rolling direction. A cylindrical-shaped specimen with the height of 10 mm and the diameter of 10 mm (ϕ 10) was prepared by following procedure. The small pieces of the plate with a size of ϕ 10 were, at first, taken from a specimen by milling, and then assembled together into the cylindrical shape, preserving the orientation of the pieces. The diffraction angle 2 θ and the distance from the sample to the detector L_2 were set to be 140° and 100 mm, respectively. The diffraction patterns were taken for 10 minutes to get sufficient intensity.

Journal of Physics: Conference Series **734** (2016) 032047
doi:10.1088/1742-6596/734/3/032047

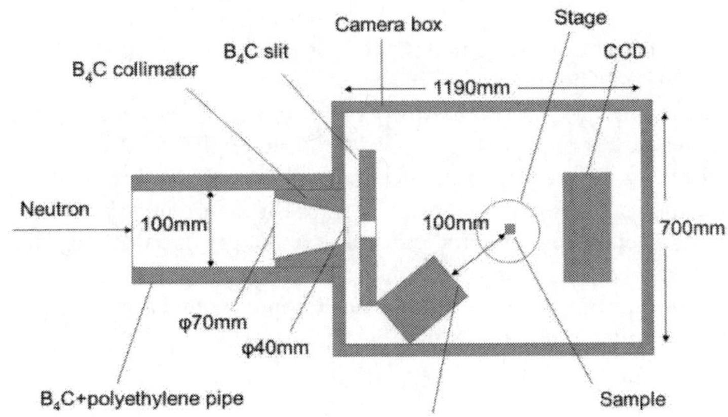

Figure 2. Schematic illustration of the experimental setup.

3. Results and discussion

Figure 3 shows the diffraction pattern in the rolling direction of the deformed specimen, in comparison with that of the undeformed specimen. It can be seen in figure 3 that the undeformed specimen may originally have strong texture even before tensile deformation because the intensity of the 110 reflection is much higher than other peaks. After applying compressive deformation by 10 %, the 110 reflection is the intensity of the 110 reflection is obviously decreased by 0.79 times, while that of the 211 reflections that are perpendicular relation to the 110 reflection is increased by 1.4 times. This is a typical texture evolution for the BCC structure, caused by plastic deformation [16]. Figure 4 shows the variation of diffraction patterns of the deformed specimen with respect to diffraction direction from RD to TD. The variation in the peak intensity shows a possibility of this measurement method for the generation of pole figures. These results suggest that the texture evolution due to the plastic deformation can be observed by RANS. TOF neutron diffraction has an advantage for efficient texture measurement by measuring multiple neutron TOF histograms in various directions of the pole figure, which are simultaneously measured by a wide area detector [17]. For RANS based on TOF neutron diffraction, therefore, it is also expected to measure the texture of material for relatively short time by using similar technique.

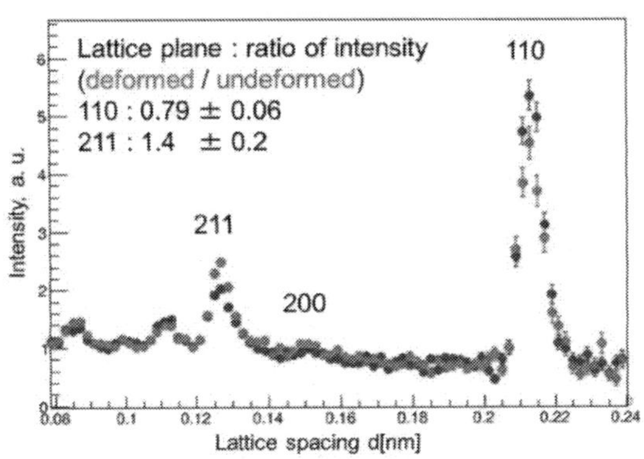

Figure 3. Comparison of the diffraction patterns of the IF steel before and after 10% compressive deformation.

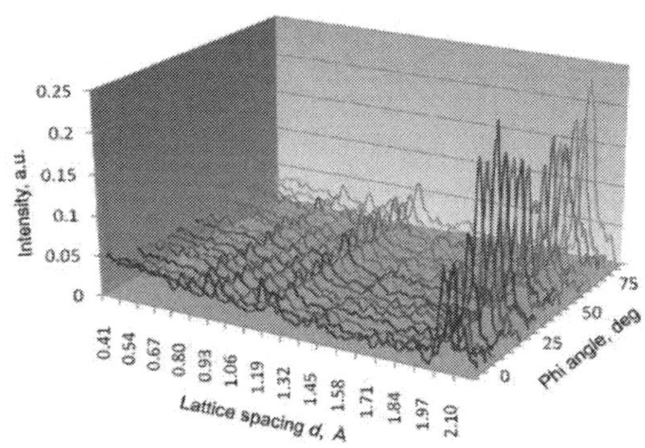

Figure 4. Variation of diffraction patterns with respect to diffraction direction.

4. Conclusions

In this study, the neutron diffraction experiments were carried out using Riken accelerator-driven compact neutron source, RANS. The diffraction pattern of IF steel samples was successfully measured by the TOF method. The texture evolution due to plastic deformation was successfully observed by measuring a change in the diffraction peak intensity. Consequently, RANS has been proved to be capable for neutron engineering diffraction aiming for the easy access measurement of texture evolution.

In the future scene of the metal forming analysis, texture measurement by a compact accelerator-based neutron source can be compared with and can validate the crystal plasticity analysis results and, at the same time, can provide valuable input information for the numerical analyses. Accurate crystal plasticity analysis can make a contribution to the macroscopic material modeling and can improve drastically the accuracy of forming simulations in industrial use.

Acknowledgments

This work has been partially supported by Photon and Quantum Basic Research Coordinated Development Program from the Ministry of Education, Culture, Sport, Science and Technology, Japan. Furthermore, this work has been supported by JSPS KAKENHI Grant Numbers 25289265 and 25420078. The author would like to thank the Iron and Steel Institute of Japan (ISIJ) Research Group for their beneficial assistance. The authors also wish to acknowledge Dr. Yujiro Ikeda of Japan Atomic Energy Agency for his beneficial advice and comments.

[1] Takamura M, Ozaki T, Miyoshi Y, Sunaga H and S. Takahashi 2011 *Proc. 10th Int. Conf. Techn. Plast. (ICTP 2011)* 591
[2] Hama T, Takuda H 2011 *Int. J. Plasticity* **27** 1072
[3] Clausen B, Brown D W, Tomé C N, Balogh L and Vogel S C 2010 *Proc. 31st Risø International Symposium on Materials Science, Challenges in materials science and possibilities in 3D and 4D characterization techniques* 11
[4] Kirstein O, Brule A, Nguyen H and Luzin V 2008 *Mater. Sci. Forum* **571-572** 213
[5] Pirling T, Bruno G and Withers P 2006 *J, Mater. Sci. Eng. A* **437** 139
[6] Hofmann M, Schneider R, Seidl G A, Rebelo-Kornmeier J, Wimpory R C, Garbe U and Brokmeier H G 2006 *Physica B* **385-386** 1035
[7] Suzuki H, Katsuyama J and Morii Y 2012 *J. Solid Mech. Mater. Eng.* **6** 574
[8] Santisteban J R, Daymond M R, James J A and Edwards L 2006 *J. Appl. Cryst.* **39** 812
[9] Harjo S, Aizawa K, Ito T, Arima H, Abe J, Moriai A, Sakasai K, Nakamura T, Nakatani T, Iwahashi T and Kamiyama T 2010 *Mater. Sci. Forum* **652** 99
[10] Wang X L, Holden T M, Stoica A D, An K, Skorpenske H D, Jones A B, Rennich G Q and Iverson E B 2010 *Mater. Sci. Forum* **652** 105
[11] Krawitz A D 2011 *Mater. Sci. Technol.* **27** 589
[12] Seki Y, Hashiguchi T, Ota H, Wang S, Taketani A and Otake Y 2014 *Physics Procedia* **60** 324
[13] Uesaka M and Kobayashi H 2016 *Reviews of Accelerator-Science and Technology* eds. A W Chao and W Chou (New Jersey: World Scientific) **8** p 181.
[14] Ikeda Y, Taketani A, Takamura M, Sunaga H, Kumagai M, Oba Y, Otake Y and Suzuki H *Nucl. Instrum. Methods Phys. Res. A* (submitted)
[15] Yamagata Y, Hirota K, Ju J, Wang S, Morita S, Kato J, Otake Y, Taketani A, Seki Y, Yamada M, Ota H, Bautista U and Jia Q 2015 *J. Radioanal. Nucl. Chem.* **305** 787
[16] Morooka S, Tomota Y, Adachi Y, Morita S and Kamiyama T 2006 *Advanced Materials Research* **15**-17 912
[17] Wenk H R, Lutterotti L and Vogel S 2003 *Nucl. Instrum. Methods Phys. Res. A* **515** 575

Journal of Physics: Conference Series **734** (2016) 032048 doi:10.1088/1742-6596/734/3/032048

Plastic deformation of high-purity α-titanium: model development and validation using the Taylor cylinder impact test

Nitin Chandola, Benoit Revil-Baudard and Oana Cazacu

Department of Mechanical and Aerospace Engineering, University of Florida, REEF, 1350 N. Poquito Rd., Shalimar, FL 32579, USA

E-mail: nitinchandola@ufl.edu, revil@ufl.edu, cazacu@reef.ufl.edu

Abstract. Results of an experimental study on the quasi-static and high-rate plastic deformation due to impact of a high-purity, polycrystalline, α-titanium material are presented. To quantify the plastic anisotropy and tension-compression asymmetry of the material, first monotonic uniaxial compression and tension tests were carried out at room temperature under quasi-static conditions. It was found that the material is transversely isotropic and displays strong strength differential effects. To characterize the material's strain rate sensitivity, Split Hopkinson Pressure Bar tests in tension and compression were also conducted. Taylor impact tests were performed for impact velocity of 196 m/s. Plastic deformation extended to 64% of the length of the deformed specimen, with little radial spreading. To model simultaneously the observed anisotropy, strain-rate sensitivity, and tension-compression asymmetry of the material, a three-dimensional constitutive model was developed. Key in the formulation is a macroscopic yield function [1] that incorporates the specificities of the plastic flow, namely the combined effects of anisotropy and tension-compression asymmetry. Comparison between model predictions and data show the capabilities of the model to describe with accuracy the plastic behavior of the α-Ti material for both quasi-static and dynamic loadings, in particular, a very good agreement was obtained between the simulated and experimental post-test Taylor specimen geometries.

1. Introduction

High-strain rates of the order of 10^4-10^5 s^{-1} can be achieved using in a Taylor-impact test, which consists of launching a solid cylindrical specimen at an elevated velocity of the order of 100-300 m.s^{-1} against a stationary rigid anvil [2]. The data obtained in such impact tests are invaluable both for development and validation of models for description of the plastic deformation in the dynamic regime. Few Taylor impact tests have been reported on titanium (Ti) materials. For example, Holt et al. [3] reported results of reverse ballistic impact tests at 225 and 294 m/s on commercially-pure Ti specimens. Metallographic examination revealed extensive deformation twinning. To account for the increase in hardening induced by twinning an incremental stress term, which depends on grain size and number of twins per grain, has been added. However, yielding was modeled using von Mises yield criterion, thus neglecting both the anisotropy and tension-compression asymmetry induced by deformation twinning. In this paper, we present a new three-dimensional model that was developed in order to model simultaneously the anisotropy, tension-compression asymmetry and strain-rate sensitivity of the plastic deformation of a high-purity α-Ti material. Furthermore, we report Taylor impact tests conducted on the same material and compare the model predictions with the measurements of the deformation of the impacted specimens. Taylor cylinder impact tests were conducted to investigate the dynamic behavior of the material (strain rates of the order of $10^5 s^{-1}$) and to validate the proposed model. The test results, namely photographs of post-test specimen profiles and measurements of their deformation are reported.

2. Constitutive modeling of impact behavior of pure Ti

The material used in this work was a high purity (99.999%) Ti that was purchased from Alpha Aesar of Johnson Matthey Electronics, Inc., (Spokane, WA). Optical microscopy showed that the material has equiaxed grains with an average grain size of about 20 μm. To quantify the plastic anisotropy and the tension-compression symmetry of the material, monotonic uniaxial compression and tension tests were carried out at room temperature under quasi-static conditions. The test results indicated that the material is quasi-isotropic in the plane of the plate, while there is a marked difference between the yield stress in the through-thickness plate direction and the average in-plane yield stress. Irrespective of orientation, the material displays strength-differential effects (stronger in compression than in tension). The Split Hopkinson Pressure Bar (SHPB) tests in tension and compression that were performed indicate that although there is an increase in flow stress with increased strain rate, the strain hardening rate or slope of the stress-strain curve exhibits little dependence on strain rate (see fig.1). This indicate that the relative activity of the plastic deformation mechanisms that are operational is very little affected by the change in strain rate.

A constitutive model was developed. Key in the formulation is the use of an yield function proposed by Cazacu et al. [1] that accounts for both anisotropy and tension/compression asymmetry. The effective stress associated with the yield criterion is:

$$\bar{\sigma} = \mathrm{B}\Big[(|\tilde{\sigma}_1| - k\tilde{\sigma}_1)^2 + (|\tilde{\sigma}_2| - k\tilde{\sigma})^2 + (|\tilde{\sigma}_3| - k\tilde{\sigma}_3)^2\Big]^{\frac{1}{2}} \tag{1}$$

where k is an internal variable, its range of variation being (-1, 1), while $\tilde{\sigma}_1$, $\tilde{\sigma}_2$ and $\tilde{\sigma}_3$ are the principal values of the transformed stress tensor $\tilde{\sigma} = [\mathbf{L}] : \mathbf{S}$ with $\mathbf{S}$ the Cauchy stress deviator applied to the material. In Eq.(2), L is a fourth-order orthotropic and symmetric tensor. In the coordinate system associated with the material symmetry axes (RD, TD and TT, respectively) and in Voigt notations, this tensor is represented by a 6x6 matrix given by

$$[\mathbf{L}] = \begin{bmatrix} L_{11} & L_{12} & L_{13} & 0 & 0 & 0 \\ L_{12} & L_{22} & L_{23} & 0 & 0 & 0 \\ L_{13} & L_{23} & L_{33} & 0 & 0 & 0 \\ 0 & 0 & 0 & L_{44} & 0 & 0 \\ 0 & 0 & 0 & 0 & L_{55} & 0 \\ 0 & 0 & 0 & 0 & 0 & L_{66} \end{bmatrix} \tag{2}$$

In Eq.(1), B is a constant defined such that the equivalent stress $\bar{\sigma}$ reduces to the tensile stress along RD. In the plastic deformation regime, the high purity Ti material investigated is transversely isotropic, the plane of isotropy being the plane of the plate (x, y) while the through-thickness plate direction (TT or z) is an axis of rotational symmetry. Therefore, $L_{11} = L_{22}$; $L_{13} = L_{23}$; $L_{55} = L_{66}$. To model the difference in hardening between tension and compression with accumulated plastic strain, all the parameters involved in the yield function were considered to evolve with the equivalent plastic strain. The model predicts that for this Ti material, at initial yielding and under 10% strain, the yield surfaces have an elliptical shape, the tension-compression asymmetry is small while above 10% strain the yield surfaces have a triangular shape, the difference in response between tension and compression being pronounced (see fig.2). The observed strain-rate sensitivity was modeled with a Johnson-Cook [4] law that was identified based on the SHPB data in tension. A user material subroutine (UMAT) was developed for the constitutive model described and implemented in the commercial implicit FE solver ABAQUS Standard.

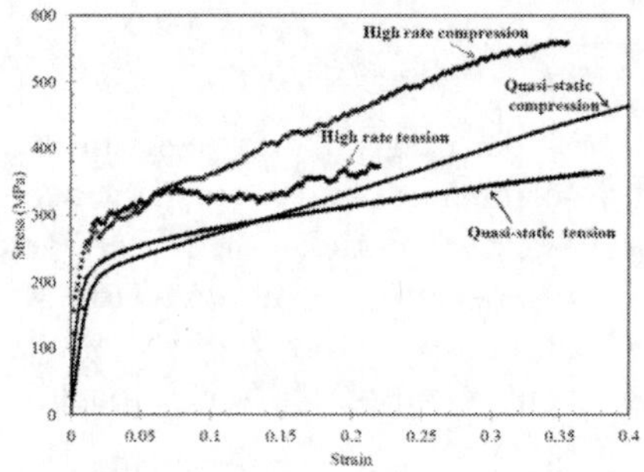

Figure 1. Experimental stress-strain response in uniaxial tension and compression along the in-plane of Ti plate for quasi-static and high-rate loadings.

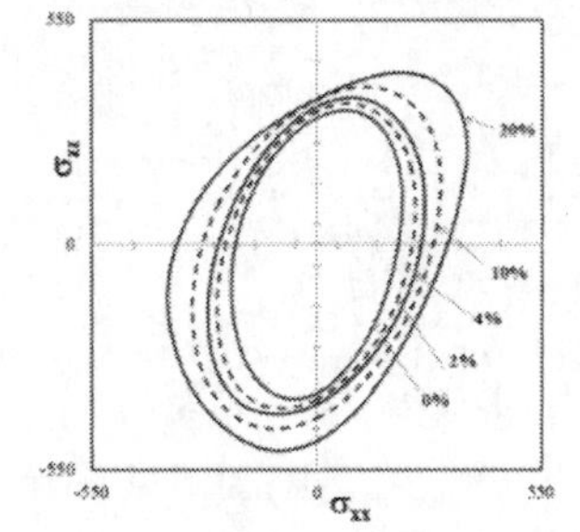

Figure 2. Projection in the $(\sigma_{xx}, \sigma_{zz})$ plane of the theoretical yield loci [1] for the Ti plate, corresponding to individual levels of equivalent plastic strain.

3. Assessment of the predictive capabilities for cold forming

High-rate impact tests (Taylor impact tests) were conducted to examine the material response at strain rates higher than those achieved using a conventional SHPB system. The Taylor impact test consists of launching a cylindrical specimen at an elevated velocity against a stationary target [2]. Similar to a uniaxial compression experiment, a Taylor impact specimen deforms plastically by contracting axially and expanding radially. However, unlike in a quasi-static compression experiment, the state of stress and level of plastic deformation within the Taylor specimen are not uniform. The highest level of deformation is observed at the impacted surface, and the least deformation is observed at the rear surface. Data concerning geometrical shape changes of the post-test specimen and measurements of the actual deformation (lateral spreading; strain profiles i.e. measurements of the deformation at different locations along the length of the specimen) are essential for the validation of the capability of structural-level models to predict dynamic behavior. Fig. 3 presents photographs of the undeformed specimen and the post-impact shape of a typical α-Ti specimen, respectively. The initial through-thickness directions is indicated by an arrow. For more details about the simulation of Taylor impact of Ti, see [5].

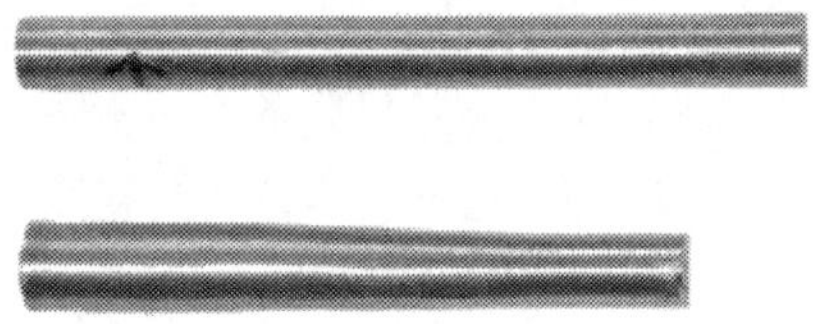

Figure 3. Photographs of a typical Ti Taylor specimen with the long-axis along IP: (a) undeformed specimen; (b) specimen after impact at velocity of 196 m/s.

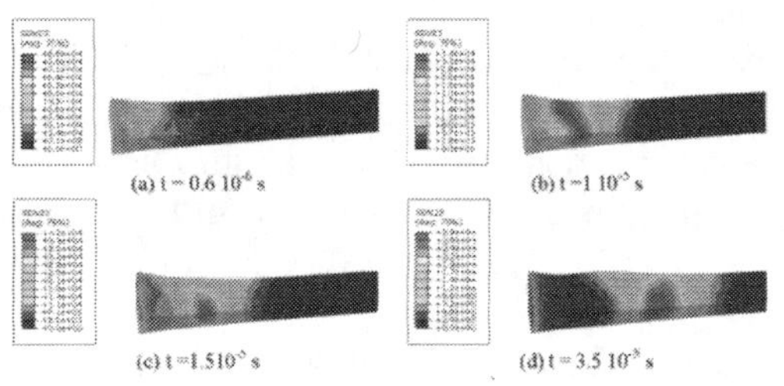

Figure 4. Isocontours of the equivalent plastic strain-rate.

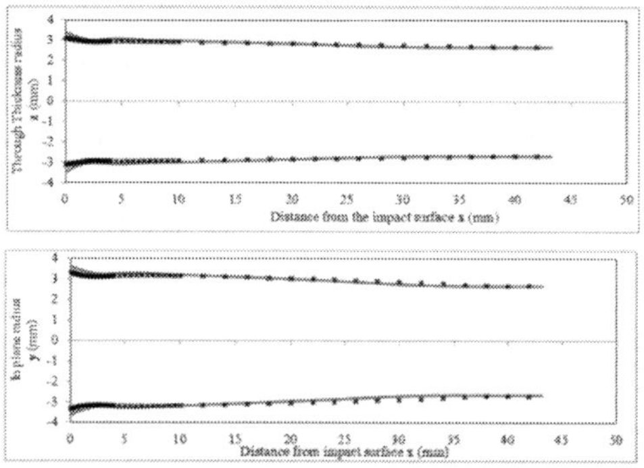

Figure 5. Comparison of the simulated and measured (symbols) minor and major outline of the specimen vs. the distance x from the impacted end.

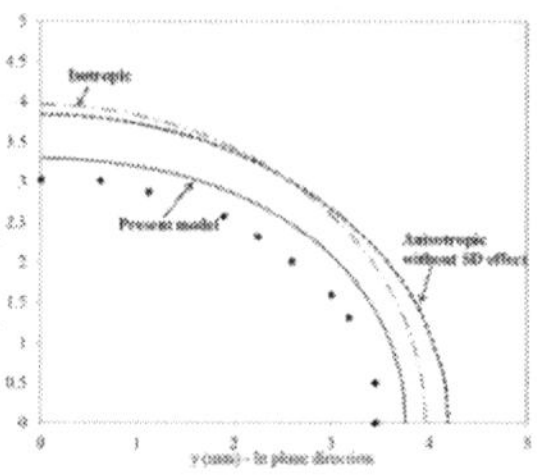

Figure 6. Comparison between the FE simulated deformation of the footprint according with the present model, von Mises yield criterion, and the anisotropic criterion without strength differential effects and the measured cross-section (symbols) after impact at 196 m/s.

Fig.4 shows the isocontours of the equivalent plastic strain rate at different times after the impact between the specimen and the anvil. First, let us note that very high plastic strain rates of the order of 10^4 s^{-1} are predicted, with the largest strain rates, of the order of 8.6 10^4 s^{-1}, being near to the impact surface. Furthermore, the largest strain rates were experienced by the material during the initial, transient phase of the deformation when non-zero plastic strain rates occur only in the vicinity of the impacted surface. After about 10^{-5}s, the zone of plastic deformation moves away from the impacted end and the region in the neighborhood of the impacted surface is not subjected to plastic strain rates anymore. Fig.5 show the strain profiles along the major (y or in-plane direction) and minor (z or through-thickness) axes of the impacted specimen, respectively. In these figures are compared the predicted (solid line) major and minor specimen profiles (solid line) with the experimental measurements (symbols) taken at several distances from the impacted end (impacted end corresponds to x = 0). Note the very good agreement between the numerical predictions and the experimental data. Furthermore, the model predicts very well the experimentally observed key features of the plastic deformation of the Ti Indeed, according to the model plastic deformation extended over approximately 64% of the deformed specimen length, with little lateral spreading. Furthermore, the plastic deformation for $0 < x < 3.5$ mm is due only to the impact while at distances x>3.5mm, the observed post-test plastic deformation is due to the plastic wave that has traveled along the length of the specimen. In Fig.6 is shown a comparison between the deformed impacted cross-section and FE simulations using von Mises yield criterion and the anisotropic yield criterion of Eq. (1) without consideration of tension-compression asymmetry, respectively. If the tension-compression asymmetry of the material is neglected, the lateral spreading of the specimen is largely overpredicted and the small ellipticity of the deformed cross-section is not captured. In contrast, the simulated cross-section according to the present model is in good agreement with the data, thus demonstrating the importance of incorporating in the constitutive description the combined influence of anisotropy and tension-compression asymmetry.

References

[1] Cazacu O, Plunkett B and Barlat F 2006 *int. J. Plasticity* **22** 1171
[2] Taylor G 1948. *Proceedings of the Royal Society of London* A **194** 289
[3] Holt W H, Mock W, Zerilli F J, Clark J B 1994 *Mech. Mater.* **17** 195
[4] Johnson G R, Cook W H 1983 *Proceedings of the 7th International Symposium on Ballistics* 541
[5] Revil-Baudard B, Cazacu O, Flater P and Kleiser G 2015 *Mech. Mater.* **80** 264

Journal of Physics: Conference Series **734** (2016) 032049 doi:10.1088/1742-6596/734/3/032049

Crack initiation observation and local stress analysis in shear fracture tests of ultra-high strength steels

Ninshu Ma[1], Kenji Takada[2] and Nao Sugimoto[2]

[1] JSOL Corporation, 2-2-4, Tosabori, Nishiku, Osaka, 550-0001, Japan
[2] Honda R&D Co. Ltd., 4630, Shimo-Takanezawa, Hagamachi, Haga-gun, Tochigi-ken, 321-3393, Japan

Abstract. To investigate the local strain and stress at the crack initiation position in shear fracture test pieces of ultra-high strength steels, a butterfly shear fracture specimen was employed. The crack initiation position and propagation direction were observed during shear fracture tests by high speed cameras and investigated through analysing the fracture surface by scanning electron microscope. Further, the finite element method was employed and the stress-triaxiality at the crack initiation position was investigated. It can be obtained that the crack initiated at the position where the stress state is close to uniaxial tensile state or plane strain state more than pure shear stress state.

1. Introduction

To make light and strong automotive bodies, ultra-high strength steels and their structure components have been increasingly employed. However, with increasing the strength of steels, the elongation and ductility become worse. These will increase the difficulties in the metal forming. In addition, these make the prediction of the crack occurrence and its propagation become more difficult. To reduce the injury to humans due to transportation accidents, the prediction accuracy on the cracking position and deformation behaviours of automotive structures under various impact loading conditions must be investigated and improved.

There have been several criteria for the prediction of cracks due to sheet metal forming which include thinning [1], forming limit diagram based on principal strains [2] and stress forming limit diagram based on principal stresses [3]. However these criteria have difficulties when loading paths and three dimensional stress states become complicated. Recent researches indicated that the fracture strain of automotive steel sheets under different stress states is not a constant and it strongly depends on the stress-triaxiality [4-8]. The fracture strain under high stress-triaxiality states has been investigated using the standard uniaxial tensile test piece or the tensile test pieces with a centre hole or notches at edges [7-8]. To measure the fracture strain under low stress-triaxiality states such as the pure shearing stress state, several shear fracture specimens were designed and their test methods were proposed [4-8]. Most researchers observed the macro shear fracture mode in these shear fracture tests and found that the macro stress state in the shear fracture tests is close to the pure stress state. They concluded that the fracture strain is smaller than the uniaxial tensile stress state [5-7]. However, the local strain and the local stress at the crack initiation position were not exactly identified. In this article, we focused the research on the local fracture strain and local stress state at the crack initiation position in the shear fracture specimens. We selected a widely used shear fracture specimen with "Butterly" shape [5] and conducted the shear fracture tests for three strength grades of 590MPa, 980MPa and 1500MPa of steel sheets. The crack initiation position during shear fracture testing was observed using high speed camera and also identified by observing fracture surface using scanning electron microscope (SEM). The local fracture strain and the stress-triaxiality at the crack initiation position of shear fracture specimens were analysed by the finite element method (FEM).

2. Butterfly shear specimen of high strength steels

The butterfly specimen for shear fracture tests is shown in Fig. 1, which was proposed by Bai and Wierzbicki [5]. The thickness in the shear zone of the butterfly specimen is 0.5mm which is much thinner than the specimen thickness (1.6mm) in other zones. The width of the shear zone varies along the length direction of the butterfly specimen to produce the shear stress state. The load in the shearing direction is applied by displacement and the vertical displacement is controlled. The three strength grades of 590MPa, 980MPa and 1500MPa steels for shear fracture tests were employed. The 590MPa and 980MPa shear specimens was made from rolled steel sheets. The 1500MPa shear specimen was cut off from a piece of the rolled sheet to which the hot stamping thermal cycle has been applied. To ensure the reliability of shear fracture tests, three pieces of specimens for each strength grade of steels were prepared and experiments were repeatedly conducted.

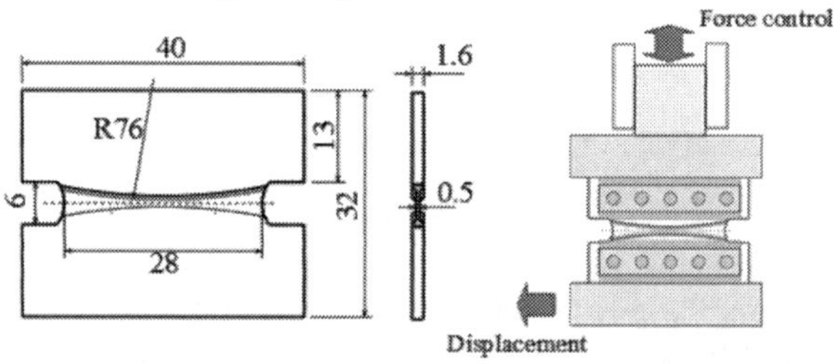

Fig. 1 Butterfly specimen and setup for shear fracture test [5].

3. Crack initiation position and propagation direction in butterfly specimen during shear tests

The crack initiation position during shear fracture tests was observed by a high speed camera and analysed using SEM photos along the fracture surface for three strength grades of steels. Fig. 2 shows macro photos of the butterfly specimen of the 1500MPa steel when the stroke is 0.0mm, 1.8mm and large enough for the final fracture, respectively. The crack initiation position is almost the same for three strength grades of steel sheets. The crack initiated at the stroke of 3.7mm, 2.5mm and 1.8mm for 590MPa, 980MPa and 1500MPa steels, respectively.

Fig. 3 shows the SEM photos along the fracture surface for the 1500MPa steel. Generally, it looks like a quasi-clearage fracture. From the photo shown in Fig. 3(b), the crack initiation position can be obviously seen around the edge of the butterfly specimen. From the river-like fracture patterns shown in Fig.3 (c), the crack propagation direction indicated by arrows in these figures.

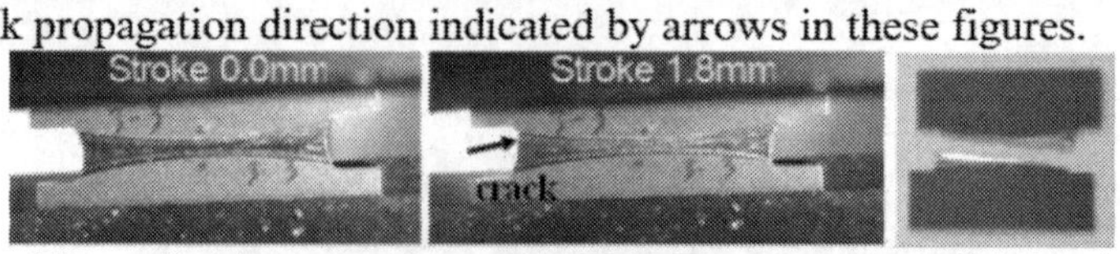

Fig.2 Macro-photos of butterfly shear specimen of 1500MPa steel at crack initiation and fracture states.

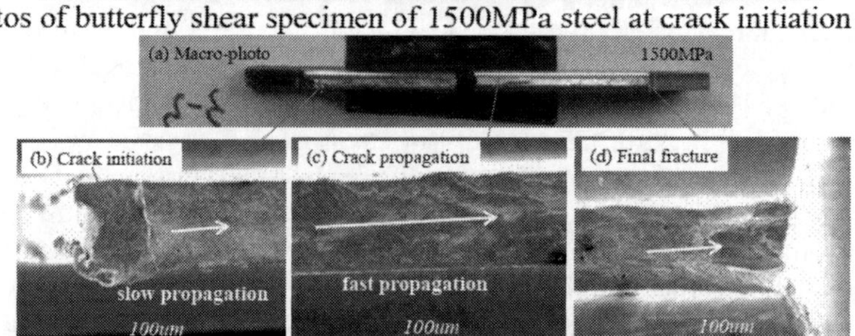

Fig.3 SEM-photos of 1500MPa shear specimen along fracture surface: (a) macro-photo, (b) around crack initiation position, (c) crack propagation zone, (d) final fracture.

4. FE analysis of fracture strain and stress-triaxiality

4.1. Mechanical properties of high strength steels

The Young's modulus and Poisson's ratio of advanced high strength steels 590MPa, 980MPa and 1500MPa are 206GPa and 0.3 respectively. The nonlinear stress and strain curves for 590MPa, 980MPa and 1500MPa materials are measured through uniaxial tensile tests [8]. The results are shown in Fig. 4 which accurately reproduced force-stroke historical curves of uniaxial tensile tests [8]. The strength coefficient A following swift hardening law for steels 590MPa, 980MPa and 1500MPa is 1012.7MPa, 1368.3MPa and 1896.2MPa, respectively. The exponential hardening coefficient n is 0.175, 0.086 and 0.058, respectively.

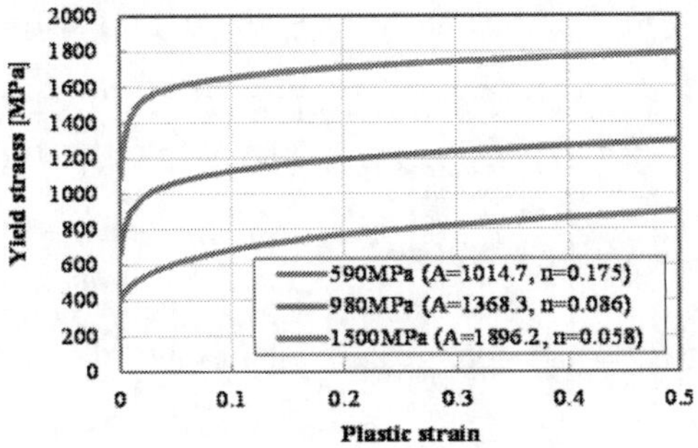

Fig. 4 Measured stress-strain curves [8].

4.2. Finite element model for butterfly shear specimens

To investigate the local stress and the local fracture strain, the nonlinear FEM software LS-DYNA was employed. Since the geometry of the butterfly shear specimen is symmetric in the thickness direction, only a half size through the specimen thickness is modelled by solid elements as shown in Fig. 5. Three elements were divided in the thickness direction of the finite element model and the minimum size of elements is about 0.1mm. The nodal displacements on the bottom surface are fixed. The vertical nodal displacement on upper surface is constrained. The horizontal nodal displacement on the upper surface is applied following the experimental data.

Fig. 5 Butterfly specimen and setup for shear fracture test.

4.3. Local fracture strain and stress-triaxiality in shear fracture tests

With the aid of FEM, the local strain and local stress were computed. The distribution of the equivalent plastic strain and the distribution of the stress-triaxiality in the shear fracture test at the crack initiated stroke (1.8mm) for the 1500MPa steel are shown in Figs. 6(a) and 6(b), respectively. The distribution of the equivalent plastic strain and the distribution of the stress-triaxiality are similar for 590MPa and 980MPa steels at the stroke 3.7mm and 2.5mm, respectively. For all shear fracture tests of three strength grades of 590MPa, 980MP and 1500MPa steels, the local plastic strain and stress-triaxiality are both largest around the edge point-A where the crack occurred.

Using the computed results during shear fracture testing, the historical changes of the plastic strain and the stress-triaxiality at the edge point-A and the inside point-B for all three strength grades of steels are plotted in Fig.7. The changing characteristics of stress-triaxiality at the edge point-A are complicated and those at the inside point-B have almost a same value of 0.05 which is close to the pure shear stress state. At the edge point-A, the stress-triaxiality is about -0.4 or -0.5 when the plastic strain is small. When the plastic strain becomes large, the stress-triaxiality increases to about 0.2~0.4. The local fracture strain for three strength grades of 590MPa, 980MPa and 1500MPa steels is 1.55, 0.88, 0.41, with the stress-triaxiality of 0.38, 0.24, 0.21, respectively. Although the plastic strain at the edge point-A is smaller than that at the inside point-B, the crack occurred at the edge point-A. This is because the stress-triaxiality at the edge point-A is larger than that at the inside point-B. From above results, we can know that the crack initiated at the position where there is higher stress-triaxility and the crack did not initiate at the pure shear stress state even the local plastic strain is larger.

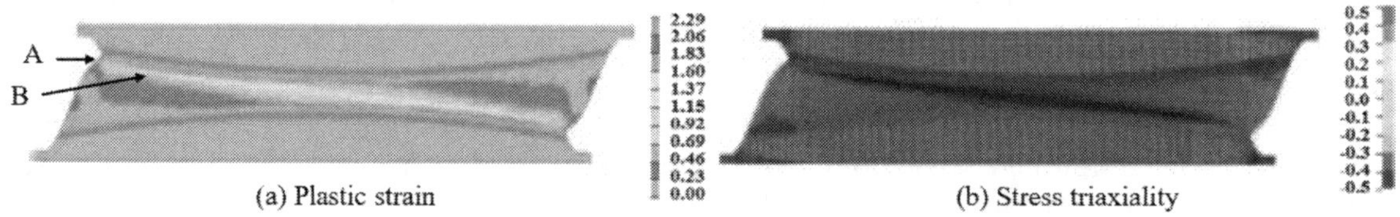

Fig. 6 Computed results of 1500MPa steel at stroke=1.8mm: (a) plastic strain, (b) stress-triaxiality.

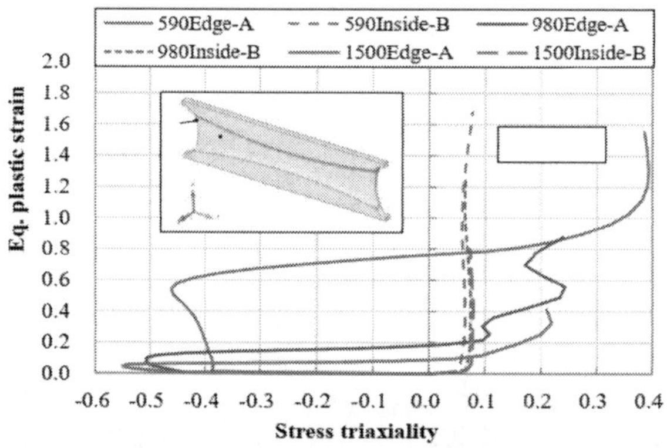

Fig.7 Local fracture strain and stress-triaxiality estimated by FEM.

5. Conclusions

(1) Shear fracture tests using butterfly specimen for three strength grades of 590MPa, 980MPa and 1500MPa steel sheets were conducted.
(2) The crack initiation position, the edge of the butterfly specimen, was observed by the high speed camera and also identified from analysis of fracture surface using scanning electric microscope.
(3) The stress-triaxiality at the crack initiation position is much higher than the pure shear stress state.
(4) Although the plastic strain at the position with the pure shear stress state is larger than the plastic strain at the edge of the butterfly specimen, the crack did not occur at the position with the shear stress state.
(5) The local fracture strain at the crack initiation position is 1.55, 0.88, 0.41 with the higher stress-triaxiality of 0.38, 0.24, 0.21 for three strength grades of 590MPa, 980MPa and 1500MPa steels, respectively.
(6) It looks that the crack propagation is along the pure shear stress zone in the butterfly specimen and detail research is necessary.

References

[1] Ma N 2013 *Sokeizai*, v54n4, pp 21-26 (in Japanese).
[2] Marciniak Z and Kuczynski K 1967 *Int. Journal of Mechanical Science,* v9, pp 609–620.
[3] Kuwabara T 2005, *Proc. NUMISHEET2005*, pp 20-39.
[4] Isik K, Soyarslan C, Richter H and Tekkaya A E 2011 *Proc. 8th European LS-DYNA Users Conf.*, Strasbourg, May, 2011.
[5] Bai Y and Wierzbicki T 2010 *Int. J. Fracture*, 161, pp 1-20.
[6] Cesar de Sa J M A 2014 *Proc. Numisheet2014*, pp 344-349.
[7] Dunand M and DirkMohr 2011 *J. Mechanics and Physics of Solids*, v59, pp1374–1394.
[8] Sato K et al. 2014 *Trans JASE*, 45-6, pp 1099-1104.
[9] Ma N et al. 2016 *Trans JASE*, 46-2, pp 603-608.

Development of measurement method of work hardening behavior in large plastic strain for sheet metal forging

Nobuo Yoshikawa, Tomohiro Yamashita, Satoshi Shirakami,

Osamu kada, Tohru Yoshida and Shunji Hiwatashi

Steel Research Laboratories, Nippon Steel and Sumitomo Metal Corporation, 20-1 Shintomi, Futtsu, Chiba 293-8511, Japan

E-mail: yoshikawa.7md.nobuo@jp.nssmc.com

Abstract. For the purpose of accuracy improvement of sheet metal forging FE analysis, we have developed a new measurement method of work hardening behavior in large plastic strain by repeatedly performing simple shear test using pre-strained steel sheet. In this method, it is possible to measure work hardening behavior more than equivalent plastic strain 2.0. In addition, it was carried out a comparison between developed method and compression test in order to verify the validity of the results by the developed method. As a result, both results were in good agreement. The validity of developed method has been verified.

Keywords: work hardening, large plastic strain, sheet metal forging, steel sheet

INTRODUCTION

The reduction in weight of automobile parts is significantly required to improve the fuel consumption of automobiles. Therefore, sheet metal forging parts which allow weight reduction by making distribution control of sheet thickness gradually increase. In addition, sheet metal forging allows for the reduction in cost by net-shape and near-net-shape forming without additional cutting and finishing operations [1].

However, sheet metal forging process is very complex because it includes both conventional press forming used tension and conventional cold forging used compression. So, it is essential for highly accurate FE analysis to quickly develop sheet metal forging process. Flow stress strongly affects the accuracy of sheet metal forging FE analysis, but extrapolated flow stress obtained by uniaxial tensile test is applied. Because it is difficult to measure the work hardening behavior in large plastic strain of a thin steel sheet using the conventional experimental procedure, such as uniaxial tensile test, hydraulic bulge test and simple shear test [2, 3]. Among these, simple shear test is difficult to cause plastic instabilities such as necking in uniaxial tensile test. However the sample is destroyed from the edge at large strain and leaded to the measurement limit [4].

Therefore, we have developed a new measurement method of work hardening behavior in large plastic strain by repeatedly performing simple shear test using pre-strained steel sheet. Furthermore, in order to verify the validity of developed method was compared with compression test.

EXPERIMENTAL PROCEDURE TO MEASURE WORKHARDENING BEHAVIOR IN LARGE PLASTIC STRAIN

Test Material

TABLE (1) shows mechanical properties of steel used in the experiment. The sample was used 440MPa grade hot-rolled steel sheet with a thickness of 2.0mm.

TABLE (1). Mechanical properties of used sample

YS /MPa	TS /MPa	u-EL /%	t-EL /%
320	445	18	35

Experimental Procedure

In order to suppress the failure from sample edges, we have devised a new method of performing repeatedly simple shear test using pre-strained steel sheet.

Simple shear test is promoted on flat sample shown in **FIGURE 1** by fixing the parts with grips that are subjected to parallel displacement. The stress is directly proportional to recorded load while the strain is obtained by measuring the inclination of scribe lines in sample surfaces.

Shear stress and shear strain was converted to equivalent stress and equivalent plastic strain by using the following equation.

$$\sigma_{eq} = \kappa\sigma_s \quad , \quad d\varepsilon_{eq}^p = \frac{1}{\kappa} d\varepsilon_s^p \qquad (1)$$

where σ_s and ε_s is shear stress and shear strain, σ_{eq} and ε_{eq} is equivalent stress and equivalent strain, κ is coefficient relating the uniaxial tensile stress and shear stress.

Pre-strain is introduced by cold rolling in order to introduce a large pre-strain as possible. Equivalent plastic strain calculated in using the following equation.

$$\varepsilon_{eq}^{p} = -\frac{2}{\sqrt{3}} \ln \frac{t}{t_0} \qquad (2)$$

where t and t_0 is initial thickness and after rolling thickness.

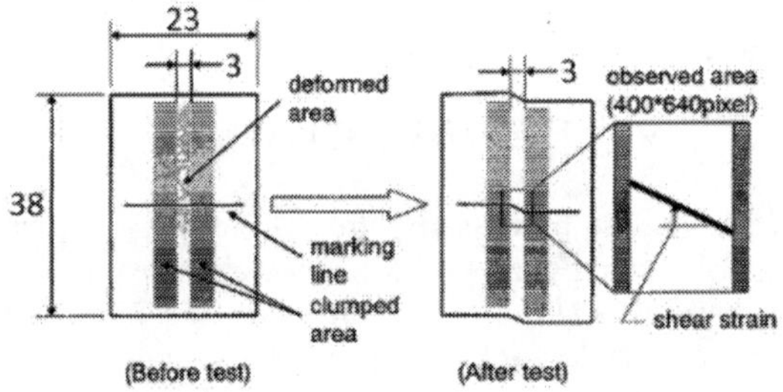

FIGURE 1. Specimen of Simple Shear Test

RESULTS AND DISCUSSION

Experimental Results

The stress-strain curves and work hardening rate obtained by uniaxial tensile test and simple shear test using non-prestrained sample are shown in **FIGURE 2** (a) and (b). Stress-strain curves of uniaxial tensile test and simple shear test were in good agreement until equivalent plastic strain 0.2 which meet plastic instability at uniaxial tensile test. From this result, conversion according to formula (1) is reasonable. In addition, the measurement limit of the simple shear test was equivalent plastic strain about 0.5, it was about twice at the uniaxial tensile test.

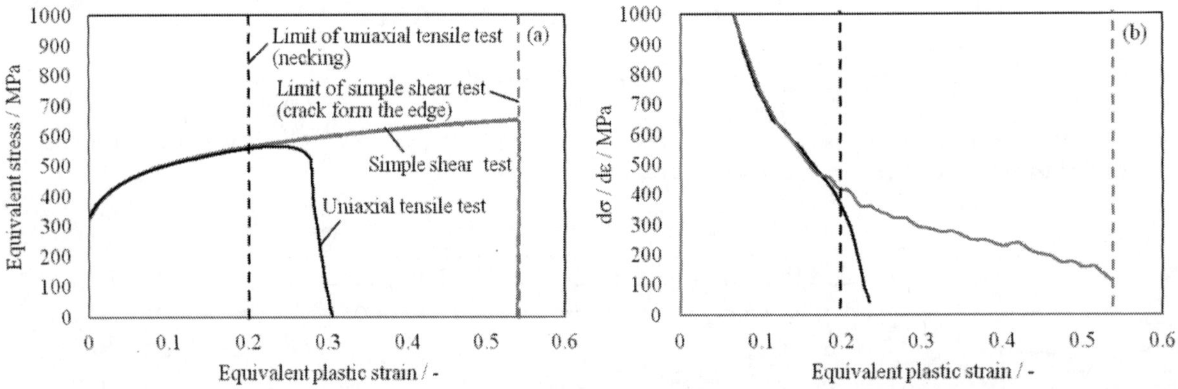

FIGURE 2. Stress - strain curves and work hardening rates by uniaxial tensile test and simple shear test.

The stress-strain curves and work hardening rate obtained by simple shear test and repeated simple shear test using non-prestrained and pre-strained sample are shown in **FIGURE 3** (a) and (b). Stress-strain curve of simple shear test and repeated simple shear test using non-prestrained were in good agreement. The measurement limit of repeated simple shear test using non-prestrained was equivalent plastic strain about 1.3, it was about six times at the uniaxial tensile test. On the other hand, Stress-strain curve of simple repeated simple shear test using non-prestrained sample and pre-strained sample were in good agreement without initial secondary deformation at pre-strained sample. Although high yield stress of repeated simple shear test using pre-strained is believed to cause cross effect that occurs because the slip system of primary deformation and secondary deformation is different, flow stress and work hardening rate is in good agreement with the non-prestrained sample in large secondly deformation. Therefore the results obtained by this method are considered to be reasonable. The measurement limit of repeated simple shear test using pre-strained was equivalent plastic strain about 2.0, it was about ten times at the uniaxial tensile test.

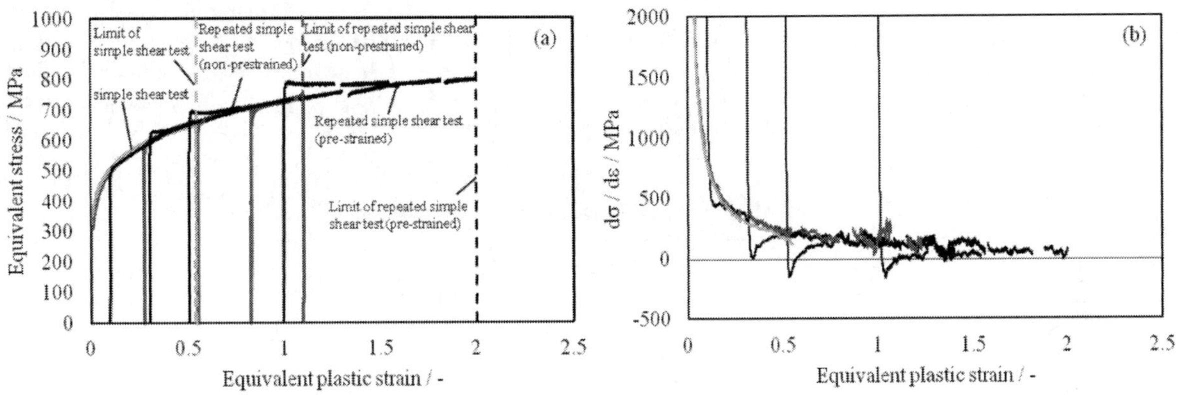

FIGURE 2. Stress - strain curves and work hardening rates by simple shear test and repeated simple shear test using non-prestrained and pre-strained sample

The stress-strain curves obtained by repeated simple shear test using pre-strained and extrapolated the uniaxial tensile test by using swift's power law are shown in **FIGURE 4**. Although both curves were in good agreement until equivalent plastic strain 0.3, it was significantly different in the more plastic strain range. Flow stress obtained by repeating simple shear test using pre-strained indicates a lower flow stress than the extrapolated curve to a tensile uniaxial test in the large strain region. The difference of flow stress occurred about 12% in the plastic strain 2.0. These differences are likely to affect the accuracy of sheet metal forging FE analysis such as forming load and material flow.

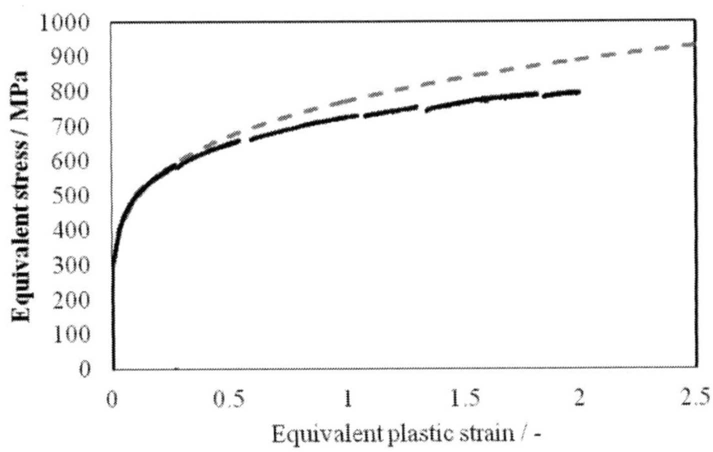

FIGURE 4. Comparison of stress - strain curves of developed method and approximate results using swift's power low.

Comparison of the Compression Test

In order to verify the validity of developed method was compared with compression test. Because it is difficult to carry out the compression test by using a thin steel sheet, cut out the sample from 540MPa grade S55C spherical annealing steel bar with a diameter of 55mm shown in **TABLE (2)**. Compression test was carried out by upsetting the cylindrical sample of a height 21mm in diameter 14 mm between grooved plates. The calculation of the equivalent stress and the equivalent plastic strain was used method proposed by Osakada et al [5].

$$\sigma_{eq} = \frac{1}{f}\frac{P}{A_0},\ \varepsilon^p_{eq} = g(e) = g(\frac{h_0 - h}{h_0}) \quad (3)$$

Where P is a test load, A_0 is the initial cross-sectional area, f is constrained coefficients, h and h_0 is height of test sample before test and after test, respectively. Relationship of the e, ε_{eq} and f was used literature values [5].

The stress-strain curves obtained by repeated simple shear test using pre-strained and compression test using non-prestrained are shown in **FIGURE 5**. It was possible to measure work hardening behavior about equivalent plastic strain 1.0 and 1.6 using repeated simple shear test and compression test, respectively. Both stress-strain curves were in good agreement within the measuring range, the validity of the developed method has been verified.

TABLE (2). Mechanical properties of used sample

YS /MPa	TS /MPa	u-EL /%	t-EL /%
365	563	17	40

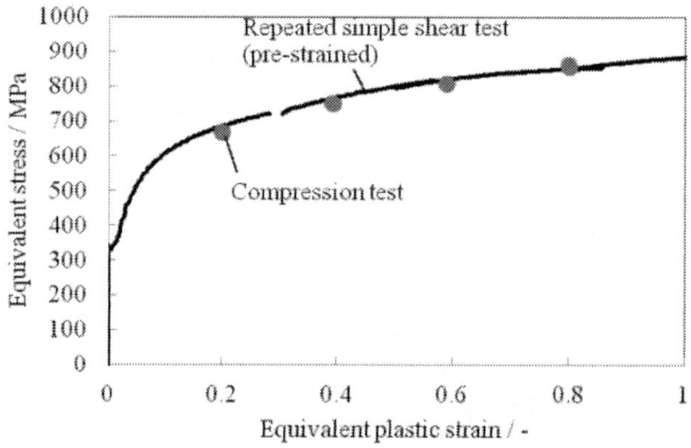

FIGURE 5. Comparison of stress - strain curves by repeated shear test and compression test.

CONCLUSION

For the purpose of accuracy improvement of sheet metal forging analysis, we have developed a measurement method of work hardening behavior in large plastic strain by repeatedly performing simple shear test using pre-strained steel sheet. The results obtained in this study are summarized as follows:

1) The newly developed method could measure work hardening behavior more than equivalent strain 2.0.
2) Both stress-strain curves were in good agreement within the measuring range, the validity of the developed method has been verified.

REFERENCES

[1] M. Merklein et al, "Bulk forming of sheet metal," ,CIRP Annals - Manufacturing Technology 61 (2012) 725-745.
[2] Johnson, W., et al.:Steel Metal Industries, 4(1965), 271.
[3] K. Miyauchi, On simple shear deformation, Sci. Papers RIKEN 81(1987) 57–67.
[4] E.F.Rauch..:J.Eng.Mater.Thechnol, 131 (2009),011107-1
[5] Osakada,K., et al.:CIRP Annals, 30-1 (1981), 135

On the modelling of strength differential and anisotropy exhibited by titanium

Niko Manopulo, Christian Raemy, Pavel Hora

ETH Zurich, Institute of Virtual Manufacturing
Tannenstrasse 3, 8092, Zurich, Switzerland

manopulo@ivp.mavt.ethz.ch

Abstract. Modelling the response of titanium alloys under plastic deformation is challenging. In addition to strong anisotropy, these materials exhibit tension-compression asymmetry as well as anisotropic hardening even under monotonous loading conditions. The present contribution aims to propose a new modelling approach for titanium materials, by a homogeneous introduction of asymmetry into existing symmetric yield functions. Furthermore, anisotropic hardening effects are tracked describing the strain dependant yield surface evolution. The results are validated using the earing profile of deep drawn cups.

1. Introduction

Given its excellent strength to weight ratio as well as its corrosion resistance, commercially pure titanium is often the material of choice, especially in aerospace and medical industries. The manufacturing of titanium parts using forming operations has, significant advantages in terms of material waste and speed over alternative methods such as machining. Forming of titanium is however far from being a trivial task. Apart from strong anisotropy, titanium alloys exhibit pronounced strength differential in tension and compression. Furthermore, given the directional dependency of twinning formation, strong anisotropic hardening is observed.

Different yield locus models have been proposed in the recent literature in order to properly capture asymmetric yielding. The CPB06 model, proposed by Cazacu et al. [1], features a homogeneous combination of the principal values of the linearly transformed stress deviator with its scaled absolute value, introducing thus an asymmetry in the positive and negative stress values. This idea has been later extended to feature more linear transformations by Plunkett et al.[2]. Nixon et al. proposed in 2010 a similar model featuring also the third invariant of the stress tensor [3]. All these models are able of capturing the yield stress asymmetry as well as R-values in tension with varying quality.

The present contribution aims to propose a generic method to construct homogeneous asymmetric yield loci based on well-established symmetric yield locus formulations. The approach is based on the recently proposed Homogeneous Anisotropic Hardening (HAH) model [1], which is primarily designed to capture Bauschinger effect. The basic framework of the HAH function has been extended to define asymmetric distortions in a finite number of predefined loading directions. The results are compared to the CPB06 model based on a cup drawing test in terms of force-displacement behavior and earing profile.

2. Proposed Model

The main form of the proposed yield function reads:

$$\Phi^q = \phi^q + \sum_k^n f_k^+ \left|\mathbf{d}_k\!:\mathbf{s} - |\mathbf{d}_k\!:\mathbf{s}|\right|^q + f_k^- \left|-\mathbf{d}_k\!:\mathbf{s} - |\mathbf{d}_k\!:\mathbf{s}|\right|^q$$

where ϕ is any homogeneous symmetric yield function. The summation over k applies asymmetric distortions to the yield locus in n different deviatoric stress directions ($\mathbf{d}_k$). The parameters $f_k^+$and f_k^- must be identified together with the parameters of ϕ. The exponent q is used to blend the distortional component with the base function preserving homogeneity.

3. Material Characterization

To fully characterize the material properties standard tensile tests as well as compression tests in rolling (RD) and transverse (TD) directions have been carried out. The compression experiments have been conducted in a dilatometer using prismatic samples of quadratic base (3mm) and 5mm height. Both tensile and compression tests were conducted in room temperature and with quasi-static strain-rates.

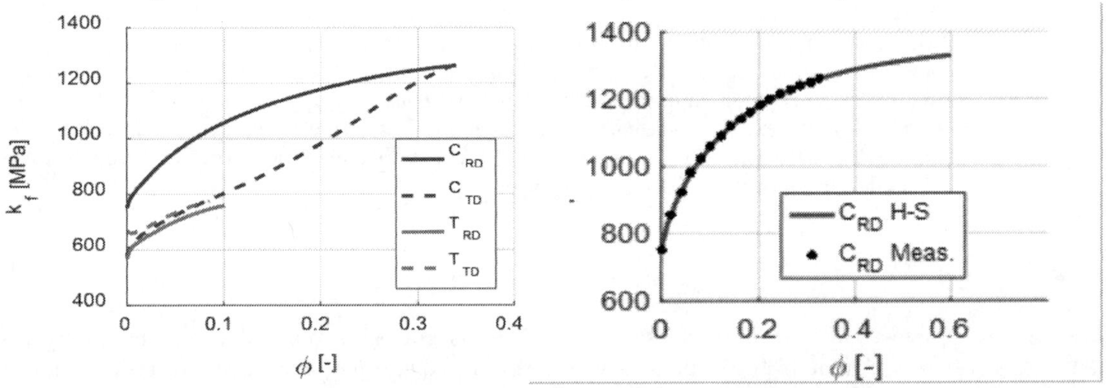

Figure 1. Flow curves in tension (T) and in compression (C) respectively in rolling (RD) and transverse (TD) directions (left). Approximation of the compression curve in rolling direction with the H-S model (right)

Figure 1 depicts the measured flow curves. Significant asymmetry can be observed in RD even for the initial yield stress, with a slightly divergent trend with increasing strain. In TD the response is nearly symmetric except for the first few percent strain. The compression test in transverse direction shows an atypical s-like hardening behavior. This is considered to be caused by the twinning of the hexagonal lattice, which hinder dislocation movement and thus increase the macroscopic yield stress.

3.1. Hardening model

The Hockett-Sherby function has been used to capture the hardening behavior:

$$k_f = B - (B - A)\exp(-m\phi^p)$$

where k_f represents flow stress, φ plastic strain, A the initial flow stress, B the saturated flow stress and m and p are hardening parameters. The function is able of accurately fitting the experimental data as depicted in Figure 1 (right).

3.2. Identification of the Yield Function

The yield function proposed in Section 2 has been identified to capture the experimental behaviour delineated in Section 3. The yld2004-18p model [1] has been used as the symmetric base function, which has been distorted in the following three deviatoric stress directions to enable asymmetry:

$$\mathbf{d}_1 = \begin{bmatrix} 2/3 & 0 & 0 \\ 0 & -1/3 & 0 \\ 0 & 0 & -1/3 \end{bmatrix}, \mathbf{d}_2 = \begin{bmatrix} -1/3 & 0 & 0 \\ 0 & 2/3 & 0 \\ 0 & 0 & -1/3 \end{bmatrix}, \mathbf{d}_3 = \begin{bmatrix} 1/3 & 0 & 0 \\ 0 & 1/3 & 0 \\ 0 & 0 & -2/3 \end{bmatrix}$$

The first two operate respectively in RD and TD whereas the third component acts over the equibiaxial stress direction. The identified yield locus, in comparison to the CPB06 yield locus can be seen in Figure 2 (left), the corresponding Lankford coefficients are depicted in Figure 2 (right). The functions have been identified by minimizing a cost-function related to yield stresses and Lankford coefficients in both RD and TD. Clearly the MRH approach provides more flexibility than required to exactly match these values. The following additional heuristic constraints have been used:

- The equibiaxial yield stress in both tension and compression has been considered to the corresponding stress in TD
- R-values in compression are assumed to be equal to the ones in tension
- R-value in 70 degrees has been assumed to be equal to the average of 45 and 90 degrees (to avoid oscillation in between

Isotropic hardening model was used to capture the evolution of the yield locus, the compression flow-curve in RD has been chosen as reference for all considered yield loci as higher equivalent plastic strain values have been reached in this case.

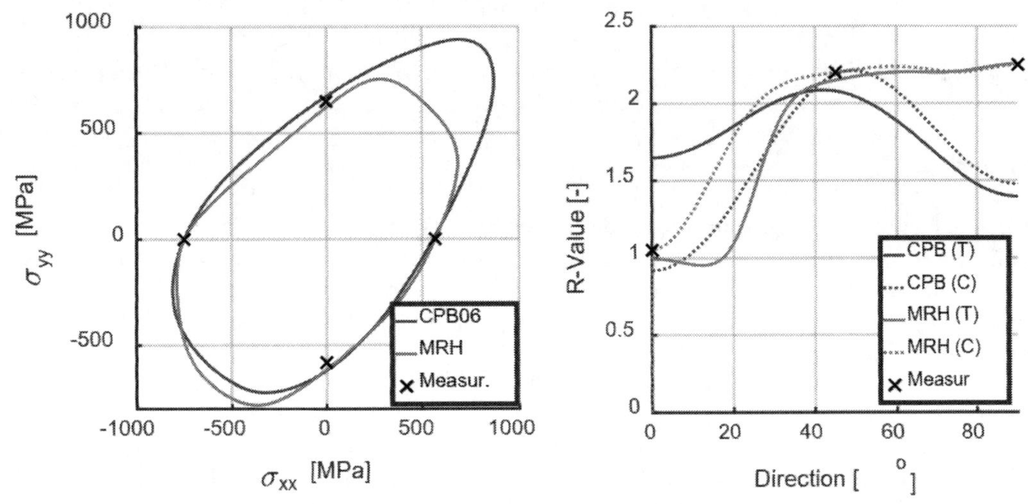

Figure 2. Radial distance of the flange edge of the deep drawn cup (left). Force-Displacement diagram (right)

4. Cup Drawing Test

The quality of the proposed model has been tested based on a cup drawing experiment. The circular blank of 110mm diameter and 0.4mm thickness has been drawn to a depth of 30mm using punch with diameter 50 mm. The blank holder force has been set to 30kN. A corresponding FE-Model has been generated in LS-Dyna featuring a quarter of the blank to exploit symmetry. The blank has been modelled with 3 layers of brick elements with selectively reduced integration. The mesh has been chosen to have a size of about 0.5mm. The Coulomb friction coefficient between the blank and the rigid tools was assumed to be 0.05, given the fact that the blank has been well lubricated.

4.1. Results

The results obtained using the different modelling approaches are depicted in Figure 3. It is seen that, as far as earing is concerned, the proposed model is well able of capturing the shape of the flange edge with especially accurate results at RD and TD. The prediction in 45 degrees is equivalent to that of Hill model. The CPB06 model underestimates draw-in in RD and overestimates it in TD. This is mostly due to the fact that CPB06 cannot capture R-values in tension and compression equally well. In fact looking at Figure 3 (right) it is observed that both MRH and CPB models more accurately capture force-displacement, given their ability to model asymmetry, whereas the Hill model overestimates the force.

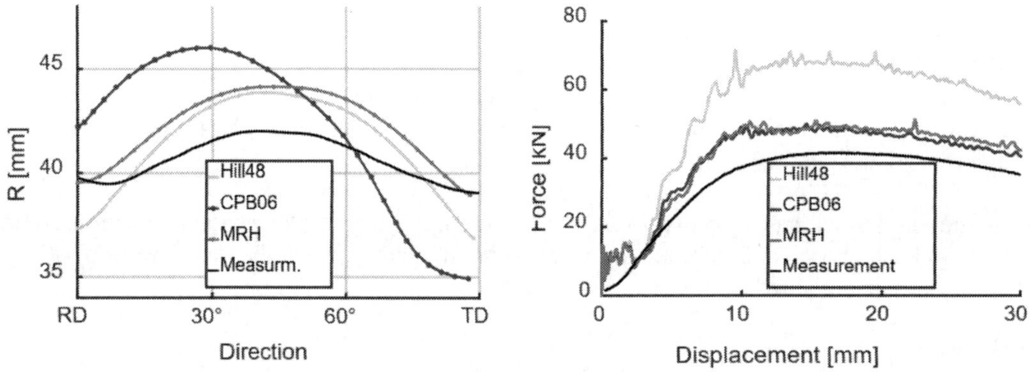

Figure 3. Radial distance of the flange edge of the deep drawn cup (left). Force-Displacement diagram (right)

5. Conclusions

A new approach has been proposed for the modelling of both asymmetry and anisotropy observed in commercially pure titanium. The model is designed to construct asymmetric functions based on well-established symmetric expressions, thus providing a generic framework in this sense. The results based on cup drawing experiments show that the proposed approach can successfully model both earing and force-displacement behaviour accurately.

References

[1] O. Cazacu, B. Plunkett, F. Barlat, Int J Plasticity **22**, 1171, 2006
[2] B. Plunkett, O. Cazacu, F. Barlat, International Journal of Plasticity, 24(5), 847-866, 2008
[3] M. E. Nixon, O. Cazacu, R. A. Lebensohn, Int J Plasticity **26 (4)**, 516-532 , 2010.
[4] Barlat F., Gracio J.J., Lee MG., Rauch E.F., Vincze G. Int J Plasticity **27**, 1309-1327, 2011.
[5] F. Barlat H. Aretz, J.W. Yoon, M.E. Karabin, J.C. Brem, R.E. Dick, Int J Plasticity **21**(5),1009-1039, 2005.

Constitutive modelling of plastic deformation and damage in anisotropic high-purity titanium and validation using ex-situ and in-situ tomography data

O Cazacu, B Revil-Baudard and N Chandola
Department of Mechanical and Aerospace Engineering, University of Florida, REEF, Shalimar, FL 32579, USA

e-mail: cazacu@reef.ufl.edu

Abstract. A study on plastic deformation and damage in titanium was conducted. The X-ray tomography data reveal that damage distribution and evolution in titanium is markedly different than for a FCC material. Theoretically, it is shown that only by modelling both the anisotropy and the tension-compression asymmetry in plastic behaviour it is possible to realistically predict titanium behaviour. For a smooth specimen under uniaxial tension, the model predicts that damage initiates at the centre of the specimen, and is diffuse; the level of damage close to failure being very low. In contrast, for a notched specimen under the same loading it is predicted that damage initiates at the outer surface of the specimen, and grows towards the centre of the specimen, which corroborates with XCMT data.

1. Introduction

Titanium materials that are predominantly hexagonal close-packed are known to display at room temperature plastic anisotropy and a strong tension–compression asymmetry. Although titanium materials are textured, modeling of their damage and failure characteristics is generally done using evolution laws for the rate of void growth or Gurson-type models. However, these models cannot realistically describe damage in Ti materials because the core hypothesis of such models is that the plastic behavior of the matrix is governed by the isotropic von Mises criterion. In this paper, we present an experimental study on plastic deformation and damage of a polycrystalline Ti material, as well as modelling of the observed behaviour using the Stewart and Cazacu model [1].

2. Plasticity-Damage couplings in polycrystalline Titanium

According to this model, the plastic strain rate is defined with respect to the stress potential φ given by:

$$\varphi(\sigma,f)=\hat{\mathrm{m}}^2\frac{\sum_{i=1}^{3}\left(\left|\hat{\sigma}_i\right|-\mathrm{k}\hat{\sigma}_i\right)^2}{\bar{\sigma}_x^{\mathrm{T}}}+\bar{\sigma}_x^{\mathrm{T}}\left(2f\cosh\left(\frac{3\sigma_m}{\mathrm{h}\bar{\sigma}_x^{\mathrm{T}}}\right)-\left(1+f^2\right)\right). \qquad (1)$$

In equation (1) k is a parameter describing the tension-compression asymmetry of the matrix; $\hat{m}$ depends on the anisotropy coefficients and k, f is the porosity, and $\hat{\sigma}_1, \hat{\sigma}_2, \hat{\sigma}_3$ are the principal values of the transformed stress tensor $\hat{\boldsymbol{\sigma}} = \mathbf{L} : \boldsymbol{\sigma}'$, where $\boldsymbol{\sigma}'$ is the deviator of the Cauchy stress tensor and **L** is a fourth-order symmetric tensor describing the orthotropy of the fully-dense material. The parameter h which controls damage evolution depends on both the anisotropy coefficients and the sign of the mean stress, σ_m. Hardening of the matrix (fully-dense material) is due to the effective plastic strain, $\bar{\varepsilon}^p$. The rate of change of the porosity is given by the evolution of existing voids and the nucleation of new ones. Evolution of existing voids is obtained from mass conservation while nucleation of new ones is due to both the plastic strain and to the mean stress. In the absence of voids (f = 0), the potential (1), reduces to that of the matrix, i.e. the quadratic form of Cazacu et al. [2] orthotropic criterion. Note that all the parameters involved in the criterion have a clear physical meaning, being related to the plastic properties of the material. Specifically, all the parameters are expressible in terms of the anisotropy coefficients (components of the tensor **L**) and/or the strength-differential parameter k. For the identification of these parameters we conducted mechanical tests in uniaxial tension on flat specimens with long axis along the rolling direction (RD) and two other in-plane directions, at 45^0 and 90^0 (TD) with respect to RD, respectively and in uniaxial compression tests (cylindrical specimens same directions). The evolution of the anisotropy in r-values with accumulated plastic deformation and the evolving difference in hardening rates between tension and compression loadings observed from the test results are indicative of texture evolution. To describe these effects, the anisotropy coefficients and k are considered to evolve with accumulated plastic deformation. Next, the capabilities of the model to predict plastic deformation and damage in the Ti material for loadings that were not used for identification are demonstrated. Key results are presented in the following. For more details, the reader is referred to [3].

3. Comparison between F.E. prediction and XCMT measurements of porosity evolution

For the smooth specimen, the XCMT scan was done ex-situ, following uniaxial tension along RD up to an axial displacement of 5.52 mm (i.e. very close to failure). In figure 2(a) are shown the reconstructed 2-D views in the (TD, TT) plane, and in the (RD, TD) plane of the Ti specimen, respectively. For comparison purposes, in figure 2(b) are shown the respective views extracted from an XCMT scan of a copper specimen, which was taken at the same axial displacement. It is very important to note that for the same axial displacement, the copper material is much damaged while titanium material shows very little damage. While a large hole/crack is seen in the middle of the copper specimen, almost no damage is observed in the Ti specimen. It means that for uniaxial tensile loading the rate of damage growth is slower in Ti than in Cu. To extract the experimental void volume fraction, a clustering technique was applied in order to identify 3-D groups of connected voxels, and thus to identify individually each pore within the matrix. A total of 2385 pores have been detected in the total volume of the specimen, which resulted into an average porosity over the specimen volume of 0.052%. However, the average porosity in the minimal cross-section is of 0.25%. Figure 2 shows a comparison between the F.E. cross-sections and isocontours of void volume fraction according to the Stewart and Cazacu [1] model and XCMT data for the same axial displacement. The F.E. predictions of the geometry of the specimen are also superposed on the different cut views It is worth noting that the model also predicts that damage is diffuse, in the minimal cross-section the maximum void volume fraction predicted being of 0.22% (see figure.2a). Likewise, for the RD-TD view (figure 2(b)) most of the voids observed by XCMT are inside the region of maximum void volume fraction predicted by the model.

Journal of Physics: Conference Series **734** (2016) 032052 doi:10.1088/1742-6596/734/3/032052

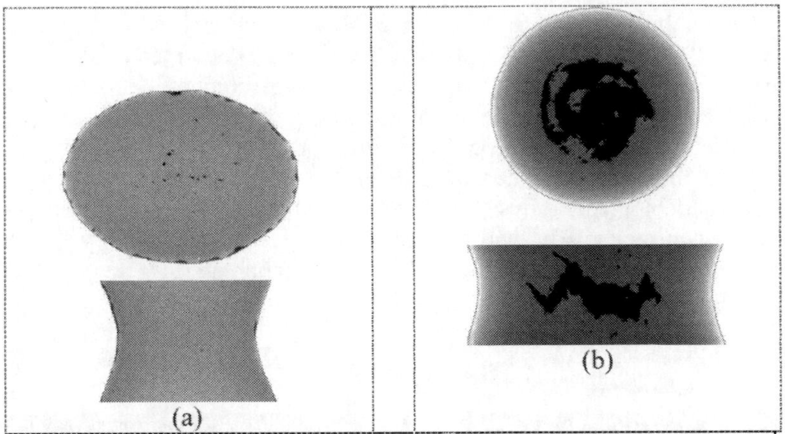

Figure 1. Post-test XCMT scans of smooth specimens subjected to uniaxial tension: (a) Ti specimen; (b) Cu specimen.

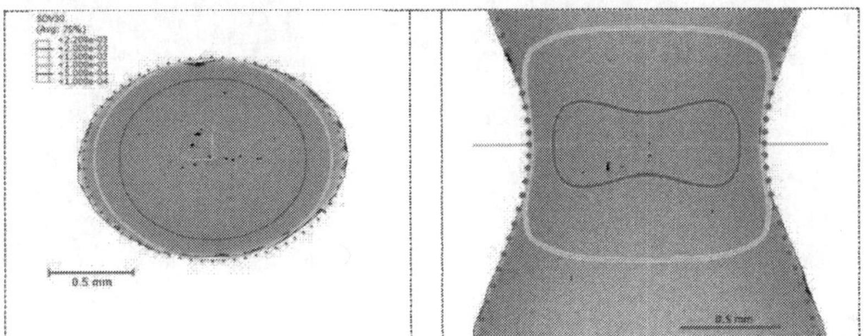

Figure 2. Comparison between F.E. cross-sections and isocontours of porosity of a smooth Ti specimen subjected to uniaxial tension along RD, according to [1]: (a) cross-section; (b) (RD-TD) section.

Note that the model correctly predicts the section geometry in all the planes. Next, we assess the predictive capabilities of the model in terms of capturing the influence of the stress triaxiality on damage and its evolution. For this purpose, in-situ XCMT in uniaxial tension was conducted on a notched axisymmetric RD specimen. The notch radius was of 0.51mm while the specimen cross-section radius was of 1.27 mm.

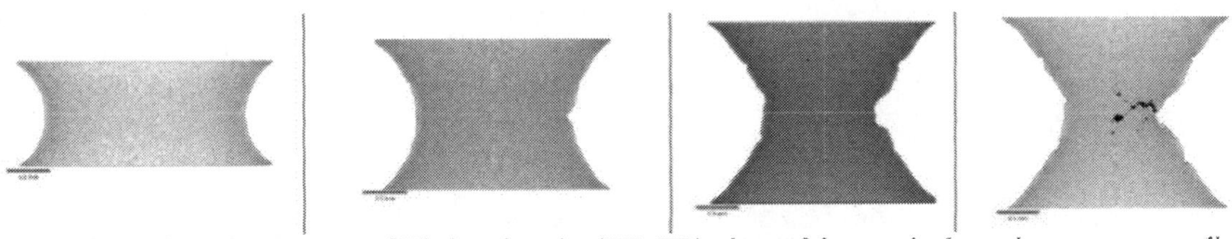

Figure 3. In-situ XCMT scans of Ti showing the (RD-TD) view of the notched specimen corresponding to displacement of 0.24 mm, 0.73 mm 1.02 mm, 1.2 mm, respectively.

Scans were taken at notch displacement of 0.24 mm, 0.73mm, 1.02 mm and 1.20 mm, respectively. As an example, in Figure 3 are shown the cross-section in the (RD, TD) corresponding to each displacement. Comparison between F.E. predictions at 1.2 mm axial displacement and model are shown in Figure 4 (b). Since according to the model, damage is driven by the plastic deformation, and the plastic anisotropy is correctly described, the model correctly predicts the location of the zones of maximum damage for the notched specimen. The importance of accounting for tension-compression asymmetry in the plastic behaviour of Ti is clearly demonstrated by comparing the predicted isocontours of the porosity in the (RD-TD) section with an anisotropic model with no tension-compression symmetry (k=0). Note that if the tension-compression asymmetry in plastic deformation is accounted for, the local distribution of the damage is totally different.

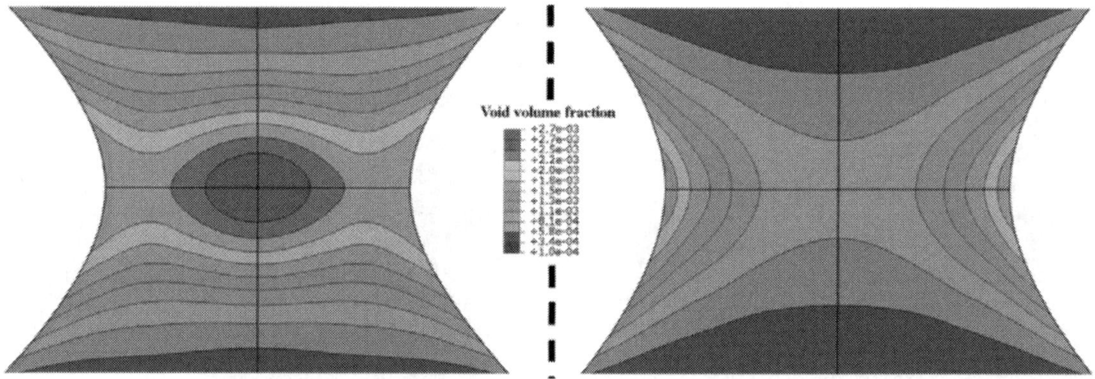

(a) (b)

Figure 4. Comparison between F.E. cross-sections and isocontours of porosity of a smooth Ti specimen subjected to uniaxial tension along RD, according to [2]: (a) cross-section; (b) (RD-TD) section.

4. Conclusions

The ex-situ and in-situ X-ray tomography measurements conducted show that damage distribution and evolution in this material is markedly different than in a typical FCC material such as copper. Stewart and Cazacu [1] anisotropic elastic/plastic damage model predicts correctly the anisotropy in plastic deformation, and its strong influence on damage distribution and damage accumulation. Specifically, for a smooth axisymmetric specimen subject to uniaxial tension, damage initiates at the center of the specimen, and is diffuse; the level of damage close to failure being very low. On the other hand, for a notched specimen subject to the same loading the model predicts that damage initiates at the outer surface of the specimen, and further grows from the outer surface to the center of the specimen, which corroborates with the in-situ tomography data.

References

[1] Stewart J B and Cazacu O 2011 *Int. J. Solids Struct.* **48** 357.
[2] Cazacu O, Plunkett B and Barlat F 2006 *Int. J. Plasticity* **22** 1171.
[3] Revil-Baudard B, Cazacu O, Flater P, Chandola N and Alves J L 2016 *J. Mech. Phys. Solids* **88** 100.

Journal of Physics: Conference Series **734** (2016) 032053 doi:10.1088/1742-6596/734/3/032053

DD3MAT – a code for yield criteria anisotropy parameters identification.

P D Barros[a], P D Carvalho[a], J L Alves[b], M C Oliveira[a] and L F Menezes[a]

[a]CEMUC, Mechanical Engineering Department, University of Coimbra, Polo II, Rua Luís Reis Santos, Pinhal de Marrocos, Coimbra, 3030-788, Portugal
[b]CMEMS, Microelectromechanical Systems Research Unit, University of Minho, Campus de Azurém, 4800-058 Guimarães, Portugal

pedro.barros@dem.uc.pt

Abstract. This work presents the main strategies and algorithms adopted in the DD3MAT in-house code, specifically developed for identifying the anisotropy parameters. The algorithm adopted is based on the minimization of an error function, using a downhill simplex method. The set of experimental values can consider yield stresses and r-values obtained from in-plane tension, for different angles with the rolling direction (RD), yield stress and r-value obtained for biaxial stress state, and yield stresses from shear tests performed also for different angles to RD. All these values can be defined for a specific value of plastic work. Moreover, it can also include the yield stresses obtained from in-plane compression tests. The anisotropy parameters are identified for an AA2090-T3 aluminium alloy, highlighting the importance of the user intervention to improve the numerical fit.

1. Introduction

Sheet metal forming processes are nowadays designed and optimized virtually using finite element analysis (FEA), which is consensually accepted as the main reason for the huge decrease in time-to-market life cycle and for notable savings in terms of money, time and effort in the design, production and process set-up of new formed parts. However, the success of finite elements solvers on the design and optimization of sheet metal formed parts is strongly dependent on their ability to accurately describe the material's mechanical behaviour. Sheet metals generally exhibit anisotropic mechanical behaviour due to the rolling process, characterized by the symmetry of the mechanical properties with respect to three orthogonal planes, i.e. orthotropic behaviour. Thus, different mechanical behaviours are expected for different loading directions and conditions. Phenomenological models are the most widely used approach to describe the elastoplastic response of metallic sheets, since they are computationally efficient when compared with microscopic models. The material's orthotropic behaviour is modelled by the yield surface, used to describe the yielding and the plastic flow of the material, based on the hardening law selected. This dual role of the yield surface requires a particular care and accuracy in its modelling and numerical implementation. The material anisotropy parameters identification must be performed in a way that a given yield criterion should reproduce the materials mechanical behaviour as close as possible. Moreover, since sheet metal forming processes are carried out with inhomogeneous deformation and under multiaxial strain paths it is important to use as much information as possible.

In this work DD3MAT in-house code is used to identify the anisotropy parameters for an AA2090-T3 aluminium alloy [1], considering three yield criteria YLD91 [2], CB2001 [3] and CPB06 [4]. The predicted in-plane distribution of the yield stresses and r-values evolution, as well as the yield surfaces, are discussed to show that a more or less accurate fit of the experimental data is achieved, depending on the selected yield criterion and on the user intervention, through the selection of the weighting factor for each test result.

2. DD3MAT – yield criteria parameters identification code.

The anisotropy parameters should be determined such that the yield criterion reproduces the material's mechanical behavior as close as possible. The most used experimental results for the identification of anisotropy parameters are the yield stresses and r-values obtained from in-plane tension, for different angles (θ) with the RD. In order to improve the description of the yield surface, it is also recommended to experimentally determine the biaxial yield stress and the biaxial anisotropy coefficient [5–7]. For the CPB06 yield criterion, uniaxial compression experimental results are also necessary for describing the strength differential effects. However, when performing compression tests for thin metallic sheets it is necessary to avoid buckling effects. This requires the use of small specimens, leading to supplementary difficulties in the acquisition and analysis of experimental results, particularly for high strain values [8]. The anisotropy parameters for the three selected yield criteria were obtained with the DD3MAT in-house code. The procedure adopted is based in an optimization problem regarding the minimization of an error function, evaluating the difference between the estimated values and the experimental ones, as follow

$$F(\mathbf{A})=\sum_{\theta=0}^{90}\left[w_{\sigma_\theta^{\mathrm{T}}}\left(\sigma_\theta^{\mathrm{T}}(\mathbf{A})/\sigma_\theta^{\mathrm{T}}-1\right)^2+w_{\sigma_\theta^{\mathrm{C}}}\left(\sigma_\theta^{\mathrm{C}}(\mathbf{A})/\sigma_\theta^{\mathrm{C}}-1\right)^2+w_{r_\theta}\left(r_\theta(\mathbf{A})/r_\theta-1\right)^2\right] \\ +w_{\sigma_b}\left(\sigma_b(\mathbf{A})/\sigma_b-1\right)^2+w_{r_b}\left(r_b(\mathbf{A})/r_b-1\right)^2 \quad (1)$$

where $\mathbf{A}$ represents the set of parameters associated with the selected yield criterion. $\sigma_\theta^{\mathrm{T}}$, $\sigma_\theta^{\mathrm{C}}$ and r_θ are the experimental yield stresses in tension, compression and r-values determined in uniaxial tension, respectively, obtained from the uniaxial tests for a specific orientation (θ) with respect to RD. σ_b is the experimental yield stress obtained from the equibiaxial tensile test, r_b is the experimental r-value obtained from the disc compression test, and $\sigma_\theta^{\mathrm{T}}(\mathbf{A})$, $\sigma_\theta^{\mathrm{C}}(\mathbf{A})$ $r_\theta(\mathbf{A})$, $\sigma_b(\mathbf{A})$ and $r_b(\mathbf{A})$ are the correspondent values predicted from the adopted yield criterion. Such procedure can be considered a generalization of the one proposed by Banabic et al., 2005 [9]. The weighting factors, $w_{\sigma_\theta^{\mathrm{T}}}$, $w_{\sigma_\theta^{\mathrm{C}}}$, w_{r_θ}, w_{σ_b} and w_{r_b} are used to balance the influence of the experimental data. Nevertheless, the selection of the weighting factors is normally a manual procedure, strongly dependent on users' expertise and knowledge. The identification procedure, defined in (1) also implies the pre-selection of an initial yield stress or the hardening law parameters, if a specific value of plastic work is defined to select the experimental values for the yield stresses.

3. Results and discussion

The anisotropy parameters for the three previously mentioned yield criteria were identified for an AA2090-T3 aluminium alloy, for which the experimental values of $\sigma_\theta^{\mathrm{T}}$, $\sigma_\theta^{\mathrm{C}}$, r_θ (at each 15° with RD), σ_b and r_b are available [1].

Figure 1 shows the evolution of the yield stress and r-values with the angle from RD, for the three yield criteria, obtained considering all weighting factors equal to 1.0 (labelled "Set 1"). As expected, YLD91 shows a less accurate fit when compared with the other yield criteria, mainly for the yield stresses. CPB06 presents a better fit for the yield stresses, when compared with the YLD91, with the added benefit of also describing the materials' behavior for compression stress states. Regarding the CB2001, both yield stresses and r-values are globally well described. Nevertheless, it is possible to see that all yield criteria fail to capture the r-value at 45° and tend to overestimate the yield stresses for angles closer to the transverse direction (TD). Therefore, a new set of parameters was identified, through the selection of the weighting factors (shown in Table 1), i.e. with user intervention, labelled "Set 2".

Figure 2 shows the evolution of the yield stress and r-values with the angle from the RD, for the three yield criteria, considering the user intervention. For the CB2001, the description is enhanced mainly for yield stresses for angles closer to TD and the predicted r-values at 45° and 90° are closer to the experimental ones. For the YLD91, the main improvement occurs for yield stresses closer to the rolling direction and the r-value at 45°. The CPB06 also benefited from the user's input in the identification procedure since the yield stresses evolution in tension and compression are globally closer to the experimental values.

Figure 3 presents the yield surfaces for the three considered yield criteria in the $\sigma_1 - \sigma_2$ plane considering, or not, user intervention. These results, together with Table 2, allows analyzing the material behavior for stress states other than uniaxial tensile stress. YLD91 presents lower experimental σ_b values, for both sets, slightly more accurate for Set 2. The CB2001 yield criterion predicts accurate σ_b and r_b values for both stets, but Set 1 presents a r_b value lower than the experimental, while it is slightly higher for Set 2. Globally, CPB06 presents the σ_b and r_b values least accurate, since the introduction of the compression yield stress in the optimization procedure reduces the relative importance of these values in the objective function. A summary of the parameters identified for the three yield criteria, considering both sets, is presented in Table 3, highlighting the parameters' interdependence.

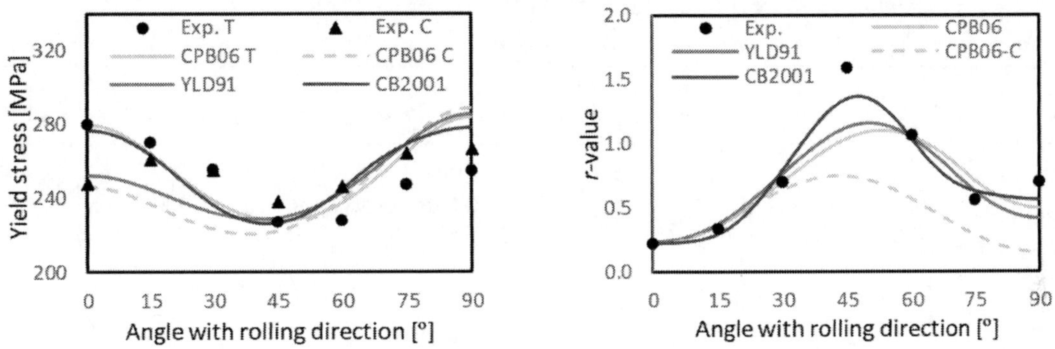

(a) (b)

Figure 1. Experimental and predicted (a) yield stresses and (b) r-values (Set 1).

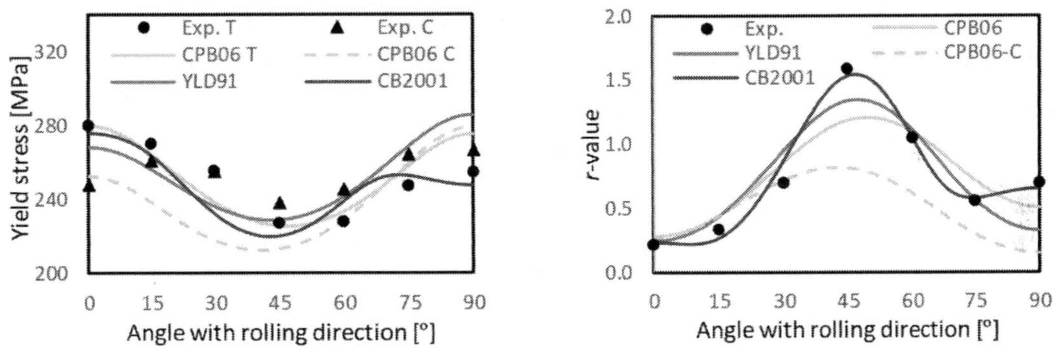

(a) (b)

Figure 2. Experimental and predicted (a) yield stresses and (b) r-values (Set 2).

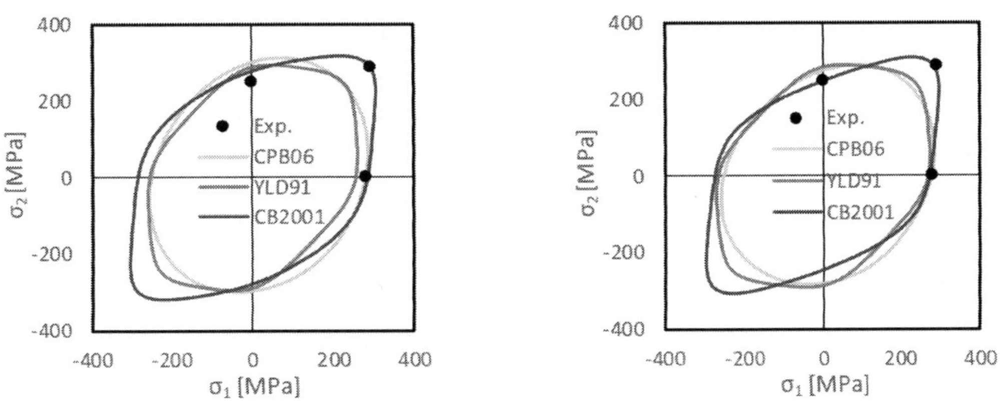

(a) (b)

Figure 3. Predicted yield surfaces in the $\sigma_1 - \sigma_2$ plane (a) Set 1 and (b) Set 2.

Table 1. Experimental and predicted yield stress and r-values for the biaxial test.

CB2001	$w_{r_{45}}=5$ $w_{r_{75}}=5$ $w_{r_{90}}=5$ $w_{\sigma_b}=20$ $w_{\sigma_{60}^{T}}=10$ $w_{\sigma_{75}^{T}}=10$ $w_{\sigma_{90}^{T}}=10$
YLD91	$w_{r_{45}}=5$ $w_{\sigma_b}=20$ $w_{\sigma_{0}^{T}}=20$ $w_{\sigma_{15}^{T}}=20$ $w_{\sigma_{75}^{T}}=20$ $w_{\sigma_{90}^{T}}=20$
CPB06	$w_{r_{45}}=7$ $w_{\sigma_b}=20$ $w_{\sigma_{90}^{T}}=30$ $w_{\sigma_{15}^{C}}=40$ $w_{\sigma_{30}^{C}}=5$ $w_{\sigma_{45}^{C}}=5$ $w_{\sigma_{60}^{C}}=5$ $w_{\sigma_{75}^{C}}=5$ $w_{\sigma_{90}^{C}}=40$

Table 2. Experimental and predicted yield stress and r-values for the biaxial test.

		Set 1			Set 2		
	Exp.	CB2001	YLD91	CPB06	CB2001	YLD91	CPB06
σ_b	289.4	289.6 (0.06%)	235.1 (-18.7%)	232.8 (-19.6%)	285.9 (-1.2%)	239.4 (-17.3%)	229.5 (-20.7%)
r_b	0.67	0.66 (-0.74%)	0.57 (-15.7%)	0.78 (+15.8%)	0.66 (-1.5%)	0.76 (+12.7%)	0.89 (+33.3%)

Table 3. Anisotropy parameters identified for the considered yield criteria.

	Set	a_1	a_2	a_3	a_4	b_1	b_2	b_3	b_4	b_5	b_{10}	c
CB2001	1	0.728	1.402	1.334	1.525	3.576	-0.838	-1.983	-0.581	-4.580	-0.567	0.608
	2	1.358	1.848	1.075	1.709	5.357	-0.623	-4.386	-3.654	-6.046	-0.882	0.857

	Set	c_1	c_2	c_3	c_6	m
YLD91	1	1.069	1.300	0.856	1.213	8
	2	1.110	1.224	0.835	1.238	8

	Set	C_{11}	C_{22}	C_{33}	C_{66}	C_{23}	C_{13}	C_{12}	k
CPB06	1	-1.010	0.475	1.199	-1.082	-0.102	0.104	-0.012	-0.057
	2	-0.818	0.449	1.232	-1.224	-0.088	-0.060	-0.819	-0.050

Note that, for metal sheets, the off plane parameters cannot be experimentally evaluated. Thus, parameters a_5, a_6 and b_k $(k=6,7,8,9,11)$, for the CB2001, c_4, c_5 for the YLD91 and C_{44}, C_{55} for the CPB06 take the corresponding isotropic values, i.e. 1.0.

4. Conclusions

The results presented show the ability of DD3MAT in-house to perform the identification of the anisotropy parameters for different yield criteria. The classical identification procedure adopted allows the user to control the importance of each test through the weighting factors. However, the users' input and knowledge is only as good as the flexibility allowed by the yield criterion considered.

Acknowledgments

The authors gratefully acknowledge the financial support of the Portuguese Foundation for Science and Technology (FCT) via the projects PTDC/EME-TME/118420/2010 and UID/EMS/00285/2013.The first author is also grateful to the FCT for the PhD grant SFRH/BD/98545/2013.

References

[1] J.W. Yoon, F. Barlat, K. Chung, F. Pourboghrat, D.Y. Yang, Int. J. Plast. 16 (2000) 1075–1104.
[2] F. Barlat, D.J. Lege, J.C. Brem, Int. J. Plast. 7 (1991) 693–712.
[3] O. Cazacu, F. Barlat, Math. Mech. Solids 6 (2001) 613–630.
[4] O. Cazacu, B. Plunkett, F. Barlat, Int. J. Plast. 22 (2006) 1171–1194.
[5] D.J. Lege, F. Barlat, J.C. Brem, Int. J. Mech. Sci. 31 (1989) 549–563.
[6] K. Pöhlandt, D. Banabic, K. Lange, in:, ESAFORM 2002, Krakow, Poland., 2002, pp. 723–727.
[7] F. Barlat, J.C. Brem, J.W. Yoon, K. Chung, R.E. Dick, D.J. Lege, F. Pourboghrat, S.H. Choi, E. Chu, Int. J. Plast. 19 (2003) 1297–1319.
[8] M. Tritschler, A. Butz, D. Helm, G. Falkinger, J. Kiese, Int. J. Mater. Form. 7 (2014) 259–273.
[9] D. Banabic, H. Aretz, D.S. Comsa, L. Paraianu, Int. J. Plast. 21 (2005) 493–512.

Application of Strain Non-Uniformity Index (SNI) Based Approach to Predict Failure in Sheet Metal Components

P P Date [1] **and K. D. Jamadar** [2]

[1] Department of Mechanical Engineering, Indian Institute of Technology Bombay, Mumbai 400076, India Web page: www.iitb.ac.in
[2] Volkswagen Nutzfahrzeug Hannover, Germany

E-mail: ppdate@iitb.ac.in, kartik.jamadar@volkswagen.de.

Abstract. Industrial components are far too complex in shape and far too large in size compared to the laboratory scale test samples used to establish failure criteria like the forming limit diagram (FLD). These have been traditionally used to predict failure, but are found to often make wrong predictions due to their sensitivity to strain path, and forming conditions like temperature. The Strain Nonuniformity Index (SNI), a single parameter determined from the spatial strain distribution, may be used to predict failure and identify locations of failure or even imminent failure. This is possible regardless of temperature or even the method of forming (cold forming, hot forming, superplastic forming etc.). The paper presents a few results demonstrating successful applications of this novel SNI based technique.

Keywords: Sheet Forming, SNI; FLD; strain distribution; failure prediction; failure location

1. Introduction

Automobile components made from sheet metal are used as skin panels, crash structures, chassis members, structural members like the A, B and C pillars and reinforcements of a variety of shapes and sizes. This generates a spatial variation of strains over the entire sheet. Often components show either unacceptably large extent of thinning, necking, and even cracking. It is important to know the potential 'failure sites' so that the tools can be designed accordingly. Actually, these failure sites are a combined outcome of the component design, shape (design) of the blank from which it was drawn, followed by tool design and the forming variables like lubrication, blank holding scheme and the press speed.

An apriori assessment about likely failure sites is often made using the forming limit diagram (FLD) so that all anticipated problems could be rectified and die rework thereby avoided. Dies are themselves very expensive, and reworking them adds significantly to the cost. At what point in punch stroke an issue might arise can be indicated using the FLD.

The FLD is however, dynamic (dependent on the material properties and prestrain) and is highly sensitive to strain path. The present work presents examples of the strain distribution based methodology (published earlier [1, 2]) for anticipating failure and identifying the potential locations of failure. This methodology applies with equal facility to different materials and is irrespective of the processing route as one relies on the strain distribution and not how it was produced.

2. Strain distribution based methodology

The quantities, namely, the Strain Non-uniformity Index (SNI) and the Constraint Factor (CF) characterise a strain distribution [1]. The maximum degree of strain non-uniformity is of interest. Hence a plane passing through the points of highest deformation, least deformation and the direction of forming would encompass this greatest non-uniformity in strain. Such a plane is called a critical plane.

In a given component one might have a multiplicity of critical planes and those with the maximum SNI are of interest in order to predict failure. Once the values of SNI in these planes exceed certain threshold value failure is deemed to be imminent / have taken place at the point of peak major strain and maximum thinning.

In the present work, strain distributions from Nakazima test samples at failure were used. Some of the samples were safe as well. A relation correlating the normalised SNI of major strain with the normalised SNI of thickness strain was obtained. Here the SNI for the strain distribution in a given plane was normalised by the peak value of strain in that plane. It was found that the correlations for different blank widths lay onto a single quadratic curve. This way, a common relation could be worked out between the normalised SNI of major strain with that of thickness strain across all blank widths of a given material.

Further, Nakazima tests performed on about 21 materials (sheet metal grade + sheet thickness defines one material) showed that the quadratic curves of several materials lay in a narrow band. Here, among the 21 materials was an aluminium alloy as well. This shows that the set of 21 quadratic curves could be replaced by a single average quadratic curve without any significant loss of accuracy.

In other words, if the quadratic curve to determine the SNI at failure for a given material is available, one uses the equation that is established by experiment. However, for a material that is not among the 21 materials studied, one can still, without introducing significant error, determine the SNI at failure. Hence the quadratic equation for an untested material need not be available for using this methodology. Availability of a quadratic curve would, however, enhance the accuracy of prediction.

Since the curve correlates the normalised SNI of major strain with that of the thickness strain, each critical plane now has an individual SNI at failure. All critical planes therefore would not fail at a single threshold SNI, which was the case in the previous studies presented. This way, the SNI of a critical plane has to exceed the threshold SNI for that plane for that instant. The threshold SNI itself is dynamic (changing with progression) and this enables a more accurate prediction of the location of failure.

The above observations are understandable since the degree of nonuniformity in thickness strain distribution is driven by the degree of nonuniformity in the major strain distribution, with the minor strain influencing the intensity of the thickness strain nonuniformity for a given non uniformity in the major strain. The constraints from the tool and the frictional conditions, together determine the distribution of the minor strains, particularly in view of minor strain being most influenced by the tool contact conditions as documented earlier by Backofen [3]. The development of a minor strain thereby influences the biaxial strain path. This is why the SNI of the thickness strain incorporates the effect of the strain path.

Application of this methodology to the results of strain distribution from AUTOFORM simulations of skin panels of a car is illustrated in Fig 1 (a and b). The locations of failure predicted using the methodology mentioned above is in line with the shop floor experience.

3. Results and discussion

AUTOFORM simulations for components A, B, and C served as a means of obtaining the strain distribution all over the component. The failure strains were represented using the FLD and locations of failure predicted. Alternately, the strain distribution over the component permitted failures to be marked based on the SNI exceeding the failure SNI established experimentally using Nakazima tests.

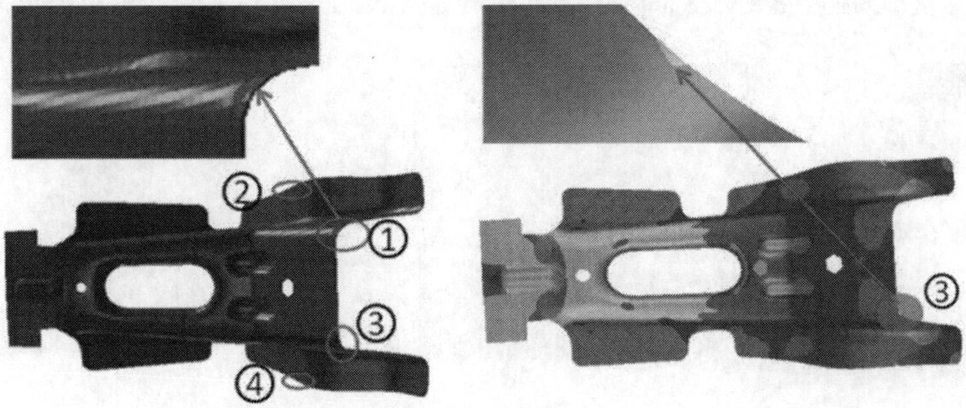

(a) (b)

Fig. 1. Failure locations in an industrial component A: (a) Using the SNI based failure criterion (b) Using the FLD. Strain data in both the cases came from AUTOFORM simulations

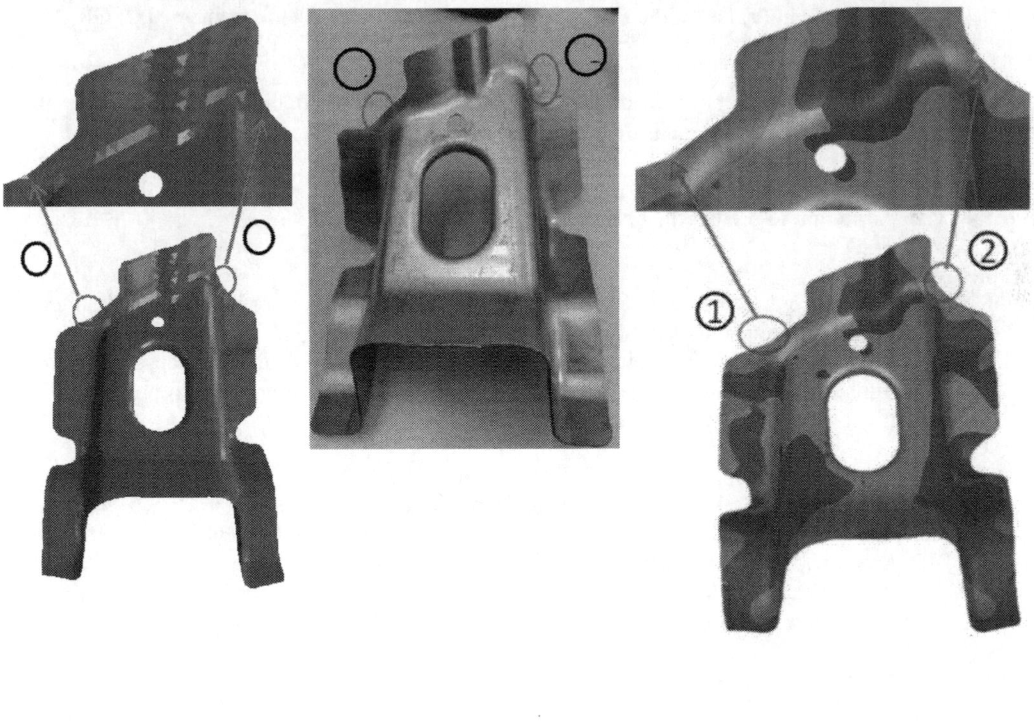

(a) (b) (c)

Fig. 2. Failure locations in an industrial component B: (a)Using the SNI based failure criterion (b) On the formed component (c) Using the FLD. Strain data in (a) and (c) came from AUTOFORM simulations

It is not necessary to use Nakazima tests to establish the threshold SNI. The component itself could have been used for the purpose. As seen in Fig. 1 (a) and (b) for component A, the strain distribution based methodology predicts failures at four locations while the FLD based predictions (for the same component simulated using AUTOFORM) shows only one location of failure. For component B, Fig. 2 (b) represents the shop floor experience. The failure locations predicted by both, the SNI as well as the FLD correlate well with the actual failure locations. A similar observation may be made from Fig. 3 pertaining to Component C.

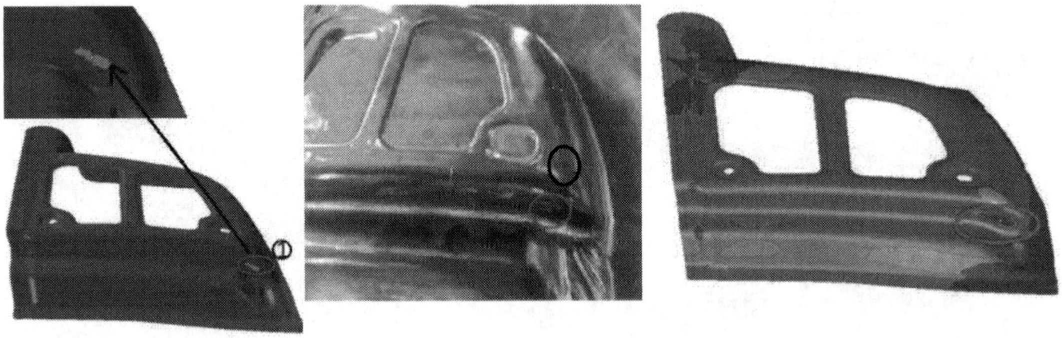

(a) (b) (c)

Fig. 3. Failure locations in an industrial component C: (a) Using the SNI based failure criterion (b) On the formed component (c) Using the FLD. Strain data in (a) and (c) came from AUTOFORM simulations

From the foregoing, it appears that predictions using the SNI based methodology seem to be as good as those made using the FLD, faring a little better occasionally.

The observations made confirm that the relatively simple strain distribution based methodology can serve as a viable criterion to predict the occurrence of failure and its location.

4. Conclusions

1) The strain distribution based failure criterion, namely the SNI is capable of predicting the occurrence of failure and its location.
2) Simplicity of establishing the failure location irrespective of the forming process, strain path, temperature and inconsistent lubrication makes the SNI a viable alternative to predict failures especially at the design stage.

5. References

[1] S. G. Desai and P. P. Date, On the quantification of strain distribution in drawn sheet metal products, Journal of Materials Processing Technology 177 (2006) 439–443.
[2] P. P. Date and K. D. Jamadar, A Strain Distribution Based Approach To Predict Failure In Sheet Metal Forming, Proc. IDDRG2016, 12-15 June 2016, Linz, Austria.
[3] W. A. Backofen, Deformation Processing, Addison-Wesley Publishers, 1972.

Acknowledgements

The authors are thankful to M/s. Volkswagen AG for funding the work and to Mr. Sandip Jadhav for the AUTOFORM simulations.

Evaluation of Bi-Material Crack Using The Finite Block Method

P. Hinneh[1, a] , P.H. Wen[1, b]

[1]School of Engineering and Materials Science, Queen Mary University of London, E1 4NS, UK

[a]p.hinneh@qmul.ac.uk , [b]p.h.wen@qmul.ac.uk

Keywords: Finite Block Method, Interface Crack, Stress Intensity Factor, Meshfree, Bi-material

Abstract. The Finite Block Method has been employed in this paper to evaluate the stress intensity factor of a bi-material plate. The complex stress intensity factor components K_1 and K_2 determined by the Finite Block Method is compared with an equivalent Finite Element Method (ABAQUS) analysis. The paper demonstrates the accuracy of the Meshfree approach by the Finite Block Method without the arduous demand of meshing around the crack surface as seen on standard FEM crack analysis. This paper also describes the application of the polygonal singular core and the collocations points around the interface crack. A computational example for various E_1/E_2 material combinations is presented.

1. Introduction

The analytical solution of interface crack is difficult to obtain as work produced by Williams [1], England [2], Hutchinson [3] and many others have shown. However, the study of cracks in monolithic material has been in existence for decades and this has led to the development of Linear Elastic Fracture Mechanics, the J-Integral method, the strain energy release rate and many other analytical solutions [6]. In an attempt to tackle the challenges presented by the mixed mode coupling and oscillatory effect of interface cracks, many numerical solutions have been presented to determine the complex stress intensity factors.

The finite element method is a popular numerical method and it has been utilised in analysing two-dimensional cracks. This has been demonstrated in recent works by Lin [4]. Building on the success of the Finite Element Method, the Extended Finite Element Method as proposed by Belytschko utilize interpolation functions, the concept of partition of unity and the discontinuity of near crack tip displacement fields which is deemed independent from the finite element mesh to analyse interface cracks [7]. The Boundary Element method is another numerical method capable of evaluating stress intensity factor for bi-material. However, the finite element and the boundary element methods are heavily mesh based models and this presents its own challenges in modeling cracks. The Finite Block Method (FBM) based on the collocation method has been developed by Wen [5] for elasticity analysis. The FBM divide's the problem domain into blocks and the governing equations are satisfied in the strong form at certain collocation points. In this paper, the finite block method using the polygon single core as developed by the authors is used to evaluate the stress intensity factor for bi-material interfacial problem.

The paper is organized as follows. In section 2 an overview of FBM is presented, followed by the evaluations of SIF using the polygonal single core. In section 3, a numerical example of a bi-material plate with center and an edge crack is presented. The research is summarized in section 5.

2. Overview of the Finite Block Method For Interface Crack

The Finite Block Method formulation as reviewed in this paper makes use of a simple square to demonstrate the mathematical formulation. In addition, the reader is advised to continue reading the following papers [5, 8] for an in-depth understanding of FBM.

From the simple square plate shown in figure 1 a smooth function $u(\xi,\eta)$ can be approximated in the domain $|\xi| \le 1$ and $|\eta| \le 1$. Where the field function is given by

$$u(\xi,\eta)=\sum_{i=1}^{M}\sum_{j=1}^{N}F(\xi,\xi_i)G(\eta,\eta_j)u_l \tag{1}$$

In this case the number of collocation points along the two axes are represented by M and N, u_l indicates the nodal value, $l=(j-1)M+i$, with uniformly distributed nodes at $\xi_i=-1+2(i-1)/(M-1)$, $i=1,2,...,M$, $\eta_j=-1+2(j-1)/(N-1)$, $j=1,2,...,N$, and two polynomial functions;

$$F(\xi,\xi_i)=\prod_{\substack{m=1\\ m\neq i}}^{M}\frac{(\xi-\xi_m)}{(\xi_i-\xi_m)},\qquad G(\eta,\eta_j)=\prod_{\substack{n=1\\ n\neq j}}^{N}\frac{(\eta-\eta_n)}{(\eta_j-\eta_n)}. \tag{2}$$

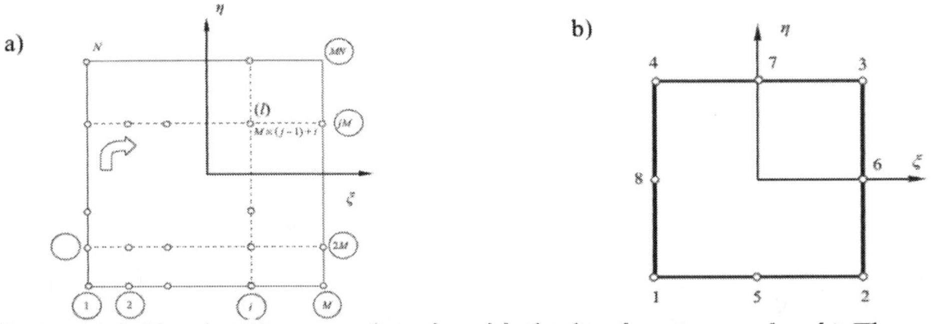

Figure 1. a) Simple 2D square domain with the local system nodes. b) The mapping geometry with the 8 seeds.

The total number of nodes is given by $Q=M\times N$. Then, the first order partial differential is determined with respect to ξ and η respectively

$$\frac{\partial u}{\partial \xi}=\sum_{i=1}^{M}\sum_{j=1}^{N}\frac{\partial F(\xi,\xi_i)}{\partial \xi}G(\eta,\eta_j)u_l\,,\quad \frac{\partial u}{\partial \eta}=\sum_{i=1}^{M}\sum_{j=1}^{N}F(\xi,\xi_i)\frac{\partial G(\eta,\eta_j)}{\partial \eta}u_l \tag{3}$$

Employing the same approach as used in the Finite Element Method and using a set of quadratic shape functions with 8 seeds, the 2D square problem in Cartesian coordinate system (x,y) is transformed into a square in the mapping domain (ξ,η) as shown in figure 1b. The quadratic shape functions are defined follows

$$N_i(\xi,\eta)=\frac{1}{4}(1+\xi_i\xi)(1+\eta_i\eta)(\xi_i\xi+\eta_i\eta-1),\quad i=1,2,3,4, \tag{4}$$

$$N_i(\xi,\eta)=\frac{1}{2}(1-\xi^2)(1+\eta_i\eta),\qquad i=5,7, \tag{5}$$

$$N_i(\xi,\eta)=\frac{1}{2}(1-\eta^2)(1+\xi_i\xi),\qquad i=6,8. \tag{6}$$

From figure 1b, a quadratic block can be mapped into a square domain using the following expression

$$x=\sum_{k=1}^{8}N_k(\xi,\eta)x_k,\quad y=\sum_{k=1}^{8}N_k(\xi,\eta)y_k\,, \tag{7}$$

where (x_k,y_k) denotes the coordinate of seed k. The first order partial differentials of function $u(x,y)$ in the Cartesian coordinate system are obtained by

$$\frac{\partial u}{\partial x}=\frac{1}{J}\left(\beta_{11}\frac{\partial u}{\partial \xi}+\beta_{12}\frac{\partial u}{\partial \eta}\right),\quad \frac{\partial u}{\partial y}=\frac{1}{J}\left(\beta_{21}\frac{\partial u}{\partial \xi}+\beta_{22}\frac{\partial u}{\partial \eta}\right)\quad , \tag{8}$$

IOP Publishing
doi:10.1088/1742-6596/734/3/032055

$$\beta_{11}=\frac{\partial y}{\partial \eta},\beta_{12}=-\frac{\partial y}{\partial \xi},\beta_{21}=-\frac{\partial x}{\partial \eta},\beta_{22}=\frac{\partial x}{\partial \xi},\ J=\beta_{22}\beta_{11}-\beta_{21}\beta_{12}. \tag{9}$$

Then, substituting (3) into (8) gives an expression interms of the nodal value of the first order partial differentials and the nodal value of displacement in the following manner

$$\frac{\partial u}{\partial x}=\frac{1}{J}\sum_{i=1}^{M}\sum_{j=1}^{N}\left[\beta_{11}\frac{\partial F(\xi,\xi_i)}{\partial \xi}G(\eta,\eta_j)+\beta_{12}F(\xi,\xi_i)\frac{\partial G(\eta,\eta_j)}{\partial \eta}\right]u_l=D_{xl}(\xi,\eta)u_l,$$
$$\frac{\partial u}{\partial y}=\frac{1}{J}\sum_{i=1}^{M}\sum_{j=1}^{N}\left[\beta_{21}\frac{\partial F(\xi,\xi_i)}{\partial \xi}G(\eta,\eta_j)+\beta_{22}F(\xi,\xi_i)\frac{\partial G(\eta,\eta_j)}{\partial \eta}\right]u_l=D_{yl}(\xi,\eta)u_l, \tag{10}$$

Using equation 10, the general expression for a 2D constitutive equation and the equilibrium equation, the nodal value of the displacement (u_x,u_y) is determined for a given boundary condition.

3. Evaluation of SIF Using The Polygonal Singular Core

In this study a single polygonal core [8] centered at the crack tip is used in determining the near crack tip stress and displacement values. From figure 2, r_0 is the radius of the circle (dash line) originating from the crack tip. This indicates the size of the singular core and $P(r_k,\theta_k)$ $(k=1,...N_b)$ is the coordinate of the collocation point on the interface between the block and the core.

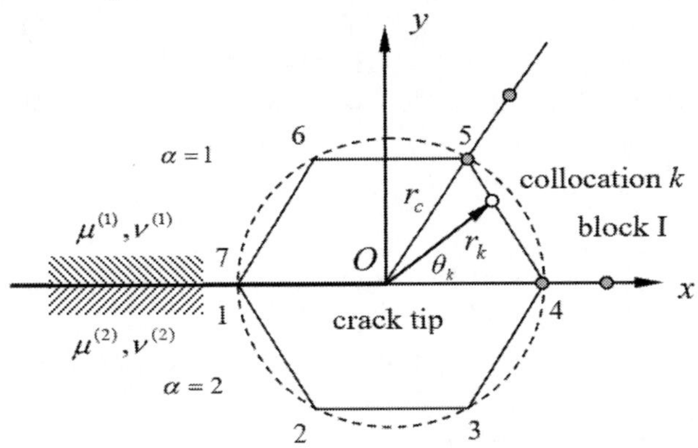

Figure 2. A single polygonal core around the crack tip. In this case a six sided core is used.

With the aid of a singular core around a straight interfacial crack, the singular stress field around the crack tip is determined by the following expression [8]

$$\sigma_y^{(\alpha)}+i\tau_{xy}^{(\alpha)}=A_{C0}^{(1)}\left(\frac{1}{2}+i\varepsilon\right)(1+\hat{\kappa})r^{-\frac{1}{2}+i\frac{\varepsilon}{2}}=\frac{K}{\sqrt{2\pi}}r^{-\frac{1}{2}+i\frac{\varepsilon}{2}} \tag{11}$$

where K, the complex stress intensity factor is defined by

$$\frac{K}{\sqrt{\pi a}}=\frac{K_1+iK_2}{\sqrt{\pi a}}=A_{C0}^{(1)}\left(\frac{1}{2}+i\varepsilon\right)(1+\hat{\kappa})\sqrt{2/a} \tag{12}$$

where $\kappa^{(\alpha)}=3-4\nu^{(\alpha)}$ for plane strain and $\kappa^{(\alpha)}=\dfrac{3-\nu^{(\alpha)}}{1+\nu^{(\alpha)}}$ for plane stress

The superscript $\alpha=1,2$ represents material 1 above and material 2 below the plate.

$$\hat{\kappa}=\frac{\kappa^{(1)}\mu^{(2)}+\mu^{(1)}}{\kappa^{(2)}\mu^{(1)}+\mu^{(2)}},\qquad \varepsilon=\ln(\hat{\kappa})/2\pi \tag{13}$$

Using the Williams' series [8] the coefficient $A_{Cn}^{(1)}$ $(n=0,1,..,N_b-1)$ is determined for each node.

4. Numerical Example

A central crack of length *2a* in a bi-material plate is studied in this example. The plate has width w = 1 and height H = 4. The uniformly distributed load σ_0 is applied on the top and bottom. Owing to the symmetry with respect to *y* axis, it is equivalent to solve the boundary value problem for a half plate with a reasonable mesh. The FEM (ABAQUS) analysis was conducted under plane strain condition with a Poison ratios $\nu_1 = \nu_2 = 0.3$ and the crack length a = 0.5. The results of SIFs versus the ratio of Young's modulus E_1/E_2 are shown in table 1 for different cases. The solutions given by ABAQUS are provided in the table for comparison.

a/w	E_1/E_2	$K_1/\sigma_0\sqrt{\pi a}$			$K_2/\sigma_0\sqrt{\pi a}$		
		Present	ABAQUS	[9]	Present	ABAQUS	[9]
0.5	1	1.190	1.188	1.1893	0.000	0.000	0
	2	1.182	1.179	1.1798	-0.056	-0.051	-0.0566
	5	1.154	1.148	1.1483	-0.110	-0.098	-0.1053
	10	1.131	1.123	1.1237	-0.132	-0.116	-0.124

Table 1 Stress intensity factors of central crack bi-material plate

5. Summary

The Finite Block Method has been utilised in this paper to study interface crack of bi-material. It is easily observed that the SIF as deduced by the FBM are in good agreement with ABAQUS and the reference.

6. References

[1] Williams, M. L. "The stresses around a fault or crack in dissimilar media."*Bulletin of the seismological society of America* 49.2 (1959): 199-204.

[2] England, A. H. "A crack between dissimilar media." *Journal of Applied Mechanics* 32.2 (1965): 400-402.

[3] Hutchinson, J.W.; " Mixed Mode Fracture Mechanics of Interfaces." in Metal-Ceramic Interfaces, Acta-Scripta Metallurgica, Proceedings Series 4 edited by M. Ruhle, A. G. Evans, M. F. Ashby and J. P. Hirth, Pergamon Press, (1990).

[4] Lin, K. Y., and J. W. Mar. "Finite element analysis of stress intensity factors for cracks at a bi-material interface." *International Journal of Fracture* 12.4 (1976): 521-531.

[5] Wen, P. H., P. Cao, and T. Korakianitis. "Finite block method in elasticity."*Engineering Analysis with Boundary Elements* 46 (2014): 116-125.

[6] V. Z. Parton (1992). Fracture Mechanics: From Theory to Practice. United Kingdom: Bell and Bain Ltd.. 64-68.

[7] Nagashima, Toshio, Youhei Omoto, and Shuichi Tani. "Stress intensity factor analysis of interface cracks using X-FEM." *International Journal for Numerical Methods in Engineering* 56.8 (2003): 1151-1173.

[8] Li M, Meng LX, Hinneh P, Wen PH. Finite block method for interface cracks. Engineering Fracture Mechanics. 2016 May 31;156:25-40.

[9] Natarajan, Sundararajan, Chongmin Song, and Salim Belouettar. "Numerical evaluation of stress intensity factors and T-stress for interfacial cracks and cracks terminating at the interface without asymptotic enrichment." Computer Methods in Applied Mechanics and Engineering 279 (2014): 86-112.

Journal of Physics: Conference Series **734** (2016) 032056 doi:10.1088/1742-6596/734/3/032056

Prediction of plastic instabilities under thermo-mechanical loadings in tension and simple shear

P.Y. Manach, L.F. Mansouri, S. Thuillier

Univ. Bretagne Sud, FRE CNRS 3744, IRDL, F-56100 Lorient, France

sandrine.thuillier@univ-ubs.fr

Abstract. Plastic instabilities like Portevin-Le Châtelier were quite thoroughly investigated experimentally in tension, under a large range of strain rates and temperatures. Such instabilities are characterized both by a jerky flow and a localization of the strain in bands. Similar phenomena were also recorded for example in simple shear [1]. Modelling of this phenomenon is mainly performed at room temperature, taking into account the strain rate sensitivity, though an extension of the classical Estrin-Kubin-McCormick was proposed in the literature, by making some of the material parameters dependent on temperature.

A similar approach is considered in this study, furthermore extended for anisotropic plasticity with Hill's 1948 yield criterion. Material parameters are identified at 4 different temperatures, ranging from room temperature up to 250°C. The identification procedure is split in 3 steps, related to the elasticity, the average stress level and the magnitude of the stress drops. The anisotropy is considered constant in this temperature range, as evidenced by experimental results [2]. The model is then used to investigate the temperature dependence of the critical strain, as well as its capability to represent the propagation of the bands. Numerical predictions of the instabilities in tension and simple shear at room temperature and up to 250°C are compared with experimental results [3]. In the case of simple shear, a monotonic loading followed by unloading and reloading in the reverse direction ("Bauschinger-type" test) is also considered, showing that (i) kinematic hardening should be taken into account to fully describe the transition at re-yielding (ii) the modelling of the critical strain has to be improved.

References

[1] Manach PY, Thuillier S, Yoon JW, Coër J, Laurent H 2014 *Int. J. Plasticity* **58** 66
[2] Coër J, Bernard C, Laurent H, Andrade-Campos A, Thuillier S 2011 *Exp. Mech.* **51** 1185
[3] Mansouri LZ, Thuillier S, Manach PY 2016 submitted for publication in *Int. J. Mech. Sc.*

Constitutive modeling of a commercially pure titanium: validation using bulge tests

Benoit Revil-Baudard[1] **and Elisabeth Massoni**[2]

[1]Department of Mechanical and Aerospace Engineering, University of Florida, REEF, 1350 N. Poquito Rd., Shalimar, FL 32579, USA
[2]MINES ParisTech, Center for Material Forming (CEMEF), UMR CNRS 7635, BP 207, 06904 Sophia-Antipolis Cedex, France

E-mail: `revil@ufl.edu`

Abstract. In this paper, mechanical tests aimed at characterizing the plastic anisotropy of a commercially pure α-titanium sheet are presented. Hemispheric and elliptic bulge tests conducted to investigate the forming properties of the material are also reported. To model the particularities of the plastic response of the material the classical Hill [1] yield criterion, and Cazacu et al. [2] yield criterion are used. Identification of the material parameters involved in both criteria is based only on uniaxial test data, while their predictive capabilities are assessed through comparison with the bulge tests data. Both models reproduce qualitatively the experimental plastic strain distribution and the final thickness of the sheet. However, only Cazacu et al. [2] yield criterion, which accounts for both the anisotropy and tension-compression asymmetry of the material captures correctly plastic strain localization, in particular its directionality. Furthermore, it is shown that accounting for the strong tension-compression asymmetry in the model formulation improves numerical predictions regarding the mechanical behavior close to fracture of a commercially pure titanium alloy under sheet metal forming processes.

1. Introduction

Titanium and its alloys have outstanding mechanical properties such as moderate weight, high ductility and high strength, exceptional corrosion resistance, and excellent high-temperature properties. In particular, in the last decade considerable efforts have been devoted to the mechanical characterization of the response of commercially purity (CP) and high-purity (HP) titanium for quasi-static uniaxial loadings [3] [4]. It was reported that CP and HP titanium materials display strength differential effects, specifically in uniaxial compression the flow stress and strength are higher than in uniaxial tension. These strength differential effects are attributed mainly to mechanical twinning. Forming of titanium and its alloys still relies on numerous trial-and-error experiments. Given the complexity of forming processes and the cost of titanium, mathematical models for predicting the plastic strain distribution, strain localization, and failure are essential for realizing the true potential of these processes. However previous studies have also shown that classic plasticity models, such as J2 plasticity, or Hill [1] yield criterion cannot accurately describe the plastic behavior of titanium materials. The use of the von Mises yield criterion is clearly inadequate because it neglects the material's plastic anisotropy. Although Hill [1]'s criterion accounts for anisotropy, it cannot predict accurately the tension-compression asymmetry of HP-titanium [3].

The main objective of this paper is to assess the capabilities of existing orthotropic models to capture the room-temperature plastic behavior of a commercially pure titanium for complex 3D loadings. The tension-compression asymmetry of the plastic flow is a key feature of the plastic behavior of HP and CP-Ti. Thus, Cazacu et al. [2] orthotropic criterion that accounts for yielding asymmetry between tension and compression associated with deformation twinning will be considered along with the classical Hill [1] criterion.

2. Constitutive modeling of pure Ti

The material used in this study is a CP α-titanium sheet of 1.6 mm thickness, called T40. To quantify the plastic anisotropy in strains and in yield and flow stresses, uniaxial tests were carried out. Based on these tests, it can be concluded that the material displays a strong orthotropy in terms of yielding and Lankford coefficients (e.g. see fig.2). Also, the strength in uniaxial tension is lower than the strength in unixial compression for ani in-plane orientations. The anisotropic properties of this material have been modeled by the classical Hill [1] yield criterion and by the Cazacu et al. [2] yield criterion, which also accounts for the observed tension-compression asymmetry. The anisotropic developed by Cazacu et al. [2] yield criterion is briefly given in what follows. The effective stress associated with the yield criterion is:

$$\bar{\sigma} = \mathrm{B}\left[(|\tilde{\sigma}_1| - k\tilde{\sigma}_1)^2 + (|\tilde{\sigma}_2| - k\tilde{\sigma})^2 + (|\tilde{\sigma}_3| - k\tilde{\sigma}_3)^2\right]^{\frac{1}{2}} \tag{1}$$

where k is an internal variable, its range of variation being (-1, 1), while $\tilde{\sigma}_1$, $\tilde{\sigma}_2$ and $\tilde{\sigma}_3$ are the principal values of the transformed stress tensor

$$\tilde{\sigma} = [\mathbf{L}] : \mathbf{S} \tag{2}$$

with $\mathbf{S}$ being the Cauchy stress deviator applied to the material. In Eq.(2), L is a fourth-order orthotropic and symmetric tensor. Modeling the anisotropy by means of this 4th order symmetric and orthotropic tensor L, ensures that the material response is invariant under any orthogonal transformation belonging to the symmetry group of the material. In the coordinate system associated with the rolling (RD), transverse (TD) and normal direction (ND) and in Voigt notations, this tensor is represented by a 6x6 matrix given by

$$[\mathbf{L}] = \begin{bmatrix} L_{11} & L_{12} & L_{13} & 0 & 0 & 0 \\ L_{12} & L_{22} & L_{23} & 0 & 0 & 0 \\ L_{13} & L_{23} & L_{33} & 0 & 0 & 0 \\ 0 & 0 & 0 & L_{44} & 0 & 0 \\ 0 & 0 & 0 & 0 & L_{55} & 0 \\ 0 & 0 & 0 & 0 & 0 & L_{66} \end{bmatrix} \tag{3}$$

In Eq.(1), B is a constant defined such that the equivalent stress $\bar{\sigma}$ reduces to the tensile stress along RD. Thus, B is expressed is expressed in terms of the anisotropy coefficients L_{ij} with i, j = 1...3 and the material parameter k as follows:

$$B = \frac{1}{[(|\mathrm{T}_1| - k\mathrm{T}_1)^a + (|\mathrm{T}_2| - k\mathrm{T}_2)^a + (|\mathrm{T}_3| - k\mathrm{T}_3)^a]^{\frac{1}{a}}} \; ; \; \mathrm{T_i} = \left(\delta_{\mathrm{ij}} - \frac{1}{3}\right) L_{ij} \tag{4}$$

The identification of the parameters involved in Eq.(1), namely k and the components of the tensor L is done based on the experimental data obtained in uniaxial tension (tensile strengths and Lankford coefficients) and in uniaxial compression tests [5], the parameter of the anisotropic form of Cazacu et al. [2] criterion (Eq. 1) has been identified. To capture the evolution of the tension-compression asymmetry and changes in plastic behavior with the plastic strain, all the material parameters involved in the expression of the yield function are considered to evolve with accumulated plastic deformation. Therefore, the shape of the yield surface evolves from an elliptical shape (low tension-compression asymmetry) for small plastic strain toward a triangular shape for large plastic strain (see Fig. 1). For comparison purposes, the Hill [1] yield criterion will also be applied to model the plastic behavior of this material. Since the plastic behavior of the titanium alloy is strongly anisotropic, the Hill [1] couldn't account correctly for both anisotropy in stresses and Lankford coefficients. Since the main goal is to assess the capabilities to correctly reproduce the plastic strain and evolution of the thickness during cold forming, only the Lankford coefficients are used in the identification procedure for Hill [1] criterion. Note that The Cazacu et al. (2006), with only one additional parameters than Hill [1] also accounts for the tension-compression asymmetry. Figure 2 compares the respective evolution of the Lankford coefficients with the angle with respect to the rolling direction for an equivalent plastic strain of 0.1. Both yield criteria capture well the in-plane strain anisotropy of the titanium material.

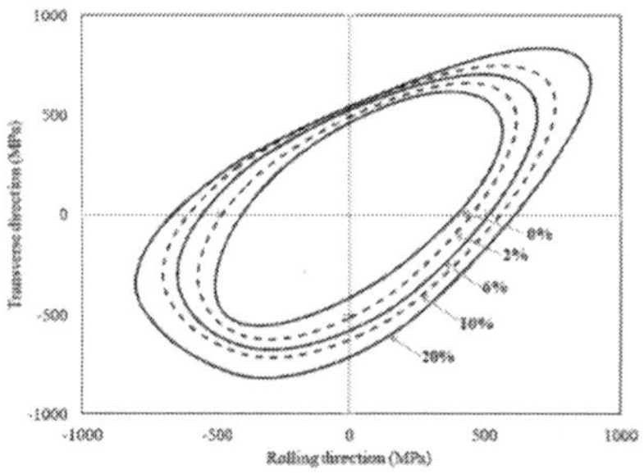

Figure 1. Yield loci for a CP α-titanium according the Cazacu et al [2] yield criterion for different level of accumulated plastic strain.

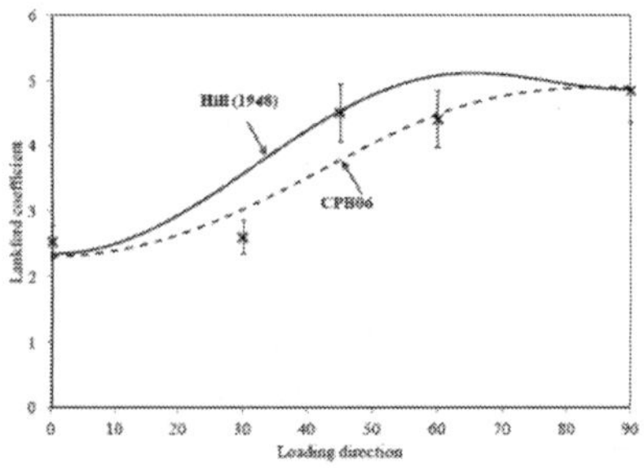

Figure 2. Evolution of the Lankford coefficient according the Hill [1] yield criterion and the Cazacu et al [2] yield function.

3. Assessment of the predictive capabilities for cold forming

Experimental bulging tests coupled with digital image correlation have been performed and revealed specific strain localization zones [4]. The results of such tests will also allow full 3-D validation of the constitutive models. Identified based on uniaxial tests, the predictive capabilities of the Hill [1] and Cazacu et al. [2] anisotropic yield criteria are assessed on experimental hemispheric and elliptical bulge tests. Both yield criteria provide good prediction for the overall equivalent plastic strain isocontours and final thickness of the metal sheet. The Hill [1] lacks accuracy in describing the plastic behavior for plastic strain close to fracture strain and cannot predict the position and orientation of the strain localization zone (see fig. 3). Only the Cazacu et al. [2] yield criterion, which accounts for the tension-compression asymmetry of the plastic behavior captures correctly the plastic strain localization and its orientation. Furthermore, the evolution of the thickness with pressure confirms that the Cazacu et al. [2] criterion better predict the plastic behavior close to fracture. To complementary illustrate these differences; in the figure 4 is plotted the evolution of the thickness at the center of the metal sheet with the pressure of the injected water for both Hill [1] and Cazacu et al. [2] yield criteria. Until the pressure reaches 21 MPa, both anisotropic yield criteria predict a similar evolution for the thickness at the center of the metal sheet. But, the Cazacu et al. [2] yield criterion is able to capture the change in the evolution rate of the thickness leading to strain localization. It is worth noting that the difference between the thicknesses predicted by the two anisotropic function increases as the tension-compression asymmetry becomes larger. Thus, for metal sheet forming, accounting of the tension-compression asymmetry of the plastic behavior provides a better description of the mechanical response of the material. By comparing the yield loci for the Hill [1] and the Cazacu et al. (2006) criteria and conclusions drawn from the bulge simulations, it is worth noting that the strain localization appears when the tension-compression asymmetry becomes significant.

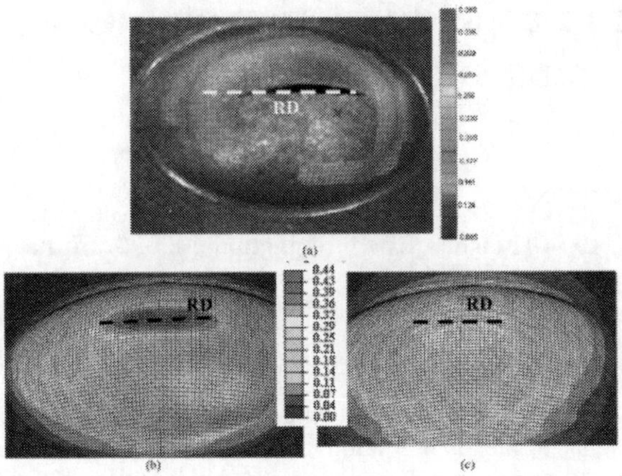

Figure 3. Isocontours of equivalent plastic strain for an elliptic bulge test of CP titanium with the RD oriented along the major axis of the ellipse: (a) DIC measure, b) the Cazacu et al [2] yield criterion, (c) Hill [1] yield criterion.

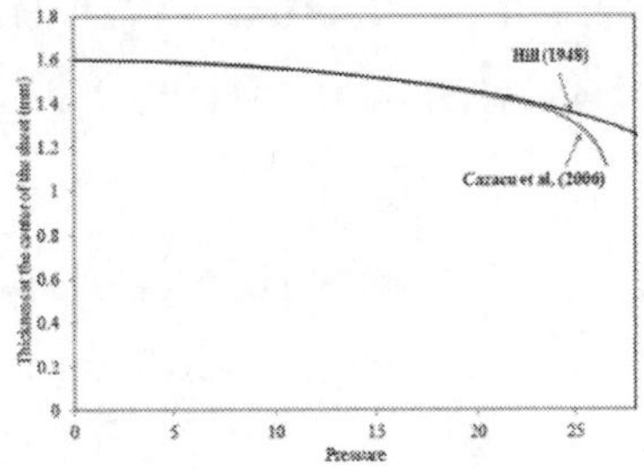

Figure 4. Evolution of the thickness of the CP titanium sheet located at the top of the bulge with pressure for an elliptic bulge test (the RD is oriented along the major axis of the ellipse): comparison between the response given by the Hill [1] yield criterion and from the Cazacu et al [2] yield function.

4. Conclusions

The prediction of the strain localization zone is directly correlated to the tension-compression asymmetry effect in three dimensional loadings, and only an anisotropic yield criterion accounting for this key feature could correctly reproduce the plastic behavior close to fracture for an CP α-titanium alloy. Use of a particular yield criterion to predict the plastic behavior of a titanium alloy during cold forming processes greatly relies on the experimental data available to identify this model and on the predictive capabilities of the model needed. If the model must reproduce the overall mechanical behavior for small to medium plastic strain, the Hill [1] yield criterion is a reliable model. Otherwise, if the plastic behavior close to rupture must be accurately described and the prediction of the strain localization zone is necessary, the Cazacu et al. [2] presents a better option.

References

[1] Hill R 1948. *Proceedings of the Royal Society of London* A **193** 281
[2] Cazacu O, Plunkett B and Barlat F 2006 *int. J. Plasticity* **22** 1171
[3] Nixon M E, Cazacu O and Lebensohn R A 2010.*int. J. Plasticity* **26** 510
[4] Revil-Baudard B 2010 *Simulation of the mechanical response of titanium alloys for cold forming* , Ph.D. Thesis Ecole Nationale Superieure des Mines de Paris, CEMEF, France.
[5] Benmhenni N, Bouvier S, Brenner R, Chauveau T and Bacroix B 2013 *Materials Science and Engineering* A **573** 222

An interpolation type anisotropic yield function and its application in sheet metal forming simulation

R Q Xiao, F Peng and X H Dong*

Shanghai Jiao Tong University, 1954 Huashan Road, Shanghai, 200030, China

E-mail: dongxh@sjtu.edu.cn

Abstract. Affected by texture, Bauschinger effect and different deformation mechanisms, plastic deformation behavior of the sheet metals are complicated, especially for HCP metals, such as magnesium and titanium. With more and more enhanced demand to describe the materials' yield behavior precisely and efficiently in numerical simulation, application of traditional continuous type yield functions encounters great challenge. So an interpolation type anisotropic yield function for plane stress is proposed. This yield function is represented by a yield surface in the polar coordinate system. The radius vector to the yield surface represents the yield stress at yielding, while the outer normal is related with the R values. The physical meaning of the parameters is directly defined. Accuracy, efficiency and flexibility can be achieved by application of such yield functions.

1. Introduction

The accuracy of yield function has essential effect in the simulation results of sheet metal forming process. During the last decades, a number of yield functions have been proposed to describe the anisotropic yield behavior of materials, most of which take the form of continuous functions. A review of these yield functions has been presented in [1]. Generally, in order to further improve the prediction accuracy and depict more strong anisotropy, the form of the yield function is bound to be more complex. In order to overcome the drawbacks of continuous yield function, Vegter and van den Boogaard [2] proposed a yield function based on the second order Bézier curve interpolation, which is adaptive in describing highly anisotropic materials without increasing the complexity of yield function. Then Peng et.al [3] proposed a new interpolation-type yield function based on the Hermite interpolation, which takes simple function form and the parameters possessing clear physical meaning, but the convexity of the yield locus may not be guaranteed. In this work, an adjusted Hermite interpolation-type yield function is proposed which consists of a base function of Hermite polynomial and an adjusting function to insure the convexity and fit additional experimental data. Yield locus calculated by the proposed yield function, some continuous yield function and experimental data are compared to verify the effectiveness of the proposed yield function.

2. The interpolation-type anisotropic yield function

2.1. *The yield locus*

For orthotropic sheet metals, the stress and plastic strain increment are usually expressed in the material coordinate system, with *x*-axis, *y*-axis and *z*-axis corresponding to the rolling, transverse and normal directions respectively. In this paper, only biaxial tension and compression loading conditions are considered, thus the yield locus in σ_x-σ_y plane is investigated. For convenience, the interpolation type yield function is expressed by polar coordinate, in which a planar stress condition is represented by a stress vector $\boldsymbol{r}(\theta)$, where r is the modulus, and θ is the angle between the stress vector and σ_x-axis. The direction of the plastic strain increment is described by β, the angle between the in-plane plastic strain increment and x-axis, which is related with the R value, i.e. the strain ratio.

As the associated flow rule being adopted, the tangent of yield locus will be normal to the direction of plastic strain increment, thus the first derivative of yield locus can be expressed by plastic strain increment direction β as follows

$$\frac{dr}{d\theta}=\frac{r(\sin\theta-\cos\theta\tan\beta)}{\sin\theta\tan\beta+\cos\theta} \tag{1}$$

2.2. *Adjusted Hermite interpolation-type yield function*

To define a segment of the yield function between the polar angle $[\theta_a,\theta_b]$, the radius r and first order derivative $(dr/d\theta)$ at its two nodes are calculated from experimentally obtained corresponding yield stress and the R value, and then substituted into the Hermite interpolation function $r(\theta)=r_H(\theta)$ [3]. However, the convexity of the Hermite interpolation-type yield function may not be guaranteed, and the yield locus between the nodes may deviate from additional experimental data. In order to solve these problems, an adjusting function $\xi(\theta)$ is added to the Hermite interpolation function (base function) as follows

$$r(\theta)=r_H(\theta)+\xi(\theta) \tag{2}$$

The adjusting function $\xi(\theta)$ is the linear combination of $\xi_i(\theta)$

$$\xi(\theta)=\sum_{i=1}^{n}\lambda_i\xi_i(\theta) \tag{3}$$

with

$$\xi_i(\theta)=\bar{\theta}^{i+1}(\bar{\theta}-1)^2 \tag{4}$$

where $\bar{\theta}=\dfrac{\theta-\theta_a}{\theta_b-\theta_a},\theta_a\le\theta\le\theta_b,i\ge1;\lambda_i$ is the undetermined parameters.

After introducing the adjusting function $\xi(\theta)$, the C_1 continuity of the yield function at the interpolation nodes is still hold for the fact that

$$\xi_i(\theta_a)=\xi_i(\theta_b)=\xi_i'(\theta_a)=\xi_i'(\theta_b)=0 \tag{5}$$

It can be proved that $\xi_i(\theta)$ is convex. Thus, if convexity of the yield locus isn't fulfilled by the base function, e.g. Hermite interpolation function, it can be insured by introducing the adjusting function in corresponding segment. Moreover, the number of adjusting function's term can be selected, which provides great flexibility to fit the experimental data between the interpolation nodes.

2.3. *Identification of the adjusting function parameters*

For an adjusting function in the interval $\theta\in[\theta_a,\theta_b]$, parameters can be determined by solving the following optimization problem

$$\min f(\boldsymbol{\lambda})=\sum_{i=1}^{n}(r(\theta_i)-r_i)^2 \tag{6}$$

$$\text{s.t. } g(\theta,\boldsymbol{\lambda})\ge0,\theta\in[\theta_a,\theta_b] \tag{7}$$

The objective function $f(\boldsymbol{\lambda})$ is the error function of yield stress vector's modulus; r_i is the yield stress vector's modulus obtained by experiment; $r(\theta_i)$ is the yield stress vector's modulus predicted by the yield function; n is the number of experimental yield points; $\boldsymbol{\lambda}=(\lambda_1, \lambda_2,\ldots, \lambda_m)$ is the vector of the adjusting function parameters; m is the number of the parameters. The constraint function $g(\boldsymbol{\lambda})$ is imposed to insure the convexity of the yield locus, which takes the following form

$$g(\theta,\boldsymbol{\lambda}) = r^2(\theta,\boldsymbol{\lambda}) + 2\left(\frac{dr}{d\theta}(\theta,\boldsymbol{\lambda})\right)^2 - r(\theta,\boldsymbol{\lambda})\left(\frac{d^2r}{d\theta^2}(\theta,\boldsymbol{\lambda})\right) \tag{8}$$

The constraint can be treated as

$$g_{\min}(\boldsymbol{\lambda}) \geq 0 \,, \theta \in [\theta_a, \theta_b] \tag{9}$$

Then the minimum $g_{\min}(\boldsymbol{\lambda})$ can be obtained by another optimization process

$$\min g(\theta,\boldsymbol{\lambda}) \tag{10}$$

$$\text{s.t.}\ \theta \in [\theta_a, \theta_b] \tag{11}$$

Therefore, a double optimization process is adopted to solve the problem. The main optimization process is to minimize the error function of yield stress vector's modulus, while the subordinate optimization process is to find the minimum $g_{\min}(\boldsymbol{\lambda})$. The two optimization process are implemented simultaneously in MATLAB. The undetermined parameter λ_i (i=1~m) are obtained from the double optimization process.

3. Application to an HCP magnesium alloy

The experiment results of an AZ31 magnesium alloy sheet [4], which show a strong tendency of tension/compression asymmetry, are adopted to compare the effectiveness of the proposed yield function with some widely-used continuous yield functions, including Hill48 [5], BBC2005 [6] and CPB2006 [7].

The proposed interpolation yield locus consists of five interpolation segments separated by 5 yield stress points, i.e., $P_1(\sigma_0^t,0)$, $P_2(\sigma_b^t,\sigma_b^t)$, $P_3(0,\sigma_{90}^t)$, $P_4(\sigma_0^c,0)$, $P_5(0, \sigma_{90}^c)$, where the superscript t and c denote the tension and compression tests respectively. Besides the R values corresponding to P_1 ~ P_5 are used for calculating $(dr/d\theta)_i$ (i=1~5).

The yield loci predicted by Hermite interpolation (denoted as Hermite) and the adjusted Hermite interpolation (denoted as Hermite(1λ)) are compared in Fig.1(a). It can be seen that the yield locus obtained by Hermite interpolation base function alone shows concave near points $P_1(\sigma_0^t,0)$ and $P_3(0,\sigma_{90}^t)$. However, the convexity is fulfilled after introducing adjusting functions with one parameter λ in the two segments of the first quadrant. The example shows that the convexity problem of Hermite interpolation may occur and the adjusting function can fix it. Moreover, the parameter λ is obtained by the aforementioned optimization process, in which the additional yield points have been used. Thus, the accuracy of yield locus in the two segments has been improved as well. In other segments, it is unnecessary to introduce the adjusting function, because the yield locus by Hermite interpolation is convex and no additional experimental data is provided, so only the Hermite interpolation base functions are adopted.

The comparison of the yield loci of normalized stress on the σ_x-σ_y plane by different yield functions and experimental data is shown in Fig.1(b). Comparing the yield loci in Fig.1(b), it can be seen that the accuracy of a yield function is determined by its function form and the number of its undetermined parameters. The Hill48 yield function can be determined by only 3 experimental data, i.e., σ_0^t, σ_{90}^t, R_0^t, thus the yield locus by Hill48 yield function is obviously deviated from the experimental data. In the identification of BBC2005 yield function, the yield stress and R values of 3 points P_1~P_3 are used, so the yield locus matches the experimental data in the first quadrant well, but the tension/compression asymmetry can't be predicted, because of its tension/compression symmetric function form. To determine the parameters of CPB2006 yield function, the yield stress and R values of 5 points $P_1 \sim P_5$ are used, as the proposed yield function. Thus, in the first quadrant the yield locus is predicted by CPB2006 yield function with a certain degree of accuracy; and the tension/compression asymmetry is described accurately, since the compression experimental data are used in parameter determination. For the proposed yield function, as mentioned above, the adjusting functions are adopted in the two segments of first quadrant to keep convex and fit the additional experimental data. Though only one adjusting parameter λ is adopted in each of the two effective segments, the accuracy is improved obviously, and further improvement can be achieved by taking more parameters in the adjusting function.

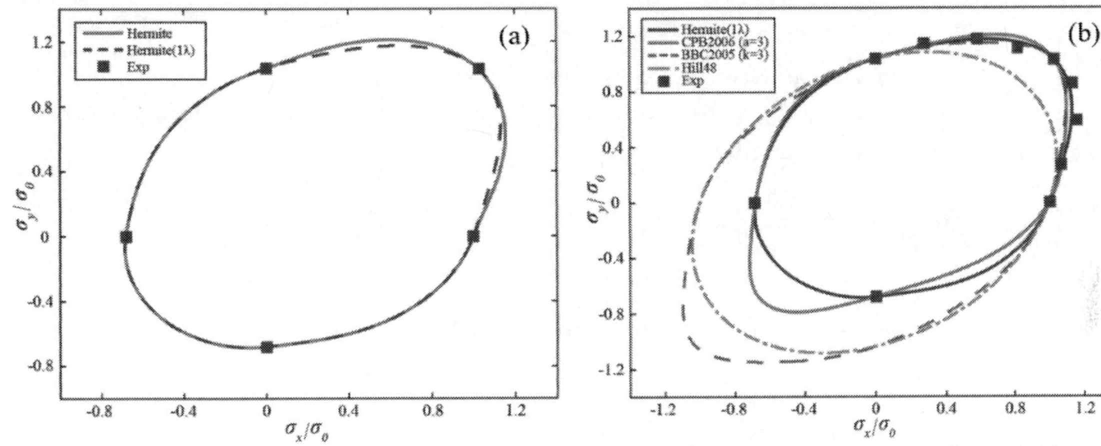

Fig 1. Comparison between experimental data and yield loci predicted by various yield functions($\bar{\varepsilon}^p$=0.004). Data after Andar M O, Kuwabara T and Steglich D [4].

4. Conclusions

In this paper, an adjusted Hermite interpolation-type yield function is proposed. The yield function is constructed in the base-adjusting function framework, the base function, such as Hermite interpolation, describes the shape of the yield locus, and the expandable adjusting function is introduced to insure convexity and fit additional experimental data. Compared with continuous type yield function, more parameters can be easily introduced into the proposed interpolation type yield function, thus more accurate yield locus can be predicted, as shown in the comparison of different predicted yield loci with the experimental data of an AZ31 magnesium alloy sheet.

Acknowledgements

The authors are grateful to the financial support of the National Natural Science Foundation of China by Grant no. 51275297.

References

[1] Banabic D 2010 *Sheet Metal Forming Processes: Constitutive Modeling and Numerical Simulation* (Berlin Heidelberg: Springer) pp 45-112
[2] Vegter H and van den Boogaard A H 2006 *Int. J. Plasticity* **22** 557-580
[3] Peng F, Dong X H, Tian X Y, Xie H Y and Liu K 2015 *Int. J. Mec. Sci* **99** 89-97
[4] Andar M O, Kuwabara T and Steglich D 2012 *Mat. Sci. Eng. A* **549** 82-92
[5] Hill R 1948 Proc. R. Soc. Lond. A **193** 281-297
[6] Banabic D, Aretz H, Comsa D S and Paraianu L 2005 *Int. J. Plasticity* **21** 493-512
[7] Cazacu O, Plunkett B and Barlat F 2006 *Int. J. Plasticity* **22** 1171-94

Non-Local Ductile Damage Formulations for Sheet Bulk Metal Forming

Steffen Beese, Stefan Loehnert, Peter Wriggers
Institute of Continuum Mechanics, Leibniz Universität, Appelstraße 11, 30167 Hannover, D
E-mail: beese@ikm.uni-hannover.de

Abstract. A ductile damage model for sheet bulk metal forming processes and its efficient and accurate treatment in the context of the Finite Element Method is presented. The damage is introduced as a non-local field to overcome pathological mesh dependency. Since standard elements tend to show volumetric locking in the bulk forming process a mixed formulation is implemented in the commercial software *simufact.forming* to obtain better results.

1. Introduction

The sheet bulk metal forming process [4] is characterized by large plastic bulk deformation of a metal sheet. This way it is possible to create functional parts, such as gears or drivers just by forming and without additional cutting processes. In order to simulate such challenging processes an accurate material model and non standard numerical techniques are required. To model the plastic behaviour the framework of elastic isomorphism, see e.g. [1, 10] is used and coupled to a continuum damage model. To circumvent the mesh depended localisation of such models a gradient enhanced non-local theory [5, 2] is used. Since the plastic deformation is nearly incompressible, standard finite elements will show volumetric locking behaviour. Therefore a strategy based on a mixed formulation is presented and implemented in the simulation environment *simufact.forming*.

2. Elasto-Plastic Material Model

Because of its general applicability and its straight forward derivation the concept of elastic isomorphism as e.g. proposed by Bertram [1] or Svendsen [10] is used for the formulation of the elasto-plastic constitutive laws. This concept states, that for every plastic deformation the elastic material response remains unchanged. If the stored elastic energy for an elastically isotropic material $\psi = \psi(\boldsymbol{C}, \mathcal{P})$ is a function of the right CAUCHY-GREEN Tensor $\boldsymbol{C}$ and the plastic Transformation $\mathcal{P}$, then this concept necessarily leads to a functional relationship of the form:

$$\psi(\boldsymbol{C}, \mathcal{P}) = \psi(\mathcal{P}^T \boldsymbol{C} \mathcal{P}) =: \psi(\boldsymbol{C}^E) \qquad \text{with} \qquad \boldsymbol{C}^E := \mathcal{P}^T \boldsymbol{C} \mathcal{P}\,. \tag{1}$$

By introducing a yield function $\Phi(\boldsymbol{M}^E, \sigma_y)$, where $\boldsymbol{M}^E$ is the elastic MANDEL stress, the elastic range is defined for all stress states by a to negative value of Φ. For $\Phi = 0$ plastic yielding takes place and $\mathcal{P}$ is assumed to evolve by means of the principle of maximum dissipation. Because of isotropic hardening the elastic range will grow during yielding. To model this effect in a physical meaningful way, we introduce an internal variable ξ which is proportional to the square root of

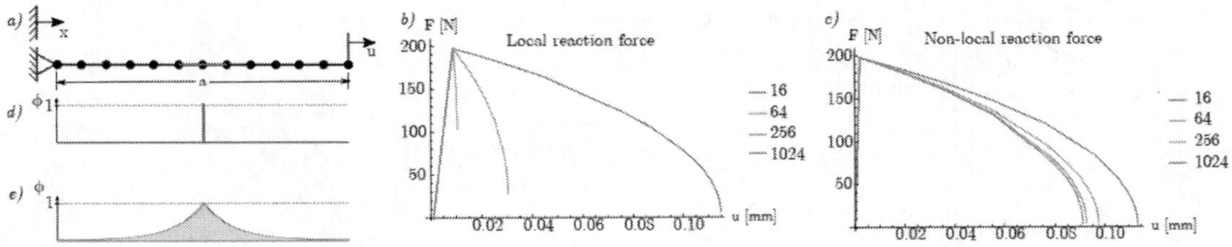

Figure 1. One dimensional Boundary Value Problem *a)* with local *b)* and non-local *c)* Reaction Forces depending on the mesh size. Final damage distribution for the local *d)* and non-local *e)* case.

the dislocation density on the micro scale of the considered steel. The evolution of this internal variable is motivated by elementary processes on the microlevel, see e.g. Stainier et.al. [9]:

$$\dot{\xi} = H(\xi\infty - \xi)\lambda\,. \tag{2}$$

As last ingredient of our model the dissipative effect of damage is considered. Here the anisotropically extended LEMAITRE damage model of Soyarslan et.al. [8] is used. Despite their original idea of splitting the stresses in positive and negative parts by means of an eigenvalue decomposition and using the complementary energy function, we use a split of the free energy function $\psi = \psi^+ + \psi^-$ depending on either the positive or negative eigenvalues of the strain like quantity $(\boldsymbol{C}^E - \mathbf{1})$. The damage potential then reads:

$$\Phi^D = \frac{S}{s+1}\left\langle\frac{-(\psi^+ + h\cdot\psi^-) - Y_0}{S}\right\rangle^{s+1}\frac{1}{(1-\tilde{\phi})^\beta}\,, \tag{3}$$

where $h \in [0,1]$ weights the compression part ψ^-. In equation (3) $\tilde{\phi} \in [0,1]$ is the local damage factor and $\{S, s, \beta\}$ are material parameters for this model. The evolution of the scalar damage variable $\tilde{\phi}$ is then obtained by an associative evolution law.

3. Non-Local Damage Model

It is well known that local softening materials exhibit a strong mesh dependency [6]. The smaller the elements become the earlier a fully degraded element will cause a singular tangent stiffness matrix. One efficient way to circumvent these issues is the application of a non-local theory. This is done by the introduction of a new field variable, the non-local damage ϕ and by coupling the local damage variable $\tilde{\phi}$ to its non-local counter part ϕ and its gradient $\nabla\phi$. Computationally this results in solving an additional equation with homogeneous NEUMANN boundary conditions in the entire domain Ω:

$$\phi - \ell^2\nabla\cdot\nabla\phi = \tilde{\phi} \quad \forall x \in \Omega \quad \text{with:} \quad \nabla\phi\cdot\mathbf{n} = 0 \quad \forall x \in \partial\Omega. \tag{4}$$

This form is either known as implicit gradient formulation [5] or as micromorphic damage model [2]. The constant ℓ represents an internal length scale which in general is determined by experiments for the considered material. In the limiting case of $\ell \to 0$, the purely local model is obtained. The benefit of additional stability has to be paid by an increase of computational costs by a factor of about 1.3, assuming the same mesh is used as for the standard local damage simulation.

The functionality of the model is best explained by the one dimensional example shown in Fig. 1. There, a clamped beam is exposed to a displacement boundary condition and the reaction

Journal of Physics: Conference Series **734** (2016) 032059 doi:10.1088/1742-6596/734/3/032059

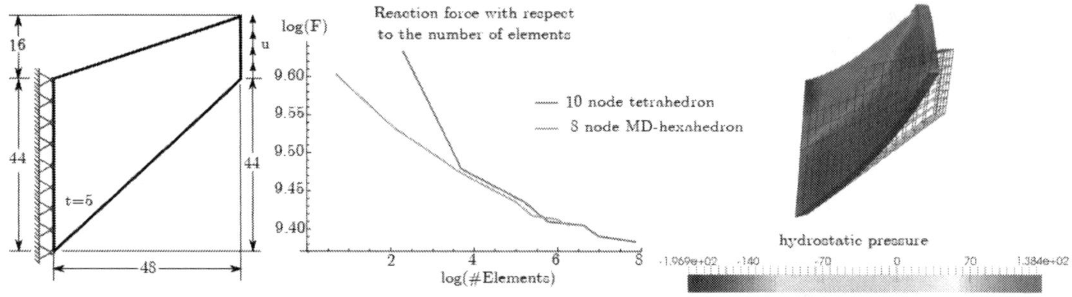

Figure 2. Setup of COOK's Membrane problem *(left)* with convergence of the reaction force *(middle)* and resulting pressure field *(right)*.

forces are recorded depending on the number of elements used for the discretisation. The yield strength is reduced by a factor of 0.99 for the two elements in the middle of the beam. While the reaction force for the non-local model converges to a stable solution the local model tends to diverge. The reason is the small domain (one integration point) where the damage is localising. In the non-local model the damage evolution is spread also to some neighbouring elements. In this example the ratio $\frac{\ell}{a}$ is 0.1.

4. Finite Element Formulation and Implementation

The balance equation for the non-local damage (4) and the linear momentum balance is solved by the finite element method. However, standard low order displacement finite elements tend to show volumetric locking behaviour when plastically incompressible materials are involved. Especially in sheet bulk metal forming processes this is often observed when functional elements are formed by large bulk deformations. There is a huge variety of element technologies to treat this issue and a very simply and quite effective technique was first reported by Simo et.al. [7]. This is based on the HU-WASHIZU variational formulation, where the additional fields $\bar{p}$ and $\bar{J}$ are introduced and the free energy function is additively split into an isochoric part $\psi_{iso}(\boldsymbol{C}^E_{iso})$ and a volumetric part $\psi vol(\bar{J} \cdot \det(\mathcal{P}))$. The HU-WASHIZU potential then reads:

$$\Pi^{HW}(\boldsymbol{u}, \bar{p}, \bar{J}) = \int_{\Omega_0} \psi iso(\boldsymbol{C}^E_{iso}) + \psi vol(\bar{J} \cdot \det(\mathcal{P})) + \bar{p}(\det(\boldsymbol{F}) - \bar{J})\, \mathrm{dV}\,, \tag{5}$$

where the last term penalises the deviation of $\bar{J}$ from the determinant of the deformation gradient $\det(\boldsymbol{F})$. By computing the variation of eq. (5) the field $\bar{p}$ is identified as the hydrostatic pressure. In order to avoid additional equations in the global equation system piecewise constant shape functions are used for the discretisation of $\{\bar{p}, \bar{J}\}$ and a static condensation is performed. The displacement field $\boldsymbol{u}$ and the non-local damage ϕ are discretised by standard low order shape function of LAGRANGE type. To show the performance of the presented mean dilatation (MD) element the classical COOK's membrane problem is solved using the hexahedral 8 node MD element and compared to the results of a higher order tetrahedral element. In figure 2 the problem setup is shown as well as the convergence behaviour and the resulting pressure distribution. Especially for coarse discretisations the presented element performs good and the convergence rate for finer meshes is comparable to the 10 node tetrahedral element.

The implementation is done in the commercial FE software *simufact.forming* as a new user element in subroutine `USELEM`. To circumvent difficulties of the built in contact formulation regarding the additional degrees of freedom needed for the field ϕ, just a weak coupling of damage the equation and the linear momentum is realized. Therefore in every load step first the

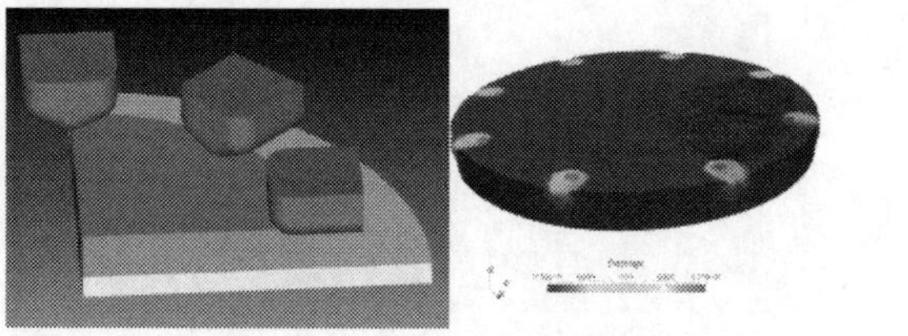

Figure 3. *simufact.forming* model and resulting damage distribution for demonstrator test problem

Figure 4. *simufact.forming* model and resulting damage distribution for deep drawing test problem

damage equation is solved and afterwards the mechanical equilibrium is calculated by means of a NEWTON-RHAPSON iteration. For post processing purposes an interface to the visualisation software *ParaView* is implemented.

5. Testcases

Two examples are considered to reflect the main load cases of sheet bulk metal forming. The first is a demonstrator problem which was first considered in [3]. In this example eight punches are placed in an axisymmetric way and they coin a round sheet of metal. In figure 3 on the left the *simufact.forming* model is shown and on the right the deformed configuration and the resulting damage distribution are shown. Compared to the local model the damage field is more smooth and the absolute value is identical at each indent, so damage is not localising in this example. The other main load case is a bending dominated problem. In figure 4 the demonstration problem of *simufact.forming* for deep drawing is shown. As expected the main region for damage accumulation is the rounding between bottom an wall of the cup. To validate the numerical model experimental tests for the two examples are in preparation.

Acknowledgments

The authors gratefully acknowledge the financial support by the German Research Foundation (DFG) within the subproject C2 of the transregional collaborative research centre (SFB TR 73) "Sheet-Bulk-Metal-Forming".

References

[1] Bertram A 2008 *Elasticity and plasticity of large deformations* (Berlin: Springer)
[2] Eringen A C 2012 *Microcontinuum field theories: I. Foundations and Solids* (New York: Springer-Verlag)
[3] Landkammer P, Loderer A, Krebs E, Söhngen B, Steinmann P, Hausotte T, Kersting P, Biermann D and Willner K 2015 *Key Engineering Materials* **639** 251-258
[4] Merklein M, Allwood J, Behrens B A, Brosius A, Hagenah H, Kuzman K, Mori K, Tekkaya A and Weckenmann A 2012 *CIRP Annals-Manufacturing Technology* **61** 725-745
[5] Peerlings R H J and De Vree J H P 1996 *International Journal for Numerical Methods in Engineering* **39** 3391-3403
[6] Pietruszczak S and Mróz Z 1981 *International Journal for Numerical Methods in Engineering* **17** 327-334
[7] Simo J, Taylor R and Pister K 1985 *Computer Methods in Applied Mechanics and Engineering* **51** 177-208
[8] Soyarslan C and Isik K 2013 In *Advanced Materials Research* **769** 205-212
[9] Stainier L, Cuitino A and Ortiz M 2002 *Journal of the Mechanics and Physics of Solids* **50** 1511-1545
[10] Svendsen B 2001 *International Journal of Solids and Structures* **38** 9579 - 9599

Error assessment in post-necking strain hardening behaviour identification of mild steel sheet

S. Coppieters[a], T. Hakoyama[b], K. Denys[a], D. Debruyne[a] and T. Kuwabara[b]

[a] Department of Materials Engineering, KU Leuven, Technology Campus Ghent, Gebroeders De Smetstraat 1, 9000 Ghent, Belgium.
[b] Division of Advanced Mechanical Systems Engineering, Institute of Engineering, Tokyo University of Agriculture and Technology, 2-24-16, Nakacho, Koganei-shi, Tokyo 184-8588, Japan.

sam.coppieters@kuleuven.be

Abstract. The information hidden in the diffuse neck of a tensile test on a thin metal sheet can be extracted using a special case of the non-linear virtual fields method yielding the so-called post-necking strain hardening behaviour. The method, however, requires a number of assumptions which are scrutinized in this paper. To eliminate experimental errors which could potentially hamper the assessment, virtual test data (i.e. strain fields at different load steps) is generated using a FE model of the tensile test. The identification strategy is then used to retrieve the reference strain hardening behaviour used in the FE simulation. This approach is used to study the necessity of incorporating rate-dependent plasticity in the identification procedure. Additionally, the necessary plane stress condition in the diffuse neck is studied.

1. Introduction

Pioneering work of Bridgman [1] resulted in a solution for the problem of diffuse necking. Bridgman considered his method a second level of approximation because the method is only concerned with the distribution of stress and strain across the diffuse neck. He also envisioned a third level of approximation, also referred to as the "complete solution" of the general problem, which takes the material state and the shape of the whole deforming specimen into account. Several researchers arrived at such complete solutions using finite-element based inverse methods. In order to avoid the shortcomings of the FE-based inverse method, a method based on the complete solution to retrieve the strain hardening behavior hidden in the diffuse necking regime was presented in [2,3]. The key point in this method is that the strain hardening behavior can be identified by minimizing the discrepancy between the internal and the external work in the region where the diffuse neck develops. The method is a special case of the Virtual Fields Method (VFM) [4]. The initial assumption [2] that an increase of the cost function caused by an error in the work hardening law is significantly larger than an increase caused by an error in the (anisotropic) yield criterion was investigated in [5]. In this contribution, two additional assumptions are scrutinized: a) the necessary plane stress condition in the diffuse neck and b) rate-independent plasticity.

2. Methodology

The Post-Necking Tensile Experiment (PNTE) is simulated using Abaqus/Standard. The computed surface strains, elongation of the monitored diffuse necking zone and the tensile force are used as input for the identification method presented in [2,3]. This approach excludes all experimental errors and enables studying the impact of the assumptions on the identified parameters. The simulation of the PNTE was performed using an appropriate material model for a mild steel sheet [6] with an initial thickness of 0.65 mm. The tensile test is simulated with a constant cross-head speed of 0.05 mm/s. The geometry of the specimen has a width of 20 mm and complies with [7]. The surface strains are extracted from a central zone (length 50 mm) around the diffuse neck. Virtual data is generated at a measurement frequency of 2Hz. The inverse identification [2,3] uses rate-independent plasticity with the von Mises yield criterion and the p-model [3] to describe isotropic hardening:

$$\sigma_{eq} = \begin{cases} K(\varepsilon_0 + \varepsilon_{eq}^{pl})^n & \textit{if } \varepsilon_{eq}^{pl} \leq \varepsilon_{\max} \\ K(\varepsilon_0 + \varepsilon_{\max})^n + Q[1 - e^{-p(\varepsilon_{eq}^{pl} - \varepsilon_{\max})}] & \textit{if } \varepsilon_{eq}^{pl} > \varepsilon_{\max} \end{cases} \quad (1)$$

where ε_{max} is the maximum uniform strain and p the post-necking hardening parameter.

2.1. *Strain rate dependency*

When uniform straining ceases the strain rate gradually increases within the diffuse neck. In [3] the strain rate in the central point of diffuse neck was measured using DIC and the maximum strain rate was about $0.02\,\frac{1}{s}$. It was, however, assumed in [3] that since very few material points violate the quasi-static upper limit of $0.01\,\frac{1}{s}$, the impact on the globally identified strain hardening behavior could be ignored. The virtual data was simulated using a plane stress (shell elements) FE model incorporating rate-dependent plasticity. The von Mises yield criterion along with a rate-dependent hardening model was used:

$$\sigma_{eq} = g(\varepsilon_{eq}^{pl})\left(\frac{\dot{\varepsilon}_{eq}^{pl}}{0.0005}\right)^m \;, \; g(\varepsilon_{eq}^{pl}) = 541(0.0036 + \varepsilon_{eq}^{pl})^{0.25} \quad (2)$$

with the exponent m equal to 0.01. Subsequently, rate-independent plasticity was used in the identification procedure to retrieve the strain hardening behavior described by the p-model. A minimum (during uniform elongation) and maximum (in the diffuse neck) plastic strain rate of $\dot{\varepsilon}_{eq}^{pl} = 0.0003\ s^{-1}$ and $\dot{\varepsilon}_{eq}^{pl} = 0.01\ s^{-1}$ was numerically predicted, respectively. Figure 1 shows the reference flow curves Eq.(2) according to those two extreme values of the plastic strain rate $\dot{\varepsilon}_{eq}^{pl}$. The p-model uses the maximum uniform strain $\varepsilon_{max} = 0.23$ obtained from the tensile test. The post-necking hardening parameter p enables to describe a Swift type (small p-value) or a saturation type (large p-value) post-necking strain hardening. It can be inferred from figure 1 that up to $2\varepsilon_{max} \approx 0.47$ the identified p-model is in close agreement with the reference hardening with $\dot{\varepsilon}_{eq}^{pl} = 0.0003\ s^{-1}$. The latter suggests that the influence of the strain rate up to $2\varepsilon_{max}$ can be safely ignored. Indeed, it can be observed from figure 2 that the change of strain rate between $\varepsilon_{eq}^{pl} = \varepsilon_{max}$ and $\varepsilon_{eq}^{pl} = 2\varepsilon_{max}$ is small. Figure 2 shows the evolution of the strain rate for different material points as a function of the elongation of the central region around the diffuse neck. Three different locations are shown: inside the diffuse neck, just outside the diffuse neck (I) and far away from the diffuse neck (II). After the onset of diffuse necking, the strain rate drops at material points far away from the diffuse neck (II). Close to the diffuse neck (II) the strain rate increases, but by approaching $\varepsilon_{eq}^{pl} = 2\varepsilon_{max}$, the strain rate eventually drops. The drop in strain rate is accompanied with further (limited) plastic deformation.

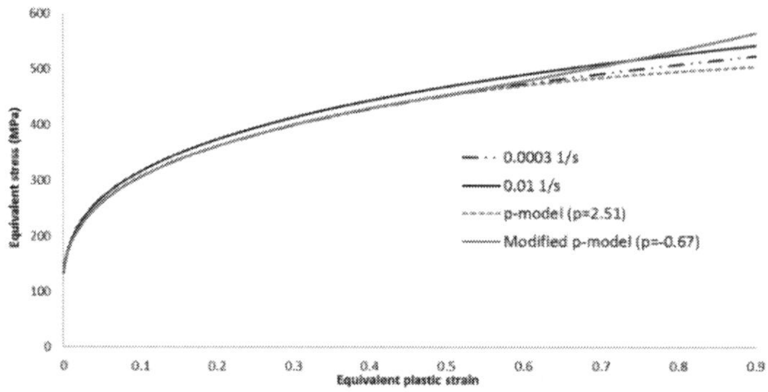

Figure 1. Reference flow curve Eq.(2) for different strain rates. Identified strain hardening models.

The latter means that during this drop the local flow stress is affected by the local instantaneous strain rate level. Clearly, the p-model is not flexible enough to capture the interplay of strain rates within the specimen affecting the local flow stress. Figure 2 also shows the difference between the virtual external work W_{ext} and the computed internal work W_{int} obtained with the identified strain hardening behaviour. It can be concluded that strain rate sensitivity hampers the identification of the rate-independent p-model. However, given that the strain rate change is moderately low between ε_{max} and $2\varepsilon_{max}$, the p-model can be modified in such a way that it responds to the strain rate effect. To do so, ε_{max} in Eq.(1) is changed to $2\varepsilon_{max}$ which is in accordance with observation that the development of transverse stresses is postponed to $\varepsilon_{eq}^{pl} = 2\varepsilon_{max}$ [8]. Figure 1 shows that the modified p-model has a negative p-value signifying a strong increase in strain hardening rate. The predicted W_{int} using the modified p-model is in close agreement with W_{ext}, see figure 2.

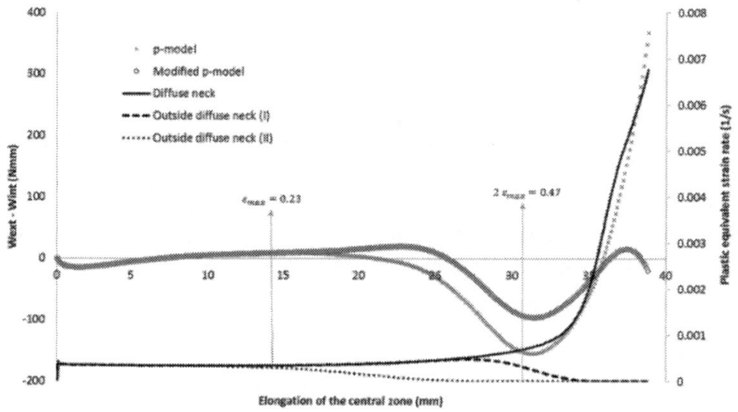

Figure 2. Discrepancy between virtual external work and computed internal work using the identified hardening models. Evolution of strain rate within the tensile specimen at 3 different locations.

2.2. Plane stress conditions

In practice, the surface strains are measured Digital Image Correlation. The volume of the specimen is assumed to remain constant so that the thickness strain can be computed using the measured surface strains. As such, out-of-plane shear strains are neglected and a plane stress state is assumed. The effect of these assumptions is studied by generating virtual data using a 3D (solid elements) FE model, thereby ignoring strain-rate sensitivity. Subsequently, the identification procedure [2,3] is used to retrieve the reference strain hardening model used in the simulation. The black solid curve in Figure 3 shows the reference strain hardening curve (labelled as *Ref Swift*). It can be inferred that the p-model identified based on the virtual data generated by the 3D FE model (labelled as *P-model (solids)*) deviates from the reference curve at large plastic strains. The latter suggests that the stress state in the diffuse neck differs from plane stress. To check this, the p-model (labelled as *P-model (plane stress)*) was also identified based on virtual data generated by a plane stress FE model. An identical deviation in the post-necking regime, however, was found. Consequently, the origin of this discrepancy cannot be attributed to a deviation from plane stress within the diffuse neck.

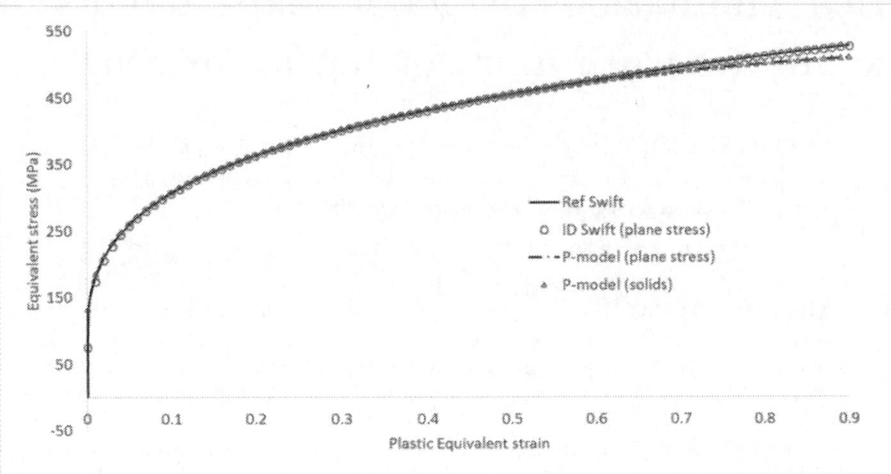

Figure 3. Influence of the stress state in the diffuse neck on the identified strain hardening behaviour.

3. Conclusions

This paper shows that strain rate effects cannot be ignored when identifying post-necking strain hardening behaviour of mild steel sheet. Plane stress conditions, however, within the diffuse neck can be safely assumed.

References

[1] P.W. Bridgman, Studies in Large Plastic Flow and Fracture, McGraw-Hill, NY, 1952.

[2] S. Coppieters, S. Cooreman, H. Sol, P. Van Houtte, D. Debruyne, Identification of the post-necking hardening behavior of sheet metal by comparison of the internal and external work in the necking zone, Journal of Materials Processing Technology, 211 (3), pp. 545-552, 2011.

[3] Coppieters, S., Kuwabara, T. (2014). Identification of Post-Necking Hardening Phenomena in D3uctile Sheet Metal. *Experimental Mechanics* (54), 1355-1371.

[4] J.-H. Kim, A. Serpantié, F. Barlat, F. Pierron, M.-G. Lee, Characterization of the post-necking hardening behaviour using the virtual fields method, International Journal of Solids and Structures, 50, pp. 3829-3842, 2013.

[5] Coppieters, S., Ichikawa, K., Debruyne, D., Kuwabara, T. (2014). Identification of Post-Necking Hardening Behaviour of Sheet Metal: Influence of the Yield Function. International Conference on Experimental Mechanics. Cambridge (UK), 7-11 july 2014.

[6] Ichikawa K., Kuwabara T., Coppieters S., Forming Simulation Considering the Differential Work Hardening Behavior of a Cold Rolled Interstitial-Free Steel Sheet, Key Engineering Materials, 611-612, pp.56-61(2014).

[7] ASTM Standard E8M-96, 1996. Standard test methods for tension testing of metallic materials [metric]. Annual Book of ASTM Standards, Vol. 03.01:76-96.

[8] Iadicola, M.A., 2011. Validation of uniaxial data beyond uniform elongation. AIP Conference Proceedings, 1383, 742-749.

Numerical simulations of biaxial experiments on damage and fracture in sheet metal forming

Steffen Gerke, Marco Schmidt and Michael Brünig

Institut für Mechanik und Statik, Universität der Bundeswehr München, Werner-Heisenberg-Weg 39, 85579 Neubiberg, Germany

E-mail: steffen.gerke@unibw.de

Abstract. The damage and failure process of ductile metals is characterized by different mechanisms acting on the micro-scale as well as on the macro-level. These deterioration processes essentially depend on the material type and on the loading conditions. To describe these phenomena in an appropriate way a phenomenological continuum damage and fracture model has been proposed.

To detect the effects of stress-state-dependent damage mechanisms, numerical simulations of tests with new biaxial specimen geometries for sheet metals have been performed. The experimental results including digital image correlation (DIC) show good agreement with the corresponding numerical analysis. The presented approach based on both experiments and numerical simulation provides several new aspects in the simulation of sheet metal forming processes.

1. Introduction

Elastic material properties, yield stress and coefficients characterizing plastic hardening are determined from one-dimensional tension tests with flat rectangular specimens. In addition tests with differently pre-notched specimens and corresponding numerical simulations have been used to study the stress dependence behavior, see [1] where only triaxialities from 0.33 and 0.6 can be reached. Lower positive stress triaxialities could be reached by one-dimensional shear specimens [1] and butterfly type specimens have been studied [2] within a special testing device. But it has to be noticed, that it is difficult to study a wide range of stress triaxialities with one-dimensionally loaded specimens. Therefore, biaxially loaded specimens are frequently used to study the yield surface of ductile metals, i.e. at strain states where no major damage can be reported. All these biaxial specimens are characterized by an extensive homogenous central region (see [3]) which does not allow the controlled study of damage behavior. Hence, new geometries of biaxially loaded specimens are proposed to examine stress state dependent damage and failure behavior.

2. Continuum damage and fracture model

The continuum damage model presented by [4] is used to predict the elastic-plastic deformations including damage. It takes into account information of the mechanisms of individual micro-defects and their interactions detected by numerical calculations on the micro-level [5]. The phenomenological approach is based on the additive decomposition of the strain rate tensor into elastic, plastic and damage parts.

In the present paper the focus is on the damage behavior: The generalized elastic-damage constitutive law is a function of elastic and damage strain tensors [4]. This allows simulation of deterioration of elastic material properties caused by formation and growth of micro-defects. In addition, onset and continuation of damage is governed by the stress-state-dependent damage criterion

$$f^{da} = \alpha I_1 + \beta\sqrt{J_2} - \sigma = 0. \tag{1}$$

where I_1 and J_2 are the first and second deviatoric stress invariants, σ is the damage threshold and α and β represent the stress-state-dependent damage mode parameters. These variables are related to different damage mechanisms and depend on the stress triaxiality $\eta = I_1/(3\sqrt{3J_2})$ as well as on the Lode parameter $\omega = (2T_2 - T_1 - T_3)/(T_1 - T_3)$ with T_1, T_2 and T_3 being the principal Kirchhoff stress components.

The increase in macroscopic strains caused by formation and growth of micro-defects leading to anisotropic damage behavior is modeled by the damage rule

$$\dot{\mathbf{H}}^{da} = \dot{\mu}\left(\bar{\alpha}\frac{1}{\sqrt{3}}\mathbf{1} + \bar{\beta}\mathbf{N} + \bar{\delta}\mathbf{M}\right). \tag{2}$$

Here $\bar{\alpha}$, $\bar{\beta}$ and $\bar{\delta}$ are stress-state-dependent kinematic variables and $\dot{\mu}$ represents the rate of the equivalent damage strain. In addition, $\mathbf{N}$ and $\mathbf{M}$ are normalized deviatoric stress tensors, see [5, 6] for details.

3. Numerical and experimental results

For biaxial testing it is preferable that the geometry of specimens can be used for testing in a large variety of stress states. With this emphasis the geometries shown in Fig. 3 are proposed and studied in detail. The geometry shown in Fig. $3_{(a)}$ (X1-specimen) has two crosswise arranged notches whereas at the geometry shown in Fig. $3_{(b)}$ (XO2-specimen) a central hole was added leading to four separated notched regions. Finally for the geometry displayed in Fig. $3_{(c)}$ (H-specimen) the notches have been arranged parallel to one of the loading axis. All specimens have been fabricated of an aluminum alloy of series 2017, from sheets of 4.0 mm thickness (see [6] for details) whereas all outer specimen dimensions are 240 mm by 240 mm. All material reduction in thickness direction, i.e. notches etc., are of 1.0 mm on each side, leaving 2.0 mm.

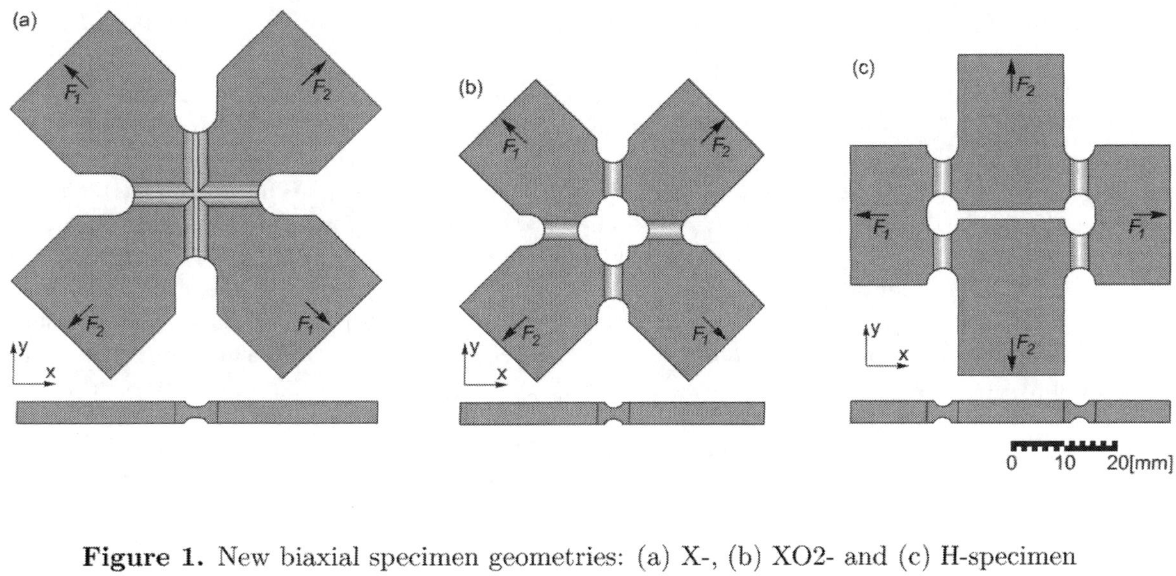

Figure 1. New biaxial specimen geometries: (a) X-, (b) XO2- and (c) H-specimen

The corresponding elastic-plastic finite element (FE) simulations have been carried out with Ansys Classic. All experiments have been performed on the biaxial test machine containing four electro-mechanically, individually driven cylinders [6]. The full 3D displacement field of the specimen surface was measured with a digital 3D image correlation (DIC) system [7].

Numerical simulations with the X1-specimen indicated major strain concentrations in the notches with maxima for load cases $F_1/F_2 = 1/-1$ and $F_1/F_2 = 1/0$ at the center of the specimen and for the load case $F_1/F_2 = 1/1$ at the ends of the notches. Based on this observation it is evident to introduce a central opening and thus four separated connectors between the specimen legs are generated. Numerical simulations with the XO2-specimen have shown that this geometry can be used in a large range of stress triaxialities from approximately $-0.1 \leq \eta \leq 0.9$.

The most promising results could be achieved with the H-Specimen. A numerical study program covering load cases from compression dominated $F_1/F_2 = -3/1$ over pure shear $F_1/F_2 = 0/1$ to tension dominated $F_1/F_2 = 3/1$ has been realized. Fig. 3 shows of the triaxiality η and the Lode parameter ω under the different loading condition, indicating a very homogeneous distribution with triaxialities in the range $-0.6 \leq \eta \leq 0.8$.

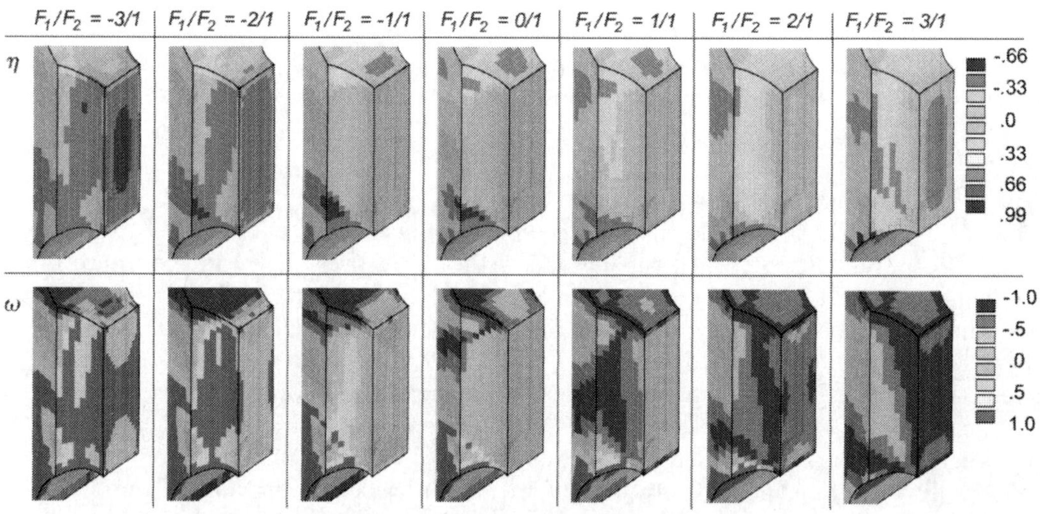

Figure 2. H-specimen: triaxiality η and Lode parameter ω under different loading conditions

Furthermore, Fig. $3_{(a)}$ compares Δu_{Ref} of all four notches within one experiment. The displacement Δu_{Ref} is experimentally taken by the difference in displacements of two reference points on opposite legs of the specimen outside the notched region while numerical results correspond to node displacements of corresponding finite elements. The behavior is almost symmetric while two crosswise arranged notches fractured. It is important to notice that this symmetric behavior is additionally forced by the clamped supports in the biaxial machine. In Fig. $3_{(b)}$ two experimental and one numerical load-displacement-curves under loading conditions $F_1/F_2 = 0/1$ are compared. Here the experimental curves of the H-specimen show only the usual experimental spreading. Furthermore it can be noted, that the experimental and numerical curves agree quite well.

In Fig. 3 the shear strains analyzed by the DIC system are compared with the corresponding numerically predicted ones at different states throughout the experiment under loading conditions $F_1/F_2 = 0/1$. At both deformation states good agreement can be observed.

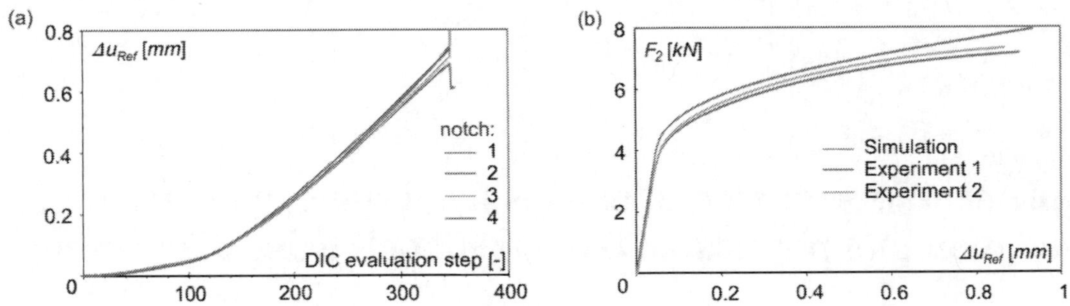

Figure 3. H-specimen: (a) comparison between notches within one experiment; (b) numerical and experimental results: load-displacement curves

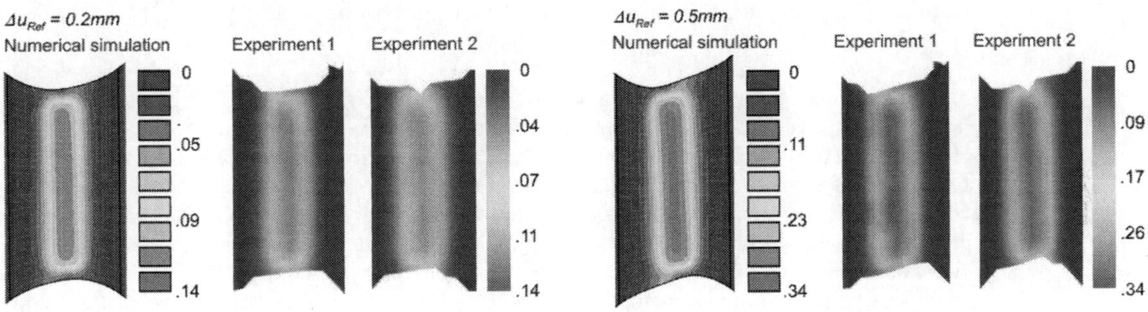

Figure 4. H-specimen: numerical predicted and experimentally observed shear strains γ_{xy}

4. Conclusions

New specimen geometries to study the stress state dependence of the damage and fracture behavior of ductile sheet metals are presented and numerically as well as experimentally studied. The geometry of the XO2- and H-specimens are characterized by four independent notched regions where damage and failure behavior can be analyzed. Thus, experiments with these new specimen geometries are recommended for experimental programs to analyze safety and lifetime of metal structures. In the future the experimentally achieved results have to be leaded back to the determination of the material parameters of the continuum damage model.

References

[1] Brünig M, Chyra O, Albrecht D, Driemeier L and Alves M 2008 *International Journal of Plasticity* **24** 1731–1755
[2] Dunand M and Mohr D 2011 *Engineering Fracture Mechanics* **78** 2919 – 2934
[3] Demmerle S and Boehler J 1993 *Journal of the Mechanics and Physics of Solids* **41** 143–181
[4] Brünig M 2003 *International Journal of Plasticity* **19** 1679–1713
[5] Brünig M, Gerke S and Hagenbrock V 2013 *International Journal of Plasticity* **50** 49–65
[6] Brünig M, Brenner D and Gerke S 2015 *Engineering Fracture Mechanics* **141** 152 – 169
[7] Brünig M, Gerke S and Schmidt M 2016 *International Journal of Fracture* **doi: 10.1007/s10704-016-0080-3**

Implementation of virtual models from sheet metal forming simulation into physical 3D colour models using 3D printing

S Junk

Department of Industrial Engineering, University of Applied Sciences Offenburg, Campus Gengenbach, Klosterstr. 14, 77723 Gengenbach, Germany

Abstract. Today the methods of numerical simulation of sheet metal forming offer a great diversity of possibilities for optimization in product development and in process design. However, the results from simulation are only available as virtual models. Because there are any forming tools available during the early stages of product development, physical models that could serve to represent the virtual results are therefore lacking. Physical 3D-models can be created using 3D-printing and serve as an illustration and present a better understanding of the simulation results. In this way, the results from the simulation can be made more "comprehensible" within a development team. This paper presents the possibilities of 3D-colour printing with particular consideration of the requirements regarding the implementation of sheet metal forming simulation. Using concrete examples of sheet metal forming, the manufacturing of 3D colour models will be expounded upon on the basis of simulation results.

1. Introduction

The Additive Manufacturing (AM) processes, also called 3D Printing (3DP), have greatly advanced over recent years. A common feature of all of these processes is that the work pieces are manufactured in layers by adding the materials. Another important feature is the direct manufacturing of the components on the basis of 3D data [1]. The bonding processes used range from the fusing of powdery materials through the adhering of powders and foils / transparencies to the polymerisation of liquid photopolymers. In this, a wide palette of various materials ranging from plaster through plastic to metals is implemented [2]. The development of the last years has also demonstrated that the use of the processes ranges from simple prototyping to production of small or medium series. Thus, the number of components built by means of AM has continued to increases over the years [3].

Forming is applied in many areas of AM. The AM processes are thus implemented to produce moulds (Rapid Tooling). On one hand, the direct application of tool data from the CAD is used here in order to drastically reduce the production time of tools without the time-consuming programing of tool paths and the tedious set-up of machine tools [4]. On the other hand, tools with complex internal forms are thereby implemented; these tools are not conventionally manufactured. One example of this is the production of tools with internal cooling passages for the hot forming of high-strength steels [5].

Additionally, AM can also be conducive to the production of sample pieces in presentation models. Because, in particular, forming dies are generally not available for the production physical samples during the development phase of formed components, one is limited to virtual models in this integral phase. Sample parts can be additively produced through AM application; these parts, for example, can serve as the basis for discussion with the development team or the client. Simple test trials can also occur with such physical models. Along with rapid availability, the haptic compilation of the model is an essential advantage because using this physical model generates a significantly better understanding of the forming process compared to observing a simple view of the virtual model on a screen.

2. Additive Manufacturing of 3D colour models

With regard to forming, it is especially interesting that not only the geometry but also a colour texture can be displayed with 3D colour printing. This is, in particular, an advantage in the presentation of simulation results. Consequently, complex geometries or the phases of a multistage forming process can be visualised through physical 3D colour prints and haptically "grasped". Three AM processes can currently be applied for the production of 3D colour models [1]. These differ from one another, particularly, in in the layer thickness and accuracy, the joining technologies implemented as well as the necessary post processing and in the construction materials used (see Table 1).

Table 1. Additive manufacturing technologies for 3D colour printing.

	Laminated Object Modelling	Binder Jetting	Polyjet Modelling
Construction material	Sheet of paper	Polymer plaster powder	Liquid photopolymer
Layer thickness	approx. 0.1 mm	approx. 0.1 mm	approx. 32 µm
Joining technology	Applying adhesive & cutting	Jetting binder	Curing using UV light
Post processing	Removing paper and infiltration	Blowing off powder and infiltration	Removing support structure using water-jet

2.1. Laminated Object Modelling

Laminated Object Modelling (LOM) is one of the oldest of a AM processes. In this process, paper is adhered in individual layers. In earlier systems, the paper was unwound from a roll and stuck together in large areas. Then the outline of the component was cut out by means of a laser. In order to reduce the equipment costs, current systems use a cutting blade rather than a laser. The paper is printed in individual sheets in colour and then adhered. Today the adhesive is selectively applied so as to avoid oversaturation and the expansion of the paper. During post-processing, the unnecessary fringe areas and support structure, which was cubed during the previous production process, are first manually removed with the aid of a graver. To increase the stability, the components can be infiltrated with epoxy resin.

2.2. Binder Jetting

In Binder Jetting (BJ) a polymer plaster powder is adhered by means of a binder. For this, a layer of polymer plaster powder is first applied with a roller. Then a liquid binder is applied with a print head so that the powder adheres. Colours as well as lettering and logos can also be applied with a second print head, similar to that in a conventional inkjet printer. In this, the colouring is only done on one of the approx. 1.5 to 2.0 mm thick edge layers; the core of the component is not coloured. Post processing in this AM process occurs in two steps. First, the 3D printed model is released, and excess powder is cleaned by means of compressed air. Because the 3D printed models, also called "green body," only exhibit weak stability, they must still be infiltrated with an epoxy resin. This infiltration significantly increases the stability of the model and also improves the brilliance of the printed colours.

2.3. Polyjet Modelling

In Polyjet Modelling (PJM), a liquid photopolymer is applied with a print head with a high number of nozzles. The applied layer is cured with UV lamps located directly next to the print head. High precision in detail is achieved through the minimal thickness of the layers. Admittedly, the production times are considerably longer for the same reason. Components that are hanging over are stabilised by support material. Material combinations with diverse features are possible (Digital Materials) through combining several different photopolymers. Moreover, levels of grey or colours as well can be created by these combinations. The support structures must be removed after the actual manufacturing process. Generally, a water jet or an alkaline bath is used for this.

3. Process chain from virtual to physical 3D model

During the production of a physical 3D colour model, the data for the virtual CAD model passes through various interfaces and software packages for the simulation and for the data processing (see Figure 1). The production process of the physical model divides itself in applying the component layer by layer and the subsequent necessary post-processing.

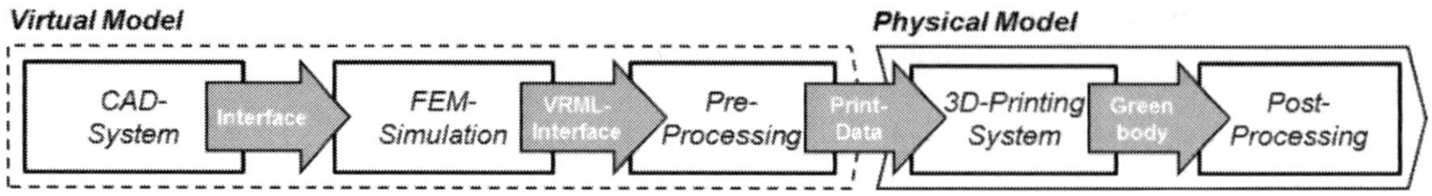

Figure 1: Process chain from the virtual model to the physical 3D colour model.

3.1. Virtual model on the basis of simulation results

The basis for every AM Process is a 3D model. For conventional applications, this is prepared using a CAD system. Alternatively, 3D scanners can also be used to digitalise existing components or tools. In this process, a time consuming Reverse Engineering is often necessary in addition to the actual scanning in order to transfer the point cloud from the 3D scan to a closed volume model. In contrast to conventional process chain, an additional process step is required in the application of simulation technology. Here the CAD data from the active surfaces of the forming tools is read into the simulation software and then the computation of various parameters takes place. After the completion of the computation, the simulation results are transferred to the data processing. The VRML format is used as the interface between forming simulation and pre-processing, because it can also transfer the texture of the surface. In addition, the positioning and orientation in the installation room of the 3D printer takes place at this point. Then the data read is sliced and transferred as print data to the 3D printer.

3.2. Physical 3D colour model

The production of the 3D colour model occurs in a layer by layer application of material. A 3D printer Projet 660 Pro from 3D Systems was used in the examples indicated in Chapter 4 as well. This printer functions according to the BJ process and has an installation room of 254 x 381 x 203 mm (L x W x H). The colour print head can represent 390,000 colours with a resolution of 600 x 540 dpi. A powdery high performance composite material consisting of plaster and polymers is used as construction material. After an idle period of approximately 90 minutes, the excess powder is vacuumed off with a hose in the installation room. Then the component is transferred to a chamber where the remaining powder is blown off with a compressed air pistol. In the second part of post-processing, the component is soaked in epoxy resin or doused with an instant adhesive to improve the stability and the brilliance of the colours.

4. Examples of implementing the 3D colour models

The following section presents several examples of the implementation of 3D colour models based on the simulation results in sheet metal forming. The simulation results can thus be read as VRML data and be transferred to the 3D printer. Specific software settings are not required for this. The file can be directly read on the post-processor.

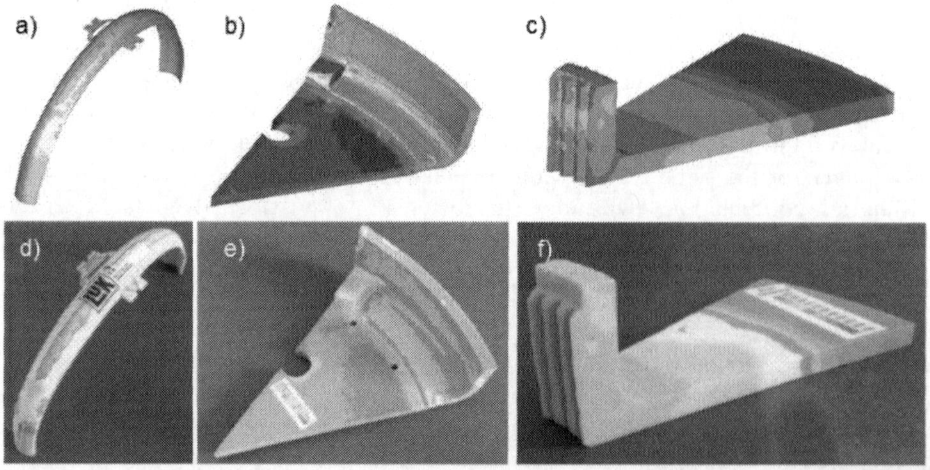

Figure 2: Simulation results and 3D colour print of washer (a,d), thick sheet (b,e) and hub (c,f).

A "curved spring washer" is an example of the use of 3D colour printing in sheet metal forming. Various forming stages are optimised and tested with FEM simulation within the framework of a development project. The measurements are 248 x 46 x 113 mm (L x W x H). A particular challenge for sheet metal forming is the minimal wall thickness of the model. Because the 3D model is very fragile when the wall thickness is less than 2 mm, secure handling during post-processing is hardly possible. Therefore, the wall thickness is scaled, i.e. increased, in the course of the data processing (see Figure 2 a, d). Additionally, the colour key and the company logo are applied.
In a second example, the simulation of the forming process of a "thick sheet" with the measurements 150 x 105 x 46 mm (L x W x H) is run as a 3D colour model (see Figure 2 b, e). The round cuttings are generated before the simulation. Because a rotationally symmetrical component is being produced here, only a part model, that is, a cutting of approximately 45°, is simulated and printed. A scaling of the wall thickness of approximately 6.5 mm was not necessary for this model. The colour key is also applied in this example.
A "hub with internal teeth" (see Figure 2 c, f) with the measurements 50 x 26 x 16 mm (L x W x H) is printed as a third example. A segment of 23.5° was also selected from this example. Because this is a very small component, the application of the key is possible but hardly readable. Therefore, an increase of the scale to 2:1 or even 3:1 in the scaling of the data processing is recommended for this example.
Three-dimensional colour printing can also be applied in representing simulation results in massive forming simulation and microstructure simulation [6]. In these applications the 3D-printing of several forming stages and the implementation of "3D-puzzle" could be demonstrated.

5. Conclusion

Currently, a variety of Additive Manufacturing Processes that enable the production of 3D colour models are available. Particular attention must be given to the selection of suitable interfaces for data processing in the application of these processes for the production of models based on simulation results taken from forming technology. An optimisation of the component, for example, in thickening the thin and thereby fragile areas, can take place during the data processing. Additionally, a scaling of especially small or large models is possible. Large models have the option to divide the into sections.
The 3D colour models produced in this manner serve the visualisation of complex geometries and detailed simulation results. A simpler "grasping" and better understanding of the geometry and the simulation results is possible with the haptic compilation. Furthermore, an extension of the 3D printing on simulation results from similar areas, such as, e.g., massive forming or even casting, and the representation of microstructure models are possible.

Acknowledgments
The author likes to thank the department of process engineering of the industrial partner LuK GmbH & Co. KG in Kappelrodeck (Germany) for the support and the providing of the simulation results.

References

[1] Gibson I, Rosen D, and Stucker B 2014 *Additive Manufacturing Technologies: 3D Printing, Rapid Prototyping, and Direct Digital Manufacturing*, Springer, New York
[2] de Beer N 2012 *Additive Manufacturing - Turning Mind into Mater*, Sierra College for Applied Competitive Technologies
[3] Wohlers T 2014 *Wohlers Report 2014; Additive Manufacturing and 3D Printing State of the Industry, Annual Worldwide Progress Report*, Wohlers Associates Inc., Fort Collins, USA
[4] Junk S, Taleb-Araghi B 2012 New Developments in Rapid Tooling Using 3D-Printing with Plaster Powders, *Proc. of Fraunhofer Direct Digital Manufacturing Conf.*, Berlin, pp.1-6
[5] Müller B, Gebauer M, Hund R, Malek R, Gerth N 2014 Metal Additive Manufacturing for tooling applications - Laser Beam Melting technology increases efficiency of dies and molds, *Proc. of Metal Additive Manufacturing Conf.*, Vienna, pp. 1-14
[6] Junk S 2015 Making virtual models „tangible“ - Additive manufacturing of simulation results using 3D colour printing, *Aachener Stahlkolloquium ASK*, Aachen

An isogeometric Reissner-Mindlin shell element for dynamic analysis considering geometric and material nonlinearities

Paul Sobota[+], Wolfgang Dornisch[*], Sven Klinkel[+]

[+] Lehrstuhl für Baustatik und Baudynamik, RWTH-Aachen University, Mies-van-der-Rohe-Str. 1, 52074 Aachen, Germany

[*] Lehrstuhl für Technische Mechanik, Technische Universität Kaiserslautern, Gottlieb-Daimler-Str., 67663 Kaiserslautern, Germany

E-mail: klinkel@lbb.rwth-aachen.de

Abstract. The present approach deals with the dynamical analysis of thin structures using an isogeometric Reissner-Mindlin shell formulation. Here, a consistent and a lumped mass matrix are employed for the implicit time integration method. The formulation allows for large displacements and finite rotations. The Rodrigues formula, which incorporates the axial vector is used for the rotational description. It necessitates an interpolation of the director vector in the current configuration. Two concept for the interpolation of the director vector are presented. They are denoted as continuous interpolation method and discrete interpolation method. The shell formulation is based on the assumption of zero stress in thickness direction. In the present formulation an interface to 3D nonlinear material laws is used. It leads to an iterative procedure at each integration point. Here, a J2 plasticity material law is implemented. The suitability of the developed shell formulation for natural frequency analysis is demonstrated in numerical examples. Transient problems undergoing large deformations in combination with nonlinear material behavior are analyzed. The effectiveness, robustness and superior accuracy of the two interpolation methods of the shell director vector are investigated and are compared to numerical reference solutions.

1. Introduction

In this contribution an isogeometric Reissner-Mindlin shell formulation for transient problems including large deformations and nonlinear material behavior is presented. Isogeometric analysis for structural dynamics was proposed in [1]. It leads to an excellent performance in the optical branch, see [2]. In the frame of plate formulations a Kirchhoff formulation is presented in [3] and a Reissner–Mindlin formulation is derived in [4]. In the context of shells, isogeometric Kirchhoff–Love shell analysis is studied in [5]. The good performance was achieved at the cost of higher computational effort. Benson et a. [6, 7] proposed isogeometric shell formulations for transient problems using explicit time integration, which are also able to deal with large deformations and nonlinear material behavior. The present approach employs an implicit time integration. The purpose of the present approach is to analyze the impact of the director interpolation. The continuous interpolation method of the shell director vector, proposed in [8], is compared to a discrete interpolation method as used e.g. in [6, 9, 10].

2. Governing equations for the Reissner-Mindlin shell

Let $\boldsymbol{x}$ be a point on the shell mid surface of the reference configuration and $\bar{\boldsymbol{x}}$ be a point of the mid surface of the current configuration. The shell continuum is defined by

$$\begin{aligned}\boldsymbol{\Phi}(\xi^1,\xi^2,\xi^3) &= \boldsymbol{x}(\xi^1,\xi^2) + \xi^3\boldsymbol{d}(\xi^1,\xi^2)\\ \bar{\boldsymbol{\Phi}}(\xi^1,\xi^2,\xi^3) &= \bar{\boldsymbol{x}}(\xi^1,\xi^2) + \xi^3\bar{\boldsymbol{d}}(\xi^1,\xi^2)\end{aligned} \tag{1}$$

Here, the quantities with an overset bar denote the values in the current configuration. $\boldsymbol{\Phi}$ represents an arbitrary point in the shell and $\boldsymbol{d}$ describes the director vector, see Fig.1. ξ^1, ξ^2 are the in-plane coordinates and ξ^3 denotes the thickness coordinate with $-h/2 \leq \xi^3 \leq h/2$, where h is the thickness of the shell. The membrane strains $\varepsilon_{\alpha\beta}$, the change of curvature $\kappa_{\alpha\beta}$

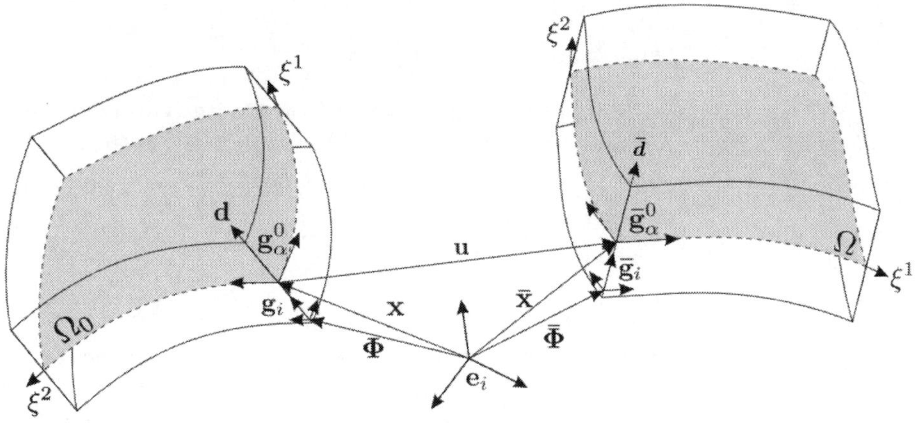

Figure 1. The reference- and current configuration of the shell continuum

and the transversal shear strains γ_α read

$$\begin{aligned}\varepsilon_{\alpha\beta} &= \frac{1}{2}\left(\bar{\boldsymbol{x}}_{,\alpha}\cdot\bar{\boldsymbol{x}}_{,\beta} - \boldsymbol{x}_{,\alpha}\cdot\boldsymbol{x}_{,\beta}\right)\\ \kappa_{\alpha\beta} &= \frac{1}{2}\left(\bar{\boldsymbol{x}}_{,\alpha}\cdot\bar{\boldsymbol{d}}_{,\beta} - \boldsymbol{x}_{,\alpha}\cdot\boldsymbol{x}_{,\beta}\right)\\ \gamma_\alpha &= \bar{\boldsymbol{x}}_{,\alpha}\cdot\bar{\boldsymbol{d}} - \boldsymbol{x}_{,\alpha}\cdot\boldsymbol{d}\,.\end{aligned} \tag{2}$$

The Greek letters indicate the local coordinate direction, it holds $\alpha,\beta \in \{1,2\}$. The vector containing all shell strains reads $\boldsymbol{\varepsilon}^T = [\varepsilon_{11}, \varepsilon_{22}, 2\varepsilon_{12}, \kappa_{11}, \kappa_{22}, 2\kappa_{12}, \gamma_1, \gamma_2]$.

3. The weak form

Let $\delta\boldsymbol{v} \in H^1$ be an admissible test function and $\boldsymbol{v}$ the displacement vector, the weak form of the equation of motion reads

$$G(\boldsymbol{v},\delta\boldsymbol{v}) = \int_{B_0} \delta\boldsymbol{v}^T\rho_0\ddot{\boldsymbol{v}}\,\mathrm{d}V + \int_{\Omega_0}\delta\boldsymbol{\varepsilon}^T\,\boldsymbol{\sigma}\,\mathrm{d}A - \int_{\Omega_0}\delta\boldsymbol{v}^T\boldsymbol{p}\,\mathrm{d}A - \int_{\Gamma_\sigma}\delta\boldsymbol{v}^T\boldsymbol{f}\,\mathrm{d}s = 0\,. \tag{3}$$

Here, $\boldsymbol{\sigma}$ represents the stress resultants which are work conjugate to the shell strains. ρ_0 is the initial mass density and $\ddot{\boldsymbol{v}}$ the acceleration. The surface load $\boldsymbol{p}$ is defined per area reference surface and the traction $\boldsymbol{f}$ is given on the boundary.

4. Interpolation methods of the director vector

4.1. Discrete interpolation method - Rotation before interpolation

Within the discrete interpolation method (DIM), see e.g. [11], the nodal basis systems in the current configuration

$$\bar{\boldsymbol{a}}_{iI} = \boldsymbol{R}\boldsymbol{a}_{iI} \tag{4}$$

are calculated by the rotation of each nodal basis system in the reference configuration at each control point I. The current director vector $\bar{\boldsymbol{d}}_I$ in each control point I equals $\bar{\boldsymbol{a}}_{3I}$. This *discrete* rotation of nodal basis systems is eponymous for the interpolation method at hand. The rotational matrix

$$\begin{aligned} \boldsymbol{R} &= \boldsymbol{1} + \frac{\sin\omega}{\omega}\boldsymbol{\Omega} + \frac{1-\cos\omega}{\omega^2}\boldsymbol{\Omega}^2 \\ \text{with } \boldsymbol{\Omega} &= \varepsilon_{ijk}\boldsymbol{\omega}_i = \begin{bmatrix} 0 & -\omega_3 & \omega_2 \\ \omega_3 & 0 & -\omega_1 \\ -\omega_2 & \omega_1 & 0 \end{bmatrix} \qquad \text{and } \omega = |\boldsymbol{\omega}| \end{aligned} \tag{5}$$

is calculated via the Rodrigues formula, where ε_{ijk} is the Levi-Civita tensor. Nodal values $\boldsymbol{\omega}_I$ have to be inserted into Eq. (5). These approximation of teh director vector through the element area is given as

$$\bar{\boldsymbol{d}}^h = \sum_{I=1}^{n_{en}} N_I \bar{\boldsymbol{d}}_I \qquad \bar{\boldsymbol{d}}^h_{,\alpha} = \sum_{I=1}^{n_{en}} N_{I,\alpha} \bar{\boldsymbol{d}}_I \,. \tag{6}$$

4.2. Continuous interpolation method - Interpolation after Rotation

The idea of the continuous interpolation method (CIM), which is described in [11], is to interpolate the rotational parameter $\boldsymbol{\omega}$ instead of the director vector itself. Instead of the interpolation of discretely rotated values as is done in Eq. (6), here the current director vector

$$\bar{\boldsymbol{d}}^h = \boldsymbol{R}\boldsymbol{d}^h = \boldsymbol{R}\sum_{I=1}^{n_{en}} N_I \boldsymbol{d}_I \tag{7}$$

is computed by first interpolating the director vector of the reference configuration and then applying the rotation $\boldsymbol{R}$ in each integration point. Thus, the condition $|\bar{\boldsymbol{d}}| = |\boldsymbol{d}|$ holds exactly. Within this approach interpolated values $\boldsymbol{\omega}^h$ and $\bar{\boldsymbol{d}}^h$ have to be used for the computation of $\boldsymbol{R}$.

5. Numerical Example - Impulsively loaded square plate

The simply supported square plate is taken from the *shell obstacle course* proposed in [12]. The material properties, loadings and dimensions are depicted in Fig. 2. The load is applied impulsively within the first time step $t_0 = 0$. Due to double-symmetry only one quarter of the system is modeled. The reference solution of the maximum deflection of the midpoint has been calculated with 25x25 elements with quartic basis functions with the CIM concept. The purpose of thi example is to investigate the effect of the different interpolation methods for the director vector, see Fig. 3. With the DIM concept, NUBRS with $p = 6$ lead to a decreasing quality of the solution (cf. Fig. 3b). Here, two different aspects are involved. The approximation of the deformed geometry increases with higher order basis functions. However, also the area of control points that influence the director vector increases. If the director vector is rotated before it is interpolated on curved geometries, the length of the director vector will in general be smaller than one and this effect increases with an increasing number of involved control points. This effect decreases with mesh refinement, because the distance and therefore the difference of the rotation angle of the involved control points becomes smaller. Note that the continuous and discrete interpolation method of the shell director vector are identical for configuration without curvature.

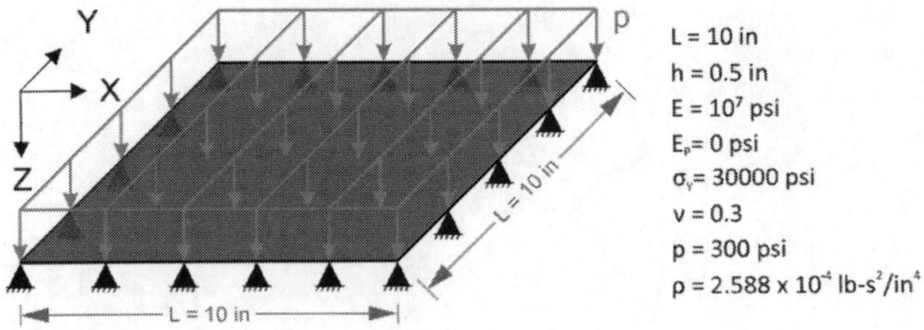

Figure 2. Impulsively loaded square plate

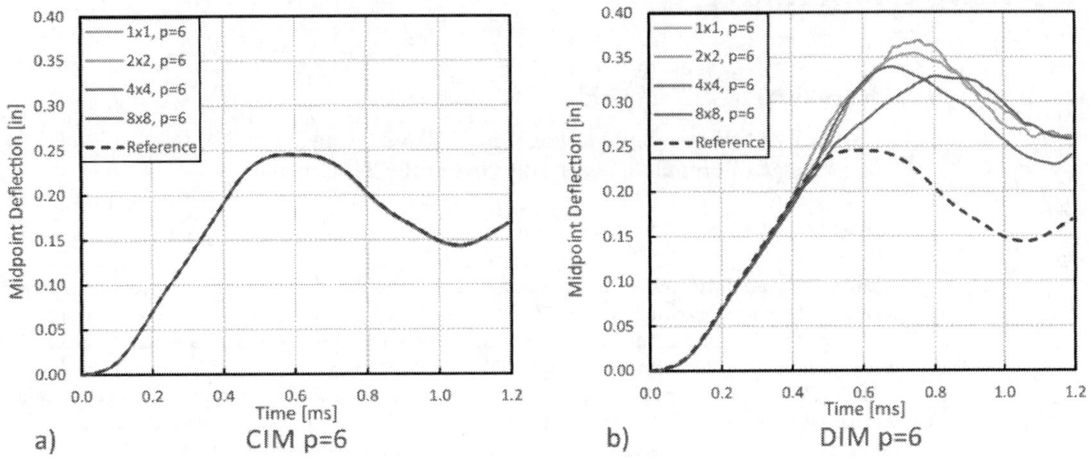

Figure 3. Comparison between interpolation methods of the shell director vector: a) Continuous Interpolation Method (CIM) for h- and k-refinement. For $p = 6$ the results for all meshes are almost indistinguishable from the reference solution; b) Discrete Interpolation Method (DIM) for h- and k-refinement.

References

[1] Cottrell J A, Reali A, Bazilevs Y and Hughes T J R 2006 *Comput. Meth. Appl. Mech. Engrg.* **195** 5257–5296
[2] Cottrell J A, Hughes T J R and Bazilevs Y 2009 *Isogeometric analysis: Toward integration of CAD and FEA* (Chichester: Wiley)
[3] Shojaee S, Izadpanah E, Valizadeh N and Kiendl J 2012 *Finite Elem. Anal. Des.* **61** 23–34
[4] Thai C H, Nguyen-Xuan H, Nguyen-Thanh N, Le T H, Nguyen-Thoi T and Rabczuk T 2012 *Int. J. Numer. Meth. Engng.* **91** 571–603
[5] Schäuble A K 2013 *Isogeometric Galerkin Finite Elements in Time and Space* (Master Thesis. Institut für Baustatik und Baudynamik, Universität Stuttgart)
[6] Benson D J, Bazilevs Y, Hsu M C and Hughes T J R 2010 *Comput. Meth. Appl. Mech. Engrg.* **199** 276–289
[7] Benson D J, Hartmann S, Bazilevs Y, Hsu M C and Hughes T J R 2013 *Comput. Meth. Appl. Mech. Engrg.* **255** 133–146
[8] Dornisch W, Müller R and Klinkel S 2016 *Comput. Meth. Appl. Mech. Engrg.* **303** 1–34
[9] Wagner W and Gruttmann F 2005 *Int. J. Numer. Meth. Engng.* **64** 635–666
[10] Adam C, Bouabdallah S, Zarroug M and Maitournam H 2015 *Comput. Meth. Appl. Mech. Engrg.* **284** 106–137
[11] Dornisch W 2015 *Interpolation of Rotations and Coupling of Patches in Isogeometric Reissner–Mindlin Shell Analysis* (Ph.D. thesis. Lehrstuhl für Baustatik und Baudynamik, RWTH Aachen)
[12] Belytschko T, Lin J I and Chen-Shyh T 1984 *Comput. Meth. Appl. Mech. Engrg.* **42** 225–251

Constitutive modelling of aluminium alloy sheet at warm forming temperatures

S Kurukuri[1], M J Worswick[1] and S Winkler[2]

[1] University of Waterloo, Waterloo, ON, N2N 3G1, Canada
[2] Dana Canada Corporation, Oakville, ON, L6K 3E4, Canada

srihari.kurukuri@uwaterloo.ca

Abstract. The formability of aluminium alloy sheet can be greatly improved by warm forming. However predicting constitutive behaviour under warm forming conditions is a challenge for aluminium alloys due to strong, coupled temperature- and rate-sensitivity. In this work, uniaxial tensile characterization of 0.5 mm thick fully annealed aluminium alloy brazing sheet, widely used in the fabrication of automotive heat exchanger components, is performed at various temperatures (25 to 250 °C) and strain rates (0.002 and 0.02 s^{-1}). In order to capture the observed rate- and temperature-dependent work hardening behaviour, a phenomenological extended-Nadai model and the physically based (i) Bergström and (ii) Nes models are considered and compared. It is demonstrated that the Nes model is able to accurately describe the flow stress of AA3003 sheet at different temperatures, strain rates and instantaneous strain rate jumps.

1. Introduction

Uniaxial tensile tests show that the work hardening in AA3003 sheet depends on the temperature and strain rate [1]. In numerical modelling of sheet metal forming, a description of material hardening is required. The conventional approach is to fit measured flow stress curves to a convenient empirical equation. The validity of these phenomenological models is often limited to situations that are comparable to the range of experiments on which they are based. Material models based on consideration of the underlying physical processes are expected to have a larger range of applicability than more commonly used empirical models [2-5]. In physically based models, the effect of microstructural-level processes is usually accounted for, perhaps in an average way, at the continuum level. In the current work, a phenomenological extended Nadai [2-4] and two physically based models due to Bergström [6] and Nes [7-9] are considered. The models are fit to measured rate- and temperature-dependent constitutive data for AA3003 brazing sheet. Their ability to capture work hardening and strain rate jump response is assessed.

2. Constitutive model

2.1. Extended Nadai model

In warm forming of aluminium sheet, the strain rate and temperature have a significant, coupled effect on the flow stress. The phenomenological model adopted in this work to capture this behavior is the Nadai hardening model with strain rate dependency [2-4]:

$$\sigma_{eq} = C(\varepsilon_{eq} + \varepsilon_0)^n \left(\frac{\dot{\varepsilon}}{\dot{\varepsilon}_0}\right)^m \tag{1}$$

Temperature dependence is introduced by allowing C, n and m be functions of temperature, T (in Kelvin). The following functions for the parameters were used in order to fit the tensile tests.

$$C(T) = C_0 + a_1\left[1 - \exp\left(a_2\frac{T-273}{T_m}\right)\right],\ n(T) = n_0 + b_1\left[1 - \exp\left(b_2\frac{T-273}{T_m}\right)\right],\ m(T) = m_0\exp\left(c\frac{T-273}{T_m}\right). \tag{2}$$

2.2 The Bergström model

In the Bergström physically based hardening model [6], the flow stress is decomposed into strain rate independent stress, $\sigma_0(T)$ and a work hardening dependent stress, $\sigma_w(\rho,T)$ due to the evolution of microstructure and is completely described by the total dislocation density ρ. The relation between the dislocation density and σ_w is given by the Taylor equation [2-4]:

$$\sigma_w = \alpha G(T) b\sqrt{\rho}, \tag{3}$$

where α is a scaling parameter of the order 1, G the elastic shear modulus and b the Burgers vector.

The evolution of dislocation density is expressed as the competition between dislocation storage and recovery by remobilization and annihilation; U represents the storage of mobile dislocations and Ω is the dynamic recovery by remobilization and annihilation. The functions U and especially Ω determine the shape of the hardening curve at different temperatures and strain rates.

$$\frac{d\rho}{d\varepsilon} = U(\rho) - \Omega(\dot{\varepsilon},T)\rho;\quad \text{with } U(\rho) = U_0\sqrt{\rho} \quad \text{and} \quad \Omega(\dot{\varepsilon},T) = \Omega_0 + C\exp\left(-\frac{mQ_v}{RT}\right)\dot{\varepsilon}^{-m}. \tag{4}$$

U_0 is the immobilization rate, Ω_0 the low temperature, high strain rate limit value of the remobilization probability, C and m are constants, Q_v the activation energy and R the gas constant. Equation (4) can be integrated analytically to obtain the total dislocation density and is expressed in terms of an incremental algorithm that can easily be implemented in FE code (Kurukuri et al., 2010):

$$\rho_{j+1} = \left[\frac{U_0}{\Omega}\left(\exp\left(\frac{1}{2}\Omega\Delta\varepsilon\right) - 1\right) + \sqrt{\rho_j}\right]^2 \exp(-\Omega\Delta\varepsilon)\cdot \tag{5}$$

The final equation for the flow stress is:

$$\sigma_{eq} = g(T)(\sigma_0 + \alpha G_{ref} b\sqrt{\rho}, \quad \text{and} \quad g(T) = 1 - C_T\exp\left(\frac{-T_1}{T}\right), \tag{6}$$

where C_T and T_1 are fitting parameters, $g(T)$ is the shear modulus, G, divided by the reference value G_{ref}. Some of the parameters in the Bergström model can be selected beforehand. The rest are determined by a least squares approximation of experimental results. The initial dislocation density ρ_0 was chosen to be $10^{11}m^{-2}$, which seems to be a reasonable value for annealed aluminium [2]. The magnitude of the Burgers vector b and the shear modulus at room temperature G_{ref} were taken from the literature.

2.3 The Nes model

The Nes model approach relies on a multi-parameter description for the microstructure evolution. At small strains the stored dislocations are arranged in a cell structure with dislocation density within the cells ρ_i and cell size δ. The cells have finite walls of higher dislocation density. At larger strains the dynamic recovery of dislocations becomes important and the cell walls collapse into sub-boundaries of well-defined misorientation φ. Extensive presentations of the model are given in [7-9]. The microstructure evolution is obtained by solving a set of differential equations describing the evolution of these parameters. Leaving out the details, these can be written as:

$$\frac{\mathrm{d}\rho_i}{\mathrm{d}\gamma}=\frac{1}{1+f(q_b^2-1)}\frac{2}{bL_{eff}}-\frac{\rho_i v_\rho}{\dot{\gamma}},\quad \frac{\mathrm{d}\delta}{\mathrm{d}\gamma}=-\frac{2\delta^2\rho_i}{\kappa_0\varphi}\frac{SL^2}{L_{eff}}+\frac{bv_\delta}{\dot{\gamma}},\quad \text{and}\quad \frac{\mathrm{d}\varphi}{\mathrm{d}\gamma}=g(\rho_i,\delta,\varphi). \tag{7}$$

The first term in these equations represents storage of dislocations (which increases work hardening), the second term represents dynamic recovery by annihilation of dislocations (which decreases work hardening). It describes the flow stress as a function of microstructure evolution parameters:

$$\tau=\tau_t+\alpha_1 Gb\left[\Gamma_1\left(\frac{q_c}{\delta\sqrt{\rho_i}}\right)\sqrt{\rho_i}+\Gamma_2\left(\frac{q_c}{\delta\sqrt{\rho_i}}\right)\frac{q_c}{\delta}\right]+\hat{\alpha}_2 Gb\left[\Gamma_2(0)\frac{1}{\delta}+\frac{1}{D}\right]. \tag{8}$$

Here, τ_t is thermal stress, due to short range interactions between mobile dislocations and intersecting stored ones, dragging of jogs, and elements in solid solution. The α_1-term is the contribution from stored dislocations and the $\hat{\alpha}_2$-term is the contribution from sub-grains and grain boundaries. G is the shear modulus, b is the Burgers vector. In applications of the model, the stress tensor at a macroscopic continuum scale is required representing contributions from many grains of various crystallographic orientation and microstructure. Hence, the flow stress and strain at continuum scale can be represented as, $\sigma_{eq}=M\tau$ and $\varepsilon_{eq}=\gamma/M$, here M is the Taylor factor. The Nes model involves a large number of parameters (nearly 40), however, most of these are pre-determined based on material constants, while some are fit through manual methods as described in [5].

3. Results

The various parameters of the models described above are obtained by fitting simultaneously with the experimental stress-strain data at temperatures of 25, 150 and 250 °C and strain rates of 0.002 and 0.02 s^{-1} using MATLAB simplex optimization algorithm. In figure 1, the simulated flow curves for the extended Nadai, Bergström and Nes models are plotted together with the experimental data. It can be seen that the Bergström and extended Nadai models are capable of describing the overall trends in material behaviour over the range of temperatures and strain rates with a reasonable accuracy. The Nes model, on the other hand, predicts the behaviour with exceptional accuracy at all strain rates and temperatures due to its strong physically motivated nature.

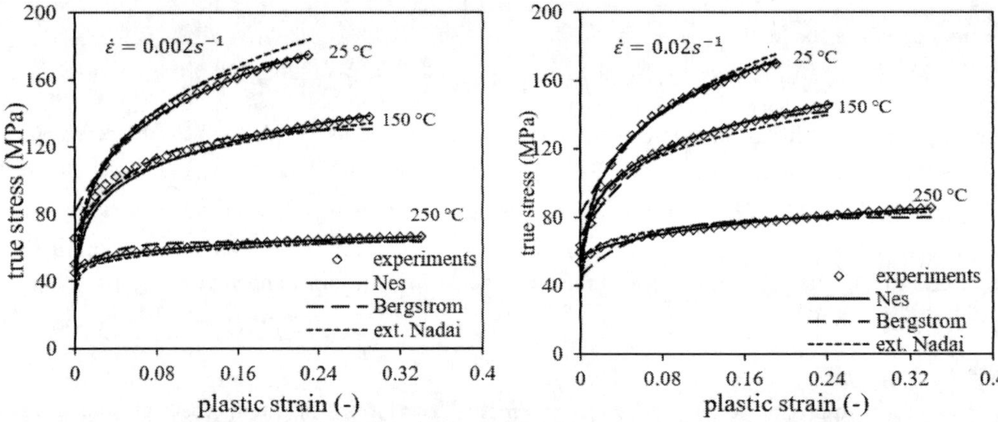

Figure 1. True stress-plastic strain curves–experiments and models.

Even though all three models considered yield reasonable results for constant strain rate conditions, the predictions are quite different if a jump in the strain rate is considered. In figure 2, the simulated stress-strain curves using three models are compared with the experiments are plotted for deformation at 150 °C and 250 °C and for strain rates of 0.002 and 0.02 s^{-1}. If strain rate changes from 0.002 to 0.02 s^{-1} or from 0.02 to 0.002 s^{-1} are applied after a strain of 15%, the extended Nadai model immediately follows the curve corresponding to a constant strain rate. With the Bergström model the constant strain rate curve is only slowly approached after continuous straining. With the Nes model, an initial jump in the flow stress is observed, followed by slowly approaching the constant strain rate curve, which is a better representation of the experiments.

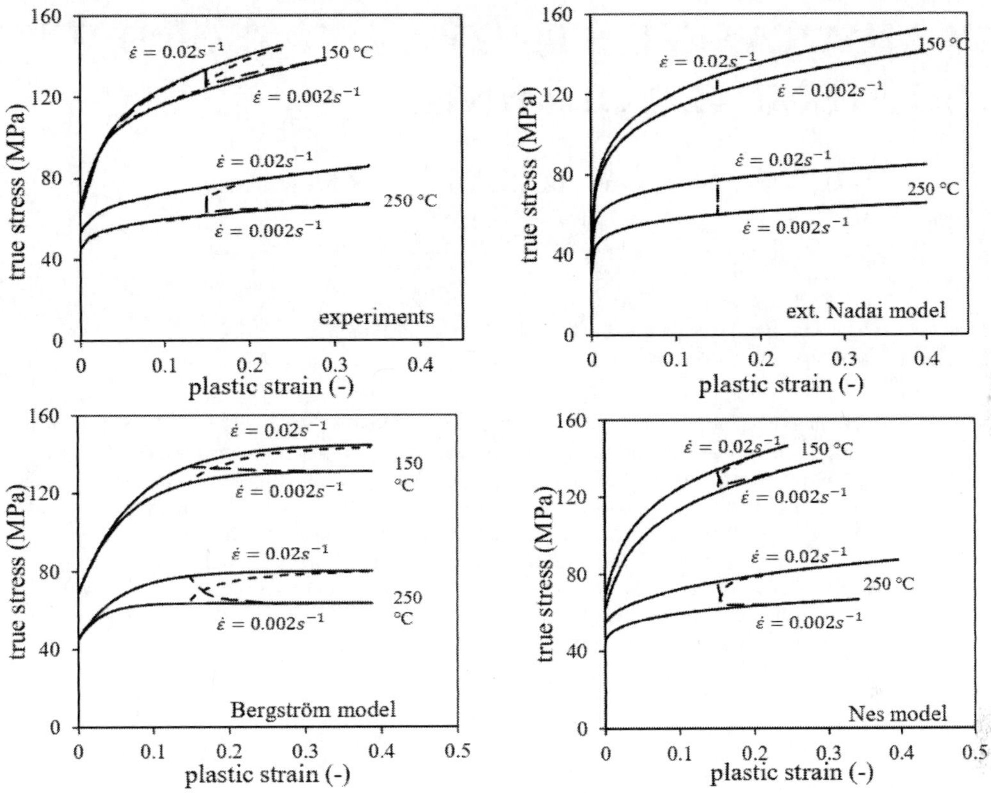

Figure 2. Flow curves with and without strain rate jumps—experiments and models.

Acknowledgements
Financial support from Dana Canada Corporation, Automotive Partnership Canada (NSERC and CFI), the Ontario Research Fund and the Canada Research Chairs Secretariat is gratefully acknowledged.

References

[1] Bagheriasl R and Worswick M 2015 *Int. J. Mater. Form.* **8** pp 229–244.
[2] Van den Boogaard A H and Huétink J 2006 *Comp. Meth. App. Mech. Engg.* **195** pp 6691–709.
[3] Kurukuri S, Van den Boogaard, A H, Miroux A and Holmedal B 2009 *J. Mater. Proc. Tech.* **209** pp 5636–5645.
[4] Kurukuri S, Miroux A, Wisselink and Van den Boogaard 2011 *Int. J. Mater. Form.* **4(2)** pp 129–140.
[5] Kurukuri S 2010 PhD Thesis, University of Twente, The Netherlands.
[6] Bergström Y 1969 Mat. Sci. Engg. **5** pp 193–200.
[7] Nes E 1998 *Prog. Mat. Sci.* **145**, pp 129–193.
[8] Marthinsen K and Nes E 2001 *Mat. Sci. Tech.* **17** pp 376–388.
[9] Marthinsen K, Nes E and Holmedal B 2004 *Mat. Sci. Tech.* **17** pp 1377–1382.

Journal of Physics: Conference Series **734** (2016) 032065 doi:10.1088/1742-6596/734/3/032065

A COMPARATIVE STUDY OF FAILURE WITH INCREMENTAL FORMING

S.H.Wu[1*], N.N.Song[2],F.M. Andrade Pires[1]
[1]Faculty of Engineering, University of Porto
Rua Dr. Roberto Frias 400, 4200-465 Porto, Portugal
*E-mail: shenghua,fpires@fe.up.pt
[2]INEGI, University of Porto
Rua Dr. Roberto Frias 400, 4200-465 Porto, Portugal
E-mail: snannan@inegi.up.pt

Abstract. Incremental forming (ISF) is an innovative flexible sheet metal forming process which can be used to manufacture complex shapes from various materials. Due to its flexibility, it has attracted more and more attention over recent decades. Localized deformation and shear through the thickness are essential characteristics of ISF. These lead to specific failure modes and formability of ISF that are different from the conventional stamping process. In this contribution, three continuum damage models (Lemaitre, Gurson, extended GTN models) are formulated and fully coupled with the finite element simulation in a commercial software ABAQUS to predict failure in incremental forming. A comparative investigation of these three damage models has been carried out to analyze both the deformation behavior and failure mechanisms.

1. Introduction

Single point incremental forming (SPIF) is the flexible process of converting a flat sheet of metal into a desired geometry without a high cost die. Due to its advantages, it has attacted more and more attentions and has been used to produce the customized complex dimensional shape products. The sheet metal in the incremental forming process not only surfers stretching and thinning deformation mode, the transverse shear through the sheet thickness also has a great influence on the formability that is restricted by geometric instabilities due to necking and strain localization, and thus the understanding of damage development and accuracy prediction of failure is of great importance. Unfortunately, the fracture is not preceded by localized necking in incremental forming which were confirmed by M. Skjoedt and Martins [1, 2].This implies that forming limit diagrams (FLDs) of conventional sheet metal forming are inapplicable to describe failure in SPIF. In order to model the plastic flow and fracture of these ductile structural metals, numerous failure criteria (including coupled and uncoupled damage approaches) are proposed. To account for the damage-behavior coupling, two approaches are widely used: the physically motivated approach initially developed by Gurson [3] and the phenomenological approach based on continuum damage mechanics (CDM) [4]. The CDM model takes the triaxiality into the thermodynamic damage force Y when it is expressed in the stress space [4].However, the third stress invariant is not considered. The original Gurson model introduces a strong coupling between plastic strain and damage, and the presence of micro voids in the formulation leads to a yield surface that depends on both the hydro-static pressure and porosity. The extended Gurson model introduced the shear effects by considering the influence of both the hydro-static pressure and Lode angle on damage evolution [5]. Although significant research efforts have been devoted to the understanding and description of the phenomenon of ductile fracture and to the formulation of constitutive models, the improper application of the models to various metal forming process may result in misleading ductile fracture predictions.

In this study, the finite element model of incremental forming fully coupled with ductile damage model are performed. The developed constitutive subroutines are implemented in the explicit scheme program. The effects of different damage models on the incremental forming have been studied.

2. Formulation of damage model

There are several damage models used to predict the failure for ductile materials. the governing equations of the constitutive models under analysis are briefly reviewed in this section.

2.1. Lemaitre damage model

The constitutive equations for ductile damage have been proposed by Lemaitre [4]. The Lemaitre damage model is a fully coupled Elasto-plasticity-damage model, which is based on a thermodynamic framework and has been extensively used in the prediction of internal damage and failure in ductile metals.

By assuming homogeneous distribution of microvoids and the hypothesis of strain equivalence, the effective stress tensor can be represented as

$$\sigma_{eff} = \frac{\boldsymbol{\sigma}}{1-D} \tag{1}$$

where σ_{eff} is the effective stress tensor, $\boldsymbol{\sigma}$ is the stress tensor for the undamaged material, in addition, the damage variable, D.

The damage energy release rate, $-Y$, corresponds to the variation of internal energy density which can be given by

$$-Y = \frac{\sigma_{eq}^2}{2E(1-D)^2}\left[\frac{2}{3}(1+\nu) + 3(1-2\nu)\left(\frac{p}{\sigma_{eq}}^2\right)\right] \tag{2}$$

For the yield function Φ, the following von-Mises type form is adopted

$$\Phi\left(\boldsymbol{\sigma}, \bar{\varepsilon}_{eq}^p, D\right) = \frac{\sigma_{eq}}{1-D} - \sigma_Y\left(\bar{\varepsilon}_{eq}^p\right) \tag{3}$$

2.2. GTN and modified GTN models

The Gurson-Tvergaard-Needleman (GTN) model, which is one of the most well known extensions of Gursons model, assumes both isotropic hardening and damage. The damage variable in this model is represented by an effective porosity f^*. The flow potential is generalized into the form:

$$\Phi(\boldsymbol{\sigma}, \bar{\varepsilon}_{eq}^p, f^*) = J_2(\boldsymbol{S}) - \frac{1}{3}\left\{1 + q_3 f^{*2} - 2q_1 f^* cosh(\frac{q_2 3p}{2\sigma_Y})\right\}\sigma_Y^2 \tag{4}$$

where,the parameters q_1, q_2 and q_3 are the introduced material parameters.

The evolution of the porosity is given by:

$$\dot{f} = \dot{f}^N + \dot{f}^G + \dot{f}^{shear} \tag{5}$$

The nucleation mechanism is driven by the plastic strain and can be represented as:

$$\dot{f}^N = \frac{f_N}{s_N\sqrt{2\pi}} exp\left[-\frac{1}{2}\left(\frac{\bar{\varepsilon}^p - \varepsilon_N}{s_N}\right)^2\right]\dot{\bar{\varepsilon}}^p \tag{6}$$

The most significant contribution to the evolution of spherical voids is the growth mechanism, which can be expressed by:

$$\dot{f}^G = (1-f)\, tr\left(\dot{\varepsilon}^p\right) = (1-f)\, \dot{\varepsilon}^p_\nu \tag{7}$$

Under shear dominated loading conditions, the distortion of voids and inter-void linking play a critical role in the evolution of the material internal degradation. Therefore, in order to improve the GTNs model predictive ability, under both zero and low levels of stress triaxialities, Xue [5] has proposed the introduction of a shear mechanism. The rate of this shear mechanism can be mathematically expressed by:

$$\dot{f}^{shear} = k_\omega f \omega\left(\boldsymbol{\sigma}\right) \frac{\boldsymbol{S} : \dot{\varepsilon}^p}{\sigma_{eq}} \tag{8}$$

3. Numerical Examples

In this section, square box application of ISF was investigated with the use of numerical treatment based on Explicit integration scheme. During the incremental forming, The tool moves along the prescribed path generated by CNC machine center to impose deformation onto a square sheet. The results obtained by performing numerical simulations with the previously described damage models, will be presented and discussed.

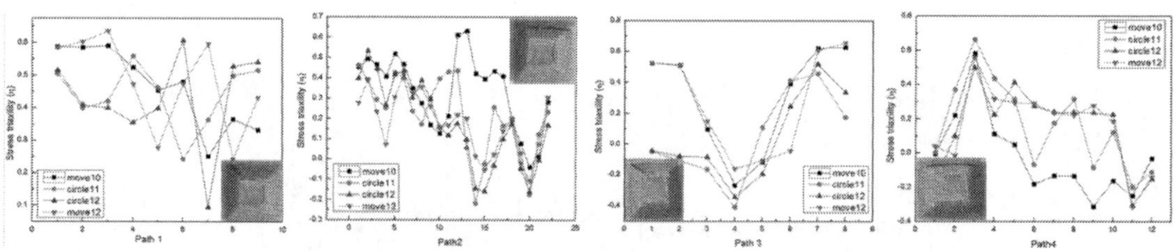

(a) Corner Path 1 (b) Edge Path 2 (c) BackCorner Path 3 (d) BackEdge Path 4

Figure 1: Triaxility distribution at different steps

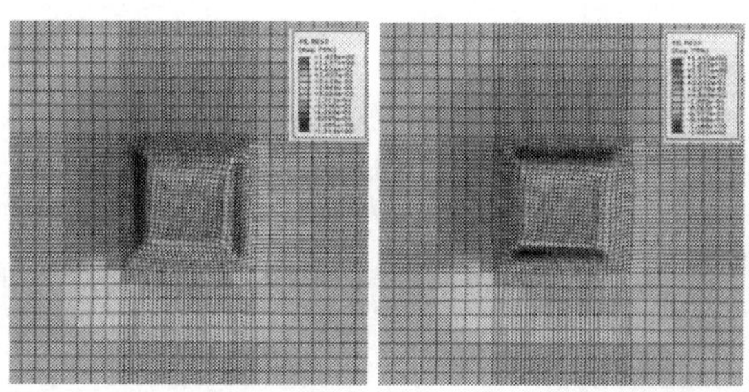

(a) PE13 (b) PE23

Figure 2: The distribution of shear plastic strain through thickness

3.1. The effect of stress triaxility
In this section, four paths: corner path(1), edge path(2), back corner path(3), and back edge path(4) and four typical steps including corner feeding, and circulating deformation are selected to study the stress triaxility distribution during the incremental forming process, which can be seen in Figure 1. From there, it is clear seen that the stress triaxility changes a lot during the incremental forming. It is out of the stress triaxility range covered by the conventional sheet metal forming process. This can also been seen in Figure 2, which demonstrates the distribution of shear plastic strain through the thickness. It means that the material during the incremental forming surfer the shear stress through the thickness.

3.2. The comparison of damage prediction
Figure 3 show the damage prediction obtained from the original Gurson model, Modified GTN model with different shear effects, and Lemaitre continuum damage model. It can be seen that with the increasing damage effect, the void fraction increases. The large damage distributed around the cup corner, the failure will occurs in the near corner. Compared with GTN model, the Lemaitre model overestimates the material failure.

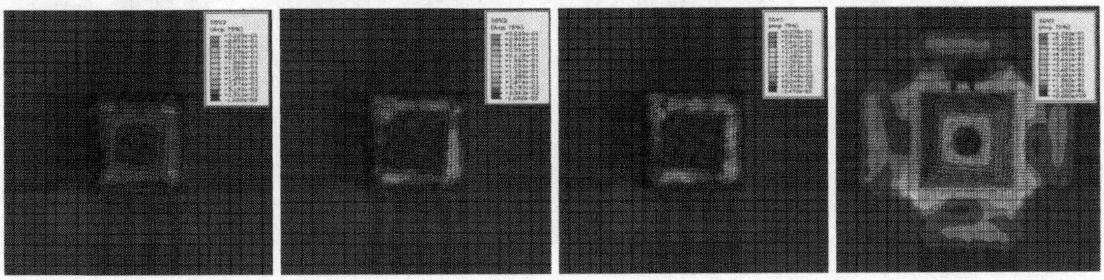

(a) GTN (b) Modified GTN ($k_\omega = 0.2$) (c) Modified GTN ($k_\omega = 0.5$) (d) Lemaitre

Figure 3: Damage prediction with different damage models

4. Conclusions
In this work, ductile damage models have been implemented in the explicit integration scheme by developing VUMAT subroutine. All of these damage models have been used to predict the failure of metallic sheets subjected to incremental forming. The stress triaxility varies at large ranges during incremental forming and The shear effect has a great influence on the failure prediction. Lemaitre continuum damage model overestimated the fracture depth of incremental forming material. This comparison provides an insight possibility to study incremental forming process and evaluate the influence of various processing parameters on the formability.

References
[1] M. Skjoedt, M.B. Silva, N. Bay, P.A.F. Martins. Formability in Multistage Single Point Incremental Forming. 7th ESMC2009, The 7th Euromech Solids Mechanics Conference, 2009.

[2] Martins P.A.F., Bayb N, Skjoedt M., Silva M.B.,Theory of single point incremental forming, CIRP Annals - Manufacturing Technology, 57, 1 p:247-252, 2008

[3] Gurson A.L., Continuum theory of ductile rupture by void nucleation and growth. Part I Yield criteria and flow rules for porous ductile media, Journal of Engineering Materials and Technology, 99, 1, p:2-15, 1977

[4] Lemaitre J., Coupled elasto-plasticity damage constitutive equations. Computer Methods Applied Mechanics and Engineering 51 p:3149, 1985.

[5] Liang Xue, Ductile fracture modeling - theory, experimental investigation and numerical verification, PhD thesis Massachusetts Institute of Technology, 2007.

[6] Kathryn Jackson, An analysis of ductile rupture in notched bars, J. Mech. Phys. Solids, 32, 6, pp. 461490, 1984.

Investigations on fracture curves in strain and stress space for advanced high strength steel forming

S Panich[1,2], K Drotleff[1], M Liewald[1] and V Uthaisangsuk[3]

[1] Institute for Metal Forming Technology, University of Stuttgart, Holzgartenstraße 17, 70174 Stuttgart, Germany

[2] Department of Production Engineering, Faculty of Engineering, King Mongkut's University of Technology North Bangkok, 1518 Pracharat 1 Road, 10800 Bangkok, Thailand

[3] Department of Mechanical Engineering, Faculty of Engineering, King Mongkut's University of Technology Thonburi, 126 Pracha Uthit Road, 10140 Bangkok, Thailand

E-mail: sansot.p@eng.kmutnb.ac.th, klaus.drotleff@ifu.uni-stuttgart.de
mathias.liewald@ifu.uni-stuttgart.de, vitoon.uth@kmutt.ac.th

Abstract. Conventional forming limit curves (FLCs) are inappropriate for describing formability for advanced high strength (AHS) steel sheets, since such steel grades experience fracture without localized necking occurrence. The aim of this work was to develop a fracture curve (FC) for the AHS steel grade DP980. The FC was determined by means of the Nakajima stretch forming test and tensile tests of various sample geometries, by which shear fracture governed. An optical strain measurement system was used to capture strain histories of deformed samples up to failure. From these results, fracture strains were gathered and plotted in a strain space. Subsequently, the strain based curve was transformed to space between stress triaxiality and plastic strain. Hereby, effects of anisotropic yield function, namely, the Hill'48 model on obtained stress fracture loci were investigated. In order to verify applicability of the determined limit curves, a Mini-tunnel part was pressed and simulated. It was found that the stress based FC do predict failure of the DP980 steel sheet more accurately than the strain based FC.

1. Introduction

Necking and fracture have been considered as major failure mechanisms in sheet metal forming. Occurrence of local necking of steel sheets leads to a rapid fracture and forming operation is consequently terminated. To predict the onset of this necking, analytical and numerical tools were developed. However, the introduction of advanced high strength (AHS) steel grades with reduced ductility brought up an issue concerning shear fracture which could not be properly predicted by the conventional forming limit curve (FLC). This formability characteristic of such steel types led to a development of new forming criterion such as the fracture curve (FC). For example, forming limit diagram has been newly presented in stress space or as a function of triaxiality that is less or not sensitive to strain path [1]. In this work, fracture forming limits with regard to the strain and stress triaxiality/plastic strain space were investigated. Initially, Nakajima stretch forming test and tensile tests of pure shear and combined loading (shear and tension) samples were performed for the AHS steel grade DP980, in which limit strains due to shear fracture were additionally obtained. Then, the determined strain based FC was transformed to stress triaxiality and effective plastic strain space under consideration of the anisotropic Hill'48 yield model. Furthermore, in order to verify applicability of the stress based fracture loci, experimental stamping test of a Mini–tunnel sample was conducted and simulated.

2. Determination of strain based FC

2.1 Nakajima stretch–forming test

For experimental determination of strain–based FC for the examined steel grade DP980, the Nakajima stretch–forming test was carried out by using the sample standard ISO 120004–2 [1,2]. The sheet samples had the same length of 200 mm, but different widths varying from 55 up to 200 mm. The steel sheet had the initial thickness of 0.97 mm. During the test, local strain histories of the deformed samples were gathered by mean of an optical strain measurement system. Basically, critical plastic strains for conventional FLC were evaluated from calculated strain distributions at the stage of localized necking, which could be characterized by considering development of local strain rates. In this work, principal in–plane plastic strains at the maximum rate of thickness reduction were identified and subsequently used as limiting strain values to fracture for the FC (strain based).

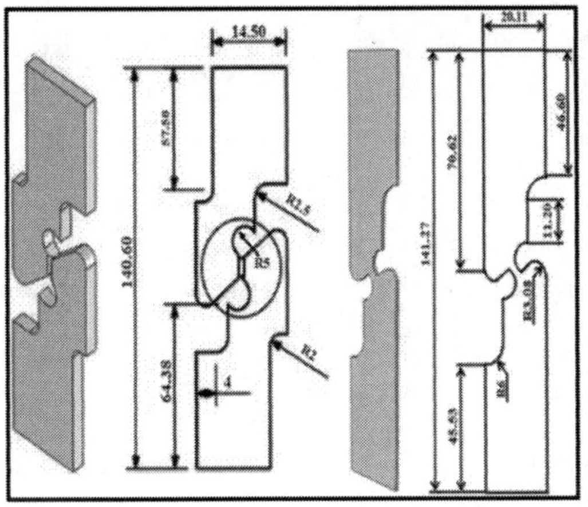

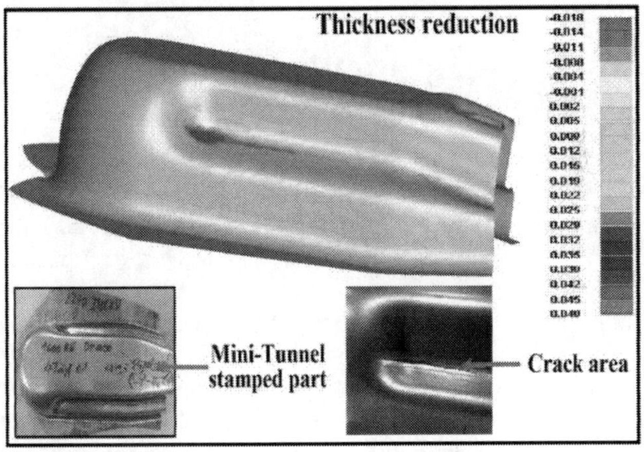

Figure 1. Pure shear/combined load sample [3] **Figure 2.** Stamped and simulated Mini–tunnel part.

2.2 Shear fracture limit test

Tensile tests of pure shear and combined loading samples were performed for obtaining fracture threshold strains due to shear failure mode. Figure 1 illustrates geometries of the used pure shear and combined loading sample [3]. In case of the pure shear sample, nearly pure shear stress governed the middle area of the sheet sample. For the combined loading sample, the pure shear sample was slightly modified to generate a stress state disclosing a combination of shear and tensile loading, as reported in Bao and Wierzbicki [4]. The optical strain measurement system was also applied for experimentally evaluating local strain distributions developing in the critical area of both specimens. Similar to the Nakazima test, the critical maximum and minimum in–plane strains were obtained at a stage before crack appearance and used as shear fracture limit strains.

3. Determination of stress based FC

3.1 Uniaxial tensile test

Uniaxial tensile tests of the investigated steel sheet were carried out by using the ASTM E8 standard specimen. The sheet samples were tested under three loading directions, 0°, 45° and 90° to the rolling direction (RD). During the tension tests, elongations within the gauge length of the test specimens regarding both sample length and width were determined. The *r*–values were then calculated by the ratio between the true width and true thickness strains measured up to 14% of the total elongation.

3.2 Transformation into stress space

Obtained forming limit strains were transformed into stress triaxiality and plastic strain space under consideration of the anisotropic Hill'48 yield function [5]. The strain ratio (ρ) is defined by $\rho = (\varepsilon_2/\varepsilon_1)$ and the average normal anisotropy is expressed as $\bar{r} = (r_0 + 2r_{45} + r_{90})/4$. The maximum and minimum principle strains could be then transformed to stress triaxiality $(\eta_{Hill'48})$ and effective plastic strain $(\bar{\varepsilon}_{Hill'48})$ space combination with Hill'48 yield function by Equation 1 and 2 [5].

$$\eta_{Hill'48} = \frac{\sqrt{1+2\bar{r}}}{3} \times \frac{1+\rho}{\sqrt{1+\left(\frac{2\bar{r}}{1+2\bar{r}}\rho\right)+\rho^2}} \tag{1}$$

$$\bar{\varepsilon}_{Hill'48} = \frac{1+\bar{r}}{\sqrt{(1+2\bar{r})}}\sqrt{\varepsilon_1^2+\varepsilon_2^2+\frac{2\bar{r}}{1+\bar{r}}\varepsilon_1\varepsilon_2} \tag{2}$$

3.3 Mini–tunnel forming test

In order to verify the applicability of the strain and stress based FC, experimental stamping test of a Mini–tunnel part was done as demonstrated in Figure 2. During the test, draw–in of the outer sample periphery into the die was controlled by means of high constant blank holder force of 1400 kN. Oil lubricant was used to minimize friction between blank and dies. The samples were pressed until crack occurred with the drawing depth of around 73 mm, which was used to validate the FE model. Subsequently, FE simulation of the Mini–tunnel part was conducted. Boundary conditions similar to the experiment were defined also for FE model. Material hardening and anisotropic behavior were described by the swift law and Hill'48 yield criterion, respectively. A friction coefficient of 0.08 was given. Punch, die and blank holder were defined as rigid analytical surfaces and steel sheet as elastic–plastic material. At the same drawing depth of the experiment, calculated stress and strain values were determined from critical areas of the formed samples.

4. Results and discussion

4.1 Fracture Forming limit in strain space

From the Nakajima stretch forming test and tensile test of shear and combined loading sample, the FC (stress based) of the examined steel grade DP980 was determined, as shown in Figure 3. The strain based fracture locus consisted of three strain branches. The first branch corresponded to the stress states between biaxial tension ($\varphi_1 = \varphi_2$) and plane strain ($\varphi_2 = 0$). The second region was the range from plane strain (φ_2=0) to uniaxial tension (φ_1 = -2φ_2). The last branch represented the stress state from uniaxial tension (φ_1 = -2φ_2) to pure shear (φ_1 = -φ_2). This third strain area showed the shear facture limit of the steel. The experimentally determined FC was then verified by the Mini–tunnel stamping. Initially, punch–displacement curve and drawing depth from the experiment and FE simulation were compared, in which the failure moment of the experiment could be identified. Up to this point of time, strain paths of major and minor principal strains were gathered from damage initiating areas on the deformed samples in the simulations. This principal strain path was calculated with regard to the Hill'48 yield criterion and was then plotted on the strain based fracture locus, as depicted in Figure 3. Obviously, the strain path up to failure in the experiment ended somewhat below the FC. It meant that the FC predicted a too high deformation state for the Mini–tunnel sample.

4.2 Estimation of fracture curve in the stress space

The strain based FC transformed to the stress triaxiality and effective plastic strain space by numerical calculations according to Equation 1 and 2 was demonstrated in Figure 4. The stress based fracture locus also showed three different branches. These three regions corresponded to the same states of stress as shown on the strain based FC. Note that the biaxial tension, plane strain, uniaxial tension and pure shear state were approximately represented by the triaxiality values η of 0.667, 0.557, 0.33 and 0, respectively. Subsequently, path of stress triaxiality and effective plastic strain obtained from the same critical area of the formed Mini–tunnel sample were evaluated and plotted up to the experimental failure on the stress based FC. It was found that the critical stress triaxiality and plastic strain values at failure from the calculated forming history of the Mini–tunnel sample were very close to the stress based FC. It meant that the stress based FC could more precisely predict material failure of the AHS steel grade DP980 than the strain based FC.

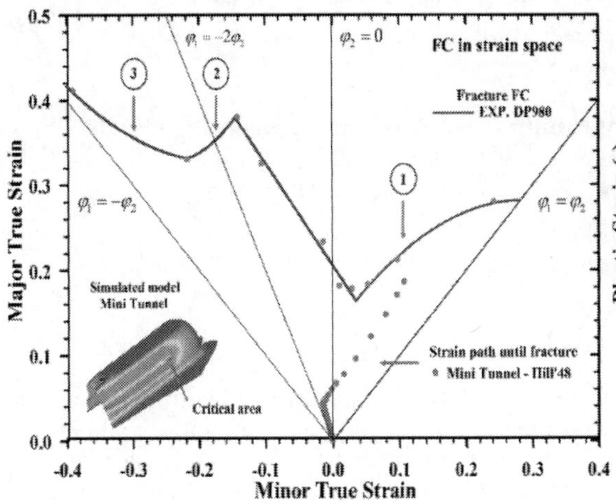

Figure 3. Strain path of the simulated Mini–tunnel sample on the strain based FC of the investigated steel

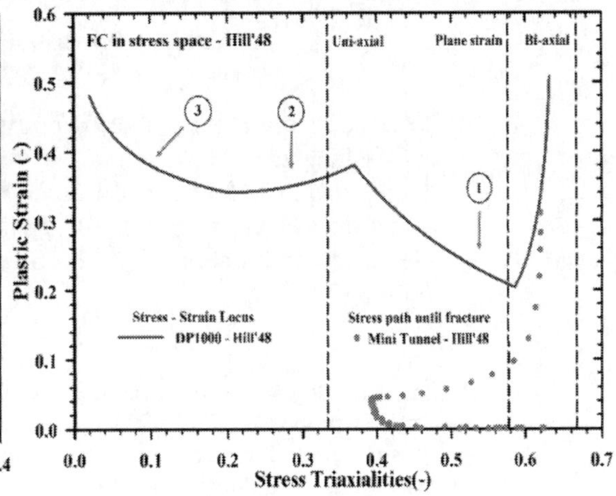

Figure 4. Stress–strain path of the simulated Mini–tunnel sample on the stress based FC of the investigated steel

5. Summary

In this work, the strain and stress based fracture curves were determined for the AHS steel grade DP980. The Nakajima test and tensile tests of pure shear and combined loading samples were performed for obtaining the strain based limit curve. Then, the strain based FC was transformed to stress triaxiality and effective strain space under consideration of the Hill'48 yield function. To verify both determined FCs stamping test of the Mini–tunnel sample was carried out and simulated. The calculated strain and stress paths were compared with the FCs. It was observed that the stress based FC do more realistically describe material formability of the examined steel than the strain based FC. In future work, this stress triaxiality space will examined on further geometries of real parts and will expanded also to Martensite steel grades up to 1150 MPa UTS using different yield functions such as Yld89 and Yld2000.

6. References

[1] Gorji M, Berisha B, Hora P and Barlat F 2015 *Int. J. Mater. Form.* 6: 1-12.
[2] Nakazima K, Kikuuma T and Hasuka K 1971 *Yawata Tech Report* 284: 678-680.
[3] Lian J, Sharaf M, Archie, F and Muenstermann S 2013 *Int. J. Damage. Mech.* 22 (2): 188-218.
[4] Bao Y and Wierzbicki T 2004 *J. Eng. Mater. Tech.* 26 (3): 314-324.
[5] Isik K, Silva M B, Tekkaya A E and Martin P A F 2014 *J. Mater. Proc. Tech.* 214: 1557-1565.
[6] Panich S, Barlat F, Uthaisangsuk, V and Jirathearanat S 2013 *Mater. Des.* 51: 756-766.
[7] Panich S, Uthaisangsuk V and Jirathearanat S 2014 *Adv. Mater. Res.* 849: 200-206.
[8] Drotleff, K and Liewald 2015 *Proc. Int. Conf. on Forming Technology Forum*: 69-74.

Plane stress yield function described by 3rd-degree spline curve and its application

Toshiro Aamaishi[1], Hideo Tsutamori[2], Eiji Iizuka[3], Kentaro Sato[3], Yuki Ogihara[3] and Yohei Matsui[2]

[1]JSOL Corporation Tosabori Daibiru Building 2-2-4 Tosabori, Nishi-ku, Osaka, 550-001 Japan
[2]Department of Mechanical Engineering, Daido University, 10-3, Takiharu-cho, Minami-ku Nagoya, 457-8530, Japan
[3]Steel Research Laboratory, JFE Steel Corporation, 1 Kokan-cho, Fukuyama, 260-0835, Japan

Abstract. In this study, a plane stress yield function which is described by 3rd-degree spline curve is proposed. This yield function can predict a material anisotropy with flexibility and consider evolution of anisotropy in terms of both r values and stresses. As an application, hole expanding simulation results are shown to discuss accuracy of the proposed yield function.

1. Introduction

Precise prediction of a material anisotropy is an important factor of accurate sheet metal forming simulation. To describe a material anisotropy, many yield functions have been proposed. For example, Hill'48 [1] quadratic model is widely used because of its simplicity. Barlat *et al.* [2] introduced yield function Yld2000-2d, which can consider biaxial stress. Vegter *et al.* [3] proposed a plane stress yield function based on second order Bézier curve. Even though this yield function also requires some experiment data of biaxial stresses to determine anisotropic material parameters, flexible expression of the yield surface enables accurate simulation. These yield functions assume the initial anisotropy is kept during plastic deformation. However, it is known that some kinds of sheet metal show evolution of anisotropy. For example, Hu [4] provided an experiment data of an aluminum sheet, which show changes of normalized flow stress and r value with increasing plastic strain.

In this paper, a 3rd-degree spline based plane stress yield function is proposed. The idea of connecting yield locus by a kind of spline curves is identical to Vegter's work. On this proposed plane stress yield function, 3rd-order Bézier curves are employed for flexibility of yield locus expression. Moreover, by using plastic strain dependent material properties, the proposed yield function can consider anisotropic hardening.

2. Yield function description

In the two-dimensional principal stress space, a stress point is represented by the vector $\{\sigma_1, \sigma_2\}^T$. Let θ $(0 \leq \theta \leq \pi/2)$ be the angle between the 1st principle axis and the rolling direction. In this study, only orthotropic materials are considered. Divide the angle from 0 to $\pi/2$ into n sections, and set

$0 \leq i \leq n$. In the area $\sigma_1 \geq 0$ and $\sigma_2 \geq 0$, a plane stress normalized yield locus is described by function $\boldsymbol{R}_i$. If $\sigma_1 \geq \sigma_2 \geq 0$ then yield locus is described by function $\boldsymbol{R}_i$. If $\sigma_1 \geq \sigma_2 \geq 0$ then

$$\boldsymbol{R}_i(t) = (1-t)^3\boldsymbol{A}_i + 3(1-t)^2t\boldsymbol{F}_i + 3(1-t)t^2\boldsymbol{G}_i + t^3\boldsymbol{C}_i \quad (1)$$

and if $\sigma_2 > \sigma_1 \geq 0$ then

$$\boldsymbol{R}_i(t) = (1-t)^3\boldsymbol{C}_i + 3(1-t)^2t\boldsymbol{H}_i + 3(1-t)t^2\boldsymbol{I}_i + t^3\boldsymbol{E}_i \quad (2)$$

where $0 \leq t \leq 1$. Note that $\boldsymbol{A}_i$, $\boldsymbol{C}_i$, $\boldsymbol{E}_i$, $\boldsymbol{F}_i$, $\boldsymbol{G}_i$, $\boldsymbol{H}_i$ and $\boldsymbol{I}_i$ are vectors of control points of 3rd-order Bézier curves as shown in Fig. 1. $\boldsymbol{B}_i$ and $\boldsymbol{D}_i$ are vectors which provide the control points of Bézier curves. Point $\boldsymbol{A}_i$ implies uniaxial tension stress, and point $\boldsymbol{E}_i$ corresponds to uniaxial tension stress that is angled at $\pi/2$ from the direction of $\boldsymbol{A}_i$. Point $\boldsymbol{C}_i$ corresponds to equi-biaxial stress that is σ_1=σ_2. The tangent at $\boldsymbol{A}_i$ corresponds to r value that is determined by uniaxial tension test, and the tangent at $\boldsymbol{C}_i$ corresponds to r value of equi-biaxial stress state that is $r_b = d\varepsilon_2^p/d\varepsilon_1^p$. Point $\boldsymbol{B}_i$ is defined as a point of intersection of two tangents. Point $\boldsymbol{D}_i$ is determined similarly. If $\sigma_1 \geq \sigma_2$, point $\boldsymbol{F}_i$ is the point of interior division ratio $(1-p{:}p)$ of line $\boldsymbol{A}_i\ \boldsymbol{B}_i$, where p is a material parameter, and point $\boldsymbol{G}_i$ is the point of interior division ratio $(1-p{:}p)$ of line $\boldsymbol{C}_i\ \boldsymbol{B}_i$. If $\sigma_1 < \sigma_2$, points $\boldsymbol{H}_i$, $\boldsymbol{I}_i$ are determined similarly.

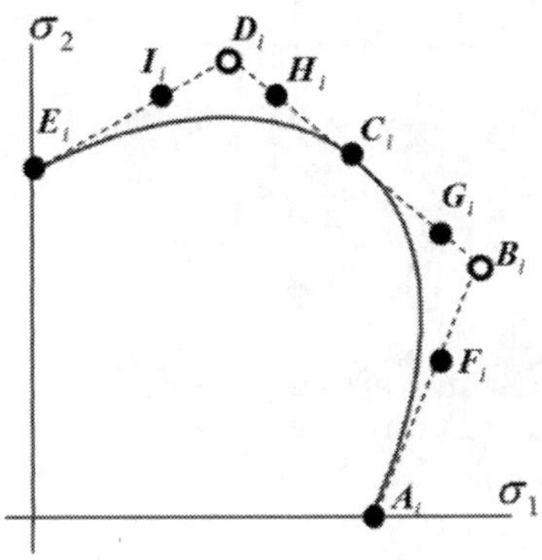

Figure 1. Yield locus and its control points

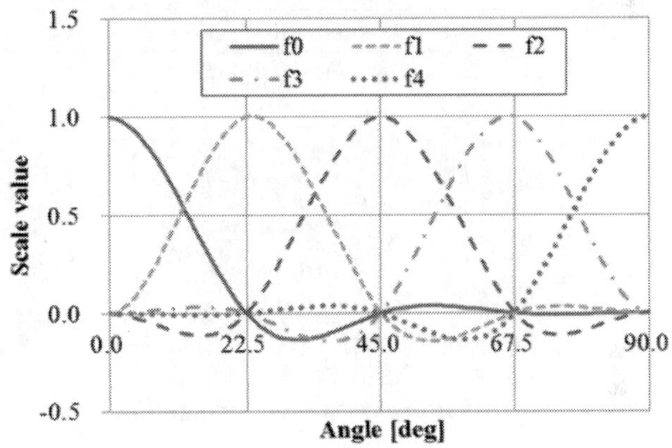

Figure 2. Spline functions to interpolate (Case n=4)

For planar anisotropic sheet metal, the yield locus and the reference strain ratio depend on the angle between the 1st principle axis and the rolling direction. The angular dependency of the reference points and strain ratio can be described by using cubic spline basis function f_i . If n=4, spline basis functions f_i are determined as in Fig. 2. By using spline function f_i , the function which describes yield locus is written as following. If $\sigma_1 \geq \sigma_2 \geq 0$ then

$$\boldsymbol{R}_i(\theta,t) = \textstyle\sum_{i=0}^{n} f_i(\theta)\{(1-t)^3\boldsymbol{A}_i + 3(1-t)^2t\boldsymbol{F}_i + 3(1-t)t^2\boldsymbol{G}_i + t^3\boldsymbol{C}_i\} \quad (3)$$

and if $\sigma_2 > \sigma_1 \geq 0$ then

$$\boldsymbol{R}_i(\theta,t) = \textstyle\sum_{i=0}^{n} f_i(\theta)\{(1-t)^3\boldsymbol{C}_i + 3(1-t)^2t\boldsymbol{H}_i + 3(1-t)t^2\boldsymbol{I}_i + t^3\boldsymbol{E}_i\} \quad (4)$$

The yield locus of uniaxial compression is assumed that it is symmetric with respect to the origin. To describe the evolution of anisotropy in terms of both stresses and r values, plastic strain dependency of the normalized yield locus needs to be considered. In this model, anisotropic hardening is described by considering plastic strain dependent control points which imply yield locus, tangents of the control points and parameter p.

3. Application

To demonstrate the applicability of the proposed yield function, hole expanding tests with a mild steel sheet and a 6000 series aluminum alloy sheet were investigated. The thickness of the mild steel sheet was 0.8mm and that of the aluminum alloy sheet was 1.0 mm. The specifications of hole expanding tests are summarized as in Table 1 and shape of blank sheets are shown in Fig.3. The mild steel specimen was 240mm × 240mm, and had the hole of 50mm diameter. Moreover, the diameter of the aluminum alloy blank was 100mm, and the initial hole diameter was 25mm. The sampling points of thickness were 1mm from the hole edge.

Table 1 Hole expanding tool specifications

	Mild Steel	Aluminum alloy
Punch diameter (mm)	100.0	40.0
Punch porofile radius (mm)	15.0	6.0
Die opening diameter (mm)	130.0	43.0
Die profile radius (mm)	15.0	6.0
Blank holding force (kN)	100.0	60.0
Punch stroke (mm)	22.1	10.0

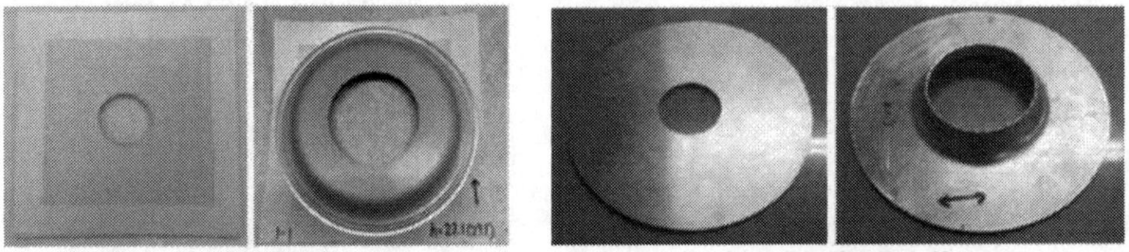

(a) Mild steel (b) Aluminum alloy
Figure.3 Pre-hole expanding and post-hole expanding blank sheet

To obtain the material characteristics, uniaxial tension tests were performed in five directions. The results are shown in Table 2 and Table 3. In hole expanding tests, the stress states on the hole edge is known to be similar to uniaxial stress states. Thus, in this study, the yield locus described by a yield function is determined by only the results of uniaxial tension tests. To determine the yield locus of proposed yield function, the angle from 0 to $\pi/2$ is divided into four sections. Thus five directions of uniaxial stresses and r value were considered as material properties. The position of equi-biaxial stress points are estimated by Hill'48 yield function. Material parameter p was determined to be the yield surface of proposed model to fit to Hill'48's yield surface.

Experiment and simulation results are shown in Fig.4. For the implementation of the proposed yield function, the user material subroutine based on the LS-DYNA's explicit time integration was employed. In the isotropic hardening case, the shape of yield locus was determined by material properties with plastic strain=20%. In the case of anisotropic hardening, evolution of the anisotropy of stresses and r values were considered by plastic strain dependent sequential curve data.

As illustrated in Fig. 4 (a), the peak of thickness decreasing direction of the mild steel is around 30 degree. The proposed model describes more accurate prediction than Hill'48. In the comparison results of the aluminum alloy in Fig.3 (b), prediction of proposed model was closer to experiment data at 90 degree. Hence the superiority of proposed model, which is based on considering uniaxial stresses and r values in five directions, is confirmed. In terms of effect of considering anisotropic hardening, there were no major differences between the isotropic hardening and the anisotropic hardening in the results. Therefore, the effect of anisotropic hardening is minor on hole expanding tests which are mainly influenced by uniaxial stress state.

Table 2 Material properties of the mild steel sheet

Angle (deg)	YS (MPa)	TS (MPa)	r value (ε_p=1%)	r value (ε_p=20%)
0.0	156.8	298.2	1.63	2.20
22.5	161.0	295.1	1.49	2.00
45.0	158.4	295.2	1.51	2.05
67.5	160.4	296.0	1.66	2.30
90.0	162.0	299.3	1.83	2.60

Table 3 Material properties of the aluminum alloy sheet

Angle (deg)	YS (MPa)	TS (MPa)	r value (ε_p=1%)	r value (ε_p=20%)
0.0	98.1	247.3	0.72	0.78
22.5	108.8	248.2	0.42	0.51
45.0	82.0	246.4	0.21	0.30
67.5	83.1	238.0	0.37	0.48
90.0	92.6	246.2	0.42	0.65

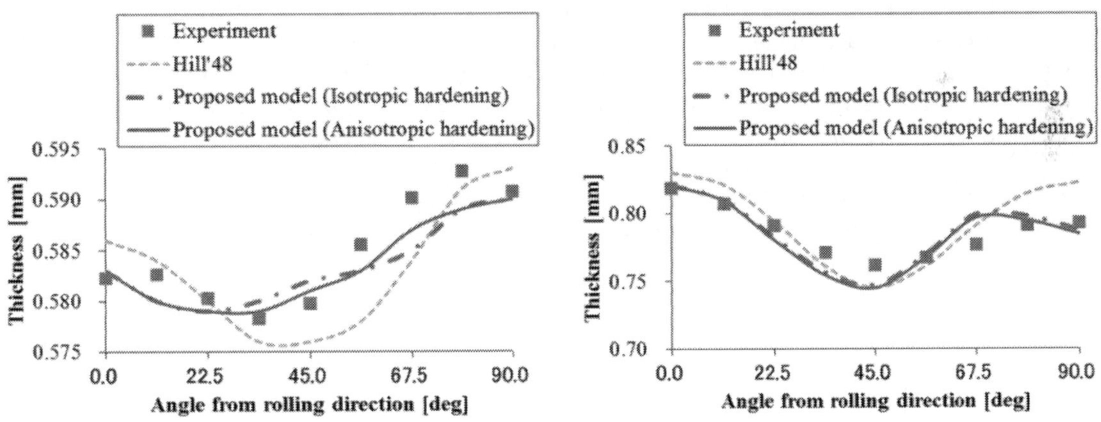

(a) Mild steel (b) Aluminum alloy

Figure 4. Comparison of thickness along the edge of the hole after hole expansion

References

[1] Hill R 1948 *Proceedings of the Royal Society of London* **A193** 281-97
[2] Barlat F, Berm J C, Yoon J W, Chung K, Dick R E, Lege D J, Porurboghrat F, Choi S H and Chu E 2003 *International Journal of Plasticity* **19** 1297-319.
[3] Vegter H and van den Boogaard A H 2006 *International Journal of Plasticity* **22** 557-80.
[4] Hu W 2007 *International Journal of Plasticity* **23** 620-39.
[5] Tsutamori H, Iizuka E, Amaishi T, Sato K, Ogihara Y and Miyamoto J 2016 *Journal of the Japan Society for Technology of Plasticity* **57** no.662 245-51.

FEM analysis of springback control with lump-punch penetration after V-bending

Takayuki Aso [1-a] **and Takashi Iizuka** [1-b]
[1]Department of Mechanical and System Engineering, Kyoto Institute of Technology, Matsugasaki Goshokaido-cho, Sakyo-ku, Japan

[a] kyotokosen94@gmail.com, [b] tiizuka@kit.ac.jp

Abstract. In actual manufacturing, some empirical methods such as the bottoming technique are generally used in order to adjust the bend angles of products. However, the problem with this is that it relies on the technique of the engineer. In this study, quantitative springback control by lump-punch penetration after V-bending is investigated with FEM analysis and experimentation. The lump at the punch tip is pushed into a bent section at the final stage of V-bending and stretches the inside surface at the bent section. The method of springback control is suggested based on the deformation state. Then, the suitability of springback control using this mechanism is investigated. It is confirmed that the springback amount is reduced by lump-punch penetration. Accordingly, it is recommended to control springback by sheet forging with a lump punch.

1. Introduction

Bend forming has been used as much as shearing in the production of transportation equipment. However, springback occurs when bend forming. Various methods are used to compensate for springback. One current method is manufacturing metal molds in which the springback amount is accounted for based on finite element method (FEM) analysis estimates [1]. Moreover, bottoming and coining are often as methods for controlling springback [2]. However, these methods strongly depend on the experience of the engineers. Therefore, a method that does not rely on experience is desired.
In the preceding study, a method in which springback is quantitatively controlled was investigated. From the experimental results, the possibility that the bend angle can be controlled by the length ratio, depending on the residual strain ratio, was suggested [3]. In this study, the control of the bend angle and the mechanism are investigated by FEM analysis.

2. Control of bend angle with lump-punch penetration after V-bending

2.1. Deformation mechanism with lump-punch penetration after V-bending

In the case of normal bend forming, when the deformation mechanism is focused, the inside surface of the bend section shrinks and the outside stretches during loading [4]. This difference in deformation is considered to be the cause of springback.
The deformation mechanism of bend forming with lump-punch penetration is shown in Figure 1. It is considered that the inside of the bend section is locally stretched, so the springback amount is reduced. In this study, because sheet forging is applied at the bend section by penetration with a lump punch after V-bending, the deformation of the inside is close to that of the outside. The method of obtaining the desired bend angle using this technique is investigated.

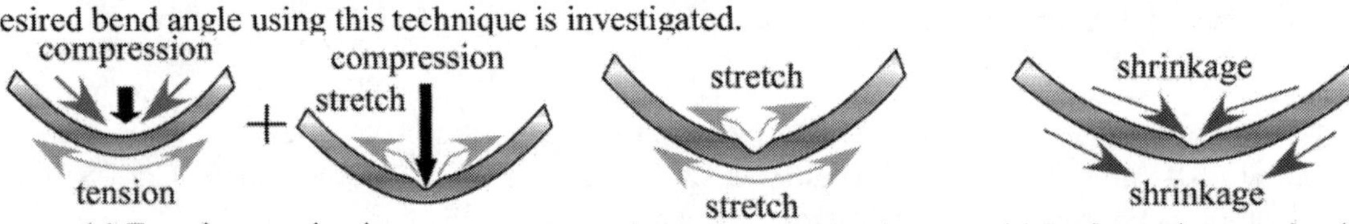

(a) Forming mechanism (b) Deformation mechanism during loading (c) Deformation mechanism during unloading

Figure 1. Deformation mechanisms of lump-punch penetration during loading and unloading.

2.2. Mold shape

Figure 2 shows the mold shape used to apply sheet forging. The punch with a lump at the tip is shown in Figure 2(a). Sheet forging is applied by pushing this lump into the sheet after V-bending. Figure 2(b) shows the die. The radius of curvature is 10 mm, and the opening angle is 90°. The specimen dimensions are shown in Figure 2(c).

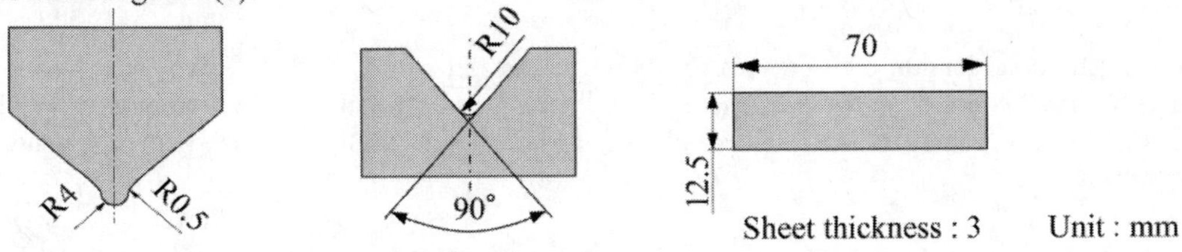

(a) Lump punch (b) Die (c) Specimen

Figure 2. Mold model.

2.3. Lump-punch penetration procedure

Figure 3 shows the analysis procedure. The punch, die, and specimen are set as shown in Figure 3(a). Then, the lump punch begins to push on the specimen, which conforms to the die. The pushing depth is from 0 mm to 1.8 mm at intervals of 0.2 mm, and the forming speed is 5 mm/min. The analysis procedure imitates the experimental procedure.

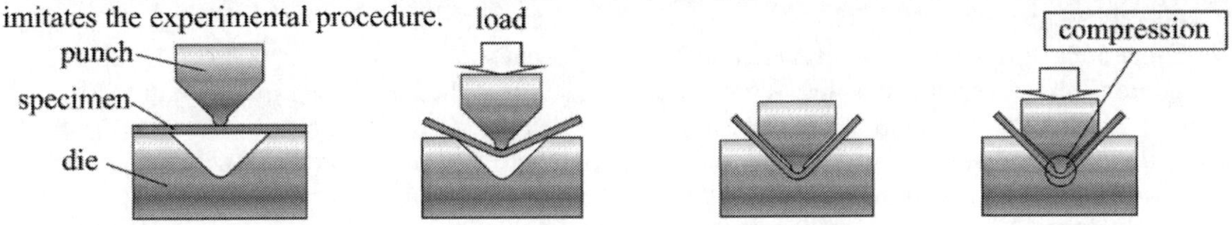

(a) setting a specimen (stroke 0mm) (b) taking a bending load (c) fitting a die (pushing volume 0mm) (d) pushing

Figure 3. Analysis procedure.

3. Analysis

3.1. Analysis conditions

In this study, Simufact Forming 13.3 was used as the analysis software. Table 1 shows the analysis conditions. In this analysis, isotropic von Mises material behavior was assumed. The material is A5052-H, and the mechanical properties were calculated using a tensile test. Moreover, the coefficient of friction was assumed to be 0.12. A symmetry plane was established at the center in the breadth direction, and the analysis was conducted using a half-symmetric model.

A sheet mesh was adopted because the number of elements in the thickness direction was kept constant during penetration shown in Figure 4. In this analysis, five mesh layers in the sheet thickness direction were used. As for the longitudinal and breadth directions, cube-shaped elements were used.

Table 1. Analysis conditions

Flow curve	$\sigma = F\varepsilon^n$
Young's modulus / GPa	71
Poason's ratio	0.33
F -value / MPa	381
n -value	0.13
Material	A5052-H
Friction	0.12

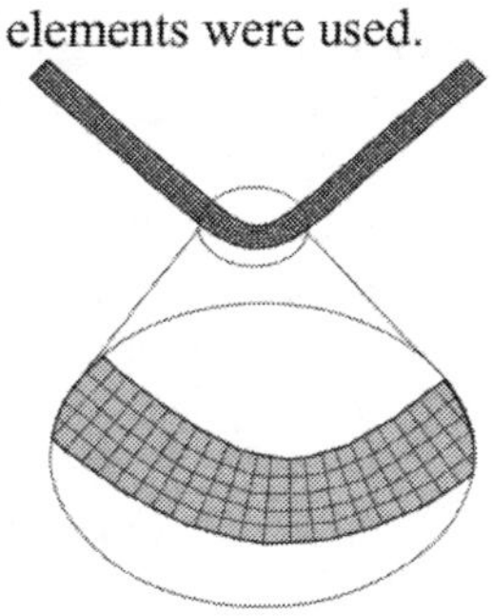

Figure 4. Mesh shape.

3.2. Shape valuation

The bend angle was measured in the CAD software after completing the analysis, as shown in Figure

5(a). The springback amount ($\Delta\theta$) was calculated from the bend angle after unloading (θ_{after}) and during loading (θ_{before}) using equation (1).

$$\Delta\theta = \theta_{after} - \theta_{before} \quad (1)$$

To measure the bend length, the nodes of the specimen model shown in Figure 5(b) were used. After forming, the nodal coordinates were extracted, and the lengths between the nodes were calculated. The bend length is the sum of these lengths. Moreover, the bend length ratio is defined as the inner length/outer length.

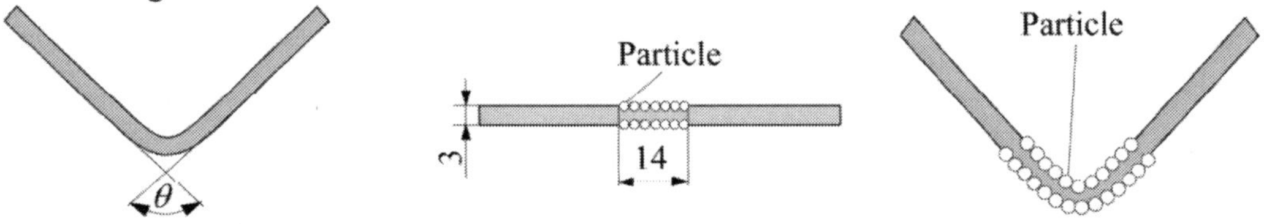

(a) Bend angle (b) Bend length

Figure 5. Shape valuation.

4. Analysis results

4.1. Effect of penetration with lump punch on bend angle and springback

The relationship between the pushing depth and the deviation of the bend angle is shown in Figure 6. It is confirmed that the deviation of the bend angle is reduced as the pushing depth increases. From this figure, there is a possibility that the bend angle can be controlled by penetration with a lump punch. There are qualitative deviations of about 2 degrees between the analysis results and the experimental results. However, the analysis tendency is qualitatively the same as the experimental result. The relationship between the pushing depth and the springback amount is shown in Figure 7. The tendency is that the springback amount decreases as the pushing depth increases. When the pushing depth is from 0.6 mm to 1.2 mm, the springback amount is 0 degrees. Then, springgo is transferred after the pushing depth reaches 1.6 mm because the specimen contacts the sidewall of the punch. At this point, although there are quantitative deviations between the analysis and experimental results, the tendencies are qualitatively the same. The inside of the bend section is stretched by sheet forging, and springback can be reduced. From these figures, springback is dominant over the bend angle after loading. Therefore, it is considered that springback needs to be controlled in order to control the bend angle.

4.2. Control of bend angle by bend length ratio

In this study, the possibility that the bend angle can be controlled by the bend length ratio is investigated. The relationship between the pushing depth and the bend length ratio (inner length/outer length) is shown in Figure 8. As for the analysis results, the bend length ratio decreases as the pushing depth increases before the pushing depth reaches 0.4 mm. Then, the tendency is that the bend length ratio increases as the pushing depth increases after the pushing depth reaches 0.6 mm. In other words, the inside of the bend section is stretched more than the outside. When the specimen reaches the die, a space exists between the specimen and the die because the curvature is different between the lump and the die. Before the pushing depth reaches 0.4 mm, the material flows in order to fill these spaces. Therefore, it is considered that the bend length ratio decreases. There are deviations between the analysis and experimental results. As for this deviation, it is considered that the error is in the experimental results. The relationship between the bend length ratio and the bend angle is shown in Figure 9. The analysis results show a different tendency from that originally hypothesized. As the bend length ratio decreases, the deviation of the bend angle also decreases when the initial penetration occurs. After further penetration, as the bend length ratio increases, the bend angle is reduced. Moreover, when the bend length is approximately 0.75 to 0.77, the deviation of the bend angle is 0 degrees. In order to control the bend angle, it is considered that not only the length ratio but also the residual strain needs to be investigated. Because the curvature radius of the tip of the punch is different from that of the die, there is space between specimen and die. The bending is caused at this space. Then, the material at this space is flowed into inside of bend section. It is considered that the bend length ratio is reduceced in the primary penetration because the material at the space between specimen and die is flowed into the inside of the bend section.

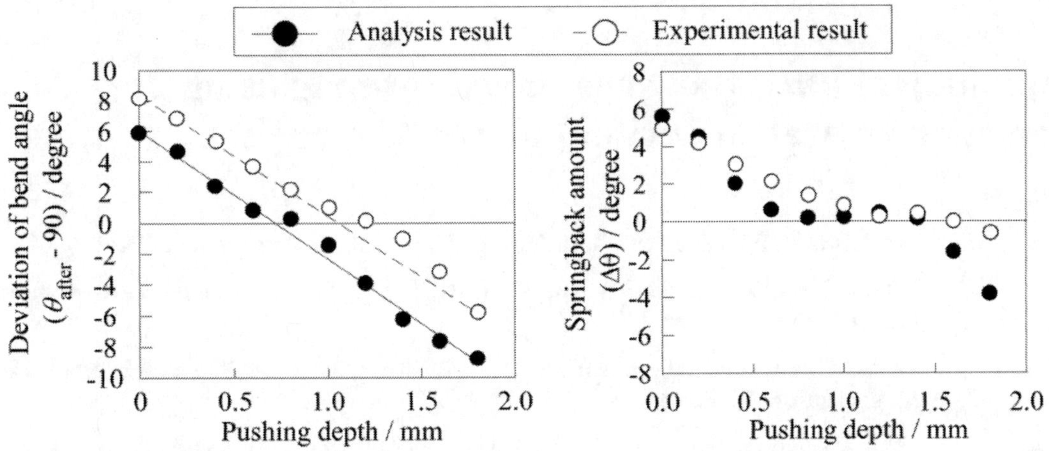

Figure 6. Relationship between pushing depth and deviation of bend angle after unloading.

Figure 7. Relationship between pushing depth and springback amount.

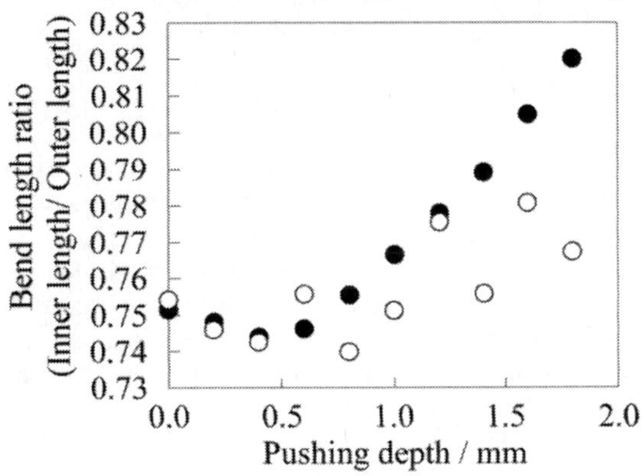

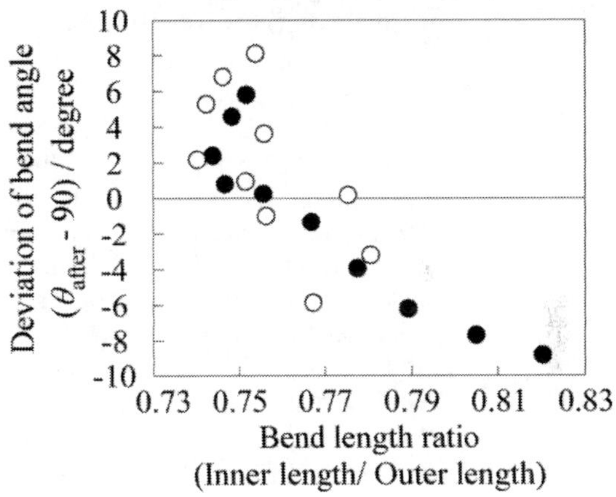

Figure 8. Relationship between pushing depth and bend length ratio.

Figure 9. Relationship between bend length ratio and deviation of bend angle.

5. Conclusion

In this study, FEM analysis of springback control with lump-punch penetration after V-bending was conducted. From the analysis results, it was determined that the bend angle and springback are reduced with lump-punch penetration, and the tendency of the analysis results is mostly the same as that of the experimental results. Additionally, from the analysis and experimental results, when the bend length ratio is almost 0.76, the deviation of the bend angle is 0°. Thus, it is considered that the bend angle can be controlled by the bend length ratio.

References

[1] Livatyali H and Altan T 2001 *J. Mater. Process. Technol* **117** 262
[2] Forcellese A, Fratini L, Gabrielli F and Micari F 1998 *J. Mater. Process. Technol* **80-81** 108
[3] Aso T, Iizuka T and Ota T 2015 *Key. Engineering. Material* **651-653** 1066
[4] Schroeder W and Calif B 1946 *T. A.S.M.E* 287

Simulation of ultra-thin sheet metal forming using phenomenological and crystal plasticity models

F Adzima[a,b], PY Manach[a], T Balan[b], L Tabourot[c], S Toutain[d], JL Diot[e]

[a] Univ. Bretagne Sud, CNRS FRE 3744, IRDL, Rue de Saint Maudé, F-56100 Lorient, France

[b] Arts et Métiers ParisTech, LEM3, UMR CNRS 7239, 4 rue Augustin Fresnel, 57078 Metz Cedex 3, France

[c] Université de Savoie, SYMME, EA 4144, 74944 Annecy-le-Vieux, France

[d] Delta Composants, 1 Rue Pierre Gilles de Genne, 72403 La Ferté-Bernard, France

[e] NovaPack SAS, 14 Rue des Glairaux, 38120 Saint Egrève, France

tudor.balan@ensam.eu

Abstract. Micro-forming of ultra-thin sheet metals raises numerous challenges. In this investigation, the predictions of state-of-the-art crystal plasticity (CP) and phenomenological models are compared in the framework of industrial bending-dominated forming processes. Sheet copper alloys 0.1mm-thick are considered, with more than 20 grains through the thickness. Consequently, both model approaches are valid on theoretical ground. The phenomenological models' performance was conditioned by the experimental database used for parameter identification. The CP approach was more robust with respect to parameter identification, while allowing for a less flexible description of kinematic hardening, at the cost of finer mesh and specific grain-meshing strategies. The conditions for accurate springback predictions with CP-based models are investigated, in an attempt to bring these models at the robustness level required for industrial application.

1. Introduction

The ongoing trend on device miniaturization has increased the demand for microparts. Micro-manufacturing technologies are being developed and optimized continuously to meet the needs of industry. Particularly, micro-forming processes can achieve high productivity and low costs while delivering good parts. However, the knowledge and the simulation tools that are successfully operated in macro-scaled forming cannot be applied to accelerate the design phase of micro-forming processes. Indeed, the employed ultra-thin sheet metals exhibit peculiarities in their deformation behavior. In the present study, crystal plasticity and phenomenological material models are implemented for the numerical simulation of industrial micro-forming processes. Their predictive capabilities are compared and the influencing parameters for accurate predictions are investigated.

2. Materials and demonstrator parts

The current study focuses on two demonstrator parts manufactured from metal sheets of CuBe2 and CuFe2P copper alloys. The gull-wing microelectronic package leads (see Fig. 1 (a)) are made of 127 µm thick CuFe2P sheets while 100 µm thick CuBe2 sheets are processed to produce the connector displayed on Fig. 1 (b). Thus, due to the part geometries, small (chip lead) and large (connector) R/T (radius/thickness) ratios situations are under investigation.

(a) (b)

Fig. 1: Demonstrator parts: (a) CuFe2P made LQFP package (small R/T ratio) and (b) CuBe2 made connector (large R/T ratio)

3. Material models and parameters

Phenomenological modeling as well as crystal plasticity based modeling are considered since there is an average of 20 grains through sheets thickness. The crystal plasticity model is based on the works of [1] and [2]. The slip rate $\dot{\gamma}^{(s)}$ on a slip system *(s)* is given by a visco-plastic law as a function of the resolved shear stress on the system $\tau^{(s)}$ and its resistance to motion $\tau_c^{(s)}$, reading

$$\dot{\gamma}^{(s)} = \begin{cases} \dot{\gamma}_0^{(s)} \; sign(\tau^{(s)}) \left|\frac{\tau^{(s)}}{\tau_c^{(s)}}\right|^n & \text{if } \left|\tau^{(s)}\right| \geq \tau_c^{(s)} \\ 0 & \text{if } \left|\tau^{(s)}\right| < \tau_c^{(s)} \end{cases} \tag{1}$$

In the previous equation, $\dot{\gamma}_0^{(s)}$ is the reference strain rate and *n* is the strain rate sensitivity coefficient. The hardening at slip system level is based on the so-called PAN (Peirce-Asaro-Needleman) model [1].

The phenomenological plasticity model is elastic-visco-plastic and accounts for material anisotropy through the yield function of Bron and Besson [3]. Mixed hardening is assumed with an isotropic hardening of Hockett-Sherby and a combined Armstrong-Frederick & Prager kinematic hardening. The phenomenological and crystal plasticity approaches were implemented in ABAQUS/Standard.

The sheets behavior was characterized through mechanical tests and a large experimental database was established. The latter consisted of tensile tests in 0, 45 and 90° to the rolling direction (R.D.), equi-biaxial loading via hydraulic bulge test, monotonic shear tests and reversed shear tests after different pre-strain levels. The crystal plasticity model was calibrated on the tensile test in the R.D. on a Representative Volume Element which grains orientations were sampled from EBSD measurements. As illustrated on Fig. 2, the CPFEM (crystal plasticity finite element method) model proved capable of providing good predictions of all experimental tests although it was calibrated on a single tensile test.

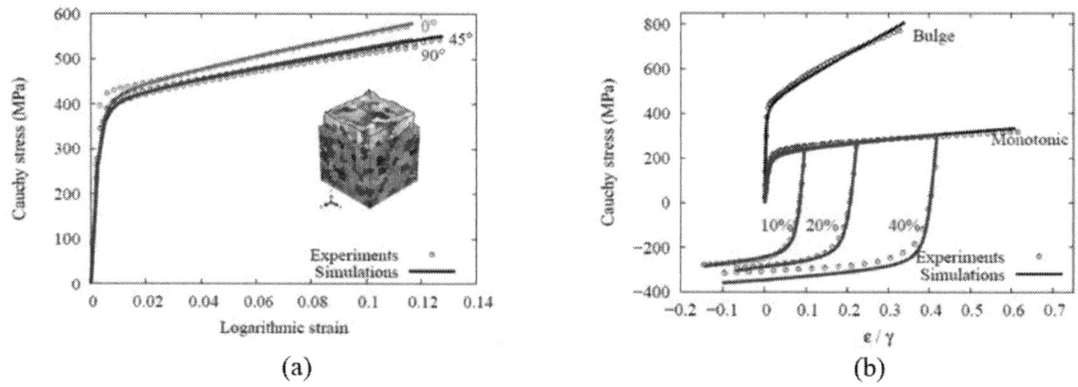

Fig. 2: Predictions by the CPFEM model of (a) tensile tests and (b) shear and bulge tests

On the other hand, the phenomenological plasticity model required a calibration procedure, implemented in SiDoLo [4], on the full database in order to describe accurately all the experimental strain paths.

4. Springback predictions and influencing parameters

The two modeling approaches (CPFEM and phenomenological) were applied to the prediction of springback in bending processes. A numerical study was conducted on several parameters of the CPFEM model. Both demonstrator parts were employed in the study.

Crystalline elastic anisotropy was found to significantly impact the CPFEM predictions of springback shapes. Indeed, assuming whether isotropic or anisotropic elasticity yielded very little difference in the forming step as anisotropic discrete slip flow drives the mechanical response in both cases. However, springback is dominated by the elastic behavior and thus affected by the type of elasticity considered. The differences that arise are illustrated on Fig. 3 which represents the profiles of a portion of the connector part.

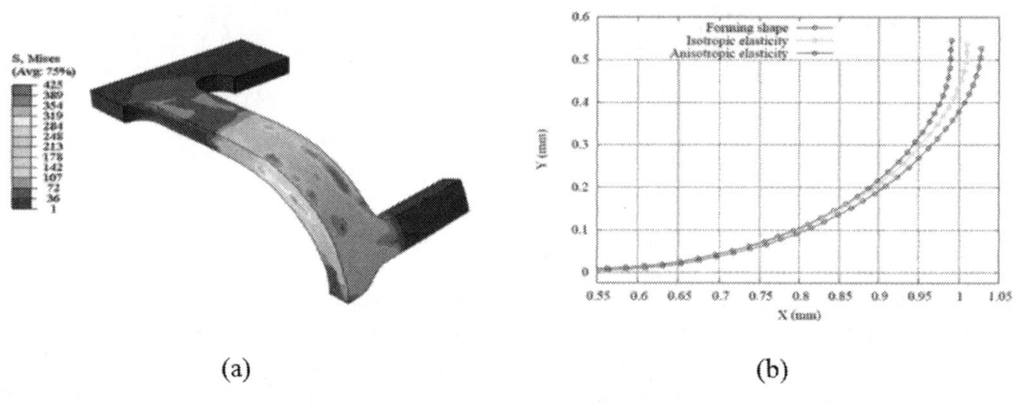

Fig. 3: Influence of elastic anisotropy on the profile (b) of a portion of the formed connector (a)

Furthermore, the ratio T/D (sheet thickness over grain diameter) which can be approximated to the number of grains in the thickness was linked to the amount of springback predicted. When this ratio and consequently the number of grains of grains in the part were increased, the predicted mechanical response converged towards the prediction of the phenomenological model. The CPFEM springback angle also grew closer to the phenomenological value and the experimental measurements (see Fig. 4).

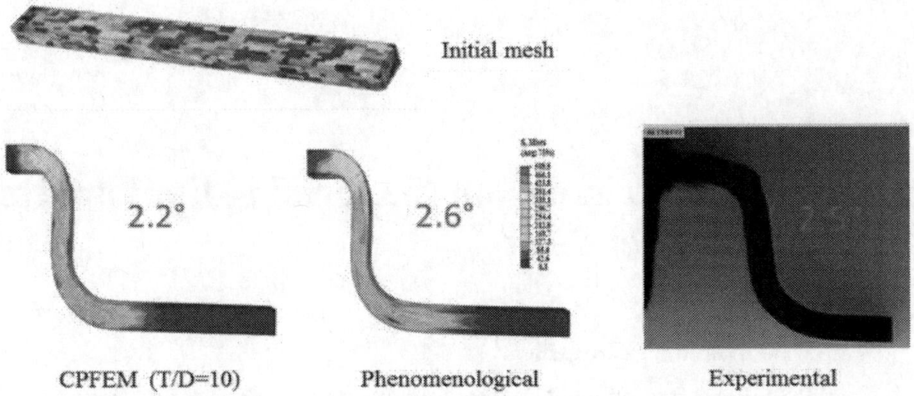

Fig. 4: Predictions and measurement of the springback angle of the chip leads

A similar trend on the evolution of springback with the T/D ratio was reported by [5] in their experimental study.

5. Summary and conclusions

The deformation behavior of copper alloys ultra-thin sheets was investigated with phenomenological plasticity and CPFEM models in the framework of micro-forming. The CPFEM model was more robust with respect to parameter identification. The springback predictions of the CPFEM model were highly sensitive to the assumption made on elasticity (isotropic or anisotropic). Moreover, increasing the number of grains in the sheet thickness led to more springback amplitude and to converging predictions with the phenomenological models in terms of stress fields and springback angle.

References

[1] Peirce, D., Asaro, R.J., Needleman, A., An analysis of nonuniform and localized deformation in ductile single crystals, Acta Metall. 30 (1982), 1087-1119

[2] Raphanel, J., Ravichandran, G., Leroy, Y., Three-dimensional rate-dependent crystal plasticity based on Runge-Kutta algorithms for update and consistent linearization, Int J Solids Struct 41 (2004) 5995-6021

[3] Bron, F. and Besson, J., A yield function for anisotropic materials Application to aluminum alloys, Int. J. Plasticity 20 (2004) 937-963

[4] Cailletaud, G., Pilvin, P., Identification and Inverse Problems: A Modular Approach, ASME Applied Mechanics Division - Publications - AMD 168 (1993) 33

[5] Jiang, C.P., and Chen, C.C., Grain Size Effect on the Springback Behavior of the Microtube in the Press Bending Process, Materials and Manufacturing Processes 27 (2012) 512-518

Journal of Physics: Conference Series **734** (2016) 032070 doi:10.1088/1742-6596/734/3/032070

Effect of the *r*-value change on the forming limit analysis for a ultra-low carbon steel sheet

T Hakoyama[1], T Kuwabara[2]

[1] JSPS Research fellow (PD), Institute of Engineering, Tokyo University of Agriculture and Technology, Koganei-shi, Tokyo, 184-8588, Japan
[2] Division of Advanced Mechanical Systems Engineering, Institute of Engineering, Tokyo University of Agriculture and Technology, Koganei-shi, Tokyo, 184-8588, Japan

E-mail: hakoyama@m2.tuat.ac.jp

Abstract. The effect of the *r*-value change on the accuracy of the forming limits predicted using the Marciniak-Kuczyński-type (M-K) forming limit analysis for a cold rolled interstitial-free (IF) steel sheet is investigated. Uniaxial tensile tests with a digital image correlation system are used to measure the *r*-value change. A tube subjected to tension-expansion loading under linear paths in the first quadrant of the stress space are performed to measure the multiaxial plastic deformation behavior and the forming limits of the test material. The observed differential hardening (DH) behavior is approximated by changing the material parameters of the Yld2000-2d yield function (Barlat et al, 2003) as functions of the reference plastic strain. The M-K analyses are performed using the *r*-value change and *r*-value constant DH model. It is concluded that the DH model considering the *r*-value change leads to the more accurate predicted forming limits.

1. Introduction

A forming limit curve (FLC) is widely used in industry. It is well known that material models have a significant effect on the prediction accuracy of FLCs [1, 2]. Therefore, yield functions that are capable of accurately reproducing the plastic deformation behavior of a test material are crucial.

One of the most popular constitutive models is isotropic hardening. However, the yield surface shape can change even in monotonic loading because of the change in texture with plastic deformation. Hill and co-authors [3, 4] proposed a new constitutive analysis method by focusing on the contours of equal plastic work (henceforth referred to as work contour) in the stress space; the change in shape of successive work contours with increasing deformation was formulated. One of the present authors developed biaxial tensile testing methods for sheet metals using cruciform specimens [5] and for tubular specimens [6, 7] to observe the differential hardening (DH) behavior of cold-rolled ultralow-carbon steel sheets [5, 7], high-strength steel sheet [8], aluminum alloy sheet [9], pure titanium sheet [6], and a magnesium alloy sheet [10].

Several authors developed constitutive models that enable us to reproduce the DH behavior in sheet forming simulations [8, 9, 11-15]. However most of them ignored the *r*-value change. Xu and Weinmann [11] calculated the FLCs of AK steel using the *r*-value change DH model based on the Marciniak-Kuczyński-type (M-K) approach [16]; however, the validity of the material models used in these analyses was not fully discussed.

Present authors investigated the effect of the DH behavior on the predictive accuracy of the forming limit for a cold-rolled IF steel sheet and concluded that the effect of the DH behavior was negligible [17]. However the r-value change was not considered. The objective of the present study is to clarify the effect of the constitutive model considering the r-value change on the predictive accuracy of the FLC and forming limit stress curve (FLSC) based on the M-K analysis for the sample used in [7, 17]. The calculated FLC and FLSC were compared with the measured ones.

2. Material modeling

2.1. *Uniaxial tensile test procedures and results*

The test material used was a 0.7 mm thick cold-rolled IF steel sheet. The work hardening characteristics and r-values measured at 0, 45 and 90° to the rolling direction (RD) are listed in table 1. Hereafter, the RD, transverse (TD) and thickness (ND) directions of the material are defined as the x-, y- and z-axes, respectively.

The digital image correlation system (ARAMIS®, Gom) was used to measure the development of the strain field for the uniaxial tensile tests. Standard uniaxial tensile specimens (JIS 13 B-type) were used. Figure 1 shows the variations in the instantaneous r-values, r', in the RD and TD with increasing the longitudinal plastic strain ε^{p}. The values of r' are almost constant for $\varepsilon^{\mathrm{p}} \le 0.1$ and $0.3 \le \varepsilon^{\mathrm{p}}$ while they gradually decrease with the increase of ε^{p} for $0.1 \le \varepsilon^{\mathrm{p}} \le 0.3$.

The variation in r' was approximated as a function of ε^{p} using the following equation:

$$r' = \frac{A_1 - A_2}{1 + \exp\left((\varepsilon^{\mathrm{p}} - A_3)/A_4 \right)} + A_2 \tag{1}$$

where A_i (i=1~4) are parameters. The approximated results were also depicted in figure 1.

Table 1. Work hardening characteristics of the test material [7].

Tensile direction (°)	$\sigma_{0.2}$ (MPa)	c^{a} (MPa)	n^{a}	α^{a}	r-value[b]
0 (RD)	164	574	0.273	0.008	2.27
45	173	574	0.272	0.008	1.77
90 (TD)	170	564	0.273	0.009	2.65

[a] Approximated using $\sigma = c(\alpha + \varepsilon^{\mathrm{p}})$ for $0.002 \le \varepsilon^{\mathrm{p}} \le 0.093$.
[b] Measured at nominal strain $\varepsilon_{\mathrm{N}} = 0.1$.

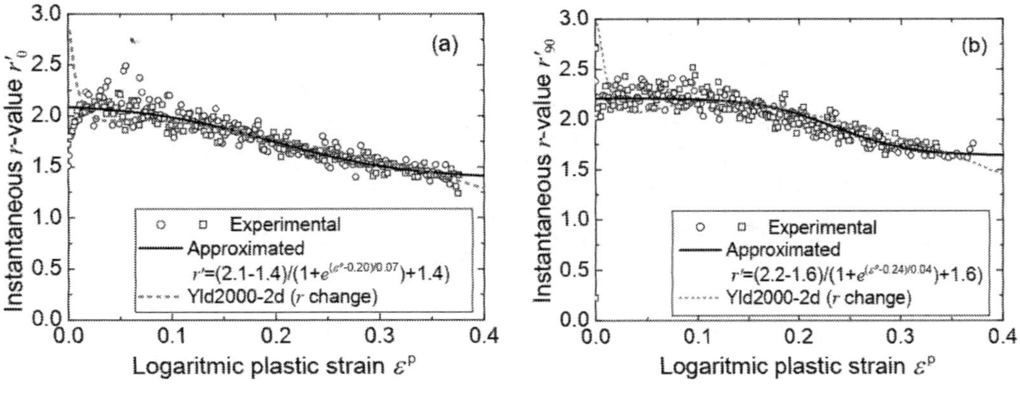

Figure 1. Variation in the instantaneous r-values, $r'_0 = d\varepsilon_y / d\varepsilon_z$ and $r'_{90} = d\varepsilon_x / d\varepsilon_z$, with increasing the logarithmic plastic strain ε^{p}.

2.2. Multiaxial tube tension-expansion forming test procedures

Multiaxial tube tension-expansion forming tests (MTET) were performed to measure the multiaxial plastic deformation behavior of the test material from initial yield to fracture using the servo-controlled multiaxial tube tension-expansion machine [7]. Tubular specimens with an inner diameter of 44.6 mm and a length of 200 mm were fabricated by bending a sheet sample into a cylindrical shape and CO_2 laser-welding the sheet edges together. Axial loads T and an internal pressure P were applied to the specimen and caused multiaxial stresses at the center of the specimen. Linear stress paths were applied to tubular specimens; the true stress ratios $\sigma_x : \sigma_y$ were chosen to be 4:1, 2:1, 4:3, 1:1, 3:4, 1:2 and 1:4. All the MTET data considered in this study are those obtained in [7].

2.3. Material modeling

The concept of the work contour in the stress space [3, 4] was introduced to evaluate the work hardening behavior of the test material under biaxial tension. The true stress-logarithmic plastic strain curve obtained from a uniaxial tension test along the RD was selected as a reference datum for work hardening; the uniaxial true stresses σ_0 and the plastic work per unit volume W_0 performed during the test up to a specified value of the uniaxial logarithmic plastic strain $\varepsilon_0^{\mathrm{p}}$. The uniaxial true stress σ_{90} in the TD and the biaxial true stress components (σ_x, σ_y) were then determined for the same plastic work as W_0. The stress points plotted in the stress space form a work contour associated with $\varepsilon_0^{\mathrm{p}}$.

Figure 2 shows the measured stress points forming work contours. Also depicted in the figure are the theoretical yield loci calculated using two DH models; one is based on constant r-values (DH-I) and the other one is based on instantaneous r-values (DH-II). The DH-I and II models were approximated by changing the anisotropic parameters α_i ($i = 1$-8) and exponent M of the Yld2000-2d yield function [18] as functions of $\varepsilon_0^{\mathrm{p}}$ using the following equations:

$$M, \alpha_i = B_1 - B_2 \exp(-B_3 \varepsilon_0^{\mathrm{p}}) + B_4 / (B_5 + \varepsilon_0^{\mathrm{p}}) \quad (2)$$

where B_i (i=1~5) are the parameters. The r-values calculated using the DH-II model are also depicted in figure 1.

3. Forming limit analyses

Figure 3 shows the forming limit strains and stresses measured using the uniaxial tensile tests, MTETs, and hydraulic bulge tests. Also depicted in the figure are the FLCs and FLSCs calculated using the M-K approach [16]. The material parameters assumed in the M-K analyses are as follows: the elastic modulus is 200 GPa, the Poisson's ratio 0.3, and the strain rate sensitivity exponent (m-value) 0.02. The strain hardening function was described as $\sigma_0 = 670(\varepsilon_0^{\mathrm{p}} + 0.01)^{0.33} - 147\varepsilon_0^{\mathrm{p}}$ (MPa) [17]. The magnitude of the initial imperfection was set to 0.997. Additional details on the calculation

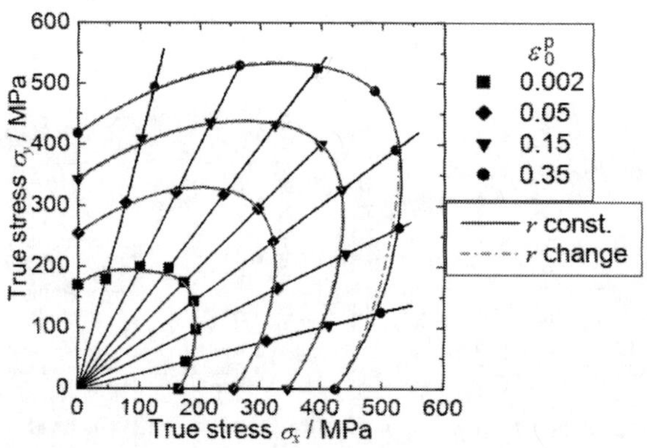

Figure 2. Measured stress points forming contours of plastic work compared with those calculated using the DH-I and II models based on the Yld2000-2d yield function.

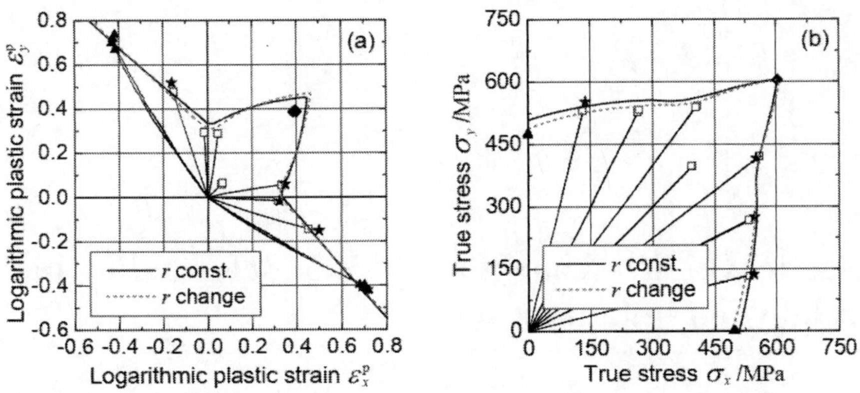

Figure 3. Measured and calculated (a) forming limit strains and (b) forming limit stresses. The measured data are the same as those in[7]; □: Fractured at a position of $\theta \le \pm 30°$ (where θ is the angle from the weld line in the circumferential direction of a tubular specimen),★: fractured at a position of $\theta > \pm 30°$, ◆: hydraulic bulge test, and ▲: Uniaxial tensile test.

procedures and constitutive equations for the M-K approach can be found in [1, 8].

The DH-II model has a better agreement with the measured FLC and FLSC than the DH-I model for the stress paths with $\sigma_y > \sigma_x$.

4. Conclusions

The DH-II model (considering the *r*-value change) had a better agreement with the measured FLC and FLSC than the DH-I model (assuming constant *r*-values).

Acknowledgments

This work was supported by Grant-in-Aid for JSPS Fellows (266118).

References

[1] Kuroda M and Tvergaard V 2000 *Int. J. Mech. Sci.* **42** 867-87
[2] Banabic D, Barlat F, Cazacu O and Kuwabara T 2010 *Int. J. Mater. Form.* **3** 165-89
[3] Hill R and Hutchinson JW 1992 *J. Appl. Mech.* **59** S1-9.
[4] Hill R, Hecker SS and Stout MG 1994 *Int. J. Solids Struct.* **31** 2999-3021.
[5] Kuwabara T, Ikeda S and Kuroda T 1998 *J. Mater. Process. Technol.* **80-81** 517-23.
[6] Ishiki M, Kuwabara T and Hayashida Y 2011 *Int. J. Mater. Form.* **4** 193-204.
[7] Kuwabara T and Sugawara F 2013 *Int. J. Plast.* **45** 103-18.
[8] Hakoyama T and Kuwabara T 2015 *From Creep Damage Mechanics to Homogenization Methods*, ed H Altenbach, T Matsuda and D Okumura (Cham: Springer) pp 67-95.
[9] Yanaga D, Takizawa H and Kuwabara T 2014 *J. Japan Soci. Technol. Plast.* **55** 55-61 (in Japanese).
[10] Andar M, Kuwabara T and Steglich D 2012 *J. Japanese Soc. Exp. Mech.* **12** 151-6.
[11] Xu S and Weinmann KJ 2000 *Int. J. Mech. Sci.* **42** 677-92.
[12] Zamiri A and Pourboghrat F 2007 *Int. J. Solids Struct.* **44** 8627-47.
[13] Aretz H 2008 *Int. J. Plast.* **24** 1457-80.
[14] Stoughton TB and Yoon JW 2009 *Int. J. Plast.* **25** 1777-817.
[15] Peters P, Manopulo N, Lange C and Hora P 2014 *Int. J. Mater. Form.* **7** 447-57.
[16] Marciniak Z and Kuczyński K 1967 *Int. J. Mech. Sci.* **9** 609–20.
[17] Hakoyama T and Kuwabara T 2015 *Key Eng. Mater.* **651-653** 9-14.
[18] Barlat F, Brem JC, Yoon JW, Chung K, Dick RE, Lege DJ, Pourboghrat F, Choi SH and Chu E 2003 *Int. J. Plast.* **19** 1297-319.

Crystal-plasticity finite-element analysis of deformation behavior in a commercially pure titanium sheet

T Hama[1], A Kobuki[1], H Fujimoto[1], and H Takuda[1]

[1]Graduate School of Energy Science, Kyoto University, Yoshida-honmachi, Sakyo-ku, Kyoto 606-8501, Japan

E-mail: hama@energy.kyoto-u.ac.jp

Abstract. A crystal-plasticity finite-element method was used to examine the deformation mechanism in a commercially pure titanium sheet. The following tension-compression asymmetry was exhibited in the stress-strain curves: the yield stress was larger under tension than under compression, whereas the work-hardening was smaller under tension than under compression. The strain hardening behaviour was predicted qualitatively well using the crystal-plasticity analysis. The simulation results suggested that the tension-compression asymmetry could be explained in terms of the difference in the activity of the twinning systems.

1. Introduction

Commercially pure (CP) α-titanium (Ti) sheets are widely used in various products due to their characteristics such as high strength, corrosion resistance, and biological compatibility [1, 2]. CP-Ti sheets exhibit deformation behavior with strong anisotropy. One of the main factors of the anisotropy is that they have the hexagonal-close-packed (hcp) structure that depicts a strong crystalline anisotropy. Specifically, the critical resolved shear stress (CRSS) as well as work-hardening differs significantly depending on the family of slip systems [3]. The twinning systems, which have the direction dependence in deformation, also play a role as the secondary plastic deformation mode [4, 5]. Moreover, rolled CP-Ti sheets have the strong basal texture in which c-axes tilt 20 to 40° from the normal direction (ND) to the transverse direction (TD) [4-9]; thus, the anisotropy is further pronounced. However, the role of each slip or twinning system on the macroscopic anisotropic deformation behavior is not yet well understood.

A crystal-plasticity model enables a direct investigation into the interaction between mesoscopic crystalline and macroscopic deformation in metals. Therefore, a crystal-plasticity analysis is useful to investigate the deformation behavior in hcp metals that exhibit strong anisotropy, and it has widely been utilized for magnesium (Mg) alloys [10-15]. There are a few past studies that examine the deformation behavior in CP-Ti using a crystal-plasticity analysis [16, 17]. However, the simulation accuracy in the previous studies seems to be insufficient; thus, the deformation mechanism is still open to discussion.

In the present study, the tension-compression asymmetry observed in the stress-strain curves in a CP-Ti JIS Grade 1 sheet is investigated using a crystal-plasticity finite-element method. First, the stress-strain curves obtained under monotonic tension and compression are compared between simulation and experimental results to verify the simulation accuracy. The mechanism that yields the tension-compression asymmetry is then investigated numerically.

2. Crystal-plasticity model

The crystal-plasticity model employed in this study follows the model used in our previous studies for Mg alloy sheets [10, 11, 15]. The crystalline slip is assumed to follow Schmid's law. The slip rate $\dot{\gamma}^{(\alpha)}$ of the α slip system is assumed to be given by a visco-plastic power law as [18]

$$\frac{\dot{\gamma}^{(\alpha)}}{\dot{\gamma}_0}=\left|\frac{\tau^{(\alpha)}}{\tau_Y^{(\alpha)}}\right|^{\frac{1}{m}}\operatorname{sign}\left(\tau^{(\alpha)}\right),\quad \dot{\tau}_Y^{(\alpha)}=\sum_\beta q_{\alpha\beta}h\left|\dot{\gamma}^{(\beta)}\right| \tag{1}$$

where $\tau^{(\alpha)}$ and $\tau_Y^{(\alpha)}$ are the Schmid's resolved shear stress and the current strength of the α slip system, respectively, $\dot{\gamma}_0$ is the reference strain rate, m is the rate sensitivity exponent, $q_{\alpha\beta}$ is the self ($\alpha=\beta$) and latent ($\alpha\neq\beta$) hardening moduli, and h is the hardening rate for which the following two laws are used.

$$h=h_0 \text{ and } h=h_0\left(1-\frac{\tau_0}{\tau_\infty}\right)\exp\left(-\frac{h_0\bar{\gamma}}{\tau_\infty}\right) \text{ with } \bar{\gamma}=\sum_\alpha\int\left|\dot{\gamma}^{(\alpha)}\right|dt. \tag{2}$$

In this study, five families of slip systems, i.e., basal {0001}<11$\bar{2}$0> slip, prismatic {10$\bar{1}$0}<11$\bar{2}$0> slip, pyramidal {10$\bar{1}$1}<11$\bar{2}$0> slip, pyramidal {11$\bar{2}$2}<11$\bar{2}$3> slip (hereafter termed as pyr <*a*+*c*>-1 slip), and pyramidal {10$\bar{1}$1}<11$\bar{2}$3> slip (hereafter termed as pyr <*a*+*c*>-2 slip), and two families of twinning systems, i.e., {10$\bar{1}$2} tensile twinning and {11$\bar{2}$2} compressive twinning, are taken into account. There are three basal, three prismatic, three pyramidal <*a*>, six pyr <*a*+*c*>-1, twelve pyr <*a*+*c*>-2, six {10$\bar{1}$2} twinning, and six {11$\bar{2}$2} twinning systems. The linear hardening (eq. (2)-1) is assumed for the twinning systems, and Voce hardening (eq.(2)-2) is assumed for the slip systems. A twinning and detwinning model proposed by the authors [10, 11, 15] is employed. The above rate-dependent crystal-plasticity model is incorporated into each Gauss point in a static finite-element method. The rate tangent modulus method [19] and the so-called generalized r_{min}-strategy are employed for the explicit time integration [20].

3. Experimental and simulation procedures

In the experiment, a monotonic tension and compression test was conducted using a rolled CP-Ti JIS Grade 1 sheet with 1.0mm thick (Kobe steel) [5]. Comb-shaped dies were employed in the experiment to suppress the occurrence of buckling during compression.

A cube that had five uniform eight-node solid elements in each direction was used as a finite-element model. The same initial crystallographic orientation was assigned to all eight Gauss integration points in an element. The initial crystallographic orientations assigned to the elements were randomly selected from a result of EBSD measurement. The interaction matrix and hardening parameters employed in the present simulation are shown in Tables 1 and 2. Note that it was assumed that latent hardening due to twinning activity was larger than that of slip activity [7, 21]. Material parameters were carefully identified with referring to stress-strain curves and evolutions of Lankford value obtained under monotonic loading. Young's modulus, Poisson's ratio, and the rate-sensitivity exponent were respectively set to E = 105 GPa, ν= 0.34, and m = 0.02, which are the typical values for CP-Ti. The reference-strain rate was taken to be γ_0=0.001s^{-1}.

4. Results and discussion

Figure 1(a) shows the absolute stress-absolute strain curves obtained from the experiment and the simulation of monotonic tension and compression in the rolling direction (RD). In the experimental results, the yield stress is larger and the work-hardening is smaller under tension than under compression. Similar

Table 1. Latent-hardening parameters $q_{\alpha\beta}$ used in the present study.

	Basal	Prism	Pyr <*a*>	Pyr <*a*+*c*>-1	Pyr <*a*+*c*>-2	{10-12}	{11-22}
Basal	0.5	0.5	0.5	0.5	0.5	0.9	0.9
Prism	0.5	0.5	0.5	0.5	0.5	0.9	0.9
Pyr <*a*>	0.5	0.5	0.5	0.5	0.5	0.9	0.9
Pyr <*a*+*c*>-1	0.5	0.5	0.5	0.5	0.5	0.9	2.0
Pyr <*a*+*c*>-2	0.5	0.5	0.5	0.5	0.5	0.9	2.0
{10-12}	0.5	0.5	0.5	0.5	0.5	0.9	0.9
{11-22}	0.5	0.5	0.5	0.5	0.5	0.9	0.9

Table 2. Identified hardening parameters, in MPa.

	Basal	Prism	Pyr <*a*>	Pyr <*a*+*c*>-1	Pyr <*a*+*c*>-2	{10-12}	{11-22}
τ_0	133	62	81	145	145	90	140
τ_∞	180	160	210	270	270		
h_0	1950	1050	580	2050	2050	350	350

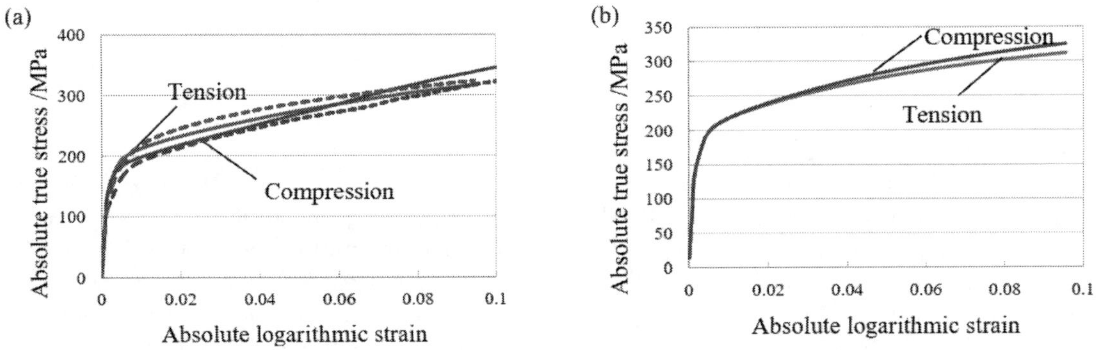

Figure 1. Stress-strain curves under tension and compression. (a) Results of experiment (dotted lines) and simulation (solid lines). (b) Simulation results obtained without considering twinning.

trends have been reported by Nixon et al. [7]. The strain hardening behavior is predicted qualitatively well in the simulation results.

The mechanism that yields the tension-compression asymmetry is examined using the evolution of relative activities (Figure 2). During tension, the activity of prismatic slip is dominant at the beginning of plastic deformation, and then the activities of pyramidal <a> slip and $\{11\bar{2}2\}$ twinning increase. These three systems are active until the end of deformation. The above activities of slip systems are similar under compression. In contrast, the twinning activity is different under compression as follows: the $\{10\bar{1}2\}$ twinning systems are active, and the $\{10\bar{1}2\}$ twinning activity under compression is larger than that of $\{11\bar{2}2\}$ twinning under tension. The differences in the twinning activities would play a significant role in the tension-compression asymmetry in the stress-strain curve because the CRSS of the $\{11\bar{2}2\}$ twinning is larger than that of the $\{10\bar{1}2\}$ twinning (Table 2) and the twinning activity is assumed to yield large latent hardening as described earlier (Table 1).

To confirm this discussion, Figure 1(b) shows the absolute stress-absolute strain curves under monotonic tension and compression in the RD obtained without considering twinning. Clearly, the yield stresses under tension and compression are almost identical and the difference in work-hardening is also reduced, exhibiting that the tension-compression anisotropy is largely owing to the differences in the twinning activities. On the other hand, the work-hardening is still slightly larger under compression than under tension. This difference may be due to the fact that the activity of pyramidal <a+c>-2 slip that has large CRSS and work-hardening rate increases gradually only under compression.

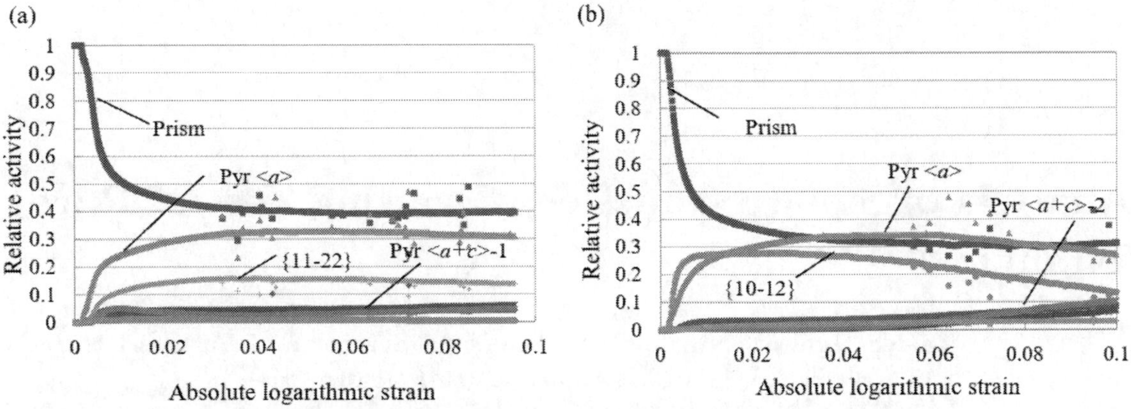

Figure 2. Relative activities obtained under monotonic (a) tension and (b) compression.

5. Conclusions

In the present study, the anisotropic deformation behavior under monotonic loading in the CP-Ti sheet was studied using the crystal-plasticity finite-element method. The tension-compression asymmetry exhibited in the stress-strain curves in the RD was predicted qualitatively well in the present simulation. The simulation results showed that the difference in the twinning activities was one of the main factors that yielded the tension-compression asymmetry.

References

[1] Ishiki M, Kuwabara T, Yamaguchi M, Maeda Y, Hayashida Y and Itsumi Y, *Trans. Jpn. Soc. Mech. Eng.* **75** (2009) 491-500.
[2] Zeng Z, Zhang Y, Jonsson S, *Mater. Des.* **30** (2009) 3105-3111.
[3] Yoshinaga H and Horiuchi R, *Trans. JIM* **5** (1963) 14-21.
[4] Hama T, Nagao H, Kobuki A, Fujimoto H, Takuda H, *Mater. Sci. Eng. A* **620** (2014) 390-398.
[5] Yi N, Hama T, Kobuki A, Fujimoto H, Takuda H, *Mater. Sci. Eng. A* **655** (2016) 70-85.
[6] Mullins S and Patchett BM, *Metall. Trans. A* **12** (1981) 853-863.
[7] Nixon ME, Cazacu O and Lebensohn RA, *Int. J. Plast.* **26** (2010) 516-532.
[8] Ishiyama S, *Titan. Japan* **54** (2006) 42-51 (in Japanese).
[9] Conrad H, Prog. *Mater. Sci.* **26** (1981) 123-403.
[10] Hama T and Takuda H, *Int. J. Plast.* **27** (2011) 1072–1092.
[11] Hama T and Takuda H, *Comput. Mater. Sci.* **51** (2012) 156-164.
[12] Graff S, Brocks W and Steglich D, *Int. J. Plast.* **23** (2007) 1957-1978.
[13] Hama T, Kitamura N and Takuda H, *Mater. Sci. Eng. A* **583** (2013), 232-241.
[14] Wang H, Wu PD, Wang J and Tome CN, *Int. J. Plast.* 49(2013), 36-52.
[15] Hama T, Tanaka Y, Uratani M and Takuda H, *Int. J. Plast.* (2016), in press.
[16] Gurao NP, Kapoor R, and Suwas S, *Acta Mater.*, 59(2011), 3431-3446.
[17] Benmhenni N, Bouvier S, Brenner R, Chauveau T, and Bacroix B, *Mater. Sci. Eng. A*, 573(2013), 222-233.
[18] Asaro RJ and Needleman A, *Acta Metall.* **33** (1985) 923-953.
[19] Pierce D, Asaro RJ and Needleman A, *Acta Metall.* **31** (1983) 1951-1976.
[20] Hama T, Nagata T, Teodosiu C, Makinouchi A and Takuda H, *Int. J. Mech. Sci.* **50** (2008) 175-192.
[21] Jain A and Agnew SR, *Mater. Sci. Eng. A* **462** (2007) 29-36.

Analysis of Square Cup Deep-Drawing Test of Pure Titanium

Takaki Ogawa[1], Ninshu Ma[1], Minoru Ueyama[2] and Yasunori Harada[2]

[1] JSOL Corporation, 2-2-4 Tosabori, Nishi-ku, Osaka 550-0001, Japan

[2] University of Hyogo, 2167 Shosya, Himeji, Hyogo 671-2280, Japan

E-mail: ogawa.takaki@jsol.co.jp

Abstract. The prediction of formability of titunium is more difficult than steels since its strong anisotropy. If computer simulation can estimate the formability of titanium, we can select the optimal forming conditions. The purpose of this study was to acquire knowledge for the formability prediction by the computer simulation of the square cup deep-drawing of pure titanium. In this paper, the results of FEM analsis of pure titanium were compared with the experimental results to examine the analysis validity. We analyzed the formability of deep-drawing square cup of titanium by the FEM using solid elements. Compared the analysis results with the experimental results such as the forming shape, the punch load, and the thickness, the validity was confirmed. Further, through analyzing the change of the thickness around the forming corner, it was confirmed that the thickness increased to its maximum value during forming process at the stroke of 35mm more than the maximum stroke.

1. Introduction

Titanium is abundant in the earth. It is the fourth in abundance next to aluminum, iron, and nickel in the earth's metal elements. Titanium is superior in strength and hardness compared to iron. Titunium is lighter than iron. The mass density is about 57% of iron. The crystal structure of α titanium is an hcp structure. Slip system of hcp structure is three and it is less than the bcc structure such as iron and the fcc structure such as aluminum. Due to this fact, the anisotropic parameter r value is large and changes greatly even in-plane of titanium sheets. From this reason, the prediction of formability of titunium is more difficult.

Computer simulation on the titanium forming process was less reported compared with the steels and aluminum. Particularly, the studies on the formability of the square cup deep-drawing of titanium are few. Ohwue et al. [1] applied finite element analysis(FEM) simulation using full model of circular titanium blanks. The occurrence of earrings in circular-shell deep-drawing tests for two different types of crystal materials titanium was investigated. On the other hand, in the sheet metal forming simulation, the studies of optimum process design are energetically carried out. Kitayama et al. [2] applied their proposed optization algorithm to the optimal variable blank holder force trajectory for the square cup deep-drawing. The objective is taken as the minimization of the deviation of the whole thickness.

If computer simulation can estimate the formability of titanium, we can use better forming conditions to improve the formability. The purpose of this study was to acquire knowledge for the formability prediction by the computer simulation of the square cup deep-drawing of pure

Journal of Physics: Conference Series **734** (2016) 032072 doi:10.1088/1742-6596/734/3/032072

titanium. In this paper, the results of FEM analsis of pure titanium were compared with the experimental results to examine the analysis validity.

2. Analysis Model and Analysis Conditions

Using the dynamic explicit method elastic-plastic FEM program LS-DYNA, 3-dimensional analysis of a square cup deep-drawing was carried out. A square cup deep-drawing test for pure titanium was performed. The specimen is JIS H 4600 TP340C with a heat treatment (pure titanium JIS2 species). Basically the analysis condition is set to be the same as this used in the deep drawing test.

The dimensions of tools are shown in Fig. 1(a). At the corner, the clearance between the die and the punch is designed to have the maximum value of 0.7 mm. The amount of maximum punch stroke is 65.5 mm. The stroke speed is 5,000 mm/s that is faster than real speed to save the calculation time of the dynamic explicit FEM. The blank holding force is 10 kN and the die is fixed. The tools are modeled as rigid bodies. The mesh size is from 0.2 mm to 5.0 mm.

The workpiece is shown in Fig. 1(b). The workpiece sheet thickness is 0.477 mm and diameter ϕ is 100 mm. Values of the material parameters used in the calculation are listed in Table 1. The mechanical properties, the relationship between stress and plastic strain obtained from the results of the tensile test.

The solid element with the reduced integration is selected to model the titanium blank. Three elements with the size of 0.16mm in the thickness direction are employed. The element size in the plane of blank is 0.5 mm. Hill's quadratic anisotropic yield function(1948) is used. We used solid element to consider the thickness contact force changes when the plate thickness increases and decreases in the forming process.

Coulomb friction for contacts between titanium blank and working tools: punch, die and blank holder was assumed to be 0.2 in the simulation.

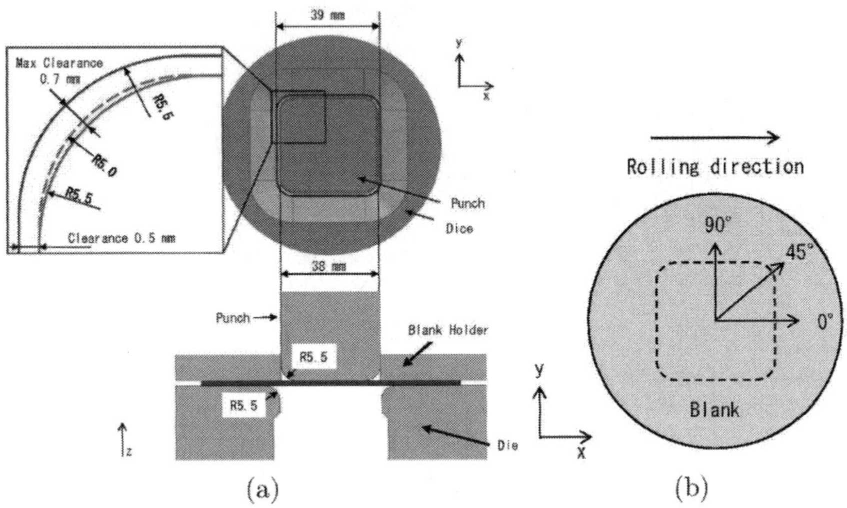

Figure 1. (a) Geometory of Tools. (b) Setting blank sheet along the rolling direction.

Table 1. Values of material parameters used in simulations.

Young's modulus	r_0	r_{45}	r_{90}	Yield stress	Poisson's ratio	Mass density
108.5 GPa	1.89	3.72	3.18	250 MPa	0.34	4.50×10^{-9} ton/mm^3

3. Analysis results

3.1. Punch load and forming shape

The relationship between punch load and punch stroke are shown in Fig. 2. The punch load increases linealy when the punch stroke is less than 10 mm, and reached the first peak when the punch stroke is about 15 mm. Then the punch load decreases gradually at the stroke of about 35 mm. Again, it is increasing until reaching the second peak at the stroke of about 40 mm , and it drops to zero at the final stroke. The result has good agreement with the experimental result.

About the formed shape, the comparison between the analysis results and the experimental results are shown in Fig. 3. The formed shapes at the punch strokes of 20.5 mm, 40.5 mm and 65.5 mm punch are respectively shown in Fig. 3(a), (b) and (c) from the top view or the side view. With increasing the punch stroke, the earing occured in the corner portion. That qualitatively matched with experiment. It can be seen that the second peak of the punch load corresponds to the certain amount earing in the corner portion.

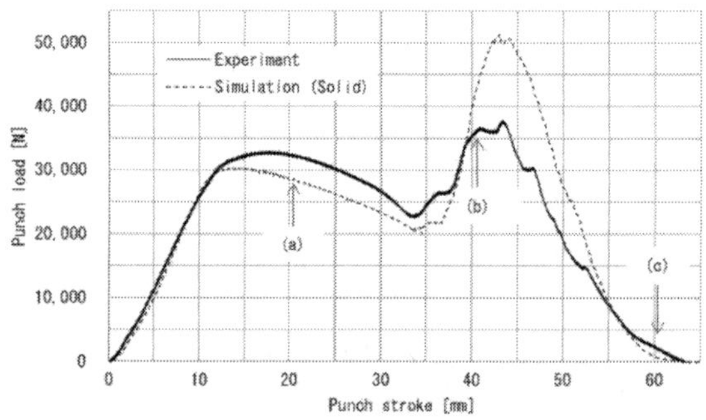

Figure 2. The relationship between punch load and stroke.

(a)20.5 mm (b)40.5 mm (c)65.5 mm

Figure 3. Comparison between simulation(left) and experiments(right).

3.2. Thickness distribution

Fig. 4 shows the relationship between the thickness and the travel distance from the center of cup along the section at the punch stroke of 20.5 mm. Fig. 4 compares the computed thickness distributions with experimental results along the rolling direction (0 degree, Fig. 4 left) and along the 45 degree from the rolling direction (Fig. 4 right). Along both the rolling direction and 45 degree from rolling direction, the thickness in the bottom of the cup remains nearly the initial thickness. Along the rolling direction, the vertical wall thickness reduced, and the blank thickness increased toward the outer periphery. On the other hand, along the 45 degree direction from the rolling direction the vertical wall thickness reduced similarly with that in the rolling direciton, but the thickness remains nearly the initial thickness toward the outer periphery.

Fig. 5 shows the relationship between the thickness and the travel distance from the center of cup along the section at the punch stroke of 40.5 mm. The vertical wall thickness along the rolling direction changes greatly. The experimental results show that the vertical wall thickness decreased and increased toward the outer along the 45 degree direction from the rolling direction. On the other hand, the computed results show that the vertical wall thicknesss increased toward outer part.

Then, Fig. 6 shows the thickness distribution of the vertical wall at the punch stroke 65.5 mm. Fig. 7 shows the hisory of thickness at the maximum thickness element. It can be seen that the thickness increases at the flange corner and decreases after passing through the die radius.

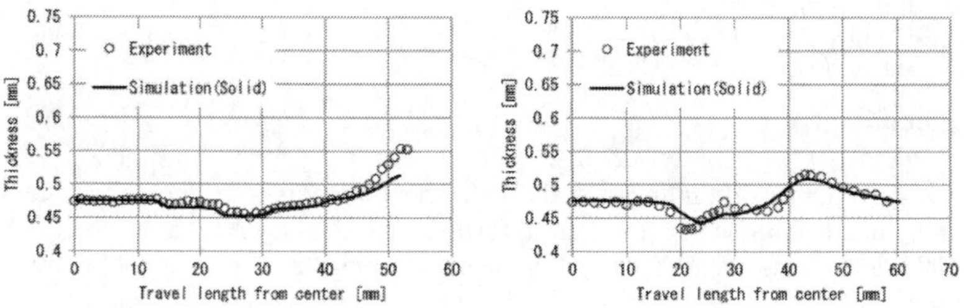

0° 45°

Figure 4. Thickness along section at the stroke of 20.5 mm.

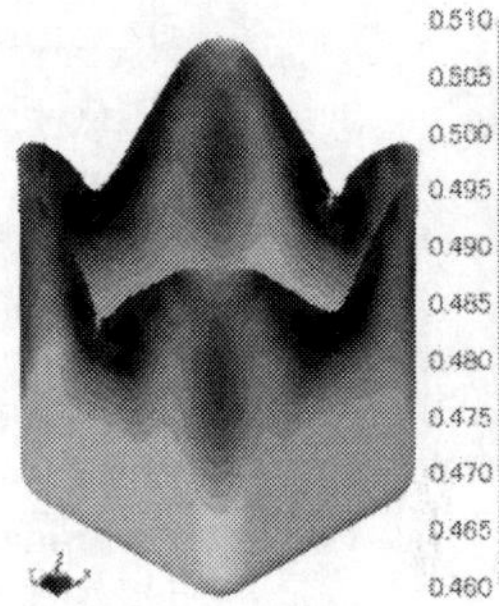

Figure 6. The thickness distribution at the punch stroke of 65.5 mm.

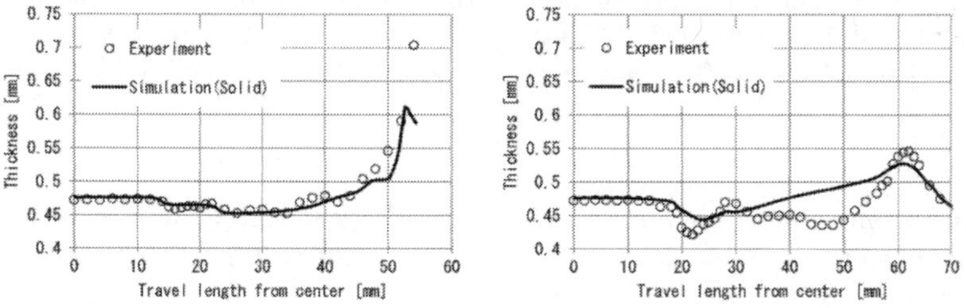

0° 45°

Figure 5. Thickness along section at the stroke of 40.5 mm.

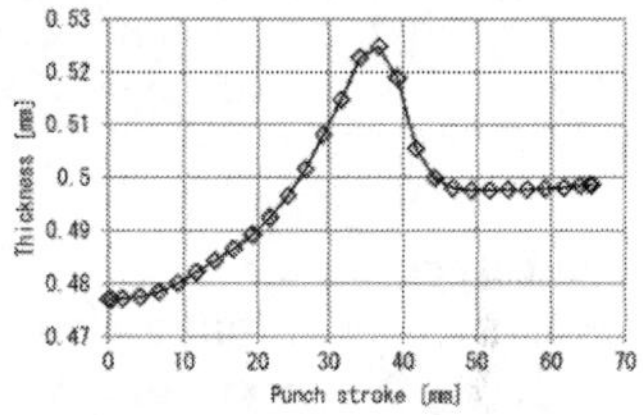

Figure 7. The hisory of thickness at the maximum thickness element.

4. Conclusions

In this study, we analyzed the formability of deep-drawing square cup of titanium by the FEM using solid elements. By comparing the analysis results with the experimental results of the forming shape, the punch load, and the thickness, the validity was confirmed. Further, through analyzing the changing history of the thickness around the formed corner, it was confirmed that the maximum thickness occured during forming process at the stroke of 35mm more than the maximum stroke.

References

[1] Tetsuro Ohwue and Yoshikazu Kobayashi. Analysis of earring in circular-shell deep-drawing of bcc and hcp sheet metals. *Procedia Engineering*, 81:887–892, 2014.

[2] Satoshi Kitayama, Kenta Kita, and Koetsu Yamazaki. Optimization of variable blank holder force trajectory by sequential approximate optimization with rbf network. *The International Journal of Advanced Manufacturing Technology*, 61:1067–1083, 2012.

Journal of Physics: Conference Series **734** (2016) 032073 doi:10.1088/1742-6596/734/3/032073

Advances in characterization of sheet metal forming limits

Thomas B. Stoughton[1], John E. Carsley[1], Junying Min[2], Jianping Lin[3]

[1]General Motors Global Research & Development Center, USA
[2]Ruhr-University Bochum, Germany
[3]Tongji University, China

Thomas.b.stoughton@gm.com

Abstract. This paper accounts for nonlinear strain path, sheet curvature, and sheet-tool contact pressure to explain the differences in measured forming limit curves (FLCs) obtained by Marciniak and Nakajima Tests. While many engineers working in the sheet metal forming industry use the raw data from one or the other of these tests without consideration that they reflect the convolution of material properties with the complex processing conditions involved in these two tests, the method described in this paper has the objective to obtain a single FLC for onset of necking for perfectly linear strain paths in the absence of through-thickness pressure and restricted to purely in-plane stretching conditions, which is proposed to reflect a true material property. The validity of the result is checked using a more severe test in which the magnitude of the nonlinearity, curvature, and pressure are doubled those involved in the Nakajima Test.

1. Introduction

Although the limitations of the forming limit diagram (FLD) concerning the role of nonlinear strain paths on the onset of necking have been known nearly since its inception, and solutions to handle these limitations had been formulated in terms of limits on the effective plastic strain and equivalently by limits on the true stress, limited recognition of the importance of these limitations was acknowledged by industry until stress-based solutions were rediscovered and more recently promoted. While there exists many equivalent representations of the stress-based solution, supported by bifurcation theory and MK Analysis, a wide-spread misunderstanding of forming limits persists as is evident in the continued ineffective use of the conventional strain-based FLD. This presentation will review the fundamental inadequacy of the strain-based FLD, in particular focusing on the inability to assess the severity of forming, i.e., the likelihood of exceeding the necking limits of the sheet material, which applies even in the forming conditions where the strain path is almost perfectly linear.

The primary contribution of this paper is to combine knowledge of the effects of 1) nonlinear strain paths, 2) sheet curvature, and 3) tool contact pressure, to explain the differences in measured forming limits obtained by Marciniak and Nakajima Tests, and define the forming limit for the onset of necking under conditions of perfectly linear in-plane stretching in the absence of a through-thickness pressure. The validity of the interpretation of the resulting FLC as a material property is demonstrated by applying the same corrections to data obtained from a 50 mm diameter punch, which doubles the magnitude of the nonlinearity, curvature, and pressure that is involved in the conventional Nakajima Test.

2. FLC Correction Procedure

In order to employ this procedure it is assumed that the history of both the strain and surface coordinates at the location of neck is recorded using Digital Image Correlation (DIC) methods. It is also assumed that the frame rate if the DIC camera system is sufficient to capture an image within a negligible strain difference from those at which onset of necking occurs. And finally, it is assumed that a method is employed to detect the true onset of necking from analysis of the strain and/or geometry data. The method adopted for this study is based on detecting a curvature change in the area of the neck, obtained by fits to the surface coordinates [1].

While the corrections employed here depend on the material model, this often-cited criticism of the stress-based approach to dealing with strain path nonlinearity is a red herring since the primary application of the FLC is in the analysis of FEM simulations, which necessarily is based on a specific material model to simulate the deformation of the metal. So the least of all concerns is the use of the material model to account for experimental processing effects on the measurement of the FLC. However, there should be no question that in both the use of the FLC and in its experimental determination, one should always use the most reliable available material model for the material in question, and that means using an advanced material model that is calibrated to uniaxial and bulge test data, even including tension-compression tests to account for kinematic effects, when these are important to the forming processes that require analysis. More details of this method can be found in the literature [2].

2.1. Correction for Nonlinear Strain Path Effect

The procedure for correcting the FLC for nonlinear strain path is most easily described and implemented in a general way in terms of the same UMAT function that is used in the FEM code. But here, instead of applying to UMAT to calculate the increments of the stress tensor and effective plastic strain for a given increment of the total strain tensor *defined by the simulation*, we use the experimental strain tensor history from the DIC strain measurements up to the onset of necking to calculate the experimental increments to the total strain tensor as input to the same calibrated UMAT. It is important to note that this procedure requires care to correctly translate these experimental DIC strain tensors into the material coordinate system in which the material model parameters are defined, to correctly utilize the features of this UMAT solution. This procedure is represented by the following two simple equations,

$$\left(\delta\vec{\sigma}(t), \delta\bar{\epsilon}_p(t)\right) = UMAT\left(\vec{\sigma}(t), \bar{\epsilon}_p(t), \delta\vec{\varepsilon}(t), MP\right) \tag{1}$$

$$\left(\vec{\sigma}(t+\delta t), \bar{\epsilon}_p(t+\delta t)\right) = \left(\vec{\sigma}(t), \bar{\epsilon}_p(t)\right) + \left(\delta\vec{\sigma}(t), \delta\bar{\epsilon}_p(t)\right), \tag{2}$$

where $\vec{\sigma}(t)$ and $\bar{\epsilon}_p(t)$ are the stress tensor and effective plastic strain at time t in the DIC record, both initialized to zero, $\vec{\sigma}(0) = 0$, and $\bar{\epsilon}_p(0) = 0$, the δ in front of these two variables represents their calculated increments obtained from the UMAT function in Eq. 1 and integrated in Eq. 2, $\delta\vec{\varepsilon}(t)$ represents the increment of the total strain at time t obtained the DIC record, and MP represents the set of Model Parameters for the selected material model. If a kinematic hardening model is used, the required back stress tensor values will also be input to the UMAT and their increments returned and integrated along with the stress tensor components and effective plastic strain. The final result of the integration for a given strain path up to the onset of necking is the stress tensor and effective plastic strain given by $\left(\vec{\sigma}, \bar{\epsilon}_p\right)_{onset}$.

2.2. Correction for the Sheet Curvature Effect

Sheet curvatures induce geometrically necessary strain gradients through the sheet thickness. Based on previous studies, onset of necking is suppressed until the conditions on every layer through the sheet thickness are able to participate in the formation of the neck geometry. This means that in the analysis of the conditions for onset of necking, the stress and strain conditions on every layer must be considered, so that the conditions on the least critical layer may be used to correctly define the FLC for in-plane stretching. Although the curvature of the sheet in the area of the neck in the Marciniak Test is small, the bending resistance of metals causes a slight crown on the surface, so that in principle, correction for the sheet curvature should be done for the Marciniak Test. However, in the case of the Nakajima Test, the curvature is dictated by the punch radius, so the strain differences between the top and bottom surface for 1 mm sheet are on the order of 2% strain, which is more than 10% of the available ductility of some metals. These gradients are even higher for thicker gauges.

Fortunately, the DIC measurement system provides surface coordinates as well as the strains, which allow measurement of the principal curvatures (κ_1, κ_2) on the convex (Outer) surface along the directions of the principal strains, which then allows calculation of the sheet thickness and computation of the principal strains on any layer through the sheet thickness by geometric constraint. These calculations are done for the Middle and Inner surfaces and the strain paths on these layers are processed through the same procedure as applied for the Outer layer using Eqs. 1-2, after transforming these calculated principal strains back to the material coordinate system. Except for the consideration of the pressure effect, to be discussed next, one can then identify the critical layer for the determination of the onset of necking under conditions of purely in-plane deformation, by determination of the layer with the lowest value of effective plastic strain, i.e., the layer with the least amount of plastic work.

2.3. Correction for the Contact Pressure Effect

From the theoretical argument that hydrostatic pressure has no effect on the plasticity of pressure-insensitive metals, it is argued that through-thickness pressure delays onset of necking by increasing both components of the biaxial stress condition above those that apply in the absence of pressure. From equilibrium conditions, the pressure P on the contacting side of a doubly curved sheet characterized by principal curvatures (κ_1, κ_2) in balance with the stress in the plane of the sheet (σ_1, σ_2), is given by

$$P = \left(1 + \tfrac{1}{2}h\kappa_2\right)h\kappa_1\sigma_1 + \left(1 + \tfrac{1}{2}h\kappa_1\right)h\kappa_2\sigma_2, \qquad (3)$$

Using this calculated pressure at the onset of necking, the biaxial stress components on the Inner layer is obtained by subtracting P from both stress components. This leads to a lower stress condition at the onset of necking for this layer and also a lower value of the effective plastic strain, which can be obtained by inverting the hardening law after computing the yield function at this lower stress state. The same correction is made for the Middle layer, but in this case $P/2$ is subtracted from the integrated stress conditions on the Middle layer, resulting in a smaller reduction of the effective plastic strain. Then, the critical condition for onset of necking is determined by the layer with the smallest value of the pressure-corrected effective plastic strain.

3. Strain-based FLC

While the described procedure results in the correct definition of the general conditions for the onset of necking, $(\vec{\sigma}, \bar{\epsilon}_p)_{onset}$, it is critical that this limit criterion be restricted to use in applications with the same material model specifically defined and calibrated in the UMAT used in the processing of the FLD experiments. However, depending on the application, the engineer may prefer to use a different material model, and it would be a serious mistake to use these stress limits or effective plastic strain limits with a different material model. In order to mitigate the consequences of using a different material model, it is recommended to convert these stress and effective plastic strain limits into a limit on the plastic strain $\vec{\epsilon}_{\text{onset}}$ defined for linear stress path. The equation for this calculation is the plastic flow rule, whose explicit equations can be extracted directly from the UMAT code. If $\bar{\sigma}_p(\vec{\sigma}, MP)$ is the plastic potential function, then the plastic strain for a linear stress path ending at the critical stress $\vec{\sigma}_{onset}$ and critical effective plastic strain $\bar{\epsilon}_{p,onset}$ is given symbolically by the following simple formula,

$$\vec{\epsilon}_{onset} = \bar{\epsilon}_{p,onset} \frac{\partial \bar{\sigma}_p(\vec{\sigma}_{onset}, MP)}{\partial \vec{\sigma}}. \qquad (4)$$

The net strain limits at the detected onset of necking involved in Marciniak (M), conventional Nakajima (N-4), and 50 mm Nakajima punch (N-2) tests on MP 980 steel without consideration of process effects are shown in Fig 1. It is obvious that these forming limits are test-dependent, and therefore do not represent a material property. The movement of the minimum of the FLC to positive minor strain that is observed in the conventional Nakajima Test (N-4), is significantly amplified in the N-2 test result, which shows that the forming limit is both translating to the right and moving upwards.

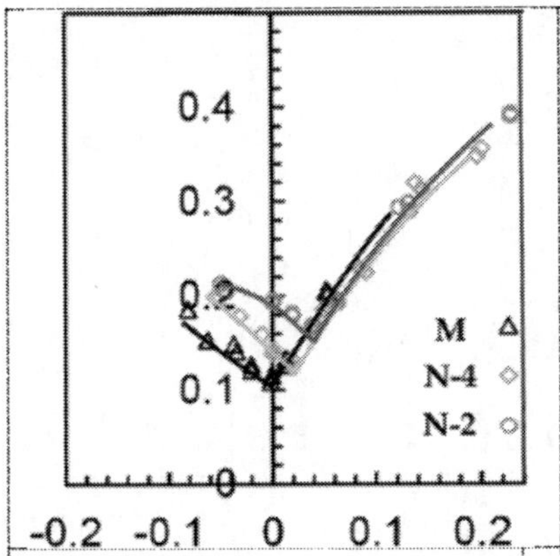

Fig 1. Onset of necking limits for MP 980 steel.

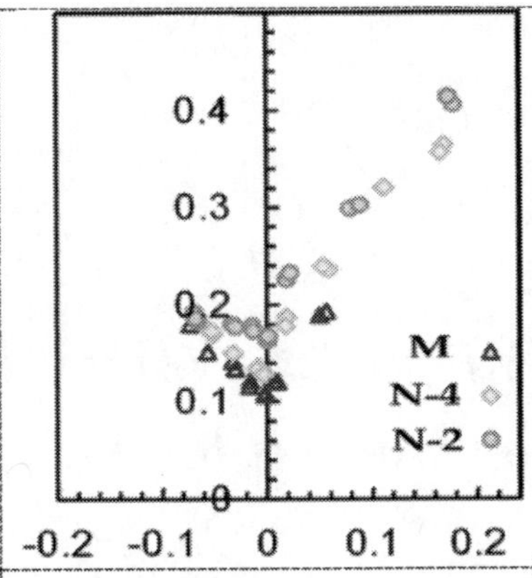

Fig 2. Correction of limits for curvature and nonlinear path.

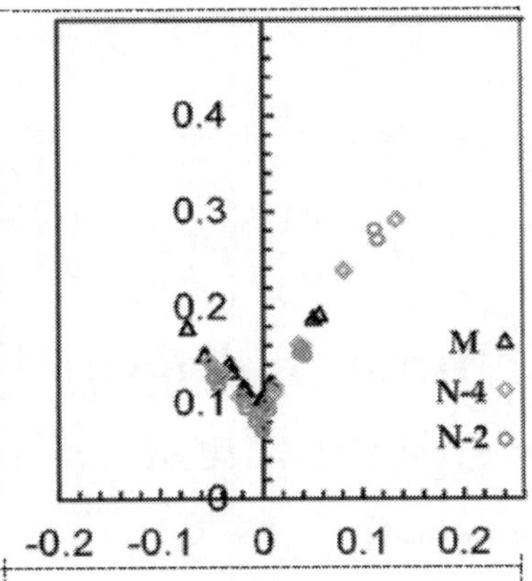

Fig 3. Additional correction for contact pressure effects.

Fig 2 shows the result of correcting for curvature and nonlinear strain path using a simple in-plane isotropic version of Hill 1948 yield function. While the result of the corrections for the N-4 test are similar to the M test, they are still systematically higher than the M test results. The non-convergence is even clearer in comparison with the same corrections for the N-2 test, which are higher still. Fig 3 shows the results of correcting for the contact pressure effects on the Inner and Middle layers of the N-2 and N-4 tests, based on an isotropic hardening law that is obtained by fitting to the experimental data. With these corrections, the forming limits obtained from the M, N-4, and N-2 tests are very close and show the same characteristic slopes on both sides of the FLD.

4. Discussion and Conclusions

Closer inspection of Fig 3 shows that the Nakajima tests are slightly but systematically below the FLC obtained from the Marciniak Test on the left hand side of the FLD. This bias can be attributed to the use of the isotropic hardening model, which leads to an over estimation of the stress and therefore the pressure corrections, particularly on the left hand side of the FLD where the strain path change is largest and the kinematic hardening effects would be expected to be largest. While such an explanation should be investigated and validated using more realistic hardening models, a key conclusion of this study is documenting the importance and role of contact pressure in the Nakajima test. This also leads to the conclusion that unless one has a fully calibrated kinematic hardening model, one should not expect to remove all the process-dependent effects in FLC data obtained from the Nakajima test. That leads to the conclusion that FLC's from the Marciniak Test are more realistic. The final conclusion is based on the observation that the very small differences in the deformation conditions involved in these tests lead to significant variations in limit strains as shown in Fig 1, which emphasizes the importance of accounting for these processes effects in the application of the FLD. As impressive as the fact that one can correct for these differences in measurement of the FLC by applying science to determine a process independent forming limit as shown in Fig 3, the benefits of utilizing this knowledge to determine formability in applications, where the complexity of strain path, the magnitude of curvature, and the contact pressure effects are orders of magnitude larger than these tests, are comparatively orders of magnitude higher.

References

[1] Min J Y, Stoughton T B, Carley J E, Lin J P, *Detection of the Onset of Localized Necking Using Digital Image Correlation Geometry Measurements Part II: Comparison with other Methods and Extension for the Nakazima Test*, Submitted to Materials and Design.

[2] Min J Y, Stoughton, T B, Carley, J E, Lin J P, *Compensation for Process-dependent Effects in the Determination of Localized Necking Limits*, Submitted to IJMS.

[3] Tharret M R, Stoughton T B, *Stretch-Bend Forming Limits of 1008 AK Steel*, SAE 2003-01-1157.

Numerical modelling of the effect of using multi-explosives on the explosive forming of steel cones

T De Vuyst[1], K Kong[2], N Djordjevic[1], R Vignjevic[3], JC Campbell[1], K Hughes[1]

[1] Brunel University London, Department of Mechanical, Aerospace and Civil Engineering, NSIRC, Granta Park, Great Abington, Cambridgeshire, CB21 6AL, UK
[2] Cranfield University, School of Aerospace, Transport and Manufacturing, Cranfield, Bedfordshire, MK43 0AL, UK
[3] Brunel University London, Department of Mechanical, Aerospace and Civil Engineering, Uxbridge, UB8 3PH, UK

E-mail: `tom.devuyst@brunel.ac.uk`

Abstract. Modelling and analysis of underwater explosive forming process by using FEM and SPH formulation is presented in this work. The explosive forming of a steel cone is studied. The model setup includes a low carbon steel plate, plate holder, forming die as well as water and C4 explosive.

The effect of multiple explosives on rate of targets deformation has been studied. Four different multi-explosives models have been developed and compared to the single explosive model. The formability of the steel plate based on forming limit failure criteria has been investigated. Aspects such as shape of plates deformation and thickness of the plate during the forming process have been examined.

The model results indicate that a multi-explosives model does not always guarantee a faster rate of target deformation without central explosive. On the other hand the model results indicate that the multi-explosives setup is capable of preventing crack failure of the steel plate during the forming process which would occur if a single explosive model was used.

1. Introduction

Explosive forming is based on releasing a high amount of energy in a short time into a surrounding environment, creating a shock wave which travels through an energy transfer medium and reaching the work piece, results in the deformation of the work piece. This process has been used in the aerospace industry since the 1960 – 70s. For example, National Aeronautics and Space Administration (NASA) has carried production work for the skin of the Apollo space rockets [1]. The main elements involved in the explosive forming process are: the explosive charge, an energy transfer medium, a forming die and the workpiece. One can distinguish two main types of explosive forming: the contact technique and stand-off technique [1]. In this paper the stand-off technique will be considered, and particularly the application of the explosive forming process to the forming of a steel cone. Water will be used as the energy transfer medium, and a C4 explosive as explosive charge. The objective of the work presented in this paper is to

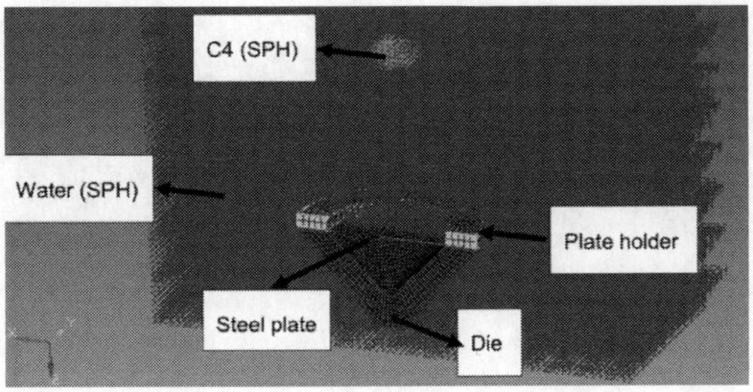

Figure 1. Overview of different parts in the numerical model

determine whether the use of multiple explosive charges offers potential benefits of the use of a single explosive charge.

2. Model Description

In Finite Element Analysis of the explosive forming process the Lagrangian, Eulerian and Arbitrary Lagrangian Eulerian spatial descriptions have been used in the past [2, 3, 4]. However, in this paper a different approach will be followed by using a coupled Finite Element - Smoothed Particle Hydrodynamics (FE-SPH) solver available in LS-Dyna [5]. This choice is based on the Lagrangian nature of the SPH method which makes tracking interfaces more straight forward than in Eulerian methods and is combined with the methods ability to deal with large deformations, hence avoiding numerical problems associated with element distortion/inversion. As stated in the previous section, the component to be formed will be a steel cone. The specific setup is based on the work presented in [6] and consists of a forming die with a base diameter of $200mm$ and angle of 90degrees, and a plate holder. The die and plate holder are modelled as a rigid body and discretised using hexahedral elements. The workpiece is a $1.6mm$ thick low carbon steel plate with a diameter of $200mm$. The plate was modelled with shell elements. Since the objective of this work is focussed on studying the potential benefit of using multiple explosive charges the material behaviour was approximated as elastic-perfectly plastic (yield stress $250MPa$). Finally to contain the water a rigid tank is defined using shell elements. The other parts in the model, the water and explosive are modelled with SPH particles. A spherical 19 gram C4 explosive charge is used throughout the paper. The water domain is modelled as a cube with dimensions $400 \times 160 \times 550mm$. The explosive is modelled using a high explosive burn model and a Jones-Wilkins-Lee equation of state (JWL EOS). The water is modelled as an inviscid fluid with a Gruneisen equation of state. The parameters for the C4 explosive and the water model were taken from [6]. The interaction between the different parts is achieved using the definition of three contact algorithms. The water particles interact with the workpiece, die and plate holder through a node to surface contact algorithm, the plate and forming die interact through a surface to surface contact algorithm, and finally the clamping of the workpiece by the plate holder is handled through a tied surface to surface contact.

Using this model five different explosive charge configurations (see Figure 2) were evaluated. Firstly one single explosive charge is placed centrally and at a distance of $150mm$ above the workpiece, secondly on top of the centrally located charge, a second charge is placed at the same distance from the workpiece, but offset laterally by $50mm$, thirdly on top of the centrally located charge, a second charge is placed at the same distance from the workpiece, but offset by $50mm$ at angle of 45degrees, fourthly on top of the centrally located charge, a second charge

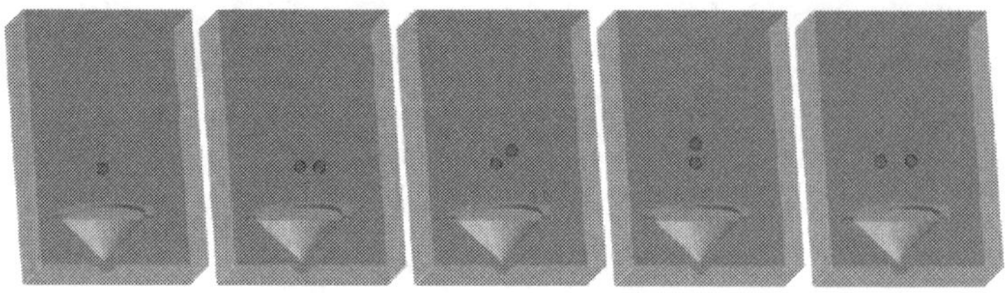

Figure 2. Overview of explosive charge configurations: from left to right configuration 1, 2, 3, 4, 5

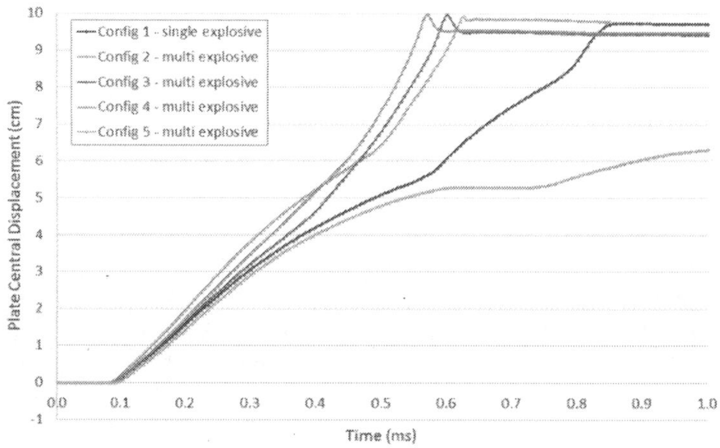

Figure 3. Comparison of displacement time histories at the plate center for the five configurations

is placed at the same distance from the workpiece, but offset vertically, $50mm$ above the first explosive, and finally two explosive charges are placed symmetrically at distance of $50mm$ from the symmetry axis and at a distance of $150mm$ from the plate. In order to assess the effect of these different explosive charge configurations on the formability of the plate, the model results are post processed based on forming limit failure criteria using a Forming Limit Diagram (FLD).

3. Results and discussion

It can be seen in Figure 3 that the multi-explosive configurations does not always result in a higher deformation rate than the single explosive charge. For the case where there is no axially located charge (configuration 5) this results in an incomplete forming operation despite the higher charge. This can be explained by the absence of a charge on the axis of symmetry which would deliver a more direct loading, and also interaction of the two pressure waves. The analysis of the FLD for the other cases shows a clear difference between the single explosive (Config. 1) and the multiple explosive (Config. 4) case, as shown in Figure 4. While the prediction for Config. 1 is one of cracks near the top of cone, Config. 4 the prediction is for wrinkling near the top of cone. In both cases the forming limit failure criteria predict cracking near the clamped area, but this could be due to the combination rigid constraints applied and the sharp corners. Considering a plot of the predicted sheet thickness (see Figure 5 the differences can be linked to different deformation patterns: the region where the sheet is thinnest in the single explosive

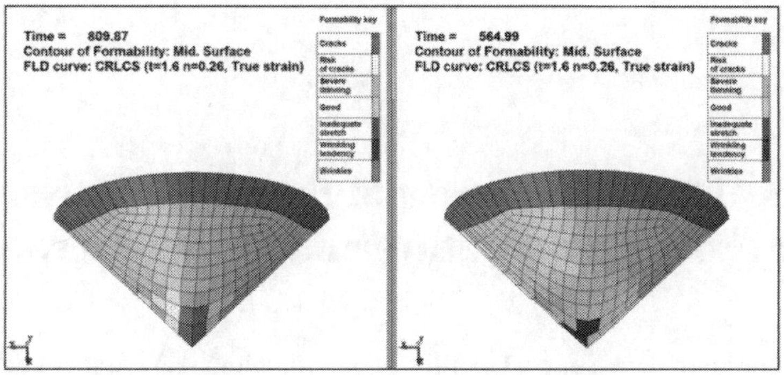

Figure 4. Comparison of Formability prediction for left: Config. 1 and right: Config. 4

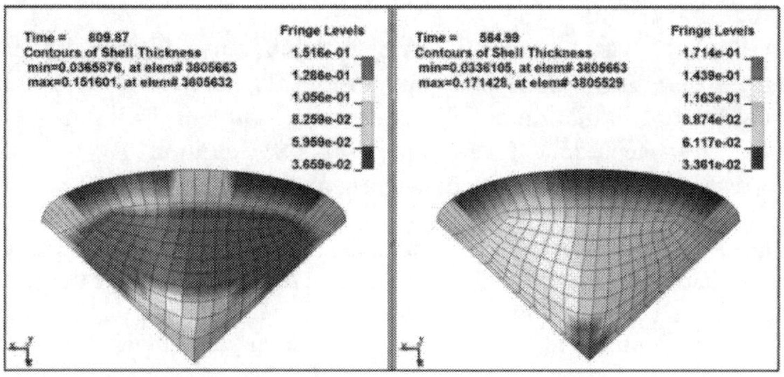

Figure 5. Comparison of predicted sheet thickness for left: Config. 1 and right: Config. 4

configuration is located near the tip of the cone, while in the multi explosive configuration (Config. 4) the thinning occurs along the flanks of the cone and the sheet is thicker near tip of the cone.

4. Conclusion

A coupled Finite Element - Smoothed Particle Hydrodynamics model has been developed to study the explosive forming process, and in particular effect of using multiple explosives on the formability of cones. The numerical results predict that a multi-explosives model does not always guarantee a faster rate of target deformation without central explosive. The model results also indicate that the multi-explosives setup is capable of preventing crack failure of the steel plate during the forming process a failure which would occur if a single explosive model was used.

References

[1] Mynors D J and Zhang B 2002, *J Mat Proc Tech,* **125-126**, 1-25.
[2] Kuroda K, Hamada H, Hamashima H and Itoh S 2007 *6th European LS-DYNA Users Conference*, Gothenburg, Sweden, 5.127-36
[3] Akbari Mousavi S A A, Riahi M and Hagh Parast A 2007 *J MAt Proc Tech* **187-188** 512-6
[4] Alipour R and Najaran F 2011 *World Academy Sci Engng Tech* **5**
[5] LS-Dyna User's Manual 2015 *LSTC*
[6] Alavi Nia A and Emami R 2010 *Steel Research Int* **81(9)**

Journal of Physics: Conference Series **734** (2016) 032075 doi:10.1088/1742-6596/734/3/032075

Modeling the ductile fracture and the plastic anisotropy of DC01 steel at room temperature and low strain rates

V Tuninetti[1,2], S Yuan[3], G Gilles[4], C F Guzmán[3], A M Habraken[3,5] and L Duchêne[3,*]

[1] Department of Mechanical Engineering, Universidad de la Frontera, Temuco, Chile
[2] Technical Department, Distribuidora Cummins Chile S.A., Quilicura, Región Metropolitana, Chile
[3] ArGEnCo Department, MS²F Division, University of Liège, Liège, Belgium
[4] Samtech SA, A Siemens Company, Digital Factory Division, Product Lifecycle Management, Simulation and Test Solutions, Angleur, Belgium
[5] Fonds de la Recherche Scientifique – FNRS, Belgium
* Corresponding author: l.duchene@ulg.ac.be

Abstract. This paper presents different extensions of the classical GTN damage model implemented in a finite element code. The goal of this study is to assess these extensions for the numerical prediction of failure of a DC01 steel sheet during a single point incremental forming process, after a proper identification of the material parameters. It is shown that the prediction of failure appears too early compared to experimental results. Though, the use of the Thomason criterion permitted to delay the onset of coalescence and consequently the final failure.

1. Introduction

There are currently several damage models available in the literature in order to describe the material degradation. The most popular micromechanical model for ductile damage is the Gurson-Tvergaard-Needleman (GTN). Gurson [1] first proposed a porous plasticity model with the void volume fraction as internal variable. The model was further improved to account for the loss of load carrying capacity associated with void nucleation and coalescence [2], hence considering the three stages of damage: void nucleation, growth and coalescence. Since then, the GTN model has been extended to cover different applications. For instance, Benzerga and Besson [3] performed the same analysis as Gurson [1] but considering an anisotropic matrix of the Hill'48 type [4]. A fully implicit integration scheme was implemented by Guzmán [5] for this anisotropic GTN model, also including the effect of the kinematic hardening.

2. Material model

The hardening of the DC01 steel is described by both Swift isotropic and Armstrong-Frederick kinematic hardening models. Swift law is formulated as follows:

$$\sigma_Y = K\left(\varepsilon_0 + \varepsilon_{eq}^p\right)^n$$

where σ_Y is the flow stress, ε_{eq}^p is the equivalent plastic strain and K, n, ε_0 are material constants identified from a tensile curve in the rolling direction. Armstrong-Frederick law is expressed as:

$$\dot{X}_{ij} = C_X\left(X_{\text{sat}}\dot{\varepsilon}_{ij}^p - \dot{\varepsilon}_{eq}^p X_{ij}\right)$$

where C_X and X_{sat} are the kinematic hardening saturation rate and value, respectively. This backstress X_{ij} is used to define the shifted stress tensor: $\tilde{\sigma}_{ij} = \sigma_{ij} - X_{ij}$.

The equivalent stress of DC01 is defined using Hill'48 criterion [4]:

$$\tilde{\sigma}_{eq} = \sqrt{\frac{1}{2}\left[F(\tilde{\sigma}_{22} - \tilde{\sigma}_{33})^2 + G(\tilde{\sigma}_{33} - \tilde{\sigma}_{11})^2 + H(\tilde{\sigma}_{11} - \tilde{\sigma}_{22})^2 + 2L\tilde{\sigma}_{23}^2 + 2M\tilde{\sigma}_{13}^2 + 2N\tilde{\sigma}_{12}^2\right]}$$

where F, G, H, L, M, N are material parameters. Note that if $F = G = H = 1$ and $L = M = N = 3$, Hill'48 yield criterion recovers the isotropic von Mises criterion.

The classical GTN yield surface is defined by:

$$F_p\ (\sigma_{ij}, X_{ij}, f, \sigma_Y) = \frac{\tilde{\sigma}_{eq}^2}{\sigma_Y^2} - 1 + 2q_1\ f \cosh\left(-\frac{3}{2}\frac{q_2\ \tilde{\sigma}_m}{\sigma_Y}\right) - (q_1\ f)^2 = 0$$

where $\tilde{\sigma}_m$ is the macroscopic mean stress and f is the void volume fraction (also called porosity), defined as the average ratio of the void volume to the total volume of the material. The damage parameters q_1 and q_2, originally equal to 1.0 in the initial Gurson model, are usually set to 1.5 and 1.0 allowing the continuum model to be in good agreement with the localization strain for cell analysis [2]. The extended Gurson model used for the simulations has been implemented by Guzmán [5], featuring Hill'48 type anisotropy of the matrix coupled with a mixed hardening law and classical nucleation and coalescence laws proposed in the GTN model. The evolution of voids is additively decomposed in the nucleation f_n and growth part f_g, hence:

$$\dot{f} = \dot{f}_n + \dot{f}_g = \frac{f_N}{S_N\sqrt{2\pi}}\exp\left[-\frac{1}{2}\left(\frac{\varepsilon_M^P - \varepsilon_N}{S_N}\right)^2\right]\dot{\varepsilon}_M^P + (1-f)\mathrm{tr}\dot{\varepsilon}_{ij}^P$$

with ε_M^p as the equivalent plastic strain in the matrix and f_N, S_N, ε_N as material parameters. Coalescence is triggered when the porosity reaches a critical value. Its effect appears as an accelerated evolution of voids through the effective porosity f^* function:

$$f^* = \begin{cases} f & \text{if } f < f_{cr} \\ f_{cr} + \frac{1/q_1 - f_{cr}}{f_F - f_{cr}}(f - f_{cr}) & \text{if } f > f_{cr} \end{cases}$$

where the failure porosity f_F and the critical coalescence porosity f_{cr} are material parameters in the classical GTN model, while the critical porosity is supposed to be reached when the following criterion is no more fulfilled in the Thomason coalescence model [6]:

$$\frac{\sigma_1}{\sigma_Y} < \left[\alpha\left(\frac{1}{\chi} - 1\right)^2 + \frac{\beta}{\sqrt{\chi}}\right](1 - \pi\chi^2)$$

where σ_1 is the maximum principal stress, α is a material parameter defined as a function of the hardening exponent n, β is equal to 1.24 and χ is the void space ratio [6].

Finally, the model was extended to account for the directional growth of voids at low triaxiality (e.g. during shearing), following the Nahshon and Hutchinson shear approach [7].

3. Parameter identification

The elasto-plastic behavior of the DC01 steel (including isotropic and kinematic hardening parameters) was characterized from experimental tensile, shear and plane strain tests along different directions in the plane of the sheet [5]. The methodology is based on using two optimization techniques: OPTIM software [8] based on inverse modeling and the classical simulated annealing [5]. The obtained parameter sets are given in Table 1.

Table 1. Material parameters for the Hill'48 yield criterion and the Swift and Armstrong-Frederick hardening models.

Hill'48	**Swift**	**Armstrong-Frederick**
$F = 0.8103$, $G = 0.9927$, $H = 1.4660$	$K = 542.49$ MPa	$C_X = 113.63$
$N = L = M = 2.9246$	$\varepsilon_0 = 0.0178$	$X_{sat} = 81.96$ MPa
	$n = 0.4328$	

The identification of the GTN model parameters does not follow the same methodology than the one used for the elasto-plastic behavior. The nature of the model, mixing macroscopic and microscopic variables, implies that the identification of the parameters should rely on various experimental measurements [5] (fractography, optical microscopy, strain field measurements on complex geometry samples). The obtained set of parameters is shown in Table 2.

Table 2. Identified set of material parameters for the GTN model.

f_0	f_N	S_N	ε_N	f_{cr}	f_F
0.0008	0.0025	0.175	0.42	0.0055	0.135

4. Single Point Incremental Forming (SPIF)

The studied SPIF process consists in the forming of a truncated cone of 30mm in depth from a circular sheet having a diameter of 182mm and a thickness of 1mm [5]. The wall angle is defined by the path of the spherical tool, whose diameter is 10mm. Various wall angles were tested in order to evaluate the formability of the material, quantified as the maximum achievable wall angle without failure and the obtained experimental value is 67° [9].

The finite element (FE) simulations were performed using the updated Lagrangian FE code Lagamine developed by the ArGEnCo Department of the University of Liège. The 8-node 3D Solid-Shell element, called RESS [10], was selected. It uses the enhanced assumed strain method with only one additional deformation mode. It involves a reduced in-plane integration scheme (with only one integration point) and an hourglass control technique, while the number of integration points along the thickness of the sheet can be adapted (a value of 3 was chosen in the present study). The SPIF process was modeled with 1492 RESS element on a 90° angle pie (using rotational boundary conditions) [5].

The goal of this study was to evaluate the accuracy of variants of the GTN damage models to predict the failure of the sheet for different wall angles. Four variants, available in our FE code, were selected:

- GTN → Classical GTN model where coalescence is initiated by the parameter f_{cr}.
- GTN + Thomason → idem but Thomason criterion is used for coalescence (the critical coalescence porosity f_{cr} is computed from Thomason criterion).
- GTN + Shear → GTN model with Nahshon and Hutchinson shear extension [7].
- GTN + Shear + Thomason → idem with Thomason coalescence criterion.

For each variant of the GTN model, several simulations were performed with increased values of the wall angle (by step of 1°) until failure is predicted. The numerical results for the maximum achievable wall angle are presented in Table 3 and compared to experimental result. For each case, a wall angle of 1° more resulted in the failure of the sheet before the end of the SPIF process.

Table 3. Experimental and numerical results of the cone test.

	GTN	**GTN + Thomason**	**GTN + Shear**	**GTN + Shear + Thomason**	**Exp.**
Maximum achievable wall angle	47°	51°	47°	51°	67°
Maximum porosity at initiation of coalescence	0.0055*	0.01357	0.0055*	0.01363	/
Maximum effective porosity reached	0.1388	0.1644	0.2004	0.1546	/

*: in these variants, by definition, coalescence is triggered when the porosity reaches f_{cr}.

The maximum values on the whole FE mesh of the porosity reached when the coalescence starts (this value is only meaningful for the Thomason coalescence criterion) and the maximum effective porosity reached at the end of the process are also presented in Table 3. It can be noted:

(i) The maximum achievable wall angles predicted by the variants of the GTN model are significantly smaller than the experimental value.

(ii) The shear extension has a very limited influence on the results.

(iii) The Thomason coalescence criterion permitted to increase the maximum achievable wall angle by delaying the onset of coalescence. Indeed, the porosity attained when Thomason criterion is no more fulfilled is way larger than the parameter f_{cr} of the classical GTN model.

(iv) The maximum effective porosity exceeded the failure limit f_F. However, such values only appear very locally for the simulations in Table 3, which was not considered as a failure of the DC01 sheet in this study.

Figure 1 presents numerical results when the failure is predicted (maximum achievable wall angle exceeded by 1°) for the GTN + Shear + Thomason variant. It is shown that the porosity reaches large values only in a zone around the final path of the tool; locally, the porosity can be significantly larger than the failure limit. According to figure 1(b), coalescence appears in a similar zone. Also, the edge effects in figure 1 prove the limitation of the rotational boundary conditions used.

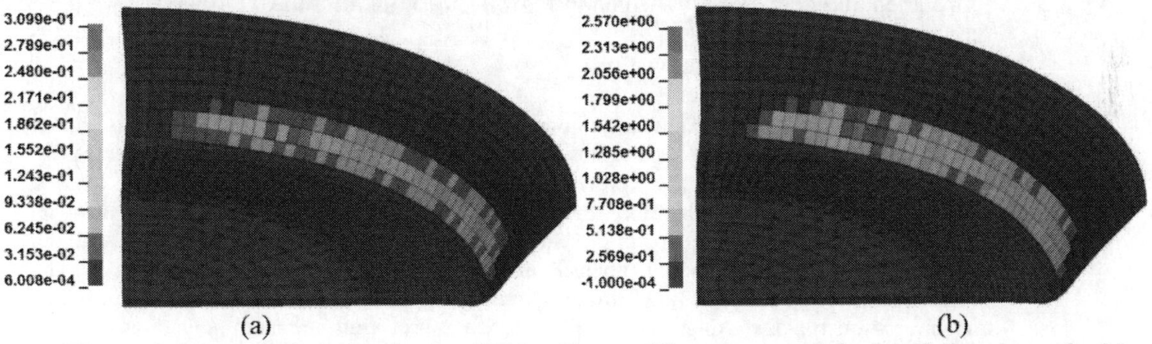

Figure 1. Numerical results for the GTN + Shear + Thomason model when failure is reached (wall angle is 52°). (a) effective porosity f^*, (b) difference between left hand side and right hand side of Thomason criterion (coalescence occurs when positive).

Acknowledgements

F.R.S. - FNRS (Belgium) and the Interuniversity Attraction Poles (IAP) Program P7/21 (Belgian Science Policy) are acknowledged for financial support.

References

[1] Gurson A L 1977 *J. Eng. Mater-T* **99** 2
[2] Tvergaard V 1989 *Advances in Applied Mechanics* ed J W Hutchinson and T Y Wu (Elsevier)
[3] Benzerga A A and Besson J 2001 *Eur. J. Mech. A-Solid* **20** 397
[4] Hill R 1948 *P. Roy. Soc. A-Math. Phy.* **193** 281
[5] Guzmán C F 2016 *Experimental and Numerical Characterization of Damage and Application to Incremental Forming* PhD thesis, Université de Liège, available from http://hdl.handle.net/2268/192884
[6] Zhang Z L, Thaulow C and Odegard J 2000 *Eng. Fract. Mech.* **67** 155
[7] Nahshon K and Hutchinson J W 2008 *Eur. J. Mech. A-Solid* **27** 1
[8] Gilles G, Hammami W, Libertiaux V, Cazacu O, Yoon J H, Kuwabara T, Habraken A M and Duchêne L 2011 *Int. J. Solids Struct.* **48** 9
[9] Behera A K 2013 *Shape feature taxonomy development for toolpath optimisation in incremental sheet forming* PhD thesis, Katholieke Universiteit Leuven
[10] Ben Bettaieb A, Velosa de Sena J I, Alves de Sousa R J, Valente R A F, Habraken A M and Duchêne L 2015 *Finite Elem. Anal. Des.* **107** 44

Contact Modelling of Large Radius Air Bending with Geometrically Exact Contact Algorithm

V Vorkov[1], A Konyukhov[2], D Vandepitte[1] and J R Duflou[1]

[1] KU Leuven, Department of Mechanical Engineering, Celestijnenlaan 300, 3001, Leuven, Belgium
[2] Karlsruhe Institute of Technology (KIT), Department of Civil Engineering, Geo and Evironmental Sciences, Otto-Ammann-Platz 9, 76131, Karlsruhe, Germany

E-mail: vitalii.vorkov@kuleuven.be

Abstract. Usage of high-strength steels in conventional air bending is restricted due to limited bendability of these metals. Large-radius punches provide a typical approach for decreasing deformations during the bending process. However, as deflection progresses the loading scheme changes gradually. Therefore, modelling of the contact interaction is essential for an accurate description of the loading scheme. In the current contribution, the authors implemented a plane frictional contact element based on the penalty method. The geometrically exact contact algorithm is used for the penetration determination. The implementation is done using the OOFEM – open source finite element solver. In order to verify the simulation results, experiments have been conducted on a bending press brake for 4 mm Weldox 1300 with a punch radius of 30 mm and a die opening of 80 mm. The maximum error for the springback calculation is 0.87° for the bending angle of 144°. The contact interaction is a crucial part of large radius bending simulation and the implementation leads to a reliable solution for the springback angle.

1. Introduction

Air bending remains one of the most popular production techniques in sheet metal forming. Recently, with the introduction of high-strength steels with limited bendability, the problem of cracking due to exceedance of deformation limits has become more pronounced. Therefore, the usage of punches with a large radius is a typical approach for the deformation reduction. The bending process with punches of big radii is dissimilar to conventional or small radius air bending. Conventional bending can be considered as a process where the loading scheme remains unchanged throughout the forming cycle and can be described as a traditional 3-point bending mechanism. In contrast with this observation, the loading scheme for large radius bending changes from 3-point bending at the onset to 4-point bending at the end of the forming cycle. Moreover, the loading scheme during the 4-point bending phase depends on the bending angle. The contact points between punch and plate move from the tip of the punch towards the die edges as the forming angle increases [1,2].

The constant change of the loading scheme poses a significant obstacle for the development of analytical models. Finite element analysis with contact interaction is a convenient way for process simulation that allows taking into account the interaction between the tool and the plate. Despite the computational power of modern computers, 3D models still require a significant amount of calculation time, particularly when contact interaction is taken into account. Thus, model reduction is advisable to keep the calculation time short, especially if the aim is to use the solution for adaptive control of forming.

Most of the commercial finite element codes use the master-slave approach for the implementation of the contact interaction [3]. In this approach one of the contacting bodies is represented as the master surface and the other contacting body is considered as the cloud of slave nodes. An efficient approach for determination of the closest distance between the master surface and slave nodes is the geometrically exact contact algorithm, which calculates the amount of contact interaction between two bodies, namely a penetration [4].

The current contribution focuses on the implementation of the frictional plane strain contact element with the geometrically exact contact algorithm as an approach for the penetration calculation.

2. Experimental procedure

Large-radius bending experiments have been conducted on a CNC press brake with a capacity of 50 tonnes. Tooling and plate dimensions are shown in **Figure 1a.** For all plates the width is 50 mm. Bending tests were performed on the high strength steel Weldox 1300. Material parameters and a detailed description of the experimental procedure can be found in reference [5].

The forming process has been recorded by a high-speed camera set-up mounted on the ram of the press brake. Videos of the bending experiments have been processed by WinTopo freeware to retrieve the formed plate profile that is subsequently measured by BricsCAD. Two bending parameters have been measured: (1) the springback angle, which is calculated as the difference between the product and the bend angle (see **Figure 1b**) and (2) the bend allowance, which is defined as the difference between the initial plate length and the sum of flange lengths for the bent part (see **Figure 1c**).

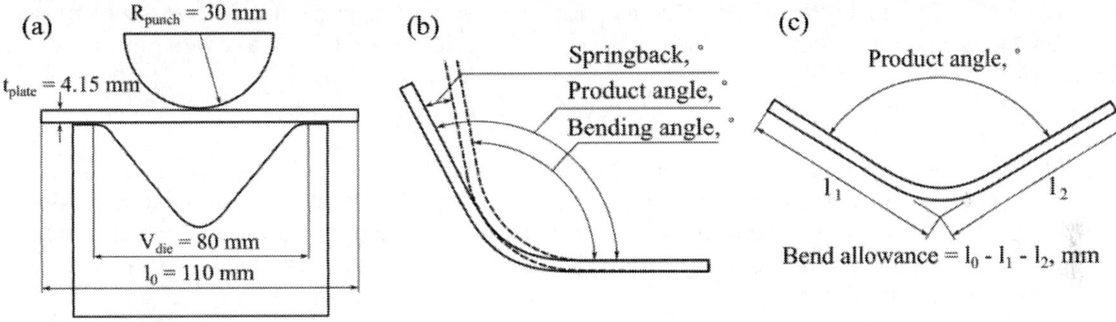

Figure 1. (a) Scheme of the experimental procedure; (b) angles and (c) bend allowance definitions.

3. Finite element approach

For finite element analysis two solutions were used: a solution based on the OOFEM open source finite element code [6]; and a model implemented in the commercial finite element code Abaqus. For the sake of comparison, the mesh and material parameters, and the friction coefficient are the same for both models. Linear elements with plane strain formulation were utilized since this situation is dominant during the bending process [7].

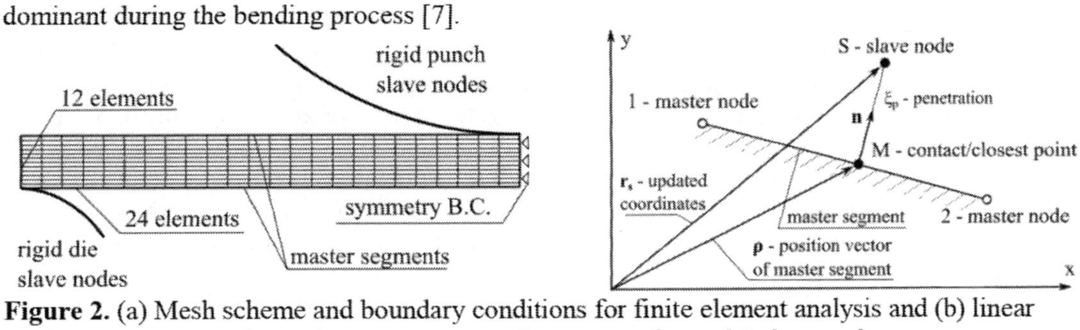

Figure 2. (a) Mesh scheme and boundary conditions for finite element analysis and (b) linear Node-To-Segment contact element geometry: 2 master nodes and 1 slave node.

The interaction between plate and tooling was modelled by means of the contact interaction with a friction coefficient of 0.3. This friction coefficient value is identified from measurement results obtained by Aerens [7] on coarse and fine materials. The contact surfaces of the forming tools (punch and die) in the plane model are represented by absolutely rigid lines. The mesh and boundary conditions are presented in **Figure 2a**. Due to the symmetry of the forming process, only half of the model has been simulated. Isotropic hardening with piecewise approximation of the stress strain curve [5] has been used for modelling the material behaviour.

3.1. OOFEM model

Plane strain elements Quad1PlaneStrain have been taken from the OOFEM element library – an element with the total Lagrangian formulation using the first Piola-Kirchoff stress and the deformation gradient. The solution has been divided into two parts along the deformation cycle: the forming phase and the springback. Both solution steps have been solved using an implicit Static Structural solver.

Despite the fact that the OOFEM library contains a large number of element types, contact elements are not available. Therefore a penalty Node-To-Segment contact element has been implemented according to the description by Konyukhov and Izi [3] with the penetration calculation implemented through the geometrically exact contact algorithm. **Figure 2b** shows the scheme of the Node-To-Segment linear contact element which is defined by two master nodes 1 and 2, and a slave node S.

3.2. Geometrically exact contact algorithm

Closest point projection is a differential geometry procedure which is formulated as the extremal problem for the determination of the shortest distance between a point and a line segment (see **Equation 1**). More information about the closest point projection procedure in a covariant form can be found in [3].

$$F(\xi) = \frac{1}{2}\|r - \rho(\xi)\| \tag{1}$$

According to the geometrically exact contact algorithm for the implemented contact element the value of the penetration is calculated in the direction of the normal vector of the element using **Equation 2** (see **Figure 2b**), where: ξ_p – penetration, $\mathbf{r}_s$ – updated coordinates, $\boldsymbol{\rho}$ – position vector of master segment, and $\mathbf{n}$ – normal vector. For the penetration the following notation is used: $\xi_p > 0$ – no contact; $\xi_p = 0$ – slave point S on contact boundary; $\xi_p < 0$ – contact with penetration.

$$\xi_p = (\boldsymbol{r}_s - \boldsymbol{\rho}) \cdot \boldsymbol{n} \tag{2}$$

3.3. Abaqus model

For the Abaqus model 4-node plane strain CPE4R plane elements were selected. The forming process in Abaqus is again split into two solution steps: the implicit solver has been used in both the forming phase and the springback phase. The mesh and material parameters as well as the friction coefficient are kept the same as in the OOFEM solution.

4. Results and discussion

For the comparison of results the springback and the bend allowance have been selected as the representative parameters for the final shape of the bent plate. **Figure 1b** and **c** show the definition of these parameters. **Figure 3** compares experimental results with OOFEM and Abaqus finite element models. Since the springback depends on the bending angle (see **Figure 1b**) and the bend allowance corresponds to the product angle **(Figure 1c),** appropriate angles are depicted in the graph.

Both finite element models predict the springback angles, with an expected trend: springback increases with increasing bending angle. When compared to experiments, the maximum error for the springback is 0.87° for OOFEM, and 1.45° for Abaqus, for the bending angle of 144°.

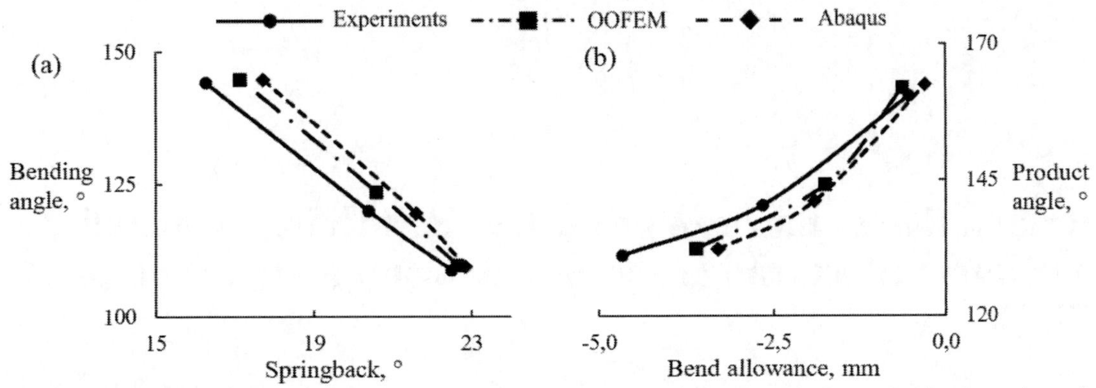

Figure 3. Comparison of experiments and finite element analyses for (a) the springback angle and (b) the bend allowance.

The bend allowance is characteristic of the final shape of the bent part, which allows to calculate the initial length of a blank before the bending process (see **Figure 1c**). This parameter primarily depends on the final or product angle and on the accuracy of the springback prediction, hence the initial error in the springback calculation leads to a significant error in the bend allowance. The maximum error for the bend allowance (the product angle of 131°) is 1.06mm for OOFEM and 1.37mm for Abaqus. The significant errors for the bend allowance mean that the prediction of the bent curvature is not accurate enough for the plane model, since this model cannot handle 3D effects, such as the anticlastic effect.

5. Conclusion

In the current contribution the 2D penalty Node-To-Segment contact element with friction has been implemented within the OOFEM environment. The penetration between the master segment and slave nodes has been calculated by means of the geometrically exact contact algorithm. The obtained solution for large-radius bending was is compared to the solution obtained by the commercial finite element code Abaqus and also to experimental data. The plane finite element model provides an appropriate estimation for the springback angle.

6. Acknowledgements

The authors would like to thank Mikael Öhman from KTH Royal Institute of Technology, Sweden for his help in the implementation of the contact algorithm using OOFEM environment.

References

[1] Vorkov V, Aerens R, Vandepitte D, Duflou JR. On the Identification of a Loading Scheme in Large Radius Air Bending. Key Engineering Materials, vol. 639, Trans Tech Publ; 2015, p. 155–162.

[2] Vorkov V, Aerens R, Vandepitte D, Duflou JR. The Multi-Breakage Phenomenon in Air Bending Process. Key Engineering Materials, vol. 611, Trans Tech Publ; 2014, p. 1047–1053.

[3] Konyukhov A, Izi R. Introduction to computational contact mechanics: a geometrical approach. John Wiley & Sons; 2015.

[4] Schweizerhof K, Konyukhov A. Covariant description for frictional contact problems. Computational mechanics 2005;35:190–213.

[5] Vorkov V, Aerens R, Vandepitte D, Duflou JR. Springback prediction of high-strength steels in large radius air bending using finite element modeling approach. Procedia Engineering 2014;81:1005–1010.

[6] Patzák B. OOFEM—an object-oriented simulation tool for advanced modeling of materials and structures. Acta Polytechnica 2012;52.

[7] Aerens R. Le pliage en l'air. CRIF, Section Construction Mécanique, MC110; 2000.

Determination of the bead geometry considering formability and stiffness effect using generalized forming limit concept (GFLC)

Wan-Gi Cha[1], Sabrina Vogel[2], Nikola Bursac[2], Albert Albers[2], Wolfram Volk[1]

[1]Institute of Metal Forming and Casting, Technische Universität München, Walther-Meißner-Str. 4, 85748, Garching, Germany.
[2]Institute of Product Engineering, Karlsruhe Institute of Technology, Kaiserstr. 10, 76131, Karlsruhe, Germany.

E-mail: wan-gi.cha@utg.de

Abstract. Beads are used in deep drawn sheet metal parts for increasing the part stiffness. Thus, reductions of sheet metal thickness and consequently weight reduction can be reached. Style guides for types and positions of beads exist, which are often applied. However, higher stiffness effects can be realized using numeric optimization. The optimization algorithm considers the two-stepped manufacturing process consisting of deep drawing and bead stamping. The formability in both manufacturing steps represents a limiting factor. Considering nonlinear strain paths using generalized forming limit concept (GFLC), acceptable geometries will be determined in simulation. Among them, the efficient geometry which has higher stiffness effects will be selected in numerical and experimental tests. These will be integrated in the optimization algorithm.

1. Introduction

With the progressive development of computer technology in the nineties, demands for lighter and stiffer metal structures have been increased. FEM-based simulation programs were used to identify improved with the help of parameter variations bead patterns for different load cases [1, 2]. The bead length was optimized based on energy, but the implementation of intersection wasn't considered [3]. A different approach is that the direction of bead stamping is determined based on the determined material distribution by topology optimization [4]. One of the first direct beading optimization was introduced that beads must be determined along the trajectories of the first principal stress [5]. Recently, this algorithm have been developed in bead optimization [6]. The existing algorithms for the optimization of beads have not taken into manufacturing processes as well as the geometry of the sheet metal part.

Within this research project, an optimization method will consider the manufacturing and geometry parameters. A failure model is used to predict formability in the bead stamping simulation after the forming simulation as deep drawing.

This paper will introduce that the cross section of bead geometry is determined by using forming simulation, which carries out the two-stepped process. In order to achieve that, the modified Nakajima test [7] is selected as example of the forming simulation to perform the various stress states as pre-forming. The bead stamping process is carried out after that. The formability of the part, which is deformed after two-stepped forming simulation, is predicted considering nonlinear strain paths by using a concept of the generalized forming limit curve (GFLC) [8]. The geometry parameter of bead profile

Journal of Physics: Conference Series **734** (2016) 032077 doi:10.1088/1742-6596/734/3/032077

varies to identify the influence on the formability and the stiffness of the part. Some experimental cases of pre-formed part is used to validate the simulation by comparison of strain in pol region.

2. Determination of bead profile

It is simulative investigated which bead profile increase effectively the stiffness of the part without the defect as instability or fracture.

2.1. Bead geometry sample

In order to investigate systematically the influences of the bead geometry on the formability and stiffness of the part, the sample of the bead profile is defined as the normalized geometry function as shown in Fig. 1. The geometry parameters consist of head radius r_h, foot radius r_f, height h, width w and edge angle α. The profile function will be implemented in the optimization program.

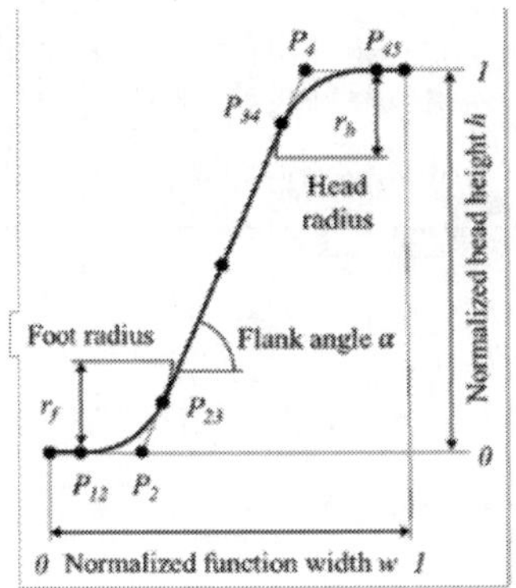

Figure 1. Sample of bead profile with parameterization.

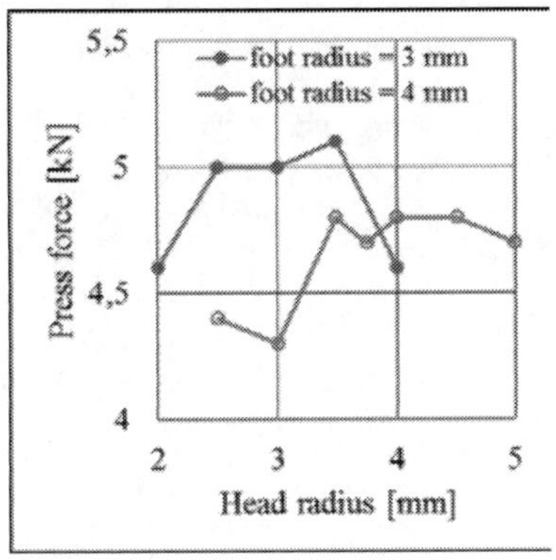

Figure 2. Influence of radii pair.

2.2. Simulation

As mentioned above, the simulation tool AutoForm [9] is used to carry out the modified Nakajima simulation and bead stamping. The materials of HC260LAD and AA6016 are used. A half-spherical punch and a blank holder of the modified Nakajima have diameter of 400 mm and inside diameter of 430 mm, respectively. The modified Nakajima and bead stamping are carried out. The small punch presses on the top of the part after bead stamping. The pressed force identifies the buckling stiffness of the parts.

Table 1. Parameter level

h [mm]	r [mm]	w [mm]	α [°]
15	5	35	35
14	4	37,5	37,5
13	3	40	40
12	2	42,5	42,5
11		45	45
10			47,5
			50

2.3. Parameter study

The defined geometry parameters are five. The influence of the relation between the head and foot radius on the stiffness of the part is investigated. The head radius varies in cases of that the foot radius is fixed as 3 and 4 mm, respectively. The Figure 3 shows a result of the influence, that the high bead forces are calculated when the radii have no big difference each other. This means that the head and foot radius can be regarded as a same radius to obtain the high stiffness. Consequently, the number of the parameters can be reduced to simplify the study.

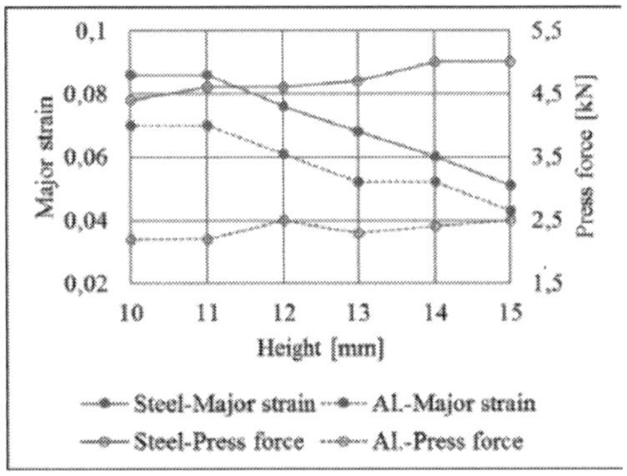

Figure 3. Influence of height.

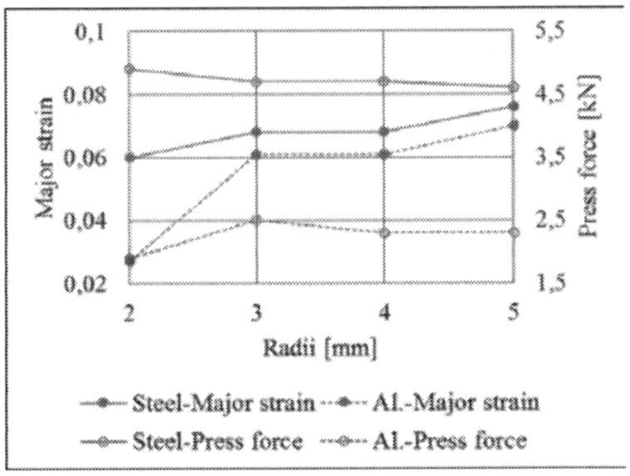

Figure 4. Influence of radii.

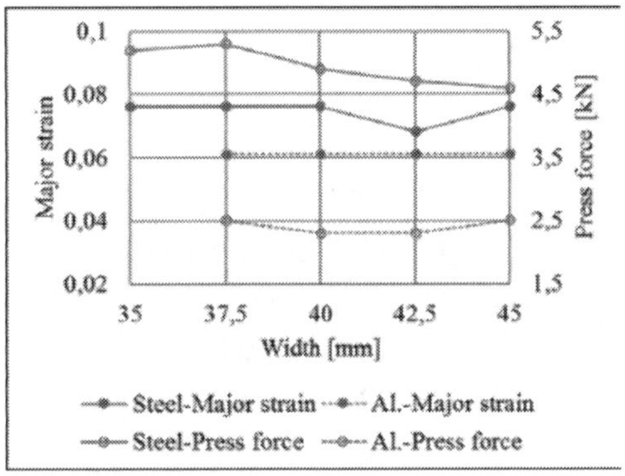

Figure 5. Influence of width.

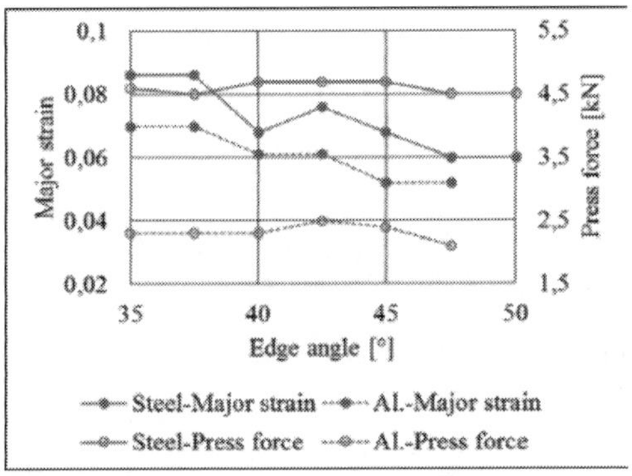

Figure 6. Influence of edge angle.

The four parameters are investigated in several levels by the precondition that the two radii are identical. The change of the parameters is shown in Table 1. The parameters in shadow are defined the reference case. Every parameter is changed to investigate the influence on the formability and the stiffness when other parameters are set the reference case as constant.

Results of the influences of height, radii, width and edge angle are shown in Figure 3-6. It is obvious that the height and the radii are proportional to the stiffness and in inverse proportion to the formability. This means, that the limited bead geometry makes also the high stiffness of part. However, cases of the width and the edge angle show a different aspect. The variation of the width has no significant influence on the formability and the stiffness. The parts have more high stiffness when the edge angles between 40 and 45° are used. For that reason, the width and the edge angle can be regarded as constant geometry parameter.

2.4. Geometry as function of major strain

The available bead profiles must be determined by depending on preforming. Therefore, the height and identical radii of profiles are simply defined as linear function of major strain from the preforming simulation (Figure 7-8).

3. Validation

In order to validate the simulation results, major strains of biaxial preformed parts are measured by ARGUS measurement system [10]. The major and minor strain values on the top are compared to that from the simulation. The biaxial preforming is carried out with strokes of 60, 70 and 80 mm. The material of AA6016 is used in this validation. The major strain from the simulation has values of 0.052,

0.063 and 0.077 in cases of strokes of 60, 70 and 80 mm, respectively. The measured major strain has values of 0.060, 0.071 and 0.083 in cases of strokes of 60, 70 and 80 mm, respectively. The simulation results have been underestimated about 10 % in comparison to the experiment. But, the differences are shown as constant. This means, that the simulation is not accurate but precise. In this work, the material parameters were not identified, but implemented material parameters from AutoForm were applied to this study. While draw beads were carried out by rectangular draw bead geometry in the experiment, they were simplified as line force in the simulation. These can make inaccurate the simulation results. The bead profile can have intentionally less dimension based on the current results after the determination from major strains.

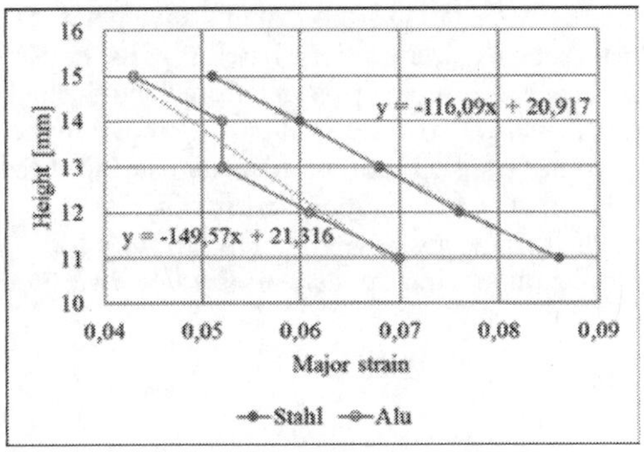

Figure 7. Height function of major strain.

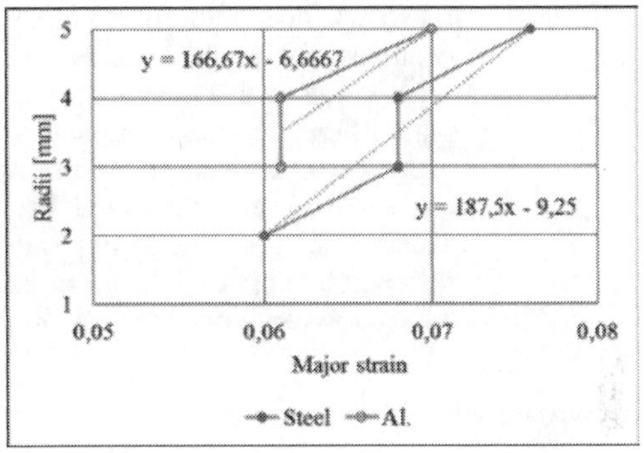

Figure 8. Radii function of major strain.

4. Conclusion

In this paper, how the limited bead profile is determined numerically and the partial validation are introduced. The bead profile can be defined as a function of major strain based on GFLC. The simulation precision was validated by comparison of the strain values from the biaxial preformed parts. Although the validation was carried out only in that case, the determination of limited bead profile from the simulation is reliable.

In the future, other material HC260LAD and other stress states of preforming are going to be examined, and the instability and failure will be confirmed in bead stamping, whether the simulation considers nonlinear strain path right by using GFLC. Moreover, the manufacturing feasibility is going to be also investigated in various bead path and the processing parameters. At the end, the bead process will be optimized under considering of part formability and manufacturing feasibility.

Acknowledgement

The authors would like to thank the German Research Foundation (DFG) for the funding of the depicted research. This project DFG VO 1487/7-1 is supported by DFG.

References

[1] Schriever T, Herrmann F and Maiwald J 1994 *RWTH Aachen FAT Schriftenreihe* **106**
[2] Herrmann F 1997 *RWTH Aachen Schriftenreihe Automobiltechnik* **4097**
[3] Luo J H and Gea H C 1998 *Finite Elements in Analysis and Design* **31** pp 55-71
[4] Yang R, Chen C and Lee C 1996 *Structural Optimization* **12** pp 217-221
[5] Klein B 1995 *Technical university of zürich* **44**
[6] Clausen O N and Pedersen C B W 2009 *WCSMO-8 (Lisbon, Portugal)*
[7] Mackensen A and Hoffmann H 2009 *MS&T 2009 (Pittsburgh, USA)* pp 2279-2289
[8] Volk V, Hoffmann H, Suh J and Kim J 2012 *CIRP Annals (Amsterdam, Holland)* **61** pp 259-262
[9] AutoForm^{plus} R6
[10] ARGUS v6.3.0-6

Effect of some process parameters on geometric errors in two point incremental forming for Al-Cu-Mg Aluminum Alloy

Xiaoqiang Li[1], Kai Han[1], Honghan Yu[1], Yongsheng Zhang [1] and Dongsheng Li[1]

[1]*Beihang University, 37 Xueyuan Road, Beijing, 100191, China*
E-mail: lixiaoqiang@buaa.edu.cn

Abstract. Two point incremental forming receives widespread study with its advantages of economy and flexibility in small batch products, such as aircraft parts. Aircraft parts, however, are rigorous in their shape errors. In this paper, one real airplane part is selected and formed with different process parameters to investigate the shape error level of part. Comparing the geometric errors caused by different process parameters, such as tool diameter, step size, feed rate and tool path, it is found that the geometric errors reduce as tool diameter increases. Meanwhile, the effect of step size is not linear. Influence law of feed rate is various with different other parameters. The bidirectional tool path, having opposite processing direction at adjacent layer, reduces the errors.

1. Introduction

Incremental sheet forming (ISF) has the advantage of flexible processing, short lead time and low costs for small-batch production. In ISF, the forming tool imposes a localized deformation on sheet and the desired part is formed by the accumulation of these localized deformations [1]. Aircraft parts, which generally are small-pitch and various, are very suitable for technical characteristics of ISF. In the meantime, aircraft part is a place where all things are accurate. For this reason, it's significance to investigate applications of this forming process in the aviation industry.

A type of ISF is called Single point incremental forming (SPIF) in which the sheet deforms under the action of forming tool without constraint of die as shown in Fig.1a. But the geometrical accuracy of the parts, manufactured by this process, is generally poor while aircraft parts are rigorous in their shape errors. Two point incremental forming (TPIF) is a particular type of ISF in which the sheet deforms under the action of forming tool with constraint of die placed under the sheet showed in Fig.1b [2].

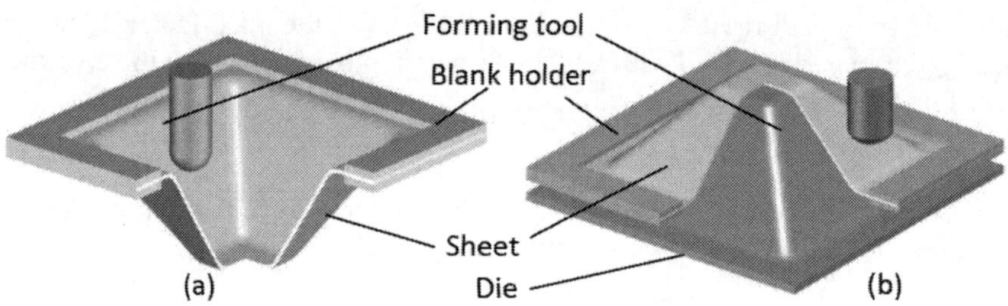

Fig.1-SPIF and TPIF (a) SPIF and (b) TPIF [1]

In previous study, it's observed that process parameters have significant impact on geometric errors [3, 4]. The objective of current study is to investigate the effect of the major process parameters, namely tool diameter, step size, feed rate and tool path, on geometric error [5, 6], with realization of a real airplane part made of Al-Cu-Mg aluminum alloy with different process parameters of positive TPIF at room temperature.

By coordinate measuring machine (CMM), 3D models of the forming part are obtained. Then, the values of geometric errors of the parts are calculated by comparing the forming shape with the desired shape. Finally, the effect of process parameters can be obtained and analyzed by comparing the values of geometric errors.

2. Experimental tests

To realize a real airplane part which 3D model is reported in Fig.2a, the part is further designed to accord with the process of ISF for its special shape as shown in Fig.2b. The part is manufactured on an incremental forming machine (Fig.3). To reduce the geometric errors of parts, TPIF function of this machine is used in present study. As shown in Fig.4, the sheet is clamped on the machine and deformed by the pressure applied from tool and positive die. To obtain the better geometric accuracy, the sheet is pre-bent with dead-weight of gripper mechanism, making sheet abut against the die.

Sheet is made of Al-Cu-Mg aluminum alloy, with 1.00 mm thickness and 600mm×400mm dimensions. The chemical composition of the material is listed in Table 1. The material of the die is wood. During the forming process, the lubricant is smeared on the surface of the part.

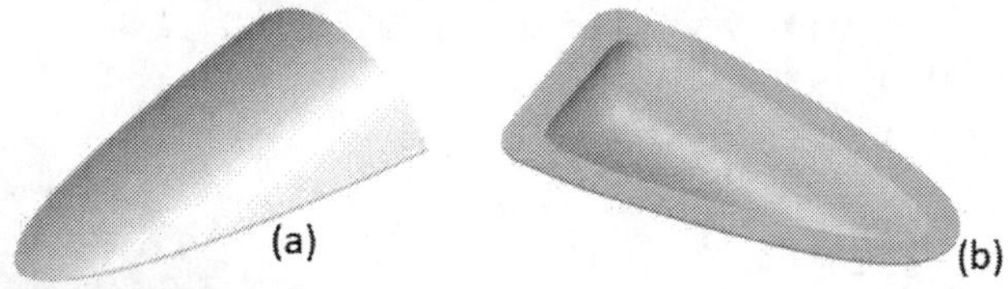

Fig.2-3D model (a) 3D model part of an airplane part and (b) the part designed

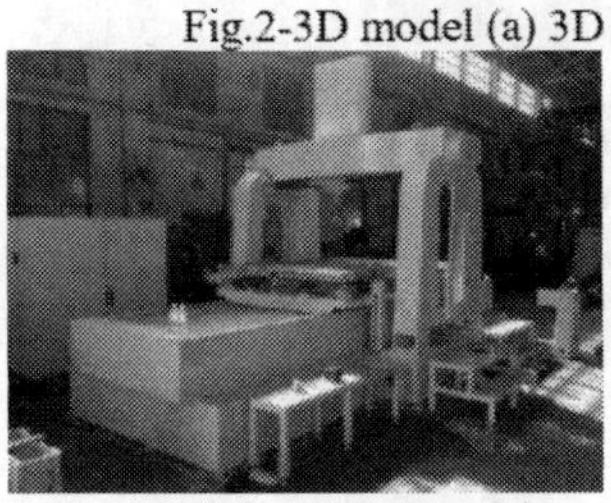

Fig.3-Incremental forming machine

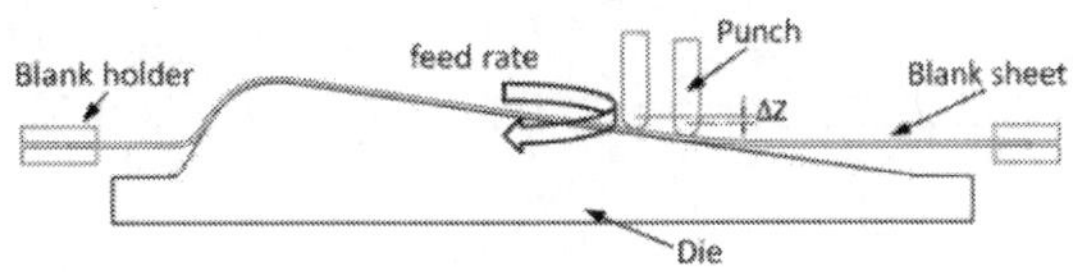

Fig.4- Positive TPIF

Table 1 chemical composition of Al-Cu-Mg aluminum alloy

Element	Cu	Mg	Mn	Fe	Si	Ti	Al
wt.%	3.8~4.9	1.2~1.8	0.3~0.9	0.3	0.2	0.1	Bal.

To study the effect of TPIF process parameters on geometric errors, the major process parameters are selected, which are tool diameter, step size, feed rate and tool path. According to the principle of a single parameter, the design of experiments is carried out and showed in Table 2. The two different tool paths are called unidirectional and bidirectional tool path showed in Fig.6a and b respectively [6]. In unidirectional tool path, the processing direction of each layer keeps same showed in Fig.6a. On the comparison, the processing direction of adjacent layer has opposite processing direction in bidirectional tool path showed in Fig.6b. All of experiments are conducted with the same thickness, dimensions and lubricant.

Table 2 Design of experiments

Exp.#	Tool diameter [mm]	Step size [mm]	Feed rate [mm/min]	Tool path
1	12 mm	0.25 mm	5000 mm/min	Unidirectional
2	12 mm	0.25 mm	8000 mm/min	Unidirectional
3	12 mm	0.25 mm	12000 mm/min	Unidirectional
4	12 mm	0.50 mm	5000 mm/min	Unidirectional
5	12 mm	1.00 mm	5000 mm/min	Unidirectional
6	10 mm	0.25 mm	5000 mm/min	Unidirectional
7	25.4 mm	0.25 mm	5000 mm/min	Unidirectional
8	12 mm	0.25 mm	5000 mm/min	Bidirectional
9	25.4 mm	0.50 mm	12000 mm/min	Unidirectional
10	25.4 mm	0.50 mm	12000 mm/min	Bidirectional

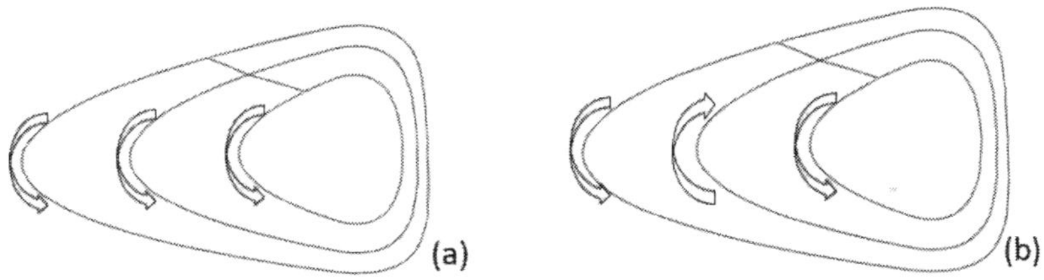

Fig.6- Tool path (a) unidirectional tool path and (b) bidirectional tool path in which arrow represents the direction of the tool path

3. Result and discussion

To evaluation of the geometric error, the point clouds of the outer surfaces are obtained by CMM, and the 3D models are generated by these clouds of points via software. In order to get discrepancies between the desired and experimental geometry, these 3D models are intercepted in three location at x=0mm, x=50mm and x=100mm respectively. As shown in Fig.7, these sections of each model are compared with the desired corresponding sections in the same location for observing shape defects of obtained shape. Then, a quantitative analysis of the obtained results, namely the value of geometric error, is carried out on the basis of the mean root square error between actual and desired profiles.

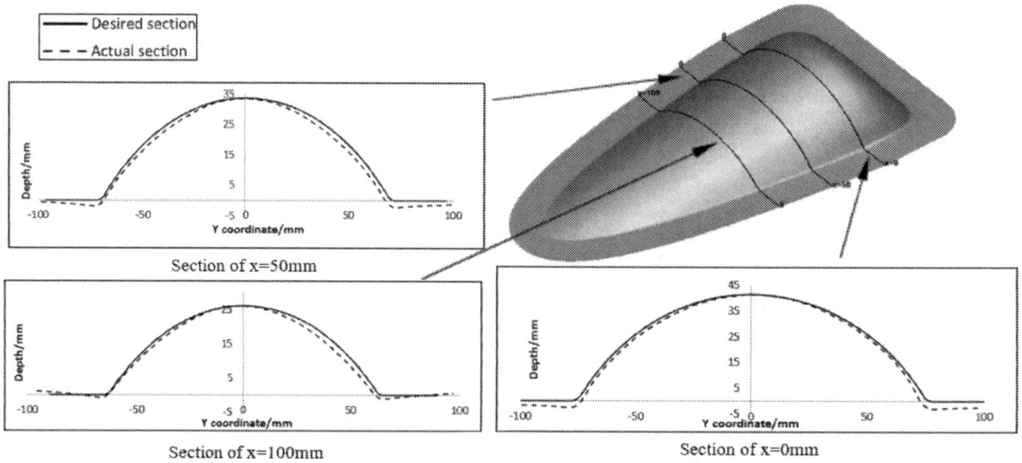

Fig.7-Sections of 3D model

3.1 Effect of tool diameter

Comparing geometric errors of various tool diameters with constant step size and feed rate of 0.25mm and 5000mm/min as shown in Fig.8, it is observed that the increment of tool diameter leads to decrease of geometric error. This phenomenon is due to the increment of tool diameter leads to increment of forming force, as a result of that the geometric error decreases as the forming force increasing [1, 2].

3.2 Effect of step size

The parts are formed at constant tool diameter and feed rate of 12mm and 5000mm/min. As shown in Fig.9, geometric errors decreases at first then increases as the step size increasing. The increment of step size leads to increment of forming force, as a result of decrease of the geometric error [2]. However, as the increment of step size, local strain rate increases, leading to the incremental yield stress. The increment of yield stress makes sheet harder to realize desired shape. Due to this reason, the geometric error decreases from 0.25mm to 0.50mm but increases from 0.5mm to 1.00mm.

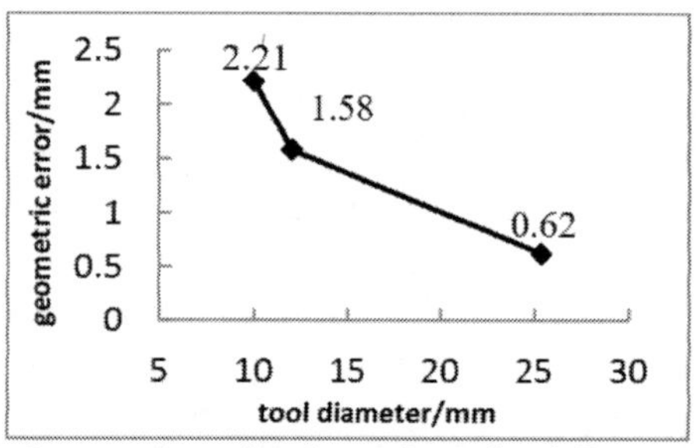

Fig.8-comparison of various values of tool diameter with step size 0.25mm, feed rate 5000mm/min, unidirectional

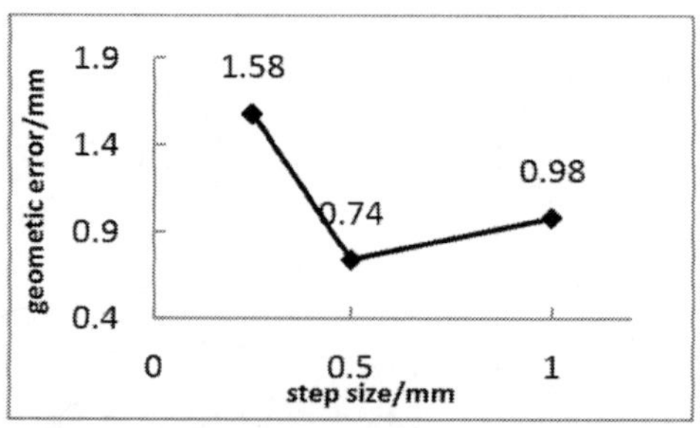

Fig.9-Comparison of various values of step size with tool diameter 12mm, feed rate 5000mm/min, unidirectional

3.3 Effect of feed rate

As showed in Fig.10, it is found that the influence law of feedrate on geometric error is various with different other parameters. When the part is formed at tool diameter and feed rate of 12mm and 5000mm/min by unidirectional tool path, the geometric error increases as feedrate increases. However, the increment of feedrate leads to the decrease of the geometric error of the part, which is formed at tool diameter and feed rate of 25.4mm and 12000mm/min by bidirectional tool path. It is confirmed that the effect of feedrate on geometric error is difference with various other process parameters

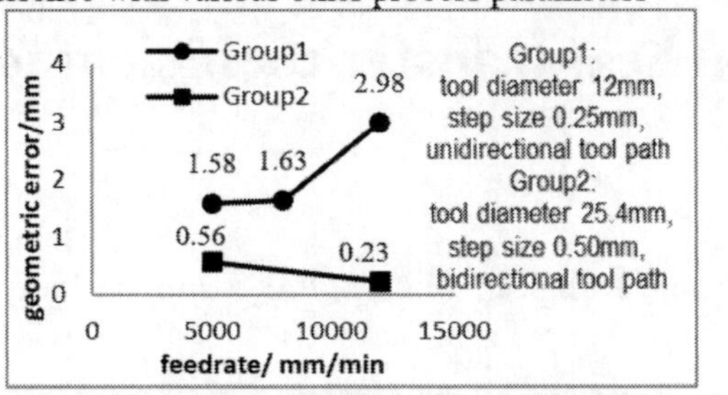

Fig.10-Comparison of various values of feed rate

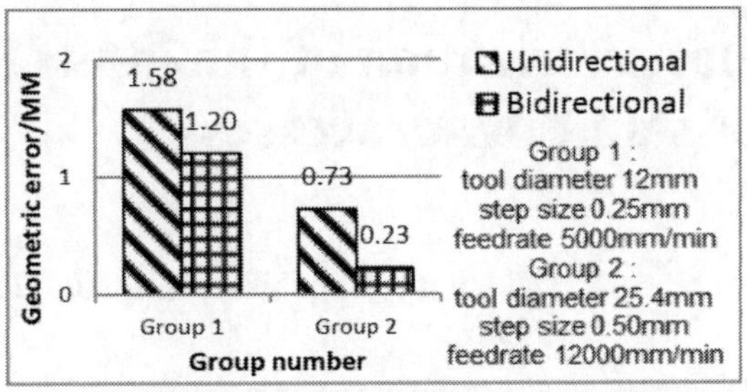

Fig.11-Comparison of geometric errors of different tool path

3.4 Effect of tool path

With different tool path and other same parameters, the parts are compared in Fig.11. It is obvious that the bidirectional tool path shows better geometric accuracy compared to unidirectional tool path. It is due to tool path of same processing direction of each layer shows worse twist, as a result of worse geometric accuracy [6]. Decreasing geometric errors effectively, bidirectional tool path can be used to improve geometric accuracy.

4. conclusion

As a preliminary study, with realization of airplane part made of Al-Cu-Mg aluminum alloy, the effect of process parameters on geometric error is obtained and analyzed. The primary conclusions are summarized as following:

1. As discussed above, it is obtained that the increment of tool diameter leads to the reduction of geometric errors.

2. The influence law of step size is not linear. Therefore, it is meaningful to study the suitable step size under the premise of insuring geometric accuracy to realize the improvement of TPIF production efficiency.

3. Influence law of geometric error affected by feed rate is various with different other parameters.

4. In this paper, bidirectional tool path has been shown to reduce the geometric errors.

References

[1] Tekkaya A E. Sheet metal forming : processes and applications[J]. Sheet Metal Forming Processes & Applications, 2012.

[2] Fiorentino A, Ceretti E, Attanasio A, et al. Analysis of forces, accuracy and formability in positive die sheet incremental forming[J]. International Journal of Material Forming, 2009, 2(1):805-808.

[3] Attanasio A, Ceretti E, Giardini C, et al. Asymmetric two points incremental forming: Improving surface quality and geometric accuracy by tool path optimization[J]. Journal of Materials Processing Technology, 2008, 197(1):59–67.

[4] Attanasio A, Ceretti E, Giardini C. Optimization of tool path in two points incremental forming[J]. Journal of Materials Processing Technology, 2006, 177(s 1–3):409–412.

[5] Jeswiet J, Micari F, Hirt G, et al. Asymmetric Single Point Incremental Forming of Sheet Metal[J]. CIRP Annals - Manufacturing Technology, 2005, 54(2):88-114.

[6] Wu SH, Reis A, Pires FMA, Santos AD, da Rocha AB, Study of tool trajectory in incremental forming. Adv Mater Res 472–475:1586–1591, 2012.

Journal of Physics: Conference Series **734** (2016) 032079 doi:10.1088/1742-6596/734/3/032079

Characterization of the interfacial heat transfer coefficient for hot stamping processes

Xi Luan, Xiaochuan Liu, Haomiao Fang, Kang Ji, Omer El Fakir and LiLiang Wang *

Department of Mechanical Engineering, Imperial College London, SW7 2AZ, UK

E-mail: liliang.wang@imperial.ac.uk

Abstract. In hot stamping processes, the interfacial heat transfer coefficient (IHTC) between the forming tools and hot blank is an essential parameter which determines the quenching rate of the process and hence the resulting material microstructure. The present work focuses on the characterization of the IHTC between an aluminium alloy 7075-T6 blank and two different die materials, cast iron (G3500) and H13 die steel, at various contact pressures. It was found that the IHTC between AA7075 and cast iron had values 78.6% higher than that obtained between AA7075 and H13 die steel. Die materials and contact pressures had pronounced effects on the IHTC, suggesting that the IHTC can be used to guide the selection of stamping tool materials and the precise control of processing parameters.

1. Introduction

Due to growing concerns about fuel economy improvement and exhaust emissions reduction, there has been a remarkable growth in the development of new forming technologies applied in the automobile and aviation industries. Weight reduction, specifically through utilizing aluminium alloys to replace steels, has become a major interest in this area. A 40% reduction in weight and 30% reduction in fuel consumption can be achieved by using aluminium alloys [1]. However, their poor ductility at room temperature restricts the widespread adoption. The most common solution to this problem is forming the alloys at elevated temperatures, e.g. using warm or hot stamping processes.

In hot stamping process, the high post-form strength potential of aluminium alloys is largely influenced by the quenching rate during the in-die quenching stage of the process [2]. The interfacial heat transfer coefficient (IHTC) is an important thermo-physical parameter characterizing the heat exchange between the blank and the forming tools. Hence, it has been widely used in both Finite Element (FE) modelling of hot forming processes and forming process parameter determination [3]. Although many studies have been conducted on the IHTC in hot forming of high strength steel [4], very few focus on hot forming of aluminium alloys, especially with different die materials. In this work, the IHTC between the aluminium alloy AA7075-T6 and two different die materials, cast iron and H13 die steel, was determined through hot stamping tests and simulations, under different contact pressures up to 17 MPa.

2. Methodology

2.1. Experiment setup

The experiments were conducted on a dedicated test rig, the IHTC-Mate, which was integrated on a Gleeble 3800 thermo mechanical simulator. The aluminium alloy AA7075-T6, a high strength alloy widely utilized in the aviation industry, was used in this study. The dimensions of the specimen were 120 mm × 10 mm with a thickness of 2 mm, while the dimensions of the contact surfaces of the die and punch were 500 mm × 250 mm. The specimen was heated by direct resistance heating, while the sliding block on the frame where the blankholders were mounted was made up of insulating material to isolate the specimen and blankholders from the rest of the rig. For each test, the heating rate, target temperature and the contact pressure in the form of the punch force were set in advance. The specimen was heated up to 490℃ with a heating rate of 10℃/s and immediately quenched after the target temperature was reached. In the quenching stage, the punch moved towards the specimen and compressed it against the die, holding it with the pre-set punch force for 10 seconds, to simulate the same process that occurs during hot stamping. The hot stamping tests were carried out at a series of contact pressures from 0 MPa to 17 MPa on the two sets of die and punch respectively. A thermocouple embedded at the centre of the test specimen provided the temperature measurement and feedback control. Another two thermocouples were inserted into the punch and blankholder and connected to a data logger. The temperature evolution was used for the calibration of the developed FE model. The surface roughness (Ra) of the die and punch were measured using a White Light Interferometry equipment (Wyko NT9100).

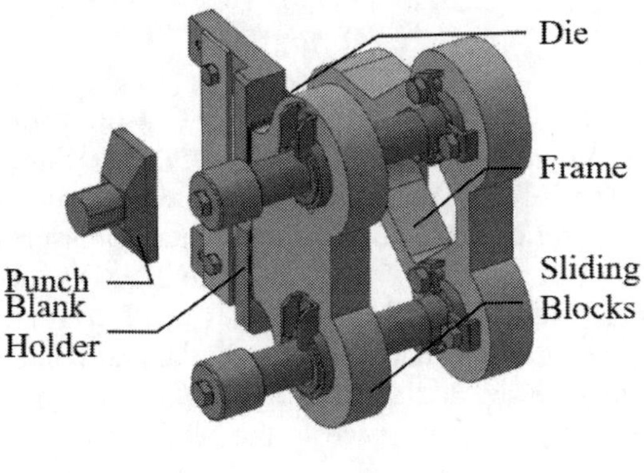

Figure 1. Structure of the dedicated test rig

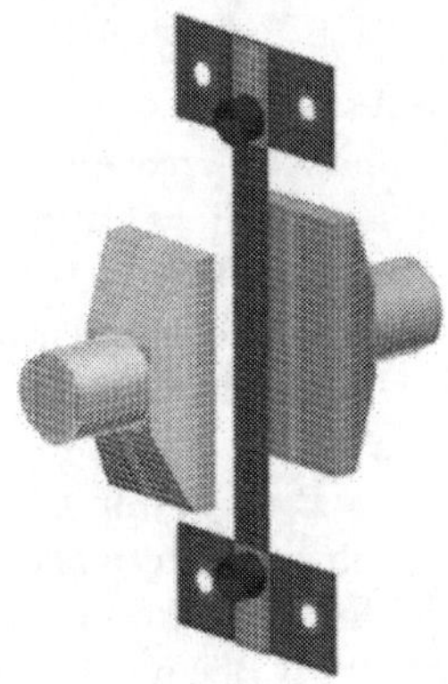

Figure 2. FE simulation model for IHTC determination

2.2. FE simulation

The FE model of the test rig was created using the commercial simulation software PAM-STAMP, as shown in Figure 2. The geometry of the blank, punch, die and blank holders generated in Solidworks were imported into PAM-STAMP and the material properties of all the parts and the process parameters were kept consistent with those used in the experiments. Within the temperature range used in the tests, the average thermal conductivities of cast iron and H13 die steel were 44.0 W/m-K and 24.4 W/m-K respectively. The initial temperature of the specimens was set to 490°C at the middle segment. The initial temperature of the die and punch were set to 30°C.

2.3. Determination of the IHTC values

Through assigning a series of IHTC values between the hot blank and the cold die and punch in PAM-STAMP, different temperature evolutions of the node at the specimen's centre during quenching were plotted and compared with the temperature evolution recorded by the data logger during the experiment on the same figure. The IHTC value and corresponding simulated temperature evolution that showed the best fitting to the experimental results was determined to be the IHTC value in the experiment. As shown in Figure 3, the IHTC value between the hot blank and cast iron was found to be 13.5 kW/m^{2}K for a contact pressure of 5 MPa, which provides the best fit to the experimental results.

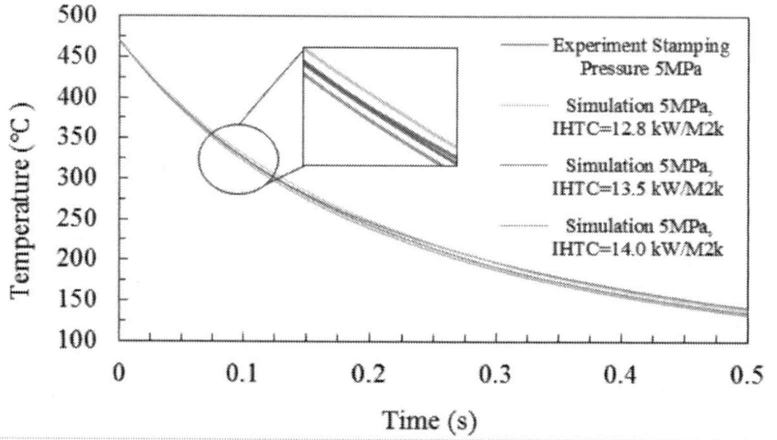

Figure 3. Temperature evolution of specimens centre with cast iron die with various pre-defined IHTC values

3. Results and discussion

Figure 4 shows that an increase in contact pressure promotes a significantly higher IHTC value within a certain pressure range. For cast iron, there is a remarkable growth of the IHTC value from 0.8 kW/m^{2}K at zero contact pressure to 15 kW/m^{2}K at a contact pressure of 13 MPa, after which a plateau in the IHTC value was observed. Similar to the trend of the IHTC for cast iron at different contact pressures, the IHTC for H13 die steel increased with increasing contact pressure from 0 MPa to 10 MPa, and subsequently plateaued at approximately 8.4 kW/m^{2}K. Higher contact pressures lead to more deformation of the asperities on the workpiece material, leading to a sharp increase in the real contact area and thus an enhanced thermal exchange, resulting in an increase in the IHTC [5]. As the real contact area approached the apparent contact area under high pressure, convergence in the IHTC value was observed. Additionally, the results suggest that thermal conductivity and surface roughness, which were different for both materials, have appreciable effects on the IHTC, which has been confirmed by previous research [6, 7].

According to research on continuous cooling transformation diagrams (CCT) of aluminium alloys [8], when the quenching rate in a forming process is lower than the alloy specific critical cooling rate, there is a tendency for some solute elements to precipitate out as coarse particles, which weakens the alloys' response to any subsequent artificial ageing. In general, AA7XXX alloys require a higher quenching rate (over 50°C/s) while AA6XXX alloys' critical quenching rate is approximately 20°C/s. For the same alloy grade, the critical quenching rate is also alloying element dependent, i.e. the higher the content of alloying elements, the greater quenching rate that will be required. By superimposing the quenching curves with the CCT diagrams, the required contact pressure for achieving the critical quenching rate for any aluminium alloy can be determined. The converged value of the IHTC determined from the experiments would also prevent an excessive contact pressure from being applied that would have no effect on the IHTC, reducing tool wear, and promoting the cost efficiency of the hot stamping process.

During the quenching stage of the hot stamping process, the hot aluminum blank is rapidly cooled down by cold dies with embedded water cooling channels. Within the typical contact pressure range encountered in aluminium alloy hot forming processes, the increase in the IHTC for cast iron is almost double that of the increase in the IHTC for H13 die steel as the contact pressure is increased, resulting in a much higher quenching rate. For a component with large dimensions and a complex geometry, the temperature distribution over the surface may be preferred to be uniform or tailored during the forming and quenching process in order to achieve the desired strength and thickness distribution [9], and this may be controlled by utilizing different die materials in different parts of the forming tool and with different contact pressures applied. By using cast iron, the design of the cooling channels may be more flexible, enabling more effective and efficient quenching, compared with conventional H13 tools in which cooling channels would need to be drilled during manufacturing. Utilizing different materials may also be beneficial for hot forming processes where blankholders are used, as they usually result in a faster than required temperature drop and premature cracking in their contact region as the central portion of the component is being formed.

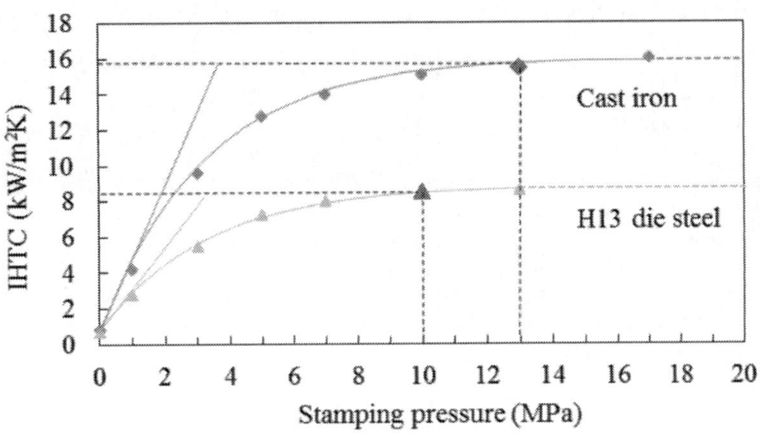

Figure 4. Effect of contact pressure on the ITHC with different die materials

4. Conclusion

In this work, the interfacial heat transfer coefficient (IHTC) between a hot aluminium alloy 7075-T6 blank and two die materials, cast iron and H13 die steel, were investigated based on hot stamping experiments in conjunction with FE simulations in PAM-STAMP. IHTC values under a series of different contact pressures were obtained, it was found that the IHTC increased with increasing contact pressure up to a plateau value. For cast iron, the IHTC converged to a value of 15 kW/m^{2}K when the contact pressure was higher than 13 MPa. For H13 die steel, the IHTC plateaued at 8.4 kW/m^{2}K at a contact pressure of 10 MPa. From the results, it can be deduced that cast iron forming tools might be more advantageous for manufacturing ultra-high strength lightweight components from AA7XXX alloys or quenching rate sensitive alloys, while H13 steel might be more suitable for manufacturing blankholders or forming aluminum alloys that are less quenching rate sensitive.

References

[1] LUTSEY N P 2010 *Institute of Transportation Studies Report* UC Davis
[2] MARTIN J W 1998 *Precipitation Hardening 2nd ed* (Britis: Butterworth Heinemann) pp79-124
[3] Bai Q *et al 2012 Int. J. of Machine Tools and Manufacture* **56** 102
[4] Caron E *et al* 2014 *Int. J. of Heat and Mass Tranfer* **71** *396*
[5] Chang Y *et al* 2016 *J of Material Processing Technology* **228** 25
[6] Kreith F *et al* 2011 *Principle of Heat Transfer 7th ed* (USA: CENGAGE Learning) pp 23-45
[7] Holman J P *et al* 2010 *Heat Transfer 10th ed* (USA: McGraw Hill) chapter 1 pp5-9
[8] Milkereit B *et al* 2012 *Material Sci. and Eng. A* **550** 87
[9] Karbasian H *et al* 2010 *J of Material Processing Technology* 210 pp 2103-2118

Foil Blanking Mechanism Research Using Rubber Tool by Finite Element Simulation and Experiment

Yang-Kai Chen, Xiao-Xing Li, Li-Hui Lang, Rui Xiao, Yu-Long Ge

School of Mechanical Engineering and Automation, Beihang University, Beijing 100191, China
E-mail: li.xiaoxing@buaa.edu.cn

Abstract. For foil blanking process, the usage of flexible tool can effectively reduce the requirement of the manufacturing and assembling precision, compared with using conventional tool. However, the blanking mechanism using rubber tool is not clear. To investigate this question, the Finite Element (FE) model of rubber and process is established using ABAQUS package. The result of FE simulation affirm that the fracture emerges as a result of shear, not tensile. Then, for titanium foil with 0.08mm thickness, the cutting experiment is executed to verify the validity of blanking mechanism and FE simulation.

1. Introduction

Micro-Electro-Mechanical System (MEMS) is attracted the attention of the scholars, recently, as its widely application, in many fields such as medical science and biological science. The mechanical parts in each MEMS are small in size, most of them are foil with micron grade thickness. High dimensional and assembly precision must be guaranteed, which go against cost and time reduction, for forming and blanking of foil sheet using rigid tools. Therefore the method substituting flexible tools for rigid mould, die or punch, are greatly used.

The effects of many parameters using rigid tools are carried out, such as relative blanking clearance, corner radius of tools and punch speed[1][2][3].

In the light of our view, one reach only is in regard of flexible blanking process. F. Takahashi[4] developed a novel method of blanking using rubberlike materials for punch instead of rigid tool. The influence on the product quality of some parameters was investigated, such as the counter pressure, punch material and punch penetration into the die. They conclude that the counter pressure at the blanking completion should range from 200 to 400MPa.

However, the mechanism is not clear. To explore this mechanism, this work carries out finite element analysis firstly. Then the tool set with rubber punch is build and achieve the experiment based on this set. The morphology of fracture surface is obtained.

2. Fracture Criterion

$$\int_0^{\varepsilon_{eq}^{**}} \frac{d\varepsilon_{eq}}{\varepsilon_{eq}^{**}} = 1 \tag{1}$$

Fracture criterions proposed by researchers mainly include constant equivalent strains, the plastic instability theory, forming limit diagram (FLD) and maximum shear stress. In this work, the criterion represented in formula (1) is applied, in which a non-linear strain path is taken into account. This comprehensive criterion can predict not only ductile fracture but also shear fracture with different expression of ε_{eq}^{**}. The validity of this criterion is verified by results of the three point bending process and the uniaxial compression test using square steel tube in Hooputra et al.[5].

3. Modelling and Experiment

3.1. Finite Element Model

Sheet blanking processes are considered as plane strain problems. The geometry model of die cutting edge, which is crucial in rubber blanking process, is established as shown in Figure 1. The thickness of sheet is 0.08mm and rubber1mm. The binder ring and male die are the same rubber.

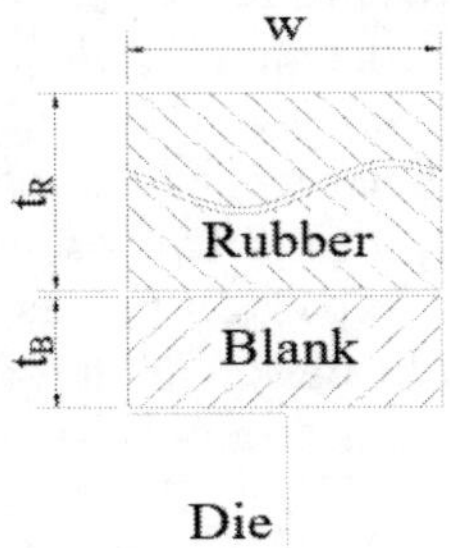

Figure 1. the geometry model of flexible blanking mould

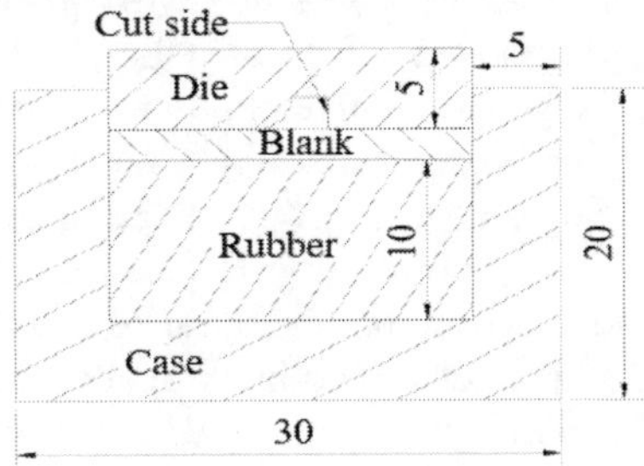

Figure 2 blanking tool

3.2. Material property

The sheet metals are 0.08mm thickness for both finite element analysis and experiment. The uniaxial tensile test is executed with Zwick(100KN) tensile testing machine. The material is TA1. The value of parameters for power exponent constitutive equation are showed in Table 1.

Table 1. Mechanical properties of material TA1 with thickness 0.08mm

Young's modulus E(GPa)	Yield strength (MPa)	Strength coefficient k(MPa)	Hardening exponent n
108	260	1004	0.44

Rubber is one of the hyperelastic materials. The mechanical properties of hyperelastic materials are described in terms of a strain energy potential W, which defines the strain energy stored in the material per unit of reference volume (volume in the initial configuration) as a function of the strain at that point in the material. The Mooney-Rivlin form, formula (2) based on isotropic materials , is applied in this article to predict the rubber's behavior on account of its validity.

$$W = C_1\left(\bar{I}_1 - 3\right) + C_2\left(\bar{I}_2 - 3\right) + \frac{1}{D_1}(J-1)^2 \tag{2}$$

where **W** is the strain energy per unit of reference volume; C_1, C_2 and D_1 are material parameters; $\bar{I}_1$ and $\bar{I}_2$ are the first and second deviatoric strain invariants; J is the elastic volume ratio.

Table 2. Mooney-Rivlin constants of rubber evaluated by ABAQUS

Hardness shore A	C_1	C_2	D_1	Poisson's ratio
60	0.123	0.0272	0	0.4

The rubber with Shore hardness 60HA and density 1.1g/mm3 is used in this text. The uniaxial tensile test is performed on Zwick(100KN) tensile testing machine and the biaxial tensile test is accomplished on biaxial tensile testing machine described in R. Xiao et al.[6]. The material parameters evaluated using the above test results in ABAQUS, is shown in Table 2.

3.3. *Blanking Experiment*

The blanking tool are shown in Figure 2. Both the width and depth of groove are 2mm. Quenching is implemented to confirm the stiffness and abrasive resistance of cutting edge. The female die is pushed down by electric pushrod with 5mm/s velocity and the maximum force of 15KN.

4. Results and Discussion

4.1. *Rubber blanking process*

The rubber blanking process is shown in Figure 3, the left one is contour map of hydrostatic pressure (MPa) and the other is contour map of the shear(MPa) stress corresponding to left one. The diagonal section line regions represent the hydrostatic pressure greater than 0MPa.

With the movement of die, the elastoplastic deformation of the sheet occur and the sheet metal enter in the groove. There is no crack shown in Figure 3(a). It is obvious that the hydrostatic pressure of die region is larger than the groove zone, demarcated by the line perpendicular to die upper surface through the point near groove edge. The maximum value meet in the area close to die top face. The greater plastic deformation is generated than conventional rigid blanking process because of whole blank under continuous increase of pressure. In this process, blending and stretching of blank fiber exist if friction between rubber and other parts is taking into account. Graph on right side indicates that the maximum shear stress takes place above the cutting edge. The decreasing proportion and increasing shear force lead to a raising shear stress. The crack occur when the shear stress comes to shearing stretch.

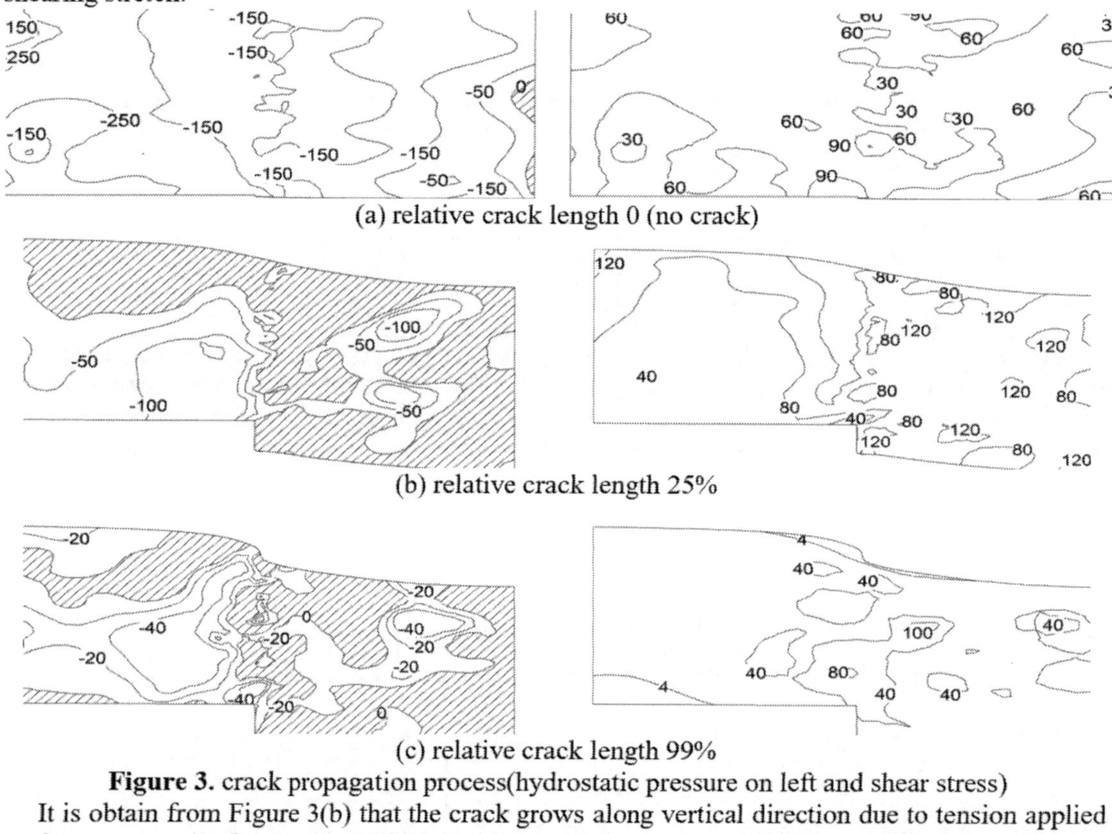

(a) relative crack length 0 (no crack)

(b) relative crack length 25%

(c) relative crack length 99%

Figure 3. crack propagation process(hydrostatic pressure on left and shear stress)

It is obtain from Figure 3(b) that the crack grows along vertical direction due to tension applied on no fracture zone by fracture part which is consistent with the contour map of hydrostatic pressure. The shear stress meet its maximum value over the crack tip from shear stress graph.

The state of stress show in Figure 3(c) represent that tensile stress is far lower than compressive stress near the crack tip according to hydrostatic pressure around. Meanwhile the isoline map of shear stress figures out the maximum shear stress appears at the crack tip and the fracture zone perpendicular to blank surface. That is to say, the shear stress play a major role in crack growth even in the end of rubber blanking process.

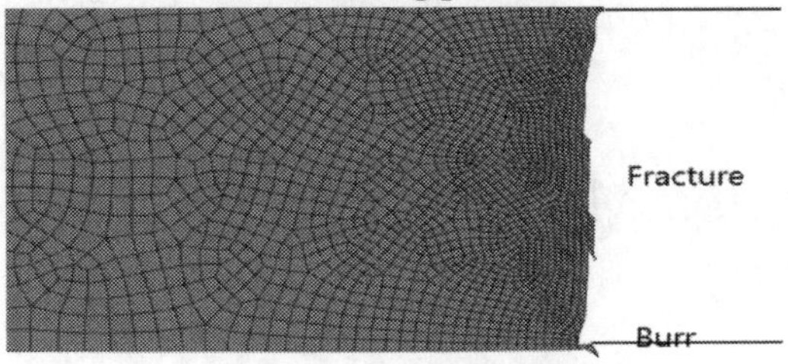

(a) finite element analysis

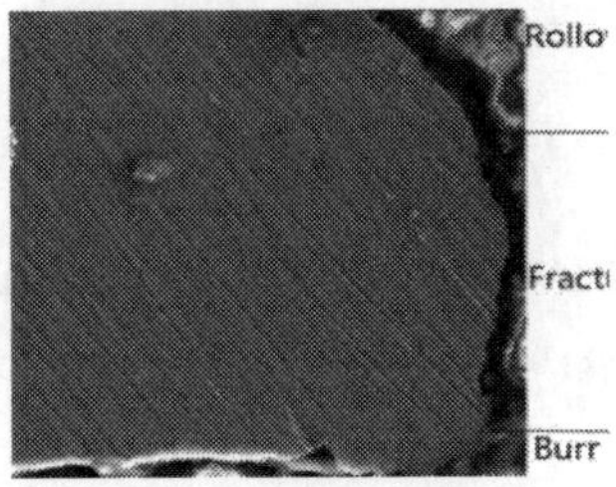

(b) experiment

Figure 4. fracture surface morphology of rubber blanking process

4.2. Fracture surface morphology

Similar to conventional blanking process, the morphology of rigid micro-blanking is composed of rollover, smooth zone, fracture zone and burr[7]. However in rubber blanking process, there are only rollover, fracture zone and burr, shown in Figure 4, which is different from rigid blanking process, observed using JSM 6010 scanning electron microscope (SEM). The first part is rollover. In the second zone, fracture zone, the maximum shear stress result in crack growth in vertical direction with rough face. The last zone is burr formed with the initialization of crack. Smooth zone is not included in rubber blanking morphology which is different to rigid blanking, because of flexible tool rubber.

5. Conclusion

In this paper, the finite element method is used for analyzing the flexible tool blanking process. Then the experiment is executed to validate the results. The followed conclusion can be drawn.

(a) The initialization and growth of crack are performed as a result of the shear stress, not the tensile stress.
(b) The direction of crack growth is perpendicular to the sheet face at the at the crack tip.
(c) The components of morphology of the shear plane are Rollover, Fracture zone and Burr without smooth zone comparing with rigid tool blanking process.

References

[1] Jie Xu, BinGuo , ChunjuWang, DebinShan. Blanking clearance and grain size effects on micro deformation behavior and fracture in micro-blanking of brass foil. International Journal of Machine Tools & Manufacture 2012(60): 27-34.

[2] Shin-Hyung Song, Woo Chun Choi. FEM Analysis on the Influence of Rounded Tool Edge on Micro-Blanking of Thin Foil with Negative Clearance. International Journal of Precision Engineering and Manufacturing 2015(16): 1101-1105.

[3] H. Marouani, A. Ben Ismail, E. Hug, M. Rachik. Numerical investigations on sheet metal blanking with high speed deformation. Materials and Design, 2009(30):3566–3571.

[4] F. Takahashi, T. Nishimura, I Suzuki. A method of blanking from amorphous alloy foils using rubber tool. Annals of the CIRP, 1991(40):315-318.

[5] H Hooputra, H Gese, H Dell and H Werner. A comprehensive failure model for crashworthiness simulation of aluminum extrusions. International Journal of Crashworthiness,2004,9(5):449-464.

[6] Rui Xiao, Xiao-Xing Li, Li-Hui Lang et al. Biaxial tensile testing of cruciform slim superalloy at elevated temperatures. Materials and Design, 2016(94):286-294.

[7] T. S. Kwak, Y. J. Kim, W.B. Bae. Finite element analysis on the effect of die clearance on shear planes in fine blanking. Journal of Materials Processing Technology, 2002:462–468

Journal of Physics: Conference Series **734** (2016) 032081 doi:10.1088/1742-6596/734/3/032081

The friction coefficient evolution of a MoS_2/WC multi-layer coating system during sliding wear

T Y Chan[1]; Y Hu[1]; Mohammad M Gharbi[2]; D J Politis[1]; L Wang[1*]

[1] Department of Mechanical Engineering, Imperial College London, London, SW7 2AZ

[2] Schuler Pressen GmbH, Bahnhofstrasse 41, 73033 Goppingen, Genmany

[*]Corresponding author: L. Wang: liliang.wang@imperial.ac.uk

Abstract. This paper discusses the evolution of friction coefficient for the multi-layered Molybdenum Disulphide (MoS_2) and WC coated substrate during sliding against Aluminium AA 6082 material. A soft MoS_2 coating was prepared over a hard WC coated G3500 cast iron tool substrate and underwent friction test using a pin-on-disc tribometer. The lifetime of the coating was reduced with increasing load while the Aluminium debris accumulated on the WC hard coating surfaces, accelerated the breakdown of the coatings. The lifetime of the coating was represented by the friction coefficient and the sliding distance before MoS_2 coating breakdown and was found to be affected by the load applied and the wear mechanism.

1. Introduction

Aluminium sheet metal forming attracts great interest from manufacturing industries in providing light weight components with high strength. However, studies suggest that the interaction between tools and Aluminium workpiece surfaces where adhesion and transferring of Aluminium on tools takes place decreases the surface quality of the finished product and damages the tool [1]. As a result, lubricant and coating are required for sheet metal forming processes especially in operations such as ironing and stamping [2].

The aim of this paper is to explore the friction and wear of an advanced coating system, where Molybdenum Disulphide (MoS_2) and WC were combined as a multilayer coating to reduce the adhesive wear in Aluminium forming process. The tribological properties of Molybdenum Disulphide are maintained at high temperatures up to 400°C, thus making it one of the most desirable lubricants to be used in warm forming. The WC, as the hard layer coating, has high resistance to thermal shock, which gives itself the ability to withstand rapid temperature fluctuations when sheet metal is in contact with the tools. The material also has the ability to withstand high compressive stresses, providing additional support to the soft MoS_2 coating.

2. Material and Experimental Method

2.1. Multilayer coating strategy

WC was applied through ultrasonic flame spray, a method where ultrasonic waves propagated and ensured precise application of coating without the need of direct contact to the material [3]. MoS_2 was applied at a later stage through film spray, which created the multilayer coatings on the G3500 cast iron substrate shown in Figure 1.

The specimen was examined through friction tests against an aluminium AA 6082 pin of radius 2.5 mm. The pin was in direct contact with the thin soft film of Molybdenum Disulphide and WC hard coating when the film broke down. The thickness of the MoS_2 film was approximately 25-30 μm and classified as a thick coating. The WC film on the other hand was 60-70 μm thick and the spraying method used allowed a more even coating distribution comparing to the MoS_2 film.

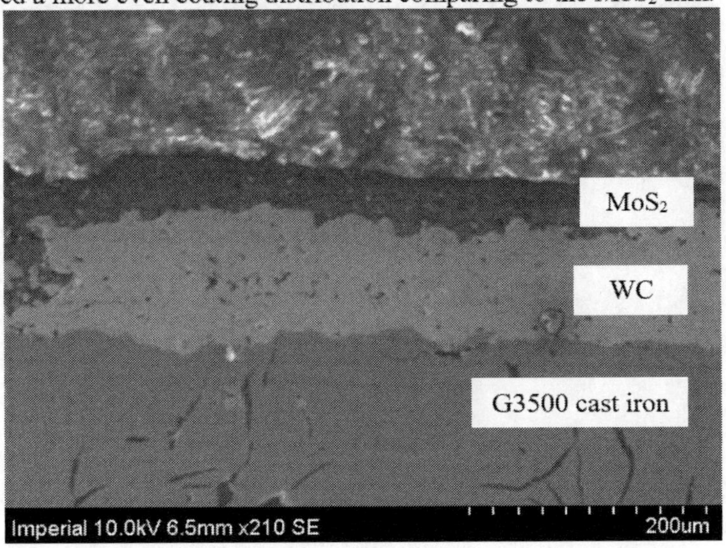

Figure 1. SEM image of Specimen Cross Section.

2.2. Friction Coefficient Test (Pin-on-disc set up) & Surface Analysis

A THT tribometer [4,5] was used to for the friction tests to assess the time, distance the pin travelled with respect to the plate and the coefficient of friction. These parameters were used to compare the evolution of friction under various circumstances.

Different loads were applied to investigate the effect on the lifetime of the coating, which was determined by the sliding distance before the occurrence of break-down. The loads applied were 1N, 3N, 7N and 9N at a sliding speed of 10 mm/s.

The optical microscope was used to analyse the topography of the specimen. It was particularly useful in observing the type of debris and wear left behind on the wear track and hence suggesting the friction mechanism taken place during the process. Moreover, Scanning Electron Microscope (SEM) was used to extract the thickness of the specimen's coating.

3. Results and Discussion

The tribological relationship was largely affected by the mechanical property of the coating and substrate as well as the size and hardness of debris in contact. Figure 2 shows the evolution of the average coefficient of friction under different loads on the MoS_2 soft coating lifetime before its breakdown.

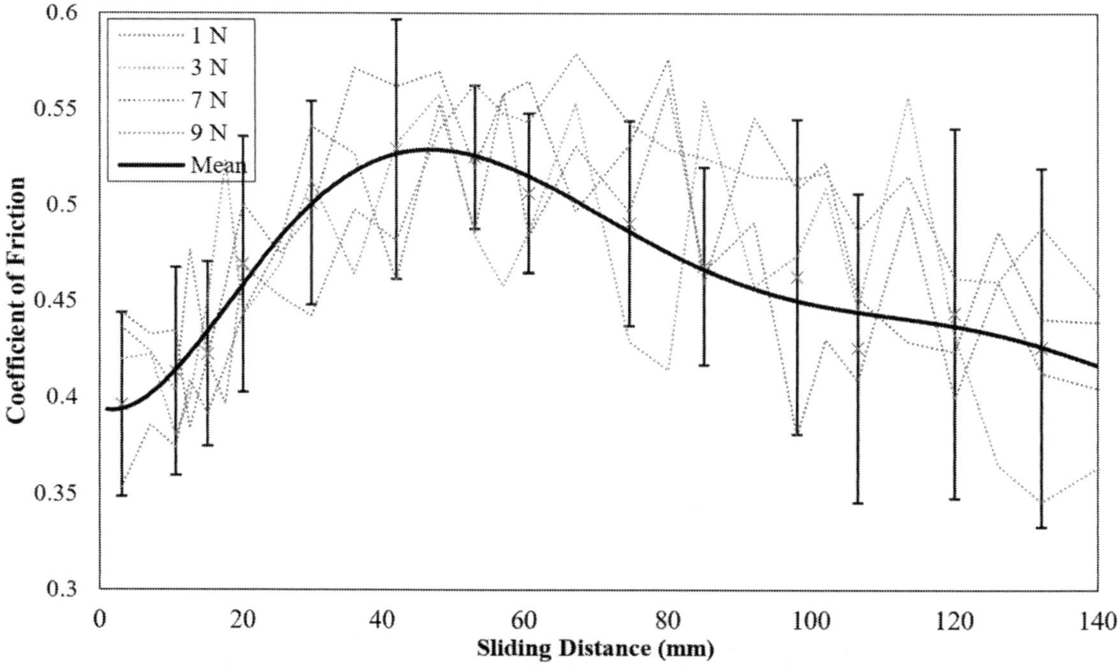

Figure 2. Mean Friction coefficient with error bars against sliding distances under 1N, 3N, 7N and 9N load.

The mean friction coefficient showed a rapid increase from 0.4 to 0.5 within a sliding distance less than 40mm. However, the effect of load on the friction coefficient evolution was not clear as the thick MoS_2 layer was penetrated very soon after the experiment started and gave a gradually increasing value of the friction coefficients. Low friction coefficient could be obtained through the introduction of multilayer coating at the initial stage. In comparison to aluminium and cast iron direct contact, the shear strength was reduced significantly through the usage of MoS_2. The low hardness of MoS_2 (400 HV) would imply a larger contact area between the aluminium slider and its surface and hence the reduction in friction through lower friction force was compromised by the increase in normal force due to the increased contact area. However, the combination of MoS_2 and thick metal layer would increase the hardness of the soft coating hardness to 1000 – 2000 HV [6]. This would result in a lower overall friction coefficient as lower shear strength was achievable without increasing the contact area.

The soft MoS_2 coating underwent rapid adhesive wear off, leading to an increase in friction coefficient within a short sliding distance. The lower platform at the beginning of figure 2 suggested the effect of MoS_2 in friction coefficient reduction. The soft coating soon experienced elastic and plastic deformation during the sliding and faced an increase in friction due to ploughing by the wore-off particles.

After reaching its peak at 40mm for 0.55, the friction coefficient experienced a gentle reduction while remaining higher (0.45) than the lower platform (0.4) experienced at the beginning. This phenomenon was observed as MoS_2 particles acted as a solid lubricant, when the soft particles were entrapped between surfaces while the ploughing forces from the MoS_2 were no longer present when particles were entrapped between the contact of the slider and hard coating. The differences in load are suggested to have little impact on the breakdown time of the MoS_2 soft coating due to the relatively little variation in the coefficient of friction. However, the increased fluctuation shown as the error bar at higher sliding distance suggested the difference in friction experienced by the WC coating. This predicted that the WC hard coating would experience breakdown while the lifetime of the WC hard coating will vary due to the difference in loads [7].

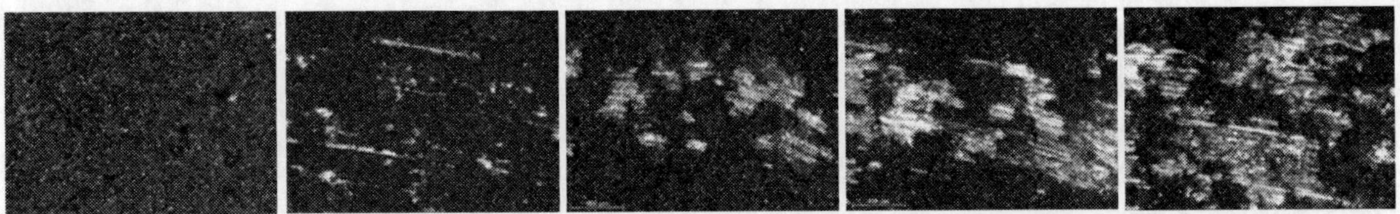

Figure 3. Specimen surfaces under vaious load: before the test, 1N, 3N, 7N, 9N (Left to right).

The wear track from Figure 3 suggested higher wear rate with increased loads. The surfaces were shown to have larger area of WC exposed as well as aluminium debris accumulation. This could be well-explained by the penetration of the soft coating due to the roughness of aluminium. Higher load increased the penetration area and resulted in higher wear rate during the friction test under the same distance travelled.

Aluminium debris was only observed along the wear track where WC coating was exposed. This proposed the benefit of MoS_2 soft coating on the reduction of material transfer, as almost no aluminium debris could be found over the presence of this soft coating. On the other hand, aluminium debris accumulated on the WC hard coating surfaces would lead to a rise in friction coefficient through a type of anchoring mechanism resisting motion. Due to the increased exposure of WC area with the increased load, an increase in aluminium debris was also shown at higher load.

The rapid wear off of the MoS_2 coating shows the ineffectiveness of the soft coating itself. To obtain more effective Aluminium sheet metal forming, application of MoS_2 spray is required for every stamping operation. This would require a lower punch force due to the lower friction achieved and would prevent the wear off of aluminium and accumulation of debris.

4. Conclusion

The MoS_2 coating on WC hard coating allows lower punch forces to be required in aluminium sheet metal forming while providing a better surface finishing. The coating prevents the accumulation of aluminium debris on the tool surfaces. However, due to the short lifetime of the soft MoS_2 coating, this requires the MoS_2 to be applied by a spraying method after each forming process, despite the fact that the MoS_2 debris would still be able to partially protect the tool surface.

5. References

[1] Riahi, A. R. & Alpas, A. T. (2007) Adhesion of AA5182 aluminum sheet to DLC and TiN coatings at 25C and 420C. *Surface & Coatings Technology.* 202 (4-7), 1055-61.

[2] Bay, N., Olsson, D. D. & Andreasen, J. L. (2008) Lubricant test methods for sheet metal forming. *Tribology International.* 41 (9-10), 844-853.

[3] Bescond, C., Lima, R. S., Gilbert, A., Kruger, S. E., Levesque, D. & Marple, B. R. (2004) Laser-ultrasonic characterization of thermal spray coatings. *Review of Progress in Quantitative Nondestructive Evaluation. 27 July-1 Aug. 2003,* USA, AIP. pp.968-75.

[4] ASTM International. (n.d.). Standard Test Method for Wear Testing with a Pin-on-Disk Apparatus. ASTM International.

[5] Paar, A. (n.d.). High Temperature Tribometer User Manual.

[6] Vierneusel, B., Tremmel, S. & Wartzack, S. (2012) Effects of deposition parameters on hardness and lubrication properties of thin MoS_2 films. *Materialwissenschaft Und Werkstofftechnik.* 43 (12), 1029-35.

[7] Ma, G., Wang, L., Gao, H., Zhang, J. & Reddyhoff, T. (2015) The friction coefficient evolution of a TiN coated contact during sliding wear. *Applied Surface Science.* 345109-115.

Numerical Modelling of Drawbeads for Forming of Aluminium Alloys

Y Joshi[1*], P Christiansen[2], I Masters[1], N Bay[2], R Dashwood[1]
[1] WMG Centre HVM Catapult, International Manufacturing Centre, University of Warwick, Coventry, CV4 7AL, United Kingdom.
[2] Department of Mechanical Engineering, Technical University of Denmark, 2800 Kgs. Lyngby, Denmark.
*Corresponding author: Phone: +442476528083, E-mail: y.k.joshi@warwick.ac.uk

Abstract. The drawbeads in stamping tools are usually designed based on experience from the forming of steel. However, aluminium alloys display different forming behaviour to steels, which is not reflected in the drawbead design for tools used for stamping aluminium. This paper presents experimental results from different semi-circular drawbead geometries commonly encountered in automotive dies and compares them to those obtained from Stoughton's analytical drawbead model and the 2D plane strain drawbead model set up using LS-DYNA. The study was conducted on lubricated NG5754 strips. The results presented are in terms of drawbead restraining force versus strip displacement, as a function of drawbead depth. The FE drawbead model agrees well with the experiments whereas the analytical model overpredicted the drawbead forces.

1. Introduction

The significance of drawbeads in automotive sheet metal forming operations has been illustrated in [1]. The drawbead restraining force, hereafter referred to as restraining force, is the braking force experienced by the strip when it is drawn through the drawbeads. It is the sum of the bending and unbending forces and sliding friction forces. Drawbeads can locally control the sheet metal flow in drawing of complex non-axisymmetric automotive components and offer a distinct advantage of imparting a higher restraining force to the blank without a significant increase in the blank size and overall blankholder force. The tensile force exerted on the blank reduces the circumferential stresses which cause wrinkling [2] and helps in achieving stretch in the part to increase its impact strength. Even though aluminium alloys, because of light-weighting initiatives, have been of particular interest in the automotive industry, the drawbead design guidelines for automotive forming are based on forming of steels [3] and may not be appropriate for forming of aluminium alloys [4].

Trial and error experiments to determine the optimum drawbead geometry for a complex automotive part and for new sheet materials is an onerous task. Therefore, several analytical and FE drawbead models were developed to predict the restraining force and the blankholder force taking into account the effect of drawbead geometry, sheet material properties, and the process parameters. Tufecki et. al. [5] analysed drawbead force models and concluded that the Stoughton drawbead force model agreed well with the drawbead strip draw experiments performed in line with the experiments conducted by Nine [6] on AKDQ steel, rimmed steel, and 2036-T4 aluminium test strips. The Stoughton model [7] attempts to not only predict the restraining force and the blankholder force but also the calculation of the effective

bending radius of the strip and hence, suits partial and full drawbead penetrations. The model uses constants which directly relate to mechanical parameters obtained from a tensile test and is a good formulation for understanding the significant factors in drawbead design.
It has been demonstrated that 2D plane strain models for drawbeads give an agreeable prediction of drawbead forces [8] if the width of the sheet is significantly larger than the thickness. Findings from an extensive review [1] also confirmed that 2D plane strain drawbead models are able to reflect the local deformation behaviour and allow the use of both linear and non-linear contact models to define dynamic friction conditions.

The aim and scope of this paper are to determine a suitable approach for predicting restraining forces required in drawbead design by comparing the Stoughton analytical restraining force model with a numerical 2D plane strain drawbead model and drawbead strip-draw experiments.

2. Research methodology

2.1 Experimental set-up

The experiments were performed on a new drawbead tool, based on Nine's design [6]. The layout can be seen in figure 1. Four drawbead depths 'd', were used; 3.9mm, 5.8mm, 8.7mm, and 11mm. The strips were drawn at 85mm/s and a blankholder force of 15kN was applied to ensure that the blankholder remain closed. Aluminium NG5754 strips, 500mm long, 50mm wide and 1.5mm thick were carefully deburred and lubricated with a synthetic forming lubricant having a viscosity of 244.5mm^2/s at 20°C, with a coat weight of 1.5±0.5gm/m^2. The strips were drawn in the rolling direction on both, roller drawbead and fixed drawbead set-ups. The roller drawbead set-up approximates frictionless conditions. Each experimental run was repeated twice to account for the measurement error.

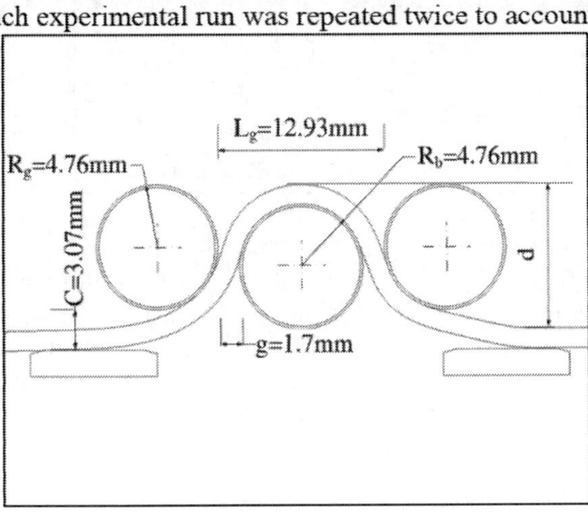

Figure 1: Drawbead geometry used in experiments and simulations

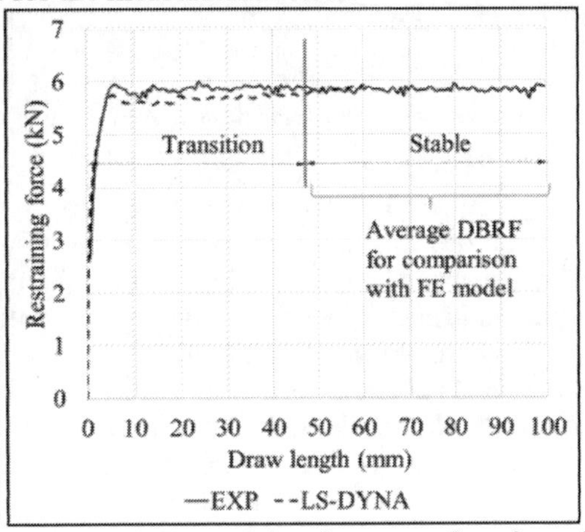

Figure 2: Restraining force as a function of draw length, NG5754, μ=0, depth=8.7mm

2.2 FE Model set-up

A drawbead model as shown in figure 1, closely representing the experimental set-up was built in LS-DYNA where the groove and the drawbead are modelled as rigid steel tools with a fine mesh and the strip had a 0.3mm square mesh with 5 elements through the thickness for better discretization. A 2D plane strain element type ELFORM 13 with 4 integration points was chosen. As the restraining force in the experiment stabilized after 50mm, figure 2, the draw length in the FE model was set to 60mm to reduce the computational time. The material flow curves used in the model were obtained from the uniaxial tensile test in the rolling direction. Voce hardening rule ($\sigma_0 = 117 + (319 - 117)(1 - e^{-12.4\epsilon}$), where σ_0 is the flow stress in MPa and ϵ is the effective plastic strain) was used to generate the flow stress curve used in LS-DYNA's MAT_24. The Coulomb friction model ($\tau = \mu p$, where τ is frictional shear stress, μ is the coefficient of friction and p is the contact pressure) was used to define friction in the FE model. For the Stoughton analytical drawbead model, a Hollomon hardening rule ($\sigma_0 = 459\ \epsilon^{0.28}\ MPa$) was used. Other material properties are given in table 1.

Table 1: NG5754 material properties

Density (kg/m^3)	Young's modulus (MPa)	Poisson's ratio	Lankford coefficient
2700	70000	0.33	0.67

3. Results and discussion

Figure 2 shows the restraining force obtained from the drawbead strip-draw experiment and the FE model. Figures 3a and 3b compare the restraining forces obtained from the drawbead strip-draw test, the 2D plane strain FE model and the Stoughton model for μ of 0 and 0.15 respectively. It can be seen that apart from the lowest and highest drawbead depths, the results from LS-DYNA agrees reasonably well with the experiments. The FE model follows the linear relationship between the restraining force and the drawbead depth as established by the experiments. However, the Stoughton drawbead model overpredicted the restraining force for both, the friction and the frictionless case. The model also showed a slightly non-linear trend.

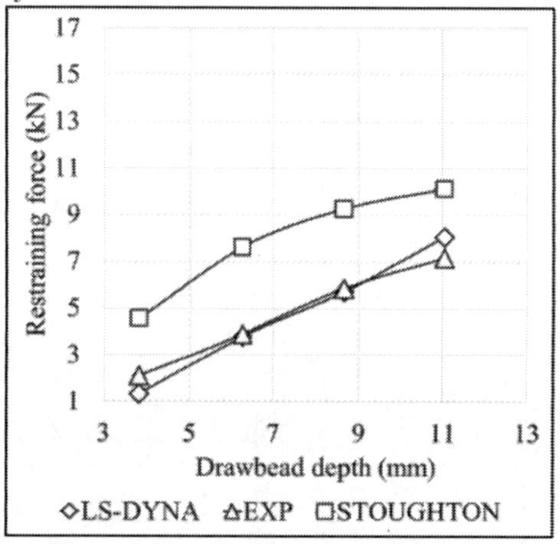

Figure 3a. Restraining force, μ=0

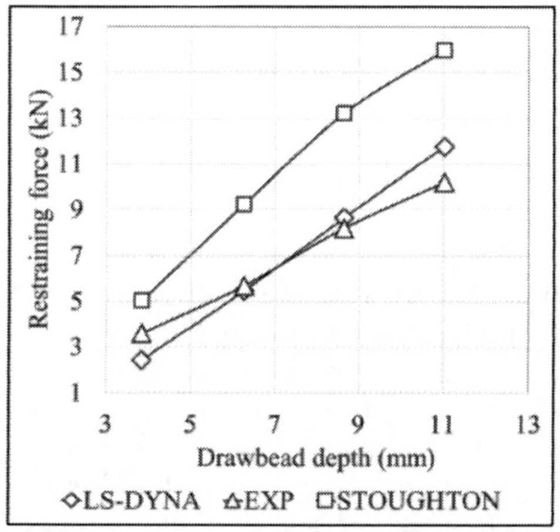

Figure 3b. Restraining force, μ=0.15

As illustrated in figures 4a and 4b, at the lowest drawbead depth of 3.9mm, there are only 3 strip nodes in contact and more than 24 nodes for the depth of 8.7mm. Therefore, the contact definition at the lowest depth is poor which may have led to smaller restraining forces simulated by LS-DYNA. At the highest depth, the FE model showed larger effective plastic strain in the strip, approximately 0.67, due to bending and unbending effect. This value is substantially higher than 0.19 which was observed at the point of instability in the uniaxial tensile test. It can be argued that LS-DYNA extrapolated the flow curve beyond effective plastic strain of 0.19 to obtain the effective stress at an effective plastic strain of 0.67 which resulted in the overprediction of the restraining force.

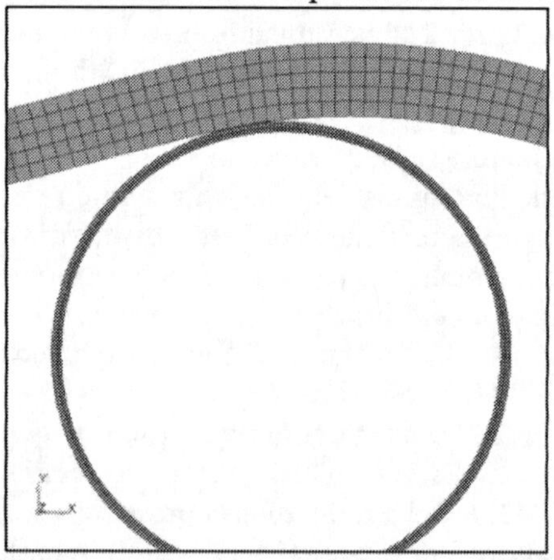

Figure 4a.Contact at d=3.9mm,μ=0.15

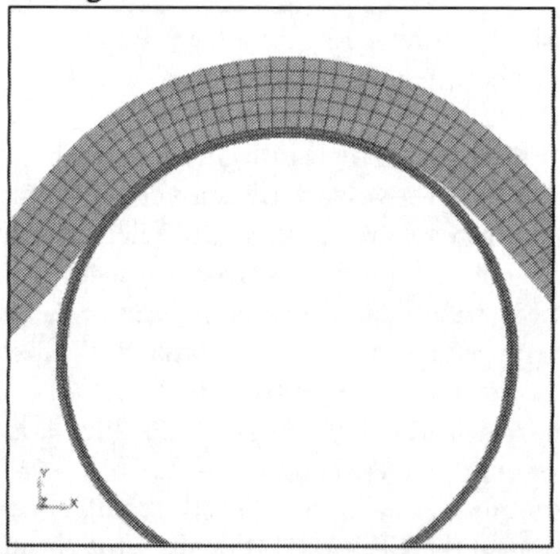

Figure 4b.Contact at d=8.7mm,μ=0.15

The coefficient of friction value was determined from the FE model by trying different values until satisfactory correlation with the results from the fixed drawbead strip-draw experiments was obtained. The coefficient of friction of 0.15 agreed well with experiments. While attempting to derive a coefficient of friction from the simulation, it was observed that no single value of the coefficient of friction gave good agreement with experimental results. This was in line with the Nine's formula [6] which gave lower coefficients of friction for increasing drawbead depth. Also, Ren [9] stated that a single coefficient of friction did not match the entire range of drawbead depths in the simulations and suggested a contact pressure based non-linear friction model which outputs lower coefficients of friction for higher drawbead depths. Both [6] and [9] contradict Coulomb's linear friction model and indicate that friction representation could be a significant source of variation in the simulated restraining forces. If this is true, then the restraining forces predicted by the FE model with μ=0, should agree with the results from roller drawbead strip-draw experiments. It can be seen from figure 3a that there is some difference between the simulated and experimental restraining forces for the frictionless case at lowest and highest depths. This difference increases in the presence of friction instead of helping in achieving a good correlation as illustrated in figure 3b. Hence, the Coulomb friction model can be appropriately applied in numerical modelling of sheet metal flow over drawbeads. However, a good contact definition must be ensured while simulating lower drawbead penetrations and applying a flow curve with larger effective plastic strains obtained from materials tests, for example, Watts-Ford test, may be beneficial.

4. Conclusion

In this paper, the Stoughton analytical drawbead model and an FE model of the drawbead strip-draw test were compared with the drawbead strip-draw experiments which were conducted on lubricated NG5754 aluminium strips. The Stoughton model significantly overpredicted the restraining forces whereas the numerical 2D plane strain drawbead model accurately predicted the restraining forces, especially for the drawbead depth ranging from 5.8mm to 8.7mm. For the both friction cases, μ=0 and μ=0.15, the FE model predicted a smaller restraining force at the lowest drawbead depth and overpredicted the restraining force at the highest drawbead depth. With proper contact definition for lower drawbead penetrations and an experimental flow curve with larger effective plastic strains, a 2D numerical plane strain model with Coulomb's friction law can reasonably predict the drawbead restraining forces.

5. References

[1] Xu S G, Bohn M L and Weinmann K J 1997 Drawbeads in Sheet Metal Stamping - A Review *SAE International*

[2] Zharkov V A 1995 *Theory and practice of deep drawing* (London: Mechanical Engineering Publications)

[3] Lange K 1986 *Handbook of Metal Forming* vol 4 (Springer-Verlag)

[4] Li R and Weinmann K J 1999 Formability in Non-Symmetric Aluminium Panel Drawing Using Active Drawbeads *CIRP Annals - Manufacturing Technology* **48** pp 209-12

[5] Tufekci S S, Wang C-T, Kinzel G L and Altan T 1994 Estimation and Control of Drawbead Forces in Sheet Metal Forming *SAE International*

[6] Nine H D 1978 *Mechanics of Sheet Metal Forming,* ed D Koistinen and N-M Wang: Springer US) pp 179-211

[7] Stoughton T B 1988 Model of drawbead forces in sheet metal forming In: *15th IDDRG,* (Dearbon, USA) p 205

[8] Carleer B D, Meinders T and Huetink H 1996 Equivalent drawbead model in finite element simulations In: *NUMISHEET '96, Third International Conference on Numerical Simulation of 3D Sheet Metal Forming Processes,* ed J K Lee, *et al.* (Dearborn, Michigan) pp 25-32

[9] Ren F, Chappuis L and Xia Z C 2009 Drawbead Restraining Force Modeling: Nonlinear Friction *SAE Int. J. Mater. Manuf.* **2** pp 530-6

Modeling of hole-expansion of AA6022-T4 aluminum sheets with anisotropic non-quadratic yield functions

Yannis P. Korkolis[1], Benjamin Brownell[1], Sam Coppieters[2] and Haobin Tian[3]

[1]U. of New Hampshire, USA

[2]KU Leuven, Technology Campus Ghent, Belgium

[3]Shanghai Second Polytechnic U., China

yannis.korkolis@unh.edu

Abstract. In the hole-expansion of anisotropic AA6022-T4 sheets, the strain around the hole is non-uniformly distributed due to the anisotropy of the material. This was examined by performing experiments with a flat-headed punch and using Digital Image Correlation (DIC). In the experiments, failure always initiated at a unique location, oriented at 45° to the Rolling Direction of the sheet. The use of DIC allowed the probing of the full-strain-field in real-time. Subsequently, the experiments were simulated in DYNAFORM using shell elements and the Yld2000-2D anisotropic non-quadratic yield function, properly calibrated for this material. In addition, the hardening curve of the material was inversely identified at large strains from the tail of the tensile test. The strain evolution is compared between the experiments and the model.

1. Introduction

The expansion of the circular hole of a thin sheet is a classical problem in plasticity [1]. Beyond its applications in assessing the performance of plasticity theories and in shedding light into the mechanics of material failure by void growth, it is also relevant to predicting splitting fractures in sheet metal forming and understanding the effect of edge preparation on sheet formability. In this study, we examine the hole-expansion of a plastically orthotropic sheet by a flat-headed punch [2], using a combination of experiments and analysis. The anisotropy and post-necking hardening response of the AA6022-T4 material of this study is first established experimentally. Subsequently, the hole-expansion experiments are described, including the use of Digital Image Correlation for real-time measurement of the strain fields. Finally, finite element simulations of hole-expansion using the commercial software DYNAFORM are presented and their predictions are compared to the experiments.

2. Material Modeling

The plastic anisotropy of the 1mm-thick AA6022-T4 aluminum alloy was probed with uniaxial tension experiments in 7 orientations to the RD, 3 plane-strain tension experiments at RD, 45° and TD and 1 disk compression experiment [3]. These experiments were fitted using the Yld2000-2D yield function [4] at a plastic work level of 27 MJ/m^3 (or 10.6% strain in RD tension), Figure 1. The predicted R-values match the anisotropy found in the experimental R-values well, both showing a minimum value at 45° to the RD, Figure 2. To calibrate the model shown in Figure 1, eight anisotropic fitting coefficients needed to be optimized and identified to describe the Yld2000-2D surface, Table 1. The governing parameters were optimized to fit the experimentally acquired stress states and R-values with an unconstrained nonlinear optimization code. Additionally, the strain hardening behavior of the material beyond the point of maximum uniform elongation (true $\varepsilon_{max} \approx 0.18$) was identified using a special case of the virtual fields method [5]. The Voce model was identified up to a plastic equivalent strain of about 0.3, see Figure 3.

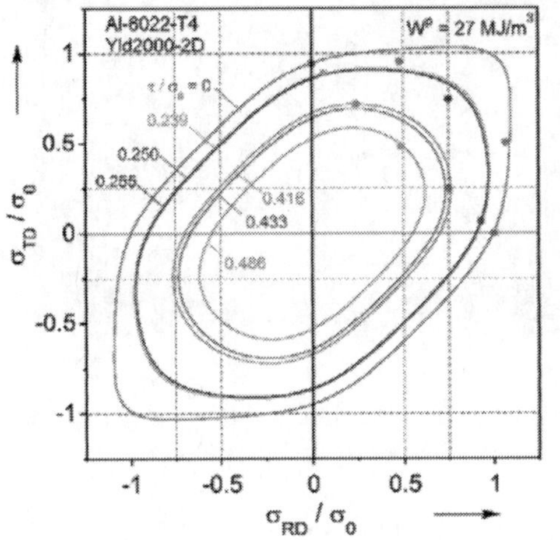

Figure 1. Normalized plane-stress yield locus (experiments and Yld2000-2D fit).

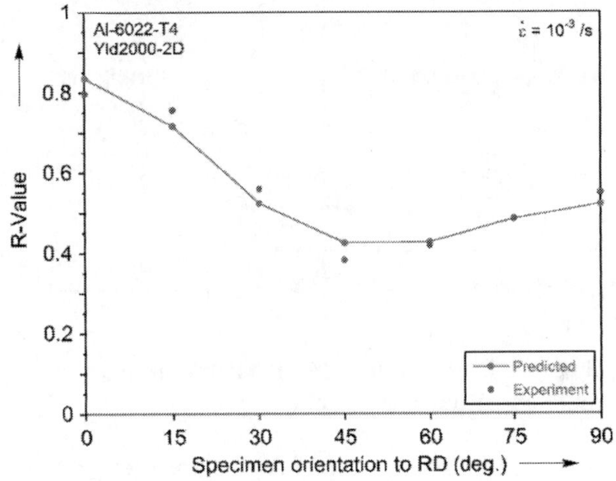

Figure 2. R-values obtained from experiments and predicted by Yld2000-2D at 27 MJ/m³.

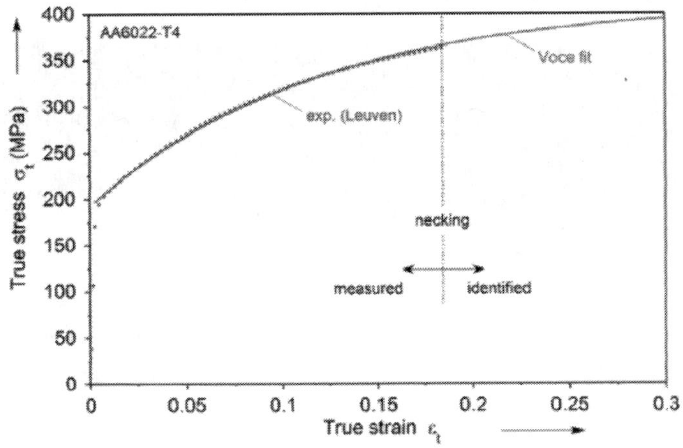

Figure 3. The true sts-stn curve of the material in the rolling direction. Inversely identified Voce law using experimental data beyond the point of maximum uniform strain (≈ 0.18).

Table 1. Anisotropic coefficients used to fit Yld2000-2D (W^p= 27 MJ/m³).

k	8
α_1	0.9751
α_2	0.9998
α_3	0.9695
α_4	1.0669
α_5	0.9981
α_6	0.9323
α_7	0.9403
α_8	1.1572

3. Hole-expansion Experiments

The hole-expansion experiments were performed on a 260kN double-action hydraulic press with 305mm of stroke and with the same punch and die geometry as for a Marciniak test, Figure 4 (see also [3]). The specimen was clamped at the periphery by a lock-bead to prevent flow. The punch was then lowered onto the sheet with a speed of 1.065mm/s, until specimen failure was observed. The cup height was approximately 19 mm at failure. A single crack was observed to propagate initially at 45° to RD of the sheet, Figure 5. To capture the evolution of strain in the specimen, a coat of matte white paint was applied to the surface of the specimen followed by a random black speckle pattern. The Correlated Solutions digital image correlation (DIC) VIC-3D software was used to analyze the images.

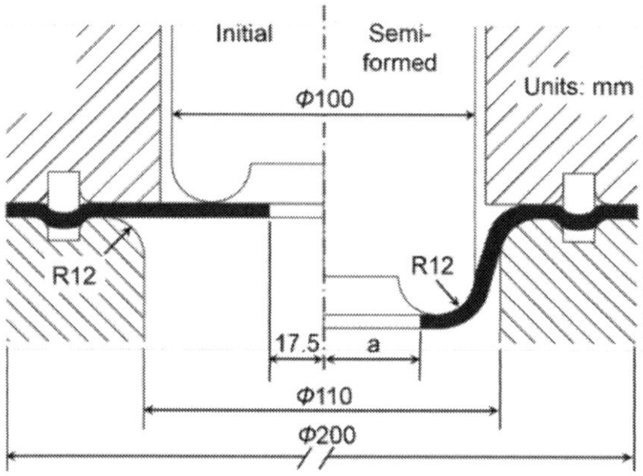

Figure 4. Punch and die geometry used for the hole-expansion experiments.

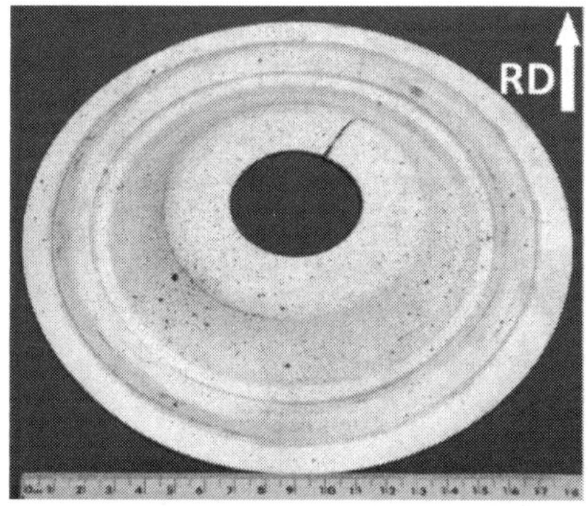

Figure 5. Specimen after experiment showing failure at 45° to the RD.

The three principal strains are shown in the last image captured prior to failure, at a punch displacement of approximately 19mm. The major and minor strains have the greatest absolute values along one of the 45 degree directions. Examining the third principal strain, i.e., the through thickness strain, it can be seen that thinning is also greatest along both 45° directions, Figure 6. That strain was determined from the two in-plane ones, using the incompressibility condition.

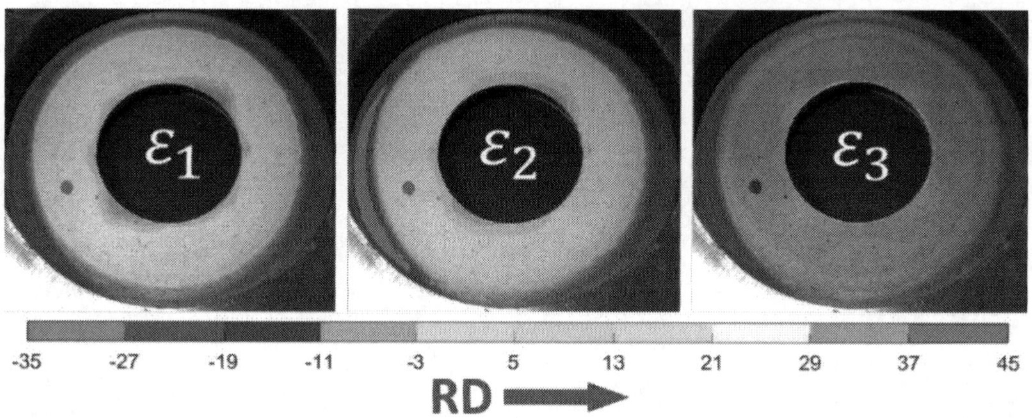

Figure 6. Principal strains shown immediately before failure is observed.

4. Hole-expansion Simulations

The hole-expansion of AA6022-T4 sheets was simulated using the explicit finite element code DYNAFORM. A quarter model was used due to the axisymmetric punch and die geometry and the fact that the R-values show 2-fold symmetry in the sheet. The mesh used 6,391 Belytschko-Tsay quadrilateral shell elements with 5 integration points through the thickness. To increase computational efficiency, the equivalent drawbead built into DYNAFORM is used to prevent the sheet flow past the drawbead. A friction coefficient of 0.17 between die and blank, punch velocity of 1000mm/s and time step size of $1.2*10^{-6}$ were used. It can be seen in Figure 8 that the experimental and predicted thickness change values are in phase with each other, and generally have comparable mean values, but in every case they have drastically different amplitudes.

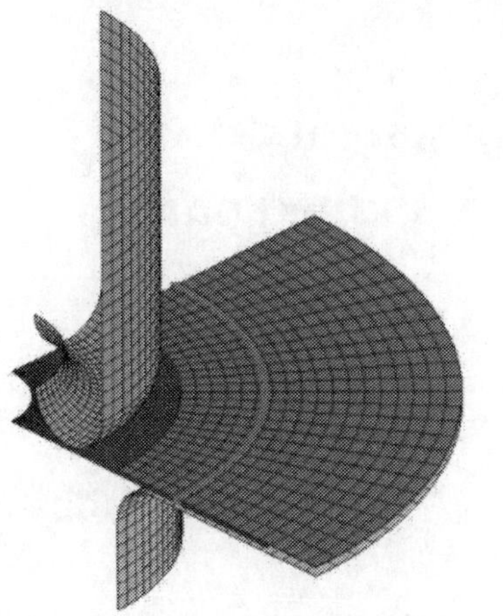

Figure 7. Finite element model used to simulate hole-expansion experiments using DYNAFORM

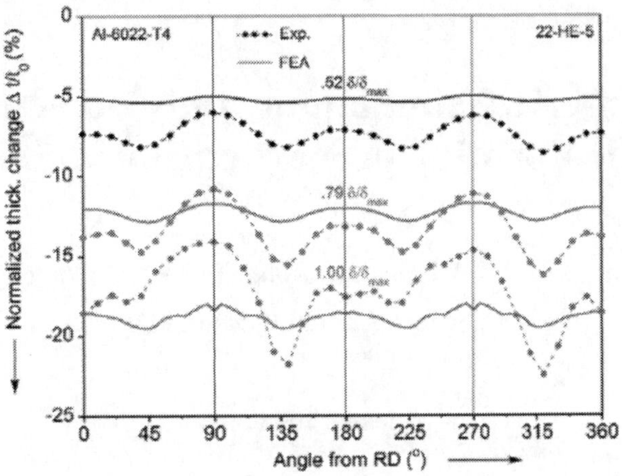

Figure 8. Evolution of experimental and FEA thickness change around the expanding hole at a fixed undeformed radius of 20mm at 3 normalized punch displacements.

5. Conclusions

The experimental data from the hole-expansion of AA6022-T4 sheets proved to be difficult to match using shell elements. This was despite the fact that the Yld2000-2D yield function, calibrated at 27 MJ/m^3, matched the plasticity experiments very well and that the post-necking hardening curve was identified with a custom, dedicated procedure. The simulations reproduced the wavelength, but not the amplitude of the thickness variation around the circumference of the hole. To further improve the predictions, solid finite element models will be considered, along with material model calibrations at higher levels of plastic work.

References

[1] Parmar, A., and Mellor, P.B. Plastic Expansion of a Circular Hole in Sheet Metal Subjected to Biaxial Tensile Stress. Int. J. Mech. Sci. 20.10 (1978): 707-20.

[2] Kuwabara, T., Hashimoto, K., Iizuka, E. and Yoon J.W. Effect of Anisotropic Yield Functions on the Accuracy of Hole Expansion Simulations. J. Mat. Proc. Tech. 211.3 (2011): 475-81.

[3] Tian, H., Brownell, B., Baral, M. and Korkolis, Y.P. Earing in Cup-Drawing of Anisotropic Al-6022-T4 Sheets. Int J Mater Form (2016): n. pag.

[4] Barlat F, Brem JC, Yoon JW, Chung K, Dick RE, Lege DJ, Pourboghrat F, Choi S-H, Chu E (2003) Plane Stress Function for Aluminum Alloy Sheets-Part I: Theory. Int J Plast 19:1297–1319.

[5] Coppieters, S., Cooreman, S., Sol, H., Van Houtte, P. and Debruyne, D. Identification of the Post-necking Hardening Behaviour of Sheet Metal by Comparison of the Internal and External Work in the Necking Zone. J. Mat. Proc. Tech. 211.3 (2011): 545-52.

Acknowledgements

This research was partially supported by the U.S. NSF GOALI grants CMMI-1031169 and CMMI-1301081. Prof. Haobin Tian acknowledges the support of National Natural Science Foundation of China (Grant No. 51405284). The AA6022-T4 sheets were provided by Dr. Edmund Chu of Alcoa.

Numerical simulation study on active and passive hydroforming process optimization of box shaped part

Y P Zeng, J L Dong, T D He, B Wang

Chengdu Aircraft Industrial (Group) Co., Ltd, Chengdu, 88 Weiyi Street, China

zengyp_aero@163.com

Abstract. Low qualified rate and inferior quality frequently occurring in the general deep drawing process of a certain box-shaped part, now use hydroforming to optimize forming process, in order to study the effect of hydroforming for improving the quality and formability, purposed five process schemes: general deep drawing, active hydroforming, passive hydroforming, general deep drawing combined with active hydroforming, passive combined with active hydroforming. Each process was simulated by finite element simulation and results were analysed. The results indicate the passive combined with active hydroforming is the best scheme which can obtain smallest thickness thinning and satisfactory formability, then optimized hydroforming pressure, blank holder force subsequently by adjust the simulation parameters. Research result proves that active/passive hydroforming is a new method for complex parts forming.

1. Introduction

Hydroforming is a kind of flexible forming technology, which use liquid pressure to force material formed as the shape of a die, better quality (excellent surface quality, less thinning of thickness, controllable wrinkling and smaller springback) and lower cost obtained at the same time in the process of hydroforming [1]. Sheet hydroforming process classified into two types: passive hydroforming and active hydroforming. The passive hydroforming which also called hydrodynamic deep drawing (HDD) [2], die filled with liquid as a pressure cavity in HDD process, it can significantly improve the forming limit compared with general deep drawing [3]. Active hydroforming combine deep drawing and bulging technology which use high pressure liquid to replace the punch, its forming limit generally less than HDD but it is more suitable for partial plastic forming and proved by production practice.

Recent years, many researchers study on the hydroforming process of box shaped part, Lang [4] investigated the pre-bulging effect in process of passive hydroforming of a box shaped part, the results show that pre-bulging height and pressure has important influences on the forming results, in paper [5]-[7], failure control methods investigated by numerical simulations and the general failure occurred in passive hydroforming process of box shaped part also discussed, in paper [8], the passive hydroforming process of different shaped box parts were studied by numerical simulations and discussed the influences of technology parameters for forming limit.

In this paper, the forming process of a box shaped part was simulated both by active and passive hydroforming, simulation results were analyzed, and proposed a new forming scheme which take the advantages of active and passive hydroforming.

2. Finite element simulation modelling

2.1. Box shaped part
The box shaped part as shown in figure 1, the maximum length of part is about 862mm, and the maximum width is about 231mm, maximum height of the part is 46mm, radius of the bottom is 7.95mm, eight reinforcing ribs equidistantly distributed on the flat surface of the part.

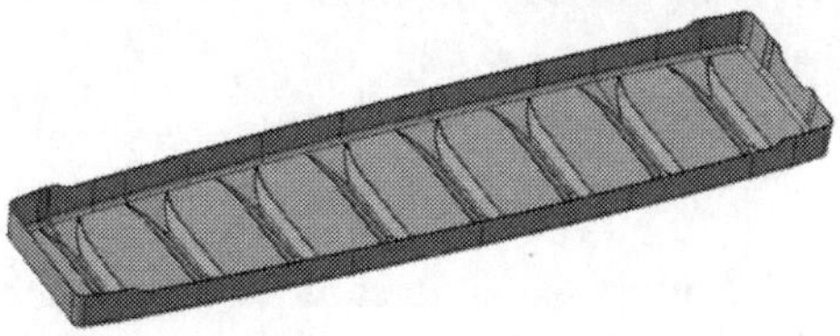

Figure 1. 3D model of box shaped part

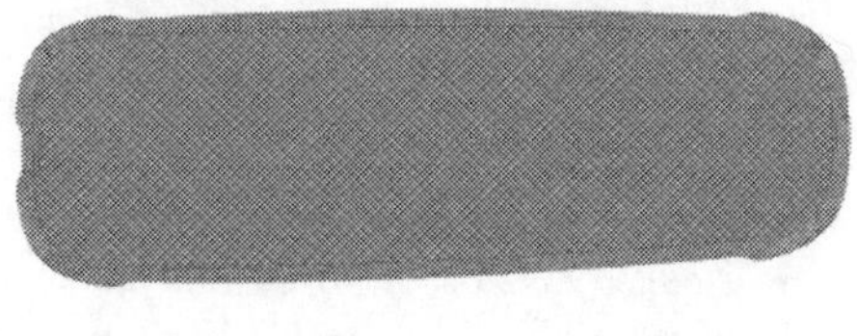

Figure 2. Optimized shape of blank

2.2. Blank information
Blank geometrical shape of box shaped part as shown in figure 2, blank profile optimized by inverse calculation of numerical simulation, and 20mm allowance add on the profile, the material used for numerical simulation is 2xxx aluminium alloy, thickness of blank is 1.6mm.

2.3. Numerical simulation modelling plan of part forming process
In order to study the results of different forming process, meanwhile, confirm the most suitable forming process for box shaped part, five simulation schemes were proposed as below:

Table 1. Forming schemes of box shaped part

Scheme number	Preforming	Final forming
1	N/A	General deep drawing
2	N/A	Passive hydroforming
3	N/A	Active hydroforming
4	General deep drawing	Active hydroforming
5	Passive hydroforming	Active hydroforming

2.4. Modelling of active and passive hydroforming
Active and passive modelling in numerical simulation software as shown in figure 3 and figure 4.

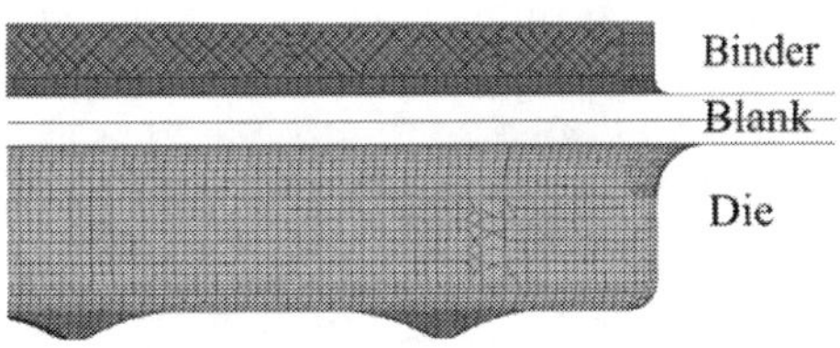

Figure 3. Active hydroforming modelling

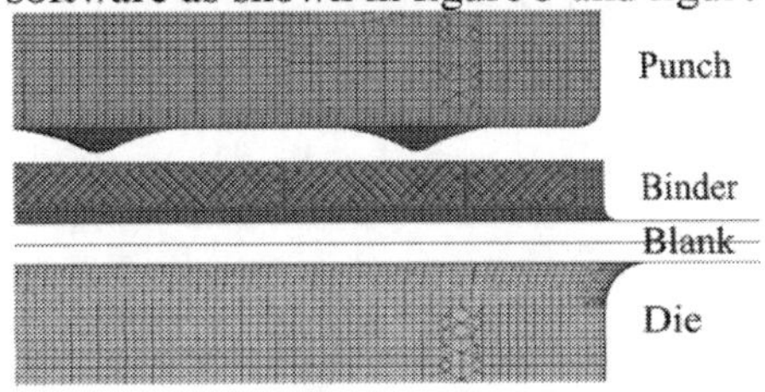

Figure 4. Passive hydroforming modelling

3. Numerical simulation analysis

3.1. General deep drawing simulation result
In the process of general deep drawing, punch and die are rigid body, blank drawn into die while the punch moving downward during forming process, the maximum stress of blank is at the position of four corners, simulation result is shown in figure 5.

Blank serious thinning and the thinnest thickness is located in the corners of the box shaped part, the minimum thickness is 0.243mm, means fracture occurred at corners, and flange winkling seriously.

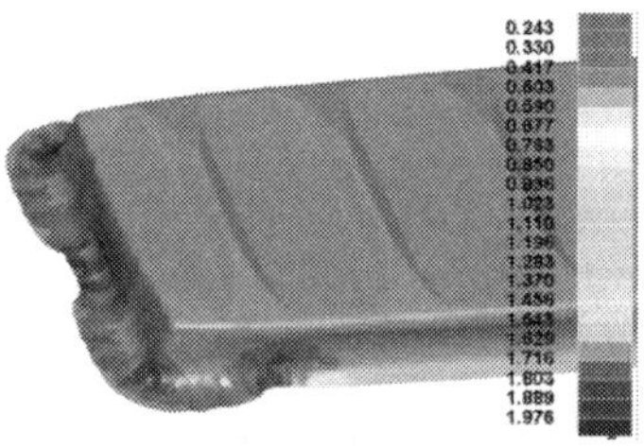

Figure 5. General deep drawing simulation

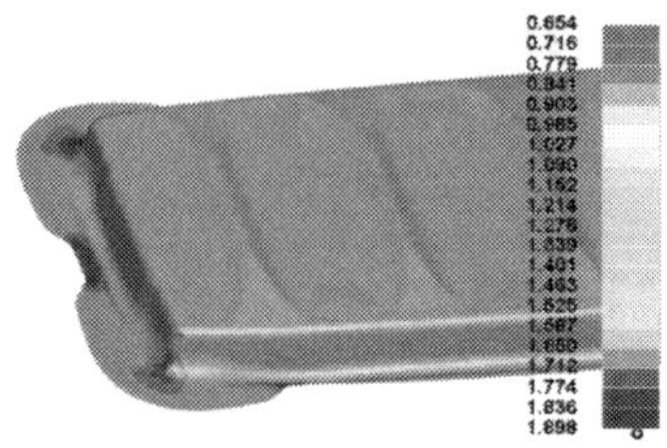

Figure 6. Passive hydroforming simulation

3.2. Passive hydroforming simulation result

In process of passive hydroforming, liquid cavity as a die, liquid pressure increased when punch moving downward, and blank formed under high pressure. In simulation, the pre-bulging pressure is 2MPa, maximum pressure is 50MPa, passive hydroforming simulation result in figure 6.

The thinnest thickness at the corner is 0.654mm, higher than the limit value in general deep drawing, but the corner still fracted due to thickness thinning over 30%, but no winkling on the flange.

3.3. Active hydroforming simulation result.

Pressurized liquid as a flexible punch in active hydroforming, at the first stage of active hydroforming, liquid deep drawing blank into die, then blank bulging and stretching by high pressure liquid at second stage. In simulation, maximum liquid pressure is 50MPa, and blank holder force changing with pressure.

Active hydroforming simulation result as below in figure 7, the thinnest thickness at the corner is 0.458mm, lower than the limit value in passive hydroforming, and small winkling on the flange.

Figure 7. Active hydroforming simulation

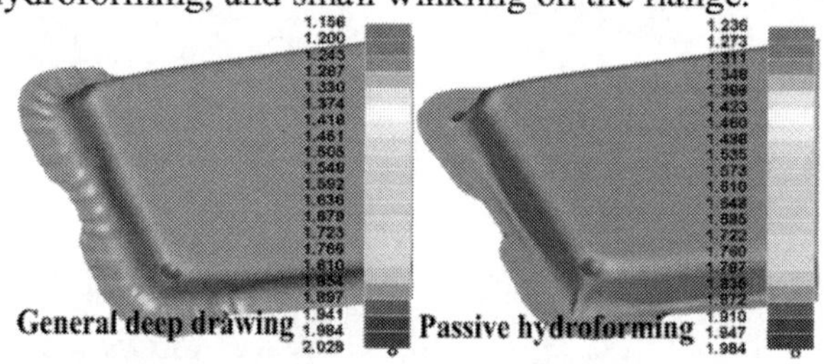

Figure 8. Simulation of two preforming processes

3.4. General deep drawing combined with active hydroforming simulation result.

In order to avoid the fracture occurred at corners, appropriate preforming is needed, in this simulation, general deep drawing as preforming process, the depth of preformed box shaped part is 46mm, and the radius of bottom is 15mm, the simulation result of general deep drawing preforming process shown in figure 8, the thinnest thickness at corner is 1.156mm, and flange wrinkling obviously.

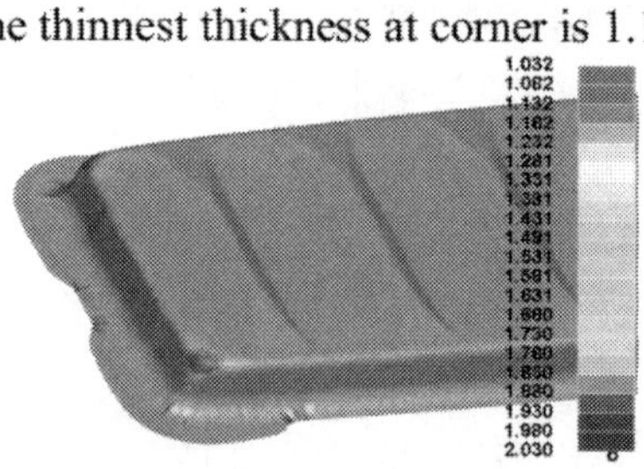

Figure 9. GDD and AH simulation

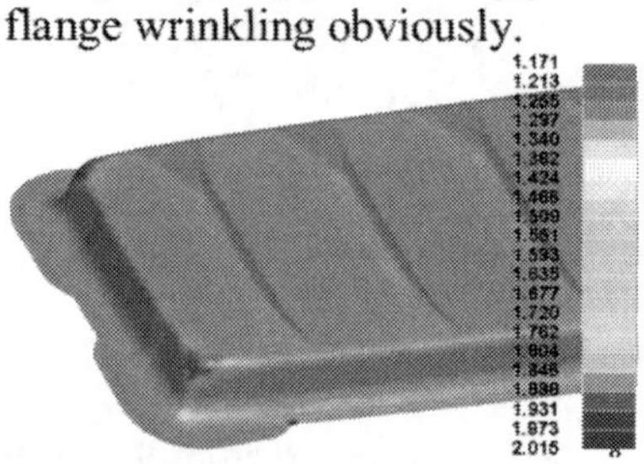

Figure 10. PH and AH simulation

General deep drawing (GDD) preformed part taken into final forming by active hydroforming (AH), in active hydroforming, the rounded corners and reinforcing ribs of box shaped part are formed as the shape of the die, and the thinnest thickness at corner decreased from 1.156mm to 1.032mm, thickness thinning is 35.5%, that means fracture occurred, and wrinkling on the flange is not eliminate.

3.5. Passive hydroforming combined with active hydroforming simulation result.

Passive hydroforming (PH) as preforming process in this simulation, geometric shape of preformed part the same as the preformed part in general deep drawing, passive hydroforming simulation result shown in figure 8, the thinnest thickness at corner is 1.236mm, and no wrinkling on the flange.

The preformed part taken into final forming by active hydroforming (AH), in final process, the thinnest thickness of formed rounded corners decreased from 1.236mm to 1.171mm, the thickness thinning is about 26.8%, which is acceptable also the best result of all the simulations, and flange of box shaped part formed in good quality.

4. Simulation optimization

Active hydroforming pressure and blank holder force are optimized to obtain the best result for box shaped part, and the results as shown in figure 11.

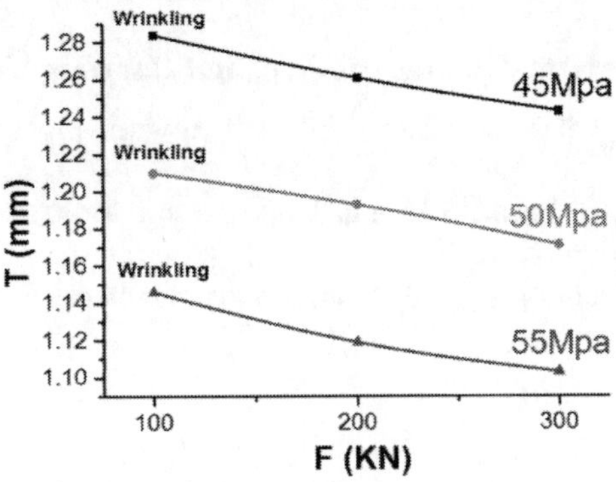

Figure 11. GDD and AH simulation

Figure 12. Hydroforming part

5. Conclusion

By comparing five forming schemes for complex box shaped part and simulation results indicated that the forming scheme combination of passive and active hydroforming is the most suitable scheme for box shaped part with small rounded corners, simulation also proved the sheet hydroforming formability is higher than general deep drawing.

References

[1] Siegert K, Haussermann M and Losch B 2000 Recent development in hydroforming technology, *J. Mater. Process. Technol.* **98** 251

[2] S H Zhang and J Danckert 1998 Development of hydromechanical deep drawing, *J Mater. Process. Technol.* **83** 14

[3] Lang L H, Wang Z R and Kang D C 2004 Hydroforming highlights: sheet hydroforming and tube hydroforming, *J Journal of Materials Processing Technology.* **151** 165

[4] Lang L H, Wang Y M, Xie Y S, Yang X Y and Xu Y Q 2012 *J. Transactions of Nonferrous Mrtals Society of China.* **22** s302-s308

[5] Lang L H, Wang Y M, Xie Y S, Xu Y Q, Yang Z H and Ma H J 2013 *J. MATERIAL SCIENCE & TECHNOLOGY.* **21** 37-43

[6] Lang L H, Xu N, Wang Y M, Yuan C, Xu Y Q, Yang Z H, Shao T W and Duan X M 2013 *J. FORGING & STAMPING TECHNOLOGY.* **38** 21-25

[7] Lang L H, Wang Y M, Xie Y S, Li H L and Hui X P 2011 *J. JOURNAL OF PLASTICITY ENGINEERING.* **18** 33-37

[8] Önder E and Tekkaya A E 2008 *J. International Journal of Machine Tools and Manufacture.* **48** 532-542

Parametric study on numerical simulation of the electromagnetic forming of DP780 steel workpiece with aluminium driver sheet

Hyeonil Park[1,2], Jinwoo Lee[1], Se-Jong Kim[1], Youngseon Lee[1] and Daeyong Kim[1,3,§]

[1]Korea Institute of Materials Science, Changwondaero 797, Changwon, Gyeongnam 642-831, Republic of Korea

[2]Department of Mechanical Engineering, Pusan National University, 2 Busandaehak-ro 63beon-gil, Busan 609-735, Republic of Korea

[3]Korea University of Science and Technology, 217 Gajeongro, Daejeon 305-350, Republic of Korea

[§] E-mail: daeyong@kims.re.kr (D. Kim)

Abstract. The purpose of this study is to investigate the influences of numerical parameters on the electromagnetic forming (EMF) simulation. The 3-dimensional coupled electromagnetic-mechanical simulations were conducted to predict the deformation behavior of the advanced high strength steel (AHSS) sheet receiving support in EMF with aluminum driver sheet. Dual phase (DP) 780 steel workpiece was formed into a hemi elliptical protrusion shape with aluminum alloy AA1050 driver sheet using a flat spiral coil actuator and open cavity die. The deformed shape of the DP780 workpiece and the computation time with respect to element size, N cycle number and time step of electromagnetic (EM) solver were analysed.

1. Introduction

There have been the demands to apply EMF process to the AHSS sheet for shaping a reinforcement rib or sharp edge on preformed automotive parts [1]. However, realizing it is not easy because low conductivity of high strength steel sheet disturbs to induce the eddy current during EMF process. One of ways to overcome this technical hurdle is using high conductivity materials as the driver sheet [2]. Because EMF process with the drive sheet is a complex problem coupled the mechanical, thermal phenomenon as well as the electromagnetism, the numerical simulation is essential tool for understanding the invisible multi-physics natures and designing the optimal processes. In EMF simulation, more iterative coupling and numerical discretization accompany inevitable increasing the numerical errors and the calculation time. Therefore the numerical parameters should be properly selected for guaranteeing reasonable prediction accuracy and computation cost. The purpose of this study is to investigate the influences of numerical parameters on the EMF simulation results for the AHSS sheet with driver sheet.

2. Numerical modeling

The workpiece and driver sheet utilized in this study are a DP780 steel sheet with a thickness of 1.4mm and an aluminum alloy AA1050 sheet with a thickness of 1.0mm, respectively. The blank size of both is 200mm x 200mm. Table 1 summarizes the mechanical and electromagnetic properties of workpiece and driver sheet.

Table 1. The mechanical and electromagnetic properties of DP780 workpiece and AA1050 driver sheet.

	Mechanical					Electromagnetic	
	Young's modulus (GPa)	Yield stress (MPa)	Tensile stress (MPa)	Uniform elongation (%)	Total elongation (%)	Electric conductivity (MS/m)	Magnetic permeability (μH/m)
DP780	210	546.9	832.4	14.51	23.89	3.50	875.0
AA1050	68.9	89.3	107.7	2.15	6.01	35.6	1.26

The EMF of DP780 workpiece with AA1050 driver sheet was performed using the flat spiral coil actuator with an open cavity die to form an aimed hemi elliptical protrusion shape analogous to the reinforcement rib as shown in Figure 1. The flat spiral coil with a cross sectional area of 30mm x 3mm was wound with 15 turns. The gap distance between the windings is 2.5mm. The open cavity having approximately 40mm x 100mm was machined at a mean radius of the flat spiral coil to avoid the dead zone. A charge voltage was chosen to be 7.3kV. Refer to the reference [3] for more details.

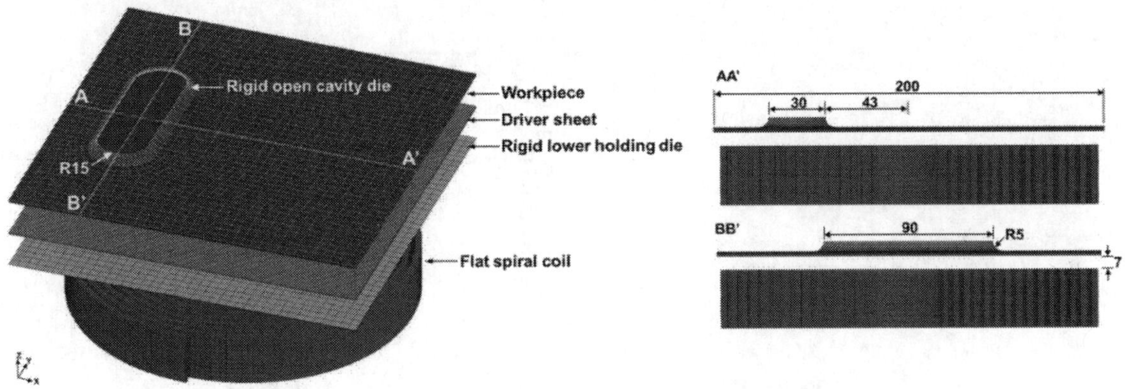

(a) (b)

Figure 1. (a) Finite element meshes of the assembly model and (b) side view of the flat spiral coil and the open cavity die (unit: mm).

Figure 1(a) shows the mesh assembly for EMF simulations. The eight-node hexahedron solid elements for the workpiece, the driver sheet and the flat spiral coil, while the four-node quadrangular shell elements for the rigid open cavity die and the lower holding die were utilized. An isotropic elasto-plasticity model was adopted for workpiece and driver sheet by neglecting their anisotropy for the sake of simplicity. The commercial software LS-Dyna was used to fully couple with electromagnetic and mechanical systems in 3-dimensional spaces [4].

The influence of three numerical parameters on the numerical simulation was investigated: (1) the element size of the workpiece and driver sheet, (2) N cycle number for electromagnetism computation frequency between EM and structural solving matrices and (3) time step of electromagnetic solver. Each parameter for the base line condition is chosen to be element size of 1.5×1.5 (mm x mm), N cycle number of 20, and time step of 2 (μs), respectively. For parametric studies, the plane element sizes were varied as 1 x 1, 1.5 x 1.5, 2 x 2, 2.5 x 2.5 and 3 x 3 (mm x mm), N cycle numbers as 1, 5, 10, 15, 20, 25, 50, 75 and 100, time steps as 0.1, 0.5, 1, 1.5, 2, 2.5, 3, 3.5 and 4 (μs). The deformed shape in the BB' cross section of the DP780 workpiece and computation time with respect to each parameter were predicted and compared.

3. Results

3.1. Influence of element size

The forming height increases and is gradually saturated as the element size decreases as shown in Figure 2(a). Figure 2(b) shows that the computation time rapidly increases as the element size approaches to 1.0mm.

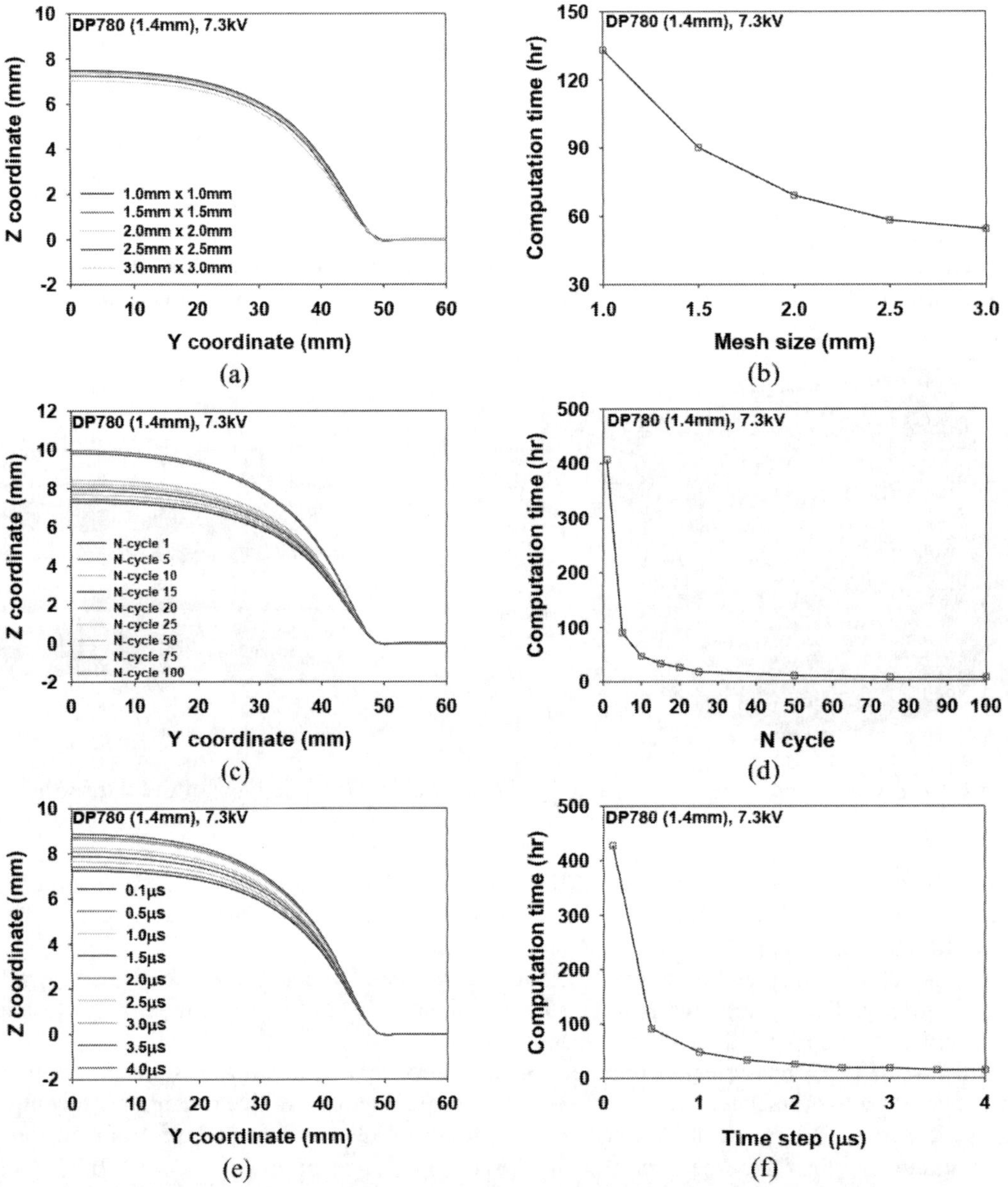

Figure 2. (a), (c) and (e) deformed shape and (b), (d) and (f) computation time with respect to mesh size ((a) and (b)), N cycle number ((c) and (d)) and time step ((e) and (f)).

3.2. Influence of N cycle number

The forming height proportionally increases until N cycle number increases up to 50 as shown in Figure 2(c). When N cycle number is more than 50, the unexpected increase of forming height is observed. Figure 2(d) shows the computation time decreases as the N cycle number increases and drastically increases when N cycle number is less than 10.

3.3. Influence of time step

The forming height increases as time step increases as shown in Figure 2(e). Figure 2(f) shows that the computation time increases as the time step decreases and drastically increases when time step is less than 1 μs.

4. Summary

The present study aims to evaluate the influence of the numerical parameters on prediction accuracy and computation cost in the numerical simulation for the EMF of the DP780 workpiece with an AA1050 driver sheet to hemi spherical protrusion shape. The deformed shape of the DP780 workpiece and the computation time with respect to element size, N cycle number and time step of EM solver were predicted. The forming height increases as the element size decreases, N cycle number and time step increase. On the other hand, the computation time increases as all parameters decrease. Further study to select proper parameter set is needed to compromise between the prediction accuracy and computation cost.

References

[1] Eguia I, Mangas A and Iturbe R 2010 Electromagnetic Forming of Longitudinal Strengthening Ribs in Roll Formed Automotive Profiles (Columbus, Ohio: Proceedings of the 4th International Conference on High Speed Forming) pp 198–207

[2] Weimar G 1963 Hochgeschwindigkeitsbearbeitung III – Umformung von Blechen und Rohren durch magnetische Kräfte *Werkstatt und Betr.* **96** 893–9

[3] Park H, Kim D, Lee J, Kim S-J, Lee Y and Moon Y H 2016 Effect of an aluminum driver sheet on the electromagnetic forming of DP780 steel sheet *J. Mater. Process. Technol.* **235** 158–70

[4] Eplattenier P L, Cook G, Ashcraft C, Burger M, Imbert J and Worswick M 2008 Introduction of an Electromagnetism Module in LS-DYNA for Coupled Mechanical-Thermal-Electromagnetic Simulations (Dortmund: Proceedings of the 3rd International Conference on High Speed Forming) pp 85–96

Acknowledgments

This study was financially supported by the Fundamental Research Program of the Korea Institute of Materials Science and R&D programs on Fusion Core Technology for Industry (No. 10040078), funded by the Ministry of Knowledge Economy, Republic of Korea.

Journal of Physics: Conference Series **734** (2016) 032086 doi:10.1088/1742-6596/734/3/032086

A Multi-Scale Modeling of Multiphase Advanced High Strength Steels Based on Crystal Plasticity and Evolutionary Yield Function

Farhang Pourboghrat, and Taejoon Park

Ohio State University, Department of Integrated Systems Engineering, Ohio State University, 210 Baker Systems, 1971 Neil Avenue, Columbus, OH 43210, United States

pourboghrat.2@osu.edu, park.2417@osu.edu

Abstract. A multi-scale modeling of multiphase advanced high strength steels was developed based on crystal plasticity and evolutionary yield function. An analytical model to directly predict the mechanical behavior of each phase was developed from crystal plasticity model and EBSD data. Considering the grain size and orientation distribution, evolution of the yield surface could be directly predicted by the analytical prediction model. The yield surface prediction results were applied to the evolutionary yield function model. Macroscopic mechanical behavior of the multiphase material could be predicted by considering the volume fraction of each phase, which enables to optimize the desired material properties during the development of the multiphase material.

1. Introduction

To reduce the amount of time and cost required for material and product design, computational approaches have been adopted to predict and evaluate the mechanical performance of materials, such as formability and springback. In particular, for the precise prediction of the mechanical behavior, advanced constitutive models are required to account for the macroscopic behavior observed from various experiments. Further endeavors have been made to develop advanced material models and to regenerate the microstructures to represent the microstructural behavior.

In this work, a multi-scale modeling of multiphase advanced high strength steels was developed based on crystal plasticity [1] and evolutionary yield function. As an alternative to the finite element approaches along with the generation of the 3D representative volume elements (RVE), an analytical model to directly predict the mechanical behavior of each phase from EBSD data was developed based on crystal plasticity. In the developed analytical prediction model, the stress-strain behavior and the evolution of the r-value and yield surface with respect to the plastic deformation could be directly predicted by considering the grain size distribution and orientation distribution obtained from the EBSD data. Then, the generated initial yield surface and its evolution were applied to calibrate the phenomenological evolutionary yield function model. A generalized J2 based non-quadratic asymmetric/anisotropic function was applied for the yield function.

By developing the analytical model, the impact that the volume fraction and material properties of each phase have on the macroscopic behavior could be directly predicted without performing finite

Journal of Physics: Conference Series **734** (2016) 032086 doi:10.1088/1742-6596/734/3/032086

element simulations. This will in turn reduce the total time required to optimize the desired material properties of a multiphase steel.

2. Development of an analytical model to directly predict yield surface based on crystal plasticity and measured EBSD data

An analytical model to directly predict the mechanical behavior of each phase from EBSD data was developed based on crystal plasticity. As schematically shown in Figure 1, the initial yield surface for each grain is calculated by considering the crystal orientation within a multiphase steel. The macroscopic yield surface is then obtained by homogenizing the yield surfaces for representative grain orientations. Figure 2 shows the comparison of the combined yield surface for BCC crystals by increasing the number of grains. In the crystal plasticity simulations, the crystal orientation for each grain was randomly generated and 24 slip systems were considered for BCC crystals (12 {110}<111> and 12 {112}). As the number of grains considered in the crystal plasticity simulation increases, the yield surface shape converges to the smooth isotropic yield surface.

Figure 3 shows the combined yield surface and its evolution generated by the developed analytical model utilizing the crystal orientations measured from EBSD data. For demonstration purposes, 1000 grains of BCC crystals were randomly selected from the EBSD data. For the generation of the initial yield surface, the number of crystal orientations within each grain was assumed, and then the yield surface evolution was calculated for different loading conditions. The work hardening rate for each slip system was obtained from the micro-pillar compression data.

Besides the calculation of the initial yield surface and its evolution, the macroscopic stress-strain behaviour, as well as the r-value evolution during the uniaxial tension were also predicted by the developed analytical model. Figure 4 shows the r-value evolutions during the uniaxial tension.

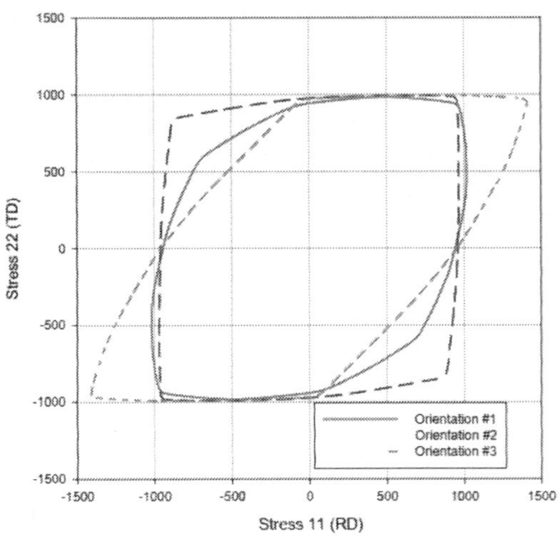

Figure 1. Calculated initial yield surfaces for individual grains of BCC crystals.

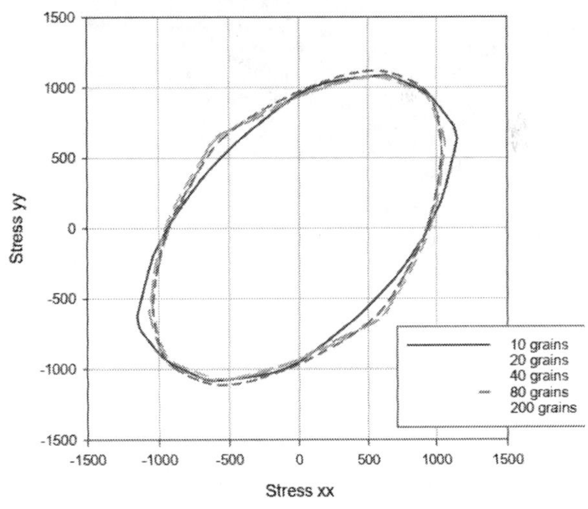

Figure 2. Comparison of the combined initial yield surfaces for BCC crystals for different grain numbers.

To evaluate the accuracy of the developed analytical model in predicting the global stress-strain curves, the crystal plasticity FE simulation was also performed to generate a uniaxial tension stress-strain curve. A hypothetical ferrite with 48 slip systems (12 {110}<111>, 12 {112} and 24 {123}<111> slip systems) was considered and the same hardening parameters were used both for the analytical crystal plasticity simulation and the CPFE simulation. A rectangular specimen shown in Figure 5 with randomly distributed BCC crystals were considered for the CPFE simulation. As shown in Figure 6, the stress-strain curve generated by the developed analytical model is almost identical to the curves obtained by the CPFE simulation.

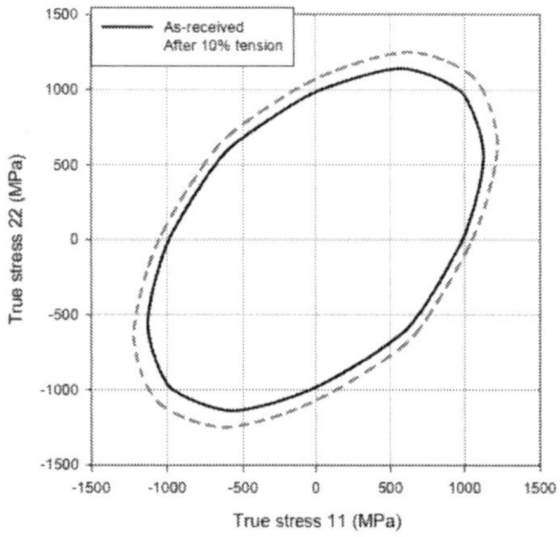

Figure 3. Calculated stress-strain curves during the uniaxial tension for different loading directions (BCC crystal).

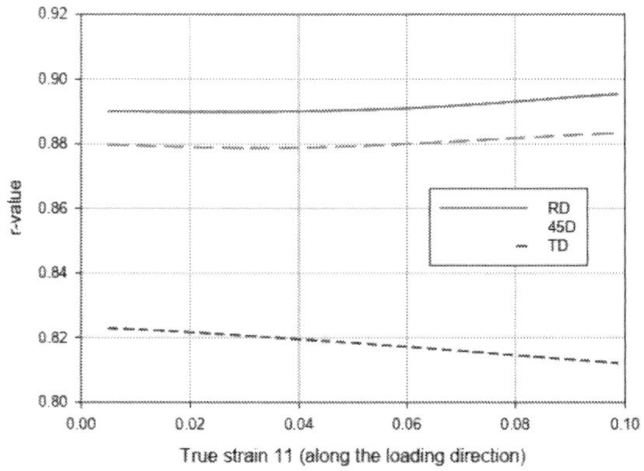

Figure 4. Calculated r-value evolutions during the uniaxial tension for different loading directions (BCC crystal).

Figure 5. A rectangular specimen with randomly distributed BCC crystals used for the CPFE simulation.

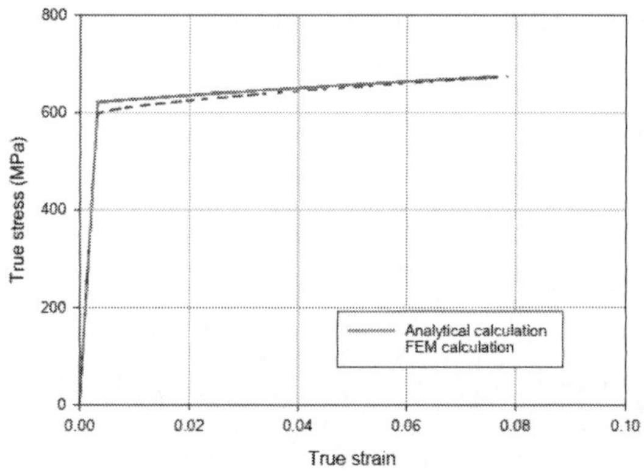

Figure 6. Comparison of the obtained stress-strain curves from the CPFE simulation and the analytical model.

3. Implementation of predicted data into the evolutionary yield function model

The calculated yield surfaces for individual phases were utilized to generate the macroscopic yield surface and its evolution for a multiphase steel. In addition to crystal orientations, the volume fraction of each phase and grain size distributions, the grain shape distributions were also utilized in the generation of the macroscopic yield surface and its evolution. The results were then implemented into the phenomenological evolutionary yield function model [2].

As a demonstration, Figures 7 and 8 show the yield surface evolution and stress-strain curves for BAO QP980 steel represented by the phenomenological evolutionary yield function model. The evolution of the yield stress for each direction were characterized for the yield function using Yld2000-2D function. The calculated yield stresses during the uniaxial tension for different tensile directions are compared with the measured yield stresses and showed reasonably good agreement, as shown in Figure 8.

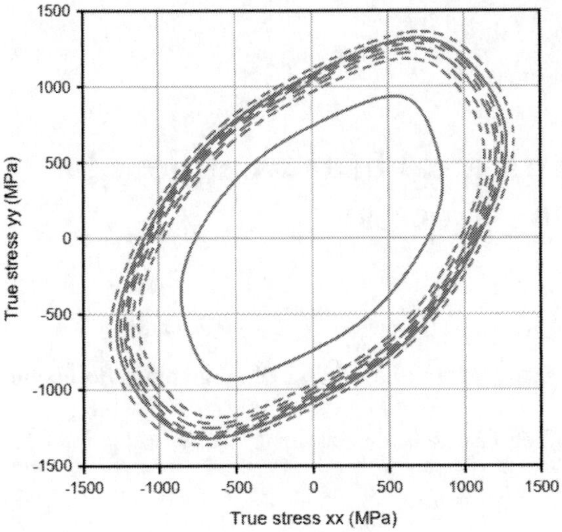

Figure 7. Contours of the yield function for BAO QP980 steel

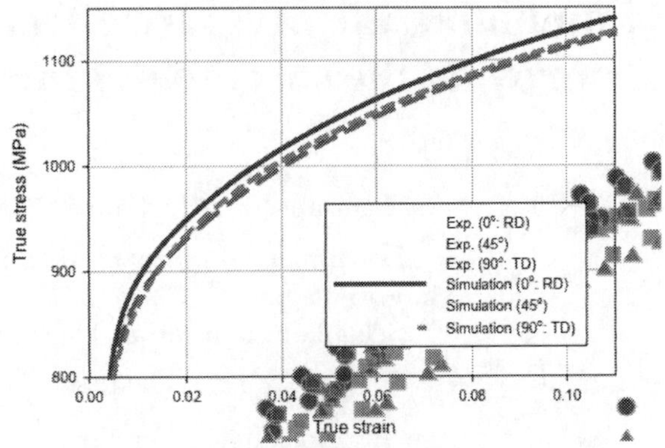

Figure 8. Comparison of the measured and simulated true stress-strain curves for the rolling, 45 and 90 degrees off tensile directions.

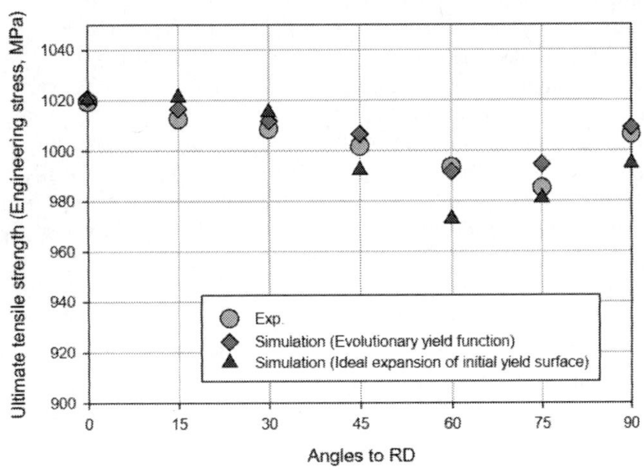

Figure 9. Comparison of the measured and simulated ultimate tensile strength distributions.

References

[1] Zamiri A and Pourboghrat F, 2010, *Int. J. Plast.* **26** 731-46
[2] Zamiri A and Pourboghrat F, 2007, *Int. J. Solid Struc.* **44** 8627-47

Combining the pressure effect with local heat treatment for improving the sheet metal forming process

G. Palumbo[1], A. Piccininni[1], P. Guglielmi[1], D. Sorgente[2], L. Tricarico[1]

1. Department of Mechanical Engineering, Mathematics and Management, Politecnico di Bari, Italy
2. School of Engineering, Università degli Studi della Basilicata, Italy

gianfranco.palumbo@poliba.it

Abstract. The present work deals with the advantages in the Hydromechanical Deep Drawing (HDD) when AA5754 Tailored Heat Treated Blanks (THTBs) are adopted. It is well known that the creation of a suitable distribution of material properties increases the process performance. When non heat-treatable alloys are considered, the THTB approach can be successfully applied to increase the Limit Drawing Ratio (LDR) by changing the peripheral zone into the annealed state starting from a cold-worked blank. If this approach is combined with the advantages of a counterpressure, even more remarkable improvements can be achieved. Due to the large number of involved parameters, the optimized design of both the local treatment and the pressure profile were investigated coupling an axial symmetric Finite Element model with the integration platform modeFRONTIER. Results confirmed the possibility of increasing the LDR from 2.0 (Deep Drawing using a blank in the annealed state) up to about 3.0 if combining the adoption of a THTB with the optimal pressure profile.

1. Introduction

The massive adoption of light alloys is becoming an unavoidable step for respecting the limitations of harmful emissions [1]. Aluminium (Al) alloys, which is a largely adopted light material, despite providing good resistance to weight ratio, present poor formability at room temperature: the adoption of such alloys thus needs deep investigations aimed at improving their attitude to be processed by stamping process. The Deep Drawing (DD) is considered an useful test to investigate such an aspect, since during the process the material undergoes a strain distribution similar to the one the blank experiences during industrial stamping processes. Several studies are reported in the literature about the possible solutions to increase the Limit Drawing Ratio (LDR), i.e. the ratio between the maximum blank diameter that can be fully drawn without rupture and the punch diameter: the adoption of a counterpressure during the DD process (Hydromechanical Deep Drawing, HDD) reduces the radial stress state in the cup wall and lowers the frictional forces in the flange region, thus having a beneficial effect [2]; an improvement can be also achieved adopting a suitable heating strategy during the stamping, for example heating the peripheral zone (lowering the yield stress and increasing the ductility) while maintaining the central region of the blank at room temperature (in order to have an high enough strength) [3]; a further strategy is to create such a local variation of material properties by means of a local heat treatments prior to the stamping process. The last approach is known as Tailored Heat Treated Blank (THTB) technology [4]. Even if plenty of researches exist concerning each of the abovementioned solutions to increase the LDR, only few works, and almost no one if the AA5xxx series is considered, investigate the advantages coming from the combination of the counterpressure with the THTB technology [5]. The present work deals with the numerical investigation of the improvement which the combination of the HDD and the THTB technology can produce. In particular, the aim is to design the local annealing treatment over the blank and, at the same time, define the optimal HDD process parameters (the pressure profile and the Blank Holder Force, BHF). Such a goal was obtained by means of a multi-objective optimization procedure which coupled an axial symmetric FE model with the integration platform modeFRONTIER.

2. The investigated alloy

The investigated material (AA5754) belongs to the group of the Al-Mg alloy, particularly suitable for automotive applications requiring high formability [6]. The mechanical behaviour was at first investigated in the annealed condition (red curve in Figure 1): tensile tests were conducted in order to determine the flow stress curve and to model it using the *Hollomon* equation ($\sigma = K\varepsilon^n$).

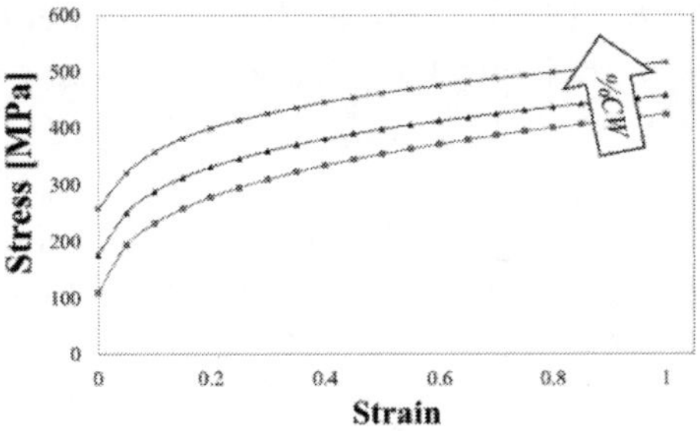

Figure 1. Flow stress curve as a function of *%CW*

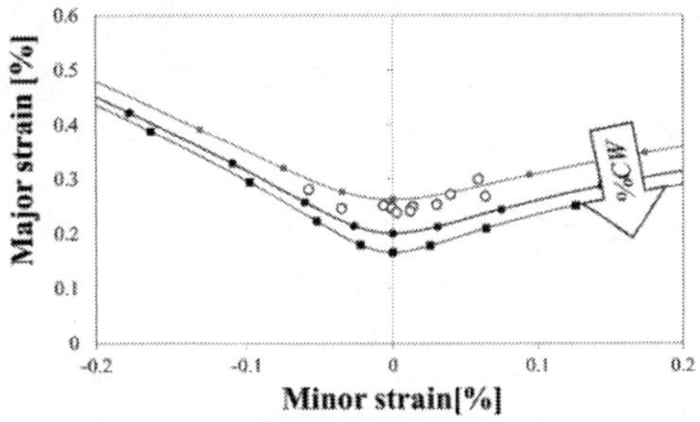

Figure 2. Comparison between the analytical and the experimental limit strains

Tests were also conducted on 5xxx alloys characterised by higher values of the cold-working (*%CW*): in such a way main mechanical parameters (yield stress, ultimate tensile strength and elongation at necking) could be modelled as a function of the parameter *%CW*. The deformative behaviour in the annealed condition was investigated by means of formability (Nakajima) tests. Tests were assisted by the non-contact Digital Image Correlation (DIC) GOM-ARAMIS system, able to acquire the strain distribution in the specimen. The experimental limit strains (white dots in Figure 2) were compared with the analytical Forming Limit Curve (FLC) calculated using flow stress data and the Modified Maximum Force Criterion (MMFC) [7]. Since a good overlapping between data was achieved, the MMFC was adopted to determine the FLCs of the investigated alloy corresponding to different values of the parameter *%CW*.

3. The numerical model and the optimization procedure

The axial symmetric numerical model in Figure 3a was created using the commercial FE code Abaqus. Tools were modelled as rigid bodies, while the blank (initial thickness 1 mm, radius variable in the range 36 - 50 mm) was deformable and meshed with 0.1 mm axisymmetric shell elements. The friction was modelled using the Coulomb law, setting the coefficient to 0.2 between the punch and the blank and to 0.05 between the blank and other tools [8]. The Von Mises yield criterion was adopted and material data described in the section 2 (in terms of both flow stress curves and FLCs) were implemented in the numerical model. The effect of the local heat treatment before the deformation was modelled setting different mechanical and deformative properties over the blank, as depicted in Figure 3b: the central part of the specimen, whose extension was managed by the parameter *L%* (it is a percentage of the radius), was modelled using material properties having a variable cold-worked condition (managed by the parameter *CW%*); the flange was always modelled in the annealed state; finally, in order to make the simulation more realistic, material properties were varied linearly (transition zone 1 mm wide) from the center to the flange.

In Figure 3c the pressure profile set in the HDD simulations is shown: it was designed in order to determine the bulging of the blank before the punch began its stroke; both the maximum pre-bulging pressure (p_{PB}) and the maximum forming pressure (p_{MAX}) were changed in the optimization loop.

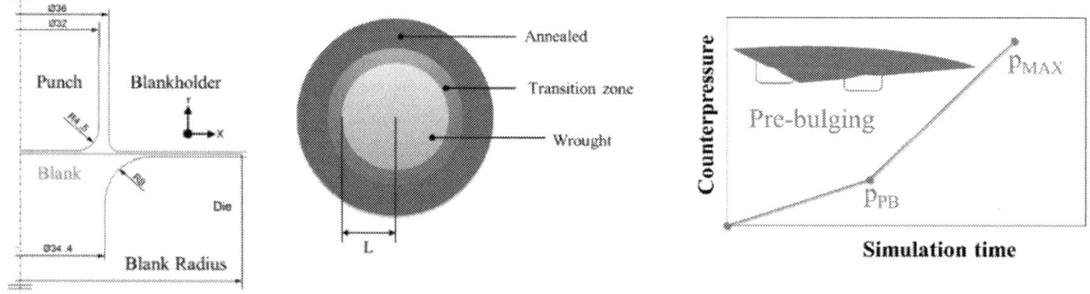

(a) (b) (c)

Figure 3. Process modelling: (a) 2D FE model; (b) distribution of material properties after the THTB; (c) pressure profile and parameters adopted for defining it

In fact, the FE model was coupled with the integration platform modeFRONTIER to carry out the optimization procedure based on a multi-objective genetic algorithm. The optimization was divided into 2 steps: in the first step, it was aimed at optimally distributing the mechanical properties over the blank (both the extension of the central blank region and the initial level of cold-working were evaluated); in the second step the pressure profile was optimized (the maximum pre-bulging pressure, the maximum forming pressure and the BHF were defined) while keeping fixed the heat treatment parameters. In particular, the second optimization step started from an initial population of 12 designs created with the Uniform Latin Hypercube algorithm [9] evolving through 20 successive generations. The objective functions in both the optimization steps were chosen to maximize the LDR and minimize the occurrence of blank rupture, monitoring the FLDCRT output variable (it is a standard output in Abaqus which takes into account how much the nodal strain is close to the Forming Limit Curve); in addition, to drive the optimization procedure toward significant results, constraints were applied to both the objective functions.

4. Results and discussions

In order to clearly describe the evolution of the input parameters during the optimization, *history charts* concerning the first optimization step are proposed in Figure 4.

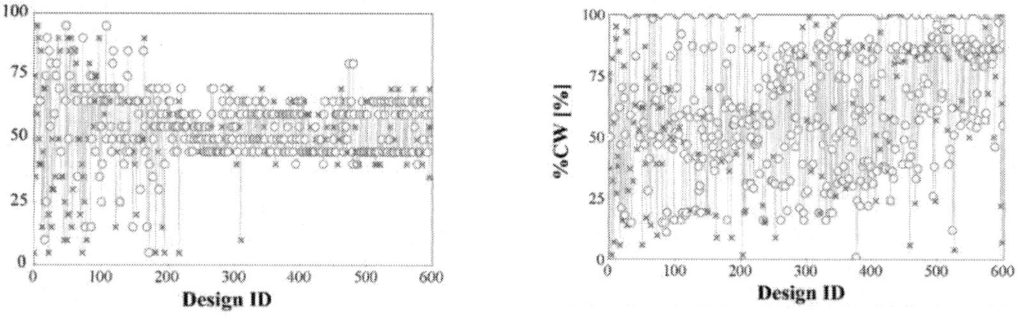

(a) (b)

Figure 4. Evolution of the values for parameters *L%* (a) and *CW%* (b) towards optimal values

In Figure 4 feasible designs (the designs respecting both the applied constraints) are indicated by empty circles, while the unfeasible ones by red crosses. The evolutions of both the parameters related to the heat treatment showed designs clusters within the last generations, highlighting that the region in contact with the punch needs a highly cold-worked condition (*%CW* about equal to 80%); at the

same time, the optimization results indicate that the annealing treatment has to be limited to the flange portion of the blank having the extension of about 35% (*L*% parameter tends to about 65%).
As concerning the second step optimization, *history charts* plotting the evolution of the pre-bulging pressure and the maximum forming pressure are shown in Figure 5.

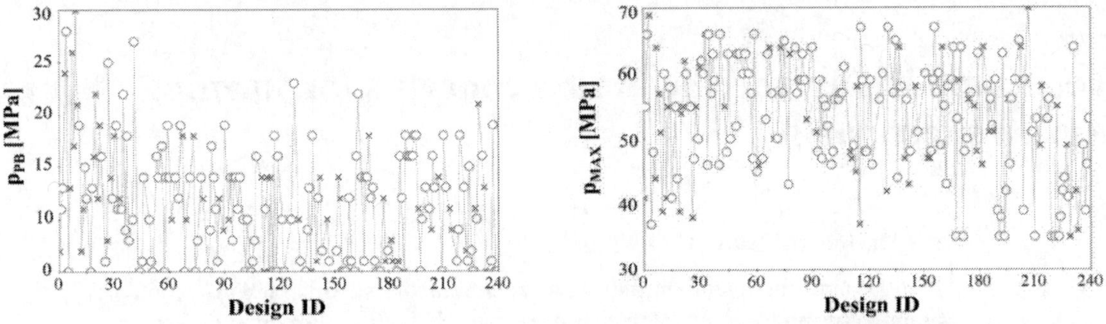

(a) (b)

Figure 5. Evolution of the values for parameters p_{PB} (a) and p_{MAX} (b) towards optimal values

The graphs indicate that to successfully draw high values of the blank radius it is necessary to pre-strain the blank under a moderate bulging (around 10 MPa) and to increase the maximum pressure up to around 60 MPa during the deep drawing. The optimized distribution of the material properties combined to a properly designed pressure profile led to an achievable LDR value equal to 2.94 (R = 47 mm), which represents an improvement of +27% if compared to classical DD process. A key role in such a result is played by the local heat treatment: in fact, at the end of the first optimization step, the LDR was already increased up to 2.8 (21% higher than the classical DD process).

5. Conclusions

The improvement of the HDD process when THTBs are adopted has been numerically demonstrated using a two-steps multi-objective optimization. The first optimization step allowed to evaluate the maximum radius of the blank able to be successfully formed when adopting the proper level of the initial work-hardening of the central region and the proper extension of such an hardened region (LDR=2.8). The second optimization step allowed to determine the maximum radius of the blank able to be successfully formed when adopting the optimal pressure profile and BHF value (LDR=2.94). Thanks to the combination of the above mentioned techniques (local treatment and counter-pressure), the DD process can be improved even if performed at room temperature, thus needing a less complex equipment and leading to a consequent higher tool saving. Future developments will be aimed at refining the FE model and at validating its results by means of experimental tests.

Acknowledgments
The authors wish to thank OMER srl (Palermo, Italy) for supporting the present research.

References

[1] Joost WJ 2012 *JOM.* **64** 1032
[2] Lang L, Li T, An D, Chi C, Nielsen KB, Danckert J 2009 *Mater. Sci. Eng.* A. **499** 320
[3] Palumbo G and Tricarico L 2007 *J. Mater. Process. Technol.* **184** 115.
[4] Geiger M, Merklein M, Vogt U 2009 *Prod. Eng.* **3** 401
[5] Piccininni A, Di Michele G, Palumbo G, Sorgente D, Tricarico L 2015 *Acta Metall. Sin. (English Lett.)* **28** 1482
[6] Hirsch J 2014 *Trans. Nonferrous Met. Soc. China* **24** 1995
[7] Mattiasson K, Sigvant M, Larsson M 2006 *IDDRG'06 Proceeding* Portugal.
[8] Halkaci HS, Turkoz M, Dilmec M 2014 *J. Mater. Process. Technol.* **214** 1638.
[9] Esteco, modeFRONTIER user manual v.4.5.3 2014

Process modelling and die design concepts for forming aircraft sheet parts

H A Hatipoğlu[1] and C O Alkaş[2]

[1]Tool Design Engineer, Turkish Aerospace Industries, Ankara, Turkey
[2]Manufacturing Engineer, Turkish Aerospace Industries, Ankara, Turkey

E-mail: hahatipoglu@tai.com.tr

Abstract. This study is about typical sheet metal forming processes applied in aerospace industry including flexform, stretch form and stretch draw. Each process is modelled by using finite element method for optimization. Tensile, bulge, forming limit and friction tests of commonly used materials are conducted for defining the hardening curves, yield loci, anisotropic constants, forming limit curves and friction coefficients between die and sheet. Process specific loadings and boundary conditions are applied to each model. The models are then validated by smartly designed experiments that characterize the related forming processes. Lastly, several examples are given in which those models are used to predict the forming defects before physical forming and necessary die design and process parameter changes are applied accordingly for successful forming operations.

1. Introduction

Sheet metal forming represents a major manufacturing effort in the aerospace industry. Typical sheet forming processes applied in aerospace industry include bending, drawing, stretching and flexforming. In all those, it is aimed to convert a flat sheet of metal into a part of desired shape without defects like fracture, excessive thinning or wrinkling. Process simulation, which is conducted by using finite element method, helps to predict the forming defects beforehand. Since there are no physical parts and tools, design and process parameter changes can be easily done leading to fast virtual try-outs. Finally, the right part is obtained the first time reducing material, workmanship and time costs.

In this study, the modelling features of such simulations will be explained and three samples will be shown in which numerical analyses were successfully applied.

2. Modelling Features

The success of the numerical analysis is highly dependent upon proper modeling of material behaviour and friction.

2.1. Material modeling

In order to capture the correct material behaviour, tensile tests (ASTM E8-04), bulge tests (ISO16808) and forming limit diagram tests (ISO 12004-2) of commonly used sheet alloys are conducted. The results are then entered to the simulation for defining the hardening curves, yield loci, anisotropic constants and forming limit curves. Barlat 89 anisotropic yield function and Swift relation are used for the yield surface and hardening curve respectively.

Journal of Physics: Conference Series **734** (2016) 032088 doi:10.1088/1742-6596/734/3/032088

Table 1 shows the obtained material parameters for Aluminum alloy 2024-O.

Table 1. Material parameters for AA 2024-O.

Elastic Modulus [GPa]	Poisson's Ratio	Density [tons/ mm^3]	Yield Strength [MPa]	Strength Coefficient (K) [MPa]	Hardening Exponent (n)	Lankford Coefficients (r_0,r_{45},r_{90})
69	0.33	2.8e-9	74	320	0.22	0.60, 0.77, 0.60

2.2. *Friction modeling*

Friction coefficients for various conditions between the tool and the sheet surfaces are determined by doing dedicated tests and comparing them with numerical conjugates. Table 2 lists some friction coefficient findings [1].

Table 2. Friction coefficients for various conditions (tool material: AA 7075, sheet material: AA 2024.

Condition	Friction Coefficient (μ)
dry	0.250
oil	0.125
oil+ nylon film	0.050

3. Samples

Three aircraft sheet parts are investigated as: rudder tip, cockpit panel and skin panel. Those parts are from various locations on the aircraft and each has its own forming process and tooling (figure 1).

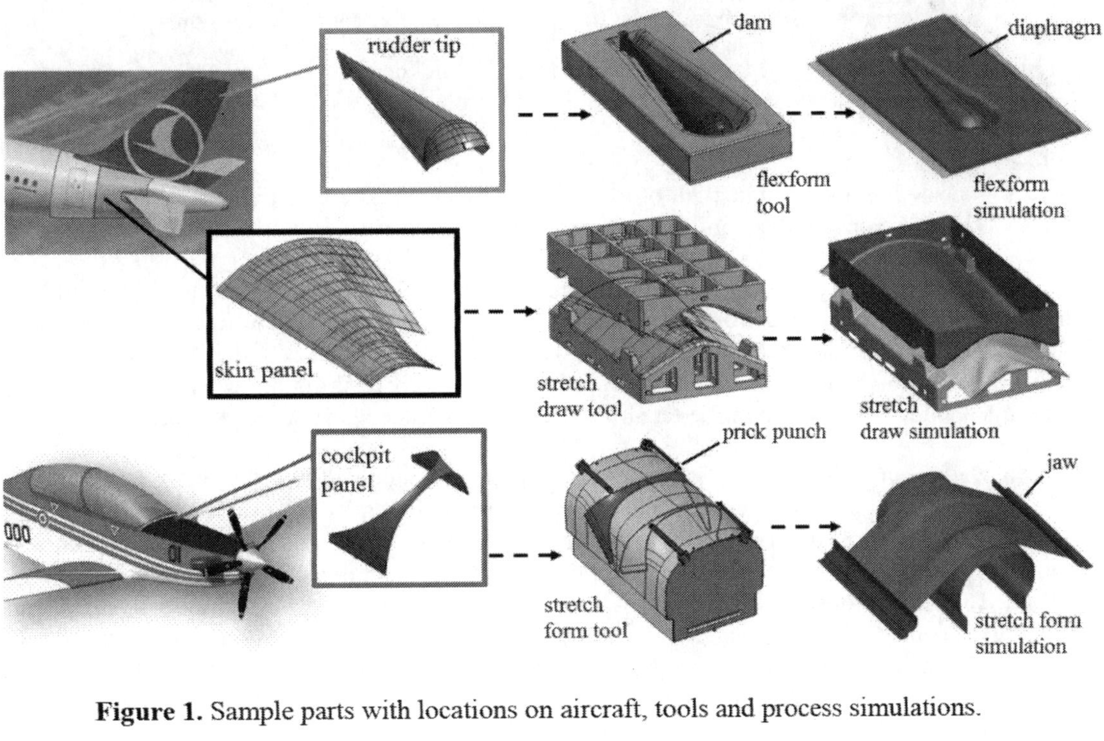

Figure 1. Sample parts with locations on aircraft, tools and process simulations.

3.1. Rudder tip

Rudder is the part which gives direction to the aircraft and rudder tip encloses the top of the rudder (figure 1). The material is AA 6061 with 1.2 mm thickness. It is rather a deep part which is formed with flexforming process.

3.1.1. Flexforming process. In this process, the sheet is formed on a single tool by fluid pressure acting through a rubber diaphragm which takes the place of the counterpart of the tool. For the rudder tip, the tool (material is AA 7075) is a punch type with a dam that prevents wrinkles on the sheet. The flat sheet is attached onto the tool with a guide pin from the tail and with indexing pads from the sides. Some lubricant with nylon film is applied between tool and sheet interface for better sliding. The final part is obtained at about 30 MPa pressure with intermediate steps of stress annealing and trimming.

3.1.2. Flexforming simulation [2]. Numerical analysis is performed before physical forming of the rudder tip. The bodies of the model are the die (rigid-shell), the sheet (plastically deformable-shell) and the rubber diaphragm (elastically deformable-solid). The process is half modelled since the part is half symmetric. The boundary conditions are guide pinned node of the sheet, hinged peripheral nodes of the diaphragm and the pressure load on the diaphragm. Explicit time integration scheme is used for solution. Based on the numerical results, forming steps of the real part are determined. Figure 2 shows the thickness distribution of the final part at the half section. The simulation error is 2% when compared with the experiment.

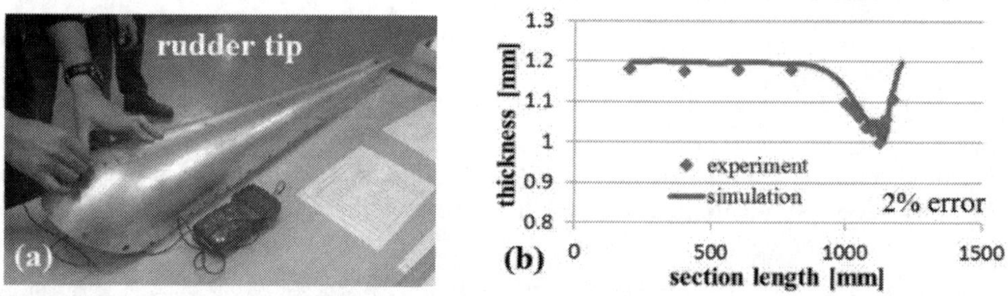

Figure 2. (a) Ultrasonic thickness measurement, (b) thickness distribution comparison between experiment and simulation.

3.2. Cockpit panel

This part is located on the front cockpit of the aircraft (figure 1). The material is Al 2024 with 1.6 mm thickness. The part has complex contoured surfaces that make it very difficult to stretch.

3.2.1. Stretch forming process. In this process, the sheet is stretched over a rigid form-die that acts by hydraulic power while gripped from the sides by jaws which are also hydraulically moved. The form die is attached onto the die table and moves vertically up and down with the die table. For the cockpit panel, the form-die (material is cast iron-GGG45) is designed to produce two parts at once. Some lubricant is applied onto the die for better sliding. Two prick punches are used per part for indexing.

3.2.2. Stretch forming simulation [3]. Numerical analysis is conducted to determine the necessary die-jaw movements for successful forming. The process is modelled utilizing quarter symmetry. Form-die and jaw are discretized with rigid shell elements, whereas sheet is with deformable ones. The boundary conditions are vertical displacement of the die, horizontal displacement of the jaw and the constrained nodes of sheet under the jaw. The problem is solved with an explicit scheme. Obtained die-jaw displacement pairs are entered to the NC control unit of the press and the part is formed successfully in a short time. After forming, the part is measured by GOM-Argus® 3-D optical strain measurement system. It can be seen from figure 3 that maximum strain difference between experiment and simulation is about 0.006 which corresponds to 8% of total strain.

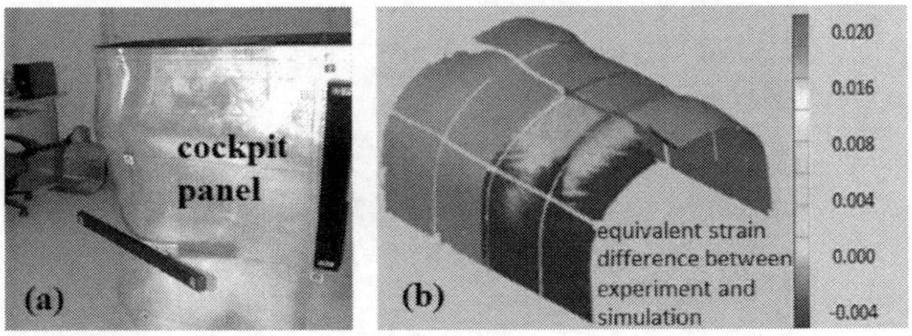

Figure 3. Optical strain measurement of cockpit panel and equivalent strain difference between experiment and simulation.

3.3. Skin panel
This part is located on the upper side of the tail barrel of the aircraft (figure 1). The material is Al 2024 with 3.2 mm thickness. Negative contours on the part geometry necessitate the usage of stretch drawing process which can be classified as a combination of stretch forming and deep drawing.

3.3.1. Stretch drawing process. The sheet is prestretched and wrapped over a lower die by the hydraulic action of flexible jaws; an upper die on the press ram is lowered to complete the required forming. The bosses on the lower die together with the cavities on the upper provide proper closing operation. The skin panel part is stretched for 3% ratio and a ram force of 150 tons.

3.3.2. Stretch drawing simulation. The process is checked with numerical analysis. Upper and lower dies are rigidly modelled with shell elements. Sheet is modelled with deformable shell elements except for the regions under flexible jaws (rigidly modelled). The boundary conditions are the movements of upper-lower die (vertically) and the jaws (horizontally). The sheet deviates from the tool surface after unloading (springback). This deviation is measured on real part with laser tracker and the result is compared with simulation. Average error is about 7% (figure 4).

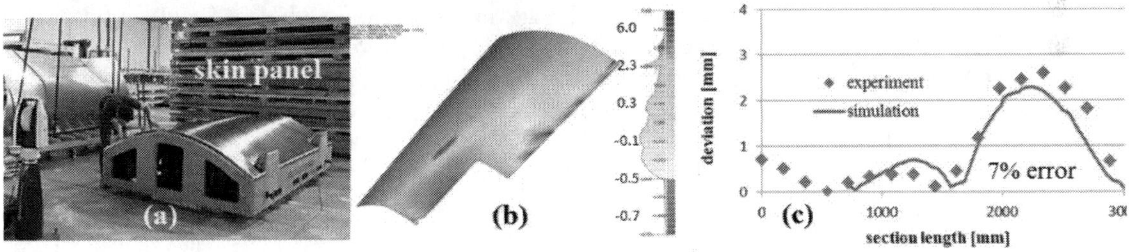

Figure 4. (a) Surface deviation measurement of skin panel with laser tracker, (b) measurement result, (c) deviation comparison between experiment and simulation.

4. Conclusion
This paper aims to show the application of numerical analysis to aircraft sheet metal forming problems. Three samples were chosen to investigate on flexforming, stretch forming and stretch drawing processes. Results prove that numerical simulation is a successful tool for the prediction of final part.

References

[1] Hatipoglu H, Polat N and Köksal A 2010 *Steel Research Int* **81** 640-642
[2] Hatipoglu H, Polat N, Köksal A and Tekkaya A E 2007 *Key Eng. Materials* **344** 469-476
[3] Hatipoglu H, Alkas C and Karadogan C 2013 *Key Eng. Materials* **554** 1451-56

Prediction of the ballistic limit of an aluminium sandwich panel

J. Campbell[1], T. De Vuyst, R. Vignjevic, K. Hughes

Brunel University London, Uxbridge, UB8 3PH, United Kingdom
[1]Corresponding author: james.campbell@brunel.ac.uk

Abstract. This paper presents research on modelling the impact of a 150g projectile on a 35mm thick aluminium sandwich panel. The objective of the work is a predictive modelling capability for the ballistic limit of the panel. A predictive modelling capability supports the design of capture and deorbit missions for large items of space debris such as satellites and rocket upper stages. A detailed explicit finite element model was built using the LSDYNA software and results were compared with experimental data for the projectile exit velocity to establish key parameters. The primary parameters influencing the model behaviour were the strength and failure of the aluminium face sheets and the friction between projectile and panel. The model results showed good agreement with experimental results for ogive nose projectiles, but overestimated the exit velocity for flat nose projectiles.

1. Introduction

The level of debris in Earth orbit is now significant and it is now necessary to reduce the debris population. A proportion of this debris consists of large items including discarded rocket upper stages and derelict satellites. To remove these large items it is proposed to launch a satellite to rendezvous with the debris, then capture and then deorbit in a controlled manner, a concept known as active debris removal [1]. Different methods for capturing derelict satellites have been proposed, with one concept based on the use of a harpoon.

The objective of this work is the development and validation of a modelling method to predict the ballistic limit of representative sandwich panels. Good knowledge of the ballistic limit is required for the harpoon capture concept, as the harpoon must penetrate the structure but must not cause significant additional damage or generate debris. Experimental data from a series of tests performed at the University of Cambridge [2] was available for comparison and the LSDYNA explicit finite element code was used to simulate the impacts.

2. Stiffened panel model

The experimental panels consisted of two 1.27mm thick aluminium 2024-T81 sheets, bonded to an aluminium honeycomb core, for a total panel thickness of 35mm. Each panel was drilled with four 6.5mm holes in the corners, on a 105mm pitch. Steel studding, passing through the panel holes, was used to hold the panel to the target support, with aluminium washers in direct contact with the panels [2]. In the experiments, 20mm diameter 150g steel projectiles with different nose shapes were fired at target panels. In this work we investigate the ogive nose and flat nose projectiles, for both normal impact and 45° impact angles.

Following previous experience with modelling the ballistic limit of aluminium plates [3] a detailed modelling approach was taken where the face sheets were meshed using solid elements (5 elements through thickness) and the honeycomb core geometry was directly modelled using shell elements. Element erosion is used to represent material failure and the projectile and supports are treated as rigid. The models for the normal and 45° impact cases are shown in figure 1. The material properties for the face sheets were derived from the typical stress strain curve for 2024-T81 alloy published in MIL-HDBK-5 [4], and the properties for the honeycomb material were based on published data for the strength of 5056 aluminum foil [5]. To verify the sheet material properties a standard tension test was modelled using solid element s of equivalent size to the face sheet model. Figure 2 shows the stress-strain curve measured from a tension test model, figure 3, overlaid on the MIL-HDBK-5 curve showing good agreement for the resultant behavior from a finite element model

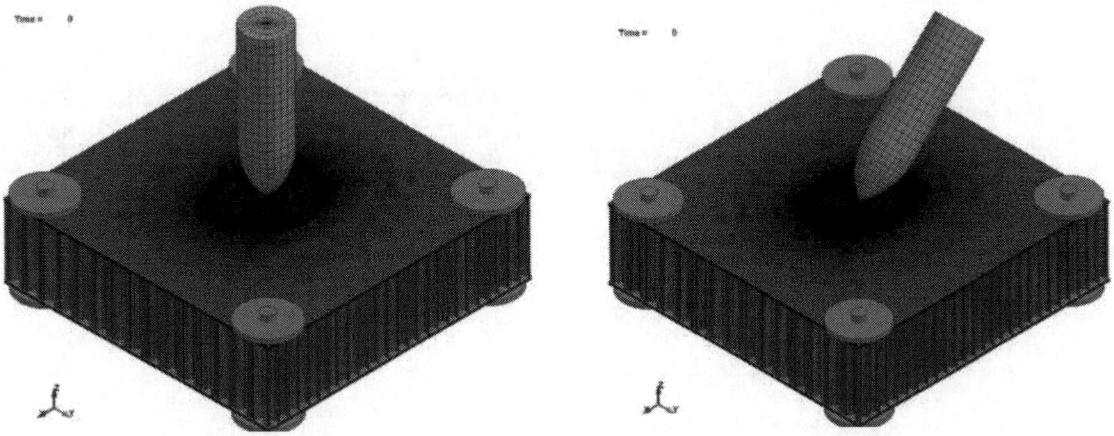

Figure 1: Finite element panel model for 0° impacts (left) and 45° impact (right). The rear face sheet mesh differs between the two models due to the different projectile exit location.

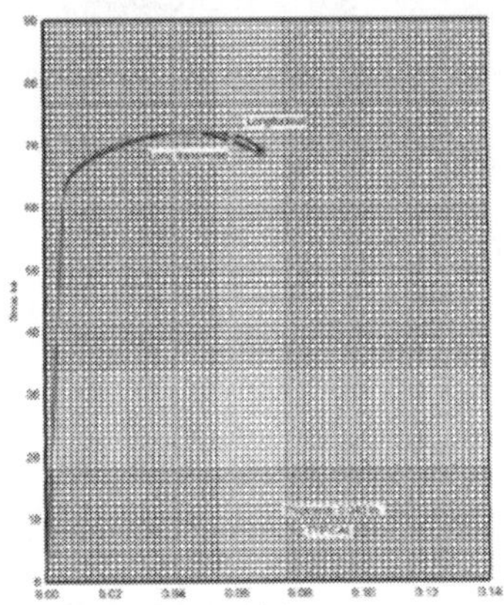

Figure 2: Stress vs strain from FE tension test model (red line) overlaid on MIL-HDBK-5 curve [4].

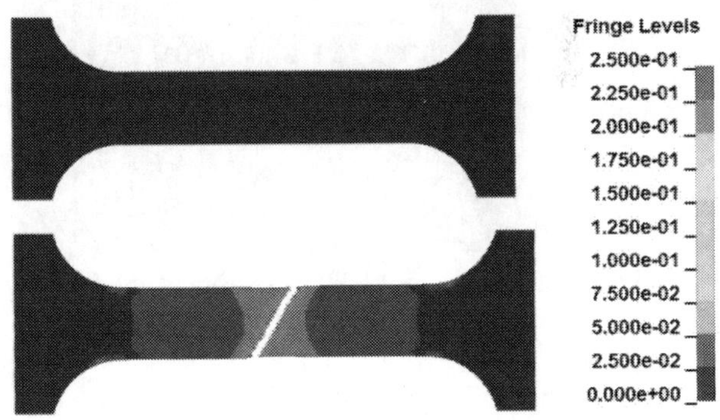

Figure 3: Initial (top) and final(bottom) view of FE tension test model used to check material stress-strain curve. Fringes of effective plastic strain.

A 70m/s impact of the ogive nose projectile was selected as the test case for sensitivity studies to investigate the influence of the sheet and honeycomb models, projectile-panel friction and the core to sheet bond strength. The friction coefficient was selected to match the experimental exit velocity.

3. Model results

For the ogive nose projectile, six analyses were run for impact velocities from 44 m/s to 98 m/s, with the only difference between each analysis being the initial velocity of the projectile. The results showed strong agreement with the experimental projectile exit velocities over the whole range of velocities considered. Figure 4 shows the comparison between the model results for the projectile exit velocity and the constant energy loss fit to the experimental results. Figure 5 shows a sequence of model results for the 70m/s analysis, illustrating the progress of the projectile penetration. Compared with the experimental results, the deformation of the rear sheet is under-predicted in the models. Analyses were also run for the 45° impact experiments, and also gave good agreement with the experimental results, including the deflection of the projectile by the rear sheet at lower impact velocities.

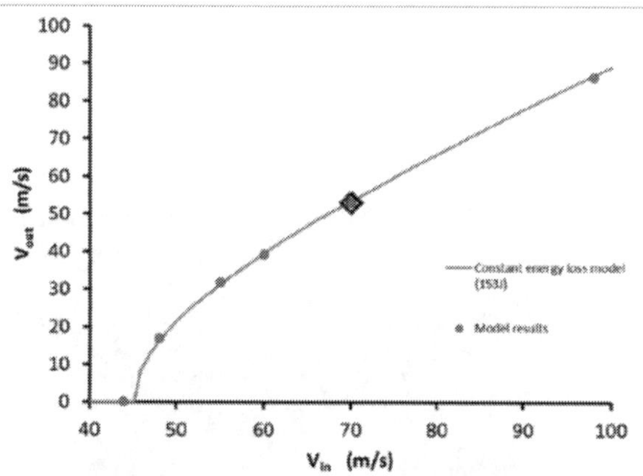

Figure 4: Comparison of numerical residual velocities with the constant penetration energy fit to the experimental results. The black diamond indicated the velocity where sensitivity studies were used to match the experimental velocity.

Figure 5: Side view of 70 m/s impact model, images at 0.5ms, 1.0ms, 1.5ms and 2.0ms.

For the flat nose projectile analyses with impact velocities of 90m/s and 74m/s were run, with all model settings otherwise identical to the ogive nose projectile analyses. The results from these are presented in figure 6, and show that the model significantly over predicts the exit velocity in both cases, although the model correctly predicts the failure modes of the front sheet (plugging) and the tearing of the rear face sheet, figure 7. From the experimental results it is clear that crushing of the honeycomb core is an important mechanism in these impacts and one that is not properly captured in the model. This is consistent with the under-prediction of the absorbed energy, as crushing of the core is not a significant mechanism for the ogive nose projectiles.

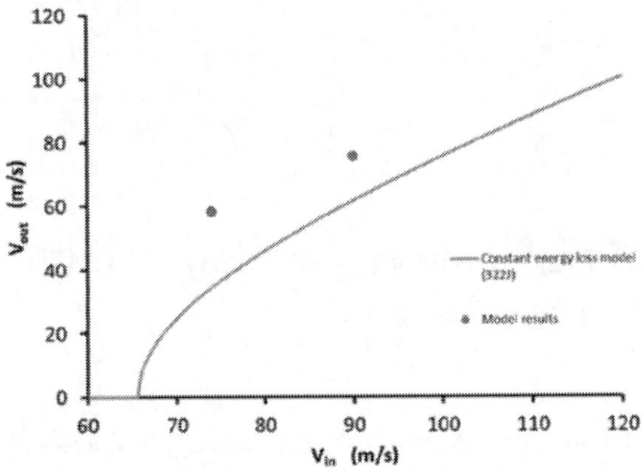

Figure 6: Residual velocity from 0° panel analyses for a flat nose projectile compared against experimental results.

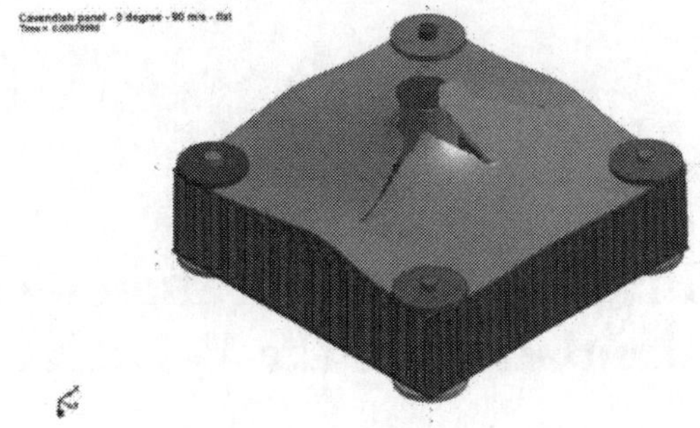

Figure 7: View of 90m/s flat nose impact analysis showing rear face sheet deformation as the projectile begins to exit the panel.

4. Conclusions

A detailed explicit finite element model of an aluminium sandwich panel has been developed. For the impact of an ogive nose projectile this model has shown excellent agreement with experimental data, including the ability to match the ballistic limit of the panel. For a flat nose projectile the model over predicts the exit velocity, as the crushing of the honeycomb core is not correctly captured. Future work will include experimental tests on the sheet material to reduce the uncertainty in material properties and investigation of how the core crush behaviour seen in the flat nose projectile impacts can be correctly modelled.

Acknowledgements
This work was funded by the European Space Agency under a contract with Airbus Defence and Space.

References
[1] Innocenti L, Soares T, Delaval J, ESA Clean Space Initiative. *In: Proc. of 6th European Conf. on Space Debris*, ESA SP-723, 2013.
[2] Taylor NE, Jardine AP Panel penetration experiments. *University of Cambridge*, August 2015.
[3] De Vuyst T, Vignjevic R, Azorin Albero A, Hughes K, Campbell JC, Djordjevic N A study of the ballistic limit of AA2024-T351 sheets impacted by spherical and cubical steel projectiles. *Submitted to: ASME Journal of Applied Mechanics,* January 2016.
[4] Military Handbook: Metallic materials and elements for aerospace vehicle structures. MIL-HDBK-5H. United States Department of Defense, December 1998.
[5] Aluminum Standards and Data, The Aluminum Association Inc., Washington DC, 1997

Friction and lubrication modelling in sheet metal forming simulations of the Volvo XC90 inner door

M Sigvant[1,2,7], J Pilthammar[1,2], J Hol[3], J H Wiebenga[3], T Chezan[4], B Carleer[5] and A H van den Boogaard[6]

[1] Volvo Cars, Dept. 81153 Stamping CAE & Die Development, Olofström, Sweden.
[2] Blekinge Institute of Technology, Valhallavägen, 371 41 Karlskrona, Sweden.
[3] TriboForm Engineering, Hengelosestraat 500, 7500 AM Enschede, The Netherlands.
[4] Tata Steel, P.O. Box 10.000, 1970 CA IJmuiden, The Netherlands.
[5] AutoForm Engineering, Emil-Figge-Str. 76-80, D-44227 Dortmund, Germany.
[6] University of Twente, P.O. Box 217, 7500 AE, Enschede, The Netherlands.
[7] Author to whom any correspondence should be addressed

E-mail address: mats.sigvant@volvocars.som

Abstract. The quality of sheet metal formed parts is strongly dependent on the friction and lubrication conditions that are acting in the actual production process. Although friction is of key importance, it is currently not considered in detail in stamping simulations. This paper presents project results considering friction and lubrication modelling in stamping simulations of the Volvo XC90 inner door. For this purpose, the TriboForm software is used in combination with the AutoForm software. Validation of the simulation results is performed based on door-inner parts taken from the press line in a full-scale production run. The project results demonstrate the improved prediction accuracy of stamping simulations.

1. Introduction
The quality of sheet metal formed parts is strongly dependent on the tribology, friction and lubrication conditions that are acting in the actual production process. These friction conditions are dependent on the tribology system, i.e. the applied sheet material, coating, tooling material, lubrication- and process conditions. Although friction is of key importance, it is currently not considered in detail in stamping simulations. The current industrial standard is to use a constant (Coulomb) coefficient of friction. At the Stamping CAE & Die Development Department at Volvo Cars, it is concluded that friction and lubrication modelling is the way forward for further improving stamping simulation accuracy. Since the results in [1] were very promising, the next step was to apply the method to stamping simulation of parts of car bodies.

2. Production Set-up
The part used in this study is the rear door inner of the right hand side of the Volvo XC90. The part is stamped with a single action die in a mechanical 3D-transfer press line at Volvo Cars Body Components, Olofström, Sweden. The die material is GGG70L and the die and the punch is chrome plated. The blankholder is only polished and then laser hardened. The blank is contour cut from a 0.7 mm thick VDA239-CR4 GI (hot-dip galvanized) sheet material with Fuchs 4107 applied as a pre-lube.

At the end of the production run, also 100 blanks of VDA239-CR4 ZM (Zinc-Magnesium coating) with the same lubricant and ordered amount were stamped. More information about the production set-up and the results from the trial are described in [2-3].

3. Numerical Models

3.1. Sheet Metal Forming Simulations

The rear door inner forming process is simulated with AutoForm^{plus} R6.0. The die, punch and blank holder surfaces have been scanned and thus also include geometrical draw-beads. The resulting deformations of the blank holder in the die structure analysis are incorporated into the sheet metal forming model by AutoForm's morphing functionality, see [4] for more details. The BBC2005 material model, see [9], is used for all simulations, using four initial yield stress, four R-values and the exponent M as input. The blank holder force is modelled with columns and the load on each column is taken from a die structural analysis. The ram speed in the simulation is taken identical to the ram speed of the mechanical press-line. The different friction models are included in the simulations using the TriboForm FEM Plug-In for AutoForm.

3.2. Friction Models

Tribological conditions in metal forming processes are dependent on local process and lubrication conditions, loading and local strain state of the sheet material as demonstrated in [5-6]. The TriboForm software allows for multi-scale modelling of a time and locally varying friction coefficient under a wide range of process conditions. The physically-based models included in TriboForm enable friction modelling in the mixed lubrication regime. This is achieved by coupling a boundary lubrication friction model [7] and a hydrodynamic friction model [8]. Information of the tribology system is required as an user input, i.e. the applied sheet material, coating and tooling material, lubrication type, lubrication amount and process conditions. This information can either be entered by the user or extracted from a database, i.e. the TriboForm Library. The friction models for GI and ZM lubrication systems assumes the following: tool surfaces are chrome plated GGG70L die material with a Ra of 0.35 and the amount of the lubricant Fuchs 4107 is 2.0 g/m^2. The lubrication amount is based on the lubrication amount measurements on the blanks and the observation that there was oil present on the forming surfaces in the die.

4. Results

4.1. Friction models

A comparison of the two lubrication systems reveals that they have a different tribological behaviour. The friction coefficient in the GI lubrication system varies more with contact pressure and relative velocity of the sheet than the ZM lubrication system. Another difference is that the friction coefficients at low contact pressures are almost twice as large for GI than for ZM. Generally, the ZM lubrication system is rather stable, i.e. similar friction coefficient independently of contact pressure and relative velocity, which is a good property of a lubrication system. More information about the two lubrication systems can be found in [3].

4.2. Draw-in predictions

The predicted draw-in values for the GI system are from a stamping simulation using a constant friction coefficient μ equal to 0.15 and a stamping simulation using the TriboForm friction model. On the top and bottom edges of the part, the predicted values are close to each other and also to measured ones. On the other two edges the difference is larger between the μ=0.15 and the TriboForm model. Also on these two edges, the predicted draw-in with the TriboForm model is very close to the measured one. The general impression is that the TriboForm model predicts the draw-in more accurately than the model using μ=0.15.

The prediction of the draw-in of the sheet edges with a TriboForm ZM lubrication model is less accurate than the GI model prediction. The ZM prediction is only accurate at front edge of the part, on all other edges the draw-in is overpredicted. On the other hand, the prediction of the size of the pre-cut holes are better for ZM lubrication system than for the GI lubrication system. Another interesting observation is that draw-in predictions with $\mu=0.15$ is more accurate than TriboForm model on all edges, except at the front edge where $\mu=0.15$ underpredicts the draw-in. These results indicate on one hand that the TriboForm ZM model predicts the friction coefficient on the part accurately, while it is underpredicting it on the addendum and on the other hand that $\mu=0.15$ is a more accurate model for ZM. More information about the draw-in predictions can be found in [3].

4.3. Major strain predictions

Figure 1 presents comparisons made between predicted and measured major strains. In areas with red colours the simulations are overpredicting the strains and in blue areas the simulations are underpredicting the strains. In black and white areas, the overestimation or underestimation is more than 0.05.

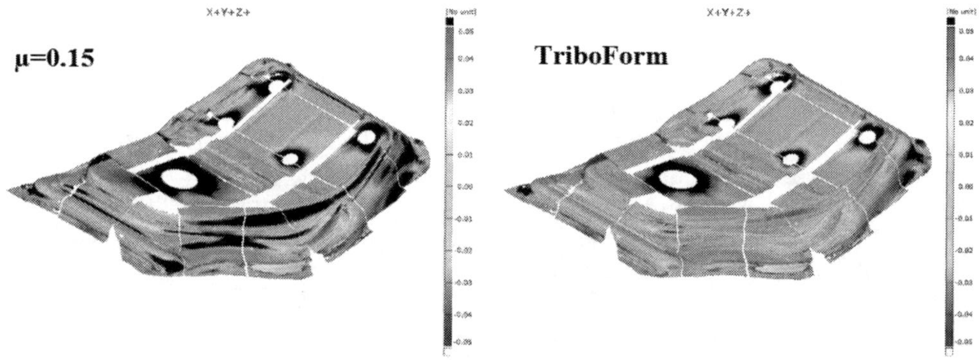

Figure 1. Difference in true major strain between simulations and experimental measurements for the GI system.

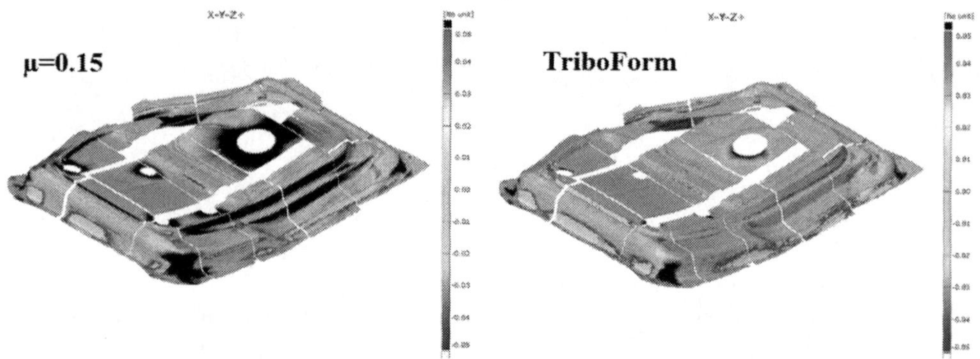

Figure 2. Difference in true major strain between simulations and experimental measurements for the ZM system.

Using a constant friction coefficient of $\mu = 0.15$ results in too large major strains in several areas, especially in vertical walls. Using the TriboForm friction model will result in lower friction coefficients in areas with high contact pressures, e.g. in radii and draw beads, and also in areas with high relative velocity between the sheet and the die surfaces. This results in a better agreement between simulated and measured major strains. The comparison for the minor strains show the same trend, i.e. that the accuracy of the sheet metal forming simulation results increases using the Triboform GI friction model.

Figure 2 shows the comparison between predicted major strains and measured strains for the ZM lubrication system. Once again the TriboForm model is predicting the strains more accurately than the $\mu = 0.15$ model. This is a very encouraging result since the draw-in was overpredicted for ZM system with the TriboForm model, which then could indicate that the TriboForm friction model for this lubrication system was less accurate.

5. Conclusions and future work

The results presented in this paper demonstrate that accounting for realistic and accurate friction and lubrication conditions bring metal forming simulations to a higher level and improve the prediction accuracy of stamping simulations.

There are several benefits of the presented approach. First of all, the TriboForm software is based on physical models with input parameters that can be efficiently collected from a database or measured with minimal effort. This enables to accurately predict the results of sheet metal forming operations before manufacturing the dies. Secondly, it enables the simulation of friction conditions for the materials and lubricants used in actual production of automotive parts and reduces the demand for experimental testing and try out. Overall, it enables Volvo Cars to further reduce lead time and development cost through the use of more accurate stamping simulations.

Although very encouraging results, there is still room for further improvements of the modelling techniques. Using the real ram velocity of the press results in strain rates in the simulation that are substantially higher than the strain rate material testing normally is performed at. Therefore, the strain rate effects of sheet material will be included in future studies and it will probably increase the accuracy even more. It is also important to have accurate models for each lubrication system used in the real die. Therefore, future implementations of the TriboForm FEM Plug-In will have the possibility to use multiple friction models.

References

[1] Sigvant M, Hol J, Chezan T and van den Boogaard A H, 2015, Friction modeling in sheet metal forming simulations: Application and validation on an U-bend product, *Proc. of FTF 2015 (Zürich, Switserland).*

[2] Sigvant M, Pilthammar J, Hol J, Wiebenga J H, Chezan T, Carleer B and van den Boogaard A H, 2016, Friction and lubrication modeling in sheet metal forming simulations of a Volvo XC90 inner door, *Proc. of IDDRG 2016 (Linz, Austria).*

[3] Sigvant M, Pilthammar J, Hol J, Wiebenga J H, Chezan T, Carleer B and van den Boogaard A H, 2016, Friction in sheet metal forming: Simulation of the Volvo XC90 inner door, *Proc.of FTF 2016 (Ohlstadt, Germany).*

[4] Pilthammar J, Sigvant M and Walter S.K., Including die and press deformations in sheet metal forming simulations, *Proc.of NUMISHEET 2016 (Bristol, UK).*

[5] Grueebler R and Hora P, 2009, Temperature dependent friction modeling for sheet metal forming. *International Journal of Material Forming, 2:251–254.*

[6] Ludwig M, Müller C and Groche P, 2010, Simulation of dynamic lubricant effects in sheet metal forming processes, *Key Engineering Materials, **438**: 171–178.*

[7] Hol J, Meinders V T, de Rooij M B and van den Boogaard A H, 2014, Multi-scale friction modeling for sheet metal forming: The boundary lubrication regime, *Tribology Int., **81**: 112–128.*

[8] Hol J, Meinders V T, Geijselaers H J M, and van den Boogaard A H, 2015, Multi-scale friction modeling for sheet metal forming: the mixed lubrication regime. *Tribology Int., **85**: 10–25.*

[9] Banabic D, Carleer B, Comsa D S, Kam E, Krasivskyv A, Mattiasson K, Sester M, Sigvant M and Zhang X, 2010, Sheet Metal Forming Processes, Constitutive Modelling and Numerical Simulation, *Springer.*

Improving stamping simulation accuracy by accounting for realistic friction and lubrication conditions: Application to the door-outer of the Mercedes-Benz C-class Coupé

J. Hol[1], J.H. Wiebenga[1], J. Stock[2], J. Wied[2], K. Wiegand[3], B. Carleer[4]

[1] TriboForm Engineering, Hengelosestraat 500, 7500 AM Enschede, the Netherlands
[2] Daimler AG, Technology Center Production Cars / Manufacturing-Equipment, 71059, Sindelfingen, Germany
[3] Daimler AG, Technology Center Production Cars / Process Development and Materials, 71059, Sindelfingen, Germany
[4] AutoForm Engineering, Emil-Figge-Str. 76-80, D-44227 Dortmund, Germany

E-mail: j.hol@triboform.com

Abstract. In the stamping of automotive parts, friction and lubrication play a key role in achieving high quality products. In the development process of new automotive parts, it is therefore crucial to accurately account for these effects in sheet metal forming simulations. Only then, one can obtain reliable and realistic simulation results that correspond to the actual try-out and mass production conditions. In this work, the TriboForm software is used to accurately account for tribology-, friction-, and lubrication conditions in stamping simulations. The enhanced stamping simulations are applied and validated for the door-outer of the Mercedes-Benz C-Class Coupé. The project results demonstrate the improved prediction accuracy of stamping simulations with respect to both part quality and actual stamping process conditions.

1. Introduction

The quality of sheet metal formed parts is strongly dependent on the tribology, friction and lubrication conditions that are acting in the actual production process. Although friction is of key importance, it is currently not considered in detail in stamping simulations. The current industrial standard is to use a constant (Coulomb) coefficient of friction. This limits the overall simulation accuracy as also demonstrated in earlier work [1].

Figure 1 demonstrates the profound influence of friction and lubrication on the accuracy of stamping simulation results. It shows the outcome of a sensitivity analysis performed using AutoForm Sigma for a uniside panel, demonstrating the influence of multiple parameters on the sheet thinning. Of most interest in the evaluation of these results are the areas that are sensitive for splits, which are located near the B-pillar and roof top (see Figure 1). In these critical zones, it is observed that friction and lubrication dominates the sheet thinning by up to 70%. This demonstrates the necessity of an accurate description of friction and lubrication conditions in stamping simulations.

This paper presents a selection of results considering friction and lubrication modeling in stamping simulations of the door-outer of the Mercedes-Benz C-class Coupé. First the overall project approach will be outlined. Next, the stamping process of the door-outer part will be introduced. A description of the project results including validation of the simulation results based on door-outer parts from a press trial is provided. Finally, the conclusions and points of future work are described.

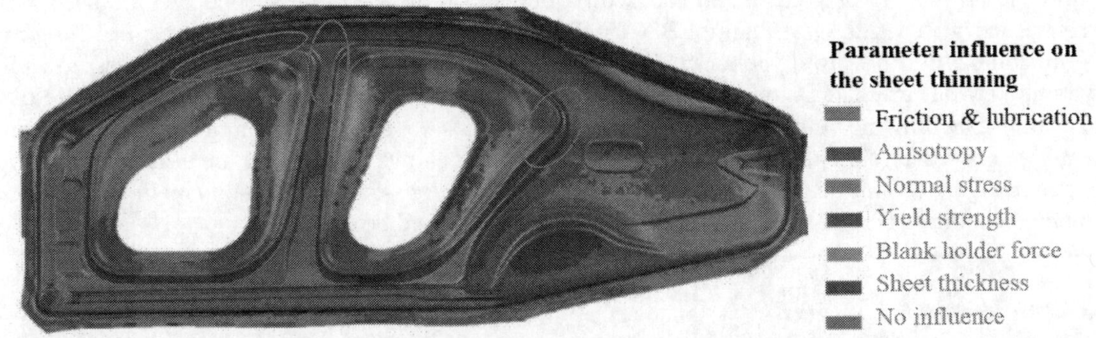

Figure 1. Friction dominates sheet thinning in encircled critical areas up to 70% for a uniside panel

2. Approach

The approach followed in this work is visualized in Figure 2 whereby the TriboForm software is used in combination with the AutoForm software.

2.1. Simulation of friction and lubrication conditions

The TriboForm software allows for modeling the friction and lubrication behavior under a wide range of process conditions. The physically-based models included in the TriboForm software enable multi-scale modeling of friction conditions in the mixed lubrication regime [2, 3]. As a user input, information of the tribological system is required, i.e. the applied sheet material, coating and tooling material, lubrication type, lubrication amount and process conditions. This information can either be entered by the user or extracted from a database, i.e. the TriboForm Library.

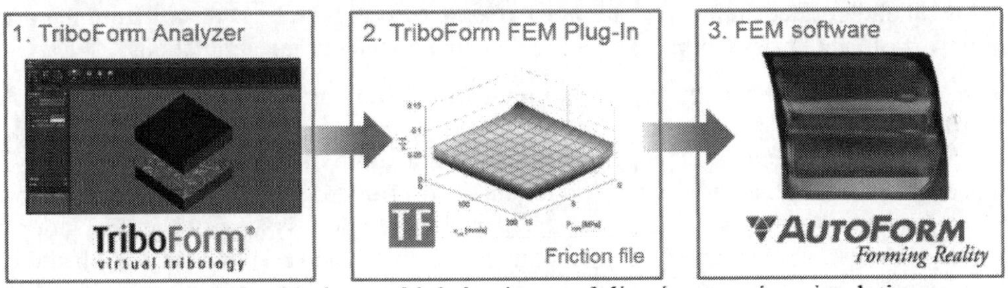

Figure 2. Approach for friction and lubrication modeling in stamping simulations

2.2. Friction model

Using the TriboForm Analyzer R1.0, a friction model is created per tribological system. The resulting friction model describes the frictional behavior as a function of the local contact pressure, relative sliding velocity, plastic strain in the sheet material and interface temperature. The resulting four dimensional friction model can be saved and exported from the TriboForm Analyzer in a friction file, see Figure 2.

2.3. Stamping simulations

Next, the friction file is imported into AutoForm^{plus} R5.2 using the TriboForm FEM Plug-In and incorporated in the door-outer simulation model. As a result, the constant coefficient of friction in AutoForm is replaced by the four dimensional friction model. Friction coefficients are computed for each node and each time increment of the FEM simulation. The local- and time-dependent friction coefficients describe the sheet, tooling and lubricant used in actual production of the door-outer part.

3. Door-outer trial and simulation results

An impression of the door-outer part is shown in Figure 3. In this work, the drawing operation OP20 is considered. The door-outer part is produced using a CR210BH-EG sheet material with a thickness of

0.7 mm and an EDT surface finish. Blanks with a dimension of 1600 mm x 1600 mm are used. The blanks are lubricated with Oest Platinol B804/3 COW-1 in varying lubrication amounts, i.e. 2.0 g/m^2, 1,2 g/m^2 and 0,5 g/m^2. A prototype tooling is used made of GG25 material. The parts are pressed using a hydraulic try-out press at Daimler AG in Sindelfingen, Germany, with a maximum stroke rate of 2 strokes/min. The drawing velocity is varied between 49 mm/s, 35 mm/s and 15 mm/s. For a decreasing drawing velocity and lubrication amount, the door-outer part shows splits in the areas as indicated and enlarged in Figure 3. This demonstrates the dependence of the door-outer part quality on the friction and lubrication conditions in the stamping process as further discussed below.

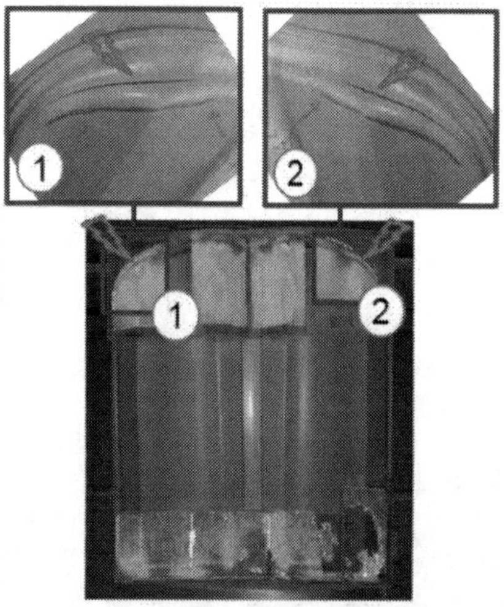

Figure 3. Critical and gridded areas of the door-outer part of the C-class Coupé

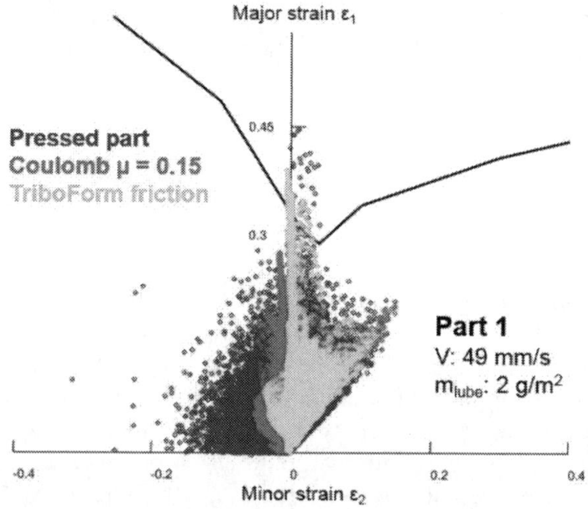

Figure 4. Comparison of the FLD results from the press trial and stamping simulations

Figure 4 shows a comparison of the Forming Limit Diagram (FLD) resulting from strain measurements performed on a pressed part (see the gridded and evaluated areas in Figure 3) and resulting from AutoForm simulations. The part shows a critical strain distribution exceeding the FLC. The simulation using a Coulomb coefficient of friction of $\mu = 0.15$ underestimates the strains in the part. Next a TriboForm friction file is used describing the actual tribological system of the door-outer part and corresponding lubrication amount of 2.0 g/m^2. AutoForm simulation results using this friction file are included in Figure 4, and show an improved prediction accuracy of the critical strain distribution with respect to the pressed part.

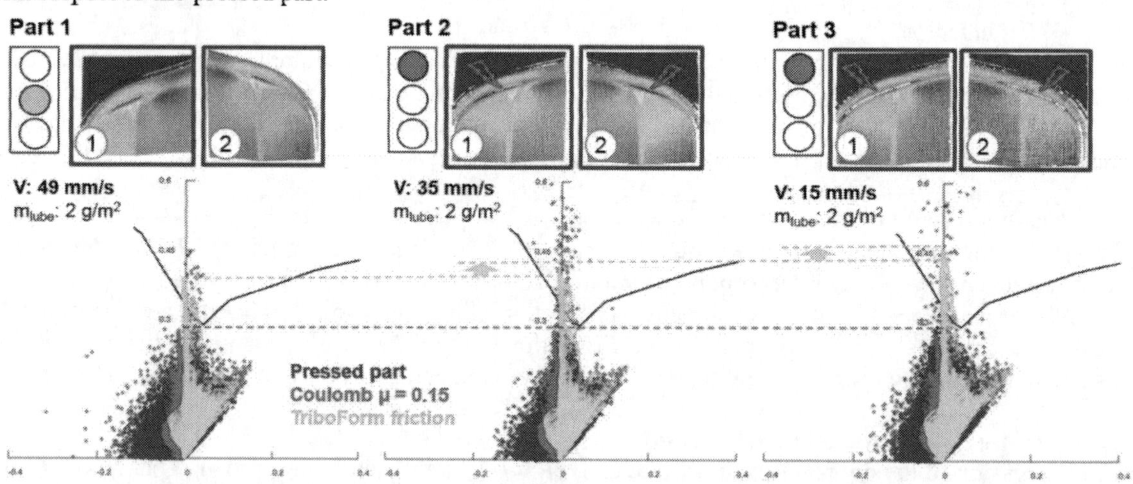

Figure 5. Comparison between press trial and simulation results: Influence of drawing velocity

A comparison between press trial parts and simulation results for decreasing drawing velocities is shown in Figure 5. The parts are pressed, and TriboForm simulations are performed, using a lubrication amount of 2.0 g/m^2. The door-outer part shows splits when the drawing velocity is reduced. This behaviour is not captured by the simulations using a constant coefficient of friction. The simulation results using the TriboForm friction file, including the friction dependency on sliding velocity, show an improved prediction accuracy of the critical strain distribution and splits with respect to the pressed parts.

A comparison between press trial and simulation results for decreasing lubrication amounts is shown in Figure 5. All parts are pressed at a drawing velocity of 49 mm/s. The door-outer part shows splits when the lubrication amount is reduced. This is not captured by the simulations using a constant coefficient of friction. The simulation results using a TriboForm friction file per lubrication amount again show an improved prediction accuracy of the critical strain distribution and splits with respect to the pressed parts.

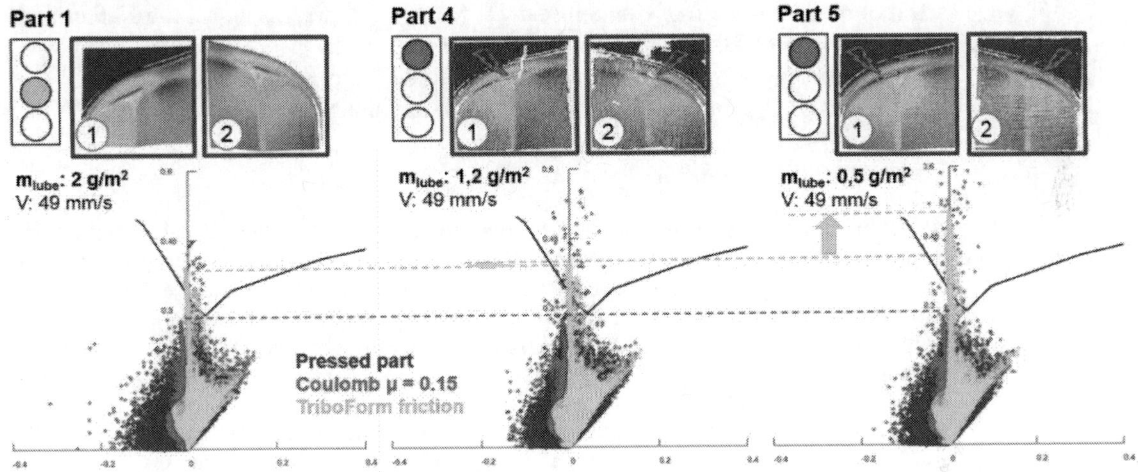

Figure 6. Comparison between press trial and simulation results: Influence of lubrication amount

4. Conclusions and future work

The results presented in this paper demonstrate the strong influence of friction and lubrication conditions on the quality of the door-outer part. Moreover, it demonstrates that accounting for realistic and accurate friction and lubrication conditions bring metal forming simulations to a higher level and improve the prediction accuracy of stamping simulations, both qualitatively and quantitatively. Future work will include further applications in series production, which should lead to the broad application of advanced friction and lubrication models in stamping simulations at Daimler AG.

References

[1] Sigvant M., Pilthammar J., Hol J., Wiebenga J.H., Chezan T., Carleer B. and van den Boogaard A.H.: Friction modeling in sheet metal forming simulations of a Volvo XC90 inner door. IDDRG conference, June 12-15, 2016, Linz, Austria.

[2] Hol J., Meinders V.T., de Rooij M.B. and van den Boogaard A.H.: Multi-scale friction modeling for sheet metal forming: The boundary lubrication regime. Tribology Int., 81: 112–128, 2014.

[3] Hol J., Meinders V.T., Geijselaers H.J.M., and vd Boogaard A.H.: Multi-scale friction modeling for sheet metal forming: the mixed lubrication regime. Tribology Int., 85: 10–25, 2015.

Acknowledgement

The authors are grateful for the financial support from Daimler AG.

Advanced friction simulation of standardized friction tests: a numerical and experimental demonstrator

J. Hol[1], J.H. Wiebenga[1], M. Hörning[2], F. Dietrich[2,3], and C. Dane[4]

[1] TriboForm Engineering, Hengelosestraat 500, 7500 AM Enschede, the Netherlands
[2] Adam Opel AG, Bahnhofsplatz, 65423 Rüsselsheim am Main, Germany
[3] PtU TU Darmstadt, Otto-Berndt-Straße 2, 64287 Darmstadt, Germany
[4] Tata Steel, P.O. Box 10.000, 1970 CA IJmuiden, the Netherlands

E-mail: j.hol@triboform.com

Abstract. For the characterization of friction conditions under sheet metal forming process conditions, different friction test set-ups are being used in industry. However, different friction tests and test set-ups are known to result in scattering friction results. In this work, the TriboForm software is utilized to numerically model the frictional behavior. The simulated coefficients of friction are experimentally validated using friction results from a standardized strip drawing friction test set-up. The experimental and simulation results of the friction behavior show a good overall agreement. This demonstrates that the TriboForm software enables simulating friction conditions for varying tribology conditions, i.e. resulting in a generally applicable approach for friction characterization under industrial sheet metal forming process conditions.

1. Introduction

The quality of sheet metal formed parts is strongly dependent on the tribology and friction conditions that are acting in the production process. These friction conditions are dependent on the utilized sheet material, tooling material and lubricant. This combination of factors is known as the tribological system [1]. Choosing the optimal tribological system in an early stage of the design process enables Adam Opel AG to design products with optimal quality at minimal time and cost investment.

To analyze the effect of the tribological system on the final part quality, there is a need to quantify the corresponding frictional behavior and include an accurate description of this behavior in Finite Element Method (FEM) simulations. One way to determine the frictional behavior for a specific tribological system is to execute friction tests. Varying experimental set-ups are available in industry to determine the frictional behavior. The question is which friction test set-up to use to determine the frictional behavior acting in the actual production process.

To answer this question, a study has been performed by the German Deep Drawing Research Group (GDDRG) [2]. In this study, a round robin test was performed by comparing the outcome of seven different friction test set-ups. In this round robin test, only the type of test set-up was altered. That is, the analyzed tribology system and test conditions were kept identical, i.e. an identical combination of sheet material, tooling material, and lubricant was used per test set-up. The outcome of the round robin test is summarized in Figure 1, showing the experimentally determined friction coefficients (μ) as a function of nominal contact pressure (P_{nom}) resulting from the seven friction test set-ups. The friction test set-up relevant for this work is the flat-bed strip draw test from the PtU Darmstadt, Germany.

Figure 1 shows that despite using an identical tribological system and identical test conditions in all seven test series, the friction coefficients range between μ = 0.03 and μ = 0.26. Also note that even the trend of the frictional behavior differs between the test set-ups. To summarize, the results from the round robin test clearly show a strong scattering in magnitude and trend of the friction results. Although not further analyzed in detail in [2], the scatter in the results can likely be related to differences in the tooling geometry and dimensions used [4,5], but also on e.g. the overall design and stiffness of the test set-ups. All these aspects differ per test set-up and hence no conclusion can be drawn on the correct magnitude and trend of the frictional behavior for this tribological system.
The ultimate goal in the design of sheet metal formed parts with A-class quality is to determine an accurate description of the frictional behavior which is acting in the real production process. This description should be generally applicable, i.e. independent of the experimental test set-up used. In this work, this is achieved by using an advanced friction modeling approach instead of an experimental approach. For this purpose, the TriboForm software is used. As a first stepping stone towards industrial application at Adam Opel AG, the outcomes of the TriboForm software are experimentally validated using the standardized strip draw friction test from the PtU Darmstadt. This will be done for different sheet surface textures and lubrication amounts, all leading to a difference in the frictional behavior.

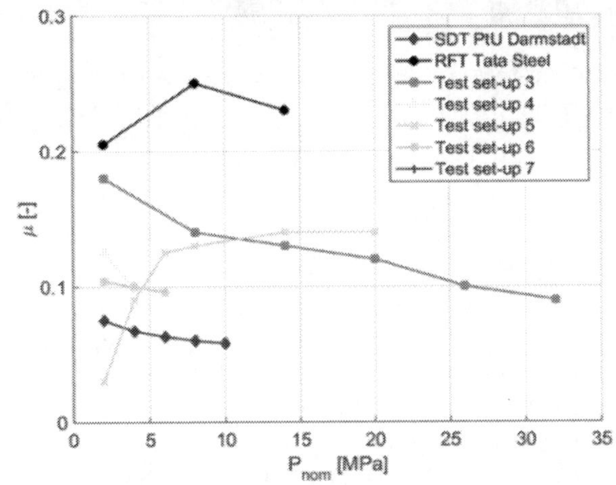

Figure 1. Experimentally determined friction coefficients resulting from seven different friction test set-ups showing a large scatter of results [2].

2. Friction modeling using the TriboForm software

Tribological conditions in metal forming processes are dependent on local process and lubrication conditions, loading and local strain state of the sheet material. The TriboForm software allows for multi-scale modelling of a time and locally varying friction coefficient under a wide range of process conditions. Information of the tribology system is required as an user input, i.e. the applied sheet material, coating and tooling material, lubrication type, lubrication amount and process conditions. This information can either be entered by the user or extracted from a database, i.e. the TriboForm Library. The friction models simulated and generated by the TriboForm software can be readily imported in FEM simulations of forming processes using the TriboForm FEM Plug-In.

3. Friction experiments

In this study, friction tests have been performed using a flat-bed strip draw test, see Figure 2. The tests have been performed at the PtU Darmstadt in Darmstadt, Germany. This friction test set-up and conditions are specified according to the VDA norm 230-213 [3]. In this study, strips of VDA239 CR4 GI sheet material with a thickness of 0.8 mm are drawn through a set of GGG70L tools at a constant relative sliding velocity of 50 mm/s while recording the friction forces. Varying sheet surface textures were considered, referred to as 'T1' and 'T2', see Figure 3. A Fuchs Anticorit PL3802 39S lubricant

was used in two different lubrication amounts, i.e. 0.5 g/m^2 ('pre-applied') and 1.2 g/m^2. The sheet material is loaded from two sides with a relatively large tooling area, which limits the maximum nominal contact pressure that can be achieved with this test set-up. In this study, a maximum pressure of 10 MPa could be achieved after which the strip of sheet material breaks.
As an input for the friction simulations, the real 3D surface topographies of the tooling and sheet surfaces are utilized. The surface topographies can be imported by the user or extracted from the TriboForm Library. The virgin sheet surface topographies of texture T1 and T2 are measured by 3D confocal microscopy and utilized in the TriboForm software for the friction simulations. Figure 3 shows a hot-dip galvanized zinc coating with a specially designed surface texture (T1) and a hot-dip galvanized zinc coating with an Electro Discharge Textured (EDT) surface finish (T2). The sheet surface textures have comparable roughness values of s_a = 1.35 μm for T1 and s_a = 1.41 μm for T2.

Figure 2. Flat-bed strip draw test, PtU Darmstadt, Darmstadt, Germany [3]

Figure 3. Sheet surface topographies: Texture 1 (T1, left) and Texture 2 (T2, right)

4. Results

4.1. Friction modelling
The basis of a TriboForm simulation model to calculate friction conditions is the projection of the tool surface topography on to the sheet surface topography. In this way, simulation models are generated in the TriboForm software for both the T1 and T2 sheet surface topography. Using the resulting two simulation models, the friction conditions are calculated by loading and sliding the tool surface over the sheet surface. The amount of lubricant on the sheet surface can be entered in the TriboForm software, i.e. 0.5 g/m^2 and 1.2 g/m^2. The friction conditions are calculated for pre-defined ranges of process conditions, i.e. nominal contact pressure, plastic strain in the sheet material, relative sliding velocity and interface temperature. In this paper, only the friction results as a function of nominal contact pressure will be discussed and experimentally validated.

4.2. Friction test results
Figures 4 and 5 display the friction test results up to a contact pressure of 10 MPa for a lubrication amount of 0.5 g/m^2 and 1.2 g/m^2 respectively. Each marker represents an average friction coefficient resulting from 5 experimental repetitions. The test results show a decreasing friction coefficient for increasing contact pressure for both the sheet surface texture T1 and T2. The hot-dip galvanized zinc coating with an EDT surface finish (T2) shows a lower overall frictional behavior compared to the hot-dip galvanized zinc coating with the specially designed surface texture (T1). Comparing Figure 4 and Figure 5, a slight reduction in the measured friction coefficient magnitude is observed for an increasing lubrication amount. Moreover, an increasing difference in frictional behavior between sheet surface texture T1 and T2 is observed for increasing lubrication amounts. This demonstrates the interaction between lubrication amount and sheet surface texture on the resulting frictional behavior.

4.3. Friction simulation results and experimental validation
TriboForm friction simulations have been performed per tribological system. The TriboForm simulation results are shown in Figures 4 and 5 up to a contact pressure of 15 MPa. Note however, that TriboForm enables the calculation of the friction behavior for much higher contact pressures as well, i.e. up to 500 MPa. The results demonstrate that the TriboForm software enables an accurate prediction of friction

coefficients for varying lubrication amounts and sheet surface textures, both in terms of the magnitude and trend of the friction results. The lower frictional behavior for the T2 sheet surface texture compared to the T1 texture is correctly predicted. For the friction results using a lubrication amount of 0.5 g/m^2 shown in Figure 4, an difference in frictional behavior for the T2 surface texture can be observed whereas the frictional behavior for the T1 texture is in good agreement. The friction results using a lubrication amount of 1.2 g/m^2 show a good agreement between simulation and experimental results for both sheet surface topographies.

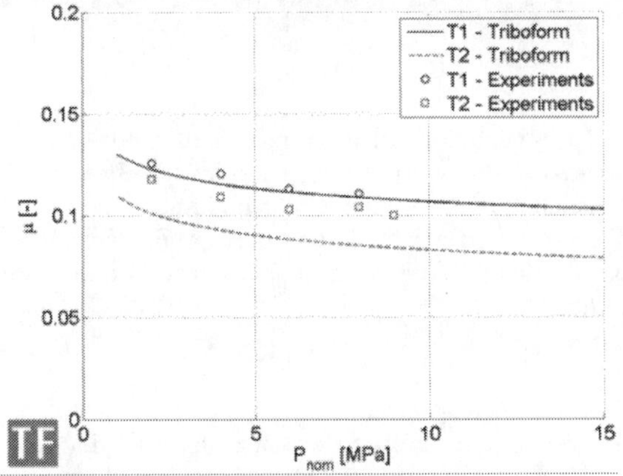

Figure 4. Experimental and simulation results for CR4 GI - T1 and T2 with 0.5 g/m^2 lubricant

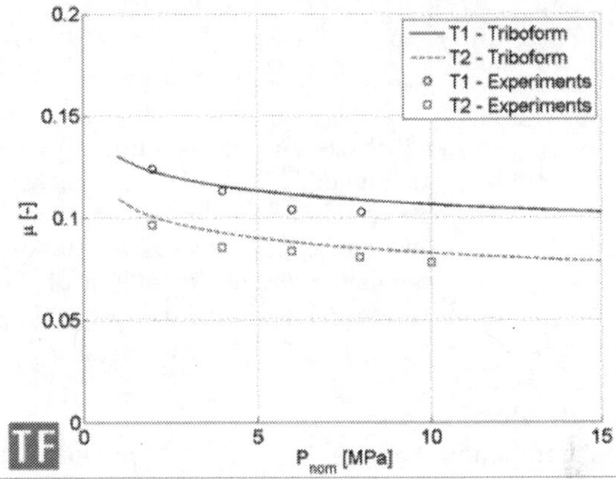

Figure 5. Experimental and simulation results for CR4 GI - T1 and T2 with 1.2 g/m^2 lubricant

5. Conclusions

The TriboForm software enables the simulation of friction coefficients resulting from standardized friction tests. Both the magnitude and the overall trend of the experimentally obtained friction results from the standardized flat-bed strip draw friction test are in good agreement. This is demonstrated for varying lubrication amounts and sheet surface topographies.

Benefits of the presented friction modelling approach are the following. First of all, the TriboForm software provides a generally applicable approach for friction characterization under industrial sheet metal forming process conditions. Secondly, it reduces the demand for experimental friction testing. Determining the same amount of information using friction tests would require a significant time and cost investment. Finally, the TriboForm simulations can be performed early in the design process and the results can be readily imported in FEM simulations of forming processes. This enables Adam Opel AG to perform more accurate metal forming simulations including the friction model corresponding to the tribological system as applied in stamping production, and design A-class quality sheet metal formed parts at minimal time and cost investment.

6. References

[1] Czichos H., Habig K.-H.: *Tribologie Handbuch: Tribometrie, Tribomaterialien, Tribotechnik*. Vieweg+Teubner Verlag, 2010.

[2] GDDRG Arbeitskreis - Arbeitsgruppe Oberflächentopographie und Umformbarkeit – Reibwert bestimmung: *VDEH Round robin friction tests*. IDDRG working group meeting, 27-2-2013.

[3] VDA norm 230-213: *Test procedures for the product classes Prelube, Prelube2, Hotmelt, Spot Lubricant*, Verband der Automobilindustrie E.V. (VDA), November 2008.

[4] Emmens W.C.: *A novel Design friction tester*. IDDRG working group, Pisa Italy 7-9 May 1991.

[5] Dane, C. and Emmens W.C.: *Influence of die dimensions on the reliability of measured coeffcients of friction*. IDDRG working group meeting, Colmar France, 22 - 24 May 1995.

Acknowledgments

The authors are grateful for the financial support from Adam Opel AG for this project.

The Role of Evolutive Elastic Properties in the Performance of a Sheet Formed Spring Applied in Multimedia Car Industry

J. Faria, J. Silva, P. Bernardo, M. Araújo and J. L. Alves

CMEMS, University of Minho, Campus of Azurém, 4800-058 Guimarães, Portugal

m.araujo@dem.uminho.pt

Abstract. The manufacturing process and the behaviour of a spring manufactured from an aluminium sheet is described and investigated in this work considering the specifications for the in-service conditions. The spring is intended to be applied in car multimedia industry to replace bolted connections. Among others, are investigated the roles of the constitutive parameters and the hypothesis of evolutive elastic properties with the plastic work in the multi-step forming process and in working conditions.

1. Introduction
The competitiveness of the automotive industry has led to a simultaneous increase of innovative design concepts, lightweight materials and a more efficient costs control. The present study aims to investigate an innovative solution for fixing a printed circuit board (PCB) under a metallic chassis (sheet metal parts) without bolted connections. This geometry must be capable of submitting the PCB under a compressive load, to ensure the ground contact between components. The finite element analysis (FEA) has been widely used by scientific and industrial communities in product design and to predict material's behaviour. FEA reduces the number of trials, reducing the cost of product development [1-2]. The problem investigated in this paper is to evaluate the usage of a new spring to replace bolted connections. The research objectives are to evaluate the effects of the manufacturing process, the evolution of the elastic properties due to the plastic work, and the hardening models.

2. The new spring: geometry description
The proposed geometry (Figure 1) is located at the cover corners of a metallic chassis. It consists in a recess with a trapezium format where the contact between the PCB and chassis will occur. The innovation presented in this geometry is focused on the geometry around the recess. This geometry consists of material removal in the shape of tabs and it was designed to have a similar behaviour to a spring. The geometry was subject of previous numerical simulations aiming to find the best geometry's shape to the tabs.

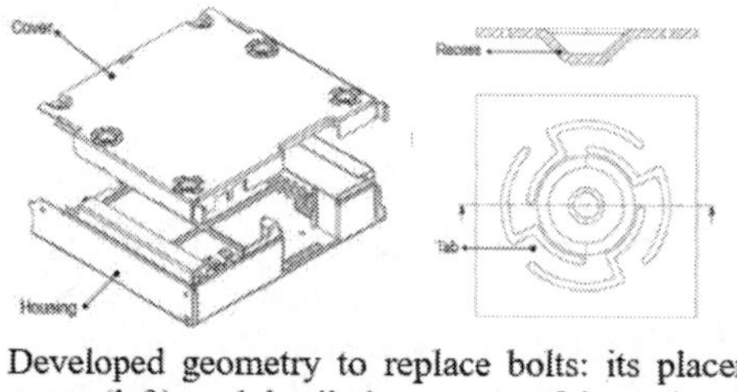

Figure 1. Developed geometry to replace bolts: its placement in the chassis top cover (left), and detailed geometry of the spring (right).

3. Numerical modelling

3.1. DD3IMP
Deep Drawing 3D IMPlicit Code (DD3IMP) is a finite element (FE) code which is the result from continuous development and optimizations aimed for the simulation of sheet metal forming processes

[3-4]. This 3-D elastoplastic FE code allows large elastoplastic strains and rotations, following a fully implicit time integration scheme. DD3IMP models the geometry of the forming tools by parametric Bézier surfaces or Nagata patches [5]. The frictional contact between the two bodies is interpreted with a classical Coulomb's law with an augmented Lagrangian approach. The frictional contact algorithm operates directly on the parametric Bézier surfaces or the Nagata patches [6].

3.2. Constitutive Models

The efforts done by scientific community to improve and create new constitutive models in FE are now providing a more efficient description of the mechanical behaviour of sheet metals [7]. The orthotropic yield criterion is currently used to model the planar anisotropy of the blank. In turn, the isotropic and kinematic hardenings describe the evolution of the flow stress and the yield surface with the plastic work.

In the present work it is used a Voce law to describe the isotropic work hardening. Kinematic hardening is computed by Lemaître & Chaboche evolution law. It is intended to manufacture the multimedia system chassis in aluminium alloy, in the grade AA5182-O [7]. This specific material has constitutive parameters to be used in Voce law [eq. (1)] and in evolution law [eq. (2)]:

$$Y = 148.5 + 192.4[1 - \exp(-9.7\,\bar{\varepsilon}^p)] \tag{1}$$

$$\dot{\mathbf{X}} = 152.7\left[\frac{26.0}{\bar{\sigma}}(\boldsymbol{\sigma}' - \mathbf{X}) - \mathbf{X}\right]\dot{\bar{\varepsilon}}^p, \quad \mathbf{X}(0) = \mathbf{0} \tag{2}$$

where $\bar{\varepsilon}^p$ is the equivalent plastic strain, Y is the flow stress, X the back stress tensor, σ' the deviatoric Cauchy stress tensor and $\bar{\sigma}$ the equivalent tensile stress of the isotropic von Mises yield criterion.

3.3. Elastic properties evolution

Most studies of sheet metal forming consider the elastic behaviour as isotropic and constant, assuming that plastic work has no effect in the elastic properties. It is considered that when a permanent plastic deformation takes place, the elastic strain will be recovered. Presently the accuracy of each study is defined through the mathematical models implemented and depends on the studied case. Alves et al. [10] point out that for some materials, taking into account the evolutionary isotropic elasticity can have a larger effect on springback prediction than taking into account the kinematic hardening.

In the present study, only macroscopic evaluations will be considered. The evolution of Young modulus as a function of the plastic strain is given throughout the law proposal by Alves et. al. [10]. This law is based on the experimental work of Yang et Al [8,9], who proposed an evolution law based on experimental measurements given by a polynomial equation, however its interval of validity is restricted to less than 25% of plastic strain. In sheet metal forming the maximum equivalent plastic strain is usually higher than 25%. Because of that, Alves et al. [10] introduced an exponential equation with saturation which describes better the evolution of the elastic properties in a wider range of plastic strains. The evolution law was implemented in DD3IMP FE code and it is given in a generic form by:

$$E = E_0 + k_E . E_0[1 - \exp(-C_E\bar{\varepsilon}^p]] \tag{3}$$

where E_0 is the value of the initial Young modulus, k_E and C_E are the parameters of the evolution law and $\bar{\varepsilon}^p$ is the equivalent plastic strain. $k_E \in [0,1]$ defines the saturation value of the Young modulus, $\lim_{\bar{\varepsilon}^p \to +\infty}\left(\frac{E_0 - E(\bar{\varepsilon}^p)}{E_0}\right) = k_E$, and C_E measures the rate of approximation to the saturation value.

4. Numerical results

4.1. Simulation Setup

The numerical study was separated in two main stages. In the first stage the stamping process was numerically simulated. This process is responsible for the major geometrical changes and plastic work field in the raw sheet metal. In the first step a blank sheet with a rectangular shape of 30 mm x 30 mm x 1.2 mm is positioned between the holder and the die. A fixed clearance is imposed between the holder and the die during all the punch displacement. The punch moves vertically giving shape to the sheet metal, and when the punch meets to the final position it moves back in the opposite direction, allowing the springback of formed part. The second stage simulates the operating conditions of the proposed geometry, in order to investigate the role of the evolutive elastic properties on the mechanical behaviour of the spring. The spring is loaded by a tool representing the PCB. As boundary conditions, a displacement of 1 mm is imposed to the tool, and the reaction force is evaluated.

In order to compare the role of either constitutive models, two main types of numerical simulations were carried out using the same boundary conditions and FE meshes. The first group only considers the assembly conditions (AC label) during the assembly of the cover in the chassis, which corresponds to pressing the spring. The second group considers all the forming history since manufacturing process (MP label) until assembly conditions (MP.AC). For each of these groups, different material laws are used. For example, "MP.AC.EPE.KH" is a simulation in which all conditions are activated, including evolutive elastic properties (EPE) and kinematic hardening (KH).

4.2. *Results and discussion*

To evaluate the influence of elastic properties evolution and kinematic hardening in numerical simulation of assembly conditions (AC), Figure 2 (a) shows the force reaction along a 1 mm displacement in four different cases. The four simulations were conducted taking into account the following phenomena: with only elastic properties evolution (AC.EPE), with only kinematic hardening (AC.KH), with both phenomena (AC.EPE.KH) and without any of them (AC).

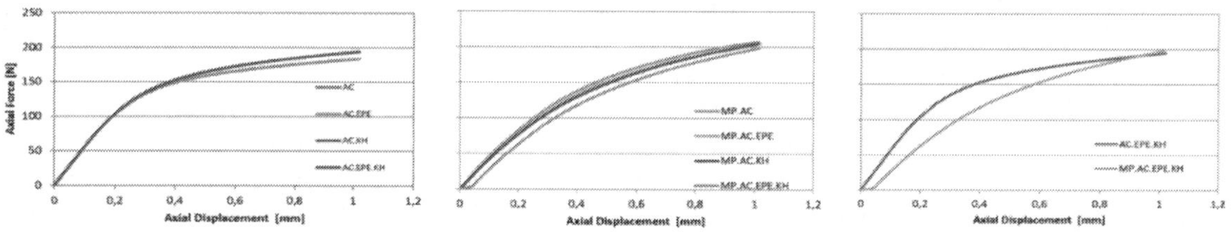

(a) Curve of the AC, AC.EPE, AC.KH and AC.EPE.KH

(b) Curves of the MP.AC, MP.AC.EPE, MP.AC.KH and MP.AC.EPE.KH

(c) Curve of the AC.EPE.KH and MP. AC.EPE.KH

Figure 2. Curve force versus displacement during the simulations of the assembling process.

In Figure 2 (a), the MP condition is not activated, i.e. the forming history is not taken into account. In this case, the material of the spring is considered as new material without any plastic work history. The AC and AC.EPE curves show a very similar behaviour. It allows concluding that elastic properties evolution does not influence the spring behaviour. When kinematic hardening was considered, the maximum force values for the same displacement of the tool (or the PCB) is lower.

The evolution of the elastic properties, i.e. of the value of the Young modulus with the plastic work has almost no influence. For the four cases on Figure 2 (a), the equivalent plastic strain is consistently low, and it occurs in well-defined zones. There is no substantial difference in the maximum plastic strain amongst the four different studied cases. In the case where the material does not consider Bauschinger effect (i.e. there is not kinematic hardening) and the elastic properties evolution, the force required to perform the same displacement is about 5% higher. When the evolution of Young modulus and the Bauschinger effect are considered, the maximum force is lower.

Figure 2 (c) indicates that there is a non-negligible difference between AC and MP.AC. For about 0.4 mm displacement the difference between the curves is of 22%, but AC.EPE.KH curve starts to saturate earlier and, at the end of 1 mm displacement, the MP.AC.EPE.KH displays a higher maximum force. The difference between the maximum force values at the end of displacement is of 3%. The maximum plastic strain takes place during the manufacturing process stage as expected. When compared with the first stage, the plastic work caused by the remaining steps is too low. In the remaining steps, deformation is concentrated in the extremities of the tabs, and mostly in the elastic regime. Figure 3 shows a comparison between the equivalent plastic strain of the cases in Figure 2 (c).

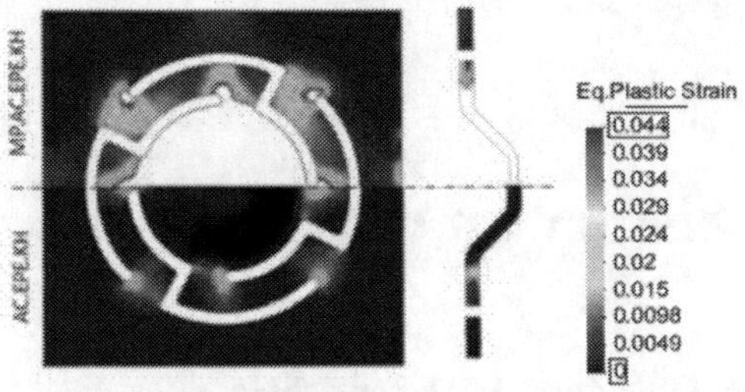

Figure 3. Comparison between the equivalent plastic strain field of the AC.EPE.KH and MP.AC.EPE.KH simulation.

5. Conclusions

The study of the mechanical behaviour and manufacturing process of a new spring to be applied in car multimedia industry was presented. The role of both constitutive modelling and evolutive elastic properties was analysed in order to determine their in-service conditions and specifications. The new spring is intended to replace several bolted connections, and thus the spring forces plays a paramount role. The analysis of the numerical results presented in the previous figures allow to conclude:

- For the investigated cases where the equivalent plastic strain is low, the evolution of the elastic properties, due to the plastic deformation, has almost no influence in the behaviour of the spring;
- The description of the Bauschinger effect (kinematic hardening) has shown different values due to cyclic loadings, and direction changes that the investigated geometry was submitted.
- The study of loading history (manufacturing process) is important to obtain reliable results during assembling and working conditions. The effect of pre-deformation on the geometry behaviour occurs, likely, due to the change of the elastic behaviour of the material. Neglecting the loading history of the material can lead to wrong predictions concerning the spring's behaviour;
- The scientific community has been shown the importance of kinematic hardening and elastic evolution properties in springback prediction of sheet metal parts. This study shows that, the loading history and evolutive elastic properties are more important than the kinematic hardening;
- Finally, a correct description of features of the mechanical behaviour has an enormous importance to accurately predict the performance of a sheet metal formed part.

6. Acknowledgements

This research was sponsored by: (a) Portugal Incentive System for Research and Technological Development. Project in co-promotion nº 36265/2013 (Project HMIExcel - 2013-2015), and (b) FCT with the reference project UID/EEA/04436/2013, by FEDER funds through the COMPETE 2020 – Programa Operacional Competitividade e Internacionalização (POCI) with the reference project POCI-01-0145-FEDER-006941.

References

[1] Neto DM et al. 2014 Materials and Design, 60:368–379, 2014.
[2] Laurent H et al. 2010 Finite Elements in Analysis and Design 46, p51–759.
[3] Oliveira MC et al. 2008 Archives of Computational Methods in Engineering, 15, n2, p113.
[4] Menezes LF et al. 2000 Journal of Materials Processing Technology, 97, p100-106.
[5] Neto DM et al. 2013 Finite Elements in Analysis and Design, vol 72, p35-46
[6] Neto DM et al. 2014 Computer Methods in Applied Mechanics and Engineering, 271, p296-320
[7] LPMTM 2001 3DS IMS 1999 000051, WP3, Task 1, University Paris 13.
[8] Yang M et al. 2004 Journal of Materials and Processing Technology 151, p232-236.
[9] Yang M et al. 2004 NUMIFORM Conference, Columbus, EUA.
[10] Alves JL et al. 2006 IDDRG'06 Conference, Porto, Portugal, p161-168.

Effects of forming history on crash simulation of a vehicle

M İ Gökler[1], U Ç Doğan[2], H Darendeliler[1]

[1]METU-BILTIR Center and Department of Mechanical Engineering, Middle East Technical University, 06800, Ankara, Turkey
[2]Robert Bosh Korea Limited Company, Korea

E-mail: gokler@metu.edu.tr

Abstract. The effects of forming on the crash simulation of a vehicle have been investigated by considering the load paths produced by sheet metal forming process. The frontal crash analysis has been performed by the finite element method, firstly without considering the forming history, to find out the load paths that absorb the highest energy. The sheet metal forming simulations have been realized for each structural component of the load paths and the frontal crash analysis has been repeated by including forming history. The results of the simulations with and without forming effects have been compared with the physical crash test results available in literature.

1. Introduction

As the vehicle fatalities increase, vehicle safety has become an important design objective besides all performance and comfort criteria of vehicles. During a collision, as much as crash energy is aimed to be absorbed by the structural components of a vehicle to provide greater protection for occupants. From the view point of passive safety, the design of the load paths is important since the components building up the load paths are the major parts absorbing the kinetic energy during crash. Since load transfer is not so easy to predict, numerical methods are used to simulate the crash behavior of vehicles [1-3].

The majority of the parts of a vehicle, especially parts which build up the main load paths, are manufactured by sheet metal forming. However, it is a known fact that sheet metal forming processes alter the mechanical properties of the materials and change the thickness due to plastic deformation. Work hardening and thickness change apparently affect the crash response significantly. Effects of forming history on the crash response of the crash relevant body parts have been discussed by several studies [4,5]. The main purpose of this study is to show that the changes in material properties and thicknesses of the parts produced by sheet metal forming process alter the crashworthiness and overall crash response of the vehicle.

2. Finite Element Model

The detailed finite element (FE) model of 2006 Ford F250 Pickup, which had been verified with tests, was downloaded from the official web site of National Crash Analysis Center (NCAC) [6] and utilized in this study. The finite element model of Ford F250 Pickup consists with a total of 726629 beam, shell and solid elements. For the numerical analyses, explicit LS-DYNA code has been selected.

The load paths in a vehicle alter according to the crash type. Different components constitute the load paths for frontal crash and side crash, whereas the load paths differ from one vehicle to another. In order to find the crash relevant parts of the particular vehicle during the frontal crash, the internal energy of each part due to crash has been investigated.

In the frontal crash simulation, the particular vehicle with a mass of 3016 kg has crashed into a rigid wall at 56 km/h. This corresponds to an initial kinetic energy of 364.90 kJ. The results of the simulation showed that 86 % of this energy has been converted to internal energy. The remaining of the kinetic energy is observed for hourglass energy, spring/damper energy, springback energy and etc.

By examining the internal energy levels of the components, the basic crash relevant parts have been found. It is observed that the load path consisting of the lower front S-rails, the connected longitudinal members to these S-rails and the engine bracket absorbed nearly 45% of the total internal energy under frontal crash. Hence, this load path, which is shown in figure 1, has been focused in this study.

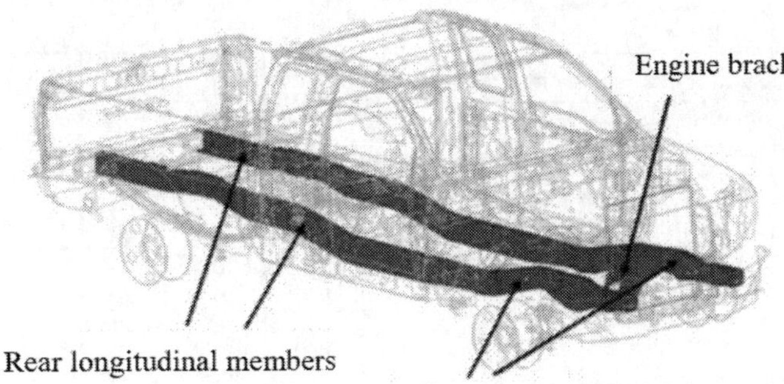

Figure 1. The first load path.

3. Forming analyses of load carrying members

To examine the forming effects on the crash response, sheet metal forming simulations have been carried out to detect the forming histories of nine crucial components building up the selected load path. In this study, it is assumed that these components are manufactured by the deep drawing process. Using the dimensions of the components, appropriate die and punch forms are modeled in order to obtain the desired geometry of the related parts. During the finite element analyses, forming tools are taken as rigid. The sheets are modeled by four node shell elements.

4. Comparison of frontal crash analysis with and without forming history

The results of the frontal crash analysis with and without forming history have been compared and presented for absorbed total energy, total displacement at the vehicle's center of gravity, the total force recorded by the load cells on the wall, the average velocity and the deceleration of the left and the right front seats.

The total absorbed energy by the main load path for the analysis without forming effects is 139.77 kJ and when the forming effects are taken into consideration its value becomes 152.67 kJ by an increase of 9.23 % as shown in figure 2.

Figure 3 shows the comparison of the resultant displacement of vehicle's center of gravity from the numerical crash analyses with and without forming history, and the crash test. The resultant displacement has decreased after the forming history has been taken into consideration, which is mainly due to the increase in energy absorption. The maximum displacement has been observed at about 0.097 s and there is a difference of about 48 mm between the results of the simulations with and without forming history. As it is seen from the figure, a good agreement is achieved with the test results, after forming histories of nine crash relevant body parts have been taken into account.

Total reaction force measured by the load cells on the wall and the values obtained by the simulations are given in figure 4 with respect to time. The maximum total force measured during the crash test is 1060 kN whereas the corresponding values obtained by the analyses with and without forming history are 1105 kN and 979 kN, respectively. It is seen that the maximum total force obtained with forming history is closer to the maximum total force measured during the crash test.

The average velocities obtained from the simulations without and with forming history for the positions of the left and right front seat accelerometers have been compared with test results and shown in figure 5. The result obtained with forming analysis shows good agreement with the test result.

Figures 6 and figure 7 show the comparison of the deceleration pulse variation from the simulations with and without forming history for the left and right seats, respectively which is very crucial since it is directly related to the injury criteria of occupants. The peak left front seat deceleration pulse has been reduced from 55.7g to 45.7g and the peak right front seat deceleration pulse has been reduced from 62.3g to 56.9g as seen from figures 6 and 7.

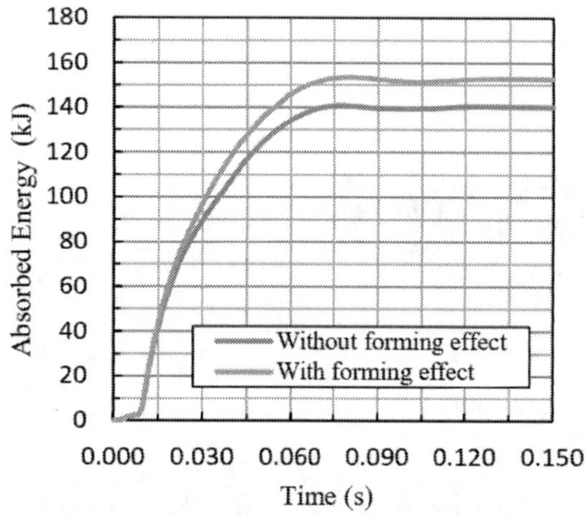

Figure 2. Change of total absorbed energy by the first load path with and without forming history.

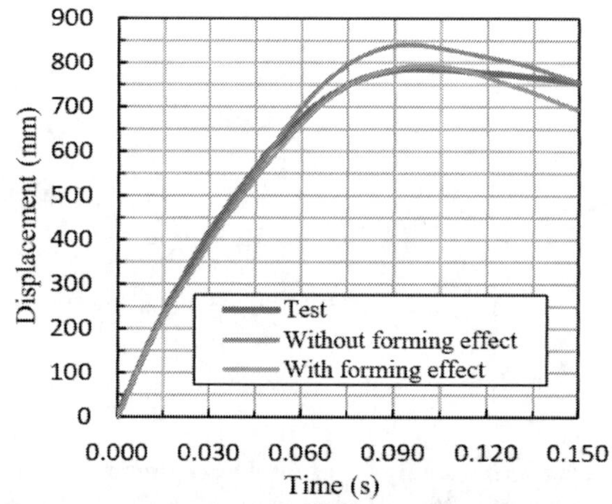

Figure 3. Comparison of simulation results and the test results [6] for resultant displacement of the vehicle center of gravity.

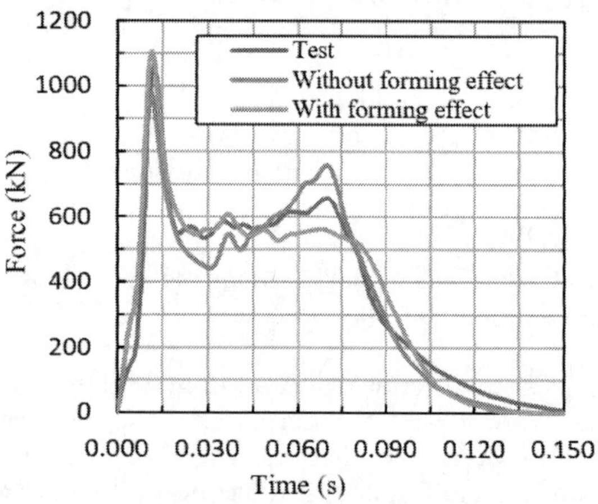

Figure 4. Comparison of total force determined by the simulations and the test results [6].

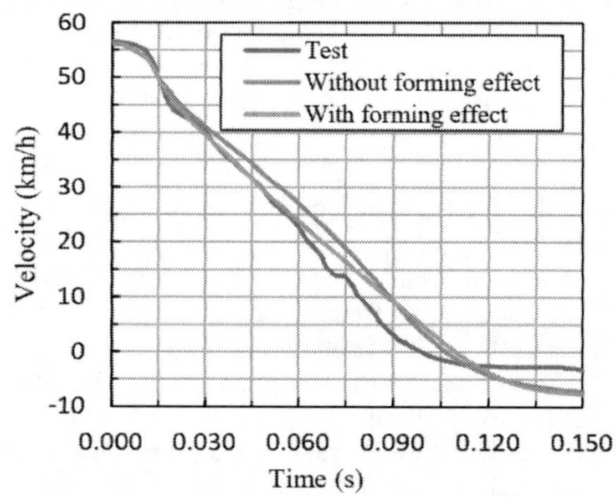

Figure 5. Comparison of average velocities of the left and right front seats obtained by the simulations and test results [6].

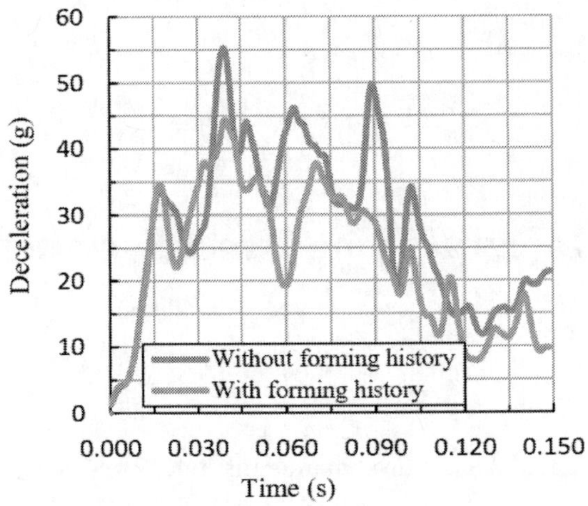

Figure 6. Variations of the left front seat deceleration for simulations with and without forming histories.

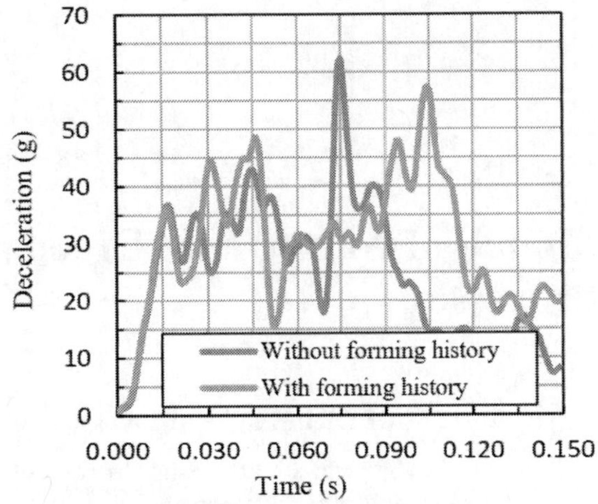

Figure 7. Variations of the right front seat deceleration for simulations with and without forming histories.

5. Conclusions

In this study, the effects of forming history on the crash simulation have been investigated. The following conclusions can be drawn from this work:

1. Consideration of the forming histories of the sheet metal components on the load path in the simulations affects the overall crash response of the vehicle.
2. Energy absorption amounts of the components of the load path change when the forming effects have been taken into consideration in the crash simulations.
3. It is found that the maximum deceleration pulses transferred to the driver's and the passenger's seats decreased when the forming effects have been used.
4. It is shown that the results obtained by the full vehicle crash simulation with the forming effects are in better agreement with the physical crash test results.
5. It is observed that the forming histories should be taken into account to get more realistic results from FE analyses.

References

[1] Deba A., Mahendrakumara M S, Chavana C, Karvea J, D. Blankenburgb D., Storenb S., 2004 Design of an aluminium-based vehicle platform for front impact safety, *Int. J. Impact Eng.* **30** 1055–79.

[2] Kwasniewski L, Bojanowski C, Siervogel J, Wekezer J W, Cichocki K 2009 Crash and safety assessment program for paratransit buses, *Int. J. Impact Eng.* **36** 235–42.

[3] Zhou Y, Lan F and Chen J 2011 Crashworthiness research on S-shaped front rails made of steel–aluminum hybrid materials *Thin-Walled Structures* **49** 291–97

[4] Lademo O G, Berstad T, Eriksson M, Tryland T, Furu T, Hopperstad O S and Langseth M 2008 A model for process-based crash simulation, *Int. J. Impact Eng.* **35** 376–88.

[5] Wang W, Xiaoyu Sun X and Wei X 2016 Integration of the forming effects into vehicle front rail crash simulation, *Int. J Crashworthiness* **1** 9-21

[6] National Crash Analysis Center, George Washington University, Washington DC, USA, http://www.ncac.gwu.edu/index.html, Last visited on June, 2009.

Models of Anisotropic Creep in Integral Wing Panel Forming Processes

A I Oleinikov[1,2] and A A Oleinikov[3]

[1] Research and Technology Center, Central Aerohydrodynamic Institute, Zhukovsky, MR, 140180, Russia
[2] Department of Aeromechanics and Flight Engineering, Moscow Institute of Physics and Technology, Zhukovsky, Russia
[3] Faculty of informational technologies, mathematics and physics, Amur State University of Humanities and Pedagogy, Komsomolsk-na-Amure, 681000, Russia

a.i.oleinikov@mail.ru

Abstract. For a sufficiently wide range of stresses the titanic and aluminummagnesium alloys, as a rule, strained differently in the process of creep under tension and compression along a fixed direction. There are suggested constitutive relations for the description of the steady-state creep of transversely isotropic materials with different tension and compression characteristics. Experimental justification is given to the proposed constitutive equations. Modeling of forming of wing panels of the aircraft are considered.

1. Introduction.

Large integral panels are widely used to extend the operating life of the airframe. The stress-strain state resulting from panel forming determines the quality, the geometrical, physical, and mechanical performance of the panel, as well as the geometry of dies and blanks. One of the key salient features of high-strength light alloys is a large difference in (heterogeneous) resistances to tension, compression, and torsion. The alloys have limited ductility; that is why the panels considered often are inappropriately classified in terms of conventional forming technologies using "instant" plastic deformations. In forming under plasticity conditions, the material life does not extend beyond the panel manufacturing stage, causing the panel to suffer intolerable damage and cracks, which results in rejection of the entire panel. Slow forming under creep, where stresses are restricted by the elastic limit vicinity, while the shape is generated by gradually accumulating creep strains, seems promising for extending the material life, minimizing damage and residual stresses, and providing high quality of panel manufacturing.

The paper deals with computations of process performance of viscoplastic forming, blank, and die tooling for large integral wing structures. The constitutive relations between the steady creep strain rates and stresses describe steady, transversely isotropic creep with different characteristics under tension and compression. Computer modeling of these forming processes involves the use of the finite element method for consecutive solutions of three-dimensional quasi-static problems of elastoplastic straining, relaxation, and springback with allowance for large displacements and turning angles, anisotropy, and different resistances of the material.

2. Constitutive Potential Model of Steady Creep of Incompressible Transverse Isotropy Materials with Different Properties under Tension and Compression.

Test data argued that the material is transversely isotropic under the steady-state creep with the isotropy plane - rolling plane. For instance, the 180°C temperature creep curves of aluminum alloy B95T2 are essentially divergent for samples in direction in the plate rolling plane, and in the direction of the normal to this rolling plane [1]. Meanwhile, the creep degree is the same for every direction, but creep coefficients are considerably different.

Let be a Cartesian rectangular system of coordinates. The strain rate of the steady-state creep depends on the current values of stresses and the temperature. If the form of this dependence does not change under the coordinate system rotation on arbitrary angle around the x_3 axis then the material is transversely isotropic (initially transverse isotropy) under the steady-state creep with the isotropy plane x_1, x_2.

We propose the following creep equations of incompressible transversely isotropic, which follow from [2],

$$\dot{\varepsilon}_{ij}^{c} = \frac{\partial \Phi}{\partial \sigma_{ij}}, \quad \Phi = \frac{f_1(\xi,\eta)}{n_1+1}\sigma_{et1}^{n_1+1} + \frac{f_2(\xi,\eta)}{n_2+1}\sigma_{et2}^{n_2+1}, \tag{1}$$

where

$$\sigma_{etm} = \sqrt{3/2\,T_m}, \quad T_m = a_m J_1^2 + b_m J_2 + c_m J_3, \quad (a_m > 0,\ b_m > 0,\ c_m > 0;\ m = 1,2), \tag{2}$$

$$\xi = \frac{J_1}{\sqrt{J}}, \quad \eta = \frac{J_4^{1/3}}{\sqrt{J}}, \tag{3}$$

where

$$J_1 = \sigma_{33} - \frac{1}{2}(\sigma_{11} + \sigma_{22}),\ J_2 = (\sigma_{11} - \sigma_{22})^2 + 4\sigma_{12}^{\ 2},\ J_3 = \sigma_{13}^{\ 2} + \sigma_{23}^{\ 2}, \tag{4}$$

$$J_4 = 4\sigma_{12}\sigma_{13}\sigma_{23} + (\sigma_{22} - \sigma_{11})(\sigma_{23}^2 - \sigma_{13}^2),\ \left(J = J_1^2 + J_2 + J_3\right) \tag{5}$$

The simplest (tensor-linear) theory follows from (1) if in these relations we do not take into account the dependence of the functions f_1 and f_2 on the parameter η and assume that the type of the stress state is adequately characterized by the following parameters ξ.

The creep curves calculated by Eqs. (1) for different constant values of tensile and compressive stresses for the same directions of the plate as those in the steady creep experiments are in reasonable agreement with the experimental curves. For comparison, we give the numerical and experimental data [1], which differ from the results obtained in the steady creep experiments. Figure 1 shows the calculated and experimental creep curves for pure torsion of a thin-walled sample tube whose centerline coincides with the transverse isotropy axis of the AK4-1T alloy in the plate at a temperature of 200°C. The following geometric parameters of the tube are used in the calculations: length 10 mm; inner and outer diameters 1.84 and 2 mm, respectively; one of the tube ends is rigidly fixed, while the other end is subjected to shear forces below the yield stress. The forces are applied rather rapidly, as compared with the time during which stress relaxation occurs and creep strains are developed, but rather slowly, as compared with the time of pulsed application where accelerative effects should be taken into account. Therefore, the configuration at the initial time (t = 0) with zero creep strains in the numerical solution is the deformed configuration of the tube made from this elastic material under the action of constant shear forces. The behavior of the sample material at the stage of force application is isotropically elastic and is described by Hooke's law; the creep of this material is transversely isotropic with different characteristics under tension and compression and is described by the flow law (1). It follows from Fig. 1 that the results calculated by Eqs. (1) ensure reasonable agreement with experimental data under pure torsion, especially under moderate stresses when the steady creep phase is realized.

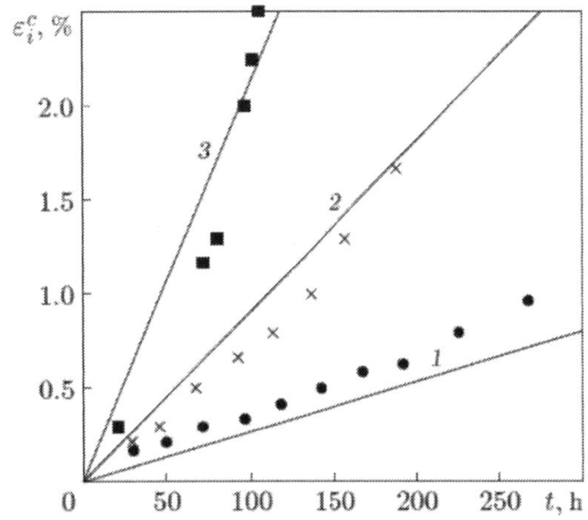

Figure 1. Calculated data (curves) and experimental data (points) on creep of the AK4-1T alloy at a temperature of 200°C under conditions of torsion of a thin-walled cylindrical sample whose centerline coincides with the transverse isotropy axis ($\sigma_i = \sqrt{3/2 s_{ij} s_{ij}}$, $\varepsilon_i^c = \sqrt{2/3 \varepsilon_{ij}^c \varepsilon_{ij}^c}$):

1) $\sigma_i = 144.4$ MPa; 2) $\sigma_i = 153$ MPa; 3) $\sigma_i = 170$ MPa.

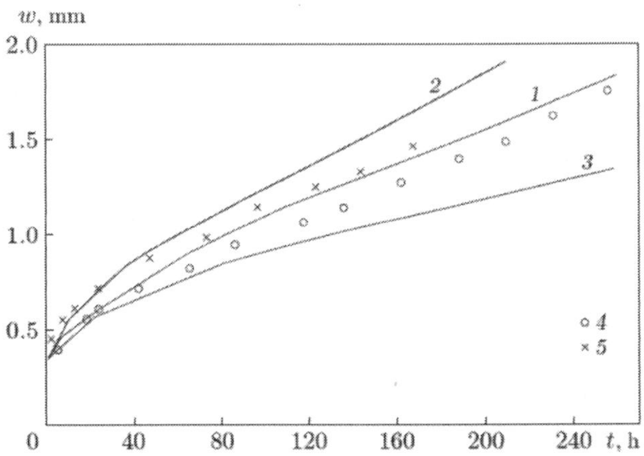

Figure 2. Deflection at the point of the plate versus time during the creep process under bending by the moments of torsion: curves are the results calculated with the use of constants (1) corresponding to tensile and compressive loading, respectively (curve 1), corresponding to tensile loading (curve 2), and corresponding to compressive loading (curve 3); points are the experimental data from experiment No. 1 and No. 2.

For the same material, we also compare the calculated and experimental curves of deflection of a square thick-walled plate with the creep process (the transverse isotropy plane of the material is parallel to the midsurface) under the action of four vertical concentrated forces P = 1850 kgf applied at the centers of the corner areas. The applied forces simulate pure bending of the plate by the moments of torsion with intensity P/2 uniformly distributed over the edges [3]. The points in Fig. 2 show the deflections obtained in the experiments with two identical plates [4]. Figure 2 also shows the time evolution of the deflection, which was calculated on a uniform grid consisting of three-dimensional eight-node hexagonal elements with the rib length of 10 mm. The following geometric parameters of the plate are specified in the calculations: length 180 mm, thickness 20 mm, and corner area size 20 × 20 mm. It follows from Fig. 3 that the results calculated by Eqs. (1), which take into account the difference in material characteristics under tension and compression are in reasonable agreement with the results of experiments on pure bending of the plate by the moments of torsion. At the same time, it is obvious that significant errors are obtained if this difference is ignored: the calculated plate deflection is overpredicted (as compared with the experimental data) if the characteristics under tension are used in all directions and underpredicted if the characteristics under compression are used. This type of plate bending can be accurately described by the constitutive relations derived in this paper for transversely isotropic materials with different characteristics under tension and compression.

2.1. *Problem of Wing Panel Forming.* In this type of forming, the blank is also subjected to bending. In contrast to the direct problems discussed above, however, the values of the surface forces and boundary displacements are unknown in the general case and have to be determined alongside with the fields of stresses, strains, and displacements. For this purpose, it is possible to use the residual boundary displacements, for instance, the displacements over the normal to the frontal face of the blank, which can be easily calculated from the specified geometric parameters of the panel.

Let us consider a case with the problem of forming at the stage of active loading being solved in the kinematic formulation. We find the sought elements of the boundary displacements. The distributions of these components at the 8th iteration are shown in Fig. 3. Using these displacements, we modeled all stages of the forming process [5].

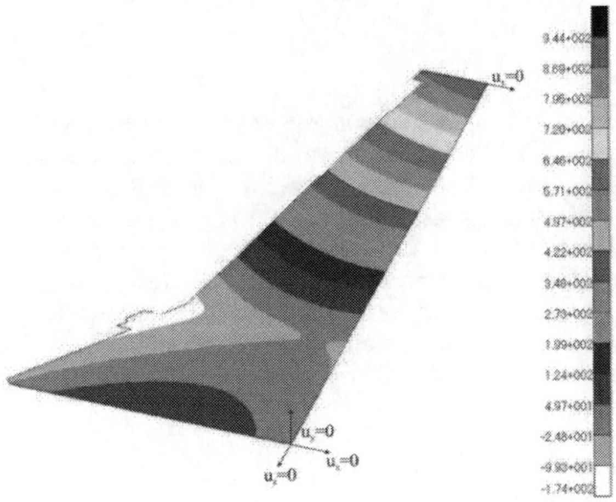

Figure 3. Distributions of the boundary displacements calculated at the 8th iteration.

References

[1] B. V. Gorev and I. Zh. Masanov, "Specific features of straining of structural sheets made of aluminum alloys in the creeping mode," Tekhnol. Mashinostr., No. 7, 13–20 (2009).

[2] A.I. Oleinikov, Models for the steady-state creep of transversely isotropic materials with different tension and compression characteristics. Sibirskii Zhurnal Industrial'noi Matematiki, 2010, Vol. XIII, No. 3, pp. 113–116. [Journal of Applied and Industrial Mathematics, 2011, Vol. 5, No. 3, pp. 406–409.]

[3] S. P. Timoshenko and S. Woinowsky-Krieger, Theory of Plates and Shells, McGraw-Hill, New York (1959).

[4] O. V. Sosnin, B. V. Gorev, and V. V. Rubanov, "Torsion of a square plate made of a material with different resistances to tension and compression at creeping," in: Strength Calculations of Ship Structures and Mechanisms (collected scientific papers), [in Russian], No. 117, Institute of Water Transport Engineering, Ministry of Inland Water Transport, Novosibirsk (1976), pp. 78–88.

[5] B. D. Annin, A. I. Oleinikov, and K. S. Bormotin, "Modeling of Forming ofWing Panels of the SSJ-100 Aircaft,"Prikl. Mekh.Tekh. Fiz. 51 (4), 155–165 (2010) [J. Appl. Mech. Tech. Phys. 51 (4), 579–589 (2010)].

Journal of Physics: Conference Series **734** (2016) 032096 doi:10.1088/1742-6596/734/3/032096

Forming parts over small radii

S K Hazra[1], D J Hughes[1], M P Pereira[2] and B F Rolfe[2]
[1] WMG, University of Warwick, Coventry CV4 7AL, UK
[2] School of Engineering, Deakin University, Australia

Abstract. Stamping simulations usually make the plane stress simplifying assumption. However, this becomes less valid when material draws around features with radius to sheet thickness ratios less than 20. Pereira, Yan & Rolfe (Wear, Vol.265, p.1687 (2008)) predicted that out-of-plane stress equivalent to material yield can occur because a line contact forms briefly at the start of the draw process. The high transient stress can cause high rates of tool wear and may cause the 'die impact line' cosmetic defect. In this work, we present residual strain results of a channel section that was drawn over a small radius. Using the neutron source at the Institut Laue-Langevin, in-plane and out-of-plane strains were measured in the channel part to show some support for the conclusions of Pereira *et. al.*

1. Introduction

Sheet metal forming is a widely used manufacturing process that converts thin, flat sheet into a desired part using a set of tooling. The process is commonly simulated using finite element (FE) models, which assume that the plastic deformation in the sheet takes place under plane stress conditions, so that out-of-plane stresses are absent. However, it is common for tools to have features (e.g. the die radius) with a radius to sheet thickness ratio (R/t) of less than 20 [1], where the plane stress assumption becomes less valid. Recently, FE models of channel sections were carried out by Groche *et. al* [2] and Pereira *et. al* [3] to study the drawing of material around a 90° die radius with an R/t ratio less than 5. Their results predicted that the plane stress assumption is inaccurate because sheet material is subject to out-of-plane contact stresses that cause it to bend and unbend as it flows around the radius. Pereira *et. al* [4] predicted that the stress is of the order of the yield stress of the material. A similar observation was made earlier by Coubrough *et. al* [5] in steady-state draw experiments with an R/t of about 12. However, Pereira *et. al* [3] also suggested that contact stresses can exceed yield stress when the flat sheet first comes into contact with the tool radius because line contact conditions are momentarily formed [3]. The resulting high stress has two important implications. First, it can cause a high rate of wear in tooling, particularly during the pressing of high strength alloys. Second, it leaves a mark known as the 'die impact line' [6], which is a visible, geometrical defect that impairs the quality of a skin panel. A panel with such a defect may be trimmed to remove the defect from the final part, leading to waste. However, with recent trends in automotive styling, these sharp features may exist on outer body panels where the material is part of the final part surface and therefore cannot be removed.

This paper tests the assumption that the stress state in a sheet material as it draws around a small radius remains in plane stress. The authors are unaware of any experimental technique that is able to measure out-of-plane stresses or strains in a sheet in-situ as it is drawing over a small radius. For this reason, the residual strain distributions of channel sections were measured because they reflect the deformation history of the material making up the part. Withers *et. al* [7] described several measurement techniques including hole drilling, neutron diffraction and X-ray diffraction techniques. In this work, the neutron diffraction technique was utilised because of the availability of the facility and its ability to measure the strain state of a material within a small near cubic 'gauge volume'.

2. Method

20mm wide strips of automotive grade aluminium, AA6111-T4, were pressed into channel sections using purpose-built tooling mounted in an Erichsen sheet metal forming tester. The samples were prepared at the labs of the School of Engineering, Deakin University, Australia using the method described by Pereira *et al.* [8]. The residual stress of the sample was measured according to ISO/TS 21432:2005. The neutron instrument used was the SALSA diffractometer located at the Institute Laue-Langevin reactor in Grenoble, France [9]. The instrument uses a monochromatic beam, delivered via a double-focusing bent Si-crystal monochromator, with a wavelength of 1.66 Å (reflecting off the [4 0 0] plane). The collimator-controlled incident beam size was (height × depth) 0.6 mm × 25 mm (FWHM) and the 2θ diffracted beam angle was 84.5° (for the aluminium [3 1 1] plane). The diffraction angle ensured that the gauge volume was close to rectangular in geometry and the size of the gauge volume ensured that a sufficiently large number of grains (~2000) were sampled within it. Note that the long axis of the gauge volume was oriented along the width of the channel sample. This ensured that the spatial resolution through the thickness of the wall was determined by the smallest dimension of the gauge volume (0.6 mm) whilst also reducing the counting time to by averaging along the width of the channel (where there is no variation is residual strain). The SALSA instrument employs a 2D Position Sensitive Detector (PSD); the data was integrated and fitted using the LAMP software to provide diffraction peak positions (2θ). The sample was positioned in the beam by securing it on a hexapod with 6-degrees of freedom. To measure the in-plane and out-of-plane strain components, the hexapod table (Fig.2) was rotated as appropriate. In particular, the out of plane sample direction was aligned along the diffraction *q*-vector, which bisects the angle between the incident and diffracted beam and represents the measured component of strain.

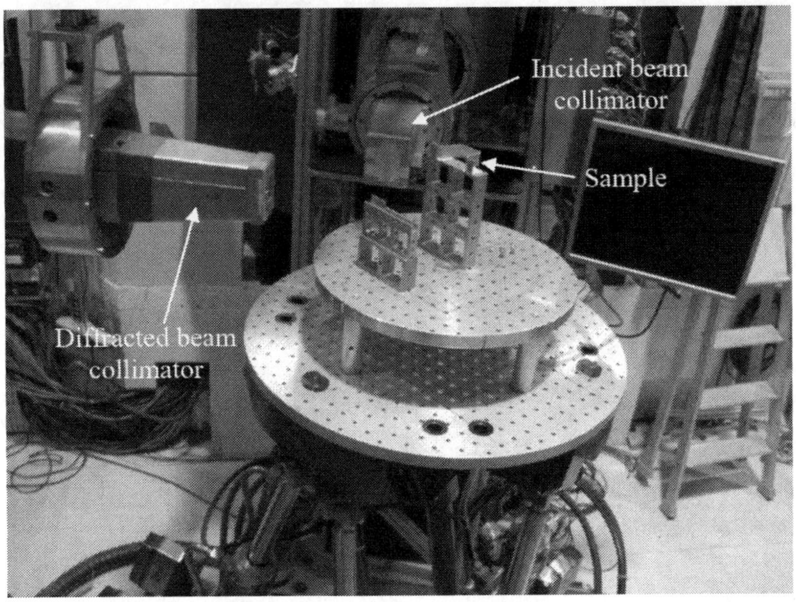

Fig.1 A photograph of the experimental setup showing the sample, hexapod, beam and detector slits

The samples were measured at 5 locations along their walls as shown in Fig.2. Location 1 corresponds to the location of the die impact line which had the appearance of a burnish. At each location, a through-thickness scan was performed.

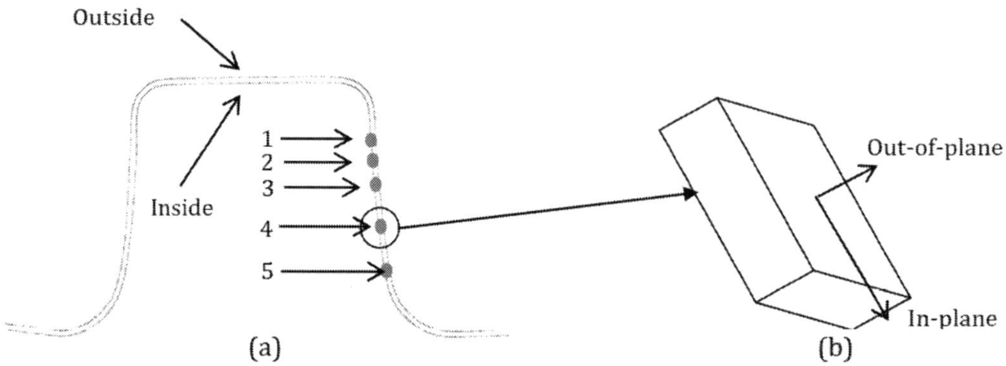

Fig.2(a) Locations of residual strain measurement (b) Orientation of out-of-plane and in-plane strains

Strains were calculated with the following relation:

$$\varepsilon = \frac{\Delta d}{d_0} = \Delta\theta \cot\theta_0$$

Where ε is the strain the in *hkl* [311] plane, d_0 is a reference lattice spacing, Δd is the change in the lattice spacing, $\Delta\theta$ is the change in the Bragg angle as a result of the change in lattice spacing and θ_0 is the Bragg angle for the reference lattice spacing. θ_0 was measured at the top of the channel where the material experienced minimal plastic deformation.

3. Results

The out-of-plane results and the in-plane results are shown in Fig.4.

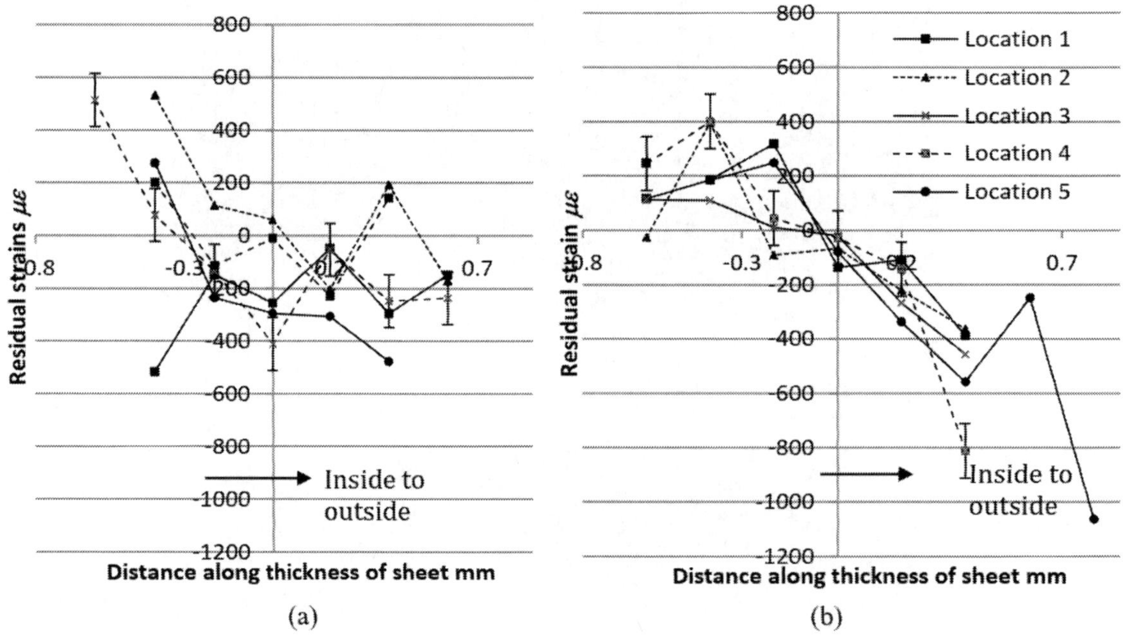

Fig.4(a) Out-of-plane strain (b) In-plane strains along the wall of the sample

At the location of the burnish (location 1), the out-of-plane strain distribution (Fig.4a) remains negative through the thickness of the sheet. In particular, it is about -520$\mu\varepsilon$ at the inside face and about -200$\mu\varepsilon$ at the outside face. Below the burnish (locations 2 to 5), the strain on the inside face is positive and becomes negative towards the outside face. The highest out-of-plane strain (531$\mu\varepsilon$) was measured on the inside face at location 2.

In-plane strains were positive in the inside face and negative in the outside face (Fig.4b). At the location of the burnish (location 1), the residual strain on the inside face was about 110$\mu\varepsilon$. The strain appears to increase slightly before decreasing to about -400$\mu\varepsilon$ at the outside face. The variation was approximately similar along the other cross-sections.

4. Discussion

Some strain measurements, such as the out-of-plane strains at location1 do not appear to balance across the thickness of the sample. This is for two reasons. First, the shear strains have not been accounted for in these measurement and second, the relatively large beam width used in contrast to the thickness of the sheet is likely to have smoothed out the actual distribution. Despite this, the measurements in Fig.4a show that the out-of-plane strain are significant compared to yield strain of AA6111-T4 (about 1500$\mu\varepsilon$). This observation supports the prediction made by Pereira *et. al* [3] that out-of-plane deformation takes place during the bending of sheet over small radii. The in-plane strains, which are the result of the stretching of the material, shows a distribution of varying strain that is consistent with bending in the sheet. The character of the distribution is likely to be the cause of the curl in the part.

5. Conclusions

The residual strains in a channel section were measured to identify out-of-plane and in-plane strains in the wall of the part. The significant out-of-plane strains that were measured are consistent with the prediction of Pereira *et. al* [3] of the existence of out-of-plane deformation during the drawing of material over small tooling radii. This observation implies that simulations over material forming over small radii should be modelled with elements that account for these strains in order to account for tool wear and the die impact line cosmetic defect. The results from this work will contribute to identifying the conditions that cause the cosmetic defects at the die impact line for automotive body panels.

Acknowledgements

The authors would like to acknowledge the beam time provided by the Institut Laue-Langevin for the duration of the experiment and the assistance provided by the scientists at the SALSA beamline.

References

[1] Bhattacharya R, Stanton M, Dargue I, Aylmore R, Williams G. Experimental Evaluation of Springback in Aluminium Alloys Using Optical Measurement and Numerical Analysis. In: Menary G, editor. The 14th International ESAFORM Conference on Material Forming; Belfast: AIP Conference Proceedings; 2011. p. 214.

[2] Groche P, Nitzsche G, Elsen A. Adhesive wear in deep drawing of aluminum sheets. CIRP Annals - Manufacturing Technology 2008;57:295 - 8.

[3] Pereira MP, Duncan JL, Yan W, Rolfe BF. Contact pressure evolution at the die radius in sheet metal stamping. Journal of Materials Processing Technology 2009;209:3532 - 41.

[4] Pereira M, Okonkwo P, Yan W, Rolfe B. Deformation and frictional heating in relation to wear in sheet metal stamping. Steel Research International 2010. p. 713-6.

[5] Coubrough GJ, Allinger MJ, Tyne CJV. Angle of contact between sheet and die during stretch–bend deformation as determined on the bending-under-tension friction test system. Journal of Materials Processing Technology. 2002;69:130-1.

[6] Marciniak Z, Duncan JL, Hu SJ. *Mechanics of Sheet Metal Forming*: Butterworth Heinemann; 2002.

[7] Withers PJ, Turski M, Edwards L, Bouchard PJ, Buttle DJ. Recent advances in residual stress measurement. International Journal of Pressure Vessels and Piping. 2008;85:118-27.

[8] Pereira M, Yan W, Rolfe B. Wear at the die radius in sheet metal stamping. Wear. 2012;274-275:355-67.

[9] Hughes DJ, Bruno G, Pirling T, Withers PJ. Scientific Review: First Impressions of SALSA: The New Engineering Instrument at ILL. Neutron News. 2006;17:28-32.

The Bendability of Ultra High strength Steels

S K Hazra[1], P Efthymiadis[1], A Alamoudi[2], R L V Kumar[1], B Shollock[1] and R Dashwood[3]
[1]WMG, University of Warwick, Coventry CV4 7AL, UK
[2]School of Engineering, University of Warwick, Coventry CV4 7AL
[3]Coventry University, Priory Street, Coventry CV1 5FB

Abstract. Automotive manufacturers have been reducing the weight of their vehicles to meet increasingly stringent environmental legislation that reflects public demand. A strategy is to use higher strength materials for parts with reduced cross-sections. However, such materials are less formable than traditional grades. The frequent result is increased processing and piece costs. 3D roll forming is a novel and flexible process: it is estimated that a quarter of the structure of a vehicle can be made with a single set of tooling. Unlike stamping, this process requires material with low work hardening rates. In this paper, we present results of ultra high strength steels that have low elongation in a tension but display high formability in bending through the suppression of the necking response.

1. Introduction

The UK manufactures over 1.6 m vehicles per annum, which is predicted to rise 9% annually. The sector accounts for 7.3% of manufacturing output and 5.2% of manufacturing employment [1]. Automotive manufacturers, however, are facing increasing demands to minimise the environmental impact of their products through lightweighting while improving safety standards. For example, the UK government passed the Climate Change Act (2008) to reduce its emissions by 80% from 1990 levels by 2050. The contradictory environmental and safety demands have led to increasing use of high strength materials such as ultra-high strength steel (UHSS) that have a high strength to weight ratio. These materials have ultimate tensile strengths above 1000MPa but low elongations, making them difficult to form with the conventional cold stamping process.

One way to adopt low formability materials is to use alternative processing routes such as roll-forming[2]. Roll forming increases formability by deforming material through incremental, localised bending [3]. Sheet material flows through a series of rollers rigidly mounted on stands that bend the sheet as it flows through the rollers. The number of stands depends on the complexity of the geometry and can range from 4 to 35. Roll forming is cheap to operate, involves high material utilisation and is environmentally friendly but is limited to processing parts with a single cross-section along its length. Even though the geometry of this cross-section can be more complex than what can be achieved by stamping, forming a fixed cross-section along the length of a part limits its appeal for processing automotive parts, which frequently requires a varying cross-section along its length.

The limitation of the method is being addressed by the recent development of the flexible roll forming process [4]. With this method, the individual rolls are mounted on actuators that can change the

position and orientation of the rollers as a sheet flows through. With correct control, parts with continually varying cross-sections along its length can be produced. Flexible roll forming, therefore, opens up the possibility of roll forming automotive components with high strength materials such as UHSS.
Recent work on roll forming has looked into the effect of residual stresses on the springback on the final part. For example, Mendiguren *et. al* [5] and Weiss *et. al* [6] tried to relate bending behaviour of dual phase steels and aluminium respectively to their springback. Hosseini *et. al* [2] developed a strain gauge-based technique to measure residual stresses of the order of yield stress in UHSS. In this work, the behaviour of a UHSS dual phase grade is investigated by comparing its bendability to its stretch performance and by characterising its through-thickness strain distribution during a bending test. The results show that the material can sustain significantly higher strain in bending compared to stretching.

2. Method

The chemical composition of the dual phase UHSS is given in Table 1.

Element	**C**	**Mn**	**Si**	**Nb**
Composition %	0.15	0.15	0.05	0.013

Table 1 Chemical composition of the dual phase grade

Stretch formability of the material was characterised with tensile tests and by measuring its forming limit curve (FLC). Tensile tests were carried out according to ISO 6892-1:2009 and FLC tests were carried out according to ISO 12004-2:2008. Tensile tests were conducted along, 45° and 90° to the grain direction. The tests were carried out on an Instron 250kN test machine, strains were measured with an optical/non-contact extensometer and loads were measured with a 100kN loadcell. FLC tests were carried on an Interlaken ServoPress 225 test machine. Strains were measured with a GOM digital image correlation (DIC) device mounted above the machine. Three repeats were carried out of each test.
Bending formability was characterised with plane strain bending tests that were carried out on a purpose built bending rig mounted on an Instron 5800 frame. To assess formability, a DIC device was used to capture strains along the thickness of the sheet (Fig.1).

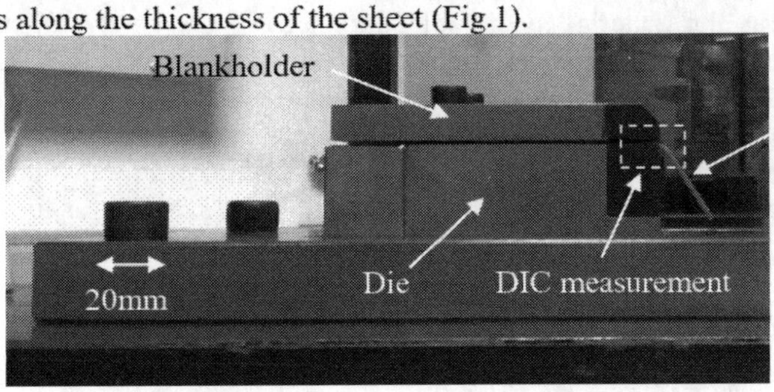

Fig.1 Photo of the bending rig mounted on an Instron 5800 frame

3. Results

The results of the tensile tests are in Table 2 and the results for the FLC tests are in Fig.2.

	0° to grain	45° to grain	90° to grain
0.2% proof stress MPa	700.4±9	715.4±6	737.4±9
Ultimate tensile strength (UTS) MPa	1046.7±0.5	1070.7±7	1080.6±12
r-value	0.38±0.02	0.49±0.03	0.46±0.03
n-value	0.14±0	0.14±0.01	0.14±0
Yield ratio = proof/UTS	0.67±0.02	0.67±0.02	0.68±0

Table 2 Summary of data for the dual phase steel used in this study

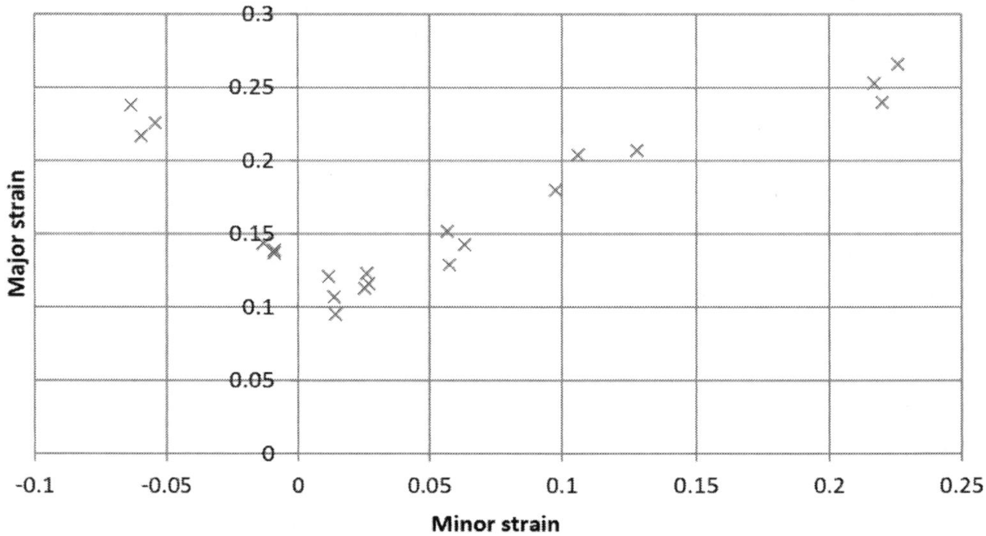

Fig.2 FLC curve for the dual phase steel

The results illustrate a material with a high strength to weight ratio (Table 1) but low formability, particularly in the plane strain region (Fig.2). Away from the plane strain path (Fig.2), formability increases so that in the biaxial region, failure takes place at around 0.25 strain. Its low formability is anticipated by a low work hardening rate of 0.14 (or high yield ratio of 0.67) in all directions with respect to its grain.

In plane strain bending, the material in the outer ligament of the bend was able to support greater strain than in the stretch mode experienced in the tensile and FLC tests (Fig.3). Compared to the plane strain deformation failure strain that was measured to be about 0.1-0.11 (Fig.2), the strain the bending sample supported was 0.16-0.17 (Fig.3). Unlike FLC samples, which necked and failed, analysis of the bending samples using scanning electron microscopy (SEM) did not reveal significant damage (Fig.4). With the current design of the bending apparatus, bending of the sample was limited to about 80°. As a result, it was not possible to explore the limits of bending formability of the material.

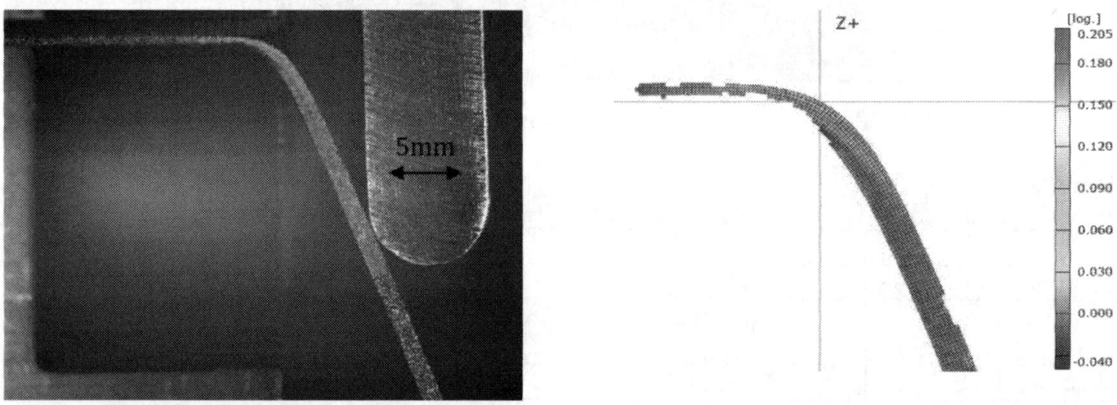

(a) Side view of the bending rig (b) Contour plot – major strain

Fig.3 Major strain contour plot in plane strain bending (a) DIC image (b) Strain result

Fig.4 SEM micrograph of the highly strained region of the bending sample

4. Conclusions

Tensile tests and plane strain bending tests were carried out on a dual phase UHSS. The results showed that the material was able to support higher major strains in bending than in stretch. The full field major strain field showed that the high strain levels were found, as expected, in the outer ligament of the sample. However, despite the higher strains in bending, little evidence of damage was found in the microstructure of the material, suggesting that it is able to sustain even higher strains.

Acknowledgements

The authors would like to thank the Engineering and Physical Sciences Research Council for financially supporting this work and Tata Steel UK, Tata Steel BV for the supply of the materials. We had many useful discussions with the staff at Tata Steel UK and Tata Steel BV, The Hadley Group, Advanced Forming Research Centre, Jaguar Land Rover and data M Sheet Metal Solutions.

References

[1] Cable V, Parry-Jones R. Driving success – a strategy for growth and sustainability in the UK automotive sector. Automotive Council UK, 2013.

[2] Hosseini SB, Sagström E, Ottosson P, Troive L. A method for measuring residual stresses in rollformed hat profiles of ultra high-strength steel. In: Kolleck R, editor. IDDRG: TOOLS AND TECHNOLOGIES FOR THE PROCESSING OF ULTRA HIGH STRENGTH STEELS; Graz, Austria: Technical University Graz; 2010. p. 245-54.

[3] Kalpakjian S, Schmid SR. *Manufacturing processes for engineering materials*. 5th ed. Singapore ; London: Pearson/Prentice Hall; 2008. xxi, 1010 p. p.

[4] Lindgren M. Experimental and Computational Investigation of the Roll Forming Process: Luleå University of Technology; 2009.

[5] Mendiguren J, Hanselman S, Atzema E, Hodgson P, Rolfe B, Weiss M. Characterization of a Dual Phase Steel Using Tensile and Free Bending Tests. In: Yoon JW, Beynon J, Hodgson P, editors. NUMISHEET 2014; Melbourne, Australia: AIP Publishing; 2013. p. 659-62.

[6] Weiss M, Rolfe B, Hodgson PD, Yang C. Effect of residual stress on the bending of aluminium. Journal of Materials Processing Technology. 2012;212:877-83.

Analysis of Strain Non-uniformity Index (SNI) for Different Geometries of Drawn Sheet Metal Parts

Sandip Jadhav[1], Ratneshwar Tekale[2], Prashant P. Date[3]

Department of Mechanical Engineering,
Indian Institute of Technology, Bombay
Powai, Mumbai-400 076, India

E-mail: [1] pdsandip@gmail.com, [2] ratneshwar.tekale@gmail.com, [3] ppdate@gmail.com,

Abstract: This paper explores the effect of the part geometry on the variation of the Strain Non-Uniformity Index (SNI) at failure in shapes drawn from a single material. Forming of different shapes, namely, a square cup, an equibiaxially stretched sample was performed experimentally as well as simulated using AUTOFORM 5.2 Plus software. Failure predictions were made using the SNI based methodology and the FLD, and compared with experimental outcomes. Forming of the cross draw (FTF benchmark), was simulated and corresponding critical SNI was established based on the failure predicted with reference to the FLD. The SNI values so obtained are discussed in light of different component shapes and draw depths during forming of various geometries.

Keywords: - Forming Limit Diagram (FLD), Strain Non-Uniformity Index (SNI), sheet metal forming, Formability

1. Introduction

Strain distribution based failure criterion based on the Strain non-uniformity index (SNI) was proposed earlier by Date et. al. [1] and has been applied to failure predictions in drawn sheet metal parts [2]. Since the strain distribution and hence the SNI is an outcome of part geometry, forming conditions and material properties, this paper examines the effect of part geometry as well as the material geometry on the evolution of the SNI over the entire forming process.

Strain Non-uniformity Index (SNI) = Peak thickness strain – Average thickness strain............ (1)

2. Work plan

In the present work, square cups of side 40mm were drawn from blanks of five different materials (four ferrous and one aluminium alloy), laser marked with circles of initial diameter of 2mm. Similarly, square blanks of side 200mm, and similar grid markings were stretched over a punch of diameter 100mm. Strain non-uniformity index (SNI) was determined based on the strain distribution taken along the diagonals for square cups and along meridians for the equibiaxially stretched sheets. SNI was also established using the strain distributions obtained from AUTOFORM simulations of different shapes, namely, square cup, equibiaxially stretched sheet, FTF Cross shape [3], FTF triangular shape, Jaguar land rover aluminium panel (Numisheet 2016 Benchmark -2) [4], and the kidney tray. In each case the evolution of the SNI with punch travel was plotted. The properties of materials used for experiments as well as for simulation are given in Table 1.

3. Results and Discussion

A good agreement with experiment, on the prediction of failure location using the FLD and the SNI based methodology is readily inferred from Fig. 1 and Fig. 2. The evolution of the SNI with punch travel for three different shapes is seen in Fig. 3 (a-c). It is found that the SNI variation with draw depth for five different materials lies in a narrow band for each of the shapes studied. Fig. 3d summarises the variation for different component geometries.

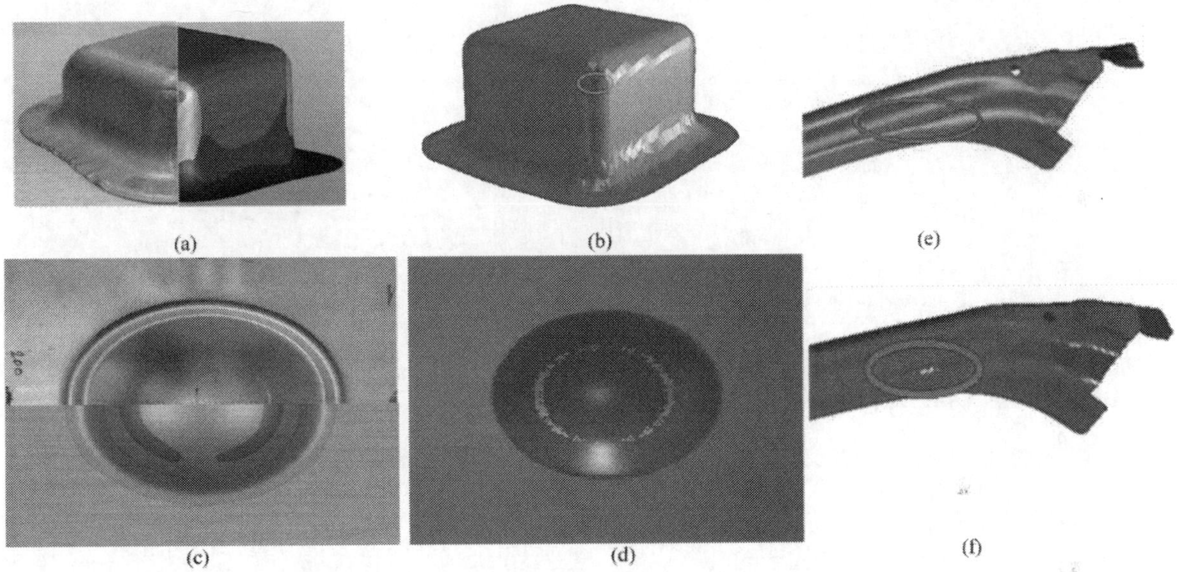

Figure. 1 Square Cup with flange (a) Simulation and experimental results, (b) SNI based failure location, (c) Equibiaxially stretched sheet simulation and experimental results, (d) Equibiaxially stretched sheet SNI based results, (e)Numisheet benchmark-2 Simulation results, (f)Numisheet benchmark-2 SNI based results

Table 1. Material Properties

Sr. No	Material Name	Yield strength (MPa)	Work Hardening Exponent	Anisotropy	Strength Co-efficient (MPa)	Sheet Thickness in (mm)
1	TL1550Z	258.8	0.194	1.24	594.2	1
2	HX260LAD	245.4	0.174	0.94	605.8	1
3	HC180BD	193.1	0.183	1.5	511.76	0.6
4	DC04	183.0	0.204	1.64	518.95	1
5	AA6016	127.4	0.241	0.59	423.69	1.25

Also seen in Figs. 3a-b are the draw depths at which failure was indicated by experiments, FLD and the SNI based criterion for square cup and equibiaxially stretched blank. It may be observed that the prediction of draw depth at failure using the SNI based criterion (circular mark) is close to the experimentally observed draw depth at failure (triangle). A square mark indicates FLD based failure prediction. In equibiaxial stretching (Fig. 3a), AA6016 shares the band with three varieties of steel.

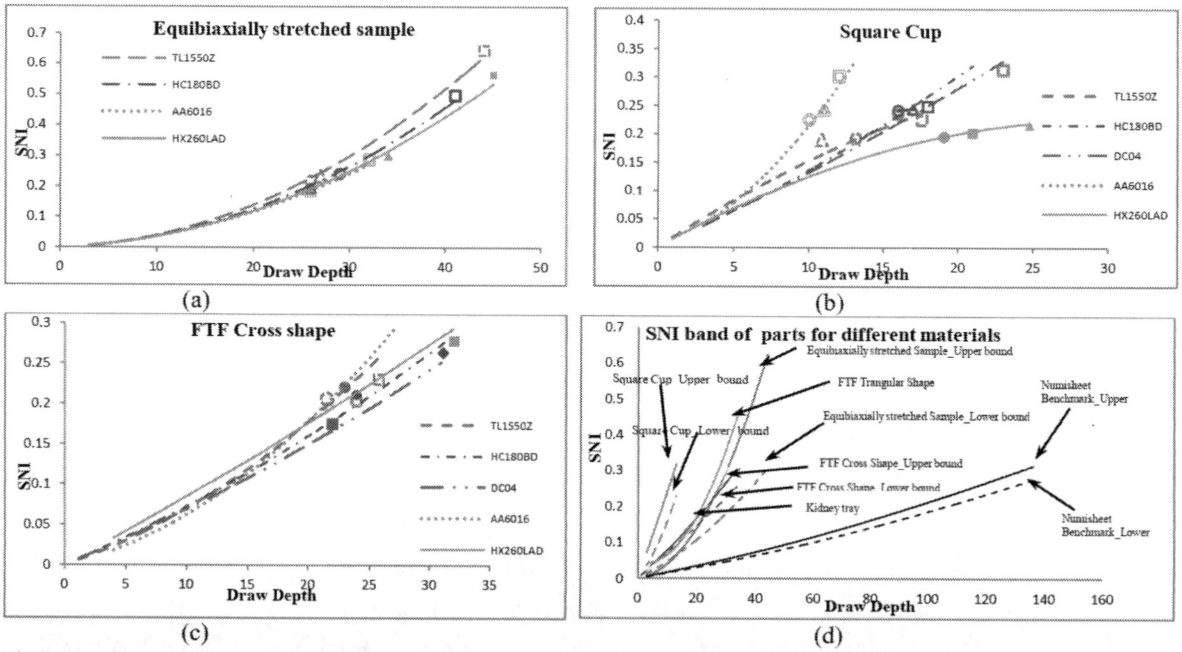

Figure 3: SNI vs. Draw Depth (o - failed as per critical SNI, △ - Failed, Experimentally, Rectangular mark indicates failed simulation) □

Table 2. Coefficients of trend lines for evolution of SNI with draw depth.

Sr.No	Part Name	Material	a	b	R^2
	Equibiaxially stretched sample[1]	TL1550Z	0.0003	0.001	0.99
a	Equibiaxially stretched sample	HC180BD	0.0003	0.0011	0.99
	Equibiaxially stretched sample	AA6016	0.0002	0.0014	0.99
	Equibiaxially stretched sample	HX260LAD	0.0002	0.0012	0.99
	Square Cup	TL1550Z	0.0002	0.0178	0.97
	Square Cup	HC180BD	0.0001	0.0122	0.96
b	Square Cup	DC04	0.00008	0.0124	0.97
	Square Cup	AA6016	0.0011	0.0102	0.99
	Square Cup	HX260LAD	0.0003	0.015	0.98
	FTF Cross Shape	TL1550Z	0.0002	0.005	0.98
	FTF Cross Shape	HC180BD	0.00008	0.0065	0.97
c	FTF Cross Shape	DC04	0.00006	0.0063	0.96
	FTF Cross Shape	AA6016	0.0003	0.0035	0.98
	FTF Cross Shape	HX260LAD	0.00003	0.0082	0.96
	Equibiaxially stretched sample_Max	All materials	0.0003	0.001	0.99
	Equibiaxially stretched sample_Min	All materials	0.00007	0.004	0.99
	Numisheet Benchmark_Max	All materials	0.000005	0.0016	0.86
	Numisheet Benchmark_Min	All materials	0.000005	0.0014	0.91
d	Square Cup_Max	All materials	0.00008	0.0235	0.82
	Square Cup_Min	All materials	0.0006	0.0101	0.92
	FTF Cross Shape_Max	All materials	0.00006	0.0072	0.97
	FTF Cross Shape_Min	All materials	0.00002	0.0069	0.95
	Kidney Tray	TL1550Z	0.0002	0.0043	0.97
	FTF Trangular Shape	TL1550Z	0.0003	0.0031	0.99

1. Highlighted to indicate results for TL1550Z material

It may be inferred from the table that the band is the narrowest for Numisheet 2016 benchmark-2, and dependence of SNI on the draw depth, almost linear. For equibiaxially stretched blanks, coefficient 'a' (and hence curvature) remains virtually unchanged over the 5 materials, while 'b' (slope at zero punch travel) is significantly different. For the TL1550Z material parameter 'a' shows no sensitivity whatsoever to the product geometry, while the value of 'b' does.

A similar trend of evolution of SNI with draw depth for a wide range of geometries made from a given single material brings out the utility of the method. A critical SNI would be attained faster in one geometry compared to another, explaining the relatively inferior formability of a given shape using a given sheet metal.

5. Conclusions

1] Failure SNI obtained by experiment is usually very close to that obtained analytically. Hence, the critical SNI obtained for the material can be used for failure prediction.
2] At the initial stage, the rate of change of SNI (which depends on the constant 'b'), depends significantly on the product shape and the material. For a given material the rate of change of SNI depends upon part geometry.
3] Geometry with sharp radii have high rate of change of SNI, which results in early failure
4] For a given part geometry, a narrow band of variation of the SNI with draw depth over several materials shows that there is no need for correlating laboratory results (using samples of a different geometry) with shopfloor experience any more. The shopfloor product itself, may be used as a sample, and evolution of SNI with punch travel established.
5] A change in material would lead to a critical SNI being achieved faster or slower depending on the coefficients of the equation.

References

[1] Date P.P., 2016 Strain distribution based approach to predict failure in sheet metal forming, IDDRG 2015, Linz, 12-15 June 2016, Linz, Austria.
[2] Date P.P., 2014 strain distribution based prediction in sheet metal components: some examples, IDDRG 2014 CETIM, 1st to 4th June 2014, Paris.
[3] Strain Prediction Benchmark - Cross-Die, 2015, 8th Forming Technology Forum Zurich.
[4] Allen M, Dutton T, Oliveira M, Hazra S, Adetoro O and Das A, Numisheet 2016, Benchmark 2 - Springback of a Jaguar Land Rover Aluminium Panel

Numerical modelling of the gas detonation process of sheet metal forming

Sandeep P. Patil, Madhur Popli, Vahid Jenkouk, and Bernd Markert
Institute of General Mechanics, RWTH Aachen University, Templergraben 64, 52062 Aachen, Germany
E-mail: patil@iam.rwth-aachen.de

Abstract. Gas detonation forming is an unconventional technique, which has the potential to form complex geometries, including sharp angles and undercuts in a very short process time. To date, most of the numerical studies on detonation forming neglect the highly dynamic pressure profile of the detonation obtained from experiments. In the present work, it is emphasised that the consideration of the actual detonation pressure as measured in the experiment is crucial. The thickness distribution and radial strain are studied using a strain-rate dependent Johnson-Cook material model. The obtained results vary significantly with change in loading rate. Moreover, the model is capable of predicting extremely sharp edges.

1. Introduction
Gas detonation forming is a high-speed forming method that uses pressure energy of a shock wave, produced as a result of the detonation of a mixture of gases like Oxygen (O_2) and Hydrogen (H_2) [1]. This forming process has many well-known advantages, e. g., higher degrees of formability for various materials in comparison to more conventional methods, the capability to form extremely complex geometries including sharp angles and fine embossing without the need of relief angles, a clean combustion, an easy automation, and less safety regulations compared to the use of solid explosives. It is worth mentioning that the costs of the process and tooling are also reduced through the simplification of the tooling requirements, and a significant potential for steel forming compared to electromagnetic forming [2].
Yasar et al. conducted both experimental and numerical investigations of aluminium cup drawing using gas detonation process [3, 4]. Also, Mokadem [5] developed a dynamic forming limit diagram for this process. Wijayathunga and Webb [6] developed a finite element model to simulate the experimental tests for the impulsive deep drawing of a brass square cup with the presence of a soft lead plug. However, the highly dynamic detonation pressure profile, which obtains in the experiment, was not considered as it is. Due to sensing limitations and in order to reduce computational cost, in Yasar et al. works, simulations were carried out with linearly increasing and decreasing load over the forming process duration (triangular loading) [3, 4]. However, the pressure curve obtained from the experiments shows that the first peak occurs in a very short time period, and this is the most influencing peak for the deformation.
The present work numerically investigates the gas detonation forming of DC04 steel cups. The 3D explicit dynamic finite element analyses are carried out using the LS-DYNA explicit solver [7]. The Johnson-Cook plasticity material model is used to study strain-rate sensitivity.

2. Numerical modelling

Figure 1 shows the schematic representation of the experimental set-up and 3D finite element model, which includes the die, the holder and the workpiece. Due to the axial symmetry of the problem, only a quarter of the whole system with symmetric boundary conditions is considered to reduce the computational time. The die and the holder are modelled using solid rigid elements (MAT_020) [7] because of their high stiffness compared to the blank as well as they are not active components during forming process. The deformation in the blank is studied by modelling it with solid and shell elements for different cases. The shell element models predict better results, hence, only shell element models will be discussed in this work. Belytschko-Tsay shell elements [7] are used to create a meshed workpiece with 5 integration points, which results into a total of 11076 elements. The die and the holder are considered to be fixed and the contact between them is defined using surface to surface segment based contact formulation [7], assuming planer segments. The loading is applied only on the free surface of the blank. The mechanical properties of DC04 steel for simulations are a elastic modulus of $E = 180$ GPa, a Poisson's ratio of $\nu = 0.3$, a density of $\rho = 7870$ kg/m^3 and a tensile strength of $\sigma = 210$ MPa.

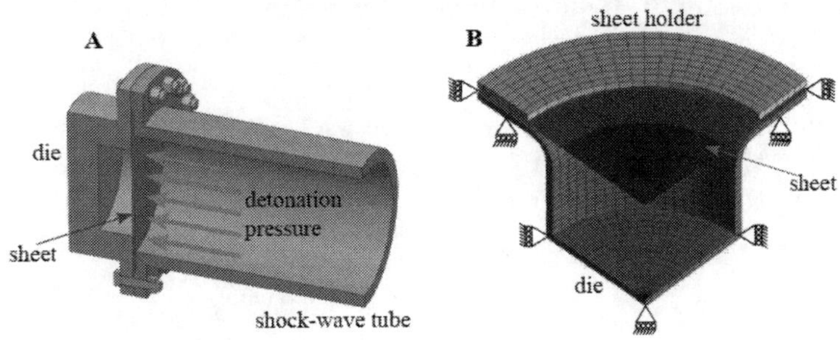

Figure 1. (A) Schematic representation of experimental set-up and (B) finite element model with symmetric boundary conditions.

The simulations are carried out with the Johnson-Cook phenomenological model [9], i. e., it is not based on traditional plasticity theory that reproduces several important material responses observed in impact and penetration of metals. In this plasticity model, the three key material responses are considered strain hardening, strain-rate effects, and thermal softening. These three effects are combined, in a multiplicative manner. The flow stresses as are expressed as:

$$\sigma_y = \left(A + B\bar{\varepsilon}^{p^n}\right)\left(1 + C\ln\frac{\dot{\bar{\varepsilon}}^p}{\dot{\varepsilon}_0}\right)\left(1 - \left[\frac{T - T_{room}}{T_{melt} - T_{room}}\right]^m\right), \quad (1)$$

where $\bar{\varepsilon}^p$ is the effective plastic strain, T_{room} the ambient temperature, T_{melt} the melting point or solidus temperature, T the effective temperature, A the yield stress, B the hardening modulus, n the strain exponent, m the temperature exponent and C the strain-rate factor. Also, $\dot{\varepsilon}_0$ represents the strain-rate for the quasi static reference loading $\dot{\varepsilon}_0 = 5.6 \times 10^{-4}s^{-1}$ [8]. The required material parameters for simulations are given in Table 1 [8].

Table 1. Values for the Johnson-Cook material model parameter

A [MPa]	B [MPa]	n	C	m
162	598	0.6	0.07	0.009

A smoothed stress-stain curve is used to simulate the deformation process. Since deformation takes place within a very short time period (a few hundred microseconds), the loading curves

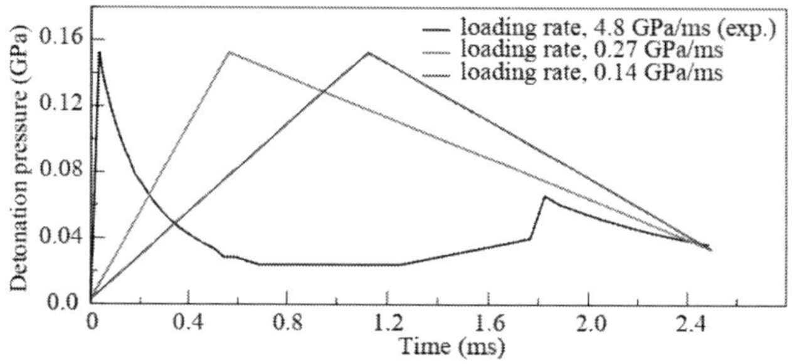

Figure 2. Load curves - different loading-rates for simulations.

are considered up to 2.5 milliseconds. One of the major factors influencing the shape as well as the distribution of strain in the formed cup is the rate at which the load is applied. Therefore, it becomes necessary to study the effects of this loading rate on the formed cup. The different loading curves as a function of time are shown in Figure 2.

3. Results and discussion

Figure 3 depicts the shape of the deformed blank with respect to loading time with a peak load of 1500 bar using the experimental loading rate. The analysis of blank shape with respect to the time highlights the fact that the whole deformation process takes place within the first peak-load and the reflected waves do not play a major role in the formation of the cup.

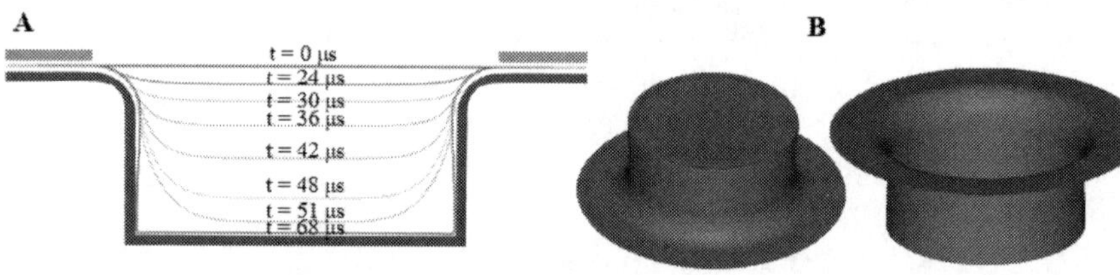

Figure 3. (A)Cup shape formation during simulation and (B) final deformed shape of the blank.

Figure 4A shows the variation of the thickness with respect to the initial radius for different loading rates. One of the main objectives of this work is to correctly predict the thickness distribution along the base and the wall of the deformed cup for different loading-rates. In the literature, thickness distribution has been studied using finite element models. However, experimental loading-rates were not considered, but simulation results were obtained using a triangular loading. Therefore, triangular loadings shows a smooth variation of the thickness and an overestimation of the magnitude (see Figure 4A). In case of experimental loading rate, thickness variation is not smooth and minimum thickness goes up to 35 % at the sharp bottom corners. The model is clearly able to predict a local maxima in the thickness value close to the 90° bend and a decrease in the thickness on side of the bend. It can be clearly observed that a decrease in the loading rate results in an increased wall thickness.
Figure 4B depicts the variation in the radial strain given by $\varepsilon_r = \frac{u_r}{r}$ as predicted by the Johnson-Cook material model for actual and triangular loading with respect to the initial radius. As expected, the radial strain is the highest in the areas where a maximum reduction in thickness is obtained. Like thickness variation results, triangular loadings show a smooth variation of the radial strain compared to the experimental loading rate.

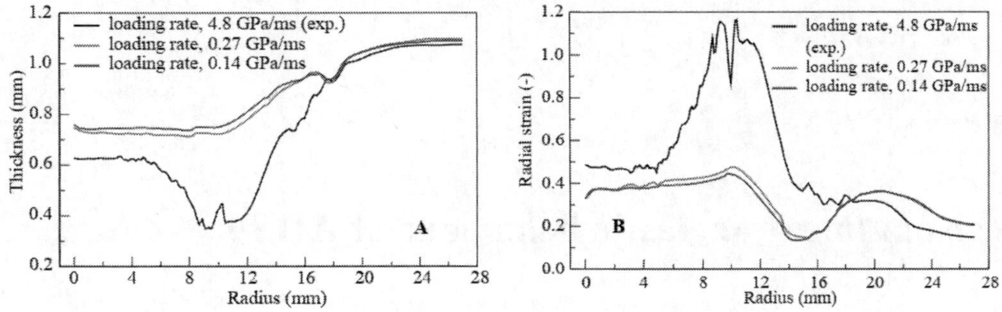

Figure 4. (A) Thickness variation in the deformed cup with respect to initial radius. (B) Radial strain in the deformed cup with respect to initial radius.

4. Conclusions and future work

Numerical simulations were carried out with the Johnson-Cook plasticity model, which can be fit over a wide range of strain, strain-rate and temperature, and hence it changes yield point. This plasticity model is the best choice to predict deformation during the forming process due to its moderate complexity and well established methods to predict the material constants. The proposed finite element model is able to predict the cup shape of the blank, including the sharp edges at the bottom corners. Moreover, the model concludes that obtained results are very sensitive to the loading-rates. Therefore, it highlights the fact that the accurate measurement of the loading rate and its consideration in the simulation process becomes very important.
Further improvements can be made to the model by systematically performing a higher number of experiments using different shaped geometries. Also, numerical studies can be performed on different shaped geometries including the temperature effects.

Acknowledgments

Authors wishing to acknowledge Olivier Herbert (SWL, RWTH Aachen University) for providing experimental pressure curves.

References

[1] Honda A and Suzuki M 1999 Sheet metal forming by using gas imploding detonation *J. Mater. Process. Technol.* **85** (1) 198–203
[2] Mynors D J and Zhang B 2002 Applications and capabilities of explosive forming *J. Mater. Process. Technol.* **125** 1–25
[3] Yasar M 2004 Gas detonation forming process and modeling for efficient spring-back prediction *J. Mater. Process. Technol.* **150** (3) 270–279
[4] Yasar M, Demirci H I and Kadi I 2006 Detonation forming of aluminium cylindrical cups experimental and theoretical modelling *Mater. Design.* **27** (5) 397–404
[5] Mokadem A E 2008 Finite element modeling of sheet metal forming using shock tube *Mechanical Design and Production* (Cairo University)
[6] Wijayathunga V and Webb D 2006 Experimental evaluation and finite element simulation of explosive forming of a square cup from a brass plate assisted by a lead plug *J. Mater. Process. Technol.* **172** (1) 139–145
[7] LS-DYNA, Ls-dyna theory manual, Livermore software Technology corporation
[8] Verleysen P, Peirs J, Van Slycken J, Faes K and Duchene L 2011 Effect of strain rate on the forming behaviour of sheet metals *J. Mater. Process. Technol.* **211** (8) 1457–1464
[9] Johnson G R and Cook W H 1983 A constitutive model and data for metals subjected to large strains, high strain rates and high temperatures *Proceedings of the 7th International Symposium on Ballistics* (The Hague, The Netherlands) **21** 541–547

Journal of Physics: Conference Series **734** (2016) 032100 doi:10.1088/1742-6596/734/3/032100

On the nonlinear anelastic behaviour of AHSS

A Torkabadi, V T Meinders and A H van den Boogaard

University of Twente, P.O. Box 217, 7500 AE Enschede, The Netherlands

a.torkabadi@utwente.nl

Abstract. It has been widely observed that below the yield stress the loading/unloading stress-strain curves of plastically deformed metals are in fact not linear but slightly curved, showing a hysteresis behaviour during unloading/reloading cycles. In addition to the purely elastic strain, extra dislocation based micro-mechanisms are contributing to the reversible strain of the material which results in the nonlinear unloading/reloading behaviour. This extra reversible strain is the so called anelastic strain. As a result, the springback will be larger than that predicted by FEM considering only the recovery of the elastic strain. In this work the physics behind the anelastic behaviour is discussed and experimental results for a dual phase steel are demonstrated. Based on the physics of the phenomenon a model for anelastic behaviour is presented that can fit the experimental results with a good accuracy.

1. Introduction

There has been an increasing interest by the automotive industry towards employing Advanced High Strength Steels (AHSS) in the past years. However the dimensional accuracy of the dual phase steel part remains an industrial issue [1]. In the past years most researches were focused on the development of novel plasticity models to give an accurate stress prediction. Nevertheless, little has been done in modeling the material behaviour during unloading upon release of the constraining force. The material behaviour upon unloading is of importance for the springback prediction considering that the experimental evidence shows that assuming a Hookean behaviour to describe the unloading is not realistic. It has been observed experimentally that the material shows a nonlinear unloading behaviour as well as reloading behaviour after being plastically deformed [2-4]. This is caused by an extra reversible strain recovered during unloading along with the pure elastic strain [5]. The root cause of this phenomenon is the short-range reversible movement of the dislocations known as anelasticity [6-8]. The dislocation structures which are impeded by the pinning points or piled up before the grain boundaries can move to a new equilibrium upon the relaxation of the lattice stress, contributing to some extra microscopic strain.

From an engineering perspective, considering that the total recovered strain during unloading governs the springback magnitude, it is essential to take into account the anelastic strain in the material models for the FEM springback simulations. In that respect, various researchers have adopted an approach attributed to E-modulus degradation [9-14]. In this approach the E-modulus of the material is made a function of the equivalent plastic strain in the simulations. Hence, the E-modulus represents the chord modulus which is measured from the experiments. The draw back with this approach is that it is assumed that all the points in the material are unloaded to the zero stress. This is not a realistic assumption in industrial forming processes. Therefore for a better springback prediction it is critical to model the nonlinear unloading behaviour. In this work by quantifying the anelastic strain, a model for describing the nonlinear unloading behaviour is proposed and the model prediction is compared with experimental data.

2. Constitutive modelling

The proposed model is based on the additive decomposition of the total strain increment into elastic $\dot{\boldsymbol{\varepsilon}}^{e}$, anelastic $\dot{\boldsymbol{\varepsilon}}^{an}$ and plastic $\dot{\boldsymbol{\varepsilon}}^{p}$ strain.

$$\dot{\boldsymbol{\varepsilon}} = \dot{\boldsymbol{\varepsilon}}^{e} + \dot{\boldsymbol{\varepsilon}}^{an} + \dot{\boldsymbol{\varepsilon}}^{p} \tag{1}$$

The rate of the elastic strain is given according to

$$\dot{\boldsymbol{\varepsilon}}^{e} = \overline{\boldsymbol{E}}^{-1}:\dot{\boldsymbol{\sigma}} \tag{2}$$

and the plastic strain rate is

$$\dot{\boldsymbol{\varepsilon}}^{p} = \dot{\lambda}\boldsymbol{N} \tag{3}$$

where $\boldsymbol{N}$ is the normal vector to the yield surface which depends on the derivative of the yield function at the end of the increment. The rate of the anelastic strain is given as

$$\dot{\boldsymbol{\varepsilon}}^{an} = \dot{\xi}\boldsymbol{N} \tag{4}$$

where ξ is the anelastic multiplier. It is assumed that the anelastic strain follows the normality condition and is co-directional with the plastic strain increment. This assumption originates from the physics of anelastic strain. The evolution of ξ is provided as [5]

$$\xi = \left(K\left(\sigma_f - \sigma_{y0}\right) + \varepsilon_{pre}^{an\ 0.5}\right)^2 \cdot g(s) \tag{5}$$

In the equation above, K is material related parameter obtained from experiment, σ_f is the flow stress of the material and can be described by a hardening law, σ_{y0} is the yield point of the material, ε_{pre}^{an} is the pre-yield anelastic strain and its value is determined from experiments. $g(s)$ is a function which describes the nonlinearity of the unloading/loading curves as a function of s. The variable s is a dimensionless stress dependent parameter which is defined as

$$s = \begin{cases} \sigma_{eq}/\sigma_f & if\ \varphi < 0\ and\ \boldsymbol{\sigma}:\dot{\boldsymbol{\sigma}} > 0 \\ 1 - \sigma_{eq}/\sigma_f & if\ \varphi < 0\ and\ \boldsymbol{\sigma}:\dot{\boldsymbol{\sigma}} < 0 \\ 1 & if\ \varphi \geq 0 \end{cases} \tag{6}$$

where σ_{eq} is the equivalent stress and φ is the yield function.
In order to describe the nonlinear behaviour of the anelastic strain, the $g(s)$ function is given as

$$g(s) = \sinh(\alpha.s^2)/\sinh(\alpha) \tag{7}$$

α is a constant whose value is determined from curve fitting to the experimental data.

There are numerous hardening models proposed in the literature relating the flow stress (σ_f) to the equivalent plastic strain (ε_{eq}^{p}). In this case, the Swift hardening law is used according to

$$\sigma_f = C\left(\varepsilon_0 + \varepsilon_{eq}^{p}\right)^n \tag{8}$$

From Equation (5), (6) and (8) the rate of the anelastic multiplier can be written as

$$\dot{\xi} = \frac{\partial\xi}{\partial\boldsymbol{\sigma}}:\dot{\boldsymbol{\sigma}} + \frac{\partial\xi}{\partial\lambda}\dot{\lambda} \tag{9}$$

Correspondingly, the rate of the anelastic strain is

$$\dot{\boldsymbol{\varepsilon}}^{an} = \left(\frac{\partial\xi}{\partial\boldsymbol{\sigma}}:\dot{\boldsymbol{\sigma}} + \frac{\partial\xi}{\partial\lambda}\dot{\lambda}\right)\boldsymbol{N} \tag{10}$$

Combining Equation (1), (2), (3) and (10) yields

$$\dot{\boldsymbol{\varepsilon}} = \overline{\boldsymbol{E}}^{-1}:\dot{\boldsymbol{\sigma}} + \dot{\xi}\boldsymbol{N} + \dot{\lambda}\boldsymbol{N} \tag{11}$$

The value of each strain is evaluated simultaneously through an implicit algorithm. The set of residual functions for every increment at the ith iteration is elaborated as

$$\boldsymbol{R}_{\sigma}^{i} = \Delta\boldsymbol{\varepsilon} - \overline{\boldsymbol{E}}^{-1}:\Delta\boldsymbol{\sigma} - \Delta\xi\boldsymbol{N} - \Delta\lambda\boldsymbol{N} \tag{12}$$

$$R_\lambda^i = -\varphi \tag{13}$$

The linearization yields

$$\begin{bmatrix} \dfrac{\partial \boldsymbol{R}_\sigma^i}{\partial \boldsymbol{\sigma}} & \dfrac{\partial \boldsymbol{R}_\sigma^i}{\partial \lambda} \\ \dfrac{\partial R_\lambda^i}{\partial \boldsymbol{\sigma}} & \dfrac{\partial R_\lambda^i}{\partial \lambda} \end{bmatrix} \begin{bmatrix} \delta\boldsymbol{\sigma} \\ \delta\lambda \end{bmatrix} = \begin{bmatrix} \boldsymbol{R}_\sigma^i \\ R_\lambda^i \end{bmatrix} \tag{14}$$

At the end of every iteration, the stress and plastic multiplier is updated by $\delta\boldsymbol{\sigma}$ and $\delta\lambda$ until the norm of the residual vector is smaller than the accepted tolerance.

3. Experimental

In order to determine the parameters for the anelastic model, cyclic uniaxial loading / unloading / reloading experiments were conducted on a DP800 steel with a thickness of 1. In such experiments the material was loaded to certain force and then unloaded to zero force. In the subsequent cycles the loading force was increased incrementally. The schematic illustration of the experimental procedure is shown in Figure 1.

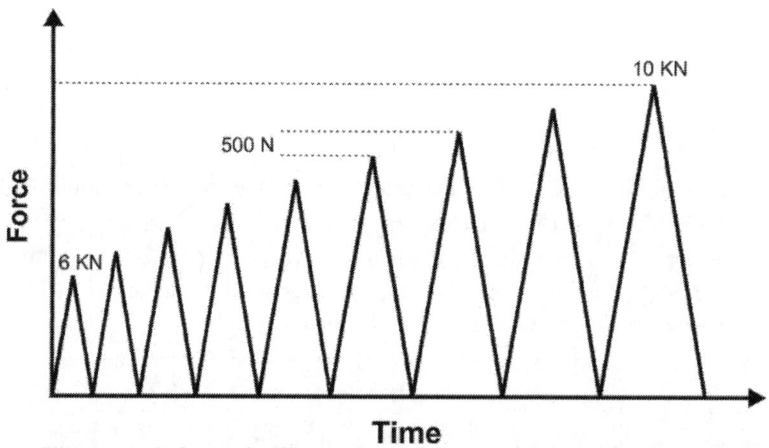

Figure 1. Schematic illustration of the experimental procedure.

From such experiments the material parameters for Swift hardening law and the anelastic model were obtained. Table 1 summarizes the material parameters obtained from experimental data to calibrate the Swift hardening and anelastic model.

Table 1. Material parameters.

$\sigma_f = C\left(\varepsilon_0 + \varepsilon_{eq}^p\right)^n$			$\xi = \left(K\left(\sigma_f - \sigma_{y0}\right) + \varepsilon_{pre}^{an\ 0.5}\right)^2 (\sinh(\alpha.s^2)/\sinh(\alpha))$			
$C\ (MPa)$	ε_0	n	$K(MPa^{-1})$	$\sigma_{y0}\ (MPa)$	ε_{pre}^{an}	α
1710	0.0072	0.28	5.1 x 10^{-5}	430	3 x 10^{-4}	0.68

4. Results and discussion

In Figure 1(a) the stress-strain curves obtained from the cyclic loading / unloading / reloading experiment and the curve predicted by the Swift + anelastic models are compared. For a better comparison, the zoomed in view of the 7th unloading/reloading cycle is shown in Figure 1(b). Besides, the unloading curves predicted by the E-modulus ($E = 210\ GPa$) and the chord modulus ($E_c = 164\ GPa$) are shown in Figure 1(b).

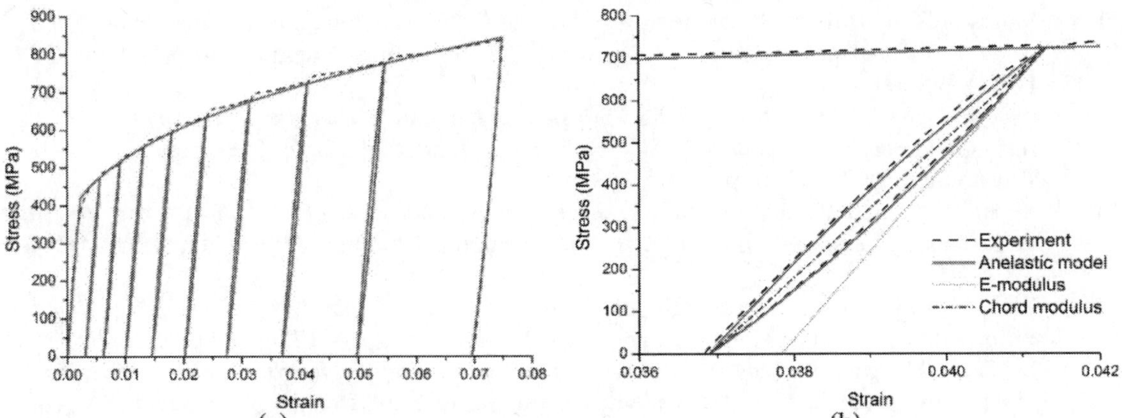

Figure 2. Stress-strain response of DP800 (a) model prediction (red) and experimental data (black); (b) the magnified view of the 7th unloading/reloading cycle.

In Figure 1(b) it can be seen that predicting the unloading strain using the E-modulus can result in a large deviation from the material behaviour. The unloading behaviour approximated by the chord modulus is closer to reality; however, it is accurate only if it is assumed that all the parts in the material are unloaded to zero stress. Yet, the unloading curve predicted by the proposed anelastic model provides a good prediction through the whole unloading curve capturing the nonlinear behaviour.

5. Conclusion

In this study a model was proposed to capture the nonlinear unloading / reloading behaviour in AHSS. The model was calibrated to the experimental data obtained from DP800 steel using cyclic unloading / reloading experiments. To that end, four parameters were fitted to the experimental data. A comparison between the experimental results and the model shows a good correlation for the whole unloading / reloading path. Hence, using the proposed model in FEM simulations of forming processes will result in better prediction of the springback behaviour of complex parts.

References

[1] Wagoner R H, Lim H, and Lee M-G, *Advanced Issues in springback.* International Journal of Plasticity, 2013. **45**(0): p. 3-20.

[2] Cleveland R M and Ghosh A K, *Inelastic effects on springback in metals.* International Journal of Plasticity, 2002. **18**(5–6): p. 769-785.

[3] Eggertsen P A, Mattiasson K, and Hertzman J, *A Phenomenological Model for the Hysteresis Behavior of Metal Sheets Subjected to Unloading/Reloading Cycles.* Journal of Manufacturing Science and Engineering, 2011. **133**(6): p. 061021-061021.

[4] Sun L and Wagoner R H, *Complex unloading behavior: Nature of the deformation and its consistent constitutive representation.* International Journal of Plasticity, 2011. **27**(7): p. 1126-1144.

[5] Torkabadi A, van Liempt P, Meinders V T, and van den Boogaard A H. *A constitutive model for the anelastic behavior of Advanced High Strength Steels.* in *COMPLAS XIII*. CIMNE.

[6] Zener C M and Siegel S, *Elasticity and Anelasticity of Metals.* The Journal of Physical and Colloid Chemistry, 1949. **53**(9): p. 1468-1468.

[7] Ghosh A K, *A physically-based constitutive model for metal deformation.* Acta Metallurgica, 1980. **28**(11): p. 1443-1465.

[8] van Liempt P and Sietsma J, *A physically based yield criterion I. Determination of the yield stress based on analysis of pre-yield dislocation behaviour.* Materials Science and Engineering: A, 2016. **662**: p. 80-87.

[9] Morestin F and Boivin M, *On the necessity of taking into account the variation in the Young modulus with plastic strain in elastic-plastic software.* Nuclear Engineering and Design, 1996. **162**(1): p. 107-116.

[10] Li X, Yang Y, Wang Y, Bao J, and Li S, *Effect of the material-hardening mode on the springback simulation accuracy of V-free bending.* Journal of Materials Processing Technology, 2002. **123**(2): p. 209-211.

[11] Yoshida F, Uemori T, and Fujiwara K, *Elastic–plastic behavior of steel sheets under in-plane cyclic tension–compression at large strain.* International Journal of Plasticity, 2002. **18**(5–6): p. 633-659.

[12] Yang M, Akiyama Y, and Sasaki T, *Evaluation of change in material properties due to plastic deformation.* Journal of Materials Processing Technology, 2004. **151**(1–3): p. 232-236.

[13] Fei D and Hodgson P, *Experimental and numerical studies of springback in air v-bending process for cold rolled TRIP steels.* Nuclear Engineering and Design, 2006. **236**(18): p. 1847-1851.

[14] Yu H Y, *Variation of elastic modulus during plastic deformation and its influence on springback.* Materials & Design, 2009. **30**(3): p. 846-850.

Joining of tubes by gas detonation forming

Vahid Jenkouk, Sandeep Patil, Bernd Markert

Institute of General Mechanics, RWTH Aachen University, Templergraben 64, 52062 Aachen, Germany

E-mail: vahid.jenkouk@ rwth-aachen.de

Abstract. For many applications, such as in structural components, it is required to join two tubes – sometimes with dissimilar material properties. Only few research studies have investigated the joining of tubular metallic components by means of high-velocity forming processes. In this paper, we present the novel process of joining of two tubes by a gas detonation pressure wave. In particular, the joining of a copper and a steel tube is discussed by means of a finite element study and a conducted experiment.

Keywords: high velocity forming, gas detonation forming, joining, metal forming

1. Introduction

Tubular parts find wide application as structural components due to their high stiffness to weight ratios. Often it is required to join two tubular components. Besides the predominantly used technique of welding, there exist a series of industrially applied forming technologies that are capable of joining two metallic parts such as clinching, riveting, hemming, rolling, spinning and hydroforming [1]. Some of the advantages of the use of forming technologies for joining are: Joining and shaping operations could be carried out simultaneously and therefore, the process chain could be significantly shortened; Negative effects of the local heat concentration on the microstructure, which may occur in welded joints, can be eliminated. Some material pairings with dissimilar properties that cannot be welded due to their difference in melting temperatures could be joined by forming [2]. Joining of dissimilar materials – e.g. soft versus high strength materials such as aluminium or copper joined with steel alloys – is of great interest for structural components or car frames as two conflicting properties such as structural stiffness and capability of energy absorption (relevant for the crash zone) could be integrated in one component.

In general, the joining by plastic deformation is performed by form-closed, force-closed and/ or metallurgical joints [3]. Several research works have investigated tube and cam joining operations that are based on the force-closed joint principle [4]. There exist several manufacturing techniques for the formation of a force-closed joint between two tubes. These joints can be created by rolling of the inner or outer tube. In this process, one tube is plastically deformed by a roller and pressed against the other. If the clearance between the tubes is properly selected, one tube will be solely elastically deformed as a consequence of the plastic deformation of the other joint partner [5]. After the removal of the forming stresses, the elastically deformed tube will impart pressure on the other tube and due to the friction the tubes will resist to motion as long as the forces, to which they are subjected, are smaller than the frictional forces. Moreover, tubes can also be joined by hydroforming [2]. The major advantages of hydroforming are the reduced processing times, the reduced amount of tooling due to the absence of tools on one side and the possibility to deform cavities in hollow shaped components. Another option for joining metallic components is electromagnetic forming (EMF). EMF is a high velocity process, which forms metals with high electrical conductivity such as aluminium by a pulsed magnetic field [6]. The advantages of the high velocity forming processes compared to hydroforming and other conventional forming technologies are their reduced processing time – milliseconds in EMF and seconds in hydroforming – and the increased formability for several materials due to the extremely high strain rates [7]. A further high velocity forming process is the gas detonation forming. Kleiner et al. have used the pressure shock wave produced by the gas charge of oxygen and hydrogen to manufacture hollow shaped tubes [8]. Experimentally they have shown that the aluminium alloys AlMgSi1 and Al99.5 exhibit higher formability when deformed by gas detonation compared to conventional forming processes.

In this work, we discuss the use of gas detonation for the joining process of two tubes. Similar to EMF, extremely high strain rates can be achieved by this forming technique. Since it is aimed to create a joint of high strength, both form-closed and force-closed joints are applied. A numerical model by means of the finite element method is established to investigate the deformation process. In a numerical study, three material pairings of a material with relatively high strength (DC04 steel alloy) and a soft material (copper) are modified and tested in virtual environment. Moreover, a physical experiment of the tube joining is presented.

2. Process principle and numerical modelling

Process principle: As illustrated in Figure 1A, similar to hydroforming joining processes, an inner tube with a thin wall thickness is inserted with a small clearance-fit into an outer tube with a larger wall thickness in order to ensure that a force-closed joint is achieved after removal of the internal pressure. The inner surface of the outer tube is grooved with the purpose of creating a form-closed joint by means of the interlocked material. For maintaining a certain geometric accuracy of the outer diameter of the outer tube, a rigid die is placed around the outer tube. The inner tube has to be subjected to the pressure wave caused by gas detonation. The principle of the gas detonation is thoroughly described in [8].

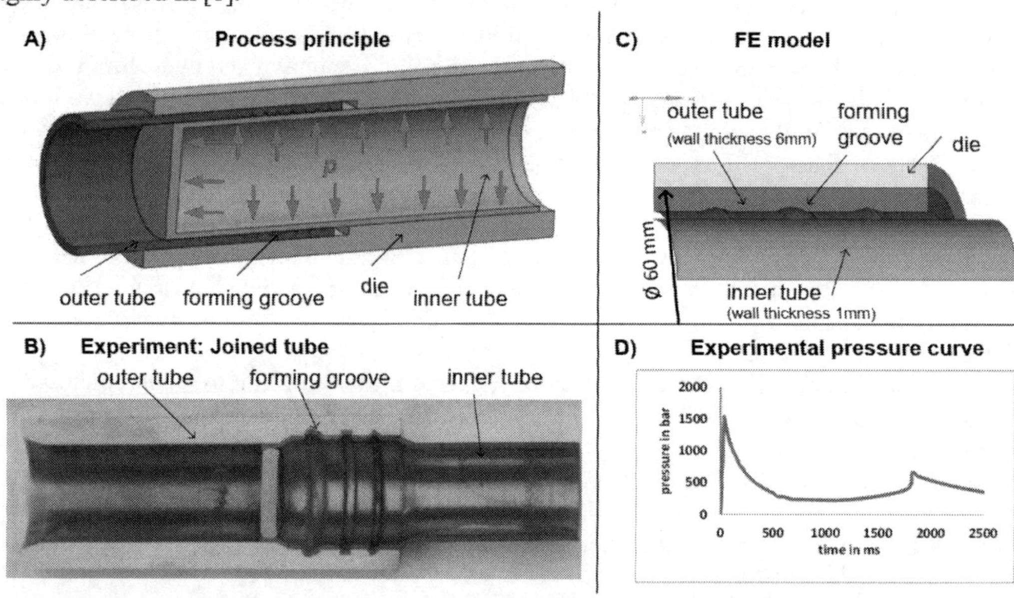

Figure 1: A) Principle of tube joining illustrated in a cross-sectional cut view. B) Cross-sectional cut of joined tubes (courtesy of SWL). C) Geometry of the finite element model. D) Plot of gas detonation pressure vs process time (courtesy of SWL).

Experimental setup: Using the described process principle, experiments with different pressure values and various groove geometries have been carried out at the Shock Wave Laboratory (SLW) of the RWTH Aachen University led by Prof Olivier. In Figure 1B a photograph of two joined tubes is shown. The inner tube is made of copper and the outer tube of steel alloy. The design of the tools is very challenging since it has to be ensured that inside the tools a technical vacuum is sustained; otherwise the metals could melt due to the extreme compression of the air. As it can be seen, the inner tube is nicely formed into the grooves of the outer tube. However, for determining the parameters required for such behaviour, a large number of trial and error experiments were required. In order to determine the appropriate parameters, the tube deformation can be modelled by means of the finite element method. For instance, appropriate pressure values need to be selected, as small pressure values may lead to partially filled grooves and high pressure values to fracture of the inner tube and excessive deformation of the outer tube. Moreover, by means of simulations the wall thickness of the tube can be predicted which strongly influences strength, durability and reliability of the components.

Finite element model: For the finite element simulation of the tube joining a quarter of the model is considered to reduce the computational cost. The outer die is modelled as rigid because of its high stiffness compared to those of the tubes. Both tubes are modelled as deformable and are assumed to be homogenous. The inner tube is modelled by 53760 Belytschko-Tsay shell elements. The outer tube is modelled by 407790 8-node linear brick elements. The outer die is fully constrained. The inner tube is subjected to pressure. The time curve of the pressure is taken from experimental data (see Figure 1D). The contacts between all components are defined by surface to surface segment based contact formulation. Because of the highly dynamic character of the process, the explicit time integration scheme of the LS-DYNA software package is used. The Johnson-Cook phenomenological material model as shown in Eq. 1 is employed in order to take into account the strain rate dependency of the used materials. The flow stress is computed as follows:

$$\sigma_y = \left(A + B\bar{\varepsilon}^{p^n}\right)\left(1 + Cln\frac{\dot{\bar{\varepsilon}}^{p^n}}{\dot{\varepsilon}_0}\right)\left(1 - \left[\frac{T - T_{room}}{T_{melt} - T_{room}}\right]^m\right) \qquad \text{Eq.1}$$

where $\bar{\varepsilon}^p$ represents the effective plastic strain, T_{room} the ambient temperature, T_{melt} the melting point, T the effective temperature, A the yield stress, B the hardening modulus, n the strain exponent, m the temperature exponent, C the strain rate factor and $\dot{\varepsilon}_0$ the strain-rate for the quasi static reference loading. The parameters and the mechanical properties of DC04 steel are taken from [7]; for copper the vales form [9] have been used. The following simulations were carried: Sim1: outer ring made of DC04 material, inner tube made of copper material; Sim2: both tubes made of DC04 material; Sim3: outer tube made of copper material and inner tube made of DC04 material.

3. Results and discussion

In Figure 2, the von Mises strain distributions of the Sim3 are depicted. It can be observed that the inner tube deforms into the grooves of the outer tube.

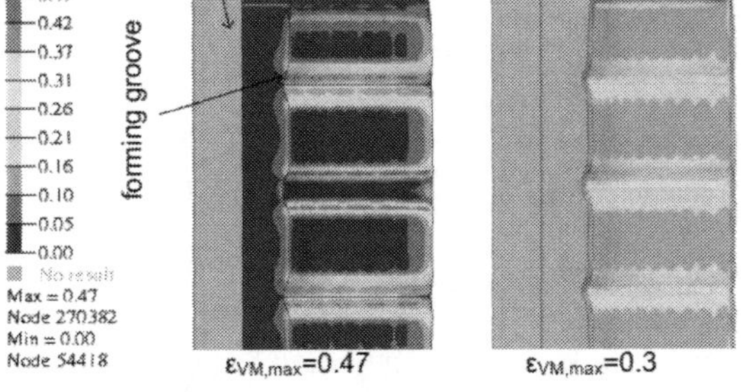

Figure 2: Exemplary simulation results of the von Mises strain distribution of Sim3 at the process time 0.5ms

	Sim1	Sim2	Sim3
Max thickness reduction tube 1 (%)	24.8	23.3	21.8
Max. strain tube 1	0.43	0.4	0.3
Max. strain tube 2	0.33	0.31	0.47
Max deformation tube 2 (mm)	0.36	0.39	0.63

Table 1: Simulation results

As expected in this case the softer outer tubes plastically deflects to a considerable amount. The maximum von Mises strain of outer tube in this case 0.47, whereas for the pairing in Sim2 (outer steel/ inner copper) the maximum value is 0.3. As expected it is not reasonable to employ a softer outer tube since its grooves even out and hence the required interlock effect is reduced.

4. Conclusion and future work

In this paper, the process of joining of two tubes by gas detonation is introduced. The feasibility of the process is shown by means of an experiment and a simulation study. In this process two joint types exist: Closed-form joint: the plastic deformation of the material of the inner tube into the grooves of the outer tube material leads to mechanical interlocking. Closed-force joint: the difference in the elastic recovery behaviour of the materials after unloading leads to a remaining pressure distribution between the tubes. As a consequence, frictional forces prevent the relative motion of the tubes.

With a FEM study, we have investigated the joining of the dissimilar metals copper and steel alloy (DC04). The study suggests that the order of the metals influences the joint strength. As expected, to place the softer material as the outer tube is not favourable since its grooves deform and flatten, and therefore the interlock effect is decreased.

The conducted experiment and our simulation study have a rather qualitative character. Therefore, there is a need for quantitative experimental and numerical studies; especially focussing on the joint strength. In particular, this requires numerical analysis of residual stresses in order to determine the joint strength resulting from force fit. Moreover, both experiments and simulations of pull-out and torsion test are suited to assess this joining technology. For determining the suited process parameters (pressure, geometric dimension of tubes and grooves, material pairing) it is also required to implement failure criteria that are capable of predicting fracture.

5. Acknowledgement

The authors would like to thank Prof Olivier, the head of the Shock Wave Laboratory (SWL) of the RWTH Aachen University, for providing the experimental data of the detonation joining.

References

[1] P. Groche, S. Wohletz, M. Brenneis, C. Pabst, and F. Resch, “Joining by forming- A review on joint mechanisms, applications and future trends,” *J. Mater. Process. Technol.*, vol. 214, no. 10, pp. 1972–1994, 2014.

[2] P. Groche and K. Tibari, “Fundamentals of Angular Joining by Means of Hydroforming,” *CIRP Ann. - Manuf. Technol.*, vol. 55, no. 1, pp. 259–262, Jan. 2006.

[3] P. Groche and M. Türk, “Smart structures assembly through incremental forming,” *CIRP Ann. - Manuf. Technol.*, vol. 60, no. 1, pp. 21–24, 2011.

[4] F. Vollertsen, “Accuracy in process chains using hydroforming,” *J. Mater. Process. Technol.*, vol. 103, no. 3, pp. 424–433, 2000.

[5] M. Hagedorn and K. Weinert, “Manufacturing of composite workpieces with rolling tools,” *J. Mater. Process. Technol.*, vol. 153, pp. 323–329, 2004.

[6] C. Weddeling, S. T. Woodward, M. Marré, J. Nellesen, V. Psyk, A. E. Tekkaya, and W. Tillmann, “Influence of groove characteristics on strength of form-fit joints,” *J. Mater. Process. Technol.*, vol. 211, no. 5, pp. 925–935, 2011.

[7] P. Verleysen, J. Peirs, J. Van Slycken, K. Faes, and L. Duchene, “Effect of strain rate on the forming behaviour of sheet metals,” *J. Mater. Process. Technol.*, vol. 211, no. 8, pp. 1457–1464, 2011.

[8] M. Kleiner, M. Hermes, M. Weber, H. Olivier, G. Gershteyn, F.-W. Bach, and A. Brosius, “Tube expansion by gas detonation,” *Prod. Eng.*, vol. 1, no. 1, pp. 9–17, Jun. 2007.

[9] G. Johnson and W. Cook, “A constitutive model and data for metals subjected to large strains, high strain rates and high temperatures,” in *Proceedings of the Seventh International Symposium on Ballistics, The Hague*, 1983, pp. 541–547.

Experimental validation of numerical model for asymmetric deep drawing of DP780 steel sheet using digital image correlation

X Xue, J Liao, G Vincze and A B Pereira

Centre for Mechanical Technology and Automation, Department of Mechanical Engineering, University of Aveiro, 3810-193, Aveiro, Portugal

Email: xin@ua.pt

Abstract. A validated numerical model for in-plane stress/strain prediction is essential in understanding the deformation behaviour of sheet metal forming process, in particular, asymmetric deep drawing of advanced high strength steel sheet. In this work, the Yld2000-2d anisotropic yield criterion integrated with the initial homogeneous anisotropic hardening model was employed to describe the complex material behaviours of DP780 steel as well as the adoption of Yoshida chord model for elastic modulus degradation. Digital image correlation technique was utilized to measure the in-plane strain and shape deviation of the developed P-channel. The validity of the numerical model was assessed by comparing the predicted strain distribution and twist springback with the measured results. It indicates that the developed numerical model based on the selected constitutive models is acceptable for the deformation analysis, although the predicted discrepancies still exist.

1. Introduction

Advanced high strength steels (AHSS) have become a favourite resource for the weight reduction of automotive industry due to their promising higher strength without compromising ductility. Compared to conventional mild steels, AHSS are prone to strain localization and a subsequent rapid fracture owing to a lower ductility. The challenges for the accurate prediction of strain and elastic-plastic deformation in sheet metal forming process should be highlighted as well as the experimental validation of the developed numerical model.

Nowadays, the use of digital image correlation (DIC) techniques is an efficient tool to collect the experimental forming data and enable the validation of simulation results by direct comparison of the global deformation. In practice, even though the material flow or the overall deformation is usually predicted quite acceptably, the significant discrepancy between the computed and measured strain distributions still exist.

In this work, a new asymmetric experimental benchmark (P-channel) is developed as well as its numerical model in case of deep drawing of DP780 steel sheet. The material constitutive models including the Yoshida chord model [1], the Yld2000-2d anisotropic yield criterion [2] and the homogeneous anisotropic hardening model [3] are adopted to describe the complex elastic-plastic behaviours such as elastic modulus degradation [4], material anisotropy, Bauschinger effect, transient hardening and permanent softening [5]. Special focus will be laid on the analyses of in-plane strain and twist springback for the validation of numerical model using digital image correlation technique.

2. Material characterization and parameter identifications

In this work, DP780 steel sheet with a thickness of 0.8 mm in the as-received state was characterized by using standard uniaxial tension tests along different directions and simple shear tests. The elastic-plastic constitutive models are defined by a yield criterion and a hardening law. The determination of elastic modulus degradation was carried out through uniaxial loading-unloading-loading (ULUL) cycle tension tests and Yoshida-Uemori model, which can be expressed by

$$E_{chord} = E_0 - (E_0 - E_a)[1 - \exp(-\zeta \bar{\varepsilon}_p)] \tag{1}$$

where E_a represents the chord modulus obtained for an infinitely large plastic pre-strain $\bar{\varepsilon}_p$ and ζ is a material parameter that determines the decrease rate of the unloading elastic modulus E_{chord}.

The anisotropic yield function Yld2000-2d was proposed by Barlat et al. [2], which is given by

$$\phi = \left|X_1^{'} - X_2^{'}\right|^a + \left|2X_2^{"} + X_1^{"}\right|^a + \left|2X_1^{"} + X_2^{"}\right|^a = 2\bar{\sigma}^a \tag{2}$$

The following linear transformation

$$\begin{aligned} \mathbf{X}' &= \mathbf{C}' \cdot \mathbf{s} = \mathbf{C}' \cdot \mathbf{T} \cdot \boldsymbol{\sigma} = \mathbf{L}' \cdot \boldsymbol{\sigma} \\ \mathbf{X}'' &= \mathbf{C}'' \cdot \mathbf{s} = \mathbf{C}'' \cdot \mathbf{T} \cdot \boldsymbol{\sigma} = \mathbf{L}'' \cdot \boldsymbol{\sigma} \end{aligned} \tag{3}$$

written in tensor notation with the stress deviator $\mathbf{s}$ and the stress tensor $\boldsymbol{\sigma}$ allow the introduction of plastic anisotropy. $\mathbf{L}'$ and $\mathbf{L}''$ provided a total of eight anisotropy coefficients α_k.

The homogeneous anisotropic hardening (HAH) model was proposed to describe the complex mechanical behaviour under stain path changes. It can be briefly presented in the following

$$\Phi(\mathbf{s}) = \left(\phi_s^q + \phi_f^q\right)^{\frac{1}{q}} = \left[\phi_s^q + f_1^q \left|\hat{\mathbf{h}}^s : \mathbf{s} + \left|\hat{\mathbf{h}}^s : \mathbf{s}\right|\right|^q + f_2^q \left|\hat{\mathbf{h}}^s : \mathbf{s} - \left|\hat{\mathbf{h}}^s : \mathbf{s}\right|\right|^q\right]^{\frac{1}{q}} = \sigma(\bar{\varepsilon}) \tag{4}$$

For the detailed descriptions of this constitutive model, the interested reader can be referred to [3]. Figure 1a shows that the decrease of the elastic modulus for DP780 is 24.0% of its initial value at a 0.09 pre-strain. The comparison of the experimental strain-stress curves and calculated results based on the adopted constitutive models, as shown in Figure 1b, indicates that the hardening model captures well the Bauschinger effect, transient hardening and permanent softening. Table 1 lists all the coefficients of material constitutive models used for simulation.

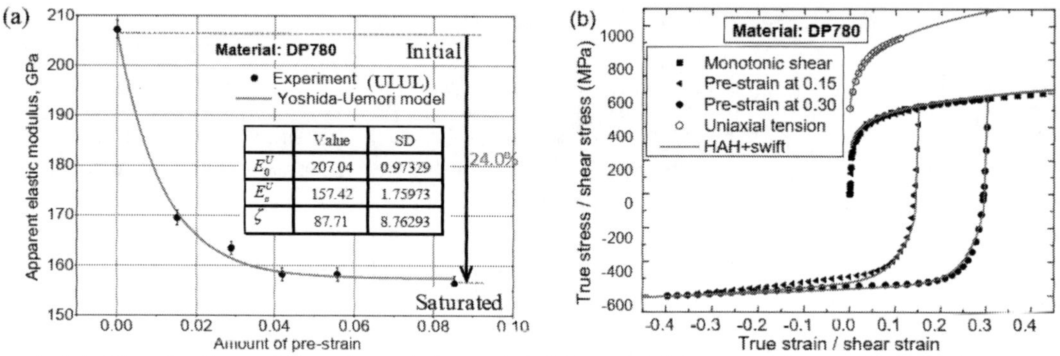

Figure 1. (a) The elastic modulus degradation of DP780 steel (b) comparison of uniaxial and shear stress-strain relations between experiment and calculation results.

Table 1. Material constitutive coefficients of DP780 steel used for simulation.

Yoshida chord model	E_0	E_a	ζ						
	207.04	157.42	87.71						
Yld2000-2d	a	α_1	α_2	α_3	α_4	α_5	α_6	α_7	α_8
	6	0.973	0.879	0.864	0.948	0.999	0.860	0.949	1.030
Initial HAH	q	k	k_1	k_2	k_3	k_4	k_5		
	2	45	50	26.2	0.3	0.9	6.5		
Reference isotropic hardening (Swift law)	ε_0	n	K	$\bar{\sigma} = K(\varepsilon_0 + \bar{\varepsilon}^p)^n$ with $\varepsilon_0 = \left(\frac{\sigma_0}{K}\right)^{1/n}$.					
	0.0015	0.151	1298.2						

3. Experimental methodologies

An asymmetric deep drawing benchmark with a variable cross-sectional shape was developed, namely P-channel. It is designed as a U-shaped channel with varied width sweeping along the longitudinal direction, as shown in Figure 2. The section widths are different and kept constant in the both ends but varied through a smooth transition in the middle. Forming this type of channel normally result in considerable deformation behaviours (e.g., twist springback) due to its asymmetric geometry and residual stress distribution. The forming equipment is a SHIMADZU universal tension-compression machine with a maximum capacity 1000 kN. The forming parameters are listed in Table 2. Inverse-install moving die-set was employed to reduce the vibration of the blank holder.

Table 2 Main process parameters of deep drawing.

No.	Parameters	Values
1	Total stroke of deep drawing, mm	40
2	Drawing velocity, mm/min	10
3	Amount of lubricant, g/m^2/side	10-12
4	The quantity of nitrogen gas spring	6
5	Initial/maximum blank holder force, kN	90/144
6	Final press contact pressure, MPa	10
7	Die clearance ratio	1.125

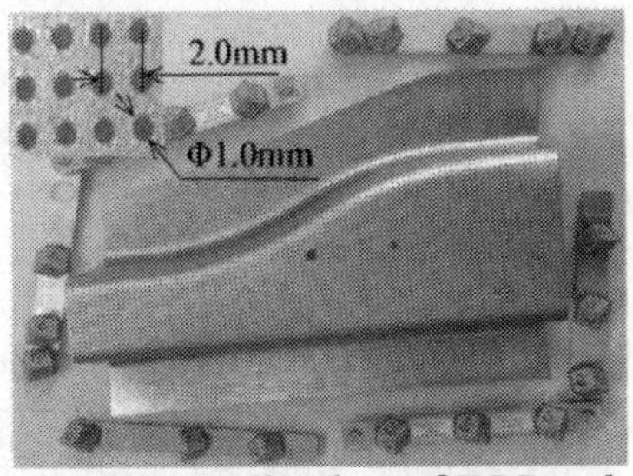

Figure 2. The application of DIC technique

Digital image correlation (DIC) technique was used to assess the deformation of the drawn parts, such as the distribution of in-plane strain, thickness reduction, and geometric shape deviation [6]. Standard grid distance of 2.0 mm and grid diameter of 1.0 mm were marked on the initial blank surface. After the forming, several images were recorded with the ARGUS 5M system, in order to acquire the points positioning from different angles. The 3D reconstruction of the final measuring object was then obtained/computed using the photogrammetric methods.

4. Result and discussion

4.1. In-plane strain

This work aims to validate the selected constitutive models on the prediction of strain and springback for the presented test case. For the selected centre cross section, the discrepancies between the DIC measured stains and the predicted results are shown in Figure 3. It can be seen that the predicted in-plane strain is underestimation at the transition area of the curved side wall compared to the DIC experimental results.

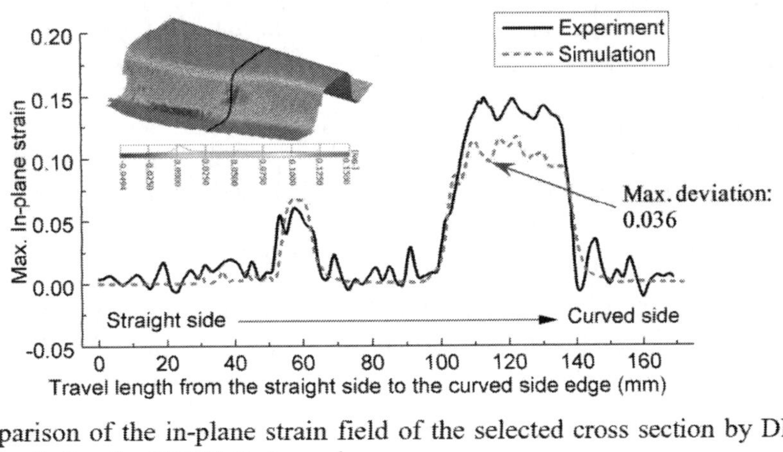

Figure 3. Comparison of the in-plane strain field of the selected cross section by DIC measurement and numerical prediction for DP780 P-channel.

Even though the simulated results for material flow are in reasonable agreement with the experiments, some significant differences exist between the computed and measured strain fields. This might be due to the point pattern of the measured object and the lack of fitting accuracy of the deformed circles because of the severe strain variations. This error deviation of in-plane strain can be up to 0.02. It also might be due to the deviation of the material description implemented into the simulation code [7]. These discrepancies should be reduced or eliminated with the development of advance deformation measurement technology and more comprehensive material constitutive models.

4.2. Preliminary result of twist springback

Figure 4 shows the displacement springback distribution along the Z-direction or 3-direction (U3) and comparison of the measured twist springback and the predicted results. It is observed that the displacement of the flange along the straight side has a significant variation while that along the curved side is almost stable, indicating the occurrence of the twisting. The relative twist angle defined as the rotation angle of the principal axis of inertia for the cross-section before and after springback [6] decreases fast initially in the narrow sections, then it starts slightly fluctuating in the sections located in the transition region, where the width of the cross section varies. After the transition region, it starts decrease again in the wide sections. The evolution of the predicted relative twist springback agrees well with the experiment. Although the predicted discrepancies still exist, the validity of the developed numerical model is acceptable for the deformation analysis.

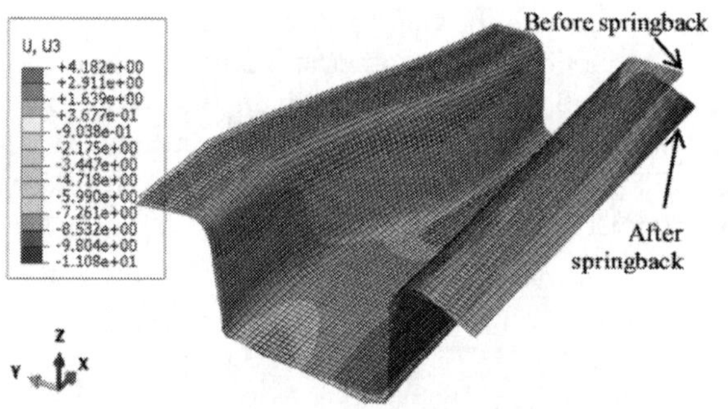

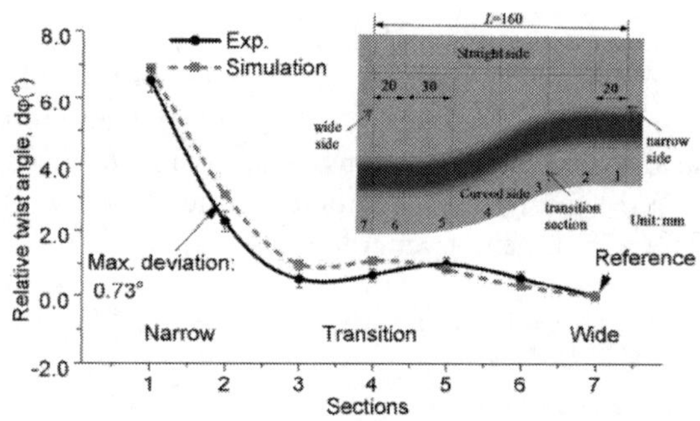

Figure 4. (a) The displacement springback distribution along the Z-direction (b) comparison of the measured relative twist angle of the DP780 P-channel and simulation results.

5. References

[1] Yoshida F, Uemori T and Fujiwara K 2002 *Int. J. Plast.* **18** 633
[2] Barlat F, Brem J C, Yoon J W and Chung K 2003 *Int. J. Plast.* **19** 1297
[3] Barlat F, Gracio J J, Lee M G, Rauch E F and Vincze G 2011 *Int. J. Plast.* **27** 1309
[4] Cleveland R and Ghosh A 2002 *Int. J. Plast.* **18** 769
[5] Yoshida F and Uemori T 2003 *Int. J. Mech. Sci.* **45**(10) 1687
[6] Xue X, Liao J, Vincze G, Sousa J A, Barlat F and Gracio J 2016 *Mater. Des.* **90** 204
[7] Xue X, Liao J, Vincze G and Barlat F 2015 *Int. J. Mater. Form.* (doi: 10.1007/s12289-015-1275-2)

Acknowledgments

The support from the Portuguese Foundation of Science and Technology (UID/EMS/00481/2013 and PTDC/EMS-TEC/0777/2012) is greatly appreciated. The authors are grateful to Prof. Myoung-Gyu Lee of Korea University for providing user-material subroutine code of the HAH model.

Industrial application and validation of forming simulation in the flexforming process

Alan Leacock[1], Dave Ling[2] and Mikael Bergkvist[3]

[1] AMFOR Group, Ulster University, BT370QB, UK
[2] AutoForm Engineering B.V. Industrieweg 2, 9th Floor NL-2921 LB Krimpen a/d IJssel
[3] Quintus Technologies, Quintusvägen 2, 721 66 Västerås, SWEDEN

ag.leacock@ulster.ac.uk

Abstract. The flex-forming process is used extensively in aerospace industry for net shape forming of sheet metal structural components. Common metals used in the manufacture of these components include 7075 and 2024 aluminium alloys; usually in an as quenched condition following solution heat treatment. While the process is commonplace, the level of manual rework remains high, driven by inherent process and material variability and the lack of upfront analysis before the manufacture of tooling. A suitable process modelling method using AutoForm is presented along with an industrial validation study for the manufacture of an aerospace frame component in 7075-W aluminium alloy. The results illustrate the importance of material model accuracy and the inclusion of through thickness compressive stresses in predicting the flange springback of the component.

1. Introduction

Flexforming is a process whereby a rubber membrane, which is pressurized with oil, is used to form a blank onto either a male or female tool, which unlike rubber pad forming applies a hydrostatic pressure. Flexforming is used extensively in the manufacture of complex low volume sheet metal components and is commonplace in aerospace applications. Aluminium aerospace components have the added complexity of requiring additional pre and post heat treatments to provide sufficient formability to form joggles/small radii and subsequently provide in-service characteristics [1].

Asnafi [2] conducted one of the most comprehensive experimental, analytical and numerical studies of the flexforming process. The author examined 2024-O and 6061-O aerospace aluminium alloys looking at both the tension and compression flange behavior, though springback was not considered within the study. Chen [3] also used CAE to model 2024-O and 7075-O straight flanged and joggled aerospace frame components. While pressure and time were shown to have little impact on the process performance, the low pressure of 400 bar and relatively stable, but weak aluminium alloys limit the use of this work for solution heat treated components. The study also illustrated the importance of additional harder 'intensifier' polyurethane pads in forming details at lower pressures of 400 bar. Chen [4] then went on to extend the work of Asnafi [2] using CAE where the primary focus of the work was to examine wrinkling in 2024-O/T3 and 7075-O with no quantifiable analysis of springback in these parts.

While these studies have provided a core fundamental source for analyzing and optimizing the flexforming process, the vast majority of aerospace components are formed in an unstable as quenched condition (W) following solution heat treatment. This paper presents a methodology for modelling the flexforming process for a 7075-W aerospace frame component without the computational cost and complexity of including rubber.

2. Material Data and Modelling Approach

The material used in this study was a 7075-O solution heat treated and naturally aged for 30 minutes before forming. A more detailed study of the post solution heat treatment behaviour of the material can be found elsewhere [5].

The material yield behaviour was modelled using the BBC2005 yield criterion and Hockett-Sherby isotropic hardening model as implemented in AutoForm. The parameters used in these models are given in Table 1 and 2 respectively

Table 1. Parameters for the BBC2005 yield criterion

Parameter	a	b	L	M	N	P	Q	R
Value	6.823	0.224	0.326	0.233	0.429	0.429	0.600	0.636

Table 2. Parameters for the Hockett-Sherby hardening representation

Parameter	S_{sat}	S_0	m	n
Value	488.780	139.325	5.410	0.827

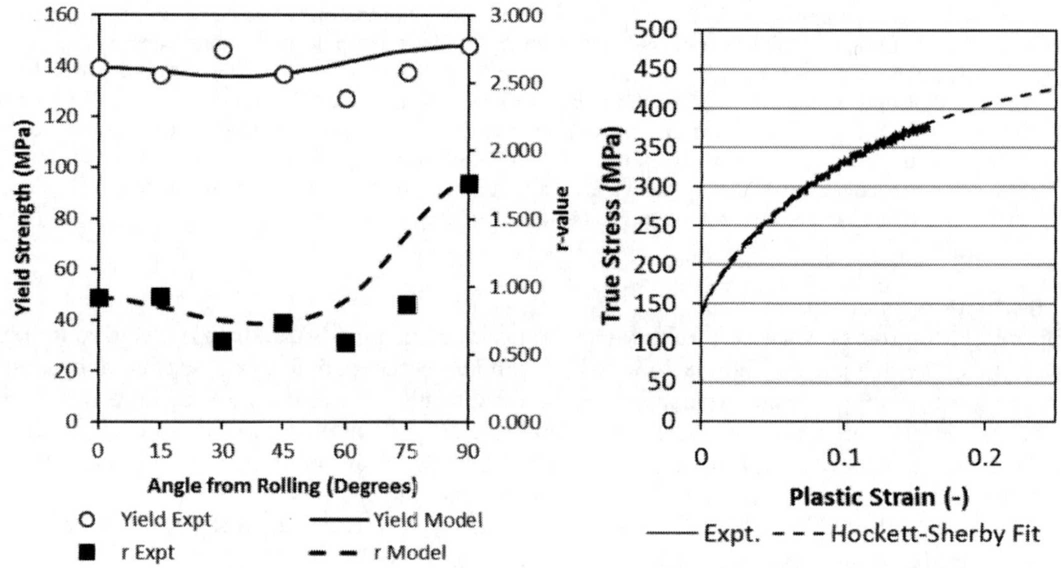

(a) (b)

Figure 1. (a) Yield strength and r-value as a function of angle from rolling, (b) Comparison between the uniaxial test data and Hockett Sherby fit in the rolling direction.

The quality of fit obtained in the model is shown graphically in Figure 1 for the uniaxial yield strength, r-values and hardening curve. The propensity for the material to exhibit serrated yielding is clearly visible in the uniaxial hardening curve shown in Figure 1(c).

3. Modelling Approach and Industrial Validation Study

In order to assess the validity of the modelling approach a small aerospace frame component shown in Figure 2 was selected. This small component was selected on the basis of the feature density contained within the form which is representative of the complexity found in the majority of aerospace frame components. There are two location holes on the part to ensure that the part remains stationary during the forming process. The blanks were first removed from the parent material using a router, deburred and subsequently solution heat treated at 465°C for 35 minutes before being stored in a freezer at -18°C until required. During the production trials the blanks were aged for 30 minutes at room temperature before being located on the aluminium tool using two pins. A Quintus Flexforming press was then used to form the components at 1000 bar. Following the forming process, the components were scanned using a Renishaw Cyclone touch-probe scanner while supported on the screw-jack fixture shown in Figure 2 (b)

The process was modelled in AutoForm^{Plus} R6 using a hydro-mechanical process setup and a limit pressure of 1000 bar. The direct pressure application was justified on the basis of the hydrostatic loading characteristic of this forming process when compared to rubber pad forming. The friction coefficient was taken to be a constant value of 0.15. Mesh refinement was controlled using an angle criterion of 22.5 degrees. The time step was set to a value of 0.3 and the convergence tolerance was 0.1. The latter value minimised instabilities in modelling the forming of flanges in net shape components. Two pin holes in the part were used to locate the part during the simulation. A post processing step was used to extract the flange angle as a function of position along the component length.

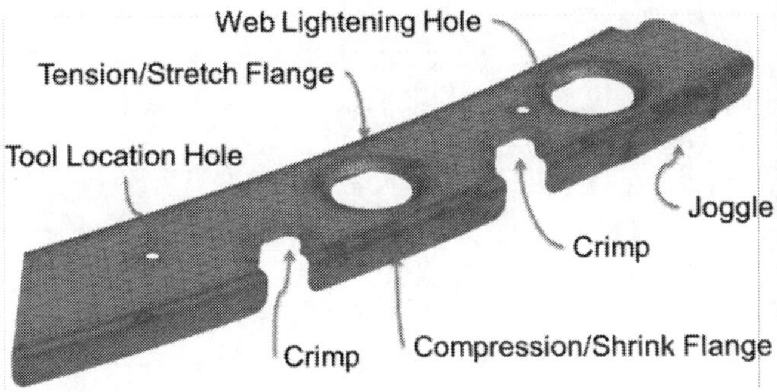

(a)

(b)

Figure 2. (a) Frame component features and (b) support fixture for measurement used in the validation study.

4. Discussion

The overall shape of the component and model output results are illustrated in Figure 3 (a) and (b). The x axis represents the distance along the length of the frame component while the zero point is referenced to the first pin hole used to locate the part during forming. There is a good correlation in the overall prediction of springback in the compression flange results, while the tension flange results tend to underestimate the degree of springback. There is also a clear influence of buckling in the compression flange that occurs during the initial pressurization that results in a variable flange angle along the part.

Following the production validation study the internal surface of the components were examined and clearly showed compression on the internal surface of the radius between the flange and the web. Localised through thickness compression in these regions is likely to influence the overall springback. The models were then re-analysed with the addition of through thickness stresses. The results of this analysis are shown in Figure 3 (b) and (c) for the compression and tension flange respectively. In both cases the inclusion of these stresses are noted to increase the simulated springback by almost 1°, improving the overall accuracy of the results.

5. Conclusions

It is possible to accurately model the flex forming process without including the rubber provided an accurate materials characterisation underpins the model, in this case through the use of the BBC 2005 material model. The interaction of the material, process and tooling must be considered, especially in the case of through thickness stresses at the bend radius.

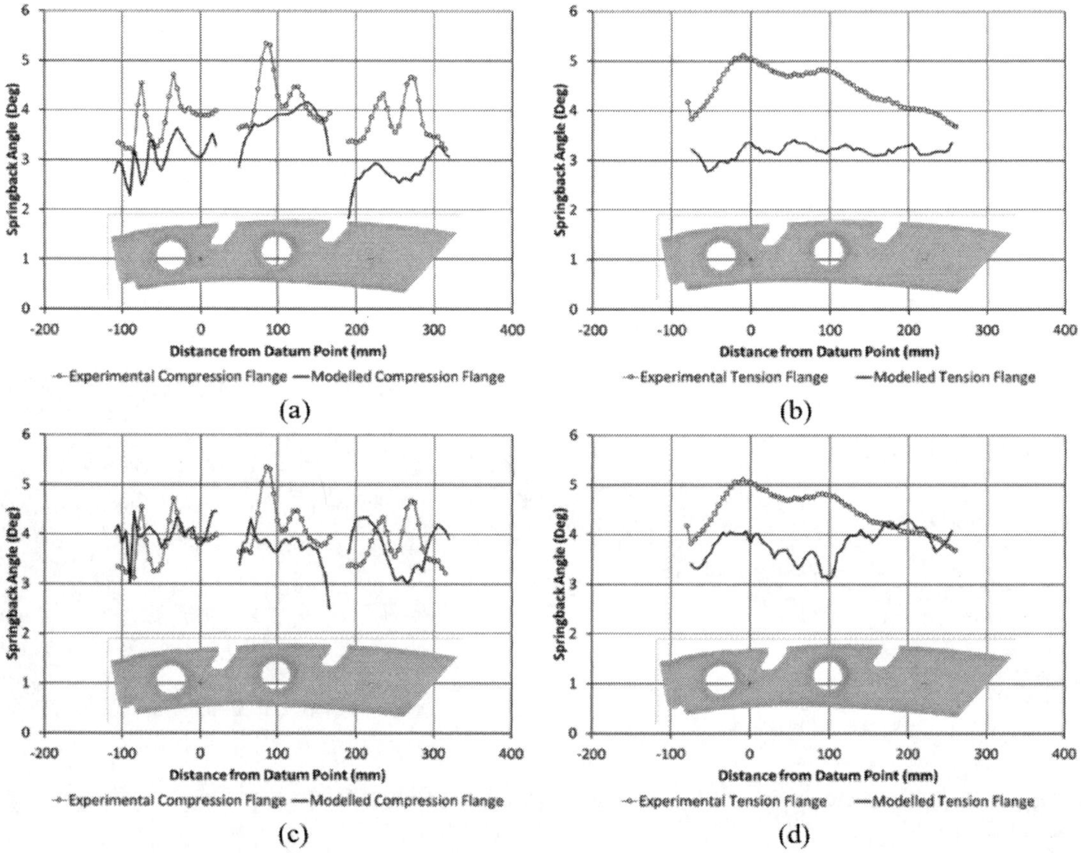

Figure 3. Springback angle for the compression flange (a), tension flange (b) and for through thickness stresses included in the compression flange (c) and tension flange (d)

References

[1] Howe C 2007 A tool design method for the compensation of springback in sheet hydroformed components. Ph.D. thesis University of Ulster.

[2] Asnafi N 1999 On stretch and shrink flanging of sheet aluminium by fluid forming *Journal of Materials Processing Technology* **96** pp 198-214

[3] Chen L et al. 2015 Experimental and simulation studies of springback in rubber forming using aluminium sheet straight flanging process *Materials and Design* **54** pp 354-360

[4] Chen L et al. 2015 Studies on wrinkling and control method in rubber forming using aluminium sheet shrink flanging process *Materials and Design* **65** pp 505-510

[5] Leacock A G et al. 2013 Evolution of mechanical properties in a 7075 Al-alloy subject to natural ageing *Materials and Design* **49** pp 160-167

EFFECT OF GEOMETRIC PARAMETERS ON FORMABILITY AND STRAIN PATH DURING TUBE HYDRFORMING PROCESS

A Omar[1], K R Harisankar[1], AsimTewari[2] and K Narasimhan[1]

[1]Department of Metallurgical Engineering and Materials Science, Indian Institute of Technology Bombay

[2]Department of Mechanical Engineering, Indian Institute of Technology Bombay
E-mail: nara@iitb.ac.in

Abstract. Forming limit diagram (FLD) is an important tool to measure the material's formability for metal forming processes. In order to successfully manufacture a component through tube hydroforming process it is very important to know the effect of material properties, process and geometrical parameters on the outcome of finished product. This can be obtained by running a finite element code which not only saves time and money but also gives a result with considerable accuracy. Therefore, in this paper the mutual effect of diameter as well as thickness has been studied. Firstly the finite element based prediction is carried out to assess the formability of seamless and welded tubes with varying thickness. Later on, effect of varying diameter and thickness on strain path is predicted using statistical based regression analysis. Finally, the mutual effect of varying material property alongwith varying thickness and diameter on constraint factor is studied.

Keywords: *Tube hydroforming; Strain path, Forming limit diagram, Finite element method*

1. Introduction

The increase emphasis on fuel consumption and reducing the carbon emission leads to an industrial demand for manufacturing lightweight component, tube hydroforming has become a widely employed manufacturing technique in automotive and aerospace sector [1]. Successfully manufacturing the component using tube hydroforming process requires optimised loading path and precise knowledge of the impact of material properties and tube geometry on any process [2]. Hence, the finite element method (FEM) has been widely used to predict the formability of tube hydroforming process and to optimise the process parameters [3]. The most widely used tool to assess the formability of a forming process is termed as "forming limit diagram" (FLD) and its concept is developed by Keeler and Beckofen [4]. Imaninejad et al.[5] has proposed the experimental methodology to generate strain path on either side of FLD by inducing different stress states as well as changing the geometrical condition to obtain FLD curve. Several researchers have since assessed the formability for different material and in one of our earlier work [6] FLD of the ERW steel tubes were drawn using the same methodology. Kang et al.[7] have studied the effect of tube diameter on hydroformability of bumper rail section and optimise the loading path to obtain the objective of reduce thinning rate. Liu et al. [8] have predicted the crack location of tailor welded tube with different thickness ratio as well as weld position.

This work deals with the prediction of the effect of varying thickness as well as diameter on the formability and strain path during tube hydroforming process. The comparisons were made with welded as well as seamless tubes to further study the impact of weld on forming characteristics.

2. Simulation Methodology

The tube hydroforming simulations were carried out on commercially available finite element based software PAM-STAMP-2G ©. The simulation methodology used in this work is discussed in detail in the work done by Nikhare et al.[9]. In order to obtain biaxial strains, the elliptical bulge methodology proposed by Li et al.[10] is used. To predict the onset of necking, a thickness gradient criterion is used [11]. In the simulation, Hill's-1948 yield theory is used and flow stress characteristic is defined by Hollomon's law. All the simulations were carried out with shell element having 1mm mesh size and same loading path. Weld size is modeled as 7.5mm as obtained in previous work [6]. The FEM model for tube hydroforming simulation is shown in figure 1.

3. Result and Discussion

3.1 Effect of thickness on FLD

The forming limit diagram for tubes of diameter =57.15 mm is plotted, as shown in figure 2, for seamless as well as welded tubes with varying thickness (t=0.6mm, 1.6mm, 2.6mm) and is based on the methodology discussed above.

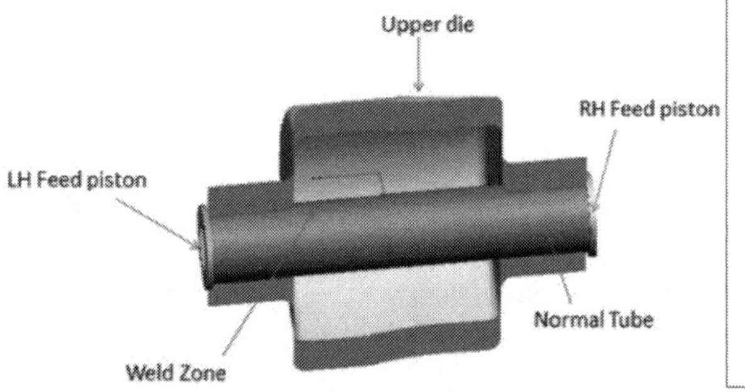

Fig.1 – CAD model for Tube hydroforming simulation

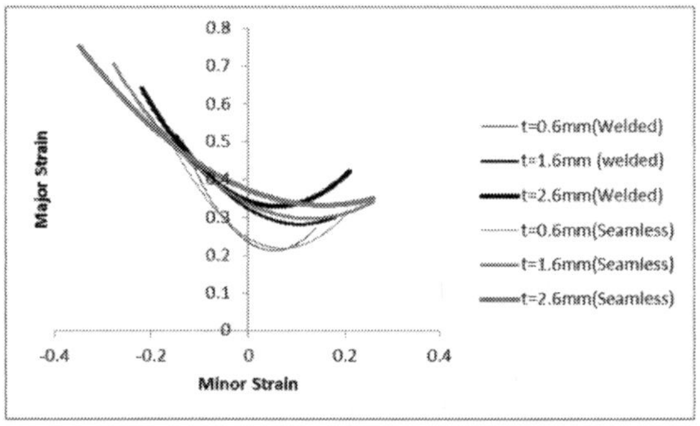

Fig.2 – Forming limit diagram for varying thickness

Table-1 The FLCo and slope values at varying thickness for welded and seamless tubes

Tube type →	Welded Tubes				Seamless Tubes			
Thickness →	0.6mm	1.6mm	2.6mm	%change	0.6mm	1.6mm	2.6mm	%change
FLCo	0.2312	0.2623	0.29	25	0.2424	0.30	0.317	32
FLC_L	-1.71	-1.56	-1.46	15	-1.64	-1.36	-1.11	32
FLC_R	0.24	0.26	0.36	50	0.28	0.18	0.125	55.3

The FLD obtained in figure 1 were linearly fitted by regression analysis to obtain the FLCo as well as slope of drawing and stretching region and these values were tabulated in table 1. From the above table it can be concluded that weld zone affect the formability in the entire region viz: drawing, plane strain and stretching region. Furthermore with increasing thickness the formability is increase with around 20-30 % in LHS of FLD. It is possible that the RHS of FLD prediction is over predicted because of less no. of data points and hence more simulation is needed to be carried out with varying oval ratio to have more point and better fit to ascertain the accuracy of prediction for biaxial region.

3.2 Effect of varying diameter and thickness on strain path for welded tubes

In order to individually predict the impact of diameter and thickness of tube on strain path, 3 different set of diameter (i.e. 28.42, 42.7, 57.15 mm) were taken with three set of thickness. 9 simulations were carried out with tube being modeled with weld and the loading path is kept same in all the simulation so as to avoid its impact on the strain path. Strain paths were analysed based on strain ratio (SR= $\varepsilon_{minor}/\ \varepsilon_{major}$) as well as coefficient of determination 'R^2' obtained from regression analysis. R^2=1 fits a straight line showing no deformation heterogeneity whereas as the heterogeneity in deformation increases R^2 decreases. Thus severe strain localisation indicates lesser R^2 value. The slope of strain path based on linear fit were obtained and tabulated below in table 2. Lesser the slope (in terms of negative value) more is the extent of drawability obtained for a given strain path.

Table-2 Strain path results for different dia. and thickness

→ Tube type	Welded Tubes		
Tube dia. and thickness	Strain Ratio= $\varepsilon_{minor}/\varepsilon_{major}$)	R^2	Slope
φ28.4t0.6	-0.17	0.89	-3.66
φ28.4t1.6	-0.32	0.82	-1.72
φ28.4t2.6	-0.36	0.78	-1.47
Φ42.7t0.6	-0.27	0.96	-2.91
Φ42.7t1.6	-0.34	0.84	-2.06
Φ42.7t2.6	-0.35	0.80	-1.49
Φ57.15t0.6	-0.22	0.96	-3.18
Φ57.15t1.6	-0.29	0.92	-2.91
Φ57.15t2.6	-0.35	0.82	-2.06

The FLD obtained in figure 1 were linearly fitted by regression analysis to obtain the FLCo as well as slope of drawing and stretching region and these values were tabulated in table 1. From the above table it can be concluded that weld zone affect the formability in the entire region viz: drawing, plane strain and stretching region. Furthermore with increasing thickness the formability is increase with around 20-30 % in LHS of FLD. It is possible that the RHS of FLD prediction is over predicted because of less no. of data points and hence more simulation is needed to be carried out with varying oval ratio to have more point and better fit to ascertain the accuracy of prediction for biaxial region.

3.3 Effect of material properties and thickness on strain distribution for welded tubes

The material properties used for this section is 0.12 and 0.187 for strain hardening exponent of base metal and 0.054 and 0.085 for weld metal. 'R' value for base metal is taken as 1.2 and 2.1 whereas for weld metal 'R' value is taken as 1. The tube diameter and thickness are taken as 28.42, 57.15 mm and 1.6, 2.6 mm respectively. The heterogeneity in thickness strain is termed as strain non-uniformity index (SNI) and the heterogeneity in surface strain distribution is termed as Constraint factor (CF), as shown in figure 3 and eqn.1, the combination of both is used for the purpose of decision making and process control [12]. The CF is defined as follows-

$$CF=[\varepsilon 2peak - \varepsilon 2pole] / [\varepsilon 1peak - \varepsilon 1pole] \;\ldots\ldots\ldots\ldots\ldots\ldots\ldots\; (1)$$

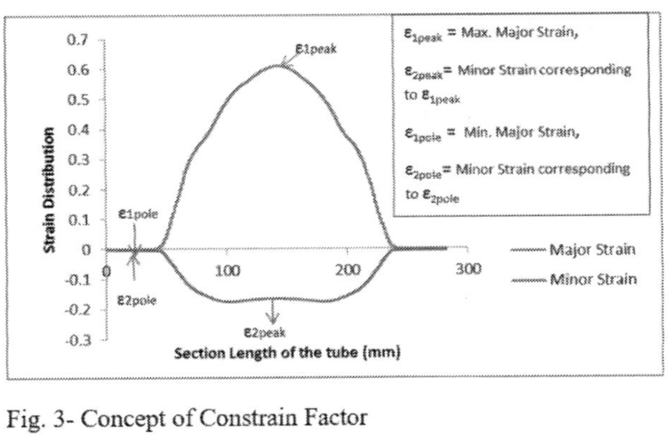

Fig. 3- Concept of Constrain Factor

Table-3 Results of ANOVA for Constraint Factor

Source	Sum of Squares	df	Mean Square	F Value	p-value Prob > F
Model	0.072	4	0.018	11.94	0.0005
A-n	9.2E-003	1	9.2E-03	6.14	**0.0307**
B-r	0.039	1	0.039	25.71	**0.0004**
C-D	0.024	1	0.024	15.84	**0.0022**
D-T	8.1E-005	1	8.1E-05	0.054	0.8207

Predicted R-Squared = 0.6438
Adjusted R-Squared = 0.7447 .

The factors which are accepted after F-testing on the model at the significance level of 5% are revealed to be plastic anisotropy 'r' followed by tube diameter and strain hardening exponent 'n'. Therefore, in order to achieve higher formability in drawing region the plastic strain ratio should be higher resisting severity in thinning.

4. Conclusion

With increase in thickness the formability of the drawing region is increased upto 15-25 % in case of welded tube whereas for seamless tube the formability is increase to 32% in same region. This result can further be strengthened by observing the effect of strain path on increasing thickness which leads to lower negative slope and higher strain ratio indicating improvement in formability. For biaxial region the increase in formability is upto 50% with increasing thickness which is quite high therefore more simulations are needed to carry out with more combination of oval ratio to acquire more data points to ascertain the reliability of the finding above. Deformation heterogeneity is assessed by co-efficient of determination 'R^2' and the results shows that for higher diameter the deformation heterogeneity reduces. Finally the mutual impact of n, r, dia. and thickness is studied on CF and it was found that r value has the most significant effect followed by tube diameter.

References

[1] Dohmann, F., Hartl, Ch, 1996. Hydroforming—a method to manufacture lightweight parts. Journal of Materials Processing Technology 60, 669–676

[2] M. Imaninejad, G. Subhash, A. Loukus , Loading path optimization of tube hydroforming process , International Journal of Machine Tools & Manufacture 45 (2005) 1504–1514

[3] B. Carleer, G. van der Kevie, L. de Winter, B. van Veldhuizen ,Analysis of the effect of material properties on the hydroforming process of tubes, Journal of Materials Processing Technology 104 (2000)pp- 158-166

[4] Keeler, S.P., Backofen, W.A., 1963. Plastic instability and fracture in sheets stretched over rigid punches. Transactions ASM 56, 25–48.

[5] Imaninejad M, Subhash G and Loukus A. Influence of end conditions during tube hydroforming of aluminium extrusions. Int J MechSci 2005; 46: 1195–1212.

[6] Ahmad Omar, AsimTewari and K Narasimhan ,Formability and microstructure evolution during hydroforming of drawing quality welded steel tube, Journal of strain analysis and engineering design, pp-1-15, July 2015,

[7] Sung-Jong Kang, Hyoung-Kwang Kim, Beom-SooKang,Tube size effect on hydroforming formability, Journal of Materials Processing Technology 160 (2005) 24–33

[8] Sung-Jong Kang, Hyoung-Kwang Kim, Beom-Soo Kang, Tube size effect on hydroforming formability, Journal of Materials Processing Technology 160 (2005) 24–33.

[9] Nikhare C and Narasimhan K. Limit strain comparison during tube and sheet hydroforming and sheet stamping processes by numerical simulation. CM 2008; 7(1): 1–8.

[10] Shuhui Li , Xianfeng Chen, Qingshuai Kong, Zhongqi Yu, Zhongqin Lin, Study on formability of tube hydroforming through elliptical die inserts, Journal of Materials Processing Technology 212 (2012) 1916–1924

[11] Kumar S, Date PP and Narasimhan K. A new criterion to predict necking failure under biaxial stretching. J Mater Process Tech 1994; 45: 583–588.

[12] S.G.Desai, P.P. Date, On the quantification of strain distribution in drawn sheet metal products, Proceedings of the 11th International Conference on Metal Forming 2006, Journal of Materials Processing Technology,Volume 177, Issues 1–3, 3 July 2006, Pages 439–443

STUDY OF FRACTURE BEHAVIOUR AND STRAIN PATH DURING TUBE HYDRFORMING PROCESS

A Omar[1], K R Harisankar[1], Asim Tewari[2] and K Narasimhan[1]

[1]Department of Metallurgical Engineering and Materials Science, Indian Institute of Technology Bombay

[2]Department of Mechanical Engineering, Indian Institute of Technology Bombay

E-mail: nara@iitb.ac.in

Abstract. Forming limit diagram (FLD) is a tool which is widely used to measure the material formability of hydroforming process. It is well known that strain based FLD is dependent on strain path which the material undergoes during the course of deformation. This work is based on understanding the deformation and fracture behaviour of as-received (AR) tube and annealed tube during hydroforming process. The strain path is being measured at different locations of the AR tube as well as annealed tube and is co-related with the microhardness value to understand the localization and fracture behaviour. The annealed tube tends to fracture suddenly from the weld line and fracture surface study reveals sudden dominant brittle mode of fracture indicating the impact of weld notch on the fracture process and thus giving lower limiting strains and hardness value.

Keywords: *Tube hydroforming; Strain path, Microhardness, ERW tube, DIC*

1. Introduction

Tube hydroforming process is an advanced manufacturing technique which is widely employed for manufacturing of tubular components in automotive, aerospace industries as well as in producing household appliances because of it being a near net shape manufacturing technique [1]. Comparing with conventional stamping techniques, tube hydroforming offers several advantages like lightweighting, increased reliability, part consolidation, high strength part and tighter dimensional tolerances [2]. In the past, several researchers have worked on obtaining the forming limit diagram for tube hydroforming with varying loading paths as well as different geometrical conditions (l/d ratio) in order to achieve variety of stress state viz: uniaxial drawing, plane strain and biaxial stretching[3,4]. Stretching side of FLC is obtained by Li et al [5] by using "elliptical bulge test" methodology to obtain higher strain ratio in biaxial region. Chen et al.[6] have also conducted elliptical bulge test on laser welded and ERW tubes and compared the FLD for the two sets of tube and reported that laser welded tube have higher formability than their ERW counterparts. In all the above reported work only final fracture is analysed with no discussion on strain paths. The strain path analysis has only been reported by Yang et al[7] to establish FLD using complex strain paths and analytical model was used to validate on SS304 stainless steel tube.

This work is an extension of previously reported work on formability of DQ ERW steel tubes [8]. In the present work the understating of fracture behaviour and its correlation with strain path will be discussed and the effect of annealing on the fracture location and strain path will be studied.

2. Experimental details

The material used in this work is ERW DQ steel tube of 57.15mm diameter and 1.6mm thickness. The weld width alongwith HAZ is estimated around 7.5mm using microhardness test [8]. The EBSD of As-received material alongwith microhardness reveals that weld material has relatively finer grain size and higher hardness and thus higher strength co-efficient than base metal. This indicates the heterogeneity across the tube circumference. For detail material characterisation of as-received tubes refer to our earlier work [8]. In an attempt to reduce the heterogeneity and homogenize the material, the as-received tube is annealed to 950 degree C for 30 minutes. The microhardness profile and EBSD (in thickness direction) alongwith the true stress strain curve shown respectively from figures 1-3 indicates that annealing process has considerably able to homogenised the material and thus able to significantly reduce the heterogeneity caused due to welding process.

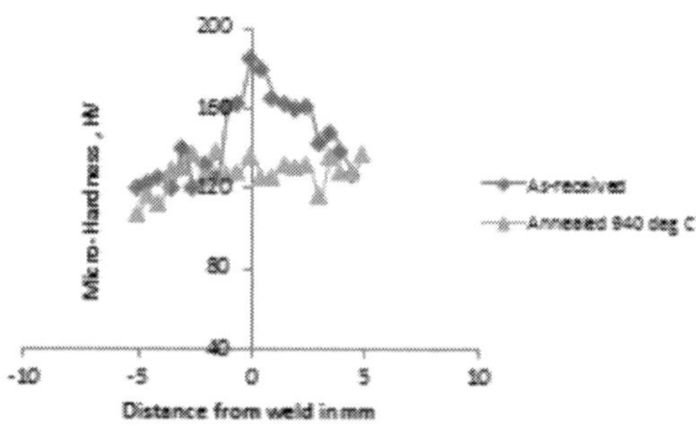

Fig.1 – Microhardness Profile

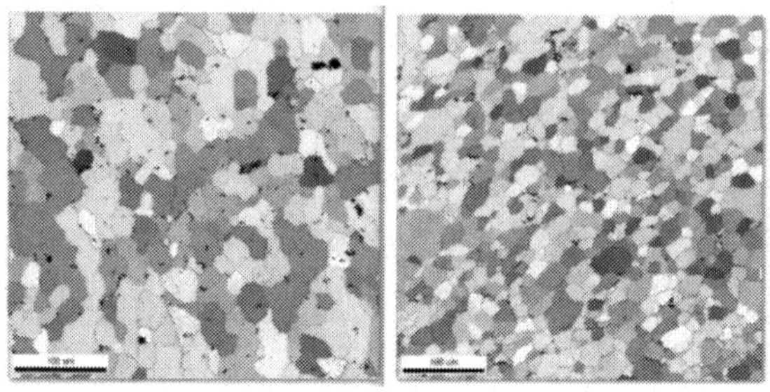

(a) (b)

Fig.2 – IPF map after annealing (a) base metal (b) weld metal

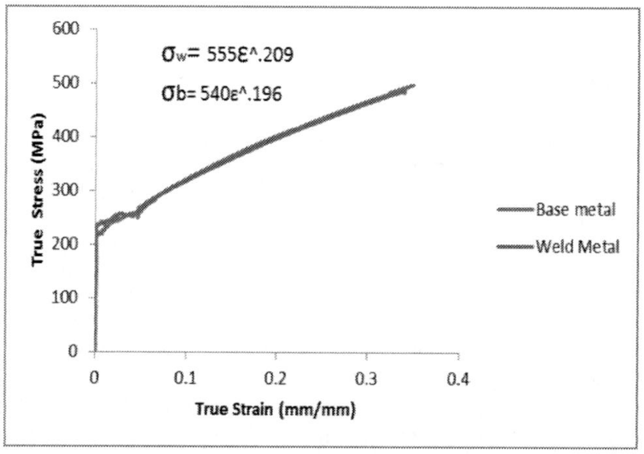

Fig.3 – True stress- true strain curve for HT sample

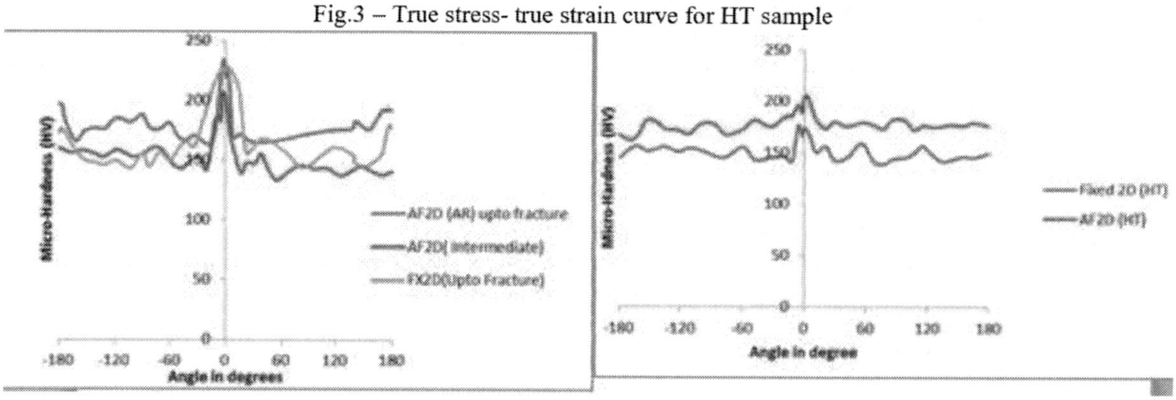

(a) (b)

Fig.4 – Microhardness profile for (a) AR and (b) HT sample after fracture (Note AF2D denoted axial feed with L/D=2 and FX2D denotes fixed condition with L/D=2).

3. Determination of Strain path

Tube hydroforming experiments were performed on a CNC hydroforming system having press capacity of 200 Ton and maximum pressure generating capacity of 1500 bar. The experiments were carried out at L/D =2 and different strain paths were generated by varying loading paths between two extremes of advanced feed and advanced pressure type. Black and white paints were sprayed on tube surface to form special speckle pattern which is used by GOM –ARAMIS digital image correlation (DIC) system to measure strains during online tube hydrobulging experiments. The strain paths were measured at the point of fracture in the base metal as well as same set of experiments were carried out with weld being kept in front of camera so as to capture the strain path near the weld i.e. opposite of the fracture position and shown in figure 5. The strain path for annealed sample is shown in figure 6. The microhardness value alongwith standard deviation is listed in table 1 giving a good indication of deformation heterogeneity across the circumference of the tube for different working conditions.

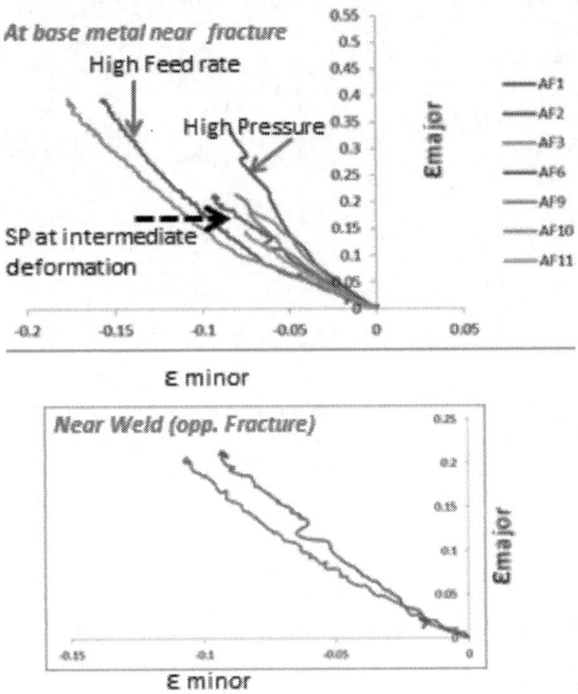

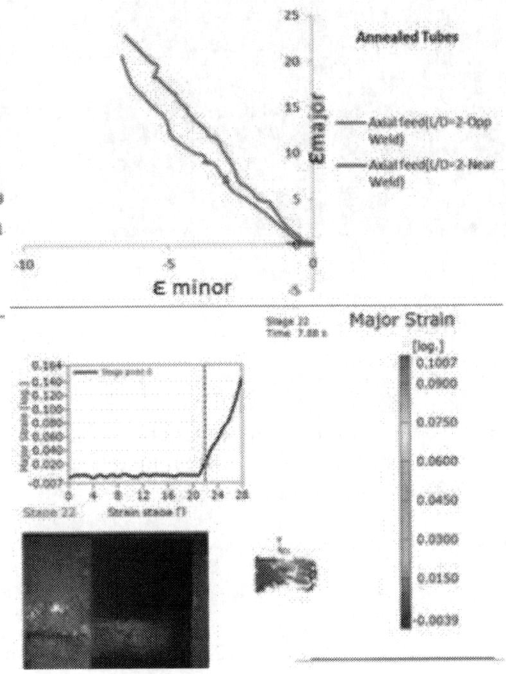

Fig.5 – Strain path for Axial feed L/D =2 (AR) Fig.6 – Strain path for Axial feed L/D =2 (HT)

It can be observed from figure 4 showing microhardness value for deformed sample of (a) AR tube and of (b) annealed tube that there is large variation throughout microhardness value for AR tube. Zero degree position is the weld position whereas the burst position is at 180 degree in case of figure 4(a) i.e. for AR tubes. One can clearly state by looking at the variation of hardness value that the point of fracture (near 180 degree) has highest hardness due to strain hardening effect and that is where strain localization takes place. The weld region already has higher hardness and strength and after hydroforming there is very little deformation occurs near weld as (almost 1/3rd of the base metal) evident from strain path comparison shown in figure 5 and therefore there is not much increase in hardness. The hardness of tube deformed till fracture in fixed condition also follows the same pattern and same fracture location but is slightly lower which is because the forming limit and hence strain hardening for fixed condition is lesser than the axial feed condition because of severe extent of thinning and strain severity owing to no material feeding in the critical region as explained in previous work [8].

On the other hand observing the microhardness value of annealed tube till fracture reveals that there is just a slight variation in hardness value throughout the circumference of the tube. In case of annealed sample the crack was observed at weld region (0 degree). Not much variation in hardness value indicates that there is very little or almost no strain localization occurring before fracture which is quite clear from strain path diagram of annealed tube as can be seen in figure 6, where strains at weld (fracture point) and at opposite to weld (i.e. base metal) is following almost the same path indicative of uniform deformation. The repeated fracture of annealed tube at weld with lesser degree of deformation showing almost no strain severity and much lower bulge height being obtained (note that bulge height for annealed sample is at max 4 mm whereas AR tube gives bulge height from 9-15 mm for different conditions indicating of that weld notch is playing a major role in fracture initiation and acts as a primary stress raiser. Fracture surface study of annealed tube presented in figure 7 also indicates the presence of large percentage of flat cleavage plane indicating the dominant mode of failure as brittle.

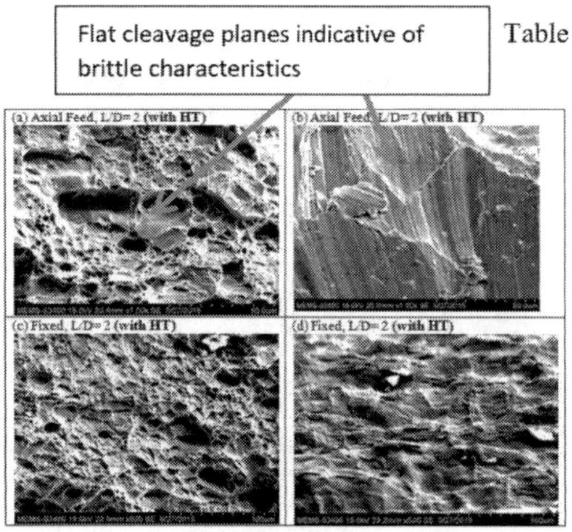

Fig.7 – Fractographs of annealed tube

Table 1 Microhardness value of different sample and its std. deviation

(Note - * indicate SD for base metal only)

Condition	HV Weld	HV HAZ	HV Base	SD
As-Received (AR)	170-180	150-160	120-135	20.9 (7.2*)
Annealed (HT)	130-140	120-130	120-140	8.96
Axial feed (AR) Intermediate deformation	~200	170-190	140-160	17.45(7.71*)
Axial Feed (AR) upto fracture	~240	210-240	180-200	18.51 (10.08*)
Axial feed HT upto fracture	~200	180-195	165-180	10.8

4. Conclusion

In this work a brief analysis of the deformation behaviour in terms of strain path and microhardness across the circumference of the tube is presented for AR tube as well HT tube in annealed condition. The changing position of fracture in case of annealed tube was studied and analysed using fracture surface. Initial investigation reveals that notch in case of HT tubes is crack initiating site and in future a detail analysis will be carried to understand the impact L/D ratio has on crack location.

References

[1] Materials, Design and Manufacturing for Lightweight Vehicles, edited by P K Mallick – pp-271. Woodhead Publishing -2010

[2] Ahmetoglu M., Altan T. (2000), Tube hydroforming: state-of-the-art and future trends, Journal of materials Processing Technology, 98 (2000) 25-33.

[3] Imaninejad M, Subhash G and Loukus A. Influence of end conditions during tube hydroforming of aluminium extrusions. Int J Mech Sci 2005; 46: 1195–1212.

[4] Chen X, Li S, Yu Z, et al. Study of experimental approaches of forming limit curve for tube hydroforming.
Int J Adv Manuf Tech 2012; 61: 87–100.

[5] Shuhui Li , Xianfeng Chen, Qingshuai Kong, Zhongqi Yu, Zhongqin Lin, Study on formability of tube hydroforming through elliptical die inserts, Journal of Materials Processing Technology 212 (2012) 1916–1924

[6] Chen X, Li S, Yu Z, et al. A theoretical and experimental study on forming limit diagram of seamed tube hydroforming. J Mater Process Tech 2011; 2011: 2012–2021.

[7] Lianfa Yang, Guolin Hu, Jianwei Liu, Investigation of forming limit diagram for tube hydroforming considering effect of changing strain path,International Journal of Advance Manufacturing Techonology (2015) 79:793-803

[8] Ahmad Omar, Asim Tewari and K Narasimhan, Formability and microstructure evolution during hydroforming of drawing quality welded steel tube, Journal of strain analysis and engineering design, pp-1-15, July 2015.

Numerical Validation of an Optimized Cooling System for Hot Stamping Die

A Zakaria , M A Abidin , M S N Ibrahim and A Senin

Universiti Kuala Lumpur Institute of Product Design and Manufacturing(IPROM) 119, Jalan 7/91, Taman Shamelin Perkasa, 3.5 Miles Cheras, 56100 Kuala Lumpur, Malaysia

dzakaria@unikl.edu.my

Abstract. Numerical analysis of hot stamping process is very complex mainly due to thermo-mechanical processes involved. Many variables such as heat transfer coefficient, density, young modulus and other thermal parameters are temperature and pressure dependent. The paper presents results of CFD analysis on the near optimized cooling system of hot stamping die for automotive structural part. By using actual parameters obtained from the industry production line, this research is aimed at comparing the performance of actual cooling system with the results obtained by CFD simulation using commercial software. The die and blank were modelled as 3D volume mesh in a closed position thus ignoring blank history data prior to stamping operation. Temperature distribution representing hardness of the simulated final part is an agreement with the QA data of the actual part thus showing viability of this method to be used in cooling system design

1. Introduction

Hot stamping or press hardening is currently the most viable process to produce automotive structural parts with tailored properties. It has many advantages over its contemporary the conventional cold stamping. First, it's increased tensile strength up to 4 times than the material as delivered. Hot stamped part has also minimum spring-back. More importantly the final properties of the part can be varied along its length. This is extremely useful property for parts require higher crashworthiness such as B-Pillar [1]

A typical hot stamping process starts with the blank heated at 950°C for about 4-10 minutes inside a furnace. It is then placed on top of a die by a robot hand. Simultaneous forming and quenching take place immediately after this. The holding time is typically 10 seconds. When the die is open the part temperature is within few hundred degrees Celsius. Finally the excess material of the hardened part is trimmed by a CNC laser cutter.

Hot stamping is very complex as it involves combined thermo-mechanical and micro-structure evolution [2]. A prior knowledge in terms of temperature dependent stress strain parameters, heat transfer between the blank and the die, coupled thermo-mechanical calculation and the material evolution of micro-structure allow engineers to predict the final properties of the part. Die design plays an important role in hot stamping to ensure the part is formed and quenched according to the desired geometry and properties. Besides having to withstand press cyclic loading, die material should able to extract the heat from the blank. This means the die must be able to absorb and evacuate energy of up to 100KW [3] by means of cooling channels. Therefore location and size of the cooling channel are critical to guarantee a complete martensitic transformation. Due to manufacturing constraints, conformal cooling channel design may not be economical. Hence the straight hole type is more popular in practice. Hu et al[4]investigated the performance of five different cooling designs namely straight hole, longitudinal CCC (conformal cooling channel), transversal CCC, parallel CCC and serpentine CCC. Each design performance index is measured by the figure of merit (FoM) concept. The longitudinal CCC design scores the highest FoM, therefore giving the highest cooling efficiency. Chen et al [5] classify cooling channel design into two: namely parallel type and serial type. In the parallel design, output channel of die insert is connected to the input of the following insert. Whereas in the case of serial design each die insert has its own input and output cooling channel. It is found that that the parallel design is more efficient.

Finite element method has been successfully implemented in simulation of hot stamping of Boron and other quenchable steels. In addition many commercial codes are now available to assist tool design engineers. However most of these codes assume the die temperature during quenching is uniform. This can only be achieved if the cooling channels are optimized. In an attempt to accurately simulate the heat transfer during the cooling process, the use of the correct heat transfer coefficient (HTC) has been proposed in the literature. It has been shown that HTC is dependent on contact pressure between the blank and die surface. However Zhang et al [6] defined this value as a function of time. The main objective of this paper is to illustrate a simple methodology based on CFD analysis for validating an optimized cooling system design by considering several critical thermal parameters obtained from a hot stamping line.

2. Methodology

A fairly complex cooling system of a hot stamping die shown in figure 1 is taken as a case study. Surface hardness distribution of the actual part shown in figure 2 was measured by a portable hardness tester.

2.1. Heat Transfer Model

Ignoring the stages prior to stamping, the die assembly in a closed position is considered in this study. This would allow, a pure CFD analysis to be carried out. Heat is released by the blank and transferred to die and cooling water by convection and conduction. If the heat loss due radiation and latent heat generated by the blank are ignored then the steady state energy balance can be written as

$$Q_b = Q_d + Q_w \tag{1}$$

Where Q_b is the heat released by the blank, Q_d is the heat absorbed by the die block and Q_w is the heat dissipated to the cooling system. To expand equation 1 further

$$h_bA(T_b\text{-}T_d) = k_dV(T_b\text{-}T_d) + h_wA(T_d\text{-}T_w) \tag{2}$$

where h_b and h_w are HTC for the blank-die and water , T are temperatures of blank , die and cooling water , A is the surface area and V is the die volume.

2.2. Numerical simulation via CFD

In order to reduce the computing times the die block, blank and cooling channels were volume meshed separately using element size of 0.05 and 0.006m respectively as shown in figure 1. Since there is no pressure involved, the HTC value used is 1300 W/m^{2}K and other parameters such as material model and boundary conditions are given in Table 1. AcuSolve solver was used to run the analysis. The results of changes the blank temperature distribution throughout the holding period of 8 seconds can be used to determine the performance of the cooling system. Temperatures of other components were also monitored at selected points.

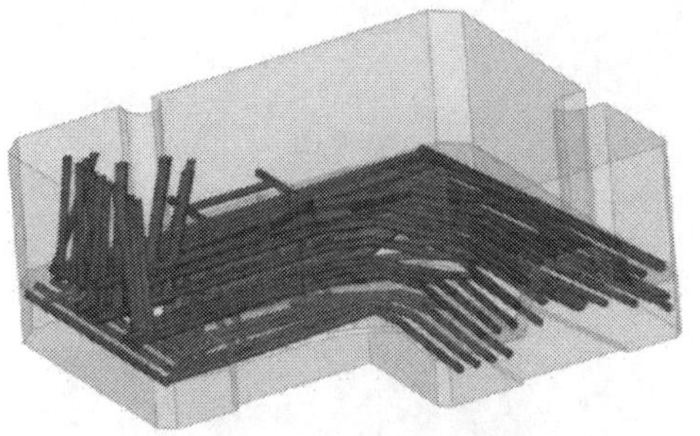

Figure 1. Cooling system

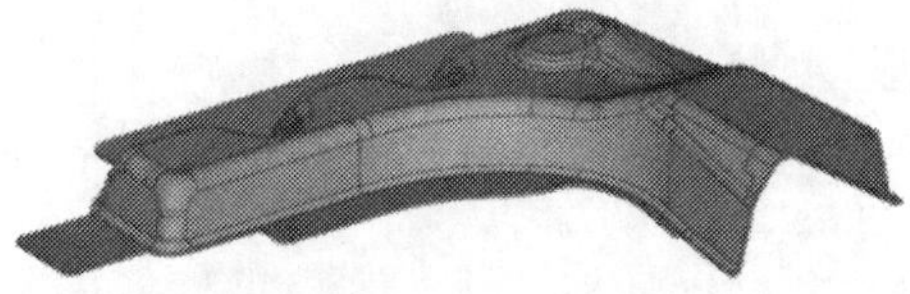

Figure 2. Final Stamped Part

Table 1. Mechanical, Thermal Properties and Boundary Condition.

	Material	ρ *Kg/m³*	K W/mK	*c* *J/kg²K*	h W/m²K	*T* *°K*
Blank	Boron	7800	35	800	1300	900
Die	HTCS	7870	66	465	-	303
Cool	water	1000	0.598	4163	3000	280

3. Results and Discussion

Temperature response after 0.1 and 8 seconds for major components is shown in figure 3 and figure 4. It depicts rapid cooling of the stamped part from initial temperature of 900°K. After 0.1 seconds the heat released by the blank causes the die and water temperatures to rise by 100°K and 10°K respectively. At the end of the holding cycle of 8 seconds average temperature of the blank is about 400°K.

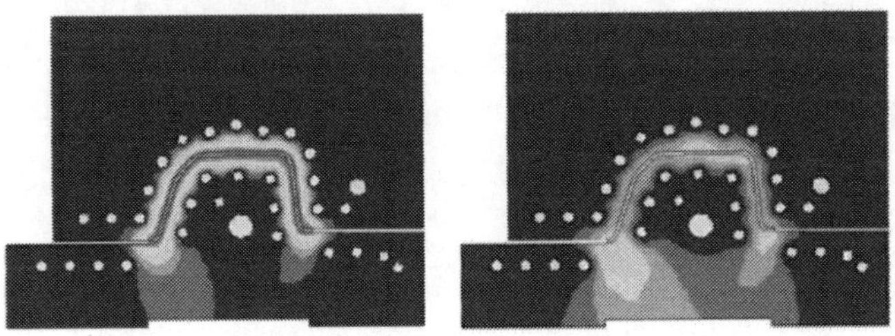

Figure 3. Blank Temperature after 0.1 and 8 seconds

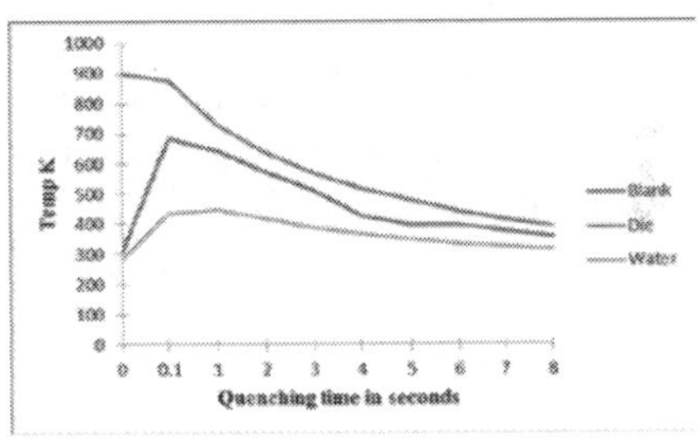

Figure 4. Temperature Response

Upon examination of the simulated part closely, it is evident the cooling is fairly uniformed. Some portions remain at high temperature of 450°K and others at about 350°K. The red spots show insufficient cooling at the areas. In fact, this non-uniformity is attributed to the manufacturing constraint stated earlier. Temperature variation of the actual part is 108°C as compared to 100°C obtained by this simulation. The best temperature distribution is at point 5 which records a maximum hardness of 45HRC. The hardness value for areas 2, 3 and 4 is slightly above 40 HRC. Detailed hardness distribution of the actual part is given in figure 4. It is inferred that the percentage of maternsite formed by the areas which cooled faster is higher than the later hence giving higher hardness value. Despite having this variation, its magnitude is considered small and acceptable by the industry standard.

No	Hardness (HRC)
1	43.5
2	40
3	40.3
4	40.8
5	45
6	44.5
7	45.6
8	42.6
9	41.9

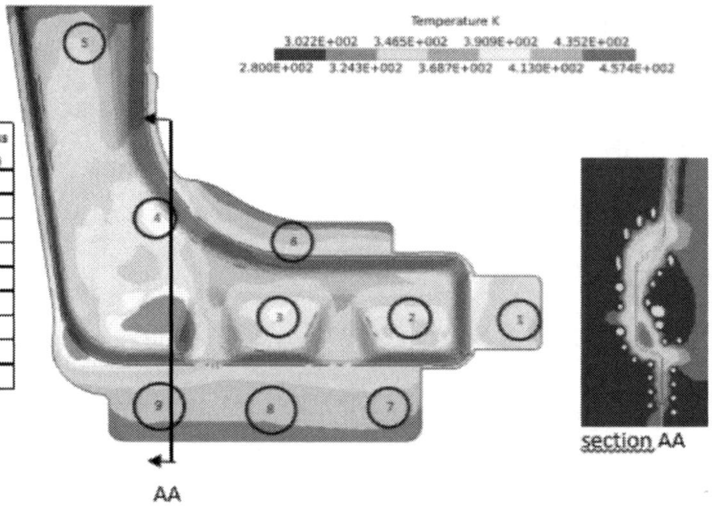

Figure 4 Temperature Pattern of Simulated Part. Cross sectional area shows insufficient cooling

4. Conclusion

It is obvious that thermal parameters play a very important role in determining the accuracy of the numerical experiment results. The simulated temperature pattern results of this CFD analysis are in a closed agreement with the actual data. In this research we have attempted to demonstrate that the proposed methodology is able to validate the actual optimized cooling system. This in turn can be used as a tool for cooling system optimization analysis.

Acknowledgement

The authors would like to acknowledge the Ministry of Science and Innovation Malaysia for funding this research. Credit should also go to our industry partner Miyazu (M) Sdn. Bhd. for providing valuable technical support.

References

[1] Maikranz-Valentin M, Weidig U, Schoof U, Becker H and Steinhoff K 2008 Components with Optimized Properties due to Advanced Thermo-mechanical Process Strategies in Hot Sheet Metal Forming, *Materials Technology* **79** No.2 92-97.

[2] Wu W, Hu P and Shen G 2015 Thermonechanical –Phase Transformation Simulation of High-Strength Steel in Hot Stamping, *Mathematical Problems in Engineering* http://dx.doi.org/10.1155/2015/982785

[3] Liu H, Bao J,Xing Z, Zhang D,Song B and Lei C 2011 Modeling and FE Simulation of Quenchable High Strength steels Sheet metal hot forming Process, *Journal of Materials Engineering and Performance* **20** 894-902

[4] Hu P, He B and Ying L 2016 Numerical Investigation on Cooling Performance of Hot Stamping Tool with Various Channel Designs, Applied Thermal Engineering **86** 338-351

[5] Chen G J, Zhang Y, Shen W, Qin L J, Deng N and Yao X C 2015 Numerical Simulation on Cooling System of Hot Stamping Mold InB-Pillar *Procs. of the 2nd International Conference (ICHSU2015)* 353-358

[6] Zhang Z, Gao P, Liu C and, Li X 2014 Experimental and simulation Study for Heat Transfer in Hot Stamping of High-strength Boron Steel, *Metallurgical and Materials Transaction B The [1]Minerals, Metal & Materials Society and ASM International* DOI:10.1007/s1163-014-0082-3

Journal of Physics: Conference Series **734** (2016) 032107 doi:10.1088/1742-6596/734/3/032107

The concept of virtual material testing and its application to sheet metal forming simulations

A Butz, J Pagenkopf, M Baiker, D Helm

Fraunhofer Institute for Mechanics of Materials IWM, 79108 Freiburg, Germany

E-mail: alexander.butz@iwm.fraunhofer.de

Abstract. A crystal plasticity based full-field microstructure simulation approach is used to virtually determine mechanical properties of sheet metals. Microstructural features like the specific grain morphology and the crystallographic texture are taken into account to predict the plastic anisotropy. A special focus is on the determination of the Lankford coefficients and on the yield surface under plane stress conditions. Compared to experimental procedures, virtual material testing allows to generate significantly more data points on the yield surface. This data is used to calibrate anisotropic elasto-plastic material models which are commonly used for sheet metal forming simulations. A numerical study is carried out to analyze the influence of the chosen points on the yield surface on the calibration procedure.

1. Introduction

An accurate description of the elasto-plastic material behavior is needed for a precise simulation of sheet metal forming processes. Here, the two most important aspects of the material models are the description of the hardening and the yield locus. To account for the anisotropy of sheet metals, an increasing number of anisotropic yield locus models is available, starting from the well-known Hill48 [1] yield locus description up to more modern models like Yld2000-2d [2]. It is known, that the simpler yield locus descriptions are usually not sufficient to model the plastic material behavior accurately, especially for recently developed advanced high strength steels and also for aluminum alloys. In the case of an associated flow rule, the curvature of the yield locus defines also the development of the plastic strain components which is of high importance for sheet metal forming simulations.

Besides the choice of an appropriate yield locus model, the identification of the model parameters is an important issue, especially when more complex yield locus models with more parameters are considered. The determination of the parameters of more sophisticated yield locus models requires a large number of experiments which can be difficult and expensive to realize. In a standard parameter identification procedure the number of experimental results (yield stresses, r-values) usually corresponds to the number of free parameters in the yield locus model. The experimental data which were considered for the parameter identification can then be precisely reproduced by the yield locus model. However, accuracy for other stress states is unclear.

Here, the concept of virtual material testing is an interesting approach. It allows to significantly extend the limited experimental data base by additional virtual experiments. These virtual data can be used as additional input for a precise parameter identification or for the assessment of yield locus models.

2. The concept of virtual material testing

The main idea of the "virtual material testing" approach is to describe the material on the micro scale (grain scale) and to calculate the macroscopic properties like flow curve, r-values or yield surface by numerical homogenization. To predict the macroscopic material behavior, it is necessary to take the most important aspects of the microstructure into account. For this reason, a so-called unit cell model is used. This model represents a sufficiently large part of the microstructure and considers the size, shape and orientation of individual grains. The elasto-plastic deformation on the grain scale is described by a single crystal plasticity material model that was implemented into the finite element software Abaqus/Standard following [3]. The material model allows the consideration different initial orientations of the crystallographic lattice for each single grain. The real texture of a sheet metal can be represented if a sufficiently large number of grains are included. Thus, the anisotropic material behavior on the macro scale with respect to yielding and r-values is a direct outcome of the applied microstructure model.

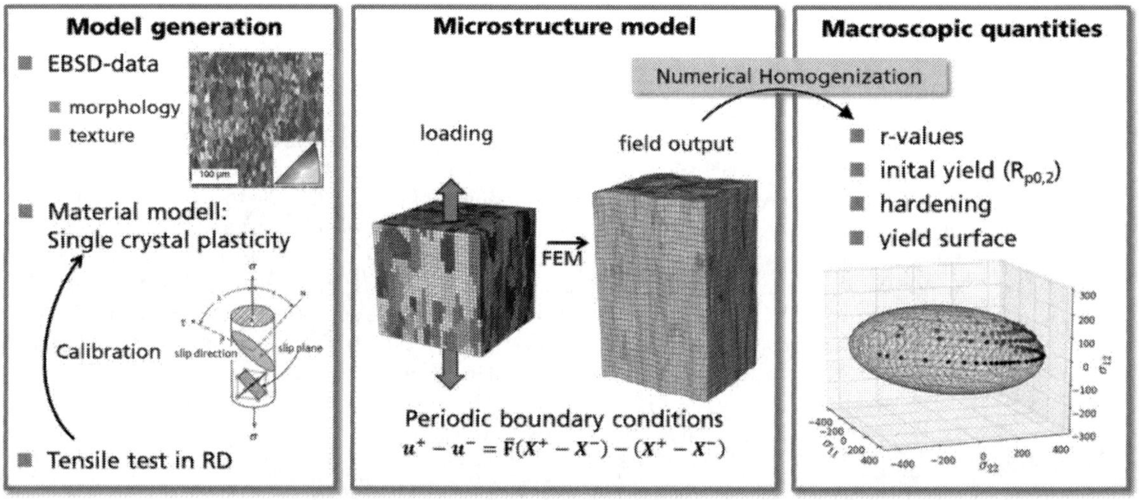

Figure 1. Concept of the "virtual material testing" approach.

Experimental data from standard material characterization are used to generate a finite element model of the microstructure and to calibrate and validate the unit cell model. Once the model is calibrated, it can be used for the analysis of arbitrary loading conditions. Figure 1 gives an overview of the applied procedure. To set up the microstructure model, an EBSD measurement and a tensile test in rolling direction of the sheet are sufficient. From the EBSD measurement, information regarding the texture and the grain morphology are obtained. The flow curve from the tensile test is used to adjust the hardening parameters of the crystal plasticity model. Details of the procedure can be found in [4].

3. Application to a mild steel DX56

The described procedure was applied to a mild steel grade DX56 with a thickness of 1.2 mm [5]. The elongated shape of the grains in rolling direction which was observed in the EBSD measurements was considered within the finite element model of the microstructure as illustrated in Figure 2. The microstructure model consists of 500 grains. By choosing an appropriate set of 500 initial orientations, the experimentally measured texture could be represented with good accuracy.

The parameters of the crystal plasticity model which describe the hardening of the slip systems were adjusted via inverse simulation with respect to the macroscopic stress-strain curve from a tensile test in rolling direction. Some results of the calibrated microstructure model are given in Figure 3. The adjusted flow curve of the calibrated model is in very good agreement with the experimental data. Furthermore, the initial yield stresses and the r-values were analyzed with respect to the orientation of the specimen. The r-values depend strongly on the orientation of the specimen which is well captured by the micromechanical model.

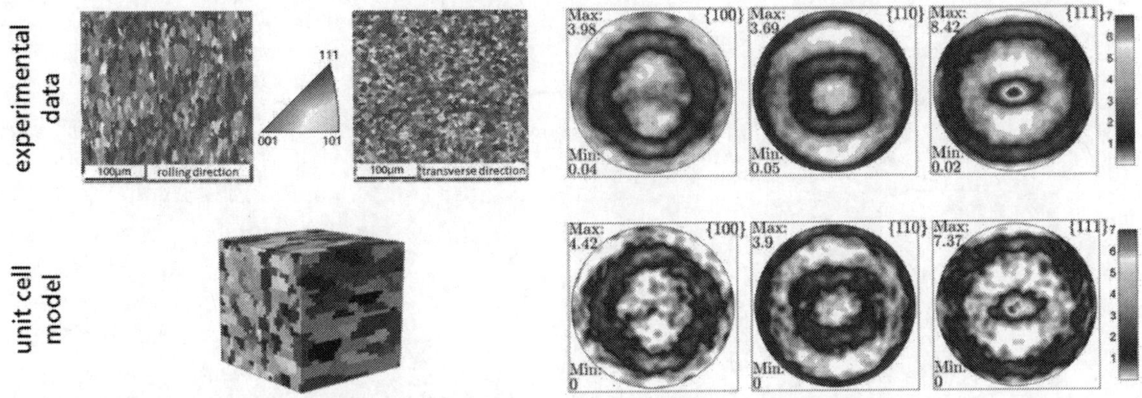

Figure 2. Generation of the microstructure model of the DX56 steel sheet: Comparison of the grain structure (left) and the crystallographic texture (right).

The calibrated model was used to calculate points on the yield surface to provide a set of yield points in the plane stress space for the identification of the model parameters of commonly used yield models. Different load cases which fulfill the assumed plane stress condition ($\sigma_{33} = \sigma_{23} = \sigma_{31} = 0$) for sheet metals were imposed to the microstructure model. It is assumed that the values of the identified parameters of the yield functions depend on the number of the yield points and their location on the yield surface. For this reason, the following numerical study was carried out: Three different sets of yield points were calculated and considered for the identification of the yield functions. In case 1, all points on the yield surface are located in the σ_{11}-σ_{22} plane in which the stress component σ_{12}=0. Case 2 considers five different levels of the component σ_{12}. Case 3 takes more points with a pronounced shear component σ_{12} into account. The yield locus models according to Hill48 [1], Yld89 [6] and Yld2000-2d [2] were adjusted to these three different sets of points on the yield surface. The results are given in Figure 4. All three cases give a good representation of the initial yield surface in the plane of σ_{12}=0. However, the yield functions predict different yield stresses and r-values as a function of the loading angle to the rolling direction. It is noted that – in contrast to the standard procedure – no r-values were considered for the identification of the parameters which partly explains the deviations to the experimental r-values. Nevertheless, the more flexible yield function Yld2000-2d is able to capture the plastic anisotropy of the DX56 if a sufficient amount of yield points with a pronounced shear stress is used for the fitting procedure (case 2 and case 3).

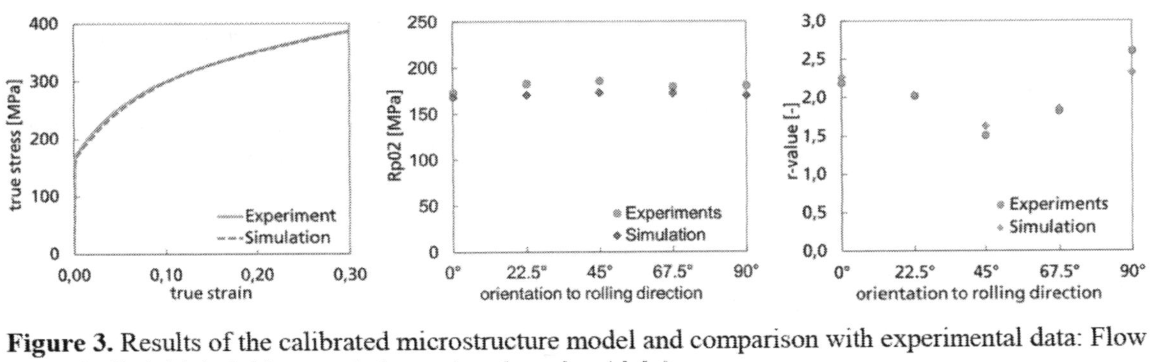

Figure 3. Results of the calibrated microstructure model and comparison with experimental data: Flow curve (left), initial yield strength (center) and r-value (right).

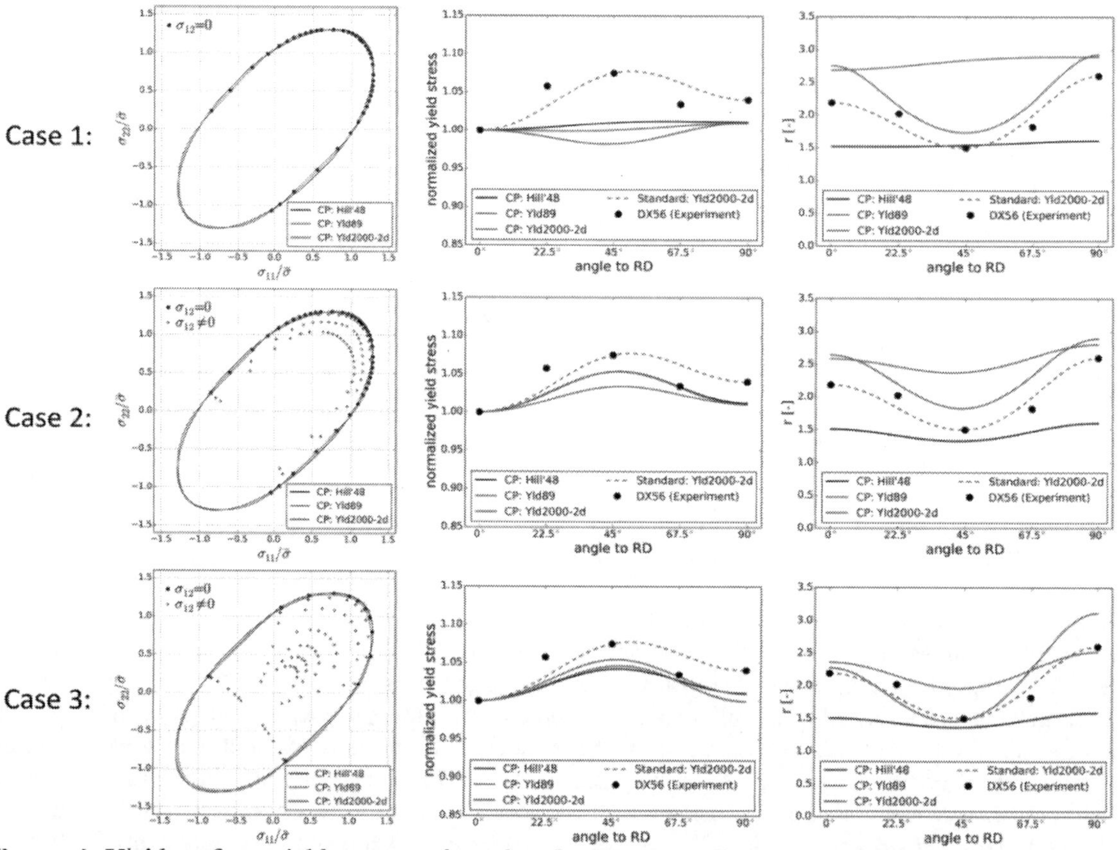

Figure 4. Yield surface, yield stress and r-value for DX56 predicted by yield functions that were fitted to different sets of numerically generated points on the yield surface. The Yld2000-2d function fitted to the experimental data (standard procedure) is shown for comparison.

4. Conclusion

The concept of virtual material testing is a promising approach for an extended characterization of sheet metals. The microstructure model used in this work gave a good representation of the experimentally available results and could be used for the determination of arbitrary points on the yield surface. Due to the high number of data points, the parameter identification problem of the yield locus models becomes highly overdetermined. As shown in the example, the identified parameters of the yield locus models depend on the number of yield points and their location on the yield surface. For this reason, further work will focus on the development of appropriate parameter identification strategies for complex yield surface models under consideration of additional, virtual material data.

Acknowledgement

The research was supported by the AiF (grant ID 17469BG) within the programme for promoting the Industrial Collective Research (IGF) of the German Ministry of Economics and Energy (BMWi).

References

[1] Hill R 1948 *Proc. R. Soc. Lond. A* **193** 281-97
[2] Barlat F et al. 2003 *Int. J. Plasticity* **19** 1297-319
[3] Kalidindi S R, Bronkhorst C A and Anand L 1992 *J. Mech. Phy. Solids* **40** 537-69
[4] Baiker M, Helm D and Butz A 2014 *Steel Res. Int.* **85** 988-98
[5] Baiker et al. 2015 EFB-Forschungsbericht Nr. 404, ISBN 978-3-86776-449-0
[6] Barlat F and Lian J 1989 *Int. J. Plasticity* **5** 51-66

Journal of Physics: Conference Series **734** (2016) 032108 doi:10.1088/1742-6596/734/3/032108

Experimental Verification of the Strain Non-Uniformity Index (SNI) based Failure Prediction

D A Dhumal [1], **Pratik Kulkarni** [2] **P P Date** [3] **and V M Nandedkar** [4]

[1, 4] Department of Production Engineering, SGGS Institute of Engineering and Technology, Nanded 431606, India
[2, 3] Department of Mechanical Engineering, Indian Institute of Technology Bombay, Mumbai 400076, India

E-mail: kajaldhumal28@gmail.com, Pratik.k.034@gmail.com, ppdate@gmail.com, vilas.nandedkar@gmail.com

Abstract. Formability of the sheet metal depends upon the uniformity of strain distribution, which depends on material properties, tooling and process parameters. Nakazima Test was conducted to study the strain distribution and establish the forming limits of AA 6016. The experimental conditions were simulated using AUTOFORM 5.2 Plus software and the failure predicted using the SNI based methodology. The failure predictions were correlated with the state of the experimentally deformed Nakazima samples, and also with the FLD based forming limits. The failure prediction from the SNI based methodology was found to correlate well with the state of the experimental Nakazima sample.

Keywords: Nakazima Test, AA6016, strain nonuniformity index (SNI), Forming Limit Curve (FLC).

1.Introduction

Inferences from spatial strain distribution over a sheet metal blank can be drawn based on several observations. Schedin and Melander [2] studied strain distributions of sheet metals using different shapes of grid elements. They also correlated the nature of strain distribution with the mechanical properties of the sheet. The concept of a strain envelop was introduced by Karima et.al [3] which refers to strain distribution plotted in the major and minor strain space. This was meant to identify process changes introduced on account of various reasons. Schedin and Melander [4] showed the effect of strain hardening exponent (n) and normal anisotropy (r) on the thickness strain distribution. Date et. al. [5] defined the strain non-uniformity index (SNI) based on strain distributions which result from complex interactions among forming parameters.

Predicting failure based on the SNI involves analysing strain distribution on a large number of critical planes. It is possible that the SNI on many critical planes attain failure at the same time. This enables predicting multiple locations of failure. However, this process is time consuming as large number of strain distributions have to be analysed.

It was therefore thought that the process of detecting onset of failure could be speeded up using the 'Global SNI', or the GSNI, which is one single value determined over the entire blank. 'Global' Strain non-uniformity index (as opposed to that for several critical planes) is simply the peak strain over the entire blank minus the average strain over the entire blank. It is computed from the strain distribution, (that obtained from a simulation, for instance) over the entire blank.

The objective of this paper is twofold, i.e., to see if a GSNI value could be identified to mark the onset of failure, so that one need not analyse strain distributions on the numerous critical planes in the safe regime of deformation, and to look for a possibility of correlating the GSNI with the critical SNI. To this end, both experiments as well as simulations were performed.

Nakazima test is the most common method for evaluating the forming limits of sheet metal thickness ranging from 0.5 mm to 3.3 mm [1]. The Nakazima test geometry, being relatively simple and standard, was used for this study.

2. Work Plan

Nakazima test was performed on AA 6016 sheet using six blank widths ranging from 75mm to 200mm. Circles of 2mm diameter with centre spscing of 3mm were laser marked on each of the blanks which were stretched to failure over a punch of 100mm diameter. The blank of 75mm width alone remained safe, as it would often tear near the bead. The press capacity was 70 tons and the blank holder force went up to 60 tons. The experimental parameters were noted and used for simulating the set of Nakazima tests using the AUTOFORM 5.2 Plus software. The properties of AA6016 used in the present work are given in Table 1. The major and minor strain distributions along with nodal and elemental data are obtained at every deformation step, and used to predict occurrence of failure based on the FLD as well as to determine the SNI and thereby predict failure.

Table 1. Mechanical properties of material

Material	Density (MPa/mm)	Young's Modulus (GPa)	Poisson's ratio (μ)	Anisotropy Parameters (orthotropic) Plastic Strain Ratio, R			Strain hardening exponent (n)
				0°	45°	90°	
AA6016	2.7 E-05	70	0.3	0.61	0.49	0.7	0.245

As the draw depth progresses, number of elements on the blank increases due to mesh refinement in adaptive meshing. The amount of non-uniformity in the strain distribution is quantified using strain non uniformity index [5]. The value of SNI for the longitudinal meridional strain distribution was obtained experimentally by grid strain analysis and analytically by an FEM simulation.

The GSNI was determined using simulation data only, while the SNI was determined analytically as well as experimentally.

3. Results and Discussion

It was found from experiments that the blanks of all six widths failed close to 30mm of dome height. The height of the dome was measured using a height gauge. Table 2 below shows the SNI, GSNI and the condition of the samples at experimental dome height of 30 mm for all 6 samples. The 'failure' in case of GSNI was determined using the FLD. For instance, for a blank of width 200mm, simulation was performed up to a punch travel of 30mm (the travel at which experimental samples failed), and were found to be 'safe' according to the FLD.

Though in simulation, the dome height could be higher than 30mm in actual experimentation the onset of crack occurred at 30 mm for all blank widths. It may be observed from Table 2 that the GSNI did not show significant variation over the six blank widths. Based on the FLD, four of the blank widths were predicted to be safe. The SNI from experiments differs with that from simulation because the actual experimental conditions cannot be reproduced in a simulation. For instance the prevalent coefficient of friction, the actual holding force, different techniques of obtaining the SNI all contribute to the difference among experimental and analytical values. The condition of the sample as predicted using simulation based SNI seems to correlate reasonably well with the experimental observations, while those using the FLD correlated exactly for two blank widths.

Table 2.SNI values and condition for experimental draw depth

Blank Dimensions	Draw Depth (mm)	Experimentation		Simulation Data, FLD Based Prediction		Simulation based SNI Prediction	
		SNI	Condition	GSNI	Condition	SNI	Condition
200 X 200	30	0.180	Failed	0.073	Safe	0.235	Risk of Split
200 X 175	30	0.210	Failed	0.084	Safe	0.234	Risk of Split
200 X 150	30	0.201	Failed	0.076	Safe	0.252	Risk of Split
200 X 125	29	0.255	Failed	0.088	Failed	0.209	Risk of Split
200 X 100	30	0.223	Failed	0.070	Failed	0.191	Risk of split
200 X 75	30	0.079	Safe	0.070	Safe	0.137	Safe

The variation of GSNI with punch travel (upto failure as per the FLD) shown in Figure 2a shows a narrow band of variation across the six blank widths studied. The red coloured rectangular blocks indicate the GSNI calculated from the simulation data at the onset of necking or cracks. From simulation it the GSNI values were found to increase with the increasing width of sample, due to a greater spatial non-uniformity of strain distribution. The narrow band describing variation over six blank widths may be represented by a single line indicating the average GSNI (Figure 2b). The extreme lines of the band lie within ± 0.02 of the average GSNI.

Figure 2 (c) indicates the correlation of experimental SNI with GSNI and the SNI determined from simulations. The correlation coefficients for the two relations are not very high on account of scatter seen in the figure. However, with respect to the experimental SNI, the simulation based SNI shows a variation of less than 0.05 from the least squares curve representing the fit. The difference in GSNI and simulation based predicted SNI is due to strain peaks being captured along the meridian of the sample, whereas calculation of the GSNI involves the strains over the entire area of the blank.

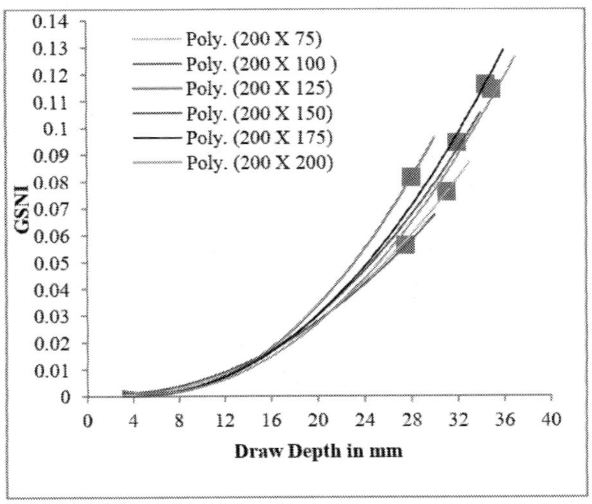

(a)

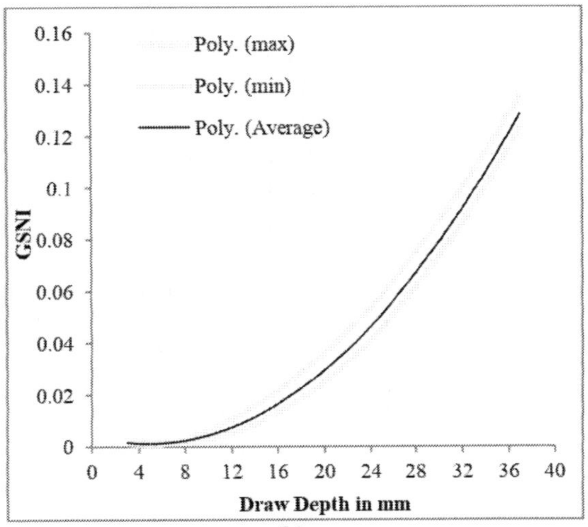

(b)

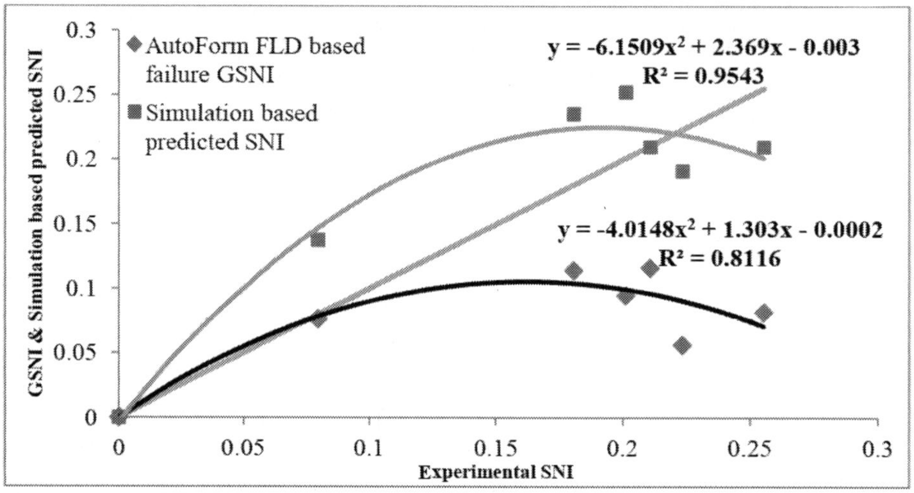

(c)

Figure 2. Draw depth vs. simulation based GSNI (a), bandwidth of GSNI along with its average (b) and correlation of experimental SNI with SNI and GSNI (c)

Also, in simulation based SNI prediction the strain values are taken along a multiplicity of critical planes, whereas the longitudinal meridional strain nonuniformity represents the SNI in the experimental Nakazima blank. The values of simulation based predicted SNI and experimental SNI are comparable with each other.

The average value of SNI at failure by experimentation (over the 6 blanks) is about 0.19 and that by simulation based SNI prediction program at failure turns out to be about 0.21, which are fairly close.

4. Conclusions

1) The relation between the experimental SNI and the analytically determined SNI and GSNI is nonlinear. The constants of the equation will depend on the material.
2) The analytically predicted SNI differs from the experimental SNI due to deviation of actual experimental conditions from those in simulation. Moreover, differences in method of obtaining the strain distribution also contribute to the difference.
3) The variation in GSNI with blank width across the 6 blank widths studied lies within a narrow band that is ± 0.02 wide.

5. References

[1] Standard test methods for determining the forming limit curve. ASTM Standards Designation: E2218 – 14 ε1 pp 1-15.
[2] Schedin E and Melander A, 1986 The evaluation of large strains from industrial sheet metal stampings with a square grid. J. Applied Metalworking 4(2) pp 143–156.
[3] Karima, Chandrasekaran N, and Tse W, 1989 Process signatures in metal stamping: basic concepts. J. Mater. Shaping Technol. 7 pp 169–183.
[4] Schedin E and Melander A, 1987 On the strain distribution during the stretch forming of low- and high-strength sheet steels. J. of Mech. Working Tech., 15 pp 181–202.
[5] Desai S. G. and Date P. P., 2006, On the quantification of strain distribution in drawn sheet metal products, Journal of Materials Processing Technology 177 pp439–443.

Experimental and Numerical Investigation into the Effect of Strain Rate Changes on Failure of AA7075 at the Elevated Temperatures

H.Zand[1], F.Zhalehfar[2], A.Vafaeesefat[3]

1. Postgraduate student, the Science and Research Branch of the Azad University
2. Corresponding author, Die Design Office, Automotive Industry Research & Innovation Centre of SAIPA
3. Professor, Dept. of Mechanical Eng., School of Engineering, Imam Hussein University
Email: Farhadzhalehfar@yahoo.com, Jalehfar.f@airic-ir.com

Abstract. This paper represents the temperature-dependent study which is performed on AA7075. This was performed through the application of deep drawing experiments at high temperature of 250, 300, 350, 400, 450 and 500◦C and different forming speeds. FEM study is applied at room and 300 and 400 0C temperature with different strain rates. Softening model is used to model the thermo-mechanical constitutive equation. Forming limit curve models (based on M-K model) were used in the analysis of simulation results in order to predict the onset of necking. Simulation results were compared with experimental data to evaluate the accuracy of onset of necking simulation and failure location prediction

1. Introduction

The 7XXX aluminum series (Al-Zn-Mg-Cu alloy system) are very attractive materials in automobile and aerospace industries [1]. But the notable drawback is their formability which is significantly lower than traditional steel alloys at room temperature .Nevertheless, the formability of them increases significantly with increasing forming temperature up to the recrystallization temperature [2, 3]. In this study, a direct hot forming process has been examined, where blanks are heated up in high temperature die, and the effects of strain rate at seven different temperatures from 25 to 500◦C, which cover both warm and hot temperatures, have been investigated. Forming speed is varied from 5 to 200 mm/min in order to evaluate the effect of strain rate on failure mechanism. FEM study has been performed where M-K based FLD failure criteria has been considered in order to predict the failure. In order to model the constitutive behavior of AA7075 at higher temperature and calculating forming limit curve, the softening model has been used.

2. Experimental procedure

Specimens of AA7075-T6 with same dimensions were considered for different strain rates and temperatures. Samples are formed in die which consists of six main parts: 1. Punch, 2. Blank Holder, 3. Heat controller equipment, 4. Heat isolation system, 5. Upper die, 6. Heat condenser system. Forming system is presented in Fig.1. Because onset of necking is investigated in this study, the

Journal of Physics: Conference Series **734** (2016) 032109 doi:10.1088/1742-6596/734/3/032109

forming test continued to the point that no drastic change in trend of required force for forming process was observed. Samples are circular blanks with 140 mm in diameter and 3.2 mm in thickness.

Fig.1. Hot forming die and its main parts.

3. Numerical Model

In order to model 3D finite element simulation, after modeling of all geometries in Fig. 1 and taking advantage of having symmetric die, three sets of uniaxial tensile tension tests have been accomplished. Anisotropic coefficients of this material along different direction proportional to rolling direction are shown in Table.1. To investigate the mesh sensitivity, some simulations based on the different size of elements vs. punch load were done and element size 0.0015 mm shows a good correlation with experimental data, thereby 0.0015 (m) size of element considered as a suitable element size for blank in simulation.

Table 1. Mechanical property of AA7075

Temperature	R_0	R_{90}	R_{45}	σ_0 - MPa
25	0.45	1.72	1.24	503
300	2.3	4.2	4.7	72
400	2.9	5.11	5.73	66.3

In order to model the softening phenomenon, a modified equation has been considered. General form of this equation is shown below,

$$\sigma_f(T,\dot{\varepsilon}) = C(T,\dot{\varepsilon})\varepsilon^{n(T,\dot{\varepsilon})}\exp(\lambda\log(T) - T_h\varepsilon) \quad (1)$$

Where λ is the fitting parameter and it is taken 10-8 for Aluminum alloys. T_h is defined as the absolute temperature T_a(K) divided by absolute melting temperature $T_m^a(K)$. Parameters C and n are improved as below:

$$C(T,\dot{\varepsilon}) = C_0 + a_1[1-\exp(a_2 \frac{T-273}{T_m})] + a_3 \log(1+\frac{\dot{\varepsilon}}{\dot{\varepsilon}_{ref}}),\ n(T,\dot{\varepsilon}) = n_0 + b_1[1-\exp(b_2 \frac{T-273}{T_m})] + b_3 \log(1+\frac{\dot{\varepsilon}}{\dot{\varepsilon}_{ref}}) \quad (2)$$

Above two equations include constants where can be calculated using the least square approximation. Calculated parameters are shown in table 2. [4]

Table 2. Fitting parameters for Softening model, T>300

Tm	Co	no	a_1	a_2	a_3	b_1	b_2	b_3
920 K	120.2 MPa	0.09	32.1 MPa	31	0.264	0.121	0.01	-0.0248

4. Results and Discussions

Fig.2 demonstrates the effect of temperature on stress-strain curve and formability at elevated temperature. As it can be seen, the alloy softens by increasing temperature, while showing the hardening behavior in higher strain rate magnitudes. As it's been argued, DRV is the main cause of softening behavior of this alloy at higher temperature below 0.4 s^{-1} strain rates. [5]

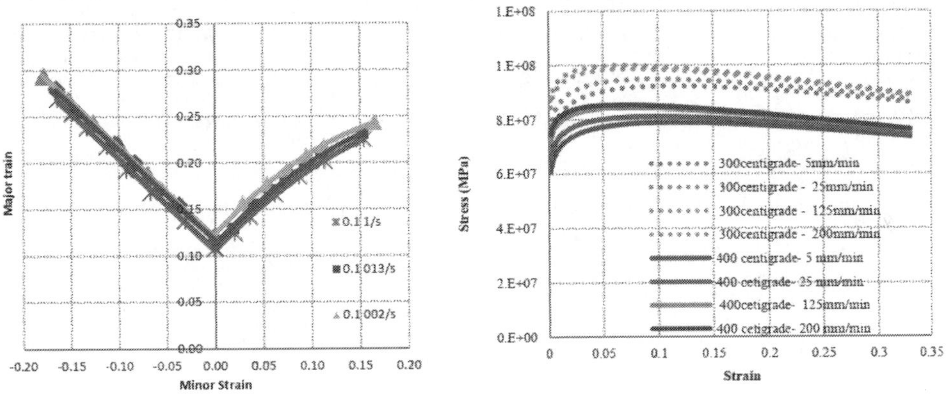

Fig. 2. Numerical assumption of Mechanical behavior of AA7075 at different strain rates

Fig.3 demonstrates the effect of temperature and strain rate on the maximum height of samples. As it can be seen, the maximum height increases markedly with an increase in temperature. Moreover, this amount decreases while the strain rate increases. In addition, the crack type changes from the brittle to ductile while temperature is increasing. Meanwhile, the failure position relocates from the top of the sample, near the punch radius, at room temperature to around die radius at higher temperatures which is correlated with the past studies [5, 6, and 7]. The effect of temperature on required force is considerable. Fig. 4 shows the force changes with temperature. Heating to higher temperatures causes significant reduction in amount of ultimate stress, and subsequently according to the Fig. 4-A, the required force will decrease in forming process. Moreover, increasing strain rate causes the slight increase in stress throughout the hardening phenomenon of material. As a result, the required force increases with the strain rate increase. Fig. 4-B illustrates the effect of temperature and strain rate variation on maximum height of samples.

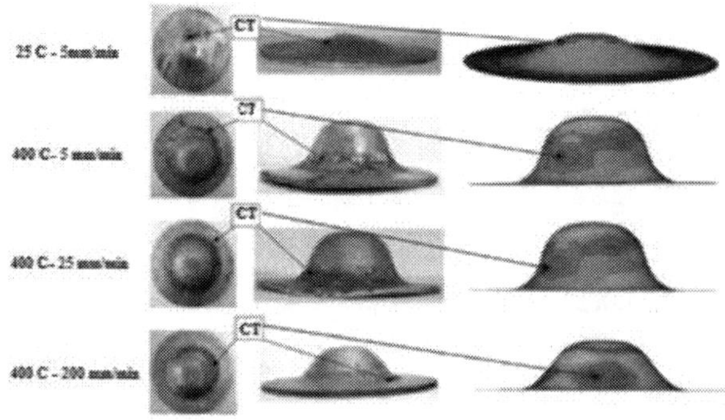

Fig.3. Samples through different strain rates. (CT: Crack Tip)

As it can be seen, increasing in temperature to below 300°C, results in a significant increase in formability of AA7075. Ironically, this trend changes after temperatures over 300°C, causing reduction to just a few millimeters at 450°C ,and drops off markedly in 500°C. The main reason of this trend can be the increasing in friction between die and blank throughout the changes of blank state from solid to semisolid condition, variation in surface topology, removal of oxide layers and wear of the tool [5]. In addition, the maximum height increases as well as the reduction of strain rate. The main reason of increase in height between room temperature and elevated temperatures is the softening of sheet under this condition. As shown, the height of samples decreases, while the strain rates rise steadily and much higher reduction is shown at higher temperatures throughout the strain rate hardening of sheet metals.

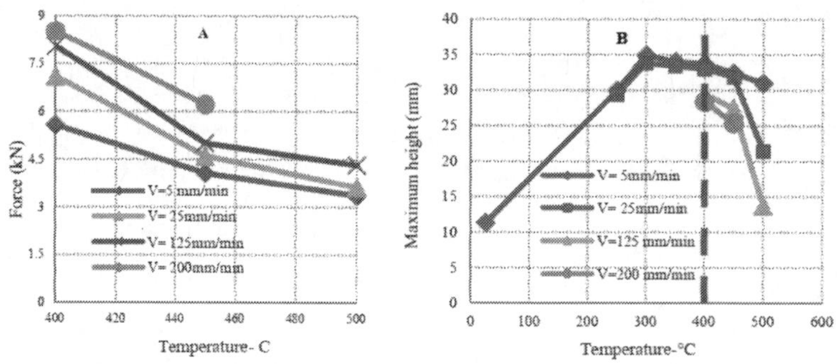

Fig.4. combined effects of temperature and forming velocity on A: required force, B: maximum height of cups

5. References

[1] Tajally.M, Emadoddin.E, 2011. Journal of Materials and Design 32, 1594-1599.
[2] Peng.P, Zhong-qin.L, Guan-long.C and Muamme.K, 2006. Trans. Nonferrous Met. SOC. China 16, 267-273.
[3] Bolt.P, Lamboo.N and Rozier.P, 2001. Journal of Materials Processing Technology, vol. 115, pp. 118–121.
[4] Toros.S, Ozturk.F, 2010. Computational Materials Science 49, 333-339.
[5] Abedroo.N, Pourboghrat.F, 2006. International Journal of Plasticity 22, 314–341.
[6] Kurukuri.S, A. H. van den Boogaard, Miroux.A, 2009. J. Materials Processing Technology vol. 209, pp. 5636–5645.
[7]Takuda.H, Mori.K, Masuda.I, Abe.Y, Matsuo.M, 2002. Journal of Materials Processing Technology,vol.120, pp.412-418.

Numisheet
Journal of Physics: Conference Series **734** (2016) 032110

IOP Publishing
doi:10.1088/1742-6596/734/3/032110

Constitutive modeling and simulation at room-temperature deformation and failure of polycrystalline Molybdenum

Geremy Kleiser[1,2], **Benoit Revil-Baudard**[2], **Oana Cazacu**[2] **and Crystal L. Pasiliao**[1]

[1] Air Force Research Laboratory, Eglin AFB, FL, USA
[2] Department of Mechanical and Aerospace Engineering, University of Florida, REEF, 1350 N. Poquito Rd., Shalimar, FL 32579, USA

E-mail: geremy.kleiser@us.af.mil, revil@ufl.edu, cazacu@reef.ufl.edu

Abstract. In this paper is presented a systematic experimental investigation of the mechanical response of polycrystalline commercially pure molybdenum (Mo). It was established that the material has ductility in tension at 10^{-5}/s and that the failure strain is strongly dependent on the orientation. A specimen taken along the rolling direction sustains large axial strains (20 %), while a specimen cut at an angle of 45° to the rolling direction could only sustain 5% strain. Irrespective of the loading orientation the yield stress in uniaxial compression is larger than in uniaxial tension. While in tension the material has a strong anisotropy in Lankford coefficients, in uniaxial compression it displays weak strain-anisotropy. An elastic-plastic orthotropic model that accounts for all the specificities of the plastic deformation of the material was developed. Validation of the model was done through comparison with data on notched specimens. Quantitative agreement with both global and local strain fields was obtained.

1. Introduction

For high-temperature applications, a promising alternative to nickel-based super alloys is the use of refractory metals. Molybdenum (Mo) is an attractive candidate due to its significant lower density than other refractory metals, but its limited ductility at room temperature is a major concern. Although experimental studies aimed at understanding the room-temperature tensile behavior were undertaken, the mechanical response in compression, the influence of loading orientation on the tension-compression asymmetry, and the influence of the strain rate on the plastic behavior have been largely unexplored. In this work, a comprehensive experimental and theoretical investigation of the quasi-static and high strain-rate, room-temperature, and mechanical response of a polycrystalline Mo material is presented. We begin by reporting the results of a series of monotonic tensile and compression tests, conducted to quantify the plastic anisotropy and the tension-compression asymmetry of the material for a large range of strain rates. Yielding is described using Cazacu et al. [1] yield criterion that accounts for both anisotropy and strength differential effects. Model validation is provided through comparison between the simulations and data on quasi-static uniaxial tension test notched specimens.

2. Constitutive modeling of impact behavior of pure Mo

The material is a high purity (99.98%) polycrystalline molybdenum plate. Electron back-scattered diffraction (EBSD) measurements show that the material has almost equiaxed grains and most of the grains within the plane of the plate had an orientation within ± 30 of the rolling direction (RD). In order to quantify the influence of the loading direction on the room-temperature mechanical response, uniaxial tensile and compression tests on smooth specimens were performed for different loading orientations according to ASTM E8 and ASTM E9, respectively. For uniaxial tensile test along RD at a strain rate of 10 $^{-2}$/s, the material displays a brittle behavior (strain at failure less than 1%). Given the strain-rate sensitivity of the brittle-to-ductile transition of refractory metals, uniaxial tension tests at a strain rate of 10^{-5}/s were conducted. For this strain rate, the material exhibits ductile behavior, but the strain at failure is strongly dependent on the loading orientation. The specimens cut along RD and at 15° from RD, sustain a tensile axial true strain of 21%, but specimens cut at θ greater than 15° show significantly lower ductility, the specimen taken at 45° to RD only admitting 4% axial strain before failure. Digital image correlation (DIC) measurements were taken in order to monitor the evolution of the local strains. To gain a better understanding of the specificities of the plastic deformation of the material studied, two cameras were disposed to independently monitor the evolution of the axial, width, and thickness strains. Camera 1 captures the evolution of axial and width strains occurring in the front of the specimen, while camera 2 monitors the evolution of the axial and thickness strains on the side of the specimen. The compression tests data showed that polycrystalline molybdenum displays tension-compression asymmetry. It was observed that irrespective of the loading orientation the yield stress in compression is larger than in tension. A constitutive model was developed. Key in the formulation is the use of an yield function proposed by Cazacu et al. [1] that accounts for both anisotropy and tension/compression asymmetry. The effective stress associated with the yield criterion is:

$$\bar{\sigma} = \mathrm{B}\Big[(|\tilde{\sigma}_1| - k\tilde{\sigma}_1)^2 + (|\tilde{\sigma}_2| - k\tilde{\sigma})^2 + (|\tilde{\sigma}_3| - k\tilde{\sigma}_3)^2\Big]^{\frac{1}{2}} \tag{1}$$

where k is an internal variable, its range of variation being (-1, 1), while $\tilde{\sigma}_1$, $\tilde{\sigma}_2$ and $\tilde{\sigma}_3$ are the principal values of the transformed stress tensor $\tilde{\sigma} = [\mathbf{L}] : \mathbf{S}$ with $\mathbf{S}$ the Cauchy stress deviator applied to the material, L being a fourth-order symmetric tensor. In Eq.(1), B is a constant defined such that the equivalent stress $\bar{\sigma}$ reduces to the tensile stress along RD. The test results in uniaxial tension and compression were used to identify the material parameters involved in the Cazacu et al. [1] yield criterion. The numerical values of these parameters corresponding to yielding are:

$$[\mathbf{L}] = \begin{bmatrix} 1.0 & 0.928 & 0.933 & 0 & 0 & 0 \\ 0.928 & 0.997 & 0.939 & 0 & 0 & 0 \\ 0.933 & 0.939 & 1.021 & 0 & 0 & 0 \\ 0 & 0 & 0 & 0.0641 & 0 & 0 \\ 0 & 0 & 0 & 0 & 0.0641 & 0 \\ 0 & 0 & 0 & 0 & 0 & 0.0641 \end{bmatrix} \quad \text{and} \quad \mathrm{k} = -0.0547 \tag{2}$$

3. Simulation of the uniaxial tensile response for smooth and notched specimens

Comparison between the experimental and simulated load vs. axial displacement curves for smooth tensile specimens oriented along RD and TD are shown in Fig.1 and Fig. 2, respectively. The experimental axial displacement corresponds to the axial displacement measured using DIC techniques. Note the excellent agreement in terms of the description of the overall response, in particular the accuracy with which it is described the influence of the loading orientation. Furthermore, for each loading orientation, the F.E. predictions of the axial, width and thickness strains isocontours (superposed on the lower part of the specimen) are compared to the DIC strain maps obtained experimentally for the same axial displacement. Irrespective of the specimen orientation, a very good quantitative agreement between the numerical predictions and the experimental measurements of the local fields was obtained.

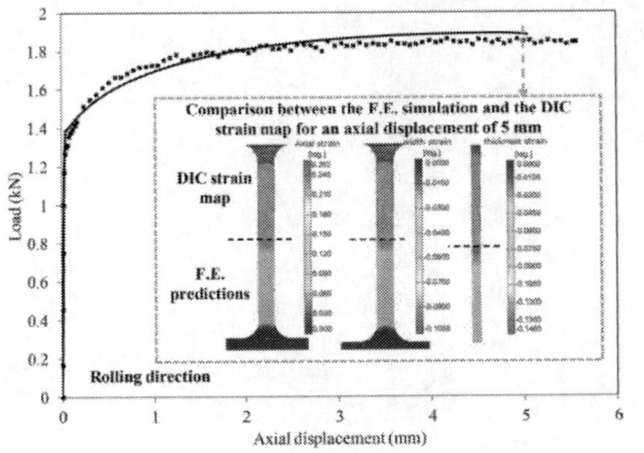

Figure 1. Comparison between the theoretical and predicted load vs. axial displacement curves in uniaxial tension on smooth specimens along RD.

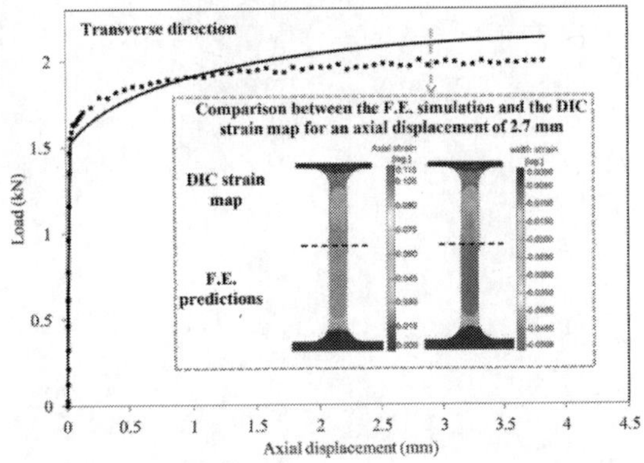

Figure 2. Comparison between the theoretical and predicted load vs. axial displacement curves in uniaxial tension on smooth specimens along TD.

Uniaxial tension tests on notched specimens have also been performed to further the understanding of the mechanical behavior of Mo when subjected to non-axisymmetric loadings and to validate the elastic-plastic model developed. To this end, F.E. simulations using the model will be compared to global and local fields on tensile tests of notched specimen using DIC techniques. Tests were conducted on RD and TD specimens. The gage length cross section of the notched specimen is a square of side of 1.5875 mm, and the notch radius is 3.175 mm. While the camera 1 allows monitoring the axial strain and the width strain developing in the front of the specimen, the camera 2 informs about the axial strain and the thickness strain occurring on the side of the specimen. Fig.3a shows the axial strain maps acquired with the two cameras for the RD specimen. Note that the axial strain is no more homogenous in the cross-section, the axial strain that develops on the front face (camera 1) is smaller than the axial strain that occurs on the side of the specimen. For quantitative comparisons with predicted results, average values of the axial strain and width strain over the line AB (camera 1) and over the line BC (camera 2) calculated from the DIC measurement were compared with the numerical predictions. As an example, for the RD specimen, comparison between the simulated and experimental measurements of the evolution with the axial displacement of the average axial strain and width strain occurring on the front of the specimen (AB) is presented in Fig.3b while the comparison of the evolution of the axial strain and thickness strain occurring on the side of the specimen is shown in Fig.3c. Note that the model correctly predicts the evolution of all the local strains (for more details, see [2]).

4. Conclusions

In this work, the experimental characterization, modeling and simulation of the plastic anisotropy and tension-compression asymmetry of polycrystalline commercially pure molybdenum at quasi-static and high-rate was presented. It was established that polycrystalline molybdenum has ductility in tension at 10^{-5}/s, and that the failure strain is strongly dependent

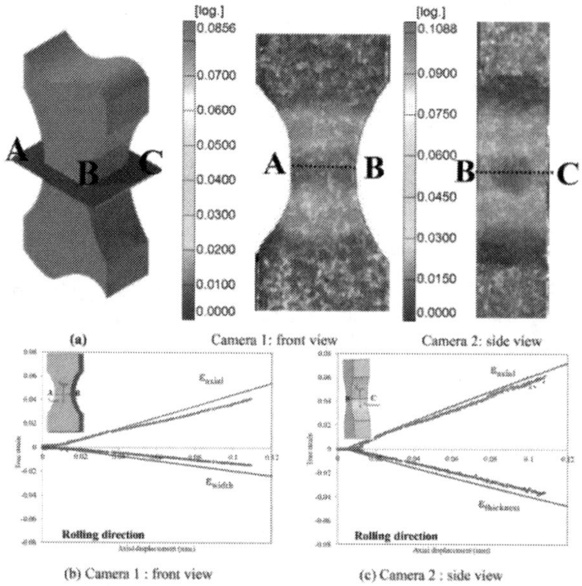

Figure 3. (a) Strains maps acquired with each camera. Comparison between model predictions (solid lines) of local strains and the average local strains at the root of the notch acquired by DIC with: (b) camera 1 (front view) and (c) camera 2 (side view), as a function of axial displacement for a notched specimen loaded along the rolling direction.

on the orientation. To accurately quantify the anisotropy in plastic deformation, DIC measurement techniques were used. The compression tests data showed that polycrystalline molybdenum displays tension-compression asymmetry, the yield stress in compression being larger than in tension. Furthermore, evaluation of the ellipticity of the deformed compression specimens allowed us to uncover that although the material exhibits a strong strain anisotropy in tension (e.g. the r-value at 45° is 0.665 while r-value in TD is close to 1). An elastic-plastic model that accounts for all the specificities of the plastic deformation of the material was developed. Key in the formulation was the use of a yield function that simultaneously accounts for anisotropy and tension compression asymmetry. The predictive capabilities of the model has been assessed for both quasi-static and impact loading. Uniaxial tension tests on notched specimens were performed, showing that the strain is no longer homogeneous along the cross-section, the axial strain that develops on the front of the specimen is smaller than the one that develops on the side. Quantitative agreement between model and data for both global and local strain fields was obtained for several orientation of the specimen. The predictive capabilities of the model has also been challenged for dynamic/impact loading. Due to the Mos limited tensile ductility, characterization of the behavior under impact is particularly challenging. Taylor impact tests were successfully conducted on this material for impact velocities in the range 140-165 m/s and the model well predicts the geometry of the final Taylor specimen, as well as the length of the plastic zone for impact loadings (For more details about the high rate deformation of Mo, see [3]).

References

[1] Cazacu O, Plunkett B and Barlat F 2006 *Int. J. Plasticity* **22** 1171
[2] Kleiser G., Revil-Baudard B., Cazacu O.,Pasiliao C.L. 2015 *Int. J. Solids Struct,* **75-76** 287
[3] Kleiser G., Revil-Baudard B., Pasiliao C.L. 2016 *Int. J. Impact Eng.*, **96** 116

Journal of Physics: Conference Series **734** (2016) 032111 doi:10.1088/1742-6596/734/3/032111

Mechanical behaviour of TWIP steel under shear loading

G. Vincze[1]*, M.C. Butuc[1] and F. Barlat[1,2]

[1]Centre for Mechanical Technology and Automation, University of Aveiro, Campus Universitário de Santiago, 3810-193 Aveiro, Portugal

[2]Graduate Institute of Ferrous Technology, Pohang University of Science and Technology, San 31 Hyoja-dong, Nam-gu, Pohang, Gyeongbuk 790-784, Republic of Korea

gvincze@ua.pt

Abstract. Twinning induced plasticity steels (TWIP) are very good candidate for automotive industry applications because they potentially offer large energy absorption before failure due to their exceptional strain hardening capability and high strength. However, their behaviour is drastically influenced by the loading conditions. In this work, the mechanical behaviour of a TWIP steel sheet sample was investigated at room temperature under monotonic and reverse simple shear loading. It was shown that all the expected features of load reversal such as Bauschinger effect, transient strain hardening with high rate and permanent softening, depend on the prestrain level. This is in agreement with the fact that these effects, which occur during reloading, are related to the rearrangement of the dislocation structure induced during the pre-deformation. The homogeneous anisotropic hardening (HAH) approach proposed by Barlat et al. (2011) [1] was successfully employed to predict the experimental results.

1. Introduction

One of the oldest materials, steel, has been continuously improved over the years, maintaining its supremacy for manufacturing even today. The Twinning Induced Plasticity (TWIP) steel is one of the new steel grades, which exhibit high tensile strength in the range 1300-1500 MPa and significant ductility with strains to failure above 0.3 (30%). This high performance is the result of the exceptional strain hardening ability induced by the TWIP effect [2-4]. The obstacles produced by twins to dislocation glide lead to the reduction of the dislocation mean free path. Recently, it was shown by Bintu et al. [5] that the strain hardening of TWIP steel is strongly dependent on twin density, which can lead to a negative strain hardening rate sensitivity parameter. In the cited work, the author changed the density of twins by varying the strain rate. In the present work, different prestrain amount are considered for the same purpose.

The large flow stress and large ductility of the TWIP steel makes this material useful in applications requiring large energy absorption. Since most applications require that the material deforms plastically under non-proportional loading, the objective of this work is to analyse the mechanical response of TWIP steel subjected to reverse loading when the prestrain is gradually increased. Moreover, this study is conducted to evaluate the capability of the homogeneous anisotropic hardening (HAH) model proposed by Barlat et al [1] to describe the material behaviour.

Journal of Physics: Conference Series **734** (2016) 032111 doi:10.1088/1742-6596/734/3/032111

2. Material and experimental details

An Fe-Mn-C-Si-Al austenitic TWIP steel supplied by POSCO (South Korea) with the chemical composition shown in Table 1 is considered in this study. The material was supplied in the form of sheet of 1.85 mm thickness. The as-received TWIP steel is austenitic at room temperature. The austenite is stabilized by a high concentration of Mn (higher than 15%). Under load, this material twins, but no phase transformation to martensite is observed. Si is added to prevent C precipitation leading to the formation of carbides.

Table 1 Chemical composition of the TWIP steel considered in the present study

C	Mn	Si	Al
0.6%	18%	0.22%	1.5%

The material was tested in simple shear using a Shimadzu Autograph Machine (Shimadzu, Japan) with maximum load capacity of 100 kN and the deformation was measured with ARAMIS - non-contact and material-independent measuring system based on digital image correlation (GOM, Germany). The simple shear device used in this work is an improvement of the previous tool developed at the University of Aveiro. This device compared to others has the particularity that it is compact enough to use it inside a thermic chamber if needed. Fig. 1 shows the simple shear device used in the present work. Moreover, due to the high strength of this material, the geometry of the specimens (Fig. 2) proposed by Guitierez-Urutia et al. [6] is adopted.

Figure 1. Simple shear device for AHSS

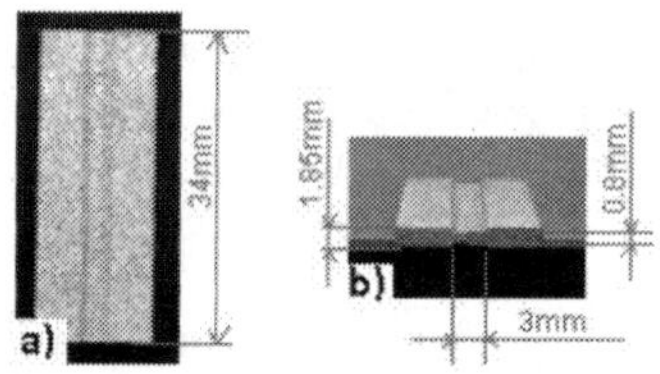

Figure 2. Sample for shear test.
a) front view; b) up view

3. Constitutive Modelling

In this study, an advanced constitutive model based on anisotropic non-quadratic yield functions and a distortional hardening framework, is used. A brief description is given below but more details can be found in [1].

3.1 Swift isotropic hardening model

The Swift model expresses the isotropic strain hardening component in the form of a the power law

$$\sigma = K(\varepsilon_0 + \bar{\varepsilon})^n \tag{1}$$

where $\bar{\varepsilon}$ is the effective plastic strain, and K, ε_0, n are material parameters.

3.2 YLD2000-2D Yield Function

The Yld2000-2d plane stress yield function introduces plastic anisotropy with two linear transformations on the Cauchy stress tensor. It is expressed in terms of the deviatoric stress components as:

$$\phi = \phi'(\tilde{S}') + \phi''(\tilde{S}'') = 2\bar{\sigma}_Y^a, \tag{2}$$

where ϕ' and ϕ'' are two isotropic functions defined by

$$\phi'(\tilde{S}') = \left|\tilde{S}'_1 - \tilde{S}'_2\right|^a, \tag{3}$$

$$\phi''(\tilde{S}'') = \left|2\tilde{S}_2'' + \tilde{S}_1''\right|^a + \left|2\tilde{S}_1'' + \tilde{S}_2''\right|^a, \tag{4}$$

while $\tilde{S}'$ and $\tilde{S}''$ are linear transformations of the effective stress tensor s, which is defined as the deviatoric part of the Cauchy stress. Hence,

$$\tilde{S}' = C' s, \qquad \tilde{S}'' = C'' s, \tag{5}$$

where C' and C'' represent the linear transformations of the effective stress tensor **s** and contain the material anisotropy coefficients.

3.3 Homogeneous anisotropic hardening (HAH)

This formulation is a combination of a conventional yield function (which can be isotropic or anisotropic) and a fluctuating component. It is described by the follows equation:

$$\Phi(\mathbf{s}) = \left(\phi_s^q + \phi_f^q\right)^{\frac{1}{q}} = \left[\phi_s^q + f_1^q\left|\hat{\mathbf{h}}^s : \mathbf{s} + \left|\hat{\mathbf{h}}^s : \mathbf{s}\right|\right|^q + f_2^q\left|\hat{\mathbf{h}}^s : \mathbf{s} - \left|\hat{\mathbf{h}}^s : \mathbf{s}\right|\right|^q\right]^{\frac{1}{q}} = \sigma(\bar{\varepsilon}) \tag{6}$$

where **s** is the stress deviator and q a constant exponent, $\hat{\mathbf{h}}$ is a normalized tensorial state variable, called the microstructure deviator; f_1 and f_2 are two functions of two state variables g1 and g2; g3 and g4 capture the permanent softening. The coefficients k, k_1 to k_5 control the evolution of the microstructure deviator and state variables g_1 to g_4, respectively(see Table 2).

Table 2. The evolutions of the state variables

$\hat{\mathbf{h}}^s : \mathbf{s} \geq 0$		$\hat{\mathbf{h}}^s : \mathbf{s} < 0$	
$\frac{dg_1}{d\bar{\varepsilon}} = k_2\left(k_3\frac{\bar{\sigma}_0}{\bar{\sigma}(\bar{\varepsilon})} - g_1\right)$	(7)	$\frac{dg_1}{d\bar{\varepsilon}} = k_1\frac{g_4 - g_1}{g_1}$	(11)
$\frac{dg_2}{d\bar{\varepsilon}} = k_1\frac{g_3 - g_2}{g_2}$	(8)	$\frac{dg_2}{d\bar{\varepsilon}} = k_2\left(k_3\frac{\bar{\sigma}_0}{\bar{\sigma}(\bar{\varepsilon})} - g_2\right)$	(12)
$\frac{dg_4}{d\bar{\varepsilon}} = k_5(k_4 - g_4)$	(9)	$\frac{dg_3}{d\bar{\varepsilon}} = k_5(k_4 - g_3)$	(13)
$\frac{d\hat{\mathbf{h}}_s}{d\bar{\varepsilon}} = k\left(\hat{\mathbf{s}} - \frac{8}{3}\hat{\mathbf{h}}_s\left(\hat{\mathbf{h}}_s : \hat{\mathbf{s}}\right)\right)$	(10)	$\frac{d\hat{\mathbf{h}}_s}{d\bar{\varepsilon}} = -k\left(-\hat{\mathbf{s}} + \frac{8}{3}\hat{\mathbf{h}}_s\left(\hat{\mathbf{h}}_s : \hat{\mathbf{s}}\right)\right)$	(14)

Table 3. Material constitutive coefficients of TWIP steel used for simulation.

Swift law	*K [MPa]*	ε_0	n						
	2398	0.12	0.685						
Yld2000-2d	a	α_1	α_2	α_3	α_4	α_5	α_6	α_7	α_8
	6	0.9441	1.0407	1.0123	0.9820	1.0322	1.0656	1.0127	0.9965
HAH	q	k	k_1	k_2	k_3	k_4	k_5		
	2	30	40	45	0.1	0.75	8		

4. Results and discussion

The experimental stress-strain curves of TWIP steel deformed in forward-reverse simple shear loading are shown in Fig. 3. For reverse loading, the amount of prestrain was gradually increased in order to vary the twin density. Besides of a strong Bauschinger effect and a transient hardening stage, it can be observed that the hardening rate decreases with increasing the prestrain amount.

In Fig. 4, the calculated and experimental shear stress-shear strain curves are presented in two graphs for the sake of clarity (two curves in each graph). For low amounts of prestrain, the model captures all the characteristics of the material behaviour very well. On the other hand, for a prestrain of 65%, the model tends to overestimate the permanent-softening behaviour. One solution is to decrease the coefficient k_4 of HAH model in order to emphasize the permanent softening, but of course this will affect the others curves. Another solution may be the replacement of the Swift model with a hardening law based on the dislocation density.

In conclusion, the HAH model can be used for a reasonably accurate modelling of TWIP steel subjected to reverse loading. In the future, it will be interesting to investigate the behaviour of TWIP steel in cross-loading conditions.

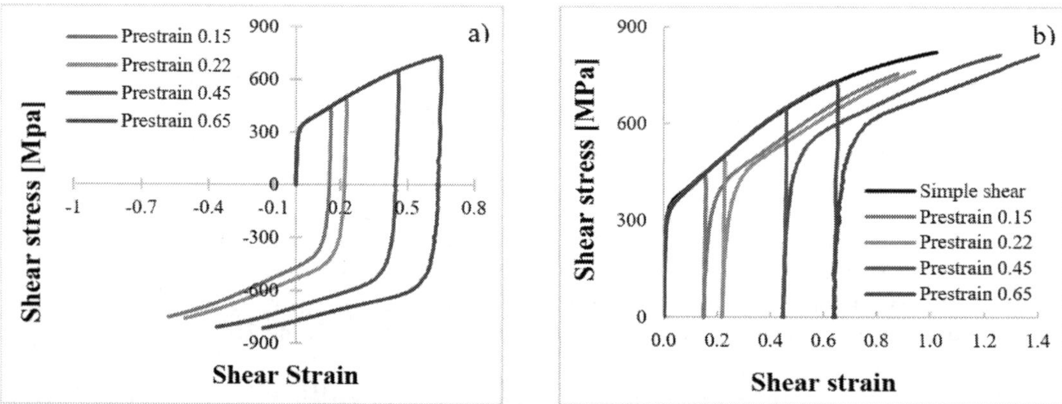

Figure 3. Experimental simple shear stress-strain curves with different prestrain amounts: a) Total strain with absolute stress and; b) cumulated strain and positive stress.

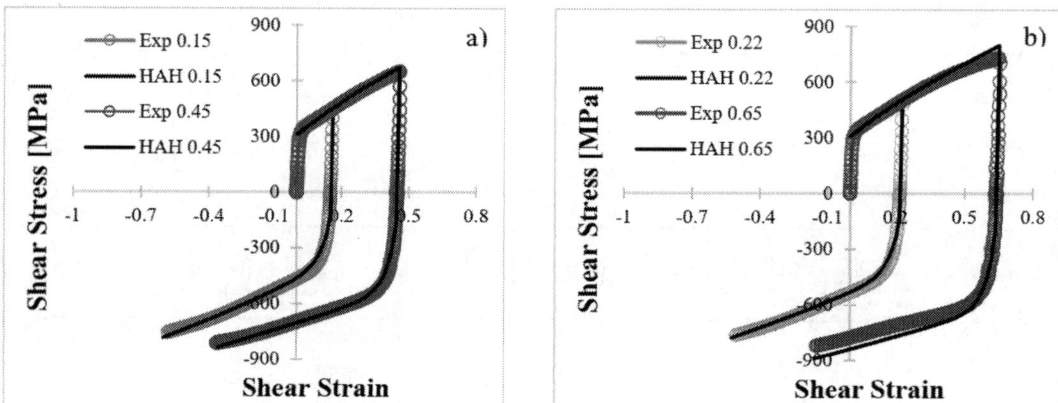

Figure 4. Calculated and experimental results with different prestrain amount: a) 15% and 45% and b) 22% and 65%

Acknowledgments

The authors acknowledge support from the Portuguese Foundation of Science and Technology (FCT) through projects PTDC/EME-PME/116683/2010 and UID/EMS/00481/2013. We thank POSCO (South Korea) for providing the studied material.

References

[1] Barlat F, Gracio J J, Lee M G, Rauch E F and Vincze G, 2011 *Int. J. Plast.* **27** 1309
[2] Barbier D, Gey N, Bozzolo N, Allain S, 2009 Journal of microscopy **235** 67
[3] U.F Kocks, 1996 Phil. Mag. **13** 541
[4] Lebedkina TA, Lebyodkin MA, Chateau JP, Jacques A, Allain S 2009 Mater. Sci. Eng.A **519** 147
[5] A. Bintu, G. Vincze, C. Picu, A. Lopes, J. Grácio, F. Barlat, 2015 Mat. Sci. Eng. A **629** 54
[6] I. Gutierrez-Urrutia, J. A. del Valle, S. Zaefferer, D. Raabe, 2010 *J Mater Sci.* **45** 6604–6610

Cold formability prediction by the modified maximum force criterion with a non-associated Hill48 model accounting for anisotropic hardening

J Lian*,1, D C Ahn2, D C Chae3, S Münstermann1 and W Bleck1

1 Steel Institute, RWTH Aachen University, Intzestraße 1, 52072 Aachen, Germany
2 Metal Forming Group, Global R&D Centre, POSCO, 180-1 Songdo-dong, Yeonsu-gu, 406-840 Incheon, South Korea
3 Stainless Steel Products Research Group, Technical Research Laboratories, POSCO, 1 Goedong-dong, Nam-gu, Pohang, Gyeongbuk 790-300, Pohang, South Korea

* junhe.lian@iehk.rwth-aachen.de

Abstract. Experimental and numerical investigations on the characterisation and prediction of cold formability of a ferritic steel sheet are performed in this study. Tensile tests and Nakajima tests were performed for the plasticity characterisation and the forming limit diagram determination. In the numerical prediction, the modified maximum force criterion is selected as the localisation criterion. For the plasticity model, a non-associated formulation of the Hill48 model is employed. With the non-associated flow rule, the model can result in a similar predictive capability of stress and r-value directionality to the advanced non-quadratic associated models. To accurately characterise the anisotropy evolution during hardening, the anisotropic hardening is also calibrated and implemented into the model for the prediction of the formability.

1. Introduction

Cold formability, normally characterised as the forming limit curve (FLC), is one of the key characteristics for the sheet metal forming industry. Among many influencing parameters, the plastic anisotropy in terms of both stress and deformation has significant impact on the cold formability behaviour [1, 2]. Various plasticity models have been developed for the accurate description of anisotropy and most of models are based on the associated flow rule [3-6]. Recently, the non-associated quadratic anisotropic models have received increasing attention, although the experimental justification of its application remains an open issue [7]. The use of these models is mainly driven by its efficiency and simplicity. Compared to the associated quadratic models, it improves the anisotropic characteristic predictive capability to a similar level to the advanced non-quadratic associated anisotropic models. During the plastic deformation process of anisotropic materials, the hardening as well as the deformation behaviour is actually evolving. These phenomena are normally referred to as anisotropic (distortional) hardening. However, only very few models consider this factor [8, 9], and most anisotropic plasticity models still use the initial/constant yield stress or r-value to determine the parameters of the yield criterion. In this study, the evolving feature of the yield surface is implemented into a non-associated Hill48 model and it is further used to give the prediction of the cold formability of a ferritic steel in conjunction with the modified maximum force criterion (MMFC) [10].

2. Modelling strategy

The Hill48 plasticity model [3] was the first and the most often employed anisotropic models for sheet metals. The yield function is defined by:

$$f = \bar{\sigma}(\boldsymbol{\sigma}) - \sigma_Y(\bar{\varepsilon}^p) \leq 0 \tag{1}$$

where $\bar{\sigma}(\boldsymbol{\sigma})$ is a function of stress components, defined in Equation (2) and $\sigma_Y(\bar{\varepsilon}_p)$ is the flow curve of specific material as a function of the equivalent plastic strain. Yield function defines the yielding process onset when $f = 0$.

$$\bar{\sigma}(\boldsymbol{\sigma}) = \sqrt{F(\sigma_{22} - \sigma_{33})^2 + G(\sigma_{33} - \sigma_{11})^2 + H(\sigma_{11} - \sigma_{22})^2 + 2L\sigma_{23}^2 + 2M\sigma_{13}^2 + 2N\sigma_{12}^2} \tag{2}$$

where *F, G, H, L, M* and *N* are material coefficients controlling the anisotropy of the yield surface. In the Hill48 plasticity model, the flow curve is obtained from the experimental true stress-true strain curve, normally the uniaxial tensile test along the rolling direction, and extrapolated to large strains. For the determination of the anisotropic coefficients of the Hill48 model, there are various possible ways, and among them two methods, the stress based method and r-value based method are the often used approaches. For details, readers are referred to Ref. [11].

One practical motivation for the non-associated Hill48 (nHill48) development is the inconsistent anisotropic prediction by the parameters calibrated based on the stress method or r-value method. For example, the model calibrated from the stress based method could properly describe the yield stress directionality but result in an inaccurate prediction on the r-value, or vice versa. The basic idea for the nHill48 formulation is that the yield function and flow potential both take the form of Hill48 model, but using the stress based method to calibrate the yield function and the r-value based method to calibrate the flow potential. To account for the anisotropic hardening, it basically requires that the yield surface/function is changing with respect to the evolution of the plastic strain. As the advantage of the nHill48 model is the straightforward anisotropic coefficient calibration without any iterative numerical procedure, the extension of it to account for the anisotropic hardening is simply to change the constant material parameters $F{\sim}N$ to a plastic strain dependent function $F(\bar{\varepsilon}^p){\sim}N(\bar{\varepsilon}^p)$.

For the prediction of FLC, the pioneering work was developed by Hill [12] for the localized necking and by Swift [13] for the diffuse necking. Based on the maximum force criterion by Swift [13], Hora [10] developed the MMFC, motived by the fact that the critical phase in the metal forming processes is not the occurrence of diffuse necking, but the localized necking, i.e. the plastic deformation is still allowed to a certain extent till a plane strain tension mode after the maximum force is reached. Therefore, the stress σ_{11} is not only a function of strain hardening but also the strain ratio β:

$$\frac{\partial\sigma_{11}}{\partial\varepsilon_{11}} + \frac{\partial\sigma_{11}}{\partial\beta}\cdot\frac{\partial\beta}{\partial\varepsilon_{11}} = \sigma_{11} \tag{3}$$

where σ_{11} and ε_{11} are, respectively, the major principal stress and strain; β is defined as the ratio between the major and minor principal strain rate, $\beta \equiv \dot{\varepsilon}_{22}/\dot{\varepsilon}_{11} = \text{const}$.

In this study, the plane stress condition is assumed. The non-associated Hill48 model is implemented together with the MMFC in an iterative manner to determine the FLC for various loading conditions. Due to the space limit, the detailed equations are not shown here.

3. Materials

In this study, a ferritic stainless steel sheet (AISI 439) with 1 mm thickness was investigated. It is a titanium stabilized, 17-18% chromium alloyed, low carbon steel, designed to resist corrosion in a variety of oxidizing environments from fresh water to boiling acids. The steel owns completely ferritic microstructure, i.e. body-centred cubic (bcc) crystal structure. Owning to its previous manufacturing processes, such as rolling and annealing, the steel shows strong anisotropy and off-γ-fibre texture component ~{554}<225> texture. For detailed texture information, readers are referred to Ref. [14].

4. Results

Tensile tests were conducted to characterise the mechanical properties according to the European standard EN 10002-1. To study the plastic anisotropy and to calibrate the material parameters of the anisotropic plasticity model, tensile tests along three typical directions, rolling direction (0°), diagonal direction (45°) and transverse direction (90°) were conducted. The engineering and true stress–strain curves for 0°, 45° and 90° are shown in Figure 1 (a) and (b), respectively. The intention of this study is to evaluate the anisotropic hardening behaviour, and therefore, the r-value results are not given due to the space limit. For more complete information, readers are referred to Ref. [11, 14].

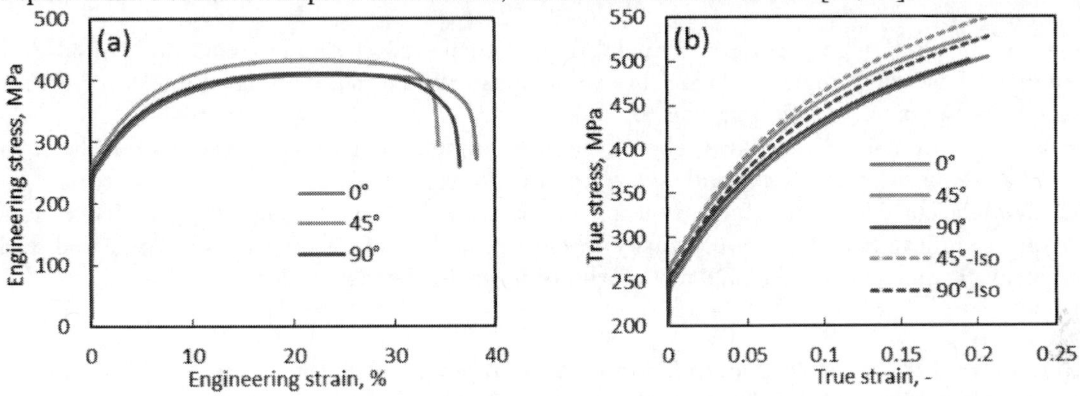

Figure 1. (a) The engineering stress–strain curves of AISI439 for different loading directions with respect to rolling direction (0°, 45°, 90°); (b) The true stress–strain curves and the anisotropic hardening behaviour along different loading directions (0°, 45°, 90°).

In Figure 1 (b), one can see that compared to the scaled isotropic hardening curve (dashed lines) calculated from 0° based on the initial anisotropic factor, the experimental flow curves for 45° and 90° are much lower at larger strains, although the agreement is very good at the beginning of the deformation. It is clearly noted that using the scaled initial anisotropic ratio results in substantial errors when the plastic deformation is getting large. Therefore, in this study, the yield surface evolution is calibrated for the non-associated Hill48 model, so that very accurate description for the anisotropic stress–strain curves at large deformation can be obtained in the subsequent formability prediction.

For the cold formability characterisation, the Nakajima tests were performed according to the standard (DIN EN ISO 12004-2). More details of the testing setup can be found in Ref. [15]. The final forming limit curves (FLCs) are presented in Figure 2 with two evaluation methods of the FLC at necking, the visual method and position dependent method.

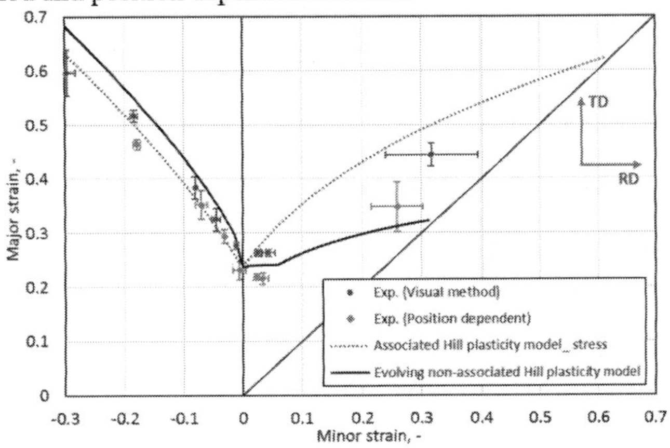

Figure 2. The FLCs measured by Nakajima tests at the instant of necking based on visual and position dependent method and the FLCs predicted by the MMFC for AISI 439.

Both the conventional associated Hill48 and the evolving nHill48 models are coupled with MMFC to perform the FLC prediction and the results are shown in Figure 2. It is noted that the parameters for the associated Hill48 are calibrated based on the stress method and only isotropic hardening is assumed. Clearly, the prediction by the associated Hill48 model overestimates the forming limit strains, especially in the biaxial tension region. On the other hand, the evolving nHill48 significantly improves the results in this region and, interestingly, it also shows a platform near the plane strain region that gives a very similar trend of the experimental results.

5. Conclusions

The aim of the study is to perform a formability prediction study by a non-associated Hill48 model accounting for the anisotropic hardening in conjunction with the modified maximum force criterion (MMFC). The following conclusions can be drawn:

- For the investigated bcc steel, significant anisotropic hardening is observed from the tensile tests along rolling, diagonal and transverse directions.
- With the consideration of the anisotropic hardening feature, the forming limit prediction results are significantly improved, especially in the biaxial tension loading region, compared to the conventional associated Hill48 model with isotropic hardening.

References

[1] Barlat F 1987 Crystallographic Texture, Anisotropic Yield Surfaces and Forming Limits of Sheet Metals *Materials Science and Engineering* **91** 55-72

[2] Lian J, Barlat F and Baudelet B 1989 Plastic Behavior and Stretchability of Sheet Metals .2. Effect of Yield Surface Shape on Sheet Forming Limit *Int. J. Plast.* **5** 131-47

[3] Hill R 1948 A Theory of the Yielding and Plastic Flow of Anisotropic Metals *Proc. R. Soc. London, Ser. A* **193** 281-97

[4] Barlat F and Lian J 1989 Plastic Behavior and Stretchability of Sheet Metals .1. A Yield Function for Orthotropic Sheets under Plane-Stress Conditions *Int. J. Plast.* **5** 51-66

[5] Barlat F, Brem J C, Yoon J W, Chung K, Dick R E, Lege D J, Pourgoghrat F, Choi S H and Chu E 2003 Plane stress yield function for aluminum alloy sheets—part I: theory *Int. J. Plast.* **19** 1297-319

[6] Barlat F, Aretz H, Yoon J W, Karabin M E, Brem J C and Dick R E 2005 Linear transfomation-based anisotropic yield functions *Int. J. Plast.* **21** 1009-39

[7] Stoughton T B 2002 A non-associated flow rule for sheet metal forming *Int. J. Plast.* **18** 687-714

[8] Stoughton T B and Yoon J W 2009 Anisotropic hardening and non-associated flow in proportional loading of sheet metals *Int. J. Plast.* **25** 1777-817

[9] Barlat F, Gracio J J, Lee M G, Rauch E F and Vincze G 2011 An alternative to kinematic hardening in classical plasticity *Int. J. Plast.* **27** 1309-27

[10] Hora P, Tong L and Reissner J 1996 A prediction method for ductile sheet metal failure. In: *NUMISHEET 1996,* ed J K Lee, *et al.* (Dearborn pp 252-6

[11] Lian J, Ahn D C, Chae D C, Münstermann S and Bleck W 2015 A Comparative Study on the Formability Prediction of Steel Sheets by Anisotropic Models Based on Associated Flow Rule and Non-Associated Flow Rule. In: *Key Eng. Mater.*, pp 150-5

[12] Hill R 1952 On Discontinuous Plastic States, with Special Reference to Localized Necking in Thin Sheets *J. Mech. Phys. Solids* **1** 19-30

[13] Swift H W 1952 Plastic Instability under Plane Stress *J. Mech. Phys. Solids* **1** 1-18

[14] Lian J, Ahn D C, Chae D C, Münstermann S and Bleck W 2015 Cold Formability of Automotive Sheet Metals: Anisotropy, Localization, Damage and Ductile Fracture. In: *Key Eng. Mater.*, pp 353-60

[15] Lian J, Liu P and Münstermann S 2012 Modeling of damage and failure of dual phase steel in Nakajima test *Key Eng. Mater.* **525-526** 69-72

Springback prediction of titanium tube bending considering Bauschinger effect and Young's modulus variation

J Ma[1], H Yang[1], H Li[1], Z J Tao[1], G J Li[2]

[1] State Key Laboratory of Solidification Processing, School of Materials Science & Engineering, Northwestern Polytechnical University, Xi'an, 710072, China
[2] Chengdu Aircraft Industry (Group) Corporation Ltd., Chengdu, 610092, China

E-mail: yanghe@nwpu.edu.cn; liheng@nwpu.edu.cn

Abstract. High strength titanium bent tubes present promising usages in advanced aircraft and spacecraft to achieve lightweight and improve overall performance. However, the high ratio of yield strength to Young's modulus results in significant springback in bending, which limits their forming accuracy. In this work, the Bauschinger effect and nonlinear unloading behavior of high strength Ti-3Al-2.5V tube are experimentally investigated. Then, to describe such behaviors, the Yoshida-Uemori (Y-U) two-surface hardening model and Chord unloading model are introduced into the elastoplastic constitutive framework and numerically implemented. Taking rotary draw (RDB) bending as a case, the springback angles are predicted and analyzed by comparison with the experimental results.

1. Introduction

Titanium alloy tubes are increasingly used for hydraulic pneumatic, fuel and environment control systems in advanced aircraft and spacecraft as bleeding components owing to their high specific strength and corrosion resistance capability [1]. As typical high strength titanium tube (HSTT), stress relieved Ti-3Al-2.5V tube can meet the aeronautic product requirements of high-pressure resistance, mass reduction and fuel economy, attracting increasing attention in aircraft manufacturing industry. During titanium tube bending, when the multi-tools are removed, more pronounced springback inevitably occurs due to its high ratio of the yield strength to Young's modulus. The springback reduces the accuracy of the bent tubular parts, and decreases the connection/sealing performance with other parts as well as the internal structure compact. Thus, it is particularly important to achieve for high precision bent tubes in the aeronautic "bleeding" tube system and the precision springback prediction should be considered as a key issue in HSTT bending.

The accurate springback prediction strongly depends on the accurate description of the Bauschinger effect and the nonlinear unloading behavior [2]. To take the Bauschinger effect into consideration in the forming-unloading deformation analysis, Yoshida and Uemori proposed a two-surface isotropic-kinematic hardening model, called Yoshida-Uemori (Y-U) model [3], which has been widely used for improving springback prediction accuracy in high strength steel sheets forming. Also, to describe the nonlinear unloading behavior, the Chord model was proposed, which is a linearly-approximated stress-strain gradient during unloading and demonstrated that the unloading modulus gradually decreases with plastic deformation [3].

However, for the titanium tubes, both of the above two behaviors still remain unclear and have been seldom found in springback studies. For improving the accuracy of springback prediction in HSTT bending, it is essential to clarify the Bauschinger effect and nonlinear unloading behaviors, and take them into consideration in springback prediction.

In this study, we attempt to experimentally clarify the Bauschinger effect and nonlinear unloading behaviors in HSTT. Then, an elastoplastic constitutive frame work, considering the above two material behaviors, was established and numerically implemented. Taking RDB as a case, the springback angles were predicted and analyzed by comparison with the experimental results.

2. Materials behaviours

2.1. Bauschinger effect

HSTT of stress relieved Ti-3Al-2.5V tube, with Φ=12 mm and t=0.9 mm, was used in this study. By uniaxial tension test, the mechanical properties, including elastic modulus, yield strength, tensile strength were obtained. To observe the Bauschinger effect in HSTT, the cyclic tension-compression test was carried out. This test was modified from the ASTM standard with a special tube specimen design for overcoming the bucking. Figure 1 shows the designed specimen with the gauge length of 2 mm and the projection width of 2 mm. To accurately measure the deformation between such a small gauge length, the DIC optional measurement instrument was applied. Figure 3 (a) shows The tension-compression stress-strain curves, indicating that the Ti-3Al-2.5V tube is strongly susceptible to the Bauschinger effect.

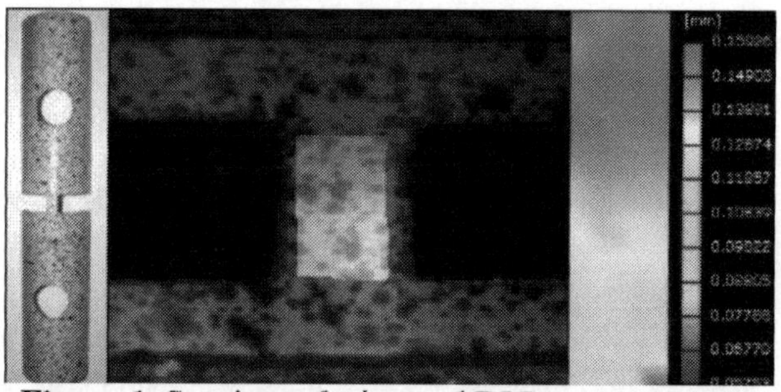

Figure 1. Specimen design and DIC measurement

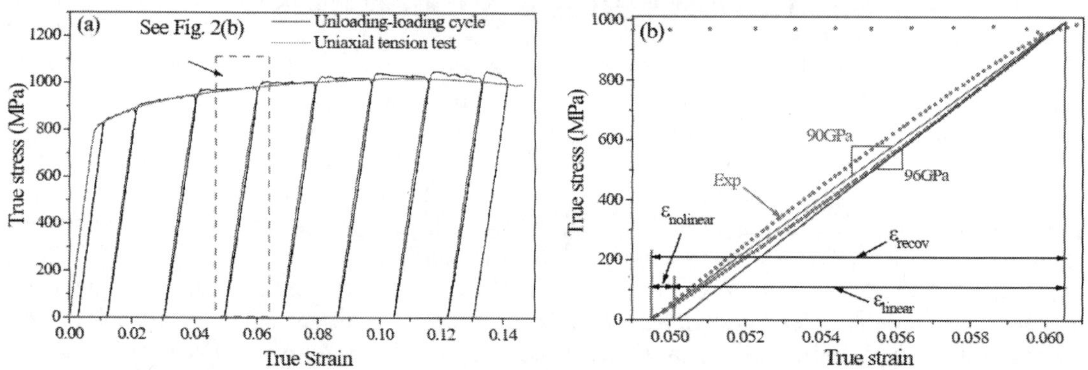

Figure 2. Stress-strain curves of HSTT under cyclic tension-unloading test

2.2. Variation of Young's modulus

Cyclic tension loading-unloading test was conducted to explore the nonlinear unloading behaviour during accommodated deformation. Figure 2 (a) illustrates The true stress-strain relationship and the Chord modulus (average Young's modulus), which are the linearly approximated stress-strain gradient during each unloading-reloading cycle at different pre-strains. It also can found from Figure 2 (b), the stress-strain curve of the forth cycle, that the unloading and reloading curves are not linear and the unloading curves always lag behind the subsequent loading curves, forming hysteresis loops that are more pronounced for larger plastic strain.

Figure 3 (b) shows the Chord modulus variations with accumulated plastic strain. It can be seen that the Chord modulus decreases dramatically when the plastic strain is less than 5% and turns to be a constant value when the plastic strain exceeds to 5%, and the percentage of the nonlinear recovery also turns to be stable (about 13.5%). Thus, it can be concluded that the unloading of HSTT exhibits a pronounced nonlinearity.

3. Elastoplastic constitutive modelling

3.1. Yoshida-Uemori model

Yoshida-Uemori model was used to evaluate the Bauschinger effect in Ti-3Al-2.3V tube. Y-U model is a two-surface isotropic-kinematic hardening model, wherein the yield surface moves within a bounding surface. The criterion for the subsequent yield surface *f* is expressed by the Eq. (1)

$$f = \phi(\sigma - \alpha) - Y = 0 \tag{1}$$

The bounding surface F is given by Eq. (2)

$$F = \phi(\sigma - \beta) - (B + R) = 0 \tag{2}$$

The Y-U model parameters (Y, C, B, b, m, R_{sat}, h) were determined simultaneously by using a FE based optimization technique. The objective function of the optimization method is defined as

$$\Gamma(x) = \sum_{t=1}^{T} \omega^t \Gamma^t(x) \tag{3}$$

where x=[Y, C, B,...] is the parameters set, $\Gamma^t(x)$ is the objective function. Thus, the parameters set of the Y-U model can be optimized. It is found from Figure 3 (a) that the predicted curves well agree with the experimental results, indicating that the calibrated Y-U model are reliable.

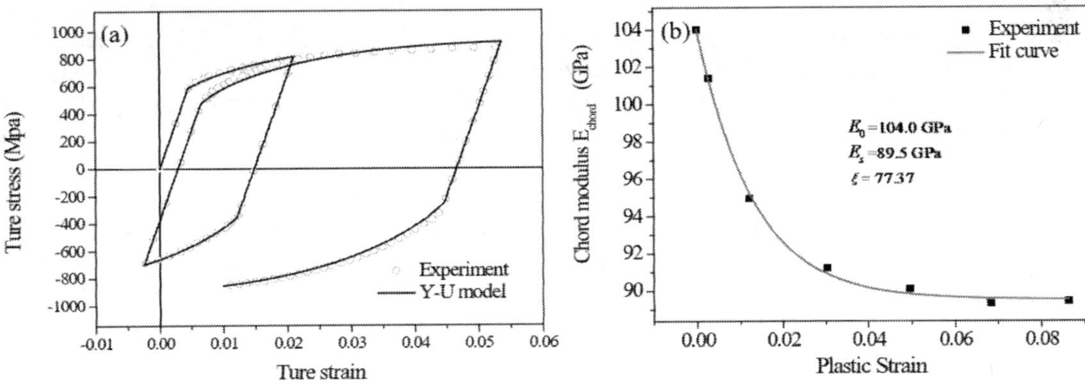

Figure 3. Comparison of experimental and calibrated results: (a) Y-U model; (b) Chord model

3.2. Chord unloading model

In order to describe the Young's modulus variation at different pre-strains, the Chord model was used in this study, as shown in Eq. (4)

$$E = E_0 - (E_0 - E_a)\{1 - \exp(-\xi \varepsilon_p)\} \tag{4}$$

where E_0 is the Young's modulus obtained by the uniaxial tension test, E_a is the converged average Young's modulus and ξ is the convergent rate. The parameters were determined by least squares method from the tension unloading-loading experimental data, as shown in Figure 3(b).

4. Springback prediction of HSTT

Taking the RDB of HSTT as a typical case, the springback was predicted and analyzed. The key modelling techniques and forming parameters can be found in Ref. [4]. The elastoplastic frame work was implemented into ABAQUS by fully implicit return mapping algorithm and stress updating algorithm. The von Mises yielding function was applied in the elastoplastic frame work. For comparison, the isotropic hardening (IH) model and constant unloading (Const. E_0) model were also used in simulation.

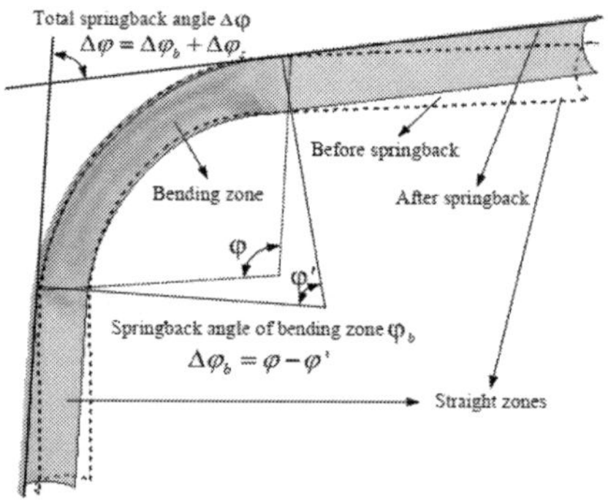

Figure 4. Schematic of springback in HSTT rotary draw bending

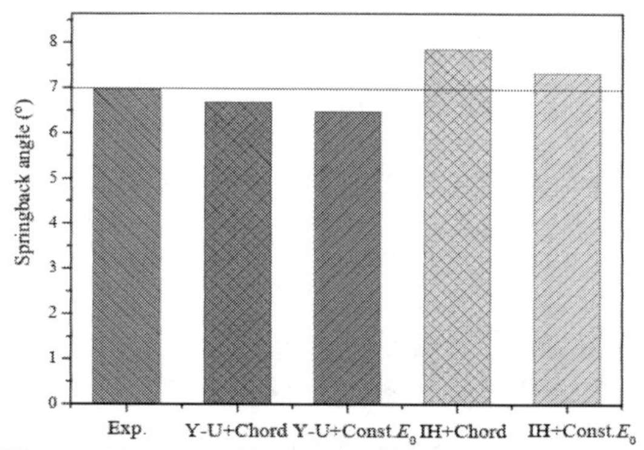

Figure 5. Comparison of springback prediction for different model combination

Figure 4 shows the schematic of springback in HSTT rotary draw bending. Figure 5 predicted springback angles under 90 ° bending. In comparison with IH + Chord *or* Const. E_0 models, consideration of the Bauschinger effect resulted in smaller springback angles. During tube bending and mandrel-pulling process, the reversal loading phenomenon occurs in some regions. Therefore, the stress at these areas becomes smaller when the Bauschinger effect is considered.

Compared with IH *or* Y-U +Const. E_0 models, the models considering the variation of Young's modulus gave larger springback angles due to the decrease of the Chord modulus at the bent tube. From the view of influence on springback prediction, the variation of Young's modulus seems to be more obvious than Bauschinger effect. Consequently, for accurate springback prediction, it is important to understand and consider the Bauschinger effect and variation of Young's modulus in HSTT rotary draw bending.

5. Conclusion

By using DIC-based cyclic tension-compression test and cyclic tension-unloading test, both the Bauschinger effect and variation Young's modulus were measured and clarified. To describe the two behaviours, the Y-U model and Chord model were established, calibrated and numerically implemented. Considering the Bauschinger effect and the variation of Young's modulus is of great importance for accurate springback prediction of HSTT bending.

Acknowledgements

The authors would like to thank the National Science Fund for Excellent Young Scholars (51522509), the National Natural Science Foundation of China (51275415) and Fundamental Research Funds for the Central Universities (3102014KYJD001).

References

[1] Boyer R R 1996 *Materials Science and Engineering: A* **213** (1) 103-114
[2] Wagoner R H, Lim H and Lee M G 2013 *International Journal of Plasticity* **45** 3-20
[3] Yoshida F and Uemori T 2002 *International Journal of Plasticity* **18** (5) 661-686
[4] Li H, Hu X, Yang H and Li L 2016 *International Journal of Plasticity* **82** 127-158

Journal of Physics: Conference Series **734** (2016) 032114 doi:10.1088/1742-6596/734/3/032114

Limit dome height test of very thin brass sheet considering the scaling effect

Jambeswar Sahu[1] and Sushil Mishra[2]

Department of Mechanical Engineering, Indian Institute of Technology Bombay, India-400076
[1]jambeswar@gmail.com, [2]sushil.mishra@iitb.ac.in

Abstract. The effect of reduction of material geometry in metal forming process is important and is more effective when material size less than 100 micron. Thin brass sheet of 30, 50, 90 micron thickness were tested by in-plane uniaxial and out-of-plane limit dome height (LDH) test to investigate the size effect. The test was carried for both as-received and annealed specimen. The microstructure of all tests was determined in details by electron back scattered diffraction (EBSD) microscopy technique. There was a clear dependency of mechanical behaviour observed on part miniaturized of both tensile and LDH test. The limiting strain is obtained maximum at plane strain path in LDH test having more misorientation. The 50 micron sheet in biaxial strain path required more load and have larger major strain, more texture and Taylor factor value.

1. Introduction

The study of part miniaturization effect is necessary for an optimized production process, because the mechanical properties of thin sheet metal are mostly different from bulk material. This lead to size effect in micro scale material geometry compared to macro scale [1, 2]. The size effect was studied by tensile testing of copper, nickel, aluminium etc. by several researcher and observed that the yield stress decreases with scaling down the material geometry [3-4]. The researchers were investigated the scaling effect on deep drawing and low formability, punch load, thickness distribution was observed [5]. The grain size versus thickness observed by Kals et al. that shows an important role in forming process [6, 7]. The effect of miniaturization and microstructural parameters like grain size, misorientation, twin fraction, Taylor factor, texture index are more in material having dimension less than 100 μm. The Taylor factor was studied by many researcher, i.e higher Taylor factor have more dislocation density and larger intragranular misorientation [8-11]. There are less studies done on the effect of microstructural parameters in case of thin sheet. In the current research work miniaturization and microstructure effect were studied by uniaxial tensile as well as limit dome height (LDH) test as tensile test does not give sufficient information about different state of stress and strain.

2. Tensile test of very thin brass sheet

A conventional tensile test machine Instron 3345 single vertical column type tensile test machine of capacity 5KN was used for tensile testing with a strain rate of 10^{-4}. The tensile curve for working material 30, 50, 90 μm thickness brass is shown in figure 1. The tensile test for each condition was carried out for three times for repeatability of result. The elongation is more in 90 μm thin sheet followed by 50 and 30 μm thin sheet. On the other hand the flow stress is more in 30 μm thin sheet followed by 50 and 90 μm thin sheet. This is lead to the miniaturization or scaling effect. The microstructure study was carried out for better understanding of the different deformation property for different thickness of specimen by FEI Quanta TM 200 Hv electron back scattered diffraction (EBSD) microscopy and analysed by TSL OIM Analysis 7.2 software. The EBSD sample was prepared by mechanical polishing followed by diamond polishing and electro polishing (electrolyte of 20% orthophosphoric acid and 80% water). The EBSD microstructure image of as received brass material of different specimen thickness is shown in figure 2 (a-c). It is observed that, the grain size in 30 μm sheet is very small even not indexed by EBSD. Which may be the cause of high flow stress and less elongation in 30 μm sheet. Brass sheet of 50 and 90 μm have 7 and 9 μm respectively, which shows comparatively less flow stress and more elongation. Due to very less elongation the material was not suitable for any for further deformation work. Therefore, the received material was annealed with 650° C for 20 minute to modify the microstructure and used for formability study by LDH test.

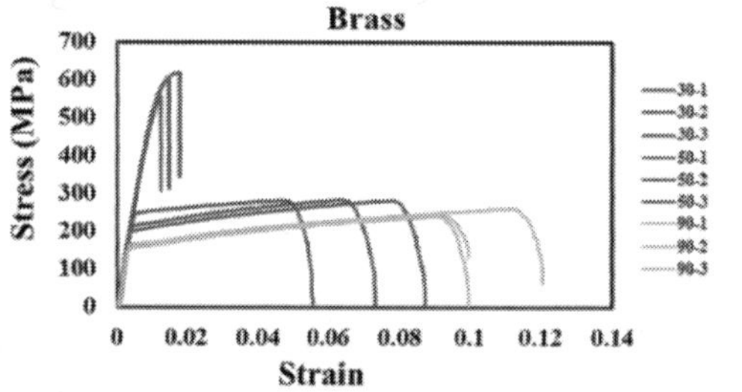

Figure 1. Tensile test of as received brass material with different specimen thickness.

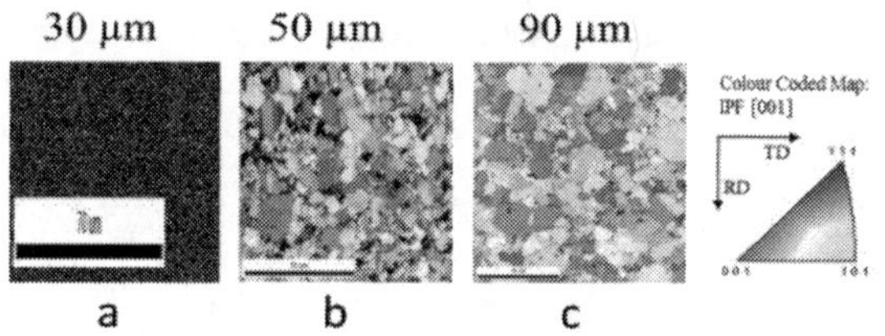

Figure 2. EBSD image (microstructure) of as received brass material of different specimen thickness.

3. Limit dome height (LDH) test of very thin brass sheet

The state of stress and strain used in sheet metal forming process are not only uniaxial condition but multiaxial conditions. Therefore, the tensile data is not sufficient to describe the forming limit and detail material behaviour in sheet metal forming process. LDH is the out-of-plane deformation method and performed in different strain path like uniaxial, plane strain and biaxial. The same machine Instron 3345 with a LDH attachment as shown in figure 3 (a) was used for LDH test. The sample dimension for biaxial, plane strain and uniaxial are scaled down and optimized shown in figure 3 (b-d) respectively. The LDH samples for different strain path were prepared by wire electric discharge machine (EDM) machine and printed with dot pattern for strain measurement. Half hemispherical punch of diameter 15mm were used for LDH test with 10^{-3} rate of punch displacement. The punch stroke versus load graph are shown in figure 4-6 for 90, 50, 30 μm thickness respectively. It is observed that the maximum load requirement decreases with miniaturization and from biaxial to uniaxial.

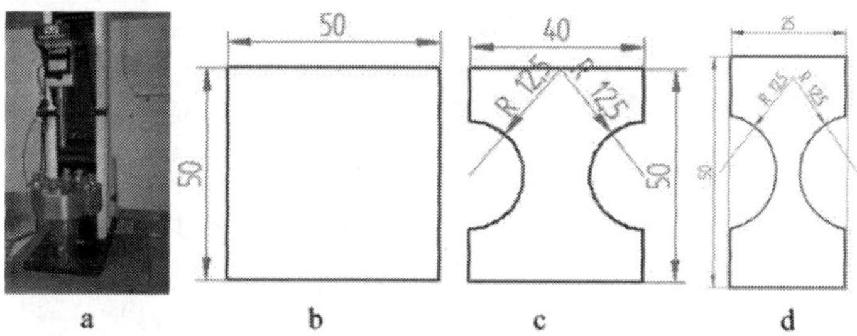

Figure 3. a) Experimental setup, b-d) Sample dimension for different strain path.

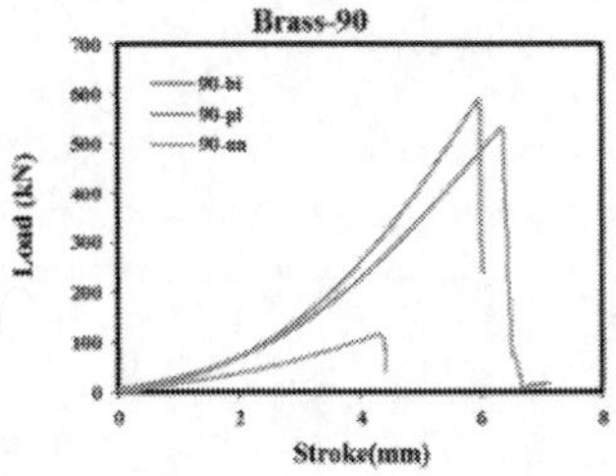

Figure 4. Punch stroke versus load plot for 90 μm thickness brass material.

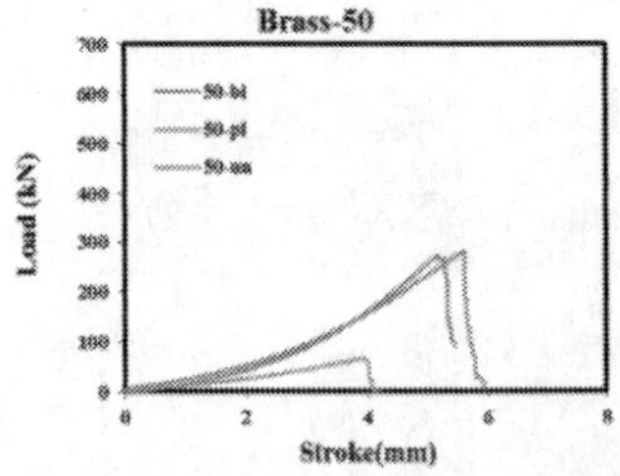

Figure 5. Punch stroke versus load plot for 50 μm thickness brass material.

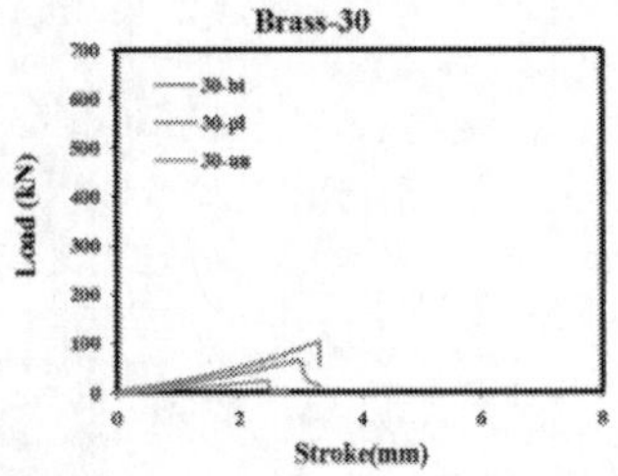

Figure 6. Punch stroke versus load plot for 30 μm thickness brass material.

The strain was measured in stereo microscope by measuring the minor and major axis before and after deformation for three strain path. The respective strain is calculated by the ratio of difference of final and initial dimension to initial dimension. The forming limit diagram (between major and minor strain) is shown in figure 7. It is observed that the maximum limiting strain is found at plane strain condition for different specimen thickness. A clear scaling/size effect is observed, that the formability (Limiting strain) decreases with decreasing thickness from 90 to 30 μm. Microstructure study was carried out to investigate the detailed deformation behaviour. Grain size was quantified in figure 8 from the TSL analysis of EBSD image shown figure 13. It is observed that the grain size decreases in uniaxial strain path but increases from uniaxial strain to biaxial strain path irrespective of thickness. The misorientation development shown in figure 9 is more on plane strain path due to the higher limiting strain in plane strain condition figure 7. Twin decay (figure 10) is more in higher thickness with deformation in all strain path but in 30 μm less decay in uniaxial and more in plane strain and biaxial. Taylor factor value is highest in 50 μm biaxial condition as shown in figure 11, so the load requirement is also more in 50 μm biaxial strain path shown in figure 5. The texture decreases with deformation (figure 12), except increasing in 50 μm biaxial strain path condition, which may be the cause of higher major strain in FLD of at 50 μm biaxial strain path (figure 7).

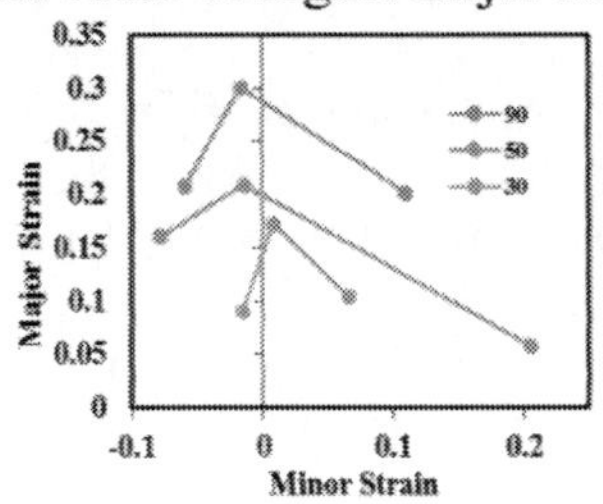

Figure 7. Forming limit diagram of brass material with different specimen thickness.

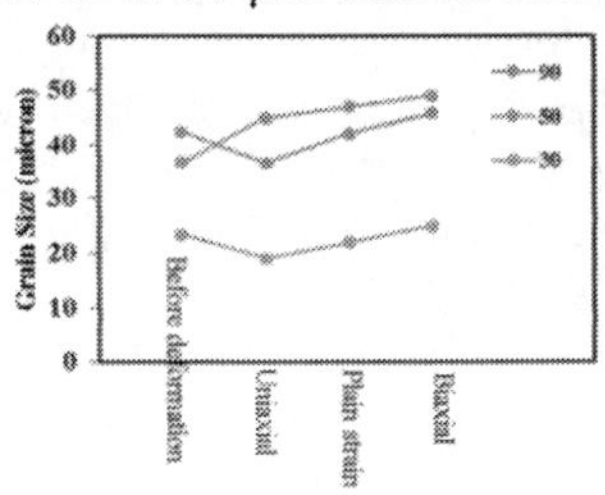

Figure 8. Grain size of different strain path with different specimen thickness.

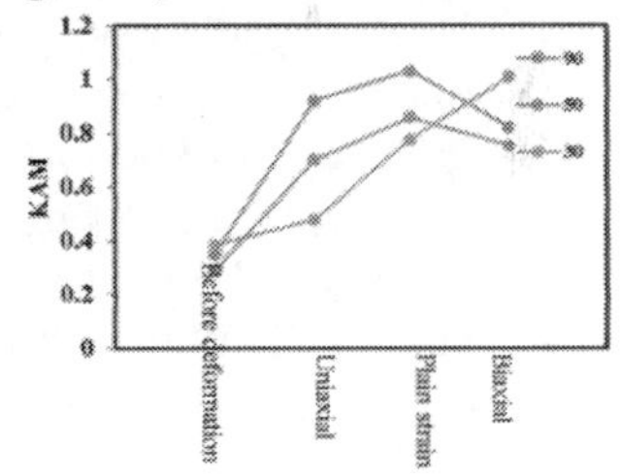

Figure 9. Kernel average misorientation development of different condition.

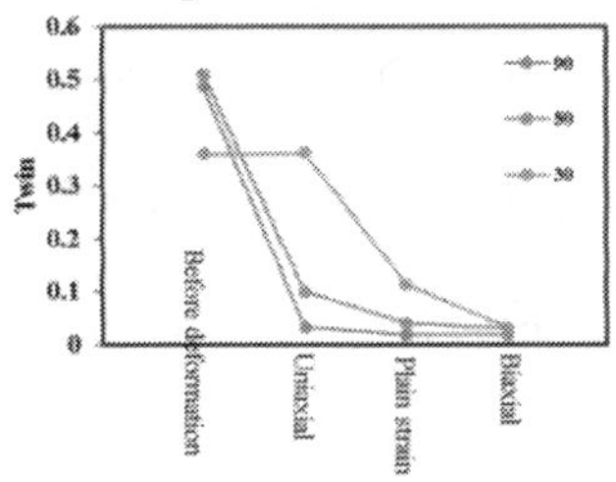

Figure 10. Twin decay of brass in different strain path with different specimen thickness.

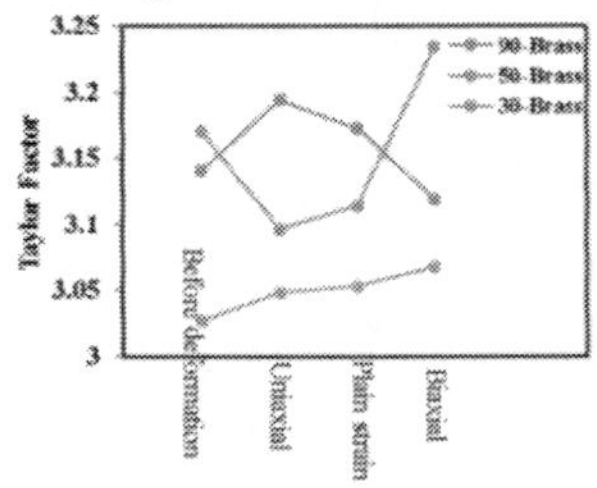

Figure 11. Taylor factor of brass in different strain path with different specimen thickness.

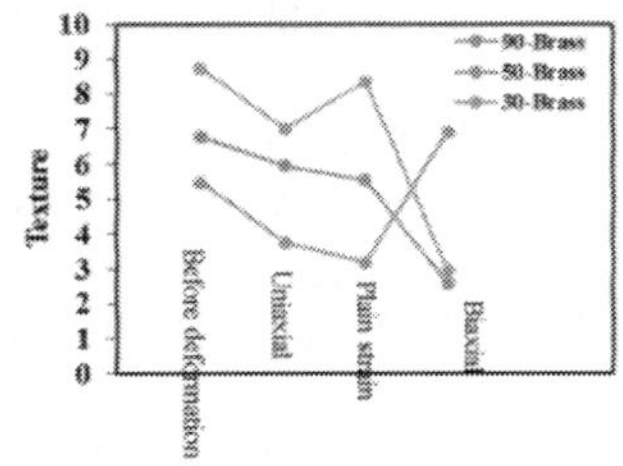

Figure 12. Texture index of brass in different strain path with different specimen thickness.

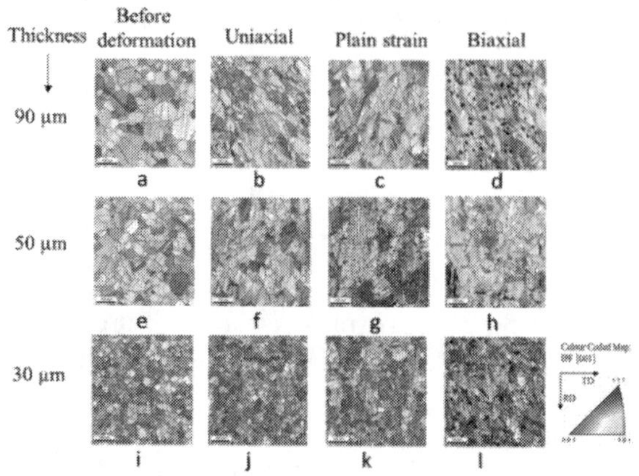

Figure 13. Tensile test of as received brass material with different specimen thickness.

4. Conclusions

Thin brass sheet of 30, 50, 90 μm thickness and comparable microstructure were tested in in-plane tensile and out-of-plane LDH test. The elongation of as-received material were found very less. Microstructure were modified by annealing for better formability. The size effect was observed in both testing that the mechanical property decreases with scaling the geometry. The limiting strain is obtained maximum at plane strain condition in LDH test for different thickness having more misorientation. The 50 micron sheet in biaxial strain path required more load and have larger major strain, more texture and Taylor factor values.

5. References

1. Hoffmannl H and Hong S 2007 Tensile Test of very thin Sheet Metal and Determination of Flow Stress Considering the Scaling Effect *Journal of CIRP Annals - Manufacturing Technology* Vol 55 (1) pp 263-266.
2. Engel U and Geiger M 1999 Specific Characteristic of Micro Sheet Metal Working *Proc. 7th Int. Conf. on Sheet Metal – SheMet '99, Nuernberg,* pp 529-536.
3. Stolken JS and Evans AG 1998 A microbend test method for measuring the plasticity length scale *Acta Mater.* Vol 46 (14) pp 5109–5115.
4. Simons G, Weippert Ch, Dual J and Villain J 2006 Size effects in tensile testing of thin cold rolled and annealed Cu foils *Journal of Materials Science and Engineering* Vol 416 (A) pp290-299.
5. Saotome Y, Yasuda K and Kaga H 2001 Microdeep drawability of very thin sheet steels *J. Mater. Process. Technol* Vol 113 pp 641–647.
6. Kals T A 1999 Fundamentals on the Miniaturization of Sheet metal working Processes Bamberg Meisenbach.
7. Kals T A and ECKSTEIN R 2000 Miniaturization in Metal Working *Journal of Material Processing Technology* Vol 103 pp 95-101.
8. Dillamore IL and Katoh H 1974 A comparision of the observed and predicted deformation texture in cubic metals *Met Sci* Vol 8 pp 21-27.
9. Hutchinson B 1999 Deformation microstructures and textures in steels *Philos. Trans. R. Soc. London* Vol 357 (A) pp 1471–1485.
10. Nesterova E, Rybin V, Zisman A and Teodosiu C. In Rybin VV, Teososiu C 2002 editors *Proceedings of physics and mechanics of large plastic strains* Russia: Saint Petersburg pp 61–74.
11. Kang JY, Bacroix B, Regle H, Oh KH and Lee HC 2007 Effect of deformation mode and grain orientation on misorientation development in a body-centered cubic steel *Acta Materialia* Vol 55 pp 4935–4946.

Journal of Physics: Conference Series **734** (2016) 032115 doi:10.1088/1742-6596/734/3/032115

Whole process modeling of joining of flareless AA 6061-T4 tube by extrusion-bulging forming using a polyurethane elastomer medium

J C Yang[1], H Li[1*], H Yang[1] and G J Li[2]

[1] State Key Laboratory of Solidification Processing, School of Materials Science, Northwestern Polytechnical University, Xi'an, 710072, China

[2] Chengdu Aircraft Industry (Group) Corporation Ltd., Chengdu 610092, China

Email: liheng@nwpu.edu.cn, yang-shuaii@126.com

Abstract. The tube joining by plastic deformation proves to be a more efficient and environmentally friendly way to achieve the tube-tube joining compared with other traditional types, such as metallurgical joining and mechanical joining. In this study, to reveal the effects of the processing parameters on the filling quality and residual contact stress, an axisymmetric finite element (FE) model of the whole joining process, including extrusion-bulging forming and unloading, was established and validated. The aluminum alloy (AA) 6061-T4 tubes, the stainless steel (ST) 15-5PH sleeve and polyurethane (PU) elastomer medium were characterized and modeled. And the implicit algorithm was adopted by comparing the prediction results between explicit and implicit FE models. The characteristics of stress distribution and plastic strain for the tube, PU elastomer and sleeve were discussed.

1 Introduction

The tubular components for "bleeding" transforming and loading carrying with enormous quantities and diversities are the aorta of aircraft [1]. The performance of the tube joints is one of the key factors affecting the whole duct systems in aircraft. Compared with other traditional types such as metallurgical joining and mechanical joining, the tube joining by plastic deformation proves to be a more efficient and environmentally friendly way to achieve the tube-tube joining [2]. How to achieve the precision and efficient joining of tubes is a crucial and urgent issue to be solved.

There are two major types of tube joining ends by plastic deformation in aerospace industry, viz., flaring joining consisting of single and double types, and flareless joining including ferrule, taper, extrusion-bulging, roller inner swaging and contraction types [3]. Compared with the flaring joining, the flareless joining presents better performance such as better sealing and vibration resistance, higher pressure resistant, lower cost and higher productivity. Among these flareless joining processes, the flareless tube joining by extrusion-bulging still contributes widely application in the manufacturing of the aerospace pipeline because of its universality and efficiency [4]. However, the whole process of extrusion-bulging forming and unloading is a strictly multiple-tool constrained process affected by so many coupling parameters such as multiple material properties, extrusion forming pressure, extrusion depth and die clearance. The forming quality is thus difficult to be controlled and the problems of "explosion, seepage, leakage" frequently occur in the tube joints. Compared with the experiments, the finite element (FE) method has become a primary tool to provide the most detailed analysis of forming processes. In this paper, based on ABAQUS platform, a FE model of the whole joining process, including extrusion-bulging forming and unloading, was established; Then, the characteristics of stress distribution and plastic strain for the tube, PU elastomer and sleeve were discussed.

2 Experiment

2.1. The mechanical properties of AA 6061-T4 tube

The uniaxial tension test of full-size tubular sections was conducted to obtain the mechanical properties as shown in the Table 1. The tube specification is 20mm × 1mm (the external diameter is 20mm and the wall thickness is 1mm). The quadratic plastic yield criteria Hill48 was applied to describe the normal anisotropic plastic deformation behavior of the material.

Table 1. The mechanical properties of AA 6061-T4 tube

Material	E/GPa	Poisson's ratio	$\delta/\%$	$\sigma_{0.2}/MPa$	σ_b/MPa	normal anisotropy index r
6061-T4	67.531	0.3	16.7	169.4	302.4	0.598

2.2. The mechanical properties of PU elastomer

The uniaxial tension and uniaxial compression were carried out and the hyperelastic material model of Ogden was adopted to describe the nonlinear elastic behaviors of the material. The PU tube specification is 17.6mm × 2.8mm (the external diameter is 17.6mm and the wall thickness is 2.8mm). Table 2 shows the mechanical properties and Table 3 provides the material constants of Ogden model.

Table 2. The mechanical properties parameters of PU elastomer

Material	$\delta/\%$	Shore Hardness	σ_b/MPa
PU elastomer	644	95	47.61

Table 3. The material constants of Ogden model

Rubber type	Mu-1	Alpha-1	D_1	Poisson's ratio
PU elastomer	5.17	2.06	0	0.4999

2.3. The mechanical properties of ST 15-5PH

The uniaxial tension test was conducted to obtain the mechanical properties, as shown in Table 4.

Table 4. The mechanical properties of ST 15-5PH sleeve

Material	E/GPa	Poisson's ratio	$\delta/\%$	$\sigma_{0.2}/MPa$	σ_b/MPa
15-5PH	183	0.26	7.4	1026.07	1042.855

3 FE modeling of whole process

A FE model of the whole process was established to simulate extrusion-bulging forming and unloading. As shown in Figure 1(a), when the pull rod is driven to produce an axial extrusion pressure, the PU elastomer tube is axially compressed to form an internal pressure, then the AA 6061-T4 tube is bulged locally into the two grooves of the ST 15-5PH sleeve; when the pull rod is unloaded, the springback occurs in the PU tube, AA tube and ST sleeve. The key techniques of the FE modeling are summarized: 1) As shown in Figure 1(b), an axisymmetric model could be used to simulate the joining process, and Figure 2 shows the detailed meshing of the five parts; 2) Figure 3 shows the axial load amplitude curve of the pull rod, and the range of maintaining pressure is from 195 MPa to 255 MPa; 3) Instead of the explicit algorithm, the implicit algorithm was adopted to model the whole process including extrusion-bulging forming and unloading.

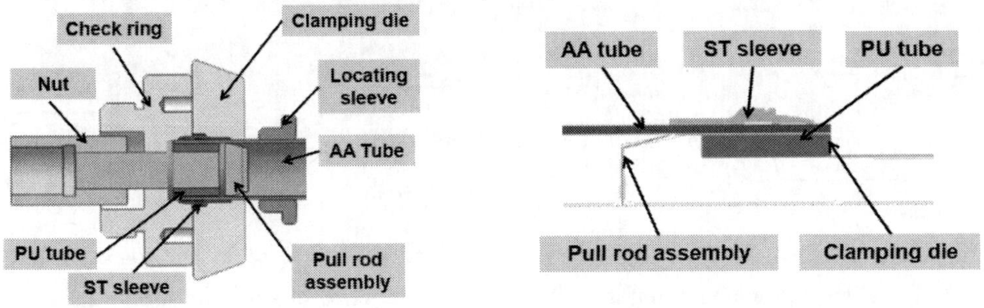

(a) 3D model before simplification (b) axisymmetric model after simplification

Figure 1. The geometrical model before and after simplification

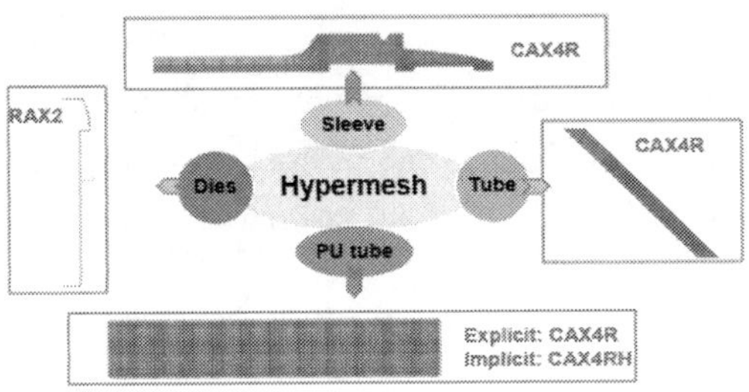

Figure 2. The hypermesh and element type of all the parts

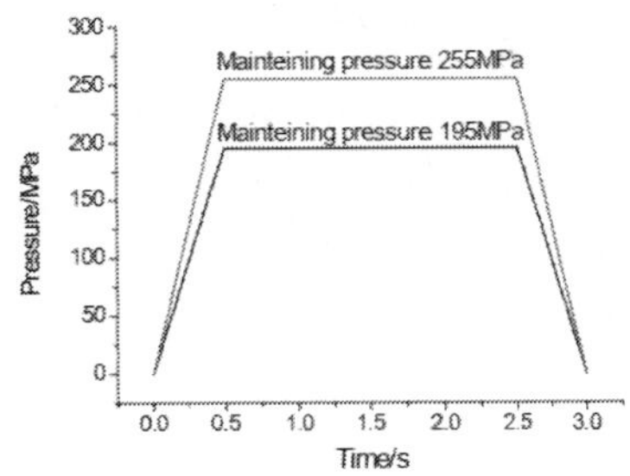

Figure 3. The axial load amplitude curve of pull rod

4 Results and discussion

The whole process simulation was carried out assigning 255 MPa to maintain the pressure of the pull rod. To obtain the residual stress, the nodes of the tube after springback were numbered and analyzed in the specified area that is shown in Figure 4. Figure 5 shows the distribution characteristics of the residual stress, plastic strain (PE) and equivalent plastic strain (PEEQ). The most reliable position appeared at the sharp corners of the sleeve's grooves where the maximum residual compression stress amounted to 292 MPa, the maximum PEEQ was 0.44 and the residual contact stress was 62 MPa. The contact stress was larger than work pressure 21 MPa. Thus a sealing line could be formed to prevent the possible leakage. The tube joining assembly is more reliable, the sealing performance of that is better, and the service performance can be also improved. The end forming experiments of the joining assembly were carried out to validate simulation results. The range of oil pressure calculated by the simulation is from 4.57 to 5.98 MPa where the eligible tube joining assembly can be fabricated. It is found that the simulation result was reasonable and coincided with experimental result.

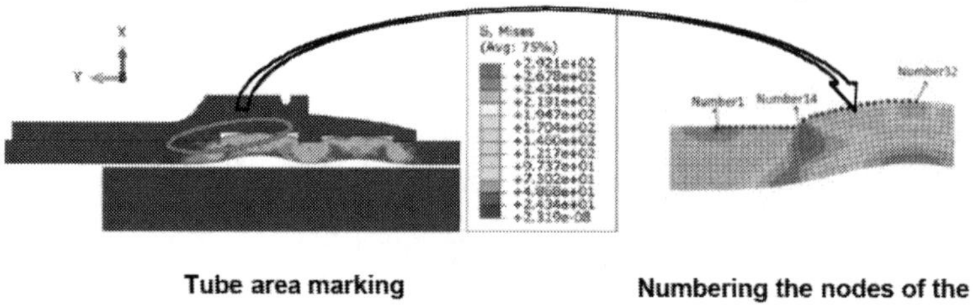

Figure 4. Numbering the special codes of the tube after springback

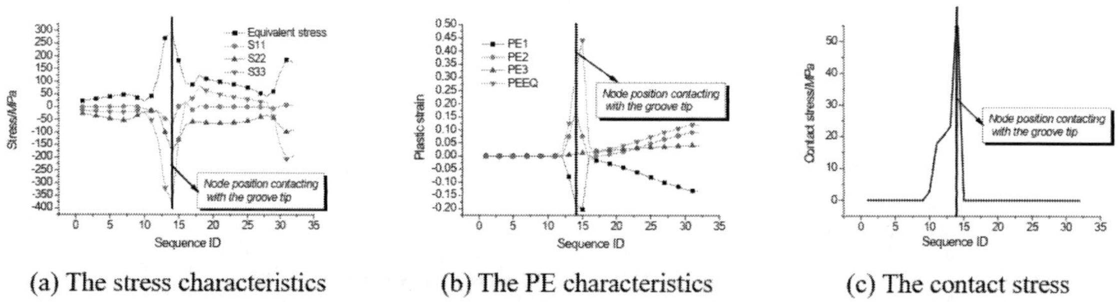

Figure 5. The distribution characteristics of stress, PE and contact stress in marking area

5 Conclusion

The stress and strain distribution characteristics of the AA 6061-T4 tube were obtained and analyzed after springback. At the sharp corners of the sleeve's grooves, the maximum compression stress amounted to 292 MPa, the maximum PEEQ was 0.44, and the maximum residual contact stress was 62 MPa that was larger than work pressure 21MPa. Thus a sealing line can be formed to prevent the leakage. By comparing the experimental result, the simulation result was reasonable and coincided with experimental result.

Acknowledgements

The authors would like to thank the National Science Fund for Excellent Young Scholars (51522509), the National Natural Science Foundation of China (51275415) and Fundamental Research Funds for the Central Universities (3102014KYJD001).

References

[1] Yang H, Li H, Zhang Z Y, Zhan M, Liu J and Li G J 2012 *Chinese Journal of Aeronautics*. **25** 1–12.

[2] Mori K-i, Bay N, Fratini L, Micari F and Tekkaya A E 2013 *CIRP Annals - Manufacturing Technol* **62** 673–694.

[3] Groche P, Wohletz S, Brenneis M, Pabst C and Resch F 2014 *Journal of Materials Processing Technology* **214** 1972.

[4] Monaghan J and Naughton P 1991 *International Journal of Machine Tools and Manufacture* **31** 283–293.

FRACTURE PREDICTION USING MODIFIED MOHR COULOMB THEORY FOR NON-LINEAR STRAIN PATHS USING AA3104-H19

Robert Dick[1,2], Jeong Whan Yoon[1,3]

Abstract. Experiment results from uniaxial tensile tests, bi-axial bulge tests, and disk compression tests for a beverage can AA3104-H19 material are presented. The results from the experimental tests are used to determine material coefficients for both Yld2000 and Yld2004 models. Finite element simulations are developed to study the influence of materials model on the predicted earing profile. It is shown that only the YLD2004 model is capable of accurately predicting the earing profile as the YLD2000 model only predicts 4 ears. Excellent agreement with the experimental data for earing is achieved using the AA3104-H19 material data and the Yld2004 constitutive model. Mechanical tests are also conducted on the AA3104-H19 to generate fracture data under different stress triaxiality conditions. Tensile tests are performed on specimens with a central hole and notched specimens. Torsion of a double bridge specimen is conducted to generate points near pure shear conditions. The Nakajima test is utilized to produce points in bi-axial tension. The data from the experiments is used to develop the fracture locus in the principal strain space. Mapping from principal strain space to stress triaxiality space, principal stress space, and polar effective plastic strain space is accomplished using a generalized mapping technique. Finite element modeling is used to validate the Modified Mohr-Coulomb (MMC) fracture model in the polar space. Models of a hole expansion during cup drawing and a cup draw / reverse redraw / expand forming sequence demonstrate the robustness of the modified PEPS fracture theory for the condition with nonlinear forming paths and accurately predicts the onset of failure. The proposed methods can be widely used for predicting failure for the examples which undergo nonlinear strain path including rigid-packaging and automotive forming.

[1] *Faculty of Science, Engineering and Built Environment, Waurn Ponds, VIC 3220, Australia*
[2] *Alcoa Technical Center, 100 Technical Drive, Alcoa Center, PA 15069, USA*
[3] *Department of Mechanical Engineering, KAIST, Daejeon 305-701, Korea*

Finite element based optimization study on hydroformed stepped tube

K R Harisankar[1], A Omar[1] and K Narasimhan[1]

[1]Department of Metallurgical Engineering and Materials Science, Indian Institute of Technology Bombay

E-mail: nara@iitb.ac.in

Abstract. Tube hydroforming process is an advanced manufacturing process in which tube is placed in between the dies and deformed with the help of hydraulic pressure. A sound tube hydroformed part depends upon die conditions, material properties and process conditions. In this work, a finite element study, along with response surface methodology (RSM) for designing the simulation, has been used to construct models with loading path, friction, anisotropic index, strain hardening exponent and tube thickness. The responses studied are the die corner radius filling and strain non-uniformity index (SNI) chosen in each step of the tube with maximum 30% thinning as stopping criteria. The factors effect and their interactions on each response were determined and analysed.

1. Introduction

The hydroforming process is widely used to form light weight, high strength and complex parts. In tube hydroforming, tube blanks are formed into a die cavity of desired shape by means of simultaneous application of hydraulic internal pressure and axial compressive loads. The main application fields have been found in automotive, aircraft and aerospace industry. Manufacturing industries are greatly benefited by this process as it reduces the number of parts per product, long term expenditure and weight [1]. Hydroformed products are mainly in the form of stepped tubes, conical tubes and box shape tubes [2].

A sound hydroformed product depends upon material properties, tube geometry, process parameters and friction. Extensive study of their effect is essential for the success of the process. The effect of material properties on the free expansion of tube was investigated by Carleer et al. [3] on different grades of steel. The effect of coefficient of friction on fracture of tube hydroforming was investigated by Zhang et al.[4] based on Gurson-Tvergaard-Needleman (GTN) ductile damage model. Furthermore, a numerical investigation was made by Alaswad et al. [5] to study the geometrical effect on bi-layered tube hydroforming.

The main aim of this study is to integrate the RSM based Design of Experiment (DoE) to construct an empirical model for corner filling and strain non-uniformity index (SNI) as a function of different process variables as well as tube thickness for a three stepped shaft to determine their mutual interaction and effect on manufacturing of such a part.

2. Finite element modeling

Finite element model was created for stepped tube hydroforming process as shown in figure 1. The tube diameter of 60 mm was selected as starting material. The model consists of a three-stepped shaft die as in figure 1. According to the dimension of the die, a tube of 60 mm diameter was selected as blank. The mechanical properties of the tube are given in the table 1 and Holloman's power law hardening model was used as the material model.

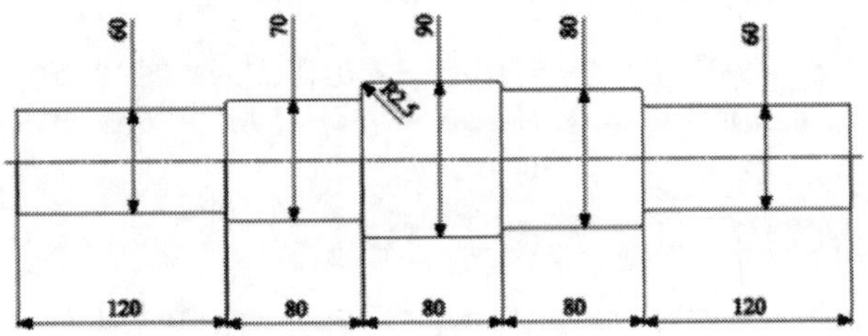

Figure 1. Dimensions of die in mm

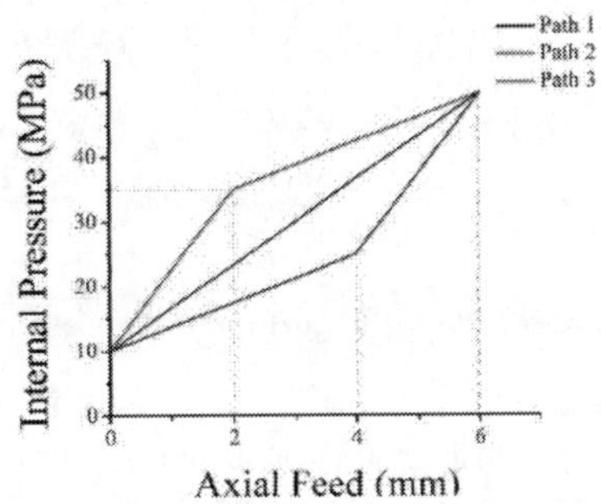

Figure 2. Loading path

3. Simulation design

The simulation methodology adopts the Box-Behnken design which consists of 46 simulation runs with different parameters [5]. Table 2 shows the independent variables and their design levels. On the basis of previous literature, the suitable loading is chosen shown in figure 2 [6]. In this study the response chosen were- (i) Corner radius (CR) - measured from die corner to tube corner and (ii) Strain Non-uniformity Index (SNI), which is the difference between Peak true thickness strain and average true thickness strain [7]. The objective of this study is to minimise both the responses.

Response Surface Methodology (RSM) was used for analysis of numerical data which was performed using a statistical software, Design Expert 9.0.6. To obtain the regression equation, a second order polynomial was fitted to the numerical data.

$$Y= b_0 + \sum b_i \chi_i + \sum b_{ii} \chi_{ii}^2 + \sum b_{ij} \chi_i \chi_j + \varepsilon \quad (1)$$

Table 1. Mechanical properties of tube

Mechanical Properties of the tube	
Elastic Modulus, E (GPa)	200
Poisson's ratio	0.3
Yield Stress, σ_s (MPa)	380
Strength coefficient K (MPa)	891

Table 2. Process variables and design levels used

Variable	Code	-1	0	1
Loading path	A	Path1	Path 2	Path 3
Coefficient of friction (μ)	B	0.05	0.1	0.15
Strain hardening exponent (n)	C	0.14	0.18	0.22
Anisotropic factor(r)	D	1.2	1.6	2
Tube thickness (t) (mm)	E	1.5	2	2.5

4. Results and discussion

The simulation was carried out using FEM based PAM-STAMP-2G© implementing to the Box-Behnken design matrix. 30% thinning was selected as the stopping criteria for all the 46 combinations of simulation. The analyses of the measured responses were carried out by using design expert software. It was found that the trend followed by each responses for all the three steps are comparatively closer and therefore the response for only one step will be discussed in this paper. The final mathematical models of each response in terms of coded factors (refer table 2) are mentioned below:

$$\text{Corner radius CR 3} = 5.51 - 0.2A + 1.45B - 0.43C - 0.71D + 0.27E - 0.017AB - 0.1AC + 0.18BD - 0.12BE + 0.047CD + 0.12A^2 + 0.14B^2 + 0.14D^2 \quad (2)$$

$$\text{SNI 2} = 0.17 - 0.000584A + 0.016B - 0.00937C - 0.00174D + 0.014E + 0.012AD - 0.00221BC + 0.011BD + 0.005075BE - 0.00225CD + 0.00475CE - 0.00474B^2 - 0.00392C^2 - 0.0055E^2 \quad (3)$$

4. 1. Effect of variables on corner radius filling(CR)

Perturbation plot for CR is shown in figure 3 and contour plot depicting the mutual interaction between μ and r on CR is shown in figure 4. Corner radius is plotted by varying only one factor in its range while maintaining other variables constant at the midpoint value of all the factors. It was found that friction is the major factor which is affecting the corner radius. As the higher friction forces resist the metal flow resulting in high corner radius, while larger anisotropic value and strain hardening exponent means easier forming and leads to low corner radius. Smaller corner radius can be produced when using advance feed type loading path (path3) since this will allow more material being pushed through the die at low pressure. However as thickness increases, corner radius found to be higher due to lesser bulge height at higher thicknesses. From figure 4 it can be noticed that smaller corner radius can be obtained at low value of friction and high value of anisotropic factor.

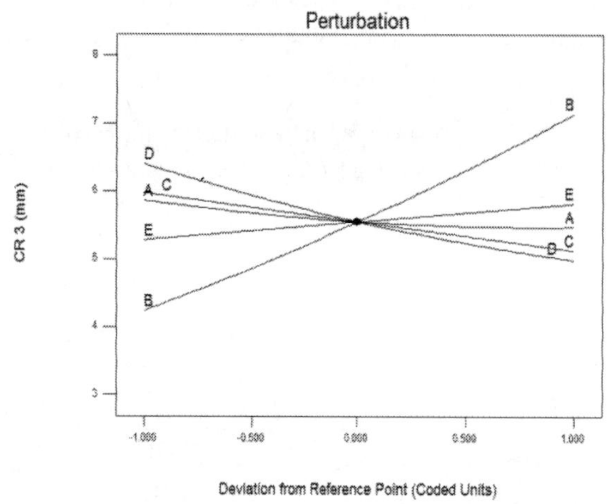

Figure 3. Perturbation of CR

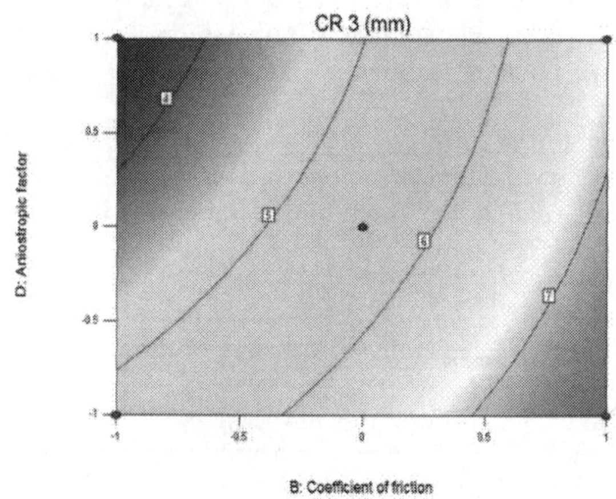

Figure 4. Contour graph of μ and r on CR

4. 2. Effect of variables on Strain Non-uniformity Index (SNI)

It can be observed from the perturbation plot of SNI as shown in figure 5 that SNI increases with the increase in coefficient of friction as well as tube thickness. This is because the material flow through the die is reduced which leads to non-uniformity in strain in the thickness direction. However, high value of anisotropic factor and strain hardening exponent will generate low value of SNI. The different types of loading paths in this study have not shown any significant effect on SNI. It is evident from the contour plot shown in figure 6 that low value of SNI is obtained at higher strain hardening exponent and lower friction coefficient.

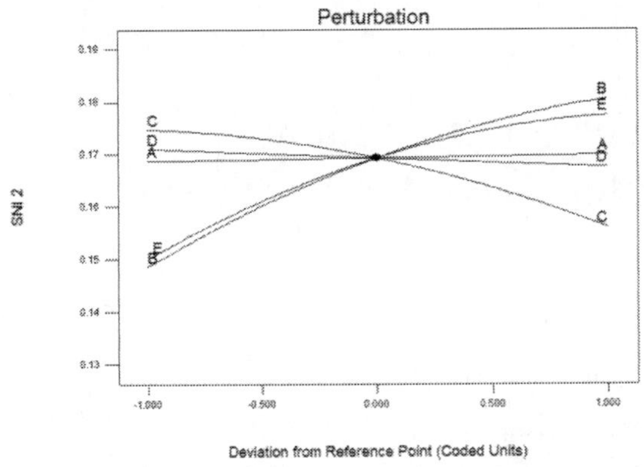

Figure 5. Perturbation of SNI

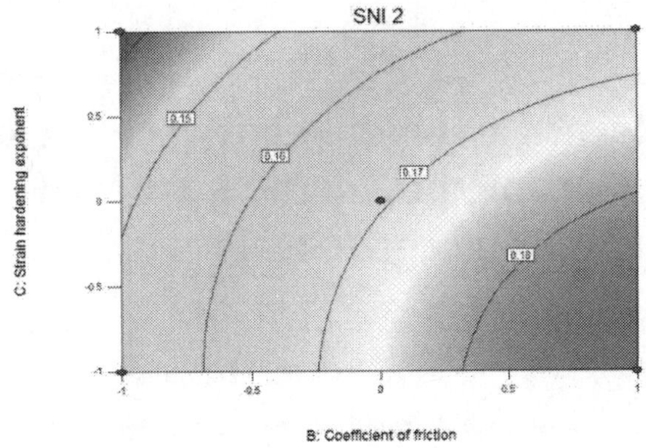

Figure 6. Contour graph of μ and n on SNI

5. Conclusion

It was found that the coefficient of friction has a very significant impact on both corner filling as well as SNI. CR and SNI have an inverse relationship on the co-efficient of friction. Material properties like anisotropic factor and strain hardening exponent have positive impact on both the responses. Furthermore, tube thickness has more impact on SNI than corner filling. But loading path does not have much impact on SNI, although it does have an influence on CR. Table 3 shows the effect of each independent variable on both responses in the descending order from left to right.

Table 3. Effect of each variable on response in descending order

Responses	**Variables (highest to lowest)**				
Corner radius	Coefficient of friction(μ)	Anisotropic factor(r)	Strain hardening exponent (n)	Loading path	Tube thickness (t)
SNI	Coefficient of friction(μ)	Tube thickness(t)	Strain hardening exponent (n)	Anisotropic factor(r)	Loading path

References

[1] Dohmann F and Hartl C 1996 Hydroforming–a method to manufacture light weight *parts J. of Mater. Process. and Technol.* **60** 669

[2] Elyasi M et al 2010 A new die design for the hydroforming of stepped tubes *Int. J. of Mater. Form.* **3** 71

[3] Carleer B et al 2000 Analysis of the effect of material properties on the hydroforming process of tubes *J. of Mater. Process. Technol.* **104** 158

[4] Zhang W W et al 2014 Analysis of corner filling behavior during tube hydroforming of rectangular section based on Gurson – Tvergaard – Needleman ductile damage model *J. of Eng. Manuf.* 1

[5] Alswad A et al 2011 Integration of finite element analysis and design of experiments to analyse the geometrical factors in bi-layered tube hydroforming *J. of Mat. and Des.* **32** 838

[6] Alaswad A et al 2011 Finite element comparison of single and bi - layered tube hydroforming processes *Int. J. of Simul.Model Pract. and Theory* **19** 1584

[7] P. P. Date 2014 Effect of mesh size on calculation of strain non – uniformity index in drawn sheet metal parts, in: *Proc. of int. conf. on technol. of plasticity* Nagoya Japan 945

Issues associated with the use of Yoshida nonlinear isotropic/kinematic hardening material model in Advanced High Strength Steels

Ming F. Shi[1], Li Zhang[2] and Xinhai Zhu[2]

[1]United States Steel Corporation, 5850 New King Court, Troy, MI 48098, U.S.A.
[2]Livermore Software Technology Corporation, 7374 Las Positas Road, Livermore, CA 94551, USA

mfshi@uss.com

Abstract. The Yoshida nonlinear isotropic/kinematic hardening material model is often selected in forming simulations where an accurate springback prediction is required. Many successful application cases in the industrial scale automotive components using advanced high strength steels (AHSS) have been reported to give better springback predictions. Several issues have been raised recently in the use of the model for higher strength AHSS including the use of two C vs. one C material parameters in the Armstrong and Frederick model (AF model), the original Yoshida model vs. Original Yoshida model with modified hardening law, and constant Young's Modulus vs. decayed Young's Modulus as a function of plastic strain. In this paper, an industrial scale automotive component using 980 MPa strength materials is selected to study the effect of two C and one C material parameters in the AF model on both forming and springback prediction using the Yoshida model with and without the modified hardening law. The effect of decayed Young's Modulus on the springback prediction for AHSS is also evaluated. In addition, the limitations of the material parameters determined from tension and compression tests without multiple cycle tests are also discussed for components undergoing several bending and unbending deformations.

1.0 Introduction

Advanced high strength steels (AHSS) are widely used in automotive body-in-white (BIW) components due to their potential weight-saving properties, while maintaining component performance in crashworthiness and structural durability. The application of AHSS in a typical vehicle will be increased from an average of 254 lbs in 2014 to an average of 483 lbs by 2025 based on a forecast by Ducker [1]. The driving force behind this forecast is the ever-increasing innovations in new steel development. Gen-3 steels aim to achieve a balance between press hardening steels (PHS) and Gen-1 dual phase steels by having higher elongation than PHS and higher strength than dual phase steels. The advantage of having higher elongation is that it is possible to cold stamp those steels using conventional stamping presses, thereby reducing fabrication cost. An additional promising feature of Gen-3 steel is its high fracture resistance compared to PHS, which makes the usage of Gen-3 steels in body-in-white (BIW) structure components attractive. Nevertheless, research has shown that for BIW applications, steels with yield strengths higher than 1.3 GPa are required for significant weight savings. However, one of the significant challenges during cold forming those steels is springback control. The ability to accurately predict springback after forming and trimming is essential in minimizing the number of die re-cuts and developing compensated dies with minimal springback and springback variation. Typically, AHSS including those Gen-3 AHSS show a significant Bauschinger effect during reverse loading and a nonlinear combined isotropic and kinematic hardening law is required to capture the Bauschinger effect of those materials. The use of the nonlinear isotropic/kinematic hardening material model proposed by Yoshida, et al [2] is useful in forming simulations where an accurate springback prediction is required. For materials such as steels showing continuous strain hardening behaviour, the Yoshida model with a modified hardening law proposed by Shi, et al [3] is often used. Many successful application cases in industrial scale automotive components using advanced high strength steels (AHSS) have been reported to give better springback predictions [4]. Several issues have been raised recently in the use of the model for higher strength AHSS including the use of two C vs. one C material parameters in the Armstrong and Frederick model (AF model), the original Yoshida model vs. Original Yoshida model with modified hardening law, and constant Young's Modulus vs. decayed Young's Modulus as a function of plastic strain. In this paper, the consistency of the material model parameters determined from the tension and compression test is verified using a one-element simulation for a 980 MPa material. Using an industrial scale automotive component, the difference in forming and springback results is addressed for two C and one C material parameters in the AF model using the Yoshida model with and without the modified hardening law. The effect of decayed Young's Modulus on springback prediction for AHSS is also evaluated. In addition, the limitations of the material parameters determined from the tension and compression tests without multiple cycle tests are discussed for those components undergoing several bending and unbending deformations.

2.0 Determinations and verifications of Yoshida model parameters

Two different types of tension and compression tests were carried out: a full cycle test and a multiple cycle test, as shown in Figure 1. The multiple cycle tests were designed to capture material hardening behaviour in loading after the material was subjected to one or more tension and compression cycles. The pure tension data were also included in Figure 1, where the pure tension data coincide with the initial tension portion of the data in the tension and compression tests, as expected.

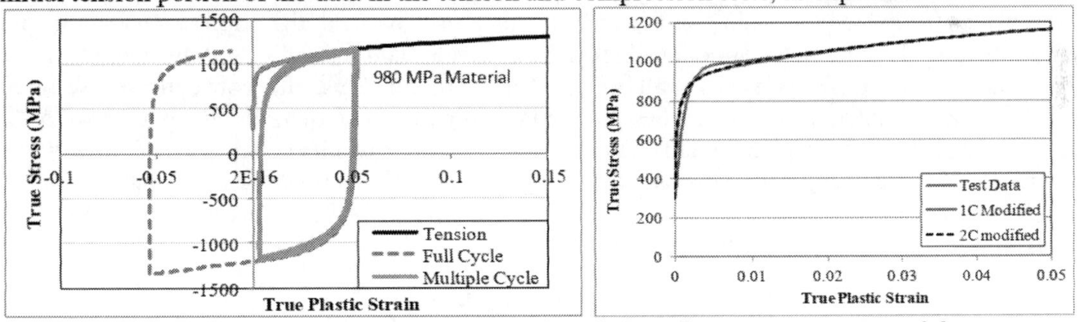

Figure 1 Tension and compression test data and data fitting of one C vs. two C models

Table 1 Material Constitutive Parameters for Original and Modified Yoshida Models

Modified	Symbols in the paper	Y (MPa)	B (MPa)	C	m	b (MPa)	h	K (MPa)	e0	N	C2
980MPa	One C Model	300	844.4	712.2	49.5	330.4	0.95	1527.3	0.010	1.13	
980MPa	Two C Model	300	835.2	657.2	48.0	323.1	0.95	1270.3	0.0535	0.93	1971.5
Original	**Symbols in the paper**	**Y (MPa)**	**B (MPa)**	**C**	**Rsat (MPa)**	**m**	**b (MPa)**	**h**	**C2**		
980MPa	One C Model	300	824.2	660.8	124.8	44.5	317.6	0.95			
980MPa	Two C Model	300	813.2	591.1	157.9	41.5	302.9	0.94	1773.2		

The material constitutive parameters were determined using an optimisation algorithm using the data shown in Figure 1(a), as shown in Table 1 for both original Yoshida model [2] and modified model

[3], where material model parameters for both one C and two C models were provided. The Two C model was selected to describe the initially higher stain hardening behaviour of AHSS. It seems from Figure 1(b) that the 2C model is better able to describe the higher strain hardening at low strain than the one C model.

The one element model was used to verify the effectiveness of the model implementation in LS-DYNA. The tension and compression test was simulated in the one element model using the parameters identified in Table 1. As shown in Figure 2, the predicted stress and strain data are consistent with the tested data from the one-element model for the original Yoshida model and the modified model. As expected, the two C model is able to capture the higher strain hardening behaviour of the material at low strains.

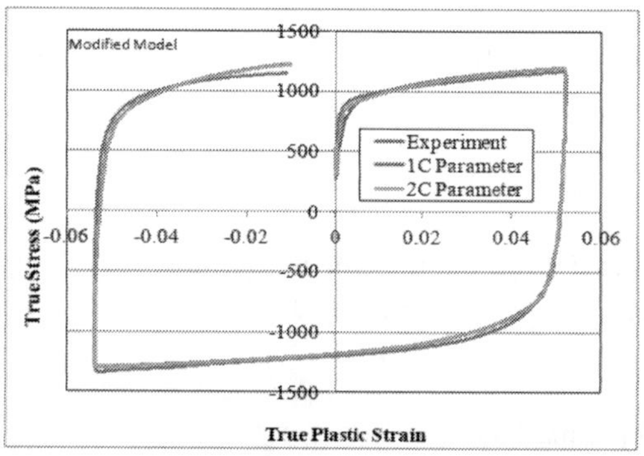

Figure 2 One-element Simulation Results

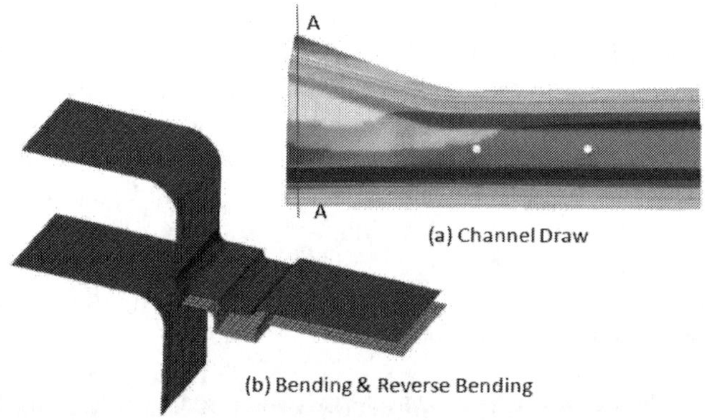

Figure 3 Components used in this study

3.0 Springback and forming strain prediction comparison

A channel draw type of industrial component was selected to evaluate the difference in the springback prediction using various models such as the one C vs. the two C models, the original Yoshida model vs. the modified Yoshida model (continuous hardening), the constant Young's modulus model vs. the decayed Young's Modulus model, as well as the Yoshida model vs. the conventional isotropic material model. As shown in Table 2, the maximum springback predictions along section A-A, including side wall curl, angle change and twist, show very similar results between one C and two C models (about 1% difference). The Yoshida model with the modified hardening law results in higher springback than the original Yoshida model while the conventional isotropic hardening model predicts significantly lower springback. As suggested by Yoshida [2], a decayed Young's Modulus (E) value may be used at various strains using the following equation:

$$E = E_0 - (E_0 - E_1)(1 - e^{-\alpha\varepsilon_p})$$

Where E_0, E_1, and α are constants and ε_p is the plastic strain. In our case where E_0 equal to 205 GPa, E_1 equal to 192 GPa, and α equal to 184, the delayed E predicted higher springback than the constant E, as shown in Table 2. However, the modified model with a constant E predicted a similar springback to the original model with delayed E. The use of a decayed E has no effect on the forming results for both models.

Table 2 Springback and thinning results for the channel and cross member

	Channel Draw					
	One C Modified	Two C Original	Two 2C odified	Isotropic	Two C Modified + Decayed E	Two C Original + Decayed E
Max Thinning (A-A) (%)	8.4	9.0	8.3	8.2	8.3	9.0
Max Thinning (part) (%)	8.7	9.7	8.7	9.4	8.7	9.7
Max Springback (A-A) (mm)	60.8	57.3	60.0	53.9	62.7	60.0

As shown in Table 2, all models predicted very similar thinning for this part and the amount of thinning is below 14% thinning limit for this material.

4.0 Effect of multiple cycle test result on forming and springback prediction
Most publications including the Numisheet 2011 benchmark use only a full cycle of the tension and compression test data to determine the Yoshida model parameters. As pointed out in [3], a multiple cycle of the tension and compression test should be included in the parameter determination, which is particularly important if a part is subject to multiple loadings and reserve loadings. The effect of multiple cycle test data on the overall forming and springback prediction can be demonstrated by a simple bending and reserve bending case, as shown in Figure 3. The modified Yoshida model parameters were also obtained excluding the multiple tension and compression test data in the parameter determination. The predicted maximum thinning and maximum springback in the z direction are respectively 13.6% and 28.0 mm using the parameters determined with all test data and 12.0% and 24.5 mm respectively using the parameters determined with the test data excluding the multiple cycle test. This case study clearly indicates that the multiple cycle test data should be included in the Yoshida model parameter determination, particularly for those parts subject to two or more loading and unloading cycle.

5.0 Conclusions
Based on the above study, the following conclusions can be made:
The two C model can capture the higher straining hardening behaviour of AHSS at a lower strain portion. Both one C and two C model predicted similar thinning but slightly different springback.
The modified Yoshida models predicted slightly higher springback than the original Yoshida model while the isotropic hardening model predicted significantly lower springback than the original and modified Yoshida models.
The use of decayed Young's Modulus resulted in higher springback prediction than the use of the constant Modulus.
The multiple cycle tension and compression test must be used in parameter determination, particularly for those parts subject to two or more loading and unloading deformations.

6.0 Reference
[1] Abraham A 2015 Great Design In Steels, www.a-sp.org
[2] Yoshida F and Uemori T 2002, Int. j. Plasticity, vol 18, p 661
[3] Shi M, Zhu X, Xia C and Stoughton T 2008 Proc. NUMISHEET p 1256
[4] Chen X, Shi M, Zhu X, Du C, Xia C, Xu S and Wang C 2009 SAE Paper 2009-01-0981

7.0 Acknowledgement
The authors are grateful to Prof. Xin Wu for providing some of Yoshida model parameters.

Analysis of earing behaviour in deep drawing of ASS 304 at elevated temperature

Amit Kumar Gupta[1*], Aditya Deole[1], Nitin Kotkunde[1], Swadesh Kumar Singh[2], and Gangadhar jella[2]

[1] Mechanical Engineering Department, BITS-Pilani Hyderabad Campus, Hyderabad-500078, Telangana, India.
[2] Mechanical Engineering Department, GRIET, Hyderabad-500090, Telangana, India

akgupta@hyderabad.bits-pilani.ac.in

Abstract. Earing tendency in a deep drawn cup of circular blanks is one the most prominent characteristics observed due to anisotropy in a metal sheet. Such formation of uneven rim is mainly due to dissimilarity in yield stress as well as Lankford parameter (*r*- value) in different orientations. In this paper, an analytical function coupled with different yield functions *viz.*, Hill 1948, Barlat 1989 and Barlat Yld 2000-2d has been used to provide an approximation of earing profile. In order to validate the results, material parameters for yield functions and hardening rule have been calibrated for ASS 304 at 250^{0}C and deep drawing experiment is conducted to measure the earing profile. The predicted earing profiles based on analytical results have been validated using experimental earing profile. Based on this analysis, Barlat Yld 2000-2d has been observed to be a well suited yield model for deep drawing of ASS 304, which also confirms the reliability of analytical function for earing profile estimation.

1. Introduction

Accurate prediction of plastic deformation in metal forming depends upon the proper selection of yield function and hardening rule [1]. Substantial work on development of anisotropic yield has been reported by researchers in the past few decades [2]. For example, Hill in 1948 extended von misses yield criterion to consider plastic anisotropy by including four additional material parameters. Although the criterion does not predict *r* value with respect to orientation from rolling direction with fair accuracy, this criterion has been extensively used in the industry due to the simplicity involved in calibrating the function using simple uniaxial tensile test [3]. Barlat and Lian proposed a non-quadratic yield function for anisotropic sheet metal and characterized *r* value accurately. But the accuracy in prediction of *r*-value and yield strength with respect to orientation from rolling direction strongly depended upon addition material parameter in the yield function which has to be evaluated iteratively [4]. Barlat et al. [5] suggested anisotropic yield function with eight constants and successfully predicted the *r*-value and yield strength directionality with excellent accuracy to confirm its applicability for metal forming.

The deformation behavior of austenitic stainless has been studied extensively to access the formability at elevated temperatures. Jong et al. constructed forming limit diagram using conventional M-K (Marciniak–Kuczyński) model and modified PMC (Parmar–Mellor–Chakrabarty) model and proved modified PMC model to be suitable since it captures the effect of surface roughening which substantially affects metal formability [6]. Stachowicz et al., for an austenitic stainless steel, investigated the effect of the forming temperature on yield strength, ultimate strength, uniform elongations and strain hardening and on spring back in V-bend test [7]. Nevertheless, it is important to note that earing is one of the important characteristics in deep drawing since it is associated with the anisotropy in the sheet. Therefore, it is necessary to estimate prediction capability of the yield function in terms of *r*-value, yield strength directionality for earing profile prediction. Yoon et al. [8] formulated an analytical function to predict earing profile by taking tool dimensions, blank dimensions, yield function and Lankford parameter into account.

In the present study, deep drawing experiment has been conducted at 250⁰ C and cup height is measured with respect to orientation from rolling direction. Anisotropic yield models, *viz.*, Hill 1948, Barlat 1989, Barlat Yld 2000-2d have been calibrated for ASS 304 to predict earing profile using analytical function suggested by Yoon. The results obtained from analytical function and simulations are compared with experimental data.

2. Analytical function for Earing profile

Yoon et al. proposed an analytical function to predict cup height with respect to orientation from rolling direction [8]. The function considers the contributions of the both of *r*-value anisotropic and yield strength anisotropy in cup height. The analytical function is given in Equation 1.

$$H^{cup}(\theta) = t_0 + R_c + \frac{R_b}{A_{\theta+90}+1}\left(d^{A_{\theta+90}} - \frac{1}{d}\right)(B_\theta)^{A_{\theta+90}} \tag{1}$$

Where, θ is orientation from rolling direction, thickness of the blank: $t_0 = 1mm$, punch corner radius: $R_b = 36\ mm$, cup corner radius: $R_c = 19\ mm$.

$$A_{\theta+90} = \frac{r_{\theta+90}}{1+r_{\theta+90}};\ B_\theta = \left(\frac{\sigma_{ref}}{\sigma^Y{}_{(\theta)}}\right)^\beta \tag{2}$$

Where, $0.5 \le \beta \le 1,\ d = \frac{R_b}{R_c}$

$$\sigma_{ref} = \left(\int_0^{2\pi} \sigma^Y{}_{(\theta)} d\theta\right) \Big/ 2\pi \tag{3}$$

In order to calculate earing profile, $\sigma^Y{}_{(\theta)}$ is substituted by yield strength obtained by anisotropic yield function for different orientations. For this purpose, yield functions namely Hill 1948, Barlat 1989 and Barlat Yld 2000-2d are calibrated for ASS 304. The required material properties for the yield criteria development have been determined using uniaxial tensile tests at 250⁰C. The calculated material properties are mentioned in Table 1. The detailed procedure was mentioned for material parameter determination by Kotkunde et al. [2]. The obtained material constants from above three yield criteria are mentioned in Table 2. Figure 1 (a), (b) and (c) shows the yield loci for ASS 304 at 250⁰C. Figure 2 (a) and (b) shows yield stress variation and Lankford coefficient variation respectively.

Table 1 Material properties for ASS 304 alloy at 250⁰C

σ_0 (MPa)	σ_{45} (MPa)	σ_{90} (MPa)	r_0	r_{45}	r_{90}
176.16	161.06	167.05	0.758	0.946	0.753

Table 2 Material Parameters for yield crietra

Yield model	Material parameters								
Hill 1948	*F*	*G*	*H*	*N*					
	1.168	1.123	0.876	3.300					
Barlat 1989	*a*	*c*	*h*	*p*	*m*				
	1.133	0.866	1.011	1.064	6				
Barlat Yld 2000-2d	a_1	a_2	a_3	a_4	a_5	a_6	a_7	a_8	*a*
	0.916	*0.996*	*0.682*	*1.007*	*0.982*	*0.794*	*1.032*	*1.376*	6

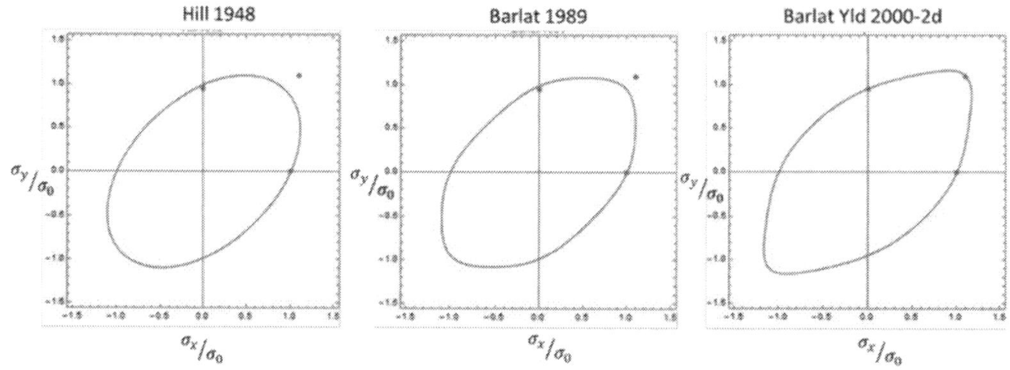

Figure 1 Yield loci (a) Hill 1948 (b) Barlat 1989 (c) Barlat Yld2000-2d

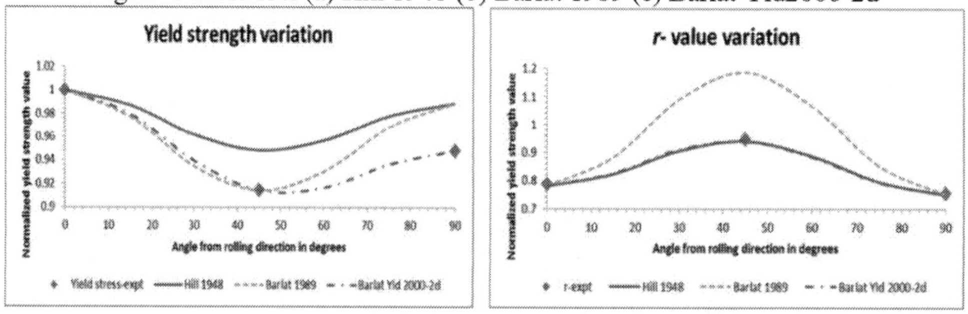

Figure 2 Variation of (a) yield stress (b) r-value with respect to orientation from rolling direction

3. Result and Discussion

Deep drawing experiments on ASS 304 have been performed at 250^{0}C with 72 *mm* blank diameter and 1 *mm* thickness sheet, and the earing profile is measured using dial gauges. The details of the setup are given in previous work by Kotkunde et al. [3]. Figure 1 (a) (b) & (c) show the yield loci plotted by Hill 1948, Barlat 1989 and Barlat Yld 2000-2d yield criteria. It is clear from the figure that yield locus plotted by Hill 1948 and Barlat 1989 yield functions are not able to predict the bi-axial yield point. But, these yield loci captures in plane yield stress asymmetry in a fairly accurate manner. However, for Barlat Yld-2000-2d yield function, the locus passes very well through uniaxial tensile yield points in rolling and transverse direction along with the biaxial yield stress point. This confirms suitability of Barlat Yld 2000-2d for representation of the yield curve for ASS 304.

It is also important to compare the prediction capability of yield stress and *r*-value directionality with experimental results. Since yield functions have a key role in evaluating the plastic strains using flow rule. Figure 2 (a) & (b) show the variation of yield stress and *r*-value with respect to orientation from rolling direction. It can be observed from Figure 4 (a) that Hill 1948 yield function does not capture yield stress variation with respect to rolling direction but it predicts *r*-value directionality accurately. In case of Barlat 1989 yield function, a good agreement is observed for yield stress and *r*-value only for two orientations and significant discrepancy with experimental data for remaining one direction. This depends on the directionality equation of yield stress or *r*-value for which parameter p is calculated iteratively. On the contrary, Barlat Yld 2000-2d calculates the yield strength and *r*-value directionality with excellent accuracy which also shows Barlat Yld 2000-2d to be suitable for calculating material response than that of the other two models.

Furthermore, the experimental cup height is compared with analytical model and it is represented in Figure 3. It is observed from Figure 3 that the analytical earing profile predicted by Hill 1948 and Barlat 1989 yield functions over-estimates cup height. On the other hand, Barlat Yld 2000-2d predicts the analytical cup height precisely with a minimum error of 1.01% and 0.96 correlation coefficient. It guarantees Barlat Yld 2000-2d to be a well suitable for calculating the cup height analytically.

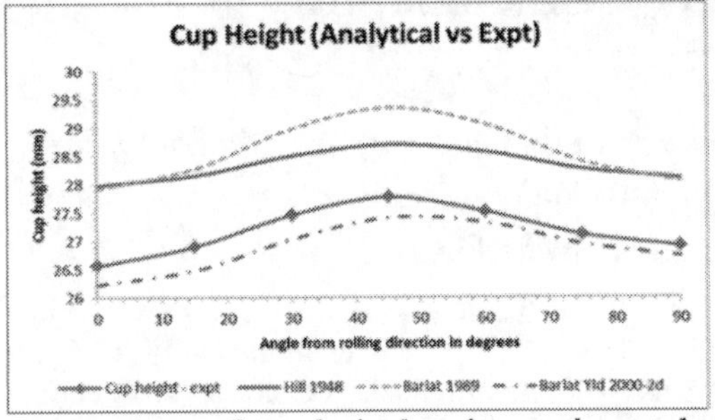

Figure 3 Comparison of Analytical and experimental earing profile

4. Conclusions

Analytical equation with three anisotropic yield functions was used in this work to provide an approximation of earing profile of deep drawn cup. The main conclusions from this study are:

Accuracy of earing profile predictions critically depends upon the proper selection of yield function. Since the anisotropy in both yield stress as well as *r*-value contributes to earing in the deep drawn cup. Hill 1948 and Barlat 1989 yield functions are not accurate enough in characterizing *r*-value and yield stress directionality. Whereas, Barlat Yld 2000-2d predicts these directionalities precisely. As a result, analytical equation coupled with Barlat Yld 2000-2d yield function provides an excellent approximation of the earing profile in agreement with experimental data.

Acknowledgment

Authors would like to thank Department of Science and Technology (DST) Government of India, for providing grant to carry our research activities at the host institute (GRIET) *SR/S3/MERC/0129/2012*.

References

[1] Kotkunde N, Krishnamurthy HN, Puranik P, Gupta AK and Singh SK 2014, *Microstructure study and constitutive modeling of Ti–6Al–4V alloy at elevated temperatures*, *Materials & Design* 2014, 54, 96-103.

[2] Kotkunde N, Deole AD, Gupta AK and Singh SK 2014, *Experimental and numerical investigation of anisotropic yield criteria for warm deep drawing of Ti–6Al–4V alloy*, Materials & Design, 63, 336-344.

[3] Kotkunde N, Deole AD, Gupta AK, Singh SK, Aditya B 2014, *Failure and formability studies in warm deep drawing of Ti–6Al–4V alloy*, Materials & Design, 60, 540-547.

[4] Barlat F and Lian K 1989, *Plastic behaviour and strechability of sheet metals Part I: a yield function for orthotropic sheets under plane stress condition*, International Journal of Plasticity, 51–66.

[5] Yoon JW, Barlat F, Dick RE, Chung K and Kang TJ 2004, *Plane stress yield function for aluminum alloy sheets—part II: FE formulation and its implementation*, International Journal of Plasticity, 20(3), 495–522.

[6] Jong H 2012, *The forming limit diagram of ferritic stainless steel sheets: Experiments and modeling*, International Journal of Mechanical Sciences, 64, 1–10.

[7] Stachowicz F and Trzepieciński T 2010, *Warm forming of stainless steel sheet*, Archives of Civil and Mechanical Engineering,10, 85–94.

[8] Yoon JW, Dick RE and Barlat F 2011, *A new analytical theory for earing generated from anisotropic plasticity*, International Journal of Plasticity, 27, 1165–1184.

On Simulation of Edge Stretchability of an 800MPa Advanced High Strength Steel

Nikky Pathak, Cliff Butcher, Michael Worswick
University of Waterloo
n3pathak@uwaterloo.ca

Abstract. In the present work, the edge stretchability of advanced high strength steel (AHSS) was investigated experimentally and numerically using both a hole expansion test and a tensile specimen with a central hole. The experimental fracture strains obtained using the hole expansion and hole tension test in both reamed and sheared edge conditions were in very good agreement, suggesting the tests are equivalent for fracture characterization. Isotropic finite-element simulations of both tests were performed to compare the stress-state near the hole edge.

1. Introduction

The hole expansion test can effectively capture the influence of different edge conditions, such as sheared, drilled or reamed, but is sensitive to the punch geometry and burr-orientation [1]. However it is unable to resolve variation in the fracture strain with material direction since the edge crack always develops in the direction with the lowest ductility. For numerical characterization, the hole expansion test is challenging since there is tool contact, a strong tendency to incite hourglass modes and initially strong through-thickness stress and strain gradients. In a study unrelated to sheared edge cracking, Dunand and Mohr [2] used a tensile specimen with a central hole (hole tensile test) for both numerical and experimental characterization of fracture in uniaxial tension. Hole tensile tests are well suited for numerical characterization to simulate edge fracture since there is no friction, tooling contact or shear burr orientation effects. Additionally, anisotropy of the fracture strains can be investigated. The objective of the present work is to directly compare the hole expansion and hole tensile tests as characterization tests to investigate edge cracking of a typical dual phase steel, DP780. Isotropic finite element models of the hole tension test and hole expansion test with a reamed edge are used to demonstrate the effectiveness of the proposed experimental methodology that does not require inverse finite-element modeling to obtain accurate engineering estimates for edge formability. This investigation will assist in the development of an alternative to the hole expansion test that is better suited for experimental and numerical characterization of edge fracture.

2. Experimental Procedure

2.1. Hole Expansion Test

The hole expansion test consists of expanding a hole in a blank with a conical punch until the onset of edge cracking. The formability metric in a hole expansion test is typically the hole expansion ratio (HER), defined as the ratio of the initial and final inner diameters of the hole as

$$HER\ (\%) = 100 \cdot \left(d_f / d_o - 1 \right) \qquad (1)$$

Hole expansion testing was conducted for both sheared and reamed (ideal) edge conditions using the ISO-16630 standard hole size of 10 mm. The procedure for the hole expansion test is outlined in [1]. The strains at the hole edge during the hole expansion test cannot be readily tracked using digital image correlation (DIC) techniques since the sheet thickness is not seen by the camera system until the hole has been expanded a sufficient amount and the hole edge rotates out-of-plane and into view of the cameras. Additionally, DIC techniques do not resolve edges so the strain measurement near the edge is very challenging. An analytical method developed by Butcher et al. [3] to determine the equivalent failure strain in a hole expansion test has been shown to be a very good estimation of the local strain in an isotropic axisymmetric simulation of a hole expansion test of DP600. The equivalent fracture strain can be estimated by measuring the inner and outer hole diameters at fracture as well as the thickness around the circumference as:

$$\varepsilon_{eq}^{macroscopic} = \frac{2}{3}(\varepsilon_c - \varepsilon_t), \qquad \varepsilon_c = \ln\left(\frac{d_{outer} + d_{inner}}{2d_o}\right), \qquad \varepsilon_t = \ln\left(\frac{t_{edge}}{t_o}\right) \tag{2-4}$$

This method is used in present study to estimate the fracture strains in a hole expansion test without using DIC techniques.

2.2. Hole Tension Test Procedure

A tensile specimen of gauge length 35 mm and width 22.5 mm was fabricated with a hole of diameter 10 mm at the center. The specimen was subjected to tension until a visible crack appeared at the edge. The majority of samples failed along the rolling direction during the hole expansion test. The transverse loading direction was therefore selected to replicate failure conditions during the hole expansion test. A 3-D DIC system by Correlated solutions Inc. was used to record the full-field strain during the test experiment using a frame rate of 4 per seconds and image size in 2448 x 2048 pixels. A vertical virtual extensometer of length 21 mm was used to measure displacement to failure. The fracture strain at the crack location was measured from the DIC analysis using a circle of size 0.25 mm.

3. Edge Stretchability

3.1. Hole Expansion Test Results

The HER for the sheared hole in the burr-up orientation was 22% and lower than the burr-down orientation, with its HER of 30%. The HER reported for the reamed edge is 51% which is significantly higher than the sheared edge because almost no damage exists at the reamed edge, corresponding to the condition of the base microstructure. The shearing process, on the other hand, introduces work-hardening and pre-nucleated voids behind the sheared edge. This pre-existing damage leads to the initiation of cracks at lower strains and consequently reduces edge stretchability. The influence of the shear burr-orientation is observed on the HER as a result of through-thickness stress and strain gradients.

3.2. Hole Tension Test

The edge stretching limit in a hole tension test is quantified by an equivalent strain obtained from DIC analysis. It is important to state that due to the inherent limitations of DIC in resolving edges, the failure strains are not true edge strains but were calculated at a distance approximately 0.1-0.3 mm away from the edge. The contour plots of equivalent strain for reamed and sheared edges just before failure are shown in Figure 1a and 1b respectively. The strain localizes near the edge during the hole tension test and therefore the edge sensitivity can be captured effectively. Figure 1c shows the plot of equivalent failure strain for the reamed and sheared holes determined from the hole tension test and hole expansion test. The equivalent strains obtained in the burr-up and burr-down positions during the hole expansion test are combined to represent an "average value" for a sheared edge condition. The equivalent strain just before fracture at the reamed edge is 0.49 and at the sheared edge is 0.27. Similar to the hole expansion test, the difference in the stretching-limit for the reamed and sheared edges during the hole tension test is evident from the values of failure strain. The displacement to failure measured by a virtual extensometer is 1.64 (±0.08) mm for the reamed edge and 0.82 mm (±0.05) for the sheared edge. The confidence intervals of failure strain for the reamed edge obtained from the two tests overlap and suggest there is no significant difference between the limit-strains of edge determined from the two tests. The average failure strains of the sheared edge determined from the hole expansion and hole tension tests matches exactly.

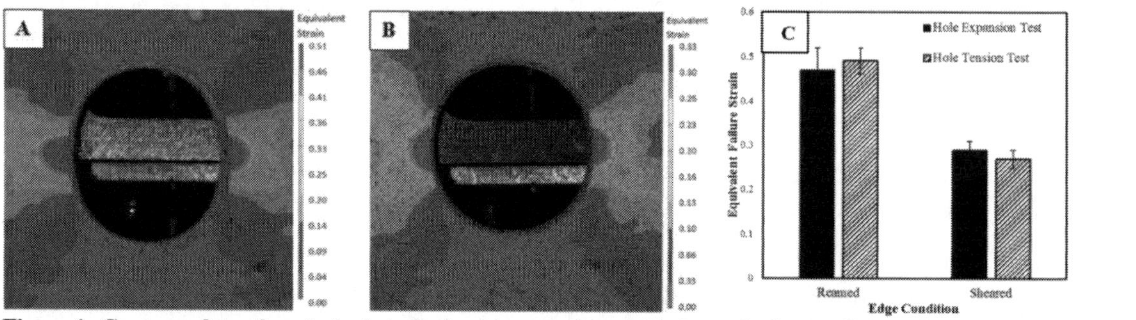

Figure 1: Contour plots of equivalent strain for (a) reamed and (b) sheared edge conditions during the hole tension test. (c) Equivalent failure strain obtained during the hole expansion and hole tension tests for the reamed and sheared edges

4. Numerical Modeling

A finite element model of the hole tension and hole expansion tests have been developed to validate the proposed methods for estimating the fracture strain. A one-eighth finite element model of the hole tension geometry and a quarter model of hole expansion geometry with a reamed edge condition were created using the commercial finite-element code, LS-DYNA. Only the reamed edge condition is considered since additional experiments are required to estimate the residual strain field in the sheared edge. The geometries were meshed using constant stress solid elements with an average equivalent length of 0.1 mm in the region near the hole edge with 8 elements through the half-thickness. A displacement boundary condition was enforced at the free-end of the hole tension specimen and an implicit solution was obtained under quasi-static loading. The displacement-to-failure was measured using the nodes that correspond to the half-length of the extensometer gauge length of 21 mm. The simulation is terminated when the experimental displacement to failure is reached. For the simulation of hole expansion test, the steel punch, binder and draw ring are modeled as rigid surfaces with a coefficient of friction of 0.10 with the blank. The simulation is terminated when the hole expansion ratio determined experimentally is achieved. The material is assumed to obey the von Mises yield criterion with the flow stress curve from [4] used to represent the material behavior in the transverse direction. No failure treatment is included within the model at present since the objective is to use the model to evaluate the transferability of the experimental hole expansion and hole tensile strains to the local strains within the material. The deformed mesh plots of hole tension and hole expansion tests are shown in Figure 2 a and b respectively.

Figure 3a shows the plot of equivalent strain versus displacement obtained by the numerical simulation of the hole tension test for a reamed edge. The equivalent strain taken at the element located at the minimum section of hole tension specimen on the hole edge. The strain accumulation with respect to displacement of the center element and the element at the hole edge is almost similar throughout the test and suggests relatively uniform deformation in a through-thickness direction. The measured strain history at the point of failure, experimentally derived from the DIC analysis, is also plotted. The failure strain derived numerically at the displacement to failure (1.64 mm) is 0.52 and lies close to the failure strain derived experimentally and hence validates the numerical model. The fracture strain of 0.51 (±0.05) from the hole expansion test would also give very good predictions for the displacement to failure.

The stress-triaxiality at the hole edges during the hole tension and hole expansion tests are shown in Figure 3b. The center element and element at the edge of expanding hole during the hole tension test are under constant uniaxial tension state throughout deformation with a constant stress-triaxiality of 1/3. This stress-state is similar to the upper element of the expanding hole during the hole expansion test. The lower edge of hole expansion test however is in compression state due to the punch contact. This mechanism generates a through-thickness stress-gradient which converges to uniaxial tension as deformation continues. Since, the DP780 reamed edge experiences fracture after the convergence of stress-gradient, the assumption of hole tension test replicating hole expansion test is appropriate. This demonstrates that the hole tension test is an efficient experiment to capture edge stretchability.

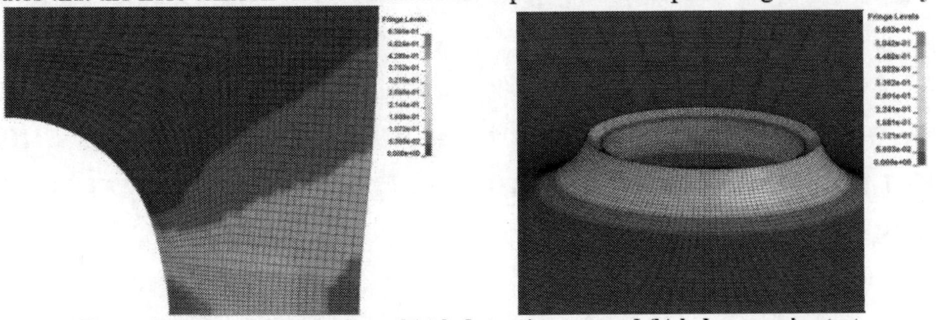

Figure 2: Deformed mesh plots of (a) hole tension test and (b) hole expansion test

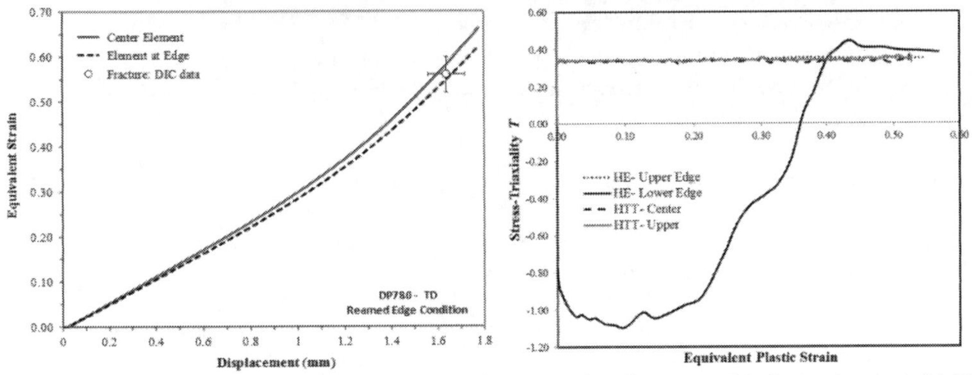

Figure 3: (a) The plot of equivalent strain versus displacement for the reamed hole tension test (b) The stress triaxiality at the edge during hole tension test

5. Conclusions

The following conclusion can be drawn from the present investigations:

1. A significant reduction in HER for the sheared edge was observed compared to the reamed edge due to the presence of pre-nucleated voids and residual strain from the shearing process.
2. Similar to the hole expansion test, the edge condition has a significant influence on the failure strain of the hole tension test.
3. An experimental method to estimate the failure strains in the hole expansion test for the reamed and sheared edge conditions is in very good agreement with the hole tensile failure strains obtained using DIC analysis.
4. Isotropic finite-element simulation shows that a nearly constant uniaxial stress state is achieved at the edge during the hole tension test, at least for the current material and ductility. A stress-gradient exists at the hole edge during the initial deformation of hole expansion test, which later converges to a uniaxial state.

References

[1] N. Pathak, C. Butcher, M. Worswick, Assessment of critical parameters influencing the edge stretchability of AHSS sheet, submitted to Journal of Materials Engineering and Performance, 2016.

[2] D. Matthieu, and D. Mohr. 'Hybrid Experimental–Numerical Analysis Of Basic Ductile Fracture Experiments For Sheet Metals'. International Journal of Solids and Structures 47.9 (2010): 1130-1143.

[3] C. Butcher, L. ten Kortenaar, M. Worswick, Experimental characterization of the sheared edge formability of boron steel, IDDRG conference 2014, Paris, France.

[4] A. Abedini, C. Butcher, D. Anderson, M. Worswick, and T. Skszek, (2013). Fracture Characterization of Automotive Alloys in Shear Loading. SAE , 2015-01-0528.

Journal of Physics: Conference Series **734** (2016) 032122 doi:10.1088/1742-6596/734/3/032122

A study on the performance of ductile failure models under different range of stress triaxiality states with experimental validation

Rui Amaral[1], Pedro Teixeira[2], Erfan Azinpour[1], Abel D Santos[1,2], J César de Sá[1,2]

[1]INEGI, Institute of Science and Innovation in Mechanical and Industrial Engineering, R. Dr. Roberto Frias, 400, 4200-465, Porto, Portugal

[2]FEUP, Faculty of Engineering, University of Porto, R. Dr. Roberto Frias, 4200-465, Porto, Portugal

ramaral@inegi.up.pt, pmcteix@gmail.com , abel@fe.up.pt, cesarsa@fe.up.pt

Abstract. In this work, experimental tests were carried out, under different loading conditions, in order to assess different ductile failure criteria, namely based on GTN, Johnson-Cook or Lemaitre models and to establish new proposals for improvement. Corresponding characterization for damage parameters is performed by an inverse analysis procedure, using reference experimental tests. Numerical simulations of a cross-shaped component are considered for the damage models, and results show a similar trend related with the experimental fracture evidence.

1. Introduction

Sheet metal forming processes are used in several manufacturing areas, including aeronautical and automotive body components. The processing may include different loading paths and operations like drawing, bending and stamping, in which the sheet metal can be subjected to large localized deformations with significant through-thickness necking, thus developing 3D stress states and promoting the fracture event of the metal blank.

The possibility to analyse larger and complex deformations to which the sheet metal can be subjected during stamping or deep-drawing operations, up to damage and fracture, is nowadays enhanced by using numerical methods (e.g., Finite Element Method) [1], which are still evolving continuously.

Under current work, the aluminium alloy AA5182-O is studied using three damage models: Johnson-Cook model, Gurson-Tvergaard-Needleman (GTN) model and Lemaitre's ductile damage model. The Johnson-Cook and GTN models are implemented in FE package Abaqus, but the corresponding damage parameters are obtained by an inverse analysis procedure using reference experimental tests. In the framework of Continuous Damage Mechanics, Lemaitre's local damage evolution model is implemented in a user material subroutine [2], which was developed to be incorporated and integrated in the FE package Abaqus.

A cross-shaped component is analysed using numerical simulation and the obtained results for the damage models are compared with the experimental ones, in order to evaluate their ability to predict damage and fracture initiation.

2. Material and damage model parameters

In this paper, a mechanical characterization of the aluminium alloy AA5182-O, with 1 mm thickness, was performed [3]. Tensile tests of AA5182-O specimens were performed for different angles relative to the rolling direction. The corresponding flow curves are shown in Fig.1.

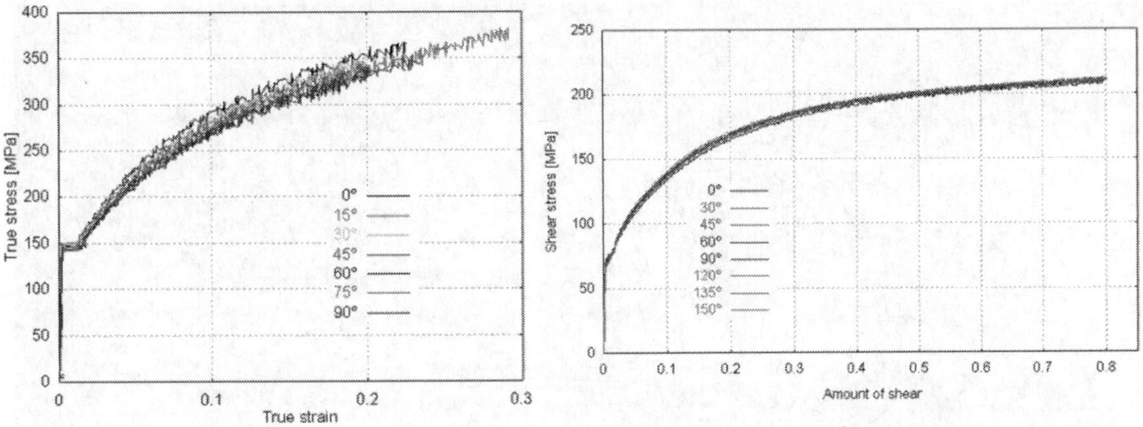

(a) (b)

Figure 1. Flow curves obtained from uniaxial tensile test (a) and from monotonic simples shear test (b) for AA5182-O, according to different directions relative to the rolling direction [3].

Table 1 contains the main mechanical properties of AA5182-O material and the parameters for hardening behaviour when modelled by Voce law.

Table 1. Mechanical properties for the aluminium alloy AA5182-O.

Property	Value	Voce Law	
Young modulus [GPa]	69		
Poisson coefficient	0.3	$Y = Y_0 + R_{sat}\left[1 - \exp\left(-C_r \cdot \bar{\varepsilon}^p\right)\right]$	
Yield stress [MPa]	149.5		
Ultimate tensile strength [MPa]	283.8	Y0 [MPa]	149
Total elongation [%]	24.4	Rsat [MPa]	208.7
r_0, r_{45}, r_{90}	0.79, 0.85, 0.7	Cr	12.1

An inverse numerical analysis, based on optimization algorithms, was used to obtain the parameters for each damage model. Corresponding parameters are presented in table 2 and they show being very similar to those identified by Sun *et al* [1], also for AA5182-O.

Table 2. AA5182-O parameters for selected damage models.

Johnson-Cook damage model				
Parameter	d_1	d_2	d_3	d_4, d_5
Value	0.024	0.38	1.5	0

GTN damage model									
Parameter	q_1	q_2	q_3	f_0	f_C	f_F	ε_N	S_N	f_N
Value	1.5	1	2.25	0.01	0.021	0.04	0.3	0.1	0.001

Lemaitre damage model [4]			
Parameter	Damage exponent - *s*	Damage denominator –*S*	Damage closure effect
Value	1.0	1.25 [MPa]	0.2

3. Results and discussion

The application of selected damage models is illustrated by means of an experimental deep drawing failure case of a cross-shaped sheet metal component. Figure 2(a) shows the selected experimental component after failure, using AA5182-O sheet with 1 mm thickness.

Figure 2. (a) Cross-shaped component with breakage. (b) Tool geometry and active parts (the punch cross geometry has 487 mm length).

A 3D FEM explicit analysis (Abaqus/Explicit) model is considered with one quarter of the experimental setup, due to symmetry. The tools are modelled as fully rigid surfaces with discretization performed by three nodded rigid elements. The blank discretization uses a double layer of deformable eight nodded solid elements with reduced integration. Regarding the material behaviour, the blank was modelled as an elasto-plastic material with hardening described by Voce Law, as already presented in section 2.

3.1. *Results*

Figures 3 and 4 present the contours of the equivalent plastic strain as well as the corresponding damage value for the three considered damage models, respectively, Johnson-Cook, GTN and Lemaitre.

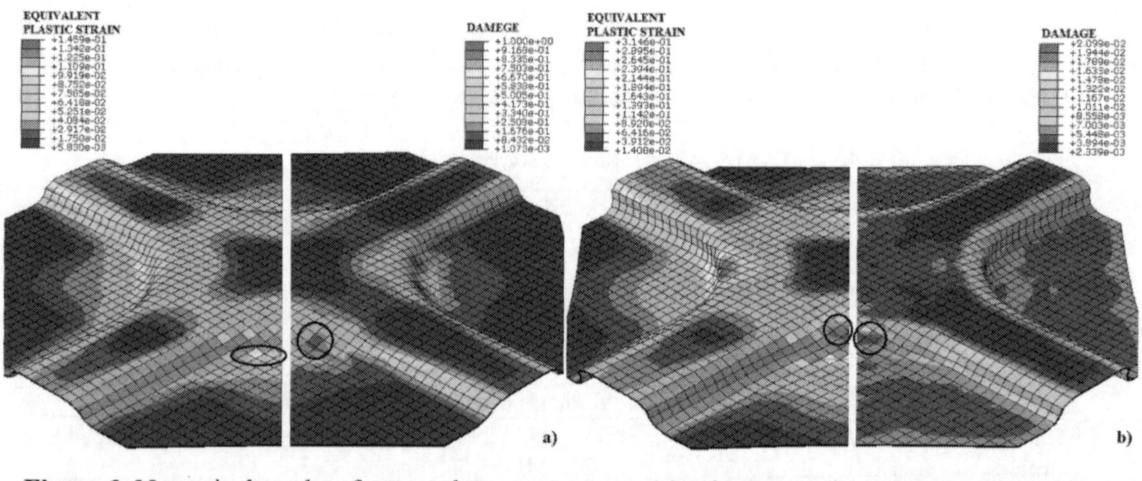

Figure 3. Numerical results of cross –shape component. (a) Johnson-Cook model; (b) GTN model.

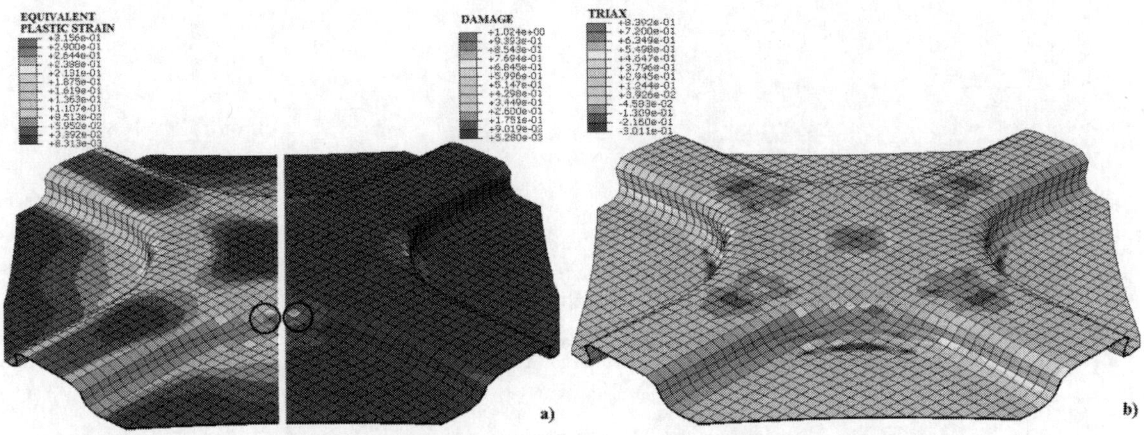

Figure 4. Numerical results of cross –shape component. (a) Lemaitre model; (b) stress triaxiality levels before failure.

It is also possible to see the prediction of fracture initiation for each model. For this experimental component, both maximum strain and damage variable have very close locations (at the vertical wall of cross-centre geometry). All damage models in this study show similar trend, which is in accordance with experimental evidence. However, the punch displacement to failure and maximum strain are different. Although GTN and Lemaitre models predict fracture initiation at the same punch displacement, Johnson-Cook model predicts failure at lower punch displacement and lower plastic strain. A reason for such difference can be due to the base of parameter identification for the studied damage models, which in turn is related with the triaxiality levels.

4. Conclusions

Three different ductile damage models, Johnson-Cook, GTN and Lemaitre were tested and compared in this paper to predict damage in sheet metal forming processes. A FE analysis of a cross-shaped component, using the parameters evaluated for the AA5182-O, has been taken as a basis of comparison between the damage models, in order to assess the accuracy of the results.

For this particular case, the simulations of GTN and Lemaitre show a good agreement with the damage location observed in the experimental component. Johnson-Cook damage model shows to overestimate the development of damage at high triaxialities. However, the range of relations between stress triaxiality and strain at failure, namely on low stress triaxiality states, is not completely incorporated for particular damage models [5], which may give directions for further developments and testing.

References

[1] Sun H T, Wang J, Shen G Z and Hu P 2013 *Int. J. Auto. Tech.* **14** 605-10
[2] Teixeira P, Santos A D, Andrade Pires F M and César de Sá J 2006 *J. Mat. Proc. Tech.* **177** 278-81
[3] 3DS report 2001 Selection and identification of elastoplastic models for the materials used in the benchmarks *Inter-regional IMS contract ''Digital Die Design Systems (3DS)''*
[4] Teixeira P, Santos A D, César de Sá J, Andrade Pires F M and Barata da Rocha A 2009 *Int. J. Mat. Form.* **2** 463-6
[5] Malcher L, Andrade Pires F M and César de Sá J 2014 *Int. J. Plast.* **54** 193–228

Acknowledgments

Author(s) gratefully acknowledge the funding of SciTech - Science and Technology for Competitive and Sustainable Industries, R&D project cofinanced by Programa Operacional Regional do Norte ("NORTE2020"), through Fundo Europeu de Desenvolvimento Regional (FEDER) and the financial support of the Portuguese Foundation for Science and Technology (FCT) under project POCI-01-0145-FEDER-016876 e PTDC/EMS-TEC/6400/2014.

Contact Modelling in Isogeometric Analysis: Application to Sheet Metal Forming Processes

Rui P.R. Cardoso[1], **O.B. Adetoro**[2] **and D. Adan**[2]

[1]Department of Mechanical, Aerospace and Civil Engineering, Brunel University London, UB8 3PH Uxbridge, London, UK

[2]Department of Engineering Design and Mathematics, University of the West of England, BS16 1QY Bristol, UK

E-mail: Rui.Cardoso@brunel.ac.uk

Abstract. Isogeometric Analysis (IGA) has been growing in popularity in the past few years essentially due to the extra flexibility it introduces with the use of higher degrees in the basis functions leading to higher convergence rates. IGA also offers the capability of easily reproducing discontinuous displacement and/or strain fields by just manipulating the multiplicity of the knot parametric coordinates. Another advantage of IGA is that it uses the Non-Uniform Rational B-Splines (NURBS) basis functions, that are very common in CAD solid modelling, and consequently it makes easier the transition from CAD models to numerical analysis. In this work it is explored the contact analysis in IGA for both implicit and explicit time integration schemes. Special focus will be given on contact search and contact detection techniques under NURBS patches for both the rigid tools and the deformed sheet blank.

1. Introduction

The constant drive for change in computational methods has created an increased demand for new computational tools as are the cases of meshless methods and, more recently, Isogeometric Analysis (IGA) based on Non-Uniform Rational B-Splines (NURBS). IGA is a powerful numerical tool for engineers and scientists as far as it allows accurate predictions of structural and fluid mechanics directly on the CAD geometries and at low computational costs. Besides many other advantages of IGA, numerical analysis directly on NURBS objects avoids the time consuming step of mesh generation, providing in this way a more competitive approach for numerical simulations in general. NURBS are widely used for geometrical modelling in many CAD software packages and, in many cases, are the standard for the interchange of many graphical formats (e.g., IGES, STEP, ACIS, STL, etc.). IGA was introduced successfully in the work of Hughes et al. [1] and then it evolved and as a result many other reference publications appeared over the term of the past few years. The isogeometric concept uses the same approximation basis functions for the geometry and for the displacement field at the NURBS's control points. The main difference from the finite element method is that the basis functions in IGA are always positive and they are not interpolatory, which causes some difficulties to handle essential boundary conditions.

In this work the contact analysis is implemented under the IGA framework and NURBS geometric approximation functions are going to be defined for contact analysis for both the rigid tools and the deformable blank sheet. The contact weak forms are developed with Lagrange multipliers for the implicit time integration scheme and with the penalty formulation for the explicit dynamic method. Special focus is given on the contact search and contact detection under IGA with NURBS objects.

2. Non-Uniform Rational B-Splines (NURBS) and IsoGeometric Analysis (IGA)

Detailed overview of B-Spline curves, B-Splines surfaces and NURBS can be found in Piegl and Tiller [2] and Cottrell et al. [3].

2.1. Knot vectors

An open knot vector is a set of non-negative parametric coordinates which are repeated $p+1$ times at the beginning and at the end of the vector (p is the order of the polynomial basis functions). For 1D:

$$\Xi = \{ \; \xi_1, \ldots, \xi_{p+1}, \ldots, \xi_{m+1}, \ldots, \xi_{m+p+1}, \; \} \tag{1}$$

where m is the number of control points or basis functions.

2.2. Control points and basis functions

For a specific local parametric coordinate ξ from an open knot vector and for a degree p of the polynomial, the basis functions are obtained recursively from the following formulae [3]:

$$N_I^p = \frac{\xi - \xi_I}{\xi_{I+p} - \xi_I} N_I^{p-1}(\xi) + \frac{\xi_{I+p+1} - \xi}{\xi_{I+p+1} - \xi_I} N_{I+1}^{p-1}(\xi), \tag{2}$$

where I is the index for the basis functions.

The Non-Uniform Rational B-Splines (NURBS) are rational polynomials obtained from a weighted linear combination of the basis functions with their control points as coefficients. In this way, a NURBS solid is constructed from a three-dimensional knot set $\Xi \times H \times Z$ and a net of control points A_{IJK} and weights W_{IJK}:

$$^{w}T^{p,q,r}(\xi, \eta, \zeta) = \frac{\sum_{k,I,J=1}^{l,n,m} N_K^r(\zeta) N_I^q(\eta) N_J^p(\xi) W_{IJK} A_{IJK}}{W}. \tag{3}$$

with:

$$W = \sum_{k=1}^{l} \sum_{I=1}^{n} \sum_{J=1}^{m} N_K^r(\zeta) N_I^q(\eta) N_J^p(\xi) W_{IJK} \tag{4}$$

being the NURBS weight function for a solid.

The rational basis functions from equation (3) are commonly used for the approximation of the kinematics of the continuum giving rise to the well-known IsoGeometric Analysis (IGA). Many works were already published on the development of different formulations for IGA making use of the NURBS approximations from equation (3). Cardoso and Cesar de Sa [4] used the enhanced assumed strain method in combination with NURBS basis functions for the increase of the subspace of incompressible deformations and, more recently, Cardoso and Cesar de Sa [5] used the Moving Least Square (MLS) approximation to project the kinematics from a fully-integrated NURBS patch into a reduced integrated space so that different kinds of locking pathologies could be alleviated or avoided. In this work, the formulation of Cardoso and Cesar de Sa [5] for a reduced integrated 3D NURBS patch is going to be utilised.

3. Contact Analysis in IGA

3.1. Contact search/detection in IGA

In Figure 1 (b) it can be seen clearly the non-interpolatory nature of NURBS in such that the control points (represented by the white circles in the figure) are not in the curve. Instead, the dark squares in the figure represent the coordinates of a point in the curve after being approximated by the basis functions from equation (3) from a knot value defined in the parametric space of the knot vectors.

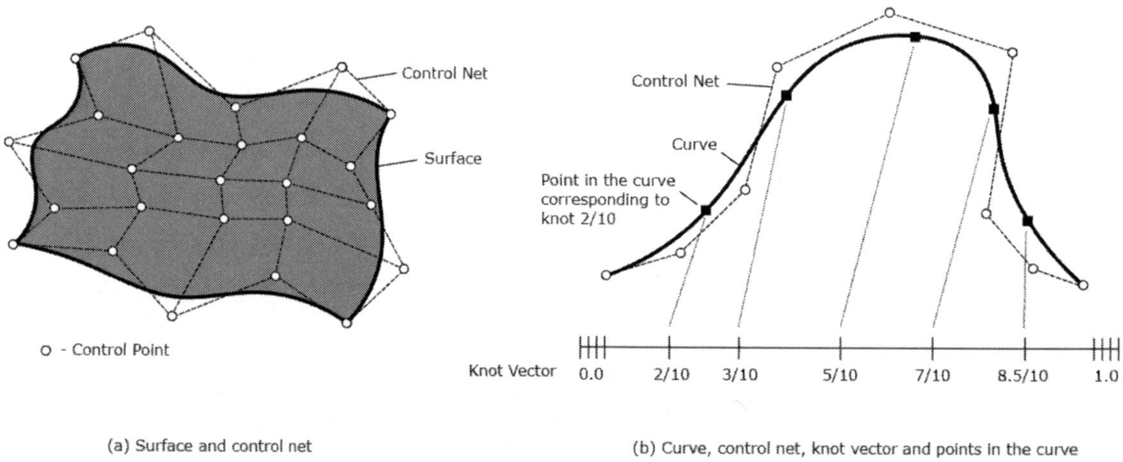

Figure 1. Definitions for control points, control net and knot vector. (a) NURBS surface and its control net. (b) 2D curve with its control points, knot vector in the parametric space and points in the curve after approximation with NURBS basis functions.

For contact search/detection is fundamental that the algorithm goes through the physical coordinates on the curve rather than through the control net. For example, the calculation of the gap in contact needs to be done from the detection of penetration between points in the curves/surfaces rather than the control points on their control nets. Due to the non-interlapolatory character of NURBS, special methods, such as point inversion [2] for the determination of the parametric knot coordinates corresponding to a minimum distance to a contacting point needs to be used.

The kinematic constraint of zero gap is imposed through the use of Lagrange multipliers in implicit time integration and through the use of a penalty parameter for the explicit time integration scheme. The stationary of the energy in its weak form for the explicit time integration becomes:

$$\delta\pi\left(\mathbf{u}\right)=\int_{\Omega}\left(\delta\boldsymbol{\epsilon}\left(\mathbf{u}\right):\boldsymbol{\sigma}\left(\mathbf{u}\right)-\delta\mathbf{u}\cdot\mathbf{b}+\delta\mathbf{u}\cdot m\ddot{\mathbf{u}}\right)d\Omega-\int_{\Gamma_t}\delta\mathbf{u}\cdot\bar{\mathbf{t}}\,d\Gamma_t+\alpha\int_{\Gamma_u}\delta\left(\mathbf{u}-\bar{\mathbf{u}}\right)d\Gamma_u=0 \quad (5)$$

where $\delta\bar{\mathbf{u}}$ is the vector with the penetration gap, which is obtained from the procedures schematically described in Figure 2. For implicit analysis, the stationary of the energy is written as:

$$\begin{aligned}\int_{\Omega}\left(\delta\boldsymbol{\epsilon}\left(\mathbf{u}\right):\boldsymbol{\sigma}\left(\mathbf{u}\right)-\delta\mathbf{u}\cdot\mathbf{b}+\delta\mathbf{u}\cdot m\ddot{\mathbf{u}}\right)d\Omega-\int_{\Gamma_t}\delta\mathbf{u}\cdot\bar{\mathbf{t}}\,d\Gamma_t+\lambda\int_{\Gamma_u}\delta\left(\mathbf{u}-\bar{\mathbf{u}}\right)d\Gamma_u &= 0\\ \delta\lambda\int_{\Gamma_u}\left(\mathbf{u}-\bar{\mathbf{u}}\right)d\Gamma_u &= 0 \quad (6)\end{aligned}$$

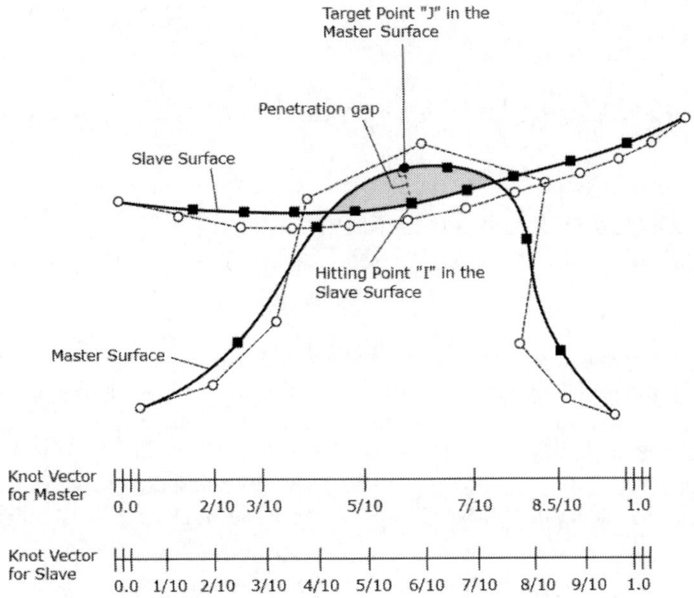

Figure 2. Contact detection between a master surface (rigid dies) and a slave surface (deformable blank sheet).

where λ are the additional Lagrange multipliers. The weak form equations for contact analysis in IGA are in fact the same as for the finite element method, however they differ in the way the kinematic quantities are interpolated/approximated.

4. Conclusions

In this work it is given a brief overview of the main challenges of developing contact formulations for IGA. The major difficulty in modelling contact in IGA is due to the non-interpolatory character of IGA, which introduces some complexities in contact search and on the determination of the penetration gap for the imposition of the kinematic constraints. The weak forms are similar to the weak forms used in the finite element method but however the approximation of the kinematic variables is performed from the use of the NURBS basis functions from equation (3).

References

[1] T.J.R. Hughes, J.A. Cottrel, and Y. Bazilevs, Isogeometric analysis: CAD, finite elements, NURBS, exact geometry and mesh refinement, Computer Methods in Applied Mechanics and Engineering, vol. 194, pp. 4135-4195, (2005).

[2] Les Piegl and Wayne Tiller, The NURBS Book. Berlin: Springer-Verlag, (1996).

[3] J. Austin Cottrell, T.J.R. Hughes, and Yuri Bazilevs, Isogeometric Analysis, Toward Integration of CAD and FEA.: Wiley, (2009).

[4] Rui P.R. Cardoso and J.M.A. Cesar de Sa, The Enhanced Assumed Strain method for the isogeometric analysis of nearly incompressible deformation of solids, International Journal for Numerical Methods in Engineering, vol. 92, pp. 56-78 (2012).

[5] Rui P.R. Cardoso and J.M.A. Cesar de Sa, The Enhanced Assumed Strain method for the isogeometric analysis of nearly incompressible deformation of solids, Computational Mechanics, vol. 53, pp. 1327-1340 (2014).

Die Starter: A New System to Manage Early Feasibility in Sheet Metal Forming

Rodrigue Narainen, Harald Porzner
ESI Group, 32605 West 12 Mile Road, Farmington Hills, MI 48334, USA

Abstract

Die Starter, a new system developed by ESI Group, allows the user to drastically reduce the number of iterations during the early tool process feasibility. This innovative system automatically designs the first quick die face, generating binder and addendum surfaces (NURBS surfaces) by taking account the full die process. Die Starter also improves the initial die face based on feasibility criteria (avoiding splits, wrinkles) by automatically generating the geometrical modifications of the binder and addendum and the bead restraining forces with minimal material usage.

This paper presents a description of the new system and the methodology of Die Starter. Some industrial examples are presented from the part geometry to final die face including automatic developed flanges, part on binder and inner binder.

Introduction

Today, sheet metal forming simulation is completely integrated into the core business of die engineering of OEM and suppliers, from early feasibility assessment and initial cost estimation to virtual try-out and final inspection. For example, the combination of ESI's Visual DIEMAKER (VDM) or CATIA DIEMAKER with PAM-STAMP provide a powerful toolset capable of working through the entire process, from tipping, part preparation, binder, addendum and analysis to trimming and flanging.

Despite the progress, early tooling feasibility process still requires multiple iterations. The modifications are based on trial and error and require skilled users. Even if best practices for die face have been integrated into die design strategies, there remain issues including:

- Die face design will vary, depending on the user
- Die face design is not necessarily optimal for a given set of requirements and constraints
- Creation of an acceptable die face solution is simply too slow, especially for complex parts.

Accordingly, there is clearly a demand for a robust, repeatable and streamlined die face engineering process that addresses early stage of feasibility assessment, when effective product design decisions must be made. Furthermore there is a critical need to greatly shorten the cycle time between part design and early tooling feasibility.
For those reasons a new approach has been proposed and developed by ESI Group and is now industrialized as a new solution, Die Starter. This new software module enables engineers to manage early feasibility in sheet metal forming by automatically generating, from minimal input parameters, a quick die face design (also known as "initial die face"), defining binder and addendum surfaces (NURBS surfaces). Based on the use of parametric geometry, modifications and improvements of initial die face can be performed quickly and automatically using a mathematical optimization algorithm procedure.

Die Starter Workflow

Die Starter is completely integrated into ESI Visual DIEMAKER (VDM, on ESI's Visual Platform) and ESI CATIA DIEMAKER (CDM, on Dassault Systèmes, CATIA V5 platform), and provides solutions for early up-front feasibility assessment.

Before running Die Starter, the part must be prepared and those steps are:

- Part and mode definition (symmetry, flanges selection,..); Tipping position
- Prepare the part to achieve a defined part outline (fill holes, rolling cylinder only if required)

Executing Die Starter for a quick die face design automatically generates the following output:

- Binder and addendum surfaces (outer and inner binder); Bead center lines and blank outline (IGES format)
- Analysis results, including thinning and strains for design check

In order to improve the initial die face geometry, a mathematical optimization technique is used. Different types of optimization are offered, specific to the binder, addendum surfaces or bead restraining forces. Performed automatically, Die Starter optimization generates the same types of data as for the initial die face step. Figure 1 illustrates the Die Starter process applied for a door inner. Once the full die face is created from Die Starter, process information and geometrical results, including draw beads restraining forces are exported to PAM-STAMP via a Quick-Link to achieve nearly automatic simulation set-up and analysis. The user can also modify or redesign their own die face based on Die Starter geometry by using Visual DIEMAKER functions. Finally, Die Starter results can be exported to any CAD system (CATIA, Unigraphics, SolidWorks…) for manual modifications.

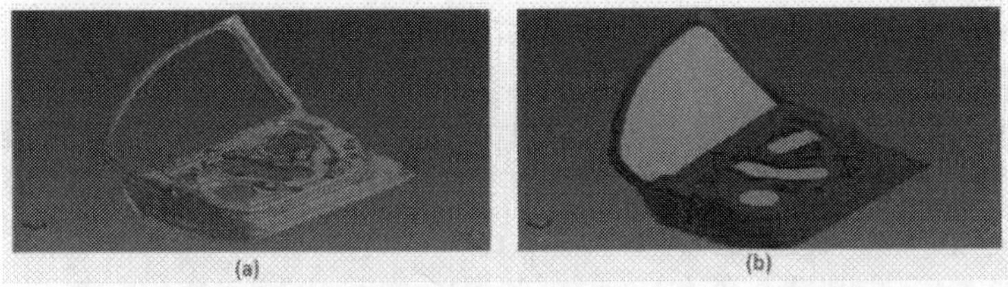

(a) Part geometry, (b) Die Starter part preparation (fill holes, smooth part outline)

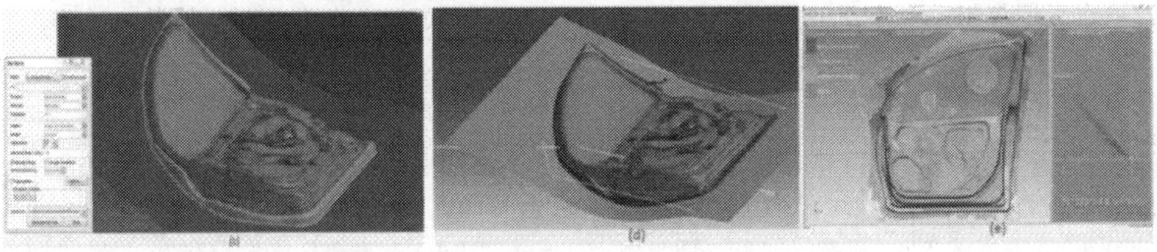

(c) Die Starter input deck and initial die face solution, (d) Improved die face after die starter optimization (binder – bead – addendum); (e) PAM-STAMP Analysis from die starter optimization results after few manual improvements.

Figure 1: Door Inner: Die Starter full process steps: (a), (b) part preparation, (c) initial die face solution, (d) final die face solution from optimization and (e) PAM-STAMP results after few manual improvements from optimized solution

Die Starter Benefit and Capabilities

The main characteristics of Die Starter are automatic initial die face then optimization and bead optimization on customized die face.

Characteristics: Customized die face:

Based on the entire die face geometry (part, binder and addendum surfaces), Die Starter will perform a quick formability analysis on the full die face and then bead optimization will provide optimum bead restraining forces (also segmentation) for improved formability.

Characteristics: Automatic initial die face:

Based on minimal part preparation, Die Starter provides within a few minutes a complete die face design that is usually acceptable and accurate. The automatic initial design is based only on geometrical and process criteria. The different characteristics are:

- Binder surfaces (NURBS surfaces) based on developable /wrapable binder criteria in most cases, including part on binder and inner binders. The binder orientation is also defined automatically. Specific double curvature binder, imposed flat binder and non- developable binder are also available.

- Addendum surfaces (NURBS surfaces) are a combination of the two principal profile concepts (standard addendum and step/bar addendum) which represent the majority of parts. An automatic global smoothing technique is used to provide clean and smooth punch and die opening lines; both critical in die face design principles. Inner addendum surfaces are also defined.
- A quick formability analysis [1-2] is automatically performed just after an initial die face design, in order to provide an estimation of the feasibility of the draw-part. This first feasibility assessment can guide the next steps, such as a direct PAM-STAMP analysis if the part appears feasible or Die-Starter optimization if thinning (high or low) needs to be corrected.

Characteristics: Automatic Optimization:

Built on a parametric surfaces engine, Die Starter has the capability to automatically modify or redefine the tooling geometry during optimization steps. The optimization uses a mathematical optimization algorithm [3 -4] combined with a formability analysis and a defined objective function with or without constraints (geometrical and mechanical).

The use of fast feasibility allows the optimization of both tooling geometry and process parameters in order to deliver at an early stage a feasible part with a stable and robust press tool process. The different optimization options are:

- Binder optimization
- Addendum optimization: Punch radius and wall addendum for type 1; Draw bar creation, punch radius and wall addendum for type 2 (type 1 and 2 are also available for inner addendum)
- Draw beads sections with associated restraining forces (also available for customized die face process)
- All of the above, individually and in combination (and all combinations between them)

Capabilities:

- Develop flanges into addendum (also binder) automatically
- Define binder and addendum (outer/inner) surfaces automatically (initial and optimization)
- Develop automatically part on binder
- Handle symmetric parts: create complete symmetrical die face (initial and optimization)
- Provide good surface quality for binder and addendum automatically. At this stage of the process, this is very important and can have a significant impact on time for creating the final die design in CAD system.
- Only few input parameters are required, related to material, process and optimization target
- A first initial die face and design check is completed in a few minutes, an optimization of an early design in few hours calculation time. Minimal human intervention is required during the Die Starter process.
- Perform quick feasibility analysis and bead optimization on a pre-defined die face
- Die face geometry (initial and optimized) from Die Starter can be loaded into any CAD system
- Die face and addendum profiles can be imported into VDM for manual modifications

Industrial Examples

This section shows some examples that ran in Die Starter for initial die face and optimization. Figure 2 shows 3 different parts that ran for initial die face. The first example (a) is a full die face of a liftgate outer with inner binder/addendum. The second example (b) is an s-rail die face with part on binder, and the third example (c) is a fully symmetrical initial die face of a hood inner calculated with a double curved binder (material saving).

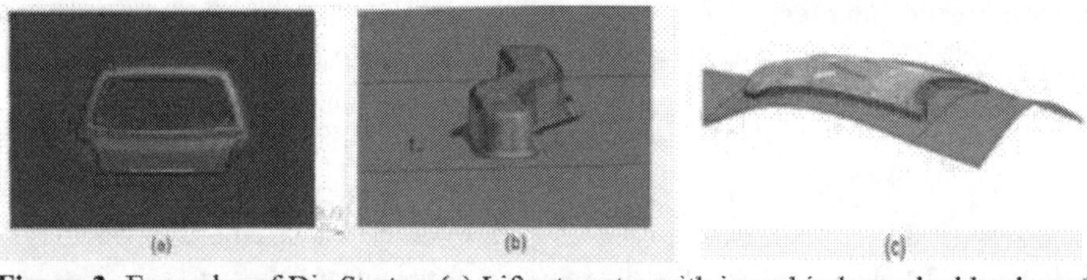

Figure 2: Examples of Die Starter: (a) Liftgate outer with inner binder and addendum. (b) S-rail initial die face with part contour on binder. (c) Hood inner fully symmetrical die face with double curved binder

Figure 3 is the initial die face of a fender outer with automatic development of flanges into the addendum (direct trimmed line from developed flange outline).

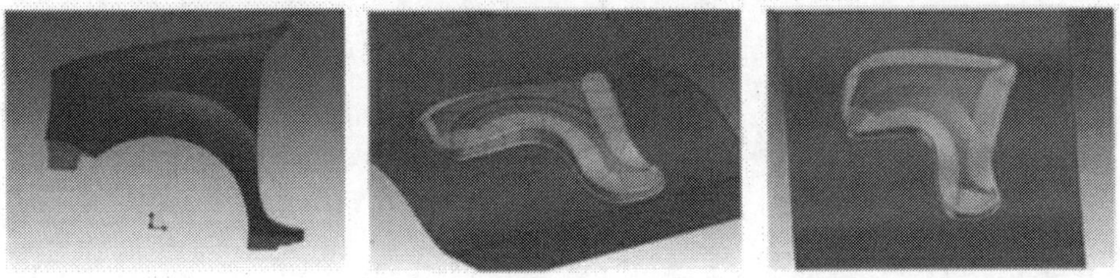

(a) (c) (d)

Figure 3: Die Starter fender initial die face with developed flanges: (a) fender part with flanges; (c), (d) initial die face with automatic developed flanges into addendum (developed trim line in orange)

The last example (figure 4) is a hood inner panel, performed for initial and optimized die face. During optimization, the rear corner areas of the addendum have automatically been changed in order to meet the maximum thinning requirement (no split after optimization). After running a validation analysis (PAM-STAMP) using the optimized die face, the feasibility of the panel is good.

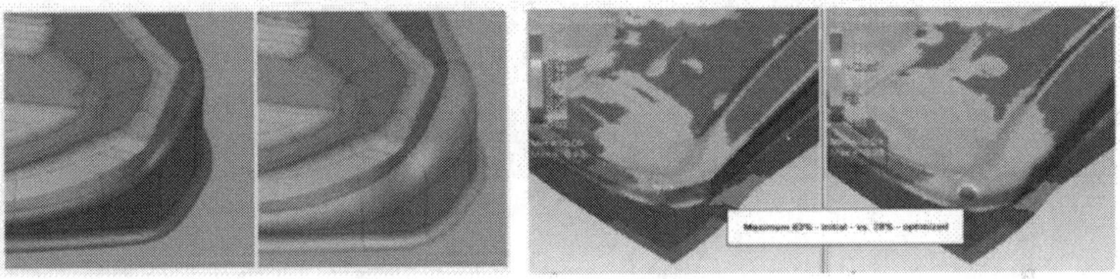

(a) (b) (c) (d)

Figure 4: Die Starter hood inner (rear corner): (a) Initial die face and (b) optimized die face, with (c), (d) thinning results comparison in PAM-STAMP

Conclusions

ESI Die Starter, a new solution module integrated into Visual DIEMAKER (built on ESI Group's Visual Platform) and CATIA DIEMAKER, has been presented in this paper. For the early stage of tool feasibility in sheet metal forming, Die Starter automatically generates a complete die face including binder and addendum surfaces, bead center lines and restraining forces associated for a prepared part. An initial (quick) die face can be created in a few minutes. A mathematical optimization method automatically modifies the initial die face in order to meet the formability specifications and requirements for minimum material usage. Die Starter solutions can be further modified in any CAD system or processed through Visual DIEMAKER/CATIA DIEMAKER to PAM-STAMP, if required.

References

[1] Guo Y.Q., Batoz J.L., EL Mouatassim M., Detraux J.M., *"On the estimation of thickness strains in thin car panels by the inverse approach"*, NUMIFORM 92, 14-18 sept 1992

[2] Guo Y.Q., Batoz J.L., Naceur H., Bouabdallah S., Mercier F., Barlet O., *"Recent developments on the analysis of sheet metal forming parts using a simplified inverse approach"*, Computers and Structures, vol. 78, 2000

[3] Powell, M.J.D., *"A fast algorithm for nonlinearly constrained optimization calculations"*, Numerical analysis, Proceedings of Biennial Conference, 1977

[4] Powell M.J.D., *"Algorithms for nonlinear constraints that use Lagrangian functions"*, Mathematical programming 14, 1978

Journal of Physics: Conference Series **734** (2016) 032125 doi:10.1088/1742-6596/734/3/032125

Isogeometric Analysis with LS-DYNA

S. Hartmann[1], **D. Benson**[2,3], **A. Nagy**[3]

[1] Dynamore GmbH, Stuttgart, Germany
[2] former: Professor at the Department of Structural Engineering,University of California, San Diego, USA
[3] Livermore Software Technology Corporation (LSTC), Livermore, CA, USA

stefan.hartmann@dynamore.de

Abstract

In the last decade numerous research has been done in the area of *Isogeometric Analysis* (IGA). The intention of this rather new technology is the wish to have a stronger integration of *Computer Aided Design* (CAD) and *Finite Element Analysis* (FEA). Its basic idea is to use the same mathematical description for the geometry as well in the design process (CAD) as in the later analysis (FEA). One of the wide spread geometry description methodology in CAD-systems is the usage of *Non-Uniform Rational B-Splines* (NURBS) as basis functions. NURBS-based finite elements have been proven to be very well suited for computational analysis leading to qualitatively more accurate results in comparison with standard finite elements based on Lagrange polynomials. Therefore continuous development of Isogeometric Analysis has been added to LS-DYNA in the last years. The present contribution will give an overview about the current possibilities in LS-DYNA to use NURBS-based CAD-geometry description for numerical simulation and outline future plans for their enhancements.

Various aspects will be addressed, i.e.:
- Available shell formulations when using NURBS-based elements
- Analysis capabilities (implicit and explicit time integration, material model database, …)
- Definition of trimmed and un-trimmed NURBS patches
- Joining of matching and non-matching, un-trimmed and trimmed NURBS patches
- Application of boundary conditions
- Treatment of contact
- Possibilities of spotweld modeling
- Mass scaling
- Parallelization and speed-up
- Pre- and Post-Processing possibilities with LS-PrePost

- Discussion of current limits and future plans for development and enhancements

Robustness and sensitivity analysis of a virtual process chain using the S-rail specimen applying random fields

T. Konrad[1,a], S. Wolff [2,b], K. Wiegand[1,c] and M. Merklein[3,d]

[1] Daimler AG, Bela-Barenyi-Str.1, 71059 Sindelfingen, Germany
[2] DYNARDO Austria GmbH, Wagenseilgasse 14, 1120 Vienna, Austria
[3] Institute of Manufacturing Technology, Friedrich-Alexander-University Erlangen-Nürnberg, Egerlandstraße 13, 91058 Erlangen, Germany

[a] tobias.konrad@daimler.com, [b] sebastian.wolff@dynardo.at,
[c] klaus.wiegand@daimler.com, [d] marion.merklein@fau.de

Abstract. An important part in robustness evaluation of production processes is the identification of shape deviations. A systematic approach is typically based on the numerical evaluation of a DoE and the application of metamodels. They provide knowledge on solver noise and sensitivities of individual model parameters. This article presents the sensitivity analysis workflow of a linked deep drawing and joining process chain. LS-DYNA®, optiSLang and SoS is used. The challenge is to separate simulative from process and material parameters of AA 6014. Spatial quantities like variations in geometry, thinning and strain have to be considered in the next process steps. At the same time the number of required virtual CAE model evaluations must be limited. The solution is based on nonlinear metamodels and random fields.

1. Introduction

Due to increasing customer requirements regarding the features of both manufactured parts and assemblies, production processes have to be robust during product life cycles. Therefore, car manufactures pay attention to produce add-on parts of highly dimensional accuracy and especially body-in-whites with equal gap dimensions [1].

In addition to these standards of quality, the necessary production efficiency and flexibility with respect to several product variants lead to a higher complexity of the entire process chain. Each step of the forming and joining process chains of add-on parts results in a shape deviation to the nominal geometry of single and assembled parts [2]. A systematic approach is important to detect shape deviations of freeform faces that have a high relevance in evaluating the dimensional accuracy of assemblies. To take input variations into account (i.e. sheet thickness and process parameters), probabilistic robustness analyses are suitable to quantify their influence on dimensional accuracy.

Whereas the sensitivities of individual model parameters of a deep drawing process are analysed in different publications [3], the analysis of linked deep drawing and joining process chains shows challenges as a consequence of overlays of input parameters. Furthermore, the number of virtual CAE model evaluations has to be kept minimizing with respect to the product development costs.

At the moment, a holistic method for analysing the robustness of process chains of add-on parts using one FE simulation software is missing. The following approach is an enhancement of the state of the art considering spatial quantities (shape deviation, thinning and strain) in the next process steps.

2. State of the Art - Random Fields

Random fields are implemented in the *Statistics on Structures (SoS)* software are effective methods for transferring the spatial quantities of the first to a subsequent second process chain. With the help of cross-correlated fields and the generation of random geometries between two process chains, the significance of statistical field properties can be estimated and evaluated [4].

Discrete geometries and result quantities of a process chain can be approximated by a random field. Subsequently computing the eigenvalues of the correlation matrix of nodal coordinate deviations, with the eigenvectors or "scatter shapes" represents their spatial correlation. Although the outcome of a probabilistic forming simulation is random for all elements, the variations of result quantities can be described through a small number of random parameters known as "random field amplitudes". These parameters are reused to generate new field variations in subsequent process steps. The advantage of using scatter shapes and random field amplitudes is to describe the variations/distribution of spatial result quantities and the opportunity to detect significant input factors in part regions [5].

State of the art scalar numbers are used to perform and analyze both robustness analyses and optimization approaches. By contrast, random fields have exact scalar values for each point in space and time. The associated field function H in space x is defined with random model parameters p:

$$y = H(\mathbf{x}, \mathbf{p}), \mathbf{x} = (x_1, x_2, x_3, t)^T, \ \mathbf{p} = (p_1, \ldots, p_n)^T \quad (1)$$

The investigated statistics of the random fields H (mean value, standard deviation) depend on the local area and are defined in equation (2) [6].

$$\textit{Mean value:} \quad \bar{H}(\mathbf{x}) = E[H(x)] \qquad \textit{Standard deviation:} \quad \sigma_{\bar{H}}^2(\mathbf{x}) = E[(H(x) - \bar{H}(\mathbf{x}))^2] \quad (2)$$

3. Methods and Setup for the Robustness Evaluation

3.1. Investigated Steps of the Forming and Joining Process Chain of the S-rail Specimen

In order to demonstrate the determination of the robustness of influence factors on the shape deviation after different forming and joining steps, the specimen S-rail were chosen [7]. The simulations are all performed with the FE solver LS-DYNA. The experimental setup and the validation of forming simulations with this specimen were part of recent benchmarks of the Numisheet 2008 conference.

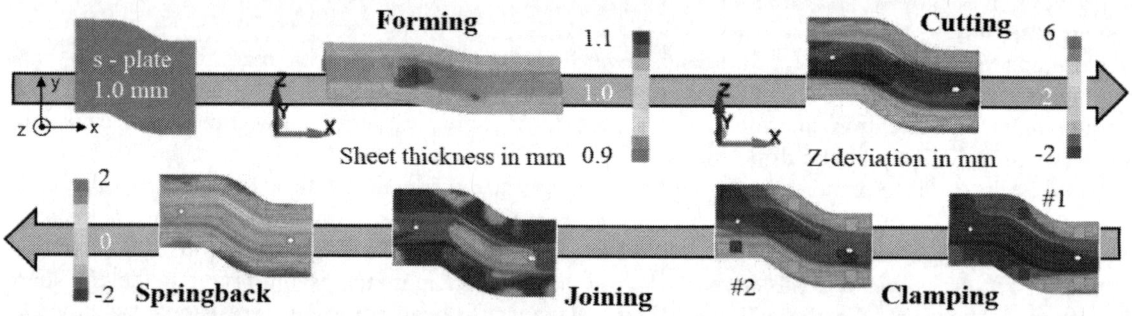

Figure 1. Forming and joining process chain of the S-rail specimen in LS-DYNA

Additionally to previous investigations, both a laser cutting operation and the subsequent steps of a joining process chain are performed: clamping, joining and springback of the joined assembly (see figure 1). Sheet thicknesses, strains and stress results of single parts can be compared both to the results of the assembly, and, after each process step of the linked forming and joining process chain. The assembly consists of two parts: upper parts (laser-cut S-rail) and plane plates with a nominal sheet thickness t_0 of 1.0 mm. The upper part is cut to a length of 270 mm and a flange length of 33 mm.

Furthermore, six clamps (clamping step #1/#2) and eight joints are set and equally distributed on both flange sides, subject to a rule for the clamping/joining sequence: "from inside to parts' outside."

3.2. Investigated Material Parameters and Input Parameters of the Robustness Analysis

In each step of this simulation process chain, the elastic-plastic material behavior of the mild aluminum alloy AA6014 (δ_0: 132 MPa, M = 6) is included. The material model and the nominal material parameters revert to the von Mises yield criterion and an isotropic material hardening law extrapolated by Vocé. In this study, Barlat YLD 2000 is implemented in the material behavior.
Based on the nominal material properties, a variation of the sheet thickness (t_0 = 1.0 mm ± 0.05) of both the upper and the lower single part is used. Due to batch differences of the mechanical properties, a variation of the elastic material behavior (E-modulus: 68420 MPa ± 342) is assumed. Furthermore, the influence of the process parameters variation of the blank-holder force (90 kN ± 1.2) and a deviation of initial plate (x-, y-) positions (± 0.5 mm) on the dimensional accuracy is investigated.

3.3. Approaches for Transferring Spatial Quantities between Multiple Process Chain Simulations

Within a linked forming and joining process chain, result quantities like actual sheet thickness, strains and stresses have to be passed between two process steps. In order to analyse the cause-effect relationships of the result quantities, different options of figure 2 are compared. Whereas using robustness analysis option 1 means simulating the whole process chains 200 times, option 2 has the advantage splitting input factors and transferring spatial quantities with random fields.

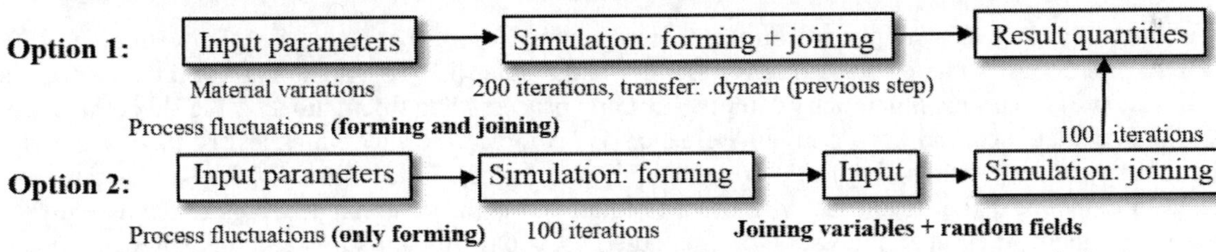

Figure 2. Approaches for robustness analyses by linking and analyzing of two process chains

4. Results of the Robustness Analysis of the Simulative Process Chain

4.1. Results of the Robustness Analysis of the Forming Process

Using robustness analysis option 2, figure 3 on the left side lists the number of necessary scatter shapes for each result quantity after the forming process. A various number of eigenvectors/scatter shapes is needed to describe each result quantity with appurtenant confidence levels. For using the correlated group of coordinate deviations, strains need five shapes and sheet thickness two shapes.

Result quantity	# scatter shapes	Expl. variation
Δx, y, z-coordinate	5	95.1701 %
Ø x-, xy-, y-, xz-, yz-, z-strain	5	96.7549 %
shell thickness	**2**	99.2426 %
Ø x-, xy-, y-, xz-, yz-, z-stress	10	90.0501 %
plastic strain/Mises σ	23 / 10	93.0/94.2 %

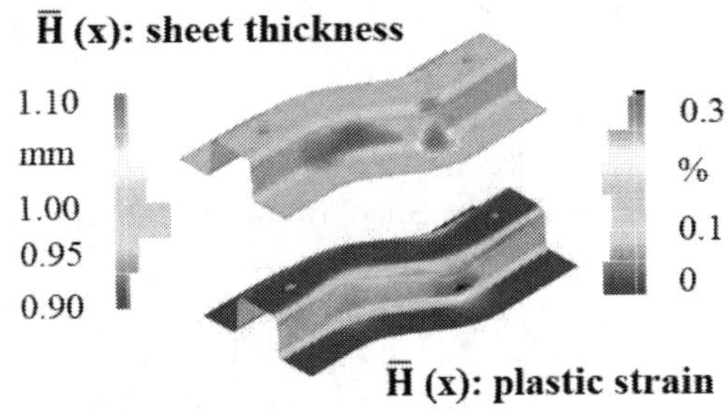

Figure 3. Number of scatter shapes and statistical data of random fields of the forming process

On the right, figure 3 additionally shows the statistical data (mean value) of two selected field quantities - $\overline{H}$ (x) of the sheet thickness and the plastic strain of 100 iterations. The spatial distribution of the sheet thickness and its relative frequency indicates that the thinning of the generated upper part is quite normally distributed, with a median about 0.97 mm. Appropriately, the spatial distribution of the plastic strain (uniform distribution) points out the highest values above the highest thickening.

4.2. Results of the Robustness Analysis of the Joining process Compared to the Forming Process

Finally, the part and assembly designers are interested in the most significant influencing parameters on the shape deviation of the single and assembled parts. Therefore, the detected parameters are compared after the forming and joining steps based on the varied parameters of the robustness analysis. Evaluated using the Coefficient of Prognosis (CoP) of the response variable coordinate deviation Δz, figure 4 contrasts with results of the forming and the joining simulations. The coordinate deviation Δz is calculated from the mean value of the selected points (11 per flange side) of figure 4.

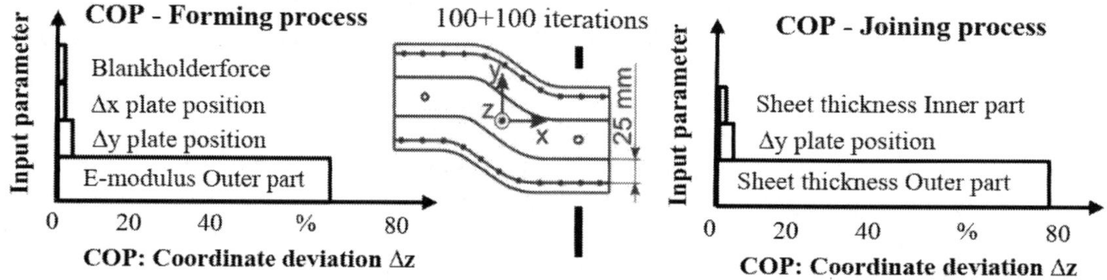

Figure 4. Results of the robustness analysis of the joining process chain using an S-rail assembly

It is found that the most important influence parameters are not the same after the forming and the joining process. At a rate of 66 %, the E-modulus of the outer part exerts the highest influence on the z-deviation of the nominal geometry after the forming process. The fluctuations of the sheet thickness, Δx- and Δy plate position exert a minor influence. By contrast, the sheet thickness of the outer part is the input parameter with the highest correlation (78 %) to the final z-deviation of the S-rail assembly. Therefore, the most significant factor in order to improve the dimensional accuracy of the assembly is the variation of the sheet thickness of the outer part. One explanation for this effect is that a comparatively small range of the sheet thickness with a deviation of ± 5 % causes the variation of the shape deviation and stress and strain values (plastic strains) of the formed parts. Higher (lower) stress values of tension-compression conditions results in a higher (lower) material thickening and finally in smaller (higher) shape deviation of formed and assembled parts.

5. Conclusion

As a consequence, spatial quantities can be analyzed after the forming steps, using random fields. In order to further analyze the cause-effect relationships with respect to spatial quantities, the random field quantities and their statistical data should be investigated after the joining step. They allow the separation of processes and improve the model knowledge and numerical simulation efficiency.

References

[1] Bohn, M.; Steinle, P., et al.: Early Stage Geometrical Deviation Optimization - An Auto-motive Example for Sheet Metal Parts, Applied Mechanics and Materials, Vol. 104, pp. 95-101, 2012.

[2] Litwa, F.; Gottwald, M.; et al.: Sensitivity analysis of several Body-in-White single part and assembly parameters using tolerance analysis, WOST, 2015, Weimar, pp.1-13.

[3] Emrich, A.: Systematik zur Robustheitsanalyse von Umformprozessen für Karosserie-komponenten aus Blech. Dissertation, Universität Stuttgart, 2013.

[4] Wolff, S.: Robustness Evaluation in Sheet Metal Forming Using Statistics on Structures (SoS) and optiSLang. In: Proceedings 32nd German ANSYS and CADFEM Users' Meeting, 2014.

[5] Bucher, C.: Computational Analysis of Randomness in Structural Mechanics: Structures and Infrastructures Book Series, 1st edition, 2009.

[6] Matthies H.; Bucher, C.: Finite elements for stochastic media problems. Comput Methods Appl. Mech. Eng. 168(1-4), 1999, pp. 3-17.

[7] Roll, K.;. Wiegand, K.; Hora, P.: Benchmark 2 – Influence of Drawbeads on the springback behavior (<<S-Rail-08>>), Numisheet 2008, Switzerland, pp. 43-57.

Journal of Physics: Conference Series **734** (2016) 032127 doi:10.1088/1742-6596/734/3/032127

Crystal Plasticity Constitutive Model for Multiphase Advanced High Strength Steels to Account for Phase Transformation and Yield Point Elongation

Taejoon Park, and Farhang Pourboghrat

Ohio State University, Department of Integrated Systems Engineering, Ohio State University, 210 Baker Systems, 1971 Neil Avenue, Columbus, OH 43210, United States

park.2417@osu.edu, pourboghrat.2@osu.edu

Abstract. A constitutive law was developed based on a rate-independent crystal plasticity to account for the mechanical behavior of multiphase advanced high strength steels. Martensitic phase transformation induced by the plastic deformation of the retained austenite was represented by considering the lattice invariant shear deformation and the orientation relationship between parent austenite and transformed martensite. The stress dependent transformation kinetics were represented by adopting the stress state dependent volume fraction evolution law. The plastic deformation of the austenite was determined to have the minimum-energy associated with the work during the phase transformation. In addition to the martensitic phase transformation, yield point elongation and subsequent hardening along with inhomogeneous plastic deformation were also represented by developing a hardening stagnation model induced by the delayed dislocation density evolution.

1. Introduction

In order to obtain the formability and strength of the sheet metals demanded for automotive industries, advanced high strength steels having multiple phases are being intensively developed. Due to the nature of the multiphase steels, it is required to develop advanced constitutive laws to account for their complex material behavior such as phase transformation and yield point elongation.

In this work, a constitutive law was developed based on a rate-independent crystal plasticity to account for the mechanical behavior of multiphase advanced high strength steels. Also, yield point elongation and subsequent hardening of the advanced high strength steels along with inhomogeneous plastic deformation were also represented by developing a hardening stagnation model induced by the delayed dislocation density evolution.

2. Development of the phase transformation model

Considering the lattice invariant shear deformation and the orientation relationship between parent austenite and transformed martensite, the martensitic phase transformation induced by the plastic deformation of the retained austenite was represented. In order to accommodate the orientation relationship, in which the corresponding closed-packed planes and directions in the parent austenite and product martensite lattices are roughly parallel to each other, the rigid body rotation was introduced to have one or more undistorted and unrotated line (invariant-line). Due to the cooperative atomic movement during the martensitic transformation, the interface between parent lattice and product lattice should be an undistorted and unrotated plane (invariant-plane). Since the combination of the rigid body rotation and Bain distortion is an invariant–line deformation, the lattice-invariant shear deformation was additionally considered for the phenomenological description of the martensitic phase transformation [1]. Based on the lattice-invariant shear deformation on planes, the macroscopic deformation gradients having 24 Kurdjumov–Sachs (KS) orientation variants were generated by decomposing the rigid body rotation and Bain distortion into the macroscopic deformation and complementary shear deformation [2, 3].

In order to consider the stress dependency of the transformation kinetics, a stress state dependent volume fraction evolution law [4] was adopted as shown in Figure 1.

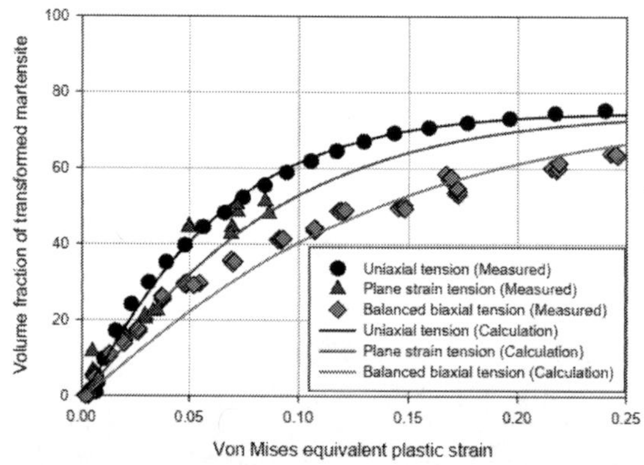

Figure 1. Measured and calibrated transformation kinetics dependent on stress state.

For the retained austenite lattice, the following additive decomposition of the strain increment was assumed,

$$d\boldsymbol{\varepsilon} = d\boldsymbol{\varepsilon}^{e} + d\boldsymbol{\varepsilon}^{slip} + d\boldsymbol{\varepsilon}^{mart} \tag{1}$$

Here, $d\boldsymbol{\varepsilon}^{e}$, $d\boldsymbol{\varepsilon}^{slip}$ and $d\boldsymbol{\varepsilon}^{mart}$ are the elastic strain increment, plastic strain increment by slip deformations and plastic strain increment by the martensitic transformation, respectively. Among the various available plastic deformations for the martensitic transformation, the optimized plastic deformation was decided to have minimum-energy associated with the work during the phase transformation:

$$d\boldsymbol{\varepsilon}^{mart} = \sum_{k=1}^{N} df_k^{\alpha'} d\boldsymbol{\varepsilon}_k^{mart} \tag{2}$$

where $\sum_{k=1}^{N} df_k^{\alpha'} = df^{\alpha'}$ and $d\boldsymbol{\varepsilon}_k^{mart}$ is the strain increment of each variant, k, obtained from the macroscopic deformation gradient during the martensitic transformation.

As for the product martensite, the strain increment was assumed to be decomposed into the elastic strain increment and plastic strain increment:

$$d\boldsymbol{\varepsilon} = d\boldsymbol{\varepsilon}^{e} + d\boldsymbol{\varepsilon}^{slip} \tag{3}$$

For a unit cell, which is a combination of parent austenite and product martensite as schematically shown in Figure 2, the Taylor type homogenization scheme was adopted. In the Taylor type scheme, which is an upper bound scheme, the parent austenite and product martensite share the same strain increment. Then, the homogenized stress can be obtained as,

$$\boldsymbol{\sigma} = (1 - f^{\alpha'})\boldsymbol{\sigma}^{\gamma} + f^{\alpha'}\boldsymbol{\sigma}^{\alpha'} \tag{4}$$

where $\boldsymbol{\sigma}^{\gamma}$ and $\boldsymbol{\sigma}^{\alpha'}$ are stresses of the retained austenite and the transformed martensite, respectively.

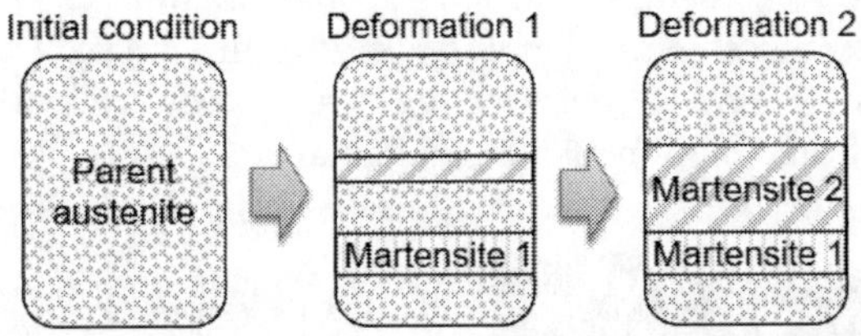

Figure 2. Schematic view of a parent austenite unit cell.

The developed constitutive law was implemented into the commercial FE code ABAQUS/Explicit with the aid of user-defined material subroutine. As a preliminary application, stress-strain response of the retained austenite, product martensite and homogenized unit cell for simple tension was calculated and the comparison results are shown in Figure 3.

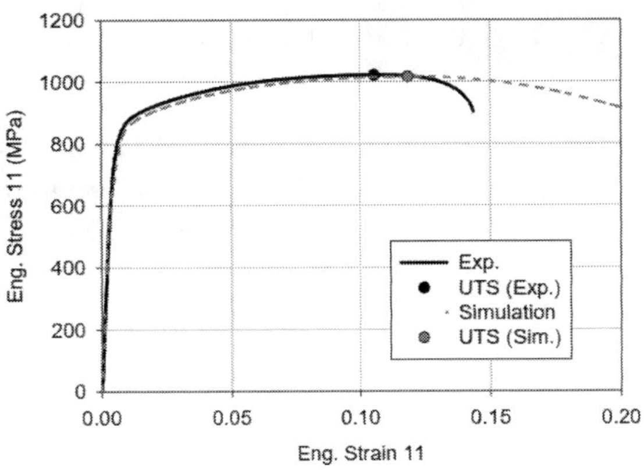

Figure 3. Comparison of the measured and calculated uniaxial tension simulation results.

3. Development of the yield point elongation model

The yield point elongation and subsequent hardening of the advanced high strength steels along with inhomogeneous plastic deformation were also represented by developing a hardening stagnation model induced by the delayed dislocation density evolution. The constitutive model was designed to account for the upper yield point due to the strong elastic interactions between the solute atoms and dislocations as well as the lower yield point due to the annihilation of the pinned dislocations. In addition to the effect of the edge dislocations on the critical resolved shear stress of the crystal slip systems, the pinned dislocation effect was added to the dislocation density:

$$\rho^{\alpha} = \rho^{\alpha}_{Edge} + \rho^{\alpha}_{Pin} \tag{5}$$

Here, ρ^{α}_{Edge} is the edge dislocation density and ρ^{α}_{Pin} is the pinned dislocation density. The edge dislocation density and the pinned dislocation density were designed to be delayed and annihilated by the interactions between the solute atoms and the dislocations.

Considering the impractical huge size of the representative volume element to represent the inhomogeneous material property distribution of the multiphase advanced high strength steels, the developed constitutive model was implemented into the phenomenological model by assigning different volume fraction for each phase within a dogbone specimen. Figures 4 and 5 show the simulated equivalent strain distribution and stress-strain curve during the uniaxial tensile test, respectively. As shown in the comparison in Figure 5, the serrated stress-strain curves as well as the yield point elongation were successfully predicted by the developed model.

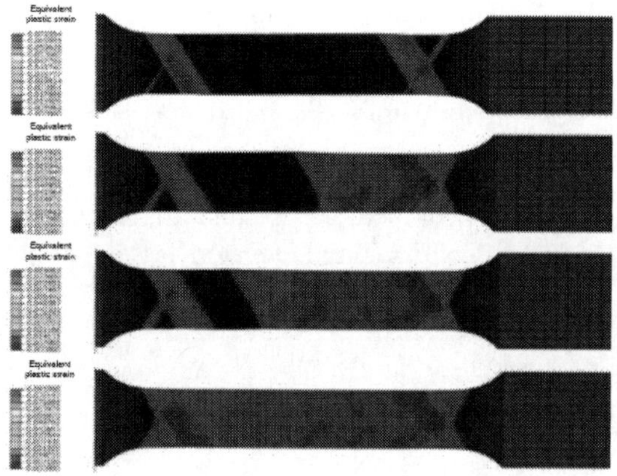

Figure 4. Equivalent plastic strain distribution with a dogbone specimen during the uniaxial tension test.

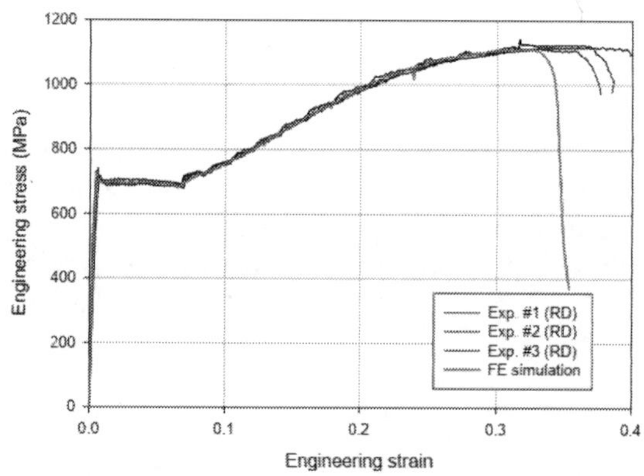

Figure 5. Comparison of the simulated and measured stress-strain curves during the uniaxial tension test.

References

[1] Nishiyama Z, Fine M E, and Wayman C M 1978, Martensitic transformation. *Academic Press.*
[2] Mackenzie J K and Bowles J S 1954 *Acta Metallurgica* **2**(1) 138-147.
[3] Wechsler M S, Lieberman D S, and Read T A 1953 *Trans. AIME* **197** 1503.
[4] Beese A and Mohr D 2011 *Acta Materialia* **59** 2589–2600.

Advanced solid elements for sheet metal forming simulation

Vicente Mataix[1], Riccardo Rossi[2], Eugenio Oñate[3], Fernando G. Flores[4]

[1 2 3]CIMNE (Centre Internacional de Mètodes Numèrics en Enginyeria), Edificio C1 Campus Nord, UPC, calle Gran Capitán s/n, 08034 Barcelona, Spain

[4] Department of Structures, Universidad Nacional de Córdoba, Casilla de Correo 916, 5000 Córdoba-Argentina, and CONICET, Argentina

E-mail: vmataix@cimne.upc.edu[1], rrossi@cimne.upc.edu[2], onate@cimne.upc.edu[3], fernando.flores@unc.edu.ar[4]

Abstract. The solid-shells are an attractive kind of element for the simulation of forming processes, due to the fact that any kind of generic 3D constitutive law can be employed without any additional hypothesis.

The present work consists in the improvement of a triangular prism solid-shell originally developed by *Flores*[2, 3]. The solid-shell can be used in the analysis of thin/thick shell, undergoing large deformations. The element is formulated in total Lagrangian formulation, and employs the neighbour (adjacent) elements to perform a local patch to enrich the displacement field. In the original formulation a modified right Cauchy–Green deformation tensor ($\bar{\mathbf{C}}$) is obtained; in the present work a modified deformation gradient ($\bar{\mathbf{F}}$) is obtained, which allows to generalise the methodology and allows to employ the **Pull-Back** and **Push-Forwards** operations.

The element is based in three modifications: (a) a classical assumed strain approach for transverse shear strains (b) an assumed strain approach for the in-plane components using information from neighbour elements and (c) an averaging of the volumetric strain over the element. The objective is to use this type of elements for the simulation of shells avoiding transverse shear locking, improving the membrane behaviour of the in-plane triangle and to handle quasi-incompressible materials or materials with isochoric plastic flow.

1. Introduction

Solid-shells have been during the last two decades[4, 5, 6, 7, 8] an important improvement in the shells simulations, providing a reliable simulations and avoiding the problematic that are associated to the kinematics hypothesis and plane stress constitutive laws related to the use of the shell element. The main advantages when using **solid-shell** elements are: a) general 3D constitutive relations; b) large transverse shear can be considered, and considering additional elements along the thickness improve this behaviour; c) there is not need to consider transitions between solid and shell elements (all the elements are solids) ;d) contact forces can be introduced directly in the geometry and in a realistic way without any additional technique, which is specially important for the consideration of friction; e) the element is **rotation-free**, avoiding the storage and computation of this variables;f) in the case where we have non-parallel boundaries this can be modelled correctly.

Most of the existing **solid-shells** are linear hexahedron[4, 5, 6, 7, 8], which have two main disadvantages; the first one is the *hourglass* effect; the second problematic is the meshing of the plane, due to the fact that meshing quadrilateral is less performant than triangles. For this reasons the triangular prisms (wedges) could be considered an interesting alternative, specially for the second problem mentioned, but this kind of geometry is not exempt of problematic, owing to the low order of interpolation of the geometry; this can be solved with the consideration of the neighbours elements, in consequence the element becomes quadratic in the plane solving this last problem.

The element has been implemented into *Kratos*[16], the in-home **FEM-Multiphycis** open-source code, implemented in C++ with parallilization capabilities. The pre/post-process of all the presented examples haven been processed with *GiD*, the *CIMNE* software for pre and post porcessing. The content of this proceeding will be the following one: 1) A resume of the theory concerning to the element is presented, focusing in what has been improved from the original element[2, 3] , 2) Some test cases are presented to show the good performance of the element. Additionally, more examples of application of the new solid-shell elements will be shown in the oral presentation in the congress.

2. Theory

For the aim of brevity we will focus in the improvements introduced in the original formulation from *Flores*[2, 3], and we will address to the original material for additional information.

2.1. Pull-Back and Push-Forward (Extension of the formulation)

The formulation presented until now is broadly the formulation already presented in the works of *Flores*[1, 2, 3], the main step forward of this work is the extension of the formulation. As we have seen the formulation presented allows us to obtain a modified left *Cauchy* tensor $\bar{\mathbf{C}}$. With this tensor we are able to obtain strains in a traditional way, *g.e* with *Green-Lagrange*, but we are unable to work with the **Pull-Back** and **Push-Forward** operations.

In order to perform these operations we need the deformation gradient $\mathbf{F}$, or in our case $\bar{\mathbf{F}}$ owing to we are working with a modified right *Cauchy* tensor $\bar{\mathbf{C}}$. Obtain one from another is not a trivial operation, and we must consider additional assumptions to obtain our modified deformation gradient. In a standard formulation to obtain the deformation gradient $\mathbf{F}$ we compute (1) from the material displacement gradient tensor $\nabla_{\mathbf{X}}\mathbf{u}$, $\mathbf{C}$ can be obtained easily from here.

$$\mathbf{F} = \nabla_{\mathbf{X}}\mathbf{u} + \mathbf{I} \rightarrow \mathbf{C} = \mathbf{F}^T\mathbf{F} \tag{1}$$

We will present now (2a) the polar decomposition of $\mathbf{F}$, which will be the key idea considered to obtain $\bar{\mathbf{F}}$. In this decomposition $\mathbf{R}$ represents the *proper orthogonal tensor* and $\mathbf{U}$ is the *right stretch tensor*. The *right stretch tensor* can be computed from the square root of the right *Cauchy* tensor. The only remaining component needed to compute the $\bar{\mathbf{F}}$ will be the modified *proper orthogonal tensor* where we will take the assumption of $\bar{\mathbf{R}} = \mathbf{R}$. The computation of this $\bar{\mathbf{F}}$ is summarized in (2b).

$$\mathbf{F} = \mathbf{R}\mathbf{U} \text{ and } \mathbf{U} = \sqrt{\mathbf{C}} \tag{2a}$$

$$\bar{\mathbf{F}} = \mathbf{R}\bar{\mathbf{U}} \begin{cases} \mathbf{C} = \mathbf{F}^T\mathbf{F} \rightarrow \mathbf{U} = \sqrt{\mathbf{C}} \rightarrow \mathbf{R} = \mathbf{F}\cdot\mathbf{U}^{-1} \\ \bar{\mathbf{U}} = \sqrt{\bar{\mathbf{C}}} \end{cases} \tag{2b}$$

With the modified $\bar{\mathbf{F}}$ we are able to compute the **Pull-Back** and **Push-Forward** operations, which allow us to to have an unified description between the *Eulerian* and *Lagrangian* tensors. In (3) we can appreciate some examples, where a **Push-Forward** by $\mathbf{F}$ of the *Lagrangian* vector $d\mathbf{X}$ to the current configuration gives the *Eulerian* vector $d\mathbf{x}$, and vice-versa with the **Pull-Back** operation. These concepts provide us a mathematically consistent method for define the time derivatives of the tensors, called **Lie-derivatives**. Considering all this, the operations can be considered for example to obtain a formulation of **hyperelastic-plastic** constitutive model based on the multiplicative decomposition of the deformation gradient, for more information consult[12].

Journal of Physics: Conference Series **734** (2016) 032128 doi:10.1088/1742-6596/734/3/032128

$$d\mathbf{x} = \mathbf{F} \cdot d\mathbf{X} \equiv \phi_* d\mathbf{X} \text{ and } d\mathbf{X} = \mathbf{F}^{-1} \cdot d\mathbf{x} \equiv \phi^* d\mathbf{x} \tag{3}$$

3. Test cases

The following test examples here presented shown the good performance and behaviour of the element. The main references of this results are taken from the original paper from *Fernando Flores*[2, 3], but many of the results are taken too from[15] a work with the most significatives benchmarks for test the shell behaviour. All the test results can be found in Figure 1.

- **Cook's membrane test:** Figures 1a and 1b represent the behaviour of the *Cook's* membrane, respectively elastic and elasto-plastic, both obtained considering an implicit scheme. The results present even less than the reference[3] locking for the elastic case and similar behaviour for the elasto-plastic case.
- **Open ended cylindrical shell test:** The results (Figure 1c) agrees with the solution from [15].
- **Slit test:** Can be seen (Figure 1d) that with the refinement of the mesh the solution tend to the correct one [15].
- **Panel test:** The panel test is a common test which presents a highly non-linear behaviour, and it is necessary an arc-length to compute. The results obtained, Figures 1e and 1e, show a clearly agreement with the reference [15].
- **Sphere test:** The Figure 1g presents the results obtained for the semi-spherical shell with a ratio $R/t = 1000$, which agrees with the results from [15].
- **Cone-shell test:** This problem is commonly considered to study the combination both plasticity and geometrical non-linearity, the results shown in Figure 1h differ from [14], but with a close behaviour.

4. Acknowledgments

I want to mention the support of the *Generalitat de Catalunya* which is funding my PhD with the grant *Agaur*.

5. References

[1] Flores F G 2013 *Mecánica Computacional* **XXXII** 63–87
[2] Flores F G 2013 *Computer Methods in Applied Mechanics and Engineering* **253** 274–286
[3] Flores F G 2013 *Computer Methods in Applied Mechanics and Engineering* **266** 81–97
[4] Dvorkin E N and Bathe K J 1984 *Engineering computations* **1** 77–88
[5] Hauptmann R and Schweizerhof K 1998 *International Journal for Numerical Methods in Engineering* **42** 49–69
[6] de Sousa R J A, Cardoso R P R, Valente R A F, Yoon J W, Grácio J J and Jorge R M N 2005 *International journal for numerical methods in engineering* **62** 952–977
[7] Klinkel S, Gruttmann F and Wagner W 2006 *Computer Methods in Applied Mechanics and Engineering* **195** 179–201
[8] Parente M, Valente R F, Jorge R N, Cardoso R and de Sousa R J A 2006 *Finite Elements in Analysis and Design* **42** 1137–1149 ISSN 0168-874X
[9] Oñate E and Flores F G 2005 *Computer Methods in Applied Mechanics and Engineering* **194** 2406–2443 ISSN 0045-7825 computational Methods for Shells
[10] Flores F G and Oñate E 2005 *Computer Methods in Applied Mechanics and Engineering* **194** 907–932 ISSN 0045-7825
[11] Flores F G and Oñate E 2011 *Finite Elements in Analysis and Design* **47** 982–990
[12] Belytschko T, Liu W K, Moran B and Elkhodary K 2014 *Nonlinear Finite Elements for Continua and Structures* 2nd ed (Wiley)
[13] Olovsson L, Unosson M and Simonsson K 2004 *Computational Mechanics* **34** 134–136 ISSN 1432-0924
[14] Klinkel S, Gruttmann F and Wagner W 2006 *Computer Methods in Applied Mechanics and Engineering* **195** 179–201 ISSN 0045-7825
[15] Sze K, Liu X and Lo S 2004 *Finite Elements in Analysis and Design* **40** 1551–1569 ISSN 0168-874X
[16] Kratos Homepage URL `http://www.cimne.com/kratos/`

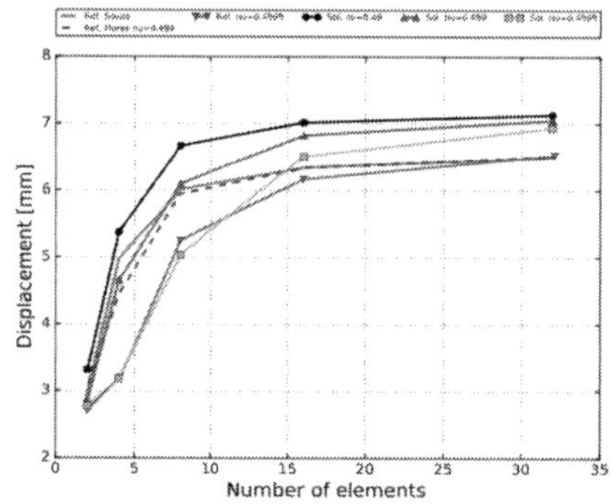

(a) Elastic Cook's membrane

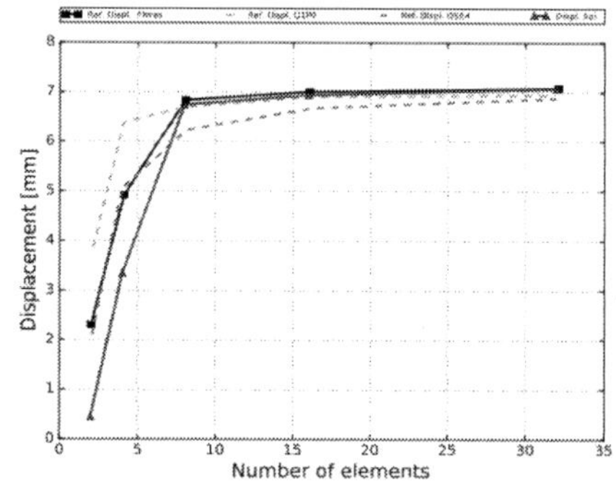

(b) Elasto-plastic Cook's membrane

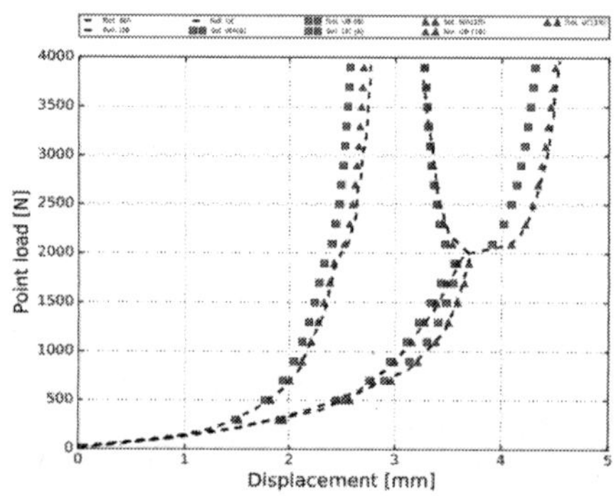

(c) Load-deflection for the open-ended cylindrical shell

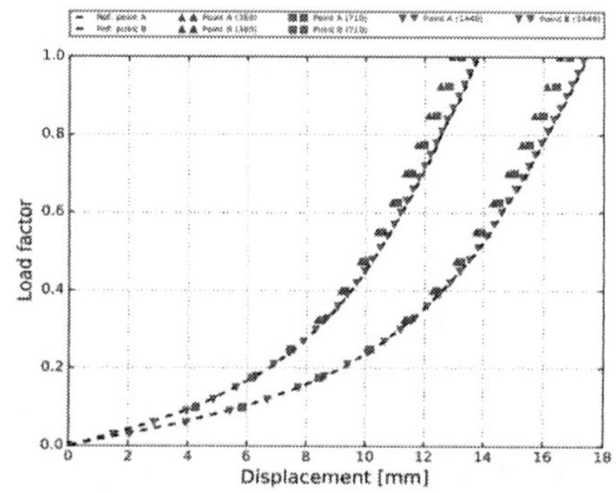

(d) Load-deflection for the slit test

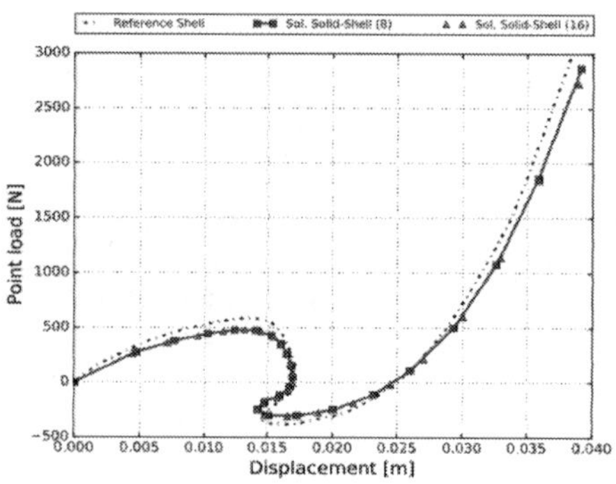

(e) Panel load-deflection $h = 6.35mm$

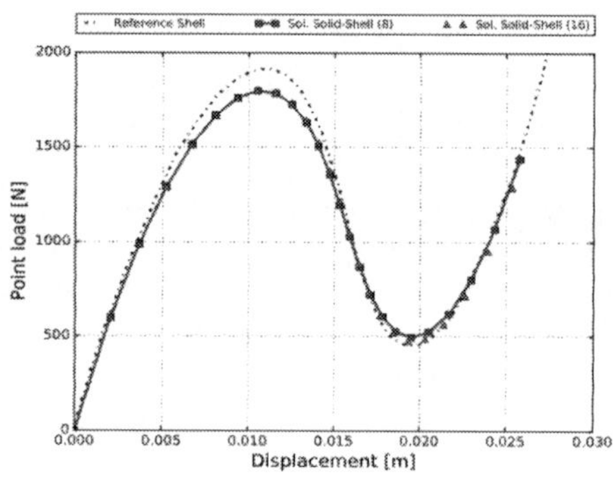

(f) Panel load-deflection $h = 12.7mm$

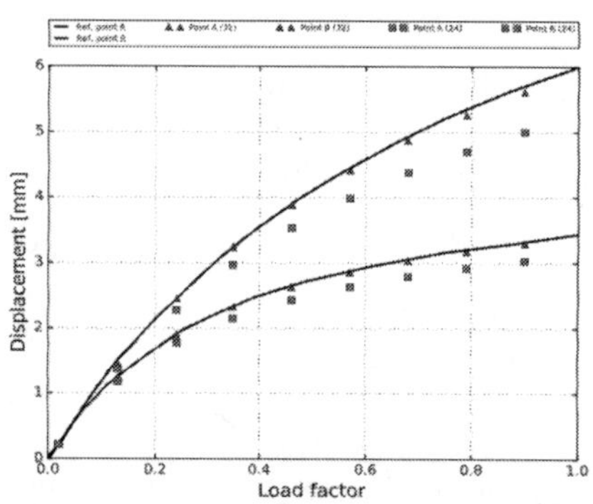

(g) Sphere load-deflection $R/t = 1000$

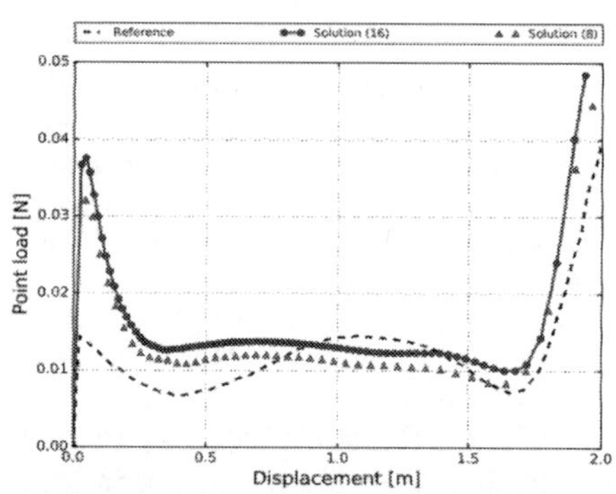

(h) Load-deflection curve for the cone-shell test

Figure 1: Resume of the tests results

A Combined Isotropic, Kinematic and Cross Hardening Model for Magnesium AZ31B-H24 under Non-linear Strain Loading Path

Yueqian Jia, Yuanli Bai*
Department of Mechanical and Aerospace Engineering
University of Central Florida, Orlando, FL 32816, USA
(*Corresponding author: bai@ucf.edu)

Abstract. A fully modularized framework was established to combine isotropic, kinematic, and cross hardening behaviors under non-monotonic loading conditions. Three sets of state variables were defined and applied to consider the effects of, a) loading history, b) twinning and de-twinning and c) different pre-strain. Experiments under two types of non-proportional loading conditions were conducted along different orientations, 1) uniaxial compression-tension reversal loading with different amounts of compressive strains, and 2) two-step uniaxial tension, known as cross-loading conditions, with different pre-strains. No apparent cross-hardening effect was observed for this material. The calibrated new hardening model, with an anisotropic CPB06ex2 yield criterion and an eMMC anisotropic fracture model, has been implemented into Abaqus/ Explicit as a user material subroutine (VUMAT). Good correlation was observed between experimental and simulation results.

1. Introduction
Magnesium alloys have been considered as a representative material of lightweight design in the automotive industry to reduce energy consumption and greenhouse gas emission. The de-twinning phenomenon occurs under compression-tension reversal loading when the subsequent tensile load is performed along the c-axis of the twinned areas, generating an unusual concave shape in the tensile stress-strain curve (Brown, Agnew et al. 2005, Lou, Li et al. 2007, Piao, Chung et al. 2012). The objectives of this paper are to, a) develop a constitutive model for magnesium alloy sheets, covering both monotonic and non-proportional loading conditions, b) conduct a sufficient set of experiments under non-linear strain path, including both reversal and cross-loading condition, and c) implement both the constitutive model and fracture model into finite element analysis to predict the non-linear historical fracture behavior.

2. Constitutive Modeling
Very strong anisotropic effects have been observed in this material, magnesium AZ31B-H24. The CPB06ex2 anisotropic yield criterion was calibrated for magnesium AZ31B-H24, which could concurrently fulfill a) different material strengths and plastic flows along different orientations, and b) the asymmetric yield strength between tension and compression, based on the work from Plunkett, Cazacu et al. (2008), Jia and Bai (2015). The modified semi-analytical Sachs model (Jia and Bai 2015) can be extended to incorporate the kinematic hardening behavior, using the state variables. Therefore, an additional item is defined, as follows

$$\bar{\sigma}_K = sign(\dot{\mu})\left[(A_{K1}\mu - A_{K2}\varepsilon_C)(\bar{\varepsilon}^p + \varepsilon_0 - \varepsilon_C)^n + X_K\sqrt{\boldsymbol{\alpha}^{dev}:\boldsymbol{\alpha}^{dev}}\right],$$

where
μ is the *ad-hoc* de-twinning state variable, evolves from 1 (no de-twinning) to 0 (de-twinning

finished). ε_C is the compressive strain in previous loading history. A_{K1} and A_{K2} are material coefficients. X_K is the loading history state variable identifying the transition area, reads,

$$X_K = \frac{1 - \dfrac{\boldsymbol{\sigma}:\boldsymbol{\alpha}}{\|\boldsymbol{\sigma}\|\|\boldsymbol{\alpha}\|}}{2},$$

$\boldsymbol{\alpha}^{dev}$ is the deviatoric part of the "dummy" back stress. The modified model becomes,

$$\overline{\sigma}_{NewSachs} = A_1(1-\chi)X_T + [A_2\chi X_T + A_3(1-X_T)](\bar{\varepsilon}^p + \varepsilon_0 - h\varepsilon_B)^n,$$

where χ indicates the fraction of the to-be-twinned grains that have already twinned, ranges from 0 to 1. It is defined as follows,

$$\chi = 1 - exp\left[-4(\frac{\bar{\varepsilon}^p - h\varepsilon_B}{\varepsilon_1})^a\right].$$

X_T is the fraction of grains expected to be twinned. ε_B is the historical accumulated plastic strain from the previous loading transitions, excluding the de-twinning process. h is an additional material coefficient attached to ε_B to control the isotropic hardening. A_1, A_2, and A_3 are simplified material coefficients. The kinematic hardening behavior during cross loading can be further compensated by an additional item $\bar{\sigma}_C$, defined as

$$\bar{\sigma}_C = -X_C^{\lambda}\sqrt{\boldsymbol{\alpha}^{dev}:\boldsymbol{\alpha}^{dev}},$$

Here, X_C is the key state variable governing the smoothness in transition during cross loading,

$$X_C = 1 - \left|\frac{\boldsymbol{\sigma}:\boldsymbol{\alpha}}{\|\boldsymbol{\sigma}\|\|\boldsymbol{\alpha}\|}\right|.$$

λ is an exponential material coefficient. Conclusively, the final constitutive model for magnesium AZ31B-H24 becomes

$$\bar{\sigma} = (\bar{\sigma}_{NewSachs} + \bar{\sigma}_K + \bar{\sigma}_C)f_{ext}(\eta, \bar{\theta}),$$

Strong anisotropic fracture has been observed in magnesium AZ31B-H24. An eMMC fracture model is applied for this material with linear transformed anisotropic equivalent plastic strain, based on the work from Jia and Bai (2015). In this paper, an additional assumption is made to incorporate the twinning/de-twinning effect as follows. The damage accumulation is temporarily deactivated, between the initialization and 90% completion of de-twinning.

3. Experimental Results and Numerical Simulation

The proposed constitutive model with CPB06ex2 anisotropic yield criterion and eMMC fracture model has been implemented into Abaqus/Explicit as a user material subroutine (VUMAT). The thickness for reversal loading test specimen was set as $2mm$ while the one for two-step tension was set as $1mm$. Three different orientations, 0°, 45°, and 90° were assigned for both models. The simulated results were then compared to the experimental ones, which were all from DIC. One can see that the new hardening model is capable to predict the compression-tension kinematic hardening behavior with pre-strain effect. The eMMC model with associated de-twinning effect could predict the fracture strains very well under 10% nominal pre-strain. The comparison of true stress-strain between experimental and simulated two-step uniaxial tension is illustrated in Figure 2, with pre-strains under different amounts and orientations. One can see that the second step true stress-strain curves with the same orientation are similar under different pre-strains. No additional hardening is apparently observed at the beginning of the second step uniaxial tension, when the loading orientation changes. In conclusion, no clear cross-hardening effect is observed for magnesium AZ31B-H24 alloy sheet in the current experimental result. The hardening curves under all scenarios are well correlated with experimental results.

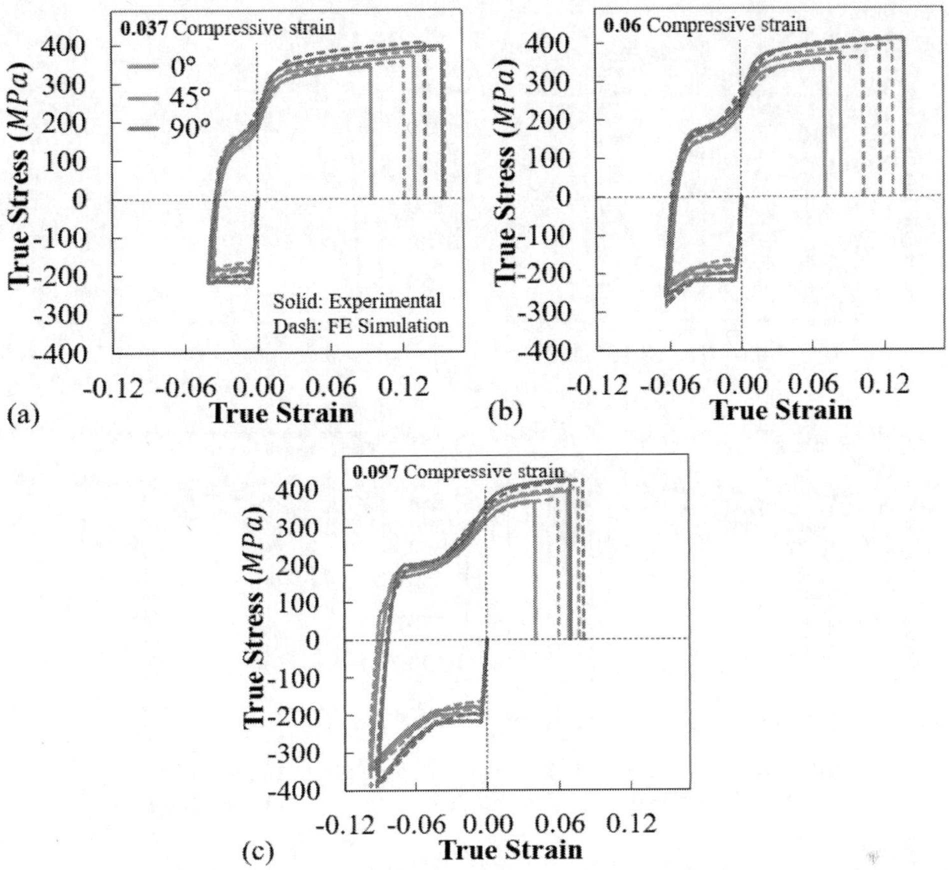

Figure 1 Comparison between tested and simulated true stress-strain curves of $2mm$ thickness AZ31B-H24 sheets up to fracture along 0°, 45° and 90° orientations, under compression-tension reversal loading with a pre-compressive strain of a) 0.037, b) 0.06, and c) 0.097.

4. Conclusions

A new Sachs-based constitutive model was developed for magnesium AZ31B-H24 alloy sheet, with a fully decoupled framework to combine isotropic, kinematic, and cross hardening behaviors. In order to verify the new constitutive model, two different types of non-proportional loading tests were conducted along different orientations, a) uniaxial compression-tension reversal loading with different pre-compressive strains, and b) two-step uniaxial tension, known as cross-loading conditions, with different pre-strains. No apparent cross-hardening effect was observed in the two-step uniaxial tension experimental results. The new model was sequentially calibrated by the experimental results from a set of different monotonic loading conditions, and a), b) above, with non-aftereffect. The calibrated new constitutive model, an anisotropic yield criterion CPB06ex2, and an eMMC anisotropic fracture model have been implemented into FE analysis to reproduce the non-proportional experiments through a user material subroutine (VUMAT) in Abaqus/Explicit. Good correlation was observed between experimental and simulation results, in both material strength and fracture behavior.

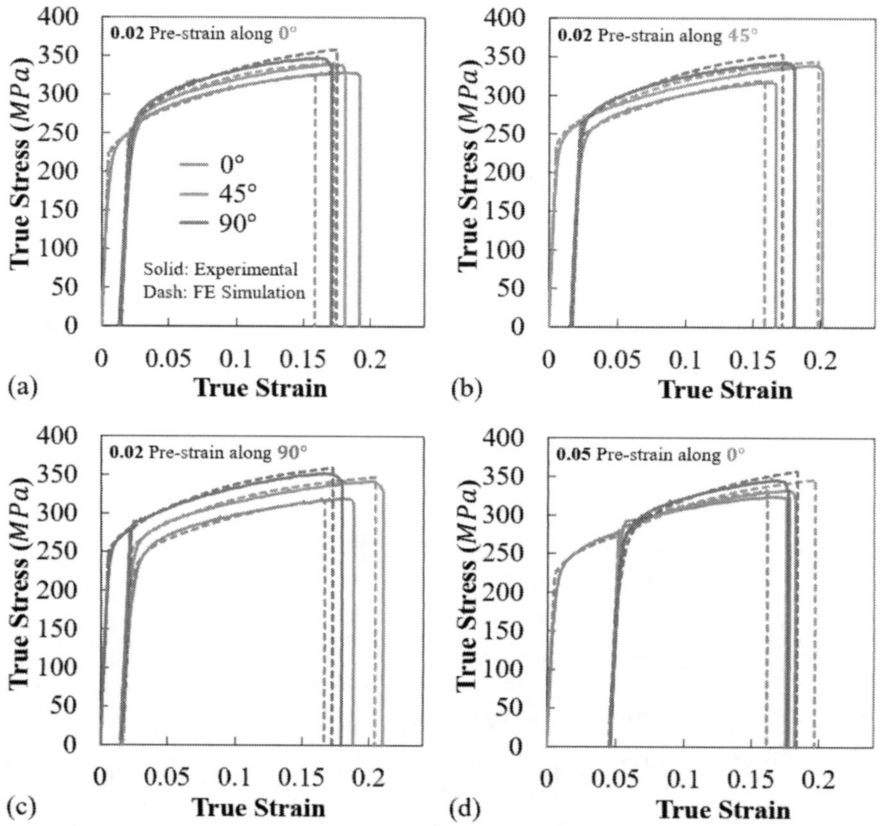

Figure 2 Comparison between tested and simulated true stress-strain curves of 1mm thickness AZ31B-H24 sheets up to fracture, under two-step uniaxial tension with a pre-strain of a), b) c) 0.02, d) 0.05, along a orientation of a), d), 0°, b), 45°, and c), 90°.

5. Acknowledgments

Partial financial supports from University of Central Florida (UCF), Volkswagen (Germany), and State Key Laboratory of Automotive Safety and Energy, Tsinghua University (Beijing, China) are greatly appreciated. Thanks are also due to the Dassault Systemes for providing Abaqus software, and the Altair Company for the continuous support with HyperMesh.

Reference

Brown, D. W., S. R. Agnew, M. A. M. Bourke, T. M. Holden, S. C. Vogel and C. N. Tomé (2005). "Internal strain and texture evolution during deformation twinning in magnesium." Materials Science and Engineering: A **399**(1–2): 1-12.

Jia, Y. and Y. Bai (2016). "Ductile fracture prediction for metal sheets using all-strain-based anisotropic eMMC model." International Journal of Mechanical Sciences, under revision.

Jia, Y. and Y. Bai (2016). "Experimental Study on the Mechanical Properties of AZ31B-H24 Magnesium Alloy Sheets under Various Loading Conditions." International Journal of Fracture: **197**(1): 25-48.

Lou, X. Y., M. Li, R. K. Boger, S. R. Agnew and R. H. Wagoner (2007). "Hardening evolution of AZ31B Mg sheet." International Journal of Plasticity **23**(1): 44-86.

Piao, K., K. Chung, M.-G. Lee and R. Wagoner (2012). "Twinning-Slip Transitions in Mg AZ31B." Metallurgical and Materials Transactions A **43**(9): 3300-3313.

Plunkett, B., O. Cazacu and F. Barlat (2008). "Orthotropic yield criteria for description of the anisotropy in tensi on and compression of sheet metals." International Journal of Plasticity **24**(5): 847-866.

Journal of Physics: Conference Series **734** (2016) 032130 doi:10.1088/1742-6596/734/3/032130

Single Point Incremental Forming to increase material knowledge and production flexibility

A.M. Habraken

University of Liège, Belgium

Nowadays, manufactured pieces can be divided into two groups: mass production and production of low volume number of parts. Within the second group (prototyping or small batch production), an emerging solution relies on Incremental Sheet Forming or ISF.

ISF refers to processes where the plastic deformation occurs by repeated contact with a relatively small tool. More specifically, many publications over the past decade investigate Single Point Incremental Forming (SPIF) where the final shape is determined only by the tool movement. This manufacturing process is characterized by the forming of sheets by means of a CNC controlled generic tool stylus, with the sheets clamped by means of a non-workpiece-specific clamping system and in absence of a partial or a full die. The advantage is no tooling requirements and often enhanced formability, however it poses a challenge in term of process control and accuracy assurance. Note that the most commonly used materials in incremental forming are aluminum and steel alloys however other alloys are also used especially for medical industry applications, such as cobalt and chromium alloys, stainless steel and titanium alloys. Some scientists have applied incremental forming on PVC plates and other on sandwich panels composed of propylene with mild steel and aluminum metallic foams with aluminum sheet metal. Micro incremental forming of thin foils has also been developed.

Starting from the scattering of the results of Finite Element (FE) simulations, when one tries to predict the tool force (see SPIF benchmark of 2014 Numisheet conference), we will see how SPIF and even micro SPIF (process applied on thin metallic sheet with a few grains within the thickness) allow investigating the material behavior. This lecture will focus on the identification of constitutive laws, on the SPIF forming mechanisms and formability as well as the failure mechanism. Different hypotheses have been proposed to explain SPIF formability, they will be listed however the lecture will be more focused on the use of SPIF to identify material parameters of well-chosen constitutive law. Results of FE simulations with damage models will be investigated to better understand the relation between the particular stress and strain states in the material during SPIF and the material degradation leading to localization or fracture.

Last but not least, as industrial world does not wait that academic scientists provide a deep and total understanding on how it works, to use interesting processes, the lecture will review some applications. Examples in fields as different as automotive guard, engine heat shield, gas turbine, electronic sensor, shower basin, medical component (patient-fitted organic shapes) and architecture demonstrate that the integration of SPIF within the industry is more and more a reality.

Note that this plenary lecture is the result of the research performed by the author in the University of Liège (Belgium) and in Aveiro (Portugal) with the team of R. de Souza during PhD theses of C. Henrard, J. Sena and C. Guzman and different research projects. It is also a synthesis of the knowledge gathered during her interactions with many research teams such as the ones of J.R. Duflou from KU Leuven in Belgium, J. Cao from Northwestern University in USA, M. Bambach in BTU Cottbus-Senftenberg in Germany, J. Jeswiet from Queen's University, Kingston, Canada who are currently working together on a state-of-the-art paper. The micro SPIF knowledge relies on contacts with S. Thibaud from the University of Franche Comte.

Meshless local integral equation method for two-dimensional nonlocal elastodynamic problems

X.J. Huang and P.H. Wen

School of Engineering and Material Science, Queen Mary University of London, London, E1 4NS, UK

E-mail: xuejiao.huang@qmul.ac.uk

Abstract. This paper presents the meshless local integral equation method (LIEM) for nonlocal analyses of two-dimensional dynamic problems based on the Eringen's model. A unit test function is used in the local weak-form of the governing equation and by applying the divergence theorem to the weak-form, local boundary-domain integral equations are derived. Radial Basis Function (RBF) approximations are utilized for implementation of displacements. The Newmark method is employed to carry out the time marching approximation. Two numerical examples are presented to demonstrate the application of time domain technique to deal with nonlocal elastodynamic mechanical problems.

1. Introduction

It is found that the distance between individual atoms or molecules cannot be ignored when it becomes obviously comparable to external scales and the discrete microstructure in materials can no longer be treated as homogeneous continuum at small size. Therefore, nonlocal elasticity theories have been initiated to model material properties exhibited at micro-, and even at nano-scale levels. Among others, Eringen [1] proposed an integral model for linear homogeneous isotropic media as

$$\sigma(\mathbf{x}) = \xi_1 \bar{\sigma}(\mathbf{x}) + \xi_2 \int_V \alpha(\mathbf{x}, \mathbf{x}', l) \bar{\sigma}(\mathbf{x}') dV(\mathbf{x}') \tag{1}$$

where σ is refered to as nonlocal stress, $\bar{\sigma}$ is the classical stress, V represents the volume of domain, $\mathbf{x}, \mathbf{x}'$ are collocation and domain integration points, l is the characteristic length. ξ_1 and ξ_2 are portion factors and $\xi_1 + \xi_2 = 1$. α is the nonlocal kernel and usually taken as $\alpha(\mathbf{x}, \mathbf{x}', l) = \lambda_0 \exp(-|\mathbf{x} - \mathbf{x}'|/l)$, $\lambda_0 = 1/2\pi l^2$.

2. LIEM for nonlocal elastodynamics

Consider a linear elastic body in two dimensional domain Ω with boundary Γ. The weak form of governing equations of motion can be written as

$$\int_{\Omega_s} (\sigma_{ij,j} + f_i - \rho \ddot{u}_i) u_i^* d\Omega = 0 \tag{2}$$

where f denotes the body force, ρ is the mass density, $\ddot{u}$ is the acceleration and u_i^* is the test function. By use of the divergence theorem, Eq.2 can be rewritten as

$$\int_{\Gamma_s} \sigma_{ij} n_j u_i^* d\Gamma - \int_{\Omega_s} (\sigma_{ij} u_{i,j}^* - f_i u_i^* + \rho \ddot{u}_i u_i^*) d\Omega = 0 \tag{3}$$

where n_j is the unit outward normal on the boundary. A unit test function $\phi_i(\mathbf{x}) = 1$ is employed and thus, local integral equations in Eq.3 becomes

$$\int_{\Gamma_s} \sigma_{ij} n_j d\Gamma + \int_{\Omega_s} (f_i - \rho \ddot{u}_i) d\Omega = 0 \tag{4}$$

Assume the local domain is enclosed by several straight lines and recall the constitutive relation of Eringin's model in Eq.1, Eq.4 can be written as

$$\int_{\Gamma_s} \sigma_{ij}(\mathbf{x}) n_j(\mathbf{x}) d\Gamma(\mathbf{x}) = \xi_1 \sum_{l=1}^{L} n_j^l \int_{\Gamma_l} \sigma_{ij}(\mathbf{x}) d\Gamma(\mathbf{x}) + \xi_2 \sum_{l=1}^{L} n_j^l \Delta_l \int_V \alpha(\mathbf{x}_i, \mathbf{x}', l) \sigma_{ij}(\mathbf{x}') dV(\mathbf{x}') \tag{5}$$

where L is number of straight lines and Δ_l is the length of the lth straight line.
The time marching approximation of Newmark mthod [2] are given as:

$$\dot{u}_{n+1} = \frac{\gamma}{\beta \Delta t}(u_{n+1} - u_n) + \frac{\beta - \gamma}{\beta} \dot{u}_n - \frac{\gamma - 2\beta}{2\beta} \Delta t \ddot{u}_n \tag{6}$$

$$\ddot{u}_{n+1} = \frac{1}{\beta \Delta t^2}(u_{n+1} - u_n) - \frac{1}{\beta \Delta t} \dot{u}_n - \frac{1 - 2\beta}{2\beta} \ddot{u}_n \tag{7}$$

in which $\dot{u}$, $\ddot{u}$ and u represent the velocity, acceleration and displacement, respectively. u_n is the displacement at $n\Delta t$, where Δt is the selected time step. Parameters γ and β are selected as $\gamma = 0.5$ and $\beta = 0.25$ in this paper.

Consider the RBF approximation in [3], and the four-point standard integral scheme is adopted to carry out the domain integrals, the following discrete equations can be derived.

$$\begin{aligned}
&\xi_1 C \sum_{k=1}^{K} u_{1(n+1)}^{(k)} \sum_{l=1}^{L} \left[\sum_{i=1}^{K} \left(E' F_{1il} n_1^l + \mu F_{2il} n_2^l\right) \alpha_{ik} + \sum_{t=1}^{T} \left(E' G_{1tl} n_1^l + \mu G_{2tl} n_2^l\right) \beta_{tk} \right] + u_{1(n+1)}^{(k)} + \\
&\xi_2 C \sum_{k'=1}^{K'} u_{1(n+1)}^{(k')} \sum_{q=1}^{V} \sum_{p=1}^{4} \sum_{l=1}^{L} \left(E' \phi_{k',1}(\mathbf{x}'_p) n_1^l(\mathbf{x}_l) + \mu \phi_{k',2}(\mathbf{x}'_p) n_2^l(\mathbf{x}_l)\right) \alpha(\mathbf{x}_l, \mathbf{x}'_p, l) \Delta_l w_p \Delta V_q + \\
&\xi_1 C \sum_{k=1}^{K} u_{2(n+1)}^{(k)} \sum_{l=1}^{L} \left[\sum_{i=1}^{K} \left(\nu E' F_{2il} n_1^l + \mu F_{1il} n_2^l\right) \alpha_{ik} + \sum_{t=1}^{T} \left(\nu E' G_{2tl} n_1^l + \mu G_{1tl} n_2^l\right) \beta_{tk} \right] + \\
&\xi_2 C \sum_{k'=1}^{K'} u_{2(n+1)}^{(k')} \sum_{q=1}^{V} \sum_{p=1}^{4} \sum_{l=1}^{L} \left(\nu E' \phi_{k',2}(\mathbf{x}'_p) n_1^l(\mathbf{x}_l) + \mu \phi_{k',1}(\mathbf{x}'_p) n_2^l(\mathbf{x}_l)\right) \alpha(\mathbf{x}_l, \mathbf{x}'_p, l) \Delta_l w_p \Delta V_q = I_1
\end{aligned} \tag{8a}$$

$$\begin{aligned}
&\xi_1 C \sum_{k=1}^{K} u_{1(n+1)}^{(k)} \sum_{l=1}^{L} \left[\sum_{i=1}^{K} \left(\nu E' F_{1il} n_2^l + \mu F_{2il} n_1^l\right) \alpha_{ik} + \sum_{t=1}^{T} \left(\nu E' G_{1tl} n_2^l + \mu G_{2tl} n_1^l\right) \beta_{tk} \right] + \\
&\xi_2 C \sum_{k'=1}^{K'} u_{1(n+1)}^{(k')} \sum_{q=1}^{V} \sum_{p=1}^{4} \sum_{l=1}^{L} \left(\nu E' \phi_{k',1}(\mathbf{x}'_p) n_2^l(\mathbf{x}_l) + \mu \phi_{k',2}(\mathbf{x}'_p) n_1^l(\mathbf{x}_l)\right) \alpha(\mathbf{x}_l, \mathbf{x}'_p, l) \Delta_l w_p \Delta V_q + \\
&\xi_1 C \sum_{k=1}^{K} u_{2(n+1)}^{(k)} \sum_{l=1}^{L} \left[\sum_{i=1}^{K} \left(E' F_{2il} n_2^l + \mu F_{1il} n_1^l\right) \alpha_{ik} + \sum_{t=1}^{T} \left(E' G_{2tl} n_2^l + \mu G_{1tl} n_1^l\right) \beta_{tk} \right] + u_{2(n+1)}^{(k)} + \\
&\xi_2 C \sum_{k'=1}^{K'} u_{2(n+1)}^{(k')} \sum_{q=1}^{V} \sum_{p=1}^{4} \sum_{l=1}^{L} \left(E' \phi_{k',2}(\mathbf{x}'_p) n_2^l(\mathbf{x}_l) + \mu \phi_{k',1}(\mathbf{x}'_p) n_1^l(\mathbf{x}_l)\right) \alpha(\mathbf{x}_l, \mathbf{x}'_p, l) \Delta_l w_p \Delta V_q = I_2
\end{aligned} \tag{8b}$$

in which $C = -\dfrac{\beta \Delta t^2}{\rho \Omega_s}$, $I_i = u_i + \Delta t \dot{u}_i + \dfrac{1 - 2\beta}{2} \Delta t^2 \ddot{u}_i$ (body force is ignored). $E' = E/(1-\nu^2)$, E is Young's modulus and ν is the Poisson's ratio and shear modulus $\mu = E/2(1+\nu)$. $k = 1, 2, \ldots, K$

and $k' = 1, 2, \ldots, K'$ are the number of nodes in the support domain centred at $\mathbf{x}$ and $\mathbf{x}'$, respectively. V is the number of sub integral domains and $w_p = 0.25$. Analytical solutions for boundary integrals of F_{jil} and G_{jtl} in closed form can be found in [4].

Traction boundary conditions are $\int_{\Gamma-\Gamma_T} t_i d\Gamma - \int_{\Gamma_T} t_i^0 d\Gamma = I_i$ for $\mathbf{x}_k$, $k = 1, 2, \ldots, M_T$, where Γ_T is the traction boundary. While displacement boundary conditions can be introduced directly as $u_i(\mathbf{x}_k) = u_i^0, k = 1, 2, \ldots, M_D$. M_T and M_D are the number of nodes located on the traction and displacement boundaries, respectively.

3. Numerical Examples

3.1. A square plane subjected to dynamic load

Consider a square plate of side $a = 1.0$ subjected to a Heaviside load $\sigma_0 H(t)$ on top and fixed along the edge at $x_2 = 0$ as shown in Fig.1. 11×11 nodes are distributed in the problem domain. Young's modulus is one unit and Poisson's ration is taken as zero. Normalized dynamic displacements on point A and B with the characteristic length $l = 0.1$ and $l = 0.2$ for different portion factors ($\xi_1 = 0.8$ / $\xi_1 = 0.5$) are plotted in Fig.2 to Fig.4, which show that the period of oscillation increases when the portion factor ξ_1 decreases or the characteristic length l rises.

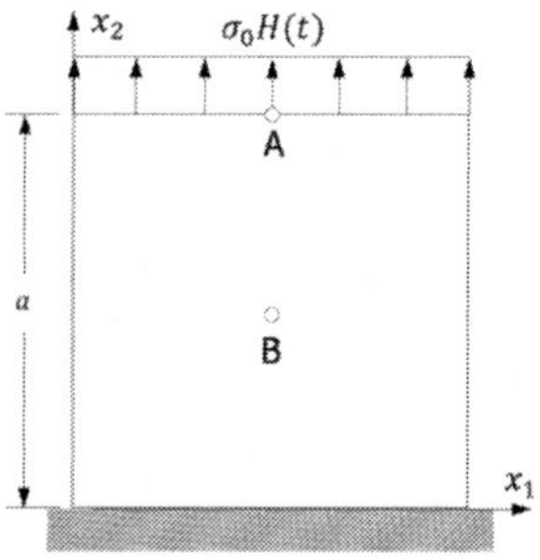

Figure 1. A square plate subjected to a Heaviside load.

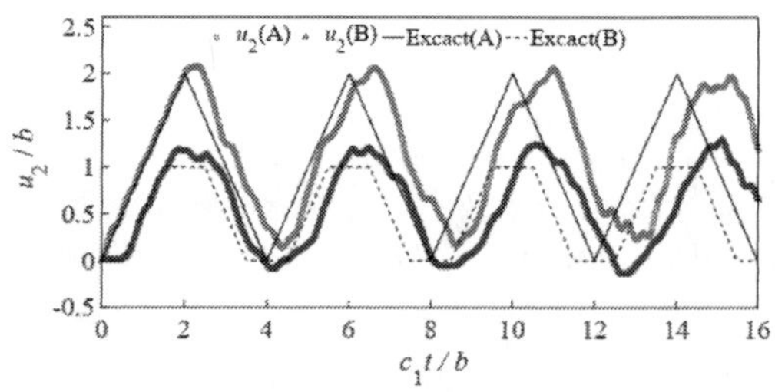

Figure 2. Normailized displacement for point A and B against normalized time c_1t/b when $\xi_1 = 0.5$, $l = 0.1$.

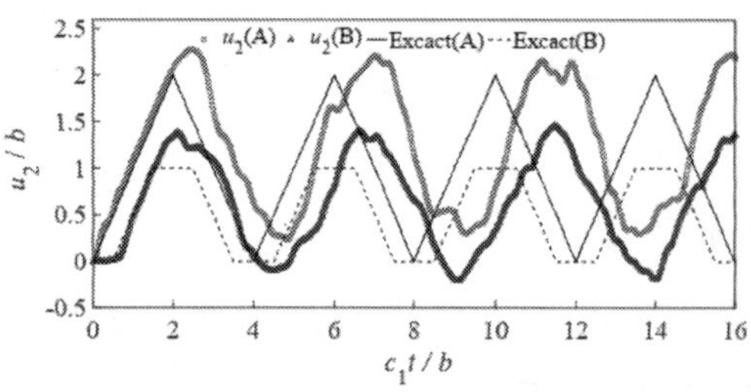

Figure 3. Normailized displacement for point A and B against normalized time c_1t/b when $\xi_1 = 0.5$, $l = 0.2$.

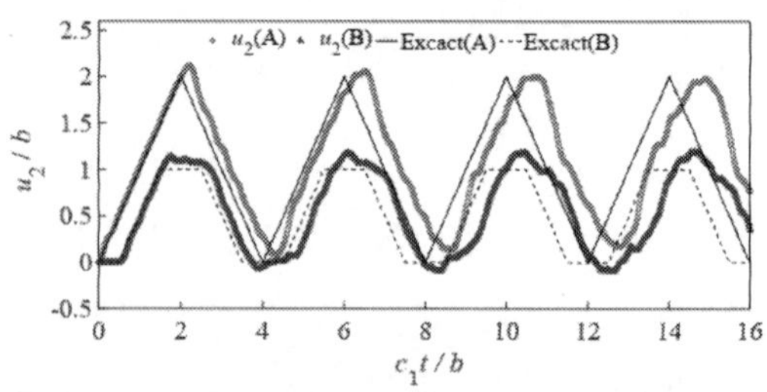

Figure 4. Normailized displacement for point A and B against normalized time c_1t/b when $\xi_1 = 0.8$, $l = 0.2$.

3.2. A ring under internal dynamic pressure

A ring under internal dynamic pressure is analysed as shown in Fig.5(a). Dimensions $a = 1cm$, $b = 2cm$. Young's modulus $E = 1.0$ and Poisson's ratio $\nu = 0.3$. Due to symmetry, only one quarter of the ring is considered. However, the contributions of strains from the whole ring

Journal of Physics: Conference Series **734** (2016) 032131 doi:10.1088/1742-6596/734/3/032131

must be taken into account in the domain integral. Nodes distributions can be seen in Fig.5(b). The normalized dynamic tangential stress at point A and B are plotted in Fig.6 and Fig.7 with $l = 0.2$ for different portion factors ($\xi_1 = 1.0$ and $\xi_1 = 0.5$). ABAQUS solutions are presented as well for comparisons when $\xi_1 = 1.0$ and good agreement have been obtained, which validates the accuracy of time domain technique to general dynamic problems.

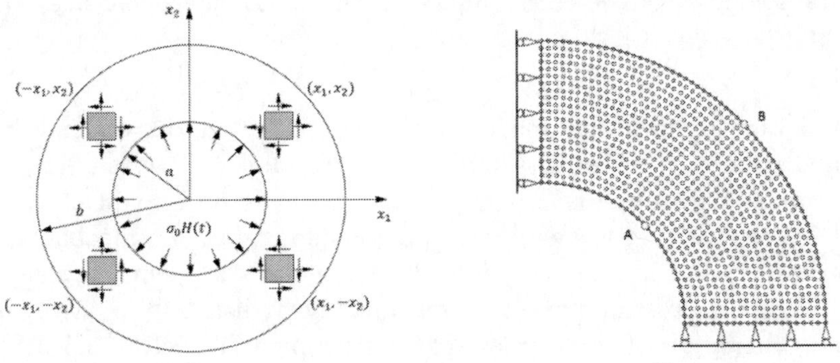

Figure 5. A ring under internal dynamic pressure :(a) geometry; (b) node distributions.

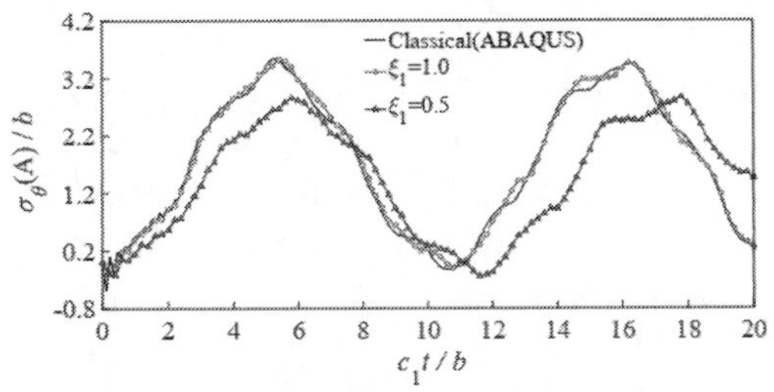

Figure 6. Normalized tangential stress against normalized time c_1t/b for point A when $l = 0.2$ for $\xi_1 = 1.0$ and $\xi_1 = 0.5$.

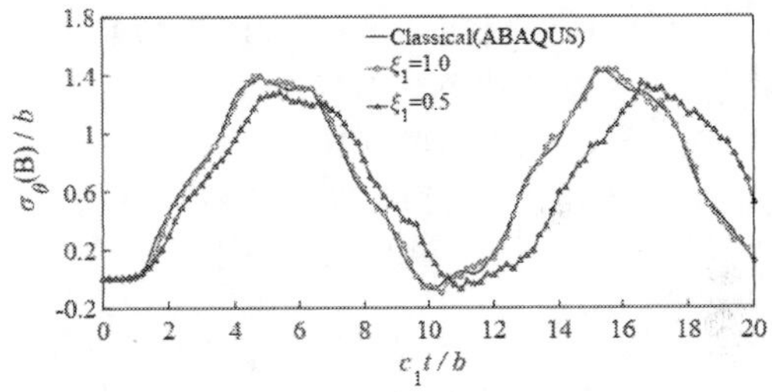

Figure 7. Normalized tangential stress against normalized time c_1t/b for point B when $l = 0.2$ for $\xi_1 = 1.0$ and $\xi_1 = 0.5$.

4. Conclusions

In this paper, LIEM formulations are presented for two-dimensional nonlocal elastodynamics. The Newmark method is selected as the approximation scheme to deal with time-dependent cases. Two numerical examples demonstrated that LIEM embedded with time domain technique is stable, convergent and accurate for solving nonlocal elastodynamic problems.

References

[1] Eringen A C and Kim B S 1974 *Mechanics Research Communications* **1** 233–237

[2] Newmark N M 1959 *Journal of the engineering mechanics division* **85** 67–94

[3] Wen P H and Aliabadi M H 2008 *Communications in Numerical Methods in Engineering* **24** 635–651

[4] Wen P and Aliabadi M 2013 *Applied Mathematical Modelling* **37** 2115 – 2126

Two-scale Modelling of material degradation and failure

Ferri M.H. Aliabadi[a]

[1] Department of Aeronautics, Imperial College London, London, SW7 2AZ, UK

[a] m.h.aliabadi@imperial.ac.uk

Keywords: transgranular, intergranular, polycrystalline, microfracture, cohesive, boundary element method.

Abstract

It is widely recognized that macroscopic material properties depend on the features of the microstructure. The understanding of the links between microscopic and macroscopic material properties, main topic of Micromechanics, is of relevant technological interest, as it may enable deep understanding of the mechanisms governing materials degradation and failure. Polycrystalline materials are used in many engineering applications. Their microstructure is determined by distribution, size, morphology, anisotropy and orientation of the crystals [1]. At temperature below 0.3-0.5 $T_{melting}$ there are no ductile or creep mechanisms and two are the main failure patterns: intergranular, where the damage follows the grain boundaries and transgranular where instead the damage goes through the grain by splitting it into two parts.

In this talk a two-scale approach to degradation and failure in polycrystalline materials will be presented. The formulation involves the engineering component level (macro-scale) and the material grain level (micro-scale). The macro-continuum is modelled using two- and three-dimensional boundary element formulation in which the presence of damage is formulated through an *initial stress approach* to account for the local softening in the neighborhood of points experiencing degradation at the micro-scale. The microscopic degradation is explicitly modelled by associating Representative Volume Elements (RVEs) to relevant points of the macro continuum, for representing the polycrystalline microstructure in the neighbourhood of the selected points. A grain-boundary formulation is used to simulate intergranular/transgranular degradation and failure in the microstructure, whose morphology is generated using the Voronoi tessellations. Intergranular/transgranular degradation and failure are modeled through cohesive and frictional contact laws. To couple the two scales, macro-strains are transferred to the RVEs as periodic boundary conditions, while overall macro-stresses are obtained as volume averages of the micro-stress field. The comparison between effective macro-stresses for the damaged and undamaged RVE allows to define a macroscopic measure of material degradation. To avoid pathological damage localization at the macro-scale, integral non-local counterparts of the strains are employed. A multiscale processing algorithm is described. Two multiscale simulations are performed to demonstrate the capability of the method.

Finally numerical simulations have been performed in order to demonstrate the validity of the model in comparison with experimental observations and literature results.

References

[1] GK Sfantos, MH Aliabadi A boundary cohesive grain element formulation for modelling intergranular microfracture in polycrystalline brittle materials, International journal for numerical methods in engineering 69 (8), 1590-1626, 2007

[2] I Benedetti, MH Aliabadi A three-dimensional cohesive-frictional grain-boundary micromechanical model for intergranular degradation and failure in polycrystalline materials. Computer Methods in Applied Mechanics and Engineering 265, 36-62, 2013
[3] GK Sfantos, MH Aliabadi Multi-scale boundary element modelling of material degradation and fracture, Computer Methods in Applied Mechanics and Engineering 196 (7), 1310-1329, 2007
[4] I. Benedetti, M.H. Aliabadi, Multiscale modeling of polycrystalline materials: A boundary element approach to material degradation and fracture *Computer Methods in Applied Mechanics and Engineering,* 265, 36–62 (2013).
[5] G. Geracia, M.H. Aliabadi Micromechanical boundary element modelling of transgranular
and intergranular cohesive cracking in polycrystalline materials, to appear.

Journal of Physics: Conference Series **734** (2016) 032133 doi:10.1088/1742-6596/734/3/032133

Non-local models for ductile failure

José César de Sá, Erfan Azinpour, Abel Santos

INEGI, Faculty of Engineering, University of Porto, Porto, Portugal

Email: cesarsa@fe.up.pt

ABSTRACT:

Ductile damage can be dealt with continuous descriptions of material, resorting, for example, to continuous damage mechanic descriptions or micromechanical constitutive models. When it comes to describe material behaviour near and beyond fracture these approaches are no longer sufficient or valid and continuous/discontinuous approaches can be adopted to track fracture initiation and propagation. Apart from more pragmatic solutions like element erosion or remeshing techniques more advanced approaches based on the X-FEM concept, in particular associated with non-local formulations, may be adopted to numerically model these problems. Nevertheless, very often, for practical reasons, some important aspects are somewhat left behind, specially energetic requirements to promote the necessary transition of energy release associated with material damage and fracture energy associated to a crack creation and evolution. Phase-field methods may combine advantages of regularised continuous models by providing a similar description to non-local thermodynamical continuous damage mechanics, as well as, a "continuous" approach to numerically follow crack evolution and branching

Acknowledgement
Authors gratefully acknowledge the funding of Project NORTE-01-0145-FEDER-000022 - SciTech - Science and Technology for Competitive and Sustainable Industries, co-financed by Programa Operacional Regional do Norte (NORTE2020), through Fundo Europeu de Desenvolvimento Regional (FEDER)

Accelerating multi-scale sheet forming simulations by exploiting local macroscopic quasi-homogeneities

J Gawad[1], Md Khairullah[1], D Roose[1] and A Van Bael[2]

[1] KU Leuven, Department of Computer Science, Celestijnenlaan 200A, 3001 Leuven, Belgium
[2] KU Leuven, Department of Materials Engineering, Kasteelpark Arenberg 44 Box 2450, 3001 Leuven, Belgium

e-mail: jerzy.gawad@cs.kuleuven.be

Abstract. Multi-scale simulations are computationally expensive if a two-way coupling is employed. In the context of sheet metal forming simulations, a fine-scale representative volume element (RVE) crystal plasticity (CP) model would supply the Finite Element analysis with plastic properties, taking into account the evolution of crystallographic texture and other microstructural features. The main bottleneck is that the fine-scale model must be evaluated at virtually every integration point in the macroscopic FE mesh. We propose to address this issue by exploiting a verifiable assumption that fine-scale state variables of similar RVEs, as well as the derived properties, subjected to similar macroscopic boundary conditions evolve along nearly identical trajectories. Furthermore, the macroscopic field variables primarily responsible for the evolution of fine-scale state variables often feature local quasi-homogeneities. Adjacent integration points in the FE mesh can be then clustered together in the regions where the field responsible for the evolution shows low variance. This way the fine-scale evolution is tracked only at a limited number of material points and the derived plastic properties are propagated to the surrounding integration points subjected to similar deformation. Optimal configurations of the clusters vary in time as the local deformation conditions may change during the forming process, so the clusters must be periodically adapted. We consider two operations on the clusters of integration points: splitting (refinement) and merging (unrefinement). The concept is tested in the Hierarchical Multi-Scale (HMS) framework [1] that computes macroscopic deformations by means of the FEM, whereas the micro-structural evolution at the individual FE integration points is predicted by a CP model. The HMS locally and adaptively approximates homogenized stress responses of the CP model by means of analytical plastic potential or yield criterion function. Our earlier work investigated simple test cases [2]. In this contribution we present a deep drawing process simulated using the HMS framework improved to exploit local quasi-homogeneities. We conclude that large performance gains (e.g. a speedup of 25) are obtained at the expense of introducing only a minor (e.g. below 1%) modelling error compared to the HMS simulation with no clustering of integration points.

References

[1] Gawad J, Van Bael A, Eyckens P, Samaey G, Van Houtte P and Roose D 2013 *Computational Materials Science* vol 66 pp 65-83
[2] Khairullah M, Gawad J, Roose D and Van Bael A 2016 *Proc. Int. European Congress on Computational Methods in Applied Sciences and Engineering* vol 7 (Crete Island)

Constitutive modelling of magnesium alloy sheets under strain path changes

Jinwoo Lee[a], Young-Seon Lee[a], Myoung-Gyu Lee[b], Daeyong Kim[a]

[a]Materials Deformation Department, Korea Institute of Materials Science, Changwon 642-831, Republic of Korea
[b]Department of Materials Science and Engineering, Korea University, Seoul 136-713, Republic of Korea

dayong@kims.re.kr

Abstract. In this study, a continuum-based approach for the description of the plastic hardening behavior of magnesium alloy sheets subjected to non-proportional strain path changes is discussed. The constitutive model is based on an anisotropic distortional yield function combining a stable component and a fluctuating component. The stable component initiates the yield criterion that characterizes the typical strength differential between tension and compression in magnesium alloys at room temperature. The evolution of the fluctuating component is reformulated based on its cubic metal counterpart to represent the deformation nature of magnesium alloys that consist of slip and twin dominant modes. The model is not formulated with a kinematic hardening rule, but it reasonably reproduces complex features of the stress-strain responses under the load reversal in magnesium alloy sheet: i.e., asymmetric hardening behavior under tension and compression, sigmoidal nature of hardening curve during monotonic compression and compression followed by tension, strong anisotropy etc.

Acknowledgments
This study was financially supported by the Fundamental Research Program of the Korea Institute of Materials Science (KIMS).

A New Thermo-Elasto-Viscoplastic Crystal Plasticity Framework to Predict the Formability of Aluminum Alloys at Elevated Temperatures

Kaan Inal[1], Edward Cyr[1], Raja K. Mishra[2]

1 Department of Mechanical and Mechatronics Engineering, University of Waterloo, Waterloo, Canada, N2L 3G1
2 General Motors Research & Development Center, Warren, MI 48090, USA

*Corresponding author: Tel: +1 519 888 4567 Ext. 38114. Fax: +1 519 885 5862. E-mail address: kinal@uwaterloo.ca.

Abstract

A new thermo-elasto-viscoplastic (TEV) crystal plasticity constitutive formulation is developed and implemented in the well-known Marciniak-Kuczynski analysis to predict the formability of aluminum alloys (AA) 5754 and 3003 at elevated temperatures. The model takes into account the temperature dependence of the single crystal elastic coefficients, single slip hardening parameters, thermal softening and slip rate sensitivity. Temperature dependent single slip hardening parameters are determined from uniaxial tension simulations at room and elevated temperatures. The new model is able to accurately predict the experimental forming limit diagrams (FLDs) without the need for further curve fitting. The effects of elastic constants and thermal softening on FLD predictions are discussed, and a new expression to represent the temperature dependence of the initial imperfection (for the M-K analysis) is developed to enable the model to successfully predict the FLDs for any temperature in the warm forming regime prior to recrystallization.

Keywords: Crystal Plasticity; Forming Limit Diagram; Temperature; M-K Analysis; aluminum alloys.

Numisheet IOP Publishing
Journal of Physics: Conference Series **734** (2016) 032137 doi:10.1088/1742-6596/734/3/032137

A shear ductile fracture criterion for metal forming

Yanshan Lou[1,3], Jeong Whan Yoon[1,2*]

[1]International Center for Innovative Manufacturing, Institute for Frontier Materials, Deakin University, Geelong Waurn Ponds, VIC 3220, Australia, y.lou@deakin.edu.au
[2]School of Mechanics, Aerospace and Systems, KAIST, 291 Daehak-ro, Ruseong-gu, Daejeon 34141, Republic of Korea, j.yoon@deakin.edu.au
[3]Institute for Forming Technology and Lightweight Construction, TU Dortmund University, Baroper Str. 303, Dortmund 44227, Germany

*Corresponding author: Jeong Whan Yoon, j.yoon@deakin.edu.au

Abstract

In this paper, an uncoupled shear ductile fracture criterion is developed for prediction of ductile fracture in sheet metal forming from shear to balanced biaxial tension. The ductile fracture criterion is calibrated by four tests of sheet metal: shear tests, uniaxial tension, plane strain tension and the Nakajima test. Specimens are designed for AA6082 T6 (t1.0) for the calibration of the proposed ductile fracture criterion. The calibrated ductile fracture criterion is then implemented into numerical simulation for the prediction of ductile fracture of the aluminum alloy. For the purpose of comparison, onset of ductile fracture is also estimated by MMC3, DF2012 and DF2014 criteria. The comparison indicates that the developed criterion can accurately predict onset of ductile fracture for all four loading conditions, but the other three criteria can only provide reasonable prediction for three tests. Accordingly, the newly developed ductile fracture criterion is suggested to be used in prediction of ductile fracture for sheet metal forming in wide loading condition from shear to the balanced biaxial tension. Discussion on calibration of this ductile fracture criterion also indicates that the ductile fracture criterion can be employed in estimation of ductile fracture for bulk metal forming processes with better predictability.

Journal of Physics: Conference Series **734** (2016) 032138 doi:10.1088/1742-6596/734/3/032138

Crystal plasticity-based modeling for predicting anisotropic behaviour and formability of metallic materials

Son Pham[1*], Youngung Jeong[2], Adam Creuziger[2], Mark Iadicola[2], Tim Foecke[2] and Anthony Rollett[3]

[1] Department of Materials, Imperial College London, London, UK
[2] Center for Automotive Lightweighting, National Institute of Standards and Technology, Gaithersburg, MD, USA
[3] Materials Science and Engineering, Carnegie Mellon University, Pittsburgh, PA, USA

*Corresponding author: son.pham@imperial.ac.uk

Abstract

Metallic materials often exhibit anisotropic behaviour under complex load paths because of changes in microstructure, e.g., dislocations and crystallographic texture. In this study, we present the development of constitutive model based on dislocations, point defects and texture in order to predict anisotropic response under complex load paths. In detail, dislocation/solute atom interactions were considered to account for strain aging and static recovery. A hardening matrix based on the interaction of dislocations was built to represent the cross-hardening of different slip systems. Clear differentiation between forward and backward slip directions of dislocations was made to describe back stresses during path changes. In addition, we included dynamic recovery in order to better account for large plastic deformation. The model is validated against experimental data for AA5754-O with path changes, e.g., Figure 1 [1].

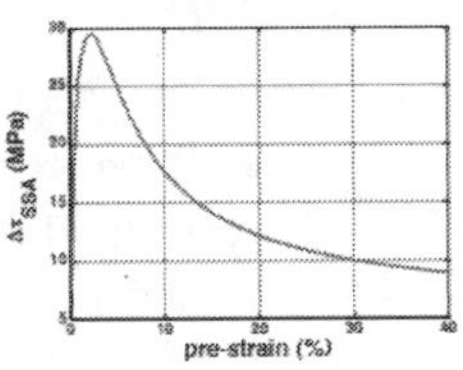

Figure 1: Modelling of the increase in the yield strength due to static ageing during path changes.

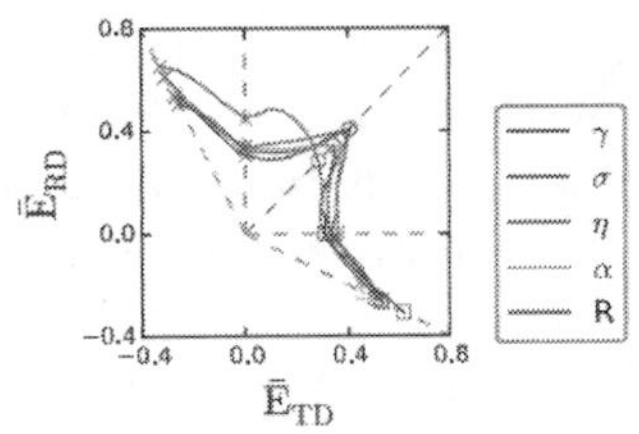

Figure 2: Strain limits of various initial textures in body-centred-cubic crystals

Another effort is to include microstructure in forming predictions with a minimal increase in computational time. This effort enables comprehensive investigations of the influence of texture-induced anisotropy on formability [2]. Application of these improvements to predict forming limits of various BCC textures, such as γ, σ, α, η and ϵ fibers and a random (R) texture. These simulations demonstrate that the crystallographic texture has significant (both positive and negative) effects on the forming limit diagrams (Figure 2). For example, the γ fiber texture, that is often sought through thermo-mechanical processing due to high r-value, had the highest forming limit in the balanced biaxial strain path but the lowest forming limit under the plane strain path among textures under consideration.

References

[1] Pham M-S, Iadicola M, Creuziger A, Hu L, Rollett AD. Thermally-activated constitutive model including dislocation interactions, aging and recovery for strain path dependence of solid solution strengthened alloys: Application to AA5754-O. Int J Plasticity 2015;75:226.
[2] Jeong YG, Pham MS, Iadicola M, Creuziger A, Foecke T. Forming limit prediction using a self-consistent crystal plasticity framework: a case study for body-centered cubic materials. Modelling and Simulation in Materials Science and Engineering 2016;24:055005.

Isogeometric Analysis of Deformation, Inelasticity and Fracture In Thin-Walled Structures

René de Borst

Department of Civil and Structural Engineering, University of Sheffield

The basic idea of isogeometric analysis (IGA) is to use splines, which are the functions commonly used in computer-aided design (CAD) to describe the geometry, as the basis function for the analysis as well. A main advantage is that a sometimes elaborate meshing process is by-passed. Another benefit is that spline basis-functions possess a higher-order degree of continuity, which enables a more accurate representation of the stress. Further, the order of continuity of the basis-functions can be reduced locally by knot insertion. This feature can be used to model interfaces and cracks as discontinuities in the displacement field.

In order to study failure-mechanisms in thin-walled composite materials, an accurate representation of the full three-dimensional stress field is mandatory. A continuum shell formulation is an obvious choice. Continuum shell elements can be developed based on the isogeometric concept. They exploit NURBS basis functions to construct the mid-surface of the shell. In combination with a higher-order B-spline basis function in the thickness direction a complete three-dimensional representation of the shell is obtained. This isogeometric shell formulation can be implemented in a standard finite element code using Bézier extraction.

.

Weak and strong discontinuities can be introduced in the B-spline function using knot-insertion to model material interfaces and delaminations rigorously as discontinuities in the displacement field. The exact representation of material interfaces vastly improves the accuracy of the through-the-thickness stress field. The ability to provide a double knot insertion enables a straightforward analysis of delamination growth in layered composite shells. Illustrative examples will be given.

.

Journal of Physics: Conference Series **734** (2016) 032140 doi:10.1088/1742-6596/734/3/032140

Strain Rate Dependant Material Model for Orthotropic Metals

Rade Vignjevic
Brunel University London, Kingston Lane,
Uxbridge UB8 3PH

Abstract

In manufacturing processes anisotropic metals are often exposed to the loading with high strain rates in the range from $10^2 s^{-1}$ to $10^6 s^{-1}$ (e.g. stamping, cold spraying and explosive forming). These types of loading often involve generation and propagation of shock waves within the material. The material behaviour under such a complex loading needs to be accurately modelled, in order to optimise the manufacturing process and achieve appropriate properties of the manufactured component.

The presented research is related to development and validation of a thermodynamically consistent physically based constitutive model for metals under high rate loading. The model is capable of modelling damage, failure and formation and propagation of shock waves in anisotropic metals. The model has two main parts: the strength part which defines the material response to shear deformation and an equation of state (EOS) which defines the material response to isotropic volumetric deformation [1]. The constitutive model was implemented into the transient nonlinear finite element code DYNA3D [2] and our in house SPH code. Limited model validation was performed by simulating a number of high velocity material characterisation and validation impact tests.

The new damage model was developed in the framework of configurational continuum mechanics and irreversible thermodynamics with internal state variables. The use of the multiplicative decomposition of deformation gradient makes the model applicable to arbitrary plastic and damage deformations.

To account for the physical mechanisms of failure, the concept of thermally activated damage initially proposed by Tuller and Bucher [3], Klepaczko [4] was adopted as the basis for the new damage evolution model. This makes the proposed damage/failure model compatible with the Mechanical Threshold Strength (MTS) model Follansbee and Kocks [5], 1988; Chen and Gray [6] which was used to control evolution of flow stress during plastic deformation. In addition the constitutive model is coupled with a vector shock equation of state which allows for modelling of shock wave propagation in orthotropic the material.

Parameters for the new constitutive model are typically derived on the basis of the tensile tests (performed over a range of temperatures and strain rates), plate impact tests and Taylor anvil tests.

The model was applied to simulate explosively driven fragmentation, blast loading and cold spraying impacts.

References

[1] Vignjevic, R., Campbell, J. C., Bourne, N. K. and Djordjevic, N. (2008), "Modeling shock waves in orthotropic elastic materials", *Jrnl of App Physics,* vol. 104, no. 4.

[2] Liu, J., (2004), *Dyna3D: A Nonlinear, Explicit, Three-Dimensional Finite Element Code for Solid and Structural Mechanics*, University of California, LLNLL, USA.

[3] Tuler, F. R. and Butcher, B. M. (1968), "A criterion for the time dependence of dynamic fracture", *Int J Fracture Mechanics,* 4 (4), pp. 431-437.

[4] Klepaczko, J. R. (1990), "Dynamic crack initiation, some experimental methods and modelling", in Klepaczko, J. R. (ed.) *Crack dynamics in metallic materials*, Springer-Verlag, Vienna, pp. 428-445.

[5] Follansbee, P. S. and Kocks, U. F. (1988), "A constitutive description of the deformation of copper based on the use of the mechanical threshold stress as an internal state variable", *Acta Metallurgica,* vol. 36, no. 1, pp. 81-93.

[6] Chen, S. and Gray, G. (1996), "Constitutive behavior of tantalum and tantalum-tungsten alloys", *Metallurgical and Materials Transactions A,* vol. 27, no. 10, pp. 2994-3006.

A comparison of deformation and failure behaviors of AZ31 and E-form Mg alloys under V-bending test

Shi-Hoon Choi*, Jaiveer Singh*, Min-Seong Kim*, Jeong-Whan Yoon**
*** Department of Printed Electronic Engineering, Sunchon National University, Sunchon 540-950, South Korea**
**** Department of Mechanical Enginnering, KAIST, Daejeon 305-701, South Korea**
shihoon@sunchon.ac.kr, jaiveer_serwa@yahoo.co.in, kimminseong@sunchon.ac.kr
j.yoon@kaist.ac.kr

Deformation and failure behaviors of magnesium (Mg) alloys (AZ31 and E-form) were investigated using V-bending test. Formability of these Mg alloys was discussed in terms of minimum bending radius. Microtexture evolution in the deformed Mg alloys was examined via electron back-scattered diffraction (EBSD) technique. Two level simulation technique which combined continuum finite element method (FEM) and crystal plasticity FEM successfully simulated the microtexture evolution in Mg alloys during V-bending test. The effect of deformation twinning on the failure in Mg alloys was also examined.

Isogeometric shape optimisation of shell structures using multiresolution subdivision surfaces

Xiao Xiao, Kosala Bandara, Fehmi Cirak

Department of Engineering, University of Cambridge, Cambridge CB2 1PZ, UK

E-mail: `f.cirak@eng.cam.ac.uk`

Abstract. We introduce the isogeometric shape optimisation of thin shell structures using subdivision surfaces. Both triangular Loop and quadrilateral Catmull-Clark subdivision schemes are considered for geometry modelling and finite element analysis. A gradient-based shape optimisation technique is implemented to minimise compliance, i.e. to maximise stiffness. Different control meshes describing the same surface are used for geometry representation, optimisation and finite element analysis. The finite element analysis is performed with subdivision basis functions corresponding to a sufficiently fine control mesh. During iterative shape optimisation the geometry is updated starting from the coarsest control mesh and proceeding to increasingly finer control meshes. The proposed approach is applied to three optimisation examples, namely a catenary, a roof over a rectangular domain, and free-form architectural shell roof. The influence of the geometry description and the used subdivision scheme on the obtained optimised curved geometries are investigated in detail.

References

[1] Bandara K and Cirak F 2016 Isogeometric shape optimisation of shell structures using multiresolution subdivision surfaces *Preprint* arXiv:1605.06288

[2] Cirak F, Scott M J, Antonsson E K, Ortiz M and Schroder P 2002 *Comput. Aided Design* **34** 137–48

[3] Cirak F, Ortiz M and Schroder P 2000 *Int. J. Numer. Meth. Eng.* **47** 2039–72

Journal of Physics: Conference Series **734** (2016) 032143 doi:10.1088/1742-6596/734/3/032143

Calibration of Gurson-type models for porous sheet metals with anisotropic non-quadratic plasticity

M. Gologanu, A. Kami, D.S. Comsa, D. Banabic

Technical University of Cluj Napoca, Romania

The growth and coalescence of voids in sheet metals are not only the main active mechanisms in the final stages of fracture in a necking band, but they also contribute to the forming limits via changes in the normal directions to the yield surface. A widely accepted method to include void effects is the development of a Gurson-type model for the appropriate yield criterion, based on an approximate limit analysis of a unit cell containing a single spherical, spheroidal or ellipsoidal void. We have recently [*1*] obtained dissipation functions and Gurson-type models for porous sheet metals with ellipsoidal voids and anisotropic non-quadratic plasticity, including yield criteria based on linear transformations (Yld91 and Yld2004-18p) and a pure plane stress yield criteria (BBC2005). These Gurson-type models contain several parameters that depend on the void and cell geometries and on the selected yield criterion. Best results are obtained when these key parameters are calibrated via numerical simulations using the same unit cell and a few representative loading conditions. The single most important such loading condition corresponds to a pure hydrostatic macroscopic stress (pure pressure) and the corresponding velocity field found during the solution of the limit analysis problem describes the expansion of the cavity.

However, for the case of sheet metals, the condition of plane stress precludes macroscopic stresses with large triaxiality or ratio of mean stress to equivalent stress, including the pure hydrostatic case. Also, pure plane stress yield criteria like BBC2005 must first be extended to 3D stresses before attempting to develop a Gurson-type model and such extensions are purely phenomenological with no due account for the out-of-plane anisotropic properties of the sheet. Therefore, we propose a new calibration method for Gurson-type models that uses only boundary conditions compatible with the plane stress requirement. For each such boundary condition we use a spectral method to solve the limit analysis problem, using as spectral basis a newly developed Mie decomposition of incompressible velocity fields for ellipsoidal cells with confocal ellipsoidal voids, extending the well-known Lee and Mear family corresponding to the spheroidal axisymmetric case [1]. We thus obtain a series of points located on the Gurson-type yield surface and their corresponding normal directions. These points are used as input to some enhanced parameter identification method developed for anisotropic yield criteria [1] to determine the calibration of the Gurson-type models. We finally use these newly calibrated Gurson-type models to evaluate forming limits for porous sheet metals.

[1] Banabic D. (Ed.), Multiscale Modeling in Sheet Metal Forming, Springer, Heidelberg-Berlin, 2016 (in press)

Trimming Simulation of Forming Metal Sheets Isogeometric Models by Using NURBS Surfaces

D. Herrero-Adan[1], Rui P.R. Cardoso[2] and O.B. Adetoro[1]
[1]Department of Engineering Design and Mathematics, University of the West of England, BS16 1QY Bristol, UK
[2]Department of Mechanical, Aerospace and Civil Engineering, Brunel University London, UB8 3PH Uxbridge, London, UK

E-mail: Daniel.Herreroadan@uwe.ac.uk

Abstract. Some metal sheets forming processes need trimming in a final stage for achieving the net-shape specification and for removing micro-cracks and irregularities. In numerical simulation, since the exact final edge location is a priori unknown in the original metal blanket, the trimming needs to be done once the forming is finished. During the forming internal stresses are generated inside the sheet. When trimming those stresses configuration is changed to achieve equilibrium as a consequence of the material removal. In this paper a novel method for simulating the trimming is presented. The part to trim is modelled using isogeometric analysis (IGA). The new surface generated is modelled with non-uniform rational B-splines (NURBS). Due to the IGA characteristics a total geometrical accuracy and an efficient residual stresses recalculation are accomplished.

Keywords. Trimming, IGA, NURBS, Metal sheet forming.

Journal of Physics: Conference Series **734** (2016) 032145 doi:10.1088/1742-6596/734/3/032145

Advances in Constitutive and Failure Models for Sheet Forming Simulation

Jeong Whan Yoon[1,2,a] and Thomas B Stoughton[3,2]

[1] *Department of Mechanical Engineering, KAIST, Daejeon 305-701, Republic of Korea*

[2] *Faculty of Science, Engineering & Built Environment, Deakin University, Geelong, Waurn Ponds, Victoria 3220, Australia*

[3] *GM R&D Center, Warren, MI 48090-9055, USA*

Non-Associated Flow Rule (Non-AFR) can be used as a convenient way to account for anisotropic material response in metal deformation processes, making it possible for example, to eliminate the problem of the anomalous yielding in equibiaxial tension that is mistakenly attributed to limitations of the quadratic yield function, but may instead be attributed to the Associated Flow Rule (AFR). Seeing as in Non-AFR based models two separate functions can be adopted for yield and plastic potential, there is no constraint to which models are used to describe each of them. In this work, the flexible combination of two different yield criteria as yield function and plastic potential under Non-AFR is proposed and evaluated. FE simulations were carried so as to verify the accuracy of the material directionalities predicted using these constitutive material models. The stability conditions for non-associated flow connected with the prediction of yield point elongation are also reviewed. Anisotropic distortion hardening is further incorporated under non-associated flow. It has been found that anisotropic hardening makes the noticeable improvements for both earing and spring-back predictions. This presentation is followed by a discussion of the topic of the forming limit & necking, the evidence in favor of stress analysis, and the motivation for the development of a new type of forming limit diagram based on the polar effective plastic strain (PEPS) diagram. In order to connect necking to fracture in metals, the stress-based necking limit is combined with a stress-based fracture criterion in the principal stress, which provides an efficient method for the analysis of necking and fracture limits. The concept for the PEPS diagram is further developed to cover the path-independent PEPS fracture which is compatible with the stress-based fracture approach. Thus this fracture criterion can be utilized to describe the post-necking behavior and to cover nonlinear strain-path. Fracture modeling in the stress space including mean stress contribution is also conducted and compared with M-C criterion. Strain-path effect on fracture criteria and the role of a damage parameter have been also discussed.

[a] Corresponding author: j.yoon@kaist.ac.kr / j.yoon@deakin.edu.au

Journal of Physics: Conference Series **734** (2016) 032146 doi:10.1088/1742-6596/734/3/032146

Non-Local Damage Modelling of Sheet Metal Forming Processes with ALE Formulation

O.B. Adetoro[1] and Rui P.R. Cardoso[2]

[1]Department of Engineering Design and Mathematics, University of the West of England, BS16 1QY Bristol, UK

[2]Department of Mechanical, Aerospace and Civil Engineering, Brunel University London, UB8 3PH Uxbridge, London, UK

E-mail: Mayo.Adetoro@uwe.ac.uk

Abstract. The modelling of material degradation due to nucleation, growth and coalescence of micro-voids is vital in sheet metal forming process due to the large deformation typically experienced by the part. Nonlocal damage modelling or nonlocal continuum is gaining a lot of interest because it is an effective approach to modelling the strain-softening, whilst avoiding the spurious localization that gives rise to strong mesh sensitivity in numerical computations. However to accurately resolve the evolving narrow bands of highly localised strain, it is necessary to use sufficient computational grids. In this paper an ALE formulation is used for modelling the localization pattern. An approach for relocating the node points is presented and explored.